10	11 1B	12 2B	13 3A	14 4A	15 5A	16 6A	17 7A	18 8A
								2 **He** 4.00260
			5 **B** 10.811	6 **C** 12.011	7 **N** 14.0067	8 **O** 15.9994	9 **F** 18.9984	10 **Ne** 20.1797
			13 **Al** 26.9815	14 **Si** 28.0855	15 **P** 30.9738	16 **S** 32.066	17 **Cl** 35.4527	18 **Ar** 39.948
28 **Ni** 58.69	29 **Cu** 63.546	30 **Zn** 65.39	31 **Ga** 69.723	32 **Ge** 72.61	33 **As** 74.9216	34 **Se** 78.96	35 **Br** 79.904	36 **Kr** 83.80
46 **Pd** 106.42	47 **Ag** 107.868	48 **Cd** 112.411	49 **In** 114.818	50 **Sn** 118.710	51 **Sb** 121.75	52 **Te** 127.60	53 **I** 126.904	54 **Xe** 131.29
78 **Pt** 195.08	79 **Au** 196.967	80 **Hg** 200.59	81 **Tl** 204.383	82 **Pb** 207.2	83 **Bi** 208.980	84 **Po** (209)	85 **At** (210)	86 **Rn** (222)
110 (269)	111 (272)	112 (277)						

64 **Gd** 157.25	65 **Tb** 158.925	66 **Dy** 162.50	67 **Ho** 164.930	68 **Er** 167.26	69 **Tm** 168.934	70 **Yb** 173.04	71 **Lu** 174.967
96 **Cm** (247)	97 **Bk** (247)	98 **Cf** (251)	99 **Es** (252)	100 **Fm** (257)	101 **Md** (258)	102 **No** (259)	103 **Lr** (260)

Chemistry and Life

An Introduction to General, Organic, and Biological Chemistry

Fifth Edition

John W. Hill
University of Wisconsin–River Falls

Stuart J. Baum
SUNY Distinguished Teaching Professor,
State University of New York, College at Plattsburgh

Dorothy M. Feigl
Saint Mary's College

PRENTICE HALL
Upper Saddle River, NJ 07458

Library of Congress Cataloging-in-Publication Data

Hill, John William.
Chemistry and life : an introduction to general, organic, and biological chemistry. — 5th ed. / John W. Hill, Stuart J. Baum, Dorothy M. Feigl.
p. cm.
Includes index.
ISBN 0-13-569294-6
1. Chemistry. I. Baum, Stuart J. II. Feigl. Dorothy M. III Title.
QD31.2.H56 1997
540—dc20 96-36303
CIP

Acquisitions Editor: *Ben Roberts*
Editor in Chief: *Paul F. Corey*
Assistant Vice President of Production and Manufacturing: *David W. Riccardi*
Executive Managing Editor: *Kathleen Schiaparelli*
Marketing Manager: *Linda Taft MacKinnon*
Marketing Assistant: *Elizabeth A. Kell*
Manufacturing Manager: *Trudy Pisciotti*
Creative Director: *Paula Maylahn*
Art Director: *Joseph Sengotta*
Interior Design and Cover Design: *Amy Rosen*
Front Cover Photography: *Braasch/Life/WoodFin Camp & Associates*
Editorial Assistant: *Ashley Scattergood*
Associate Editor: *Mary Hornby*
Photo Research Administrator: *Melinda Reo*
Photo Editor: *Lori Morris-Nantz*
Photo Researcher: *Cindy Lee Overton*
Art Manager: *Gus Vibal*
Art Studio: *Precision Graphics* and *Thompson Steele Production Services*
Copyediting, Text Composition, and Electronic Page Makeup: *Thompson Steele Production Services*

Printed in the United States of America

10 9 8 7 6 5 4 3 2 1

ISBN 0-13-569294-6

Prentice-Hall International (UK) Limited, *London*
Prentice-Hall of Australia Pty. Limited, *Sydney*
Prentice-Hall Canada Inc., *Toronto*
Prentice-Hall Hispanoamericana, S.A., *Mexico*
Prentice-Hall of India Private Limited, *New Delhi*
Prentice-Hall of Japan, Inc., *Tokyo*
Simon & Schuster Asia Pte. Ltd., *Singapore*
Editora Prentice-Hall do Brasil, Ltda., *Rio de Janeiro*

Contents

Preface

Our world has been transformed by science and technology. The impact of science on the quality of human life is profound. Yet, to beginning students, the scientific disciplines that daily influence their lives often seem mysterious and incomprehensible. Those of us who enjoy the study of science, however, find it a fascinating and rewarding experience precisely because it can provide reasonable explanations for seemingly mysterious phenomena.

Hydrogen bonds in ice.

Chemistry and Life has been written in that spirit. Apparently obscure phenomena are explained in an informal, readable style. We assume that the student has little or no chemistry background and clearly explain each new concept as it is introduced. Chemical principles and biological applications are carefully integrated throughout the text, with liberal use of drawings, diagrams, and photographs.

Our selection of topics and choice of examples make the text especially appropriate for students in health and life sciences, but it is also suitable for anyone seeking to become a better-informed citizen of our technological society. The text provides ample material for a full year's course. The 11 Special Topics cover optional material for added flexibility. They may be omitted or assigned as outside reading without loss of continuity. We have also included many short essays that focus on interesting applications of topics presented in the text. We have consciously increased the sophistication of chemical understanding as the student progresses through the chapters.

Changes in the Fifth Edition

All the text has been updated to reflect the latest scientific knowledge. In addition, we have responded to suggestions of users and reviewers of the fourth edition and used our own writing and teaching experience to make the following changes:

Organization

For this fifth edition, we have included detailed contents on the opening page of each chapter. At the end of each chapter, we have added a list of Key Terms and a chapter Summary. The key terms are boldfaced in the text and are defined in the Glossary (Appendix IV).

A microscopic view of soap action.

We have added a greater number of photos, figures, and margin notes to clarify and enliven the text discussion. Many sections have undergone extensive rewriting, especially in the chapters on lipids and proteins. The discussion of the VSEPR theory and the shapes of molecules has been taken out of Chapter 4; it now makes up new Special Topic B.

The health-related topics from the fourth edition have been retained, and we have included several new essays. For example, arteriosclerosis, lead poisoning, development of cancer, diabetes, hemophilia, blood doping, and blood types.

Exercises

In addition to the many examples and practice exercises in the body of each chapter, we have three kinds of end-of-chapter exercises.

- Review Questions are intended to provide a qualitative measure of your understanding of the chapter.
- The Problems are arranged by topic; they test your mastery of the material and—where pertinent—of problem-solving techniques introduced in the chapter. These problems are arranged in matched pairs.
- The Additional Problems are not grouped by type. Some are intended to be a bit more challenging; they often require a synthesis of ideas from more than one chapter. Others, however, are not any more difficult than those arranged by topic. Rather, they pursue an idea further than is done in the text, or they introduce new ideas.

Many of the worked out examples have been revised to improve the pedagogy.

Supplements

For the Student

- **Student Study Guide with Solutions,** by Marvin L. Hackert of the University of Texas at Austin and Roger K. Sandwick of the State University of New York at Plattsburgh. This student-friendly manual contains chapter summaries, additional examples and problems, and numerous self-tests (with answers). Solutions correspond to the odd-numbered problems in the text. (0-13-574666-3; 57466-5)
- **Chemistry and Life in the Laboratory: Experiments in General, Organic, and Biological Chemistry,** by Victor L. Heasley and Val. J. Christensen of Point Loma Nazarene College and Gene E. Heasley of Southern Nazarene University. This manual contains 36 experiments that cover the same general topics as the text. Laboratory instructions are clear and thorough and the experiments are well written and imaginative. This revision includes more information on issues of safety and disposal. All experiments have been thoroughly class tested. (0-13-597725-8; 59772-4)
- **Allied Health Chemistry: A Companion,** by Tim Smith and Diane Vukovich, both of the University of Akron. This student companion teaches the basic mathematics inherent in the course by methods that are friendly and hands-on. (0-13-470460-6; 47046-8)
- **Prentice Hall/The New York Times Themes of Times.** Through this unique program, adopters of the *Chemistry and Life* are eligible to receive our *New York Times* supplement for their students. This newspaper-format resource uses current chemistry-related articles to emphasize the importance and relevance of chemistry in everyday life. (Free in quantity to qualified adopters through your local representative.)

For the Instructor

- **Instructor's Manual to the Laboratory Manual,** by Heasley et al. (0-13-597741-X; 59774-0)
- **Instructor's Manual,** by Roger K. Sandwick of the State University of New York at Plattsburgh, and **Test Bank,** prepared by Aninna Carter of Adirondack Community College. This instructor's resource contains solutions to the problems that are not answered in the text. The test bank has been revised extensively and now contains over 1100 multiple-choice questions. (0-13-594096-7; 57409-5)

- **PH Custom Test WIN** (0-13-574633-7; 57463-2) **PH Custom Test MAC** (0-13-574641-8; 57464-0). Electronic versions of the *Chemistry and Life* test bank which contains over 1100 multiple-choice questions.
- **Transparencies:** 125 full-color transparency acetates selected by the text authors.

Acknowledgments

JWH would like to thank his colleagues at the University of Wisconsin–River Falls for so many ideas that made their way into his other texts—some of which appear in this one—and to Kathy Sumter, Program Aide in the Department of Chemistry. He is especially indebted to Ina Hill and Cynthia Hill for library research, typing, and unfailing support throughout this project, and to Mike Davis for his encouragement and for sharing his love of learning.

We greatly appreciate the substantial support and guidance from the editorial staffs at Prentice Hall and Thompson Steele. Ben Roberts, Irene Nunes, Andrea Fincke, and Craig Kirkpatrick all played significant roles in converting the manuscript into a book of which we are all proud.

Both of us would like to thank our students, who have challenged us to be better teachers, and the reviewers of this and our other books, who have challenged us to be better writers.

Hugh A. Akers
Lamar University

Dave Becker
Oakland Community College

Robert L. Clark
North Idaho College

Lawrence Duffy
University of Alaska–Fairbanks

William H. Flurkey
Indiana State University

Wesley O. Fritz
College of DuPage

James A. Golen
University of Massachusetts

Marc M. Greenberg
Colorado State University

James C. Howard
Middle Tennessee State University

James T. Johnson
Sinclair Community College

Lidija Kampa
Kean College of New Jersey

Lauren E. H. McMills
Ohio University

W. Robert Midden
Bowling Green State University

Mary O'Sullivan
Indiana State University

Rhonda Scott-Ennis, *University of Wisconsin–River Falls*

Michael A. Serra
Youngstown State University

Michael J. Strauss
University of Vermont

Ron Swisher
Oregon Institute of Technology

Rod S. Tracey
College of the Desert

Linda A. Wilson
Middle Tennessee State University

No book—or other educational device—can replace a good teacher; thus we have designed this book as an aid to the classroom teacher. The only valid test of this or any text is in a classroom. We would greatly appreciate receiving comments and suggestions based on your experience with this book.

J. W. H.
S. J. B.
D. M. F.

To the Student

What is chemistry?

Chemistry is such a broad, all-encompassing area of study that people almost despair in trying to define it. Indeed, some have taken a cop-out approach by defining chemistry as "what chemists do." But that won't do; it's much too narrow a view.

Chemistry is what we all do. We bathe, clean, and cook. We put chemicals on our faces, hands, and hair. Collectively, we use tens of thousands of consumer chemical products in our homes. Professionals in the health and life sciences use thousands of additional chemicals as drugs, antiseptics, or reagents for diagnostic tests.

Your body itself is a remarkable chemical factory. You eat and breathe, taking in raw materials for the factory. You convert these supplies into an unbelievable array of products, some incredibly complex. This chemical factory—your body—also generates its own energy. It detects its own malfunctions and can regenerate and repair some of its component parts. It senses changes in its environment and adapts to these changes. With the aid of a neighboring facility, this fabulous factory can create other factories much like itself.

Everything you do involves chemistry. Your read this sentence; light energy is converted to chemical energy. You think; protein molecules are synthesized and stored in your brain. All of us do chemistry.

Chemistry affects society as well as individuals. Chemistry is the language—and the principal tool—of the biological sciences, the health sciences, and the agricultural and earth sciences.

Chemistry has illuminated all of the natural world, from the tiny atomic nucleus to the immense cosmos. We believe that a knowledge of chemistry can help you. We have written this book in the firm belief that beginning chemistry can be related immediately to problems and opportunities in the life and health sciences. And we believe that this can make the study of chemistry interesting and exciting, especially to nonchemists.

For example, an "ion" is more than a chemical abstraction. Enough mercury ions in the wrong place can kill you, but the right number of calcium ions in the right place can keep you from bleeding to death. "$PV = nRT$" is an equation, but it is also the basis for the respiratory therapy that has saved untold lives in hospitals. "Hydrogen bonding" is a chemical phenomenon, but it also helps to account for the fact that a dog has puppies while a cat has kittens and a human has human babies. There are hundreds of similar fundamental and interesting applications of chemistry to life.

A knowledge of chemistry has already had a profound effect on the quality of life. Its impact on the future will be even more dramatic. At present we can control diabetes, cure some forms of cancer, and prevent some forms of mental retardation because of our understanding of the chemistry of the body. We can't *cure* diabetes or cure *all* forms of cancer or *all* mental retardation, because our knowledge is still limited. So learn as much as you can. Your work will be enhanced and your life enriched by your greater understanding.

Be prepared. Something good might happen to you—and to others because of you.

1 Matter and Measurement

Accurate measurements are important in science and in medicine. Measurement of height and weight are meaningful first steps in any physical examination. Other measurements, often more sophisticated than these, are also essential in many medical procedures.

This book is called *Chemistry and Life*. What does chemistry have to do with life? What is chemistry? For that matter, what is life?

The last question is more than rhetorical. Progress in science, technology, and medicine has blurred the distinction between life and death. Is a person whose heart has stopped beating necessarily dead? Is a person whose vital functions are being maintained by machine truly alive? In this book we won't even attempt to supply a definitive answer to the question "What is life?" We'll simply note the critical significance of this question for our society.

How about the first question: "What does chemistry have to do with life?" A chemist would answer, "Just about everything." The human body, for example, is the most extraordinarily complicated, most elegantly designed, and most efficiently operated chemical laboratory there is. Our attempts to answer this first question will fill most of this text.

That leaves us with the middle question, "What is chemistry?"—which is the subject of this first chapter. We shall see how science in general and chemistry in particular have developed from earlier human endeavors. Our study will include a consideration of the methods of science and the manner of its progress. Finally, we shall develop some basic concepts necessary to our study of chemistry and its relationship to life.

1.1 Science and the Human Condition

We are taught in elementary school that people have three basic needs: food, clothing, and shelter. Certainly those three things—if adequate in quantity and quality—are enough to keep us alive. Most of us, however, would agree that there are two more requirements for a *good* life: reasonably good health and some chance for happiness.

In early human societies, nearly all human effort was dedicated to the hunting and gathering of food, the making of clothing, and the provision of shelter. Our early ancestors had no knowledge of the biological and chemical basis of illness, and they could do little about their health except pray and make sacrifices to their gods. With the coming of civilization, some people began to have enough leisure time to turn their thoughts to the human condition and to the natural world around them. Over the centuries, what we now call science grew out of their speculations. As this scientific study of the material universe progressed, the responsibility for adding to the growing body of knowledge was divided among various disciplines, and one of these disciplines was chemistry.

Modern chemistry's roots are planted in alchemy, a kind of mystical chemistry that flourished in Europe during the Middle Ages (about 500 to 1500 C.E.). Modern chemists inherited from the alchemists an abiding interest in human health and the quality of life. Alchemists not only searched for a philosopher's stone that would turn cheaper metals into gold but also sought an elixir that would confer immortality on those exposed to it. Alchemists never achieved these goals, but they discovered many new chemical substances and perfected techniques, such as distillation and extraction, that are still used today.

It was a Swiss physician, Theophrastus Bombastus von Hohenheim (1493–1541), who urged all chemists to turn away from their attempts to make gold and to seek instead medicines with which to treat disease. Possessed of a monstrous ego, von Hohenheim (who preferred the self-chosen name Paracelsus) alienated many of his contemporaries. His followers, however, were numerous enough to ally forever the science of chemistry with the art of medicine.

By the seventeenth century, a changed attitude, characterized by a reliance on experimentation, had been adopted by astronomers, physicists, physiologists, and philosophers. It was this change in orientation that signaled the emergence of chemistry from alchemy. The English philosopher Francis Bacon (1561–1626) had visions of these new scientific methods endowing human life with new inventions and wealth.

By the middle of the twentieth century, it appeared that science and its application in technology had made the dreams of Bacon and von Hohenheim a reality. Many diseases—such as smallpox, polio, and plague—had been virtually eliminated. Fertilizers, pesticides, and scientific animal breeding had increased and enriched our food supply. Transportation was swift, and communication was nearly instantaneous. Nuclear energy seemed to promise an unlimited source of power for our every need. New materials—plastics, fibers, metals, and ceramics—were developed to improve our clothing and shelter.

The Alchemist, a painting done by the Dutch artist Cornelis Bega around 1660, depicts a laboratory of the seventeenth century.

Indeed, it seemed that, despite its sometimes less than honorable intentions, science could do no wrong. For example, during World War I, when the German armies' supply of ammonia (which they needed to make nitrate explosives) was cut off, a process invented by Fritz Haber (1868–1934) provided them with an alternative supply. Haber's work probably lengthened the war, but it is far more significant for its influence on modern agriculture. Ammonia and nitrates are the stuff of which fertilizers are made, and fertilizers are essential to modern high-yield farming. In fact, most of the ammonia made by the Haber process today goes into fertilizers.

Much of twentieth-century technology has grown out of scientific discoveries, and technological developments are used by scientists as tools for even more discoveries. These developments in science and technology are, to a considerable extent, the cornerstone of what we mean by the "modern" world.

The explosive potential of ammonium nitrate in contact with organic materials was dramatically revealed in an enormous explosion aboard a cargo ship in Texas City, Texas in 1947. Nearly 600 people were killed and several thousand were injured. Explosive mixtures of ammonium nitrate fertilizer and fuel oil were used in the terrorist attacks on the World Trade Center in New York in 1993 and on the federal building in Oklahoma City in 1995.

1.2 Problems in Paradise

If during the first half of the twentieth century science was viewed as humankind's savior, during the latter half it is sometimes viewed as quite the opposite. Those anesthetics that made surgery painless for the patient have caused female anesthesiologists, surgeons, and surgical nurses to suffer a high percentage of miscarriages compared with other health personnel. Fertilizer runoff from farms has polluted streams, and insecticides have had a devastating effect on some wildlife. On occasion, industrial workers making modern products for our use have died from diseases caused by the chemicals they worked with, and chemical waste dumps may threaten the health of us all.

One solution to these problems would be simply to throw out science. But do we really wish to return to surgery without anesthetics? Most of us don't. We need scientists, for it is they who will search for safer anesthetics, for approaches to increased agricultural production compatible with the natural environment, and for analytical techniques that will ensure healthful working conditions for industrial personnel.

The simple fact is that chemistry and its products, both good and bad, are so intimately involved in determining the quality of life that to ignore the subject is to court disaster. It will take an educated, informed society to ensure that science is used for the betterment of the human condition.

1.3 The Way Science Works—Sometimes

Textbooks often define science as a "body of knowledge," and it is frequently taught as a finished work rather than an ever-changing approach to learning. Science is organized into concepts. For example, even though we will often speak of the structure of an atom as if it were readily observable, atomic structure is merely a convenience that successfully describes many observable facts in a metaphorical way. It is not the "body of facts" that characterizes science but the *organization* given to those facts. To be useful, concepts must have predictive value. If the atomic theory is to be useful, it should enable scientists to predict how matter will behave.

The most distinguishing characteristic of science is its use of processes or methods. The making of observations and the cataloging of facts are bare, though necessary, beginnings to these intellectual processes. Scientists must be able to make careful measurements, but they must also be able to grasp the central theme of these observations. They must recognize the variables and be able to note the effect of changing one variable at a time. Scientists must be able to sort out the useful aspects of information and ignore the irrelevancies. Perhaps basic to these intellectual processes is the ability to formulate testable hypotheses. Even an educated guess is of little value to scientists unless an experiment can be devised to test its validity. In fact, if a hypothesis cannot be tested, the question is generally considered to lie outside the realm of science.

Science is not totally different from other disciplines. For example, creativity is central to both science and the humanities. Science does not involve cold logic to the exclusion of other human characteristics. Albert Einstein recognized that there was no *logical* path to some of the laws that he formulated. Even he relied on intuition based on experience and understanding.

It is important to realize that there is no single "scientific method" that, when followed, produces guaranteed results. Scientists observe, gather facts, and make hypotheses, but somewhere along the way they test their hunches and their organization of facts by experimentation. Scientists, like other human beings, use intuition and may generalize from a limited number of facts. Sometimes they are wrong. One of the strengths of science lies in the fact that results of experiments are published in scientific journals. These results are read—and often checked—by other scientists in all parts of the world. To become an accepted part of the "body of knowledge," the results must be reproducible. Scientists also extend each other's work, sometimes to the point that we see a "bandwagon" effect. One breakthrough sometimes results in the unleashing of vast quantities of new data and leads to the development of new concepts. For example, early in the nineteenth century it was thought that certain chemical substances, called *organic* compounds, could be produced only by living tissue, such as someone's liver or the leaf of a plant. These substances were in contrast to other materials, labeled *inorganic,* which could be prepared by a chemist in a laboratory. In 1828, Friedrich Wöhler (1800–1882), a German chemist, succeeded in making an organic compound from an inorganic one in the laboratory. The belief that such a compound could not be prepared in this manner was so strong that Wöhler did the same thing over and over again to assure himself that he had really done the "impossible." When he finally published his work, other chemists quickly repeated it and then proceeded to make hundreds of thousands of organic compounds. That bandwagon is still rolling today, with chemists making hormones, vitamins, and even genes.

Thus, contrary to an often-expressed popular notion, scientific knowledge is not absolute. Science is cumulative, and the "body of knowledge" is dynamic and constantly changing. Old concepts or even old "facts" are discarded as new tools, new questions, and new techniques reveal new data or generate new concepts. To truly understand what science is, one has to observe what the entire worldwide community of scientists has done over a period of several years rather than look over the shoulder of a single scientist for a few days.

1.4 What Is Chemistry? Some Fundamental Concepts

Chemistry is the study of the composition, structure, and properties of matter and of changes that occur in matter. What is matter? It is the stuff of which things are made. One way to think about matter is that particles or objects of matter occupy space, and that no two objects can occupy the same space at the same time. Wood, sand, people, water, and air are all examples of matter. Light is not matter; it is a form of energy. The main concern of chemists is the tiny, submicroscopic building blocks of matter known as atoms and molecules. **Atoms** are the smallest units that we associate with the chemical behavior of matter. **Molecules** are larger units composed of groups of atoms. Ultimately, samples of matter are what they are because of the atoms or molecules that form them.

Mass is a measure of the quantity of matter that an object contains. The greater the mass of a thing, the more difficult it is to change its velocity. You can easily deflect a tennis ball coming toward you at 30 meters per second (m/s), but you would have difficulty stopping a cannonball of the same size moving at the same speed. A cannonball has more mass than a tennis ball of equal size.

The mass of an object does not vary with location. An astronaut has the same mass on the moon as on Earth (Figure 1.1). **Weight,** on the other hand, measures a force. On Earth, it measures the force of attraction between our planet and the mass in question. On the moon, where gravity is one-sixth that on Earth, an astronaut weighs only one-sixth as much as on Earth. Weight varies with gravity; mass does not. In this text, we investigate processes that occur on Earth, and we therefore use mass and weight interchangeably.

◀ **FIGURE 1.1**
Astronaut John W. Young leaps from the lunar surface, where the force of gravity is only one-sixth that on Earth.

Example 1.1

On Mars, gravity is one-third that on Earth.

a. What would be the mass on Mars of a person who has a mass of 55 kilograms (kg) on Earth?

b. What would be the weight on Mars of a person who weighs 126 kg on Earth?

Solution

a. The person's mass would be the same (55 kg) as on Earth; the quantity of matter does not depend on location.

b. The person would weigh only 42 kg; the force of attraction between planet and person is only one-third that on Earth.

Practice Exercise

On Planet X, the force of gravity is 2.4 times that on Earth. (a) What would be the mass of a standard 1.00-kg object on Planet X? (b) A man who weighs 198 lb on Earth would weigh how much on Planet X?

To distinguish between samples of matter, we can compare their *properties* (see Figure 1.2). When a sample exhibits a **physical property,** such as color, odor, or hardness, its atomic or molecular building blocks do not change. In exhibiting a **chemical property,** matter undergoes a *chemical* change—the original substance is replaced by one or more new substances. The burning of sulfur in air results in a chemical change. Sulfur, which is made up of one type of atom, and oxygen from air, which is made up of another type of atom, combine to form sulfur dioxide, which is comprised of molecules that have sulfur and oxygen atoms in the ratio 1:2.

When ice melts, solid water is replaced by liquid water. The process of melting is a *physical* change, and the temperature at which it occurs—the melting point—is a physical property. Although it may be difficult at times to decide whether a change is physical or chemical, the key lies in what happens to the composition of the matter involved. *Composition* refers to the types of atoms that are present and their relative proportions. With few exceptions, a chemical change results in a change in composition, whereas a physical change does not. Table 1.1 lists some examples of physical and chemical properties.

FIGURE 1.2 ▶
Copper (left) and ethyl alcohol (right) are easily distinguished by their properties. Copper is a solid; ethyl alcohol is a liquid. Copper is opaque and has a red-brown color. Ethyl alcohol is transparent and colorless. Also, ethyl alcohol burns whereas copper does not.

Table 1.1 Some Examples of Physical and Chemical Properties

Physical Properties		Chemical Properties
Temperature	Mass	Iron rusts.
Color	Odor	Coal undergoes combustion.
Boiling point	Solubility	Silver tarnishes.
Heat capacity	Hardness	Cement hardens.
Electrical conductance	Density	Dynamite explodes.

Example 1.2

Which of the following events involve chemical changes and which involve physical changes?

a. You trim your nails.
b. Lemon juice converts milk to curds and whey.
c. Water boils.
d. Water is broken down into hydrogen gas and oxygen gas.

Solution

a. Physical change: the composition of the nail is not changed by clipping.
b. Chemical change: the compositions of curds and whey are different from the composition of the milk.
c. Physical change: liquid water and invisible water vapor formed when liquid water boils have the same composition; the water merely changes from a liquid to a gas.
d. Chemical change: new substances, hydrogen and oxygen, are formed.

Practice Exercise

Which of the following events involve chemical changes and which involve physical changes?

a. Liquid alcohol vaporizes from an open container.
b. A piece of lithium metal burns in air to form a white powder called lithium oxide.
c. A dull saw is sharpened with a file.

There are three familiar *states of matter:* solid, liquid, and gas. A **solid** object ordinarily maintains its shape and volume regardless of its location. A **liquid** occupies a definite volume, but assumes the shape of the occupied portion of its container. If you have a 12-oz soft drink, you have 12 oz whether the soft drink is in a can, in a bottle, or, through some slight mishap, on the floor—which demonstrates another property of liquids. Unlike solids, liquids flow readily. A **gas** maintains neither shape nor volume. It expands to fill completely whatever container or space it occupies. Gases can be easily compressed. For example, enough air for many minutes of breathing can be compressed into a steel tank for underwater diving. We shall take up the topic of the states of matter in more detail in Chapters 7 and 8.

1.5 Elements, Compounds, and Mixtures

Figure 1.3 shows several other ways that chemists classify matter. A **substance** has a definite, or fixed, composition that does not vary from one sample to another. The composition of a **mixture** is variable. Pure gold (24-karat gold) consists

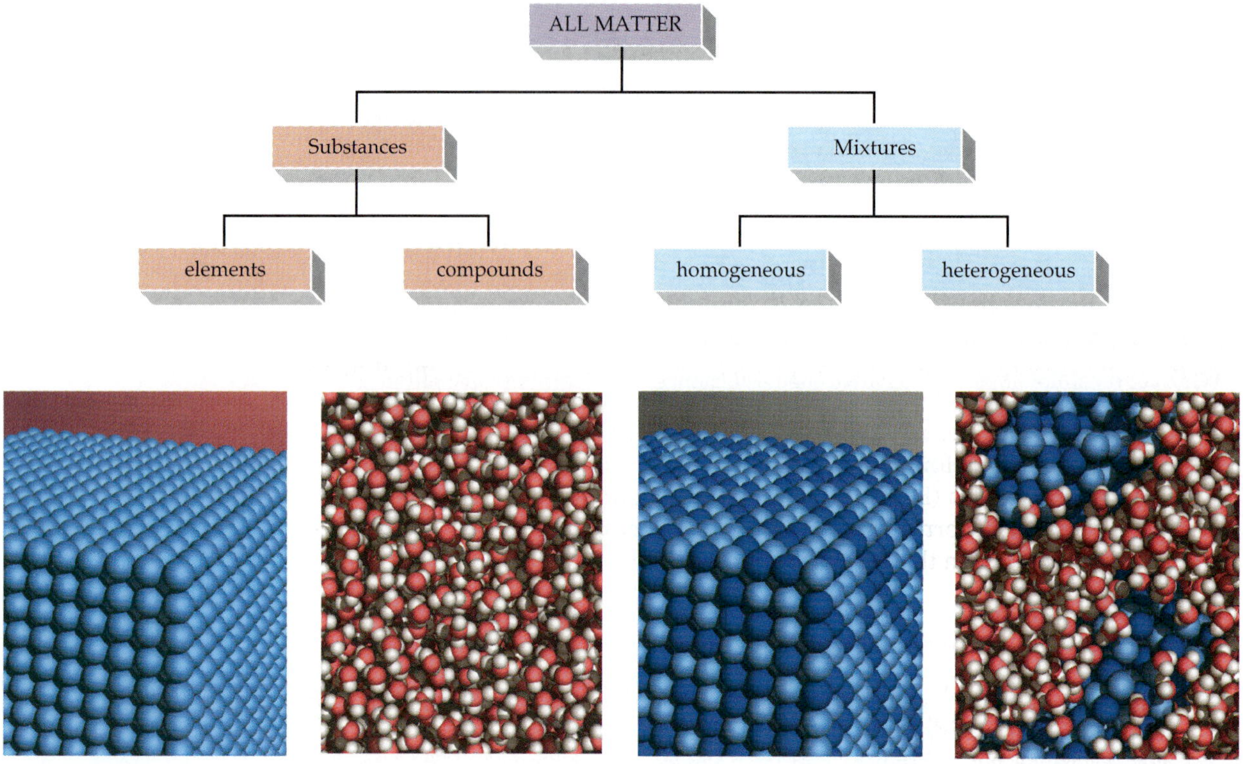

▲ **FIGURE 1.3**

Scheme for classifying matter, as explained in the text. The "molecular-level views" are of gold (an *element*), water (a *compound*), 12-karat gold (a *homogeneous mixture* of silver and gold), and 12-karat gold in water, a *heterogeneous mixture* of particles.

entirely of one type of atom; it is a substance. All samples of water are comprised of molecules consisting of two hydrogen atoms and one oxygen atom; water is a substance. On the other hand, a saline solution—a solution of salt in water—is a mixture. The proportions of salt and water can vary from sample to sample.

Any given saline solution is a **homogeneous mixture:** it has the same composition and properties—the same "saltiness"—throughout the solution. By contrast, a **heterogeneous mixture** varies in composition and/or properties from one part of the mixture to another. A mixture of ice and water is heterogeneous. Although the composition is the same regardless of whether we examine a piece of the ice or a sample of the water, the physical properties of the two are different. In a sand–water mixture, both the composition and properties vary within the mixture.

All *substances* are either elements or compounds. An **element** is composed entirely of a single type of atom. Because bulk matter cannot be made exclusively from any particles smaller than atoms, elements are the simplest of all substances. At present, 112 elements are known, but many are quite rare. Among the familiar elements are oxygen, nitrogen, carbon, sulfur, iron, copper, aluminum, silver, and gold. A **compound** is made up of atoms of two or more elements, with the different kinds of atoms joined in fixed ratios. In contrast to the limited number of elements, the possible number of compounds is essentially limitless. Currently, more than 15 million compounds are known. Water, carbon dioxide, sodium chloride (table salt), sucrose (cane sugar), and iron oxide (rust) are all compounds.

Because elements and compounds are so fundamental to the study of chemistry, we find it useful to refer to them by symbols and formulas. A **chemical symbol** is a one- or two-letter designation derived from the name of an element. Most symbols are based on English names; a few are based on the Latin name of the element or one of its compounds (see Table 1.2). The first letter of a symbol is capitalized and the second is always lowercase. (It makes a difference. For example, Hf is the symbol for the element hafnium, whereas HF represents the compound hydrogen fluoride.) Compounds are designated by combinations of chemical symbols called *formulas*. We'll postpone until Chapter 4 the writing of chemical formulas, which is somewhat more difficult than the writing of symbols.

The names and symbols of all the elements are listed on the inside front cover of this book.

A chemical symbol in a formula stands for one atom of the element. If more than one atom is to be indicated in a formula, a subscripted number is used after the symbol. For example, the formula Cl_2 represents two atoms of chlorine, and the formula CCl_4 stands for one atom of carbon and four atoms of chlorine. Table 1.2 gives actual formulas (Br_2, P_4, S_8, and so on) for some of the elements as they occur in the elemental form. You need not concern yourself with such formulas now; they are included in the table for future reference.

The modern definition of an element is based on its *atomic number.* This definition is given and explained in Chapter 2.

1.6 Energy and Energy Conversion

Energy is the capacity for doing work. **Work** must be done to make something happen that wouldn't happen by itself. Getting out of bed, building a house, and mining coal all require energy. Eating requires energy; that forkful of spaghetti would never make it to your mouth by itself. Energy is the basis for change in the material world. When something moves or breaks or cools or shines or grows or decays, energy is involved.

Energy exists in two basic forms: potential energy and kinetic energy. **Potential energy** is energy of position or arrangement; it is the energy associated with forces of attraction or repulsion between objects. **Kinetic energy** is energy of motion. Figure 1.4 illustrates these two types of energy. Water at the top of a dam has potential energy by virtue of its gravitational attraction to Earth's center. The

◀ **FIGURE 1.4**

Shasta Dam in California, where the potential energy of the water in the reservoir is converted to kinetic energy (and then to electricity) as the water falls 480 feet through pipes to the electric power station at the bottom of the dam.

Table 1.2 Names, Symbols, and Physical Characteristics of Some Common Elements

Name (Latin name)	Symbol	Selected Properties
Aluminum	Al	Light, silvery metal
Argon	Ar	Colorless gas
Arsenic	As	Grayish white solid
Barium	Ba	Silvery white metal
Beryllium	Be	Steel gray, hard, light solid
Boron	B	Black or brown powder; several crystal forms
Bromine	Br	Reddish brown liquid (Br_2)
Calcium	Ca	Silvery white metal
Carbon	C	Soft, black solid (graphite) or hard, brilliant crystal (diamond)
Chlorine	Cl	Greenish yellow gas (Cl_2)
Copper (Cuprum)	Cu	Light reddish brown metal
Fluorine	F	Pale yellow gas (F_2)
Gold (Aurum)	Au	Yellow, malleable metal
Helium	He	Colorless gas
Hydrogen	H	Colorless gas (H_2)
Iodine	I	Lustrous black solid (I_2)
Iron (Ferrum)	Fe	Silvery white, ductile, malleable metal
Lead (Plumbum)	Pb	Bluish white, soft, heavy metal
Lithium	Li	Silvery white, soft, light metal
Magnesium	Mg	Silvery white, ductile, light metal
Mercury (Hydrargyrum)	Hg	Silvery white, liquid, heavy metal
Neon	Ne	Colorless gas
Nickel	Ni	Silvery white, ductile, malleable metal
Nitrogen	N	Colorless gas (N_2)
Oxygen	O	Colorless gas (O_2)
Phosphorus	P	Yellowish white, waxy solid or red powder (P_4)
Plutonium	Pu	Silvery white, radioactive metal
Potassium (Kalium)	K	Silvery white, soft metal
Silicon	Si	Lustrous gray solid
Silver (Argentum)	Ag	Silvery white metal
Sodium (Natrium)	Na	Silvery white, soft metal
Sulfur	S	Yellow solid (S_8)
Tin (Stannum)	Sn	Silvery white, soft metal
Uranium	U	Silvery, radioactive metal
Zinc	Zn	Bluish white metal

water has the capacity to do work, but as long as it remains behind the dam it does none. When the water is allowed to flow through a pipe to a lower level, some of its potential energy is converted to kinetic energy. Water rushing through the pipe can be made to turn the blades of a turbine (a water wheel), which in turn can rotate a coil of wire in an electrical generator, producing electricity. The net result is that some of the potential energy originally stored in the water is converted to electrical work.

Kinetic energy depends on mass and velocity. The bigger an object is and the faster it is moving, the more kinetic energy it has and the more work it can do. Mathematically, kinetic energy equals one-half the mass times the square of the velocity.

$$KE = \frac{1}{2}mv^2$$

Energy is often classified in other ways. These classifications are based on some characteristic of the energy being considered, such as its source. It is easier to discuss some of these types of energy by indicating their significance in the overall pattern of energy flow on Earth.

The source of nearly all our energy is the sun. *Solar energy* radiates through space. A small portion of the sun's radiant energy reaches Earth, where a part of it is converted to *heat energy*. This heat causes water to evaporate and then rise to form clouds. The water in the clouds has potential energy. As it falls as rain and then flows in rivers, the potential energy is converted to kinetic energy. When the river water is held behind a dam, it again has potential energy that can be used to power a mill or to generate electricity (see again Figure 1.4).

Some of the solar energy striking the Earth is absorbed by green plants, which use a complicated chemical process called photosynthesis to convert solar energy into *chemical energy*. The chemical energy stored by plants—now and in ages past—is used by animals and humans for food and fuel. Nearly all of the vast quantities of energy being used in our modern civilization came originally from the sun by way of green plants. Plants of the current age are harvested by foresters and farmers. Those of ancient ages are reaped as fossil fuels—coal, petroleum, and natural gas.

Another form of energy is *nuclear energy*. This type of energy is not derived from the sun. Rather, it was stored in the Earth's crust when the solar system was formed some 4 or 5 billion years ago. We recover it for use (or, as many people believe, misuse) when we mine uranium and build nuclear reactors or atomic bombs.

Example 1.3

What kind of energy—potential or kinetic—does each of the following possess?

a. a thrown softball
b. a softball resting at the edge of a table
c. gasoline

Solution

a. A thrown softball has kinetic energy; it is moving through the air.
b. A softball at the edge of a table has potential energy by virtue of its position; it could release energy by falling.
c. Gasoline contains chemical energy, a form of potential energy; the energy is released when the gasoline burns.

Example 1.4

What energy changes occur during each of the following events?

a. Fuel oil is burned.
b. A softball falls off the edge of a table.
c. Sunlight falls on your back.

Solution

a. Chemical energy is converted to heat energy.
b. Potential energy is converted to kinetic energy.
c. Radiant energy is converted to heat energy.

1.7 Electric Forces

To deal with energy transformations, chemistry often borrows fundamental concepts from its neighboring discipline, physics. One such concept is force. A *force* is a push or a pull that sets an object in motion, or stops a moving object, or holds an object in place. *Gravity* is a force. Objects—including us—are held to the surface of the Earth by gravity, the attraction of Earth's mass for other masses. The weight of an object is the force of gravity that exists between the object and the Earth.

Electric forces are extremely important in chemistry. Some particles of matter are neutral, and some bear an electric charge, either positive (+) or negative (−). No one can explain exactly what an electric charge is. We simply accept the fact that a particle with a "charge" can exert a force—that is, can push or pull another particle that also has a "charge." The particles do not have to be touching to attract or repel one another. For this reason, we say that charged particles have *force fields* around them. Even at a distance they attract and repel one another, although these forces get weaker as the particles get farther apart. Particles with *like* charges (both positive or both negative) repel one another. Those with *unlike* charges (one positive and one negative) attract one another (Figure 1.5).

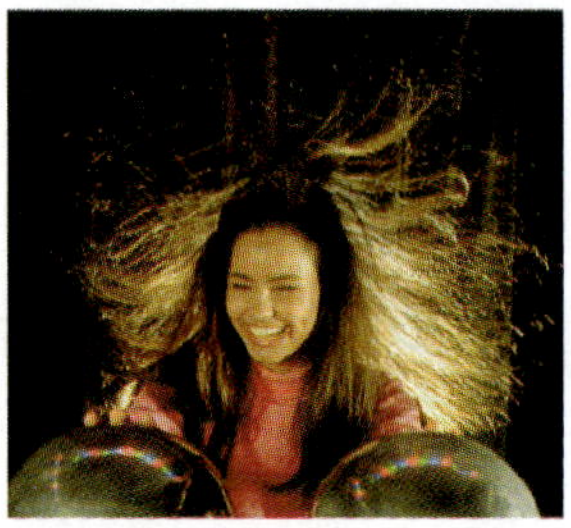

An electrostatic generator causes an electric charge on this woman's hair. The hairs all have like charges and repel one another.

This phenomenon of charged substances is not unfamiliar to you. Anyone who has pulled clothes from an automatic dryer on a cold winter day has probably seen what commercials like to call "static cling"—pieces of clothing sticking to one another. The "cling" is a result of the attraction of *unlike* charges. If, on the other hand, you brush your hair vigorously on a cold day, your hair may become "unmanageable" (another term used often in advertisements). Instead of lying flat against your head, the hair sticks out in all directions, each strand seemingly trying to get away from all the other strands. That is exactly what is happening; the strands have *like* charges and thus repel one another.

1.8 Measurement: The Modern Metric System

Accurate measurement of such quantities as mass, volume, time, and temperature is essential to the compilation of dependable scientific "facts." Such facts may be

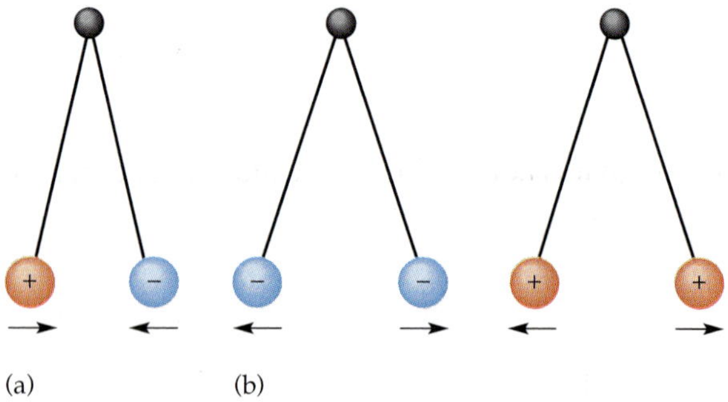

FIGURE 1.5 ▶
Diagram illustrating (a) the mutual attraction of particles with unlike charges and (b) the mutual repulsion of particles with like charges.

used by a chemist interested in basic research, but similar information is of critical importance in every science-related field. Certainly we are all aware that measurements of both temperature and blood pressure are routinely made in medicine. It is also true that modern medical diagnosis depends on a wide variety of other measurements, including careful chemical analyses of blood and urine.

Use of the metric system in the United States has become more common in recent years. For example, the contents of most bottled beverages are now given in metric units. Metric measurements are also common in sporting activities, especially track and field.

The system of measurement used by most scientists is an updated metric plan called the **International System of Units,** or **SI** (from the French *Système International*). Indeed, SI has been adopted worldwide for everyday use. Even the United States is committed to change to SI, but conversion is voluntary and has proceeded rather slowly so far. To date, the United States, Liberia, and Burma are the only countries that still use the old English system (pounds, inches, and degrees Fahrenheit).

The beauty of SI is that it is based on the decimal system. This makes conversion from one unit to another rather simple. SI has only a few basic units. For example, the basic unit of length is the **meter** (m), a distance only slightly greater than a yard. The SI unit of mass is the **kilogram** (kg), a quantity slightly greater than two pounds. All other units for length, mass, and volume can be derived from these basic units. For example, area can be measured in square meters (m^2) and volume in cubic meters (m^3)

A disadvantage of the basic SI units is that they are often of awkward magnitude. We seldom work with kilogram quantities in the laboratory. More convenient units can be derived by the use (or, in the case of the kilogram, deletion) of prefixes (Table 1.3). For example, in the laboratory we often work with grams (g) or milligrams (mg) of materials.

$$1 \text{ kg} = 1000 \text{ g} = 10^3 \text{ g}$$

$$1 \text{ mg} = 0.001 \text{ g} = 10^{-3} \text{ g}$$

Chemists can now detect masses in the microgram (μg), the nanogram (ng), and even the picogram (pg) range. You should learn the more common prefixes (those in red and blue in Table 1.3) right away.

To measure lengths much larger than the meter, such as distances along a highway, we often use the kilometer (km).

$$1 \text{ km} = 1000 \text{ m} = 10^3 \text{ m}$$

In the laboratory, it is often most convenient to use lengths smaller than the meter, such as the centimeter (cm) and the millimeter (mm).

$$1 \text{ cm} = 0.01 \text{ m} = 10^{-2} \text{ m}$$

$$1 \text{ mm} = 0.001 \text{ m} = 10^{-3} \text{ m}$$

The SI unit of volume is the cubic meter (m^3) but two units that are more commonly used in the laboratory are the cubic centimeter (cm^3) and the cubic decimeter (dm^3).

$$1 \text{ cm}^3 = (0.01)^3 \text{ m}^3 = 10^{-6} \text{ m}^3$$

$$1 \text{ dm}^3 = (0.1)^3 \text{ m}^3 = 10^{-3} \text{ m}^3$$

Although it is not an SI unit, the old metric unit *liter* is also commonly used. A **liter (L),** which is slightly larger than a quart, is the same as one cubic decimeter, or 1000 cubic centimeters.

The units liter and milliliter are commonly used for volumes of liquids and gases; the units cubic centimeter and cubic decimeter are more often used for solids.

$$1 \text{ L} = 1 \text{ dm}^3 = 1000 \text{ cm}^3$$

The milliliter (mL) is the same as a cubic centimeter: 1 mL = 1 cm^3.

Table 1.3 Approved Numerical Prefixes*

Exponential Expression	Decimal Equivalent	Prefix	Pronounced	Symbol
10^{12}	1,000,000,000,000	tera-	TER-uh	T
10^{9}	1,000,000,000	giga-	GIG-uh	G
10^{6}	1,000,000	mega-	MEG-uh	M
10^{3}	1,000	kilo-	KIL-oh	k
10^{2}	100	hecto-	HEK-toe	h
10	10	deka-	DEK-uh	da
10^{-1}	0.1	deci-	DES-ee	d
10^{-2}	0.01	centi-	SEN-tee	c
10^{-3}	0.001	milli-	MIL-ee	m
10^{-6}	0.000,001	micro-	MY-kro	μ
10^{-9}	0.000,000,001	nano-	NAN-oh	n
10^{-12}	0.000,000,000,001	pico-	PEE-koh	p
10^{-15}	0.000,000,000,000,001	femto-	FEM-toe	f
10^{-18}	0.000,000,000,000,000,001	atto-	AT-toe	a
10^{-21}	0.000,000,000,000,000,000,001	zepto-	ZEP-toe	z

*The most commonly used prefixes are shown in color.

Conversions within the SI system are much easier than those using the more familiar units of pounds, feet, and pints. This is best shown by examples.

Example 1.5

Convert 0.742 kg to grams.

Solution

$$0.742\ \cancel{\text{kg}} \times \frac{1000\ \text{g}}{1\ \cancel{\text{kg}}} = 742\ \text{g}$$

Note: If you are not familiar with the use of conversion factors or would like to review the mathematics of conversions, see Special Topic A immediately following this chapter.

Practice Exercise

Convert 16.3 mg to grams.

Example 1.6

Convert 0.742 lb to ounces.

Solution

$$0.742\ \cancel{\text{lb}} \times \frac{16\ \text{oz}}{1\ \cancel{\text{lb}}} = 11.9\ \text{oz}$$

Practice Exercise

Convert 24.5 ounces to pounds.

Example 1.7

Convert 1247 mm to meters.

Solution

$$1247\ \cancel{\text{mm}} \times \frac{1\ \text{m}}{1000\ \cancel{\text{mm}}} = 1.247\ \text{m}$$

Practice Exercise

Convert 2.53 meters to centimeters.

Example 1.8

Convert 1247 in. to yards.

Solution

$$1247\ \cancel{\text{in.}} \times \frac{1\ \text{yd}}{36\ \cancel{\text{in.}}} = 34.64\ \text{yd}$$

Practice Exercise

Convert 6.75 ft to inches.

In conversions involving pounds, ounces, inches, and yards, you multiply and divide by numbers such as 16 or 36. In metric conversions, you multiply by 10 or 100 or 1000, and so on; you need only shift the decimal point.

In the United States, it is sometimes necessary to convert a measurement from one system to the other. If you need to learn how to do this, such conversions are discussed in Special Topic A following this chapter. It is useful to have some idea of the relative sizes of comparable units. Some comparisons are shown in Figure 1.6. Other comparisons easily remembered are that the U.S. ten-cent coin, the

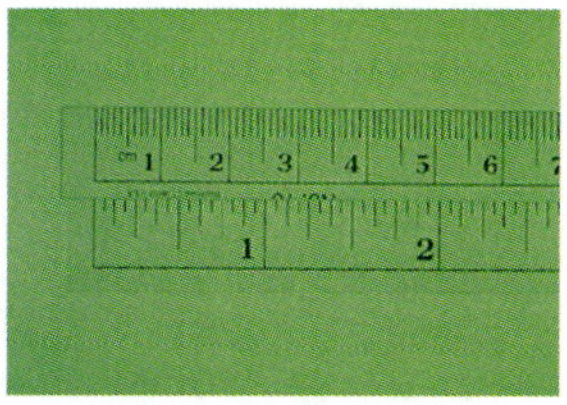

Customary and metric measures of length.
1 in. = 2.54 cm

◀ **FIGURE 1.6**

A comparison of metric and customary units of measure. The beaker at the left contains 1 kg of candy and the one next to it contains 1 lb (1 kg = 2.2046 lb). The green ribbon is 1 cm wide around a stick 1 m long (1 m = 1.0936 yd and 1 cm = 0.3937 in.). The flasks each hold 1 L when filled to the mark. The flask on the right and the bottle behind it each contain 1 L of orange juice. The flask on the left and the carton behind it each contain 1 qt of milk (1 L = 1.057 qt).

dime, is about 1 mm thick, and the U.S. five-cent piece, the nickel, has a mass of about 5 g.

$$1.0 \text{ m} = 39 \text{ in. or } 1.1 \text{ yd}$$

$$1.0 \text{ L} = 1.1 \text{ qt}$$

$$1.0 \text{ kg} = 2.2 \text{ lb}$$

1.9 Measuring Energy: Temperature and Heat

The SI unit for temperature is the **kelvin** (K), but for much of their work scientists use the **Celsius scale.** On this scale, the freezing point of water is 0 °C and its boiling point is 100 °C. The scale between these two reference points is divided into 100 equal divisions, each a Celsius degree. (The Celsius scale used to be called the centigrade scale because of this 100-degree interval.)

The Kelvin scale is called an **absolute scale** because its zero point is the coldest temperature possible, or absolute zero. (This fact was determined by theoretical considerations and has been confirmed by experiment.) The zero point on the Kelvin scale, 0 K, is equal to −273.15 °C, often rounded to −273 °C. A kelvin is the same size as a degree Celsius, so the freezing point of water on the Kelvin scale is 273 K. The Kelvin scale has no negative temperatures, and we don't use a degree sign with the K. To convert from degrees Celsius to the Kelvin scale, simply add 273 to the Celsius temperature.

$$\text{K} = {}^\circ\text{C} + 273$$

Example 1.9

The boiling point of water is 100 °C. What is the boiling point of water on the Kelvin scale?

Solution

$$100\ {}^\circ\text{C} + 273 = 373 \text{ K}$$

Practice Exercise

How would normal body temperature (37 °C) be expressed in kelvins?

Some books use the following equations to interconvert Fahrenheit and Celsius.

$${}^\circ\text{F} = \frac{9}{5}({}^\circ\text{C}) + 32$$

$${}^\circ\text{C} = ({}^\circ\text{F} - 32)\frac{5}{9}$$

In the United States, weather reports and cooking recipes still use the **Fahrenheit scale.** This scale defines the freezing temperature of water as 32 °F and the boiling point of water as 212 °F. Figure 1.7 compares the Fahrenheit, Celsius, and Kelvin scales. Some additional temperature equivalents are found in Appendix I.

The difference between the freezing point of water and the boiling point of water is 100° on the Celsius scale and 180° on the Fahrenheit scale. It follows that a Celsius degree is 1.8 times as large as a Fahrenheit degree. Remember also that the Fahrenheit freezing point of water is 32 degrees above zero. Thus, Fahrenheit temperatures (°F) and Celsius temperatures (°C) can be converted to each other by using the following equations.

$${}^\circ\text{F} = 1.8({}^\circ\text{C}) + 32$$

$${}^\circ\text{C} = ({}^\circ\text{F} - 32)0.555$$

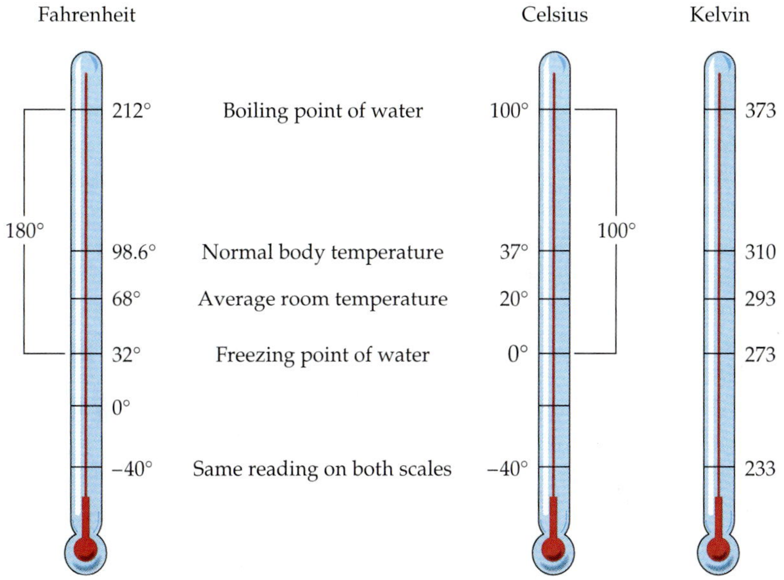

◀ **FIGURE 1.7**
A comparison of the Fahrenheit, Celsius, and Kelvin temperature scales.

Example 1.10

Before the concern for energy conservation, many people kept their homes at a temperature of 72 °F. What is that on the Celsius scale?

Solution

$$\begin{aligned} °\text{C} &= (°\text{F} - 32)0.555 \\ &= (72 - 32)0.555 \\ &= 22 \end{aligned}$$

Practice Exercise

A frozen pizza is to be baked at 425 °F. What is this temperature on the Celsius scale?

Example 1.11

Convert the temperature −20 °C to the Fahrenheit scale.

Solution

$$\begin{aligned} °\text{F} &= 1.8(°\text{C}) + 32 \\ &= 1.8(-20) + 32 \\ &= -4 \end{aligned}$$

Practice Exercise

Ethyl alcohol boils at 78.5 °C. What is that on the Fahrenheit scale?

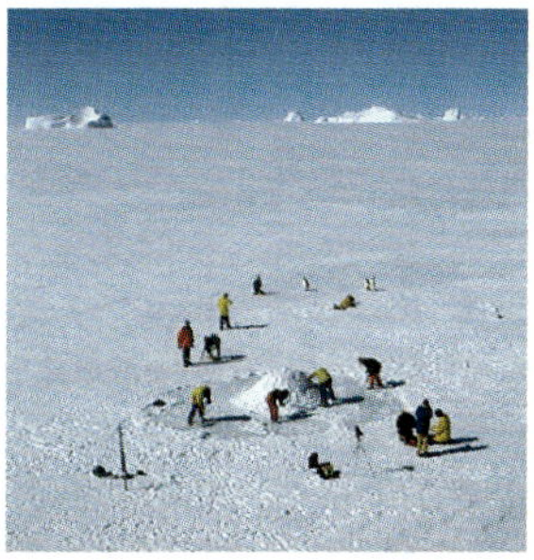

The lowest temperature ever recorded on Earth was −128.6 °F in Antarctica. What is that temperature in degrees Celsius and in kelvins?

Scientists often need to measure amounts of heat energy. You should not confuse heat with temperature. *Heat* is a measure of quantity, that is, of how much energy a sample contains. *Temperature* is a measure of intensity, that is, of how energetic each particle of the sample is. A glass of water at 70 °C contains less heat than a bathtub of water at 60 °C. The particles of water in the glass are more energetic, on the average, than those in the tub, but there is far more water in the tub, and its total heat content is greater.

The SI unit of heat is the *joule* (J), but we will use the more familiar *calorie.*

$$1 \text{ cal} = 4.184 \text{ J}$$

$$1000 \text{ cal} = 1 \text{ kcal} = 4184 \text{ J}$$

A **calorie** (cal) is the amount of heat required to raise the temperature of 1 g of water 1 °C. There is a more precise definition, but this version will do for our purposes.

The specific heat of a substance actually varies somewhat with temperature, and some of the calculations we do here yield only approximate answers. The most important specific heat value that we use in these calculations is that of water. At about room temperature the specific heat of water is 1.00 cal/(g · °C), and over the temperature range from 0 °C to 100 °C it remains within 1% of this value.

Some substances have greater heat capacities than others. The **specific heat** of a substance is the amount of heat required to raise the temperature of 1 g of the substance by 1 °C. The definition of the calorie indicates that the specific heat of water is 1 cal/(g · °C). Water is a poor conductor of heat. Metals are good heat conductors. A sample of water must absorb much more heat to raise its temperature than must a metal sample of similar mass. The metal sample may become red hot after absorbing a quantity of heat that will make the water sample only lukewarm.

We can use the following equation, in which ΔT is the change in temperature, to calculate the quantity of heat absorbed or released by a system.

$$\text{Heat absorbed or released} = \text{mass} \times \text{specific heat} \times \Delta T$$

Example 1.12

How much heat, in calories and kilocalories, does it take to raise the temperature of 225 g of water from 25.0 °C to 100.0 °C?

Solution

The specific heat of water is $\frac{1.00 \text{ cal}}{(\text{g})(°\text{C})}$. The temperature change is (100.0 − 25.0) °C = 75.0 °C, and the quantity of water to be heated is 225 g.

$$\begin{aligned} \text{Heat absorbed or released} &= \text{mass} \times \text{specific heat} \times \Delta T \\ &= 225 \text{ g } H_2O \times \frac{1.00 \text{ cal}}{(\text{g } H_2O)(°\text{C})} \times 75.0\ °\text{C} \\ &= 16{,}900 \text{ cal} \\ &= 16.9 \text{ kcal} \end{aligned}$$

Practice Exercise

How much heat, in calories and kilocalories, does it take to raise the temperature of 814 g of water from 18.0 °C to 100.0 °C?

Example 1.13

What mass of water, in kilograms, can be heated from 5.5 °C to 55.0 °C by 217,000 cal of heat?

Solution

$$\text{mass } H_2O = \frac{\text{heat absorbed}}{\text{specific heat} \times \Delta T}$$

$$= \frac{217{,}000 \text{ cal}}{1.00 \text{ cal/(g } H_2O) \text{ (°C)} \times 49.5 \text{ °C}} = 4380 \text{ g} = 4.38 \text{ kg}$$

Practice Exercise

What temperature change occurs when a 475-g sample of water absorbs 2.44 kcal of heat?

For measuring the energy content of foods, the *large* **Calorie** (note the capital C, or *kilocalorie* (kcal), is sometimes used. A dieter might be aware that a banana split contains 1500 Calories. If the same dieter realized that this means 1,500,000 calories, giving up the banana split might be easier.

1.10 Density

An important property of matter, particularly in scientific work, is *density*. When one speaks of lead as "heavy" or aluminum as "light," one is referring to the density of the metal. **Density** is defined as the quantity of mass per unit of volume:

$$d = \frac{m}{V}$$

Rearrangement of this equation gives

$$m = d \times V \quad \text{and} \quad V = \frac{m}{d}$$

These equations are useful for calculations. Densities are usually reported in grams per milliliter (g/mL) or grams per cubic centimeter (g/cm^3).

The brass weight and the pillow have the same mass. Which has the greater density?

Example 1.14

What is the density of iron if 156 g of iron occupies a volume of 20.0 cm^3?

Solution

$$d = \frac{m}{V} = \frac{156 \text{ g}}{20.0 \text{ cm}^3} = 7.80 \text{ g/cm}^3$$

Practice Exercise

What is the density of gold if a cube that measures 2.00 cm on an edge weighs 154.4 g?

Example 1.15

What is the mass of a liter of gasoline if its density is 0.660 g/mL?

Solution

$$m = d \times V = \frac{0.660 \text{ g}}{1 \cancel{\text{mL}}} \times 1 \cancel{\text{L}} \times \frac{1000 \cancel{\text{mL}}}{1 \cancel{\text{L}}} = 660 \text{ g}$$

Practice Exercise

What is the mass of 50.0 mL of grenadine, which has a density of 1.32 g/mL?

Mass does not vary with temperature, but volume does, and so does density. At around room temperature we can generally assume the density of water to be 1.00 g/mL.

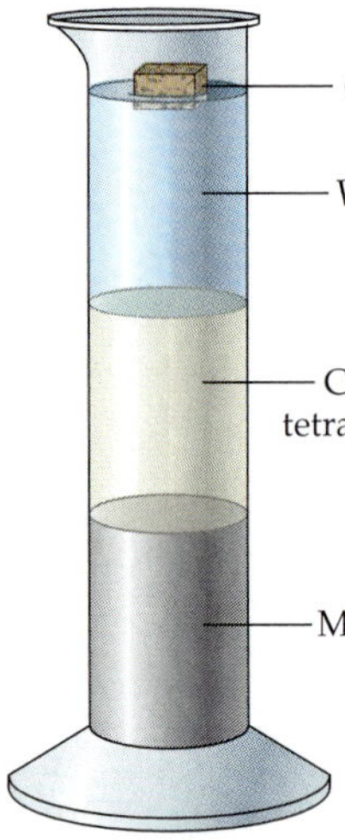

▲ **FIGURE 1.8**
Liquid carbon tetrachloride (d = 1.54 g/mL) floats on liquid mercury (d = 13.6 g/cm^3). Water (d = 1.00 g/mL), which does not mix with carbon tetrachloride, floats on it. Finally, a cork (d = 0.22 g/cm^3) floats on the water.

Example 1.16

What volume is occupied by 461 g of mercury? (The density of mercury is 13.6 g/mL.)

Solution

$$V = \frac{m}{d} = 461\,\cancel{g} \times \frac{1\text{ mL}}{13.6\,\cancel{g}} = 33.9\text{ mL}$$

Practice Exercise

What volume would be occupied by a 10.0-g piece of aluminum? (The density of aluminum is 2.70 g/mL.)

The density of water is 1 g/mL, or 1 g/cm^3 (remember that 1 mL = 1 cm^3). This nice round number for the density of water is not an accident. The metric system was originally set up in such a way as to ensure that this was the case.

Ethylene glycol (the principal ingredient in many antifreeze solutions), water, and ethanol are all colorless liquids. They can be distinguished by their densities. At 20 °C, the density of ethylene glycol is 1.114 g/mL, that of water is 0.998 g/mL, and that of ethanol is 0.789 g/mL. If you combine two liquids that are *immiscible* (that do not mix to form solutions), the liquid of lower density will float on top of the liquid of higher density (Figure 1.8).

A term related to density is *specific gravity*. **Specific gravity** is the ratio of the mass of any substance to the mass of an equal volume of water. The specific gravity of water itself, therefore, is 1. Mercury (the silvery liquid in thermometers) has a specific gravity of 13.6. That means it has a density 13.6 times as great as water's. The specific gravity of methyl alcohol is 0.791; thus, methyl alcohol is less dense than water. Because it is the ratio of two values, specific gravity is a number without units, whereas density is reported in units of mass per volume. And because the density of water is 1.00 g/mL, the specific gravity of a substance is numerically the same as its density.

$$\text{Specific gravity of a substance} = \frac{\text{density of the substance}}{\text{density of water}}$$

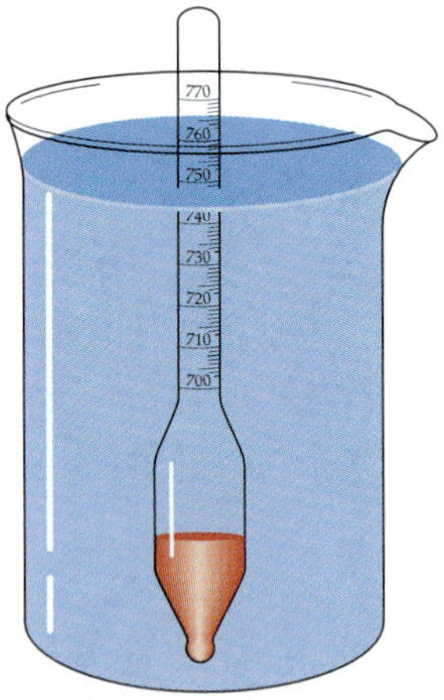

▲ **FIGURE 1.9**
The hydrometer shown here measures specific gravities over a range of 0.700 to 0.770.

Example 1.17

The density of chloroform is 1.5 g/mL. What is its specific gravity?

Solution

$$\frac{1.5\,\cancel{g/mL}}{1.0\,\cancel{g/mL}} = 1.5$$

Practice Exercise

The specific gravity of a sample of battery acid is 1.39. What is its density?

The specific gravity of a liquid is frequently measured by a device called a *hydrometer*. The hydrometer is placed in the liquid. How far it dips down into the liquid is determined by the density of the liquid (Figure 1.9). The stem of the hydrometer is calibrated in such a way that the specific gravity can be read directly at the surface of the liquid. Hydrometers can be used to measure the strength of the "battery acid" in your car, the percentage of antifreeze in the radiator, sugar content in maple syrup, dissolved solids in urine, alcohol content of wine, and many other properties of solutions that are related to specific gravity.

For those who are interested and those whose work might require it, measurement is discussed further in Appendix I, and a variety of conversion tables are collected there for convenient reference.

Fat Floats: Body Density and Fitness

How fit are you physically? If you are an athlete, or if you simply want to get in better physical condition through exercise, your first step might well be a fitness evaluation. One part of the evaluation is the determination of percent body fat, and this is best done by measuring body density. The density of human fat is 0.903 g/mL, whereas that of water is 1.00 g/mL. If placed in water, fat floats. The higher the proportion of body fat in a person, the more buoyant the person is in water. Body density is determined from body mass (weight) and volume. Body volume is determined by weighing a person when the person is submerged in water (Figure 1.10). Once body density is determined, the percent body fat is estimated from a graph or from a table of compiled data. The average percent body fat is about 16% for an adult male and 25% for an adult female. Male athletes in superb condition will have less than 7% body fat, and females will have less than 12% body fat.

▲ **FIGURE 1.10**

It is easy to determine a person's mass, but what about a person's volume? When submerged in water, a body displaces its own volume of water. The difference in a person's weight in air and when submerged in water corresponds to the mass of displaced water. This mass, divided by the density of water, gives the volume of displaced water and hence—with an appropriate correction for the volume of air in the lungs—the person's volume.

Summary

The body of knowledge we call **science** is an ever-changing collection of information that has been gathered over centuries and is still being gathered today. Scientists observe events and catalog data. They then use their observations and data, along with intellectual creativity and intuition, to formulate **testable hypotheses.**

If a hypothesis holds up in numerous experiments, it becomes a **theory** but is still subject to being discarded as soon as one experiment proves it wrong. For example: Prior to 1828, a prevalent theory held that no materials made by living organisms could be manufactured in a test tube. In that year, however, one chemist in one laboratory manufactured one compound known to be made by living organisms, and so this theory had to be thrown out.

Matter is anything that has mass and occupies space, and **chemistry** is the study of all the changes that take place in matter.

Mass is a measure of how much matter an object contains, whereas **weight** is a measure of how much force Earth's gravity exerts on the object. Mass and weight are *related* terms—the more mass in an object, the stronger the pull of gravity on it—but they are not the same thing. Mass is a constant for any given object, but the object's weight changes as the distance between the object and Earth's center changes. An apple has the same mass whether it is lying on the ground or riding in a space capsule 100 miles above Earth. However, its weight measured as it lies on the ground is different from its weight measured in the space capsule.

Every type of matter has its own characteristic **chemical properties,** which describe how that particular matter reacts with all other types of matter, and its own characteristic **physical properties,** which describe its physical appearance and behavior. Example: How a substance reacts (or doesn't react) with oxygen is a chemical property; melting point and color are physical properties. A *chemical change* is one that involves a change in chemical composition; iron metal reacting with oxygen from the air to form the iron oxide we call rust is a chemical change. A *physical change* is one that does not entail any change in chemical composition; solid iron melting in a blast furnace is a physical change.

Matter exists in three states: **solid, liquid,** or **gas.**

A pure **substance** is one that has a definite, unchanging chemical composition; water and sodium chloride (table salt) are examples. A **mixture** is a collection of pure substances that can have various compositions; a sodium chloride–water mixture, for instance, can be 10% sodium chloride, 90% water or 50% sodium chloride, 50% water or any other combination that adds to 100%. Mixtures can be either **homogeneous,** which means the various parts of the mixture are evenly distributed throughout, or **heterogeneous,** in which case the components are not evenly distributed. For example, salt dissolves in water to form—after thorough mixing—a homogeneous solution. A water–sand mixture is heterogeneous; no matter how much it is shaken, the sand remains visible as a separate component.

An **element** is a fundamental substance that cannot be broken down into simpler substances and retain the chemical and physical properties of the original substance. A **compound** is any substance in which two or more elements are chemically combined in fixed proportions. All matter is made up of elements, either one element alone (iron, oxygen) or two or more elements combined. The compound "water" is made of the elements hydrogen and oxygen; the compound "table sugar" is made up of the elements carbon, hydrogen, and oxygen.

Potential energy is the energy an object has because of its position or its configuration; **kinetic energy** is the energy an object has whenever it is moving. Examples: A bicyclist poised at the top of a long downgrade has potential energy because of her position; once she begins coasting down the hill, she has kinetic energy because she is moving. The squeezed-down spring in a jack-in-the-box with the lid closed has potential energy because of its compressed configuration; once the lid is removed, the spring has kinetic energy as it springs back to its relaxed position.

A force is a push or pull on any object. The force of **gravity** is the pull exerted by Earth on all objects at or near its surface. Electric forces, which are forces that two electrically charged bodies exert on each other, can be positive or negative. Two like electric forces ($+/+$ or $-/-$) repel each other; two unlike electric forces ($+/-$) attract each other.

The units used to measure quantities in scientific work are from the **International System of Measurement,** abbreviated **SI.** In this system, the **meter** is the basic unit of length and the **kilogram** is the basic unit of mass. The **liter** is a unit of volume, and the **kelvin** is the SI unit of temperature. SI units are used even in nonscientific work in most parts of the world. In the United States and a few other countries, nonscientific measurements are made in non-SI units, such as feet instead of meters and gallons instead of liters.

There are three temperature scales in use today: the *Kelvin scale* (also called the **absolute scale**), the **Celsius scale,** and the **Fahrenheit scale.** The relationships between the Celsius and Kelvin scales are

$$\text{K} = {}^\circ\text{C} + 273$$

$$^\circ\text{C} = \text{K} - 273$$

The relationships between the Celsius and Fahrenheit scales are

$$^\circ\text{F} = 1.80(^\circ\text{C}) + 32$$

$$^\circ\text{C} = 0.555(^\circ\text{F} - 32)$$

Heat and temperature are not the same thing. **Temperature** is a measure of how much internal energy, or thermal energy, a body has; the greater the amount of thermal energy in a body, the higher its temperature. **Heat** is a measure of how much thermal energy *moves from one body to another.* When there is no transfer of thermal energy going on, we cannot talk about heat. Example: A cup of tea at 100 °C has more internal energy than a glass of water at 20 °C; in other words, the temperature of the tea is higher than the temperature of the water. Left sitting on the counter, the tea cools as some of its thermal energy moves to the air; in other words, heat is transferred from tea to air.

The SI unit of heat is the *joule*, and one non-SI heat unit is the **calorie,** defined as the amount of heat needed to raise the temperature of 1 g of water by 1 °C. These two units are related to each other by the equations

$$1 \text{ cal} = 4.184 \text{ J}$$

$$1000 \text{ cal} = 1 \text{ kcal} = 4184 \text{ J}$$

The **specific heat** of a substance is the heat required to raise the temperature of 1 g of the substance by 1 °C. When the same amount of heat is added to a low-specific-heat substance and a high-specific-heat substance, the temperature of the former increases more than the temperature of the latter.

Density is mass per unit volume:

$$d = m/V$$

By definition, the units of density are a mass unit divided by a volume unit, usually g/cm^3 or g/mL.

The **specific gravity** of any substance is the ratio of the mass of a given volume of that substance to the mass of the same volume of water. Because it is a ratio, specific gravity has no units (they cancel). The specific gravity of mercury, for example, is calculated as follows, where we arbitrarily choose 1 L as our volume:

1 L mercury has mass 13.6 kg
1 L water has mass 1 kg

Specific gravity of mercury is therefore

$$\frac{13.6 \text{ kg}}{1 \text{ kg}} = 13.6$$

Key Terms

absolute scale (1.9)
atom (1.4)
calorie (cal) (1.9)
Calorie (Cal) (1.9)
Celsius scale (1.9)
chemical property (1.4)
chemical symbol (1.5)
chemistry (1.4)
compound (1.5)
density *(d)* (1.10)
element (1.5)
energy (1.6)
Fahrenheit scale (1.9)
gas (1.4)
heterogeneous mixture (1.5)
homogeneous mixture (1.5)
International System of Units (SI) (1.8)
kelvin (K) (1.9)
kilogram (kg) (1.8)
kinetic energy (1.6)
liquid (1.4)
liter (L) (1.8)
mass (1.4)
meter (m) (1.8)
mixture (1.5)
molecule (1.4)
physical property (1.4)
potential energy (1.6)
solid (1.4)
specific gravity (1.10)
specific heat (1.9)
substance (1.5)
weight (1.4)
work (1.6)

Review Questions

1. Define each of the following terms.
  a. chemistry **b.** matter
  c. mass **d.** weight
  e. energy **f.** calorie
  g. density **h.** specific gravity

2. What is a hypothesis? How are hypotheses tested?

3. Which of the following are examples of matter?
  a. air **b.** anger
  c. the human body **d.** gasoline
  e. prayer **f.** an idea

4. Explain the difference between mass and weight.

5. How do physical and chemical properties differ?

6. How do physical and chemical changes differ?

7. How do gases, liquids, and solids differ in their properties?

8. What are the names and symbols of the SI base units for length, mass, and temperature?

9. Which is the most compressible form of water: ice, liquid, or steam?

10. What is the difference between a substance and a mixture?

11. All samples of glucose, a simple sugar, consist of 8 parts (by mass) oxygen, 6 parts carbon, and 1 part hydrogen. Is glucose a substance or a mixture?

12. What is an element?

13. What is a chemical compound?

14. What is the difference between kinetic energy and potential energy?

15. Describe what happens to two particles with like charges when they are brought close together. What happens to particles with unlike charges when they are brought close together?

16. What is the basic unit of length in the SI system? What are the derived units for area and volume?

Problems

A *word of advice:* You cannot learn to work problems by reading them or watching your teacher work them any more than you could become a basketball player solely by reading about basketball skills or attending a basketball game. Plan to work through most of these practice problems.

Some Fundamental Concepts

17. Two samples are weighed under identical conditions in a laboratory. Sample A weighs 143 g and sample B weighs 286 g. Does sample B have twice the mass of sample A? Explain your answer.

18. Sample A, which is on the moon, has the same mass as sample B on Earth. Do the two samples weigh the same? Explain your answer.

19. Which of the following describes a chemical change, and which describes a physical change?
a. Sheep are sheared and the wool is spun into yarn.
b. A lawn grows thicker after being fertilized and watered.
c. Milk turns sour when left out of the refrigerator for many hours.

20. Which of the following describes a chemical change, and which describes a physical change?
a. Silkworms feed on mulberry leaves and produce silk.
b. An overgrown lawn is manicured by mowing it with a lawn mower.
c. Molten lava from an erupting volcano flows down the side of a mountain and solidifies.

Elements, Compounds, and Mixtures

21. Which of the following represent elements and which do not? Explain.
a. C **b.** CO **c.** Cl

22. Which of the following represent elements and which do not? Explain.
a. $CaCl_2$ **b.** Na **c.** KI

23. Which of the following are substances and which are mixtures?
a. Helium gas used to fill a balloon.
b. Maple syrup collected from a maple tree.
c. Smog formed in the early morning air.

24. Which of the following are substances and which are mixtures?
a. Vinegar made from wine.
b. Salt used to deice roads.
c. Mercury used in a thermometer.

25. Which of the following mixtures are homogeneous and which are heterogeneous?
a. high octane gasoline **b.** iced tea

26. Which of the following mixtures are homogeneous and which are heterogeneous?
a. Italian salad dressing **b.** white wine

27. Without consulting Table 1.2, name the elements with the following symbols.
a. He **b.** N **c.** F
d. K **e.** Fe **f.** Cu

28. Without consulting Table 1.2, name the elements with the following symbols.
a. Mg **b.** Si **c.** S
d. Br **e.** P **f.** Sn

29. Without consulting Table 1.2, give the symbols for the following elements.
a. hydrogen **b.** carbon **c.** oxygen
d. zinc **e.** iodine **f.** mercury

30. Without consulting Table 1.2, give the symbols for the following elements.
a. aluminum **b.** phosphorus **c.** sodium
d. chlorine **e.** calcium **f.** cobalt

Metric Measurement

31. Change each of the following measurements to one in which the unit has an appropriate SI prefix.
a. 4.54×10^{-3} g **b.** 3.76×10^{-2} m
c. 6.34×10^{-6} g

32. Change each of the following measurements to one in which the unit has an appropriate SI prefix.
a. 1.09×10^{-6} L **b.** 9.01×10^{-3} cm^3 **c.** 7.77×10^3 m

33. How many millimeters and how many centimeters are there in 1 m?

34. For each of the following, indicate which is the larger unit.
a. mm or cm **b.** kg or g **c.** dL or μL

35. For each of the following, indicate which is the larger unit.
a. L or cm^3 **b.** cm^3 or mL

36. For each of the following, indicate which is the larger unit.
a. in. or m **b.** lb or g **c.** L or gal

37. How many meters are there in each of the following?
a. 50 km **b.** 25 cm

38. How many millimeters are there in each of the following?
a. 1.5 m **b.** 16 cm

39. How many deciliters are there in each of the following?
a. 1.0 L **b.** 20 mL

40. How many liters are there in each of the following?
a. 2056 mL **b.** 47 kL

41. Make the following conversions.
a. 15,000 mg to g **b.** 0.086 g to mg

42. Make the following conversions.
a. 0.149 L to mL **b.** 47 mL to L

43. Make the following conversions.
  a. 1.5 L to mL b. 18 mL to L
44. How many milliliters are there in 1 cm^3? In 15 cm^3?

Energy: Temperature and Heat

45. Which has the greater kinetic energy: a sprinter or a long-distance runner, assuming that the two weigh the same?
46. Which has the greater kinetic energy: a cannonball or a bullet, both fired at the same speed?
47. Which has the greater kinetic energy: a bicyclist traveling at 15 mi/hr or an automobile traveling at 40 mi/hr?
48. Which has the greater kinetic energy: a 110-kg football tackle moving slowly across the field or an 80-kg halfback racing quickly down the field?
49. Which has the greater potential energy: a diver on the 1-m board or the same diver on the 10-m platform?
50. Which has the greater potential energy: an elevator stopped at the twentieth floor or one stopped at the twelfth floor?
51. For each of the following, indicate which is the larger unit.
  a. °C or °F b. cal or Cal
52. Order the temperatures from coldest to hottest: 0 K, 0 °C, 0 °F.
53. Convert the following to degrees Fahrenheit.
  a. 37 °C b. −100 °C c. 273 °C
54. Which is the higher temperature: 100 °C or 100 °F?
55. Convert the following to degrees Celsius.
  a. 98 °F b. 5 °F c. −5 °F
56. How many calories are there in each of the following?
  a. 2.75 kcal
  b. 0.74 Cal
57. How many calories are required to raise the temperature of 50 g of water from 20 °C to 50 °C?
58. How many calories are required to raise the temperature of 13 g of water from 15 °C to 95 °C?
59. How much heat would be released by 2.0 kg of water cooling from 90 °C to 20 °C?
60. How much water can be heated from 20 °C to 50 °C by 800 cal of energy?

Density

61. A 25.0-mL sample of liquid bromine has a mass of 78.0 g. What is the density of the bromine?
62. What is the density of a salt solution if 50.0 mL has a mass of 57.0 g?
63. What is the mass, in grams, of 30.0 mL of grenadine, which has a density of 1.32 g/mL?
64. What is the mass, in kilograms, of 2.75 L of the liquid glycerol, which has a density of 1.26 g/mL?
65. What is the volume of a 898-kg piece of cast iron (d = 7.76 g/cm^3)?
66. What is the volume of a 253-g sample of bromoform, which has a density of 2.90 g/mL?
67. Some metal chips having a volume of 3.29 cm^3 are placed on a piece of paper and weighed. The combined mass is found to be 18.43 g. The paper itself weighs 1.21 g. Calculate the density of the metal.
68. A glass container weighs 48.462 g. A sample of 4.00 mL of antifreeze solution is added, and the container plus the antifreeze weigh 54.51 g. Calculate the density of the antifreeze solution.
69. The density of a normal urine sample is 1.02 g/mL. What is its specific gravity?
70. The specific gravity of an antifreeze solution is 1.1044. What is its density?

Additional Problems

71. What is the volume of 5.79 mg of gold (d = 19.3 g/cm^3)? If the gold is hammered into gold leaf of uniform thickness with an area of 44.6 cm^2, what is the thickness of the gold leaf?
72. A box with a base 0.80 m on a side and a height of 1.20 m is filled with 3.2 kg of expanded polystyrene packing material. What is the bulk density, in g/cm^3, of the packing material? (The bulk density includes the air between the pieces of polystyrene foam.)
73. Which of the following items would be most difficult to lift into the bed of a pickup truck: (1) a 100-lb bag of potatoes, (2) a 15-gal plastic bottle filled with water, (3) a 3.0-L flask filled with mercury (d = 13.6 g/cm^3)?
74. A rectangular block of gold-colored material measures 3.00 cm × 1.25 cm × 1.50 cm and has a mass of 28.12 g. Can the material be gold? The density of gold is 19.3 g/cm^3.

Special Topic A
Unit Conversions

You will find chemistry problems in several places in this text. Related problems are common in other sciences and in medicine. For example, we sometimes need to convert a measurement from one unit to another. Here we describe a useful method, called the *unit-conversion method,* for solving this and many other kinds of problems. The method uses units, such as L, cm/ft, mi/h and g/cm^3, as aids in setting up and solving problems. The general approach is to multiply the known quantity (and its unit or units!) by one or more conversion factors such that the answer is obtained in the desired unit(s).

$$\begin{matrix}\text{Known quantity} \\ \text{and unit(s)}\end{matrix} \times \begin{matrix}\text{conversion} \\ \text{factor(s)}\end{matrix} = \begin{matrix}\text{answer in} \\ \text{desired units}\end{matrix}$$

The method is best learned by practice. We urge you to learn it now and thereby save yourself a lot of time and wasted effort later.

A.1 Conversions Within a System

Quantities can be expressed in a variety of units. For example, you can buy beverages by the 12-ounce can or by the 2-liter bottle. If you wish to compare prices, you must be able to convert from one unit to another. Such a conversion changes the number and the unit, but it does not change the quantity. Your actual weight, for example, remains unchanged whether it is expressed in pounds or kilograms.

You know that multiplying a number by 1 doesn't change its value. Multiplying by a fraction equal to 1 also leaves the value unchanged. A fraction is equal to 1 when the numerator is equal to the denominator. For example, we know that

$$1 \text{ ft} = 12 \text{ in.}$$

Therefore,

$$\frac{1 \text{ ft}}{12 \text{ in.}} = 1$$

Similarly,

$$\frac{12 \text{ in.}}{1 \text{ ft}} = 1$$

If we wish to convert from inches to feet, we can do so by choosing one of these fractions as a **conversion factor.** Which one do we choose? The one that gives us an answer with the right unit! Let's illustrate by means of an example.

A person's height is the same whether it is measured in meters or in feet; it is simply expressed differently.

Example A.1

My bed is 72 inches long. What is its length in feet? You know the answer, of course, but let's see how the answer is obtained by using unit conversions.

Solution

We need to multiply 72 inches by one of the fractions that relate inches to feet. Which one? The known quantity is 72 inches.

$$72 \text{ in.} \times \text{conversion factor} = ? \text{ ft}$$

For the conversion factor, choose the fraction that, when inserted into the equation, cancels the unit *inch* and leaves the unit *feet*.

$$72 \cancel{\text{in.}} \times \frac{1 \text{ ft}}{12 \cancel{\text{in.}}} = 6.0 \text{ ft}$$

Now let's try the other conversion factor.

$$72 \text{ in.} \times \frac{12 \text{ in.}}{1 \text{ ft}} = 864 \text{ in.}^2/\text{ft}$$

Absurd! How can a bed measure 864 in.^{2}/feet? You should have no difficulty choosing between the two possible answers.

Note that conversion factors are not usually given in problems. Rather, they are obtained from tables such as those in Appendix I. For the following examples, please refer to those tables for conversion factors that you don't already know.

It is possible (and frequently necessary) to manipulate units in the denominator as well as in the numerator of a problem. Just remember to use conversion factors in such a way that the unwanted units cancel.

Example A.2

At a track meet, a runner completes the mile in 4.0 minutes. How fast is this in miles per hour?

Solution

The time for 1 mile is 4.0 minutes, or 1 mile per 4.0 minutes, or

$$\frac{1\ \text{mi}}{4.0\ \text{min}}$$

This is the known quantity and unit. The unit desired is miles per hour.

$$\frac{1\ \text{mi}}{4.0\ \text{min}} \times \text{conversion factor} = ?\ \frac{\text{mi}}{\text{h}}$$

To change minutes to hours, we need the equivalence

$$1\ \text{h} = 60\ \text{min}$$

Now arrange this equivalence as the fractions

$$\frac{1\ \text{h}}{60\ \text{min}} \text{ and } \frac{60\ \text{min}}{1\ \text{h}}$$

Choose for your conversion factor the fraction that cancels the unit *minute* and leaves the unit *hour* in the denominator.

$$\frac{1\ \text{mi}}{4.0\ \cancel{\text{min}}} \times \frac{60\ \cancel{\text{min}}}{1\ \text{h}} = \frac{15\ \text{mi}}{1\ \text{h}} \left(\text{or } \frac{15\ \text{mi}}{\text{h}}\right)$$

Problems may involve the use of several conversion factors, as in the following example.

Example A.3

If your heart beats at a rate of 72 times per minute and your lifetime will be 70 years, how many times will your heart beat during your lifetime?

Solution

Two equivalences are given in the problem.

$$72\ \text{beats} = 1\ \text{min}$$
$$1\ \text{lifetime} = 70\ \text{years}$$

Three others that you will need you can recall from memory.

$$1\ \text{year} = 365\ \text{days}$$
$$1\ \text{day} = 24\ \text{h}$$
$$1\ \text{h} = 60\ \text{min}$$

Start now with the factor 72 beats per 1 minute (the known quantity and unit) and apply conversion factors as needed to get an answer in beats/lifetime (the desired unit).

$$\frac{72\ \text{beats}}{1\ \cancel{\text{min}}} \times \frac{60\ \cancel{\text{min}}}{1\ \cancel{\text{h}}} \times \frac{24\ \cancel{\text{h}}}{1\ \cancel{\text{day}}} \times \frac{365\ \cancel{\text{days}}}{1\ \cancel{\text{year}}} \times \frac{70\ \cancel{\text{years}}}{1\ \text{lifetime}}$$
$$= 2{,}600{,}000{,}000\ \text{beats/lifetime}$$

Most problems that you will have to deal with will be much simpler than Example A.3.

Now let's do some conversions within the metric system. They are much easier.

Example A.4

How many milliliters are there in a 2.0-liter bottle of soda?

Solution

From memory or from the tables in Appendix I you find that

$$1\ \text{L} = 1000\ \text{mL}$$

$$2.0\ \cancel{\text{L}} \times \frac{1000\ \text{mL}}{1\ \cancel{\text{L}}} = 2000\ \text{mL}$$

Notice that we picked the conversion factor that allowed us to cancel liters and obtain an answer in the desired unit, milliliters.

Example A.5

In the United States, the usual soft drink can holds 360 milliliters. How many such cans could be filled from one 2.0-liter bottle?

Solution

The problem tells us that

$$1\ \text{can} = 360\ \text{mL}$$

Using that equivalence, we can calculate the answer.

$$2.0\,\cancel{L} \times \frac{1000\,\cancel{mL}}{1\,\cancel{L}} \times \frac{1\text{ can}}{360\,\cancel{mL}} = 5.6\text{ cans}$$

Example A.6

How many 325-milligram aspirin tablets can be made from 875 grams of aspirin?

Solution

The problem tells us that

$$1\text{ tablet} = 325\text{ mg}$$

From memory or the tables, we have

$$1\text{ g} = 1000\text{ mg}$$

$$875\,\cancel{g} \times \frac{1000\,\cancel{mg}}{1\,\cancel{g}} \times \frac{1\text{ tablet}}{325\,\cancel{mg}} = 2690\text{ tablets}$$

A.2 Conversions Between Systems

To convert from one system of measurement to another, you need a conversion table such as the one in Appendix I. Let's plunge right in and work some examples.

Example A.7

You know that your weight is 140 pounds, but the job application form asks for your weight in kilograms. What is it?

Solution

From the table we find

$$1.00\text{ lb} = 0.454\text{ kg}$$

So your weight is

$$140\,\cancel{lb} \times \frac{0.454\text{ kg}}{1.00\,\cancel{lb}} = 63.6\text{ kg}$$

Example A.8

A recipe calls for 750 milliliters of milk, but your measuring cup is calibrated in fluid ounces. How many ounces of milk will you need?

Solution

$$750\,\cancel{mL} \times \frac{1.00\text{ fl oz}}{29.6\,\cancel{mL}} = 25.3\text{ fl oz}$$

Example A.9

A doctor puts you on a diet of 5500 kilojoules (kJ) per day. How many kilocalories (food Calories) is that?

Solution

$$5500\,\cancel{kJ} \times \frac{1000\,\cancel{J}}{1\,\cancel{kJ}} \times \frac{1.000\text{ kcal}}{4184\,\cancel{J}} = 1315\text{ kcal}$$

In Example A.10, we find that *two* units in the original measured quantity must be replaced. We solve the problem in two ways, and compare the two approaches.

Example A.10

A sprinter runs a 100.0-meter dash in 11.00 seconds. What is her speed in kilometers per hour?

Solution

First, note that the measured quantity is a speed, expressed by the ratio

$$\frac{100.0\text{ m}}{11.00\text{ s}}$$

Our task is to convert this speed to kilometers per hour. We must convert from meters to kilometers in the numerator and from seconds to hours in the denominator. We need this set of equivalent values to formulate conversion factors:

$1\text{ km} = 1000\text{ m}$ $\quad 1\text{ min} = 60\text{ s}$ $\quad 1\text{ h} = 60\text{ min}$

In one approach we arrange all the necessary conversion factors in a single straight-line setup that yields a final answer with the desired units.

$$?\,\frac{\text{km}}{\text{h}} = \frac{100.0\,\cancel{m}}{11.00\,\cancel{s}} \times \frac{1\text{ km}}{1000\,\cancel{m}} \times \frac{60\,\cancel{s}}{1\,\cancel{min}} \times \frac{60\,\cancel{min}}{1\text{ h}}$$

$$= 32.73\text{ km/h}$$

In an alternative approach, we do the conversions in the numerator and in the denominator separately and then combine the results.

$$\text{Numerator: } 100.0\,\cancel{m} \times \frac{1\text{ km}}{1000\,\cancel{m}} = 0.1000\text{ km}$$

$$\text{Denominator: } 11.00\,\cancel{s} \times \frac{1\,\cancel{min}}{60\,\cancel{s}} \times \frac{1\text{ h}}{60\,\cancel{min}} = 3.056 \times 10^{-3}\text{ h}$$

$$\text{Division: } \frac{0.1000\text{ km}}{3.056 \times 10^{-3}\text{ h}} = 32.72\text{ km/h}$$

(The two answers differ slightly due to rounding.)

Practice Exercise

Convert 90.0 kilometers per hour to meters per second.

A.3 Density Problems

Many types of problems can be solved using unit conversions. Let's try some that involve density.

Example A.11

The mass of 325 milliliters of methyl alcohol, a liquid, is 257 grams. What is the density of methyl alcohol?

Solution

Densities of liquids are usually expressed in grams per milliliter (Section 1.10), so we arrange the data such that the answer will come out in those units.

$$\frac{257\ \text{g}}{325\ \text{mL}} = 0.791\ \text{g/mL}$$

Practice Exercise

Calculate the density of liquid mercury if a 345-gram sample is found to occupy a volume of 25.4 milliliters.

Example A.12

What is the mass of 58.7 milliliters of methyl alcohol? The density of methyl alcohol is 0.791 grams per milliliter.

Solution

Another way to describe the density is through the equation

$$1\ \text{mL methyl alcohol} = 0.791\ \text{g methyl alcohol}$$

We can then formulate two conversion factors between the mass and volume of methyl alcohol.

$$\text{(a)}\quad \frac{0.791\ \text{g methyl alcohol}}{1\ \text{mL methyl alcohol}} \quad \text{and} \quad \text{(b)}\quad \frac{1\ \text{mL methyl alcohol}}{0.791\ \text{g methyl alcohol}}$$

To convert a volume to a mass, we use density as a conversion factor—that is, we use form (a).

$$?\ \text{kg} = 58.7\ \text{mL methyl alcohol} \times \frac{0.791\ \text{g methyl alcohol}}{1\ \text{mL methyl alcohol}}$$

$$= 46.4\ \text{g methyl alcohol}$$

Practice Exercise

An experiment calls for 8.65 grams of carbon tetrachloride ($d = 1.59$ g/mL). What is the volume of such a mass? [*Hint:* To convert from mass to volume, we use the inverse of the density, or form (b) of the conversion factor in Example A.12.]

Key Term

conversion factor (A.1)

Review Questions

1. What is a conversion factor?
2. How must the numerator and denominator of a conversion factor be related?
3. Arrange the following in order of increasing length (shortest first):
   (1) a 1.21-m chain, (2) a 75-in. board,
   (3) a 3-ft-5-in. rattlesnake, (4) a yardstick.
4. Arrange the following in order of increasing mass (lightest first):
   (1) a 5-lb bag of potatoes,
   (2) a 1.65-kg cabbage,
   (3) 2500 g of sugar.
5. How would you describe a young man who is 160 cm tall and has a mass of 94 kg?
6. A woman who weighs 38.5 kg is most likely to be which of the following?
   (a) a basketball center (b) a gymnast
   (c) a discus thrower

Problems

Conversions Within a System

7. Carry out the following conversions.
   a. 50.0 km to meters b. 546 mm to meters
   c. 98.5 kg to grams d. 47.9 mL to liters
   e. 578 μg to milligrams f. 237 mm to centimeters
8. Carry out the following conversions.
   a. 87.6 μg to kilograms
   b. 1.00 m to micrometers
   c. 0.0962 km/min to meters per second
   d. 55 mi/h to kilometers per minute
9. Carry out the following conversions. (You may refer to the tables in Appendix I when necessary.)
   a. 413 in. to yards
   b. 86.2 oz to pounds
   c. 64.0 fl oz to quarts
   d. 12.6 ft/s to miles per hour

10. Carry out the following conversions. (You may refer to the tables in Appendix I when necessary.)
   a. 4.53 ft to inches
   b. 8.08 lb to ounces
   c. 6.13 qt to fluid ounces
   d. 70.06 mi/h to feet per second

Conversions Between Systems

(You may refer to the tables in Appendix I when necessary.)

11. Carry out the following conversions.
   a. 16.4 in. to centimeters
   b. 4.17 qt to liters
   c. 1.61 kg to pounds
   d. 9.34 g to ounces
12. Carry out the following conversions.
   a. 2.05 fl oz to milliliters
   b. 105 lb to kilograms
   c. 143 cm to feet
   d. 775 mL to quarts
13. Football player Nate Newton weighs 320 lb. What is his mass in kilograms?
14. Basketball player Shaquille O'Neal is 7.3 ft tall. What is his height in meters? In centimeters?
15. The speed limit in rural Ontario is 90 km/h. What is this speed in miles per hour?
16. The speed of light is 186,000 mi/sec. What is this speed in meters per second?
17. Milk costs \$3.89/gal or \$1.15/L. Which is cheaper?
18. State your height in feet and your weight in pounds. Convert these values to centimeters and kilograms.

Density

(Use unit conversions to solve the following density problems.)

19. A 50.0-mL sample of the liquid hexane, a solvent used to extract oil from soybeans, has a mass of 33.0 g. What is the density of hexane?
20. What is the density of a salt solution if 75.0 mL has a mass of 87.5 g?
21. What is the mass, in grams, of 30.0 mL of the liquid glycerol, a moisturizing agent for foods, which has a density of 1.26 g/mL?
22. What is the mass, in grams, of 125 mL of castor oil, a laxative, which has a density of 0.0962 g/mL?
23. What is the volume of a 475-g piece of copper? The density of copper is 8.94 g/cm^3.
24. What is the volume of a 5.79-g piece of silver? The density of silver is 10.5 g/cm^3.

Additional Problems

25. Adult male Hooker's sea lions are 250 to 350 cm long and weigh 300 to 450 kg. Convert these measurements to inches and pounds.
26. In its nonstop, round-the-world trip, the aircraft *Voyager* traveled 25,102 mi in 9 days, 3 min, and 44 s. Calculate the average speed of *Voyager* in mi/h.
27. The furlong is a unit used in horse racing, and the units *chain* and *link* are used in land surveying. There are 8 furlongs in 1 mi, 10 chains in one furlong, and 100 links in one chain. Calculate the length of one link, in inches (1 mi = 5280 ft; 1 ft = 12 in.)
28. In the United States, land area is commonly measured in acres: 640 acre = 1 mi^2. In most of the rest of the world, land area is measured in hectares: 1 hectare = 1 hm^2. [1 hectometer (hm) = 100 m]. Which is the larger area, the acre or the hectare? Write a conversion factor that relates the acre and the hectare. (1 mi = 5280 ft; 1 m = 39.37 in.)
29. In scientific work, densities usually are expressed in g/cm^3. In engineering work, the unit lb/ft^3 is often used. The density of water at 20 °C is 0.998 g/cm^3. What is its density in lb/ft^3?
30. A square of aluminum foil (d = 2.70 g/cm3) is 5.10 cm on a side and has a mass of 1.762 g. Calculate the thickness of the foil, in millimeters.

2 Atoms

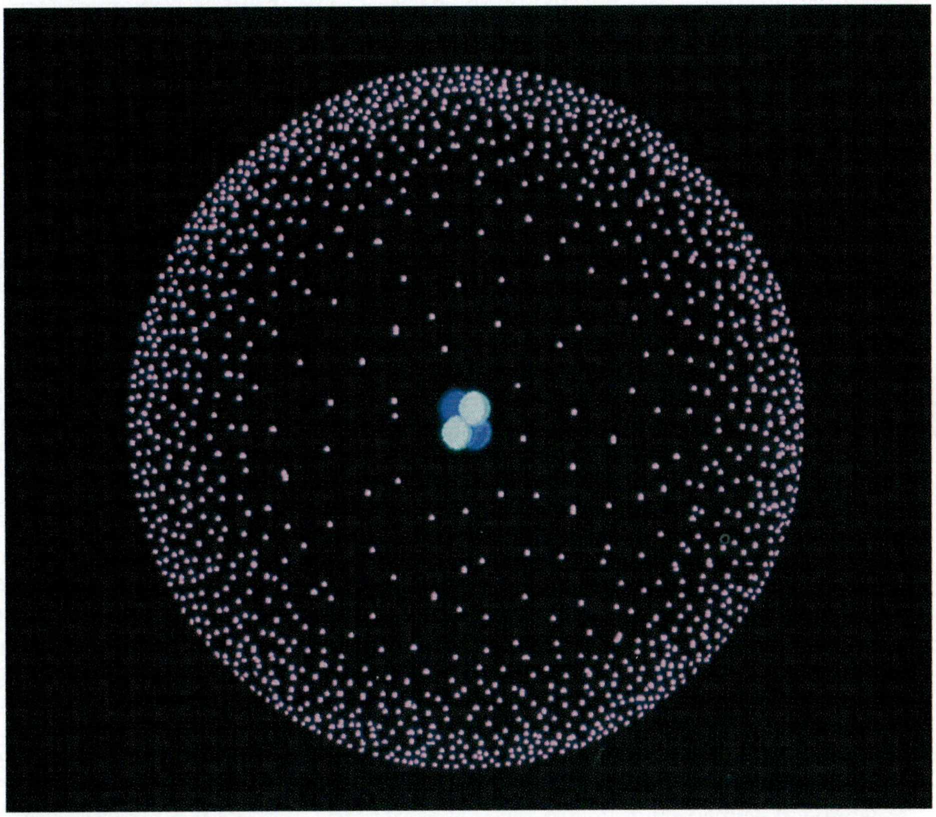

A computer-generated graphic of the helium atom. Two protons and two neutrons are contained within the nucleus, and these are surrounded by a "cloud" of electrons.

Atom—the smallest chemically obtainable portion of an element

An *atom* is the smallest characteristic particle of an element. This concept has been around for thousands of years. For most of that time, however, it was not a popular idea. The development of chemistry as a science was largely a matter of accepting the existence of atoms and then working to define the properties of the atom more and more precisely.

As originally proposed by both ancient Hindu and Greek scholars, the concept of the atom was as philosophical as it was practical. Both groups spoke of earth, water, fire, and air as elements composed of eternal and unchanging atoms in perpetual motion. For the Hindus, it was Brahma, the soul of the universe, who set the atoms in motion. Some Greek philosophers believed that atoms were in motion because they were falling toward the center of the universe. The atomistic view of matter fell into disfavor when the greatest of the Greek philosophers, Aristotle, elected to side with those who believed that matter was continuous, that is, infinitely divisible. According to Aristotle and his followers, there was no particle of matter so small that it could not be subdivided into still smaller pieces. It is a tribute to Aristotle that his view prevailed for 2000 years, even though it was wrong.

A gradual accumulation of data that could not be explained within Aristotle's philosophy finally forced a reevaluation of his views. The reassessment and ultimate rejection of Aristotle's concept of the nature of matter was a major victory for experimental science. From that point onward, scientific theories were accepted or rejected on the basis of their ability to explain experimental data and not because of their beauty or elegance or even their appeal to common sense.

Why do we care about the structure of atoms? Because it is the arrangement of the parts of atoms that determines the properties of different kinds of matter. Only by understanding atomic structure can we know how atoms combine to make the many different substances in nature and, even more importantly, how we can modify materials to meet our needs more precisely. A knowledge of atomic structure is even essential to your health. Many medical diagnoses are based on the analysis of body fluids such as blood and urine. Many such analyses depend on knowledge of how the structure of atoms is changed when energy is absorbed.

John Dalton (1766–1844) was not a good experimenter—perhaps because he was color blind. However, he skillfully used the results of the experiments of many others in formulating his atomic theory.

2.1 Dalton's Atomic Theory

The man whose scientific theory did explain all the available experimental data was John Dalton, an English schoolteacher. Dalton accepted the atomistic view of matter and presented the modern atomic theory. The theory is called "modern" because it was based on the best evidence available in Dalton's time (the early nineteenth century) rather than on ideas formulated in Aristotle's time (fourth century B.C.E.). The most important points of Dalton's atomic theory are:

1. All matter is composed of extremely small particles called atoms. [Dalton assumed atoms to be indivisible. This isn't quite so, as we see in the case of radioactivity (Chapter 3).]
2. All atoms of a given element are alike, but atoms of one element differ from the atoms of any other element. (Dalton assumed that all the atoms of a given element have the same mass. This too is now known to be incorrect, as we see in the next section).
3. Compounds are formed when atoms of different elements combine in fixed proportions. (The numbers of each kind of atom in a compound usually form a simple ratio. For example, the ratio of carbon atoms to oxygen atoms is 1:1 in carbon monoxide and 1:2 in carbon dioxide.)
4. A chemical reaction involves a *rearrangement* of atoms. No atoms are created or destroyed or broken apart in a chemical reaction.

What did Dalton mean when he said that all atoms of a given element are the same? Were the atoms of an element all the same color or the same shape? Rather—and this was Dalton's most important insight—he proposed that the mass of an atom identifies it uniquely. Atoms are extremely minute. In the nineteenth century, it was impossible to determine actual masses of atoms. Indirect measurements, however, could indicate their relative masses. Dalton set up a table of relative masses for the elements based on hydrogen, the lightest element. He believed (incorrectly) that an oxygen atom had a mass seven times the mass of a hydrogen atom.

Relative masses are now expressed in terms of the **atomic mass unit,** often abbreviated amu. The official SI unit, however, is u. We say "atomic weights are expressed in amu," but we say "the atomic weight of carbon is 12.011 u."

Dalton was wrong in several ways. He was wrong about the mass of oxygen atoms and about all atoms of the same element being identical. He was even wrong about atoms being indivisible and indestructible. Yet Dalton is still regarded, and rightly so, as one of the giants of modern chemistry. Other scientists took up his ideas and modified them, and a new era in chemistry began. Science is not simply an accumulation of correct bits of information; it is the gradual development of a model for our universe, a model that enables us to understand the workings of nature. Ideas are important if they help us understand. Some ideas have to be corrected or modified as our understanding increases, but they are nonetheless valuable. Without them it might not be possible to advance to a higher level of understanding. This situation is not unique to chemistry. An accepted medical treatment of 50 (to say nothing of 100) years ago is likely to be unacceptable today. Yet such now outdated treatment did once keep people alive and helped medical researchers develop better, safer, more effective procedures.

Marie Curie (1867–1934), a Polish-born scientist who worked in France, discovered the elements radium and polonium and gave the phenomenon of radioactivity its name.

2.2 The Nuclear Atom

Although Dalton erred in some details, he started us along the right path. If one begins with the premise that atoms exist, then it is possible to devise experiments to find out more about these atoms. In the century following the work of Dalton, many such experiments were conducted. Data continued to accumulate, and many of the results could not be explained by Dalton's atomic theory without modification. An international array of scientists were discovering that the atom was not indivisible. It could still be regarded as the smallest characteristic unit of an *element,* but it certainly was not the smallest particle of matter.

The studies of a German physicist, Eugen Goldstein, and an English physicist, William Crookes, demonstrated that atoms could be torn apart. In an apparatus called a discharge tube (Figure 2.1), electric energy was used to kick out small pieces from atoms of the metal making up one electrode. The tiny *subatomic* particles were shown to be negatively charged and could be detected as a beam of particles (called a **cathode ray**) that crossed from one electrode in the apparatus to the other. No matter what element the electrode was made of, identical negatively charged subatomic particles were kicked out to form the cathode ray. These particles were named **electrons.**

Cathode rays travel in straight lines in the absence of an external applied field.

Electron—a subatomic particle with a negative charge and essentially no mass

When a small amount of gas was admitted into the tube, the cathode-ray particles struck the atoms of the gas, knocking more electrons loose from these atoms (Figure 2.2). Those atoms that lost negatively charged electrons were left with a positive charge. Although all the negatively charged particles were

◀ **FIGURE 2.1**

Cathode rays are deflected in a magnetic field. Although these rays are invisible, we can detect them because they produce a green fluorescence as they strike a screen coated with zinc sulfide. The beam of cathode rays begins at the cathode on the left. It is deflected by the field of a magnet that is located slightly behind the anode. The deflection we observe is that expected of negatively charged particles.

FIGURE 2.2 ▶

Positive rays are also produced in cathode-ray tubes that contain some residual gas. Cathode rays (blue) stream toward the anode (+). They collide with gas atoms and knock electrons from the atoms to produce positively charged ions. These ions are attracted to the cathode (−), but some pass through the holes in the cathode and appear as a stream of positive particles (red) on the other side.

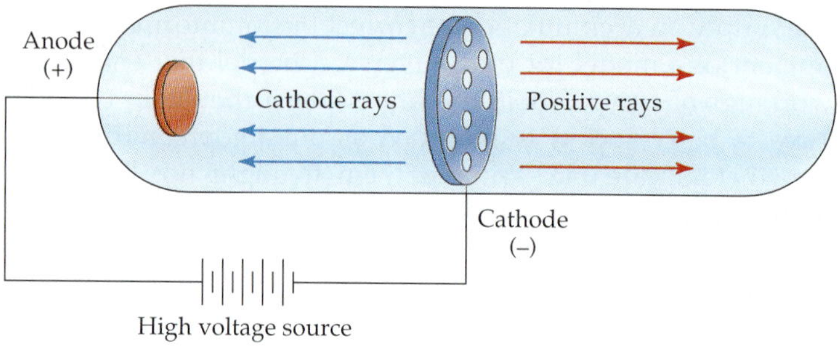

Proton—a subatomic particle with a positive charge; 1837 times as massive as an electron

Certain atoms naturally decay and release subatomic particles.

Radioactivity—the emission by atoms of subatomic particles

Neutrons are similar in mass to protons but have no charge.

identical, the positively charged particles differed depending on what gas was in the tube. For example, helium produced more massive positively charged particles than did hydrogen. The lightest positive particles found were those from hydrogen and were given a special name, **protons.** The charges on the proton and the electron are the same size (although opposite in sign), but the proton was found to be 1837 times as massive as the electron. Thus, the smallest positive particle isolated was many times heavier than the negatively charged electron.

A French physicist, Antoine Henri Becquerel, discovered that some atoms fall apart all by themselves. Atoms of the element uranium, for example, emitted "rays," some of which proved to be subatomic particles that were a fraction of the size of the original atoms (see Section 3.2). Some of the emitted particles were negatively charged and were shown to be identical to the electrons formed in a cathode-ray tube. Uranium also emitted positively charged particles much smaller than the uranium atom itself but much more massive than the electrons. A Polish-born chemist, Marie Sklodowska Curie, named these phenomena **radioactivity** (see Chapter 3).

A British scientist, Ernest Rutherford, used radioactive atoms as "atomic guns," with the tiny emitted particles playing the role of bullets (Figure 2.3). By shooting such "bullets" at other atoms, he discovered that the target atoms were mostly empty space. Most of the bullets passed right through the target atoms. Some of the bullets were deflected, however, suggesting that there was a concentrated bit of "solid" material in the atom.

Ernest Rutherford (1871–1937) was born in New Zealand but worked in Canada and England. He developed the theory of the nuclear atom.

A student of Rutherford, James Chadwick, discovered a subatomic particle unlike those previously found. This particle, called a **neutron,** has about the same mass as the proton, but it is neutral—neither positively nor negatively charged.

Early in the twentieth century, a new model of the atom was proposed to account for all these new data. According to this model there are three main types of subatomic particles: protons, neutrons, and electrons. A proton and a neutron have nearly the same mass, 1.007276 u and 1.008665 u, respectively. This is like saying that two different people weigh 100.7 lb and 100.9 lb; the difference is so small that it can be ignored. For many purposes, we can assume that the masses of the proton and the neutron are 1 amu each. The proton has a charge equal in magnitude but opposite in sign to that of an electron. This charge on a proton is written as 1+. The electron has a charge of 1− and a mass of 0.000549 u. The electrons in an atom contribute so little to its total mass that their mass is usually disregarded—electrons are treated as if they have zero mass.[1] These properties of the subatomic particles are summarized in Table 2.1.

---

[1] To emphasize how small subatomic particles are, we list here their masses in grams: proton = 1.673×10^{-24} g (0.000000000000000000000001673 g); neutron = 1.675×10^{-24} g; electron = 9.107×10^{-28} g. (Appendix II reviews exponential notation.)

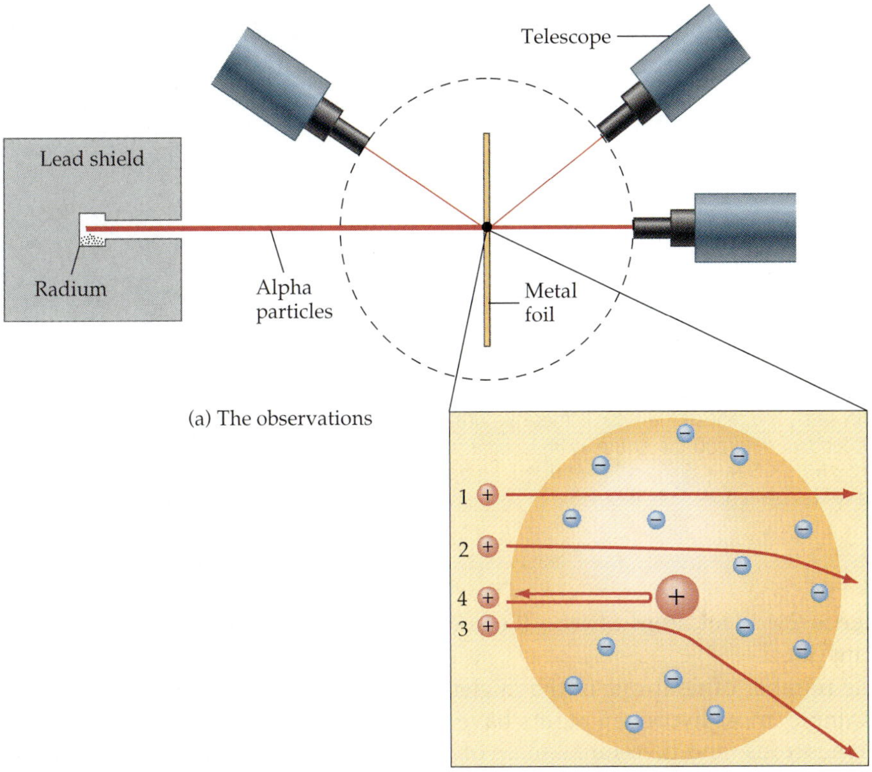

◀ **FIGURE 2.3**
Alpha particles are scattered by thin metal foil. (a) The observations. (1) Most of the alpha particles pass through the foil undeflected. (2) Some alpha particles are deflected slightly as they penetrate the foil. (3) A few (about 1 in 20,000) are greatly deflected. (4) A similar small number do not penetrate the foil at all; rather they are reflected back toward the source. (b) Rutherford's interpretation. If the atoms of the foil have a massive, positively charged nucleus and light electrons outside the nucleus, we can see that: (1) most alpha particles can pass through the atom undeflected; (2) an alpha particle is deflected slightly as it passes near an electron; (3) an alpha particle is deflected significantly as it passes close to the atomic nucleus; and (4) an alpha particle bounces back toward the source when it approaches the nucleus head on.

Table 2.1 Subatomic Particles

Particle	Symbol[a]	Approximate Mass (u)	Charge	Location in Atom
Proton	p^+	1	1+	Nucleus
Neutron	n	1	0	Nucleus
Electron	e^-	0	1−	Outside nucleus

[a]The superscripts are often omitted. We will keep them to remind us of the charge of each particle.

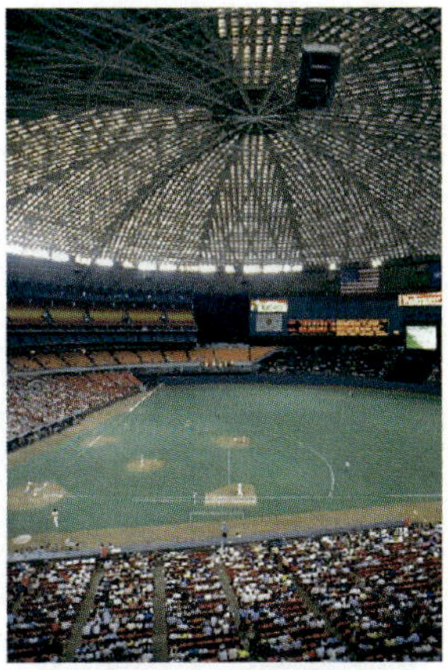

If an atom were as large as this stadium, the nucleus would be the size of a pea at the center of the stadium.

When combined to form an atom, the more massive particles (protons and neutrons) are crowded into the **nucleus,** a tiny volume of space at the center of the atom (Figure 2.4). The electrons are scattered throughout the remaining volume of the atom. To visualize how "empty" an atom is, picture a balloon that stands 10 stories high. If the balloon were an atom, its nucleus would be the size of a BB. The rest of the space in the balloon-atom is the domain of the electrons, whose mass is smaller than that of the nucleus by a factor of thousands.

Atomic number = number of protons. All the atoms of a given element have the same number of protons. For neutral atoms, the atomic number also gives the number of electrons.

The number of protons in the nucleus of an atom of any element determines the **atomic number** of that element. An *element* is a substance in which all the atoms have the same atomic number. This number determines the kind of atom—that is, the identity of the element. Dalton said that the mass of an atom determines the element. We now say it is not the mass but the number of protons that determines the element. For example, an atom with 26 protons (one whose atomic

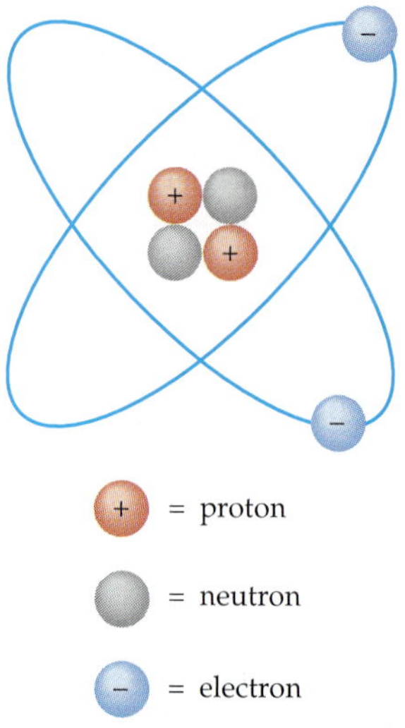

FIGURE 2.4 ▶

The nuclear model of an atom—illustrated here by helium—shows that an electrically neutral atom has the same number of electrons outside the nucleus as it has protons in the nucleus. This drawing shows the electrons much closer to the nucleus than they really are. The actual case is more accurately represented by imagining the entire atom to be a room measuring 5 m × 5 m × 5 m and the nucleus to be the period at the end of this sentence.

number is 26) is an atom of iron (Fe). An atom with 92 protons is an atom of uranium (U).

The number of neutrons in the nuclei of atoms of a given element may vary. For example, most hydrogen atoms have a nucleus consisting of a single proton and no neutrons (and therefore a mass of 1 u). About one hydrogen atom in 6700, however, does have a neutron as well as a proton in the nucleus. This heavier hydrogen is called *deuterium* and has a mass of 2 u. Both kinds are hydrogen atoms (any atom with atomic number 1—that is, with one proton—is a hydrogen atom). Atoms that have this sort of relationship—the same number of protons but different numbers of neutrons—are called **isotopes** (Figure 2.5). A third, rare isotope of hydrogen is *tritium,* which has two neutrons and one proton in the nucleus (and thus a mass of 3 u). Most, but not all, elements exist in nature in isotopic forms. This fact also requires a major modification of Dalton's original theory. He said that all atoms of the same element have the same mass. We now know that most elements have several isotopes—that is, atoms with different numbers of neutrons and therefore different masses.

An element's isotopes have the same number of protons but different numbers of neutrons.

The number of electrons in an atom equals the number of protons. Therefore, positively and negatively charged particles balance in an atom. The atom as a whole is neutral.

FIGURE 2.5 ▶

The isotopes of hydrogen

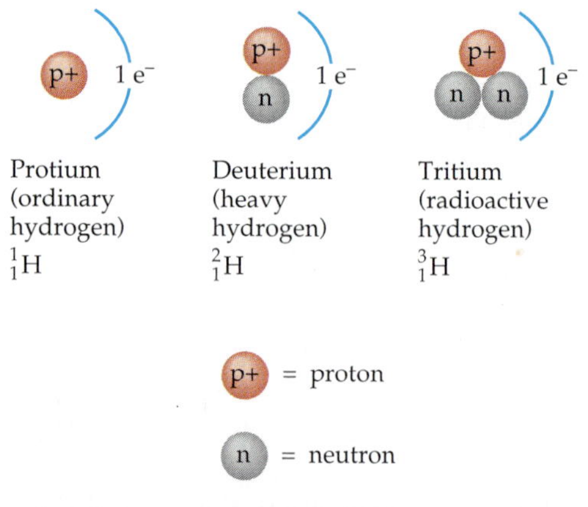

2.3 Nuclear Arithmetic

To represent the different isotopes of the elements, symbols with subscripted and superscripted numbers are used. In the generalized symbol

$$^{A}_{Z}\mathrm{X}$$

the subscript Z represents the *atomic number*—that is, the number of protons. The superscript A represents the **mass number**—that is, the number of protons plus the number of neutrons. From the symbol

neutrons + protons → $^{35}_{17}\mathrm{Cl}$ ← protons

we know that the number of protons in a chlorine atom is 17 and that the number of neutrons is $35 - 17 = 18$.

Atomic number Z = number of protons

Mass number A = number of protons + number of neutrons

The mass number is sometimes called the *nucleon number*. (Protons and neutrons, collectively are called nucleons.)

Example 2.1

Write the symbol for an isotope with a mass number of 58 and an atomic number of 27, and identify the element.

Solution

In the general symbol $^{A}_{Z}\mathrm{X}$, A is the mass number, which is 58 in this case.

$$^{58}_{Z}\mathrm{X}$$

Z is the atomic number, or 27.

$$^{58}_{27}\mathrm{X}$$

From the periodic table we can determine that the element with atomic number 27 is cobalt.

$$^{58}_{27}\mathrm{Co}$$

Practice Exercise

Write the symbol for an isotope with a mass number of 90 and an atomic number of 42.

Example 2.2

Indicate the number of protons, neutrons, and electrons in a neutral atom of the isotope $^{235}_{92}\mathrm{U}$.

Solution

The atomic number gives the number of protons and electrons in a neutral atom of the isotope.

$$\text{Atomic number} = \text{protons} = \text{electrons} = 92$$

The number of neutrons is obtained by subtracting the atomic number from the mass number.

Mass number	235
Atomic number	− 92
Number of neutrons	143

Practice Exercise

Indicate the number of protons, neutrons, and electrons in a neutral atom of the isotope ${}^{90}_{38}\mathrm{Sr}$.

Example 2.3

Which of the following (a) are isotopes of the same element, (b) have identical mass numbers, and (c) have the same number of neutrons? We are using the letter X as the symbol for all elements so that the symbol does not identify the element.

$${}^{16}_{8}\mathrm{X} \qquad {}^{16}_{7}\mathrm{X} \qquad {}^{14}_{7}\mathrm{X} \qquad {}^{14}_{6}\mathrm{X} \qquad {}^{12}_{6}\mathrm{X}$$

Solution

a. ${}^{16}_{7}\mathrm{X}$ and ${}^{14}_{7}\mathrm{X}$ are isotopes of the element nitrogen (N). ${}^{14}_{6}\mathrm{X}$ and ${}^{12}_{6}\mathrm{X}$ are isotopes of the element carbon (C).

b. ${}^{16}_{8}\mathrm{X}$ and ${}^{16}_{7}\mathrm{X}$ have the same mass number. The first is an isotope of oxygen, and the second is an isotope of nitrogen. ${}^{14}_{7}\mathrm{X}$ and ${}^{14}_{6}\mathrm{X}$ also have the same mass number. The first is an isotope of nitrogen, and the second is an isotope of carbon.

c. ${}^{16}_{8}\mathrm{X}$ ($16 - 8 = 8$ neutrons) and ${}^{14}_{6}\mathrm{X}$ ($14 - 6 = 8$ neutrons) have the same number of neutrons.

Practice Exercise

Which of the following are isotopes of the same element?

$${}^{90}_{37}\mathrm{X} \qquad {}^{90}_{38}\mathrm{X} \qquad {}^{88}_{37}\mathrm{X} \qquad {}^{88}_{36}\mathrm{X} \qquad {}^{93}_{38}\mathrm{X}$$

A particular isotope can be identified by its nuclear symbol—for example, ${}^{35}_{17}\mathrm{Cl}$ or ${}^{37}_{17}\mathrm{Cl}$. Because the atomic number of an element is implied by its chemical symbol (the element with the symbol Cl always has the atomic number 17),

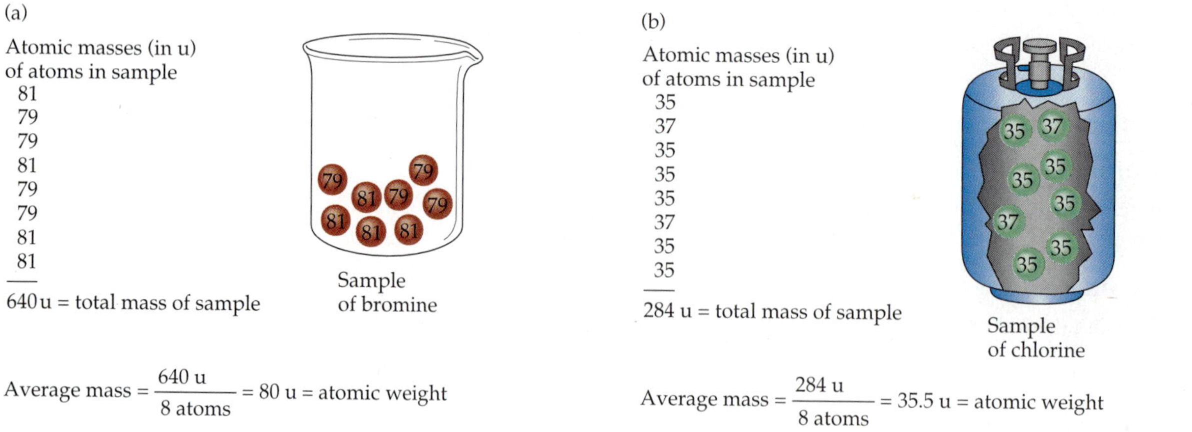

▲ **FIGURE 2.6**

Atomic weights of elements are averages of the masses of the atoms in representative samples of the elements. In a sample of bromine (a), about half the atoms have a mass of 79 u and half a mass of 81 u; the atomic weight of bromine is therefore about 80 u. (The actual value listed in the periodic table, 79.9 u, indicates that there is slightly more ${}^{79}\mathrm{Br}$ than ${}^{81}\mathrm{Br}$.) In a sample of chlorine (b), about three-fourths of the atoms have a mass of 35 u and one-fourth a mass of 37 u; the atomic weight of chlorine is therefore about 35.5 u. The average mass is much closer to that of the ${}^{35}\mathrm{Cl}$ because there is more of that isotope in the sample.

sometimes simplified forms such as ^{35}Cl and ^{37}Cl are used. More often isotopes are identified by the name of the element followed by the mass number, such as chlorine-35 and chlorine-37.

Naturally occurring carbon consists of a *mixture* of two isotopes. The much more abundant isotope is $^{12}_{6}C$. Scientists now use the *pure* isotope carbon-12, which is assigned a mass of *exactly* 12 atomic mass units, as the atomic weight standard. Based on this standard, we can define an *atomic mass unit* (abbreviated amu and having the unit u) as exactly 1/12 the mass of a carbon-12 atom. The other natural carbon isotope is $^{13}_{6}C$, with a mass of 13.00335 u. When carbon atoms participate in a chemical reaction, even though a mixture of isotopes is involved, it's more convenient to think in terms of a hypothetical, "average" atom. This "average" carbon atom would have a mass somewhere between 12.00000 u and 13.00335 u. However, this mass must be "weighted" toward the mass of the more abundant carbon-12 isotope. The weighted average of the masses of the naturally occurring carbon atoms is 12.011 u. The **atomic weight*** of an element, then, is the weighted average of the masses of the naturally occurring isotopes of that element. Figure 2.6 illustrates an approximation of how the atomic weights of chlorine and bromine are derived from the masses of their individual atoms.

2.4 The Bohr Model of the Atom

Have you ever watched colored fireworks or thrown chemicals into a fireplace to color the flames? If you have, then you have seen a phenomenon that long puzzled scientists and triggered another modification of our concept of the atom. Light is pure energy. Light of different color is light of different energy. For example, blue light packs more energy than red light. If you throw compounds containing lithium into a fire, you see the flames colored red. With sodium compounds (such as ordinary table salt), the light is yellow-orange; with strontium, red; with potassium, lavender (Figure 2.7). Why? That is what early-twentieth-century

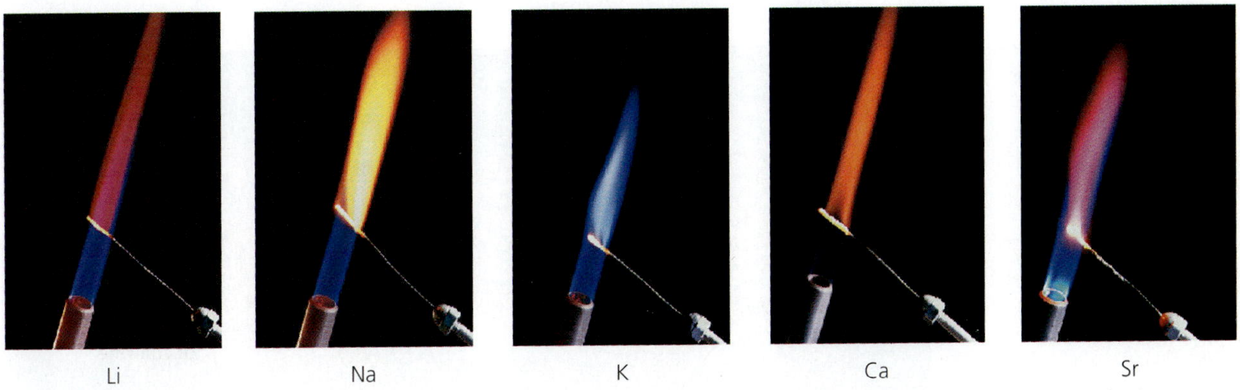

▲ **FIGURE 2.7**
Certain chemical elements can be identified by the characteristic colors that their compounds impart to flames. Five examples are shown.

* The term **"atomic mass"** is more appropriate than **"atomic weight."** However, the term "atomic weight" is so widely used by chemists that it is likely to remain in service for many years.

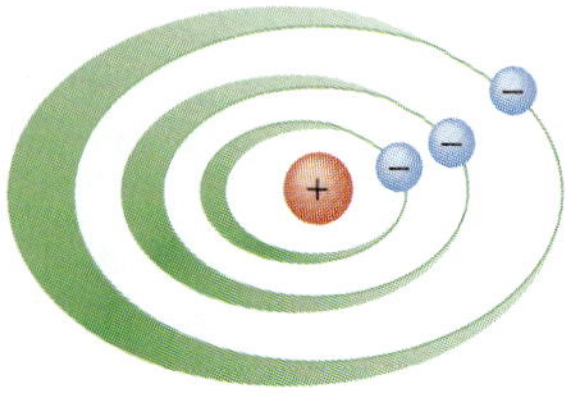

▲ **FIGURE 2.8**
Bohr visualized the atom as planetary electrons circling a nuclear sun.

scientists asked themselves. A Danish physicist, Niels Bohr, provided the answer. He said that electrons cannot be located just anywhere outside the nucleus. They must move around only in well-defined paths or orbits (Figure 2.8). An electron in one of these orbits has a characteristic amount of energy (a certain amount of kinetic energy because of its motion around the orbit and a certain amount of potential energy because it is located a certain distance from the nucleus). If an electron changes from one orbit to another, its energy also changes. If it moves from a higher-energy orbit to a lower-energy orbit, it loses energy, and that lost energy appears as light. The color of the light depends on the difference in energy between the two orbits. Atoms in which all electrons are in their lowest possible **energy levels** are said to be in the *ground state.* If one or more electrons occupy higher energy levels than in the ground state, the atom is in an *excited state.*

Energy levels—regions outside the atomic nucleus in which electrons have different energies

Bohr also said that any orbit within an atom can hold only a certain number of electrons (Figure 2.9). We shall simply state Bohr's findings in this regard. The maximum number of electrons in a given energy level is given by the formula $2n^2$, where n is the energy level being considered. For the first energy level ($n = 1$), the maximum population is $2(1^2)$, or 2. For the second energy level ($n = 2$), the maximum number of electrons is $2(2^2)$, or 8. For the third level, the maximum number is $2(3^2)$, or 18.

Example 2.4

What is the maximum number of electrons in the fifth energy level?

Solution

For the fifth level, $n = 5$, so we have

$$2 \times 5^2 = 2 \times 25 = 50$$

Practice Exercise

What is the maximum number of electrons in the fourth energy level?

The colors of fireworks result from changes in electron energy levels in atoms.

Imagine building up atoms by adding one electron to the proper energy level as each proton is added to the nucleus. Electrons will go to the lowest energy level available. For hydrogen, with a nucleus of only one proton, the one electron goes into the first energy level. For helium, with a nucleus of two protons (and two neutrons), two electrons go into the first energy level. According to Bohr, the maximum population of the first energy level is two electrons; thus, that level is filled in the helium atom. With lithium, which has three electrons, the extra electron goes into the second energy level. This process of adding electrons is continued until the second energy level is filled with eight electrons, as in the neon atom (which has a total of 10 electrons, two in the first level and eight in the second). This buildup is diagrammed in Figure 2.10. One could continue to build atoms (in one's imagination) in this manner and, with a few modifications, build up the entire collection of known elements.

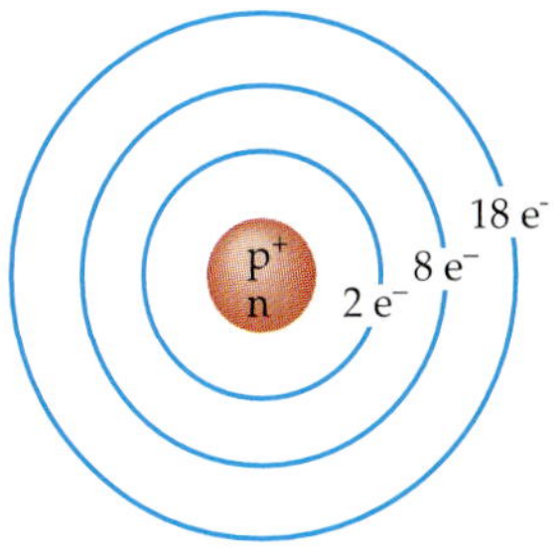

▲ **FIGURE 2.9**

A Bohr diagram for atoms of an element pictures specified numbers of electrons in distinct energy levels.

Example 2.5

Draw a Bohr diagram for a fluorine (F) atom.

Solution

Fluorine is element number 9; it has nine protons and nine electrons. Two of the electrons go into the first energy level, and the remaining seven go into the second level.

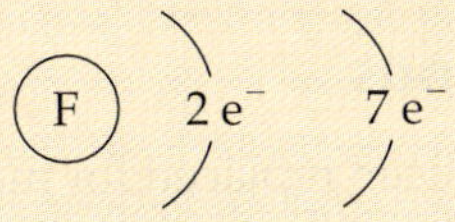

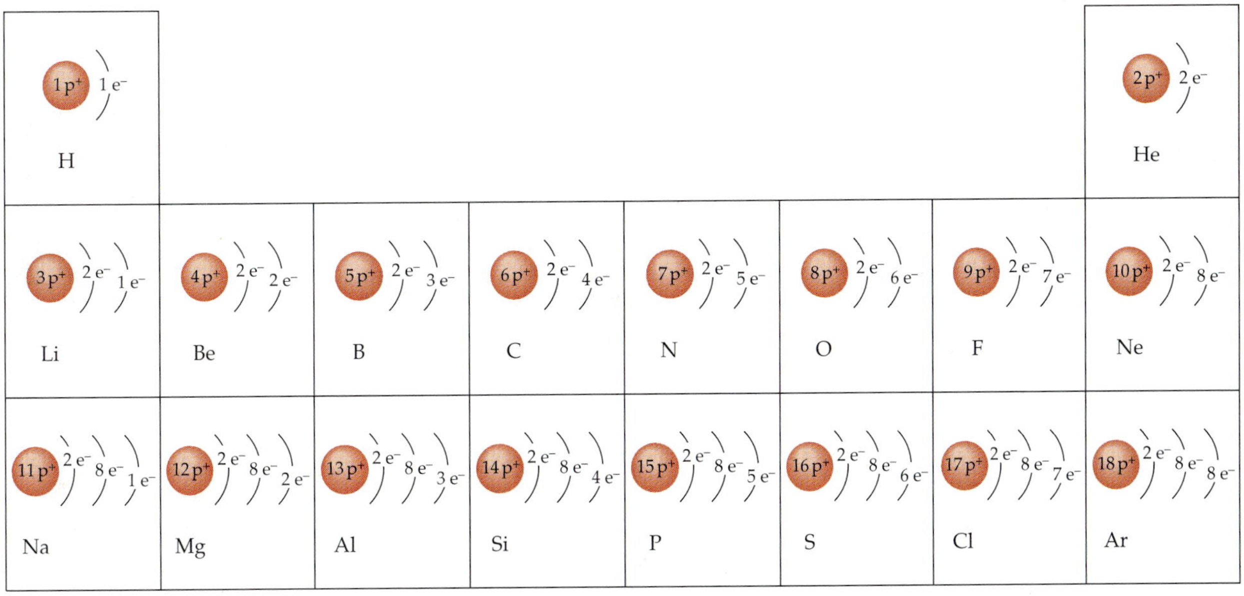

▲ **FIGURE 2.10**

Bohr diagrams of the first 18 elements.

Practice Exercise

Draw a Bohr diagram for a beryllium (Be) atom.

Example 2.6

Draw a Bohr diagram for a sodium (Na) atom.

Solution

Sodium is element number 11. It has 11 protons and 11 electrons. Two of the electrons can go into the first energy level. Of the nine remaining electrons, eight go into the second energy level. That leaves one electron for the third energy level.

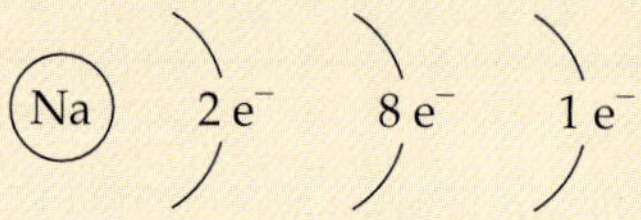

Practice Exercise

Draw a Bohr diagram for an aluminum (Al) atom.

Sometimes the Bohr configuration is written without the arcs to indicate energy levels, as follows.

Na	2, 8, 1
Ar	2, 8, 8
K	2, 8, 8, 1
F	2, 7

2.5 Electron Configurations

The Bohr model of the atom has been replaced for many purposes by more sophisticated models that explain more data in greater detail than does the simpler planetary model of Bohr. This presents us with something of a quandary. We can more accurately interpret the nature of matter only by using a model that is more difficult to understand. Fortunately, however, we need not understand the sophisticated mathematics that generates these models to make use of some of the results.

Quantum mechanics is a highly mathematical discipline used in the late 1920s by the Austrian physicist Erwin Schrödinger to develop elaborate equations that describe the properties of electrons in atoms. The results provide a measure of the probability of finding an electron in a given volume of space. The definite planetary orbits of the Bohr model are replaced by shaped volumes of space in which electrons move. The term **orbital** (replacing Bohr's *orbits*) is used for this new description of the location of electrons.

Orbital—the defined volume of an atom where an electron exists

Suppose you had a camera that could photograph electrons (there is no such thing, but we are just supposing) and you left the shutter open while an electron zipped about the nucleus. When you developed the picture, you would have a record of where the electron had been. Doing the same thing with an electric fan

FIGURE 2.11 ▶

Charge-cloud representations of atomic orbitals.

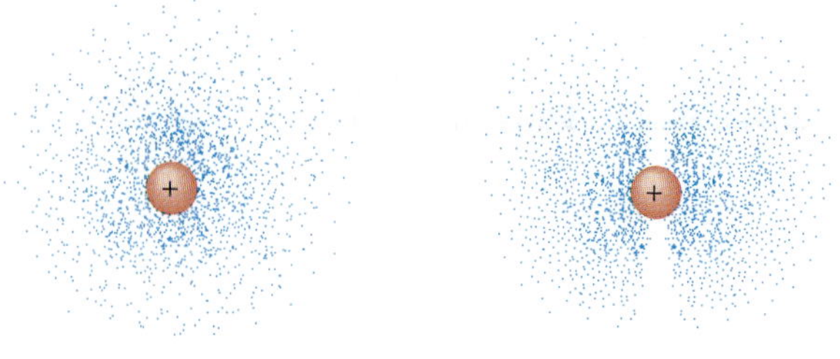

An *s* orbital   A *p* orbital

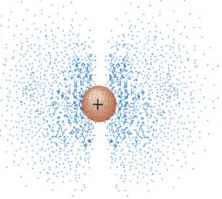

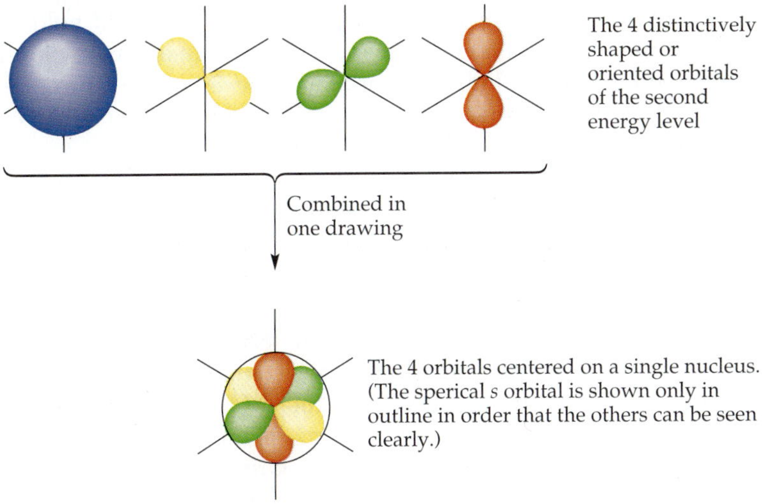

The 4 distinctively shaped or oriented orbitals of the second energy level
Combined in one drawing
The 4 orbitals centered on a single nucleus. (The sperical s orbital is shown only in outline in order that the others can be seen clearly.)

FIGURE 2.12
In these drawings of electron orbitals, the nucleus is located at the intersection of the axes. The eight electrons of the second energy level are distributed among these four orbitals, with two electrons in each orbital.

that was turned on would give you a picture in which the blades of the fan would look like a disk of material. The blades move so rapidly that their photographic image is blurred. Similarly, electrons in the first energy level of an atom would appear in our imaginary photograph as a fuzzy ball (Figure 2.11). The fuzzy ball (frequently referred to as a *charge cloud* or *electron cloud*) is the rough equivalent of an orbital.

Like Bohr, Schrödinger concluded that only two electrons could occupy the first energy level in an atom. In the quantum mechanical atom, this level is referred to as the 1*s* orbital (which is spherical—that is, shaped like a ball). Also like Bohr, Schrödinger stated that the second energy level of an atom could contain a maximum of eight electrons. However, Schrödinger concluded that these eight electrons must be located in four different orbitals. Each individual orbital could contain a maximum of two electrons. One of the orbitals of the second energy level is spherical in shape and is referred to as the 2*s* orbital. The remaining three orbitals of the second level have identical dumbbell shapes (see again Figure 2.11). They differ in their orientation in space (i.e., the direction in which they point; see Figure 2.12). As a group these orbitals are called the 2*p* orbitals, and to distinguish them from one another they are individually referred to as the $2p_x$, $2p_y$, and $2p_z$ orbitals. Electrons in these three orbitals all lie at the same energy level and possess slightly more energy than the electrons in the 2*s* orbital. Thus, the quantum mechanical atom distinguishes between main energy levels (1, 2, 3, and so on) and sublevels (2*s* and 2*p*, for example). See Tables 2.2 and 2.3.

An orbital can hold no more than two electrons.

Table 2.2 Energy Levels and Orbitals

Energy Level	Orbital(s)
1	*s*
2	*s*, *p*
3	*s*, *p*, *d*

Table 2.3 Electrons in Orbitals

Orbital Type	Orientation	Number of Orbitals	Electrons per Orbital	Total Number of Electrons
s	Spherical	1	2	2
p	Perpendicular	3	2	6
d	Perpendicular	5	2	10

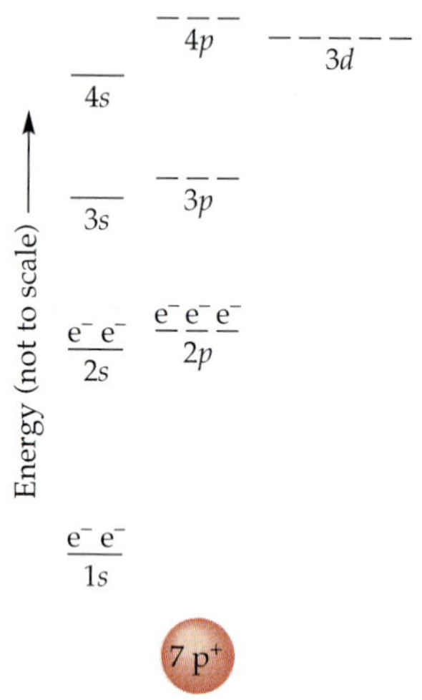

▲ **FIGURE 2.13**
Energy-level diagram for a nitrogen atom.

The third energy level of Bohr's model, and of Schrödinger's, could hold 18 electrons. In the quantum mechanical atom, however, the third main energy level is divided into three sublevels totaling nine orbitals: one 3*s* orbital, three 3*p* orbitals, and five 3*d* orbitals. Each higher main energy level adds another sublevel. The orbitals in each new sublevel have their own special shapes.

Just as one could build up atoms for all the elements by fitting electrons into the orbits of the Bohr model of the atom, so it is possible to construct a table of the elements by placing electrons into the orbitals of the quantum mechanical model. The energy-level diagram for nitrogen (which has seven electrons) is shown in Figure 2.13. The quantum mechanical description of the nitrogen atom focuses on a more detailed description of its **electron configuration** (arrangement): $1s^22s^22p^3$. The superscripts indicate the total number of electrons contained in a particular energy sublevel. For example, in nitrogen there are a total of three electrons in the 2*p* orbitals. Table 2.4 lists the electron configurations of the first 20 elements. Notice that the orbitals fill in order of increasing energy: first the 1*s*, then the 2*s*, 2*p*, 3*s*, and so on. The *d* orbitals introduce a minor complication because the 3*d* orbitals turn out to be at a higher energy level than the 4*s* orbitals. Figure 2.14 illustrates the order in which orbitals are filled.

Electron configuration—the specific ordering of electrons in an atom

To check whether you understand the fundamentals of orbitals and electron configurations, work through the following examples.

Example 2.7

Write the electron configuration for carbon, atomic number 6.

Table 2.4 Electron Configurations for Atoms of the First 20 Elements

Name	Atomic Number	Electron Structure
Hydrogen	1	$1s^1$
Helium	2	$1s^2$
Lithium	3	$1s^22s^1$
Beryllium	4	$1s^22s^2$
Boron	5	$1s^22s^22p^1$
Carbon	6	$1s^22s^22p^2$
Nitrogen	7	$1s^22s^22p^3$
Oxygen	8	$1s^22s^22p^4$
Fluorine	9	$1s^22s^22p^5$
Neon	10	$1s^22s^22p^6$
Sodium	11	$1s^22s^22p^63s^1$
Magnesium	12	$1s^22s^22p^63s^2$
Aluminum	13	$1s^22s^22p^63s^23p^1$
Silicon	14	$1s^22s^22p^63s^23p^2$
Phosphorus	15	$1s^22s^22p^63s^23p^3$
Sulfur	16	$1s^22s^22p^63s^23p^4$
Chlorine	17	$1s^22s^22p^63s^23p^5$
Argon	18	$1s^22s^22p^63s^23p^6$
Potassium	19	$1s^22s^22p^63s^23p^64s^1$
Calcium	20	$1s^22s^22p^63s^23p^64s^2$

Solution

The atomic number indicates that there are six electrons in the carbon atom. The first energy level can accommodate only two of these electrons in its 1*s* orbital.

$$1s^2 \ldots$$

The remaining four electrons must occupy orbitals in the second main energy level. The lowest energy orbital of the second level is the 2*s* orbital, which can accommodate two electrons.

$$1s^2 2s^2 \ldots$$

The remaining two electrons must go to the 2*p* orbitals, which can handle up to six electrons if necessary.

$$1s^2 2s^2 2p^2$$

Practice Exercise

Write out the electron configuration for fluorine.

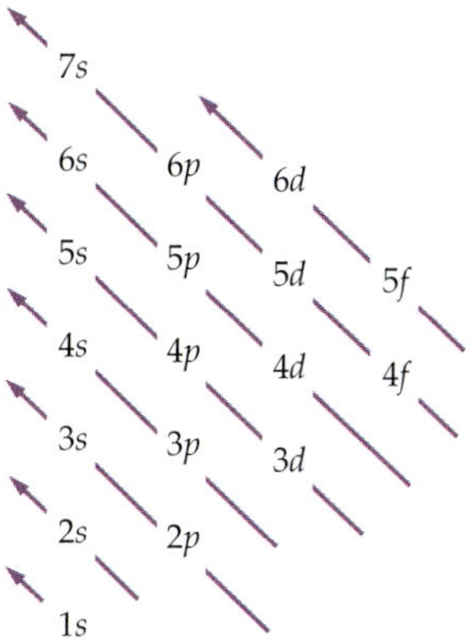

▲ **FIGURE 2.14**
An order-of-filling chart for determining the electron configurations of atoms. The first energy level has only *s* orbitals, the second has *s* and *p* orbitals, the third has *s*, *p*, and *d* orbitals, and so on.

Example 2.8

Write the electron configuration for phosphorus, atomic number 15.

Solution

We have 15 electrons to distribute. Two go into the first main energy level ($1s^2$), eight go into the second ($2s^2 2p^6$), and five go into the third ($3s^2 3p^3$). In summary, the configuration of phosphorus is

$$1s^2 2s^2 2p^6 3s^2 3p^3$$

Practice Exercise

Write out the electron configuration for chlorine.

When an atom gains or loses electrons, it becomes charged and is referred to as an **ion.** Thus, if an atom of lithium were to lose an electron, its electron configuration would be $1s^2 2s^0$. It would then be Li^+. In the same way, if an atom of fluorine were to gain an electron, it would become F^- with the electron configuration $1s^2 2s^2 2p^6$. We discuss ions further in Chapter 4.

Some **ions** are simply electrically charged atoms. Other ions consist of a group of atoms that act as a unit (see Section 4.12).

Quantum mechanics offers detailed descriptions of the electron structure of atoms. However, we shall make most use of the picture it paints of electrons as clouds of negative charge. Sometimes the shape of the cloud (that is, the orbital) is presented simply in outline. (Figure 2.12 uses shaded drawings to present the combined orbitals of the second energy level.) Later we shall see how the shape and orientation of the electron cloud can determine the shapes of what we call molecules.

2.6 The Periodic Table

As our picture of the atom becomes more detailed, we find ourselves in a dilemma. With more than 100 elements to deal with, how can we keep all this information straight? One way is by using the **periodic table of the elements.** (See

The periodic table of the elements provides a wealth of information.

Table 2.5 Mendeleev's Periodic Table

Reihen	Gruppe I R^2O	Gruppe II RO	Gruppe III R^2O^3	Gruppe IV RH^4 RO^2
1	H = 1			
2	Li = 7	Be = 9,4	B = 11	C = 12
3	Na = 23	Mg = 24	Al = 27,3	Si = 28
4	K = 39	Ca = 40	-- = 44	Ti = 48
5	(Cu = 63)	Zn = 65	-- = 68	-- = 72
6	Rb = 85	Sr = 87	?Yt = 88	Zr = 90
7	(Ag = 108)	Cd = 112	In = 113	Sn = 118

the inside front cover of this book.) The periodic table neatly tabulates information about atoms. It records how many protons and electrons the atoms of a particular element contain. It permits us to calculate the number of neutrons in the most common isotope for most elements. It even stores information about how electrons are arranged in the atoms of each element. The most extraordinary thing about the periodic table is that it was largely developed before anyone knew there were protons or neutrons or electrons in atoms.

Not long after Dalton presented his model for the atom (an indivisible particle whose mass determined its identity), chemists began preparing listings of elements arranged according to their atomic weights. While working out such tables of elements, these scientists observed patterns among the elements. For example, it became clear that elements that occurred at specific intervals shared similarities in certain properties. Among the approximately 60 elements known at that time, the second and ninth showed similar properties, as did the third and tenth, the fourth and eleventh, the fifth and twelfth, and so on.

Dmitri Ivanovich Mendeleev (1834–1907). Mendeleev invented the periodic table while trying to systematize the properties of the elements for presentation in a chemistry textbook. His highly influential text lasted for 13 editions—five of them after his death.

In 1869, Dmitri Ivanovich Mendeleev, a Russian chemist, published his periodic table of the elements. Mendeleev prepared his table by taking into account both the atomic weights and the periodicity of certain properties of the elements. The elements were arranged primarily in order of increasing atomic weight. In a few cases, Mendeleev placed a slightly heavier element before a lighter one. He did this only when it was necessary in order to keep elements with similar chemical properties in the same column. For example, he placed tellurium (atomic weight = 128) ahead of iodine (atomic weight = 127) because tellurium resembled sulfur and selenium in its properties, whereas iodine was similar to chlorine and bromine.

Mendeleev left a number of gaps (blue) in his table (Table 2.5). Instead of looking upon those blank spaces as defects, he boldly predicted the existence of elements as yet undiscovered. Furthermore, he even predicted the properties of some of these missing elements. In succeeding years, many of the gaps were filled in by the discovery of new elements, and their properties were often quite close to those Mendeleev had predicted. It was the predictive value of this great innovation that led to wide acceptance of Mendeleev's table.

It is now known that properties of an element depend mainly on the number of electrons in the outermost energy level of the atoms of the element (Chapter 4).

Sodium atoms have one electron in their outermost energy level (the third). Lithium atoms have a single electron in their outermost level (the second). The chemical properties of sodium and lithium are similar. The atoms of helium and neon have filled outer electron energy levels, and both elements are inert— that is, they do not undergo chemical reactions readily. Apparently, not only are similar chemical properties shared by elements whose atoms have similar electron *configurations* (arrangements) but also certain configurations appear to be more stable (less reactive) than others. This is a point we shall explore in Chapter 4.

In Mendeleev's table, the elements were arranged by atomic weights for the most part, and this arrangement revealed the periodicity of chemical properties. Because the number of electrons determines the element's chemical properties, that number determines the order of the periodic table. In the modern periodic table, the elements are arranged according to atomic number. Remember, this number indicates both how many protons and how many electrons there are in a neutral atom of the element. The modern table, arranged in order of increasing atomic number, and Mendeleev's table, arranged in order of increasing atomic weight, parallel one another because an increase in atomic number is generally accompanied by an increase in atomic weight. In only a few cases (noted by Mendeleev) do the weights fall out of order. Atomic weights do not increase in precisely the same order as atomic numbers because *both protons and neutrons* contribute to the mass of an atom. It is possible for an atom of lower atomic number to have more neutrons than one with a higher atomic number. An atom with a lower atomic number can therefore have a greater mass than an atom with a higher atomic number. Thus the atomic mass of Ar (no. 18) is more than that of K (no. 19), and Te (no. 52) has a mass greater than that of I (no. 53); see the periodic table.

The modern periodic table (Figure 2.15) has vertical columns called **groups** (or families) of elements. All elements in a given group have the same number of electrons in the outermost electron shell and therefore have similar properties. **Periods** are the horizontal rows of elements. Each period indicates the opening of the next main energy level. For example, sodium opens the third period; the outermost electron of a sodium atom is the first to be placed in the *third* energy level. Because each period begins a new energy level, the size of atoms increases from top to bottom within a group. And because electrons are easier to remove when farther from the nucleus, the larger atoms have lower ionization energies. (*Ionization energy* is the energy required to remove an electron from an atom.)

Columns in the periodic table are called groups or families.

Rows in the periodic table are called periods.

Example 2.9

Write out the sublevel notation for the electrons in the highest main energy level for strontium (Sr) and arsenic (As).

Solution

Strontium is in Group 2A and the fifth period of the periodic chart. Its outer electron configuration is $5s^2$. (Each period of the periodic table corresponds to the filling, or at least partial filling, of a main energy level.) Arsenic is in Group 5A and the fourth period. Its five outer (valence) electrons have the configuration denoted by $4s^24p^3$.

Practice Exercise

Write out the sublevel notation for the electrons in the highest main energy level for rubidium (Rb) and selenium (Se).

1 Group 1A																	18 Group 8A
1 H 1.00794	2 Group 2A											13 Group 3A	14 Group 4A	15 Group 5A	16 Group 6A	17 Group 7A	2 He 4.002602
3 Li 6.941	4 Be 9.01218											5 B 10.811	6 C 12.011	7 N 14.0067	8 O 15.9994	9 F 18.998403	10 Ne 20.179
11 Na 22.98977	12 Mg 24.305	3 Group 3B	4 Group 4B	5 Group 5B	6 Group 6B	7 Group 7B	8 Group ←	9 Group 8B	10 Group →	11 Group 1B	12 Group 2B	13 Al 26.98154	14 Si 28.0855	15 P 30.97376	16 S 32.066	17 Cl 35.453	18 Ar 39.948
19 K 39.0983	20 Ca 40.078	21 Sc 44.9559	22 Ti 47.88	23 V 50.9415	24 Cr 51.9961	25 Mn 54.9380	26 Fe 55.847	27 Co 58.9332	28 Ni 58.6934	29 Cu 63.546	30 Zn 65.38	31 Ga 69.723	32 Ge 72.59	33 As 74.9216	34 Se 78.96	35 Br 79.904	36 Kr 83.80
37 Rb 85.4678	38 Sr 87.62	39 Y 88.9059	40 Zr 91.22	41 Nb 92.9064	42 Mo 95.94	43 Tc (98)	44 Ru 101.07	45 Rh 102.9055	46 Pd 106.42	47 Ag 107.8682	48 Cd 112.41	49 In 114.818	50 Sn 118.710	51 Sb 121.757	52 Te 127.60	53 I 126.9045	54 Xe 131.29
55 Cs 132.9054	56 Ba 137.33	57 La 138.9055	72 Hf 178.49	73 Ta 180.9479	74 W 183.84	75 Re 186.207	76 Os 190.23	77 Ir 192.22	78 Pt 195.08	79 Au 196.9665	80 Hg 200.59	81 Tl 204.383	82 Pb 207.2	83 Bi 208.9804	84 Po (209)	85 At (210)	86 Rn (222)
87 Fr (223)	88 Ra 226.0254	89 Ac 227.0278	104 Unq (261)	105 Unp (262)	106 Unh (263)	107 Uns (262)	108 Uno (265)	109 Une (266)									

Metals ← → Nonmetals

58 Ce 140.12	59 Pr 140.9077	60 Nd 144.24	61 Pm (145)	62 Sm 150.36	63 Eu 151.96	64 Gd 157.25	65 Tb 158.9254	66 Dy 162.50	67 Ho 164.9304	68 Er 167.26	69 Tm 168.9342	70 Yb 173.04	71 Lu 174.967
90 Th 232.0381	91 Pa 231.0359	92 U 238.0289	93 Np 237.0482	94 Pu (244)	95 Am (243)	96 Cm (247)	97 Bk (247)	98 Cf (251)	99 Es (252)	100 Fm (257)	101 Md (258)	102 No (259)	103 Lr (260)

▲ **FIGURE 2.15**
The modern periodic table.

Elements are also divided into two main classes by a heavy, stepped, diagonal line. Those to the left of the line, except for hydrogen, are **metals,** elements that have a characteristic luster and are generally good conductors of heat and electricity. Most metals are *malleable,* which means that they can be hammered into thin sheet or foil. Also, most metals are *ductile;* they can be drawn into wire. Except for mercury, which is a liquid, all the metals are solid at room temperature.

The metals copper (top) and gold (bottom) are malleable and ductile.

Elements to the right of the stepped line are **nonmetals,** elements that lack metallic properties. For example, nonmetals generally are poor conductors of heat and electricity. At room temperature, several nonmetals, such as oxygen, nitrogen, fluorine, and chlorine, are gases. Others, such as carbon, sulfur, phosphorus, and iodine, are brittle solids. Bromine is the only nonmetal that is a liquid at room temperature.

Some of the groups of elements have special names. Group 1A elements (except hydrogen, a nonmetal) are called **alkali metals.** Group 2A elements are the **alkaline earth metals.** Group 7A is known as the **halogens,** and Group 8A is made up of the **noble gases.** The alkali metals (Group 1A) are strongly metallic and the halogens (Group 7A) are distinctly nonmetallic. In the middle of the table we see intermediate behavior. Group 4A has carbon (a nonmetal), at the top and two metals, tin and lead, at the bottom. The noble gases are exceptionally reluctant to

Unsettled Issues Concerning the Periodic Table

The periodic table has been around for well over a century, and in that time many variations have been proposed. Yet there still is no single table accepted by all chemists. A particular state of confusion results from differing use of the letters A and B. In the United States, A is used to designate the representative (nontransition) elements and B the transition elements. In Europe, the groups to the left of Fe/Ru/Os have typically been designated as A groups, and those to the right of Ni/Pd/Pt as B groups. In order to eliminate the confusion over the use of A and B, the International Union of Pure and Applied Chemistry (IUPAC; see Section 13.5) has recommended that the groups simply be numbered from 1 to 18. (These numbers are shown above the A/B group designations in Figure 2.15). There are advantages to each system. The A/B system is used in this text. It is helpful, for example, in that the A-group numbers are equal to the numbers of outer-shell electrons in atoms of the representative elements. There is also a lack of international agreement on the names of elements 104 through 109, and elements 110, 111, and 112 have not yet been named. These synthetic elements are rarely encountered and will not be considered further here. Although chemists disagree over which *form* of the periodic table is *most* useful, there is no disagreement over the great utility of the table.

enter into chemical reactions. This lack of reactivity is a reflection of their electron configurations, as we shall see in Chapter 4.

The B group elements are generally known as the **transition elements;** those of the A groups are called the *main group* or **representative elements.** The periodic table on the inside front cover depicts the physical states of the elements. Special Topic D (which follows Chapter 12) discusses the chemistries of some of the members of these various groups.

2.7 Which Model to Use?

For our purposes, we will use whichever model of the atom is more helpful for understanding a particular concept. In trying to evaluate the behavior of gases, chemists often use Dalton's hard, indivisible (billiard ball) model. In discussing how atoms combine to form molecules, we will use the Bohr model. To explain the shapes of molecules (very important in explaining the actions of drugs, for instance), we will use an extension of the Bohr theory called the *valence shell electron pair repulsion* (VSEPR) theory (Special Topic B, page 107), for the most part. Occasionally we will employ the orbital atom theory. Don't let this inconsistent use disturb you. Remember that a model or a theory is used to explain phenomena. When a better model or theory is invented, the old one is sometimes discarded or pushed into a secondary role, but occasionally, it can be used to clarify some point that a newer, more complicated model only obscures.

Summary

Dalton's atomic theory states that

1. Elements are made up of small, indivisible atoms.
2. All atoms of a given element are identical.
3. Atoms of different elements combine to form compounds.
4. Chemical reactions involve change in the composition of compounds, not change in the atoms making up those compounds.

Atoms are not indivisible but rather are made up of three kinds of subatomic particles: protons, electrons, and neutrons. Each **proton** carries a positive electric charge of +1 and has a mass of approximately 1 u. Each **neutron** is electrically neutral and has a mass of approximately 1 u. Each **electron** carries a negative electric charge of −1 and has a mass of 0.000549 u. This mass is so small relative to the mass of the proton and the neutron that we usually speak of electrons as having zero mass. In any atom, the number of protons equals the number of electrons, and so the atom is electrically neutral.

An atom can be pictured as a sphere. The protons and neutrons are clustered tightly together in a relatively small volume at the center of the sphere. This proton–neutron cluster is called the **nucleus.** The electrons are scattered outside the nucleus in all the remaining volume of the sphere.

Every element has its own unique **atomic number** Z, which tells us the number of protons in the nucleus of each atom of the element. In addition, the **mass number** A of any atom tells us the number of protons plus neutrons in the atom. Example: An atom containing six protons, six neutrons, and six electrons has an atomic number of 6 and a mass number of 6 + 6 = 12. Note that there is no "electron number" for an atom. Such a number is unnecessary because the number of electrons is the same as the number of protons.

Although all atoms of a given element must contain the same number of protons, the number of neutrons can vary from one atom to another of the same element. Atoms containing the same number of protons but different numbers of neutrons are **isotopes** of each other. Isotopes are represented by the notation ${}^{A}_{Z}X$, where X stands for the abbreviation of the element's name. Example: An atom containing six protons ($Z = 6$) and six neutrons ($A = 12$) and a second atom containing six protons ($Z = 6$) and eight neutrons ($A = 14$) are both atoms of the element carbon (because $Z = 6$) and are isotopes of each other (different numbers of neutrons). The two are represented by the symbols ${}^{12}_{6}C$ and ${}^{14}_{6}C$.

The relative masses of atoms are given in units called **atomic mass units (amu).** The isotope of carbon containing six neutrons, ${}^{12}_{6}C$, is arbitrarily assigned a mass of exactly 12 u, and the masses of all other elements are measured relative to this standard carbon mass.

In the Bohr model of the atom, the electrons are confined to well-defined orbits that circle the nucleus. Because this picture so closely resembles planets circling a central sun, this model is alternately referred to as the planetary model. Each orbit represents a certain **energy level** for the electrons in it. In any given orbit, an electron has a certain (constant) amount of energy. Whenever an electron jumps to another orbit, the energy of the electron changes.

Normally, an atom has all its electrons in the lowest possible energy level. Such an arrangement is called the **ground state** of the atom. When an atom has absorbed energy to boost one or more electrons to a higher energy level, the atom is said to be in an **excited state.** An electron gains energy as it is promoted to a higher energy level; it releases energy as it returns to a lower energy level. Each main energy level (main shell) can hold only a certain number of electrons. That number is given by the formula $2n^2$, where n is the energy level.

Today's view of the atom uses a quantum-mechanical model in which the well-defined orbits of the Bohr model are replaced by fuzzy "volumes of space" occupied by the electrons. These fuzzy areas are called **orbitals,** and each orbital can hold at most two electrons. Further, the main energy levels (shells) are divided into sublevels (subshells). The sublevels are designated by the letters *s*, *p*, *d*, and *f*. The first shell has only one orbital, designated 1*s*, where the 1 designates the first energy level and the *s* describes a type of orbital that is spherical in shape. The second shell has two subshells, the 2*s* with one spherical orbital and the 2*p* with three dumbbell-shaped orbitals oriented perpendicular to each other.

As with the Bohr model, each energy level in the quantum-mechanical model can hold $2n^2$ electrons. The way these electrons are distributed in any atom is called the **electron configuration** for that atom. Example: The electron configuration for the element neon, $1s^22s^22p^6$, tells us that the ten electrons of neon are distributed two in the 1*s* orbital, two in the 2*s* orbital, and six evenly distributed among the three 2*p* orbitals.

An atom that loses or gains one or more electrons acquires a net electric charge (because now the number of negative electrons is no longer balanced by the number of positive protons). Such a charged species is called an **ion.**

The **periodic table of the elements** organizes all the known elements into rows and columns. The chart is called *periodic* because that word implies a pattern that repeats at regular intervals. The columns are called either **groups** or *families,* and all the members of one family contain the same number of electrons in the outermost energy level. This similarity means that all members of a family have similar chemical properties. The rows of the periodic table are called **periods,** and the beginning of each new period (new row) indicates an additional energy level in the atoms.

Key Terms

alkali metal (2.6)
alkaline earth metal (2.6)
atomic mass (2.3)
atomic mass unit (u) (2.1)
atomic number (*Z*) (2.2)
atomic weight (2.3)
cathode ray (2.2)
electron (2.2)
electron configuration (2.5)
energy level (2.4)
group (2.6)
halogen (2.6)
ion (2.5)
isotope (2.2)
mass number (*A*) (2.3)
metal (2.6)
neutron (2.2)
noble gas (2.6)
nonmetal (2.6)
nucleus (2.2)
orbital (2.6)
period (2.6)
periodic table of the elements (2.6)
proton (2.2)
radioactivity (2.2)
representative element (2.6)
transition element (2.6)

Review Questions

1. What is the distinction between the atomistic view and the continuous view of matter?

2. If foods were described as atomistic or continuous, which designation would you use for each of the following?
- **a.** hard-boiled eggs
- **b.** hot dogs
- **c.** mashed potatoes
- **d.** milk
- **e.** peas
- **f.** scrambled eggs

3. Outline the main points of Dalton's atomic theory.

4. How did the discovery of radioactivity contradict Dalton's atomic theory?

5. Give the distinguishing characteristics of the proton, the neutron, and the electron.

6. Should a proton and an electron attract or repel one another?

7. Should a neutron and a proton attract or repel one another?

8. What is the atomic nucleus? What subatomic particles are found in the nucleus?

9. What subatomic particles are found outside the nucleus?

10. Compare Dalton's model of the atom with the nuclear model of the atom.

11. The nucleus of an atom with a mass number of 23 has 11 protons.
- **a.** How many electrons are there in the neutral atom?
- **b.** How many neutrons are there in the nucleus of the atom?

12. What are isotopes, and what is meant by the mass number of an isotope?

13. Explain what tabulated atomic weight values, such as those found on the inside front cover, represent.

14. List some characteristic properties of metals, and indicate where these elements are located in the periodic table.

15. The two principal isotopes of lithium are lithium-6 and lithium-7. The atomic weight of lithium is 6.9 u. Which is the predominant isotope of lithium?

16. Out of every five atoms of boron, one has a mass of 10 u and four have a mass of 11 u. What is the atomic weight of boron? Use the periodic table only to check your answer.

17. If three electrons were added to the outermost energy level of a phosphorus atom, the new electron configuration would resemble that of what element?

18. If two electrons were removed from the outermost energy level of a magnesium atom, the new electron configuration would resemble that of what element?

19. What is meant by the ground state of an atom? The excited state?

20. When light is emitted by an atom, what change has occurred within the atom?

21. Which has absorbed more energy, an atom in which an electron has moved from the second energy level to the third energy level, or an otherwise identical atom in which an electron has moved from the first energy level to the third?

22. Consider the quantum mechanical notation $2p^6$.
- **a.** How many electrons does this notation represent?
- **b.** What is the general shape of the orbitals described?
- **c.** How many orbitals are included?

Problems

Dalton's Atomic Theory

23. To the nearest atomic mass unit, an atom of calcium has a mass of 40 u and an atom of vanadium has a mass of 50 u. Are these findings in agreement with Dalton's atomic theory? Explain your answer.
24. To the nearest atomic mass unit, an atom of calcium has a mass of 40 u and an atom of potassium has a mass of 40 u. Are these findings in agreement with Dalton's atomic theory? Explain your answer.
25. To the nearest atomic mass unit, one atom of calcium has a mass of 40 u and another calcium atom has a mass of 44 u. Are these findings in agreement with Dalton's atomic theory? Explain your answer.
26. An atom of uranium-235 is struck by a neutron and splits into two smaller atoms. Are these findings in agreement with Dalton's atomic theory? Explain your answer.

The Nuclear Atom

27. Indicate how many electrons and how many protons there are in a neutral atom of each of the following elements. (You may use the periodic table.)
    a. calcium b. sodium
    c. fluorine d. argon
28. Indicate how many electrons and how many protons there are in a neutral atom of each of the following elements. (You may use the periodic table.)
    a. beryllium b. nitrogen
    c. iron d. uranium
29. Indicate the number of protons and the number of neutrons in atoms of the following isotopes.
    a. ^{62}Zn b. ^{241}Pu
    c. ^{99}Tc d. ^{99}Mo
30. Indicate the number of protons and the number of neutrons in atoms of the following isotopes.
    a. ^{11}B b. ^{154}Sm
    c. ^{81}Kr d. ^{121}Te
31. Which of the following pairs of symbols represent isotopes?
    a. $^{70}_{33}E$ and $^{70}_{34}E$ b. $^{57}_{28}E$ and $^{66}_{28}E$
    c. $^{186}_{74}E$ and $^{186}_{74}E$ d. $^{7}_{3}E$ and $^{3}_{2}E$
    e. $^{22}_{11}E$ and $^{11}_{6}E$
32. Use the symbolism $^{A}_{Z}E$ to represent each of the following isotopes. (You may refer to the periodic table.)
    a. boron-8 b. carbon-14
    c. uranium-235 d. cobalt-60

The Bohr Model

33. According to Bohr, what is the maximum number of electrons in the third energy level?
34. If the third energy level contains two electrons, what is the total number of electrons in the atom?
35. Draw Bohr diagrams for each of the following elements.
    a. silicon b. nitrogen c. sulfur
36. Draw Bohr diagrams for each of the following atoms.
    a. helium b. chlorine c. magnesium

Electron Configurations

37. Give the electron configurations (using quantum mechanical notation) for the elements in Problem 35.
38. Which elements have the following electron configurations?
    a. $1s^22s^2$
    b. $1s^22s^22p^3$
    c. $1s^22s^22p^63s^23p^1$
39. Give the atomic numbers of the elements described in Problem 38.
40. None of the following electron configurations is reasonable. Explain why in each case.
    a. $1s^22s^23s^2$
    b. $1s^22s^22p^23s^1$
    c. $1s^22s^22p^62d^5$

The Periodic Table

41. Indicate the group and period in which each of the following elements is found. Classify each as a metal or nonmetal. (You may use the periodic table.)
    a. C b. Ca c. Cd d. Cl
    e. B f. Ba g. Bi h. Br
42. Indicate the group and period in which each of the following elements is found. Classify each as a metal or nonmetal. (You may use the periodic table.)
    a. S b. Sn c. Sm d. Sr
    e. Ta f. Tc g. Ti h. Tl
43. Identify the element in each of the following descriptions. (You may use the periodic table.)
    a. Group 3A, Period 4 b. Group 1B, Period 4
    c. Period 5, halogen
44. Identify the element in each of the following descriptions. (You may use the periodic table.)
    a. Group 4A, nonmetal b. Group 7B, Period 5
    c. Period 2, alkali metal
45. Which of the following elements are halogens?
    a. Ag b. At c. As
46. Which of the following elements are alkali metals?
    a. K b. Y c. W
47. Which of the following elements are noble gases?
    a. Fe b. Ne c. Ge d. He e. Xe
48. Which of the following elements are transition metals?
    a. Ti b. Tc c. Te
49. How many electrons are in the outermost energy level of the halogens?
50. How many electrons are in the outermost energy level of Group 2A elements?

Additional Problems

51. Fill in the blanks for the six elements listed.

	Atomic No.	No. of Protons	No. of Electrons	No. of Neutrons	Mass No.
Pb	———	———	———	126	208
Sr	———	38	———	50	———
N	7	———	———	———	14
Cr	———	———	24	———	52
Ag	47	———	———	60	———
As	———	———	33	42	———

52. Indicate the group number or numbers of the following families.
a. alkali metals **b.** transition metals
c. halogens **d.** alkaline earth metals

53. Which atom is larger?
a. K or Rb **b.** Ne or Na **c.** Zn or Zr

54. Which atom will ionize more easily?
a. Li or Cs **b.** N or Sb **c.** Ar or Ac

55. What is the difference between the electron configurations of oxygen (O) and fluorine (F)?

56. What is the difference between the electron configurations of fluorine (F) and sulfur (S)?

57. Elements are defined on a theoretical basis as being composed of atoms that share the same atomic number. On the basis of this theory, would you think it possible for someone to discover a new element that would fit between magnesium (atomic number 12) and aluminum (atomic number 13)?

58. Referring only to the periodic table, indicate what similarity in electron structure is shared by fluorine (F) and chlorine (Cl). What is the difference between their electron structures?

59. Use quantum mechanical notation to describe the electron configuration of the atom represented in the following Bohr diagram.

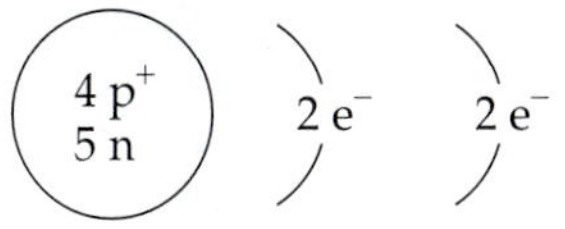

60. The following Bohr diagram is supposed to represent a neutral atom of an element. The diagram is incorrectly drawn. Identify the error.

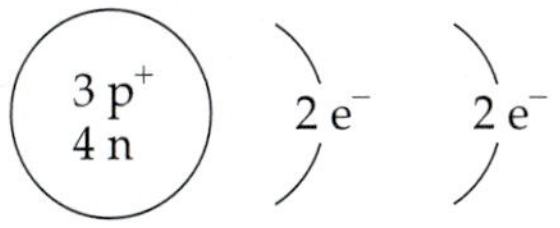

3 Nuclear Processes

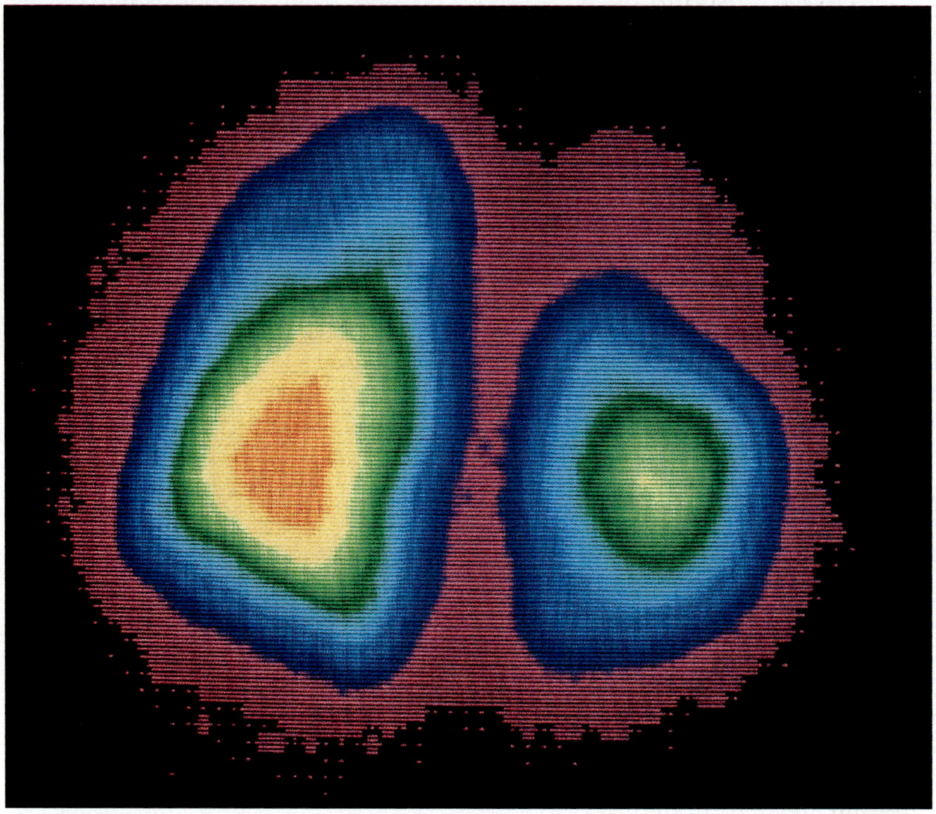

Knowledge of the properties of atomic nuclei makes it possible to obtain images that correspond to the distribution of various biochemical reactions in the body.

Chemists often focus only on the outer electrons of atoms, for it is usually only the outer electrons that are involved in chemical reactions. We start our discussion of chemical reactions in the next chapter and devote much of this text to the study of these reactions. For now, though, let us turn our attention to that tiny speck of matter called the nucleus. The volume of an entire atom is about 10,000 times that of its nucleus, yet it is the nucleus that holds the power that has become the symbol of our age.

Nuclear power confronts us with a great paradox. Although nuclear power unleashed in wrath can destroy cities and perhaps civilizations, controlled nuclear power can provide the energy necessary to run our cities and maintain our civilization. Yet even the peaceful uses of nuclear power are controversial. As citizens of the nuclear age, we have difficult decisions to make. Nuclear bombs may kill, but nuclear medicine saves lives. Diseases once regarded as incurable can be diagnosed and treated effectively with radioactive isotopes. Applications of nuclear chemistry to biology, industry, and agriculture have improved the human condition significantly. The use of radioisotopes in biological and agricultural research has led to increased crop production, which provides more food for a hungry world.

3.1 Discovery of Radioactivity

In Chapter 2, we discussed the existence of isotopes. Isotopes usually are not considered in ordinary *chemical* reactions. Such reactions involve the outer electrons of atoms, and differences in the numbers of neutrons buried deep in the hearts of atoms would not be expected to have a major effect. In *nuclear reactions,* however—reactions that do involve the heart of the atom—isotopes are of utmost importance, as we shall see.

The discovery of the first nuclear reactions was triggered by the discovery of X-rays, which are not nuclear phenomena. In 1895, a German scientist, Wilhelm Konrad Roentgen, found that he could produce a form of radiation that passed right through solid materials. This mysterious radiation was like visible light in that it was a form of pure energy. Also, like visible light, it resulted from electrons moving from one energy level to another. It was unlike visible light in that you could not see it. For want of a better name, Roentgen called the invisible, penetrating radiation *X-rays*. **X-rays** are high-energy radiation. They contain more energy than even the most energetic visible light. It is their high energy that gives X-rays such penetrating power.

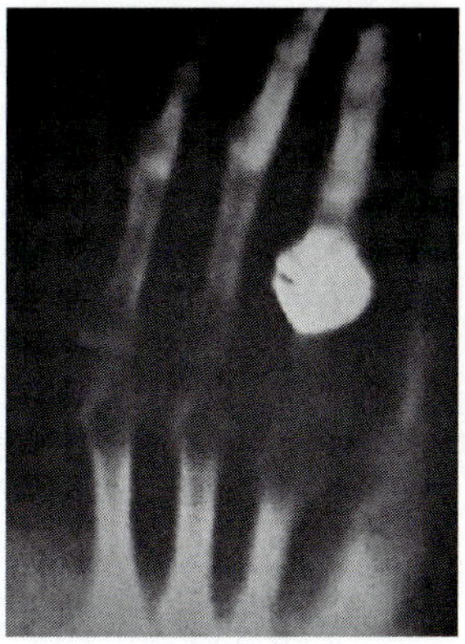

▲ **FIGURE 3.1**
The first-ever X-ray of a human being was made by Wilhelm Roentgen (1845–1923) shortly after his discovery of X-rays in 1895. It shows the hand of his wife, with the ring she was wearing.

The medical community immediately recognized the significance of the penetrating power of X-rays. The X-ray picture shown in Figure 3.1 was taken in February 1896, within two months of the publication of Roentgen's discovery. The round black dots are gunshot pellets. This picture was used to establish their positions for removal by surgery.

This phenomenon fascinated other scientists, also. One, a French physicist named Antoine Henri Becquerel (1852–1908), had been studying a different phenomenon called fluorescence. Fluorescent compounds, after being exposed to strong sunlight, continue to glow even when taken into a dark room. Becquerel wondered if there were any substances that would give off X-rays when they fluoresced. In trying to answer this question, Becquerel tested a large number of materials—among them, uranium compounds. Becquerel found that uranium did emit invisible, penetrating rays. These rays, however, proved to have nothing to do with fluorescence or with X-rays. Becquerel had discovered a totally new phenomenon.

X-rays—high-energy radiation

At this point, Marie and Pierre Curie and Ernest Rutherford entered the picture. We first mentioned these scientists in our discussion of models for the atom. Marie Curie named the phenomenon discovered by Becquerel **radioactivity.** The Curies went on to discover several new radioactive elements, including radium.

3.2 Types of Radioactivity

Scientists soon realized that the radiation emanating from uranium and other radioactive elements was of three types. When this radiation was passed through a strong magnetic or electric field, one portion was deflected in one direction, another portion was deflected in the opposite direction, and a third was not deflected at all. Rutherford named these portions *alpha* (α) *rays, beta* (β) *rays,* and *gamma* (γ) *rays,* respectively (Figure 3.2).

Only the **gamma rays** are radiation in the same sense that visible light and X-rays are radiation—that is, pure radiant energy. Gamma rays have the highest energy and are the most penetrating form of lightlike radiation yet discovered.

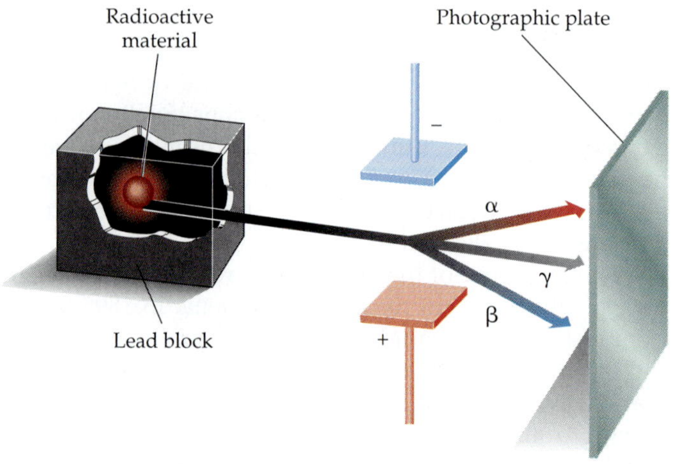

FIGURE 3.2 ▶
Three types of radiation emanate from radioactive material enclosed in a lead block. When the radiation is passed through an electric field, it splits into three beams. One beam is attracted to the negative plate; it is composed of positively charged alpha (α) particles. The second beam is attracted to the positive plate; it is a beam of beta (β) particles. The third beam, called gamma (γ) rays, is not deflected by an electric field.

Alpha (α) particle
$^{4}_{2}He$

Beta (β^{-}) particle
$^{0}_{-1}e$

Both alpha rays and beta rays were found to be streams of tiny particles. An **alpha (α) particle** has a mass four times that of a hydrogen atom and a charge twice that of a proton. These properties make alpha particles identical to helium nuclei ($^{4}_{2}He$). Beta rays were shown to be streams of electrons. Thus a **beta (β^{-}) particle** ($^{0}_{-1}e$) has a mass only 1/1837 that of a proton and has a negative charge. As a form of radiant energy, gamma rays have no mass and no charge.

We can describe in detail the forms of radioactivity, but we must still answer the question of what radioactivity is. What is the phenomenon that produces radioactivity? The answer is that some nuclei are unstable as they occur in nature; they undergo **radioactive decay**. Radium atoms with a mass number of 226, for example, break down spontaneously, giving off alpha particles (Figure 3.3). Because alpha particles are identical to helium nuclei, this process can be summarized by the equation

$$^{226}_{88}Ra \longrightarrow \overbrace{^{4}_{2}He + ^{222}_{86}Rn}^{226}_{88}$$

The new element, with two fewer protons, is identified by its atomic number (86) as radon (Rn). Note that the mass number of the starting material must equal the total of the mass numbers of the products. The same is true for atomic numbers. The symbol $^{4}_{2}He$ for the alpha particle is preferred to the symbol α because it allows us to check the balance of mass and atomic numbers more readily.

Tritium nuclei are also unstable. Tritium is one of the heavy isotopes of hydrogen mentioned in Section 2.2. Like all hydrogen nuclei, the tritium nucleus contains one proton. Unlike the most common isotope of hydrogen, however, the tritium nucleus contains two neutrons, and its mass is therefore 3 u($^{3}_{1}H$). Tritium decomposes by *beta decay*. Because a beta particle is identical to an electron, this process can be written as

$$^{3}_{1}H \longrightarrow \overbrace{^{0}_{-1}e + ^{3}_{2}He}^{3}_{1}$$

The product isotope is identified by its atomic number as helium.

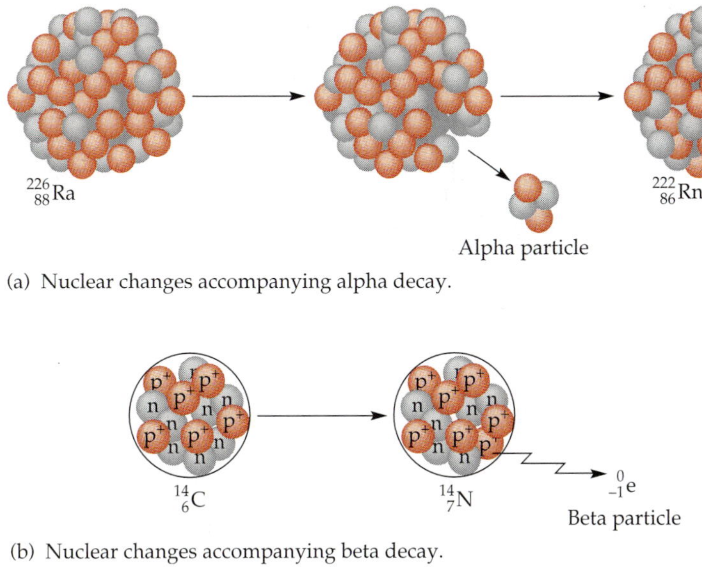

FIGURE 3.3

Nuclear emission of (a) an alpha particle and (b) a beta particle.

How can the original nucleus, which contains only a proton and two neutrons, emit an electron? We can envision one of the neutrons in the original nucleus splitting into a proton and an electron.

$$^{1}_{0}n \longrightarrow {}^{1}_{1}p + {}^{0}_{-1}e$$

The new proton is retained by the nucleus (therefore, the atomic number of the product increases by 1), and the almost massless electron or beta particle is kicked out (the product nucleus has the same mass as the original). (See again Figure 3.3 for a second example of beta decay.)

The third type of radioactivity is *gamma decay*. No particle is emitted; gamma rays are pure radiant energy. Gamma emission involves no change in atomic number or mass. The process is analogous to the emission of light from an atom. Visible light is emitted when an electron changes from a higher to a lower energy level. In gamma emission, a nucleus in a higher energy state drops to a lower energy state. Because this type of emission involves no particle, no equation is needed. Gamma decay is particularly useful when **radioisotopes** (radioactive isotopes) are used for diagnostic purposes in medicine. We discuss such uses in Section 3.9.

The emission of an alpha or beta particle is often accompanied by the emission of a gamma ray.

The major types of radioactive decay and the resulting nuclear changes are summarized in Table 3.1.

Example 3.1

Plutonium-239 emits an alpha particle when it decays. What new element is formed?

Solution

$$^{239}_{94}Pu \longrightarrow {}^{4}_{2}He + ?$$

Mass and charge are conserved. The new element must have a mass of 239 − 4 = 235 and a charge of 94 − 2 = 92. The nuclear charge (atomic number) of 92 identifies the element as uranium (U)

$$^{239}_{94}Pu \longrightarrow {}^{4}_{2}He + {}^{235}_{92}U$$

Practice Exercise

Fermium-250 undergoes alpha decay. What new isotope is formed?

Example 3.2

Protactinium-234 undergoes beta decay. What new element is formed?

Solution

$$^{234}_{91}\text{Pa} \longrightarrow {}^{0}_{-1}\text{e} + ?$$

The new element still has the mass number of 234. It must have a nuclear charge of 92 in order for the total charge to be the same on each side of the equation. The nuclear charge identifies the new atom as another isotope of uranium (U).

$$^{234}_{91}\text{Pa} \longrightarrow {}^{0}_{-1}\text{e} + {}^{234}_{92}\text{U}$$

Practice Exercise

Selenium-85 undergoes beta decay. What new isotope is formed?

Table 3.1 Major Types of Radioactive Decay

The Radiation					The Emitting (Absorbing) Nucleus	
Type	Symbol	Mass No.	Charge	Penetrating Power	Change in Mass Number	Change in Atomic Number
Alpha	α or $^{4}_{2}$He	4	2+	Slight	Decreases by 4	Decreases by 2
Beta	β^- or $_{-1}^{0}$e	0	1-	Intermediate	No change	Increases by 1
Gamma	γ	0	0	Great	No change	No change
Positron	β^+ or $_{+1}^{0}$e	0	1+	(Note 1)	No change	Decreases by 1
Electron capture	(E.C.)	0	—	(Note 2)	No change	Decreases by 1

(1) The penetrating power of a positron is quite limited because when a positron comes into contact with an electron, the two particles annihilate each other and are replaced by gamma radiation.

(2) Electron capture is a process in which the nucleus absorbs an electron from an inner electronic shell, usually the first or second. When an electron from a higher quantum level drops to the level vacated by the captured electron, an X-ray is released; E.C. is always accompanied by X-radiation. Once inside the nucleus, the captured electron combines with a proton to form a neutron.

3.3 Penetrating Power of Radiation

Radioactive materials can be dangerous because the radiation emitted by decaying nuclei can damage living tissue. The ability of the radiation to inflict injury depends, in part, on its penetrating power.

Alpha radiation—least penetrating

Gamma rays—most penetrating

All other things being equal, the more massive the particle, the less its penetrating power. Of alpha, beta, and gamma radiation, alpha particles are the least penetrating. These are streams of helium nuclei, each particle with a mass number of 4. Beta particles are more penetrating than alpha particles. The electrons that make up the stream of beta particles are assigned a mass number of 0. These particles are not really massless, but they are very much lighter than alpha particles. Gamma rays are high-energy radiation and, like light, truly have no mass. This is the most penetrating form of nuclear radiation.

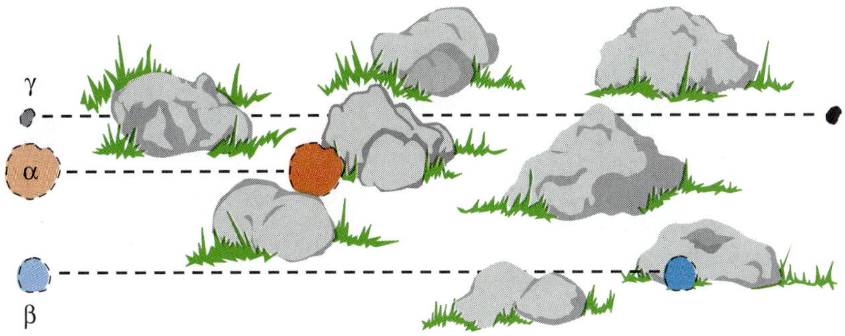

◀ **FIGURE 3.4**

Shooting radioactive particles through matter is like rolling rocks through a field of boulders; the larger rocks are stopped more quickly.

That the biggest particles make the least headway may seem contrary to common sense. Consider that penetrating power reflects the ability of the radiation to make its way through a sample of matter. It is as if you were trying to roll some rocks through a field of boulders. The alpha particle acts as if it were a boulder itself. Because of its size, it cannot get very far before it bumps into and is stopped by other boulders. The beta particle acts like a small stone. It can sneak between boulders and perhaps ricochet off one or another until it has made its way farther into the field (Figure 3.4). The gamma ray can be compared to a grain of sand that can get through the smallest openings: although the sand grain may brush against some of the boulders, it can, in general, make its way through most of the field without being stopped.

We have said that penetrating power is determined by the mass of the particle, all other things being equal. But all other things are not always equal. The faster a particle moves or the more energetic the radiation is, the more penetrating power it has.

If a radioactive substance is *outside* the body, alpha particles of low penetrating power are least dangerous; they are stopped by the outer layer of skin. Beta particles also usually are stopped before reaching vital organs. Gamma rays readily pass through tissues; an external gamma source can be quite dangerous. When the radioactive source is *inside* the body, the situation is reversed. The nonpenetrating alpha particles can do great damage. All such particles are trapped within the body, which must then absorb all the energy released by the particle. Alpha particles inflict all their damage in a very small area because they do not travel far. Beta particles distribute the damage over a somewhat larger area because they travel farther. Tissue may recover from limited damage spread over a large area; it is less likely to survive concentrated damage.

People working with radioactive materials can do several things to protect themselves (Table 3.2). The simplest is to move away from the source, because

Table 3.2 Protection from Nuclear Radiation

1. Distance: The more distant the source, the greater the safety.
2. Sample size: The smaller the radiating sample, the greater the safety.
3. Radiation: The less penetrating the radiation, the greater the safety. Thus, for external sources safety increases in the order $\gamma < \beta < \alpha$.
4. Half-life: The longer the half-life, the greater the safety.
5. Time: Generally, the shorter the time of exposure, the greater the safety.
6. Frequency: The fewer the exposures, the greater the safety.

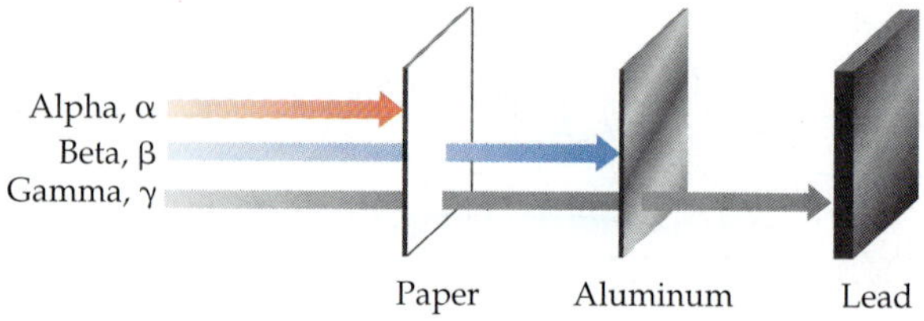

FIGURE 3.5 ▶
The relative penetrating powers of alpha, beta, and gamma radiation. Alpha particles are stopped by a sheet of paper and beta particles by a sheet of aluminum foil. It takes a block of lead several centimeters thick to stop gamma rays.

intensity of radiation decreases with distance from the source. Workers can also be protected by shielding. A sheet of paper will stop most alpha particles A block of wood or a thin sheet of aluminum will stop beta particles. But it takes several meters of concrete or several centimeters of lead to stop gamma rays (Figure 3.5).

3.4 Radiation Measurement

Several units of measurement are associated with the phenomenon of radioactivity. The *rate* at which nuclear disintegration occurs in a particular sample is measured in **curies** (Ci), named in honor of Marie Curie. A sample whose activity is hundreds or thousands of curies might be used as a source of externally applied radiation for the treatment of cancer. A sample with an activity of 10 mCi can be taken internally for diagnostic purposes by an adult, whereas a sample administered internally to a child might be measured in microcuries.

Primitive organisms such as yeast, bacteria, and viruses have much greater ability to withstand radiation than do mammals.

The *effect of radiation on matter* can be measured in a number of ways. The **roentgen (R)** is a measure of the ability of a source of X-rays or gamma rays to ionize an air sample. The **rad** (*r*adiation *a*bsorbed *d*ose) measures the amount of energy absorbed by any form of matter from any ionizing radiation. Alpha, beta, and gamma rays, as well as X-rays, are forms of **ionizing radiation**—that is, they cause the formation of ions from neutral particles. In the body, ionizing radiation most often interacts with water molecules. The reactive particles formed from water attack other molecules essential to proper cell function. Ionizing radiation damages living tissue by this route.

▲ FIGURE 3.6
Film badges are worn by people who work around radioactive materials. Radiation clouds the photographic film.

Some cells are more susceptible to radiation than others. Those that are being constantly and rapidly replaced are affected most. These include the intestinal mucosa, germ cells, embryonic cells, blood cells, and the organs responsible for producing blood cells, such as the bone marrow. Damage to reproductive cells will show up as abnormalities in the descendants of affected persons.

The **rem** (*r*oentgen *e*quivalent, *m*an) is a measure of the relative biological damage produced by a particular dose of radiation. For our purposes, the roentgen and the rad are about equivalent because 1 R generates about 1 rad of energy when absorbed in muscle tissue. It is estimated that a whole-body exposure of about 500 rads would kill most of us. The International Commission on Radiological Protection recommends that adults whose occupations expose them to ionizing radiation limit their exposure to 5 rem in any one year.

Techniques for measuring radiation range from the simple to the sophisticated. Individuals who work with radioactive materials wear *film badges* on their pockets or at their waist or as rings (Figure 3.6). The film in these badges reacts to radiation from radioactive isotopes or X-ray sources just as photographic film

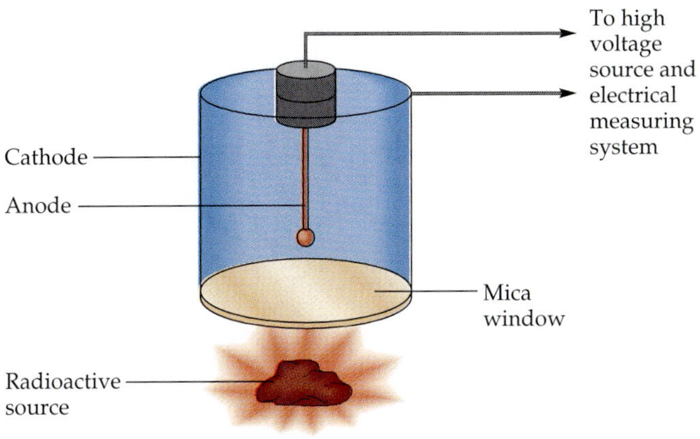

FIGURE 3.7

A Geiger counter. The schematic at left shows that radiation enters the tube through the mica window. Ions produced by the radiation cause an electrical discharge through the gas (usually argon) in the tube. Each pulse of electric current is counted as it passes through the circuit.

reacts to visible light. A certain dose of radiation will cause the film in the badge to become exposed, alerting the wearer to the potential danger. Sophisticated electronic devices, such as the *Geiger counter* (Figure 3.7), measure the ionizing effects of radiation and translate them into an observable signal (a meter reading or a clicking sound or a flashing light). Other detectors provide a permanent visual record of the intensity of the radiation. Such detectors play an important role in medical diagnosis (see Section 3.9).

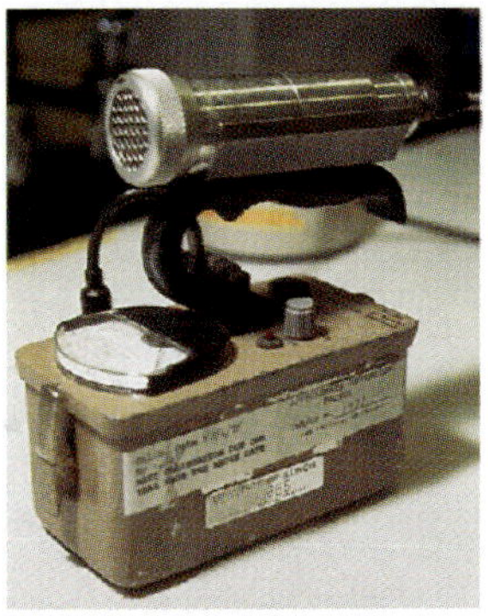

A typical geiger counter.

3.5 Half-Life

Radioactivity results when nuclei decay. We cannot predict when a particular nucleus will decay, but we can accurately predict the *rate of decay* of large numbers of radioactive atoms. (Life insurance companies cannot tell exactly when you will die, but their business depends on being able to predict how many of their clients will die over a particular period of time.)

A radioactive isotope can be characterized by a quantity called its **half-life.** The half-life of a radioactive isotope is that period of time in which one-half of the original number of atoms undergo radioactive decay. Suppose, for example, that you had 16 billion atoms of the radioactive hydrogen isotope $^{3}_{1}H$. The half-life of this isotope is 12.3 years. This means that in 12.3 years, one-half, or 8 billion, of the original 16 billion atoms would undergo radioactive decay. In another 12.3 years, half of the remaining 8 billion atoms would decay. That is, after two half-lives, 4 billion atoms or one-fourth of the original number of atoms would remain unchanged. Two half-lives do not make a whole! The concept of half-life is shown graphically in Figure 3.8.

It is impossible to say exactly when *all* the $^{3}_{1}H$ will have decayed. For practical purposes one may assume that nearly all the radioactivity is gone after about 10 half-lives. For the tritium sample considered here, 10 half-lives would be 123 years, at which time only a small fraction of 1% of the original atoms would still be present.

Much of the concern over nuclear power centers on the long half-lives of some isotopes that can be released in a nuclear accident or that are simply left as byproducts of the normal operation of a nuclear reactor. In the first instance, the fear

Half-life ($t_{1/2}$)—the time in which one-half of the radioactive atoms present undergo decay. The half-life of an element can be very long (millions of years) or extremely short (tiny fractions of a second). The half-life of uranium-238 is 4.5 billion years; that of boron-9 is 8×10^{-19} s.

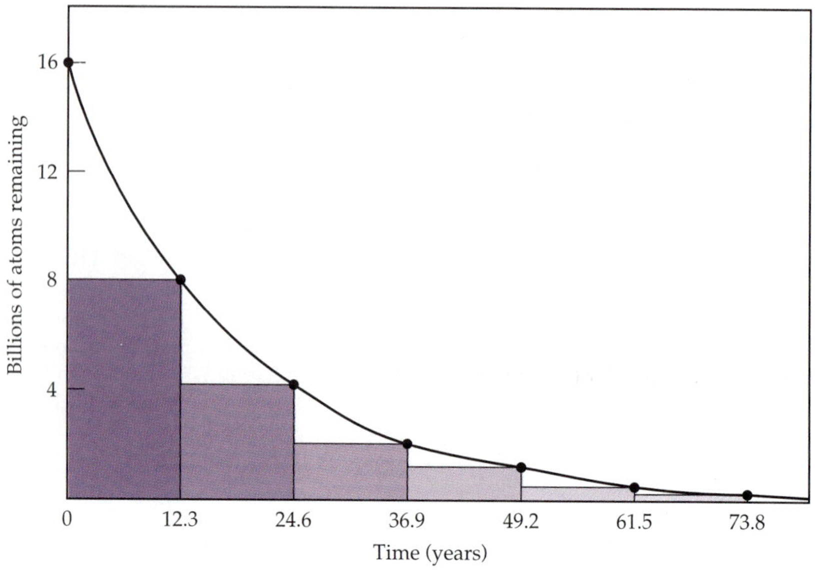

FIGURE 3.8 ▶
The radioactive decay of $^{3}_{1}H$. Each colored block represents one half-life.

is that large areas could be rendered uninhabitable for thousands of years if an explosion released long-lived radioisotopes into the atmosphere. Even with no accidents, normal operation of a reactor produces nuclear wastes that must be safely stored for thousands of years.

After 10 half-lives, the activity is $1/2^{10} = 1/1024$ of the original value. Generally, we can say that one-thousandth of the original activity remains.

We can calculate the fraction of the original isotope that remains after a given number of half-lives from the relationship

$$\text{Fraction remaining} = \frac{1}{2^n}$$

where n is the number of half-lives.

Example 3.3

You obtain a new 400-mg sample of cobalt-60, which has a half-life of 5.25 years. How much cobalt-60 remains after 15.75 years (three half-lives)?

Solution

The fraction remaining after three half-lives is

$$\frac{1}{2^n} = \frac{1}{2^3} = \frac{1}{(2 \times 2 \times 2)} = \frac{1}{8}$$

The amount of cobalt-60 remaining is (1/8)(400 mg) = 50 mg.

Practice Exercise

You have 1.224 mg of freshly prepared gold-189, which has a half-life of 30 min. How much of the gold-189 sample remains after five half-lives?

Example 3.4

You obtain a 20.0-mg sample of mercury-190, which has a half-life of 20 minutes. How much of the mercury-190 sample remains after 2 hours?

Solution

There are 120 minutes in 2 hours. There are 120/20 = 6 half-lives in 2 hours. The fraction remaining after six half-lives is

$$\frac{1}{2^n} = \frac{1}{2^6} = \frac{1}{(2 \times 2 \times 2 \times 2 \times 2 \times 2)} = \frac{1}{64}$$

The amount of mercury-190 remaining is (1/64)(20.0) mg = 0.313 mg.

Practice Exercise

A sample of 16.0 mg of nickel-57, with a half-life of 36.0 hours, is produced in a nuclear reactor. How much of the nickel-57 sample remains after 7.5 days?

3.6 Radioisotopic Dating

The half-lives of certain isotopes can be used to estimate the ages of rocks and archaeological artifacts. Uranium-238 decays with a half-life of 4.5 billion years. The initial products of this decay are also radioactive, and breakdown continues until an isotope of lead (lead-206) is formed. By measuring the relative amounts of uranium-238 and lead-206, chemists can estimate the age of a rock. Some of the older rocks on the Earth have been found to be from 3.0 to 4.5 billion years old. Moon rocks and meteorites have been dated at a maximum age of about 4.5 billion years. Thus, the age of the Earth is generally estimated to be about 4.5 billion years.

The dating of artifacts usually involves a radioactive isotope of carbon. Carbon-14 is formed in the upper atmosphere by the bombardment of ordinary nitrogen atoms by neutrons from cosmic rays.

$$^{14}_{7}\mathrm{N} + {}^{1}_{0}\mathrm{n} \longrightarrow {}^{14}_{6}\mathrm{C} + {}^{1}_{1}\mathrm{H}$$

This process leads to a steady-state concentration of carbon-14 in atmospheric CO_2. Living plants and animals incorporate this isotope into their own cells. When they die, however, the incorporation of carbon-14 ceases, and the carbon-14 in the plants and animals decays—with a half-life of 5730 years—to nitrogen-14. Thus, we merely need to measure the carbon-14 activity remaining in an artifact of plant or animal origin to determine its age. For example, a sample that has half the carbon-14 activity of new plant material is 5730 years old; it has been dead for one half-life. Similarly, an artifact with 25% of the carbon-14 activity of new plant material is 11,460 years old; it has been dead for two half-lives.

Carbon-14 dating, as outlined here, assumes that the formation of the isotope was constant over the years. This is not quite the case. However, for the most recent 7000 years or so, carbon-14 dates have been correlated with those obtained from the annual growth rings of trees. Calibration curves have been constructed from which accurate dates can be determined. Generally, the carbon-14 method is reasonably accurate for dating objects up to about 50,000 years old. Objects older than 50,000 years have too little of the isotope left for accurate measurement.

Charcoal from the fires of an ancient people, dated by determining the carbon-14 activity, is used to estimate the ages of other artifacts found at the same

Table 3.3 Several Isotopes Useful in Radioactive Dating

Isotope	Half-Life (years)	Useful Range	Dating Applications
Carbon-14	5730	500 to 50,000 years	Charcoal, organic material
Tritium (^3_1H)	12.3	1 to 100 years	Aged wines
Potassium-40	1.3×10^9	10,000 years to the oldest Earth samples	Rocks, the Earth's crust, the moon's crust
Rhenium-187	4.3×10^{10}	4×10^7 years to oldest samples in the universe	Meteorites
Uranium-238	4.5×10^9	10^7 years to the oldest Earth samples	Rocks, the Earth's crust

archaeological site. Carbon-14 dating also has been used to detect forgeries of supposedly ancient artifacts.

Perhaps you have heard of the Shroud of Turin, which is an old piece of linen cloth, about 4 meters long, bearing a faint human likeness (Figure 3.9). Since about 1350 C.E. it had been alleged to be part of the burial shroud of Christ. However, carbon-14 dating studies in 1988 by three different nuclear laboratories indicated that the flax used in making the cloth was not grown until sometime between 1260 and 1890 C.E. Therefore the cloth could not possibly have existed at the time of Christ. Unlike the Dead Sea Scrolls, which were shown by carbon-14 dating to be authentic records from a civilization that existed about 2000 years ago, the Shroud of Turin was exposed as a hoax.

▲ **FIGURE 3.9**
The Shroud of Turin, a linen cloth over 4 m long that shows the faint image of a man. The image is best seen in a photographic negative, as shown here. The shroud was thought by some to be the burial shroud of Jesus Christ.

Tritium, the radioactive isotope of hydrogen, also is useful for dating. Its half-life of 12.3 years makes it useful for dating items up to about 100 years old. An interesting application is the dating of brandies. These alcoholic beverages are quite expensive when aged for periods of time ranging from 10 to 50 years. Tritium dating can be used to check the truth of advertising claims about the most expensive brandies.

Many other isotopes are useful for estimating the ages of objects and materials. Several of the more important ones are listed in Table 3.3.

Example 3.5

A piece of fossilized wood has a carbon-14 activity that is one-eighth that of new wood. How old is the artifact? The half-life of carbon-14 is 5730 years.

Solution

The carbon-14 has gone through three half-lives.

$$\frac{1}{8} = \left(\frac{1}{2}\right)^3 = \frac{1}{2} \times \frac{1}{2} \times \frac{1}{2}$$

It is therefore about $3 \times 5730 = 17{,}190$ years old.

Practice Exercise

How old is a piece of charcoal that has a carbon-14 activity one-sixteenth that of new wood? The half-life of carbon-14 is 5730 years.

3.7 Artificial Transmutation and Induced Radioactivity

The forms of radioactivity encountered thus far occur in nature. The helium we use to fill balloons has accumulated over billions of years from the alpha decay of radioactive elements in the Earth's crust (Figure 3.10). It is possible to bring about nuclear reactions not encountered in nature. Such reactions are referred to as **artificial transmutations** because in the process one element is changed into another. These reactions are brought about by bombardment of stable nuclei with alpha particles, neutrons, or other subatomic particles. These particles, given sufficient energy, penetrate the target nucleus and trigger a nuclear reaction.

Artificial transmutation:
Artificial—not occurring in nature
Transmutation—the changing of one element into another

Ernest Rutherford studied the bombardment of the nuclei of a variety of light elements with alpha particles. One such experiment, in which he bombarded nitrogen nuclei, resulted in the production of protons.

$$^{14}_{7}\text{N} + ^{4}_{2}\text{He} \longrightarrow ^{17}_{8}\text{O} + ^{1}_{1}\text{H}$$

(Recall that the hydrogen nucleus is a proton; hence the alternative symbol for the proton is $^{1}_{1}\text{H}$.) Notice that the sum of the mass numbers on the left of the equation equals the sum of the mass numbers on the right of the equation. The atomic numbers are also balanced.

Irène Curie (daughter of Marie and Pierre) and her husband, Frédéric Joliot, studied the bombardment of aluminum nuclei with alpha particles. The reaction yielded neutrons and an isotope of phosphorus.

$$^{27}_{13}\text{Al} + ^{4}_{2}\text{He} \longrightarrow ^{30}_{15}\text{P} + ^{1}_{0}\text{n}$$

Much to their surprise, the target continued to emit particles after the bombardment was halted. The phosphorus isotope was radioactive. It emitted particles equal in mass but opposite in charge to the electron.

$$^{30}_{15}\text{P} \longrightarrow ^{0}_{+1}\text{e} + ^{30}_{14}\text{Si}$$

The $^{0}_{+1}\text{e}$ particle is called a **positron.** Once again the question arises: if the nucleus contains only protons and neutrons, where does this particle come from?

Previously we accounted for a beta particle (an electron) popping out of a nucleus by saying that a neutron changed into a proton and an electron. Perhaps a

Frédéric (1900–1958) and Irène (1897–1956) Joliot Curie discovered artificially induced radioactivity in 1934. (Frédéric changed his name to Joliot Curie when he married Irène in order to perpetuate the Curie name.) They received the Nobel prize in chemistry in 1935.

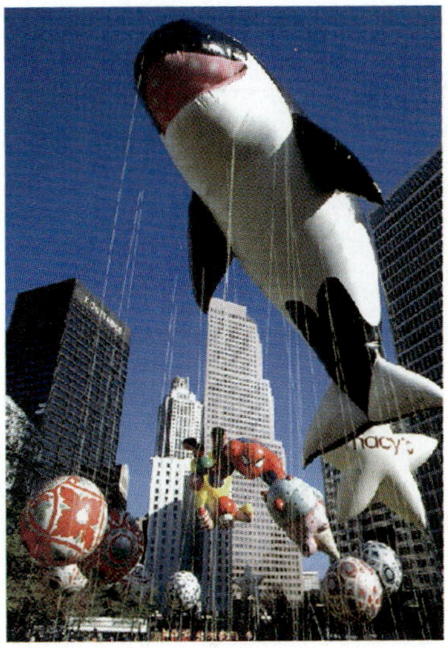

FIGURE 3.10

The balloons in this parade are filled with helium gas, a product of the alpha decay of radioactive elements.

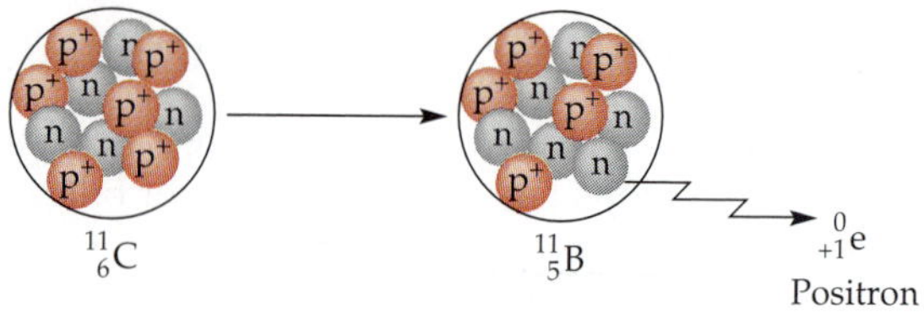

FIGURE 3.11 ▶
Nuclear emission of a positron, $^{0}_{+1}e$.

similar event can account for the appearance of a positron. Imagine a proton in the nucleus changing into a neutron and a positron. (Remember that a proton is the same as a hydrogen nucleus.)

$$^{1}_{1}H \longrightarrow {}^{0}_{+1}e + {}^{1}_{0}n$$

The equation balances nicely. When the positron is emitted, the original radioactive nucleus suddenly has one less proton and one more neutron than it had before. Therefore, the mass of the product nucleus is the same, but its atomic number is 1 less than that of the original nucleus. Figure 3.11 presents another example of nuclear decay involving positron emission.

Example 3.6

When potassium-39 is bombarded with neutrons, chlorine-36 is produced. What other particle is emitted?

Solution

$$^{39}_{19}K + {}^{1}_{0}n \longrightarrow {}^{36}_{17}Cl + ?$$

To balance the equation, a particle with a mass number of 4 and an atomic number of 2 is required—that is, an alpha particle.

$$^{39}_{19}K + {}^{1}_{0}n \longrightarrow {}^{36}_{17}Cl + {}^{4}_{2}He$$

Practice Exercise

Technetium-97 is produced by bombarding molybdenum-96 with a deuteron (hydrogen-2 nucleus). What other particle is emitted?

Many of the isotopes used in nuclear medicine undergo decay by **electron capture (E.C.)** (see again Table 3.1, p.58) For example, iodine-125, which is used in medicine to diagnose pancreatic function and intestinal fat absorption, decays by electron capture.

$$^{125}_{53}I + {}^{0}_{-1}e \longrightarrow {}^{125}_{52}Te$$

(followed by X-radiation)

Note that electron capture has the same effect on the nucleus as positron emission. Obviously, though, the X-rays emitted as a result of E.C. differ from the gamma rays formed during positron annihilation.

Example 3.7

Carbon-10, a radioactive isotope, emits a positron when it decays. Write an equation for this process.

Solution

$$^{10}_{6}C \longrightarrow {}^{0}_{+1}e + ?$$

To balance the equation, a particle with a mass number of 10 and an atomic number of 5 (boron) is required.

$$^{10}_{6}C \longrightarrow {}^{0}_{+1}e + {}^{10}_{5}B$$

Practice Exercise

Gold-188 decays by positron emission. Write a balanced nuclear equation for this process.

Nuclear medicine depends on the availability of a broad range of radioisotopes, and many of these are artificially produced. Later in this chapter we will look into some aspects of nuclear medicine.

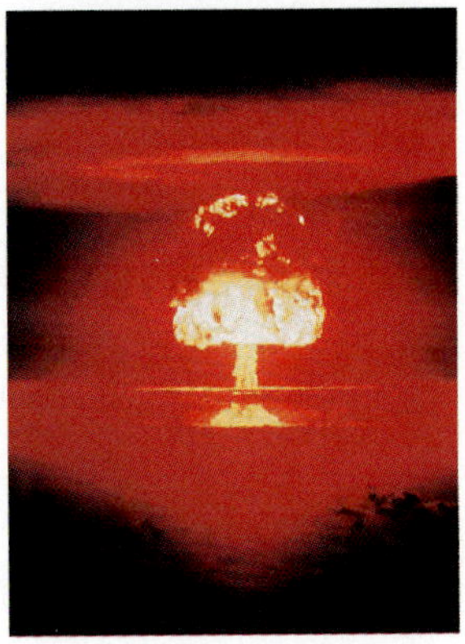

The nuclear age is often characterized by the mushroom-shaped cloud that accompanies a nuclear explosion.

3.8 Fission and Fusion

In the nuclear reaction called **nuclear fission,** a large unstable nucleus is bombarded with relatively slow-moving neutrons. In the resulting nuclear reaction, the large nucleus does not simply emit a small particle; it breaks apart, leaving two medium-sized nuclei and releasing more neutrons. A typical fission reaction is

$$^{235}_{92}\text{U} + {}^{1}_{0}\text{n} \longrightarrow {}^{90}_{38}\text{Sr} + {}^{143}_{54}\text{Xe} + 3\,{}^{1}_{0}\text{n}$$

Because it was bombardment with neutrons that triggered the reaction in the first place, the product neutrons can go on to cause more of the large nuclei to fission (break apart). Neutrons produced by this second wave of reactions will trigger more reactions, and so on. Thus nuclear fission is a chain reaction (Figure 3.12). Each reaction in the chain releases energy. If the chain of reactions is carried out in a controlled manner, the energy released can be used to generate electric power, such as is done in a nuclear reactor (Figure 3.13). The reaction can also be carried out in such a way that all the energy is released in one gigantic explosion. What one then has is a bomb—a nuclear bomb. It is not possible for a nuclear reactor to produce a nuclear-bomblike explosion because of the special conditions required for an effective bomb. The products of fission, however, whether from a bomb or a reactor, are radioactive. When the bomb explodes, these products are thrown into the atmosphere and eventually reach the ground as dangerous "fallout." In a reactor, these products must be periodically removed and transferred to storage facilities.

In nuclear fission, the nucleus breaks apart to form two nuclei.

Another important nuclear reaction is called nuclear fusion. In **nuclear fusion,** two smaller nuclei are combined (fused) into a larger nucleus, a process accompanied by the release of vast amounts of energy. A typical fusion reaction is illustrated in Figure 3.14.

At present, more than 20% of the electric power generated in the United States is produced by nuclear power plants.

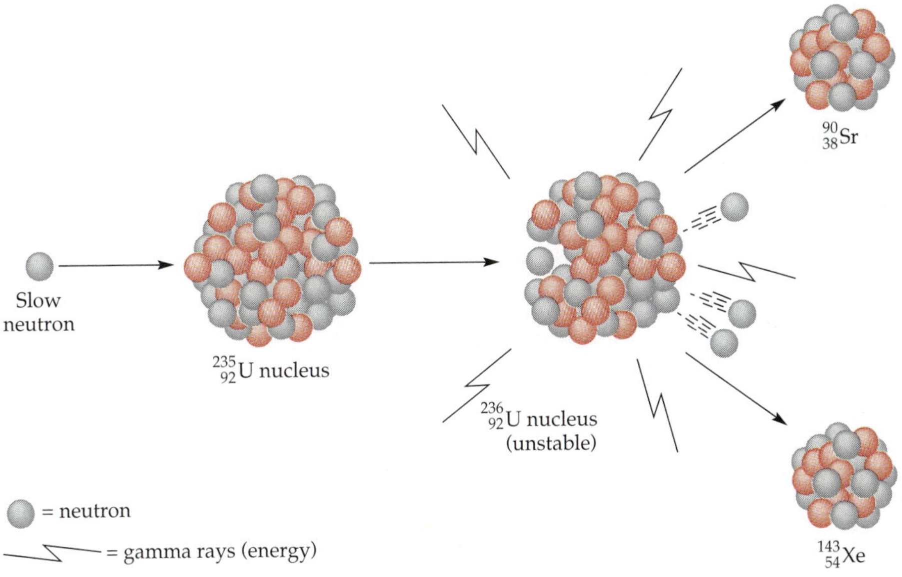

FIGURE 3.12
A uranium-235 nucleus is split when it is struck by a relatively slow-moving neutron. An unstable uranium-236 nucleus is formed, but it breaks into two fragments with the release of several neutrons. These neutrons can induce the fission of other uranium-235 nuclei. Fission of 1 g of uranium-235 yields 22,600 kW of energy.

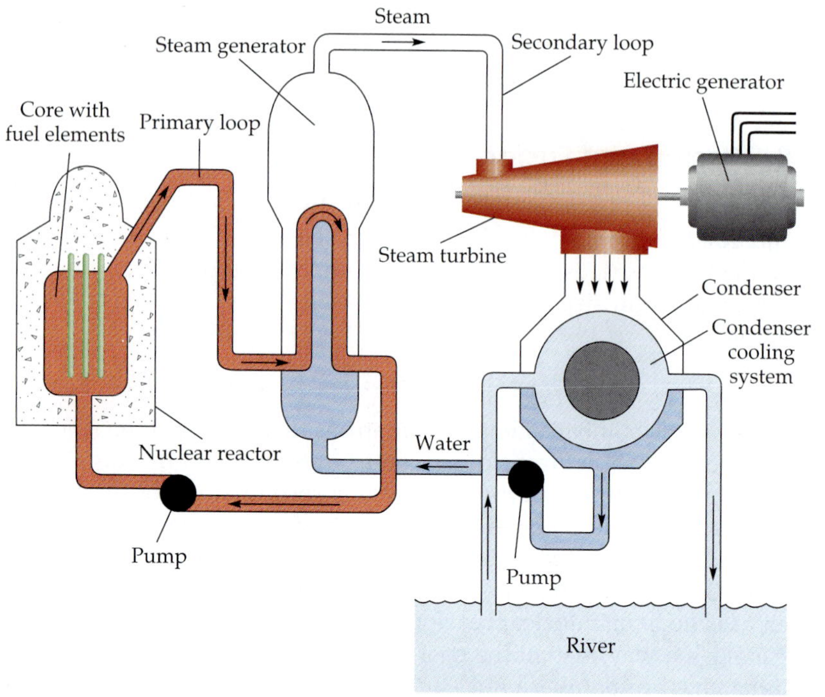

FIGURE 3.13 ►
Schematic diagram of a nuclear power plant used to generate electricity. The energy released during fission is used to generate steam, which turns a turbine that generates electricity.

In nuclear fusion, two or more nuclei are combined to form one larger nucleus.

Our sun is powered by the *fusion* of atomic nuclei, and its fuel supply—mostly ^1_1H—will last for billions of years. On Earth, scientists have unleashed the extraordinary energy of uncontrolled fusion reactions in hydrogen bombs. However, there are daunting challenges in developing a fusion energy source. Before they will fuse, the nuclei of deuterium and tritium must be forced extremely close together. And because the positively charged nuclei repel one another so strongly, close approach requires enormously high temperatures. At the required temperatures, gases are completely ionized into a mixture of atomic nuclei and electrons known as *plasma*. A temperature of over 40,000,000 °C is necessary to initiate self-sustaining fusion—a nuclear reaction that releases more energy than it took to get it started. Another requirement is that the plasma be confined at a very high density long enough for the fusion to occur. Moreover, this confinement must be done without allowing the plasma to contact the walls of the reactor, where it would immediately lose heat and thus its capability to fuse. One method is to confine the plasma in a magnetic field (Figure 3.15).

FIGURE 3.14 ►
The fusion reaction that is most promising for development of a controlled nuclear fusion reactor is the deuterium-tritium (DT) reaction. A deuterium (^2_1H) nucleus fuses with a tritium (^3_1H) nucleus to form a helium nucleus plus a neutron, and a considerable amount of energy is released.

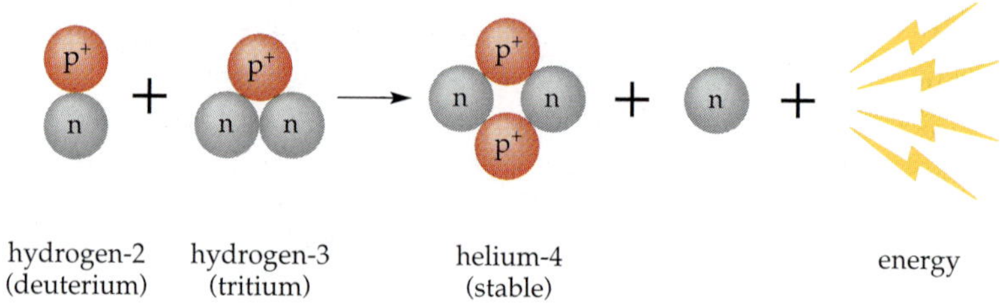

$$^2_1H + ^3_1H \longrightarrow ^4_2He + ^1_0n$$

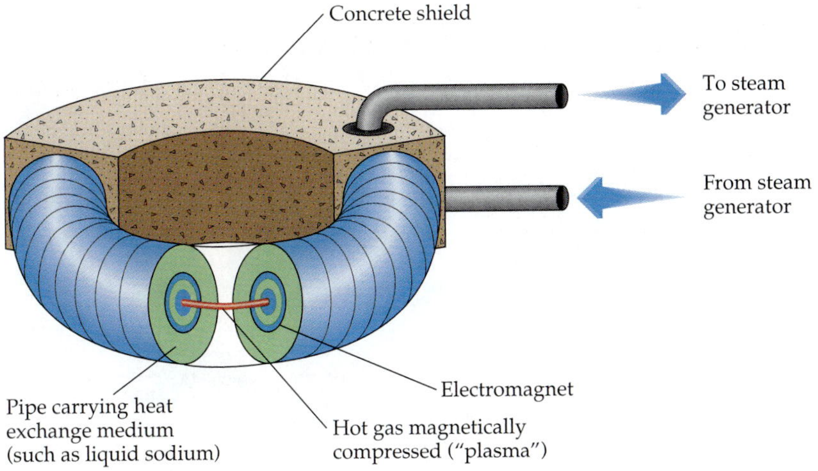

FIGURE 3.15
A nuclear fusion device. Nuclear fusion requires that plasma be confined at exceedingly high temperatures and pressures. A giant donut-shaped electromagnet called a *tokamak* is one device designed to contain the plasma.

Nuclear fusion, if perfected, will offer distinct advantages over nuclear fission for power generation. The two most important advantages are (1) greater energy production per fusion event than per fission event and (2) the fact that the radioactive waste produced in nuclear fusion is much more limited and of relatively short half-life. The half-life of tritium, for example, is only 12.3 years. Long-term storage of nuclear waste is not required.

3.9 Nuclear Medicine

Nuclear medicine involves two distinct uses of radioisotopes: therapeutic and diagnostic. In radiation therapy, an attempt is made to treat or cure disease, such as cancer, with radiation. Cancer is not one disease but many. Some forms are particularly susceptible to radiation therapy. Radiation is carefully aimed at the cancerous tissue, and exposure of normal cells is minimized (Figure 3.16). If the cancer cells are killed by the destructive effects of the radiation, the malignancy is halted. But persons undergoing radiation therapy often get quite sick from the treatment. Nausea and vomiting are the usual symptoms of radiation sickness. (Remember that the intestinal mucosa is particularly susceptible to radiation.) Thus, the aim of radiation therapy is to destroy the cancerous cells before too much damage is done to healthy tissue. Radiation is most lethal to rapidly reproducing cells, and this is precisely the characteristic of cancer cells that allows the therapy to be successfully applied.

Radioisotopes are also used for diagnostic purposes, to help provide information about the type or extent of illness. Radioactive iodine-131 ($^{131}_{53}I$) is used to determine the size, shape, and activity of the thyroid gland as well as to treat cancer located in this gland and to control a hyperactive thyroid. First, the patient drinks a solution of potassium iodide, KI, incorporating iodine-131. The body concentrates iodide in the thyroid. Large doses are used for treatment of thyroid cancer; the radiation from the isotope concentrates in the thyroid cancer cells even if the cancer has spread to other parts of the body. For diagnostic purposes, however, only a small amount is needed. A detector is set up so that readings are translated into a permanent visual record showing the differential uptake of the isotope. The "picture" that results is called a *photoscan*, and it can pinpoint the location of a tumor in that area of the body.

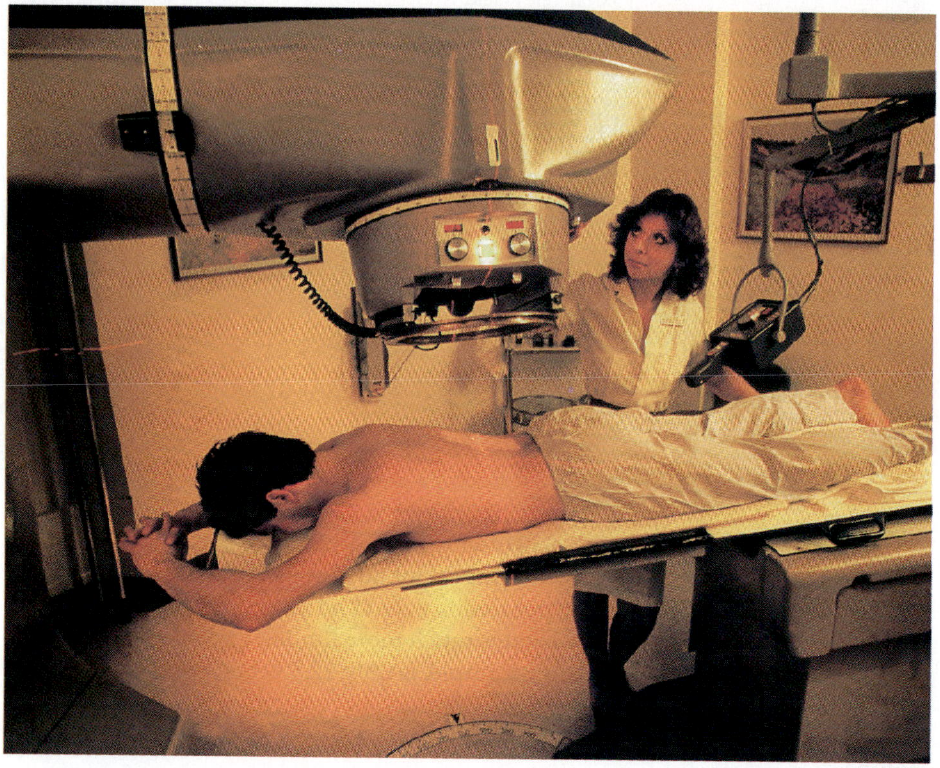

FIGURE 3.16 ▶

A cobalt-60 unit for radiation therapy.

The radioisotope most widely used in medical imaging is gadolinium-153. This isotope is used to determine bone mineralization. Its popularity is an indication of the large number of people, mostly women, who suffer from *osteoporosis* (reduction in the quantity of bone) as they grow older. Gadolinium-153 gives off two characteristic radiations, a gamma ray and an X-ray. A scanning device compares these radiations after they pass through bone. Bone densities are then determined by differences in absorption of the rays.

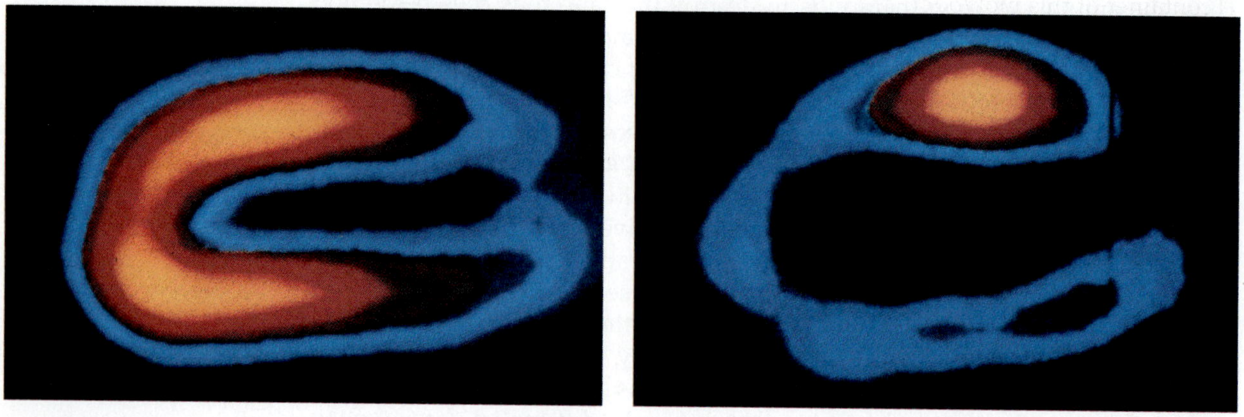

▲ FIGURE 3.17

Blood flow patterns in a healthy heart (left) and a damaged heart (right). The highlighted images from a technetium-99^m compound indicate regions receiving adequate blood flow.

Table 3.4 Some Radioisotopes and Their Application in Medicine

Isotope	Name	Radiation	Uses
^{51}Cr	Chromium–51	E.C., γ	Determination of volume of red blood cells and total blood volume
^{57}Co	Cobalt–57	E.C., γ	Determination of uptake of vitamin B_{12}
^{60}Co	Cobalt–60	$\beta^- \gamma$	Radiation treatment of cancer
^{153}Gd	Gadolinium–153	E.C., γ	Determination of bone density
^{131}I	Iodine–131	β^-,γ	Detection of thyroid malfunction; measurement of liver activity and fat metabolism; treatment of thyroid cancer
^{192}Ir	Iridium–192	E.C., γ	Radiation treatment for breast cancer
^{59}Fe	Iron–59	β^-, γ	Measurement of rate of formation and lifetime of red blood cells
^{32}P	Phosphorus–32	β^-	Detection of skin cancer or cancer of tissue exposed by surgery
^{238}Pu	Plutonium–238	α, γ	Power pacemakers in patients having irregular heartbeat
^{226}Ra	Radium–226	α, γ	Radiation therapy for cancer
^{24}Na	Sodium–24	β^-, γ	Detection of constrictions and obstructions in the circulatory system
^{99m}Tc	Technetium–99^{m}	γ	Imaging of brain, thyroid, liver, kidney, lung, and cardiovascular system
^{3}H	Tritium	β^-	Determination of total body water

Technetium-99^{m} is used in a variety of diagnostic tests (Figure 3.17). The *m* stands for *metastable,* which means that this isotope will give up some energy to become a more stable version of the same isotope, (same atomic number, same atomic weight). The energy it gives up is the gamma ray needed to detect the isotope

$$^{99m}_{43}\text{Tc} \longrightarrow {}^{99}_{43}\text{Tc} + \gamma$$

Notice that the decay of technetium-99^{m} produces no alpha or beta particles, which could cause unnecessary damage to the body. Technetium-99^{m} also has a short half-life (about 6 hours), which means that the radioactivity does not linger in the body long after the scan has been completed. With this short a half-life, use of the isotope must be carefully planned. In fact, the isotope itself is not what is purchased. Technetium-99^{m} is formed by the decay of molybdenum-99.

$$^{99}_{42}\text{Mo} \longrightarrow {}^{99m}_{43}\text{Tc} + {}^{\ \,0}_{-1}\text{e} + \gamma$$

A container of this molybdenum isotope is obtained, and the decay product, technetium-99^{m}, is "milked" from the container as needed.

Table 3.4 is a list of some radioisotopes in common use in medicine. The list is necessarily incomplete. Even this abbreviated tabulation should give you an idea of the importance of radioisotopes in medicine. The claim that nuclear science has saved many more lives than nuclear bombs have destroyed is not an idle one.

In 1977, Rosalyn Yalow shared the Nobel prize in medicine for her work with radioisotopes. The technique she developed, known as radioimmunoassay (RIA), is extremely sensitive and can be used to detect substances not detectable by other methods. RIA uses a radionuclide that selectively binds to a biochemical compound, such as a drug or hormone.

3.10 Medical Imaging

Medical imaging provides a means of looking at internal organs without resorting to surgery. The history of medical imaging dates back to the discovery of X-rays at the turn of the century. Penetrating X-rays were used almost immediately to visualize skeletal structure. Radiation from an external X-ray source passes through the body (except where it is absorbed by more dense structure, such as bone) and exposes film, thus providing a picture that distinguishes the more dense structures from the less dense tissue. Softer tissue can be visualized by introducing

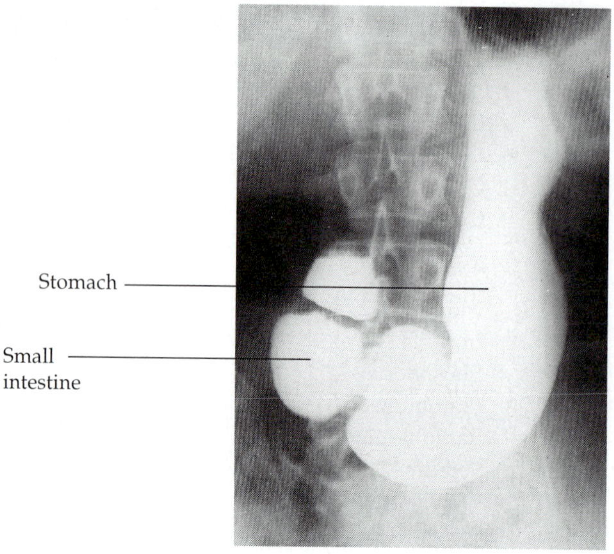

FIGURE 3.18 ▶

Barium sulfate ($BaSO_4$) is insoluble in water and opaque to X-rays. The salt can be swallowed by the patient and an X-ray photographic outline of the stomach taken.

material that absorbs X-rays into the area to be studied. For example, compounds of barium have been used to view portions of the digestive tract (Figure 3.18).

Radioisotopes can also be used to visualize internal organs. In this technique, the source of the radiation is inside the body, and the radiation (usually gamma rays) is detected as it emerges from the body.

Both X-ray technology and nuclear imaging have been coupled with computer technology to provide versatile and powerful imaging techniques. In *computed tomography* (referred to as CT or, sometimes, CAT scanning), many X-ray readings are obtained, processed by a computer, and then displayed. The resulting pictures present cross-sectional slices of a portion of the body. A series of these pictures gives a three-dimensional view of organs such as the brain (Figure 3.19). CT scans are widely used for detecting tumors and cancer.

Positron emission tomography (PET) can be used to measure dynamic processes occurring in the body, such as blood flow or the rate at which oxygen or glucose is being metabolized. PET scans are used to pinpoint the area of the brain damage that triggers severe epileptic seizures and to detect tiny blockages in blood vessels

FIGURE 3.19 ▶

This image was made using computer tomography (CT), a scanning technique that uses X-rays rather than radioisotopes. It shows a view, through the skull, of a pituitary tumor.

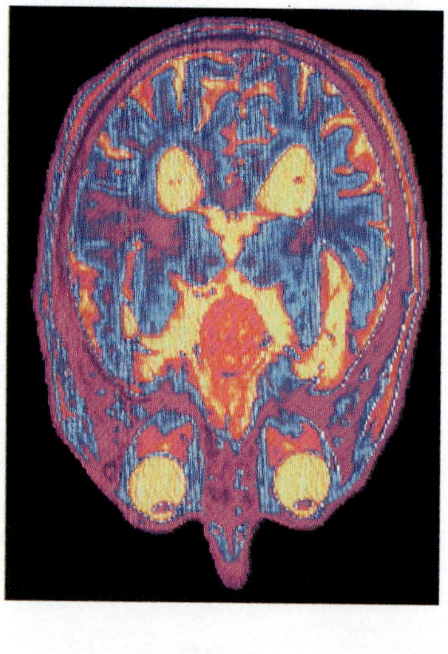

and hard-to-spot tumors. Compounds incorporating positron-emitting isotopes, such as fluorine-18, are inhaled or injected prior to the scan. Before the emitted positron can travel very far in the body, it encounters an electron (in any ordinary matter there are numerous electrons), and two gamma rays are produced.

$$^{18}_{9}\text{F} \longrightarrow {}^{18}_{8}\text{O} + {}^{0}_{+1}\text{e}$$

$$^{0}_{+1}\text{e} + {}^{0}_{-1}\text{e} \longrightarrow 2\,\gamma$$

The gamma rays exit the body in exactly opposite directions and are recorded by detectors positioned on opposite sides of the patient. If the recorders are set so that two simultaneous gamma rays must be "seen," gamma rays resulting from natural background radiation are ignored. A computer is then used to calculate the point within the body at which the annihilation of the positron and electron occurs, and an image of that area is produced (Figure 3.20).

Positron emission tomography can reveal metabolic changes that occur in the brain, which depends on glucose for most of its energy. Changes in how this sugar is metabolized or used by the brain may signal a disease such as cancer, epilepsy, Parkinson's disease, or schizophrenia.

Both X-rays and nuclear radiation are ionizing radiations, so there is always some tissue damage involved. Modern techniques keep this damage to a minimum. Other imaging techniques use nonionizing radiation.

Perhaps you are familiar with *ultrasonography*. In this technique, high-frequency sound waves are bounced off tissue, and the echo of the sound wave is recorded. Once again, a computer is used to process the data and produce an image of the tissue. (The technique is related to sonar detection of submarines.) Because no ionizing radiation is involved, ultrasonography is used extensively in obstetrics to follow fetal development (Figure 3.21). A 1984 conference on the use of ultrasonography (sponsored by the National Institutes of Health) concluded, however, that even this technique should not be used casually. The conference recommended that an ultrasonogram be obtained only if there is a good medical reason for doing so (for example, to evaluate fetal growth in mothers who have diabetes or hypertension).

Another nonionizing technique is *magnetic resonance imaging* (MRI). This technique depends on the fact that some nuclei behave as if they were little magnets; placed in a strong magnetic field (e.g., between the poles of a much more powerful

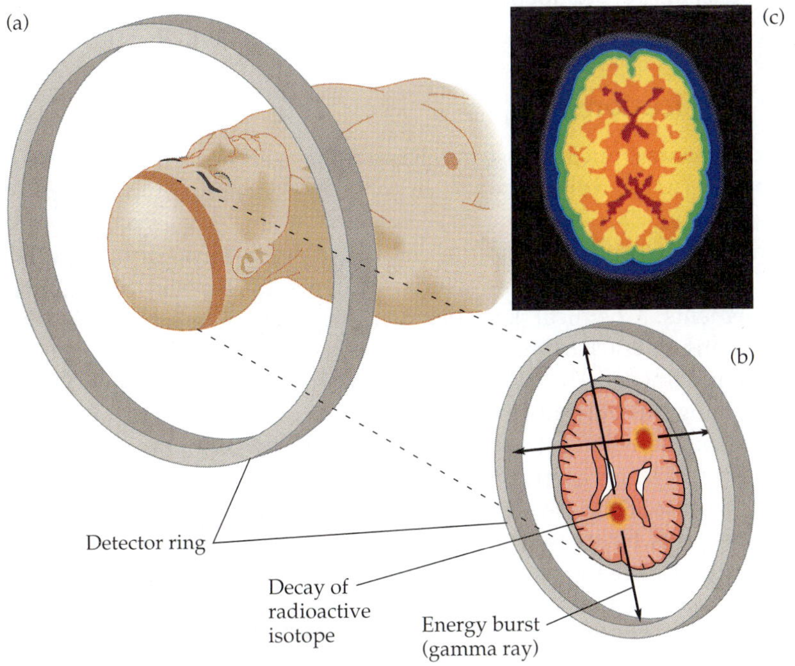

FIGURE 3.20

Positron emission tomography (PET) uses a positron-emitting radioisotope incorporated in a compound such as glucose, a simple sugar, which is given to a patient (a). The geometry of the X-rays emitted when the positron is annihilated by an electron provides an image of a "slice" of the head (b). The numbers of decays at each location are indicated by colors on a television screen (c).

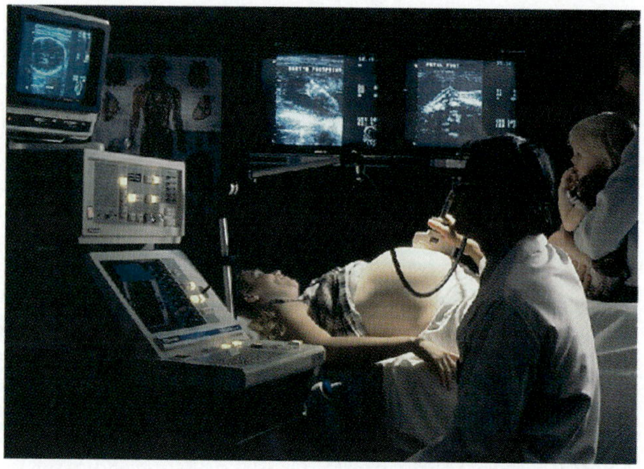

FIGURE 3.21 ▶
An ultrasonogram test on a woman who is 8 months pregnant.

magnet), the nuclei line up in a certain manner. When supplied with the right amount of energy, the nuclei absorb the energy and flip over in the magnetic field. The energy required to flip the nuclei is provided by nonionizing radiation. The absorption of the energy by the nuclei can be detected by appropriate equipment, and an image of the tissue in which the nuclei reside is produced. (Again, a computer is used to process the image.) The MRI technique has demonstrated its potential for providing not only images of organs but also information about the metabolic activity in particular tissues. MRI is particularly adept at detecting small tumors, blockages in blood vessels, damage to vertebral discs, and problems in joints—especially of the knees.

MRI is similar to CT scanning except that radio waves are used instead of X-rays.

Most of these computer-based technologies are quite expensive, ranging into the millions of dollars for a single installation. The great advantage of these techniques is that they provide information that could otherwise be obtained only by subjecting the patient to the risks of surgery. In some instances, the information provided by these techniques is not available by any other means.

3.11 Other Applications

In the last several decades, nuclear research has emphasized peaceful rather than military applications. Aside from the uses already mentioned, many other applications have been found.

Elements that do not appear in nature can be prepared through transmutation. Examples of synthetic elements include elements 43, 61, and 85 and the transuranium elements. Chemical and biochemical reaction mechanisms (the ways in which atoms interact to make new products) can often be determined only with radioactive tracers. The calcium uptake in metabolism has been found to be 90% in the young but only 40% in older animals and humans. Polycythemia vera (formation of too many red blood cells) has been studied with the aid of iron-59. It was found that 10 times as much iron as the body can use is assimilated by patients with this disease. Through similar studies it has also been learned that the uptake of trace elements by trees is quite pronounced in the winter. Thus, zinc moves up the trunk of a tree at the rate of about 2 ft/day. The effectiveness of industrial lubricants can be measured by monitoring the concentrations of metal

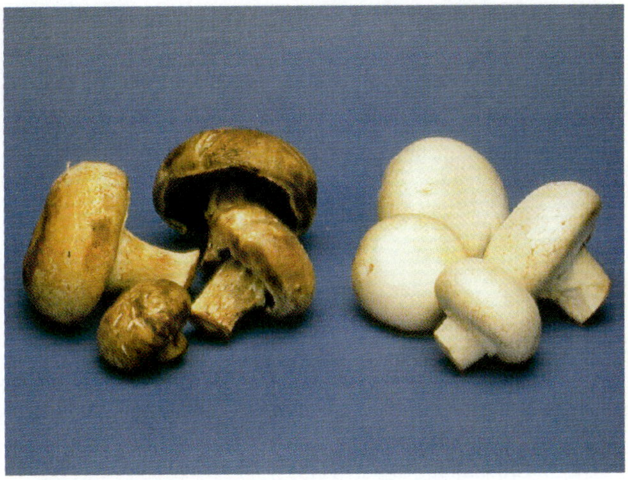

◀ **FIGURE 3.22**
Gamma irradiation delays the decay of mushrooms. Those on the right were irradiated; those on the left were not.

residues in the oils. By tagging the metal with radioactive isotopes, concentrations as low as 10^{-19} g/L can be detected.

Radioisotopes are also used as sources for the irradiation of foodstuffs as a method of preservation (Figure 3.22). The radiation destroys microorganisms that cause food spoilage. Irradiated food shows little change in taste or appearance. Some people are concerned about possible harmful effects of chemical substances produced by the radiation, but there is no good evidence of harm to laboratory animals fed irradiated food, nor are there any known adverse effects in humans in countries where irradiation has been used for several years. There is no residual radioactivity in the food after the sterilization process.

Radioisotopes have been used extensively in basic scientific research. The mechanism of photosynthesis was worked out in large part by using carbon-14 as a tracer. For example, to determine how plants make the sugar glucose from carbon dioxide and water, the plant is exposed to radioactive carbon dioxide. The compounds formed from these starting materials and their order of formation are then followed by determining which new compounds become radioactive, and in what sequence. Using data from radioactive tracer experiments, scientists determine metabolic pathways in plants, animals, and humans.

3.12 The Nuclear Age Revisited

We live in an age in which fantastic forces have been unleashed. The threat of nuclear war was a constant specter from World War II until the collapse of the Soviet Union in 1991. Nuclear bombs have been used to destroy cities—and men, women, and children. Science and scientists have been greatly involved in it all. Would the world be a better place had we not discovered the secrets of the atomic nucleus? Consider this: More lives have been saved through nuclear medicine than have been destroyed by nuclear bombs. Also, nuclear power, with all its attendant problems, still may be one of our best hopes for a plentiful energy supply until well into the twenty-first century. Imagine yourself to be Dalton. Could this gentle Quaker, whose formal education ended when he was 11 years old, have anticipated the chain of scientific developments that would follow from his original speculations on the nature of matter (Figure 3.23)?

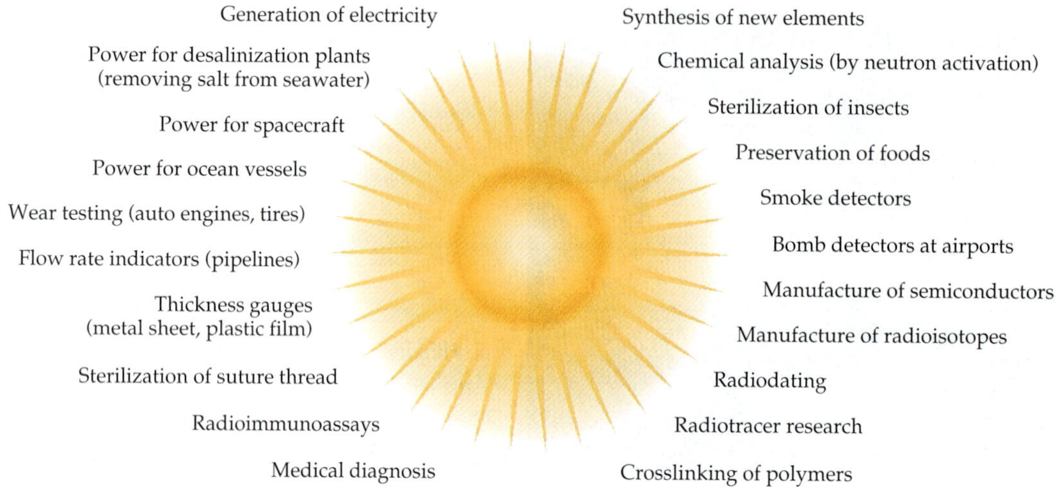

▲ **FIGURE 3.23**
Some peacetime uses of nuclear energy.

Summary

In some atoms, the cluster of protons and neutrons in the nucleus is unstable, and such atoms emit radiation as they try to reach a more stable nuclear configuration. This phenomenon is called **radioactivity,** and the isotopes that undergo the process are called **radioisotopes.** There are three kinds of radiation involved: **alpha particles, beta particles,** and **gamma rays.**

An alpha particle (α) has a mass equal to four times the mass of a hydrogen atom and carries a charge of $+2$, making it identical to the nucleus of a helium atom. For this reason, the symbol for the alpha particle is often written ${}^{4}_{2}\text{He}$. A beta particle (β) is an electron and so is nearly massless and carries a charge of -1. The symbol for the electron, ${}^{0}_{-1}\text{e}$, is often used to represent a beta particle. Gamma rays are radiation rather than particles and so have no mass and carry no charge.

An alpha decay has the general form

$$ {}^{A}_{Z}\text{X} \longrightarrow {}^{A-4}_{Z-2}\text{Y} + {}^{4}_{2}\text{He} $$

Example:

$$ {}^{238}_{92}\text{U} \longrightarrow {}^{234}_{90}\text{Th} + {}^{4}_{2}\text{He} $$

Uranium-238 Thorium-234 Alpha particle

A beta decay has the general form

$$ {}^{A}_{Z}\text{X} \longrightarrow {}_{Z+1}^{A}\text{Y} + {}^{0}_{-1}\text{e} $$

Example:

$$ {}^{32}_{15}\text{P} \longrightarrow {}^{32}_{16}\text{S} + {}^{0}_{-1}\text{e} $$

Phosphorus-32 Sulfur-32 Beta particle

In beta decay, a neutron in the starting material breaks up into a proton plus an electron:

$$ {}^{1}_{0}\text{n} \longrightarrow {}^{1}_{+1}\text{p} + {}^{0}_{-1}\text{e} $$

It is this newly formed nuclear electron that is emitted from the atom. The electrons outside the nucleus have nothing to do with beta decay.

In gamma decay, no particles leave the nucleus, and so the starting material and the product are the same. The nucleus drops to a lower energy as a gamma ray (γ) is given off.

Gamma rays are the most penetrating of the three types of nuclear emissions. Alpha particles are the least penetrating.

The **curie** (Ci) is the unit used to measure the rate at which a radioisotope decays. The relationship is

$$ 1\ \text{Ci} = 3.7 \times 10^{10}\ \text{decays/second} $$

The unit called the **roentgen** measures the ability of a radioactive source of gamma rays to ionize air. The **rad** measures how much energy is absorbed by an object when that object is exposed to radiation. A dose of 1 rad is the amount of radiation that causes 1 kg of a substance to absorb 0.01 J of energy.

Because different types of radiation cause different amounts of damage to living tissue, exposure levels are measured not in rads but in a unit called the **rem.**

Alpha particles, beta particles, and gamma radiation are all **ionizing radiation,** which means they can knock orbital electrons out of neutral atoms and thereby create ions.

The **half-life** of any radioactive isotope is the length of time it takes for half the atoms in a sample of that isotope to decay. Example: An isotope has a half-life of two days, and a certain sample contains 1000 atoms of the isotope. After two days, there are 500 isotope atoms remaining; after four days, there are 250; after six days, there are 125.

Atoms whose nuclei are stable in their natural setting can be made to undergo radioactive decay. The forced decay—called **artificial transmutation**—is accomplished by bombarding the stable nuclei with alpha particles, neutrons, or other particles. The artificial transmutation of aluminum-27 led to the discovery of a subatomic particle called the **positron** (symbol ${}^{0}_{+1}e$), which has the same mass as the electron, but carries a charge of +1:

$$\underset{\text{Aluminum-27}}{{}^{27}_{13}\text{Al}} + \underset{\text{Alpha particle}}{{}^{4}_{2}\text{He}} \longrightarrow \underset{\text{Phosphorus-30}}{{}^{30}_{15}\text{P}} + {}^{1}_{0}\text{n}$$

$$\underset{\text{Phosphorus-30}}{{}^{30}_{15}\text{P}} \longrightarrow \underset{\text{Silicon-30}}{{}^{30}_{14}\text{Si}} + \underset{\text{Positron}}{{}^{0}_{+1}\text{e}}$$

The positron forms in the nucleus during the decay; it is produced when one of the nuclear protons transmutes to a neutron plus a positron:

$${}^{1}_{+1}\text{p} \longrightarrow {}^{1}_{0}\text{n} + {}^{0}_{+1}\text{e}$$

In **nuclear fission,** a large, unstable nucleus splits into two medium-sized nuclei. Fission is initiated when a sample of the large-nucleus isotopes is bombarded with neutrons from an external source. The products of the reaction are two medium-sized nuclei and more neutrons:

$$\underset{\text{Uranium-235}}{{}^{235}_{92}\text{U}} + {}^{1}_{0}\text{n} \longrightarrow \underset{\text{Strontium-90}}{{}^{90}_{38}\text{Sr}} + \underset{\text{Xenon-143}}{{}^{143}_{54}\text{Xe}} + 3\,{}^{1}_{0}\text{n}$$

The three product neutrons can then each hit one uranium-235 nucleus, causing the fission to happen three more times. The ever-growing number of fission events is called a **chain reaction.** When such a reaction is controlled, the result is a usable source of nuclear power; when the reaction is allowed to proceed uncontrolled, the result is a nuclear explosion.

The opposite of nuclear fission is **nuclear fusion,** a reaction in which two smaller nuclei come together (they *fuse*) to form a larger nucleus:

$$\underset{\substack{\text{Deuterium}\\\text{(Hydrogen-2)}}}{{}^{2}_{1}\text{H}} + \underset{\substack{\text{Tritium}\\\text{(Hydrogen-3)}}}{{}^{3}_{1}\text{H}} \longrightarrow \underset{\text{Helium-4}}{{}^{4}_{2}\text{He}} + {}^{1}_{0}\text{n}$$

As with fission, a controlled fusion is a source of nuclear power; an out-of-control fusion is a thermonuclear explosion. Scientists have not yet come up with the technology necessary to produce large-scale controlled fusion.

Radioisotopes are used in medicine to treat some types of cancer and as a diagnostic tool. The most useful element for the latter is technetium-99^{m}, which emits only gamma radiation.

Key Terms

alpha (α) particle (3.2)
artificial transmutation (3.7)
beta (β) particle (3.2)
curie (Ci) (3.4)
electron capture (E.C.) (3.7)
gamma ray (3.2)
half-life (3.5)
ionizing radiation (3.4)
nuclear fission (3.8)
nuclear fusion (3.8)
positron (β^{+}) (3.7)
rad (3.4)
radioactive decay (3.2)
radioactivity (3.1)
radioisotope (3.2)
rem (3.4)
roentgen (R) (3.4)
X-ray (3.1)

Review Questions

1. Define or identify each of the following.
   a. alpha particle
   b. artificial transmutation
   c. rem
   d. electron capture
   e. roentgen
   f. half-life
   g. ionizing radiation
   h. radioisotope
   i. rad
2. What is the main difference between a gamma ray and an X-ray? How are they similar?

3. What type of radiation is emitted when a nucleus undergoes the following changes in its atomic number?
   a. an increase of one unit
   b. a decrease of two units
4. What two radioactive processes produce a decrease of one unit in the atomic number of the nucleus?
5. When a nucleus emits a gamma ray, what changes occur in the mass number and atomic number of the nucleus?
6. When a nucleus emits a neutron, what changes occur in the mass number and atomic number of the nucleus?
7. Describe briefly the meaning of each of the following terms.
   a. radioactive tracer
   b. curie
8. Briefly compare nuclear fission and nuclear fusion.
9. Why is the rem a more satisfactory unit for measuring radiation dosage than the rad?
10. What are the chief advantages of nuclear fusion over nuclear fission as a power source?
11. What is the basic principle that underlies radiation processing of foods?
12. Which subatomic particles are responsible for carrying on the chain of reactions that are characteristic of nuclear fission?
13. Why are such high temperatures required for nuclear fusion reactions?
14. What dangers are there in using nuclear fission to generate power?
15. List two ways in which workers can protect themselves from the radioactive materials with which they work.
16. A pair of gloves would be sufficient to shield the hands from which type of radiation: the heavy alpha particles or the massless gamma rays?
17. Heavy lead shielding is necessary as protection from which type of radiation: alpha, beta, or gamma?
18. Plutonium is especially hazardous when inhaled or ingested because it emits alpha particles. Why would alpha particles cause more damage to tissue than beta particles?
19. Explain how radioisotopes can be used for therapeutic purposes.
20. Which radioisotope has been used extensively for treatment of overactive or cancerous thyroid glands?
21. Describe the use of a radioisotope as a diagnostic tool in medicine.
22. What are some of the characteristics that make technetium-99^{m} such a useful radioisotope for diagnostic purposes?
23. What form of radiation is detected in CT scans?
24. What form of radiation is detected in PET scans?
25. What is the advantage of using nonionizing radiation for medical imaging?
26. Name two imaging techniques that do not use ionizing radiation.

Problems

Nuclear Symbols

27. Supply a name for each of the following.
   a. ${}^{4}_{2}He$ b. β^-
   c. ${}^{1}_{0}n$ d. ${}^{2}_{1}H$
28. Supply a symbol for each of the following.
   a. gamma ray b. tritium
   c. positron d. carbon-14

Nuclear Equations

29. Lead-209 undergoes beta decay. Write a balanced nuclear equation for this reaction.
30. Thorium-225 undergoes alpha decay. Write a balanced nuclear equation for this reaction.
31. Sulfur-31 decays by positron emission. Write a balanced nuclear equation for this reaction.
32. A radioactive isotope decays by alpha emission to produce bismuth-211. What was the original element?
33. Write a balanced nuclear equation for the emission of a neutron by bromine-87.
34. Write a balanced nuclear equation for the emission of a proton by magnesium-21.
35. When magnesium-24 is bombarded with a neutron, a proton is ejected. What new isotope is formed? [*Hint*: Write a balanced nuclear equation.]
36. When nitrogen-14 is bombarded with an alpha particle, a proton is ejected. What new isotope is formed? [*Hint*: Write a balanced nuclear equation.]
37. Complete the following nuclear equations.
   a. ${}^{10}_{5}B + {}^{1}_{0}n \longrightarrow ? + {}^{1}_{1}H$
   b. ${}^{121}_{51}Sb + ? \longrightarrow {}^{121}_{52}Te + {}^{1}_{0}n$
   c. ${}^{59}_{27}Co + {}^{1}_{0}n \longrightarrow {}^{56}_{25}Mn + ?$
38. Complete the following nuclear equations.
   a. ${}^{154}_{62}Sm + {}^{1}_{0}n \longrightarrow ? + 2\,{}^{1}_{0}n$
   b. $? + {}^{4}_{2}He \longrightarrow {}^{133}_{57}La + 4\,{}^{1}_{0}n$
   c. ${}^{246}_{96}Cm + {}^{13}_{6}C \longrightarrow {}^{254}_{102}No + ?$
39. Write an equation to represent each of the following nuclear processes.
   a. The reaction of two deuterons to produce helium-3.
   b. The production of ${}^{243}_{97}Bk$ by the α-particle bombardment of ${}^{241}_{95}Am$.
   c. The bombardment of ${}^{121}_{51}Sb$ by α particles to produce ${}^{124}_{53}I$, followed by its radioactive decay by positron emission.

40. What is the new nucleus formed in each of the following processes?
   a. Lead-196 goes through two successive E.C. processes.
   b. Bismuth-215 decays through two successive β^- emissions.
   c. Protactinium-231 decays through four successive α emissions.

Half-Life

41. The half-life of iodine-131 is 8.04 days. We follow the activity of a sample of iodine-131 while it falls to one-eighth of its initial value. How long will this take?
42. A 128-mg sample of technetium-99^m, half-life of 6.0 h, is used in a medical procedure. How much of the technetium-99^m sample remains after 24 h?
43. The half-life of molybdenum-99 is 67 h. How much time passes before a sample with an activity of 160 disintegrations per minute has decreased to an activity of 5.0 disintegrations per minute.
44. Krypton-81^m is used for lung ventilation studies. Its half-life is 13 s. How long will it take the activity of this isotope to reach one-fourth of its original value?

Additional Problems

45. What changes would you look for to establish whether a particular radioactive decay was by γ emission or electron capture? Explain.
46. One proposal for reducing nuclear wastes is to use neutron bombardment to convert long-lived radioisotopes to ones with shorter half-lives. For example, technetium-99, a major by-product of nuclear weapons production, has a half-life of 210,000 years. Neutron bombardment converts it to technetium-100, which decays with a 16-s half-life to ruthenium-100. Write nuclear equations for these reactions.
47. In 1994, German scientists claimed the synthesis of an isotope of element $Z = 110$. When bombarded with nickel-62 nuclei, a lead-208 nucleus yields a nucleus with mass number 269 plus one neutron. Only five atoms of the new element were formed, but it was identified by its decay pattern of two positrons and three successive α decays ending with nobelium-257. Write nuclear equations for the synthesis of element 110 and its subsequent decay reactions.
48. Ionizing radiation can be used to measure and control the thickness of paper, textiles, or rubber sheeting produced in a continuous process. The intensity of the beam of radiation is diminished as it passes through the material. The counting rate when the radiation strikes a Geiger Müller tube is directly related to the thickness. Which type of radiation—α, β, or γ—do you think is best suited to this application. Explain.
49. The activity of a radiation source is 500 Ci. The activity of another source is 10 mCi. To be used properly, one source is taken into the body and the other remains outside the body during treatment. Which is likely to be the internal source, and which the external?
50. Patient A takes internally an iodine-131 sample with an activity of 150 mCi. Patient B takes internally a dose of 15 μCi. In which patient is the iodine-131 being used to treat a malignancy, and in which patient is the isotope being used for imaging the thyroid gland? Both patients are adults.
51. About 2 mCi of thallium-201 is given by intravenous administration for imaging the heart. It is estimated that the total body radiation dose in humans is about 0.07 rad per mCi of thallium-201. How does the radiation dose used in this procedure compare with the lethal dose for humans?
52. Radioactive radon gas in homes is thought to be a lung cancer risk. One set of reactions involves two successive alpha decays from radon-222. What products are formed?
53. Selenium-82 undergoes a rare reaction in which two beta particles are emitted. What is the product?
54. C. E. Bemis and colleagues at Oak Ridge National Laboratory confirmed the synthesis of element 104, the half-life of which was only 4.5 s. Only 3000 atoms of the element were created in the tests. How many atoms were left after 4.5 s? After 9.0 s?
55. Abandoned salt mines are often cited as good places to store nuclear waste. What are the pros and cons of such disposal?
56. Discuss the impact of nuclear science on the following.
   a. war and peace
   b. medicine
   c. our energy needs
57. Cesium is one of the components of nuclear fallout. Why is it a particularly dangerous threat to the environment?
58. Nuclear wastes typically need to be stored for 20 half-lives to be safe. This translates into hundreds of years of storage time. (For instance, cesium-137 and strontium-90 have half-lives of about 30 years.) Would the shooting of such wastes into outer space be a responsible solution?

4 Chemical Bonds

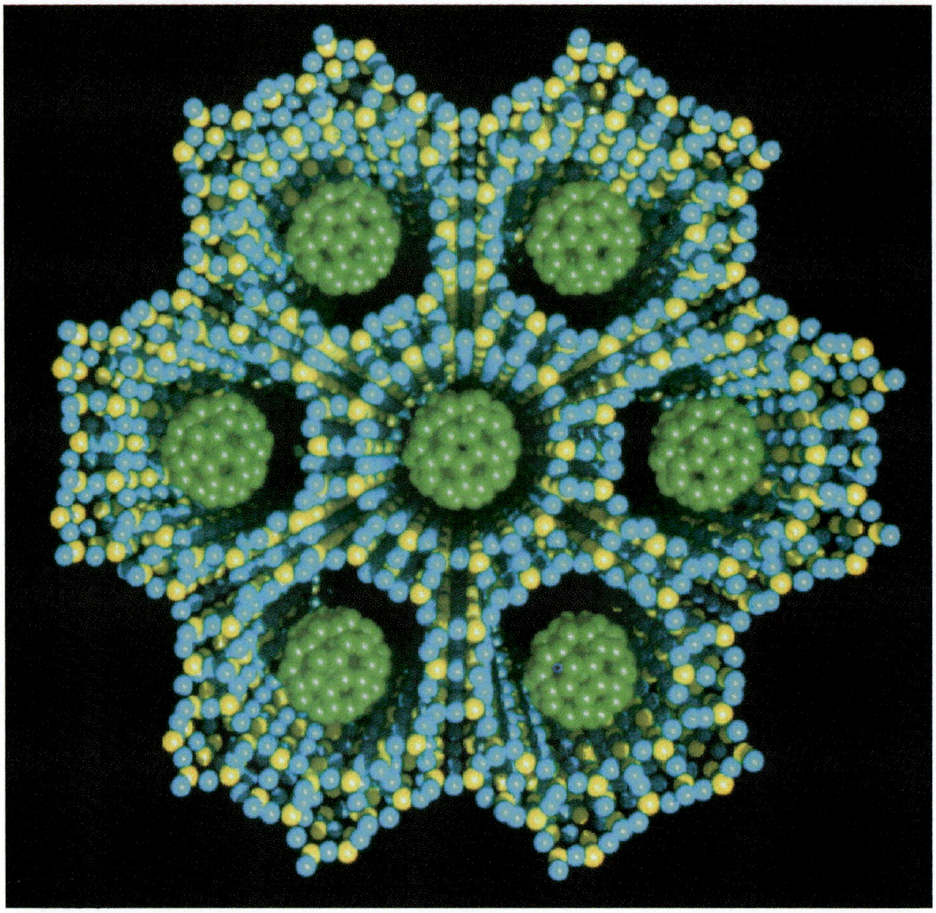

A computer graphics representation of buckminsterfullerene, a novel form of carbon (green spheres) fitting inside the pores of a zeolite crystal. Chemical bonds hold atoms in place in molecules and crystals, giving structure to matter.

There are slightly more than 100 chemical elements. There are millions of chemical compounds. To form these compounds, atoms of different elements must be held together in specific combinations. **Chemical bonds** are the forces that maintain these arrangements. Chemical bonding also plays a major role in determining the state of matter. At room temperature, water is a liquid, carbon dioxide is a gas, and table salt is a solid—all because of differences in chemical bonding. Bonds even determine the physical shape and flexibility of molecules—that is, whether they are spherical or flat, rigid or wobbly.

As scientists developed an understanding of the nature of chemical bonding, they gained the ability to manipulate the structure of compounds. Dynamite, birth control pills, synthetic fibers, and a thousand other products were fashioned in chemical laboratories and have dramatically changed the way we live. We are now entering an era that promises (some would say threatens) even greater change.

The DNA molecule—the chemical basis of heredity—carries its genetic message in its bonds. Two DNA molecules in turn are joined to each other by a special kind of bonding (see Section 23.3). Whether an organism is fish, fowl, hippopotamus, or human is determined by the arrangement of the bonds in DNA. Scientists already have the ability to rearrange these bonds, and this ability has given them limited control over the structure of living matter. Genetic engineering now enables scientists to custom-tailor certain organisms.

4.1 The Art of Deduction: Stable Electron Configurations

In our discussion of the atom and its structure (Chapter 2), we followed the historical development of some of the more important atomic concepts. Some of the nuclear concepts (Chapter 3) were approached in the same manner. We could continue to look at chemistry in this manner, but that would require several volumes of print—if we got very far—and perhaps more of your time than you care to spend. We won't abandon the historical approach entirely, but we will emphasize that other aspect of scientific enterprise: deduction.

The art of deduction works something like this.

1. **Fact** The noble gases, such as helium, neon, and argon, are inert (i.e., they undergo few, if any, chemical reactions).
2. **Theory** The inertness of the noble gases is due to their electron structures.
3. **Deduction** If other elements could alter their electron structures to become more like the noble gases, they would become less reactive.

To illustrate, let's look at an atom of the element sodium (Na). It has 11 electrons—two in the first energy level, eight in the second, and one in the third. If the atom gave up an electron, it would have the same electron configuration as an atom of the noble gas neon (Ne).

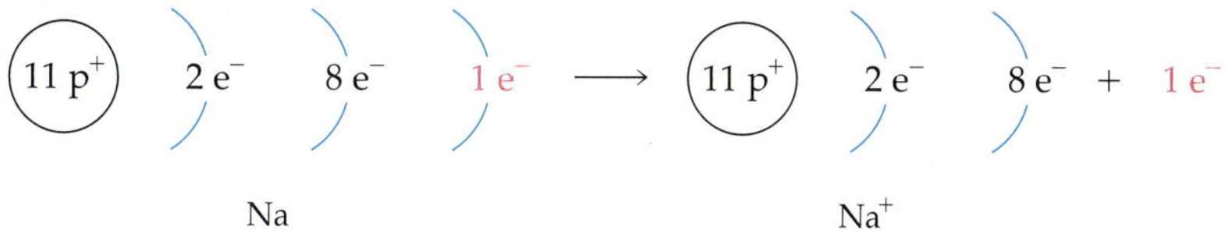

Recall that neon has the structure

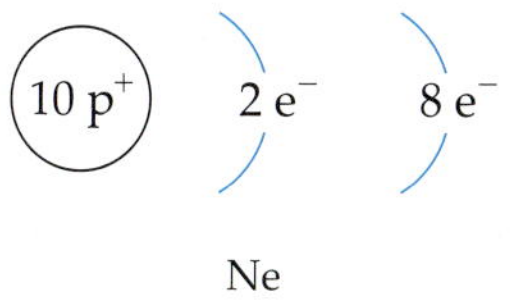

If a chlorine atom (Cl) gained an electron, it would have the same electron configuration as argon (Ar).

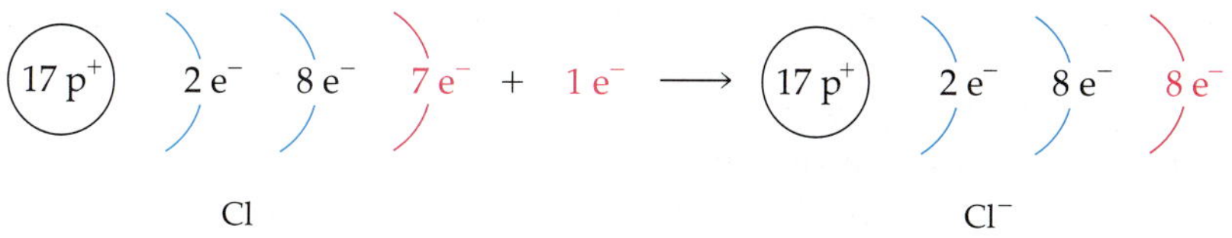

Chlorine does not *become* argon. The chloride ion and argon atom have the same electron configuration, but the chloride ion has 17 protons in its nucleus and a net charge of 1−. The argon atom has 18 protons in its nucleus and no net charge; it is electrically neutral. Nor does sodium become neon by giving up an electron. The sodium ion has the same electron configuration as neon, but the nuclei differ in the number of protons.

The electron configuration of the argon atom is

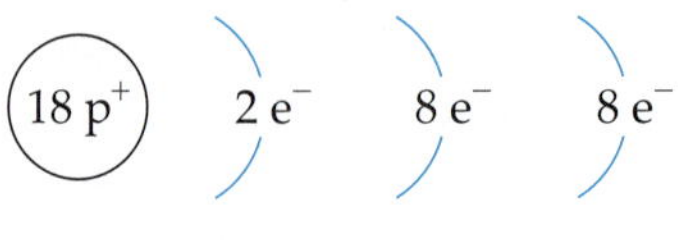

Ar

Some *ions* are simply electrically charged atoms. Other ions consist of a group of atoms that act as a unit. Positively charged ions are called *cations* and negatively charged ions are called *anions* (see Section 12.1).

The sodium atom, having lost an electron, becomes positively charged. It has 11 protons (11+) and only 10 electrons (10−). It is written Na^+ and is called a *sodium ion*. The chlorine atom, having gained an electron, becomes negatively charged. It has 17 protons (17+) and 18 electrons (18−). It is written Cl^- and is called a *chloride ion*. Note that a positive charge, as in Na^+, indicates that one electron has been lost per atom. Similarly, a negative charge, as in Cl^-, indicates that one electron has been gained per atom. **Ions** are charged units—that is, units in which the number of electrons is *not* equal to the number of protons.

Example 4.1

What is the charge on an aluminum atom that has lost three electrons?

Solution

The neutral aluminum atom has 13 electrons and 13 protons (its atomic number is 13). The ion has 13 protons (13+) and 10 electrons (10−). The net charge on the aluminum ion is 3+. The symbol is Al^{3+}.

Practice Exercise

What is the charge on a sulfur atom that has gained two electrons?

4.2 Lewis (Electron Dot) Structures

The core electron configurations of sodium and chlorine atoms do not change when these atoms form ions. It is convenient to let the chemical symbol represent the core or kernel of an atom—that is, the nucleus plus the inner electrons. The **valence electrons**—those in the outer shell—are represented by dots placed around the symbol. To illustrate, we can reduce the equations of the preceding section to

$$\text{Na}\cdot \longrightarrow \text{Na}^+ + 1e^-$$

and

$$:\underset{\cdot\cdot}{\overset{\cdot\cdot}{\text{Cl}}}\cdot + 1e^- \longrightarrow :\underset{\cdot\cdot}{\overset{\cdot\cdot}{\text{Cl}}}:^-$$

Representations in which the symbol of an element stands for the core of the atom and dots stand for its valence electrons are called **Lewis structures** or *electron dot structures*.

It is especially easy to write Lewis structures for the main group elements. The number of valence electrons for each of these elements is equal to the group number (Table 4.1).

A useful practice in placing dots around a chemical symbol is to place lone dots on each of the four sides of the symbol before pairing any of the dots.

Symbolism is a convenient, shorthand way to convey a lot of information in compact form. It is the chemist's most efficient and economical form of communication. Learning this symbolism is much like learning a foreign language. Once you master a certain basic "vocabulary," the rest is easier.

Table 4.1 Lewis Structures of Selected Main Group Elements

1A	2A	3A	4A	5A	6A	7A	Noble Gases
$\mathrm{H}\cdot$							$\mathrm{He}:$
$\mathrm{Li}\cdot$	$\cdot\mathrm{Be}\cdot$	$\cdot\dot{\mathrm{B}}\cdot$	$\cdot\underset{\cdot}{\dot{\mathrm{C}}}\cdot$	$:\underset{\cdot}{\dot{\mathrm{N}}}\cdot$	$:\underset{\cdot}{\ddot{\mathrm{O}}}\cdot$	$:\underset{\cdot\cdot}{\ddot{\mathrm{F}}}\cdot$	$:\underset{\cdot\cdot}{\ddot{\mathrm{Ne}}}:$
$\mathrm{Na}\cdot$	$\cdot\mathrm{Mg}\cdot$	$\cdot\dot{\mathrm{Al}}\cdot$	$\cdot\underset{\cdot}{\dot{\mathrm{Si}}}\cdot$	$:\underset{\cdot}{\dot{\mathrm{P}}}\cdot$	$:\underset{\cdot}{\ddot{\mathrm{S}}}\cdot$	$:\underset{\cdot\cdot}{\ddot{\mathrm{Cl}}}\cdot$	$:\underset{\cdot\cdot}{\ddot{\mathrm{Ar}}}:$
$\mathrm{K}\cdot$	$\cdot\mathrm{Ca}\cdot$				$:\underset{\cdot}{\ddot{\mathrm{Se}}}\cdot$	$:\underset{\cdot\cdot}{\ddot{\mathrm{Br}}}\cdot$	$:\underset{\cdot\cdot}{\ddot{\mathrm{Kr}}}:$
$\mathrm{Rb}\cdot$	$\cdot\mathrm{Sr}\cdot$				$:\underset{\cdot}{\ddot{\mathrm{Te}}}\cdot$	$:\underset{\cdot\cdot}{\ddot{\mathrm{I}}}\cdot$	$:\underset{\cdot\cdot}{\ddot{\mathrm{Xe}}}:$

Example 4.2

Without referring to Table 4.1, give Lewis structures for calcium, oxygen, and phosphorus.

Solution

Calcium is in Group 2A, oxygen is in Group 6A, and phosphorus is in Group 5A. The Lewis structures are therefore

$$\cdot\mathrm{Ca}\cdot \qquad :\underset{\cdot}{\dot{\mathrm{O}}}: \qquad :\underset{\cdot}{\dot{\mathrm{P}}}\cdot$$

Practice Exercise

Without referring to Table 4.1, give Lewis structures for each of the following elements.
a. Ar **b.** Sr **c.** F **d.** N **e.** K **f.** S

4.3 Sodium Reacts with Chlorine: The Facts

Sodium is a highly reactive metal. It is soft enough to be cut with a knife. When freshly cut, it is bright and silvery, but it dulls rapidly because it reacts with oxygen in the air. In fact, it reacts so readily in air that it is usually stored under oil or kerosene. Sodium reacts violently with water also, producing so much heat that it melts. A small piece will form a spherical bead after melting and race around on the surface of the water as it reacts.

Chlorine is a greenish yellow gas. It is familiar as a disinfectant for swimming pools and city water supplies. (The actual substance added may be a compound that reacts with water to form chlorine.) Who hasn't been swimming in a pool that had "so much chlorine in it that you could taste it"? Chlorine is extremely irritating to the respiratory tract. In fact, chlorine was used as a poison gas in World War I.

When a piece of sodium is dropped into a flask containing chlorine gas, a violent reaction ensues. A white solid that is quite unreactive is formed. It is a familiar compound—sodium chloride (table salt) (Figure 4.1).

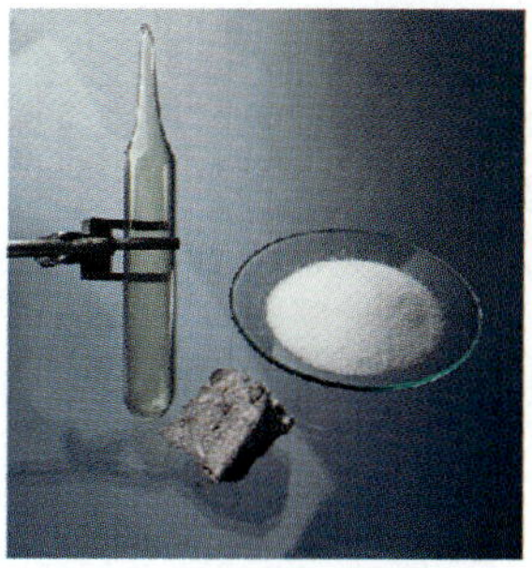

Sodium is a soft, silvery, reactive metal. Chlorine is a poisonous green gas. Sodium chloride is a crystalline white solid used to enhance the flavor of food.

FIGURE 4.1 ▶

The reaction of sodium and chlorine. Sodium metal and chlorine gas provide striking visual evidence of their reaction to produce the ionic substance sodium chloride.

4.4 The Sodium–Chlorine Reaction: Theory

Actually, chlorine gas is composed of Cl_2 molecules. Each atom of the molecule receives an electron from a sodium atom. Two sodium ions and two chloride ions are formed.

$$Cl_2 + 2\,Na \longrightarrow 2\,Cl^- + 2\,Na^+$$

The sodium–chlorine reaction illustrates a basic tendency of nature. More reactive (more energetic) substances tend to become less reactive (less energetic) substances, releasing energy in the process. A sodium atom becomes less reactive by *losing* an electron. A chlorine atom becomes less reactive by *gaining* an electron. What happens when sodium atoms come into contact with chlorine atoms? A chlorine atom gets an electron from a sodium atom. Using Lewis structures, the equation for this reaction is written

$$\mathrm{Na}\cdot + :\underset{\cdot\cdot}{\ddot{\mathrm{Cl}}}\cdot \longrightarrow \mathrm{Na}^+ + :\underset{\cdot\cdot}{\ddot{\mathrm{Cl}}}:^-$$

The two ions formed from sodium and chlorine atoms have opposite charges. They are strongly attracted to one another. Remember, however, that in even a tiny grain of salt, there are billions and billions of each kind of ion. These ions arrange

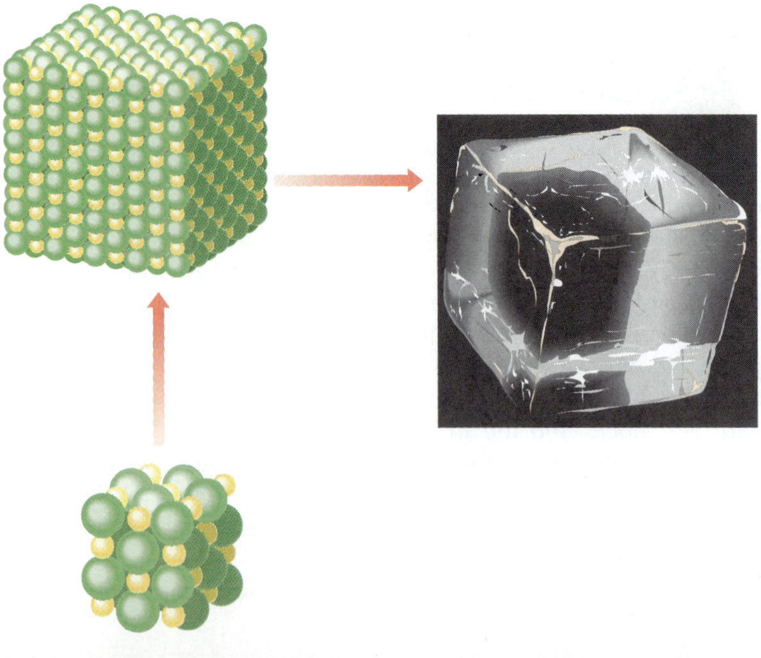

FIGURE 4.2 ▶

In a crystal of sodium chloride, ions are arranged in a regular fashion. Each Na^+ ion (small sphere) is surrounded by six Cl^- ions (large spheres). In turn, each Cl^- ion is surrounded by six Na^+ ions. This arrangement repeats itself many, many times, ultimately resulting in a crystal of sodium chloride (right).

themselves in an orderly fashion. The arrangements are repeated many times in all directions—front and back, left and right, top and bottom—to make a crystal of sodium chloride (Figure 4.2). Each sodium ion attracts (and is attracted by) six chloride ions (the ones to its front and back, its top and bottom, and its two sides). Each chloride ion attracts (and is attracted by) six sodium ions. The forces holding the crystal together (the attractive forces between positive and negative ions) are called **ionic bonds.**

A **crystal** is a solid substance with a regular arrangement of its constituent particles. The solid as a whole has a well-defined, regular shape.

Scientists sometimes use different models to represent the same system. The model employed in Figure 4.2 is a space-filling model showing the relative sizes of the sodium and chloride ions. Sometimes a ball-and-stick model is employed to better show the geometry of the crystal (Figure 4.3). From this model it is easy to see the cubic arrangement of the ions. In the crystal as a whole, for each sodium ion there is one chloride ion; thus, the ratio of ions is 1:1, and the simplest formula for the compound is NaCl. The symbols Na and Cl, written together, stand for the compound sodium chloride. The formula is also used to represent one sodium ion and one chloride ion.

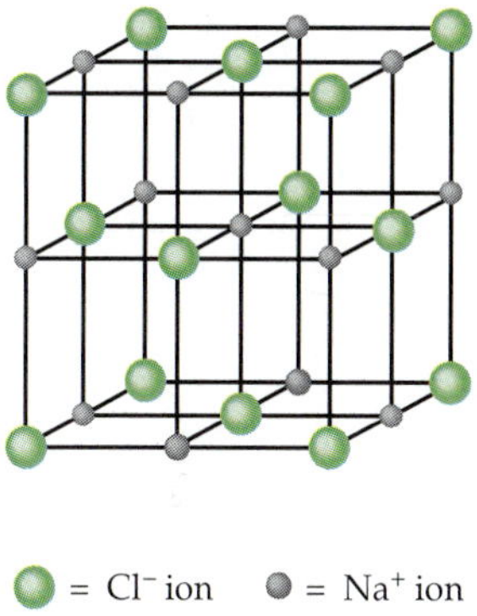

▲ **FIGURE 4.3**
A ball-and-stick model of a sodium chloride crystal.

4.5 Ionic Bonds: Some General Considerations

As we might expect, potassium, a metal in the same family as sodium, also reacts with chlorine. The reaction yields potassium chloride (KCl).

$$K\cdot + \cdot\underset{\cdot\cdot}{\overset{\cdot\cdot}{Cl}}: \longrightarrow K^+ + :\underset{\cdot\cdot}{\overset{\cdot\cdot}{Cl}}:^-$$

And potassium reacts with bromine, a reddish brown liquid in the same family as chlorine, to form a stable white crystalline solid called potassium bromide (KBr).

$$K\cdot + \cdot\underset{\cdot\cdot}{\overset{\cdot\cdot}{Br}}: \longrightarrow K^+ + :\underset{\cdot\cdot}{\overset{\cdot\cdot}{Br}}:^-$$

Sodium also reacts with bromine to form sodium bromide.

Example 4.3

Use Lewis structures to show the transfer of electrons from sodium atoms to bromine atoms to form ions with noble gas configurations.

Solution

Sodium has one valence electron, and bromine has seven. Transfer of the single electron from sodium to bromine leaves each with a noble gas configuration.

$$Na\cdot + \cdot\underset{\cdot\cdot}{\overset{\cdot\cdot}{Br}}: \longrightarrow Na^+ + :\underset{\cdot\cdot}{\overset{\cdot\cdot}{Br}}:^-$$

Practice Exercise

Use Lewis structures to show the transfer of electrons from lithium atoms to fluorine atoms to form ions with noble gas configurations.

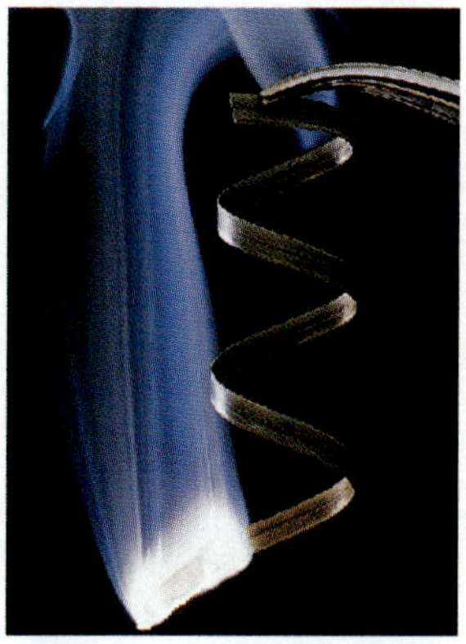

Magnesium metal burns in air by combining with oxygen gas to produce a brilliant white light and a "smoke" of solid, white magnesium oxide.

Magnesium, a Group 2A metal, is harder and less reactive than sodium. Magnesium reacts with oxygen, a Group 6A element (a colorless gas), to form another stable white crystalline solid called magnesium oxide (MgO).

$$\cdot Mg\cdot + \cdot\underset{\cdot}{\overset{\cdot\cdot}{O}}: \longrightarrow Mg^{2+} + :\underset{\cdot\cdot}{\overset{\cdot\cdot}{O}}:^{2-}$$

Magnesium must give up two electrons and oxygen must gain two electrons for each to have the same configuration as the noble gas neon.

An atom such as oxygen, which needs two electrons to complete a noble gas configuration, may react with potassium atoms, each of which has only one electron to give. In this case, two atoms of potassium are needed for each oxygen atom. The product is potassium oxide (K_2O).

$$
\begin{array}{ccc}
\text{K·} & & \text{K}^+ \\[4pt]
& + \quad \text{·}\ddot{\underset{\cdot\cdot}{\text{O}}}\text{:} \quad \longrightarrow & \quad + \quad :\ddot{\underset{\cdot\cdot}{\text{O}}}\text{:}^{\,2-} \\[4pt]
\text{K·} & & \text{K}^+
\end{array}
$$

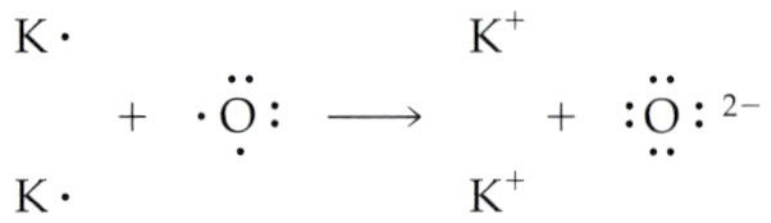

By this process, each potassium atom achieves the argon configuration. Oxygen again assumes the neon configuration.

Example 4.4

Use Lewis structures to show the transfer of electrons from magnesium atoms to nitrogen atoms to form ions with noble gas configurations.

Solution

$$
\begin{array}{ccccc}
\text{·Mg·} & & & \text{Mg}^{2+} & \\[4pt]
& \text{·}\ddot{\underset{\cdot}{\text{N}}}\text{·} & & & :\ddot{\underset{\cdot\cdot}{\text{N}}}\text{:}^{\,3-} \\[4pt]
\text{·Mg·} \;\; + & & \longrightarrow \;\; \text{Mg}^{2+} \;\; + & \\[4pt]
& \text{·}\ddot{\underset{\cdot}{\text{N}}}\text{·} & & & :\ddot{\underset{\cdot\cdot}{\text{N}}}\text{:}^{\,3-} \\[4pt]
\text{·Mg·} & & & \text{Mg}^{2+} &
\end{array}
$$

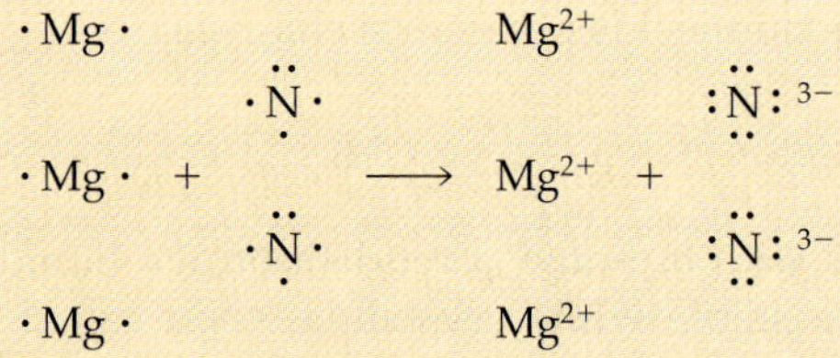

Each of the three magnesium atoms gives up two electrons (a total of six), and each of the two nitrogen atoms acquires three electrons (a total of six). Notice that the total positive and total negative charges on the products are equal (6+ and 6−). Magnesium reacts with nitrogen to produce magnesium nitride (Mg_3N_2).

Practice Exercise

Use Lewis structures to show the transfer of electrons from aluminum atoms to oxygen atoms to form ions with noble gas configurations.

Generally speaking, metallic elements in Groups 1A, 2A, and 3A of the periodic table react with nonmetallic elements in Groups 5A, 6A, and 7A to form stable crystalline ionic solids. The metals tend to give up electrons to the nonmetals. The ions so formed have noble gas electron configurations. The crystalline ionic solids are held together by the attraction of oppositely charged ions—that is, by ionic bonds.

4.6 Names of Simple Ions and Ionic Compounds

Names of simple monatomic (single-atom) positive ions (cations) are derived from those of their parent elements by addition of the word *ion*. A sodium atom (Na), upon losing an electron, becomes a sodium ion (Na^+). A magnesium atom (Mg), upon losing two electrons, becomes a magnesium ion (Mg^{2+}). Names of simple monatomic negative ions *(anions)* are derived from those of their parent elements by changing the usual ending to -*ide* and adding the word ion. A chlor*ine* atom

Table 4.2 Symbols and Names for Some Monatomic Ions

Group	Element	Name	Symbol for Ion
1A	Hydrogen	Hydrogen ion[a]	H^+
	Lithium	Lithium ion	Li^+
	Sodium	Sodium ion	Na^+
	Potassium	Potassium ion	K^+
2A	Magnesium	Magnesium ion	Mg^{2+}
	Calcium	Calcium ion	Ca^{2+}
3A	Aluminum	Aluminum ion	Al^{3+}
5A	Nitrogen	Nitride ion	N^{3-}
6A	Oxygen	Oxide ion	O^{2-}
	Sulfur	Sulfide ion	S^{2-}
7A	Fluorine	Fluoride ion	F^-
	Chlorine	Chloride ion	Cl^-
	Bromine	Bromide ion	Br^-
	Iodine	Iodide ion	I^-
1B	Copper	Copper(I) ion (cuprous ion)	Cu^+
		Copper(II) ion (cupric ion)	Cu^{2+}
	Silver	Silver ion	Ag^+
2B	Zinc	Zinc ion	Zn^{2+}
8B	Iron	Iron(II) ion (ferrous ion)	Fe^{2+}
		Iron(III) ion (ferric ion)	Fe^{3+}

[a] Does not exist independently in aqueous solution.

(Cl), upon gaining an electron, becomes a chlor*ide* ion (Cl^-). A sulf*ur* atom (S), upon gaining two electrons, becomes a sulf*ide* ion (S^{2-}).

Names and symbols for several important monatomic ions are given in Table 4.2. Note that the charge on an ion of a Group 1A element is 1+ (usually written simply as +). The charge on an ion of a Group 2A element is 2+, and that on an ion of a Group 3A element is 3+. You can calculate the charges on the negative ions in Table 4.2 by subtracting 8 from the group number. For example, the charge on an oxide ion (oxygen is in Group 6A) is $6 - 8 = -2$. The charge on a nitride ion (nitrogen is in Group 5A) is $5 - 8 = -3$. The periodic relationship of these monatomic ions is shown in Figure 4.4.

There is no simple way to determine the most likely charge on ions formed from elements in B subgroups. Indeed, you may have noticed that these elements can form ions with different charges. In such cases, chemists use Roman numerals with the names, to indicate the charge. Thus, *iron(II) ion* means Fe^{2+}, and *iron(III) ion* means Fe^{3+}.

An older system of naming ions is occasionally used, based on the Latin stems and using "ous" and "ic" endings. The "ous" indicates the lower of two possible charges and "ic" denotes the higher charge: Fe^{2+} is *ferrous* ion and Fe^{3+} is *ferric* ion. See similar names for the two copper ions in Table 4.2.

Compounds such as sodium chloride, potassium bromide, magnesium oxide, and potassium oxide are called **ionic compounds.** The constituent units of these compounds are charged particles—ions. Yet the compound as a whole is electrically neutral. We can use this principle of electrical neutrality to determine the combining ratio of ions. Potassium ions (K^+) combine with bromide ions (Br^-) in a ratio of 1:1. The formula KBr expresses this ratio and represents the compound potassium bromide. The combining ratio (1:1) and the ionic charges are understood.

Let's try another example. One calcium ion (Ca^{2+}) combines with *two* chloride ions (Cl^-). This ratio is expressed in the formula $CaCl_2$. In this formula, the ionic charges are understood. As with a coefficient of 1 in algebra, a subscript of 1 is

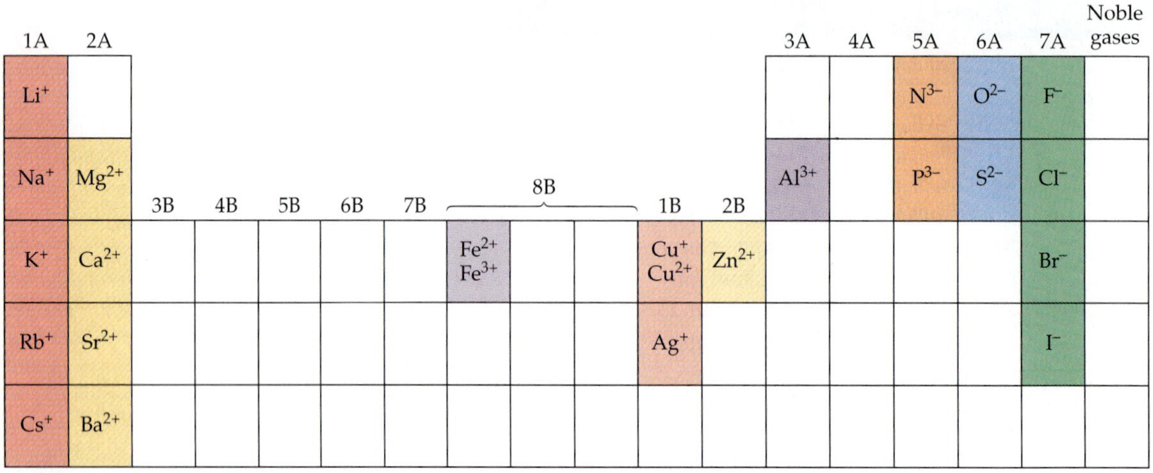

▲ **FIGURE 4.4**
The periodic relationships of some monatomic ions.

understood, so in the formula $CaCl_2$ the subscript 1 for calcium ion is understood whereas the 2 for chloride is written explicitly (not Ca_1Cl_2 but $CaCl_2$). Thus, $CaCl_2$ not only gives us the combining ratio (1:2) but also stands for the ionic compound calcium chloride. It is a shorthand way of writing 1 Ca^{2+} and 2 Cl^-.

You can use the charges on the ions in Table 4.2 to determine formulas for compounds of these elements. You can use the periodic table to predict the charges on ions formed from subgroup A elements, with Group 4A being the dividing line between positive and negative ions.

Example 4.5

What is the formula for sodium sulfide?

Solution

First, write the symbols for the ions (positive ion first). Sodium is in Group 1A; the charge on the sodium ion is 1+. Sulfur is in Group 6A, and the charge on the sulfide ion is $6 - 8 = -2$. The symbols are Na^+ and S^{2-}. The smallest number into which both charges can be evenly divided—that is, the *least common multiple* (LCM)—is 2. The least common multiple simply indicates the smallest number of electrons that can be evenly exchanged between the two elements. The subscript for each symbol can be determined by division of its charge (without the plus or minus) into the least common multiple. This step determines how many atoms of each element are needed to supply or accept the smallest common number of electrons. For Na^+,

$$\frac{2\ (\text{LCM})}{1\ (\text{charge})} = 2$$

For S^{2-},

$$\frac{2\ (\text{LCM})}{2\ (\text{charge})} = 1$$

Thus, we have the formula Na_2S_1 (2 Na^+ and 1 S^{2-}), or Na_2S.

Practice Exercise

What is the formula for magnesium bromide?

Example 4.6

What is the formula for aluminum oxide?

Solution

The symbols are Al^{3+} and O^{2-} (Al is in Group 3A, and O is in Group 6A). The LCM is 6. For Al^{3+},

$$\frac{6}{3} = 2$$

For O^{2-},

$$\frac{6}{2} = 3$$

The formula is therefore Al_2O_3 (2 Al^{3+} and 3 O^{2-}).

Practice Exercise

What is the formula for calcium nitride?

Another way to look at this is the cross-over method: the charge number for one ion becomes the subscript for the other. Thus, in Example 4.6, the charge on the aluminum ion is 3, which becomes the subscript for oxygen in aluminum oxide; and the charge on oxygen is 2, which becomes the subscript for the aluminum ion.

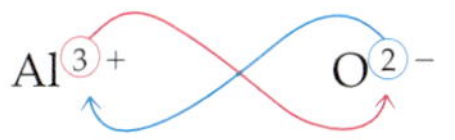

Al_2O_3

Thus, two aluminum ions have six positive charges, and three oxide ions have six negative charges. The resulting compound, Al_2O_3, is therefore neutral—as all compounds are.

Note that the cross-over method works because it is based on the transfer of electrons and the conservation of charge. Two aluminum atoms lose three electrons each (that's six electrons lost), and three oxygen atoms gain two electrons each (that's six electrons gained). Electrons lost equal electrons gained, and all is well.

What Is a Low-Sodium Diet?

People with high blood pressure are usually advised to follow a low-sodium diet. Just what does this mean? Surely they are not being advised to reduce their consumption of sodium itself. Sodium is an extremely reactive metal that reacts violently with moisture; eating it wouldn't be safe. The concern is really with sodium *ions*, Na^+. The average adult in the United States consumes too much Na^+, most of it from sodium chloride, common table salt. It is not uncommon for some individuals to eat 6 or 7 g of sodium chloride a day, most of it in prepared foods. Many snack foods, such as potato chips, pretzels, and corn chips, are especially high in salt. The American Heart Association recommends that adults limit their salt intake to about 3 g per day.

We cannot overemphasize the difference between ions and the atoms from which they are made. A metal atom and its cation are as different as a whole peach (atom) and a peach pit (ion). The names and symbols look a lot alike, but the species themselves are quite different. Unfortunately, the situation is confused because people talk about needing "iron" to perk up "tired blood" and "calcium" for healthy teeth and bones. What they really mean is iron(II) *ions* (Fe^{2+}) and calcium *ions* (Ca^{2+}). You wouldn't think of eating iron nails to get "iron." Nor would you eat highly reactive calcium metal. Although persons who are not chemists do not always make careful distinctions, we try to use precise terminology here.

Some manufacturers enrich their cereals with very fine specks of iron filings. These iron filings dissolve in the acidic environment of the stomach, producing iron ions.

Naming these binary ionic compounds is simple. First write the name of the positive ion and then the name of the negative ion. (The word ion is not used. It is understood in each case.)

Example 4.7

What is the name of the compound Na_2S?

Solution

Find the constituent ions in Table 4.2. They are sodium ion (Na^+) and sulfide ion (S^{2-}). The compound is sodium sulfide.

Practice Exercise

What is the name of the compound Li_2O?

Example 4.8

What is the name of the compound FeS?

Solution

There are two kinds of iron ions. Because sulfur exists as the S^{2-} ion and one iron ion is combined with it, the iron ion in this compound must be Fe^{2+}. The name of FeS is iron(II) sulfide.

Practice Exercise

What is the name of the compound $CuBr_2$?

Example 4.9

What is the name of the compound $FeCl_3$?

Solution

Because the charge on the chloride ion is 1− and three of these ions are combined with one iron ion, the iron ion must be Fe^{3+}. The name of $FeCl_3$ is iron(III) chloride.

Practice Exercise

What is the name of the compound Fe_2O_3?

Ionic compounds generally exist as crystalline solids. However, many of them are soluble in water. Ionic compounds are found dissolved in all natural waters—including the water in the cells of our bodies, where they are involved in such critical functions as the transmission of nerve impulses.

4.7 Covalent Bonds: Shared Electron Pairs

One might expect a hydrogen atom, with its one electron, to acquire another electron and assume the helium configuration. Indeed, hydrogen atoms do just that in the presence of atoms of a reactive metal such as lithium—that is, a metal that readily gives up an electron.

$$\text{Li}\cdot + \text{H}\cdot \longrightarrow \text{Li}^+ + \text{H}:^-$$

But what if there are no other kinds of atoms around? What if there are only hydrogen atoms? One atom can't gain an electron from another because hydrogen atoms all have an equal attraction for electrons. They can compromise, however, by *sharing a pair* of electrons.

$$\text{H}\cdot + \cdot\text{H} \longrightarrow \text{H}:\text{H}$$

By sharing electrons, the two hydrogen atoms form a hydrogen molecule. The bond formed by a shared pair of electrons is called a **covalent bond.**

H:H
covalent bond (shared pair of electrons)

A **molecule** is a group of atoms that are chemically bonded together. Molecules are represented by chemical formulas. The symbol H represents an atom of hydrogen; the formula H_2 represents a *molecule* of hydrogen, which is composed of two hydrogen atoms.

A shared single pair of electrons is called a **single bond.**

Consider next the case of chlorine. A chlorine atom will pick up an extra electron from any atom that can readily give one up. But, again, what if the only thing around is another chlorine atom? Chlorine atoms, too, can attain a more stable arrangement by sharing a pair of electrons.

$$:\overset{\cdot\cdot}{\underset{\cdot\cdot}{\text{Cl}}}\cdot + \cdot\overset{\cdot\cdot}{\underset{\cdot\cdot}{\text{Cl}}}: \longrightarrow :\overset{\cdot\cdot}{\underset{\cdot\cdot}{\text{Cl}}}:\overset{\cdot\cdot}{\underset{\cdot\cdot}{\text{Cl}}}:$$

Each chlorine atom in the chlorine molecule has eight electrons around it, an arrangement like that of the noble gas argon. This stable *octet* of electrons is the arrangement characteristic of all the noble gases except helium. Most of the covalently bonded atoms that we shall consider, except hydrogen, follow the **octet rule:** they seek an arrangement that will surround them with eight electrons. The shared pair of electrons in the chlorine molecule is another example of a covalent bond; these two electrons are called a **bonding pair.** The other electrons that stay on one atom and are not shared are called *nonbonding pairs* or **lone pairs.**

Molecular models of H_2 (top) and Cl_2 (bottom).

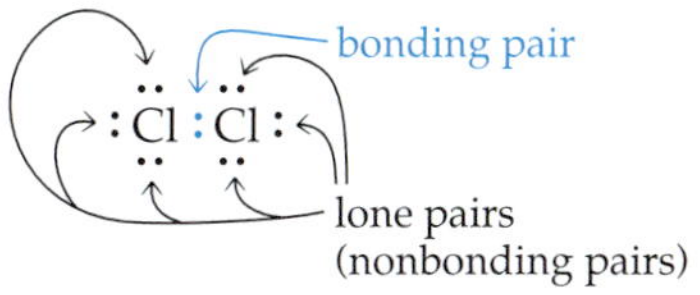

For simplicity, the hydrogen molecule is often represented as H_2 and the chlorine molecule as Cl_2. In each case, the covalent bond between the atoms is understood. Sometimes the covalent bond is indicated by a dash, H—H and Cl—Cl. Nonbonding pairs of electrons often are not shown.

4.8 Multiple Covalent Bonds

Atoms can share more than one pair of electrons. In carbon dioxide, for example, the carbon atom shares two pairs of electrons with each of the two oxygen atoms.

$$:\overset{\cdot\cdot}{\text{O}}::\text{C}::\overset{\cdot\cdot}{\text{O}}: \quad (\text{or } \text{O}{=}\text{C}{=}\text{O})$$

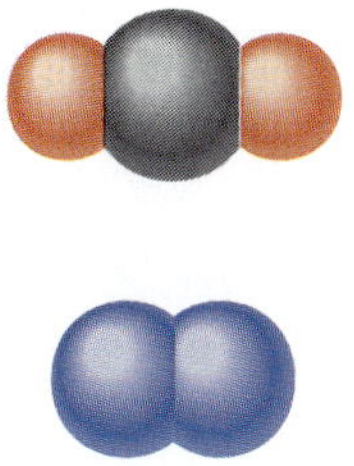

Molecular models of CO_2 (top) and N_2 (bottom).

Note that each atom has an octet of electrons about it as a result of this sharing. We say that the atoms are joined by a **double bond,** a covalent linkage in which the two atoms share two pairs of electrons.

Atoms also can share three pairs of electrons. Nitrogen is in Group 5A, and it has five valence electrons. It could share a pair of electrons with another nitrogen atom and would then look like this:

$$:\ddot{N}\cdot \ \ + \ \ \cdot\ddot{N}: \ \longrightarrow \ :\ddot{N}:\ddot{N}: \quad (incorrect\ structure)$$

To satisfy the octet rule, each nitrogen atom shares three pairs of electrons with the other.

$$:N:::N: \quad (or \ :N\!:\!:\!N: \ or \ N\equiv N\)$$

Covalent bonds often are represented as dashes. We can therefore show the three kinds of bonds as follows.

Cl—Cl O=C=O N≡N

The atoms are joined by a **triple bond,** a covalent linkage in which two atoms share three pairs of electrons. Each nitrogen also has a lone pair of electrons. Note that we could have drawn the lone pair of electrons above or below the atomic symbol. Such a drawing would represent the same molecule.

4.9 Unequal Sharing: Polar Covalent Bonds

So far, we have seen that atoms combine in two different ways. Some that are quite different in electron structure (from opposite ends of the periodic table) react by the complete transfer of an electron from one atom to another to form an ionic bond. Atoms that are identical combine by sharing one or more pairs of electrons to form covalent bonds. Now let's consider bond formation between atoms that are different, but not different enough to form ionic bonds.

Hydrogen and chlorine react to form a colorless gas called hydrogen chloride. We use the individual atoms in order to focus on the sharing of electrons to form a covalent bond.

$$H\cdot \ + \ \cdot\ddot{C}l: \ \longrightarrow \ H:\ddot{C}l: \quad (or\ H—Cl)$$

Both hydrogen and chlorine need an electron to achieve a noble gas configuration, so they share a pair and form a covalent bond.

Example 4.10

Use Lewis structures to show the formation of a covalent bond between (a) two fluorine atoms and (b) a fluorine atom and a hydrogen atom.

Solution

$$:\ddot{F}\cdot \ + \ \cdot\ddot{F}: \ \longrightarrow \ :\ddot{F}:\ddot{F}: \quad \text{Bonding pair}$$

$$H\cdot \ + \ \cdot\ddot{F}: \ \longrightarrow \ H:\ddot{F}: \quad \text{Bonding pair}$$

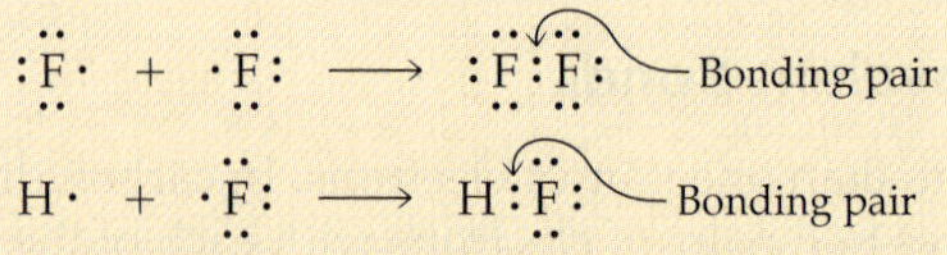

Practice Exercise

Use Lewis structures to show the formation of a covalent bond between (**a**) two bromine atoms, (**b**) a hydrogen atom and a bromine atom, and (**c**) an iodine atom and a chlorine atom.

One might reasonably ask why the hydrogen molecule and the chlorine molecule react at all. Have we not just explained that they themselves were formed to provide a more stable arrangement of electrons? Yes, indeed, we did say that. But there is stable, and there is more stable. The chlorine molecule represents a more stable arrangement than two separate chlorine atoms. Nevertheless, given the opportunity, a chlorine atom would rather form a bond with a hydrogen atom than with another chlorine atom.

Both hydrogen and chlorine consist of diatomic molecules; the reaction is more accurately represented by the scheme

$$\mathrm{H{:}H} + \mathrm{{:}\ddot{\underset{..}{Cl}}{:}\ddot{\underset{..}{Cl}}{:}} \longrightarrow 2\,\mathrm{H{:}\ddot{\underset{..}{Cl}}{:}}$$

For the sake of convenience and simplicity, the reaction of hydrogen (molecule) and chlorine (molecule) to form hydrogen chloride often is reduced to

$$H_2 + Cl_2 \longrightarrow 2\,HCl$$

Molecules of hydrogen chloride consist of one atom of hydrogen and one atom of chlorine. These unlike atoms share a pair of electrons. Sharing does not mean sharing equally, though. Chlorine atoms have a greater attraction for a shared pair of electrons than do hydrogen atoms; chlorine is said to be more *electronegative* than hydrogen. The shared electrons are held more tightly by the chlorine atom, and this results in the chlorine end of the molecule being more negative than the hydrogen end (see Section 4.13).

In a covalent bond between two atoms with equal electronegativities, there is an equal sharing of an electron pair, and the electrons are not drawn any closer to one atom than to the other. A bond of this type is a **nonpolar covalent bond.** The H—H and Cl—Cl bonds are nonpolar. In a covalent bond between atoms of different electronegativities, there is an unequal sharing of an electron pair, and the electrons are drawn closer to the atom of higher electronegativity. Such a bond is a **polar covalent bond.** The H—Cl bond is a polar bond. The polar covalent bond is *not* an ionic bond. In an ionic bond, complete electron transfer occurs from the metal to the nonmetal. In a polar covalent bond, the atom at the positive end of the bond (H in HCl) still has some share in the bonding pair of electrons.

Let's picture the electron-pair bond between two atoms as a cloud of negative electric charge that encompasses both atoms. (Modern quantum theory permits us to do this, just as we described atomic orbitals as electron clouds; see Section 2.5.) In a *nonpolar* bond such as H—H, the electron cloud density is greatest between the two hydrogen nuclei but is otherwise uniformly distributed (Figure 4.5). In a *polar* bond such as H—Cl, although the greatest electron density is still found between the two nuclei, the cloud is strongly displaced toward the more electronegative chlorine.

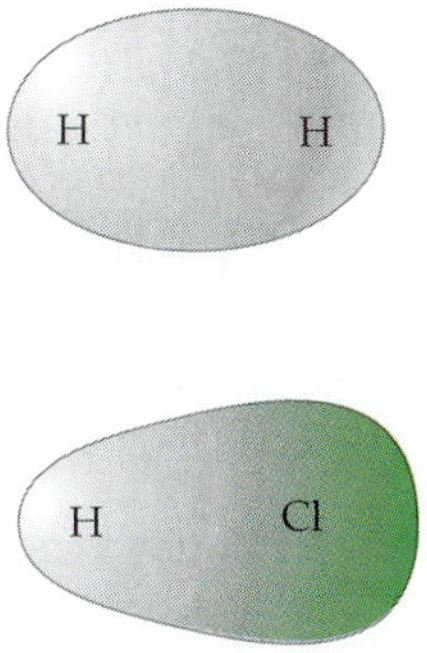

▲ **FIGURE 4.5**

Nonpolar and polar covalent bonds. In H—H, there is an even distribution of electronic charge density between the atoms. In H—Cl, electron charge density is displaced toward the Cl atom.

To indicate the polar nature of a bond, we use the representation

$$\overset{\delta+}{\mathrm{H}}\text{—}\overset{\delta-}{\mathrm{Cl}}$$

The $\delta+$ and $\delta-$ (read "delta plus" and "delta minus") signify that one end (H) is partially positive and one end (Cl) partially negative. The term *partial* charge signifies something less than the full charges of the ions that would result from complete electron transfer. This unequal sharing of electrons has a marked effect on the properties of a compound.

The gas hydrogen chloride dissolves readily in water. The aqueous solution formed is called hydrochloric acid (sometimes muriatic acid). This acid is used for, among other things, cleaning toilet bowls and removing excess mortar from new brick buildings. Hydrochloric acid is also the well-known "stomach acid." Acids are defined and further discussed in Chapter 10.

Molecular models of H_2O (top), NH_3, and CH_4 (bottom).

Ammonia (NH_3) is a gas at room temperature. Vast quantities of it are compressed into tanks and used as fertilizer. Ammonia is quite soluble in water. It forms a basic solution (Chapter 10). Such aqueous preparations are familiar household cleaning solutions.

Methane (CH_4) is the simplest of the hydrocarbons, a group of organic compounds discussed in detail in Chapter 13. It is the principal component of natural gas.

4.10 Polyatomic Molecules: Water, Ammonia, and Methane

To obtain an octet of electrons, an oxygen atom must share electrons with two hydrogen atoms, a nitrogen atom must share electrons with three hydrogen atoms, and a carbon atom must share electrons with four hydrogen atoms. In general, many nonmetals often form a number of covalent bonds equal to 8 minus the group number. Oxygen, which is in Group 6A, forms $8 - 6 = 2$ covalent bonds in most compounds. Nitrogen, in Group 5A, forms $8 - 5 = 3$ covalent bonds in most of its compounds. Carbon, in Group 4A, forms $8 - 4 = 4$ covalent bonds in most carbon compounds, including the great host of organic compounds. These simple rules will enable you to write formulas for many molecules.

Water is one of the most familiar chemical substances. Electrolysis experiments (Section 12.4) and other ample evidence indicate that the molecular formula for water is H_2O. In order to be surrounded by an octet, an oxygen atom must therefore bond with two hydrogen atoms.

$$\cdot\underset{\cdot}{\ddot{O}}: \; + \; 2H\cdot \; \longrightarrow \; \begin{array}{c} H:\underset{\cdot\cdot}{\ddot{O}}: \\ H \end{array}$$

This arrangement completes the octet in the outer energy level of oxygen, giving it the neon electron configuration. It also completes the outer energy level of the hydrogen atoms, each of which now has the helium electron configuration.

We have chosen to represent the water molecule with the hydrogen atoms arranged at an angle rather than on a straight line with the oxygen atom. This bent arrangement is necessary to explain the *polar* nature of the water molecule (see Section B.2 in Special Topic B).

An atom of the element nitrogen has five electrons in its outer energy level. It can assume the neon configuration by sharing three pairs of electrons with *three* hydrogen atoms. The result is the compound ammonia.

$$\cdot\underset{\cdot}{\ddot{N}}\cdot \; + \; 3H\cdot \; \longrightarrow \; \begin{array}{c} H:\underset{\cdot\cdot}{\ddot{N}}:H \\ H \end{array}$$

In ammonia, the bond arrangement is that of a tripod, with a hydrogen atom at the end of each "leg" and the nitrogen atom with its lone pair of electrons sitting at the top. This pyramidal shape is necessary to explain the polar nature of the ammonia molecule (Section B.2).

An atom of carbon has four electrons in its outer energy level. It can assume the neon configuration by sharing pairs of electrons with four hydrogen atoms, forming the compound methane.

$$\cdot\underset{\cdot}{\dot{C}}\cdot \; + \; 4H\cdot \; \longrightarrow \; \begin{array}{c} H \\ H:\underset{\cdot\cdot}{\ddot{C}}:H \\ H \end{array}$$

The methane molecule has a *tetrahedral* shape (Section B.2).

4.11 Names for Covalent Compounds

Many covalent compounds have common and widely used names. Examples are water (H_2O), methane (CH_4), and ammonia (NH_3). For other compounds, the prefixes *mono-*, *di-*, *tri-*, and so on, are used to indicate the number of atoms of each element in the molecule. A list of these prefixes for up to 10 atoms is given in Table 4.3.

Simply use the prefixes to indicate the number of each kind of atom. For example, the compound N_2O_4 is called *dinitrogen tetroxide.* (The *a* often is dropped from tetra- and other prefixes when it precedes another vowel.) We often leave off the mono- prefix (NO_2 is nitrogen dioxide), but do include it to distinguish between two compounds of the same pair of elements (CO is carbon monoxide; CO_2 is carbon dioxide).

Table 4.3
Prefixes That Indicate the Number of Atoms of an Element in a Compound

Prefix	Number of Atoms
Mono-	1
Di-	2
Tri-	3
Tetra-	4
Penta-	5
Hexa-	6
Hepta-	7
Octa-	8
Nona-	9
Deca-	10

Example 4.11

What is the name of SCl_2? Of SF_6?

Solution

With one sulfur atom and two chlorine atoms, SCl_2 is sulfur dichloride. With one sulfur atom and six fluorine atoms, SF_6 is sulfur hexafluoride.

Practice Exercise

What is the name of BrF_3? Of BrF_5?

Example 4.12

Give the formula for carbon tetrachloride.

Solution

The name indicates one carbon atom and four chlorine atoms. The formula is CCl_4.

Practice Exercise

Give the formula for dinitrogen pentoxide.

Example 4.13

Give the formula for tetraphosphorus hexoxide.

Solution

The name indicates four phosphorus atoms and six oxygen atoms. The formula is P_4O_6.

Practice Exercise

Give the formula for tetraphosphorus triselenide. (The symbol for selenium is Se.)

4.12 Polyatomic Ions

Many compounds contain both ionic and covalent bonds. Sodium hydroxide, commonly known as lye, consists of sodium ions (Na^+) and hydroxide ions (OH^-). The hydroxide ion contains an oxygen atom covalently bonded to a hydrogen atom, plus an "extra" electron. Whereas the sodium atom becomes a cation by giving up an electron, the hydroxide group becomes an anion by gaining an electron.

$$e^- + \cdot\ddot{\underset{\cdot\cdot}{O}}\cdot + \cdot H \longrightarrow [:\ddot{\underset{\cdot\cdot}{O}}:H]^-$$

The formula for sodium hydroxide is NaOH; for each sodium ion there is one hydroxide ion.

Table 4.4 Some Common Polyatomic Ions

Charge	Name	Formula
1+	Ammonium ion	NH_4^+
	Hydronium ion	H_3O^+
1−	Hydrogen carbonate (bicarbonate) ion	HCO_3^-
	Hydrogen sulfate (bisulfate) ion	HSO_4^-
	Acetate ion	$CH_3CO_2^-$ (or $C_2H_3O_2^-$)
	Nitrite ion	NO_2^-
	Nitrate ion	NO_3^-
	Cyanide ion	CN^-
	Hydroxide ion	OH^-
	Dihydrogen phosphate ion	$H_2PO_4^-$
	Permanganate ion	MnO_4^-
2−	Carbonate ion	CO_3^{2-}
	Sulfate ion	SO_4^{2-}
	Monohydrogen phosphate ion	HPO_4^{2-}
	Oxalate ion	$C_2O_4^{2-}$
	Dichromate ion	$Cr_2O_7^{2-}$
3−	Phosphate ion	PO_4^{3-}

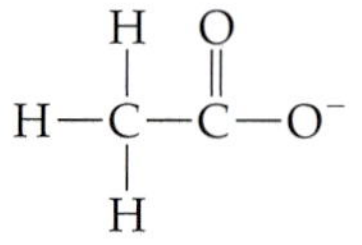

Acetate ion

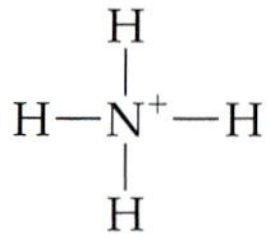

Ammonium ion

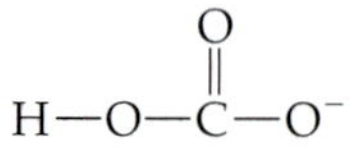

Hydrogen carbonate ion
(Bicarbonate ion)

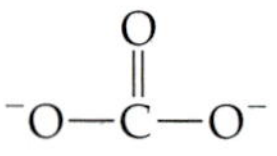

Carbonate ion

Nitrite ion

▲ **FIGURE 4.6**
Polyatomic ions have both covalent bonds (dashes) that hold the atoms together as a group and ionic charges (+ or −) that allow them to interact with other ions to form compounds.

There are many groups of atoms that (like the hydroxide ion) remain together through most chemical reactions. **Polyatomic ions** are charged particles containing two or more covalently bonded atoms (Figure 4.6). A list of common polyatomic ions is given in Table 4.4. You can use these ions, in combination with the monatomic ions in Table 4.2, to determine formulas for compounds that contain polyatomic ions.

Example 4.14

What is the formula for sodium sulfate?

Solution

First, write the formula for the ions.

$$Na^+ \qquad SO_4^{2-}$$

Crossing over,

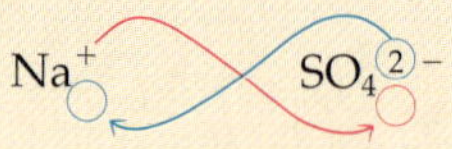

we get

$$Na_2^+ (SO_4)_1^{2-}$$

Then, dropping the charges, we have

$$Na_2(SO_4)_1 \quad \text{or simply} \quad Na_2SO_4$$

Practice Exercise

What is the formula for potassium phosphate?

Example 4.15

What is the formula for ammonium sulfide?

Solution

The ions are

$$NH_4^+ \qquad S^{2-}$$

Crossing over, we get

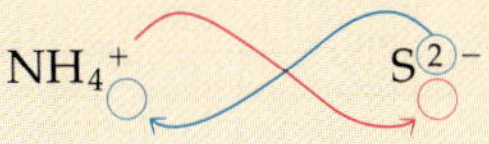

giving us

$$(NH_4)_2^+S_1^{2-}$$

Dropping the charges gives

$$(NH_4)_2S$$

The parentheses with a subscript 2 indicate that the entire ammonium unit is taken twice; there are two nitrogen atoms and $4 \times 2 = 8$ hydrogen atoms.

Practice Exercise

What is the formula for calcium acetate?

Example 4.16

What is the name of the compound NaCN?

Solution

The ions are

$$Na^+ \qquad CN^-$$

The name is sodium cyanide.

Practice Exercise

What is the name of the compound $CaCO_3$?

Example 4.17

What is the name of KH_2PO_4?

Solution

The ions are

$$K^+ \qquad H_2PO_4^-$$

The name is potassium dihydrogen phosphate.

Practice Exercise

What is the name of $K_2Cr_2O_7$?

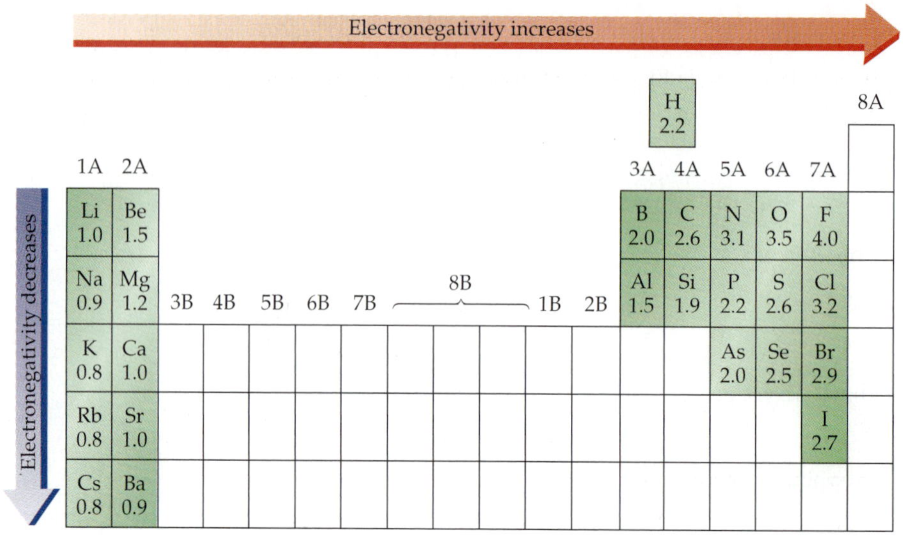

FIGURE 4.7 ▶ Electronegativities of some common elements.

4.13 Electronegativity

Before we look at the structures of any more molecules, let's take a closer look at the concept of *electronegativity*. The **electronegativity** of an element is a measure of the tendency of its atoms in molecules to attract electrons to themselves. *The greater the electronegativity of an atom in a molecule, the more strongly the atom attracts the electrons in a covalent bond.* The elements at the upper right in the periodic table—all nonmetals—are precisely those whose atoms tend to gain electrons and form negative ions; they have the greatest electronegativities. The elements on the left—metals—tend to give up electrons to become positive ions; they have the smallest electronegativities. We can use the periodic table to predict trends in electronegativities. *Within a period, elements become more electronegative from left to right. Within a group, electronegativity decreases from top to bottom.* Thus, chlorine is less electronegative than fluorine and sulfur is less electronegative than oxygen. Hydrogen does not fit well into this scheme. It is often placed in Group 1A because, like the alkali metals, it has one electron in its outer electron level. However, hydrogen, a nonmetal, is not much like the alkali metals. Alternatively, it could be placed in Group 7A because, like fluorine, it is just one electron short of having the configuration of a noble gas. However, hydrogen isn't nearly as electronegative as fluorine. If you look at a number of versions of the periodic table, you'll see that chemists still haven't decided what to do with hydrogen. We have placed it at the top, away from everything, in Figure 4.7. Its electronegativity fits that position. It will take electrons from an atom that gives them up readily (lithium, for example), and it will shift electrons to a more electronegative element (chlorine, for example).

4.14 Rules for Writing Lewis Structures

Recall that electrons are transferred (Section 4.5) or shared (Section 4.7) in ways that leave most atoms with an octet of electrons in the outermost energy level. In this section we describe the procedure we can follow in writing Lewis structures for molecules. First we must put the atoms of the molecules in their proper places.

The *skeletal structure* of a molecule tells us the order in which the atoms are attached to one another. In the absence of experimental evidence, the following rules help us to devise likely skeletal structures.

1. Hydrogen atoms form only one bond; they are shown at the end of a sequence of atoms. Hydrogen often is bonded to carbon, nitrogen, or oxygen.

$$
\begin{array}{cccc}
& \text{H} & & \text{H} \\
& | & & | \\
\text{H}-\text{C}-\text{H} \quad & \text{H}-\text{N}-\text{H} \quad & \text{H}-\text{O} \quad & \text{H}-\text{C}-\text{O}-\text{H} \\
& | & | & | \\
& \text{H} & \text{H} \quad \text{H} & \text{H}
\end{array}
$$

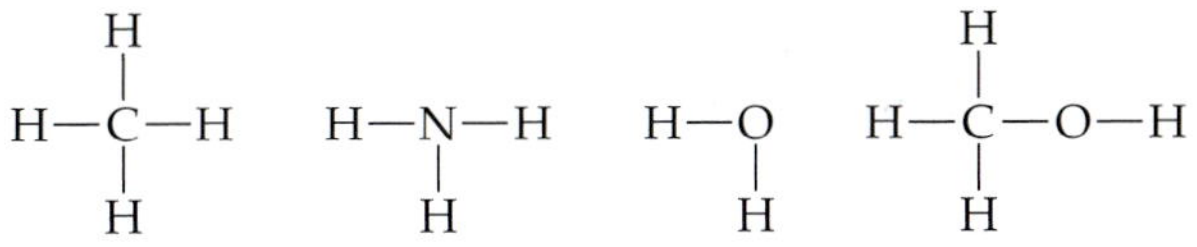

2. Polyatomic molecules and ions often consist of a central atom surrounded by more electronegative atoms. (Hydrogen is an exception; it is always on the outside, even when bonded to a more electronegative element.)

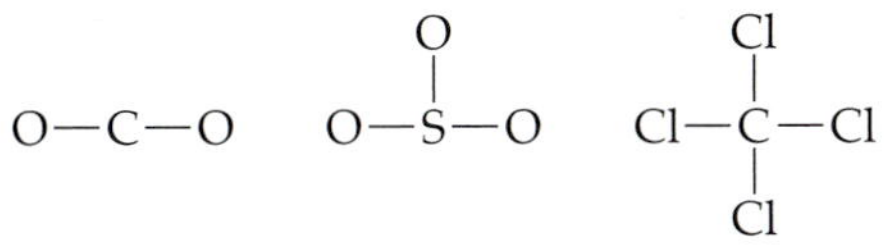

$$
\begin{array}{cccc}
& \text{O} & \text{Cl} & \\
& | & | & \\
\text{O}-\text{C}-\text{O} \quad & \text{O}-\text{S}-\text{O} \quad & \text{Cl}-\text{C}-\text{Cl} \quad & (\textit{incomplete structures}) \\
& & | & \\
& & \text{Cl} &
\end{array}
$$

After a skeletal formula for a polyatomic molecule or ion has been chosen, we can use the following steps to write an electron dot formula.

1. Calculate the total number of valence electrons. The total for a molecule is the sum of the valence electrons for all the atoms. For a polyatomic anion, add the number of negative charges. For a polyatomic cation, subtract the number of positive charges.

 Examples:

 N_2O_4 has $(2 \times 5) + (4 \times 6) = 34$ valence electrons.

 NO_3^- has $5 + (3 \times 6) + 1 = 24$ valence electrons.

 NH_4^+ has $5 + (4 \times 1) - 1 = 8$ valence electrons.

2. Write the skeletal structure, and connect bonded pairs of atoms by a dash (one electron pair).
3. Place electrons about outer atoms so that each (except hydrogen) has an octet.
4. Subtract the number of electrons assigned so far from the total calculated in Step 1. Any electrons that remain are assigned in pairs to the central atom(s).
5. If a central atom has fewer than eight electrons after Step 4, a multiple bond is likely. Move one or more lone pairs from an outer atom to the space between the atoms to form a double or triple bond. A deficiency of two electrons suggests a double bond, and a shortage of four electrons is indicative of a triple bond or two double bonds to the central atom.

Example 4.18

Give the Lewis structure for methanol, CH_4O.

Solution

1. The total number of valence electrons is $4 + (4 \times 1) + 6 = 14$.

2. The skeletal structure in which all the hydrogen atoms are on the outside and the least electronegative atom, carbon, is most central is

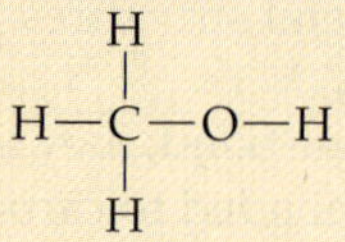

3. Now, counting each bond as two electrons gives 10 electrons. The four remaining electrons are placed (as two lone pairs) on the oxygen atom.

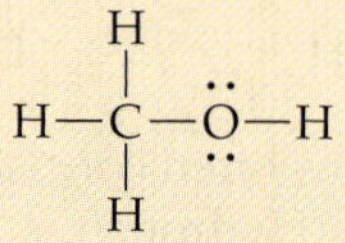

(The remaining steps are not necessary; both carbon and oxygen have an octet of electrons.)

Practice Exercise

Give the Lewis structure for ethyl chloride, C_2H_5Cl.

Example 4.19

Give the Lewis structure for nitrogen trifluoride, NF_3.

Solution

1. There are $5 + (3 \times 7) = 26$ valence electrons.
2. The skeletal structure is

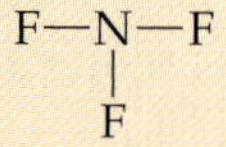

3. Place three lone pairs on each fluorine atom.

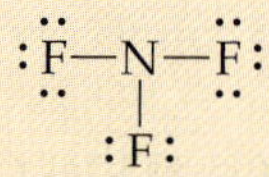

4. We have assigned 24 electrons. Place the remaining two as a lone pair on the nitrogen atom.

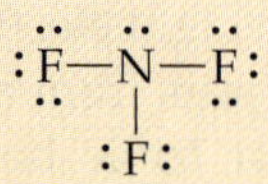

(Each atom has an octet; Step 5 is not needed.)

Practice Exercise

Give the Lewis structure for oxygen difluoride, OF_2.

Example 4.20

Give the Lewis structure for the BF_4^- ion.

Solution

1. There are $3 + (4 \times 7) + 1 = 32$ electrons.
2. The skeletal structure is

$$
\begin{array}{c}
\text{F} \\
| \\
\text{F}-\text{B}-\text{F} \\
| \\
\text{F}
\end{array}
$$

3. Place three lone pairs on each fluorine atom.

$$
\begin{array}{c}
:\!\ddot{\text{F}}\!: \\
| \\
:\!\ddot{\text{F}}-\text{B}-\ddot{\text{F}}\!: \\
| \\
:\!\ddot{\text{F}}\!:
\end{array}
$$

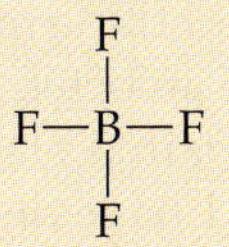

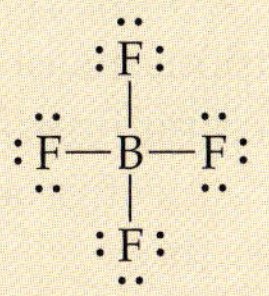

4. We have assigned 32 electrons. None remains to be assigned.

Practice Exercise

Give the Lewis structure for the PH_4^+ ion.

Example 4.21

Give the Lewis structure for carbon dioxide, CO_2.

Solution

1. There are $4 + (2 \times 6) = 16$ valence electrons.
2. The skeletal structure is

$$\text{O}-\text{C}-\text{O}$$

3. Place three lone pairs on each oxygen atom.

$$:\!\ddot{\text{O}}-\text{C}-\ddot{\text{O}}\!:$$

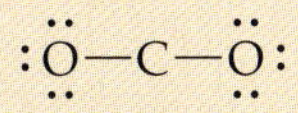

4. We have assigned 16 electrons. None remains to be placed.
5. The carbon atom has only four electrons. It needs to form two double bonds in order to have an octet. (It is not reasonable to expect that carbon would form a triple bond to one of the oxygen atoms. Why?) Move a lone pair from each oxygen atom to the space between the atoms to form a double bond on each side of the carbon atom.

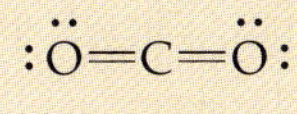

$$:\!\ddot{\text{O}}\!=\!\text{C}\!=\!\ddot{\text{O}}\!:$$

Practice Exercise

Give the Lewis structure for nitryl fluoride, NO_2F.

4.15 Exceptions to the Octet Rule

Many molecules made of atoms of the main group elements have electron structures that follow the octet rule. There are many exceptions, however. The exceptions fall into three main groups. Each type is readily identified by some structural characteristic.

Molecules with odd numbers of valence electrons obviously cannot satisfy the octet rule. Examples of such molecules are nitrogen monoxide (NO, also called nitric oxide), with 5 + 6 = 11 valence electrons; nitrogen dioxide (NO_2), with 17 valence electrons; and chlorine dioxide (ClO_2), which has 19 outer electrons. Obviously one of the atoms in each of these molecules will have an odd number of electrons and therefore cannot have an octet.

Any atom or molecule with an odd number of electrons must have one unpaired electron. Filled energy levels and sublevels have all their electrons paired, with two electrons in each orbital (Section 2.5). We need only consider valence electrons to determine whether or not an atom or molecule is a free radical. Electron dot formulas for NO, NO_2, and ClO_2 show that one atom of each has an unpaired electron; that atom obviously does not have an octet of electrons in its outer energy level.

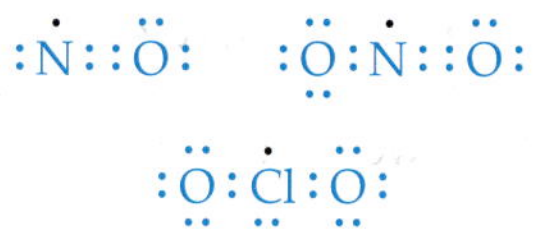

Atoms and molecules with unpaired electrons are called **free radicals.** Most free radicals are highly reactive and have only transitory existence as intermediates in chemical reactions. An example is the chlorine atom that is formed from the breakdown of chlorofluorocarbons in the stratosphere and that leads to the depletion of the ozone shield (Section 13.9). Some free radicals are quite stable, however. The nitrogen oxides are major components of smog. Chlorine dioxide is made in vast quantities and is used for bleaching flour, paper, and other materials.

Boron atoms have three valence electrons; fluorine atoms have seven. When boron reacts with fluorine, it shares those electrons with three fluorine atoms to form boron trifluoride.

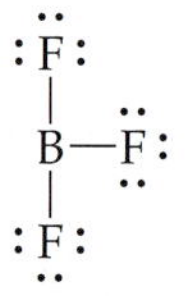

This structure uses all 24 of the valence electrons; there are none left to put on the central boron atom. Experimental evidence indicates that this structure is consistent with the reactivity of BF_3 toward molecules with unshared pairs. For instance, BF_3 reacts readily with ammonia to form BF_3NH_3, a compound in which all atoms (except hydrogen) have octets of electrons.

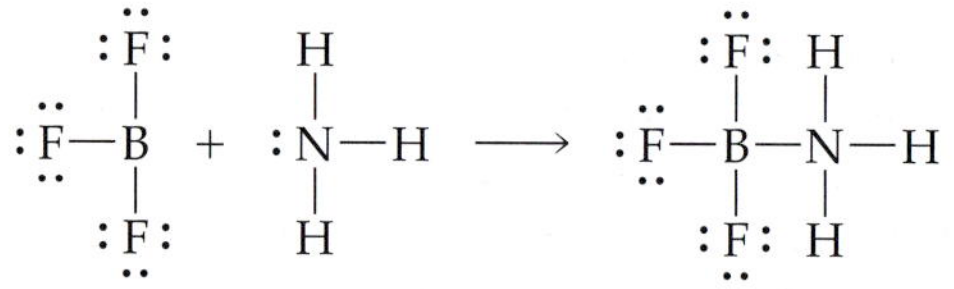

The second-period elements carbon, nitrogen, oxygen, and fluorine, nearly always obey the octet rule. (The odd-electron compounds are obvious exceptions.) The valence electron level of the second-period elements holds a maximum of eight electrons ($2s^22p^6$). The third main energy level can hold up to 18 electrons ($3s^23p^63d^{10}$). Third-period elements therefore can violate the octet rule by having more than eight electrons in the valence level. These so-called **expanded octets** are evident in the following compounds.

It is a good idea to use the octet rule except in cases where it obviously doesn't apply: when there is an odd number of electrons, when there are too few electrons to make an octet, and (third period and beyond) where there are obviously more than eight electrons that must be in the valence level.

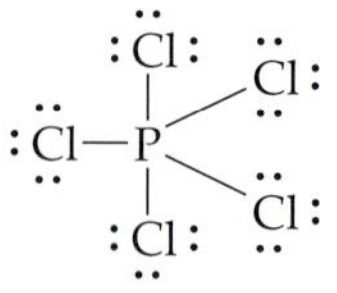

Phosphorus pentachloride

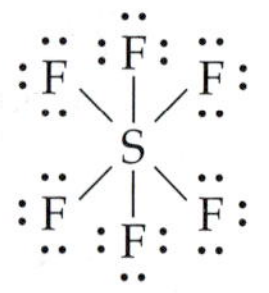

Sulfur hexafluoride

Elements in the third period and beyond, then, are not limited to an octet. Yet many of their compounds still follow the octet rule.

Summary

Chemical bonds are the forces that hold atoms together in molecules.

The electrons in the outermost energy level of any atom are that atom's **valence electrons,** and the outermost energy level is called the valence shell. An atom that *gains* one or more valence electrons becomes an *anion* and carries a net negative charge. One that *loses* one or more valence electrons is a *cation* and carries a net positive charge.

Valence electrons are the only ones shown in **Lewis structures:**

$$\overset{\cdot}{Ca}\cdot \qquad\qquad :\overset{\cdot\cdot}{\underset{\cdot}{Br}}:$$

| Lewis structure | Lewis structure |
| for the calcium atom | for the bromine atom |

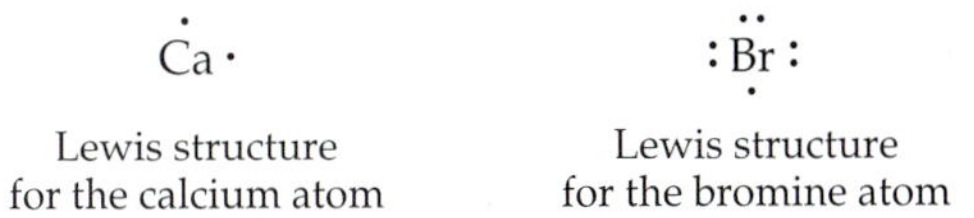

For main-group elements, the number of valence electrons shown in a Lewis structure is equal to the group number shown on the periodic table.

Reactive substances tend to undergo chemical reactions that make them less reactive. This is so because reactivity is a measure of the energy an atom has, and everything in nature tends to seek its state of lowest energy. The elements belonging to the noble-gas family (extreme right of periodic table) are unreactive because each has a filled outermost energy level, a configuration representing the lowest possible energy state. The **octet rule** states that an element participating in a chemical reaction seeks to end up with a filled valence energy level.

A sodium atom has one valence electron and tends to give it up easily. A chlorine atom contains seven valence electrons and readily accepts the lone sodium valence electron. The result is a sodium cation, Na^+, that has a filled outermost energy level containing eight electrons and a chloride anion, Cl^-, that also has a filled outermost level containing eight electrons. The positive charge and negative charge on these ions attract each other, and, as a result, Na^+ and Cl^- ions become associated with each other in a crystal. The forces holding these positive and negative species together in the crystal are called **ionic bonds,** and any compound held together by ionic bonds is an **ionic compound:**

$$Li\cdot \ + \ \cdot\overset{\cdot\cdot}{\underset{\cdot\cdot}{F}}: \ \longrightarrow \ Li^+ \ :\overset{\cdot\cdot}{\underset{\cdot\cdot}{F}}:^-$$

| Lithium atom | Fluorine atom | Lithium fluoride, an ionic compound of lithium cations and fluoride anions arranged in a crystal |

$$\overset{\cdot}{Ca}\cdot \ + \ :\overset{\cdot\cdot}{O}: \ \longrightarrow \ Ca^{2+} \ :\overset{\cdot\cdot}{\underset{\cdot\cdot}{O}}:^{2-}$$

| Calcium atom | Oxygen atom | Calcium oxide, an ionic compound of calcium cations and oxide anions arranged in a crystal |

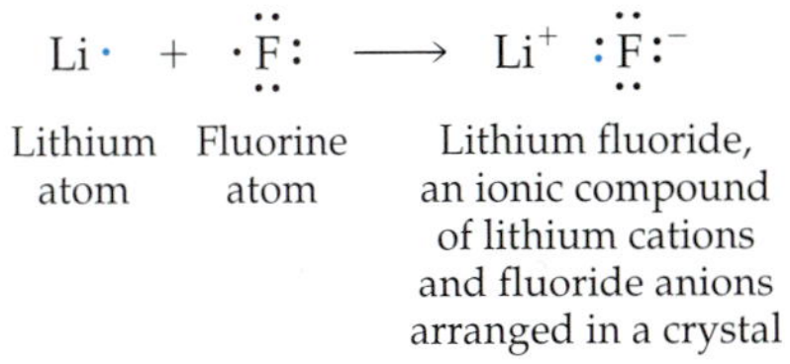

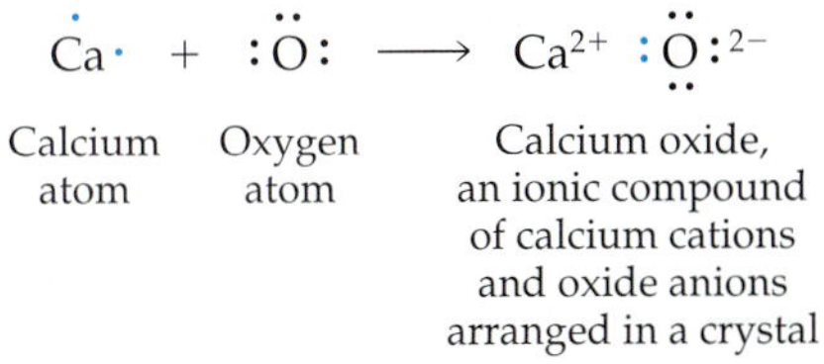

The chemical bond formed when two or more atoms share electrons is called a **covalent bond.** The resulting compound is called a *covalent compound:*

$$H\cdot \ + \ H\cdot \qquad H:H$$

| Hydrogen atom | Hydrogen atom | H_2, a covalent compound |

$$:\overset{\cdot\cdot}{F}: \ + \ \cdot\overset{\cdot\cdot}{F}: \qquad :\overset{\cdot\cdot}{F}:\overset{\cdot\cdot}{F}:$$

| Fluorine atom | Fluorine atom | F_2, a covalent compound |

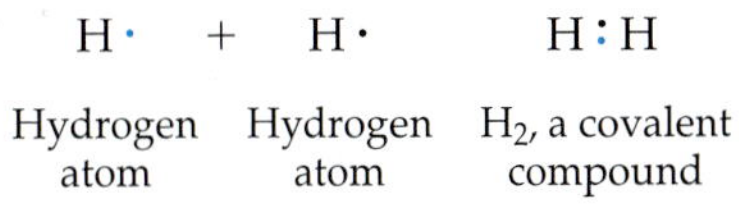

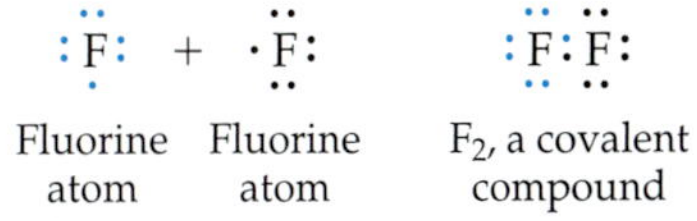

The electrons participating in the covalent bond are called a **bonding pair,** and all the other electrons are referred to as **lone pairs.**

As with ionic bonding, the octet rule drives the formation of covalent bonds. A covalent bond can be **single** (one bonding pair of electrons), **double** (two bonding pairs), or **triple** (three bonding pairs).

An atom's tendency to attract electrons to itself is called its **electronegativity.** Ignoring the noble gases because they tend not to form compounds, electronegativity increases from left to right across a row in the periodic table and decreases from top to bottom down a column.

When the atoms sharing a covalent bond are identical, as in H_2 or N_2, the two electronegativity values are the same. As a result, electrons are shared equally by the two atoms, and the covalent bond is **nonpolar.** When a covalent bond is formed from two atoms having different electronegativity values, the electrons are pulled closer to the more electronegative atom. Now the covalent molecule has a positive end and a negative end, and the covalent bond is **polar:**

$$\overset{\delta+}{H} \ :\overset{\cdot\cdot}{\underset{\cdot\cdot}{Cl}}:\,^{\delta-}$$

The number of covalent bonds formed by a nonmetal atom is generally eight minus the family number of the atom. Example: oxygen is in Group 6A and so forms $8 - 6 = 2$ covalent bonds, as in H_2O; silicon is in Group 4A and so forms $8 - 4 = 4$ covalent bonds, as in SiH_4.

Covalently bonded species that carry a net charge are called **polyatomic ions.** Such ions can form ionic compounds with an ion carrying an opposite charge. Example:

$$NH_4^+ + Cl^- \longrightarrow NH_4Cl$$

We consider three **exceptions to the octet rule.** In all three cases, stable covalent compounds are formed even though not all atoms in the molecule have eight electrons in the valence energy level:

1. Molecules in which the total number of valence electrons is an odd number:

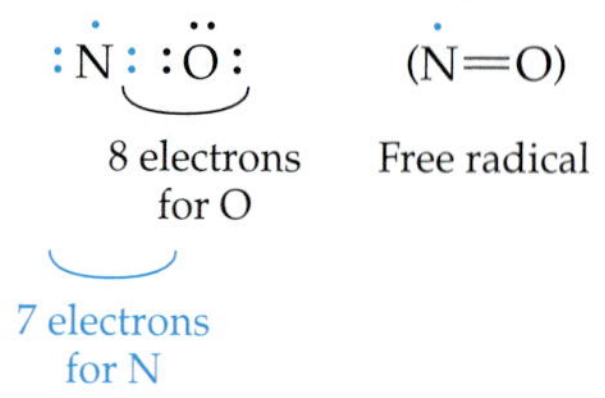

(Any molecule containing such an unpaired electron is called a **free radical.**)

2. Molecules in which the total number of valence electrons is too low to allow all atoms to have a filled valence shell:

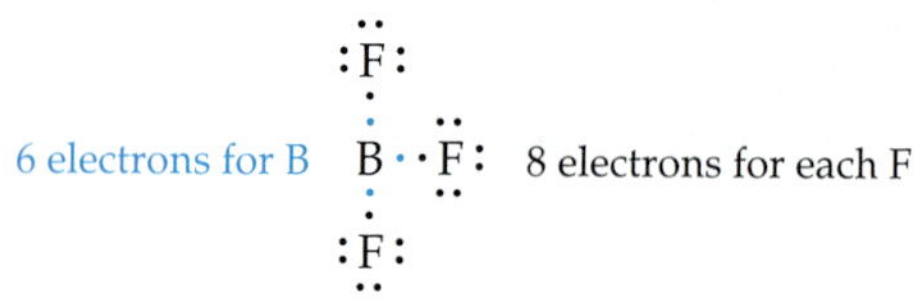

3. Molecules involving elements from periods 3 through 7, all of which may have more than eight electrons in the valence energy level.

Key Terms

bonding pair (4.7)
chemical bond (introduction)
covalent bond (4.7)
crystal (4.4)
double bond (4.8)
electronegativity (4.13)
expanded octet (4.15)
free radical (4.15)
ion (4.1)
ionic bond (4.4)
ionic compound (4.6)
Lewis structure (4.2)
lone pair (4.7)
molecule (4.7)
nonpolar covalent bond (4.9)
octet rule (4.7)
polar covalent bond (4.9)
polyatomic ion (4.12)
single bond (4.7)
triple bond (4.8)
valence electron (4.2)

Review Questions

1. Which group of elements in the periodic table is characterized by especially stable electron configurations?
2. What is the structural difference between a sodium atom and a sodium ion? How does sodium metal differ from sodium ions (in sodium chloride, for example) in chemical reactivity?
3. What is the structural difference between a sodium ion and a neon atom? What is the similarity between them?
4. What are the structural differences among chlorine atoms, chlorine molecules, and chloride ions? How do they differ in chemical reactivity?
5. Why is hydrogen so difficult to place in the periodic table?
6. Why does neon tend not to form chemical bonds?
7. How many covalent bonds do each of the following usually form? You may refer to the periodic table.
   a. H **b.** C **c.** O
   d. F **e.** N **f.** Br
8. List three main types of compounds that are exceptions to the octet rule. Give an example of each.
9. Draw Bohr diagrams for each of the following.
   a. K^+ **b.** S^{2-} **c.** F^- **d.** Al^{3+}.
10. Draw Bohr diagrams for each of the following.
   a. Mg^{2+} **b.** Cl^- **c.** Li^+ **d.** N^{3-}
11. What is the charge on simple ions formed from **(a)** atoms of Group 2A elements and **(b)** atoms of Group 7A elements?
12. What is the charge on simple ions formed from **(a)** atoms of Group 1A elements and **(b)** atoms of Group 6A elements?
13. Classify the following bonds as ionic or covalent. For those bonds that are covalent, indicate whether they are polar or nonpolar.
   a. KF **b.** IBr **c.** MgS **d.** F_2
14. Classify the following bonds as ionic or covalent. For those bonds that are covalent, indicate whether they are polar or nonpolar.
   a. NO **b.** CaO **c.** NaBr **d.** Br_2

Problems

Lewis Structures: Elements

15. Write Lewis structures for each of the following elements. You may use a periodic table.
   a. sodium **b.** oxygen
   c. fluorine **d.** aluminum
16. Write Lewis structures for each of the following elements. You may use a periodic table.
   a. carbon
   b. potassium
   c. magnesium
   d. chlorine

Lewis Structures: Ions and Ionic Bonding

17. Using Lewis structures, show the formation of an ion from
    a. barium
    b. bromine
18. Using Lewis structures, show the formation of an ion from
    a. aluminum
    b. sulfur
19. Use Lewis structures to show the transfer of electrons (a) from calcium to bromine atoms and (b) from magnesium to sulfur atoms. In each case, form ions with noble gas electron configurations.
20. Use Lewis structures to show the transfer of electrons (a) from aluminum to sulfur atoms and (b) from magnesium to phosphorus atoms. In each case, form ions with noble gas electron configurations.
21. Draw Lewis structures for
    a. Ca and Ca^{2+}
    b. S and S^{2-}
    c. Rb and Rb^{+}
    d. P and P^{3-}
22. Draw Lewis structures for
    a. Al and Al^{3+}
    b. Br and Br^{-}
    c. Mg and Mg^{2+}
    d. O^{2-} and Ne

Lewis Structures: Ionic Compounds

23. Give Lewis structures for
    a. sodium fluoride
    b. potassium chloride
    c. sodium oxide
    d. calcium chloride
    e. magnesium bromide
24. Give Lewis structures for
    a. potassium fluoride
    b. magnesium iodide
    c. potassium sulfide
    d. sodium nitride
    e. aluminum oxide

Naming Ions and Ionic Compounds

25. Name
    a. Na^{+}
    b. Mg^{2+}
    c. Al^{3+}
    d. Cl^{-}
    e. O^{2-}
    f. N^{3-}
26. Name
    a. K^{+}
    b. Ca^{2+}
    c. Zn^{2+}
    d. Br^{-}
    e. Li^{+}
    f. S^{2-}
27. Name
    a. Fe^{3+}
    b. Cu^{2+}
    c. Ag^{+}
28. Name
    a. Fe^{2+}
    b. Cu^{+}
    c. I^{-}
29. Give symbols for
    a. bromide ion
    b. calcium ion
    c. potassium ion
    d. iron(II) ion
30. Give symbols for
    a. sodium ion
    b. aluminum ion
    c. oxide ion
    d. copper(II) ion
31. Name
    a. NaBr
    b. $CaCl_2$
    c. $FeCl_2$
    d. LiI
    e. K_2S
    f. CuBr
32. Name
    a. KCl
    b. $MgBr_2$
    c. CuI_2
    d. CaS
    e. $FeCl_3$
    f. Al_2O_3
33. Name
    a. CO_3^{2-}
    b. HPO_4^{2-}
    c. MnO_4^{-}
    d. OH^{-}
34. Name
    a. NO_3^{-}
    b. SO_4^{2-}
    c. $H_2PO_4^{-}$
    d. HCO_3^{-}
35. Give formulas for
    a. ammonium ion
    b. hydrogen sulfate ion
    c. cyanide ion
    d. nitrite ion
36. Give formulas for
    a. phosphate ion
    b. hydrogen carbonate ion
    c. dichromate ion
    d. oxalate ion
37. Give formulas for
    a. magnesium sulfate
    b. sodium hydrogen carbonate
    c. potassium nitrate
    d. calcium monohydrogen phosphate
38. Give formulas for
    a. calcium carbonate
    b. potassium dihydrogen phosphate
    c. magnesium cyanide
    d. lithium hydrogen sulfate
39. Give formulas for
    a. iron(II) phosphate
    b. potassium dichromate
    c. copper(I) iodide
    d. ammonium nitrite
40. Give formulas for
    a. iron(III) oxalate
    b. sodium permanganate
    c. copper(II) bromide
    d. zinc monohydrogen phosphate

Lewis Structures: Covalent Bonds and Molecules

41. Use Lewis structures to show the sharing of electrons between two iodine atoms to form an iodine molecule. Label all electron pairs as bonding or lone pairs.
42. Use Lewis structures to show the sharing of electrons between a hydrogen atom and a fluorine atom. Label the ends of the molecule with symbols that indicate polarity.
43. Use Lewis structures to show the sharing of electrons between the indicated atoms to form a molecule in which each atom (except hydrogen) has an octet of valence electrons.
    a. P and H
    b. C and F
44. Use Lewis structures to show the sharing of electrons between the indicated atoms to form a molecule in which each atom (except hydrogen) has an octet of valence electrons.
    a. Si and H
    b. N and Cl
45. Give Lewis structures that follow the octet rule for
    a. CH_4O
    b. CH_2O
    c. NOH_3
    d. N_2H_4
    e. COF_2
    f. PCl_3
46. Give Lewis structures that follow the octet rule for
    a. NF_3
    b. C_2H_2
    c. C_2H_4
    d. CH_5N
    e. H_2SiO_3
    f. HCN

Naming Covalent Compounds

47. Give formulas for
    a. dinitrogen monoxide
    b. tetraphosphorus trisulfide
    c. phosphorus pentachloride
    d. sulfur hexafluoride
48. Give formulas for
    a. oxygen difluoride
    b. dinitrogen pentoxide
    c. phosphorus tribromide
    d. tetrasulfur tetranitride
49. Name
    a. CS_2 b. N_2S_4 c. PF_5 d. S_2F_{10}
50. Name
    a. CBr_4 b. Cl_2O_7 c. P_4S_{10} d. I_2O_5

Electronegativity

51. If atoms of the two elements in each set below are joined by a covalent bond, which atom will more strongly attract the electrons in the bond?
    a. N and S b. B and Cl c. As and F
52. Using only the periodic table (inside front cover), indicate which element in each set is more electronegative.
    a. Br or F b. Br or Se c. Cl or As
53. Without referring to figures or tables in the text, arrange each of the following sets of atoms in order of increasing electronegativity.
    a. B, F, N b. As, Br, Ca c. C, O, Ga
54. Without referring to figures or tables in the text, arrange each of the following sets of atoms in order of increasing electronegativity.
    a. I, Rb, Sb b. Cs, Li, Na c. Cl, P, Sb

Exceptions to the Octet Rule

55. Give a Lewis structure for
    a. NO b. BeI_2 c. PCl_5
56. Give a Lewis structure for
    a. BCl_3 b. PF_5 c. $AlBr_3$

Polyatomic Ions and Their Compounds

57. Give Lewis structures that follow the octet rule for
    a. ClO^- b. ClO_2^-
    c. HPO_4^{2-} d. BrO_3^-
58. Give Lewis structures that follow the octet rule for
    a. CN^- b. IO_4^-
    c. PO_3^{3-} d. HSO_4^-
59. Name
    a. KNO_2 b. LiCN c. NH_4I
    d. $NaNO_3$ e. $KMnO_4$ f. $CaSO_4$
60. Name
    a. $NaHSO_4$ b. $Al(OH)_3$ c. Na_2CO_3
    d. $KHCO_3$ e. NH_4NO_2 f. $Ca(HSO_4)_2$
61. Name
    a. Na_2HPO_4 b. $(NH_4)_3PO_4$
    c. $Al(NO_3)_3$ d. NH_4NO_3
62. Name
    a. Li_2CO_3 b. $Na_2Cr_2O_7$
    c. $Ca(H_2PO_4)_2$ d. $(NH_4)_2C_2O_4$

Additional Problems

63. Fill in this table assuming that elements W, X, Y, and Z are all main-group elements. Use the first column (W) as an example.

Element	W	X	Y	Z
Group Number	7A	1A	______	______
Lewis Symbol	$:\overset{\cdot}{\underset{\cdot\cdot}{W}}:$	______	$\cdot Y\cdot$	______
Charge on Ion	1−	______	______	2−

64. Consider the hypothetical elements X, Y, and Z with the Lewis symbols:

$$:\overset{\cdot\cdot}{\underset{\cdot\cdot}{X}}\cdot \quad :\overset{\cdot\cdot}{\underset{\cdot}{Y}}\cdot \quad :\overset{\cdot}{\underset{\cdot}{Z}}\cdot$$

    a. To which group in the periodic table would each belong?
    b. Write the Lewis structure for the simplest compound of each element with hydrogen.
    c. Write Lewis structures for the ions formed when X and Y react with sodium.
65. Potassium is a soft silvery metal that reacts violently with water and ignites spontaneously in air. Your doctor recommends you take a potassium supplement. Would you take potassium metal? What would you take?
66. Explain what is wrong with the statement "A crystal of ordinary salt is comprised of an enormous number of NaCl molecules in a highly ordered three-dimensional network."
67. Two different molecules have the formula C_2H_6O. Write Lewis structures for the two molecules.
68. Draw acceptable Lewis structures for *two* substances with the formula S_2F_2.
69. Chlorine dioxide is used to bleach flour. Give the formula for chlorine dioxide.
70. Tetraphosphorus trisulfide is used in the tips of "strike anywhere" matches. Give the formula for tetraphosphorus trisulfide.
71. The gas phosphine (PH_3) is used as a fumigant to protect stored grain and other durable produce from pests. Phosphine is generated in situ by adding water to aluminum phosphide or magnesium phosphide. Give formulas for the two phosphides.

Special Topic B
Molecular Shapes and Properties

The shapes of molecules are of considerable importance. Molecular shape often can be predicted by a simple procedure called the VSEPR theory. Shape can determine polarity and other properties. Of utmost importance are the shapes of biologically active molecules (Chapters 19 to 23), which can carry out their vital functions only if they have the right groups of atoms in the right places. In this special topic, we examine the shapes and properties of some simple molecules.

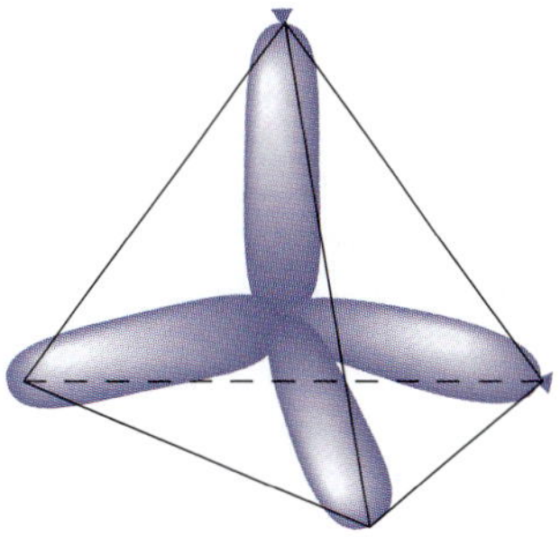

Valence shell electron pair repulsion can be pictured using balloons. The lobes of the balloons represent electron pairs. Like electron pairs, the lobes are directed toward the corners of a tetrahedron.

B.1 The VSEPR Theory

Although molecules are represented in two dimensions on paper, they have three-dimensional shapes. We use Lewis structures as a part of the process of predicting molecular shape. The shapes that we consider in this book are shown in Figure B.1.

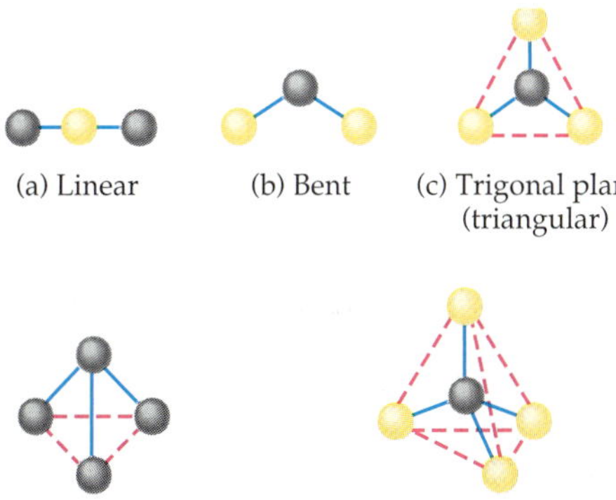

▲ **FIGURE B.1**

Shapes of molecules. In a *linear* molecule (a), all the atoms are along a line; the bond angle is 180°. The *bent* or angular molecule (b) has an angle less than 180°. Connecting the three outer atoms of the *triangular* molecule (c) with imaginary lines forms a triangle with an atom at the center, an arrangement called *trigonal planar.* Imaginary lines connecting all four atoms of a *pyramidal* molecule (d) form a three-sided pyramid. Connecting the four outer atoms of a *tetrahedral* molecule (e) with imaginary lines produces a tetrahedron with an atom at the center. (A tetrahedron is a four-sided figure in which each side is a triangle.)

The **valence shell electron pair repulsion (VSEPR) theory** often is used to predict the arrangement of atoms about a central atom. The basis of the VSEPR theory is that electron pairs will arrange themselves about a central atom in a way that minimizes repulsion between the like-charged particles. This means that they will get as far apart as possible. Table B.1 uses ball-and-stick molecular models to show the geometric shapes associated with the arrangement of two, three, or four electron groups about a central atom.

The farthest apart two substituent atoms can get is the distance between the opposite sides of the central atom at an angle of 180°. Three groups assume a triangular arrangement about the central atom, forming angles of separation of 120°. Four groups form a tetrahedral array around the central atom, with a separation of about 109.5°.

You can determine the shapes of many molecules (and polyatomic ions; see Section 4.12) by following these simple rules.

1. Draw a Lewis structure. In the structure, indicate a shared electron pair (**bonding pair, BP**) by a line. Use dots to indicate any **lone pairs (LPs)** of electrons.
2. To determine the shape, count the number of atoms *and* lone pairs attached to the *central atom.*

Table B.1 Bonding and the Shapes of Molecules

Number of Bonded Atoms	Number of Lone Pairs	Number of Sets	Molecular Shape	Examples			
2	0	2	Linear	$BeCl_2$	$HgCl_2$	CO_2	$BeCl_2$
3	0	3	Trigonal planar (triangular)	BF_3	$AlBr_3$	CH_2O	BF_3
4	0	4	Tetrahedral	CH_4	CBr_4	$SiCl_4$	CH_4
3	1	4	Pyramidal	NH_3	PCl_3		NH_3
2	2	4	Bent (angular)	H_2O	H_2S	SCl_2	H_2O
2	1	3	Bent (angular)	SO_2	O_3		SO_2

Note that a multiple bond counts as only *one set*. Examples are

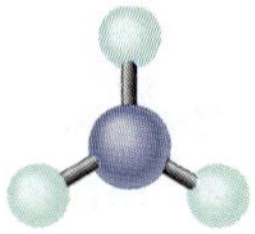

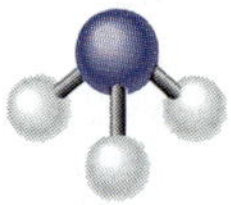

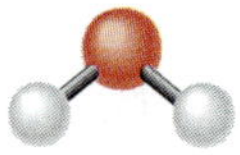

$$H-C\equiv N:$$

Two sets
(2 atoms)

$$:\overset{..}{O}=C=\overset{..}{O}:$$

Two sets
(2 atoms)

$$\overset{\overset{..}{O}:}{\underset{|}{\overset{||}{H-C-H}}}$$

Three sets
(3 atoms)

$$[:\overset{..}{\underset{..}{O}}-\overset{..}{N}=\overset{..}{\underset{..}{O}}:]^-$$

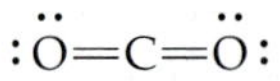

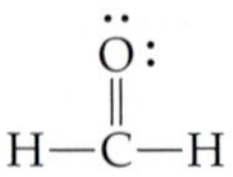

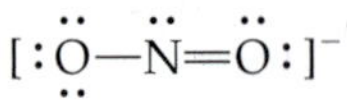

Three sets
(2 atoms, 1 LP)

$$\overset{\overset{..}{O}:}{\underset{|}{H-\underset{|}{O}:}}\overset{|}{H}$$

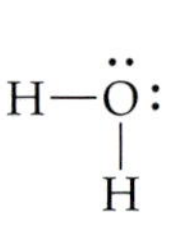

Four sets
(2 atoms, 2 LPs)

$$\overset{\overset{..}{}}{H-\overset{|}{\underset{|}{N}}-H}\overset{|}{H}$$

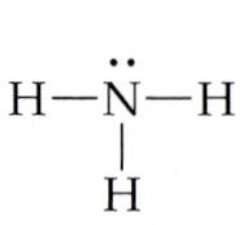

Four sets
(3 atoms, 1 LP)

$$\overset{\overset{H}{|}}{H-\overset{|}{\underset{|}{C}}-H}\overset{|}{H}$$

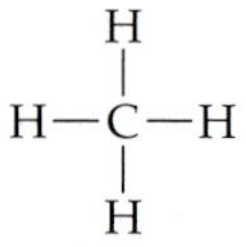

Four sets
(4 atoms)

3. Determine the number of electron sets and draw a shape *as if* all were bonding pairs.
4. Sketch that shape, placing the electron pairs as far apart as possible (see Table B.1). If there is *no* lone pair, that is the shape of the molecule. If there *are* lone pairs, ignore them, leaving the bonding pairs exactly as they were. (This may

seem strange, but it stems from the fact that *all* the sets determine the shape but only the arrangement of bonded atoms is considered in the shape of the molecule.)

Example B.1

What is the shape of the BH_3 molecule?

Solution

1. The Lewis structure is

$$\begin{array}{c} H \\ | \\ B\!-\!H \\ | \\ H \end{array}$$

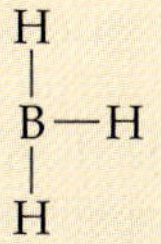

2. There are three electron sets to consider. The three sets get as far apart as possible, forming a triangular arrangement of the sets.

$$\begin{array}{c} H \diagdown \\ \diagup B\!\!-\!\!H \\ H\diagup \ \ 120° \end{array}$$

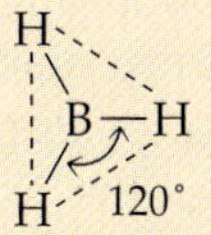

3. All the sets are bonding pairs; the molecular shape is trigonal planar, the same as the arrangement of the electrons.

Practice Exercise

What is the shape of the BeH_2 molecule?

Example B.2

What is the shape of the SCl_2 molecule?

Solution

1. The Lewis structure is

$$\begin{array}{c} :\!\ddot{S}\!-\!\ddot{C}l\!: \\ | \\ :\!\ddot{C}l\!: \end{array}$$

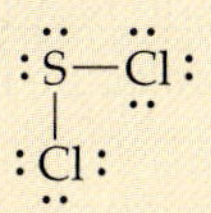

2. There are four electron sets on the sulfur atom to consider, two chlorine atoms and two lone pairs. The four sets get as far apart as possible, forming a tetrahedral arrangement of the sets about the central atom.

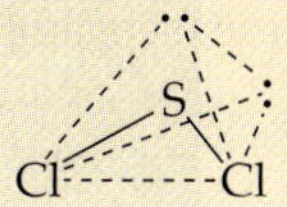

3. Ignore the lone pairs; the molecular shape is bent or angular, with a bond angle of 109.5°.

Practice Exercise

What is the shape of the PH_3 molecule?

B.2 Polar and Nonpolar Molecules

In Section 4.9 we discussed polar and nonpolar bonds. Diatomic molecules are polar if their bonds are polar and nonpolar if their bonds are nonpolar.

$$\begin{array}{cc} \overset{\delta+ \ \ \delta-}{H\!-\!Cl} & Cl\!-\!Cl \\ \text{Polar} & \text{Nonpolar} \end{array}$$

For molecules with three or more atoms, we also must consider the orientation of the bonds to determine whether or not the molecule as a whole is polar.

We should expect the bonds in water to be polar because oxygen is more electronegative than hydrogen. (Like chlorine, oxygen is to the right in the periodic table, where electronegativity increases as one moves from left to right.) Just because a molecule contains polar bonds, however, does not mean that the molecule as a whole is polar. If the atoms in the water molecule were in a linear arrangement, the two polar bonds would cancel one another.

$$\overset{\delta+ \ \ \delta- \ \ \delta+}{H\!-\!O\!-\!H} \quad \textit{(incorrect structure)}$$

Instead of one end of the molecule being positive and the other end negative, the electrons would be pulled toward the right in one bond and toward the left in the other. Overall there would be no net dipole. By a **dipole,** we mean a molecule that has a positive end and a negative end.

But water *does* act like a dipole. If you place a sample of water between two electrically charged plates, the water molecules align themselves, one end attracted toward the positive plate and the other end toward the negative plate. To act like a dipole, the molecule must be bent so that the bonds do not cancel one another out.

$$\delta+H:\overset{\cdot\cdot}{\underset{\overset{|}{H}{\delta+}}{O}}:\delta- \quad \text{or} \quad \overset{\delta+H}{}\diagdown\overset{O\delta-}{\underset{\overset{|}{H}\delta+}{}}$$

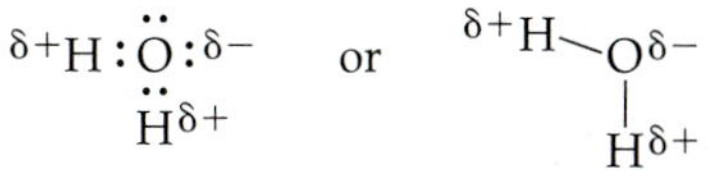

Such molecules would align themselves between charged plates as shown in Figure B.2.

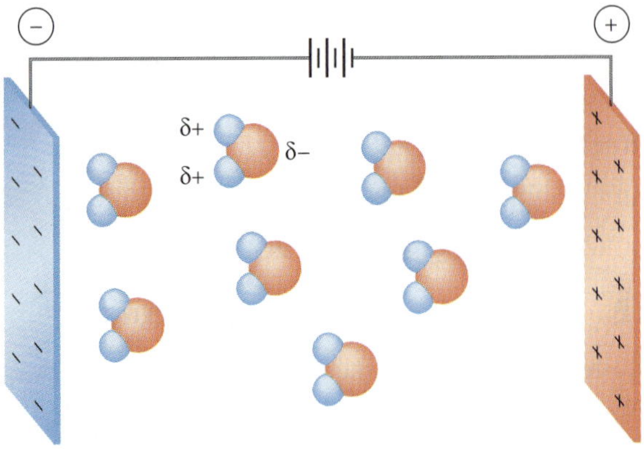

▲ FIGURE B.2

Polar molecules are aligned in an electric field, with the positive end of the molecule preferentially pointing toward the negative plate and the negative end of the molecule directed toward the positive plate.

Dipoles often are represented by an arrow with a plus at the tail end.

$$\overset{\longmapsto}{H-Cl}$$

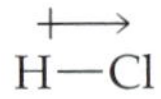

The positive end of the arrow is obvious; the head of the arrow indicates the negative end of the dipole. Using these arrows, we see that carbon dioxide is nonpolar despite its polar bonds.

$$\overset{\longleftarrow+\ +\longmapsto}{O=C=O}$$

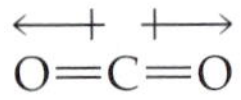

The two dipoles cancel one another out. In water, both dipoles point toward the oxygen, resulting in a net dipole toward that end of the molecule.

$$\underset{H\quad H}{\diagup^{O}\diagdown} \quad \text{or} \quad \underset{H\quad H}{\overset{\uparrow}{\underset{}{\diagup\!O\!\diagdown}}}$$

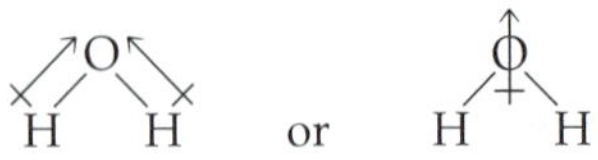

The shape of the water molecule can be accounted for by a modification of the VSEPR theory. According to VSEPR, the two bonds and two lone pairs should form a tetrahedral arrangement.

$$\underset{H\cdots H}{\overset{\cdots\overset{\cdot\cdot}{O}\cdot\cdot}{\diagup\diagdown}}$$

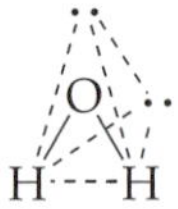

Ignoring the lone pairs, the molecular shape has the atoms in a bent arrangement, with a bond angle of 109.5° (the tetrahedral angle).

The predicted bond angle of 109.5° for water is a bit larger than the measured angle of 104.5°. The difference is explained by the fact that the lone pairs occupy a greater volume than do the bonding pairs. These larger orbitals push the smaller BPs closer together.

In ammonia there are three BPs and one lone pair about the nitrogen atom. The VSEPR theory predicts a tetrahedral arrangement with bond angles of 109.5°. The actual bond angles are 107°, close to the theoretical value. Presumably, the unshared pair of electrons occupies a greater volume than does a shared pair, pushing the latter slightly closer together. The arrangement is therefore a tripod with a hydrogen atom at the end of each leg and the nitrogen atom with its unshared pair sitting at the top (Figure B.3). Each nitrogen–hydrogen bond is somewhat polar, making the ammonia molecule polar.

In the methane molecule there are four pairs of electrons on the central carbon atom. Using the VSEPR theory, we would expect a tetrahedral arrangement and bond angles of 109.5°. The actual bond angles are 109.5°, in perfect agreement with theory (Figure B.4). All four electron pairs are shared with hydrogen atoms; thus, all four pairs occupy identical volumes. Each carbon–hydrogen bond is slightly polar, but the methane molecule as a whole is symmetrical. The slight bond polarities cancel out, leaving the methane molecule nonpolar.

FIGURE B.3 ▶

The ammonia molecule. In the drawing (a), solid lines indicate covalent bonds, and the dashed lines outline the tetrahedron. The lone pair of electrons is ignored in determining the *pyramidal* shape of the molecule. The photograph (b) shows a space-filling model of an ammonia molecule.

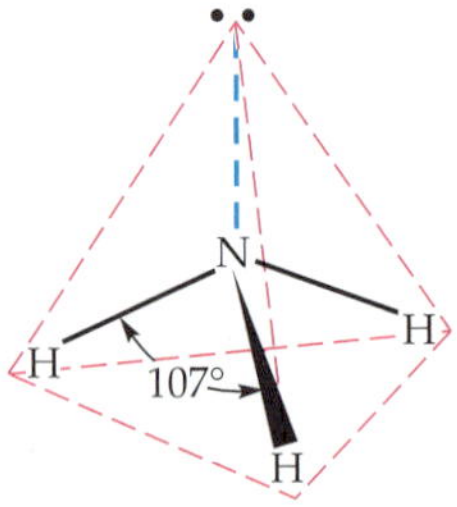

(a)

(b)

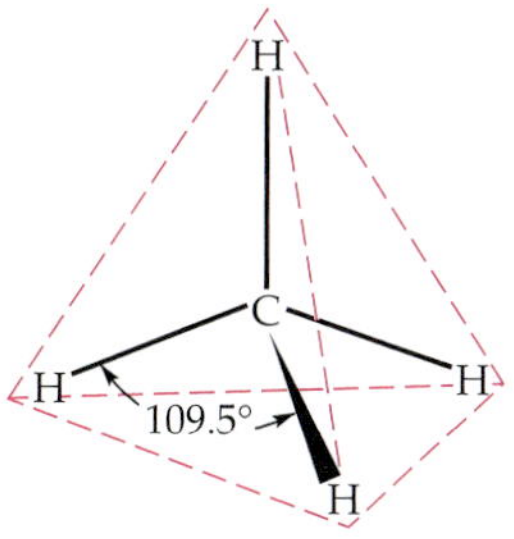

(a)

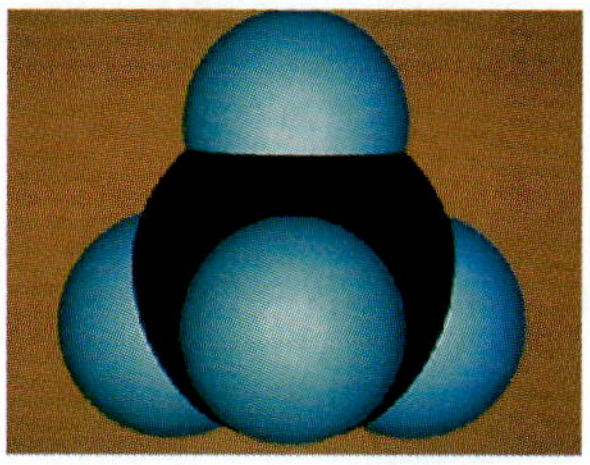

(b)

◀ **FIGURE B.4**

The methane molecule. In the drawing (a), solid lines indicate covalent bonds, and the dashed lines outline the tetrahedron. All bond angles are 109.5°. The photograph (b) shows a space-filling model of a methane molecule.

Many of the properties of compounds—such as melting point, boiling point, and solubility—depend on the polarity of the molecules of the compound.

Key Terms

bonding pair (BP) (B.1)
dipole (B.2)
lone pair (LP) (B.1)
valence shell electron pair repulsion (VSEPR) theory (B.1)

Review Questions

1. All diatomic molecules are linear. Why must this be so?
2. Is it possible for a molecule consisting of three or more atoms to be linear? Explain.
3. In VSEPR theory, the four electron pairs about the central nitrogen atom in ammonia are arranged tetrahedrally, yet we say that the shape of the ammonia molecule is pyramidal. Explain.
4. In VSEPR theory, electron pairs are designated as bonding pairs (BPs) or lone pairs (LPs). Which of the electron pair repulsions are the strongest? Explain.
5. Solutions of iodine chloride, ICl, are used as disinfectants. Are the molecules of ICl ionic, polar covalent, or nonpolar covalent?
6. What approximate bond angle is expected in triatomic molecules having the following electron pair arrangements about the central atom?
   a. linear
   b. trigonal planar
   c. tetrahedral
7. Is it possible to have a *linear* molecule in which the electron pair arrangement is *tetrahedral?* If so, give an example.
8. How does VSEPR theory treat electron pairs in a multiple bond (double or triple)?
9. Must every chemical bond have a bond dipole moment? Must every molecule have a resultant dipole moment? Explain.
10. Water is polar. Explain why this fact proves that the H_2O molecule must have a bent shape.

Problems

VSEPR Theory

11. Use the VSEPR theory to predict the shape of each molecule.
    a. beryllium chloride ($BeCl_2$)
    b. boron chloride (BCl_3)
12. Use the VSEPR theory to predict the shape of each molecule.
    a. arsine (AsH_3)
    b. carbon tetrafluoride (CF_4)
13. Use the VSEPR theory to predict the shape of each molecule.
    a. oxygen difluoride (OF_2)
    b. silicon tetrachloride ($SiCl_4$)
14. Use the VSEPR theory to predict the shape of each molecule.
    a. silane (SiH_4) b. hydrogen selenide (H_2Se)
15. Use the VSEPR theory to predict the shape of each molecule.
    a. nitrogen trichloride (NCl_3)
    b. sulfur dichloride (SCl_2)
16. Use the VSEPR theory to predict the shape of each molecule.
    a. phosphorus trifluoride (PF_3)
    b. dichlorodifluoromethane (CCl_2F_2)

Polar and Nonpolar Molecules

17. Classify the covalent bonds as polar or nonpolar.
    a. H—O b. N—Cl
    c. C—C
18. Classify the covalent bonds as polar or nonpolar.
    a. Si—Cl b. Cl—Cl
    c. P—Cl
19. Use the symbols $\delta+$ and $\delta-$ to indicate partial charges, if any, on the bonds in Problem 17.
20. Use the symbols $\delta+$ and $\delta-$ to indicate partial charges, if any, on the bonds in Problem 18.

21. The molecule BeF_2 is linear. Is it polar or nonpolar? Explain.
22. The molecule SF_2 is bent. Is it polar or nonpolar? Explain.
23. Use differences in electronegativity values to arrange each of the following sets of bonds in order of increasing polarity.
    a. Cl—F, F—F, H—F
    b. H—Br, H—F, H—H
24. Use differences in electronegativity values to arrange each of the following sets of bonds in order of increasing polarity.
    a. H—C, H—F, H—H, H—N, H—O
    b. C—Br, C—C, C—Cl, C—F, C—I
25. Use the symbols $\delta+$ and $\delta-$ to indicate partial charges, if any, on the bonds in Problem 23.
26. Use the symbols $\delta+$ and $\delta-$ to indicate partial charges, if any, on the bonds in Problem 24.

Additional Problems

27. Explain why SO_2 is a polar molecule whereas SO_3 is not.
28. Draw a charge cloud picture for the HF molecule. Use the symbols $\delta+$ and $\delta-$ to indicate the polarity of the molecule.
29. Predict whether or not each of the following species is probable. For any species that seems improbable, tell why it is so.
    a. a linear water molecule, H_2O
    b. a planar molecule, SO_3
    c. a planar molecule, PH_3.
30. Predict the molecular shape of each of the following.
    a. PCl_3
    b. ClO_4^-
    c. OCN^-
31. Predict the molecular shape of each of the following.
    a. PH_4^+
    b. NI_3
    c. Cl_2CO
32. Consider the following statement. The greater the electronegativity difference between the atoms in a molecule, the greater the resultant dipole moment of that molecule. Is this a valid statement? Explain.
33. Predict whether or not each of the following species is probable. For any species that seems improbable, tell why it is so.
    a. a tetrahedral molecule, $GeCl_4$
    b. a bent molecule, HCN

5 Chemical Reactions

The seeming "magic" of chemistry is demonstrated in this reaction in which two colorless liquids combine to produce a yellow solid.

The complex chemical processes in the living cell involve changes in energy as well as changes in chemical composition. Some reactions provide the energy that keeps the cell alive and well. Other reactions, vital to life processes, require an input of energy.

Chemical reactions proceed at various rates. Some are explosively fast; others are exceedingly slow. Rates are affected by a number of factors. Perhaps the most important of these factors, for living organisms, are complex molecules called enzymes that accelerate reaction rates enormously. Yet, strange as it may seem, the enzymes are still there unchanged *after* doing their jobs (see Chapter 22).

Chemical reactions also proceed to different extents. In some, the reactants are converted entirely to products; these reactions are said to go to *completion.* In others, the products react, re-forming the original starting materials. These reactions, outside the cell, come to equilibrium. In a living cell, equilibrium would be deadly. The cellular processes must go to completion—or nearly so. Products can become reactants in the body, but not under equilibrium conditions.

In this chapter, we will examine energy changes, reaction rates, and equilibria. For the most part, we will deal with simple, nonliving systems. The principles developed, however, will be exceedingly important in later chapters, where we will deal with the more complex chemistry of living cells.

5.1 Balancing Chemical Equations

Chemistry is a study of matter and the changes it undergoes. More than that, it is a study of the energy that brings about those changes, or that is released when those changes occur. So far, we have discussed the symbols and formulas that have been invented to represent elements and compounds. Now that we have learned the letters (symbols) and words (formulas) of our chemical language, we are ready to write sentences (chemical equations). A **chemical equation** is a shorthand way of describing chemical change, using symbols and formulas to represent the elements and compounds that are involved in the change.

Carbon reacts with oxygen to form carbon dioxide. In chemical shorthand, this reaction is written

$$C + O_2 \longrightarrow CO_2$$

The plus sign (+) indicates the addition of carbon to oxygen (or vice versa) or a mixing of the two in some manner. The arrow (→) is often read "yields." Substances on the left of the arrow are **reactants,** or *starting materials.* Those on the right are the **products** of the reaction. The conventions here are like those we used in writing nuclear equations (Section 3.2). Now, however, the nucleus will remain untouched. Chemical reactions involve only electrons.

Chemical equations have meaning on the atomic and molecular levels. The equation

$$C + O_2 \longrightarrow CO_2$$

means that one atom of carbon (C) reacts with one molecule of oxygen (O_2) to produce one molecule of carbon dioxide (CO_2).

Sometimes the physical states of the reactants and products are indicated. The initial letter of the state is written immediately following the formula. Thus (g) indicates a gaseous substance, (l) a liquid, and (s) a solid. The label (aq) indicates an aqueous solution—that is, a water solution. Using these labels, our equation becomes

$$C(s) + O_2(g) \longrightarrow CO_2(g)$$

We can't represent all chemical reactions as simply as we can the reaction between carbon and oxygen to form carbon dioxide. For the reaction of hydrogen and oxygen to form water, we might first write

$$H_2(g) + O_2(g) \longrightarrow H_2O(l) \ \textit{(not balanced)}$$

However, this representation is not consistent with the law of conservation of mass. Two oxygen atoms are shown among the reactants, as O_2, but only one appears among the products, in H_2O. For the equation to represent the chemical event correctly, it must be *balanced.* First, to balance oxygen atoms we place the coefficient 2 in front of the formula for water.

$$H_2(g) + O_2(g) \longrightarrow 2\,H_2O(l) \ \textit{(oxygen balanced, hydrogen not balanced)}$$

This coefficient means that two molecules of water, and therefore two oxygen atoms, are involved. A coefficient preceding a formula multiplies everything in the formula, and a coefficient of 1 is understood when no other number appears.

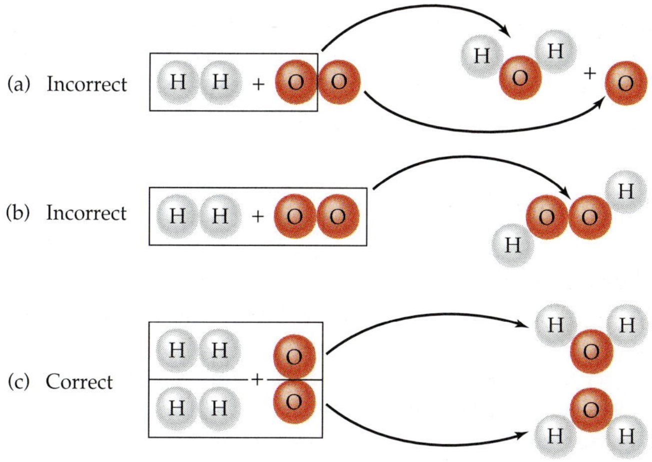

FIGURE 5.1

Balancing the chemical equation for the reaction between hydrogen and oxygen to form water. (a) *Incorrect.* There is no atomic oxygen (O) as a product. *Extraneous products cannot be introduced simply to balance an equation.* (b) *Incorrect.* The product of the reaction is water (H_2O), not hydrogen peroxide (H_2O_2). *A formula can't be changed simply to balance an equation.* (c) *Correct.* An equation can be balanced only through the use of *correct formulas and coefficients.*

In the last equation on page 114, the coefficient 2 not only increases the number of oxygen atoms on the right to two but also increases the number of hydrogen atoms to four. The equation, however, is still not balanced. To balance the numbers of hydrogen atoms, we place the coefficient 2 in front of H_2 on the left.

$$2\,H_2(g) \;+\; O_2\,(g) \;\longrightarrow\; 2\,H_2O(l)\ \textit{(balanced)}$$

Now there are enough hydrogen atoms on the left. In fact, there are four H atoms and two O atoms on each side of the equation. The law of conservation of mass is obeyed. Figure 5.1 illustrates two common pitfalls in the process of balancing equations, as well as the correct method.

Although simple equations can be balanced by trial-and-error, there are a couple of strategies that are often helpful. (1) If an element occurs in just one compound on each side of the equation, try balancing that element *first.* (2) Balance any reactants or products that exist as the *free* element *last.* Perhaps the most important step in any strategy is to check an equation to ensure that it is indeed balanced. Remember that for each element, the same number of atoms of the element must appear on each side of the equation; atoms are conserved in chemical reactions.

Example 5.1

Balance the following equation.

$$Fe \;+\; O_2 \;\longrightarrow\; Fe_2O_3$$

Solution

Begin by balancing the oxygen atoms. The least common multiple of 2 and 3 is 6. We need *three* O_2 and *two* Fe_2O_3

$$Fe \;+\; 3\,O_2 \;\longrightarrow\; 2\,Fe_2O_3\ \textit{(not balanced)}$$

We now have four atoms of iron on the right side. We can get four on the left by placing the coefficient 4 in front of Fe.

$$4\,Fe \;+\; 3\,O_2 \;\longrightarrow\; 2\,Fe_2O_3\ \textit{(balanced)}$$

(Checking that the equation is balanced, we count four Fe atoms and six O atoms on each side.)

Practice Exercise

Balance the following equation.

$$P_4 + H_2 \longrightarrow PH_3$$

Example 5.2

Balance the following equation.

$$CH_4 + O_2 \longrightarrow CO_2 + H_2O$$

Solution

Oxygen appears as the free element on the left. Let's leave oxygen for last and balance the other two elements first. Carbon is already balanced; there is one C atom on each side. To balance hydrogen, we need the coefficient 2 before H_2O.

$$CH_4 + O_2 \longrightarrow CO_2 + 2\,H_2O \text{ (}\textit{not balanced}\text{)}$$

Now, if we count oxygen atoms, we find two on the left and four on the right. If we place a 2 in front of O_2, the O atoms balance.

$$CH_4 + 2\,O_2 \longrightarrow CO_2 + 2\,H_2O \text{ (}\textit{balanced}\text{)}$$

(Checking that the equation is balanced, we count one C atom, four H atoms, and four O atoms on each side.)

Practice Exercise

Balance the following equations.

a. $Mg + B_2O_3 \longrightarrow B + MgO$
b. $NO_2 + H_2O \longrightarrow HNO_3 + NO$
c. $H_2 + Fe_2O_3 \longrightarrow Fe + H_2O$

Example 5.3

Balance the following equation.

$$H_2SO_4 + NaCN \longrightarrow HCN + Na_2SO_4$$

Solution

Notice that the SO_4 and CN groups remain unchanged in the reaction. For purposes of balancing the equation, we can treat each group as a whole rather than breaking it down into its constituent atoms. To balance hydrogen atoms, we place a 2 before HCN.

$$H_2SO_4 + NaCN \longrightarrow 2\,HCN + Na_2SO_4 \text{ (}\textit{not balanced}\text{)}$$

To balance the sodium atoms, we put a 2 in front of the NaCN.

$$H_2SO_4 + 2\,NaCN \longrightarrow 2\,HCN + Na_2SO_4 \text{ (}\textit{balanced}\text{)}$$

To check, we note that in the final balanced equation there are two H atoms, two Na atoms, one SO_4 group, and two CN groups on each side of the equation.

Practice Exercise

Balance the following equations.

a. $H_3PO_4 + Ca(OH)_2 \longrightarrow Ca_3(PO_4)_2 + H_2O$
b. $CaO + P_4O_{10} \longrightarrow Ca_3(PO_4)_2$
c. $Al(OH)_3 + H_2SO_4 \longrightarrow Al_2(SO_4)_3 + H_2O$

We have made the task of balancing equations deceptively easy by considering simple reactions, but it is more important at this point for you to understand the principle than to be able to balance complicated equations. You should know what is meant by a balanced equation and be able to handle simple systems.

5.2 Volume Relationships in Chemical Equations

Chemists generally cannot work with individual atoms and molecules. Even the tiniest speck of matter that we can see contains billions of atoms. John Dalton postulated that atoms of different elements had different masses. Therefore, equal masses of different elements would contain different numbers of atoms. Consider the analogous situation of golf balls and ping-pong balls. A kilogram of golf balls contains a smaller number of balls than a kilogram of ping-pong balls. One could determine the number of balls in each case simply by counting them. For atoms, however, such a straightforward method is not available. It was in the experiments of a French chemist and the mind of an Italian scientist that approaches to the problem of numbering atoms were found.

In 1809, the French scientist Joseph Gay-Lussac (1778–1850) published experimental results summarized in a relationship known as the **law of combining volumes.** The law says that *when gases measured at the same temperature and pressure are allowed to react, the volumes of gaseous reactants and products are in small whole-number ratios.* For example, at 100 °C, two volumes of hydrogen unite with one volume of oxygen to produce two volumes of steam (water vapor), as suggested by Figure 5.2. The combining ratio is 2 : 1 : 2.

In another experiment (Figure 5.3), Gay-Lussac found that if hydrogen is permitted to react with nitrogen to form ammonia, the combining volumes are three of hydrogen with one of nitrogen to give two of ammonia (3 : 1 : 2).

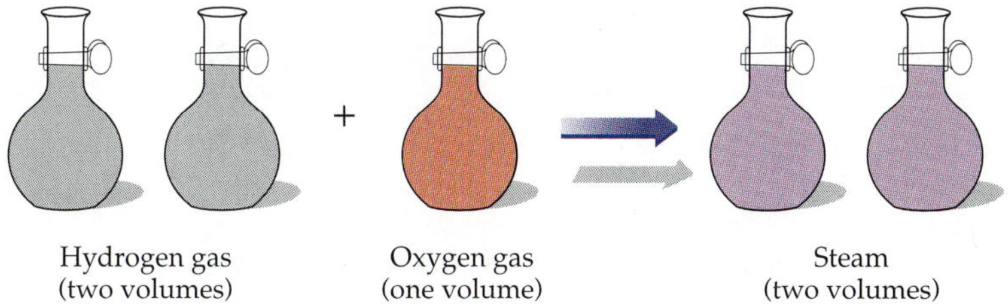

FIGURE 5.2

Gay-Lussac's law of combining volumes. When measured at the same temperature and pressure, two volumes of hydrogen gas react with one volume of oxygen gas to yield two volumes of steam.

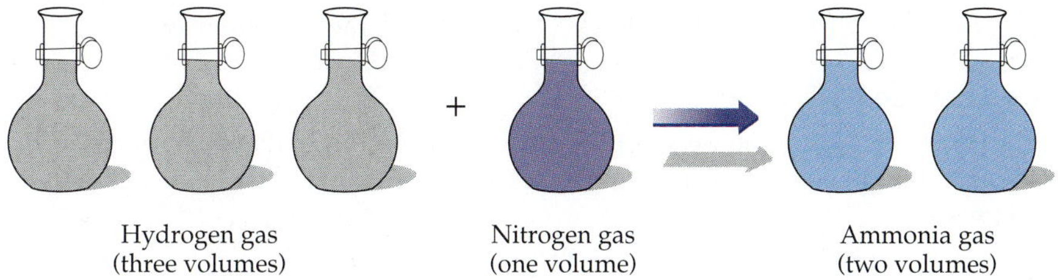

FIGURE 5.3

Gay-Lussac's law of combining volumes. When measured at the same temperature and pressure, three volumes of hydrogen gas react with one volume of nitrogen gas to yield two volumes of ammonia gas.

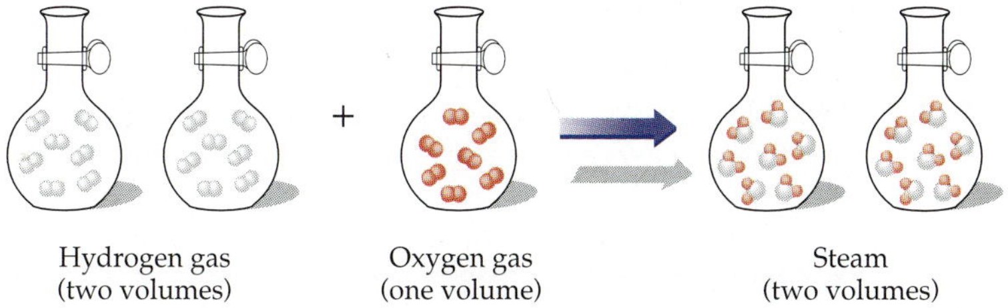

FIGURE 5.4

Avogadro's explanation of Gay-Lussac's law of combining volumes. Equal volumes of each of the gases contain the same number of molecules.

Amadeo Avogadro (1776–1856) did not live to see his ideas accepted by the scientific community. Acceptance finally came in 1860 at a scientific conference at which Stanislao Cannizzaro (1826–1910) effectively communicated Avogadro's ideas from half a century earlier.

Gay-Lussac thought there must be some relationship between the *numbers* of molecules and the *volumes* of gaseous reactants and products. But it was the Italian chemist Amadeo Avogadro who first explained the law of combining volumes. **Avogadro's hypothesis,** based on shrewd interpretation of experimental facts, was that *equal volumes of all gases (at the same temperature and pressure) contain the same number of molecules* (Figure 5.4).

The equation for the combination of hydrogen and oxygen to form water (steam) is

$$2\,H_2(g) \quad + \quad O_2(g) \quad \longrightarrow \quad 2\,H_2O(g)$$

The coefficients of the molecules are the same as the combining ratio of the gas volumes, 2:1:2 (see Figure 5.2). Similarly, the formation of ammonia is described in the equation

$$3\,H_2(g) \quad + \quad N_2(g) \quad \longrightarrow \quad 2\,NH_3(g)$$

The coefficients are identical to the factors of the combining ratio (see Figure 5.3). The equation says that a nitrogen molecule reacts with three hydrogen molecules to produce two ammonia molecules. It also indicates that if you had 1 million nitrogen molecules, you would need 3 million hydrogen molecules to produce 2 million ammonia molecules. The equation provides the combining ratios. If identical volumes of gases contain identical numbers of molecules, then, according to the equation, one volume of nitrogen reacts with three volumes of hydrogen to produce two volumes of ammonia.

Example 5.4 is a straightforward application of the law of combining volumes. Example 5.5 demonstrates that, even if solids or liquids are involved in a reaction, the law still applies, as long as the substances involved in the calculation are *gases*.

Example 5.4

What volume of oxygen is required to burn 0.556 L of propane, if both gases are measured at the same temperature and pressure?

$$C_3H_8(g) \quad + \quad 5\,O_2(g) \quad \longrightarrow \quad 3\,CO_2(g) \quad + \quad 4\,H_2O(g)$$

Solution

The equation indicates that *five* volumes of $O_2(g)$ are required for every volume of $C_3H_8(g)$. Thus,

$$?\,L\,O_2(g) = 0.556\,L\,C_3H_8(g) \times \frac{5\,L\,O_2(g)}{1\,L\,C_3H_8(g)} = 2.78\,L\,O_2(g)$$

Practice Exercise

Using the equation in Example 5.4, calculate the volume of $CO_2(g)$ that is produced when 0.492 L of propane is burned if the two gases are compared at the same temperature and pressure.

Example 5.5

Calculate the volume of methane that must decompose to produce 10.0 liters of hydrogen in the following reaction, if the two gases are compared at the same temperature and pressure.

$$CH_4(g) \quad \longrightarrow \quad C(s) \quad + \quad 2\,H_2(g)$$

Solution

$$10.0\ \text{L}\ H_2 \times \frac{1\ \text{L}\ CH_4}{2\ \text{L}\ H_2} = 5.00\ \text{L}\ CH_4$$

Practice Exercise

Using the equation in Example 5.4, calculate the volume of $O_2(g)$ that must react to form 10.0 liters of steam, if the two gases are compared at the same temperature and pressure.

It was also Avogadro who first suggested that certain elements such as hydrogen, oxygen, and nitrogen were made up of diatomic molecules. If these substances were monatomic, the equations would be

$$2\ H(g) + O(g) \longrightarrow H_2O(g)$$
(*incorrect*)

$$3\ H(g) + N(g) \longrightarrow NH_3(g)$$
(*incorrect*)

Note that these equations would give the wrong ratios of combining volumes. The first, for example, shows a ratio of 2:1:1, rather than the observed volume ratio of 2:1:2. On the other hand, if hydrogen, oxygen, and nitrogen molecules are diatomic, we get the observed ratios.

5.3 Avogadro's Number: 6.02×10^{23}

The periodic table (inside front cover) gives *relative* atomic weights for the various elements. It isn't possible to weigh individual atoms. What we do is to weigh *equal numbers* of atoms or molecules of different substances and use the ratio of masses as relative weights. Avogadro's hypothesis, which has been verified repeatedly through the years, gives us a way to do that by measuring the masses of equal volumes of gases.

Once the relative weights are known, it is possible to plan reactions such that no materials are wasted. If we could (we can't) weigh out 12 u of carbon and 16 u of oxygen, we would have one atom of each, and we could make one molecule of carbon monoxide (CO). If we weigh out 12 g of carbon and 16 g of oxygen, we still have the proper *ratio* of atoms. Gram quantities are easily weighed. From these amounts of reactants, we could make 28 g of CO, with none of the reactants left over. Similarly, if we wished to make carbon dioxide (CO_2), we could weigh out 12 g of carbon and 32 g of oxygen to make 44 g of CO_2 with no left over reactants.

Avogadro had no way of knowing how many molecules there were in a given volume of gas. Scientists since his time have determined the number of atoms in various weighed samples of substances. The numbers are enormously large, even for tiny samples. In defining atomic weights (Section 2.3), the mass of a carbon-12 atom is defined as exactly 12 u. The number of carbon-12 atoms in a 12-g sample of carbon-12 is called **Avogadro's number.** The value of Avogadro's number has been determined experimentally to be 6.0221367×10^{23}. For most purposes the number is rounded to 6.02×10^{23}.

Avogadro's number is an enormous number, so large that it staggers the imagination. If you had 6×10^{23} dollars, for example, you could spend a billion dollars every second for as long as you lived, and you would still have more than 99.999% of your money left to pass on to your children. Or if 6×10^{23} snowflakes fell evenly all across the United States, the blanket of snow would completely cover up every building in the country, including the tallest skyscrapers. See Figure 5.5 for further illustrations as to how gigantic Avogadro's number is.

5.4 Molecular Weights and Formula Weights

We use the term "average" when speaking of the mass of an individual molecule because two molecules of a compound may have different isotopes of one or more of their constituent elements.

As we learned in Chapter 2, each element has a characteristic atomic weight. Because chemical compounds are made up of two or more elements, the masses that we associate with compounds are combinations of atomic weights. For a molecular substance, the **molecular weight** is the average mass of a molecule of a substance relative to that of a carbon-12 atom. More simply, it is the sum of the weights of the atoms represented in a molecular formula. For example, because the formula O_2 specifies two O atoms per molecule of oxygen, the molecular

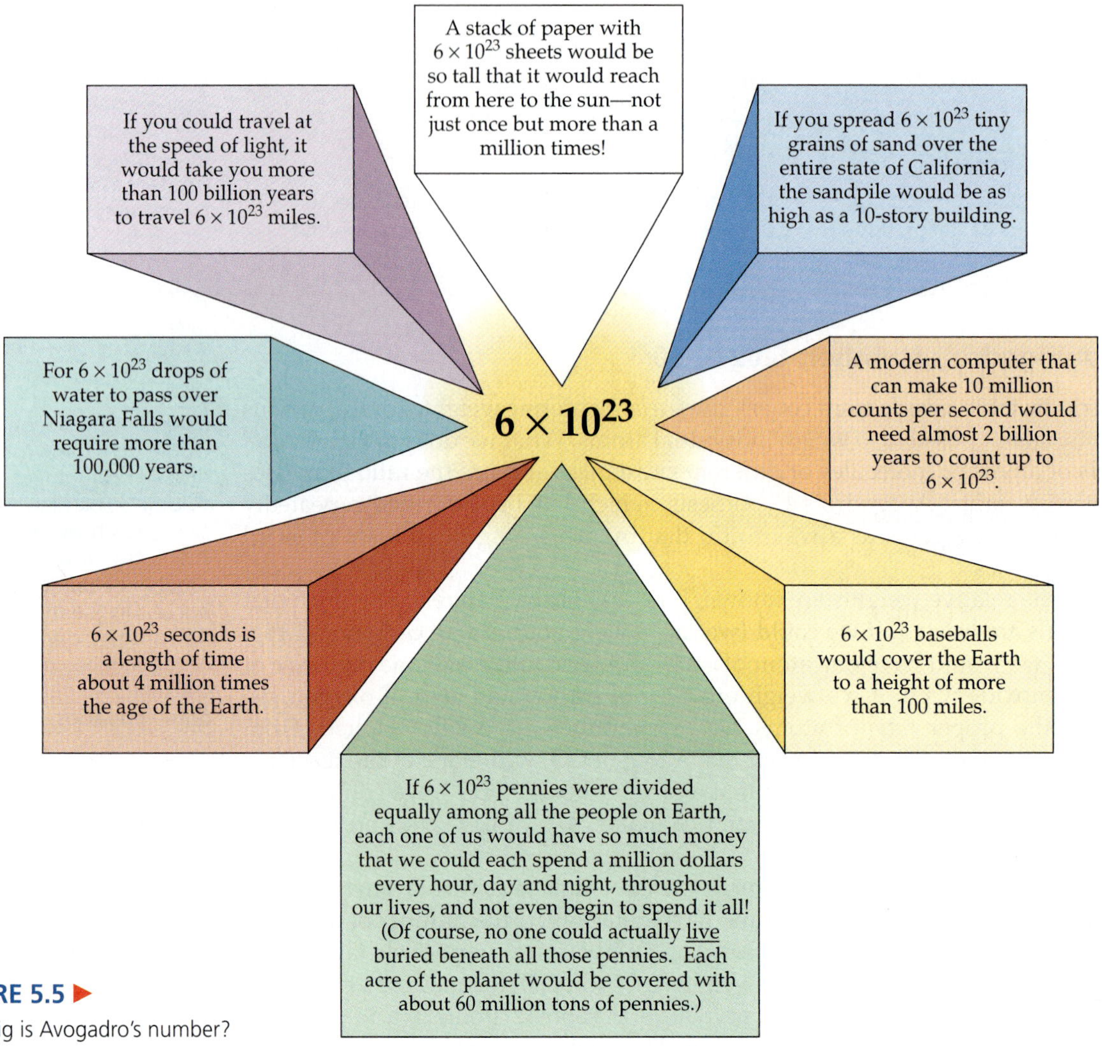

A stack of paper with 6 × 10^23 sheets would be so tall that it would reach from here to the sun—not just once but more than a million times!
If you could travel at the speed of light, it would take you more than 100 billion years to travel 6 × 10^23 miles.
If you spread 6 × 10^23 tiny grains of sand over the entire state of California, the sandpile would be as high as a 10-story building.
For 6 × 10^23 drops of water to pass over Niagara Falls would require more than 100,000 years.
6 × 10^23
A modern computer that can make 10 million counts per second would need almost 2 billion years to count up to 6 × 10^23.
6 × 10^23 seconds is a length of time about 4 million times the age of the Earth.
6 × 10^23 baseballs would cover the Earth to a height of more than 100 miles.
If 6 × 10^23 pennies were divided equally among all the people on Earth, each one of us would have so much money that we could each spend a million dollars every hour, day and night, throughout our lives, and not even begin to spend it all! (Of course, no one could actually live buried beneath all those pennies. Each acre of the planet would be covered with about 60 million tons of pennies.)

FIGURE 5.5 ▶
How big is Avogadro's number?

weight of oxygen (O_2) is twice the atomic weight of oxygen.

$$2 \times \text{atomic weight of O} = 2 \times 15.9994 \text{ u} = 31.9988 \text{ u}$$

The molecular weight of carbon dioxide (CO_2) is the sum of the atomic weight of carbon and twice the atomic weight of oxygen.

$$\begin{aligned} 1 \times \text{atomic weight of C} &= 1 \times 12.011 \text{ u} &&= 12.011 \text{ u} \\ 2 \times \text{atomic weight of O} &= 2 \times 15.9994 \text{ u} &&= \underline{31.99988 \text{ u}} \\ \text{Molecular weight of } CO_2 & &&= 44.010 \text{ u} \end{aligned}$$

Example 5.6

Calculate the molecular weight of sulfur dioxide, SO_2, an irritating gas formed when sulfur is burned.

Solution

We think about the problem in the following way. Add the atomic weight of sulfur to twice the atomic weight of oxygen. However, if we use an electronic calculator, we need only write down the final answer, 64.065 u. That is, we have no need to record the numbers 32.066 and 15.9994.

$$\begin{aligned} 1 \times \text{atomic weight of S} &= 1 \times 32.066\ \text{u} = 32.066\ \text{u} \\ 2 \times \text{atomic weight of O} &= 2 \times 15.9994\ \text{u} = \underline{31.9988\ \text{u}} \\ \text{Molecular weight of } SO_2 &= 64.065\ \text{u} \end{aligned}$$

Practice Exercise

Calculate the molecular weight of each of the following compounds.

a. $C_6H_4Cl_2$

b. $C_2H_4Cl_2$

c. H_3PO_4

The term "molecular weight" is not always appropriate. For ionic compounds, such as NaCl and K_2O, in which individual molecules do not exist, we use the term *formula unit*. **Formula weight** is the average mass of a formula unit relative to that of a carbon-12 atom. In short, the formula weight is the sum of the weights of the atoms or ions represented by the formula.

A **formula unit** is simply the atoms or ions specified by the formula. A formula unit of Al_2O_3 is two aluminum atoms and three oxygen atoms.

Example 5.7

Calculate the formula weight of ammonium sulfate, $(NH_4)_2SO_4$, a fertilizer commonly used by home gardeners.

Solution

To determine a formula weight, we add the atomic weights of the constituent elements. In the summation below, remember that everything within the parentheses must be multiplied by 2.

$$\begin{aligned} 2 \times \text{atomic weight of N} &= 2 \times 14.0067\ \text{u} = 28.0134\ \text{u} \\ 8 \times \text{atomic weight of H} &= 8 \times 1.00794\ \text{u} = 8.06352\ \text{u} \\ 1 \times \text{atomic weight of S} &= 1 \times 32.066\ \text{u} = 32.066\ \text{u} \\ 4 \times \text{atomic weight of O} &= 4 \times 15.9994\ \text{u} = \underline{63.9976\ \text{u}} \\ \text{Formula weight of } (NH_4)_2SO_4 &= 132.141\ \text{u} \end{aligned}$$

Practice Exercise

Calculate the formula weight of each of the following compounds.

a. K_2CO_3

b. $K_2Cr_2O_7$

c. $NaB(C_6H_5)_4$

5.5 Chemical Arithmetic and the Mole

We buy socks by the pair (2 socks), eggs by the dozen (12 eggs), pencils by the gross (144 pencils), and paper by the ream (500 sheets). A dozen is the same *number* whether we are counting a dozen melons or a dozen oranges. But a dozen oranges and a dozen melons do not *weigh* the same. If a melon weighs three times as much as an orange, then a dozen melons will weigh three times as much as a dozen oranges.

Chemists count atoms and molecules by the *mole*. (A single carbon atom is much too small to see, but a *mole* of carbon atoms will fill a tablespoon.) A mole of

Even a sample of carbon as small as a pencil-mark period at the end of a sentence contains about 10^{18} C atoms—that is, about 1,000,000,000,000,000,000 C atoms.

FIGURE 5.6 ▶
One mole of each of four familiar substances: salt (left), sugar (top), copper (right), and carbon (center front). Each dish contains Avogadro's number of formula units of the substance it contains. There are 6.02×10^{23} formula units of NaCl, 6.02×10^{23} molecules of sugar ($C_{12}H_{22}O_{11}$), 6.02×10^{23} atoms of copper, and 6.02×10^{23} carbon atoms in the respective samples.

carbon and a mole of magnesium both contain the same number of atoms. But a magnesium atom weighs twice as much as a carbon atom, so a mole of magnesium will weigh twice as much as a mole of carbon.

According to the SI definition, a **mole** (abbreviated *mol*) is an amount of substance that contains the same number of elementary units as there are atoms in 12 g of carbon-12. That number is 6.02×10^{23}, known as Avogadro's number. The elementary units may be atoms (such as S or Ca), molecules (such as O_2 or CO_2), ions (such as K^+ or SO_4^{2-}), or any other kind of formula unit. A mole of NaCl, for example, contains 6.02×10^{23} NaCl formula units, which means that it contains 6.02×10^{23} Na^+ ions and 6.02×10^{23} Cl^- ions. Figure 5.6 is a photograph of one mole of each of several chemical substances.

The **molar mass** of a substance is the mass of 1 mol of that substance. The molar mass is numerically equal to the atomic weight, molecular weight, or formula weight, but it is expressed in the unit *g/mol.* The atomic weight of sodium is 22.99 u; its molar mass is 22.99 g/mol. The molecular weight of carbon dioxide is 44.01 u; its molar mass is 44.01 g/mol. The formula weight of magnesium chloride is 95.21 u; its molar mass is 95.21 g/mol. We can use these facts, together with the basic definition of the number of elementary units in a mole, to write the following relationships.

$$1 \text{ mol Na} = 22.99 \text{ g Na}$$
$$1 \text{ mol } CO_2 = 44.01 \text{ g } CO_2$$
$$1 \text{ mol } MgCl_2 = 95.21 \text{ g } MgCl_2$$

In turn, these relationships supply the conversion factors that we need to make conversions between mass in grams and amount in moles, as illustrated in the following examples.

Example 5.8

How many grams of Na are there in 0.250 mol of Na?

Solution

Sodium has an atomic weight of 22.99 u and a molar mass of 22.99 g Na/mol Na.

Therefore:

$$0.250 \text{ mol Na} \times \frac{22.99 \text{ g Na}}{1 \text{ mol Na}} = 5.75 \text{ g Na}$$

Practice Exercise

Calculate the mass, in grams, of each of the following.

a. 55.5 mol H_2O

b. 0.0102 mol $C_4H_{10}O$

c. 2.45 mol C_2H_6

Example 5.9

Calculate the number of moles of CO_2 present in a 225-g sample of the gas.

Solution

In this case we need the molar mass of a molecular substance, CO_2. We determined its molecular weight, 44.010 u, on page 120. Its molar mass is therefore 44.01 g CO_2/mol CO_2. Also, we see that in converting from a mass in grams to an amount in moles we must use the *inverse* of the molar mass to get the proper cancellation of units.

$$225 \text{ g CO}_2 \times \frac{1 \text{ mol CO}_2}{44.01 \text{ g CO}_2} = 5.11 \text{ mol CO}_2$$

Practice Exercise

Calculate the amount, in moles, of each of the following.

a. 3.71 g Fe

b. 76.0 g phosphoric acid (H_3PO_4)

c. 165 g C_4H_{10}

5.6 Mole and Mass Relationships in Chemical Equations

Chemical equations not only represent ratios of atoms and molecules but also give us mole ratios. The equation

$$\text{C} + \text{O}_2 \longrightarrow \text{CO}_2$$

tells us that one atom of carbon reacts with one molecule (two atoms) of oxygen to form one molecule of carbon dioxide. The equation also indicates that 1 mol (6.02×10^{23} atoms) of carbon reacts with 1 mol (6.02×10^{23} molecules) of oxygen to yield 1 mol (6.02×10^{23} molecules) of carbon dioxide. Since the molar mass in g/mol of a substance is numerically equal to the formula weight of the substance in atomic mass units, the equation also tells us (indirectly) that 12.0 g (1 mol) of carbon reacts with 32.0 g (1 mol) of oxygen to yield 44.0 g (1 mol) of CO_2 (Figure 5.7).

Example 5.10

Nitrogen monoxide (nitric oxide) combines with oxygen to form nitrogen dioxide according to the equation

$$2\,\text{NO} + \text{O}_2 \longrightarrow 2\,\text{NO}_2$$

State the molecular, molar, and mass relationships indicated by this equation.

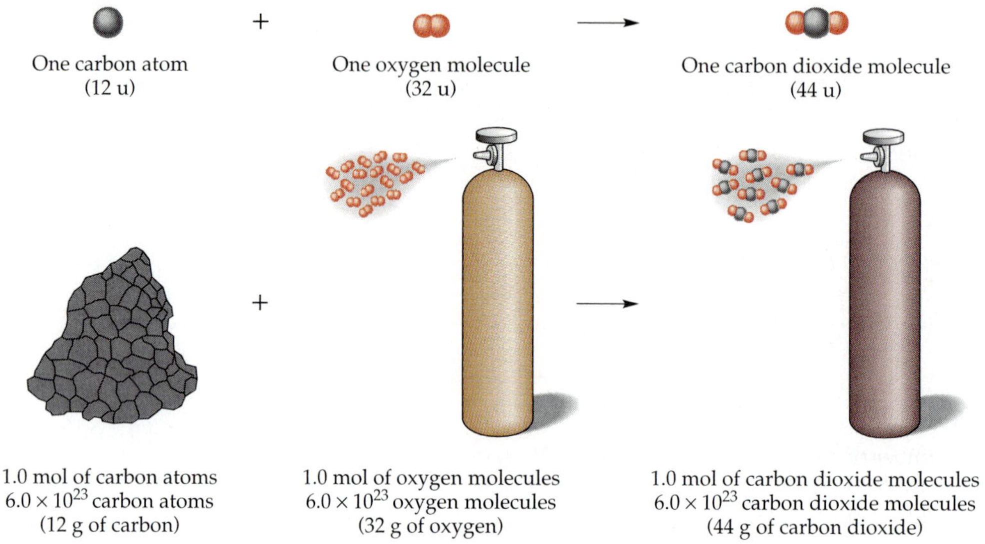

▲ **FIGURE 5.7**

We cannot weigh single C atoms or O_2 or CO_2 molecules, but we can weigh large numbers of these entities.

Solution

Molecular: 2 molecules of NO react with 1 molecule of O_2 to form 2 molecules of NO_2. Molar: 2 mol of NO react with 1 mol of O_2 to form 2 mol of NO_2. Mass; 60.0 g of NO react with 32.0 g of O_2 to form 92.0 g of NO_2.

Practice Exercise

Hydrogen sulfide burns in air to produce sulfur dioxide and water according to the equation

$$2\,H_2S \;+\; 3\,O_2 \;\longrightarrow\; 2\,SO_2 \;+\; 2\,H_2O$$

State the molecular, molar, and mass relationships indicated by this equation.

Now consider the combustion of propane in air to form carbon dioxide and water.

$$C_3H_8 \;+\; 5\,O_2 \;\longrightarrow\; 3\,CO_2 \;+\; 4\,H_2O$$

The coefficients allow us to make statements such as "1 mol of C_3H_8 reacts with 5 mol of O_2," "3 mol of CO_2 is produced for every 1 mol of C_3H_8 that reacts," and "4 mol of H_2O is produced for every 3 mol of CO_2 produced." Moreover, we can turn these statements into conversion factors known as stoichiometric factors. A **stoichiometric factor** relates the amounts of any two substances involved in a chemical reaction, on a *mole* basis. In the examples that follow, stoichiometric factors are shown in color. (Conversion factors are explained in Special Topic A.)

Example 5.11

When 0.105 mol of propane is burned in a rich supply of oxygen, how many moles of oxygen are consumed?

$$C_3H_8 \;+\; 5\,O_2 \;\longrightarrow\; 3\,CO_2 \;+\; 4\,H_2O$$

Solution

The equation tells us that 5 mol of O_2 is required to burn 1 mol of C_3H_8. This gives us the stoichiometric factor 5 mol O_2/1 mol C_3H_8 to use as a conversion factor.

$$0.105 \text{ mol } C_3H_8 \times \frac{5 \text{ mol } O_2}{1 \text{ mol } C_3H_8} = 0.525 \text{ mol } O_2$$

Practice Exercise

For the combustion of propane in Example 5.11: (a) How many moles of carbon dioxide are formed when 0.529 mol of C_3H_8 is burned? (b) How many moles of water are produced when 76.2 mol of C_3H_8 is burned? (c) How many moles of carbon dioxide are produced when 1.010 mol of O_2 is consumed?

Although the mole is essential in calculations based on chemical equations, we cannot measure out molar amounts directly. We have to relate them to quantities that we can measure. We usually express mass in grams, in which case we can follow the five-step approach outlined below.

Step 1. Write a balanced equation for the reaction.
Step 2. Determine the molar masses of the substances involved in the calculation.
Step 3. Use the molar mass of the substance for which information is given to convert grams of the *given* substance, to moles of that substance.
Step 4. Obtain a stoichiometric factor from the balanced equation to convert from moles of the given substance to moles of the substance about which information is sought, the *desired* substance.
Step 5. Use the molar mass of the desired substance to convert moles to grams of that substance.

We illustrate this five-step approach in Figure 5.8 and Example 5.12, where we also show how the steps easily can be combined into the preferred single setup method.

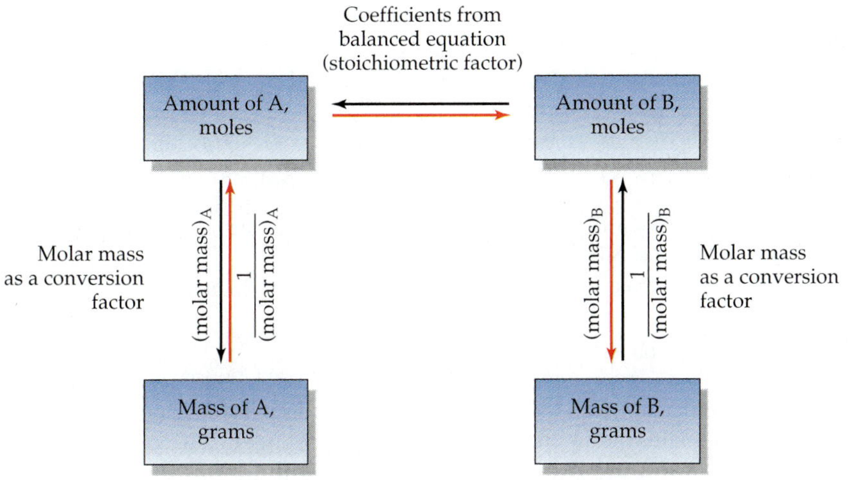

FIGURE 5.8
A chemical equation relates *moles* of reactants and products in a chemical reaction. For *mass* relationships, we must convert the mass of A to moles of A, relate moles of A to moles of B, then convert moles of B to a mass of B.

Example 5.12

Calculate the mass of oxygen needed to react with 10.0 g of carbon in the reaction that forms carbon dioxide.

Solution

Step 1. The balanced equation is

$$C + O_2 \longrightarrow CO_2$$

Step 2. The molar masses are $2 \times 15.9994 = 31.9988$ g/mol for O_2 and 12.011 g/mol for C.

Step 3. We convert the mass of the given substance, carbon, to an amount in moles.

$$10.0 \text{ g C} \times \frac{1 \text{ mol C}}{12.011 \text{ g C}} = 0.833 \text{ mol C}$$

Step 4. We use coefficients from the balanced equation to establish the stoichiometric factor that relates the amount of oxygen to that of carbon.

$$0.833 \text{ mol C} \times \frac{1 \text{ mol } O_2}{1 \text{ mol C}} = 0.833 \text{ mol } O_2$$

Step 5. We convert from moles of oxygen to grams of oxygen.

$$0.833 \text{ mol } O_2 \times \frac{31.9988 \text{ g } O_2}{1 \text{ mol } O_2} = 26.7 \text{ g } O_2$$

We can also combine all of the steps outlined above into a single setup.

$$10.0 \text{ g C} \times \frac{1 \text{ mol C}}{12.011 \text{ g C}} \times \frac{1 \text{ mol } O_2}{1 \text{ mol C}} \times \frac{31.9988 \text{ g } O_2}{1 \text{ mol } O_2} = 26.6 \text{ g } O_2$$

(The slightly different answers are due to rounding in the intermediate steps.)

Practice Exercise

Calculate the mass of oxygen needed to react with 0.334 g of nitrogen in the reaction that forms nitrogen dioxide.

Example 5.13

Ammonia, NH_3, a common fertilizer, is made by causing hydrogen and nitrogen to react at a high temperature and pressure. What mass of ammonia, in grams, can be made from 60.0 g of hydrogen?

Solution

Step 1. We must write a chemical equation

$$N_2 + H_2 \longrightarrow NH_3 \text{ (\textit{not balanced})}$$

and balance it.

$$N_2 + 3\,H_2 \longrightarrow 2\,NH_3 \text{ (\textit{balanced})}$$

Step 2. The molar masses are $2 \times 1.008 = 2.016$ g/mol for H_2 and $14.01 + 3 \times 1.008 = 17.03$ g/mol for NH_3.

Step 3. We convert the mass of the given substance, hydrogen, to an amount in moles.

$$60.0 \text{ g } H_2 \times \frac{1 \text{ mol } H_2}{2.016 \text{ g } H_2} = 29.8 \text{ mol } H_2$$

Step 4. We use coefficients from the balanced equation to establish the stoichiometric factor that relates the amount of ammonia to that of hydrogen.

$$29.8 \text{ mol } H_2 \times \frac{2 \text{ mol } NH_3}{3 \text{ mol } H_2} = 19.9 \text{ mol } NH_3$$

Step 5. We convert from moles of ammonia to grams of ammonia.

$$19.9 \text{ mol } NH_3 \times \frac{17.03 \text{ g } NH_3}{1 \text{ mol } NH_3} = 339 \text{ g } NH_3$$

As is usually the case, all of the steps outlined above can be combined into a single setup.

$$60.0 \text{ g } H_2 \times \frac{1 \text{ mol } H_2}{2.016 \text{ g } H_2} \times \frac{2 \text{ mol } NH_3}{3 \text{ mol } H_2} \times \frac{17.03 \text{ g } NH_3}{1 \text{ mol } NH_3} = 338 \text{ g } NH_3$$

Practice Exercise

The decomposition of potassium chlorate ($KClO_3$) produces potassium chloride and oxygen gas. How many grams of oxygen can be made from 2.47 g of potassium chlorate?

5.7 Structure, Stability, and Spontaneity

We saw at the beginning of Chapter 4 that some electron configurations are more stable than others. Sodium *ions* and chloride *ions*, arranged in a crystal lattice, are less reactive than sodium *atoms* and chlorine *molecules*. When sodium metal and chlorine gas are mixed, a vigorous reaction ensues.

When heat energy is released during a chemical reaction, heat often is listed as a product in the chemical equation. When energy is supplied to keep a chemical reaction going, heat can be listed as a reactant in the equation.

$$2\,Na + Cl_2 \longrightarrow 2\,NaCl$$

A great deal of energy is produced as heat and light during this reaction. Sometimes this energy is listed as one of the products in the chemical equation.

$$2\,Na + Cl_2 \longrightarrow 2\,NaCl + \text{energy}$$

Chemical reactions that result in the release of heat are said to be **exothermic.** The burning of methane is an exothermic reaction, as is the burning of gasoline or coal (see Figure 5.10, p. 129). In each case, chemical energy is converted into heat energy. There are other reactions, such as the decomposition of water, for which energy must be supplied. If the energy is supplied as heat, such reactions are said to be **endothermic.**

$$2\,H_2O + \text{energy} \longrightarrow 2\,H_2 + O_2$$

The energy released in an exothermic reaction is related to the *chemical energy* present in the reactants. This energy is a form of *potential energy.* Potential energy is released (as heat or light, for example) when reactants are brought together in a way that allows them to achieve more stable electron arrangements.

To say that a reaction is exothermic does not necessarily mean that it is instantaneous. For example, coal (carbon) has a lot of chemical energy. Carbon reacts with oxygen to form carbon dioxide with the release of considerable heat. But coal doesn't react very rapidly with oxygen at ordinary temperatures. In fact, one can store a pile of coal in air indefinitely without perceptible change. Before coal will react with oxygen to release its stored energy, it must be heated to a temperature of several hundred degrees. The coal must be supplied with a certain amount of energy (called the energy of activation; Section 5.8) before it will begin to burn steadily. Once this energy of activation is supplied, the heat evolved in the reaction will keep the coal burning brightly, and the energy eventually produced will exceed the amount of energy required to start the reaction. Overall, you get more

Coal burns in a highly exothermic reaction. The heat released can be used to convert water to steam that can turn a turbine to produce electricity.

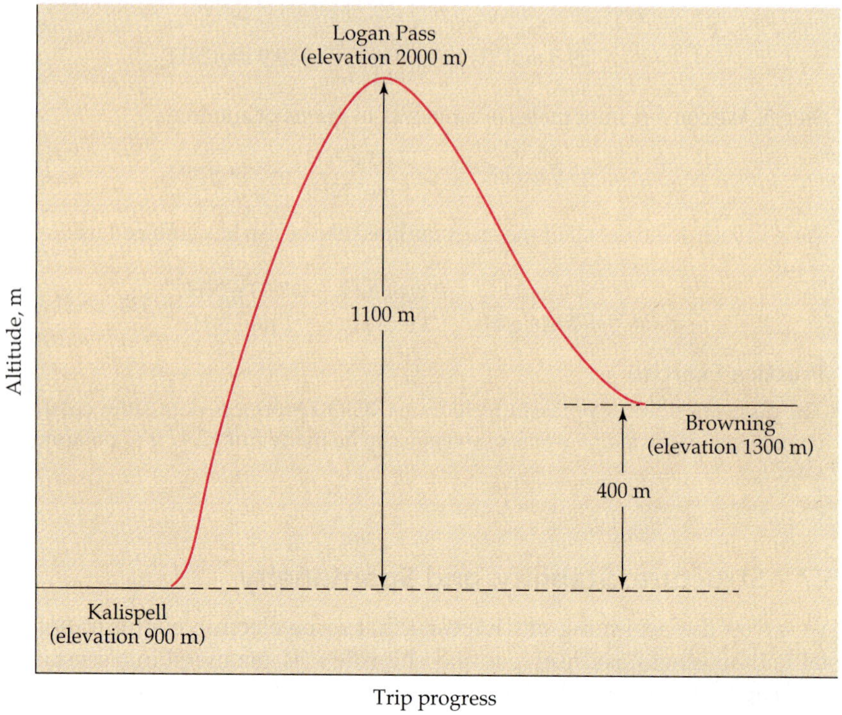

FIGURE 5.9 ▶ An analogy to the profile of a reaction and of activation energy. If you were in Kalispell, Montana (the reactants), and wanted to go to Browning (the products), you could drive the scenic Going-to-the-Sun Highway (the reaction profile) through Glacier National Park. First you have to climb 1100 m (the activation energy) to the continental divide at Logan Pass, but then it is downhill the rest of the way.

energy out than you put in. The reaction is exothermic. In an endothermic reaction, more energy goes in than comes out.

We will return to the burning of coal in a moment, but first let's consider an analogy involving a more familiar situation. If you were in Kalispell, Montana (elevation 900 m) and wished to travel to Browning (elevation 1300 m), you could choose to drive the scenic Going-to-the-Sun Highway through Glacier National Park, which crosses the continental divide at Logan Pass (elevation 2000 m). First, you would have to climb 1100 m, but then it would be downhill the rest of the way (Figure 5.9).

The reaction of coal (carbon) with oxygen can be explained in much the same way. The reactants, carbon and oxygen, lie in a rather high potential energy valley (Figure 5.10). A certain amount of energy has to be put into the system for it to get over the top of the "mountain." Once the reaction is under way, the heat released more than compensates for the energy needed to get the reaction going in the first place.

5.8 Reversible Reactions

As we shall see in Chapter 24, living cells take in oxygen and "burn" glucose to obtain energy, a process called *respiration.*

$$\underset{\text{Glucose}}{C_6H_{12}O} + 6\,O_2 \longrightarrow 6\,CO_2 + 6\,H_2O + \text{energy}$$

Green plants carry out the reaction in the opposite direction (an endothermic reaction called **photosynthesis**), using energy from sunlight to convert carbon dioxide and water to glucose and oxygen.

$$6\,CO_2 + 6\,H_2O + \text{energy} \longrightarrow C_6H_{12}O_6 + 6\,O_2$$

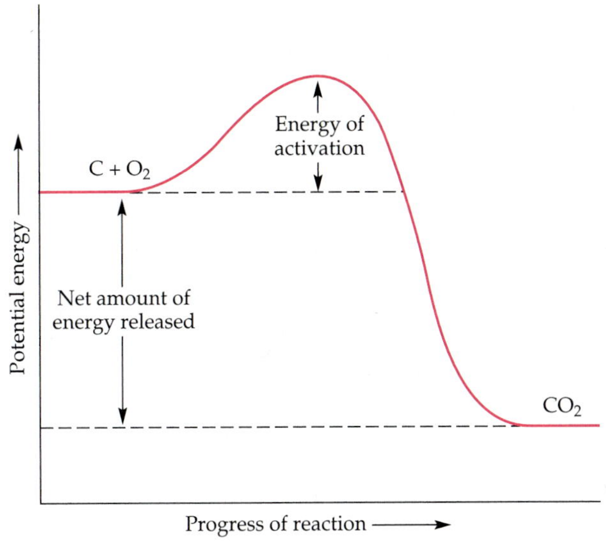

FIGURE 5.10

To get from reactants (carbon and oxygen) to product, we must put some energy (the energy of activation) into the system. (The actual reaction is much more complicated than is indicated by this simplified reaction profile.)

All life on our planet is based on these opposing reactions. But what makes it go one way in one case and the opposite way in another? One obvious answer is energy. Since energy is released when glucose is oxidized, the reactants (glucose and oxygen) must be at a higher energy level than the products (carbon dioxide and water). As we have indicated previously, this does not mean that the path is straight downhill from reactants to products. Indeed, if glucose and oxygen are mixed at ordinary temperatures outside a cell, no perceptible reaction occurs. As with the reaction of coal and oxygen, a certain amount of energy, the energy of activation, must be supplied before the reaction takes place. The potential energy diagram for this reaction (Figure 5.11) strongly resembles that for the coal and oxygen reaction (see Figure 5.10).

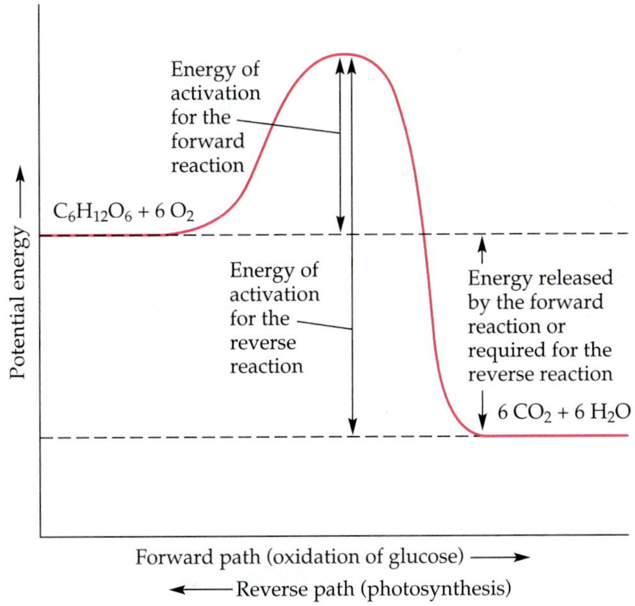

FIGURE 5.11

A reaction profile for respiration (the forward reaction) and for photosynthesis (the reverse reaction).

For **reversible reactions,** the energy hill, or the barrier to reaction, may be approached from either side. That is, we can read the diagram in Figure 5.11 from left to right or from right to left. With the burning of glucose, we are considering an exothermic reaction in which higher-energy reactants are converted to lower-energy products. When read in the reverse direction, the diagram presents the energy changes that occur in the endothermic reaction called photosynthesis. The climb to the top of the barrier is longer from one side than from the other. The **energy of activation** (also called *activation energy*) for a chemical reaction is the minimum energy needed to get the reaction started. It is the difference in energy between the level the reactants are on and the top of the energy hill. When the reactants are in the valley at the left (as in the forward reaction), the climb to the top is not so long. When the reactants are in the valley to the right (as in the reverse reaction), the climb to the top is much longer. The potential energy diagram shown in Figure 5.11 is much simplified. Just as one seldom crosses a mountain by going straight up one side and straight down the other, so reactions seldom proceed by a smooth, one-hump potential energy change.

5.9 Reaction Rates: Collisions, Orientation, and Energy

Before atoms, molecules, or ions can react, they must first get together: they must collide. Second, for all except the simplest particles, they must come together in the proper orientation. Third, the collision must provide a certain minimum energy, the energy of activation.

First, let's consider the frequency of collision, which is influenced by two factors: concentration and temperature. The more concentrated the reactants, the more frequently the particles will collide, simply because there are more of them in a given volume. Molecules also move faster at higher temperatures; thus, an increase in temperature also increases the frequency of collision. The effects of concentration and temperature on reaction rates will be discussed in more detail in following sections.

For the moment, let's take a closer look at the effect of orientation. It might not be obvious that orientation is important. Sometimes, although rarely, it is not important. Consider a situation in which you want to knock someone down. You can tackle from the front, back, or side and still accomplish your objective. If,

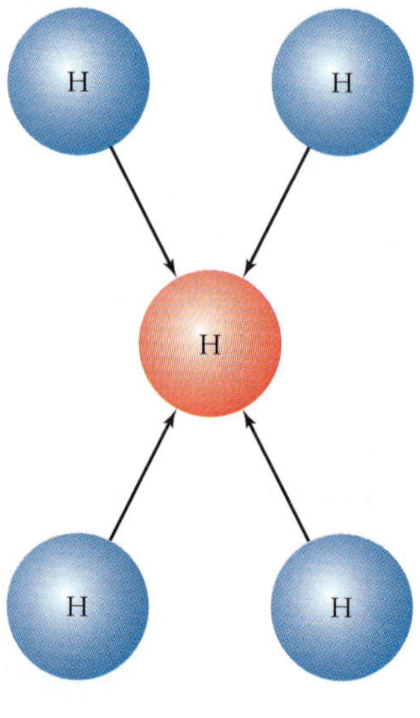

In the reaction between two hydrogen atoms to form a hydrogen molecule, orientation of the two atoms is unimportant. No matter from which direction the oncoming hydrogen atom (blue) approaches the "target" hydrogen atom (red), its "view" of the impending collision is the same. There is no preferred orientation for the collision.

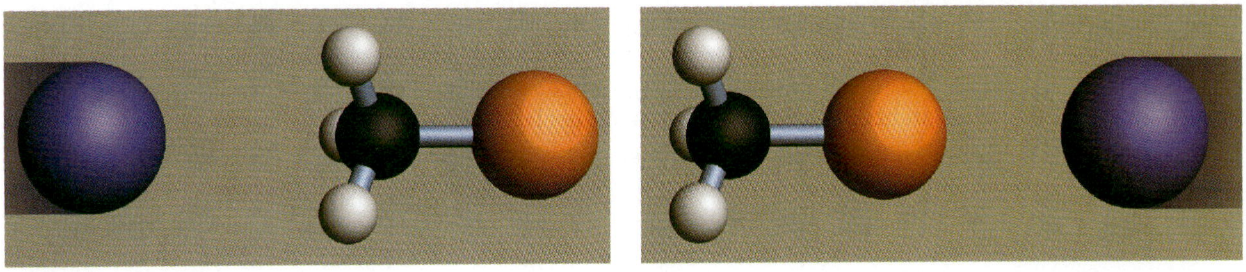

(a) (b)

▲ **FIGURE 5.12**

The importance of orientation of colliding molecules is illustrated in the reaction of iodide ion with methyl bromide to produce methyl iodide and bromide ion. The I^- ion in (a) collides with the C atom of CH_3Br, a favorable collision for reaction to occur. The I^- ion in (b) collides with the Br atom of CH_3Br, an unfavorable collision for chemical reaction.

however, a kiss is the objective, then an orientation in which the participants are face-to-face works best. For chemical reactions, there are also a few instances in which orientation is not important. If two hydrogen atoms are to react to form a hydrogen molecule,

$$2\,H \longrightarrow H_2$$

then orientation is unimportant. The quantum mechanical view of hydrogen atoms depicts them as symmetrical (spherical) electron clouds. Front, back, top, and bottom all look the same. Therefore, all orientations of two hydrogen atoms are identical. If such atoms come into contact with one another, they can react.

In most cases, however, orientation of the colliding molecules is a factor. Consider the reaction of iodide ion with methyl bromide (CH_3Br) to produce methyl iodide (CH_3I) and bromide ion.

$$I^- + CH_3Br \longrightarrow CH_3I + Br^-$$

As shown in Figure 5.12, a collision of an I^- ion with the C atom of the CH_3Br can be effective in leading to a reaction. A collision of an I^- ion with the Br atom of CH_3Br cannot. We shall point out another example of proper orientation shortly.

The third factor, that collisions must provide a certain minimum energy, is a good deal more subtle. There is a temperature effect, which will be discussed in some detail in the next section. Certain substances, called catalysts, may substantially lower this activation energy, thus increasing the reaction rate. The effect of catalysts will be discussed more thoroughly in Section 5.11.

5.10 Reaction Rates: The Effect of Temperature

Reactions generally take place faster at higher temperatures. For example, coal (carbon) reacts so slowly with oxygen (from the air) at room temperature that the change is imperceptible. However, when coal is heated to several hundred degrees, it reacts at a much more rapid rate. The heat evolved in the reaction keeps the coal burning smoothly. The effect of temperature on the rates of chemical reactions is explained by the kinetic-molecular theory. We shall consider this theory in more detail in Chapter 7. For the moment, we will cite one of its postulates, which states that at high temperatures molecules move more rapidly. Thus, they collide more frequently, providing a greater chance for reaction. At high temperatures, the rapidly moving molecules also strike one another harder. The harder these collisions are, the more energy involved. These harder collisions are more likely to supply the activation energy needed to break chemical bonds and get the reaction going. Consider the reaction that takes place between hydrogen gas and chlorine gas. In order for hydrogen and chlorine to react, bonds between hydrogen atoms and between chlorine atoms must be broken. In general, energy is absorbed when chemical bonds are broken, and energy is released when chemical bonds are formed.

$$H{-}H + Cl{-}Cl \longrightarrow 2\,H{-}Cl + \text{heat}$$

This is an exothermic reaction. Once it has started, the energy released by the formation of hydrogen–chlorine bonds more than compensates for that required to break H–H and Cl–Cl bonds. There is a net conversion of chemical energy to heat energy.

In endothermic reactions, the energy released by bond formation is less than that required to break the necessary bonds. For these reactions, energy must be supplied continuously from an external source, or the process will stop. A typical endothermic reaction is the decomposition of the salt potassium chlorate ($KClO_3$) to give oxygen and another salt, potassium chloride (KCl).

$$2\,KClO_3(s) + \text{heat} \longrightarrow 2\,KCl(s) + O_2(g)$$

The potassium chlorate must be heated continuously. If the source of heat is removed, the reaction quickly subsides.

We make use in our daily lives of our knowledge of the effect of temperature on chemical reactions. For example, we freeze foods to retard the chemical reactions that lead to spoilage. On the other hand, if we want to cook our food more rapidly, we turn up the heat. The chemical reactions that occur in our bodies generally do so at a constant temperature of 37 °C (98.6 °F). A few degrees' rise in temperature (fever) leads to an increase in respiration, pulse rate, and other physiological reactions. A drop in body temperature of a few degrees slows these same processes considerably, as is exemplified by the slowed metabolism of hibernating animals. This phenomenon is used in some surgical procedures. In some cases of heart surgery, the body temperature of the patient is lowered to about 15 °C (60 °F). Ordinarily, the brain is permanently damaged when its oxygen supply is interrupted for more than 5 min. But at the lower temperature, metabolic

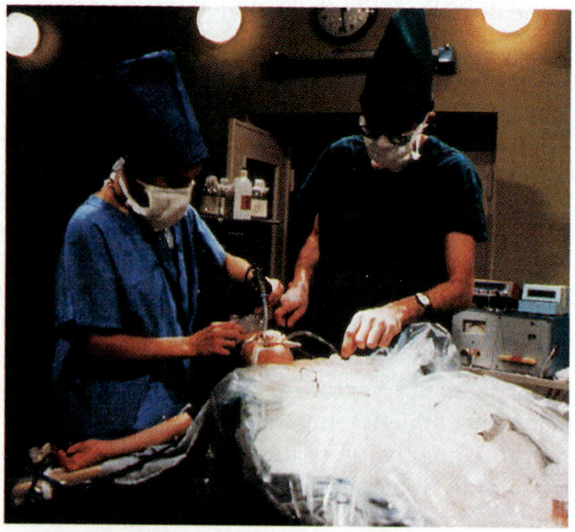

FIGURE 5.13 ▶

Heart surgery can be performed with the patient's temperature lowered to slow metabolic processes and thus minimize the possibility of brain damage.

processes slow considerably, and the brain can survive much longer periods of oxygen deprivation. The surgeon can stop the heartbeat, perform an hour-long surgical procedure on the heart, and then restart the heart and bring the patient's temperature back to normal (Figure 5.13).

It must be said that despite these examples the manipulation of reaction rates in living systems through changes in temperature is severely restricted. Increasing temperature, for example, may deactivate (render inactive) enzymes (Section 5.11) that mediate cellular chemistry and are essential to life. We kill germs by heat sterilization in autoclaves. For living cells, there is often a rather narrow range of optimum temperatures. Both higher and lower temperatures can be disabling, if not deadly.

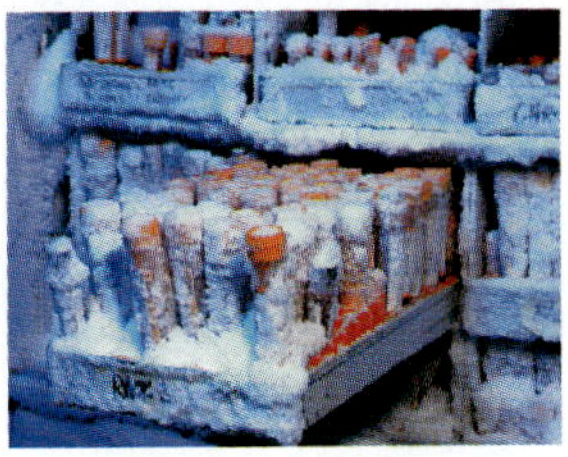

Blood plasma is usually frozen to slow its decomposition and extend its shelf life. Organs harvested for transplant are transported in ice-packed "coolers" for the same reason.

5.11 Reaction Rates: Catalysis

Recall the decomposition of potassium chlorate to oxygen and potassium chloride (Section 5.10). For oxygen to be produced at a useful rate, the potassium chlorate must be heated to over 400 °C. However, if we add a small amount of manganese dioxide (MnO_2), we can get the same rate of oxygen evolution by heating the reactant to just 250 °C. Further, after the reaction is complete, the manganese dioxide can be completely recovered, unchanged. A substance that, like manganese dioxide, changes the rate of a chemical reaction without itself being changed is called a **catalyst.** In general, catalysts act by lowering the activation energy required for the reaction to occur (Figure 5.14). If activation energy is lower, then the collisions of more slowly moving molecules will be sufficient to supply that energy. The lower temperature required for a catalyzed reaction reflects this. This energy of activation is lowered because the catalyst changes the *path* of the reaction. To return to our analogy of the trip from Kalispell to Browning, it is possible to take an alternate route. U.S. Highway 2 crosses the continental divide through Marias Pass. This route involves a climb of only 700 m, compared with 1100 m via Logan Pass (Figure 5.15). This alternate route is analogous to that provided by a catalyst in a chemical reaction.

Catalysts are of great importance in the chemical industry. A reaction that would otherwise be so slow as to be impractical can be made to proceed at a reasonable rate with the proper catalyst.

Catalysts are even more important in living organisms, where raising the temperature by 100 °C is not a feasible way of increasing the rate of critical reactions. If we raised our body temperature by 100 °C, we'd boil our blood, among other things. Biological catalysts, called **enzymes,** mediate nearly all the chemical reactions that take place in living systems. But these catalysts themselves may

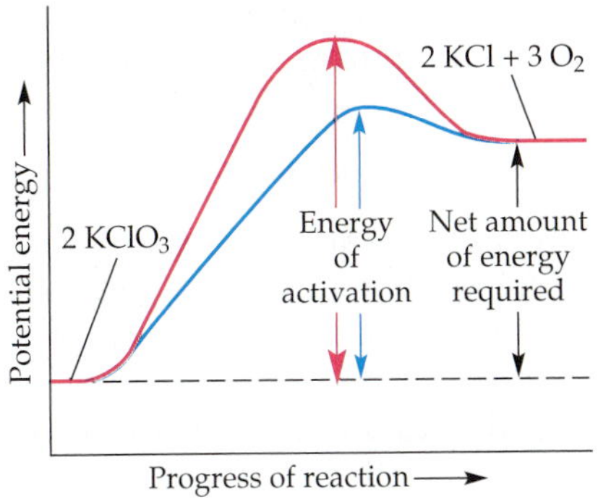

FIGURE 5.14

In the decomposition of potassium chlorate, a catalyst acts to lower the energy of activation (blue arrow and curve) as compared with that for the uncatalyzed reaction (red arrow and curve).

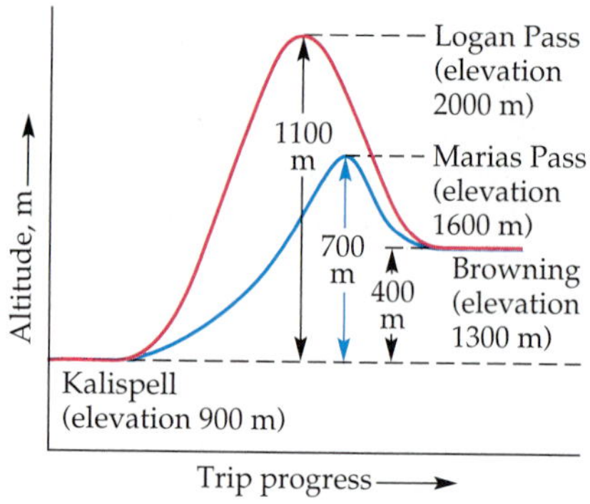

FIGURE 5.15 ▶ An analogy to the profile of a reaction and of activation energy for a catalyzed reaction. To return to our analogy of the trip from Kalispell to Browning (Figure 5.9), you could take an alternate route. U.S. Highway 2 crosses the continental divide at Marias Pass (catalyzed reaction). This route involves a climb of only 700 m (lower activation energy) compared with 1100 m via Logan Pass (uncatalyzed reaction). This alternate route is analogous to the alternate pathway provided by a catalyst in a chemical reaction.

suffer irreversible damage when living cells are subjected to excessive temperature. Once a catalyst is deactivated, the reactions that require the catalyst no longer proceed at a rate that maintains life. In addition to serving as examples of catalysts, enzymes offer a striking illustration of the importance of orientation in chemical reactions. These compounds are huge molecules, each possessing what is called an active site. The reactants must come into contact with this active site if the enzyme is to catalyze the reaction. If the reactants do not collide at the active site, no reaction occurs. So important are enzymes to life that we discuss them more fully in Chapter 22.

5.12 Reaction Rates: The Effect of Concentration

Another factor affecting the rate of a chemical reaction is the concentration of reactants. The more reactant molecules there are in a volume of space, the more collisions will occur. The more collisions there are, the more reactions will occur. For example, you can light a wood splint and then blow out the flame. The splint will continue to glow as the wood reacts slowly with the oxygen in the air. If the glowing splint is placed in pure oxygen, the splint will burst into flame, indicating a much more rapid reaction. This more rapid reaction can be interpreted in terms of the concentration of oxygen. Air is about one-fifth oxygen. The concentration of O_2 molecules in pure oxygen is therefore about five times as great as in air. The caution against smoking in a hospital room where a patient is in an oxygen tent is not merely a concession to the sensitivity of nonsmokers. It is meant to prevent disaster (Figure 5.16).

▲ **FIGURE 5.16** The No Smoking sign isn't simply an attempt to avoid exposing the patient to the hazards of second-hand smoke. It is a warning about the increased danger of fire in an oxygen-rich atmosphere.

For reactions in solution, the concentration of a reactant can be increased if more of it is dissolved. One of the first studies of reaction rate was done by Ludwig Wilhelmy in 1850. He studied the rate of reaction of sucrose (cane or beet sugar) with water. The products are two simpler sugars, glucose and fructose (see Chapter 19).

$$\text{Sucrose} + H_2O \xrightarrow{HCl} \text{Glucose} + \text{Fructose}$$

(We will not be concerned with the chemical formulas for these sugars at this time. The hydrochloric acid serves as a catalyst for the reaction. Chemists generally write formulas or names of catalysts above the arrow, because the catalyst is recovered unchanged.) Wilhelmy found that the rate of the reaction was proportional to the concentration of sucrose. If he dissolved 0.001 mol of sucrose in 1 L of water, the sucrose reacted at a certain rate. If he doubled the concentration of sucrose (i.e., if he dissolved 0.002 mol of sucrose in 1 L of water), the reaction rate was doubled.

In general, when the temperature is constant and the catalyst (if any) is present in a fixed amount, the rate of the reaction can be related quantitatively to the amounts of reacting substances. The relationship is not necessarily a simple one, however.

Consider again the photosynthesis reaction.

$$6\,CO_2 \;+\; 6\,H_2O \;+\; \text{energy} \longrightarrow C_6H_{12}O_6 \;+\; 6\,O_2$$

For this reaction to occur in one step, six carbon dioxide molecules and six water molecules would all have to come together at the same instant in an effective collision. Such an event is highly unlikely. In fact, even three-body collisions are rare. Most reactions proceed through a series of steps involving collisions of only two molecules each. Some steps may involve only the breaking of a bond in a single molecule. The photosynthesis reaction involves an extraordinarily complex web of intermediate steps, but each individual step is fairly simple. Virtually all of these steps require enzymes as catalysts.

The step-by-step process, or **mechanism,** by which a reaction takes place is determined by a study of the rate of the reaction as various factors, such as temperature, concentrations, and catalysts, are changed. Such studies are important in chemistry, for chemists want to get as much product as possible with a minimum expenditure of materials and energy. Investigations of mechanisms are also important in the chemistry of living cells. Some types of cancer are thought to be induced by chemicals. A lot of research is under way to work out the mechanism by which the chemicals act—or are acted upon—in the induction of cancer. A knowledge of the mechanisms of various metabolic reactions has enabled scientists to determine how certain poisons work, and this knowledge has led to effective treatments in many instances. Certain diseases of genetic origin involve the disruption of a single step in the mechanism of a reaction essential to health. Again, this information has permitted physicians to deal successfully with what might otherwise have been a fatal defect.

5.13 Equilibrium in Chemical Reactions

For reversible reactions—those that proceed in both a forward and a backward direction—the concept of rate requires further consideration. In such reactions, the reverse reaction cannot occur until some product molecules (that is, product with respect to the forward reaction) have been formed. Even then, because of the effect of concentration on rate, the reverse reaction will be slow until sufficient product has been formed to increase the chance of these molecules getting together to re-form the original reactants. Similarly, as the reactants change to products, there will be fewer collisions between reactant molecules, because their number becomes depleted. In isolated systems (ones that are cut off from outside influences), the rates of the forward and reverse reactions eventually become equal, and a condition called **equilibrium** is established (Figure 5.17). At equilibrium there *appears* to be no further reaction. In this case appearances are deceiving. Reactants are still changing to products, and products are still changing back to reactants. It is just that these changes are occurring at precisely the same rate. For every reactant molecule lost through the forward reaction, one is gained through the reverse reaction. Once equilibrium is established, we can measure no change in the concentrations of reactants or products. Because molecules are still reacting,

Time from start of reaction	Rate of forward reaction / Rate of reverse reaction	Reaction mixture	Concentration	
		● Reactants ○ Products	Reactants	Products
0			20	0
10			12	8
20			8	12
30			6	14
40			6	14
50			6	14

At equilibrium {30, 40, 50}

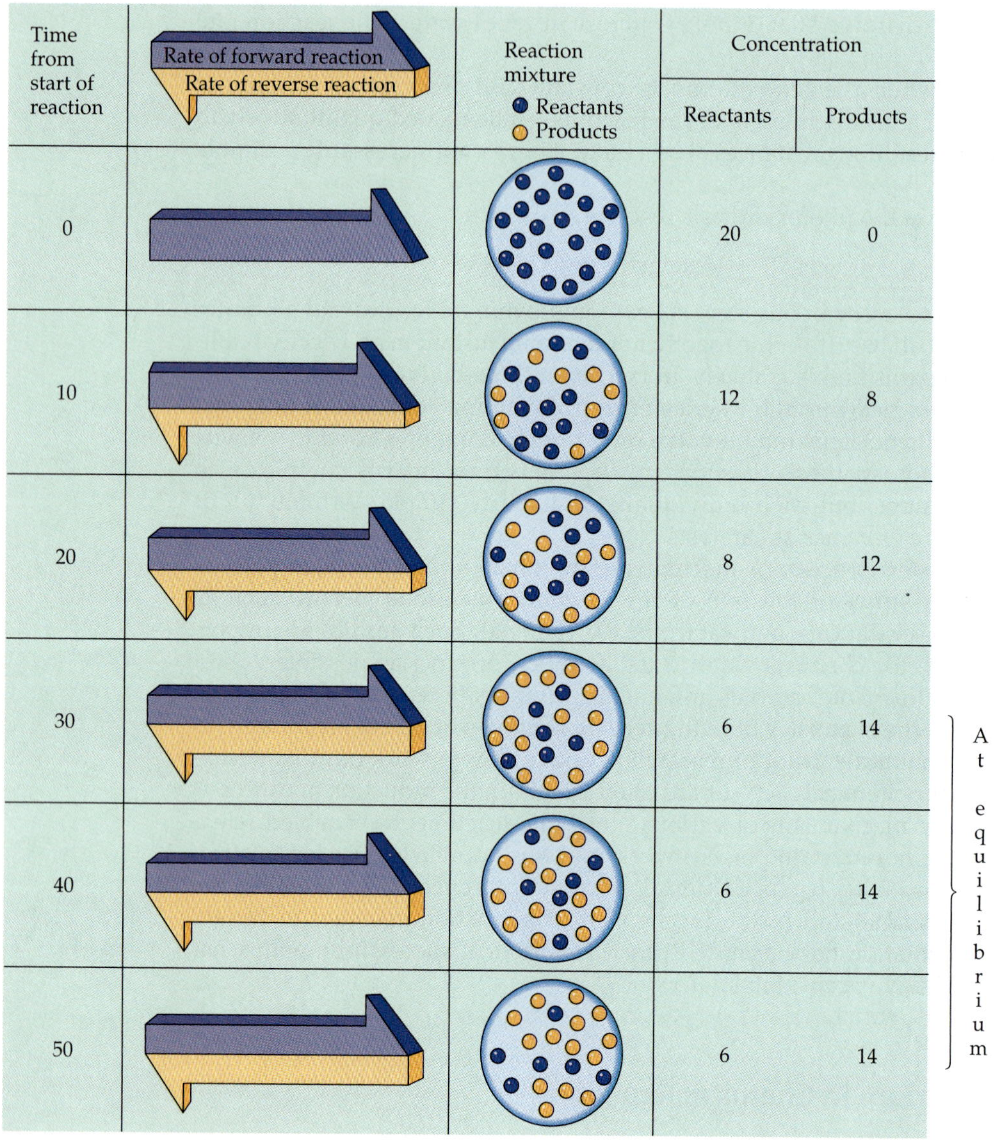

▲ **FIGURE 5.17**

Progress of a reaction toward equilibrium.

even though their concentrations don't change, we say that equilibrium is a dynamic situation.

Equilibrium is established for reversible reactions in isolated systems, those for which external conditions such as temperature and pressure do not change. What happens if those external conditions do change? For the answer to this question, we shall once again use as an example the reaction of nitrogen and hydrogen to form ammonia.

$$N_2(g) \; + \; 3\,H_2(g) \; \rightleftharpoons \; 2\,NH_3(g) \; + \; heat$$

The double arrow indicates that this is a reversible reaction. All of the compounds involved in this reaction are gases. From the work of Gay-Lussac and Avogadro, we know that the coefficients given in the equation reflect the combining volumes of the gases. The reactants occupy four volumes and the products occupy two. In Chapter 7, we study the effects of temperature and pressure on the volume of a gas. For the moment, let us just say that increased pressure tends to reduce the volume of a gas. If the pressure of the equilibrium system we are considering is increased, the equilibrium will be disrupted, and the rate at which N_2 reacts with H_2 to form NH_3 will increase; that is, the rate of the forward reaction will increase (Figure 5.18). Increasing the rate of the forward reaction has the effect of changing H_2 and N_2, which occupy four volumes, to NH_3, which occupies only two volumes. The increased pressure therefore has the effect of changing reactants that occupy four volumes to products that occupy only two. Thus, the total volume of the system will decrease. If the pressure is held constant at the new, higher value, equilibrium will again be established. But in the new equilibrium the concentration of ammonia (NH_3) will be higher than it was before the pressure changed. We sometimes describe this change by saying that the equilibrium has been shifted to the right. Thus, for the reversible reaction of nitrogen and hydrogen to form ammonia, an increase in pressure shifts the equilibrium to the right.

Observations of changes that occur in an equilibrium system when factors such as concentration, pressure, and temperature are changed were made in the nineteenth century. These effects were summarized in 1884 by a French chemist, Henri Louis Le Châtelier. His rule, still called **Le Châtelier's principle,** can be stated as follows: If a stress is applied to a system in equilibrium, the system rearranges in such a direction as to minimize the stress. If heat is added to the $N_2—H_2—NH_3$ system, the reaction will proceed to the left, using up heat. If more nitrogen gas is added to the system, the reaction will go to the right, using up nitrogen. If additional hydrogen gas is introduced, the system will shift to the right, to use up hydrogen. Additional ammonia will cause the reaction to proceed to the left, using up ammonia.

The equilibrium can also be shifted by removal of one of the substances. Removal of ammonia will cause hydrogen and nitrogen to react, forming more ammonia. Removal of hydrogen will cause ammonia to break down and form more hydrogen (and, incidentally, more nitrogen). It should be noted, however, that a catalyst will not shift an equilibrium system. Catalysts change the rate of

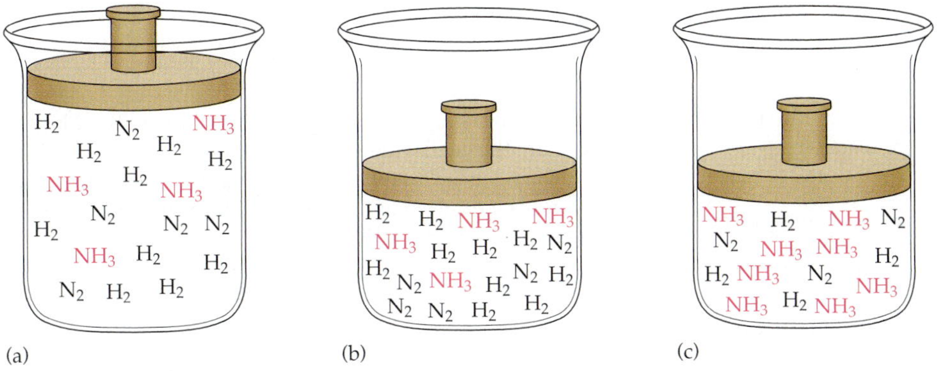

▲ **FIGURE 5.18**

Le Châtelier's principle illustrated. (a) System at equilibrium with 10 H_2, 5 N_2, and 4 NH_3, a total of 19 molecules. (b) The same molecules forced into a smaller volume, creating a stress on the system. (c) Six H_2 and 2 N_2 have been converted to 4 NH_3. A new equilibrium has been established with 4 H_2, 3 N_2, and 8 NH_3, a total of 15 molecules. The stress imposed by the smaller volume is partially relieved by the reduction in the total number of molecules.

both the forward and reverse reactions. They do not change the position of the equilibrium, and the equilibrium concentrations of reactants and products are not altered.

Example 5.14

What effect, if any, will each of the following changes have on the equilibrium in the reaction

$$2\,CO(g) + O_2(g) \rightleftharpoons 2\,CO_2(g) + \text{heat}$$

a. adding CO
b. removing O_2
c. cooling the reaction mixture
d. increasing the pressure
e. adding a catalyst

Solution

a. The reaction shifts to the right to use up the added CO.
b. The reaction shifts to the left to replace the O_2 that is removed.
c. The reaction shifts to the right to replace the lost heat.
d. The reaction shifts to the right to relieve the pressure by converting three molecules ($2\,CO + 1\,O_2$) into two molecules ($2\,CO_2$).
e. A catalyst has no effect on the position of equilibrium.

Practice Exercise

How will the addition of $Cl_2(g)$ to the system below (isolated, constant temperature, and at equilibrium) affect the equilibrium concentration of CO?

$$CO(g) + Cl_2(g) \rightleftharpoons COCl_2(g)$$

In Chapter 8, we discuss equilibria between liquid and solid phases and between liquid and vapor phases. In Chapter 9, we encounter equilibria involving solutions, and in Special Topic C we illustrate some equilibrium calculations. The concept of equilibrium is quite important to our study of the chemistry of life.

Summary

A **chemical equation** is a shorthand way of showing the changes that occur in any chemical reaction:

$$\text{reactant } 1(g) + \text{reactant } 2(l) + \ldots \longrightarrow \text{product } 1(s) + \text{product } 2(g) + \ldots \longrightarrow$$

where the parenthetical letters indicate physical state: (g) = gaseous state, (l) = liquid state, (s) = solid state.

Every chemical equation must be **balanced,** which means that, for every element shown in the equation, the number of atoms on the left must equal the number on the right.

Example:

$3\,H_2$	+	N_2	$\longrightarrow$	$2\,NH_3$
6 H atoms		2 N atoms		2 N atoms 6 H atoms

The 3 before H_2, the (implicit) 1 before N_2, and the 2 before NH_3 are the **coefficients** of the equation.

Gay-Lussac's **law of combining volumes** states that, for gases measured at the same temperature and pressure, the numbers in the ratio

$$\text{vol}_{\text{reactant1}} : \text{vol}_{\text{reactant2}} : \text{vol}\text{---} : \text{vol}_{\text{product1}} : \text{vol}_{\text{product2}} : \text{vol}\text{---}$$

are always small, whole numbers. Example: the formation of two volumes of $NH_3(g)$ always involves the combining of three volumes of $H_2(g)$ with one volume of $N_2(g)$.

Avogadro reasoned that Gay-Lussac's law is true because gas volume is a macroscopic indicator of number of gas molecules present. Gay-Lussac always measured three volumes of H_2 plus one volume of N_2 yielding two volumes of NH_3 because every N_2 molecule must combine with three H_2 molecules in order to form two molecules of NH_3. That the volume of H_2 gas must always be three times the volume of N_2 gas led to **Avogadro's hypothesis:** at the same temperature and pressure, equal volumes of any two gases contain equal numbers of gas molecules.

Avogadro could not count the molecules in a given volume. All he could do was reason that equal volumes contain equal numbers of molecules. Scientists since his time, however, have determined experimentally that, **x** grams of any element, where **x** is the atomic weight of the element listed in the periodic table, contains 6.02×10^{23} atoms of the element. Example:

atomic weight of H is 1.0 u,
therefore 1.0 g of H contains 6.02×10^{23} H atoms
atomic weight of C is 12.0 u,
therefore 12.0 g of C contains 6.02×10^{23} C atoms
atomic weight of O is 16.0 u,
therefore 16.0 g of O contains 6.02×10^{23} O atoms

This number 6.02×10^{23} is called **Avogadro's number.**

The **mole** is the basic SI unit for amount of substance. It is defined as the amount of substance containing the number of basic units equal to the number of atoms in exactly 12 g of carbon-12. Because there are 6.02×10^{23} atoms in 12 g of carbon-12, there are 6.02×10^{23} basic units in 1 mol of any substance:

1 mol chemistry books = 6.02×10^{23} chemistry books
1 mol CO_2 = 6.02×10^{23} CO_2 molecules
1 mol NaCl = 6.02×10^{23} NaCl formula units

The **molar mass** of any substance is the mass in grams of 1 mol of that substance. The molar mass of any substance is calculated from the atomic weights of the atoms in its formula. To do so, we use the fact that, for any element, the atomic weight in grams contains Avogadro's number of atoms. Example:

1 mol CO_2 is 6.02×10^{23} CO_2 molecules, which is made up of

	6.02×10^{23} C atoms		= 12 g C
plus	$2(6.02 \times 10^{23})$ O atoms	2(16)	= 32 g O
	molar mass of CO_2		44 g/mol

1 mol sucrose, $C_{12}H_{22}O_{11}$ is 6.02×10^{23} sucrose molecules, which is made up of

	$12(6.02 \times 10^{23})$ C atoms	12(12)	= 144 g C
plus	$22(6.02 \times 10^{23})$ H atoms	22(1)	= 22 g H
plus	$11(6.02 \times 10^{23})$ O atoms	11(16)	= 176 g O
	molar mass of sucrose		342 g/mol

The coefficients in chemical equations represent moles as well as atoms and molecules. The equation $3\ H_2 + N_2 \rightarrow 2\ NH_3$, for instance, tells us that three moles of H_2 combines with one mole of N_2 to form two moles of NH_3. Knowing the numbers of moles, we can calculate amounts of reactants and products:

Molar mass of H_2 is 2 g/mol; 1 mol has a mass of 2 g,
3 mol has a mass of 6 g
molar mass of N_2 is 28 g/mol; 1 mol has a mass of 28 g
molar mass of NH_3 is 17 g/mol; 1 mol has a mass of 17 g,
2 mol has a mass of 34 g

$$6\text{ g }H_2 + 28\text{ g }N_2 \longrightarrow 34\text{ g }NH_3$$

Chemical reactions that give off heat are **exothermic reactions**; those that require an input of heat in order to proceed are **endothermic reactions.** The **energy of activation** of any chemical reaction is the amount of energy needed to get the reaction started. A reaction can be exothermic and still require some initial input of activation energy, as exemplified by the burning of coal. The only requirement for a reaction to be exothermic is that the amount of heat given off must exceed the amount of activation energy added.

The rate at which a chemical reaction proceeds depends on (1) frequency of collisions between reactants, (2) how colliding reactants are aligned with each other, and (3) how much energy each collision produces. Because collision frequency is directly proportional to reactant concentrations and to temperature, reaction rate is also directly proportional to these two variables.

A **catalyst** is any chemical substance that increases the rate of a chemical reaction but is in no way changed by the reaction. A catalyst works by lowering the activation energy needed to get a reaction going.

A **reversible reaction** is one that can proceed in either direction. The reaction reactants $\longrightarrow$ products is called the forward direction, and the reaction products $\longrightarrow$ reactants is called the reverse direction. At first, a reversible reaction is mostly forward because concentration determines reaction rate. High concentrations of reactants means the forward reaction proceeds; low product concentrations means that the reverse reaction does not happen to any measurable extent. As more and more reactant is used up (and so more and more product formed), the forward reaction slows down and the reverse reaction speeds up. At the point where $\text{rate}_{\text{forward}}$ equals $\text{rate}_{\text{reverse}}$ **equilibrium** is established. At this point, there are no changes in product concentration or in reaction concentration. Both the forward reaction and the reverse reaction are still taking place. However, every reactant molecule consumed in the forward reaction is replaced by a reactant molecule produced in the reverse reaction. Every product molecule produced in the forward reaction takes the place of a product molecule used up in the reverse reaction.

Once equilibrium is established, it continues until some external agent changes the reaction conditions. **Le Châtelier's** principle lets us predict how any such change affects equilibrium: Any stress put on an equilibrium system causes the system to react in such a way as to reduce the stress. Here "stress" means a change in product or reactant concentration, a change in temperature, or a change in pressure. Example:

equilibrium system	$3\ H_2 + N_2 \rightleftharpoons NH_3$
stress	addition of H_2
result	forward reaction speeds up in order to consume more H_2; equilibrium shifted to right
stress	removal of H_2
result	reverse reaction speeds up in order to produce more H_2; equilibrium shifted to left

Key Terms

Avogadro's hypothesis (5.2)
Avogadro's number (5.3)
catalyst (5.11)
chemical equation (5.1)
endothermic (5.7)
energy of activation (5.8)
enzyme (5.11)
equilibrium (5.13)
exothermic (5.7)
formula unit (5.4)
formula weight (5.4)
law of combining volumes (5.2)
Le Châtelier's principle (5.13)
mechanism (5.12)
molar mass (5.5)
mole (5.5)
molecular weight (5.4)
photosynthesis (5.8)
product (5.1)
reactant (5.1)
reversible reaction (5.8)
stoichiometric factor (5.6)

Review Questions

1. Define or illustrate
   a. endothermic
   b. exothermic
   c. energy of activation
   d. catalyst
   e. reversible reaction
   f. equilibrium
2. Explain the difference between the *atomic weight* of nitrogen and the *molecular weight* of nitrogen. Explain how each is determined from data in the periodic table.
3. What is Avogadro's number and how is it related to the quantity called one mole?
4. What is Avogadro's hypothesis? How did it explain Gay-Lussac's law of combining volumes?
5. What are the molecular weight and the molar mass of carbon dioxide? Explain how each is determined from the formula, CO_2.
6. How do the law of combining volumes and Avogadro's hypothesis indicate that hydrogen gas is composed of diatomic molecules rather than individual atoms?
7. How many oxygen molecules are there in 1.00 mol of O_2? How many oxygen atoms are there in 1.00 mol of O_2?
8. How many calcium ions and how many chloride ions are there in 1.00 mol of $CaCl_2$?
9. What is the purpose of balancing a chemical equation?
10. Explain the meaning of the equation
$$CH_4(g) + 2\ O_2(g) \longrightarrow CO_2(g) + 2\ H_2O(g)$$
at the molecular level. Interpret the equation in terms of moles. State the mass relationships conveyed by the equation.
11. Translate the chemical equations into words.
    a. $2\ H_2(g) + O_2(g) \longrightarrow 2\ H_2O(g)$
    b. $2\ KClO_3(s) \longrightarrow 2\ KCl(s) + 3\ O_2(g)$
    c. $2\ Al(s) + 6\ HCl(aq) \longrightarrow 2\ AlCl_3(aq) + 3\ H_2(g)$
12. Refer to the following reaction diagram.

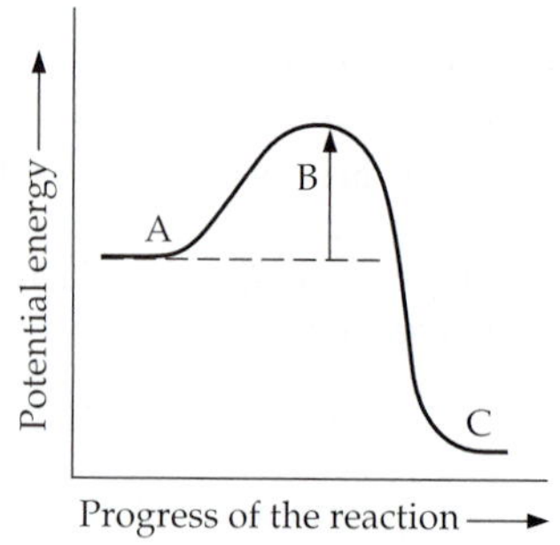

    a. Which letter in the diagram refers to the products?
    b. Which letter refers to the activation energy?
    c. Which letter refers to the reactants?
13. Is the reaction diagrammed in Question 12 endothermic or exothermic? Explain.

14. If the reaction diagrammed in Question 12 were reversible, would the reverse reaction be endothermic or exothermic? Explain.
15. If all other conditions are held constant, what is the effect of increasing the temperature on the rate of a reaction?
16. If all other conditions are held constant, what is the effect of increasing the concentration of a reactant on the rate of a reaction?
17. If all other conditions are held constant, what is the effect of adding a catalyst on the rate of a reaction?
18. Explain how the orientation of reactant molecules affects the rate of a reaction.
19. What is meant by the *mechanism* of a reaction?
20. Why does a catalyst affect the rate of a reaction?

Problems

Molar Volume

21. What is the molar volume of each of the gases?
    a. He b. H_2 c. C_2H_6
22. What is the mass of a molar volume of each of the gases in Problem 21?

Molecular Formulas and Formula Units

23. How many hydrogen atoms are indicated in one formula unit of
    a. NH_4NO_3 b. CH_3OH
    c. $CH_3CH_2CH_3$ d. C_6H_5COOH
24. How many hydrogen atoms are indicated in one formula unit of
    a. $(NH_4)_2HPO_4$ b. $Al(C_2H_3O_2)_3$
25. How many oxygen atoms are indicated in one formula unit of
    a. $Al_2(C_2O_4)_3$ b. $Ca_3(PO_4)_2$
26. How many atoms of each kind (Al, C, H, and O) are indicated by the notation 2 $Al(C_2H_3O_2)_3$?

Balancing Chemical Equations

27. Indicate whether the following equations are balanced. (You need not balance the equation; just determine whether it is balanced as written.)
    a. $Mg + H_2O \longrightarrow MgO + H_2$
    b. $FeCl_2 + Cl_2 \longrightarrow FeCl_3$
    c. $F_2 + H_2O \longrightarrow 2\,HF + O_2$
    d. $Ca + 2\,H_2O \longrightarrow Ca(OH)_2 + H_2$
    e. $2\,LiOH + CO_2 \longrightarrow Li_2CO_3 + H_2O$
28. Indicate whether the following equations are balanced as written.
    a. $2\,KNO_3 + 10\,K \longrightarrow 6\,K_2O + N_2$
    b. $2\,NH_3 + O_2 \longrightarrow N_2 + 3\,H_2O$
    c. $4\,LiH + AlCl_3 \longrightarrow 2\,LiAlH_4 + 2\,LiCl$
    d. $SF_4 + 3\,H_2O \longrightarrow H_2SO_3 + 4\,HF$
    e. $4\,BF_3 + 3\,H_2O \longrightarrow H_3BO_3 + 3\,HBF_4$
29. Balance the following equations.
    a. $Cl_2O_5 + H_2O \longrightarrow HClO_3$
    b. $V_2O_5 + H_2 \longrightarrow V_2O_3 + H_2O$
    c. $Al + O_2 \longrightarrow Al_2O_3$
    d. $Sn + NaOH \longrightarrow Na_2SnO_2 + H_2$
    e. $PCl_5 + H_2O \longrightarrow H_3PO_4 + HCl$
30. Balance the following equations.
    a. $TiCl_4 + H_2O \longrightarrow TiO_2 + HCl$
    b. $C_4H_{10} + O_2 \longrightarrow CO_2 + H_2O$
    c. $WO_3 + H_2 \longrightarrow W + H_2O$
    d. $Al_4C_3 + H_2O \longrightarrow Al(OH)_3 + CH_4$
    e. $Al_2(SO_4)_3 + NaOH \longrightarrow Al(OH)_3 + Na_2SO_4$
31. Balance the following equations.
    a. $Na_3P + H_2O \longrightarrow NaOH + PH_3$
    b. $Cl_2O + H_2O \longrightarrow HClO$
    c. $CH_3OH + O_2 \longrightarrow CO_2 + H_2O$
    d. $Zn(OH)_2 + H_3PO_4 \longrightarrow Zn_3(PO_4)_2 + H_2O$
    e. $C_3H_8 + O_2 \longrightarrow CO_2 + H_2O$
32. Balance the following equations.
    a. $Ca_3P_2 + H_2O \longrightarrow Ca(OH)_2 + PH_3$
    b. $Cl_2O_7 + H_2O \longrightarrow HClO_4$
    c. $MnO_2 + HCl \longrightarrow MnCl_2 + Cl_2 + H_2O$
    d. $Fe + O_2 \longrightarrow Fe_3O_4$
    e. $C_5H_{12} + O_2 \longrightarrow CO_2 + H_2O$

Molecular Weights and Formula Weights

33. Calculate the molecular weight or formula weight of
    a. C_6H_5Br b. H_3PO_4 c. $K_2Cr_2O_7$
34. Calculate the molecular weight or formula weight of
    a. $C_2H_5NO_2$ b. $Na_2S_2O_3$ c. $(NH_4)_3PO_4$

Molar Masses

35. Calculate the mass, in grams, of
    a. 0.00500 mol MnO_2 b. 1.12 mol CaH_2
    c. 0.250 mol $C_6H_{12}O_6$
36. Calculate the mass, in grams, of
    a. 4.61 mol $AlCl_3$ b. 0.615 mol Cr_2O_3
    c. 0.158 mol IF_5
37. Calculate the amount, in moles, of
    a. 98.6 g HNO_3 b. 9.45 g CBr_4
    c. 9.11 g $FeSO_4$ d. 11.8 g $Pb(NO_3)_2$
38. Calculate the amount, in moles, of
    a. 16.3 g SF_6 b. 25.4 g $Pb(C_2H_3O_2)$
    c. 35.6 g $FeCl_3$ d. 75.3 g $Co(ClO_3)_2$

Volume Relationships in Chemical Equations

39. Consider the following equation.

$$2\,C_4H_{10}(g) + 13\,O_2(g) \longrightarrow 8\,CO_2(g) + 10\,H_2O(g)$$

a. How many liters of $H_2O(g)$ are formed when 0.529 L of $C_4H_{10}(g)$ is burned? Assume both gases are measured under the same conditions.

b. How many liters of $O_2(g)$ are required to burn 16.1 L of $C_4H_{10}(g)$? Assume both gases are measured under the same conditions.

40. Consider the following equation.

$$C_2H_4(g) + 3\,O_2(g) \longrightarrow 2\,CO_2(g) + 2\,H_2O(g)$$

a. How many liters of $CO_2(g)$ are formed when 2.93 L of $C_2H_4(g)$ is burned? Assume both gases are measured under the same conditions.

b. How many liters of $O_2(g)$ are required to form 0.370 L of $CO_2(g)$? Assume both gases are measured under the same conditions.

Mole Relationships in Chemical Equations

41. Consider the reaction for the combustion of octane.

$$2\,C_8H_{18} + 25\,O_2 \longrightarrow 16\,CO_2 + 18\,H_2O$$

a. How many moles of CO_2 are produced when 2.09 mol of octane is burned?

b. How many moles of oxygen are required to burn 4.47 mol of octane?

42. Consider the reaction for the combustion of octane.

$$2\,C_8H_{18} + 25\,O_2 \longrightarrow 16\,CO_2 + 18\,H_2O$$

a. How many moles of H_2O are produced when 2.81 mol of octane is burned?

b. How many moles of CO_2 are produced when 4.06 mol of oxygen is consumed?

Mass Relationships in Chemical Equations

43. What mass of ammonia, in grams, can be made from 440 g of H_2?

$$N_2 + H_2 \longrightarrow NH_3 \ (\textit{not balanced})$$

44. What mass of hydrogen, in grams, is needed to react completely with 892 g of N_2?

$$N_2 + H_2 \longrightarrow NH_3 \ (\textit{not balanced})$$

45. What mass of oxygen, in grams, can be prepared from 24.0 g of H_2O_2?

$$H_2O_2 \longrightarrow H_2O + O_2 \ (\textit{not balanced})$$

46. Toluene and nitric acid are used in the production of trinitrotoluene (TNT), an explosive.

$$\underset{\text{toluene}}{C_7H_8} + HNO_3 \longrightarrow \underset{\text{TNT}}{C_7H_5N_3O_6} + H_2O \ (\textit{not balanced})$$

What mass of nitric acid, in grams, is required to react with 454 g of C_7H_8?

47. Use the equation in Problem 46 to calculate the mass of TNT that can be made from 829 g of C_7H_8.

48. What mass of quicklime (calcium oxide), in kilograms, can be made when 4.72×10^9 g of limestone (calcium carbonate) is decomposed by heating?

$$CaCO_3(s) \longrightarrow CaO(s) + CO_2(g)$$

49. What mass of nitric acid, in grams, can be made from 971 g of ammonia?

$$NH_3 + O_2 \longrightarrow HNO_3 + H_2O \ (\textit{not balanced})$$

50. In an oxyacetylene welding torch, acetylene (C_2H_2) burns in pure oxygen with a very hot flame.

$$C_2H_2 + O_2 \longrightarrow CO_2 + H_2O \ (\textit{not balanced})$$

What mass of oxygen, in grams, is required to react with 52.0 g of C_2H_2?

Le Châtelier's Principle

51. According to Le Châtelier's principle, what effect will increasing the temperature have on the following equilibria?

a. $H_2(g) + Cl_2(g) \rightleftharpoons 2\,HCl(g) + \text{heat}$

b. $2\,CO_2(g) + \text{heat} \rightleftharpoons 2\,CO(g) + O_2(g)$

c. $3\,O_2(g) + \text{heat} \rightleftharpoons 2\,O_3(g)$

52. According to Le Châtelier's principle, what effect will decreasing the concentration of $O_2(g)$ have on the following equilibria?

a. $3\,O_2(g) \rightleftharpoons 2\,O_3(g)$

b. $2\,CO_2(g) \rightleftharpoons 2\,CO(g) + O_2(g)$

c. $2\,NO(g) + O_2(g) \rightleftharpoons 2\,NO_2(g)$

Additional Problems

53. Aluminum chloride, used as a topical astringent (a substance that causes contraction of pores and thus retards perspiration), can be made by the reaction of aluminum metal with hydrogen chloride gas. The equation (not balanced) is

$$Al(s) + HCl(g) \longrightarrow AlCl_3(s) + H_2(g)$$

How much $AlCl_3$ can be made from an aluminum beverage can that has a mass of 3.51 g?

54. Indicate whether the following equations are balanced as written.

a. $2\,Sn + 2\,H_2SO_4 \longrightarrow 2\,SnSO_4 + SO_2 + 2\,H_2O$

b. $3\,Cl_2 + 6\,NaOH \longrightarrow 5\,NaCl + NaClO_3 + 3\,H_2O$

55. Propane burns in air to form carbon dioxide and water. The equation is

$$C_3H_8 + 5\,O_2 \longrightarrow 3\,CO_2 + 4\,H_2O$$

State the molecular, molar, and mass relationships indicated by this equation.

56. The poison gas phosgene reacts with water in the lungs to form hydrogen chloride and carbon dioxide. The equation is

$$COCl_2 + H_2O \longrightarrow 2\,HCl + CO_2$$

State the molecular, molar, and mass relationships indicated by this equation.

57. At temperatures above 300 °C, silver oxide, Ag_2O, decomposes to produce metallic silver and oxygen gas. A 2.95-g sample of *impure* silver oxide yields 0.183 g of oxygen. Assuming that $Ag_2O(s)$ is the only source of oxygen, what is the mass percent of Ag_2O in the sample?

58. A piece of aluminum foil that measures 12.3 cm × 14.3 cm × 2.2 mm reacts with an excess of hydrochloric acid. What mass of hydrogen is produced? The density of aluminum is 2.70 g/cm³.

$$Al(s) + HCl(aq) \longrightarrow AlCl_3(aq) + H_2(g) \text{ (not balanced)}$$

59. A coal-burning power plant burns 228 trainloads of western subbituminous coal per year. Each train is comprised of 115 cars, and each car carries 90.5 metric tons of coal (1 metric ton = 1000 kg). If the coal is 64.3% carbon, what mass of carbon dioxide, in metric tons, is produced by the plant each year?

60. The following is a side reaction in the manufacture of rayon fibers from wood pulp.

$$3\,CS_2 + 6\,NaOH \longrightarrow 2\,Na_2CS_3 + Na_2CO_3 + 3\,H_2O$$

How many grams of Na_2CS_3 is produced in the reaction of 88.0 mL of liquid CS_2 (density = 1.26 g/mL)?

61. Aspartame (see Table 19.1) is an artificial sweetener about 160 times as sweet as sucrose. Aspartame breaks down in acidic solution to produce (among other products) methanol, which is fairly toxic. The equation is

$$\underset{\text{Aspartame}}{C_{14}H_{18}N_2O_5} + 2\,H_2O \xrightarrow{H^+} C_4H_7NO_4 + C_9H_{11}NO_2 + \underset{\text{Methanol}}{CH_4O}$$

a. A typical can of soda contains about 40 g of sucrose. How much aspartame is required to obtain the same level of sweetness?

b. How much methanol is formed by the complete breakdown of that much aspartame?

c. How many cans of soda with decomposed aspartame would you have to drink to get the approximate lethal dose of 25 g of methanol?

62. Joseph Priestley discovered oxygen in 1774 by heating "red calx of mercury," mercury(II) oxide. The calx decomposed to the elements. The equation (not balanced) is

$$HgO \longrightarrow Hg + O_2$$

How much oxygen is produced by the decomposition of 10.8 g of HgO?

63. How much iron can be converted to the magnetic oxide of iron (Fe_3O_4) by 8.80 g of pure oxygen? The equation (not balanced) is

$$Fe + O_2 \longrightarrow Fe_3O_4$$

64. Laughing gas (dinitrogen monoxide, N_2O, also called nitrous oxide) can be made by heating ammonium nitrate with great care. The equation (not balanced) is

$$NH_4NO_3 \longrightarrow N_2O + H_2O$$

How much N_2O can be made from 4.00 g of ammonium nitrate?

65. Small amounts of hydrogen gas often are made by the reaction of calcium metal with water. The equation (not balanced) is

$$Ca + H_2O \longrightarrow Ca(OH)_2 + H_2$$

How many grams of hydrogen are formed by the reaction of water with 0.413 g of calcium?

66. For many years the noble gases were called the "inert gases" because it was thought that they formed no chemical compounds. Neil Bartlett made the first noble gas compound in 1962. Xenon hexafluoride is made according to the equation (not balanced)

$$Xe + F_2 \longrightarrow XeF_6$$

How many grams of fluorine are required to make 0.112 g of XeF_6?

67. Phosphine gas (used as a fumigant to protect stored grain) is generated by the reaction of water with magnesium phosphide. The equation (not balanced) is

$$Mg_3P_2 + H_2O \longrightarrow PH_3 + Mg(OH)_2$$

How much magnesium phosphide is needed to produce 134 g of PH_3?

6 Oxidation and Reduction

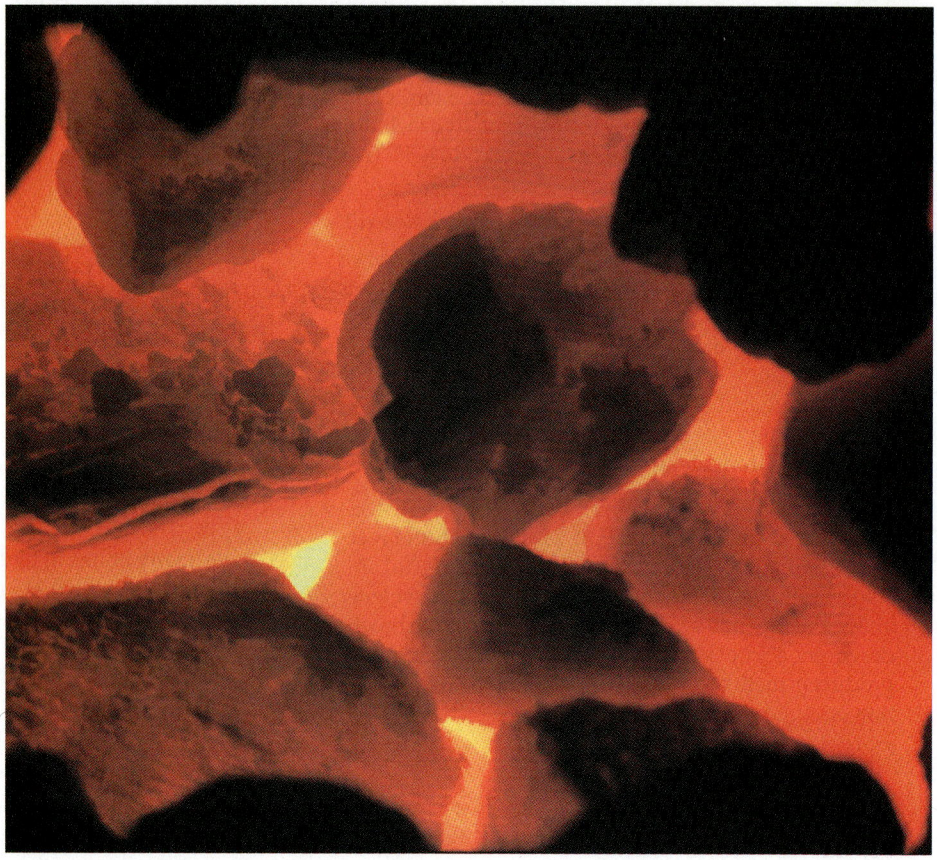

When coal burns, releasing its chemical energy as heat, the carbon in the coal is oxidized to carbon dioxide as oxygen from the air is reduced to water.

Chemical reactions can be classified in several ways. In this chapter, we consider an important group of chemical reactions called *oxidation–reduction reactions.* The two processes—oxidation and reduction—always occur together. You can't have one without the other. When one substance is oxidized, another is reduced (Figure 6.1). For convenience, however, we may choose to talk about only a part of the process—the oxidation part or the reduction part.

Chemists often refer to oxidation–reduction reactions by turning the words around and abbreviating the term as "redox" reactions; redox is a little easier to say.

Our cells obtain energy to maintain themselves by oxidizing foods. Green plants, using energy from sunlight, produce food by the reduction of carbon dioxide (photosynthesis—see Section 6.8). We win metals from their ores by reduction, then lose them again to corrosion as they are oxidized. We maintain our technological civilization by oxidizing fossil fuels (coal, natural gas, and petroleum) to obtain the chemical energy that was stored in these materials eons ago by green plants.

Reduced forms of matter—food, coal, gasoline—are high in energy (Figure 6.2). Oxidized forms—carbon dioxide and water—are low in energy. Let's

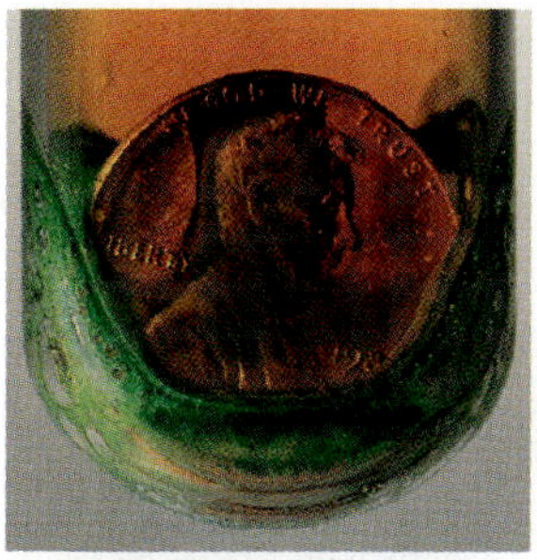

▲ **FIGURE 6.1**
Oxidation and reduction always occur together. In this photograph, a copper penny reacts with nitric acid. The copper metal is oxidized to copper(II) ions (Cu^{2+}), producing a green-blue solution. The nitric acid, HNO_3, is reduced to NO_2 gas, as indicated by the red-brown fumes.

(a) (b) (c)

▲ **FIGURE 6.2**
Some reduced forms of matter. The energy in food (a) and fossil fuels (b) and (c) is released when these materials are oxidized.

examine the processes of oxidation and reduction in some detail, in order that we might better understand the chemical reactions that keep us alive and enable us to maintain our civilization.

6.1 Oxygen: Abundant and Essential

Oxygen is surely one of the most important elements on Earth. It is one of the two dozen or so elements essential to life. Oxygen occurs in the atmosphere as the uncombined free element in the form of O_2 molecules. Overall, it makes up about 21% by volume of the air we breathe. In addition, oxygen forms compounds with nearly all the other elements. Combined with hydrogen in the remarkable compound water, oxygen makes up about 89% by mass of the oceans, seas, rivers, and lakes that cover three-quarters of Earth's surface. Oxygen makes up 45.5% by

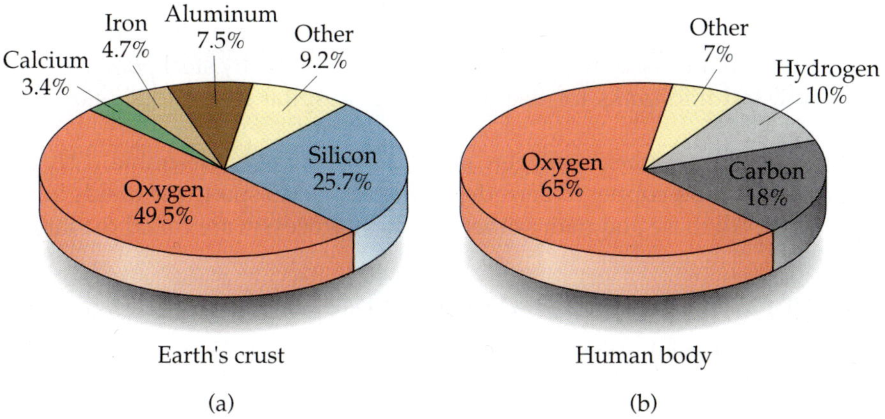

◀ **FIGURE 6.3**
Percent by mass of elements in (a) Earth's crust (including the atmosphere and the oceans, lakes, and streams) and (b) the human body.

▲ **FIGURE 6.4**
Cooking, breathing, and burning fuel all involve oxidation.

mass of Earth's solid crust, where it is combined with silicon (sand is mainly SiO_2) and many other elements.

Oxygen is found in most of the compounds that are important to living organisms. Foodstuffs—carbohydrates, fats, and proteins—all contain oxygen. The human body is approximately 65% water (by mass). Because water is 89% oxygen by mass, and many other compounds in your body also contain oxygen, about two-thirds of your body mass is oxygen (Figure 6.3).

The atmosphere is about 21% oxygen by volume. [The rest is mainly nitrogen (N_2), which is rather unreactive.] The oxygen in the air is taken into our lungs, passes into our bloodstreams, is carried to our body tissues, and combines with the food we eat. This is the process that provides us with all our energy. Fuels such as wood, gasoline, and coal also need oxygen to burn and release their stored energy (Figure 6.4). Combustion of such fossil fuels currently supplies about 86% of the energy that turns the wheels of civilization.

Pure oxygen is obtained by liquefying air and then letting the nitrogen and argon boil off. (Nitrogen boils at −196 °C, argon at −186 °C, and oxygen at −183 °C.) Nearly 50 billion pounds of oxygen is produced annually in the United States, but most of it is used directly by industry, much of it by steel plants. About 1% is compressed into tanks for use in welding, hospital respirators, and other purposes.

Not all that atmospheric oxygen does is immediately desirable. Oxygen causes iron to rust and copper to corrode. It causes food to spoil, and it aids in the decay of wood. All these and many other chemical processes are called oxidation.

6.2 Chemical Properties of Oxygen: Oxidation

When iron rusts, it combines with oxygen from the air to form a reddish brown powder.

$$4\,Fe + 3\,O_2 \longrightarrow 2\,Fe_2O_3$$

The chemical name for rust is iron(III) oxide (Fe_2O_3). Sometimes it is called by an older name, ferric oxide. Many metals react with oxygen to form metal oxides.

Most nonmetals also react with oxygen to form oxides. For example, carbon and sulfur burn in oxygen to form carbon dioxide and sulfur dioxide, respectively.

$$C + O_2 \longrightarrow CO_2$$

$$S + O_2 \longrightarrow SO_2$$

Example 6.1

Magnesium combines readily with oxygen when ignited in air. Write the equation for the reaction.

Solution

Magnesium is an element with the symbol Mg. Oxygen occurs as diatomic molecules (O_2). The two react to form MgO. The reaction is

$$Mg + O_2 \longrightarrow MgO\ (\textit{not balanced})$$

To balance the equation, we need two oxygen atoms (2 MgO) on the right and then 2 Mg on the left.

$$2\,Mg + O_2 \longrightarrow 2\,MgO$$

Practice Exercise

When ignited, zinc burns in air with a bluish-green flame to form zinc oxide (ZnO). Write the equation for the reaction.

Oxygen also reacts with many compounds. Methane, the principal ingredient in natural gas, burns in air to produce carbon dioxide and water.

$$CH_4 + 2\,O_2 \longrightarrow CO_2 + 2\,H_2O$$

Hydrogen sulfide, a gaseous compound with a rotten-egg odor, burns, producing water and sulfur dioxide.

$$2\,H_2S + 3\,O_2 \longrightarrow 2\,H_2O + 2\,SO_2$$

In each example, oxygen combines with each of the elements in the compound to form their oxides.

Example 6.2

Carbon disulfide is highly flammable; both elements combine with oxygen. What products are formed? Write the equation.

Solution

Carbon disulfide is CS_2. The products are carbon dioxide and sulfur dioxide. The balanced equation is

$$CS_2 + 3\,O_2 \longrightarrow CO_2 + 2\,SO_2$$

Practice Exercise

When heated in air, lead sulfide (PbS) combines with oxygen to form lead(II) oxide (PbO) and sulfur dioxide. Write a balanced equation for this reaction.

The clouds of gas in interstellar space are mostly hydrogen, by far the most abundant element in the universe.

6.3 Hydrogen: A Reactive Lightweight Element

Hydrogen is another important element. It is combined with oxygen in the vital compound water. By mass, hydrogen makes up only about 0.9% of Earth's crust (including the atmosphere and oceans). However, because hydrogen is the lightest of all the elements, hydrogen atoms are quite abundant. In numbers of *atoms* in Earth's crust, hydrogen ranks third (15.1%), after oxygen (53.3%) and silicon (15.9%). If we look beyond our home planet, it has been estimated that hydrogen atoms make up 89% of the atoms of the sun, and that 85 to 95% of the atoms in the atmospheres of the outer planets (Jupiter, Uranus, Saturn, and Neptune) are hydrogen atoms. In the universe as a whole, about 90% of all atoms are hydrogen—and the rest are mainly helium.

Unlike oxygen, hydrogen is seldom found in a free uncombined state on Earth. Most of it is combined with oxygen in water. Some is combined with carbon in petroleum and natural gas, which are mixtures of *hydrocarbons.* Nearly all compounds derived from plants and animals contain combined hydrogen.

Small amounts of elemental hydrogen can be made for laboratory use by reacting zinc with hydrochloric acid.

$$Zn(s) \quad + \quad 2\,HCl(aq) \longrightarrow ZnCl_2(aq) \quad + \quad H_2(g)$$

Because hydrogen does not dissolve in water, it can be collected by water displacement (Figure 6.5). Commercial quantities of hydrogen are obtained as byproducts of petroleum refining or by reaction of natural gas with steam. About 200 million kg of hydrogen is produced each year in the United States, at least two-thirds of it being used to make ammonia.

Hydrogen is a colorless, odorless gas and the lightest of all substances. Its density is only one-fourteenth that of air, and for this reason it was once used to fill lighter-than-air craft. Unfortunately, hydrogen can be ignited by a spark, which is what occurred in 1937 when the German airship *Hindenburg* was destroyed in a disastrous fire and explosion as it was landing in New Jersey. The use of hydrogen in airships was discontinued after that, and the dirigible industry

Hot air balloons, as their name implies, are buoyed by hot air, which is less dense than the ambient air.

FIGURE 6.5 ▶

Hydrogen gas is prepared in the laboratory by the reaction of zinc with hydrochloric acid. The gas bubbles from the reaction flask and is trapped in the inverted bottle. When the reaction starts, the bottle is filled with water, but the hydrogen gas pushes the water out as it collects in the bottle.

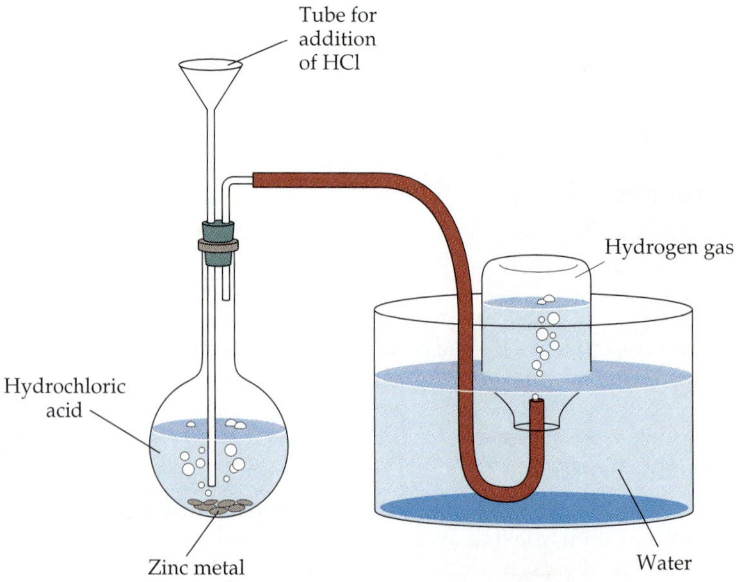

▲ **FIGURE 6.6**

Hydrogen is the most buoyant gas, but it is highly flammable. The disastrous fire in the German zeppelin *Hindenburg* led to the replacement of hydrogen by nonflammable helium, which buoys the Fuji blimp.

never recovered. Today the few airships that are still in service are filled with nonflammable helium; such airships are used mainly in advertising (Figure 6.6).

A stream of pure hydrogen will burn quietly in air with an almost colorless flame but when a mixture of hydrogen and oxygen is ignited by a spark or a flame, a violent reaction occurs (Figure 6.7). The product in both cases is water.

$$2\,H_2 \;+\; O_2 \;\longrightarrow\; 2\,H_2O$$

Hydrogen has such strong attraction for oxygen that it can remove oxygen atoms from many metal oxides to yield the free metal. For example, when hydrogen is passed over heated copper oxide, metallic copper and water are formed.

$$CuO \;+\; H_2 \;\longrightarrow\; Cu \;+\; H_2O$$

Certain metals, such as platinum and palladium, have unusual affinity for hydrogen, being able to *absorb* large volumes of the gas. Palladium can absorb up to 900 times its own volume of hydrogen! It is interesting to note that hydrogen and oxygen can be mixed at room temperature with no perceptible reaction. But if a piece of platinum gauze is added, the gases react violently at room temperature. The platinum acts as a catalyst; it lowers the activation energy for the reaction (Section 5.11). The heat from the initial reaction heats up the platinum, making it glow; it then ignites the hydrogen–oxygen mixture, causing an explosion.

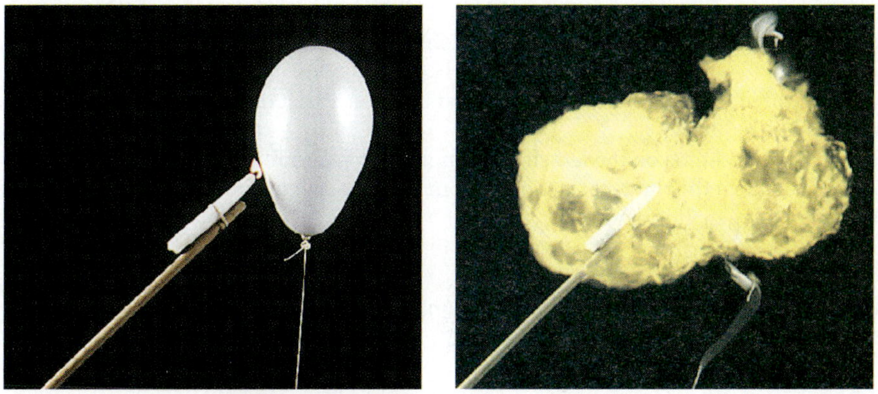

(a) (b)

◀ **FIGURE 6.7**

(a) A candle is held near a balloon filled with a mixture of hydrogen gas and oxygen gas. (b) The $H_2(g)$ ignites violently, combining with $O_2(g)$ to form water.

Platinum and palladium, as well as nickel, are often used as catalysts for reactions involving hydrogen. The metals have greatest catalytic activity when they are finely divided and have lots of active surface area. Hydrogen adsorbed on the surface of these metals is more reactive than ordinary hydrogen gas.

6.4 Oxidation and Reduction: Some Definitions

Oxidation–reduction reactions include all combustion processes, most metabolic reactions in living organisms, the extraction of metals from their ores, the manufacture of countless chemicals, and many of the reactions occurring in our natural environment.

The term "oxidation" originated to describe reactions in which a substance combines with oxygen. The opposite process, the removal of oxygen, was described by the term "reduction." We still find it convenient, especially in organic chemistry and biochemistry (Chapter 13 and following), to use three simple definitions of each of these terms (Figure 6.8).

1. Oxidation is a gain of oxygen atoms.
   Reduction is a loss of oxygen atoms.

Metals are oxidized when they corrode. Iron combines with oxygen to form iron(III) oxide, the familiar iron rust.

$$4\,Fe(s) + 3\,O_2(g) \longrightarrow 2\,Fe_2O_3(s)$$

The iron atoms gain oxygen atoms; iron is oxidized.

When lead dioxide is heated at high temperatures, it decomposes as follows.

$$2\,PbO_2 \longrightarrow 2\,PbO + O_2$$

The lead dioxide loses oxygen, so it is reduced.

Example 6.3

In each of the following, is the reactant undergoing oxidation or reduction? (These are not complete chemical equations.)

a. $Pb \longrightarrow PbO_2$

b. $SnO_2 \longrightarrow SnO$

c. $KClO_3 \longrightarrow KCl$

d. $Cu_2O \longrightarrow 2\,CuO$

Solution

a. Pb *gains* oxygen atoms (it has none on the left and two on the right); it is oxidized.

b. Sn *loses* an oxygen atom (it has two on the left and only one on the right); it is reduced.

c. There are three oxygen atoms on the left and none on the right. The compound *loses* oxygen; it is reduced.

d. The two copper atoms on the left share a single oxygen atom; they have half an oxygen each. On the right, each copper atom has an oxygen atom all its own. Cu has *gained* oxygen; it is oxidized.

Practice Exercise

In each of the following, is the reactant undergoing oxidation or reduction? (These are not complete chemical equations.)

a. $3\,Fe \longrightarrow Fe_3O_4$

b. $NO \longrightarrow NO_2$

c. $Cr_2O_3 \longrightarrow CrO_3$

d. $C_3H_6O \longrightarrow C_3H_6O_2$

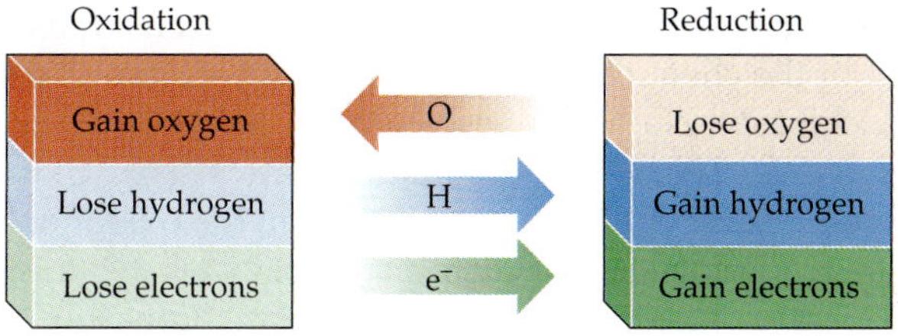

FIGURE 6.8 Three different definitions of oxidation and reduction.

2. Oxidation is a loss of hydrogen atoms.
   Reduction is a gain of hydrogen atoms.

Methyl alcohol, when passed over hot copper gauze, forms formaldehyde and hydrogen gas.

$$CH_4O \longrightarrow CH_2O + H_2$$

Since the methyl alcohol loses hydrogen, it is oxidized in this reaction.

Methyl alcohol can be made by reaction of carbon monoxide with hydrogen.

$$CO + 2\,H_2 \longrightarrow CH_4O$$

Since the carbon monoxide has gained hydrogen atoms, it has been reduced.

Example 6.4

In each of the following, is the reactant undergoing oxidation or reduction? (These are not complete chemical equations.)

a. $C_2H_6O \longrightarrow C_2H_4O$ **b.** $C_2H_2 \longrightarrow C_2H_6$

Solution

a. There are six hydrogen atoms in the compound on the left and only four in the one on the right. The compound *loses* hydrogen atoms; it is oxidized.

b. There are two hydrogen atoms in the compound on the left and six in the one on the right. The compound *gains* hydrogen atoms; it is reduced.

Practice Exercise

In each of the following, is the reactant undergoing oxidation or reduction? (These are not complete chemical equations.)

a. $C_6H_6 \longrightarrow C_6H_{12}$ **b.** $C_3H_6O \longrightarrow C_3H_4O$

3. Oxidation is a loss of electrons.
   Reduction is a gain of electrons.

When magnesium metal reacts with chlorine, magnesium ions and chloride ions are formed.

$$Mg + Cl_2 \longrightarrow Mg^{2+} + 2\,Cl^-$$

Just remember Leo the lion.

LEO says *GER*

*L*oss of *E*lectrons is *O*xidation.

*G*ain of *E*lectrons is *R*eduction.

Since the magnesium atom loses electrons, it is oxidized; and since the chlorine atoms gain electrons, they are reduced.

These definitions based on the loss or gain of electrons lead us to a much broader definition of oxidation and reduction. First, however, we introduce a concept that assists us in formulating this comprehensive definition.

In a general way, **oxidation state (O.S.)** refers to the number of electrons transferred or shared in the formation of the chemical bonds in a substance. For example, in the formation of sodium chloride, Na atoms lose electrons and Cl atoms gain them, forming the ions Na^+ and Cl^-. Na is in the oxidation state +1 and Cl is in the oxidation state −1. In the compound $CaCl_2$, chlorine is also in the oxidation state −1, existing as Cl^- ions. The oxidation state of calcium, however, is +2; it is present as Ca^{2+} ions. The total of the oxidation states of the atoms (ions) in a formula unit of $CaCl_2$ is $+2 - 1 - 1 = 0$.

In the formation of a molecule, no electrons are actually transferred; they are shared. We can, however, *arbitrarily* assign oxidation states. In a binary molecule we assign a negative oxidation state to the more electronegative element. This means that in H_2O the oxidation state of H should be positive and that of O should be negative. We assign H the oxidation state +1. We also require that the total of the oxidation states of the three atoms in H_2O be *zero*. This means that the oxidation state of O must be −2 that is, $+1 + 1 - 2 = 0$. In the H_2 molecule the H atoms are identical and must have the same oxidation state. The sum of these oxidation states must be *zero*, so the oxidation state of each H atom must also be 0.

We can use the following rules to assign oxidation states in the great majority of compounds. A few exceptions to the rules are listed in the marginal notes. The rules are listed *by priority*, with the highest priority listed first. If two rules are in conflict, use the rule with the higher priority, and this should generally take care of exceptions. Some examples are listed for each rule, and all the rules together are applied in Example 6.5.

1. *The total of the oxidation states (O.S.) of all the atoms in a neutral molecule, an isolated atom, or a formula unit is 0.*
   [Examples: The O.S. of the Fe atom is 0. The sum of the O.S. of all the atoms in each of the molecules Cl_2, S_8, and $C_6H_{12}O_6$ is 0, and the sum of the O.S. of the ions in $MgBr_2$ is 0.]
2. *In their compounds, the Group 1A metals all have an O.S. of +1, and the Group 2A metals have an O.S. of +2.*
   [Examples: The O.S. of Na in Na_2SO_4 is +1, and that of Ca in $Ca_3(PO_4)_2$ is +2.]
3. *In its compounds, hydrogen has an O.S. of +1.*
   [Examples: The O.S. of H is +1 in HCl, H_2O, NH_3, and CH_4.
4. *In its compounds, oxygen has an O.S. of −2.*
   [Examples: The O.S. of O is −2 in CO, CH_3OH, and $C_6H_{12}O_6$.]
5. *In their binary (two-element) compounds with metals, Group 7A elements have an O.S. of −1, Group 6A elements have an O.S. of −2, and Group 5A elements have an O.S. of −3.*
   [Examples: The O.S. of Br is −1 in $CaBr_2$, that of S is −2 in Na_2S, and that of N is −3 in Mg_3N_2.]

The principal exception to rule 3 is when H is bonded to an element that is less electronegative than itself, as in metal hydrides.

The principal exception to rule 4 is when oxygen is bonded to itself, as in peroxides—e.g., H_2O_2.

Example 6.5

What is the oxidation state (O.S.) of each element in the following?

a. I_2 **b.** Cr_2O_3 **c.** $AlCl_3$
d. Na_2SO_4 **e.** CaH_2

Solution

a. The sum of the O.S. for the two I atoms is 0 (rule 1). The two I atoms are identical, so the O.S. of each I atom must be 0.

b. The O.S. of O is −2 (rule 4), and the total for *three* O atoms is −6. The total of the O.S. for all atoms must be 0 (rule 1). This means that the total O.S. of *two* Cr atoms is +6, and that of one Cr atom is +3.

c. The O.S. of Cl is −1 (rule 5), and the total for *three* Cl atoms is −3. The total of the O.S. for the molecule must be 0 (rule 1), and therefore the *one* Al atom must have an O.S. of +3.

d. The O.S. of Na is +1 (rule 2), and for the *two* Na atoms, +2. The O.S. of O is −2 (rule 4), and the total for *four* O atoms is −8. The total for the formula unit must be 0 (rule 1). The O.S. of S must therefore be +6.

e. The O.S. of Ca is +2 (rule 2). The total for the formula unit must be 0 (rule 1), and so the total O.S. for the *two* H atoms must be −2. Even though the O.S. of H is usually +1 (rule 3), here it must be −1. Rule 2 takes priority over rule 3.

Practice Exercise

What is the oxidation state (O.S.) of each element in the following?

a. Al_2O_3 **b.** P_4 **c.** $NaMnO_4$
d. H_2O_2 **e.** CH_3F **f.** $CHCl_3$

(The assignment of oxidation states in CH_3F and $CHCl_3$ demonstrates the variability of the oxidation state of carbon in organic compounds.)

▲ **FIGURE 6.9**
The thermite reaction is an exceptionally vigorous redox reaction.

We can use the concept of oxidation states to help us identify oxidation–reduction reactions. For example, the spectacular reaction pictured in Figure 6.9, called the *thermite* reaction, is used to produce liquid iron for welding large iron objects.

$$2\,Al(s) + Fe_2O_3(s) \longrightarrow 2\,Fe(l) + Al_2O_3(s)$$

Even by the limited definitions we gave at the start of this section, we can call this an *oxidation–reduction* reaction. Al is *oxidized* to Al_2O_3; aluminum atoms take on or *gain* oxygen atoms. Fe_2O_3 is *reduced* to Fe; iron(III) oxide *loses* oxygen atoms.

Now let's assign oxidation states to the elements involved in the thermite reaction. These are the small numbers written above the chemical symbols in the equation.

$$2\,\overset{0}{Al}(s) + \overset{+3}{Fe}{}_2\overset{-2}{O}{}_3(s) \longrightarrow 2\,\overset{0}{Fe}(l) + \overset{+3}{Al}{}_2\overset{-2}{O}{}_3(s)$$

In this reaction, the oxidation state of Al atoms *increases* from 0 to +3, and the oxidation state of Fe atoms *decreases* from +3 to 0.

> In an *oxidation–reduction* reaction, the oxidation state of one or more elements *increases*—an **oxidation** process—and the oxidation state of one or more elements *decreases*—a **reduction** process. Oxidation and reduction must always occur together.

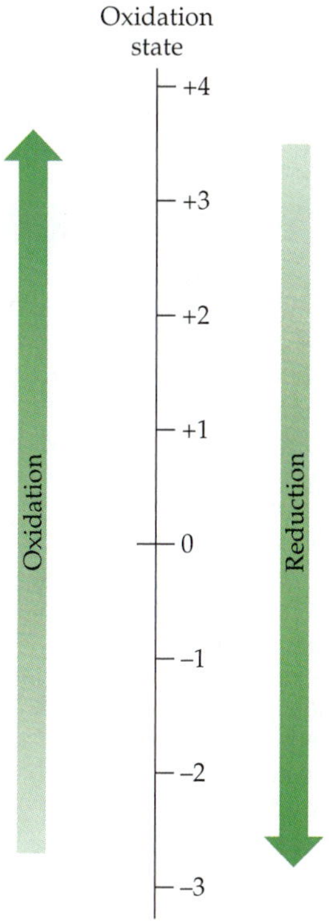

An increase in oxidation state means a loss of electrons and is therefore oxidation. A decrease in oxidation state means a gain of electrons and is therefore reduction.

The reaction pictured in Figure 6.10 is strikingly different in appearance from the thermite reaction, but the expanded definition identifies this also as an oxidation–reduction reaction. Oxidation states are noted in the equation below.

$$\overset{0}{Mg}(s) + \overset{+2}{Cu^{2+}}(aq) \longrightarrow \overset{+2}{Mg^{2+}}(aq) + \overset{0}{Cu}(s)$$

We can also visualize a redox reaction as two **half-reactions** that occur simultaneously. In one of these, the oxidation half-reaction, Mg atoms *lose* two electrons and are *oxidized* to Mg^{2+} ions.

$$\textit{oxidation:}\; Mg(s) \longrightarrow Mg^{2+}(aq) + 2\,e^-$$

▲ **FIGURE 6.10**
The reaction between magnesium and copper(II) ions is a redox reaction. In the top photograph, a coil of Mg(s) is added to a solution of $CuSO_4$(aq). After several hours, the solution is no longer blue because all the Cu^{2+}(aq) has been displaced from solution, leaving a deposit of red-brown copper metal, some unreacted magnesium, and a clear, colorless solution of $MgSO_4$(aq).

In the other half-reaction, the reduction half-reaction, Cu^{2+} ions *gain* two electrons and are *reduced* to Cu atoms.

$$\textit{reduction:}\ Cu^{2+}(aq) + 2\,e^- \longrightarrow Cu(s)$$

Example 6.6

The following are supposed half-reactions (not balanced). In each case, state whether the reactant is undergoing oxidation, reduction, or neither.

a. $Zn(s) \longrightarrow Zn^{2+}(aq)$
b. $Fe^{3+}(aq) \longrightarrow Fe^{2+}(aq)$
c. $CaCO_3(s) \longrightarrow Ca^{2+}(aq) + CO_2(g)$
d. $AgNO_3(aq) \longrightarrow Ag(s)$

Solution

a. To form a 2+ ion, zinc *loses* two electrons

$$Zn(s) \longrightarrow Zn^{2+}(aq) + 2\,e^-$$

The oxidation state of zinc increases from 0 to +2. Zinc is *oxidized.*

b. To go from a 3+ ion to a 2+ ion, iron must *gain* an electron.

$$Fe^{3+}(aq) + e^- \longrightarrow Fe^{2+}(aq)$$

The oxidation state of iron decreases from +3 to +2. Iron is *reduced.*

c. Let's start by assigning oxidation states to each of the elements.

$$\overset{+2}{Ca}\overset{+4}{C}\overset{-2}{O_3}(s) \longrightarrow \overset{+2}{Ca}{}^{2+}(aq) + \overset{+4}{C}\overset{-2}{O_2}(g)$$

None of the atoms change in oxidation state; neither oxidation nor reduction is involved. (Note that we do not have to balance the relationship.)

d. In this case we should recognize that $AgNO_3$(aq) is an ionic compound with Ag^+(aq) and NO_3^-(aq) ions. In going from Ag^+(aq) to Ag(s), silver *gains* an electron.

$$Ag^+(aq) + e^- \longrightarrow Ag(s)$$

Silver *decreases* in oxidation state from +1 to 0. Silver is *reduced.*

Practice Exercise

The following are supposed half-reactions (not balanced). In each case, state whether the reactant is undergoing oxidation, reduction, or neither.

a. $Cu^{2+}(aq) \longrightarrow Cu(s)$
b. $Sn^{2+}(aq) \longrightarrow Sn^{4+}(aq)$
c. $CuO(s) \longrightarrow Cu^{2+}(aq) + H_2O$
d. $Cu(s) \longrightarrow CuSO_4(aq)$

Now let's consider some complete reactions and determine whether or not they are oxidation–reduction reactions.

Example 6.7

Does the following equation represent an oxidation–reduction reaction?

$$Mn^{2+}(aq) + O_2(g) + H^+(aq) \longrightarrow MnO_2(s) + H_2O\ (\textit{not balanced})$$

Solution

By designating oxidation states in the equation

$$\overset{+2}{Mn}{}^{2+}(aq) + \overset{0}{O_2}(g) + \overset{+1}{H}{}^+(aq) \longrightarrow \overset{+4}{Mn}\overset{-2}{O_2}(s) + \overset{+1}{H_2}\overset{-2}{O}\ (\textit{not balanced})$$

we see that the O.S. of Mn increases from +2 to +4; that is, Mn^{2+} is *oxidized* to MnO_2. The O.S. of O decreases from 0 to −2; O_2 is *reduced* to H_2O. The equation *does* represent an oxidation–reduction reaction.

Practice Exercise

Does the following equation represent an oxidation–reduction reaction?

$$I_2(s) + 5\,Cl_2(g) + 6\,H_2O \longrightarrow 2\,IO_3^-(aq) + 12\,H^+(aq) + 10\,Cl^-(aq)$$

6.5 Oxidizing and Reducing Agents

As we have said, oxidation and reduction always go hand in hand; you can't have one without the other. When one substance is oxidized, another is reduced. For example, in the reaction

$$CuO(s) + H_2(g) \longrightarrow Cu(s) + H_2O(g)$$

the copper goes from O.S. +2 to O.S. 0; the copper(II) oxide is *reduced*. At the same time, the hydrogen goes from O.S. 0 to O.S. +1; the H_2 is *oxidized*. The substance that is *oxidized* (H_2)—because it causes some other substance (CuO) to be reduced—is called a **reducing agent.** Similarly, the substance that is *reduced* (CuO)—because it causes another substance (H_2) to be oxidized—is called an **oxidizing agent.**

Even though the changes in O.S. occur in Cu and H atoms, we do not refer to the *atoms* as the oxidizing or reducing agents. Rather, the *substances* in which these atoms are found—that is, CuO and H_2, respectively—are given these labels.

Reduction:
copper oxide is being reduced;
CuO is the oxidizing agent.

$$CuO + H_2 \longrightarrow Cu + H_2O$$

Oxidation:
hydrogen is being oxidized;
H_2 is the reducing agent.

OXIDATION is electron DRAIN,
While REDUCTION is electron GAIN;
Forever linked, they must have one another,
For one cannot occur without the other.

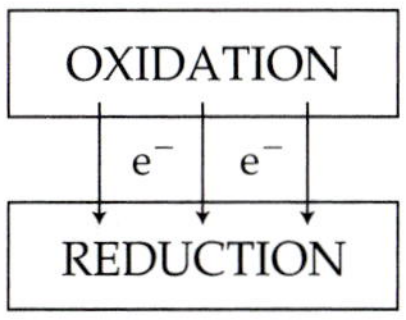

Example 6.8

Circle the oxidizing agents and underline the reducing agents in the following reactions.

a. $2\,C(s) + O_2(g) \longrightarrow 2\,CO(g)$
b. $N_2(g) + 3\,H_2(g) \longrightarrow 2\,NH_3(g)$
c. $SnO(s) + H_2(g) \longrightarrow Sn(s) + H_2O(g)$
d. $Mg(s) + Cl_2(g) \longrightarrow MgCl_2(s)$

Solution

a. Using the first definition (page 150), we can see that C gains oxygen and is *oxidized;* C(s) is the *reducing agent.* Oxygen goes from O.S. 0 in $O_2(g)$ to O.S. −2 in CO; O_2 is *reduced* and is therefore the *oxidizing agent.*

$$2\,\underline{C}(s) + \boxed{O_2(g)} \longrightarrow 2\,CO(g)$$

b. Using the second definition (page 151), we can see that N gains hydrogen and is *reduced;* it is the *oxidizing agent.* Hydrogen goes from O.S. 0 in $H_2(g)$ to O.S. +1 in $NH_3(g)$; H_2 is *oxidized* and is therefore the *reducing agent.*

$$N_2(g) + 3\,H_2(g) \longrightarrow 2\,NH_3(g)$$

c. Tin goes from O.S. +2 in SnO(s) to O.S. 0 in Sn(s). SnO is *reduced;* it is the *oxidizing agent.* Hydrogen goes from O.S. 0 in $H_2(g)$ to O.S. +1 in H_2O. H_2 is *oxidized;* it is the *reducing agent.*

$$SnO(s) + H_2(g) \longrightarrow Sn(s) + H_2O(g)$$

d. Magnesium goes from O.S. 0 in Mg(s) to +2 in $MgCl_2(s)$. Mg(s) is *oxidized;* it is the *reducing agent.* Chlorine goes from O.S. 0 in $Cl_2(g)$ to O.S. −1 in $MgCl_2(s)$. $Cl_2(g)$ is *reduced;* it is the *oxidizing agent.*

$$Mg(s) + Cl_2(g) \longrightarrow MgCl_2(s)$$

Practice Exercise

Circle the oxidizing agents and underline the reducing agents in the following reactions.

a. $Se + O_2 \longrightarrow SeO_2$
b. $CH_3C{\equiv}N + 2\,H_2 \longrightarrow CH_3CH_2NH_2$
c. $V_2O_5 + 2\,H_2 \longrightarrow V_2O_3 + 2\,H_2O$
d. $2\,K + Br_2 \longrightarrow 2\,K^+ + 2\,Br^-$

6.6 Some Common Oxidizing Agents

Oxygen itself is the most common oxidizing agent. Making up one-fifth of the air, it oxidizes the wood in our campfires and the gasoline in our automobiles. It even "burns" the food we eat to give us the energy to move and to think. Large amounts of purified oxygen are used in hospital respirators and welding torches, but most of it goes into steel furnaces.

Another common oxidizing agent is hydrogen peroxide (H_2O_2). Pure hydrogen peroxide is a syrupy liquid. It is available (in the laboratory) as a dangerous 30% solution that has powerful oxidizing power or as a 3% solution sold in stores for various uses around the home. It has the advantage of being converted to water in most reactions.

An oxidizing agent often used in the laboratory is potassium dichromate ($K_2Cr_2O_7$), which is used frequently to oxidize alcohols to compounds called *aldehydes* and *ketones* (Chapter 15). For the oxidation of ethyl alcohol (found in alcoholic beverages) to acetaldehyde, the reaction is

$$8\,H^+(aq) + \underset{\text{Dichromate ion (orange)}}{Cr_2O_7^{2-}(aq)} + 3\,C_2H_5OH(aq) \longrightarrow \underset{\text{Chromium(III) ion (green)}}{2\,Cr^{3+}(aq)} + 3\,C_2H_4O(aq) + 7\,H_2O$$

Potassium permanganate is a black, shiny, crystalline solid. It dissolves in water to form deep purple solutions. This purple color disappears as the permanganate is reduced (remember, if permanganate is an oxidizing agent, it must be reduced). So potassium permanganate is often used as a test for oxidizable substances. For example, potassium permanganate is used to oxidize iron from Fe^{2+} to Fe^{3+}. The amount of iron(II) ion in a sample can be determined by its reaction with permanganate. One can add permanganate solution, which is deep purple, to a sample of iron(II) ion. As the permanganate is reduced, it is changed to manganese(II) ion, and the purple color disappears (Figure 6.11). When all the iron(II) ion has been oxidized, the addition of more permanganate will not be accompanied by the loss of the purple color because there will be no iron(II) ion left to reduce the permanganate. Thus, one can measure just how much iron(II) ion there is in the sample by keeping track of how much permanganate is reduced—that is, by measuring how much permanganate is added until the purple color remains. The equation for this reaction is

$$\underset{\text{Permanganate ion (purple)}}{MnO_4^-(aq)} + 5\,Fe^{2+}(aq) + 8\,H^+(aq) \longrightarrow \underset{\text{Manganese(II) ion (pale pink)}}{Mn^{2+}(aq)} + 5\,Fe^{3+}(aq) + 4\,H_2O$$

Permanganate solutions can also be used to oxidize oxalic acid (a poisonous compound found in rhubarb leaves), sulfur dioxide (SO_2), and many other compounds.

Other common oxidizing agents are the halogens—fluorine (F_2), chlorine (Cl_2), bromine (Br_2), and iodine (I_2). Bromine, for example, oxidizes phosphorus to form phosphorus tribromide.

$$P_4(s) + 6\,Br_2(l) \longrightarrow 4\,PBr_3(l)$$

Many antiseptics are mild oxidizing agents. (Antiseptics are compounds applied to living tissue to kill microorganisms or prevent their growth.) For example, a 3% solution of hydrogen peroxide is often used to treat minor cuts, and tincture of iodine has long been a household antiseptic. Ointments for treating acne often contain 5 to 10% of benzoyl peroxide, a powerful antiseptic and also a skin irritant. It causes old skin to slough off and be replaced by new, fresher-looking skin. When used on areas exposed to sunlight, however, benzoyl peroxide may promote skin cancer.

A tincture is a solution made up in alcohol.

Disinfectants are substances that are applied to nonliving tissue to kill microorganisms.

Oxidizing agents are also used as disinfectants. A good example is chlorine, which is used to kill disease-causing microorganisms in drinking water.

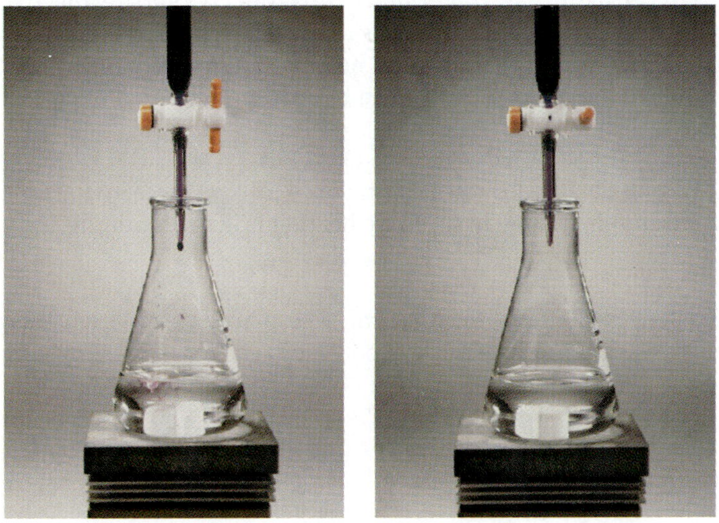

(a) (b)

FIGURE 6.11

Purple permanganate ions (MnO_4^-) are reduced to Mn^{2+} ions by iron(II) ions (Fe^{2+}). (a) The buret contains $MnO_4^-(aq)$ and the flask contains $Fe^{2+}(aq)$. (b) As the permanganate solution is added to the solution of iron(II) ions, the purple color disappears.

Swimming pools are often chlorinated with calcium hypochlorite [$Ca(OCl)_2$]. Since calcium hypochlorite is alkaline, it also raises the pH of the water. (When a pool becomes too alkaline, the pH is lowered by adding hydrochloric acid. Swimming pools are usually maintained at pH 7.2 to 7.8, see Section 11.3.)

Bleaches are oxidizing agents, too. Bleaches remove unwanted color from fabrics or other material. Nearly any oxidizing agent would do the job, but some might be unsafe, or harmful to fabrics, or perhaps too expensive. Laundry bleaches are usually sodium hypochlorite (NaOCl) as an aqueous solution (in products such as Purex and Clorox) or calcium hypochlorite [$Ca(OCl)_2$], known as bleaching powder. The powder is usually preferred for large industrial operations, such as the whitening of paper or fabrics. It is also used in hospitals to disinfect bedding and clothes. Nonchlorine bleaches often contain sodium perborate (a loose combination of $NaBO_2$ and H_2O_2).

For lightening hair color, the bleaches are usually 6% or 12% solutions of hydrogen peroxide, which oxidizes the dark pigment (melanin) in the hair to colorless products. Hydrogen peroxide can also be used to lighten certain paints by oxidizing sulfides (S^{2-}) to sulfates (SO_4^{2-}). When lead-based paints are exposed to air containing hydrogen sulfide (H_2S), they turn black because of the formation of lead sulfide (PbS). Hydrogen peroxide oxidizes the black sulfide to white sulfate.

Hydrogen peroxide can be used to restore the once-white areas of old paintings that have darkened by the reaction of white lead compounds (paint pigments) with sulfur compounds. The darkened pigments (black PbS) are converted to white $PbSO_4$ by the hydrogen peroxide.

$$\underset{\text{black}}{PbS(s)} + 4\,H_2O_2(aq) \longrightarrow \underset{\text{white}}{PbSO_4(s)} + 4\,H_2O$$

Stain removal is more complicated than bleaching. A few stain removers are oxidizing agents, but some are reducing agents, some are solvents or detergents, and others act in still other ways. Some stains require specific stain removers.

6.7 Some Reducing Agents of Interest

In every reaction involving oxidation, the oxidizing agent gets reduced, and so the substance being oxidized is acting as a reducing agent. But let us consider here reactions in which the purpose of the reaction is reduction.

Most metals occur in nature as compounds. In order to prepare the free metals, the compounds must be reduced. Metals are often freed from their ores with coal or coke (elemental carbon obtained by heating coal to drive off volatile matter). Tin oxide is one of the many ores that can be reduced with coal or coke.

$$SnO_2(s) + C(s) \longrightarrow Sn(s) + CO_2(g)$$

Sometimes a metal can be obtained by heating its ore with a more active metal. Chromium oxide, for example, can be reduced by heating it with aluminum.

$$Cr_2O_3(s) + 2\,Al(s) \longrightarrow Al_2O_3(s) + 2\,Cr(s)$$

Hydrogen is an excellent reducing agent that can free many metals from their ores, but it is generally used to produce more expensive metals, such as tungsten.

$$WO_3(s) + 3\,H_2(g) \longrightarrow W(s) + 3\,H_2O$$

Hydrogen can be used to reduce many kinds of chemical compounds. Ethylene, for example, can be reduced to ethane.

$$C_2H_4(g) + H_2(g) \xrightarrow{Ni} C_2H_6(g)$$

(Nickel is used as a catalyst in this reaction.) Hydrogen also reduces nitrogen, from the air, in the industrial production of ammonia.

$$N_2(g) \quad + \quad 3\,H_2(g) \xrightarrow{\text{Fe}} 2\,NH_3(g)$$

(An iron catalyst is used in this case.)

Perhaps a more familiar reducing agent, by use if not by name, is the developer used in black and white photography. Photographic film is coated with a silver *salt* (Ag^+Br^-). Silver ions that have been exposed to light react with the developer, a reducing agent (such as the organic compound hydroquinone), to form metallic silver.

$$\underset{\text{Hydroquinone}}{C_6H_4(OH)_2(aq)} \quad + \quad 2\,Ag^+(aq) \longrightarrow C_6H_4O_2(aq) \quad + \quad \underset{\text{Silver metal}}{2\,Ag(s)} \quad + \quad 2\,H^+(aq)$$

Silver ions not exposed to light are not reduced by the developer. The film is then treated with "hypo," a solution of sodium thiosulfate ($Na_2S_2O_3$), which washes out unexposed silver bromide to form the negative. This leaves the negative dark where the metallic silver has been deposited (where it was originally exposed to light) and transparent where light did not strike it. Light is then shined through the negative onto light-sensitive paper to make the positive print. Figure 6.12 shows positive and negative prints.

In food chemistry, reducing agents are sometimes referred to as **antioxidants.** Ascorbic acid (vitamin C) can prevent the browning of fruit (such as sliced apples or pears) by inhibiting air oxidation. Whereas vitamin C is water soluble, tocopherol (vitamin E) is a fat-soluble antioxidant. Both of these vitamins are believed to retard various oxidation reactions that are potentially damaging to vital components of living cells. (See Special Topic J.)

▲ **FIGURE 6.12**

A photographic negative (left) and a positive print (right).

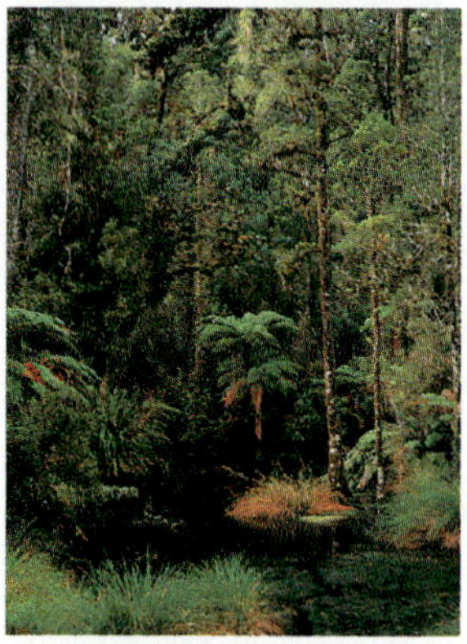

▲ **FIGURE 6.13**

Photosynthesis occurs in these green plants. The chlorophyll pigments that catalyze the photosynthetic process make much of Earth's land area green against the blue of the oceans.

6.8 Oxidation, Reduction, and Living Things

Perhaps the most important oxidation–reduction processes are photosynthesis and respiration (Section 5.8). Bread and many of the other foods we eat are largely made up of carbohydrates (Chapter 19). If we represent carbohydrates with the simple example glucose ($C_6H_{12}O_6$), we can write the overall equation for their metabolism as follows.

$$C_6H_{12}O_6 \;+\; 6\,O_2 \;\longrightarrow\; 6\,CO_2 \;+\; 6\,H_2O \;+\; \text{energy}$$

The carbohydrate is oxidized in the process.

Meanwhile, plants use carbon dioxide and water to produce carbohydrates. The energy needed comes from the sun and the process is called **photosynthesis** (Figure 6.13). The chemical equation is

$$6\,CO_2 \;+\; 6\,H_2O \;+\; \text{energy} \;\longrightarrow\; C_6H_{12}O_6 \;+\; 6\,O_2$$

During photosynthesis, carbon dioxide is reduced to glucose, a carbohydrate.

The carbohydrates produced by photosynthesis are the ultimate source of all our food, since fish, fowl, and other animals either eat plants or eat other animals that eat plants (Figure 6.14). Note that the photosynthesis process not only makes carbohydrates but also yields free elementary oxygen, O_2. In other words, photosynthesis does not just provide all the food we eat; it also provides all the oxygen we breathe.

We can see that green plants carry out the oxidation–reduction reaction that makes possible all life on Earth. Animals can only oxidize the foods that plants provide. We might therefore consider crop farming a process of reduction. Energy captured in cultivated plants, whether the plants are used directly or are fed to animals, is the basis for human life.

▲ **FIGURE 6.14**

The food we eat is oxidized to provide energy for our activities. That energy, which comes ultimately from the sun, is trapped by green plants through photosynthetic reactions that reduce carbon dioxide to carbohydrates.

Summary

Oxidation can be thought of in several ways, depending on the particular chemical reaction being analyzed:

1. the gain of oxygen atoms, as in the oxidation of metallic zinc: $2\ Zn + O_2 \longrightarrow 2\ ZnO$
2. the loss of hydrogen atoms, as in the oxidation of methyl alcohol: $CH_3OH \longrightarrow CH_2O + H_2$
3. the loss of electrons, as in the oxidation of magnesium metal to its cation: $Mg \longrightarrow Mg^{2+} + 2\ e^-$

Reduction, too, can be thought of in several ways:

1. the loss of oxygen atoms, as in the reduction of lead oxide: $PbO + H_2 \longrightarrow Pb + H_2O$
2. the gain of hydrogen atoms, as in the reduction of carbon monoxide to methyl alcohol: $CO + 2\ H_2 \longrightarrow CH_3OH$
3. the gain of electrons, as in the reduction of a magnesium cation to the metal: $Mg^{2+} + 2\ e^- \longrightarrow Mg$

The most fundamental definition of oxidation is based on oxidation states. The **oxidation state** is related to the number of electrons that the atom transfers or shares in forming bonds. Oxidation entails an *increase* in oxidation state and reduction, a *decrease* in oxidation state.

Oxidation and reduction always occur together. Whenever one reactant in a chemical reaction is oxidized, some other reactant in the reaction is reduced. The overall reaction is referred to as a *redox reaction*.

An **oxidizing agent** is any substance that causes some other substance to be oxidized; an oxidizing agent is always *reduced* in a redox reaction. A **reducing agent** is any substance that causes some other substance to be reduced; a reducing agent is always *oxidized* in a redox reaction. Example: In the reaction $CuO + H_2 \longrightarrow Cu + H_2O$, the CuO loses its oxygen atom and so is reduced; this reduction is caused by the reducing agent H_2. The H_2 gains an oxygen atom to become H_2O, and so it is oxidized; the oxidation is caused by the oxidizing agent CuO.

Oxygen is the most frequently encountered oxidizing agent. Others are potassium dichromate ($K_2Cr_2O_7$), hydrogen peroxide (H_2O_2), potassium permanganate ($KMnO_4$), and the halogens (F_2, Cl_2, Br_2, I_2). Two frequently used reducing agents are hydrogen gas and carbon.

Antiseptics, compounds used to control microorganism growth on living tissue, are oxidizing agents. *Bleaches,* used to remove color from cloth, hair and other substances, are also oxidizing agents, usually solutions of sodium hypochlorite (NaOCl).

The redox reaction most important to living organisms is *photosynthesis*, the process by which green plants convert solar energy to chemical energy locked up in the chemical bonds of sugar molecules:

$$6\ CO_2 + 6\ H_2O + \text{sunlight} \longrightarrow C_6H_{12}O_6 + 6\ O_2$$

The CO_2 is reduced because it gains H atoms; water, in the presence of sunlight, is the reducing agent. The H_2O is oxidized because it loses H atoms to become O_2; the CO_2 is the oxidizing agent.

Key Terms

antioxidant (6.7)
half-reaction (6.4)
oxidation (6.4)
oxidation state (O.S.) (6.4)
oxidizing agent (6.5)
photosynthesis (6.8)
reducing agent (6.5)
reduction (6.4)

Review Questions

1. Define oxidation and reduction in terms of oxygen atoms gained or lost. Give examples.
2. Define oxidation and reduction in terms of hydrogen atoms gained or lost. Give examples.
3. Define oxidation and reduction in terms of electrons gained or lost. Give examples.
4. Define oxidation and reduction in terms of a change in oxidation state. Give examples.
5. List four common oxidizing agents. Give one use for each.
6. List three common reducing agents. Give one use for each.
7. Name some oxidizing agents used as antiseptics and disinfectants.
8. Relate the chemistry of photosynthesis to the chemistry that provides the energy for your heartbeat.
9. List several bleaching agents. Give one use for each of them.
10. What is the usual oxidation state of hydrogen in its compounds? What is the usual oxidation state of oxygen in its compounds? What are some exceptions?
11. What happens to the oxidation state of one of its elements when a compound is oxidized and when it is reduced?
12. Are there any circumstances under which an oxidation half-reaction can occur unaccompanied by a reduction half-reaction? Explain.

Problems

Reactions of Oxygen

13. Complete and balance the following equations.
    a. $C + O_2 \longrightarrow$ b. $CH_4 + O_2 \longrightarrow$
    c. $N_2 + O_2 \longrightarrow$ d. $C_3H_8 + O_2 \longrightarrow$
14. Complete and balance the following equations.
    a. $S + O_2 \longrightarrow$ b. $CS_2 + O_2 \longrightarrow$
    c. $H_2 + O_2 \longrightarrow$ d. $C_6H_{12}O_6 + O_2 \longrightarrow$

Oxidation States

15. Indicate the oxidation state of the underlined element in each of the following.
    a. $\underline{Cr}$ b. $\underline{Cl}O_2$ c. $K_2\underline{Se}$ d. $\underline{Te}F_6$
16. Indicate the oxidation state of the underlined element in each of the following.
    a. $Ca\underline{Ru}O_3$ b. $\underline{N}H_2OH$ c. $\underline{C}_2H_6$ d. $H\underline{C}OOH$

Recognizing Redox Reactions

17. In the following supposed half-reactions (not balanced), indicate whether oxidation or reduction, or neither, is involved.
    a. $ClO_2(g) \rightarrow HClO_3(aq)$ b. $Mn^{2+}(aq) \rightarrow MnO_2(s)$
    c. $HOBr(aq) \rightarrow Br_2(l)$ d. $SbH_3(g) \rightarrow Sb(s)$
18. In the following supposed half-reactions (not balanced), indicate whether oxidation or reduction, or neither, is involved.
    a. $V_2O_5(aq) \longrightarrow V(s)$
    b. $P_4(s) \longrightarrow H_3PO_4(aq)$
    c. $CrO_3(s) \longrightarrow H_2CrO_4(aq)$
    d. $CH_3CH_2OH(aq) \longrightarrow CO_2(g)$
19. In the reaction
    $$Cu(s) + 2\,H_2SO_4(aq) \longrightarrow CuSO_4(aq) + 2\,H_2O + SO_2(g)$$
    is the $H_2SO_4(aq)$ oxidized, reduced or neither? Explain.
20. Is the reaction
    $$5\,H_2O_2(aq) + 2\,MnO_4^-(aq) + 6\,H^+(aq) \longrightarrow 2\,Mn^{2+}(aq) + 8\,H_2O + 5\,O_2(g)$$
    a redox reaction? Explain.

Oxidizing Agents and Reducing Agents

21. Circle the oxidizing agent and underline the reducing agent in these reactions.
    a. $4\,Al + 3\,O_2 \longrightarrow 2\,Al_2O_3$
    b. $2\,SO_2 + O_2 \longrightarrow 2\,SO_3$
22. Circle the oxidizing agent and underline the reducing agent in these reactions.
    a. $Cl_2 + 2\,KBr \longrightarrow 2\,KCl + Br_2$
    b. $C_2H_4 + H_2 \longrightarrow C_2H_6$
23. Circle the oxidizing agent and underline the reducing agent in these reactions.
    a. $Fe + 2\,HCl \longrightarrow FeCl_2 + H_2$
    b. $CS_2 + 3\,O_2 \longrightarrow CO_2 + 2\,SO_2$
24. Circle the oxidizing agent and underline the reducing agent in these reactions.
    a. $2\,AgNO_3 + Cu \longrightarrow Cu(NO_3)_2 + 2\,Ag$
    b. $CuCl_2 + Fe \longrightarrow FeCl_2 + Cu$
25. In the following reactions, which substance is oxidized and which is reduced?
    a. $2\,HNO_3 + SO_2 \longrightarrow H_2SO_4 + 2\,NO_2$
    b. $2\,CrO_3 + 6\,HI \longrightarrow Cr_2O_3 + 3\,I_2 + 3\,H_2O$
26. In the following reactions, which substance is oxidized and which is the oxidizing agent?
    a. $H_2CO + H_2O_2 \longrightarrow H_2CO_2 + H_2O$
    b. $5\,C_2H_6O + 4\,MnO_4^- + 12\,H^+ \longrightarrow 5\,C_2H_4O_2 + 4\,Mn^{2+} + 11\,H_2O$
27. To test for an iodide ion (for example, in iodized salt), a solution is treated with chlorine to liberate iodine. The reaction is
    $$2\,I^- + Cl_2 \longrightarrow I_2 + 2\,Cl^-$$
    Which substance is oxidized? Which is reduced?
28. Molybdenum metal, used in special kinds of steel, can be manufactured by the reaction of its oxide with hydrogen. The reaction is
    $$MoO_3 + 3\,H_2 \longrightarrow Mo + 3\,H_2O$$
    Which substance is reduced? Which is the reducing agent?
29. Green grapes are exceptionally sour due to a high concentration of tartaric acid. As the grapes ripen, this compound is converted to glucose.
    $$\underset{\text{Tartaric acid}}{C_4H_6O_2} \longrightarrow \underset{\text{Glucose}}{C_6H_{12}O_6}$$
    Is the tartaric acid being oxidized or reduced?
30. The dye indigo (used to color blue jeans) is formed from indoxyl by exposing it to air.
    $$\underset{\text{Indoxyl}}{2\,C_8H_7ON} + O_2 \longrightarrow \underset{\text{Indigo}}{C_{16}H_{10}N_2O_2} + 2\,H_2O$$
    What substance is oxidized? What is the oxidizing agent?
31. Acetylene (C_2H_2) reacts with hydrogen to form ethane (C_2H_6). Is the acetylene oxidized or reduced? Explain your answer.
32. Unsaturated vegetable oils react with hydrogen to form saturated fats. A typical reaction is
    $$C_{57}H_{104}O_6 + 3\,H_2 \longrightarrow C_{57}H_{110}O_6$$
    Is the unsaturated oil oxidized or reduced? Explain.
33. Vitamin C (ascorbic acid) is thought to protect our stomachs from the carcinogenic effect of nitrite ions by converting the ions to NO gas.
    $$NO_2^- \longrightarrow NO$$
    Is the nitrite ion oxidized or reduced? Is ascorbic acid an oxidizing agent or a reducing agent?
34. In the preceding reaction (Problem 33), ascorbic acid is converted to dehydroascorbic acid.
    $$C_6H_8O_6 \longrightarrow C_6H_6O_6$$
    Is ascorbic acid oxidized or reduced in this reaction?

Additional Problems

35. When the water pump failed in the nuclear reactor at Three Mile Island in 1979, zirconium metal reacted with the very hot water to produce hydrogen gas.

$$Zr + 2\,H_2O \longrightarrow ZrO_2 + 2\,H_2$$

What substance was oxidized in this reaction? What was the oxidizing agent?

36. Use the rules on page 152 to calculate the oxidation state of S in sodium peroxodisulfate, $Na_2S_2O_8$. Actually, *two* of the oxygen atoms are involved in a peroxide linkage (—O—O—; that is what the prefix "peroxo" signifies) and have an O.S. of −1. Make a new calculation to show that the O.S. of S is really +6.

37. Calculate the oxidation state of O in CsO_2. (This is another exception to rule 4 that occurs in compounds called superoxides.)

38. When phosphorus, P_4, is heated with water it forms both phosphine, PH_3, and phosphoric acid, H_3PO_4, in a type of reaction called a disproportionation. Calculate oxidation states for the three substances and propose a definition for the word *disproportionation.*

39. Cyanide wastes can be detoxified by the addition of chlorine to a basic solution.

$$NaCN(aq) + NaOH(aq) + Cl_2(g) \longrightarrow NaOCN(aq) + NaCl(aq) + H_2O$$

Following the addition of acid to make the solution less basic, a further reaction occurs.

$$NaOCN(aq) + NaOH(aq) + Cl_2(g) \longrightarrow NaCl(aq) + H_2O + NaHCO_3(aq) + N_2(g)$$

Balance the two equations and identify the oxidizing agent and reducing agent in each equation.

40. Incineration of a chlorine-containing toxic waste such as a polychlorinated biphenyl (PCB) produces CO_2 and HCl.

$$C_{12}H_4Cl_6 + O_2 \longrightarrow CO_2 + HCl + H_2O$$

Balance the equation for this combustion reaction. Comment on the advantages and disadvantages of incineration as a method of disposal of such wastes.

7 Gases

The properties of gases make it possible for humans to fly in hot air balloons.

Humans have walked on the surface of Earth's barren, airless moon. Spacecraft have photographed the dusty desolation of Mercury, the planet nearest the sun, from a few kilometers away. Robotic probes have dropped through clouds of sulfuric acid and a thick blanket of carbon dioxide to land on the hot, inhospitable surface of Venus. Other space probes have descended through the thin, dusty atmosphere of Mars in a vain search for life on its surface. A probe has been dropped through the crushing, turbulent atmosphere of Jupiter, and spacecraft have examined the atmospheres of Saturn, Uranus, and Neptune. Pluto has also been studied, but only from vast distances. It is now clear that Earth, a small island of green and blue in the vastness of space, is uniquely equipped to serve the needs of the living organisms that inhabit it.

The life-support system of Spaceship Earth consists in part of a thin blanket of gases called the atmosphere. Although other planets in our solar system have atmospheres, Earth's atmosphere appears to be unique in its ability to support life.

The atmosphere is composed of about 5.2×10^{15} metric tons of air spread over a surface area of 5.0×10^{8} km^2.

It is difficult to measure just how deep the atmosphere is. It does not end abruptly, but gradually fades as the distance from the surface of the Earth increases. We do know, however, that 99% of the atmosphere lies within 30 km of the surface of the Earth—a thin layer of air indeed (like the peel of an apple, only relatively thinner).

Earth's atmosphere is divided into layers (Figure 7.1). The layer nearest Earth, the troposphere, harbors nearly all living things and nearly all human activity. The next region, the stratosphere, is where we find the ozone layer that shields living creatures from deadly ultraviolet radiation.

Air is so familiar, and yet so nebulous, that it is difficult to think of as matter. But it is matter—matter in the gaseous state. All gases, air included, have mass and occupy space. Like other forms of matter, gases obey certain physical laws. In this chapter, we will examine some of those laws and see how they are related to certain vital processes—such as breathing.

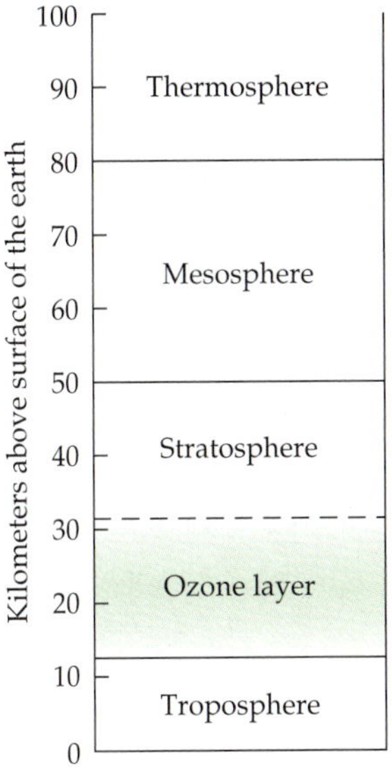

▲ **FIGURE 7.1**
The altitudes of the several layers of the atmosphere are only approximate. For example, the height of the troposphere varies from about 8 km at the poles to 16 km at the equator. The ozone layer is shown in green.

7.1 Air: A Mixture of Gases

Air is a mixture of gases. By volume, dry air consists of about 78% N_2, 21% O_2, and 1% Ar. Water vapor can make up as much as 4% of humid air (Section 7.10). There are a number of minor constituents of air, the most important of which is carbon dioxide. The concentration of CO_2 in air has increased from about 275 parts per million (ppm) in 1900 to its present value of about 360 ppm. It most likely will continue to rise as more and more fossil fuels (coal, oil, and natural gas) are burned. The composition of dry air is summarized in Table 7.1.

Air is a mixture of gases, but what are gases? Perhaps they are best understood in terms of the kinetic-molecular theory.

7.2 The Kinetic-Molecular Theory

The states of matter—solid, liquid, and gas—are obviously different from one another (Chapter 1). Chemists use models to explain these differences. The model used to explain the behavior of gases is the **kinetic-molecular theory.** The basic postulates of this theory are:

1. All matter is composed of tiny, discrete particles called molecules.
2. Molecules are in rapid, constant motion and move in straight lines.
3. The molecules of a gas are small compared with the distances between them.
4. There is little attraction between molecules of a gas.
5. Molecules collide with one another, and energy is conserved in these collisions—although one molecule can gain energy at the expense of another.
6. Temperature is a measure of the *average* kinetic energy of the gas molecules.

The kinetic-molecular theory treats gases as collections of individual particles in rapid motion (*kinetic* derives from the Greek word for motion). The particles of nitrogen gas, for example, are molecules (N_2); those of argon gas (Ar) are atoms. The distances between the particles are quite large compared with the dimensions of the particles themselves. Therefore, unlike solids and liquids, gases can be readily compressed. (According to the theory, the individual particles of a solid or

Table 7.1 Composition of Dry Air (Near Sea Level)

Component	Percent by Volume
Nitrogen (N_2)	78.084
Oxygen (O_2)	20.946
Argon (Ar)	0.934
Carbon dioxide (CO_2)	0.0360

Plus traces of: neon (Ne), helium (He), methane (CH_4), krypton (Kr), hydrogen (H_2), dinitrogen monoxide (N_2O), xenon (Xe), ozone (O_3), sulfur dioxide (SO_2), nitrogen dioxide (NO_2), ammonia (NH_3), carbon monoxide (CO), iodine (I_2)

Although gas molecules do not settle under the force of gravity in containers of ordinary size, they do so in that largest of containers—the atmosphere. The density of air decreases with increasing altitude above Earth's surface.

liquid are in contact with one another. They can't be pushed closer together because they are already touching.)

The particles of a gas are in such rapid motion, and gases have such low densities, that gravity seems to have little effect on them. They move up and down and sideways with ease and will not fall to the bottom of a container (as a liquid will). Any container of gas is completely filled. By *filled* we do not mean that gas molecules are packed tightly, but rather that the gas is distributed throughout the container's entire volume. A particle moves along a straight path until it strikes something (another particle or the walls of the container). Then it may bounce off at an angle and travel from the point of collision along a straight path until it hits something else. These collisions are *perfectly elastic;* there is no tendency for the collection of particles to slow down and eventually stop. Two particles that are about to collide have a certain combined kinetic energy. After the collision, the sum of their kinetic energies has not changed. One of the particles may have been slowed down by the collision, but the other will have been speeded up just enough to compensate (Figure 7.2).

The kinetic-molecular theory explains what it is we are measuring when we measure temperature. According to the theory, temperature is just a reflection of the kinetic energy of the gas particles. The higher the kinetic energy—that is, the faster the particles are moving—the higher the temperature of the sample. On the

FIGURE 7.2 ▶

According to the kinetic-molecular theory, molecules of a gas are in constant, random motion. They move in straight lines and undergo collisions with each other and with the walls of the container.

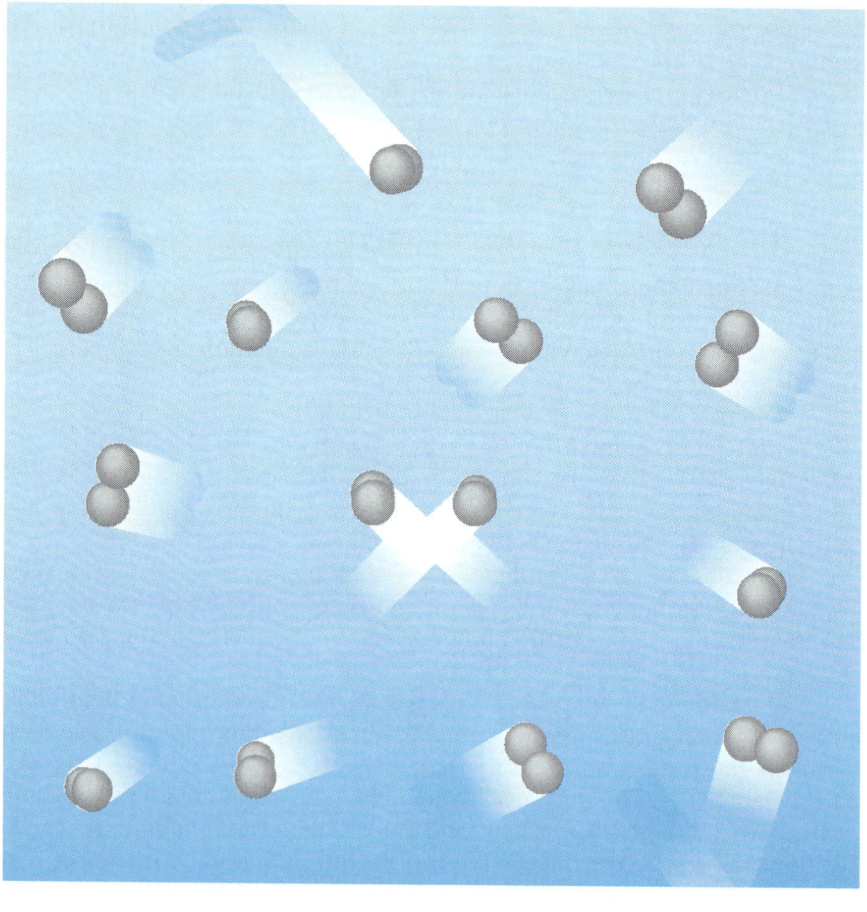

average, the particles of a cold sample are moving more slowly than the particles of a hot sample. In any single sample, some particles are moving faster than others. Temperature reflects the *average* kinetic energy of the particles.

As a particle strikes a wall of its container, it gives the wall a little push. (If you were hit with a baseball or a brick, you would feel a push.) The sum of all these tiny pushes over a given area of the wall is what we call *pressure.*

7.3 Atmospheric Pressure

Molecules in the air are constantly bouncing off our skin. They are so tiny, however, that we don't feel their individual impacts. However, when we go up in altitude rapidly—by driving up a mountain or riding an express elevator to the top of a tall building—our ears may "pop." This popping sensation is caused by an unequal air pressure on the two sides of our ear drums. Because the density of air decreases at the higher altitudes, there are fewer molecules of air outside our ear drums pushing in than on the inside pushing out. The popping stops as soon as the pressure inside the ear decreases and becomes equal to that at the higher altitude.

Pressure is force per unit area—that is, force divided by the area over which the force is exerted.

$$\text{Pressure} = \frac{\text{force}}{\text{area}} = \frac{F}{A}$$

In the SI system, force is expressed in *newtons* (*N*) and area in square meters (m^2). (A newton is a force that will give a 1-kg mass an acceleration of 1 meter per second, or m/s.) The derived SI unit for pressure is therefore the newton per square meter, commonly called a **pascal.**

$$1 \text{ Pa} = 1 \text{ N/m}^2$$

The pascal is such a small unit that the kilopascal (kPa) is often used instead.

The pressure of the atmosphere is measured by a device called a **barometer.** A simple type, known as a mercury barometer, was invented in 1643 by the Italian scientist Evangelista Torricelli (1608–1647). The mercury barometer consists of a long glass tube, closed at one end, filled with mercury, and inverted in a shallow dish that also contains mercury (Figure 7.3). Suppose the tube is 1 m long. Some of the mercury in the tube drains into the dish, but *not all* of it. The mercury drains out only until the pressure exerted by the mercury remaining in the tube exactly balances the pressure exerted by the atmosphere on the surface of the mercury in the dish. The mercury column tends to flow out under the influence of gravity, and the atmospheric pressure tends to push the mercury back into the tube. At some point these two opposing tendencies reach a stalemate.

Mercury is a dense liquid. On average, at sea level, a column of mercury about 760 mm high will be supported by air pressure. The pressure that is exerted by a column of mercury 760 mm high is called 1 **atmosphere (atm).** The pressure unit, 1 **millimeter of mercury (mmHg),** is often called a *torr* (after Torricelli).

$$1 \text{ atm} = 760 \text{ mmHg} = 760 \text{ torr}$$

The relationship of the atmosphere to SI units is

$$1 \text{ atm} = 101{,}325 \text{ Pa} = 101.325 \text{ kPa}$$

For approximate work, it is helpful to remember that 1 atm is about 100 kPa.

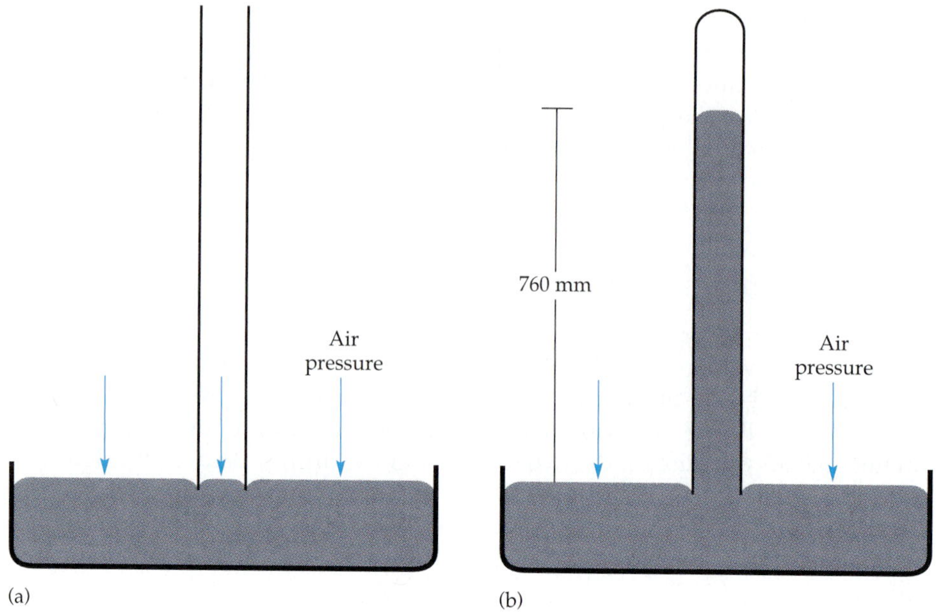

FIGURE 7.3 ▶
Measurement of air pressure with a mercury barometer. (a) The mercury levels are equal inside and outside the *open-end* tube because the tube is open to the atmosphere and filled with air. (b) A column of mercury 760 mm high is maintained in the *closed-end* tube. The space above the mercury is devoid of air and contains only a trace of mercury vapor.

Several other pressure units are widely used. Weather reports in the United States often include atmospheric pressure in *inches of mercury* (*in.Hg*).

$$1 \text{ atm} = 29.921 \text{ in.Hg}$$

Engineers generally use *pounds per square inch* (*lb/in.*^{2} *or psi*) for practical applications such as steam pressure in boilers and turbines.

$$1 \text{ atm} = 14.696 \text{ lb/in.}^2$$

Respiratory therapists use the unit *centimeters of water,* a useful scale for measuring small differences in pressure.

$$1 \text{ mmHg} = 1.36 \text{ cmH}_2\text{O}$$

Robert Boyle (1627–1691) published *The Sceptical Chymist* in 1661 in which he argued that theories are no better than the experiments on which they are based. Gradually, this point of view was accepted, and Boyle's text marked a turning point for the importance of experimentation. His experiments on air helped to found modern science.

7.4 Boyle's Law: The Pressure–Volume Relationship

A simple gas law, discovered by the Irish chemist Robert Boyle in 1662, describes the relationship between the pressure and volume of a gas. **Boyle's law** states that *for a given amount of gas at a constant temperature, the volume of a gas varies* inversely *with its pressure.* That is, in a closed container of gas when the pressure increases, the volume decreases; when the pressure decreases, the volume increases.

Think of gases as pictured in the kinetic-molecular theory (Figure 7.4). A gas exerts a particular pressure because the gas molecules bounce against the container walls with a certain frequency and speed. If the volume of the container is expanded while the amount of gas remains fixed, the number of molecules per unit volume of gas decreases. The frequency with which molecules strike a unit area of the container walls decreases, and the gas pressure decreases. Thus, as the volume of a gas is increased, its pressure decreases.

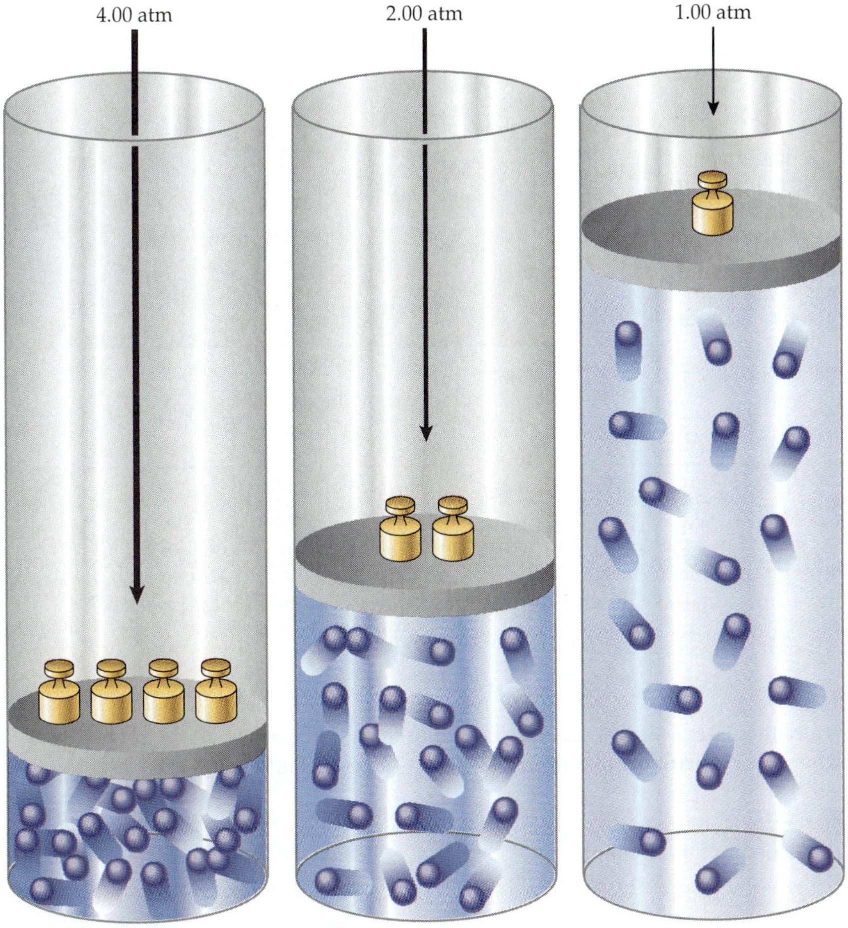

◀ **FIGURE 7.4**
A kinetic-molecular theory view of Boyle's law. As the pressure is reduced from 4.00 atm to 2.00 atm and then to 1.00 atm, the volume of the gas doubles and then doubles again.

Mathematically, for a given amount of gas at a constant temperature, Boyle's law is written

$$V \propto \frac{1}{P}$$

where the symbol $\propto$ means "is proportional to." This may be changed to an equation by inserting a proportionality constant, a.

$$V = \frac{a}{\mathrm{P}}$$

Multiplying both sides of the equation by P, we get

$$PV = a$$

Another way to state Boyle's law, then, is that for a given amount of gas at a constant temperature, the product of the pressure and volume is a constant. This is an elegant and precise, if somewhat abstract, way of summarizing a lot of experimental data. If the product $P \times V$ is to be constant, then if V increases P must decrease, and vice versa. This relationship is demonstrated in Figure 7.5 by a pressure–volume graph.

Boyle's law has a number of practical applications perhaps best illustrated by some examples. In Example 7.1, we see how to *estimate* an answer. Sometimes an estimation is all we need. Even when we want a quantitative answer, however, the estimate helps us to determine whether or not our answer is reasonable.

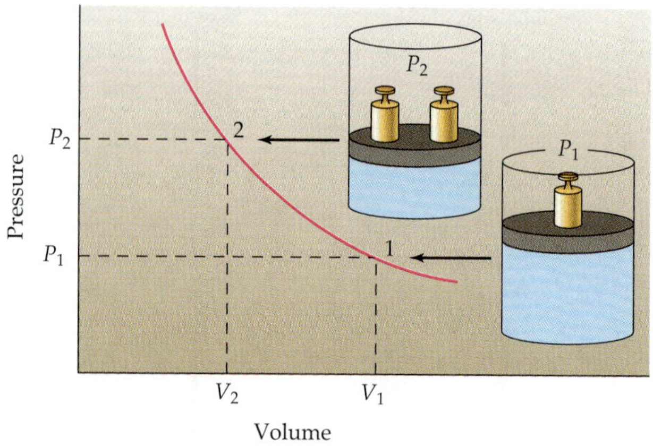

FIGURE 7.5 ▶ A graphical representation of Boyle's law. As the pressure on the gas is increased, its volume decreases. When the pressure is doubled ($P_2 = 2 \times P_1$), the volume of the gas decreases to one-half its original value ($V_2 = 1/2 \times V_1$). The pressure–volume product is a constant ($PV = a$)

Example 7.1

A gas is enclosed in a cylinder fitted with a piston. The volume of the gas is 2.00 L at 0.524 atm. The piston is moved to increase the gas pressure to 5.15 atm. Which of the following is a reasonable value for the volume of the gas at the greater pressure?

0.20 L 0.40 L 1.00 L 16.0 L

Solution

The pressure increase is almost tenfold. The volume should drop to about one-tenth of the initial value. We estimate a volume of 0.20 L. (The calculated value is 0.203 L.)

Practice Exercise

A gas is enclosed in a 10.2-L tank at 1208 mmHg. Which of the following is a reasonable value for the pressure when the gas is transferred to a 30.0-L tank?

25 lb/in.^{2} 300 mmHg 400 mmHg 3600 mmHg

Gases are usually stored under high pressure, even though they will be used at atmospheric pressure. This allows a large quantity of gas to be stored in a small volume.

Examples 7.2 and 7.3 illustrate quantitative calculations using Boyle's law. Note that in these applications any units can be used for pressure and volume, as long as the same units are used throughout a calculation. As long as we use the *same* sample of a confined gas at a constant temperature, the product of the initial volume (V_1) times the initial pressure (P_1) is equal to the product of the final volume (V_2) times the final pressure (P_2). Thus, the following useful equation representing Boyle's law can be written.

$$V_1P_1 = V_2P_2$$

Example 7.2

A cylinder of oxygen has a volume of 2.00 L. The pressure of the gas is 1470 psi at 20 °C. What volume will the oxygen occupy at standard atmospheric pressure (14.7 psi), assuming no temperature change?

Solution

It is most helpful to first separate the initial from the final condition.

Initial	Final	Change
$P_1 = 1470$ psi	$P_2 = 14.7$ psi	↓
$V_1 = 2.00$ L	$V_2 = ?$	↑

Then use the equation $V_1P_1 = V_2P_2$ and solve for the desired volume or pressure. In this case, we solve for V_2.

$$V_2 = \frac{V_1P_1}{P_2}$$

$$V_2 = \frac{(2.00\ \text{L})(1470\ \cancel{\text{psi}})}{14.7\ \cancel{\text{psi}}} = 200\ \text{L}$$

Practice Exercise

A sample of air occupies 73.3 mL at 98.7 atm and 0 °C. What volume will the air occupy at 4.02 atm and 0 °C?

Because the final pressure in Example 7.2 is *less than* the initial pressure, we expect the final volume to be *larger than* the original volume and it is.

Boyle's Law and Breathing

The pressure–volume relationship helps to explain the mechanics of breathing. When we breathe in (inspire), the diaphragm is lowered and the chest wall is expanded, increasing the volume of the chest cavity (Figure 7.6). According to Boyle's law, the pressure inside the cavity must decrease. Outside air enters the lungs because it is at a higher pressure than the air in the chest cavity. When we breathe out (expire), the diaphragm rises and the chest wall contracts, decreasing the volume of the chest cavity. The pressure is increased and some air is forced out.

During normal inspiration, the pressure inside the lungs drops about 3 mmHg below atmospheric pressure. During expiration, the internal pressure is about 3 mmHg above atmospheric pressure. About half a liter of air is moved in and out of the lungs in this process, and this normal breathing volume is referred to as the *tidal volume*. The *vital capacity* is the maximum volume of air that can be forced from the lungs and ranges from 3 to 7 L, depending on the individual. A pressure inside the lungs 100 mmHg greater than the external pressure is not unusual during such a maximum expiration.

The lungs are never emptied completely, however. The space around the lungs is maintained at a slightly lower pressure than are the lungs themselves, causing the lungs to be kept partially inflated by the higher pressure within them. If a lung, the diaphragm, or the chest wall is punctured, allowing the two pressures to equalize, the lung will collapse. Sometimes a damaged lung is collapsed intentionally to give it time to heal. The lung reinflates after the opening is closed.

People were breathing long before Boyle formulated his law, but it is satisfying to understand how the process works. We get more than just satisfaction out of science, however. An understanding of the pressure–volume relationship has enabled us to keep people alive. When paralysis prevents people from being able to breathe, they can be kept alive by artificial respirators. The iron lung, which kept many polio victims alive during the 1950s, is a sealed chamber connected to a compressor and bellows (Figure 7.7). The pressure in the chamber is varied rhythmically. When the bellows moves out, the pressure in the chamber is reduced. The pressure around the nose and mouth (outside the tank) is greater than the pressure on the chest (inside the tank), so air flows in and fills the lungs. When the bellows is moved in, pressure in the tank increases, and air is expelled from the lungs.

The iron lung, designed to enclose the patient completely (except for the head), is cumbersome and uncomfortable. It has been replaced by respirators that enclose the chest only. In fact, the whole area of respiratory therapy has become far more sophisticated in recent years. Specialists in this area are an indispensable part of the medical team.

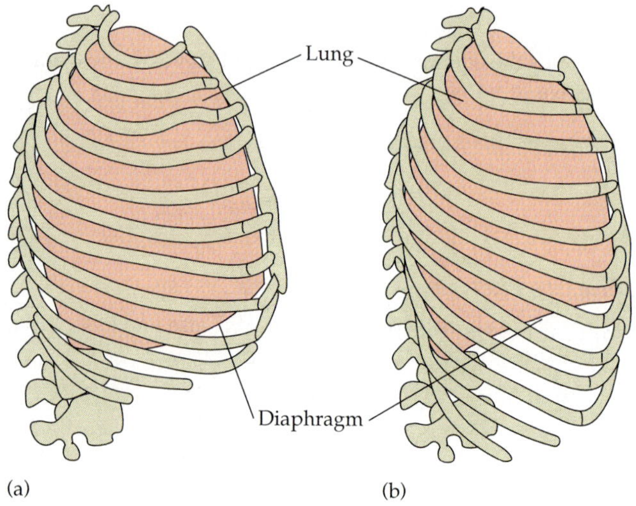

FIGURE 7.6 ▶

The mechanics of breathing. (a) Inspiration. The diaphragm is pulled down and the rib cage is lifted up and out, increasing the volume of the chest cavity. (b) Expiration. The diaphragm is relaxed; the rib cage is down, and the volume of the chest cavity decreases.

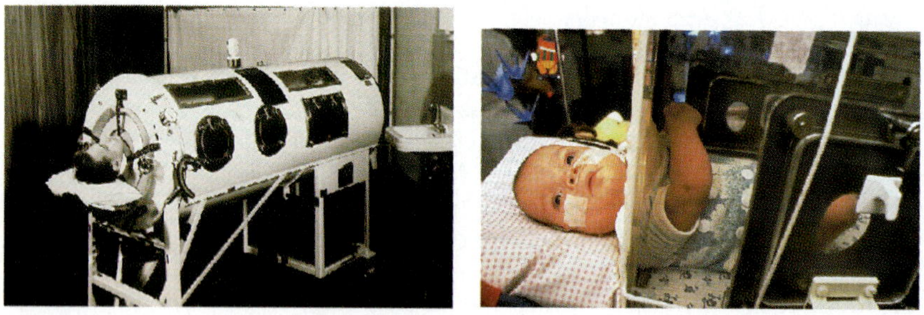

(a) (b)

▲ FIGURE 7.7

An iron lung uses changes in pressure to force air into and out of the lungs. (a) The older version enclosed the entire body except for the head. (b) Baby in a modern iron lung.

Example 7.3

A space capsule is equipped with a tank of air that has a volume of 0.10 m^3. The air is under a pressure of 110 atm. After a space walk, during which the cabin pressure drops to zero, the cabin is closed and filled with the air from the tank. What will be the final pressure if the volume of the capsule is 11 m^3?

Solution

Initial	Final	Change
$V_1 = 0.10\ \text{m}^3$	$V_2 = 11\ \text{m}^3$	↑
$P_1 = 110\ \text{atm}$	$P_2 = ?$	↓

In this case we solve for P_2.

$$P_2 = \frac{V_1 P_1}{V_2}$$

$$P_2 = \frac{(0.10\ \cancel{\text{m}^3})(110\ \text{atm})}{11\ \cancel{\text{m}^3}} = 1.0\ \text{atm}$$

As expected, the final pressure is less than the initial pressure.

Practice Exercise

A sample of nitrogen gas occupies 80.0 mL at 1.00 atm pressure. At what pressure will the nitrogen gas occupy 60.0 mL, assuming the temperature is constant?

7.5 Charles's Law: The Temperature–Volume Relationship

In 1787, the French physicist Jacques Charles (1746–1823) studied the relationship between volume and temperature of gases. He found that when a fixed quantity of gas is cooled at constant pressure, its volume decreases. When the gas is heated, its volume increases. Temperature and volume vary directly; that is, they rise or fall together. But this law requires a bit more thought. If a quantity of gas that occupies 1.00 L is heated from 100 °C to 200 °C at constant pressure, the volume does not double but only increases to about 1.27 L. The relationship between temperature and volume is not as tidy as it may seem on first impression.

Zero pressure or zero volume really means zero—no pressure or volume to be measured. Zero degrees Celsius (0 °C) means only the freezing point of water. This zero point is arbitrarily set, much like mean sea level is the arbitrary zero used to measure altitudes on Earth. Temperatures below 0 °C are often encountered, as are altitudes below sea level.

Charles noted that for each Celsius degree rise in temperature, the volume of a gas expands by 1/273 of its volume at 0 °C. For each Celsius degree drop in temperature, the volume of a gas decreases by 1/273 of its value at 0 °C. If we plot volume against temperature we get a straight line (Figure 7.8). We can *extrapolate* the line beyond the range of measured temperatures to the temperature at which the volume of the gas would become zero. This temperature is −273.15 °C. In 1848, William Thomson (Lord Kelvin) made this temperature the zero point on an *absolute* temperature scale, which is now called the **Kelvin scale.** The unit of temperature on this scale is the kelvin (K) (Section 1.9).

Before a gas ever reaches this temperature, however, it liquefies—and then the liquid freezes—so this is an exercise for the imagination.

A modern statement of **Charles's law,** then, is that *the volume of a fixed amount of a gas at a constant pressure is directly proportional to its Kelvin temperature.* Mathematically, this relationship is expressed as

$$V \propto T$$

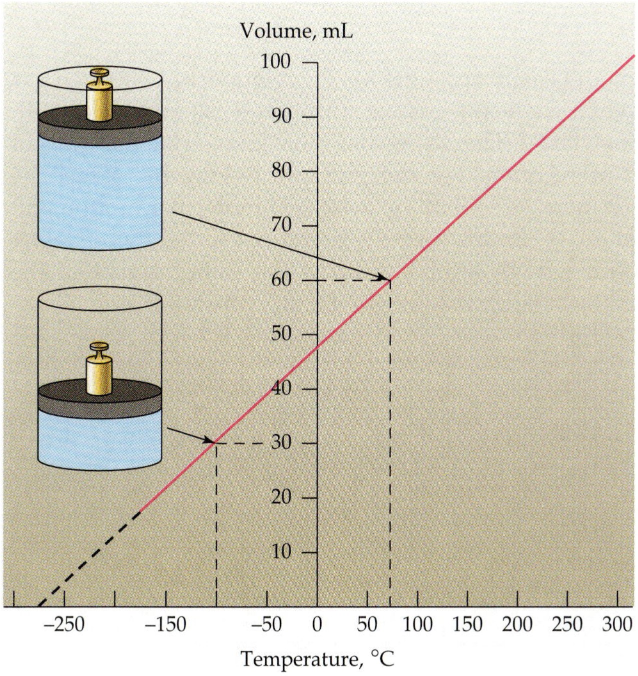

◀ **FIGURE 7.8**

Charles's law relates gas volume to temperature at constant pressure. When the gas shown has been cooled to about 70 °C, its volume is 60 mL. In the temperature interval from about 70 °C to −100 °C, the volume drops to 30.0 mL. The volume continues to fall as the temperature is lowered. The extrapolated line intersects the temperature axis (corresponding to a volume of zero) at about −270 °C.

(a) (b)

◀ **FIGURE 7.9**
A dramatic illustration of Charles's law. (a) Liquid nitrogen (boiling point, −196 °C) cools the balloon and its contents to a temperature far below room temperature. (b) As the balloon warms to room temperature, the volume of air increases proportionately (about fourfold).

In the form of an equation, this becomes

$$V = bT \quad \text{or} \quad \frac{V}{T} = b \text{ (with constant pressure and mass)}$$

where b is a constant. To keep V/T equal to a constant value, when the temperature increases, the volume must also increase. When the temperature decreases, the volume must also decrease (Figure 7.9).

For all gas law calculations involving temperature, absolute temperatures in kelvins must be used (not °C or °F).

As long as we are using the *same* sample of trapped gas at a constant pressure, the initial volume (V_1) divided by the initial *absolute* temperature (T_1) is equal to the final volume (V_2) divided by the final *absolute* temperature (T_2). The following equation is useful in solving Charles's law problems.

$$\frac{V_1}{T_1} = \frac{V_2}{T_2}$$

The kinetic-molecular model easily accounts for the relationship between gas volume and temperature. When we heat a gas we supply the gas molecules with energy and they begin to move faster. These speedier molecules strike the walls of their container harder and more often. For the pressure to stay the same, the volume of the container must increase so that the increased molecular motion will be distributed over a greater space. In this way, the pressure exerted by the faster molecules in the larger volume (high temperature) is the same as that of the slower-moving molecules in the smaller volume (low temperature).

Example 7.4

A balloon indoors, where the temperature is 27 °C, has a volume of 2.00 L. What will its volume be outdoors, where the temperature is −23 °C? (Assume no change in the gas pressure.)

Solution

First, convert both temperatures to the Kelvin scale

$$T(\text{K}) = T(°\text{C}) + 273$$

The initial temperature is

$$27 + 273 = 300\ \text{K}$$

and the final temperature is

$$-23 + 273 = 250\ \text{K}$$

Initial	**Final**	**Change**
$t_1 = 27\ °\text{C}$	$t_2 = -23\ °\text{C}$	↓
$T_1 = 300\ \text{K}$	$T_2 = 250\ \text{K}$	↓
$V_1 = 2.00\ \text{L}$	$V_2 = ?$	↓

Solving the equation

$$\frac{V_1}{T_1} = \frac{V_2}{T_2}$$

for V_2, we have

$$V_2 = \frac{V_1 T_2}{T_1}$$

$$V_2 = \frac{(2.00\ \text{L})(250\ \cancel{\text{K}})}{300\ \cancel{\text{K}}} = 1.67\ \text{L}$$

As we expected, the volume decreased because the temperature decreased.

Practice Exercise

A sample of oxygen gas occupies a volume of 2.10 L at 25 °C. What volume will this sample occupy at 150 °C? (Assume no change in pressure.)

Example 7.5

What would be the final volume of the balloon in Example 7.4 if it were measured where the temperature was 47 °C? (Assume no change in pressure.)

Solution

The initial temperature is

$$27 + 273 = 300\ \text{K}$$

The final temperature is

$$47 + 273 = 320\ \text{K}$$

Initial	**Final**	**Change**
$t_1 = 27\ °\text{C}$	$t_2 = 47\ °\text{C}$	↑
$T_1 = 300\ \text{K}$	$T_2 = 320\ \text{K}$	↑
$V_1 = 2.00\ \text{L}$	$V_2 = ?$	↑

In this case, since the temperature increases, the volume must also increase.

$$V_2 = \frac{V_1 T_2}{T_1}$$

$$V_2 = \frac{(2.00\ \text{L})(320\ \cancel{\text{K}})}{300\ \cancel{\text{K}}} = 2.13\ \text{L}$$

Practice Exercise

At what Celsius temperature will the initial volume of oxygen in the preceding practice exercise occupy 0.750 L? (Assume no change in pressure.)

7.6 Avogadro's Law: The Molar Volume of a Gas

In Section 5.2, we saw that Amadeo Avogadro proposed an important hypothesis to explain Gay-Lussac's law of combining volumes. His hypothesis was that equal numbers of molecules of different gases compared at the same temperature and pressure occupy equal volumes. Let's now restate Avogadro's hypothesis in the form generally called **Avogadro's law:** *At a fixed temperature and pressure, the volume of a gas is directly proportional to the number of molecules of gas or to the number of moles of gas, n.* If we double the number of moles of gas at a fixed T and P, the volume of the gas doubles. Because the mass of a gas is proportional to the number of moles, doubling the *mass* of a gas also doubles its volume. Mathematically, we can state Avogadro's law as

$$V \propto n \quad \text{or} \quad V = cn \text{ (where } c \text{ is a constant)}$$

When we use Avogadro's hypothesis to compare different gases, the gases must be at the same temperature and pressure. A convenient temperature/pressure combination for such comparisons is 0 °C (273 K) and 1 atm (760 torr), known as **standard conditions of temperature and pressure (STP).**

Standard conditions of temperature and pressure (STP)

$T = 273$ K

$P = 1$ atm

Suppose that in comparing different gases we use STP as the fixed temperature and pressure and Avogadro's number as the number of molecules present. Avogadro's hypothesis states that under these conditions, 1 mol (6.022×10^{23} molecules) of *all* gases should occupy the same volume. By experiment, this **molar volume of a gas** at *STP* is found to be 22.428 L of H_2, 22.404 L of N_2, 22.394 L of O_2, 22.360 L CH_4, and so on. To *three* significant figures, we can state that

$$1 \text{ mol gas} = 22.4 \text{ L gas (at STP)}$$

Figure 7.10 pictures a volume of 22.4 L and relates it to some familiar objects. The 22.4-L container would hold 28.0 g of N_2, 32.0 g of O_2, or 44.0 g of CO_2.

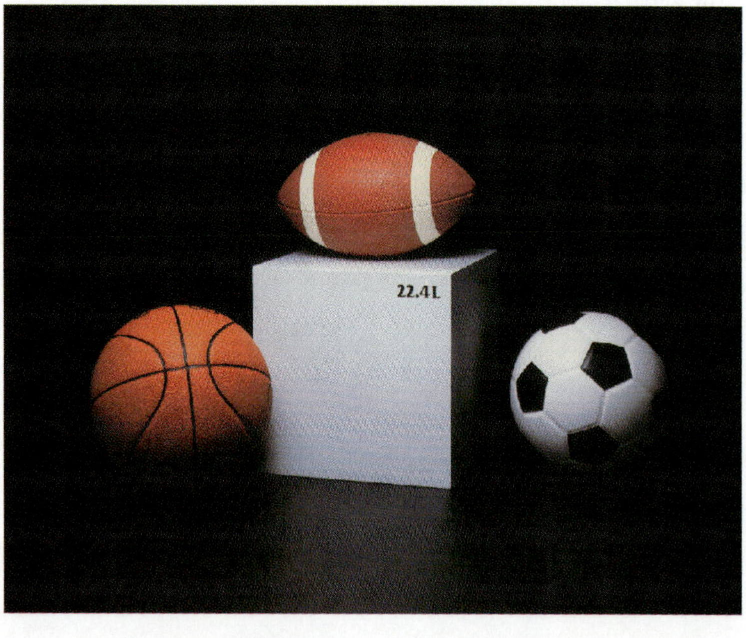

FIGURE 7.10 ► The molar volume of a gas visualized. The wooden cube has the same volume as 1 mol of gas at STP: 22.4 L. By contrast, the basketball has a volume of 7.5 L; the soccer ball, 6.0 L; and the football, 4.4 L.

Example 7.6

Calculate the volume occupied by 4.11 g of methane gas at STP.

Solution

We must convert the mass of gas to an amount in moles, and then use the molar volume relationship as a conversion factor to go from the amount of gas to its volume at STP. We can do all of this in a single setup.

$$4.11 \text{ g } CH_4 \times \frac{1 \text{ mol } CH_4}{16.04 \text{ g } CH_4} \times \frac{22.4 \text{ L } CH_4}{1 \text{ mol } CH_4} = 5.74 \text{ L } CH_4$$

Practice Exercise

Solid carbon dioxide, called "dry ice," is useful in maintaining frozen foods because it vaporizes to $CO_2(g)$ rather than melting to a liquid. How many grams of dry ice can be produced from 5.00 L of $CO_2(g)$ measured at STP?

7.7 The Combined Gas Law

From the three simple gas laws, it seems reasonable that the volume of a gas (V) should be directly proportional to the Kelvin temperature (T) and to the amount of gas (n), and inversely proportional to the pressure (P). That is,

$$V \propto \frac{nT}{P}$$

Or, expressed as an equation rather than a proportionality,

$$\frac{PV}{nT} = \text{constant}$$

This **combined gas law** is most useful in situations where a fixed quantity of gas is described under an initial set of conditions, the gas is subjected to a change, and a question is asked about the final set of conditions. In these cases we write

$$\frac{V_1P_1}{T_1} = \frac{V_2P_2}{T_2}$$

Example 7.7

A balloon is partially filled with helium on the ground at 27 °C and 740 mmHg pressure. Its volume is 10.0 m³. What would the volume be at a higher altitude where the pressure is 370 mmHg and the temperature is −23 °C?

Solution

The temperature decreases from 27 °C to −23 °C, or from

$$27 + 273 = 300 \text{ K}$$

to

$$-23 + 273 = 250 \text{ K}$$

Initial	Final	Change
t_1 = 27 °C	t_2 = −23 °C	↓
T_1 = 300 K	T_2 = 250 K	↓
P_1 = 740 mmHg	P_2 = 370 mmHg	↓
V_1 = 10.0 m³	V_2 = ?	

Helpful hint: Regardless of which variable is to be determined, it is recommended that you solve the equation for the unknown *before* substituting values into the equation. Rearranging a few letters takes less time than rearranging and rewriting complex terms.

Solving the combined gas law equation for V_2, we have

$$V_2 = \frac{V_1P_1T_2}{P_2T_1}$$

Substituting the given values (and using a calculator) yields

$$V_2 = \frac{(10.0\ \text{m}^3)(740\ \cancel{\text{mmHg}})(250\ \cancel{\text{K}})}{(370\ \cancel{\text{mmHg}})(300\ \cancel{\text{K}})} = 16.7\ \text{m}^3$$

Practice Exercise

A sample of helium occupies 38.4 mL at 40 °C and 680 mmHg. What volume will the helium occupy at 80 °C and 720 mmHg?

Example 7.8

What is the volume at STP of a sample of carbon dioxide whose volume at 25 °C and 4.00 atm is 10.0 L?

Solution

Initial	Final	Change
$t_1 = 25\ °\text{C}$	$t_2 = 0\ °\text{C}$	
$T_1 = 298\ \text{K}$	$T_2 = 273\ \text{K}$	↓
$P_1 = 4.00\ \text{atm}$	$P_2 = 1.00\ \text{atm}$	↓
$V_1 = 10.0\ \text{L}$	$V_2 = ?$	

$$V_2 = \frac{V_1P_1T_2}{P_2T_1}$$

$$V_2 = \frac{(10.0\ \text{L})(4.00\ \cancel{\text{atm}})(273\ \cancel{\text{K}})}{(1.00\ \cancel{\text{atm}})(298\ \cancel{\text{K}})} = 36.6\ \text{L}$$

Practice Exercise

A sample of hydrogen gas has a volume of 1.10 L at –40 °C and 0.520 atm. What will be its volume at STP?

Densities of gases are usually reported in the literature in units of grams per liter at STP. We can use the molar mass of a gas and its density at STP to calculate its molar volume. For example, the molar mass of $N_2(g)$ is 28.0 g/mol and its density at STP is 1.25 g/L. Dividing, we get

$$\frac{28.0\ \cancel{\text{g}}/\text{mol}}{1.25\ \cancel{\text{g}}/\text{L}} = 22.4\ \text{L/mol}$$

For $O_2(g)$ the density at STP is 1.43 g/L. The molar mass is 32.0 g/mol. Dividing, we get

$$\frac{32.0\ \cancel{\text{g}}/\text{mol}}{1.43\ \cancel{\text{g}}/\text{L}} = 22.4\ \text{L/mol}$$

Of course, these are just the values that we expect for the molar volume of a gas at STP. Since we know the molar volume, we can calculate the density of a gas at STP, as illustrated in Example 7.9.

Example 7.9

Calculate the density of $CO_2(g)$ at STP.

Solution

The molar mass of CO_2 is 44.0 g/mol. Dividing by the molar volume gives the density.

$$\frac{44.0\ \text{g}/\cancel{\text{mol}}}{22.4\ \text{L}/\cancel{\text{mol}}} = 1.96\ \text{g/L}$$

(The experimental value is 1.98 g/L.)

Practice Exercise

Calculate the density of $H_2(g)$ at STP.

7.8 The Ideal Gas Law

So far we have done calculations in which the quantity of a gas does not change. As we saw above (in our discussion of molar volume), equal volumes of gases at the same temperature and pressure contain equal numbers of moles. Thus, we can write a gas law that takes into account varying quantities of gas. This relationship is called the **ideal gas equation** or **ideal gas law.**

$$PV = nRT$$

In this equation, the pressure is in atmospheres, the volume in liters, and the temperature in kelvins. The number of moles of the gas is given by n. The constant R, which has a value of

$$0.0821\ \frac{\text{L} \cdot \text{atm}}{\text{mol} \cdot \text{K}}$$

is called the **universal gas constant.**

This R value is read as 0.0821 liter-atmosphere per mole-kelvin.

The ideal gas law can be used to calculate any of the four quantities—P, V, n, or T—if the other three are known.

Example 7.10

Use the ideal gas law to calculate the volume occupied by 1.00 mol of nitrogen gas at 244 K and 1.00 atm pressure.

Solution

$$V = \frac{nRT}{P} = \frac{1.00\ \cancel{\text{mol}}}{1.00\ \cancel{\text{atm}}} \times \frac{0.082\ \text{L} \cdot \cancel{\text{atm}}}{\cancel{\text{mol}} \cdot \cancel{\text{K}}} \times 244\ \cancel{\text{K}} = 20.0\ \text{L}$$

Practice Exercise

Determine the volume occupied by 0.200 mol of nitrogen gas at 25 °C and 0.980 atm.

Example 7.11

Use the ideal gas law to calculate the pressure exerted by 0.50 mol of oxygen in a 15-L container at 303 K.

Solution

$$P = \frac{nRT}{V} = \frac{0.50\ \cancel{\text{mol}}}{15\ \cancel{\text{L}}} \times \frac{0.0821\ \cancel{\text{L}} \cdot \text{atm}}{\cancel{\text{mol}} \cdot \cancel{\text{K}}} \times 303\ \cancel{\text{K}} = 0.83\ \text{atm}$$

Practice Exercise

Determine the pressure exerted by 0.0330 mol of oxygen in an 18.0-L container at 40 °C.

7.9 Henry's Law: The Pressure–Solubility Relationship

Henry's law: Solubility of a gas is directly proportional to the pressure of the gas at the surface of the liquid.

In the 1760s, Joseph Priestley invented soda water by dissolving carbon dioxide gas in water. No doubt you have noticed the hissing sound and the formation of bubbles when you opened a bottle of soda pop. Carbon dioxide is dissolved in the liquid, and the bottle is capped under pressure. William Henry, a close friend of John Dalton, spent a great deal of time studying the solubility of gases in liquids. In 1801 he summarized his findings in the law we know as **Henry's law:** The solubility of a gas in a liquid at a given temperature is directly proportional to the pressure of the gas at the surface of the liquid. To get back to the bottle of soda pop: when the bottle was capped under pressure, a certain amount of carbon dioxide was dissolved in the soda pop. When you opened the bottle, the pressure was *reduced* (the hissing sound you heard was pressure being released), and the solubility of the carbon dioxide was *reduced.* (The bubbles of gas you noticed were carbon dioxide escaping from solution).

The pressure–solubility relationship is also used in therapy. In cases of carbon monoxide poisoning (see Section 28.9), the victim is placed in a hyperbaric (high-

A Painful Application of Henry's Law: Divers' Bends

Underwater divers who surface too quickly may experience the painful and dangerous condition known as "the bends."

Divers who go deeply into the sea must take their own supply of air. We breathe about 800 L of air per hour. To have enough air for an hour or so of underwater exploring, the air must be compressed. The compressed air is much more soluble in blood and other body fluids than is air at normal pressures. The divers must be careful to return to the surface slowly or spend considerable time in a decompression chamber where the pressure is gradually lowered. If they don't, excess N_2(g) comes out of solution as tiny bubbles that block blood flow in capillaries and impair nerve transmission. The diver will experience decompression sickness, commonly called *the bends.* The bends are characterized by severe pains in joints and muscles. The person may faint or suffer deafness, paralysis, or even death. People who work in deep mines or tunnels where air pressure is increased to keep water from infiltrating have similar problems.

If the return to normal pressure is slow enough, the excess gases leave the blood gradually. Excess O_2(g) can be used in metabolism, and excess N_2(g) can be removed to the lungs and expelled by normal breathing. About 20 minutes of slow decompression is needed for each atmosphere of pressure above normal that the person experiences. [At the surface, the pressure is 1 atm (14.7 lb/in.^{2}). At a depth of 200 ft, the pressure is 90 lb/in.^{2}.]

Bends can also be avoided by using a helium–oxygen mixture as a substitute for compressed air. Helium is considerably less soluble in blood than is nitrogen, and fewer bubbles form as the diver ascends.

pressure) chamber. This chamber is a device that supplies oxygen at pressures of 3 or 4 atm. More oxygen is forced into the tissues at these pressures to compensate for the lack of oxygen that accompanies carbon monoxide poisoning.

Hyperbaric chambers are also used to treat infections by anaerobic bacteria (bacteria that live without oxygen). Gangrene is one such disease. The organisms that cause gangrene cannot survive in an oxygenated atmosphere. If sufficient oxygen can be forced into the diseased tissues, the infection can be arrested.

7.10 Dalton's Law of Partial Pressures

Although John Dalton is most renowned for his atomic theory (Section 2.1), he had wide-ranging interests, including meteorology. In trying to understand the weather, he did a number of experiments on water vapor in the air. In one experiment, he found that if he added water vapor at a certain pressure to dry air, the pressure exerted by the air would increase by an amount equal to the pressure of the water vapor. Based on this and other experiments, Dalton concluded that each of the gases in a mixture behaves independently of the other gases. Each gas exerts its own pressure. The total pressure of the mixture is equal to the sum of the *partial pressures* exerted by the separate gases (Figure 7.11).

Mathematically, **Dalton's law of partial pressures** is stated as

$$P_{\text{total}} = P_1 + P_2 + P_3 + \cdots$$

where the terms on the right side refer to the partial pressures of gases 1, 2, 3, and so on.

Gases such as oxygen, nitrogen, and hydrogen are nonpolar. They are only slightly soluble in water and are usually collected over water by the technique of displacement (Section 6.3). Such gases always contain water vapor, and the total pressure in the collection vessel is that of the gas plus that of water vapor.

The vapor pressure of water depends on the temperature of the water. (The **vapor pressure** of a substance is the partial pressure exerted by the molecules of the substance that are in the gas phase above the liquid phase of the substance.) The hotter the water, the higher the vapor pressure. If a gas is collected over water,

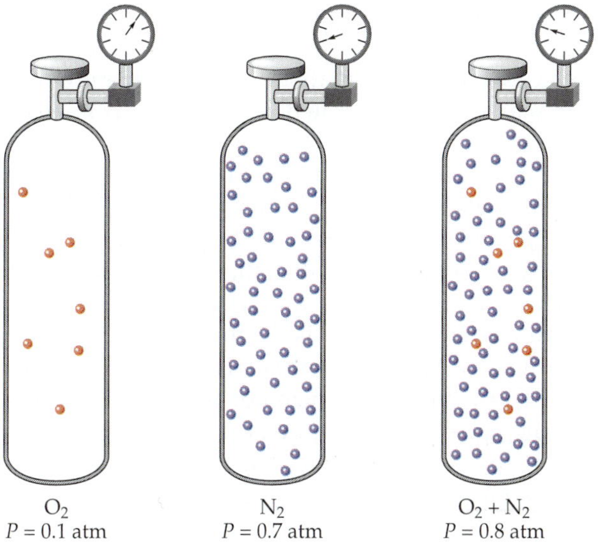

FIGURE 7.11
Dalton's law of partial pressures states that the pressure of a mixture of gases is equal to the sum of the partial pressures of the individual gases.

Table 7.2 Water Vapor Pressure at Various Temperatures

Temperature (°C)	Water Vapor Pressure (mmHg)
0	5
10	9
20	18
30	32
40	55
50	93
60	149
70	234
80	355
90	526
100	760

we can make use of vapor pressure tables (Table 7.2) to calculate the pressure due to the gas alone. We need only subtract the vapor pressure of the water, as determined from the table, from the value for the total pressure within the collection vessel.

Example 7.12

Oxygen is collected over water at 20 °C. The total pressure inside the collection jar is 740 mmHg. What is the pressure due to the oxygen alone?

Solution

From Table 7.2 we find that the vapor pressure of water at 20 °C is 18 mmHg. Because the total pressure is equal to 740 mmHg, we have

$$P_{total} = P_{O_2} + P_{H_2O}$$
$$P_{O_2} = P_{total} - P_{H_2O}$$
$$P_{O_2} = 740\ \text{mmHg} - 18\ \text{mmHg} = 722\ \text{mmHg}$$

Practice Exercise

Hydrogen gas is collected over water at 20 °C. The total pressure inside the collection jar is set at the barometric pressure of 738 mmHg. If the volume of the gas is 246 mL, what mass of hydrogen is collected?

Humidity is a measure of the amount of water vapor in the air. **Relative humidity** compares the actual amount of water vapor in the air with the maximum amount the air could hold at the same temperature. If the temperature is 20 °C and the vapor pressure of water in the atmosphere is 12 mmHg, the relative humidity is

$$\frac{12\ \cancel{\text{mm Hg}}}{18\ \cancel{\text{mm Hg}}} \times 100 = 67\%$$

The 18 mmHg in the denominator is the partial pressure of water at 20 °C obtained from Table 7.2. When the relative humidity is 100%, the air is saturated with water vapor. (Note that 100% relative humidity does not mean that the air is 100% water vapor, just that it is holding as much water as it can. At 20 °C and 100% relative humidity, only about two or three molecules in every 100 molecules of air are water.)

Cool air can hold less water vapor than warm air. As the temperature falls during the night, the atmosphere may become saturated. Water vapor condenses from the air as dew.

The **heat index** relates temperature to relative humidity. Higher humidity inhibits evaporation of sweat, which normally cools your body, thus making you feel hotter.

Respiratory therapists must concern themselves with the humidity of the gases they administer to patients. Normally, as we breathe, the inspired air is saturated with moisture as it passes through the nose and respiratory passages. Oxygen as it comes from a tank is quite dry. If oxygen is administered over a long period of time, it must be humidified to prevent it from irritating the mucous linings of the nasal passages and the lungs. If the oxygen or mixture of gases is conducted through the nose, the therapist may assist the normal body processes by imparting about 30% humidity to the inspired gases. If the breathing mixture is conducted directly to the trachea (bypassing the nose), the therapist saturates the gas mixture with water vapor.

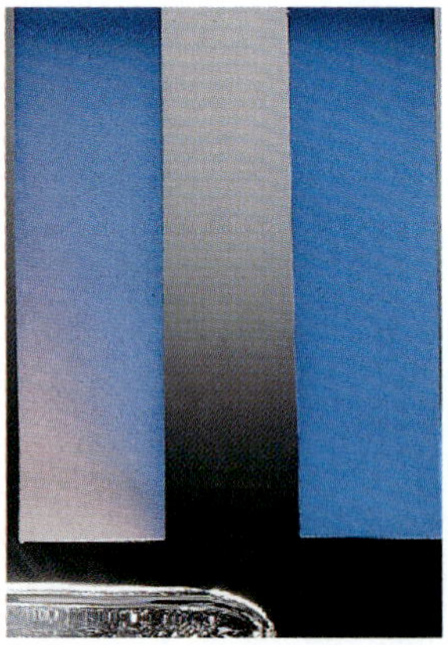

A crude measure of relative humidity can be made with strips of filter paper impregnated with an aqueous solution of cobalt(II) chloride and allowed to dry. In dry air the strip is blue, the color of *anhydrous* $CoCl_2$ (right). In more humid air, the strip acquires the red color of the *hexahydrate*, $CoCl_2 \cdot 6H_2O$ (left). (Hydrates and anhydrous salts are discussed in Section 9.4).

7.11 Partial Pressures and Respiration

When we breathe in, the inspired air becomes moistened and warmed to our body temperature of 37 °C. The air is drawn into our lungs, where it enters a highly branched system of tubes that end in tiny air sacs called alveoli (Figure 7.12). These thin-walled pouches are surrounded by blood vessels that are part of a circulatory system serving every cell in the body.

Inspired air is rich in oxygen (P_{O_2} = 150 mmHg) and poor in carbon dioxide (P_{CO_2} = 0.2 mmHg). The fluid in our cells is poor in oxygen (P_{O_2} = 6 mmHg) and rich in carbon dioxide (P_{CO_2} = 50 mmHg). Our cells use up oxygen in metabolic reactions designed to produce energy. Carbon dioxide accumulates in the cells as a waste product of these reactions. To maintain life, we must transfer the oxygen in the inspired air to our cells. At the same time, the carbon dioxide waste in our cells must be transferred to our lungs and then exhaled to the atmosphere. The transfer of both gases occurs through the process of **diffusion.** In diffusion, gases flow from regions of higher concentration to regions of lower concentration (Figure 7.13). In our bodies, oxygen makes its way to the cells through a pressure gradient—that is, in a series of steps in which oxygen diffuses from areas where its concentration is higher into areas where its concentration is lower. By the same method, carbon dioxide moves in the opposite direction. It makes its way from the cells, where its partial pressure is high, to the atmosphere, where its partial pressure is low. Figure 7.14 shows the steps in the gradient for both gases. Thus, given the mechanical action of the chest and diaphragm (Section 7.4) to get air into and out of the lungs, gases go "downhill" from higher to lower partial pressure.

Diffusion—the flow of gas from a region of higher concentration to a region of lower concentration

Normally, the carbon dioxide level in the blood (not the oxygen level) acts as a trigger for the breathing process. To oversimplify, when carbon dioxide levels build up, we take a breath; if they get too low, we don't. Thus, one of the concerns of a therapist is the partial pressure of carbon dioxide in the blood. Under certain

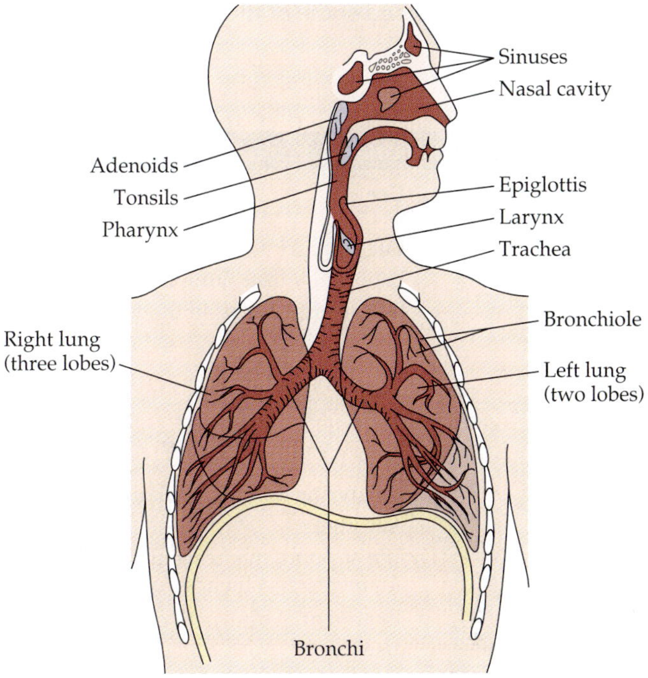

FIGURE 7.12

The respiratory system, showing the route of air through the nose, pharynx (throat), larnyx (voice box), and trachea (windpipe), into the bronchi and bronchioles (bronchial tubes) and ending in the alveoli (air sacs).

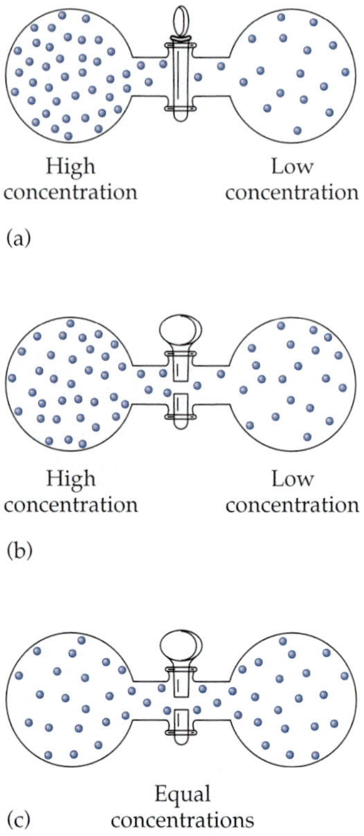

FIGURE 7.13 ▶

A gas (air in the lungs, for example) flows from an area of high concentration to an area of low concentration. (a) With the stopcock closed, no flow is possible. (b) With the stopcock open, there is a net flow of gases from the area of higher concentration on the left to the area of lower concentration on the right. (c) When the two concentrations become equal, net flow stops.

unusual conditions, it is possible for the level of carbon dioxide to fall so low that it fails to trigger the breathing mechanism. Even with access to a plentiful supply of oxygen, the person suffocates because he or she simply stops breathing.

Several commercially supplied gases are used in medicine, the majority in respiratory therapy. Table 7.3 lists these gases and some of their uses.

Table 7.3 Ten Compressed Gases Used in Medicine

Gas	Chemical Formula	Use
Air	N_2 and O_2 (mixture)	Life support
Carbon dioxide	CO_2	Laboratory tests, lung function tests
Carbon dioxide–oxygen mixtures	CO_2 and O_2	Diagnosis, inhalation therapy
Cyclopropane	C_3H_6	Anesthetic
Helium	He	Laboratory analyses
Helium–oxygen mixtures	He and O_2	Inhalation therapy, diagnostic tests
Nitrogen	N_2	Diagnostic testing, inhalation therapy
Nitrous oxide	N_2O	Anesthetic
Oxygen	O_2	Life support, medical emergencies, adjunct to anesthetics
Oxygen–nitrogen mixtures	O_2 and N_2	Treatment of obstructive lung diseases

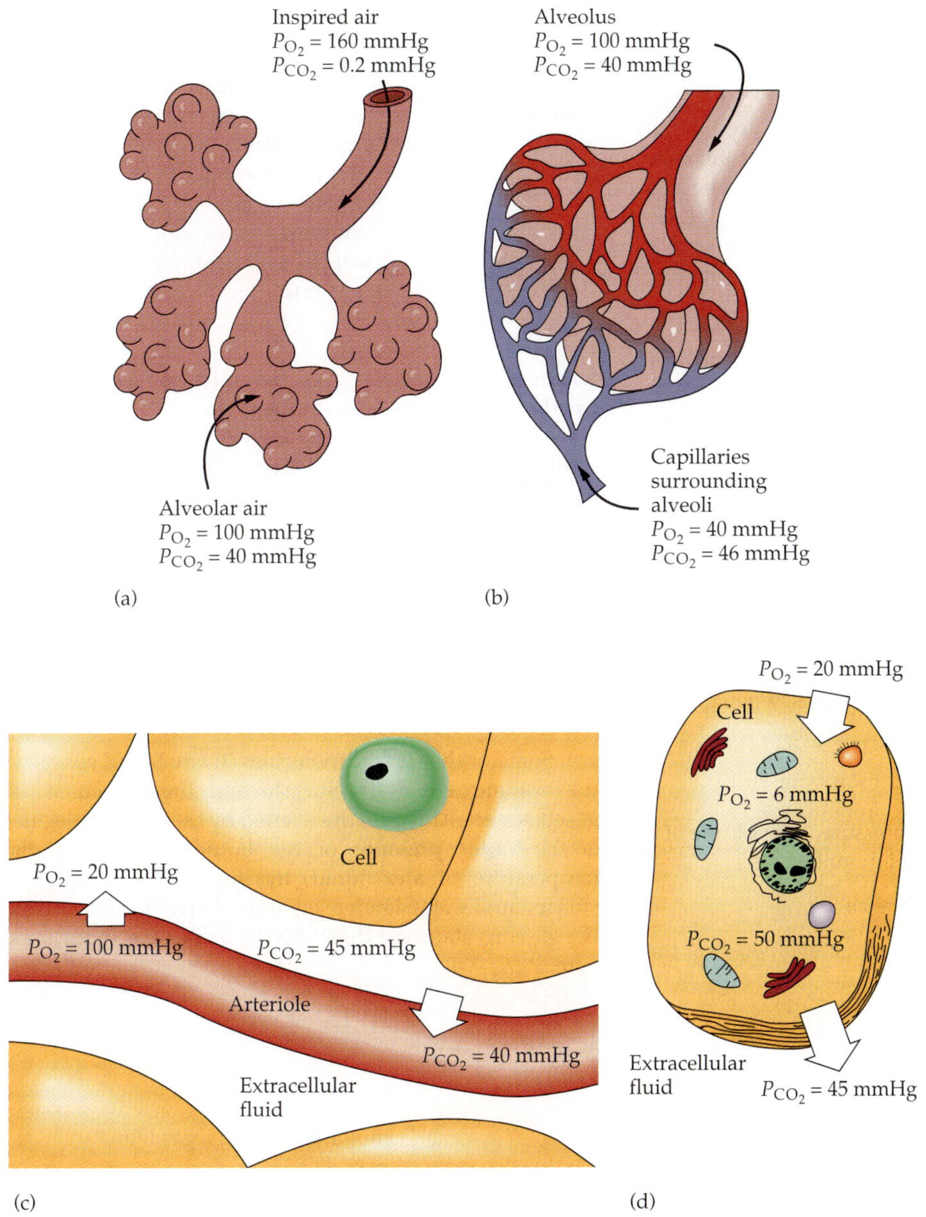

▲ FIGURE 7.14

(a) $O_2(g)$ flows from the inspired air into the alveolar air, and $CO_2(g)$ flows in the opposite direction. (b) $O_2(g)$ diffuses from an arteriole into the extracellular fluid; $CO_2(g)$ flows in the opposite direction. (c) $O_2(g)$ diffuses from an arteriole into the extracellular fluid; $CO_2(g)$ flows in the opposite direction. (d) $O_2(g)$ diffuses into a cell from the extracellular fluid; $CO_2(g)$ moves in the opposite direction. In each case, the flow is from an area of high partial pressure to a region of low partial pressure.

Summary

The air making up Earth's atmosphere is a gaseous mixture of nitrogen, oxygen, argon, trace constituents, and water vapor.

The kinetic-molecular theory of gases pictures any gas as a collection of fast-moving particles constantly bouncing off one another and off the walls of their container. The faster the particles move, the higher the *temperature* of the gas. The *pressure* of the gas is the combined effect of all the forces exerted on the walls of the container by all the moving gas particles.

At sea level, *atmospheric pressure,* the pressure exerted by air on any surface it touches, is 760 mm of mercury (mmHg). The air-pressure unit called the **atmosphere** is equivalent to 760 mmHg, and standard pressure at sea level is 760 mmHg = 1 atm. The definition of pressure is "force per unit area," and two other units for expressing pressure are that used by engineers and others, *pounds per square inch (psi),* and the SI unit, the **pascal,** defined as a force of 1 newton exerted on an area of 1 square meter: 1 Pa = 1 N/m^2.

Boyle's law describes how, at constant temperature, the volume of any gas varies inversely with the gas pressure. When the gas pressure is increased, the gas volume decreases; when the pressure is decreased, the volume increases:

$$V \propto \frac{1}{P} \quad \text{or} \quad V_1P_1 = V_2P_2 \qquad \text{(constant temperature)}$$

Charles's law describes how, at constant pressure, the volume of any gas varies directly with gas temperature. When the gas is heated, its volume increases; when the gas is cooled, its volume decreases:

$$V \propto T \quad \text{or} \quad \frac{V_1}{T_1} = \frac{V_2}{T_2} \qquad \text{(constant pressure)}$$

At a fixed temperature and pressure, the volume of a gas is directly proportional to the amount of gas:

$$V \propto n \quad \text{or} \quad V = cn$$

These three relationships are combined in the **combined gas law:**

$$\frac{V_1P_1}{T_1} = \frac{V_2P_2}{T_2}$$

A gas is said to be at **standard temperature and pressure (STP)** when its pressure is 1 atm and its temperature is 0 °C. The **molar volume of a gas** is the volume occupied at STP by 1 mol of any gas and is equal to 22.4 L.

The combined gas law is made to apply to any amount of gas by factoring into the equation the quantity *n,* the number of moles of gas in the sample. This relationship is called the **ideal gas law:**

$$PV = nRT$$

where the constant *R,* called the **universal gas constant,** has the value 0.082 L·atm/mol·K.

Henry's law states that, at constant temperature, the solubility of a gas in a liquid is directly proportional to the gas pressure at the liquid surface.

When a number of gases are confined to the same container, each gas behaves as if it were alone in the container. In other words, the pressure exerted by any component of a gas mixture is independent of the pressures exerted by all other components. This is **Dalton's law of partial pressures:**

$$P_{\text{total}} = P_1 + P_2 + \cdots$$

Some of the surface molecules in any liquid escape to the volume above the surface and thereby enter the gaseous state. The pressure exerted by these gas molecules is the **vapor pressure** of the liquid. The higher the temperature of any liquid, the higher the number of surface molecules having enough energy to escape into the gaseous state and therefore the higher the vapor pressure: $P_{\text{vapor}} \propto T$.

Key Terms

atmosphere (atm) (7.3)
Avogadro's law (7.6)
barometer (7.3)
Boyle's law (7.4)
Charles's law (7.5)
combined gas law (7.7)
Dalton's law of partial pressures (7.10)
diffusion (7.11)
heat index (7.10)
Henry's law (7.9)
ideal gas equation (7.8)
ideal gas law (7.8)
Kelvin scale (7.5)
kinetic-molecular theory (7.2)
millimeters of mercury (mmHg) (7.3)
molar volume of a gas (7.6)
pascal (7.3)
pressure (7.3)
relative humidity (7.10)
standard conditions of temperature and pressure (STP) (7.6)
universal gas constant (*R*) (7.8)
vapor pressure (7.10)

Review Questions

1. Which layer of the atmosphere lies nearest the surface of the Earth? Which contains the ozone layer?
2. List the three major components of dry air and give the approximate (nearest whole number) volume percent of each.
3. Explain what is meant by the relative humidity of an air sample.
4. Define or explain the following terms.
   a. combined gas law
   b. Henry's law
   c. mmHg
   d. tidal volume

e. vital capacity
f. vapor pressure
g. diffusion

5. Why is atmospheric pressure greater at sea level than at the top of a high mountain?
6. Give a molecular-level description of a gas.
7. Why don't the molecules of a gas settle to the bottom of their container?
8. Give a kinetic-molecular explanation of the origin of gas pressure.
9. What does a mercury barometer measure? How does it work?
10. Why is mercury (rather than water or another liquid) used as the fluid in barometers?
11. State Boyle's law in words and as a mathematical equation.
12. Use the kinetic-molecular theory to explain Boyle's law.
13. What is the advantage of storing gases under high pressure—for example, oxygen used in respiration therapy?
14. State Charles's law in words and in the form of a mathematical equation.
15. Use the kinetic-molecular theory to explain Charles's law.
16. Why must an absolute temperature scale rather than the Celsius scale be used for Charles's law calculations?
17. How is the Kelvin scale related to the Celsius scale?
18. What effect will the following changes have on the volume of a fixed amount of gas?
    a. an increase in pressure at constant temperature
    b. a decrease in temperature at constant pressure
    c. a decrease in pressure coupled with an increase in temperature
19. What effect will the following changes have on the pressure of a fixed amount of gas?
    a. an increase in temperature at constant volume
    b. a decrease in volume at constant temperature
    c. an increase in temperature coupled with a decrease in volume
20. According to the kinetic-molecular theory, (**a**) what change in temperature is occurring if the molecules of a gas begin to move more slowly, on average? (**b**) what change in pressure results when the walls of the container are being struck less often by molecules of the gas?
21. Container A has twice the volume but holds twice as many gas molecules as container B at the same temperature. Use the kinetic-molecular theory to compare the pressures in the two containers.
22. For each of the following, indicate in which container the gas would be expected to have the higher density.
    a. Containers A and B have the same volume and are at the same temperature, but the gas in A is at a higher pressure.
    b. Containers A and B are at the same pressure and temperature, but the volume of A is greater than that of B.
    c. Containers A and B are at the same pressure and volume, but the gas in A is at a higher temperature.
23. What are the standard conditions of temperature and pressure for gases? Why is it useful to define such conditions?
24. What is meant by the *molar volume* of a gas? What is the value of the molar volume of a gas at STP?
25. State the ideal gas law in words and in the form of a mathematical equation.
26. State Dalton's law of partial pressures in words and in the form of a mathematical equation.
27. Describe how Dalton's law of partial pressures is applied to gases collected over water.
28. Interpret what we mean by temperature in terms of the kinetic-molecular theory.
29. When the cap is removed from a bottle of soda pop, carbon dioxide gas escapes. Explain why this occurs.
30. Would it be a good idea to sterilize the water to be used in a fish bowl by boiling? Explain.
31. During inhalation, does the chest cavity expand or contract? Is the pressure inside the lungs decreased or increased? What happens during exhalation?
32. When air is inspired it becomes fully saturated with water vapor as it passes through the trachea on its way to the lungs. What is the approximate partial pressure of water vapor in the air in the alveoli? (*Hint:* At what temperature will the air be?)
33. In which net direction, cells to lungs or lungs to cells, does oxygen move? What about carbon dioxide?
34. Why does oxygen flow from the alveoli to the pulmonary capillaries? Why does carbon dioxide flow in the reverse direction?

Problems

Pressure

35. Carry out the following conversions between pressure units.
    a. 0.985 atm to mmHg
    b. 849 mmHg to atm
    c. 721 mmHg to atm
36. Carry out the following conversions between pressure units.
    a. 4.00 atm to mmHg
    b. 642 mmHg to kPa
    c. 105.7 kPa to mmHg
37. What is the height of a mercury column that exerts a pressure of 213 mmHg?

38. Calculate the height of a mercury column that exerts a pressure of 4.36 atm.

Boyle's Law

39. A sample of helium occupies 521 mL at 1572 mmHg. Assume that the temperature is held constant, and determine (**a**) the volume of the helium at 752 mmHg and (**b**) the pressure, in mmHg, if the volume is changed to 315 mL.
40. A decompression chamber used by deep-sea divers has a volume of 10.3 m^3 and operates at an internal pressure of 4.50 atm. What volume, in m^3, would the air in the chamber occupy if it were at 1.00 atm pressure, assuming no temperature change?
41. Oxygen used in respiratory therapy is stored at room temperature under a pressure of 150 atm in gas cylinders with a volume of 60.0 L.
   a. What volume would the gas occupy at a pressure of 750.0 mmHg? Assume no temperature change.
   b. If the oxygen flow to the patient is adjusted to 8.00 L per minute, at room temperature and 750.0 mmHg, how long will the tank of gas last?
42. The pressure within a 2.25-L balloon is 1.10 atm. If the volume of the balloon increases to 7.05 L, what will be the final pressure within the balloon, if the temperature does not change?

Charles's Law

43. A gas at a temperature of 100 °C occupies a volume of 154 mL. What will the volume be at a temperature of 10 °C, assuming no change in pressure?
44. A balloon is filled with helium. Its volume is 5.90 L at 26 °C. What will be its volume at −78 °C, assuming no pressure change?
45. A 567-mL sample of a gas at 305 °C and 1.20 atm is cooled at constant pressure until its volume becomes 425 mL. What is the new gas temperature?
46. A sample of gas at STP is to be heated at constant pressure until its volume triples. What is the new gas temperature?

Avogadro's Law and Molar Volume

47. Which of the following gas samples at STP contains the greatest number of molecules?
   a. 5.0 g of H_2 **b.** 50 L of SF_6
   c. 1.0×10^{24} molecules of CO_2
48. How many molecules are present in 475 mL of $CO_2(g)$ at STP?
49. What is the volume occupied by 0.837 g of xenon gas at STP?
50. What is the mass of 498 L of neon gas at STP?

The Combined Gas Law

51. A sealed can with an internal pressure of 721 mmHg at 25 °C is thrown into an incinerator operating at 755 °C. What will be the pressure inside the heated can, assuming the container remains intact during incineration?
52. A fixed amount of He exerts a pressure of 775 mmHg in a 1.05-L container at 26 °C. To what value must the temperature be changed to change the gas pressure to 725 mmHg? Assume the volume of gas remains constant.
53. If a fixed amount of gas occupies 2.53 m^3 at a temperature of −15 °C and 191 mmHg, what volume will it occupy at 25 °C and 1142 mmHg?
54. What volume will 575 mL of gas, measured at 23 °C and 725 mmHg, occupy at STP?
55. If the gas present in 4.65 L at STP is changed to a temperature of 15 °C and a pressure of 756 mmHg, what will be the new volume?
56. What volume will 498 mL of a fixed amount of gas, measured at 27 °C and 722 mmHg, occupy at STP?

The Ideal Gas Law

57. Calculate the volume, in liters, of 1.12 mol of $H_2S(g)$ at 62 °C and 1.38 atm.
58. Calculate the volume, in liters, of 0.00600 mol of a gas at 31 °C and 661 mmHg.
59. Calculate the pressure, in atmospheres, of 4.64 mol of CO(g) in a 3.96-L tank at 29 °C.
60. Calculate the pressure, in mmHg, of 0.0108 mol of $CH_4(g)$ in a 0.265-L flask at 37 °C.
61. How many moles of Kr(g) are there in 2.22 L of the gas at 698 mmHg and 45 °C?
62. How many grams of CO(g) are there in 745 mL of the gas at 784 mmHg and 36 °C?

Gas Densities

63. Calculate the density, in grams per liter, for each of the following gases at STP.
   a. CO
   b. AsH_3
   c. Ar
   d. N_2
64. What is the molar mass of a gas that has a density of 2.57 g/L at STP?

Dalton's Law of Partial Pressures

65. Oxygen is collected over water at 30 °C and a barometric pressure of 742 mmHg. What is the partial pressure of $O_2(g)$ in the container?
66. A container holds oxygen at a partial pressure of 0.25 atm, nitrogen at a partial pressure of 0.50 atm, and helium at a partial pressure of 0.20 atm. What is the pressure inside the container?
67. A container is filled with equal numbers of nitrogen, oxygen, and carbon dioxide molecules. The total pressure in the container is 750 mmHg. What is the partial pressure of nitrogen in the container?

68. The pressure of the atmosphere on the surface of Venus is about 100 atm. Carbon dioxide makes up about 97% of the atmospheric gases. What is the partial pressure of carbon dioxide in the atmosphere of Venus?

69. Atmospheric pressure on the surface of Mars is about 6.0 mmHg. The partial pressure of carbon dioxide is 5.7 mmHg. What percent of the Martian atmosphere is carbon dioxide?

Additional Problems

70. The interior volume of the Hubert H. Humphrey Metrodome in Minneapolis is 1.70×10^{10} L. The Teflon-coated fiberglass roof is supported by air pressure provided by 20 huge electric fans. How many moles of air are present in the dome if the pressure is 1.02 atm at 18 °C?

71. A hyperbaric chamber is an enclosure containing oxygen at higher-than-normal pressures used in the treatment of certain heart and circulatory conditions. What volume of $O_2(g)$, from a cylinder at 25 °C and 151 atm, is required to fill a 4.20×10^3-L hyperbaric chamber to a pressure of 3.00 atm at 17 °C?

72. Typically, when a person coughs, he or she first inhales about 2.0 L of air at 1.0 atm and 25 °C. The epiglottis and the vocal chords then shut, trapping the air in the lungs, where it is warmed to 37 °C and compressed to a volume of about 1.7 L by the action of the diaphragm and chest muscles. The sudden opening of the epiglottis and vocal chords releases this air explosively. Just prior to this release, what is the approximate pressure of the gas inside the lungs?

73. *Eleodea* is a green plant that carries out photosynthesis under water.

$$6\,CO_2(g) + 6\,H_2O(l) \longrightarrow C_6H_{12}O_6(aq) + 6\,O_2(g)$$

In an experiment, some *Eleodea* produce 122 mL of $O_2(g)$, collected over water at 743 mmHg and 21 °C. What mass of oxygen is produced? What mass of glucose ($C_6H_{12}O_6$) is produced concurrently?

74. In an attempt to verify Avogadro's hypothesis, small quantities of several different gases were weighed in 100.0-mL syringes. Masses were determined to the nearest 0.1 mg on an analytical balance. The following masses were obtained: 0.0080 g of H_2, 0.1112 g of N_2, 0.1281 g of O_2, 0.1770 g of CO_2, 0.2320 g of C_4H_{10}, and 0.4824 g of CCl_2F_2. Within experimental error, are these results consistent with Avogadro's hypothesis? Explain.

75. A novel energy storage system involves storing air under high pressure. (Energy is released when the air is allowed to expand.) How many cubic feet of air, measured at standard atmospheric pressure (14.7 psi), can be compressed into a 19-million-ft^3 underground cavern at a pressure of 1070 pounds per square inch (psi)?

76. Use the ideal gas equation to calculate the pressure exerted by 1.00 mol of $CO_2(g)$ when it is confined to a volume of 2.50 L at 298 K.

77. Calculate the temperature in °C of 0.78 mol of oxygen if its volume is 68 L and its pressure is 0.37 atm.

78. A sample of intestinal gas was collected and found on analysis to consist of 44% CO_2, 38% H_2, 17% N_2, 1.3% O_2, and 0.003% CH_4. (The percentages do not add to exactly 100% because of rounding.) What is the partial pressure of each gas if the total pressure in the intestine is 820 mmHg?

79. If the P_{H_2O} in air is 12 mmHg on a day when the temperature is 20 °C, what is the relative humidity?

80. Two flasks are connected. Flask A contains only oxygen at a pressure of 460 mmHg. Flask B has oxygen at a partial pressure of 320 mmHg and nitrogen at a partial pressure of 240 mmHg.
   a. Which direction will the net flow of oxygen take?
   b. Which direction will the net flow of nitrogen take?

81. A person at rest breathes about 80 mL of air per second. How long does it take to breathe 22.4 L of air?

8 Liquids and Solids

Giant icebergs are composed of water in the solid state. If they could be easily transported, icebergs could be melted to provide abundant liquid water anywhere on Earth.

In Chapter 4, we considered the subject of bonding. Our primary concern then was how atoms combine to form molecules and why elements react to form compounds. Now we're going to expand our consideration of bonding, and this time we will be looking for an answer to this question: What makes a substance a solid rather than a liquid or a liquid rather than a gas? Some force of attraction holds the particles of solids and liquids in contact with one another. The particles of a gas fly about at random; those of liquids or solids cling together. Gas particles interact so little with one another that a collection of them retains neither a specific volume nor a specific shape. But particles of a liquid are held together with sufficient force that a collection of them has a specific volume. And particles of a solid are so rigidly held together that not only the volume but also the shape of a given sample is fixed (Figure 8.1). Are these mysterious forces of attraction important? They are if appearances are important to you. You are built of solids and liquids. If you think you've got it all together, you should thank the special forces of attraction that give shape and volume to the condensed forms of matter.

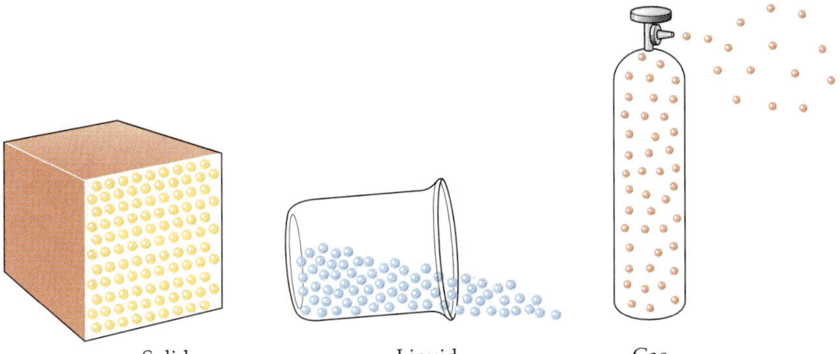

◀ **FIGURE 8.1**

Solids, liquids, and gases compared. (a) In solids, the particles (atoms, molecules, or ions) are close together and rigidly held in place; solids retain both shape and volume. (b) In liquids, the particles are close together but free to move over and around each other; liquids retain volume but not shape. (c) In gases, the particles are far apart and moving at random; gases maintain neither shape nor volume.

In this chapter, we're also going to consider changes in the state of matter—what happens when a solid is converted to a liquid or a liquid to a gas. Is this important? Well, consider perspiration. From the amount of advertising directed against this lowly liquid, you would think it was an unnecessary annoyance. However, were it not for the conversion of this liquid to a vapor on skin surfaces, we would find it difficult just to survive in the temperate and tropical zones of our planet, let alone carry out vigorous physical activity.

8.1 Sticky Molecules: Intermolecular Forces

In Chapter 4 we focused on the forces, called *chemical bonds,* that bind atoms to one another *within* molecules. These bonds determine the geometric shapes and resulting chemical properties of molecules. There are also attractive forces that exist *between* molecules—**intermolecular forces**—that are quite important. Intermolecular forces determine the *physical* properties of liquids and solids. In fact, if there were no intermolecular forces, there would be no liquids or solids—everything would be a gas.

Intermolecular forces pertain to the forces of attraction *between* a molecule and neighboring molecules. *Intramolecular forces* pertain to those *within* a molecule due to chemical bonding.

In our physical model of gases (Chapter 7), we visualize speedy, energetic molecules undergoing frequent collisions and never coming to rest or clumping together. Intermolecular forces are relatively unimportant in gases, because the molecules are far apart. The ideal gas law (Section 7.8) actually assumes that they do not exist at all. If intermolecular forces are sufficiently strong, however, molecules cluster together into the liquid or solid state. In our discussion of liquids and solids in subsequent sections, we will consider the different kinds of intermolecular forces and their relative strengths. However, let's first consider some fundamental phenomena involving liquids and solids.

We can gain a pretty good initial understanding of liquids and solids with this simple observation: *Molecules have a tendency to remain apart from each other.* Intermolecular forces of attraction are most likely to overcome this tendency at *low* temperatures, where the molecules have relatively low energies, and at *high* pressures, where molecules are forced close together. A substance is likely to exist as a gas at high temperatures and at low pressures, where molecules are far apart. The solid state exists at low temperatures and at moderate to high pressures, where molecules are most closely packed. The liquid state—a sort of in-between state—exists at intermediate temperatures and moderate to high pressures.

The adjectives "high" and "low" are relative terms. A temperature of 1000 °C is far above the boiling point of water (100 °C), but is rather low in relation to the melting point of iron (1530 °C).

Before studying intermolecular forces in detail, we can make some generalizations. First, all ionic compounds are solids at room temperature. It is possible to obtain them as liquids (by melting them), but generally only at high temperatures.

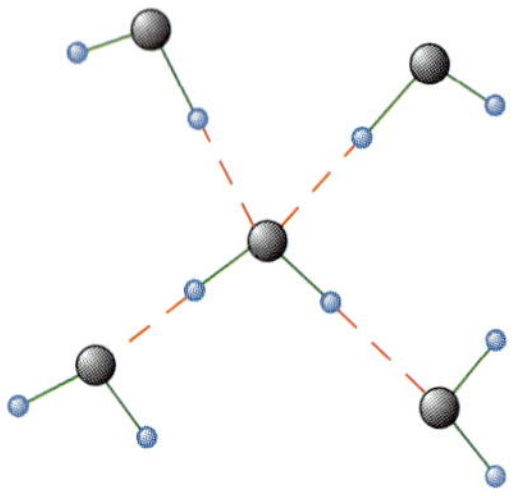

*Inter*molecular and *intra*molecular forces compared. (*Inter* is Latin for "between" and *intra* is Latin for "within.") In the hypothetical case pictured here, *intra*molecular forces—chemical bonds—are shown as solid green lines. Forces between molecules—*inter*molecular forces—are indicated by broken red lines.

Second, nearly all metals (mercury is a notable exception) are solids at room temperature. They, too, can be melted—some at fairly low temperatures, but others only at high temperatures.

It would be nice to have a third generalization to cover molecular substances. However, molecular substances exist in all three physical states at room temperature. Nitrogen (N_2) and carbon dioxide (CO_2) are gases, water (H_2O) and octane (C_8H_{18}) are liquids, and sulfur (S_8) and glucose ($C_6H_{12}O_6$) are solids. The physical state of a molecular substance is determined both by molecular weight (the mass of the molecules) and by the type of force between molecules. If one of these two variables can be eliminated (or held constant), simple generalizations are possible. For example, the Group 7A elements all exist as nonpolar diatomic molecules. Intermolecular forces are similar; thus, any variation in properties can be attributed to variations in molecular weight. And we do find such a trend. Fluorine (F_2), with a molecular weight of 38 u, and chlorine (Cl_2), with a molecular weight of 71 u, are gases at room temperature. Bromine (Br_2), with a molecular weight of 160 u, is a liquid, and iodine (I_2), with a molecular weight of 254 u, is a solid. A similar trend can be found for compounds that are subject to similar intermolecular forces, as is the case for the following compounds of carbon and the halogens.

Compound	CF_4	CCl_4	CBr_4	CI_4
Molecular weight (u)	88	154	332	520
Physical state at 20 °C	Gas	Liquid	Solid	Solid

As we shall see, the types of intermolecular forces are usually of overriding importance if we compare molecules that are subject to dissimilar forces.

8.2 Ionic Bonds as Forces Between Particles

There really is no such thing as a "molecule" of a solid ionic compound. So, there are really no intermolecular forces in ionic compounds. As we saw in Section 4.5, there are simply interionic attractions in which each ion is simultaneously attracted to several ions of the opposite charge. These interionic attractions—ionic bonds—extend throughout an ionic crystal.

We can use the following generalization to compare the strengths of various kinds of interionic attractions: *The attractive force between a pair of oppositely charged ions increases as the charges on the ions increase and as ionic size decreases.* Ionic solids melt if enough thermal energy is supplied to break down the crystalline lattice.

Example 8.1

Which member of each of the following pairs of ionic solids has the higher melting point?

a. MgO or NaCl **b.** $CaBr_2$ or $CaCl_2$

Solution

a. Mg^{2+} and O^{2-} ions have higher charges than do Na^+ and Cl^-. The forces between the doubly charged ions of MgO are greater than those between the singly charged ions of NaCl. By the generalization stated above, MgO should have the *higher* melting point.

b. In comparing $CaBr_2$ and $CaCl_2$, the cation (Ca^{2+}) is the same in both. Also, both anions carry a charge of −1. However, the Br^- ion is a larger ion than the Cl^- ion. Consequently, we expect the interionic attractions in $CaBr_2$ to be somewhat weaker than those in $CaCl_2$. We expect $CaBr_2$ to have a lower melting point than $CaCl_2$.

Practice Exercise

Which member of each of the following pairs of ionic solids has the higher melting point?

a. KCl or KI **b.** HgS or AgCl

8.3 Dipole Forces

The attractive forces between small molecules are not nearly as great as those between oppositely charged ions. Hydrogen chloride is a gas at room temperature. It melts at −112 °C and boils at −85 °C. We know that in the HCl molecule, the hydrogen atom and chlorine atom are joined by a covalent bond (Section 4.9), but what sort of force holds one HCl molecule to another in the liquid or solid state?

Remember that the HCl molecule is a dipole (Section B.2); it has a positive end and a negative end. HCl is a *polar* substance. Two dipoles brought close together will attract one another. The dipoles attempt to align themselves with the positive end of one dipole directed toward the negative ends of neighboring dipoles and vice versa. These **dipole forces** exist throughout the liquid or solid (Figure 8.2). Even if attractive forces between permanent dipoles are fairly weak, they are greater than those in a nonpolar substance of about the same molecular weight (MW). We see this effect in the series: nitrogen, nitrogen monoxide, and oxygen. The NO molecule is polar and has the highest boiling point.

Attractions between polar molecules result from forces between centers of *partial* charge, whereas the much stronger attraction between ions involves fully charged particles.

	N_2	NO	O_2
MW	28.0 u	30.0 u	32.0 u
bp	−196 °C	−152 °C	−183 °C

Example 8.2

Which member of each of the following pairs of substances has the higher boiling point?

a. hydrogen bromide, HBr (MW 80.92 u); or krypton, Kr (MW 83.30 u)

b. silane, SiH_4 (MW 32.09 u); or phosphine, PH_3 (MW 34.00 u)

Solution

a. HBr is polar. Krypton, a monatomic gas, is nonpolar. The molecular weights are comparable; that of Kr is only slightly higher than that of HBr. We expect the polar substance to have greater intermolecular forces than the nonpolar substance. HBr has the higher boiling point.

b. Although the Si—H bonds are slightly polar, the tetrahedral SiH_4 molecule as a whole is nonpolar (compare methane, Section B.2). P—H bonds are polar and the pyramidal PH_3 molecule is polar. The molecular weights are comparable; that of PH_3 is only slightly higher than that of SiH_4. We expect PH_3 to have the higher boiling point.

Practice Exercise

Which member of each of the following pairs of substances has the higher boiling point?

a. hydrogen chloride, HCl (MW 36.47 u); or fluorine, F_2 (MW 38.00 u)

b. silane, SiH_4 (MW 32.09 u); or hydrogen sulfide, H_2S (MW 34.08 u)

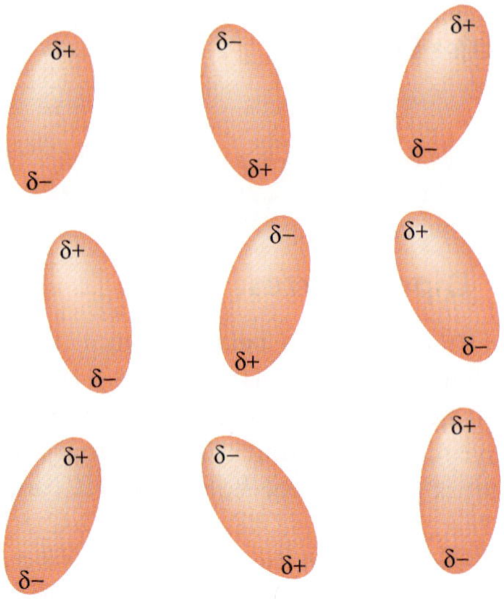

FIGURE 8.2 ▶

Dipole forces illustrated. Molecular motion prevents a perfect alignment of the dipoles. Nevertheless, the dipoles do maintain a general arrangement leading to the attractions $\delta+ \ldots \delta-$.

8.4 Hydrogen Bonds

Suppose we knew the boiling points of hydrogen sulfide (H_2S), hydrogen selenide (H_2Se), and hydrogen telluride (H_2Te) and were asked to predict the boiling point of water (H_2O).

	H_2O	H_2S	H_2Se	H_2Te
MW	18.02 u	34.08 u	80.98 u	129.63 u
bp	?	−60.33 °C	−41.3 °C	−2 °C

All are bent, polar molecules, but molecular weight increases in a regular fashion. We might therefore expect the boiling points to vary in a similar manner. If we plot boiling point versus molecular weight, we could well predict a boiling point of about −68 °C for water (Figure 8.3). However, our prediction would be wrong. Water has a much higher boiling point, 100 °C. Our incorrect prediction suggests that there is an *additional* kind of intermolecular force in water that is not found in the other molecules. There is indeed such a force, and it is known as a *hydrogen bond*.

A **hydrogen bond** between molecules is an intermolecular force in which a hydrogen atom covalently bonded to a nonmetal atom in one molecule is *simultaneously* attracted to a nonmetal atom of a neighboring molecule. The strongest hydrogen bonds are formed if the nonmetal atoms are *small* and *highly electronegative*. Only nitrogen, oxygen, and fluorine routinely fit this requirement.

Think of a hydrogen bond in this way: In a covalent bond an electron cloud joins a hydrogen atom to another atom—oxygen, for example. The electron cloud is much denser at the oxygen end of the bond. The bond is polar, with a partial positive charge ($\delta+$) on the H atom and a partial negative charge ($\delta-$) on the O atom. This leaves the hydrogen nucleus somewhat exposed, and an oxygen atom of a neighboring molecule can approach the exposed hydrogen nucleus rather closely and share some of its "electron wealth" with the hydrogen atom. Hydrogen bonding in water is illustrated in Figure 8.4, where we follow the customary convention of representing hydrogen bonds by dotted lines.

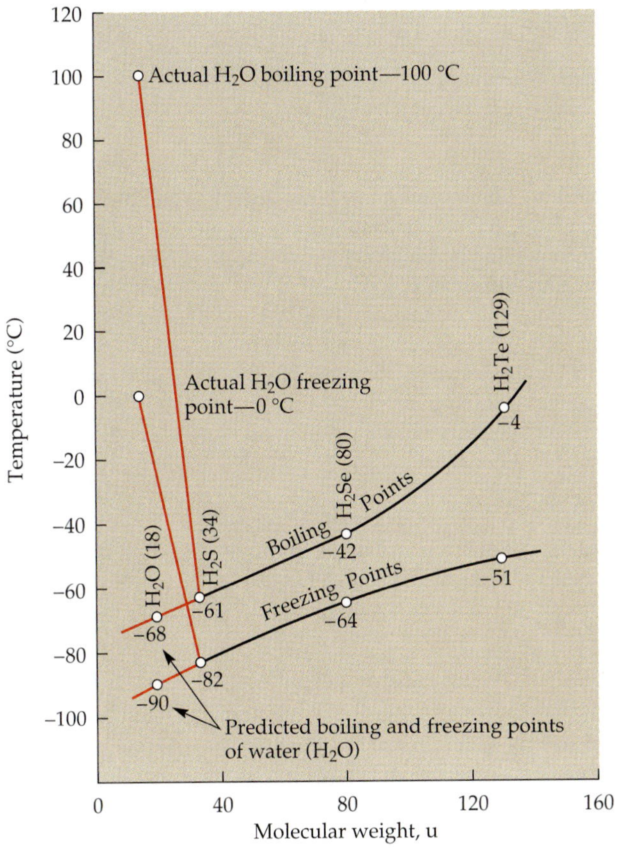

FIGURE 8.3

Boiling points and freezing points as a function of molecular weights. In order for ice to melt or liquid water to boil, hydrogen bonds must be broken in addition to overcoming the dipolar forces to achieve a change of state. The boiling point and freezing point of water are therefore much higher than would otherwise be expected for a substance composed of such small molecules.

Water has both an unusually high boiling point and an unusually high melting point for a substance with such a low molecular weight. Both these unusual values are a result of intermolecular hydrogen bonds between water molecules in the liquid and solid states.

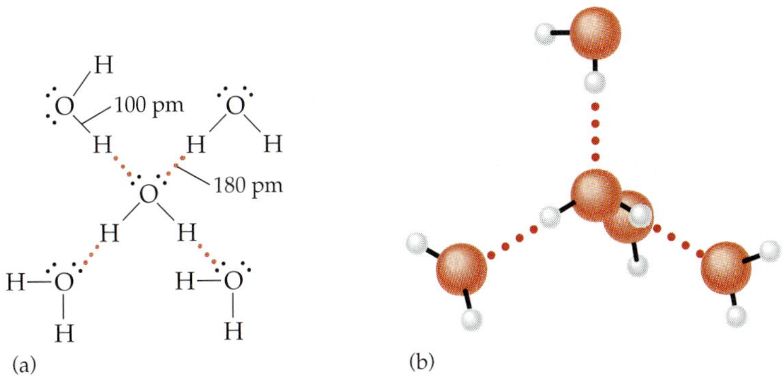

FIGURE 8.4

Hydrogen bonds in water. As suggested through (a) Lewis structures and (b) ball-and-stick models, each water molecule is linked to four others through hydrogen bonds. Each H atom lies along a line that joins two O atoms. The shorter distances (100 picometers) correspond to O—H covalent bonds, and the longer distances (180 pm) to the hydrogen bonds.

Example 8.3

For each of the following substances, comment on whether hydrogen bonding is an important intermolecular force: N_2, HI, HF, $(CH_3)_2O$, CH_3OH.

Solution

N_2: *No.* N is a highly electronegative atom, but we can't have hydrogen bonds without H atoms.

HI: *No.* H atoms are present, but they must be bonded to small, highly electronegative nonmetal atoms. Iodine atoms are quite large.

HF: *Yes.* Both of the requirements stated above are met: H is bonded to a highly electronegative nonmetal.

$(CH_3)_2O$: *No.* Both H and a highly electronegative nonmetal (oxygen) are present, but the H atoms are bonded to *carbon*, not oxygen.

CH_3OH: *Yes.* Both H and highly electronegative O are present, and one of the H atoms is bonded to O.

Practice Exercise

For each of the following substances, comment on whether hydrogen bonding is an important intermolecular force: NH_3, CH_4, C_6H_5OH, H_2S, H_2O_2.

The hydrogen bond may, at this point, seem merely an interesting piece of chemical theory, but its importance to life and health is immense. The structure of proteins, chemicals essential to life, is determined, in part, by hydrogen bonding. And the hereditary traits that one generation passes on to the next are dependent on an elegant application of hydrogen bonding (see Chapters 21 and 23).

8.5 Dispersion Forces

Because unlike charges attract, it is easy to understand interionic forces and intermolecular forces between dipolar molecules. But how can we explain the fact that nonpolar substances can also exist in liquid and solid states? Helium atoms are non-polar, yet helium condenses to a liquid at temperatures below about 5 K. Some kind of intermolecular force must hold the He atoms close enough together to keep the helium in the liquid state. What is the source of this force?

To answer this question, we first must realize that the electron cloud pictures we used in Section 4.9 represent *average* situations only. On average, the electron charge density associated with helium's two 1*s* electrons is evenly distributed about the nucleus. However, at any given instant, purely by chance, the electron charge density may become uneven. The normally nonpolar atom becomes momentarily polar; an *instantaneous* dipole is formed. This transitory dipole, in turn, can displace electrons in a neighboring helium atom, also producing a dipole. One dipole *induces* the other, and the newly formed dipole is therefore called an *induced* dipole. Taken together, these two events lead to an intermolecular force of attraction (Figure 8.5). The force of attraction between the instantaneous dipole and the induced dipole is known as a **dispersion force** (also called a *London* force, named for Fritz London, a professor of chemical physics at Duke University who offered a theoretical explanation of these forces in 1928.)

Dispersion forces are weak for small, nonpolar molecules such as H_2, N_2, and CH_4. These forces can be substantial between large molecules. The properties of polymers such as polyethylene (Section 13.13) are determined to a large degree by dispersion forces between long chains of repeating $\text{+}CH_2CH_2\text{+}_n$ units. (In this formula, *n* is several hundred or even several thousand.)

Recall that at room temperature, fluorine (F_2) and chlorine (Cl_2) are gases, bromine (Br_2) is a liquid, and iodine (I_2) is a solid. Dispersion forces are greater for larger molecules than for smaller ones. Larger atoms have larger electron clouds.

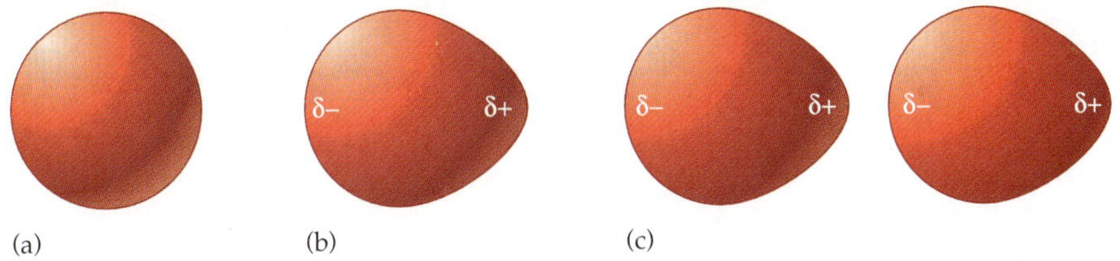

▲ **FIGURE 8.5**

Dispersion forces explained. (a) *Unpolarized molecule.* The electron charge distribution is symmetrical. (b) *Instantaneous dipole.* A momentary displacement of electron charge density (to the left) produces an instantaneous dipole. (c) *Induced dipole.* The instantaneous dipole on the left induces charge separation in the molecule on the right, making it also a dipole. The attraction between the two dipoles is an intermolecular force called a dispersion force.

Their outermost electrons are farther from the nucleus than those of smaller atoms. These remote electrons are more loosely bound and can shift toward another atom more readily than the tightly bound electrons in a smaller atom. This makes molecules with larger atoms more *polarizable* than small ones. Iodine molecules are attracted to one another more strongly than bromine molecules are attracted to one another. Bromine molecules have greater dispersion forces than chlorine molecules, and chlorine molecules, in turn, have greater dispersion forces than fluorine molecules. If you look at the periodic table, you will see that this is the order in which these elements (called, as a group, the halogens) appear in Group 7A, with iodine the largest and fluorine the smallest. Dispersion forces, to a large extent, determine the physical properties of nonpolar compounds.

It should be noted that dispersion forces may be important even when other types of forces are present. Even though we might think of such forces as individually weaker than dipolar attractions or ionic bonds, in substances composed of large molecules the cumulative effect of dispersion forces may be considerable. For large ions, such as silver (Ag^+) and iodide (I^-), dispersion forces play a significant role, even though interionic forces also exist. Dispersion forces play a major role in the presence of some dipolar forces. In hydrogen chloride (HCl), for example, dipolar forces may contribute as little as 15% to the intermolecular attraction; the rest is due to dispersion forces.

The viscosity of motor oils is indicated by a number assigned by the Society of Automotive Engineers (SAE). SAE 40 motor oil (left) is more viscous than SAE 10 oil (right).

8.6 The Liquid State

Molecules of a liquid are much closer together than those of a gas. Consequently, liquids can be compressed only slightly. The molecules are in constant motion, but their movements are greatly restricted by neighboring molecules. One liquid can diffuse into another, but such diffusion is much slower than in gases because of the restricted molecular motion of liquids.

The shape of the molecules that make up a liquid has an effect on an important property of the liquid—its **viscosity,** or resistance to flow. Liquids with low viscosity—"thin" liquids—generally consist of small, symmetrical molecules with weak intermolecular forces. Viscous liquids, on the other hand, are generally made up of large or unsymmetrical molecules with fairly strong intermolecular forces (Figure 8.6). Viscosity generally decreases with increasing temperature. Increased kinetic energy partially overcomes the intermolecular forces. Cooking

Viscosity—a measure of the resistance of a liquid to flow; the higher the viscosity, the slower the liquid's rate of flow

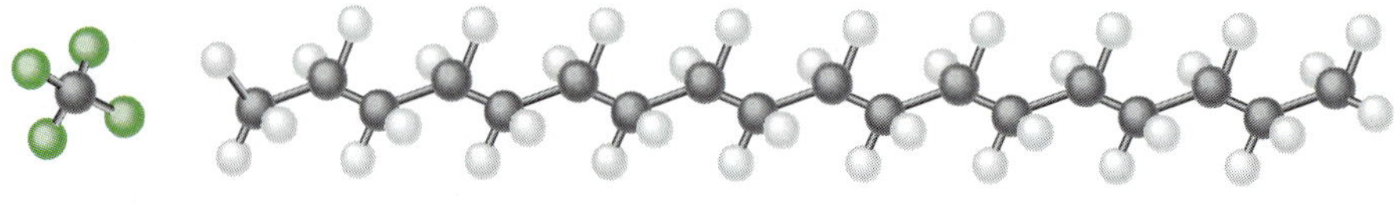

▲ **FIGURE 8.6**

(a) Carbon tetrachloride (CCl_4) consists of relatively small symmetrical molecules with fairly weak intermolecular forces. It has a low viscosity. (b) Octadecane ($C_{18}H_{38}$) is made up of long-chain molecules with comparatively strong intermolecular forces. It has a higher viscosity than carbon tetrachloride.

oil, for example, as it pours from the bottle is thick and "oily." After it's been heated in a frying pan, it becomes thinner and more watery—that is, more like water in its consistency.

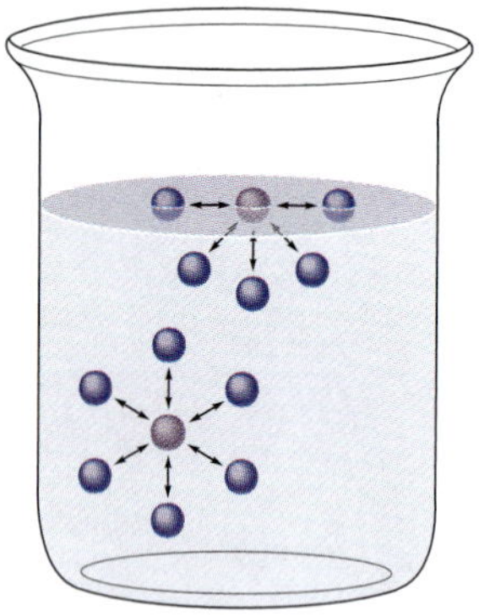

▲ **FIGURE 8.7**

Surface tension explained. Molecules in the body of a liquid are attracted equally in all directions. Those at the surface, however, are pulled downward and sideways but not upward.

Another property of liquids is **surface tension.** A clean glass can be slightly overfilled with water before it spills over. A small needle, carefully placed, can be made to float horizontally on the surface of water—even though steel is several times denser than water. A variety of insects can walk or skate across the surface of a pond with ease. These phenomena indicate something quite unusual about the surface of a liquid. There is a special force or tension there that resists disruption by the needle or water bug.

These surface forces can be explained by intermolecular forces. A molecule in the center of a liquid is pulled equally in all directions by the molecules surrounding it. A molecule on the surface, however, is attracted only by molecules at its sides and below it (Figure 8.7). There is no corresponding upward attraction. These unequal forces tend to pull inward at the surface of the liquid and cause it to contract, much as a stretched sheet of rubber would tend to contract. A small amount of liquid will "bead" to minimize its surface area, and a drop will be spherical for the same reason. The smaller the surface area, the smaller the number of molecules subjected to the unequal pull. Soaps and other detergents act in part by lowering surface tension, enabling water to spread out and wet a solid surface (see Section 20.5).

8.7 From Liquid to Gas: Vaporization

The rate of evaporation of a particular liquid depends on the temperature of the liquid and the amount of the exposed surface area.

The molecules of a liquid are in constant motion, some moving fast, some more slowly. Occasionally one of these molecules has enough kinetic energy to escape from the liquid's surface and become a molecule of vapor. If a liquid, such as water, is placed in an open container, it will soon disappear through *evaporation*. The vapor molecules disperse throughout the atmosphere, and eventually all the liquid molecules escape as they enter the vapor state. On the other hand, if the liquid is placed in a closed container, it does not go away. Some of the liquid is converted to vapor, but the vapor molecules are trapped within the container. Eventually the air above the liquid becomes saturated, and evaporation seems to stop. This vapor exerts a partial pressure (the *vapor pressure*—Section 7.10) that is constant at a given temperature.

It may well appear that nothing further is happening, but molecular motion has not ceased. Some molecules of liquid are still escaping into the vapor state. Vapor molecules in the space above the liquid occasionally strike the liquid's surface, are captured, and thus return to the liquid state. To begin with, there are lots of liquid molecules and no vapor molecules. So, at first, conversion of liquid to vapor (**vaporization**) is taking place, but conversion of vapor to liquid (**condensation**) is not. As more molecules pass into the vapor state, the rate of condensation increases. Eventually, the rate of condensation equals the rate of vaporization. This equilibrium condition appears static but is, in fact, dynamic; two opposing processes are taking place at the same rate. This situation is analogous to that encountered in the case of reversible chemical reactions (Section 5.8).

As the temperature of a liquid is raised, more molecules have enough energy to escape from the liquid state. The rate of vaporization increases. At equilibrium, the rates of vaporization and condensation are again equal, but the pressure exerted by the vapor is higher at the higher temperature. The vapor pressures of liquids increase with temperature.

Now let's see what's going on at the molecular level (Figure 8.8). As soon as molecules appear in the vapor state, some of them strike the liquid surface, are captured, and return to the liquid state. Condensation and vaporization both occur at the same time. We can represent these two opposing simultaneous processes by arrows pointing in opposite directions.

$$\text{Liquid} \underset{\text{condensation}}{\overset{\text{vaporization}}{\rightleftharpoons}} \text{Vapor}$$

If a liquid is placed in an open container, the escape of molecules of the liquid is opposed by atmospheric pressure. If the liquid is heated, the vapor pressure will increase. Continued heating will eventually result in a vapor pressure equal to atmospheric pressure. At that temperature, the liquid will begin to boil. Vaporization will take place not only at the surface of the liquid but also in the body of the liquid, with vapor bubbles forming and rising to the surface. The **boiling point** of a liquid is the temperature at which its vapor pressure becomes equal to atmospheric pressure. Since the latter varies with altitude and weather conditions, boiling points of liquids do also (Figure 8.9). The cooking of foods requires that they be supplied with a certain amount of energy. When water boils at 100 °C, an egg can be placed in the water and soft-boiled in 3 minutes. At reduced

The surface tension of the water keeps the water strider from breaking through the surface. Notice how the water surface is indented but not penetrated by the insect's feet.

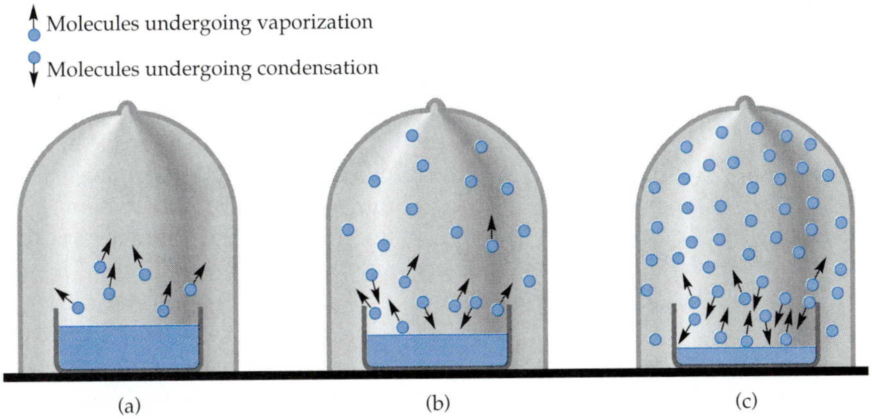

▲ **FIGURE 8.8**

Liquid-vapor equilibrium and vapor pressure illustrated. (a) Vaporization of a liquid begins. (b) Condensation begins as soon as the first vapor molecules appear, although in this illustration the rate of condensation is still less than the rate of vaporization. (c) The rates of vaporization and condensation have become equal. The maximum number of molecules that can be accommodated in the vapor state has been reached. The partial pressure exerted by these molecules—the vapor pressure of the liquid—remains constant.

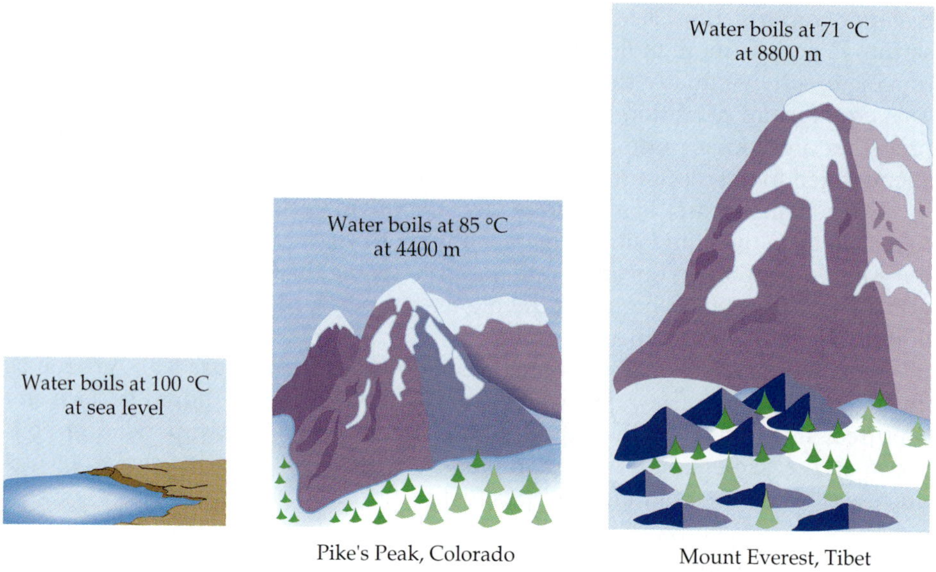

FIGURE 8.9 ▶

The boiling point of water decreases with altitude because the atmospheric pressure decreases with altitude.

Table 8.1 Boiling Points of Pure Water at Various Pressures

Pressure (mmHg)	Temperature (°C)
707	98
760	100
816	102
875	104
938	106
1004	108
1075	110
1283	115
1535	121

Table 8.2 Boiling Points of Various Compounds at 1 Atm

Compound	Boiling Point (°C)
Diethyl ether (anesthetic)	34.6
Acetone (solvent)	56.2
Methyl alcohol (wood alcohol)	64.5
Ethyl alcohol (grain alcohol)	78.3
Water	100.0
Mercury	356.6

atmospheric pressure, water boils at a lower temperature, and it contains less heat energy with which to cook the egg. It would take longer to prepare a soft-boiled egg on top of Mount Everest.

The boiling point is increased when external pressure is increased. Autoclaves and pressure cookers are based on this principle. We can achieve a higher temperature at the higher pressures attained in these closed vessels. (Heat added to a liquid at atmospheric pressure will merely convert liquid to vapor. No increase in temperature occurs until all the liquid has vaporized.) The chemical reactions involved in the cooking of a tough piece of meat proceed more rapidly at the temperature that can be attained in a pressure cooker. Bacteria (even resistant spores) are killed more rapidly in an autoclave, not directly by the increased pressure but rather by the higher temperatures attained. Table 8.1 gives the temperatures attainable with pure water at various pressures.

The boiling point of a liquid is a useful physical property, often used as an aid in identifying compounds. Since boiling point varies with pressure, it is necessary to define the **normal boiling point,** that temperature at which a liquid boils under standard pressure (1 atm, or 760 mmHg). Alternatively, one can specify the pressure at which the boiling point was determined. For example, the *Handbook of Chemistry and Physics* lists the boiling point of antipyrine (a pain reliever and fever reducer) as 319^{741}. This means that the compound boils at 319 °C under a pressure of 741 mmHg. Table 8.2 gives the normal boiling points of some familiar liquids.

Liquids can be purified by a process called **distillation.** Imagine a mixture of water and some nonvolatile material—that is, some material that will not vaporize readily. If the mixture is heated until it boils, the water will vaporize, but the nonvolatile material will not. The water vapor can then be condensed back to the liquid state and collected in a separate container. In such a distillation, the water is separated from the other component of the mixture and thereby purified.

Even if a mixture contains two or more volatile components, purification by distillation is possible. Consider a mixture of two components, one of which is somewhat more volatile than the other (e.g., water and alcohol). At the boiling

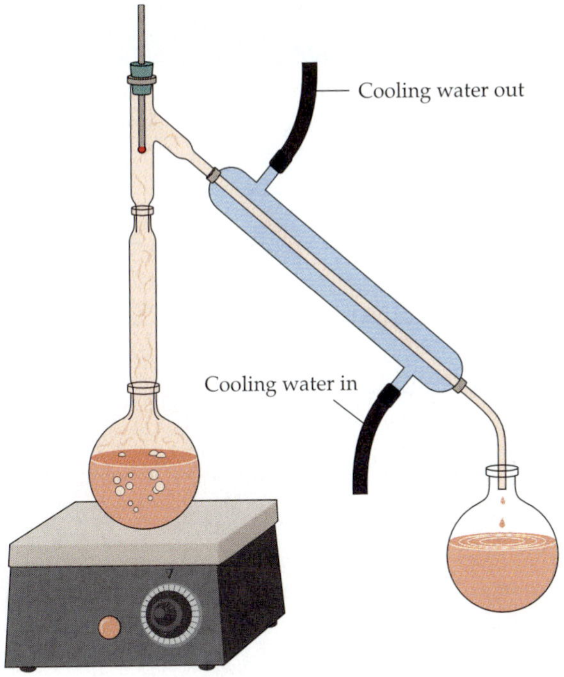

FIGURE 8.10

A distillation apparatus. A mixture is heated in the flask at the left. The vapors formed travel up the vertical column, and are then condensed in the cooled tube angled downward toward the right. The condensed liquid is then collected in the flask at the right.

point of such a mixture, both components will contribute some molecules to the vapor. The more volatile component (alcohol in this example), because it is more easily vaporized, will have more of its molecules in the vapor state than will the less volatile component. When the vapor is condensed into another container, the resulting liquid will be richer in the more volatile component than the original mixture was. Thus, a purer sample of the more volatile component will have been produced. Figure 8.10 shows a typical distillation apparatus.

Heat is required for the conversion of a liquid to a vapor. A liquid evaporating at room temperature absorbs heat from its surroundings. Most of us are familiar with this cooling effect of evaporation. Even on a warm day, we feel cool after a swim, because the water evaporating from our skin removes heat. Volatile liquids, such as ethyl chloride (C_2H_5Cl), which boils at 12.5 °C, can be used as local anesthetics. Rapid evaporation from the skin removes enough heat to freeze a small area, rendering it insensitive to pain. Alcohol rubs also act to cool the skin by their evaporation.

The amount of heat required to vaporize a given amount of liquid can be measured. The quantity of heat required to vaporize 1 mol of a liquid at a constant pressure is called the **molar heat of vaporization.** The heat of vaporization is characteristic of a given liquid. It depends to a large extent on the types of intermolecular forces in the liquid. Water, with molecules strongly associated through hydrogen bonding, has a heat of vaporization of 9.7 kcal/mol. Methane, with molecules held together by weak dispersion forces only, has a heat of vaporization of only 0.232 kcal/mol. Heats of vaporization of several liquids are given in Table 8.3.

Given the molar heat of vaporization, we can calculate the heat of vaporization in calories per gram.

Table 8.3 Heats of Vaporization (at the Normal Boiling Point) of Several Liquids

Compound	Molar Heat of Vaporization (cal/mol)	Heat of Vaporization (cal/g)
Diethyl ether ($C_2H_5OC_2H_5$)	6,200	84
Benzene (C_6H_6)	7,300	94
Methyl alcohol (CH_3OH)	8,400	260
Water	9,700	540
Mercury	14,200	71

Example 8.4

The molar heat of vaporization of ammonia (NH_3) is 5200 cal/mol. What is the heat of vaporization in calories per gram?

Solution

The molar mass of ammonia is 17.0 g/mol. Therefore, the heat of vaporization is

$$\frac{5200 \text{ cal}/\cancel{\text{mol}}}{17.0 \text{ g}/\cancel{\text{mol}}} = 306 \text{ cal/g}$$

Practice Exercise

The molar heat of vaporization of chloroform ($CHCl_3$) is 7050 cal/mol. What is the heat of vaporization in calories per gram?

Example 8.5

How much heat is required to vaporize 400 g of water at its boiling point?

Solution

The heat of vaporization of water is 540 cal/g.

$$\frac{540 \text{ cal}}{1.00 \cancel{\text{g}}} \times 400 \cancel{\text{g}} = 216{,}000 \text{ cal or } 216 \text{ kcal}$$

Practice Exercise

How much heat is required to vaporize 400 g of chloroform at its boiling point? (Use your answer from the preceding practice exercise.)

When a vapor condenses to a liquid, it gives up the same amount of heat energy as was absorbed in converting the liquid to a vapor. A refrigerator operates by alternately vaporizing and condensing a fluid. The heat required to vaporize the fluid is drawn from the refrigerated compartment. The heat is released to the outside atmosphere when the fluid is condensed back to the liquid state.

8.8 The Solid State

Solids resemble liquids in that the particles (atoms, molecules, or ions) in them are close together, making them virtually incompressible. But these two physical states differ significantly in the motion of their particles. In the liquid state, particles are in constant (if somewhat restricted) motion. In solids, there is little motion other than vibration about a fixed point. Consequently, diffusion in solids is generally extremely slow. An increase in temperature will increase the vigor of the vibrations in a solid. If the vibrations become violent enough, the solid will melt (Section 8.9).

Most solids are *crystalline.* In a scientific sense, a *crystal* is a piece of a solid substance that has plane surfaces, sharp edges, and a regular geometric shape. The atoms, ions, or molecules that make up the crystal are assembled in a regular, repeating manner extending in three dimensions throughout the crystal. We can figure out the entire structure of a crystal from just a tiny portion of it, called a **unit cell.** We can generate the entire crystal by stacking unit cells much as we stack toy building blocks. Solids that lack this ordered arrangement, of which glass is an example, are called *amorphous solids.*

We all deal with repeating patterns at one time or another, whether we are stringing beads or laying floor tiles. Usually, though, we need to consider only two dimensions (length and width). To describe crystals, we need to work with *three*-dimensional patterns. The framework on which we outline the pattern is called a **crystal lattice.** With 14 different lattices we could describe all crystalline solids, but we will limit our discussion to three of the *cubic* type.

The simplest unit cell, the **simple cubic,** has structural particles (atoms, ions, or molecules) only at its corners, but most crystal structures are better described by a unit cell with more structural particles. The **body-centered cubic (bcc)** structure has an additional structural particle at the center of the cube. The **face-centered cubic (fcc)** structure has an additional particle at the center of each face. These three unit cells are shown in Figure 8.11. The common metals Fe, K, Na, and W have a bcc crystal structure, and Al, Cu, Pb, and Ag have an fcc structure.

Crystalline solids can also be classified by the types of intermolecular forces holding the particles together. The four main classes are ionic, molecular, covalent network, and metallic.

Ionic solids have ions at definite points in the lattice. A typical ionic solid is sodium chloride (NaCl), which we discussed in Section 4.4. In the lattice, each

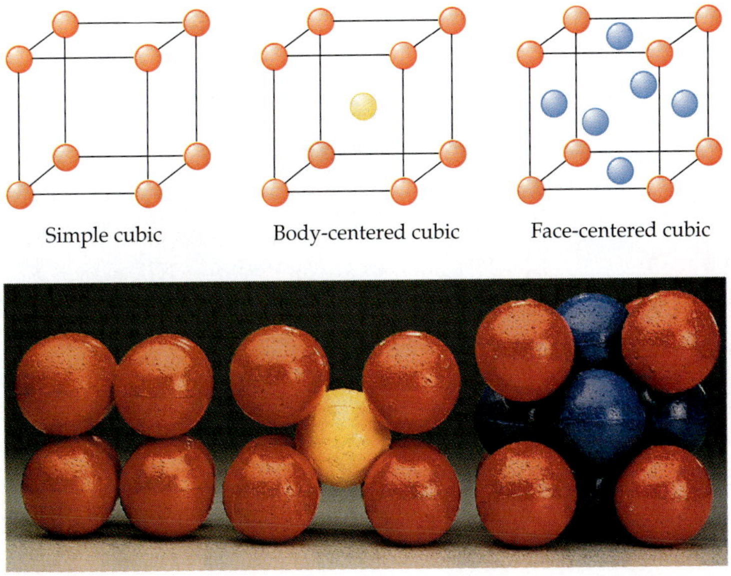

◀ **FIGURE 8.11**
Three types of cubic crystal structures. Only the centers of spheres (atoms) are shown in the top row. The space-filling models in the bottom row show that certain of the spheres are in direct contact.

chloride ion is surrounded by six sodium ions, and each sodium ion is surrounded by six chloride ions. Interionic forces are very strong; ionic solids consequently have high melting points and low vapor pressures, and are quite hard.

Molecular crystals have discrete covalent molecules at the lattice points. These are held together by rather weak dispersion forces, as in crystalline iodine, by dispersion forces plus dipolar forces, as in iodine chloride (ICl), or by hydrogen bonds, as in ice. Molecular solids often have lower melting points than ionic solids. Ice is an exceptional molecular solid; the H_2O molecules are strongly associated by hydrogen bonding (Section 8.10).

Covalent network crystals have atoms at the lattice points. These are joined into extensive networks by covalent bonds; each crystal is in essence one large molecule. Covalent network solids are generally extremely hard and nonvolatile and melt (with decomposition) at very high temperatures. Diamond is a familiar example. Carbon atoms occupy the lattice points. Each is covalently bonded to four other carbon atoms. Silicon carbide (SiC), also called Carborundum, is another familiar compound with an extensive network of covalent bonds.

At the lattice sites in *metallic solids* are positive ions. These ions are formed when metal atoms, such as silver (Ag), lose their outer electrons. The electrons thus released are distributed throughout the lattice, almost like a fluid. These electrons, which can move freely about the lattice, make metals good conductors of heat and electricity. Some metals, such as sodium and potassium, are fairly soft and have low melting points. Others, such as magnesium and calcium, are hard and have high melting points. The extra valence electrons in calcium and magnesium atoms seem to lead to stronger forces between atoms. As a class, metals are malleable; that is, they can be shaped under the influence of pressure or heat. They can be rolled into bars, pressed into sheets, or extruded into wire.

Table 8.4 lists some characteristics of crystalline solids, and Figure 8.12 illustrates some examples.

▲ **FIGURE 8.12**

Examples of three types of crystalline substances: (a) ionic, (b) covalent, and (c) metallic.

Table 8.4 Some Characteristics of Crystalline Solids

Particles in Crystal	Principal Attractive Force Between Particles	Melting Point	Electrical Conductivity of Liquid	Characteristics of the Crystal	Examples
		IONIC CRYSTALS			
Positive and negative ions	Electrostatic attraction between ions (very strong)	High	High	Hard, brittle, most dissolve in polar solvents	NaCl, CaF_2, K_2S, MgO
		COVALENT NETWORK CRYSTALS			
Atoms	Covalent bonds (very strong)	Generally do not melt	—	Very hard, insoluble	Diamond (C), SiC, AlN
		METALLIC CRYSTALS			
Positive ions plus mobile electrons	Metallic bonds (strong)	Most are high	Very high	Most are hard, malleable, ductile, good electrical and thermal conductors, insoluble unless a reaction occurs	Cu, Ca, Al, Pb, Zn, Fe, Na, Ag
		MOLECULAR CRYSTALS			
		Hydrogen-bonded			
Molecules with H on N, O, or F	Hydrogen bonds (intermediate)	Intermediate	Very low	Fragile, soluble in other H-bonding liquids	H_2O, HF, NH_3, CH_3OH
		Polar			
Polar molecules (no H bonds)	Electrostatic attraction between dipoles (rather weak)	Low	Very low	Fragile, soluble in other polar and many nonpolar solvents	HCl, H_2S, $CHCl_3$, ICl
		Nonpolar			
Atoms or nonpolar molecules	Dispersion forces only (weak)	Very low	Extremely low	Soft, soluble in nonpolar or slightly polar solvents	Ar, H_2, I_2, CH_4, CO_2, CCl_4

8.9 From Solid to Liquid: Melting (Fusion)

Solids can be changed to liquids; that is, they can be *melted.* The solid is heated, and the heat energy is absorbed by the particles of the solid. The energy causes the particles to vibrate in place with more and more vigor until, finally, the forces holding the particles in a particular arrangement are overcome. The solid has become a liquid. The temperature at which this happens is called the **melting point** of that solid. A high melting point is one indication that the forces holding a solid together are very strong.

The amount of heat required to convert 1 mol of a solid to a liquid at the melting point is called the **molar heat of fusion.** Generally, the heat of fusion is much less than the heat of vaporization (Figure 8.13). The heat of fusion is the amount of energy that will disrupt the crystal lattice but still leave the particles in contact

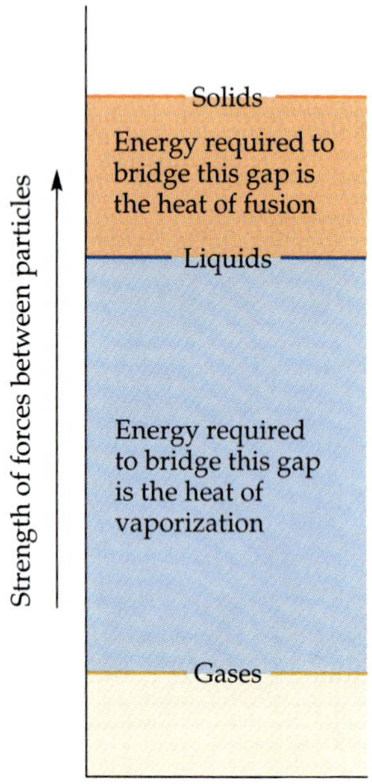

▲ **FIGURE 8.13**

The heat of fusion of a substance is usually only a fraction of its heat of vaporization.

with one another and under the influence of their mutual attraction. A much larger amount of energy is required to vaporize the liquid because, in vaporization, the attraction between particles must be completely overcome (or very nearly so). Table 8.5 gives some representative heats of fusion.

Example 8.6

The heat of fusion of water is 80 cal/g. How much heat is required to melt 400 g of ice?

Solution

$$400\ \cancel{g} \times \frac{80\ \text{cal}}{1.0\ \cancel{g}} = 32{,}000\ \text{cal or } 32\ \text{kcal}$$

Practice Exercise

The heat of fusion of sodium chloride is 120 cal/g. How much heat is required to melt 400 g of sodium chloride?

Example 8.7

The molar heat of fusion of naphthalene ($C_{10}H_8$) is 4610 cal/mol. What is the heat of fusion in calories per gram?

Solution

The molar mass of naphthalene is 128 g/mol. The heat of fusion is therefore

$$\frac{4610\ \text{cal}/\cancel{\text{mol}}}{128\ \text{g}/\cancel{\text{mol}}} = 36.0\ \text{cal/g}$$

Table 8.5 Heats of Fusion (at the Melting Point) of Several Solids

Compound	Melting Point (°C)	Heat of Fusion (cal/mol)	Heat of Fusion (cal/g)
Ammonia (NH_3)	−78	1620	95
Water (H_2O)	0	1440	80
Benzene (C_6H_6)	−6	2550	33
Copper (Cu)	1083	3110	49
Sodium chloride (NaCl)	804	7000	120
Tungsten (W)	3410	8050	43

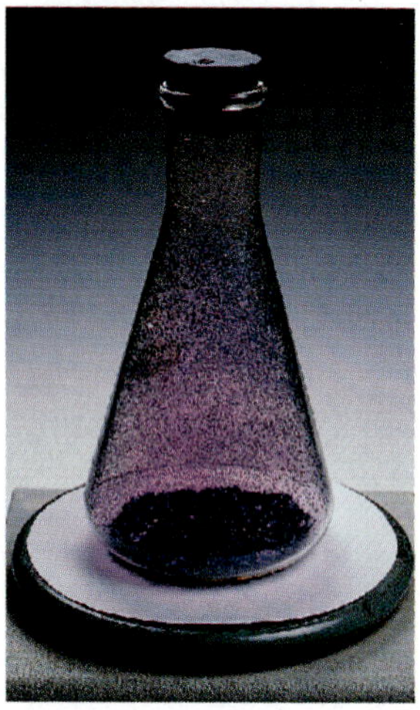

▲ **FIGURE 8.14**

Sublimation of iodine. At atmospheric pressure and a temperature of 70 °C—well below its melting point of 114 °C—iodine has a high vapor pressure. As indicated by the purple iodine vapor $I_2(g)$ and the crystals of $I_2(s)$ on the cooler walls of the flask, iodine sublimes; that is, it goes directly from the solid state to the vapor state.

Sublimation

Although few solids are as volatile as some familiar liquids, solids do vaporize, even if only to a limited extent. The passage of molecules directly from the solid to the vapor state (Figure 8.14) is called **sublimation.** The reverse process, the condensation of a vapor to a solid, is generally called *deposition*. A dynamic equilibrium is reached when the rates of sublimation and deposition become equal.

People living in cold climates are especially familiar with the phenomenon of sublimation. Snow disappears from the ground and ice from the windshield of an automobile even though the temperature stays below 0 °C. This occurs through sublimation, not melting; there is no liquid water at any point in the process. The vapor pressure of ice at 0 °C is 4.58 mmHg.

8.10 Water: A Most Unusual Liquid

Now that we have laid something of a theoretical foundation, let's look more closely at a very special liquid, water. Next to air, water is the most familiar substance on Earth. (It is the only common liquid on the surface of our planet.) Even so, it is a most unusual compound. At room temperature, it is the only liquid compound with a molar mass as low as 18 g/mol. The solid form of water (ice) is less dense than the liquid, a relatively rare situation. The consequences of this peculiar characteristic for life on this planet are immense. Ice forms on the surfaces of lakes when the temperature drops below freezing, and this ice insulates the lower layers of water, enabling fish and other aquatic organisms to survive the winters of the temperate zones. If ice were denser than liquid water, it would sink to the bottom as it formed. This would permit the new surface water to freeze and, in its turn, sink to the bottom. The repetition of this process would eventually result in a lake frozen from top to bottom. Even the deeper lakes of the northern latitudes would freeze solid in winter. Life in the northern lakes and rivers would be quite different from what it is now, if indeed there were life in those waters at all.

Water is the only substance on Earth to exist in large amounts in all three physical states: solid (ice caps); liquid (oceans); and gas (steam from geysers).

Nearly all other solids are more dense than their liquids. Consider a lead ball and molten lead, an iron ball and molten iron, etc.

The same property—the relative densities of ice and liquid water—has dangerous consequences for living cells. Since ice has a lower density than liquid water, 1 g of ice occupies a larger volume than 1 g of water. As ice crystals form in living cells, the expansion ruptures and kills the cells. The slower the cooling, the larger the crystals of ice and the more damage to the cell. The frozen food industry takes into account this property of water. Food is "flash frozen," that is, frozen so rapidly that the ice crystals are kept small and do minimum damage to the cellular structure of the food.

Liquid water has a higher density than most other familiar liquids. As a consequence, liquids that are less dense than water and insoluble in water float on its surface. A familiar problem in recent years has been the gigantic oil spills that occur when a tanker ruptures or when an offshore well gets out of control. The oil, floating on the surface of the water, covers the feathers of waterfowl and the coats of sea animals, such as the otter and the seal. The oil is often washed onto beaches, where it does considerable ecological and aesthetic damage (Figure 8.15). If oil were denser than water, it would sink to the bottom. The problem would be of a different nature, although not necessarily less acute.

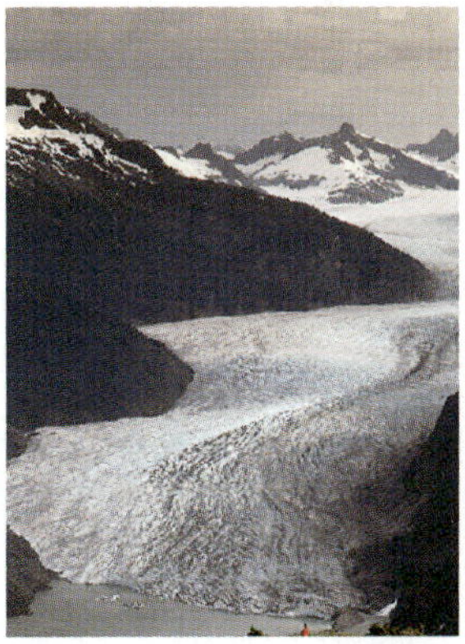

About 2% of Earth's water is in the form of ice.

◄ FIGURE 8.15
Oil coats the water of Prince William Sound, Alaska, where the *Exxon Valdez* ran aground on Bligh Reef in 1989, spilling 42 million liters of oil. One liter of oil can create a slick 2.5 hectares (6.2 acres) in size.

Table 8.6 Specific Heats of Some Common Substances

Substance	Specific Heat [cal/(g · °C)]
Water (liquid)	1.00
Water (solid)	0.50
Water (gas)	0.47
Ethyl alcohol	0.59
Wood	0.42
Aluminum	0.22
Glass	0.12
Iron	0.11
Copper	0.09
Silver	0.06
Gold	0.03

Another unusual property of water is its high specific heat (Section 1.9). It takes 1 cal of heat to raise the temperature of 1 g of water 1 °C. That's almost 10 times as much energy as is required to raise the temperature of the same amount of iron 1 °C. The specific heats of a number of common materials are given in Table 8.6.

The reason cooking utensils are made of iron, copper, aluminum, or glass is that these materials have low specific heats. Thus, they heat up quickly. The reason the handle of a frying pan is usually made of wood or plastic is that these materials have high specific heats. When they are exposed to heat, their temperatures increase more slowly.

The high specific heat of water means not only that much energy is required to raise the temperature of water but also that much heat is given off by water for even a small drop in temperature. The vast amounts of water on the surface of the Earth thus act as a giant thermostat to moderate daily temperature variations. We need only consider the extreme temperature changes on the surface of the waterless moon to appreciate this important property of water. The temperature of the moon varies from just above the boiling point of water (100 °C) to about −175 °C, a range of 275 °C. In contrast, temperatures on Earth rarely fall below −50 °C (−58 °F) or rise above 50 °C (122 °F), a range of only 100 °C.

Example 8.8

How much heat is required to increase the temperature of 12 g of water from 10 °C to 30 °C?

Solution

The specific heat of liquid water is 1 cal/(g · °C). The temperature change is 30 °C − 10 °C = 20 °C.

$$12\,\cancel{g} \times 20\,\cancel{°C} \times \frac{1\text{ cal}}{(1\,\cancel{g})(1\,\cancel{°C})} = 240\text{ cal}$$

Practice Exercise

How much heat is required to increase the temperature of 1.00 mol of water from 20 °C to 30 °C?

Example 8.9

How much energy is required to change 10 g of ice at −10 °C to steam at 100 °C?

Solution

This problem should be worked in parts. First calculate the energy required to raise the temperature of 10 g of ice from −10 °C to 0 °C, a change of 10 °C. The specific heat of ice is 0.5 cal/(g · °C).

$$10\,\cancel{g} \times 10\,\cancel{°C} \times \frac{0.5\text{ cal}}{(1\,\cancel{g})(1\,\cancel{°C})} = 50\text{ cal}$$

Next, using the heat of fusion of water (80 cal/g), calculate the energy required to melt the ice at 0 °C.

$$10\,\cancel{g} \times \frac{80\text{ cal}}{1\,\cancel{g}} = 800\text{ cal}$$

Then calculate the heat required to change the temperature of the water from 0 °C to 100 °C. The specific heat of water is 1 cal/(g · °C), and the temperature change is 100 °C.

$$10\,\cancel{g} \times 100\,\cancel{°C} \times \frac{1\text{ cal}}{(1\,\cancel{g})(1\,\cancel{°C})} = 1000\text{ cal}$$

Now calculate the amount of heat required to change the water (at 100 °C) to steam (at 100 °C). The heat of vaporization for water is 540 cal/g.

$$10\,\cancel{g} \times \frac{540\text{ cal}}{1\,\cancel{g}} = 5400\text{ cal}$$

Finally, total all the calculated values.

To raise the temperature of the ice from −10 °C to 0 °C	50 cal
To change the ice to liquid water	800 cal
To raise the temperature of the water from 0 °C to 100 °C	1000 cal
To change the water to steam	5400 cal
Total	7250 cal

Note that almost 75% of the total energy is used in vaporizing the water.

Practice Exercise

How much energy is required to change 100 g of ice at −10 °C to liquid water at 15 °C?

(a)

(b)

▲ FIGURE 8.16
Hydrogen bonds in ice. (a) Oxygen atoms are arranged in layers of distorted hexagonal rings. Hydrogen atoms lie between pairs of O atoms, closer to one (covalent bond) than to the other (hydrogen bond). (b) At the macroscopic level, this structural pattern is revealed in the hexagonal shapes of snowflakes.

Still another way in which water is unusual is that it has an exceptionally high heat of vaporization; that is, a large amount of heat is required to evaporate a small amount of water. This is of enormous importance to animals. Large amounts of body heat, produced as a by-product of metabolic processes, can be dissipated by the evaporation of small amounts of water (perspiration) from the skin. The heat of vaporization of this water is obtained from the body, and the body is cooled. Conversely, when steam condenses, considerable heat is released. For this reason, steam causes serious burns when it contacts the skin. We previously mentioned that water's high specific heat modifies the climate. Water's high heat of vaporization also contributes to the climate-modifying effect of lakes and oceans. A large portion of the heat that would otherwise warm up the land is used instead to vaporize water from the surface of a lake or the sea. Thus, in summer it is cooler near a large body of water than in interior land areas.

All of these fascinating properties of water result from the unique structure of the water molecule (Section B.2). Recall that the water molecule is polar. In the liquid state, water molecules are strongly associated by hydrogen bonding (see again Figure 8.4). These strong attractive forces account for the high heat of vaporization of water. They must be overcome if vaporization is to take place, and this is why a large amount of energy must be supplied to water for it to convert from liquid to vapor.

In the liquid state, water molecules are quite close together but randomly arranged. When water freezes, its molecules take on a more ordered arrangement with large hexagonal holes (Figure 8.16). This three-dimensional structure extends out for billions and billions of molecules. The holes account for the fact that ice is less dense than liquid water. The hexagonal arrangement allows water to assume forms of exquisite beauty as snowflakes.

Our bodies are about 65% water. Chemical reactions in living cells take place in water solutions. The importance of solutions is such that we devote the next chapter to the subject.

Most of the heat energy that water absorbs is used to break hydrogen bonds, not to increase the kinetic energy of the H_2O molecules.

Summary

Intermolecular forces are those acting among the particles making up any gas, liquid, or solid. These forces are usually negligible in gases but of prime importance in liquids and solids.

The strongest force between particles is the *interionic force,* which holds together positive and negative ions in a solid. The intermolecular forces called *dipole interactions* are important in solids and liquids made up of polar molecules.

A special type of dipole interaction is the **hydrogen bond,** found in molecules having a hydrogen bonded to a fluorine, oxygen, or nitrogen atom. The latter three atoms attract the electrons of the FH, OH, or NH bond, with the result that the F, O, or N carries a partial negative charge and the H carries a partial positive charge. In any liquid or solid in which hydrogen bonds are present, the partially positive H from one molecule attracts the partially negative part of another molecule of the solid or liquid, and this attraction is what is called a hydrogen bond.

The intermolecular forces holding nonpolar liquids and solids together are **dispersion forces.** These transient forces are a result of the fact that the electrons surrounding any nucleus are not stationary. At any moment, the moving electrons in the covalent bond of a nonpolar molecule may be clustered at one side of the molecule, forming a momentary dipole. This momentary dipole then causes a neighboring nonpolar molecule to also become a momentary dipole. The sum total of all the momentary dipole interactions are the dispersion forces.

Viscosity is a liquid's resistance to flow. High-viscosity liquids, which flow sluggishly, are generally made up of large, nonsymmetrical molecules experiencing strong intermolecular forces. Low-viscosity liquids, which flow freely, are generally made of small, symmetrical molecules that experience only weak intermolecular forces.

The molecules in the surface layer of any liquid experience only lateral and downward intermolecular forces; there are no such forces in the upward direction. This imbalance of forces causes **surface tension.**

Vaporization is the change of state in which a liquid becomes a gas. The change of state in the opposite direction, gas to liquid, is **condensation.** The amount of heat needed to vaporize one mole of any liquid is the **molar heat of vaporization** for that liquid.

For a liquid in a closed container and at a given temperature, there is a balance between vaporization and condensation at the liquid surface. When the liquid is heated, its vapor pressure increases, and so the rate of vaporization increases. As the number of gas molecules increases, the rate of condensation also increases, until a new vaporization–condensation equilibrium is reached. This adjustment of equilibrium happens at every increase in temperature until the vapor pressure of the liquid equals atmospheric pressure. At this point, molecules in the interior of the liquid are also vaporized, and the liquid begins to boil. The **boiling point** is therefore the temperature at which a liquid's vapor pressure equals atmospheric pressure. The **normal boiling point** is the temperature at which a liquid boils when the air pressure on the liquid surface is 1 atm.

Solids in which the particles are arranged in a regular array called a lattice are *crystalline solids.* The particles of a crystalline solid may be ions (an *ionic solid),* molecules (a *molecular solid*), covalently bonded atoms (a *covalent-network solid*) or positively charged metal ions (a *metallic solid*). Solids in which the particles are arranged randomly are *amorphous solids.*

Fusion is the change of state in which a solid becomes a liquid. The change of state in the opposite direction, liquid to solid, is solidification. (*Melting* is another term for fusion, and *freezing* is another term for solidification.) The temperature at which fusion occurs is the **melting point** of a solid, and the amount of heat needed to melt one mole of a solid is that solid's **molar heat of fusion.**

Key Terms

body-centered cubic (bcc) (8.8)
boiling point (8.7)
condensation (8.7)
crystal lattice (8.8)
dipole forces (8.3)
dispersion forces (8.5)
distillation (8.7)
face-centered cubic (fcc) (8.8)
hydrogen bond (8.4)
intermolecular forces (8.1)
melting point (8.9)
molar heat of fusion (8.9)
molar heat of vaporization (8.7)
normal boiling point (8.7)
simple cubic (8.8)
sublimation (8.9)
surface tension (8.6)
unit cell (8.8)
vaporization (8.7)
viscosity (8.6)

Review Questions

1. Define or illustrate each of the following terms.
  a. surface tension
  b. condensation
  c. ionic crystal
  d. metallic solid
  e. covalent-network solid

f. hydrogen bond
g. melting point
h. molecular solid
i. molar heat of fusion
j. molar heat of vaporization

2. How do liquids and solids differ from gases in their compressibility, spacing of molecules, and intermolecular forces?
3. In what ways are liquids and solids similar? In what ways are they different?
4. List four types of interactions between particles in the liquid and solid states. Give an example of each type.
5. What is the difference between an *intra*molecular force and an *inter*molecular force?
6. Explain how oxygen can be liquefied if the temperature is lowered sufficiently, even though O_2 molecules are nonpolar.
7. What types of interactions exist between molecules of each of the following?
   a. H_2 b. NO c. HF
8. List three distinctive properties of water.
9. Use the kinetic-molecular theory (Section 7.2) to explain the properties of liquids and solids.
10. What is the distinction between the terms *vaporization* and *boiling?*
11. What is the distinction between the terms *boiling point of a liquid* and *normal boiling point of a liquid?*
12. Use intermolecular forces to explain the phenomena of surface tension.
13. What do we mean when we say that a liquid is *viscous?*
14. How does a pressure cooker work?
15. How is the heat of vaporization of a liquid related to intermolecular forces in the liquid?
16. How does temperature affect the vapor pressure of a liquid? Explain.
17. How does temperature affect the viscosity of a liquid? Explain.
18. What is the difference between an *instantaneous* and an *induced* dipole?
19. What is a dispersion force?
20. What is the difference in meaning of the terms *polar molecule* and *polarizability* of a molecule?
21. Why does a polar liquid generally have a higher normal boiling point than a nonpolar liquid of the same molecular weight?
22. State the principal reasons why CH_4 is a gas at room temperature whereas H_2O is a liquid.
23. Of the substances NI_3, BF_3, and CH_3CH_3, only one is a gas at STP. Which do you think it is, and why?
24. Describe the general way in which ionic charges and sizes affect the melting point of an ionic solid.
25. What is meant by the term *crystal lattice?*
26. Why does it take longer to boil an egg at high altitude than at sea level? Does it take longer to fry an egg at high altitude? Explain.
27. Why does ice float on liquid water?
28. Why does steam at 100 °C cause more severe burns than liquid water at the same temperature?
29. Which has a higher boiling point, methane (CH_4) or ethane (C_2H_6)? Why?
30. Which has a higher boiling point, ethane or methanol? Why?

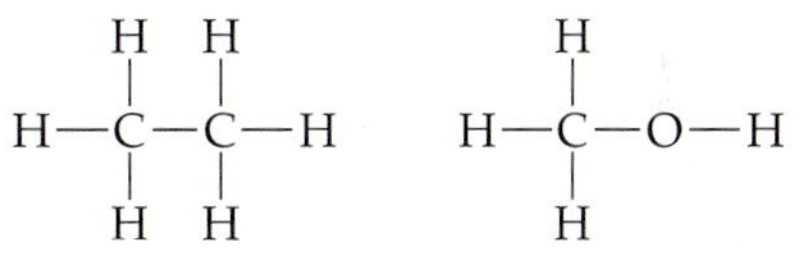

Ethane Methanol

31. Which noble gas, neon (Ne) or xenon (Xe), has the higher boiling point? Why? (Both exist as monatomic gases.)
32. In which process is energy absorbed by the material undergoing the change of state?
   a. melting or freezing
   b. condensation or vaporization
33. Label each arrow with the term that correctly identifies the process described.

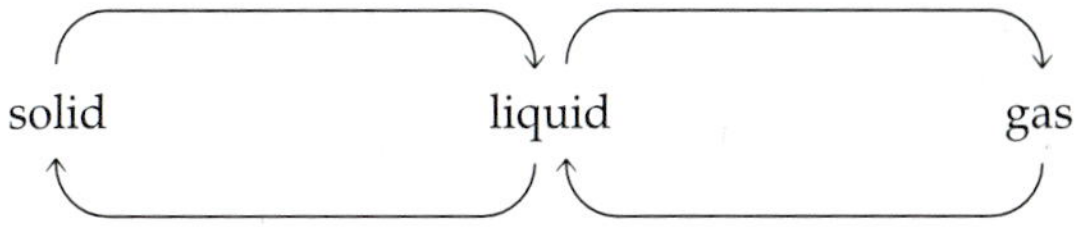

34. What is the value of the vapor pressure of a liquid at the normal boiling point of that liquid?

Problems

Heat of Vaporization

35. The heat of vaporization of bromine (Br_2) is 45 cal/g. What is the molar heat of vaporization of bromine?
36. The heat of vaporization of ammonia (NH_3) is 327 cal/g. What is the molar heat of vaporization of ammonia?
37. The molar heat of vaporization of acetic acid ($C_2H_4O_2$) is 5.81 kcal/mol. How much heat is required to vaporize 1.00 g of acetic acid?
38. The molar heat of vaporization of acetone (C_3H_6O) is 7.23 kcal/mol. How much heat is required to vaporize 5.80 g of acetone?
39. How many kilocalories of heat are required to raise the temperature of 25.0 g of H_2O from 18.0 °C to 60.0 °C? The specific heat of $H_2O(l)$ is 1.00 cal/(g · °C).

40. How many kilocalories of heat are released when 42.4 g of $CH_3OH(l)$ at 55.0 °C is cooled to 15.5 °C? The specific heat of $CH_3OH(l)$ is 0.6 cal/(g · °C).

Heat of Fusion

41. The molar heat of fusion of acetone (C_3H_6O) is 2.58 kcal/mol. How much heat is required to vaporize 7.75 g of acetone?

42. The heat of fusion of silver is 25 cal/g. Calculate the molar heat of fusion of Ag.

43. How many kilocalories of heat are required to melt a 355-g block of ice? (Refer to Table 8.5.)

44. How many kilocalories of heat are released when 275 mL of water freezes at 0 °C? (Refer to Table 8.5.)

Intermolecular Forces

45. Which of the following would you expect to have the *lower* boiling point: carbon disulfide (CS_2) or carbon tetrachloride (CCl_4)? Why?

46. Which of the following would you expect to have the *lower* boiling point: phosphine (PH_3) or arsine (AsH_3)? Why?

47. Which of the following would you expect to have the *higher* boiling point: propane (C_3H_8) or ethanol (C_2H_5OH)? Why?

48. Which of the following would you expect to have the *higher* boiling point: water (H_2O) or carbon monoxide (CO)? Why?

49. Arrange the following substances in the expected order of *increasing* boiling point: H_2S, H_2Se, H_2Te. Give the reasons for your ranking.

50. Arrange the following substances in the expected order of *increasing* boiling point: H_2O, HCl, CH_4. Give the reasons for your ranking.

Additional Problems

51. Describe the principal features of the crystal lattices known as simple cubic, bcc, and fcc.

52. The following data are given for CCl_4: normal melting point, −23 °C; normal boiling point, 77 °C; density of liquid, 1.59 g/mL; heat of fusion, 0.78 kcal/mol. How much heat must be absorbed to convert 10.0 g of solid CCl_4 to liquid at −23 °C?

53. In which of the following compounds would hydrogen bonding be an important intermolecular force?

a.

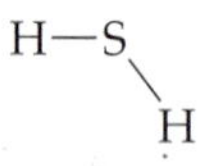

b.

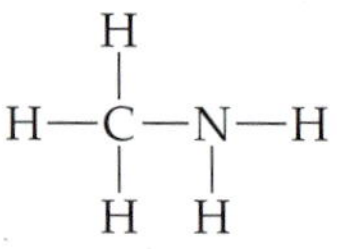

c.

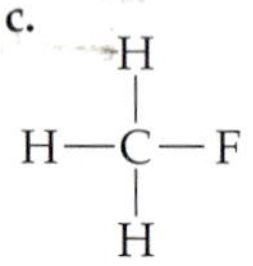

d.

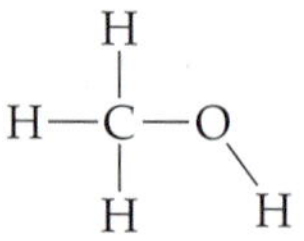

e.

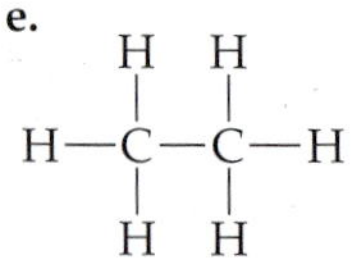

f.

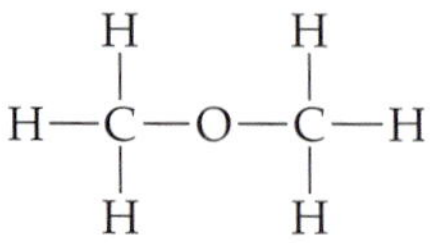

54. There is a rule of thumb that says that for many liquids the molar heat of vaporization is approximately 21 times the normal boiling point in kelvins. Use this rule to calculate the molar heat of vaporization for benzene, which has a boiling point of 353 K. How does your calculated value compare with the experimental value given in Table 8.3?

55. The specific heat of silver is 0.06 cal/(g · °C). That of gold is 0.03 cal/(g · °C). Which metal will be hotter (i.e., will be at a higher temperature) if both absorb the same amount of heat?

56. The molar heats of vaporization of ethanol (C_2H_6O) and ethyl acetate ($C_4H_8O_2$) are 9.39 kcal/mol and 7.77 kcal/mol, respectively. Which liquid has the stronger intermolecular forces?

57. How much energy will be expended in changing 100 g of ice at −5 °C to steam at 100 °C?

58. How much heat is required to convert 10.0 g of ice at −12.0 °C to steam at 130 °C?

59. To obtain water each day, a bird in winter eats 5 g of snow at 0 °C. How many kcal (food Calories) of energy does it take to melt this snow and warm the liquid to the bird's body temperature of 40 °C?

60. To obtain water on a winter hike, a woman decides to eat snow. How many extra kcal (food Calories) of food would she have to take in each day to raise the 1500 g of snow that she needs from −10 °C to 0 °C, melt it, then raise the liquid water from 0 °C to her body temperature of 37 °C?

9 Solutions

Fish, plants, and other marine organisms reside within a water solution that contains all the necessary ingredients to support life.

You are a solution of sodium ions, potassium ions, calcium ions, bicarbonate ions, chloride ions, glucose, amino acids, fatty acids, glycerol, fats, proteins, acetylcholine, and lots of other substances. Well, you aren't all in solution, or else you would wash away in the shower. But, except for a few solid and semisolid parts, such as skin and muscle and bone, you are mostly water. The rest of you is dissolved in that water.

Almost all living systems are made up of thin chemical "soups" in contact with membranes and small cellular parts called organelles. The membranes and organelles are made up of complex chemicals called lipids (fats and fatlike substances), carbohydrates (sugars, starches, and cellulose), proteins, and nucleic acids (DNA and RNA). Life processes occur in solutions and at the interfaces between solutions and semisolid organelles and membranes. Although the chemistry of these processes is now being rapidly unraveled, it is still poorly understood. Therefore, in this chapter we will deal mainly with simpler solutions.

9.1 Solutions: Definitions and Types

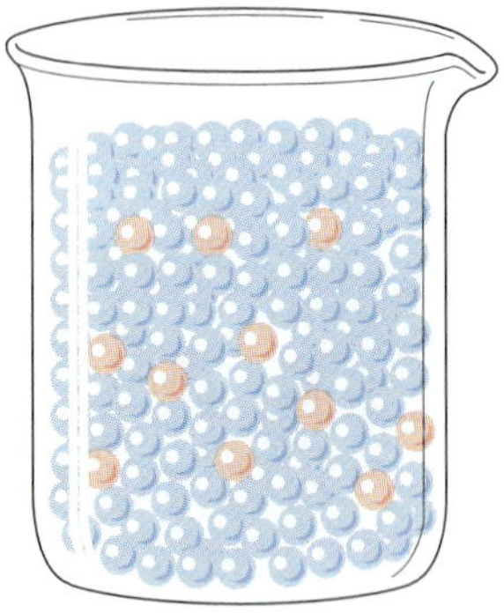

▲ **FIGURE 9.1**
In a solution, the solute molecules (pink) are randomly distributed among the solvent molecules (blue).

Put a teaspoonful of sugar in a cup of water and stir until the sugar has all dissolved. Taste the sweetened water from one side of the cup and then from the other. Use a straw to taste it from the center and the bottom. If you mixed it thoroughly, the water has the same degree of sweetness throughout; it is *homogeneous*. You could add more sugar to make the solution sweeter or less sugar to make it less sweet. You could boil away the water and recover the solid sugar because the water and sugar have not reacted chemically.

A **solution** is a homogeneous mixture of two or more substances. A solution of sugar in water does not consist of tiny particles of solid sugar dispersed among droplets of liquid water. Rather, individual sugar molecules are randomly distributed among water molecules in much the same way that pink marbles can be distributed among blue ones. Because a solution is *homogeneous*, the composition and physical and chemical properties are identical in all parts of it.

The components of a solution are the **solute,** the substance being dissolved, and the **solvent,** the substance doing the dissolving (Figure 9.1). The solute is usually the component present in lesser amounts, and the solvent is usually present in the greatest amount. There are many solvents: Hexane dissolves grease. Ethanol dissolves many drugs. Isopentyl acetate, a component of banana oil, is a solvent for the glue used in making model airplanes. Water is no doubt the most familiar solvent, dissolving as it does many common substances such as sugar, salt, and ethanol. A solvent need not be a liquid. Air is a solution of O_2, argon, water vapor, and other gases in $N_2(g)$. Steel is a solution of carbon in iron—a solid in a solid. The most common types of solutions are listed in Table 9.1. We limit our discussion in this chapter to **aqueous solutions,** those in which water is the solvent.

Table 9.1 Types of Solutions

Solute	Solvent	Solution	Example
Gas	Gas	Gas	Air (O_2 in N_2)
Gas	Liquid	Liquid	Club soda (CO_2 in H_2O)
Liquid	Liquid	Liquid	Wine (alcohol in H_2O)
Solid	Liquid	Liquid	Saline solution (NaCl in H_2O)
Solid	Solid	Solid	14-karat gold (Ag in Au)

Carbonated water is a solution of $CO_2(g)$ in water. The carbonated water is bottled under pressure; when the bottle is opened, bubbles of $CO_2(g)$ escape.

9.2 Qualitative Aspects of Solubility

We say that sugar is *soluble* in water. Just what does this mean? Can we dissolve a teaspoonful of sugar in a cup of water? Can we dissolve 10 teaspoonfuls, or 100 teaspoonfuls? We know from everyday experience that there is a limit to the amount of sugar we can dissolve in a given volume of water. Nevertheless, we still find it convenient to say that sugar is soluble in water, for we can dissolve an appreciable amount.

Some substances can be mixed in all proportions. Water and alcohol are familiar examples. We say that such substances are completely **miscible.** For most "soluble" substances, though, there is a limit to the amount that will dissolve in a given solvent. For others, which we call *insoluble*, that limit is near zero. Put an iron nail in a beaker of water. There is no apparent change. We say that iron is insoluble in water. Even insolubility is relative, however. If we had a method sensitive enough, we would find that some iron had dissolved. The amount might well be regarded as insignificant, and that is the sense in which the term *insoluble* is used. We will find terms such as "soluble" and "insoluble" useful, but they are imprecise and must be used with care.

Two other imprecise but sometimes useful terms are "dilute" and "concentrated." A **dilute solution** is one that contains a little bit of solute in lots of solvent. A **concentrated solution** is one in which lots of solute is dissolved in a relatively small amount of solvent. If we dissolve a few milliliters of ethylene glycol, or antifreeze (Chapter 14), in several liters of water, the dilute solution is quite "thin"—little changed in appearance from that of pure water. However, if we dissolve 1 L of ethylene glycol in 1 L of water, the concentrated solution is rather syrupy—similar to pure ethylene glycol in consistency.

The terms "dilute" and "concentrated" are used in a quantitative way for solutions of acids and bases. We specify their meanings in this context in Section 11.1.

9.3 Solubility of Ionic Compounds

In Chapter 8, we saw that the unique structure of water results in relatively strong forces between water molecules. Further, we examined the different types of forces that exist between identical molecules in pure liquids. Now let's look at the types of forces that exist between the solute and solvent molecules in solutions. The solubility of a given solute depends on the relative attraction between particles in the pure substances and in the solution.

Water is a good solvent for compounds of the Group 1A elements: most lithium, sodium, and potassium compounds are quite soluble in water. Examples are sodium chloride (NaCl), sodium sulfate (Na_2SO_4), potassium phosphate (K_3PO_4), and lithium bromide (LiBr). Further, nearly all nitrate salts are soluble, as are compounds that incorporate the ammonium ion. Silver nitrate ($AgNO_3$), mercury(II) nitrate [$Hg(NO_3)_2$], aluminum nitrate [$Al(NO_3)_3$], ammonium chloride (NH_4Cl), ammonium sulfate [$(NH_4)_2SO_4$], and ammonium phosphate [$(NH_4)_3PO_4$] are examples. Why do these compounds dissolve in water? In essence, three things must happen. The attractive forces holding the ions of the solute together must be overcome. Similarly, the attractive forces holding at least some of the water molecules together must be overcome. Finally, the solute and solvent molecules must interact; that is, they must attract one another. *Hydration* is the process in which water molecules surround the solute ions.

It works this way. Water molecules surround the crystal. Those that approach a negative ion align themselves so that the positive ends of their dipoles point toward the ion. With a positive ion the process is reversed, and the negative end of the water dipole points toward the ion. Still, the attraction between a dipole and an ion is not as strong as that between two ions. To compensate for their weaker attractive power, several molecules surround each ion, and in this way the many *ion–dipole* interactions overcome the *ion–ion* interactions (Figure 9.2).

In some solids, the forces holding the ions together are so strong that they cannot be overcome by the hydration of the ions. Many solids in which both ions are doubly or triply charged are essentially insoluble in water. Examples are calcium carbonate (Ca^{2+} and CO_3^{2-}), aluminum phosphate (Al^{3+} and PO_4^{3-}), and barium sulfate (Ba^{2+} and SO_4^{2-}). The large electrostatic forces between the ions hold the particles together despite the attraction of solvent molecules. Table 9.2 summarizes the solubilities of a variety of ionic compounds.

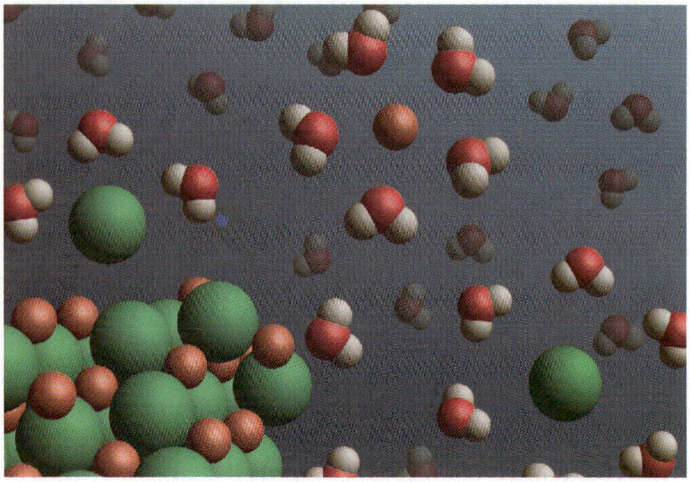

FIGURE 9.2 ▶
Ion–dipole forces in the dissolving of an ionic crystal. Water dipoles attract ions in the crystal lattice, causing them to enter the solution. The ion–dipole forces persist within the solution as well; the ions are *hydrated*.

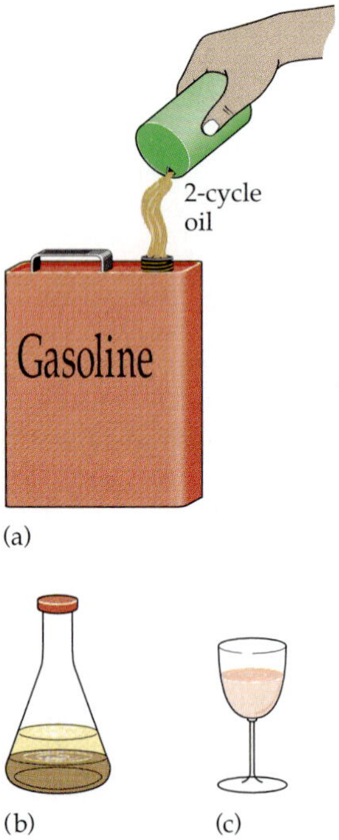

▲ **FIGURE 9.3**
"Like dissolves like." (a) Lawn mowers with two-cycle engines are fueled and lubricated with a solution of nonpolar lubricating oil in nonpolar gasoline. (b) In Italian dressing, polar vinegar and nonpolar olive oil are mixed. The two liquids do not form a solution; they separate on standing.
(c) Wine is a solution of polar ethyl alcohol in polar water.

Table 9.2 Solubilities of Solid Ionic Compounds in Pure Water[a]

	NO_3^-	CH_3COO^-	Cl^-	SO_4^{2-}	OH^-	S^{2-}	CO_3^{2-}	PO_4^{3-}
NH_4^+	S	S	S	S	N	S	S	S
Na^+	S	S	S	S	S	S	S	S
K^+	S	S	S	S	S	S	S	S
Ba^{2+}	S	S	S	I	S	D	I	I
Ca^{2+}	S	S	S	P	P	P, D	I	I
Mg^{2+}	S	S	S	S	I	D	I	I
Cu^{2+}	S	S	S	S	I	I	I	I
Fe^{2+}	S	S	S	S	I	I	I	I
Fe^{3+}	S	N	S	P	I	D	N	I
Zn^{2+}	S	S	S	S	I	I	I	I
Pb^{2+}	S	S	P	I	I	I	I	I
Ag^+	S	P	I	I	N	I	I	I
Hg_2^{2+}	S, D	P	I	I	N	I	I	I
Hg^{2+}	S	S	S	D	N	I	N	I

[a] S = is soluble in water; P = is partially soluble in water; I = is insoluble in water; D = decomposes; N = does not exist as an ionic solid.

9.4 Solubility of Covalent Compounds

An old but helpful chemical rule states that *like dissolves like*. This means that nonpolar (or only slightly polar) solutes dissolve best in nonpolar solvents, and polar solutes dissolve best in polar solvents (Figure 9.3). The rule works well for nonpolar substances. Fats, oils, and greases (nonpolar or only slightly polar) dissolve well in nonpolar solvents such as toluene, C_7H_8. The forces that hold nonpolar molecules together are generally weak. Thus, the amount of energy needed to pull apart molecules of pure solute and to disrupt the attractive forces between molecules of pure solvent is small. And this energy can be balanced by the energy released through the interaction of solute and solvent molecules, although this too is slight.

The important thing for water solubility of covalent compounds is the ability of water to form *hydrogen bonds* to the solute molecules. Thus, molecules containing a high proportion of nitrogen or oxygen atoms will dissolve in water because these are the usual elements that can form hydrogen bonds (Section 8.4). One example is methyl alcohol (CH_3OH), which is completely miscible with water. Methylamine (CH_3NH_2) is also quite soluble in water. Figure 9.4 gives the structures of these molecules (and others mentioned in this section) and shows how they interact with water by forming hydrogen bonds.

There need not be a hydrogen atom on the oxygen (or nitrogen) atom of the solute molecules. Acetaldehyde (CH_3CHO) is completely miscible with water, even though none of the hydrogen atoms in the molecule is attached to the oxygen (see Figure 9.4). The hydrogen bonds depicted in the drawing incorporate hydrogen atoms covalently bonded to the oxygen of the water molecules. We will have many occasions to discuss the importance of hydrogen bonding in later chapters of this text.

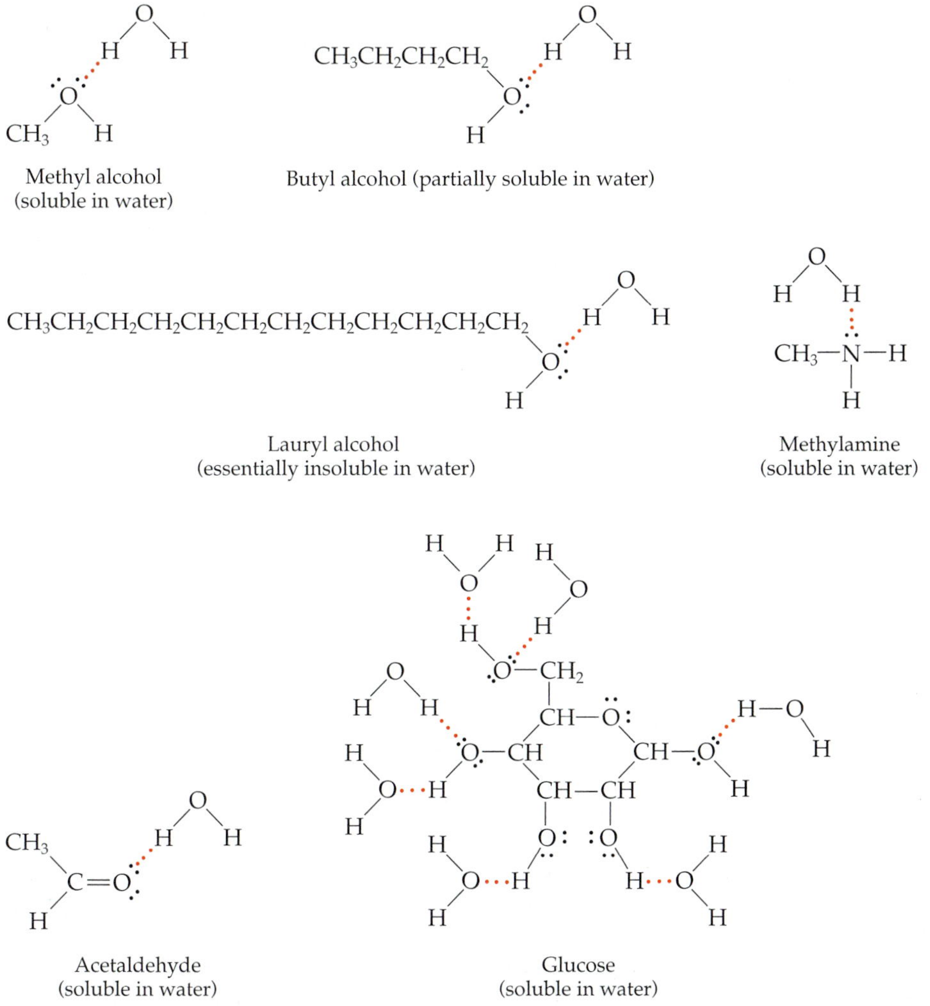

▲ **FIGURE 9.4**

Water can form hydrogen bonds to molecules that contain N or O atoms. Molecules with four or fewer C atoms per N or O atom are usually soluble in water.

Terminology of Aqueous Systems

The importance of water as a solvent is reflected in the number of terms that have been coined to describe systems involving water. For example, the general term for the interaction of solvent with solute is *solvation*, but there is a special term for the interaction of water with a solute—*hydration*.

Certain compounds, such as calcium sulfate, tend to hold on to some water molecules even when they crystallize from solution. These compounds with their bound water molecules are called **hydrates.** The formulas for these crystals are written in such a way as to indicate the number of attached water molecules. Plaster of Paris is $(CaSO_4)_2 \cdot H_2O$ (one water molecule for every two calcium sulfate units). If more water is available, $CaSO_4 \cdot 2H_2O$ is formed (now there are two water molecules for every $CaSO_4$ unit). When a plaster cast is formed to immobilize a broken bone, the first hydrate is converted to the second. The powdery plaster of Paris changes to the rigid, protective material of the cast.

If a hydrate is heated strongly enough, the bound water can be driven off to produce the **anhydrous** compound—that is, the compound without water. Some hydrates lose their bound water simply on standing in dry air. Such compounds are said to be **efflorescent.** Other compounds form hydrates by picking up water from the atmosphere. These are described as **hygroscopic.** And finally, some compounds are so good at pulling water molecules from the air that they eventually dissolve in the water thus accumulated. These compounds are said to be **deliquescent.**

Some fairly complex molecules, such as those of the sugars, are quite soluble in water. Glucose ($C_6H_{12}O_6$) contains six carbon atoms, but its six oxygen atoms permit it to hydrogen bond to many water molecules, thereby making it water-soluble.

Hydrogen bonding, then, is the important factor in water solubility. Polarity alone is not enough. Methyl chloride (CH_3Cl) and methyl alcohol (CH_3OH) have about the same polarity, yet methyl chloride is essentially insoluble in water while methyl alcohol is completely miscible with water. Methyl chloride does not engage in hydrogen bonding, and methyl alcohol does. A few polar substances, such as hydrogen chloride (HCl), dissolve in water because they react chemically to form ions; these are discussed in subsequent chapters.

9.5 Dynamic Equilibria

For most substances, there is a limit to how much solute can be dissolved in a given volume of solvent. This limit varies with the nature of the solute and that of the solvent. Solubilities are often expressed in terms of grams of solute per 100 g of solvent. Since solubility varies with temperature, it is necessary to indicate the temperature at which the solubility is measured. For example, at 20 °C, 100 g of water will dissolve up to 109 g of sodium hydroxide (NaOH). At 50 °C, 145 g of NaOH will dissolve in 100 g of water. In a shorthand method, the solubility of sodium hydroxide is expressed as 109^{20} and 145^{50} (the 100 g of water is understood).

The solubility of sodium chloride, or common table salt (NaCl), is 36 g per 100 g of water at 20 °C. Suppose we place 40 g of NaCl in 100 g of water. What happens? Initially, many of the sodium (Na^+) ions and chloride (Cl^-) ions leave the surface of the crystals and wander about at random through the solvent. Some of the ions in their wanderings return to a crystal surface. These ions can even be trapped there, becoming once more a part of the crystal lattice. As more and more salt dissolves, the number of "wanderers" that return to be trapped once again in the solid state increases. Eventually (when 36 g of NaCl has dissolved), the number of ions leaving the surface of the undissolved crystals just equals the number returning. A condition of **dynamic equilibrium** is established. The *net* amount of sodium chloride in solution remains the same despite the fact that there is still a lot of activity as ions come and go from the surface of the crystals. The net amount of undissolved crystals also remains constant (in this example, 4 g), although individual crystals may change in shape and size as ions leave one part of the crystal to enter solution while others are deposited at another part of the lattice. Some small crystals may even disappear as others grow larger, yet the net amount of undissolved salt does not change. The rate of dissolution just equals the rate of regrowth.

A solution that contains all the solute that it can at equilibrium and at a given temperature is said to be a **saturated solution.** One that contains less than this amount is an **unsaturated solution.** A solution with 24 g of NaCl in 100 g of water at 20 °C is unsaturated because it could dissolve 12 g more at that temperature.

Equilibrium is established at a given temperature. If the temperature changes, solute will dissolve or separate until equilibrium is established at the new temperature. Consider once more a sodium hydroxide solution. If we add 145 g of NaOH to 100 g of water at 20 °C, 109 g of the NaOH will dissolve, leaving 36 g as undissolved solute. If the solution is then warmed, more solute will dissolve. Finally, at 50 °C all 145 g of NaOH is in solution.

Most solid compounds are increasingly soluble as the temperature is raised (Figure 9.5). This should not be surprising. As the temperature goes up, the motion of all the particles is increased. More ions are knocked loose from the lattice and go into solution. Further, it is more difficult for the crystal to recapture the ions that return to its surface, because they are moving at higher speeds. There are a few exceptions to this general rule of increased solubility at higher temperatures. Note that the solubility of NaCl changes little over the indicated range of temperatures, and the solubility of $Ce_2(SO_4)_3$ decreases.

If a saturated solution (with excess solid solute present) is cooled, more solute precipitates until the equilibrium is once again established at the lower

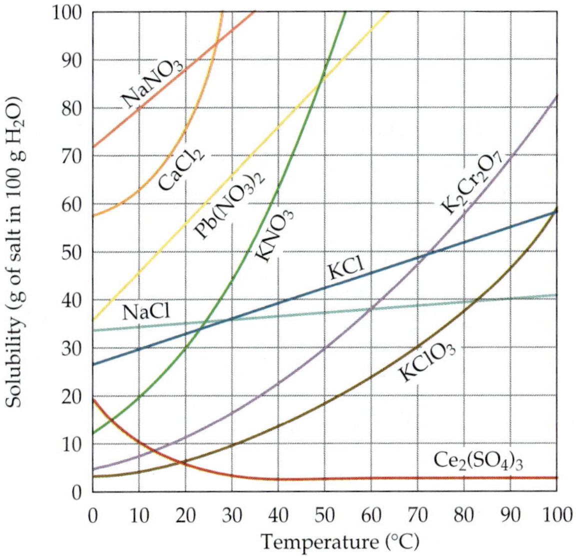

FIGURE 9.5
The solubilities of several salts in water as a function of temperature.

Some Supersaturated Solutions in Nature

Supersaturated solutions are not just laboratory curiosities; they occur naturally. Honey is one example; a principal solute is the sugar glucose. If honey is left to stand, the glucose crystallizes. We say, not very scientifically, that the honey has "turned to sugar." Supersaturated sucrose (cane sugar) solutions are fairly common in cooking. Jellies are one example. Sucrose often crystallizes from jelly that has been standing for a long time.

Some wines have high concentrations of potassium hydrogen tartrate, $KHC_4H_4O_6$. When chilled, the solution may become supersaturated. After a time, crystals may form and settle out if the wine is stored in the consumer's refrigerator. Modern wineries solve this problem—and render the wine less acidic—by a process known as cold stabilization. The wine is chilled to near 0 °C, a temperature below that commonly found in refrigerators. Tiny seed crystals of $KHC_4H_4O_6$ are added to the supersaturated wine. Crystallization is complete after a period of time, and the excess crystals are filtered off. At one time, winemaking was the principal source of potassium hydrogen tartrate, the "cream of tartar" used in baking.

temperature. For example, consider a saturated solution of lead nitrate at 90 °C. For each 100 g of water, there is 120 g of $Pb(NO_3)_2$ dissolved. When the solution is cooled to 20 °C, the solution at equilibrium can contain only 54 g of $Pb(NO_3)_2$. The excess, 66 g, will precipitate, increasing the amount of undissolved solute.

A **precipitate** is an insoluble or nearly insoluble solid that separates from a solution.

Now consider what would happen if one started to cool a saturated solution of lead nitrate with *no* excess solute present. Would lead nitrate precipitate? It might. Then again, it might not. There is no equilibrium—no crystals to capture the wandering ions. One might well be able to cool the solution to 20 °C without precipitation. Such a solution, containing solute in excess of that which it could contain if it were at equilibrium, is said to be a **supersaturated solution.** This system is not stable because it is not at equilibrium. Solute may precipitate when the solution is stirred or if the inside of the container is scratched with a glass rod.

FIGURE 9.6

Seeding a supersaturated solution. Addition of a seed crystal induces rapid crystallization of excess solute from a supersaturated solution.

Addition of a "seed" crystal of solute will nearly always result in the sudden precipitation of all the excess solute. Equilibrium is established rather rapidly when there is a crystal to which the ions can attach themselves (Figure 9.6).

9.6 Solutions of Gases in Water

Solutions of gases in water are more common than you might think. First, there is the familiar case of carbonated beverages. All are solutions of CO_2(g) in water, sometimes with added flavors and sweeteners. Other examples include blood, which contains dissolved O_2(g) and CO_2(g); formalin, an aqueous solution of formaldehyde gas (HCHO) used as a biological preservative; and a variety of household cleaners that are aqueous solutions of NH_3(g). In addition, *all* natural waters contain dissolved O_2(g) and N_2(g) and traces of other gases.

Unlike most solid solutes, gases generally become *less* soluble in water as the temperature increases. The gas molecules acquire more kinetic energy as the temperature increases and are "driven off" from solution.

The world's major ocean fisheries are located in *cold* regions such as the Bering Sea and the Grand Banks of Newfoundland.

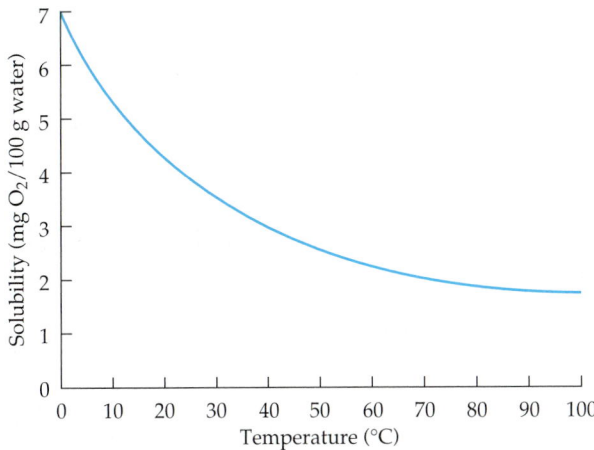

The water solubility of oxygen at atmospheric pressure and 20 °C is only about 0.0043 g/100 g H_2O, but even this limited amount is essential to aquatic life. Fish depend on dissolved air for O_2(g). Moreover, the fact that the solubility *decreases* with temperature (Figure 9.7) explains why many fish (trout, for example) cannot survive in warm water—they don't get enough oxygen. At 30 °C, the amount of dissolved O_2(g) in water is only about one-half of what is found at 0 °C.

At constant temperature, the solubility of a gas in water is directly proportional to the pressure of the gas in equilibrium with the aqueous solution (Section 7.9). The higher the pressure, the more gas that will dissolve in a given volume of water. Figure 9.8 shows how the solubility of oxygen in water at 25 °C varies with pressure. A moderate pressure of CO_2(g) above the beverage in a soft drink bottle keeps a lot of the gas dissolved in the water. When the bottle is opened, this pressure is released and dissolved CO_2(g) escapes, causing the familiar fizzing.

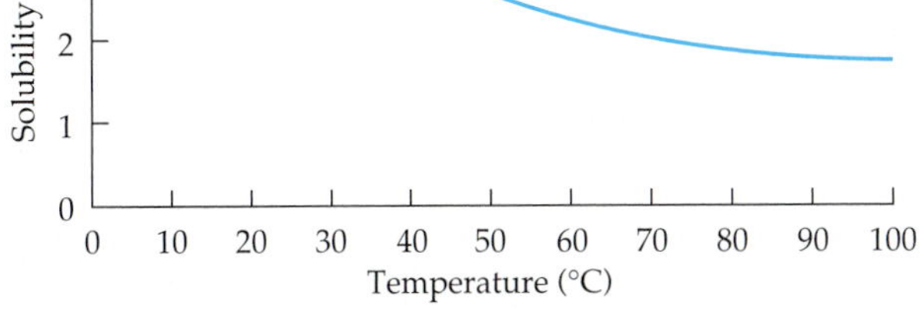

▲ FIGURE 9.7

The solubility of oxygen gas at 1 atm pressure as a function of temperature.

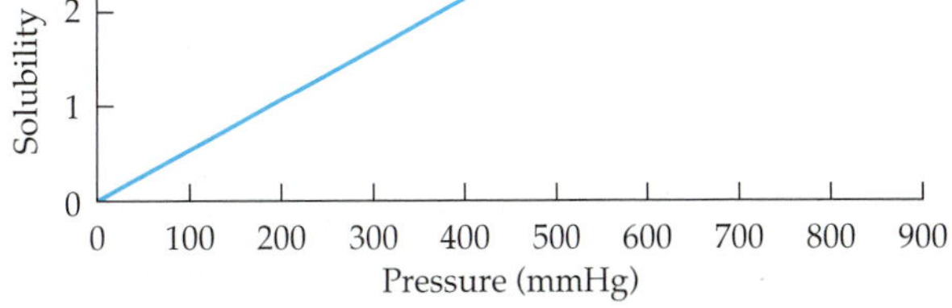

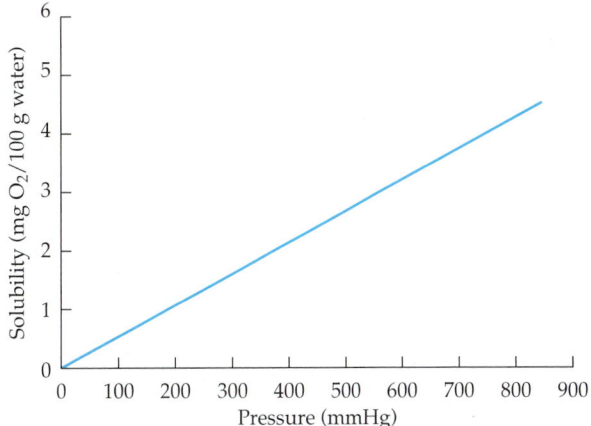

▲ FIGURE 9.8

The solubility of oxygen gas at 25 °C as a function of pressure.

9.7 Molarity

Most reactions of interest to us, including those in our bodies, take place in solution. A good cook may well get by with concentrations expressed as "a pinch of salt in a pint of water," but scientific work generally requires more precise measurement of amounts. We have already discussed the measurement of solubility in grams of solute per 100 g of solvent. Since substances enter into chemical reactions according to certain *molar* ratios, we often must measure the amount of solute in moles. The quantity of solution is usually measured in liters or milliliters. The concentration unit that chemists use most is *molarity*. The **molarity (M)** is the amount of solute, in moles, per liter of solution.

$$\text{Molarity (M)} = \frac{\text{moles of solute}}{\text{liters of solution}}$$

Example 9.1

Calculate the molarity of a solution made by dissolving 3.50 mol of NaCl in enough water to produce 2.00 L of solution.

$$\text{Molarity} = \frac{\text{moles of solute}}{\text{liters of solution}} = \frac{3.50 \text{ mol NaCl}}{2.00 \text{ L solution}} = 1.75 \text{ M NaCl}$$

We read "1.75 M NaCl" as "1.75 molar NaCl."

Practice Exercise

Calculate the molarity of a solution made by dissolving 0.750 mol of H_3PO_4 in enough water to produce 775 mL of solution.

We cannot determine moles of a substance directly; we usually work with a given mass and divide by the molar mass of the substance, as illustrated in Example 9.2.

Example 9.2

What is the molarity of a solution in which 333 g of potassium hydrogen carbonate is dissolved in enough water to make 10.0 L of solution?

Solution

First, prepare the setup that would convert from mass of $KHCO_3$ to moles of $KHCO_3$.

$$333 \cancel{\text{g } KHCO_3} \times \frac{1 \text{ mol } KHCO_3}{100.1 \cancel{\text{g } KHCO_3}} = 3.33 \text{ mol } KHCO_3$$

Now use this value as the numerator in the defining equation for molarity. The solution volume, 10.0 L, is the denominator.

$$\text{Molarity} = \frac{3.33 \text{ mol } KHCO_3}{10.0 \text{ L solution}} = 0.333 \text{ M } KHCO_3$$

Practice Exercise

Calculate the molarity of each of the following solutions.

a. 18.0 mol of H_2SO_4 in 2.00 L of solution
b. 3.00 mol of KI in 2.39 L of solution
c. 0.206 mol of HF in 752 mL of solution
d. 0.522 g of HCl in 0.592 L of solution
e. 4.98 g of $C_6H_{12}O_6$ in 224 mL of solution
f. 10.5 g of C_2H_5OH in 24.7 mL of solution

Frequently we need to know the mass of solute required to prepare a given volume of solution of a given molarity. In such calculations we can use molarity as a conversion factor between moles of solute and liters of solution. Thus, in Example 9.3, "6.67 M NaOH" means 6.67 mol of NaOH per liter of solution, expressed as the conversion factor 6.67 mol NaOH/1 L soln.

Example 9.3

How many grams of NaOH are required to prepare 0.500 L of 6.67 M NaOH?

Solution

First we calculate the moles of NaOH.

$$0.500\ \cancel{\text{L soln}} \times \frac{6.67\ \text{mol NaOH}}{1\ \cancel{\text{L soln}}} = 3.34\ \text{mol NaOH}$$

Then we use the molar mass to calculate the grams of NaOH.

$$3.34\ \cancel{\text{mol NaOH}} \times \frac{40.01\ \text{g NaOH}}{1\ \cancel{\text{mol NaOH}}} = 133\ \text{g NaOH}$$

Practice Exercise

How many grams of potassium hydroxide are required to prepare each of the following solutions?

a. 2.00 L of 6.00 M KOH
b. 100.0 mL of 1.00 M KOH
c. 10.0 mL of 0.100 M KOH
d. 33.0 mL of 2.50 M KOH

Quite often, solutions of known molarity are available. How would you calculate the volume you would need in order to get a certain number of moles of solute? We can again rearrange the definition of molarity to obtain

$$\text{Liters of solution} = \frac{\text{moles of solute}}{\text{molarity}}$$

Example 9.4

How many liters of 12 M HCl solution would one have to take to get 0.48 mol of HCl?

Solution

$$\text{Liters of HCl solution} = \frac{0.48\ \text{mol HCl}}{12\ \text{M}} = \frac{0.48\ \cancel{\text{mol HCl}}}{12\ \cancel{\text{mol HCl}}/\text{L}} = 0.040\ \text{L}$$

We would need 0.040 L (40 mL) of the solution to have 0.48 mol.

Practice Exercise

How many liters of 15 M aqueous ammonia (NH_3) solution do you need to get 0.45 mol of NH_3?

▲ **FIGURE 9.9**

The 1.000 M NaOH solution in this flask was made by dissolving 1.000 mol of NaOH (40.01 g of NaOH) in water, and then carefully diluting to a final volume of 1.000 L.

Remember that molarity is moles per liter of *solution*, not per liter of solvent. To make a liter of a 1 M solution, we weigh out 1 mol of solute and place it in a volumetric flask, which is a standard piece of laboratory glassware designed to contain a precisely specified volume of liquid (Figure 9.9). Enough water is added to dissolve the solute, and then more water is added to bring the volume up to the mark that indicates 1 L of *solution*. (Simply adding 1 mol of solute to 1 L of water would, in most cases, give more than 1 L of solution.)

9.8 Percent Concentrations

For many practical applications, we often express solution compositions in percentage composition. Then, if we require a precise quantity of solution, we simply measure out a mass or volume. If both the solute and solvent are liquids, **percent by volume** is often used, because liquid volumes are so easily measured.

$$\% \text{ by volume} = \frac{\text{volume of solute}}{\text{volume of solution}} \times 100\%$$

Ethanol used for medicinal purposes is generally of a grade referred to as USP (an abbreviation of United States Pharmacopoeia, the official publication of standards for pharmaceutical products). USP ethanol is 95% C_2H_5OH, by volume.

Example 9.5

What is the percent by volume of a solution made by dissolving 235 mL of ethanol in enough water to make exactly 500 mL of solution?

Solution

$$\% \text{ by volume} = \frac{235 \text{ mL ethanol}}{500 \text{ mL solution}} \times 100 = 47.0\%$$

Practice Exercise

What is the percent by volume of a solution made by dissolving 11.7 mL of ethanol in enough water to make 25.0 mL of solution?

Example 9.6

Describe how to make 775 mL of a 40.0% by volume solution of acetic acid.

Solution

$$40.0\% = \frac{\text{mL acetic acid}}{775 \text{ mL solution}} \times 100$$

Rearranging, we get

$$\text{mL acetic acid} = \frac{40.0 \times 775 \text{ mL}}{100} = 310 \text{ mL acetic acid}$$

Take 310 mL of acetic acid and add enough water to make 775 mL of solution.

Practice Exercise

Describe how to make 67.5 mL of a 33.0% by volume solution of acetic acid.

Many commercial solutions are supplied with the concentration expressed as a **percent by mass.** For example, sulfuric acid is sold as a solution that is 35.7% H_2SO_4 for use in storage batteries, 77.7% H_2SO_4 for the manufacture of phosphate fertilizers, and 93.2% H_2SO_4 for pickling steel. Each of these figures is a percent by mass: 35.7 g of H_2SO_4 per 100 g of sulfuric acid solution, and so on.

$$\% \text{ by mass} = \frac{\text{mass solute}}{\text{mass of solution}} \times 100\%$$

Example 9.7

What is the percent by mass of a solution of 5.0 g of NaCl dissolved in 495 g (495 mL) of water?

Solution

$$\% \text{ by mass} = \frac{\text{mass NaCl}}{\text{total mass}} \times 100\%$$

$$= \frac{5.0 \text{ g}}{5.0 \text{ g} + 495 \text{ g}} = 1.0\% \text{ NaCl}$$

Practice Exercise

What is the percent by mass of a solution of 9.0 g of H_2O_2 dissolved in 335 g (335 mL) of water?

Example 9.8

How would you prepare 750 g of an aqueous solution that is 2.5% NaOH by mass?

Solution

We know that the mass of solution is to be 750 g. We can determine the required mass of NaOH by using the definition of percent composition by mass. The required mass of water will be 750 g minus this mass of NaOH.

$$\% \text{ NaOH} = \frac{\text{mass NaOH}}{\text{total mass}} \times 100\% = \frac{\text{mass NaOH}}{750 \text{ g}} \times 100\% = 2.5\% \text{ NaOH}$$

$$\text{Mass NaOH} = \frac{2.5 \times 750 \text{ g}}{100} = 19 \text{ g}$$

$$\text{Mass } H_2O = 750 \text{ g solution} - 19 \text{ g NaOH} = 731 \text{ g } H_2O$$

To make 750 g of solution, weigh out 19 g of NaOH and add it to 731 g of water.

Practice Exercise

How would you prepare 275 g of an aqueous solution that is 5.5% glucose by mass?

Another percentage unit widely used in medicine is that of **mass/volume percent.** For example, a solution of sodium chloride used in intravenous injections has the composition 0.89% (mass/vol) NaCl; that is, it contains 0.89 g of NaCl per 100 mL of solution. A volume of 100 mL—one-tenth of a liter—is also a *deciliter.* If the mass of solute is expressed in *milligrams,* the mass/volume concentration unit is *milligrams per deciliter* (mg/dL). A blood cholesterol reading of 187, for example, means 187 mg cholesterol/dL blood. The use of milligrams per deciliter avoids the sometimes cumbersome use of decimal numbers. For example, in mass/volume percent the cholesterol reading just cited is 0.187%.

Mass/volume percent and mass/mass percent are nearly the same for *dilute aqueous* solutions, because their densities are approximately 1.00 g/mL. That is, 100 mL of such a solution weighs 100 g.

For solutions that are extremely dilute, concentrations are often expressed in parts per million (ppm), parts per billion (ppb), or even parts per trillion (ppt). For example, in fluoridated drinking water, fluoride ion is maintained at about 1 ppm. A typical level of the contaminant chloroform in municipal drinking water taken from the lower Mississippi River is 8 ppb.

Consider these comparisons. One cent in $10,000 is one ppm, and one cent in $10,000,000 is one ppb. Five dollars is about one ppt of the current U.S. national debt. A single individual in the city of San Diego represents about one ppm, and a single individual in the People's Republic of China, about one ppb.

Setting Environmental Standards

Current concern over clean air, the purity of drinking water, and soil contamination centers on trace amounts of potentially dangerous compounds. For example, benzene has been shown to produce leukemia symptoms in laboratory animals and humans. The U.S. Supreme Court has dealt with the question of whether the concentrations of benzene in the air breathed by workers should be limited to 10 ppm or 1 ppm. The Court decided that industries could not be required to lower the concentration from 10 to 1 ppm unless the higher concentration was *proven* to be dangerous.

As technology becomes more sophisticated, our ability to detect minute quantities of materials increases. This increase in the sensitivity of analytical techniques raises important questions for which there are no set answers. When a substance is first detected in the ppb or ppt range, is it a new contaminant in the environment, or has it been there all along at levels that were previously undetectable? What is the relationship between the level at which a substance can be detected and the level at which it is injurious to the health of individuals or the environment?

A special aspect of percent concentrations is that the mass of the solute needed doesn't depend on what the solute is. A 10% by mass solution of NaOH contains 10 g of NaOH per 100 g of total solution. Similarly, 10% HCl and 10% $(NH_4)_2SO_4$ and 10% $C_{110}H_{190}N_3O_2Br$ each contain 10 g of the specified solute per 100 g of solution. For molar solutions, however, the mass of solute in a solution of specified molarity is different for different solutes. A liter of a 0.1 M solution requires 4 g (0.1 mol) of NaOH, 3.7 g (0.1 mol) of HCl, 13.2 g (0.1 mol) of $(NH_4)_2SO_4$, and 166 g (0.1 mol) of $C_{110}H_{190}N_3O_2Br$.

9.9 Colligative Properties of Solutions

Solutions have higher boiling points and lower freezing points than the corresponding pure solvents. The antifreeze in automobile cooling systems is there precisely because of these effects. If water alone were used as the engine coolant, it would boil away in the heat of summer and freeze in the depths of northern winters. Addition of antifreeze to the water raises the boiling point of the coolant and also prevents the coolant from freezing solid when the temperature drops below 0 °C. Salt is thrown on icy sidewalks and streets because the salt lowers the freezing point of the ice. The ice melts because the outdoor temperature is no longer low enough to maintain it as a solid.

The extent to which freezing points and boiling points are affected by solutes is related to the number of solute particles present in solution. The higher the concentration of solute particles, the more pronounced the effect. **Colligative properties** of solutions are those properties, like boiling point elevation and freezing point depression, that depend on the number of solute particles present in solution. For living systems, perhaps the most important colligative property is osmotic pressure. Osmosis is a phenomenon we shall discuss in detail in the next section.

Before we discuss osmosis, however, we must first consider a rather subtle aspect of solute concentration. See if you can answer the following questions. How many solute particles are there in 1 L of a 1 M glucose ($C_6H_{12}O_6$) solution? How many in 1 L of a 1 M sodium chloride (NaCl) solution? How many in 1 L of a 1 M calcium chloride ($CaCl_2$) solution? The answer "should" be 6×10^{23}, right? All of the solutions contain 1 mol of their respective solute compounds. However, the question did not ask for the number of *formula units;* it asked for the number of *solute particles.* Glucose is a covalent compound; its atoms are all firmly tied together in molecules. In the glucose solution, each solute particle is a glucose molecule, and there *are* 6×10^{23} of these. But in the sodium chloride solution, each formula unit of NaCl consists of a separate sodium ion (Na^+) and chloride ion (Cl^-) in solution. When sodium chloride dissolves in water, individual ions are carried off into solution by solvent molecules. So 6×10^{23} formula units of NaCl produce 12×10^{23} particles in solution—6×10^{23} sodium ions plus 6×10^{23} chloride ions. Each calcium chloride unit produces three particles in solution—one calcium ion plus two chloride ions. Thus, the effect of a 1 M NaCl solution on colligative properties is twice that of a 1 M glucose solution. A calcium chloride solution has about three times the effect of a glucose solution of the same molarity.

When colligative properties (specifically osmotic pressure, Section 9.10) are discussed, concentration is often reported in terms of *osmolarity* (osmol/L). An **osmole (osmol)** is a mole of solute particles. A 1 M NaCl solution contains 2 osmol of solute per liter of solution; a 1 M $CaCl_2$ solution contains 3 osmol/L. The osmolarity of a 1 M glucose solution is 1 osmol/L. The concentration of body fluids is typically reported in milliosmols per liter (mosmol/L).

$$\underset{1\text{ M}}{C_6H_{12}O_6} \xrightarrow[\text{dissociate}]{\text{does not}} \underset{1\text{ osmol/L}}{C_6H_{12}O_6}$$

$$\underset{1\text{ M}}{NaCl} \xrightarrow[\text{dissociate}]{\text{does}} \underbrace{\underset{1\text{ osmol/L}}{Na^+} + \underset{1\text{ osmol/L}}{Cl^-}}_{2\text{ osmol/L}}$$

$$\underset{1\text{ M}}{CaCl_2} \xrightarrow[\text{dissociate}]{\text{does}} \underbrace{\underset{1\text{ osmol/L}}{Ca^{2+}} + \underset{2\text{ osmol/L}}{2\,Cl^-}}_{3\text{ osmol/L}}$$

9.10 Solutions and Cell Membranes: Osmosis

Everyday experience tells us that we can separate coffee grounds from brewed coffee by passing the mixture through filter paper. However, the filter paper does not remove the caffeine from the brewed coffee. Paper is *permeable* to water and other solvents and to solutes in solution. We also know that some materials are *impermeable* to liquids and solutions as well as to solids. Water and solutions do not pass through the metal walls of cans nor through the glass walls of jars and bottles. Are there, perhaps, materials with intermediate properties? Materials that will pass solvent molecules but not solute molecules? Materials that are permeable to some solutes but not others? The answer is yes. **Semipermeable membranes** are films of a material containing a network of submicroscopic holes or pores through which small solvent molecules can pass but which severely restrict the flow of solute particles. These may be natural materials of animal or vegetable origin, such as pig's bladder and parchment, or synthetic materials such as cellophane. Cell membranes, the lining of the digestive tract, and the walls of

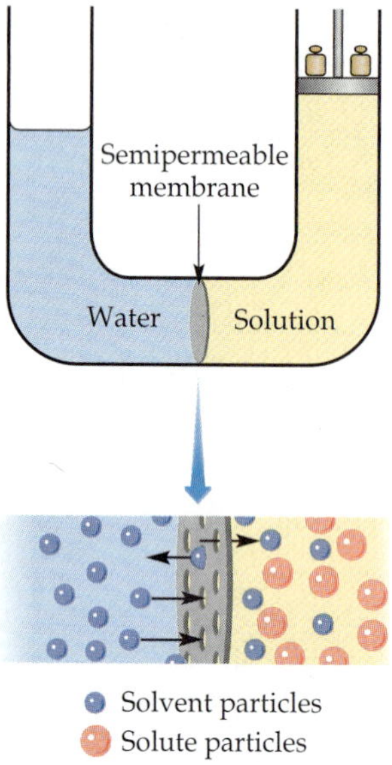

▲ **FIGURE 9.10**
The sieve model of osmosis holds that the semipermeable membrane has pores large enough to permit the passage of water and other small molecules but too small to allow the passage of larger molecules.

For a 20% sucrose solution, osmotic flow could lift a column of solution to a height of 150 m. The solution has an osmotic pressure of about 15 atm.

blood vessels are all semipermeable; they allow certain substances to go through while holding others back.

If a semipermeable membrane separates a compartment containing pure water from one containing a sugar solution, an interesting thing happens. The volume of liquid in the compartment containing sugar increases, and the volume in the pure water compartment decreases.

Use Figure 9.10 as a reference. On both sides of the membrane all molecules are moving about at random, occasionally bumping against the membrane. If a water molecule happens to hit one of the pores, it passes through the membrane into the other compartment. When the much larger sugar molecule strikes a pore, it bounces back instead of through. The more sugar molecules there are in solution (i.e., the more concentrated the solution), the smaller the chance that a water molecule will strike a pore. Thus, in our example, there will be a *net* flow of water from the left compartment into the right compartment. This net diffusion of water through a semipermeable membrane is called **osmosis.** During osmosis, there is always a *net* flow of solvent from the more dilute (or pure solvent) area to the compartment in which the solution is more concentrated (Figure 9.11). The net diffusion of water would be *from* a 5% sugar solution into a 10% sugar solution.

As the liquid level in the right compartment builds up and that in the left compartment drops, the increased quantity of fluid in the right compartment exerts pressure that makes it more difficult for additional water molecules to enter that compartment. (See Section 7.3 on how a barometer works.) Eventually the buildup of pressure is sufficient to prevent further *net* flow of water into that compartment. The rates at which water molecules move back and forth are equal.

Instead of waiting for the liquid level to build up and stop the new flow of water, we can apply an external pressure to the compartment containing the more concentrated solution and accomplish the same thing. The precise amount of pressure needed to prevent the net flow of solvent from the dilute solution to the concentrated one is called the **osmotic pressure.** The magnitude of the osmotic pressure depends only on the concentration of solute particles—that is, on the osmolarity of the solution. The more particles (the higher the osmolarity), the greater the osmotic pressure. You can think of osmotic pressure or osmolarity as a measure of the tendency of a solution to draw solvent into itself.

Examples of osmosis are found in living organisms everywhere. Cells are much like semipermeable bags filled with solutions of ions, small and large

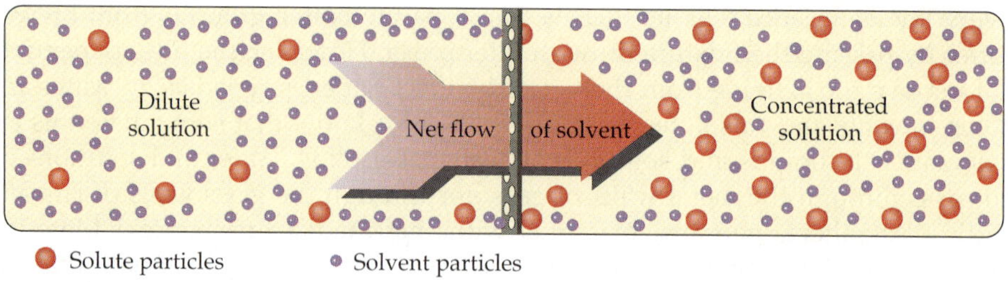

▲ **FIGURE 9.11**
Net flow through a semipermeable membrane occurs spontaneously in only one direction, from the compartment containing the more dilute solution (or pure solvent) into the region that has the more concentrated solution. Remember that ordinarily the terms *dilute* and *concentrated* refer to concentration of the *solute*. The net flow of *solvent* is from the sector in which the solvent is more concentrated to the region where it is less concentrated.

molecules, and still larger cell components. If we place red blood cells in pure water, a net osmotic flow of water into the cells causes them to burst. On the other hand, if we place the cells in a solution that is 0.89% NaCl (mass/vol), there is no net flow of water through the cell membranes and the cells are stable. The fluids inside the cells have the same osmotic pressure as the sodium chloride solution. A solution having the same osmotic pressure as body fluids is an **isotonic solution.** If the concentration of NaCl in a saline solution is greater than 0.89%, a net flow of water *out* of the cells causes them to shrink (Figure 9.12). The saline solution is a **hypertonic solution;** it has a higher osmotic pressure than red blood cells. If the concentration of NaCl is less than 0.89%, water flows *into* the cells. The solution is a **hypotonic solution;** it has a lower osmotic pressure than red blood cells.

One modern application of osmosis, called *reverse osmosis,* is based on *reversing* the normal net flow of water molecules through a semipermeable membrane. That is, by applying to a solution a pressure *exceeding* the osmotic pressure, water can be driven from a solution into pure water. In this way, pure water can be extracted from brackish water, seawater, or industrial wastewater. The success of reverse osmosis requires the use of membranes that can withstand high pressures. Reverse osmosis is widely used in ships at sea and in water-poor nations of the Middle East.

A cucumber is made into a pickle by placing it in a concentrated salt solution. Water flows from the cucumber into the salt solution.

Example 9.9

What are the osmolarities of the two isotonic solutions (glucose and sodium chloride) mentioned in the medical applications essay (page 230)?

Solution

The concentration of glucose is 0.31 M. Glucose is a molecular substance; it does not ionize in aqueous solution. The osmolarity of a 0.31 M glucose solution is therefore 0.31 osmol/L (or 310 mosmol/L). NaCl is ionic; each formula unit provides two particles (1 Na^+ and 1 Cl^-) in aqueous solution. The osmolarity of a 0.16 M NaCl solution is therefore 0.32 osmol/L (or 320 mosmol/L).

Practice Exercise

What is the osmolarity of each of the following solutions?

a. 0.50 M $CaCl_2(aq)$ **b.** 0.15 M $KNO_3(aq)$ **c.** 1.32 M $C_6H_{12}O_6(aq)$

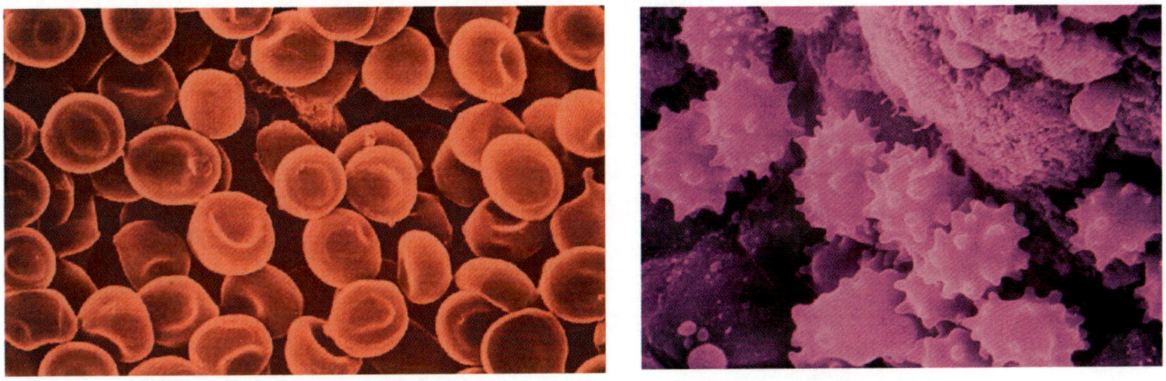

(a) (b)

▲ **FIGURE 9.12**

(a) Normal human red blood cells. (b) After they are placed in a hypertonic solution, the cells are wrinkled and shriveled.

Medical Applications of Osmosis

The rupture of a cell by a *hypotonic* solution is called *plasmolysis*. If the cell is a red blood cell, the more specific term is *hemolysis*. The shrinkage of a cell in a *hypertonic* solution, called *crenation*, can lead to the death of a cell.

In replacing body fluids intravenously, it is important that the replacement fluid be isotonic. Otherwise, hemolysis or crenation may result and the patient's well-being may be jeopardized. As we have already described, an 0.89% NaCl (mass/vol) solution, called physiological saline, is isotonic with the fluid in red blood cells. The "D5W" so often referred to by television's doctors and paramedics is a 5.5% solution of glucose (also called dextrose, D) in water (W). It also is isotonic with the fluid in red blood cells. The 0.89% NaCl (mass/vol) is about 0.16 M, and the 5.5% glucose solution is approximately 0.31 M.

There are limits to intravenous feeding. There is a limit to how much water a patient can handle in a day—about 3 L. If an isotonic solution of 5.5% glucose is used, 3.0 L of this solution supplies only about 160 g of glucose, yielding an energy value of about 640 kcal (640 food Calories) per day. This is woefully inadequate. Even a resting patient requires about 1400 kcal per day. And for a person suffering from serious burns, for example, requirements may be as high as 10,000 kcal per day. With carefully formulated solutions containing other vital nutrients as well as glucose, the feeding of a patient can be increased to about 1200 kcal per day. This still falls short of the requirements of many seriously ill people.

One answer to this problem is to use solutions that are about six times as concentrated as isotonic solutions. But instead of being administered through a vein in an arm or a leg, this solution is infused directly through a tube into the superior vena cava, a large blood vessel leading to the heart (Figure 9.13). The large volume of blood flowing through this vein quickly dilutes the solution to levels that do not damage the blood. With this technique, patients have been given 5000 kcal per day and have even gained weight.

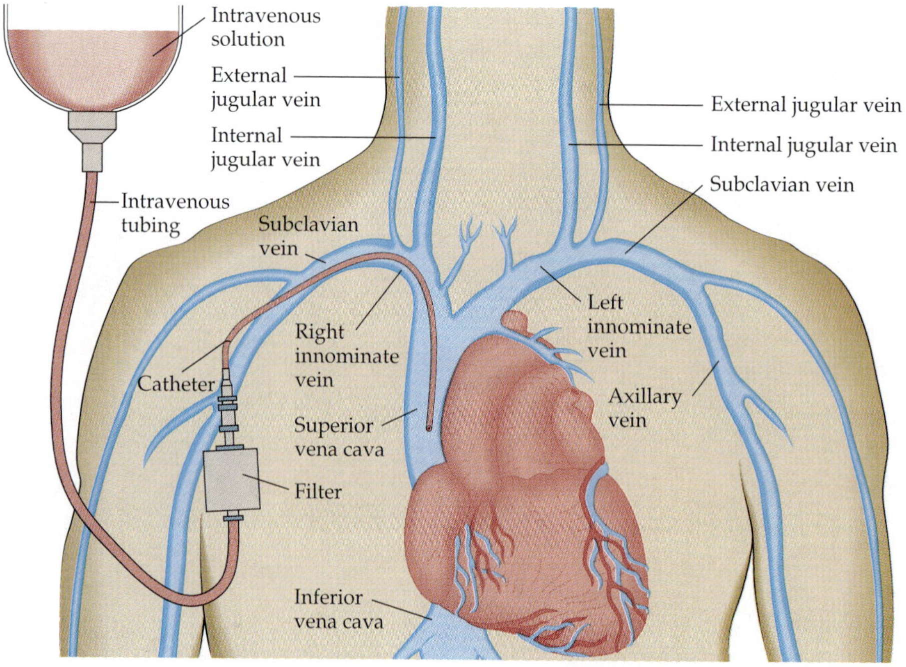

FIGURE 9.13 ▶ Concentrated nutrient solution is sometimes infused directly into the superior vena cava. The solution is quickly diluted to near-isotonic concentration by the large volume of blood that flows through the vein.

9.11 Colloids

The particles in a solution—atoms, ions, or molecules—are of submicroscopic size. Once the solute and solvent are thoroughly mixed, the *solute does not settle out;* molecular motion keeps the particles randomly distributed. The mixture is *homogeneous.* For example, the sugar in a bottle of apple juice does not settle to the bottom, and the last drop is just as sweet as, but no sweeter than, the first.

If we try to dissolve sand (silica, SiO_2) in water, the two substances may momentarily appear to be mixed, but the sand rapidly settles to the bottom of the container. The temporary dispersion of sand in water is called a *suspension.* We can separate the sand and water by pouring the suspension into a funnel fitted with filter paper; water passes through the paper and sand remains behind. The mixture of water and sand is obviously *heterogeneous;* part of it is clearly sand with one set of properties and part of it is water with another set of properties.

Is there no halfway point between true solutions, with particles of atomic, ionic, or molecular size, and suspensions, with gross chunks of insoluble matter? Yes, there is something else: *colloids.*

Even though silica (SiO_2) is insoluble in water, it's possible to prepare a dispersion of silica in water with up to 40% SiO_2 by mass that is stable for years. Such a dispersion does not involve ordinary grains of silica, nor is the suspended silica of ionic or molecular size. The dispersion is called a colloidal mixture. Figure 9.14 gives another example of a colloidal mixture.

A material is called a colloid, not because of the kind of matter in a dispersion, but because of the extent of its subdivision. True solutions have solute and solvent particles less than about 1 nm in diameter. Ordinary suspensions have particle dimensions of about 1000 nm or more. A **colloid** has dispersed matter with one or more dimensions (length, width, or thickness) in the range from about 1 nm to 1000 nm.

Recall that 1 nm = 1×10^{-9} m. Simple optical microscopes are not able to resolve particles smaller than about 1000 nm. Colloidal particles generally cannot be seen under a microscope.

◀ **FIGURE 9.14**

The red colloidal hydrous iron(III) oxide, $Fe_2O_3 \cdot xH_2O$, on the left was made by adding concentrated $FeCl_3(aq)$ to boiling water. (The word *hydrous* indicates an indefinite amount of water, as symbolized by the x in the formula.) When a few drops of $Al_2(SO_4)_3(aq)$ are added, the colloidal particles rapidly coagulate into a precipitate of $Fe_2O_3 \cdot xH_2O$ (right).

Table 9.3 Types of Colloidal Dispersions

Type	Particle Phase	Medium Phase	Example
Foam	Gas	Liquid[b]	Whipped cream
Solid foam	Gas	Solid	Floating soap
Aerosol	Liquid	Gas	Fog, hair spray
Liquid emulsion	Liquid	Liquid	Milk, mayonnaise
Solid emulsion	Liquid	Solid	Butter
Smoke	Solid	Gas	Fine dust or soot in air
Sol[a]	Solid	Liquid	Starch solutions, jellies
Solid sol	Solid	Solid	Pearl

[a] Sols that set up in semisolid, jellylike form are called gels.

[b] By their very nature, gas mixtures always qualify as solutions. The size of particles in gas mixtures and their homogeneity fulfill the requirements of a true solution.

(a) The aqueous solution is colored by its solute, potassium permanganate [$KMnO_4(aq)$]. After passing through the filter paper, the solution is still colored; the $K^+(aq)$ and $MnO_4^-(aq)$ ions pass through the pores of the paper. (b) A suspension of sand in water can be separated by filtration. The suspended sand particles are retained by the filter paper; a colorless filtrate passes through the paper.

There are eight different kinds of colloids, based on the physical state of the particles themselves (the dispersed phase) and of the "solvent" (the dispersion medium). These are listed, together with examples, in Table 9.3. Much of living matter consists of colloidal particles, mainly in the form of sols and emulsions. Substances with high molecular weights, such as starches and proteins, generally form colloidal dispersions rather than true solutions in water.

As you might expect, the properties of colloidal dispersions are different from those of true solutions and suspensions (Table 9.4). Colloids often appear milky or cloudy. Even those that appear clear reveal their colloidal nature by scattering a beam of light passed through them. This phenomenon, first studied by John Tyndall in 1869, is known as the **Tyndall effect** (Figure 9.15). The spectacular sunsets often seen in the desert are caused, at least in part, by the preferential scattering of blue light as sunlight passes through dust-laden air (an aerosol). The transmitted light is deficient in the color blue; it appears a reddish color.

In most colloidal dispersions, the particles are charged. This charge is often due to the adsorption of ions on the surface of the particle. A given colloid will preferentially adsorb only one kind of ion (either positive or negative) on its surfaces; thus, all the particles of a given colloidal dispersion bear like charges. Since like charges repel, the particles tend to stay away from one another. They cannot come together and form particles large enough to settle out. These colloids can be made to coalesce and separate out by addition of ions of opposite charge, particularly those doubly or triply charged. Aluminum salts, which contain Al^{3+}

Table 9.4 Properties of Solutions, Colloids, and Suspensions

Property	Solution	Colloid	Suspension
Particle size	0.1–1.0 nm	1–1000 nm	>1000 nm
Settles on standing?	No	No	Yes
Filter with paper?	No	No	Yes
Separate by dialysis?	No	Yes	Yes
Homogeneous?	Yes	Borderline	No

(a)

(b)

▲ **FIGURE 9.15**

The Tyndall effect. (a) The flashlight beam is invisible as it passes through a true solution (left). However, it is visible as it passes through colloidal iron(III) oxide (right). (b) Light passing through fog, a colloidal dispersion of water in air, also illustrates the Tyndall effect.

ions, are great for breaking up colloids in which the particles are negatively charged. (See again Figure 9.14.)

Some colloids are stabilized by addition of a material that provides a protective coating. Oil is ordinarily insoluble in water, but it can be emulsified by soap. The soap molecules form a negatively charged layer about the surface of each tiny oil droplet. These negative charges keep the oil particles from coming together and separating out (see Section 20.5). In a similar manner, bile salts emulsify the fats we eat, keeping them dispersed as tiny particles that can be more efficiently digested. Milk is an emulsion in which fat droplets are stabilized by a coating of casein, a protein. Casein, soap, and bile salts are examples of **emulsifying agents,** substances that stabilize emulsions.

9.12 Dialysis

To live, an organism must take in food and get rid of toxic wastes. The nutrients necessary to life must enter into cells, and the wastes must leave the cells. Generally, then, cell membranes must permit the passage not only of water molecules but also of other small molecules and ions. At the same time, it is important that large molecules and colloidal particles not be lost from the cell. Membranes that pass small molecules and ions while holding back large molecules and colloidal particles are called *dialyzing membranes.* The process is called **dialysis**. It differs from osmosis in that osmotic membranes pass only solvent molecules. In dialysis, molecules and ions always diffuse from areas of higher concentration to areas of lower concentration.

Dialyzing membranes are used in laboratories to purify colloidal dispersions by removing smaller molecules and ions. The mixture to be purified is placed in a bag made of a dialyzing membrane. The bag is placed in a container of pure water (Figure 9.16). The ions and small molecules pass out through the membrane, leaving the colloidal particles behind. Pure water is continuously pumped past the bag, and the unwanted small particles are carried away. The dialyzing membrane

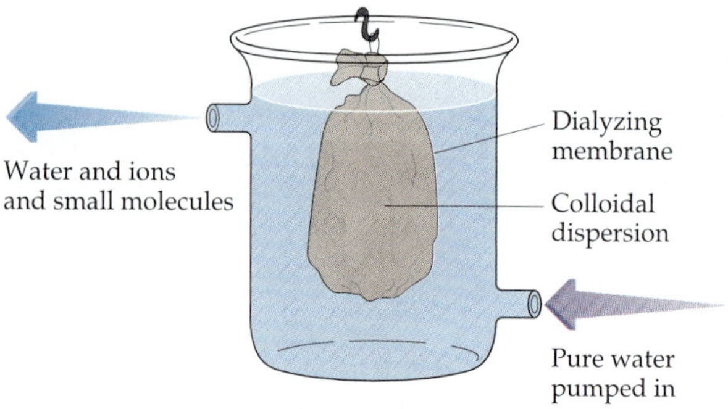

FIGURE 9.16 ▶
A simple apparatus for dialysis.

may be an animal bladder, or it may be an artificial bladder made from cellophane or from collodion, a semisynthetic plastic made by treatment of cellulose with nitric acid, alcohol, and camphor.

The kidneys are a complex dialyzing system responsible for the removal of certain potentially toxic waste products from the blood. By first gaining an understanding of the function of living kidneys, scientists have been able to construct artificial ones. Artificial kidneys are more elaborate in structure than the simple apparatus shown in Figure 9.16, but their principle of operation is the same. We discuss the operation of the kidneys in Section 28.12.

Many substances—medications, poisons, anesthetics—act by changing the permeability of membranes. Methyl mercury, a powerful poison that acts on the nervous system, is thought to act by making the membranes of nerve cells leakier than normal. A person going into shock has leaky capillaries that allow proteins and other colloidal particles as well as fluids to escape into the spaces between cells. If untreated, the cells die from lack of oxygen and nutrients. As research increases our understanding of cellular membranes, we will undoubtedly gain a better understanding of drug action and of poisoning, health, and disease.

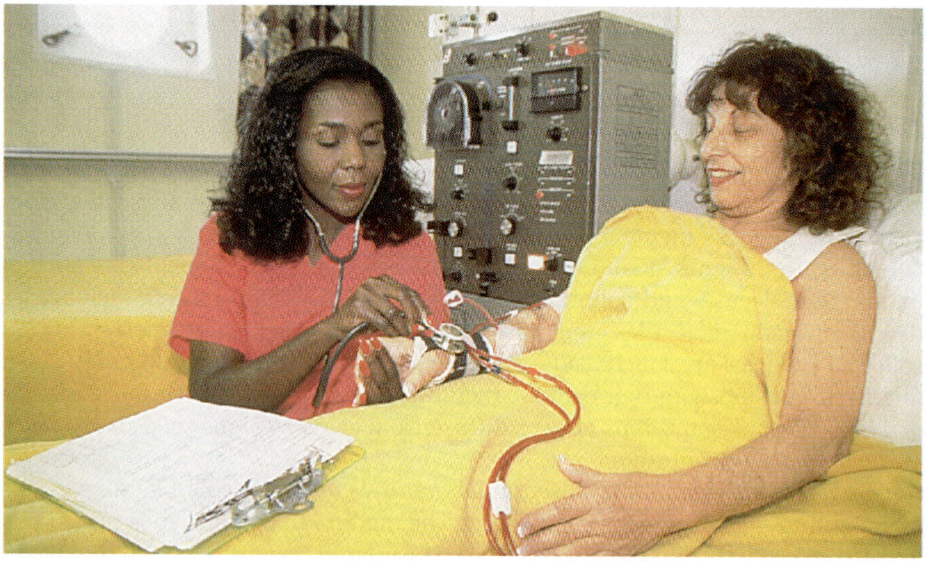

In renal dialysis, elaborate machinery substitutes for human kidneys that are no longer capable of removing toxic waste products from the blood.

Summary

A **solution** is a homogeneous mixture of two or more chemical species (atoms, molecules, or ions). In any solution, there is usually a large amount of one species, the **solvent,** and lesser amounts of all other species, the **solutes.** An **aqueous solution** is one in which the solvent is water.

A substance is **soluble** in a solvent if some appreciable amount of the substance dissolves in the solvent. Any substance that has unlimited solubility in a solvent is **miscible** with that solvent. Any substance that does not dissolve significantly in a solvent is **insoluble** in that solvent. In general, like dissolves like. Polar solutes are soluble in polar solvents and insoluble in nonpolar solvents. Nonpolar solutes are soluble in nonpolar solvents and insoluble in polar solvents.

Compounds of Group 1A elements, nitrate compounds, and ammonium compounds are soluble in water. The ionic bonds in these compounds are weak enough to be overcome by the attractive force exerted by the polar water molecules. Once the bonds are broken, water molecules surround each positive and negative ion in a process called *hydration.* Many compounds containing doubly- or triply-charged ions are insoluble in water.

Solubility is measured in grams of solute per 100 g of solvent at a specified temperature. A solution containing all the solute it can hold at equilibrium at a given temperature is **saturated.** One that contains a lesser amount of the solute is **unsaturated.** In a **supersaturated** solution, the amount of solute is greater than the amount in a saturated solution of the same solute.

One way of expressing solute concentration is **molarity:**

$$\text{molarity} = \frac{\text{moles of solute}}{\text{liters of solvent}}$$

A related concentration unit is *osmolarity:*

$$\text{osmolarity} = \frac{\text{osmoles of solute}}{\text{liters of solvent}}$$

where an **osmole** is 1 mol of solute particles. Example: 1 mol of NaCl dissolved in water gives 1 osmol of Na^+ ions plus 1 osmol of Cl^- ions; therefore the NaCl concentration is 1 molar but 2 osmolar.

Concentrations can be expressed as percentages. When both solute and solvent are liquids, the concentration is given as **percent by volume:**

$$\text{percent by volume} = \frac{\text{volume of solute}}{\text{volume of solution}} \times 100$$

When the solute is a solid, the concentration is given as **percent by mass:**

$$\text{percent by mass} = \frac{\text{mass of solute}}{\text{mass of (solute + solvent)}} \times 100$$

A **colligative property** of a solution is one governed by the number of solute particles present in the solution—in other words, governed by the osmolarity of the solution.

Any barrier that allows solute and solvent particles to pass is *permeable.* Any barrier that does not allow either to pass is *impermeable.* A barrier that allows some solute particles to pass but not others, or that lets solvent particles pass but not solute particles, is *semipermeable.*

The passage of water through a semipermeable membrane is **osmosis.** The direction of flow is from the side having the higher water concentration (which means the side having the lower solute concentration) to the side having the lower water concentration (and thus the higher solute concentration).

As water passes through a semipermeable membrane, the pressure in the compartment it enters builds up. At some point, this pressure becomes high enough to prevent any more water molecules from entering, so that the number of water molecules passing into the compartment is the same as the number passing out. The pressure at which this equilibrium occurs is the **osmotic pressure** of the solution. This amount of pressure applied externally to the solution prevents osmosis. Osmotic pressure is directly proportional to osmolarity.

In living organisms, the solutions surrounding cells are characterized according to concentration. Solutions that have the same osmolarity as the solution inside the cell are **isotonic.** Those that have a lower osmolarity than the cell solution are **hypotonic,** and those that have a higher osmolarity than the cell solution are **hypertonic.**

An insoluble substance mixed as uniformly as possible in a solvent is a **suspension.** Example: sand in water. A suspension is unstable, with the insoluble material settling out soon after mixing is stopped. Suspensions form when the particles of the insoluble substance are relatively large, about 1000 nm or greater in diameter. When the particles being suspended are smaller (ranging from 1.0 nm to 1000 nm in diameter), mixing causes a *colloidal dispersion* (also called a **colloid**) to form. Example: corn starch in water. The particles in a colloidal dispersion do not settle out when mixing is stopped, and the dispersion is often cloudy.

A colloidal dispersion is stabilized when the dispersed particles are coated with a stabilizing material called an **emulsifying agent.** Example: When clothes are laundered, the emulsifying agent soap coats soil particles (mainly oils) and thereby keeps them dispersed in water.

Key Terms

anhydrous (9.4)
aqueous solution (9.1)
colligative property (9.9)
colloid (9.11)
concentrated solution (9.2)
deliquescent (9.4)
dialysis (9.12)
dilute solution (9.2)
dynamic equilibrium (9.5)
efflorescent (9.4)
emulsifying agent (9.11)
hydrate (9.4)
hygroscopic (9.4)
hypertonic solution (9.10)
hypotonic solution (9.10)
isotonic solution (9.10)
mass/volume percent (9.8)
miscible (9.2)
molarity (M) (9.7)
osmole (osmol) (9.9)
osmosis (9.10)
osmotic pressure (9.10)
percent by mass (9.8)
percent by volume (9.8)
precipitate (9.5)
saturated solution (9.5)
semipermeable membrane (9.10)
solute (9.1)
solution (9.1)
solvent (9.1)
supersaturated solution (9.5)
Tyndall effect (9.11)
unsaturated solution (9.5)

Review Questions

1. Define or explain—and, where possible, illustrate—the following terms.
a. solution **b.** solvent
c. solute **d.** dilute solution
e. concentrated solution

2. Define or explain—and, where possible, illustrate—the following terms.
a. aqueous **b.** soluble
c. insoluble **d.** miscible
e. precipitate

3. Define or explain—and, where possible, illustrate—the following terms.
a. percent by mass **b.** percent by volume
c. molarity **d.** mass/volume percent

4. Define or explain—and, where possible, illustrate—the following terms.
a. unsaturated solution **b.** saturated solution
c. supersaturated solution

5. Define or explain—and, where possible, illustrate—the following terms.
a. colligative property
b. semipermeable membrane
c. osmosis
d. osmotic pressure
e. isotonic solution
f. hypotonic solution
g. hypertonic solution
h. osmolarity

6. Define or explain—and, where possible, illustrate—the following terms.
a. suspension **b.** colloid
c. Tyndall effect **d.** dialysis
e. emulsifying agent

7. Define or explain—and, where possible, illustrate—the following terms.
a. crenation **b.** plasmolysis **c.** hemolysis

8. Define or explain—and, where possible, illustrate—the following kinds of compounds.
a. anhydrous **b.** hydrate **c.** efflorescent
d. hygroscopic **e.** deliquescent

9. In a dynamic equilibrium involving a saturated solution, describe the two processes for which the rates are equal.

10. Precipitation is induced in a supersaturated solution by the addition of a seed crystal. When no more solid crystallizes, is the solution saturated, unsaturated, or supersaturated? Explain.

11. Arrange the following in order of increasing concentration: 1% by mass, 1 mg/dL, 1 ppb, 1 ppm, 1 ppt.

12. Benzene, C_6H_6, is a nonpolar solvent. Would you expect NaCl to dissolve in benzene? Explain.

13. Motor oil is nonpolar. Would you expect motor oil to dissolve in water? In benzene (C_6H_6)? Explain.

14. When the cap is removed from a bottle of soda pop, carbon dioxide gas escapes. Explain why this occurs.

15. Why are the world's major ocean fisheries located in cold areas?

16. Contrast a true solution and a colloid in terms of (**a**) the size of the solute particles; (**b**) the nature of the distribution of solute and solvent particles; (**c**) the color and clarity of the solution; (**d**) the Tyndall effect.

17. Compare and contrast a colloidal dispersion and a suspension.

18. Is it possible to have a colloidal dispersion of one gas in another? Explain.

19. Compare and contrast osmosis and dialysis with respect to the kinds of particles that pass through the semipermeable membrane.

20. If two containers of a gas at different pressures are connected, the net diffusion of gas is from the *higher* to the lower pressure. The net diffusion in osmosis is from the solution of *lower* to that of higher concentration. How can you explain this apparent difference?

21. What is meant by the term *reverse osmosis?* Give an example of its use.

22. The text describes an isotonic solution of NaCl as being about 0.16 M whereas an isotonic solution of glucose is about 0.31 M. Explain why the concentrations of these isotonic solutions are not the same.

23. What would be the effect on red blood cells if they were placed in (a) 5.5% NaCl(aq) (b) 0.92% glucose(aq)?
24. Use the kinetic-molecular theory to explain why NaCl dissolves in water.
25. Use the kinetic-molecular theory to explain why most solid solutes become more soluble in water with increasing temperature but gases generally become less soluble.
26. Ethyl alcohol, CH_3CH_2OH, is soluble in water; ethyl chloride, CH_3CH_2Cl, is insoluble in water. Explain.
27. Without referring to Table 9.2, indicate which of the following compounds you would expect to be soluble in water. Explain each answer. You may use the periodic table.
    a. NaBr
    b. $Ca(NO_3)_2$
    c. $BaCO_3$
    d. $FePO_4$
28. Without referring to Table 9.2, indicate which of the following compounds you would expect to be soluble in water. Explain each answer. You may use the periodic table.
    a. $(NH_4)_2CO_3$
    b. RbCl
    c. $PbSO_4$
    d. LiOH

Problems

Molarity

29. Calculate the molarity of each of the following solutions.
    a. 6.00 mol of HCl in 2.50 L of solution
    b. 0.00700 mol of Li_2CO_3 in 10.0 mL of solution
30. Calculate the molarity of each of the following solutions.
    a. 2.50 mol of H_2SO_4 in 5.00 L of solution
    b. 0.200 mol of C_2H_5OH in 18.4 mL of solution
31. Calculate the molarity of each of the following solutions.
    a. 8.90 g of H_2SO_4 in 100.0 mL of solution
    b. 439 g of $C_6H_{12}O_6$ in 1.25 L of solution
32. Calculate the molarity of each of the following solutions.
    a. 44.3 g of KOH in 125 mL of solution
    b. 2.46 g of $H_2C_2O_4$ in 750.0 mL of solution
33. How many grams of solute are needed to prepare each of the following solutions?
    a. 2.00 L of 1.00 M NaOH
    b. 10.0 mL of 4.25 M $C_6H_{12}O_6$
34. How many grams of solute are needed to prepare each of the following solutions?
    a. 250 mL of 2.50 M $K_2Cr_2O_7$
    b. 20.0 mL of 0.0100 M $KMnO_4$
35. What volume of 6.00 M NaOH is required to contain 1.25 mol of NaOH?
36. What volume of 2.50 M NaOH is required to contain 1.05 mol of NaOH?
37. What volume of 0.0250 M $KMnO_4$ must one take to get 8.10 g of $KMnO_4$?
38. What volume of 4.25 M $C_6H_{12}O_6$ must one take to get 205 g of $C_6H_{12}O_6$?

Percent Concentration

39. What is the volume percent concentration of each of the following solutions?
    a. 35.0 mL of water in 725 mL of an ethanol–water solution
    b. 78.9 mL of acetone in 1.55 L of an acetone-water solution
40. What is the volume percent concentration of each of the following solutions?
    a. 58.0 mL of water in 625 mL of an ethanol–water solution
    b. 79.1 mL of methanol in 755 mL of a methanol-water solution
41. What is the mass percent concentration of each of the following solutions?
    a. 4.12 g of NaOH in 100.0 g of water
    b. 5.00 mL of ethanol ($d = 0.789$ g/mL) in 50.0 g of water
42. What is the mass percent concentration of each of the following solutions?
    a. 175 mg of NaCl/g solution
    b. 275 mL of methanol ($d = 0.791$ g/mL)/kg water
43. Describe how you would prepare 775 g of an aqueous solution that is 10.0% NaCl *by mass*.
44. Describe how you would prepare 125 g of an aqueous solution that is 5.50% KOH *by mass*.
45. Describe how you would prepare exactly 2.00 L of an aqueous solution that is 2.00% acetic acid *by volume*.
46. Describe how you would prepare exactly 500 mL of an aqueous solution that is 30.0% isopropyl alcohol *by volume*.
47. Describe how you would prepare 250.0 mL of an aqueous solution of $MgSO_4$ that is 1.5% (mass/vol).
48. Describe how you would prepare 2.00 L of an aqueous solution of $AlCl_3$ that is 2.15% (mass/vol).
49. On average, glucose makes up about 0.10% of human blood, by mass. What is the approximate concentration in mg/dL?
50. A vinegar sample has a density of 1.01 g/mL and contains 5.88% acetic acid by mass. What mass of acetic acid is contained in a 1-L bottle of the vinegar?

Saturated, Unsaturated, and Supersaturated Solutions

51. Use data from Figure 9.5 to determine (a) whether a solution containing 48 g of KNO_3 per 100 g of water at

40 °C is saturated, unsaturated, or supersaturated, and (**b**) the approximate temperature to which a mixture of 60 g of KNO_3 and 100 g of water must be heated so that the KNO_3 is completely dissolved.

52. Use data from Figure 9.5 to determine (**a**) the mass percent of $KClO_3$ in a saturated aqueous solution at 50 °C, and (**b**) the molarity of saturated $Ce_2(SO_4)_3(aq)$ at 50 °C.

Osmolarity and Colligative Properties

53. How many solute particles does each formula unit of each of the following compounds give in aqueous solution? How many osmol are there in one mole of each compound?
a. KCl **b.** CH_3OH **c.** $(NH_4)_2SO_4$

54. How many solute particles does each formula unit of each of the following compounds give in aqueous solution? How many osmol are there in one mole of each compound?
a. $CaBr_2$ **b.** NaOH **c.** $Al(NO_3)_3$

55. For each pair of solutions at 25 °C, indicate which member has the higher osmotic pressure.
a. 0.1 M $NaHCO_3$, 0.05 M $NaHCO_3$
b. 1 M NaCl, 1 M glucose

56. For each pair of solutions at 25 °C, indicate which member has the higher osmotic pressure.
a. 1 M NaCl, 1 M $CaCl_2$ **b.** 1 M NaCl, 3 M glucose

57. How many moles of each of the following are needed to provide 1.0 osmol?
a. KCl **b.** CH_3OH **c.** $(NH_4)_2SO_4$

58. How many moles of each of the following are needed to provide 1.0 osmol?
a. $CaBr_2$ **b.** NaOH **c.** $Al(NO_3)_3$

59. For each pair of solutions at 25 °C, indicate which member has the higher osmotic pressure.
a. 1.0 osmol/L NaCl, 1.0 osmol/L $CaCl_2$
b. 1.0 osmol/L $CaCl_2$, 2.0 osmol/L glucose ($C_6H_{12}O_6$)

60. For each pair of solutions at 25 °C, indicate which member has the higher osmotic pressure.
a. 1.0 osmol/L $NaHCO_3$, 0.50 osmol/L $NaHCO_3$
b. 2.0 osmol/L CH_3OH, 1.0 osmol/L $CaCl_2$

Additional Problems

61. Can you think of any way in which solute can be crystallized from an unsaturated solution without changing the solution temperature? Explain.

62. Are there any exceptions to the general rule that a supersaturated solution can be made to deposit excess solute by cooling? Explain.

63. The solubility of $O_2(g)$ in water is 4.43 mg O_2/100 g H_2O at 20 °C when the gas pressure is maintained at 1 atm. What is the molarity of the saturated solution?

64. Consider 1.0 mol of each of the following: $Al_2(SO_4)_3$, CH_3COOH, CH_3OH, $MgBr_2$, and NaCl. Which would lower the freezing point of 1.0 L of water the most? The least? Explain.

65. Pickles are made by soaking cucumbers in a salt solution. Which has the higher osmotic pressure, the salt solution or the liquid in the cucumber? Explain.

66. Two aqueous solutions of glucose ($C_6H_{12}O_6$) are separated by a semipermeable membrane. Solution A has 3.00% glucose (mass/volume) and solution B is 0.10 M glucose. In which direction will a net flow of water occur, from A to B or from B to A?

67. Aluminum sulfate is commonly used to coagulate or precipitate colloidal suspensions of clay particles in municipal water treatment plants. Why do you suppose this electrolyte is more effective than sodium chloride?

68. An aqueous solution is prepared by dissolving 11.3 mL of CH_3OH (d = 0.793 g/mL) in enough water to produce 73.5 mL of solution. What is the percent of CH_3OH, expressed as (**a**) volume percent and (**b**) mass-volume percent?

69. You have a stock solution of 6.0 M HCl. How many moles of HCl are there in the following amounts of solution?
a. 1.0 L **b.** 100 mL **c.** 1.0 mL

70. On the average, glucose ($C_6H_{12}O_6$) makes up about 0.10% by weight of human blood. How much glucose is there in 1.0 kg of blood?

71. A cyanide solution, made by adding 1.0 lb of sodium cyanide (NaCN) to 1.0 ton (2000 lb) of water, is used to leach gold from its ore. What is the concentration of the cyanide solution in each of the following units of concentration?
a. percent by mass **b.** ppm
c. g/kg **d.** mol/L

72. Complete the following table.

Concentration of Solute (g/L)	Molarity (mol/L)	Molar Mass of Solute (g/mol)
98	1.0	—
32	0.5	—
0.74	0.01	—
—	0.1	26
—	0.025	80
120	—	40
17	—	68

10 Acids and Bases I

Most foods are either neutral or slightly acidic. We use bases—called antacids—to counter acid indigestion.

All of us practice a complicated chemistry, from digesting food and shedding tears to taking antacids and baking bread. Central to much of this chemistry are two special kinds of compounds called *acids* and *bases*. We eat them and drink them; we use vinegar (acetic acid) in cooking and in salad dressings and drink beverages made tart by citric acid or phosphoric acid. We treat excess stomach acid (hydrochloric acid) with bases such as aluminum hydroxide and magnesium hydroxide. Our bodies produce and consume acids and bases. We clean with ammonia and lye (sodium hydroxide), two familiar bases. Some of these and other acids and bases used around the home are shown in Figure 10.1.

The four tastes are related to acid–base chemistry. Acids taste *sour* and bases taste *bitter*. Salts, compounds formed when acids react with bases, taste *salty*. *Sweet* tastes are related to acid–base chemistry in a more subtle way. To taste sweet, a compound must have a hydrogen-bond donor group (acidic), a hydrogen-bond acceptor group (basic), and a nonpolar group, all arranged in just the proper geometry to fit the sweet-taste receptor.

Acid rain is an environmental problem in industrialized countries. Bitter, undrinkable alkaline (basic) water is often all that is available in areas with dry climates. Much of our understanding of air and water pollution depends on our knowledge of acids and bases.

(a) (b)

FIGURE 10.1 ▶

(a) Some common acids: toilet bowl cleaner, vinegar, aspirin, beer, and fruit juices. (b) Some common bases: drain cleaners, oven cleaner, quinine water, and so on.

According to the Biblical account, the Israelites, in their journey from Egypt to Canaan, came upon the bitter waters at Marah (Exod. 15:23). Although the writers may not have meant to, they thus recorded for posterity the existence of a base.

In this chapter, we consider some of the more qualitative properties of acids and bases—what they are, their names, how they are produced, and how they react chemically. In Chapter 11, we consider some of the more quantitative aspects of acid–base chemistry.

10.1 Acids and Bases: Definitions and Properties

Historically, acids and bases have been classified according to some distinctive properties. *Acids* are substances that, when dissolved in water,

- taste sour, if diluted with enough water to be tasted safely.
- produce a prickling or stinging sensation on the skin.
- turn the color of the indicator dye litmus from blue to red.
- react with many metals, such as magnesium, zinc, and iron, to produce ionic compounds and hydrogen gas.
- react with bases, thereby losing their acidic properties.

Concentrated or dilute acids used in the laboratory are quite corrosive; they should never be tasted.

Bases are substances that, when dissolved in water,

- taste bitter, if diluted with enough water to be tasted safely.
- feel slippery or soapy on the skin.
- turn the color of the indicator dye litmus from red to blue.
- react with acids, thereby losing their basic properties.

Many medicines taste bitter because they contain *alkaloid* (baselike) compounds. Quinine, caffeine, and the antihistamines are familiar examples.

Litmus is a dye obtained from lichens. Litmus paper is often used to determine whether a material is acidic or basic. The soil sample on the right turns blue litmus red and is acidic. The soil sample on the left turns red litmus blue and is basic.

In 1887, the Swedish chemist Svante Arrhenius proposed that an **acid** is a molecular substance that breaks up in aqueous solution into H^+ ions and anions. The acid is said to ionize. He viewed a **base** as a substance that produces OH^- in aqueous solution, either because it contains OH^-, such as an ionic hydroxide, or because it can ionize to produce OH^-. He proposed that the essential reaction between an acid and a base, *neutralization*, is the combination of H^+ and OH^- to form water. The cation originally associated with the OH^- and the anion associated with the H^+ give rise to an ionic compound, a **salt.**

How can you tell when a compound is an acid? You can tell by dissolving some of the compound in water. Since tasting may be hazardous (some acids are poisonous, and nearly all are corrosive unless they have been highly diluted), you can stick a piece of blue litmus paper into the solution. If the paper turns red, the compound is an acid.

Many foods are acidic. Vinegar contains about 5% acetic acid. Citrus fruits and many fruit-flavored drinks contain citric acid. If a food tastes sour, most likely it contains one or more acids. Lactic acid is formed in sour milk and is responsible for the tart taste of yogurt. Phosphoric acid is used to impart tartness to beer and some forms of soft drinks.

The Arrhenius theory has its limitations. We now know, for example, that a simple free proton does not exist in water solution because H^+ has such a high positive charge density that it immediately seeks out a negative charge. It finds a lone pair of electrons on the O atoms of an H_2O molecule and attaches itself to form a **hydronium ion, H_3O^+.**

By a "proton" we mean an ionized hydrogen atom—that is, H^+.

$$H-\overset{..}{O}: \quad + \quad H^+ \quad \longrightarrow \quad \left[\overset{\textstyle H}{\underset{\overset{..}{}}{H-O-H}} \right]^+$$

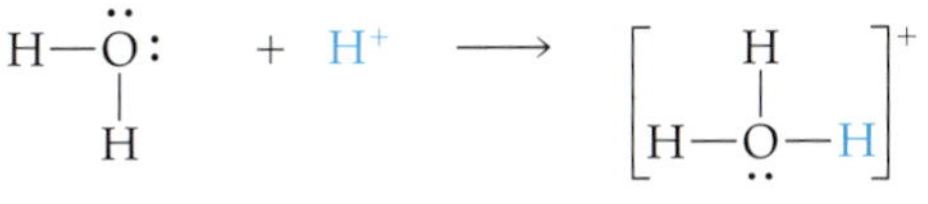

Water Hydronium ion

The H^+ ion is probably associated with several H_2O molecules—for example, four H_2O molecules in the ion $H(H_2O)_4^+$ or $H_9O_4^+$. For most purposes, however, we can use the simple hydronium ion (H_3O^+) and ignore any further hydration. The hydronium ion is quite reactive; it can readily transfer a proton to another molecule or ion. Thus, to talk about protons when we mean their sources (hydronium

▲ **FIGURE 10.2**

Ionization of HCl as a Brønsted-Lowry acid. The red arrow represents proton transfer from the HCl molecule to the H_2O molecule. Water molecules accept protons; water is a Brønsted-Lowry base in this reaction.

ions) is permissible as long as we understand that we are using a simplification of the real situation. In water, then, the properties of acids are those of the hydronium ion. It is the hydronium ion that turns litmus red, tastes sour, and reacts with active metals and bases.

The Arrhenius theory is also limited in that it applies only to reactions in aqueous solution, and it does an inadequate job of explaining where the OH^- comes from in the ionization of bases such as ammonia, NH_3. Like many scientific theories, it has been supplanted by a better one based on newer data.

The shortcomings of the Arrhenius theory were largely overcome by a theory proposed independently, in 1923, by J.N. Brønsted in Denmark and T. M. Lowry in Great Britain. In their theory, an acid is a **proton donor** and a base is a **proton acceptor.**

The theory describes the ionization of hydrogen chloride in this way:

$$\underset{\text{acid}}{HCl(g)} + \underset{\text{base}}{H_2O} \longrightarrow H_3O^+(aq) + Cl^-(aq)$$

The main feature of the Brønsted-Lowry notation, also suggested by Figure 10.2, is that the overall reaction consists of a proton transfer from an acid to a base. In this case, H_2O acts as a base by accepting a proton from HCl, an acid.

Hydrochloric acid has one ionizable H atom per molecule; it is a **monoprotic acid.** Nitric acid (HNO_3) and hydrocyanic acid (HCN) are also monoprotic acids. Sulfuric acid (H_2SO_4) and carbonic acid (H_2CO_3) each have two ionizable H atoms; they are **diprotic acids.** Phosphoric acid (H_3PO_4) has three ionizable H atoms; it is a **triprotic acid.** H_2CO_3, H_2SO_4, and H_3PO_4 collectively are known as polyprotic acids. **Polyprotic acids** are acids that have more than one ionizable H atom per molecule. You should not assume that all the hydrogen atoms in a molecule are acidic, meaning that they are released in water solutions. For example, none of the hydrogens of methane (CH_4) is given up in aqueous solution. Only one of the hydrogen atoms in acetic acid ($C_2H_4O_2$) is acidic. For this reason, the formula for acetic acid is frequently written CH_3COOH or $HC_2H_3O_2$ to emphasize that only one proton is released. You will gain experience in determining which hydrogen atoms in a molecule are acidic as you learn more about molecular structure.

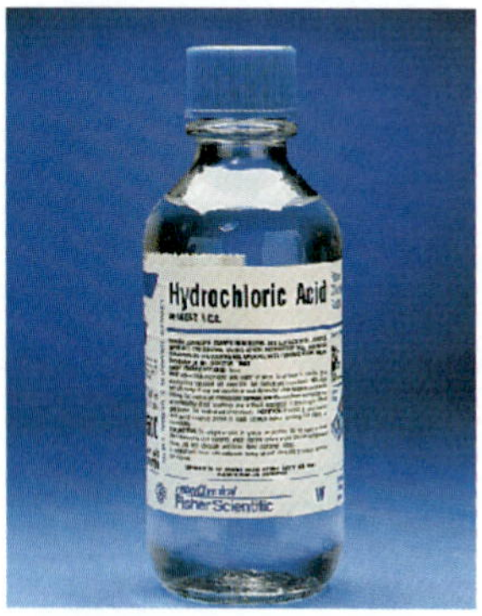

Hydrogen chloride is a gas, usually represented as HCl(g). Hydrochloric acid, represented as HCl(aq), is an aqueous solution of hydronium ions (H_3O^+) and chloride ions (Cl^-).

10.2 Strong and Weak Acids

Acids that are completely ionized in water solution are called **strong acids.** To represent the ionization of a strong acid, such as hydrochloric acid, we can write the equation

$$HCl(g) + H_2O \longrightarrow H_3O^+(aq) + Cl^-(aq)$$

The reaction is complete; essentially no HCl molecules remain. There are only $H_3O^+(aq)$ and $Cl^-(aq)$ ions in the hydrochloric acid solution. In 0.0010 M HCl(aq), for example, the molar concentrations (indicated by square brackets, []) are

$$[H_3O^+] = 0.0010\ M; \quad [Cl^-] = 0.0010\ M; \quad [HCl] = 0$$

The vast majority of acids, because they are only partially ionized in aqueous solution, are called **weak acids.** One important difference between weak acids and strong acids is in how we represent them in solution. The ionization of a weak acid in aqueous solution is a *reversible* reaction, represented in an equation with a double arrow. The ionization of hydrogen fluoride, a weak acid, is represented by the following equation.

$$HF(aq) + H_2O \rightleftharpoons H_3O^+(aq) + F^-(aq)$$

When this reversible reaction reaches equilibrium, most of the HF molecules remain nonionized in 1 M HF(aq), fewer than 1% of the molecules ionize. Because the predominant solute species are hydrofluoric acid molecules, an aqueous solution of hydrofluoric acid is best represented as HF(aq), *not* as $H_3O^+(aq) + F^-(aq)$.

To calculate concentrations in an aqueous solution of a weak electrolyte, such as $[H^+]$, $[F^-]$, and [HF] in HF(aq), requires a kind of calculation that we consider in Special Topic C.

So, you may wonder, how can you tell if a substance is an acid and, if so, whether it's a strong or weak acid? Chemists generally identify *ionizable* hydrogen atoms by writing formulas in one of two ways.

1. *We write a molecular formula with ionizable H atoms first.* HNO_3, H_2SO_4, and H_3PO_4 are acids with one, two, and three ionizable H atoms, respectively. Methane, CH_4, has four H atoms, but they are *not* ionizable; CH_4 is *not* an acid. From its name alone we know that acetic acid is an acid. When its formula is written as $HC_2H_3O_2$, we see that it has four H atoms, but only one that is ionizable, and it is the H atom that is shown first.
2. *In organic chemistry, we often use formulas that show the ionizable hydrogen atoms last.* In Chapter 16, we consider a family of acids called *carboxylic acids.* The most familiar of these is acetic acid, probably the most familiar of all the weak acids. We write the formula as CH_3COOH. This formula shows that the three nonionizable H atoms are bonded to a C atom. The fourth H atom is bonded to an O atom, and it is this one that is ionizable. Other water-soluble carboxylic acids include formic acid (HCOOH), propionic acid (CH_3CH_2COOH), and butyric acid ($CH_3CH_2CH_2COOH$). In each of these, only the H atom on the O atom is ionizable.

The simplest way to tell whether an acid is strong or weak is to note that the *common* strong acids are limited to those listed as strong in Table 10.1. Unless you are given information to the contrary, assume that any other acid is a weak acid.

Strong acids, when moderately concentrated, can cause serious damage to skin and flesh. They eat holes in clothes made of natural fibers such as cotton, silk, and wool. Strong acids also destroy most synthetic fibers such as nylon, polyester, and acrylics. Care always should be taken to prevent spills on skin and clothing. In very dilute solutions, strong acids can be harmless. Hydrochloric acid, for example, is produced in dilute solutions in our stomachs, where it aids in the

Table 10.1 Common Strong Acids

Name	Formula
Hydrochloric acid	HCl(aq)
Hydrobromic acid	HBr(aq)
Hydriodic acid	HI(aq)
Nitric acid	$HNO_3(aq)$
Sulfuric acid	$H_2SO_4(aq)$[a]
Perchloric acid	$HClO_4(aq)$

[a] Only the first ionization of H_2SO_4 is complete; HSO_4^- is only partly ionized and is classified as a weak acid.

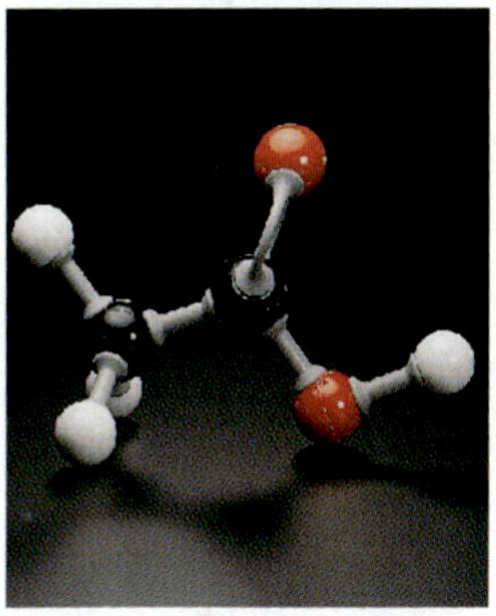

A ball-and-stick model of an acetic acid molecule. Only the H atom bonded to the O atom (lower right) is ionizable. The three H atoms attached to the C atom (left) are not ionizable.

The term *strong* does not refer to the amount of solute in the solution. Whether an acid is strong or weak, a solution with a relatively large amount of acid solute in a given volume of solution is called a *concentrated acid.* A solution with relatively little acid solute in a given volume of solution is called a *dilute acid* (see Section 11.1).

digestion of certain foodstuffs. Even there, however, under certain conditions, it can cause problems—as anyone with an ulcer can attest. Concentrated weak acids can cause problems too. Although it is pleasant to eat salads with vinegar as a condiment, concentrated solutions of acetic acid are corrosive to the skin and are especially destructive to the lining of the digestive tract.

10.3 Names of Some Common Acids

In Chapter 4 you learned the names of some common anions. Now we can use the following scheme to relate the names of acids to the anions produced when the acids ionize.

Acid name		**Anion name**
hydro___ic acid	⟶	*___ide*
hypo___ous acid	⟶	*hypo___ite*
___ous acid	⟶	*___ite*
___ic acid	⟶	*___ate*
per___ic acid	⟶	*per___ate*

To get the molecular formula for the acid, use the number of H atoms that is the same as the number of H^+ ions required to balance the electric charge of the anion. That is, one H^+ is needed to balance the 1− charge on the Cl^- ion, so we add one H atom in HCl. Similarly, we need two H atoms in H_2SO_4, and three H atoms in H_3PO_4. The scheme for naming acids is illustrated by the examples in Table 10.2. Note that most of the names have the same stem in the acid as in the anion. A few have extra letters—for example, the *ur* in sulfuric and sulfurous and the *or* in phosphoric. These you should learn as special cases if you haven't done so already.

Table 10.2 Names of Some Common Acids and Their Salts

Formula of Acid	Name of Acid	Sodium Salt	
		Formula	Name
HCl	*Hydro*chlor*ic* acid	NaCl	Sodium chlor*ide*
HClO	*Hypo*chlor*ous* acid	NaClO	Sodium *hypo*chlor*ite*
$HClO_2$	Chlor*ous* acid	$NaClO_2$	Sodium chlor*ite*
$HClO_3$	Chlor*ic* acid	$NaClO_3$	Sodium chlor*ate*
$HClO_4$	*Per*chlor*ic* acid	$NaClO_4$	Sodium *per*chlor*ate*
HNO_2	Nitr*ous* acid	$NaNO_2$	Sodium nitr*ite*
HNO_3	Nitr*ic* acid	$NaNO_3$	Sodium nitr*ate*
$HC_2H_3O_2$[a]	Acet*ic* acid	$NaC_2H_3O_2$	Sodium acet*ate*
H_2S	*Hydro*sulfur*ic* acid	Na_2S	Sodium sulf*ide*
H_2SO_3[b]	Sulfur*ous* acid	Na_2SO_3	Sodium sulf*ite*
H_2SO_4[b]	Sulfur*ic* acid	Na_2SO_4	Sodium sulf*ate*
H_3PO_4[b]	Phosphor*ic* acid	Na_3PO_4	Sodium phosph*ate*

[a] Acetic acid is also represented as CH_3COOH (see Section 10.2) and sodium acetate as CH_3COONa.
[b] Table 4.4 also lists anions of these acids in which only one of the H atoms is replaced by Na (one or two in the case of H_3PO_4).

Example 10.1

The formula for phosphite ion is PO_3^{3-}. What is the formula for phosphorous acid?

Solution

It takes three H^+ ions to balance the electric charge on the PO_3^{3-} ion. We therefore need three H atoms in the molecular acid. The formula for phosphorous acid is H_3PO_3.

Practice Exercise

The formula for arsenate ion is AsO_4^{3-}. What is the formula for arsenic acid?

Example 10.2

The formula for selenic acid is H_2SeO_4. What is the formula for selenious acid? For sodium selenate? For sodium selenite?

Solution

From Table 10.2 we see the general pattern that an acid with a name ending in *-ous* has one less O atom than an acid with a name ending in *-ic*. The formula for selenious acid is therefore H_2SeO_3. Sodium selenate is a salt of selenic acid; its formula is Na_2SeO_4. Sodium selenite is a salt of selenious acid; its formula is Na_2SeO_3.

Practice Exercise

The formula for iodic acid is HIO_3. What is the formula for hypoiodous acid? For sodium iodate? For sodium iodite? For sodium hypoiodite?

10.4 Some Common Bases

The Arrhenius definition of a base we gave in Section 10.1 is that of a substance which produces hydroxide ions, OH^-, in aqueous solution. Many bases are ionic compounds that contain Group 1A or Group 2A cations along with hydroxide ions. Each has enough OH^- ions to balance the charge of the accompanying cation, and they are named like other ionic compounds:

$NaOH$ = sodium hydroxide (commonly called *lye*)
KOH = potassium hydroxide
$Ca(OH)_2$ = calcium hydroxide (commonly called *slaked lime*)

Soluble ionic hydroxides such as NaOH are obviously bases. Moreover, because they are completely ionic, they are **strong bases.**

$$NaOH(s) \xrightarrow{H_2O} Na^+(aq) + OH^-(aq)$$

Some compounds produce OH^- ions by *reacting* with water, not just by dissolving in it. Such substances are also bases. The equation below shows that gaseous ammonia dissolves in water and reacts with water to produce an equilibrium mixture of ions and molecules.

$$NH_3(g) + H_2O(l) \rightleftharpoons NH_4^+(aq) + OH^-(aq)$$

As in the case of acetic acid, most of the ammonia molecules in the aqueous

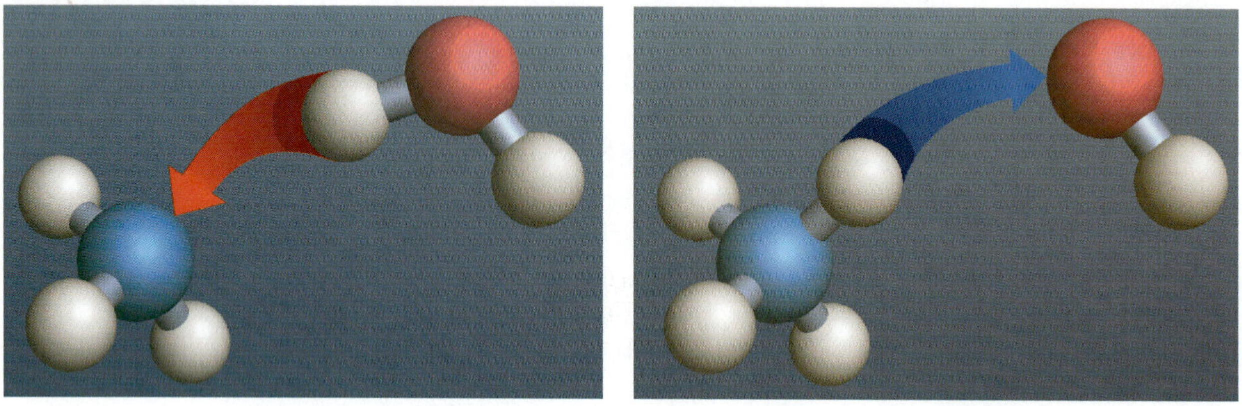

$$NH_3 + H_2O \rightleftharpoons NH_4^+ + OH^-$$

▲ **FIGURE 10.3**

Ionization of NH_3 as a Brønsted-Lowry base. The red arrow represents proton transfer in the forward reaction and the reverse reaction is indicated by the blue arrow. Only relatively few ammonia molecules react, however; most remain unchanged in solution as NH_3 molecules. There are therefore relatively few hydroxide ions and ammonium ions; ammonia is a weak base.

Many familiar household cleaning agents contain ammonia, an ingredient easily identified by its characteristic odor. Plain ammonia solutions are about 3.5% NH_3 (mass/vol). Sudsy ammonia solutions contain detergents and often have less NH_3.

solution remain nonionized. Ammonia is therefore a **weak base.**[1] Most molecular substances that act as bases are *weak* bases.

How can an ammonia molecule act as a base—a proton acceptor? Recall that the N atom of ammonia has a lone pair of electrons. The lone pair can be used to attach a proton. When a proton leaves a water molecule, it leaves behind the electron pair that joined it to the O atom. The water molecule becomes a negatively charged hydroxide ion (see also Figure 10.3).

$$\underset{\text{Base}}{H:\overset{..}{\underset{H}{N}}:} \; + \; \underset{\text{Acid}}{\textcircled{H}:\overset{..}{\underset{..}{O}}:} \; \rightleftharpoons \; \underset{\substack{\text{Ammonium}\\\text{ion}}}{\left[H:\overset{..}{\underset{H}{N}}:H\right]^+} \; + \; \underset{\substack{\text{Hydroxide}\\\text{ion}}}{\left[:\overset{..}{\underset{..}{O}}:H\right]^-}$$

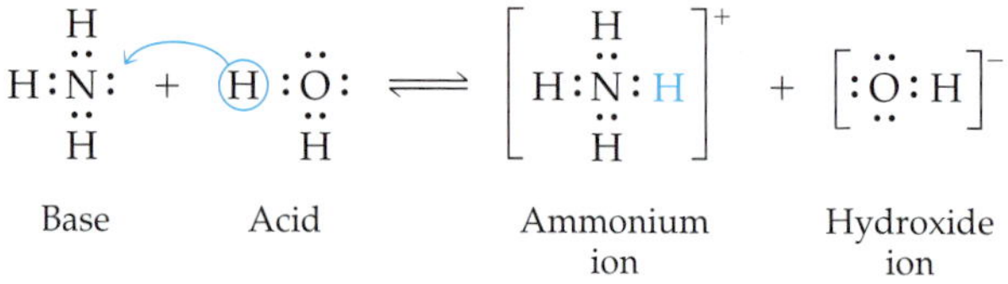

The concept of a base also includes not only hydroxide ion and molecules such as ammonia, but also ions such as carbonate (CO_3^{2-}), bicarbonate (HCO_3^-), and phosphate (PO_4^{3-}) (see Section 10.6). For example, carbonate ions react with water to yield a basic solution.

$$CO_3^{2-}(aq) \; + \; H_2O \; \rightleftharpoons \; HCO_3^-(aq) \; + \; OH^-(aq)$$

The family of organic compounds called **amines** (Chapter 17) have molecules in which one or more of the H atoms of NH_3 are replaced by a hydrocarbon group. Some examples are *methylamine,* CH_3NH_2; *dimethylamine,* $(CH_3)_2NH$; *trimethylamine,* $(CH_3)_3N$; and *ethylamine,* $CH_3CH_2NH_2$. The amines with one to four carbon atoms are water soluble, and, like ammonia, they are *weak* bases.

$$CH_3NH_2(aq) \; + \; H_2O \; \rightleftharpoons \; CH_3NH_3^+(aq) \; + \; OH^-(aq)$$

How can we recognize a base and how do we know whether the base is weak

[1] In Arrhenius's time, chemists generally believed that a substance must *contain* OH groups to be a base; $NH_3(aq)$ was thought to be NH_4OH (ammonium hydroxide). Even though there is no compelling evidence for the existence of NH_4OH, this formula is still often seen as a representation of $NH_3(aq)$.

or strong? If an ionic compound contains OH^- ions, expect it to be a base; if it is water-soluble, expect it to be a strong base. NaOH and KOH are strong bases. On the other hand, methanol (CH_3OH) is *not* a base. The OH group is not present as OH^-; it is covalently bonded to the C atom. Similarly, acetic acid (CH_3COOH) is *not* a base. It does not contain OH^-, nor does it produce $OH^-(aq)$ in water; rather, it produces $H^+(aq)$ and is therefore an *acid*. To identify a weak base, you usually need a chemical equation for the ionization reaction. However, you can identify many by using these facts: *There are only a few common strong bases, listed in Table 10.3. The most common weak bases are ammonia and the amines.* Most bases are molecular substances and do not contain hydroxide ion. They produce it by reacting with water.

The idea of an acid as a proton donor and a base as a proton acceptor greatly expands our concept of acids and bases, but the broader concept still *includes* the more limited definitions. It's just that the broader concept is useful in a greater variety of situations (Figure 10.4).

Table 10.3 Strong Bases

Name	Formula
Alkali metal hydroxides	
Lithium hydroxide	$LiOH(aq)$
Sodium hydroxide	$NaOH(aq)$
Potassium hydroxide	$KOH(aq)$
Rubidium hydroxide	$RbOH(aq)$
Cesium hydroxide	$CsOH(aq)$
Alkaline earth hydroxides	
Calcium hydroxide	$Ca(OH)_2(aq)$
Strontium hydroxide	$Sr(OH)_2(aq)$
Barium hydroxide	$Ba(OH)_2(aq)$

▲ **FIGURE 10.4**
An acid is a proton donor. A base is a proton acceptor.

10.5 Acidic and Basic Anhydrides

Many acids are made by the reaction of nonmetal oxides with water. For example, sulfur trioxide reacts with water to form sulfuric acid.

$$SO_3 + H_2O \longrightarrow H_2SO_4$$

Similarly, carbon dioxide reacts with water to form carbonic acid.

$$CO_2 + H_2O \longrightarrow H_2CO_3$$

This is a general reaction of nonmetal oxides.

$$\text{Nonmetal oxide} + H_2O \longrightarrow \text{Acid}$$

Nonmetal oxides are called **acid anhydrides.** *Anhydride* means "without water." These reactions explain why rainwater is acidic, particularly when the rain forms in air that is polluted with sulfur oxides (see pages 251 and 312).

Example 10.3

Give the formula for the acid formed when sulfur dioxide reacts with water.

Solution

Simply write the equation for the reaction.

$$SO_2 + H_2O \longrightarrow H_2SO_3$$

Practice Exercise

Give the formula for the acid formed when dinitrogen pentoxide (N_2O_5) reacts with water. (*Hint: Two* molecules of acid are formed.)

Many common bases can be made from metal oxides. For example, calcium oxide (lime) reacts with water to form calcium hydroxide (slaked lime).

$$CaO + H_2O \longrightarrow Ca(OH)_2$$

Another example is the reaction of lithium oxide with water to form lithium hydroxide.

$$Li_2O + H_2O \longrightarrow 2\,LiOH$$

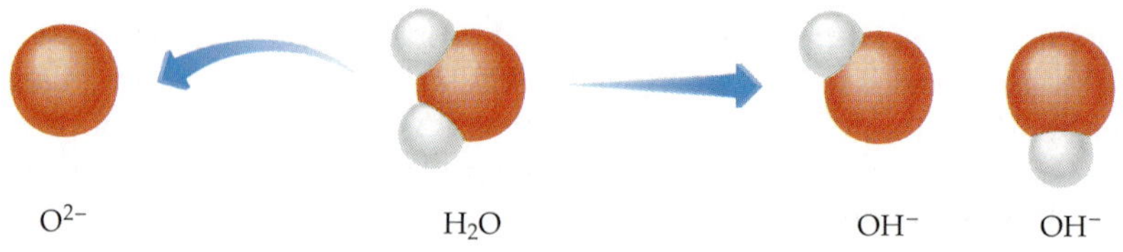

▲ **FIGURE 10.5**
Metal oxides are basic because the oxide ion reacts with water to form two hydroxide ions.

In general, metal oxides react with water to form bases (Figure 10.5). These metal oxides are called **basic anhydrides.**

$$\text{Metal oxide} \; + \; H_2O \longrightarrow \text{Metal hydroxide}$$

Example 10.4

What base is formed by the addition of water to barium oxide (BaO)?

Solution

Simply write the equation for the reaction.

$$BaO \; + \; H_2O \longrightarrow Ba(OH)_2$$

Practice Exercise

What base is formed by the addition of water to potassium oxide (K_2O)? (*Hint:* Two moles of base are formed for each mole of potassium oxide.)

10.6 Neutralization Reactions

In the reaction of an acid and a base, called **neutralization,** the identifying characteristics of the acid and base cancel out or neutralize each other. The acid and base are converted to an aqueous solution of an ionic compound, called a salt. If we use conventional formulas for the acid and base, we can write what we might call a "complete formula" equation[2] for a neutralization reaction as follows:

$$HCl(aq) \; + \; NaOH(aq) \longrightarrow NaCl(aq) \; + \; H_2O(l)$$

But this "complete formula" equation is not the best way to show what happens in the neutralization. We can better do that by indicating the actual ions and molecules present in solution; we write the equation in *ionic* form.

$$\underset{\text{(acid)}}{H_3O^+(aq) + \cancel{Cl^-(aq)}} + \underset{\text{(base)}}{\cancel{Na^+(aq)} + OH^-(aq)} \longrightarrow \underset{\text{(salt)}}{\cancel{Na^+(aq)} + \cancel{Cl^-(aq)}} + \underset{\text{(water)}}{H_2O(l)}$$

When we eliminate "spectator" ions—those that just "look on" and that appear unchanged in the ionic equation—we find that the equation reduces to the even more informative **net ionic equation.**

$$H_3O^+(aq) \; + \; OH^-(aq) \longrightarrow 2\,H_2O(l)$$

---

[2] The "complete formula" equation is often called a "molecular" equation, but this term is misleading. Many of the formulas written in such equations—e.g., NaCl(aq)—represent formula units, not actual molecules.

The essence of a neutralization reaction, then, is that H_3O^+ ions from an acid and OH^- ions from a base combine to form water. If the spectator ions form a soluble salt, they remain in solution. If the water is evaporated, the soluble salt is left as a solid. Example 10.5 provides additional examples of the different types of equations described here.

Example 10.5

Barium nitrate is used to produce a green color in fireworks. It can be made by the reaction of nitric acid with barium hydroxide. Write **(a)** "complete formula," **(b)** ionic, and **(c)** net ionic equations for this neutralization reaction.

Solution

a. Write chemical formulas for the substances involved in the reaction, and balance the equation.

$$HNO_3(aq) + Ba(OH)_2(aq) \longrightarrow Ba(NO_3)_2(aq) + H_2O(l) \text{ (not balanced)}$$

$$2\,HNO_3(aq) + Ba(OH)_2(aq) \longrightarrow Ba(NO_3)_2(aq) + 2\,H_2O(l) \text{ (balanced)}$$

b. Now, represent the strong acid, strong base, and soluble salt with the formulas of their ions and the nonelectrolyte water with its molecular formula.

$$2\,H_3O^+(aq) + 2\,NO_3^-(aq) + Ba^{2+}(aq) + 2\,OH^-(aq) \longrightarrow Ba^{2+}(aq) + 2\,NO_3^-(aq) + 4\,H_2O(l)$$

c. Cancel the spectator ions (Ba^{2+} and NO_3^-) in the above equation.

$$2\,H_3O^+(aq) + 2\,OH^-(aq) \longrightarrow 4\,H_2O(l)$$

or, more simply,

$$H_3O^+(aq) + OH^-(aq) \longrightarrow 2\,H_2O(l)$$

Practice Exercise

Calcium hydroxide is used to neutralize a waste stream of hydrochloric acid. Write **(a)** "complete formula," **(b)** ionic, and **(c)** net ionic equations for this neutralization reaction.

10.7 Reactions of Acids with Carbonates and Bicarbonates

Add vinegar to baking soda, and you get a vigorous fizzing action. Some antacid preparations are designed to effervesce. What causes the fizz? Generally, it is the evolution of CO_2 by the reaction of a carbonate or a bicarbonate salt with an acid. Carbonates and bicarbonates are salts of carbonic acid (H_2CO_3). This diprotic acid is a very weak one; it holds on to its protons quite tightly. (Notice the apparent contradiction: it's the weak acids that hold tightly to their protons and the strong acids that don't.) Conversely, carbonate ions and bicarbonate ions pick up protons readily to form carbonic acid. Further, carbonic acid is quite unstable. It decomposes to carbon dioxide and water.

$$H_2CO_3(aq) \longrightarrow H_2O + CO_2(g)$$

This reaction is not related to the acidity of carbonic acid. It simply indicates that the compound, which happens to be an acid, is not stable. You can't purchase a

The leavening action of baking soda. When acidified, here with citric acid from a lemon, baking soda ($NaHCO_3$) reacts to produce carbonic acid (H_2CO_3), which decomposes to carbon dioxide and water. The carbon dioxide gas produces a "lift" when dough is baked.

bottle of carbonic acid; it is too unstable to be isolated and bottled. It can be formed in solution, but as soon as it is formed, it tends to decompose.

If sodium bicarbonate is dissolved in water, a solution of sodium ions and bicarbonate ions is formed. If hydrochloric acid is added, the bicarbonate ions come into contact with the hydronium ions from the acid and immediately grab a proton to form carbonic acid.

$$\cancel{Na^+(aq)} + HCO_3^-(aq) + H_3O^+(aq) + \cancel{Cl^-(aq)} \rightarrow H_2CO_3(aq) + \cancel{Na^+(aq)} + \cancel{Cl^-(aq)} + H_2O$$

But carbonic acid is unstable and immediately decomposes, forming carbon dioxide and water. The gaseous carbon dioxide bubbles out of solution (the fizz). If the remaining solution is evaporated, sodium chloride is left. Even if acid is poured on solid sodium bicarbonate, carbon dioxide is released. The complete formula equation is

$$NaHCO_3(\text{s or aq}) + HCl(aq) \longrightarrow NaCl(aq) + CO_2(g) + H_2O$$

The net ionic equation is simpler.

$$HCO_3^-(aq) + H_3O^+(aq) \longrightarrow \begin{matrix} H_2CO_3(aq) \\ + \\ H_2O \end{matrix} \longrightarrow H_2O + CO_2(g)$$

Similarly, if hydrochloric acid is added to sodium carbonate, bubbles of carbon dioxide and a solution of sodium chloride are formed.

$$Na_2CO_3(\text{s or aq}) + 2\,HCl(aq) \longrightarrow 2\,NaCl(aq) + CO_2(g) + H_2O$$

The net ionic equation shows that the doubly negative carbonate ion picks up two protons to form carbonic acid. The latter breaks down to form carbon dioxide and water.

$$CO_3^{2-}(aq) + 2\,H_3O^+(aq) \longrightarrow \begin{matrix} H_2CO_3(aq) \\ + \\ 2\,H_2O \end{matrix} \longrightarrow H_2O + CO_2(g)$$

Acid Rain

Limestone and marble are mainly calcium carbonate ($CaCO_3$). Both are important building stones. Marble is also used in statues, monuments, and sculptures. The calcium carbonate in it is readily attacked by acids in the atmosphere or in rain. And the atmosphere has been made increasingly acidic in this century, mainly by the burning of sulfur-containing fossil fuels. The sulfur combines with oxygen to form sulfur dioxide.

Burning of sulfur-containing coal in electric power plants results in the release of large quantities of acidic sulfur oxides into the atmosphere. The oxides ultimately are converted to sulfuric acid and fall as acid rain.

$$S + O_2 \longrightarrow SO_2$$

Some of the sulfur dioxide reacts further with oxygen to form sulfur trioxide.

$$2\,SO_2 + O_2 \longrightarrow 2\,SO_3$$

The sulfur trioxide then reacts with water to form sulfuric acid.

$$SO_3 + H_2O \longrightarrow H_2SO_4$$

The sulfuric acid, in the form of an aerosol mist or diluted by rainwater, furnishes the hydronium ions that dissolve the marble and limestone (Figure 10.6).

$$CaCO_3(s) + 2\,H_3O^+ \longrightarrow Ca^{2+}(aq) + CO_2(g) + 3\,H_2O$$

The acid mists and acidic rainwater also attack and dissolve many metals. Damage to our buildings, automobiles, and other structures and machines from air pollution amounts to billions of dollars per year. Even then the story is not complete. These acid pollutants are also damaging to human health, as we shall see in Section 10.9.

It should be noted that natural rainwater is slightly acidic (pH ≈ 5.6). Water falling through the air dissolves carbon dioxide and forms carbonic acid.

$$H_2O + CO_2 \longrightarrow H_2CO_3$$

This mild acid has little effect compared with the strong acids formed in polluted air.

▲ **FIGURE 10.6**
Marble statues are slowly eroded by the action of acid rain on the marble (calcium carbonate).

The English chemist Robert Smith coined the term *acid rain* after studying the rainfall in London. (See also page 312.) He found that the air, heavily polluted from the burning of coal, produced rain that was abnormally acidic. The term "acid rain" has persisted, and today it usually refers to rain having a pH less than 5.6. During thunderstorms, the pH of rainwater can be much lower due to nitric acid formed by lightning.

10.8 Antacids: A Basic Remedy

The stomach secretes hydrochloric acid as an aid in the digestion of food. The stomach lining is normally protected from the corrosive effects of the acid by a mucosal lining. Holes can develop in the lining, however, that allow the acid to attack the underlying tissue. These lesions, called *ulcers,* frequently bleed and are often quite painful. Over the years, ulcers have been treated by restrictive diets and by preparations called **antacids,** basic compounds that neutralize the stomach acid. Today, we know that most ulcers are caused by bacteria (see page 262), and the preferred treatment is antibiotics. Nevertheless, antacids are still widely used

Table 10.4 Some Common Antacids

Commercial Product	Antacid Ingredient(s)
Alka-Seltzer	$NaHCO_3$, citric acid, aspirin
Amphojel	$Al(OH)_3$
Baking soda	$NaHCO_3$
DiGel	$CaCO_3$
Maalox	$Al(OH)_3$, $Mg(OH)_2$
Milk of magnesia	$Mg(OH)_2$
Mylanta	$Al(OH)_3$, $Mg(OH)_2$
Rolaids	$AlNa(OH)_2CO_3$
Rolaids Sodium-Free*	$CaCO_3$, $Mg(OH)_2$
Tums	$CaCO_3$

*Sodium-containing antacids are not recommended for people with hypertension (high blood pressure).

Drugs such as Zantac, Tagamet, Pepcid AC, and Carafate are used to treat people with ulcers. These drugs inhibit the release of hydrochloric acid by the stomach.

to treat heartburn, a condition in which stomach acid gets up into the esophagus, which has no protective lining. The antacid neutralizes the acid and relieves the burning sensation.

There are many brands of antacids available in the United States. They are often aggressively advertised, and sales approach a billion dollars annually. Despite the variety of brand names, antacids contain only a few different ingredients. Calcium carbonate, aluminum hydroxide, magnesium hydroxide, and sodium bicarbonate are typical components (Table 10.4). Let's look at some popular antacids from the standpoint of acid–base chemistry.

Sodium bicarbonate ($NaHCO_3$), commonly called baking soda, is an old standby antacid. It is probably safe and effective for occasional use by most people. Overuse can make the blood too alkaline, a condition called *alkalosis* (see Section 11.7). Sodium bicarbonate is not recommended for those with hypertension (high blood pressure), because high concentrations of sodium ion tend to aggravate the condition. The antacid in Alka-Seltzer is sodium bicarbonate. This popular remedy also contains citric acid and aspirin. When Alka-Seltzer is placed in water, the reaction of bicarbonate ions with hydronium ions from the citric acid produces the familiar fizz.

$$HCO_3^-(aq) + H_3O^+(aq) \longrightarrow CO_2(g) + 2\,H_2O$$

Alka-Seltzer can be dangerous because the aspirin is sometimes harmful to people with ulcers and other stomach disorders.

Calcium carbonate, commonly called precipitated chalk ($CaCO_3$), is another common antacid ingredient. It is safe in small amounts, but regular use can cause

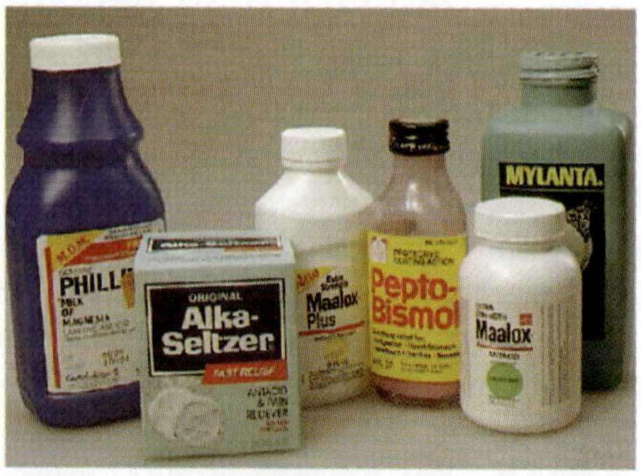

A great variety of antacid formulations are available to the consumer. The main ingredients comprise a short list of basic compounds.

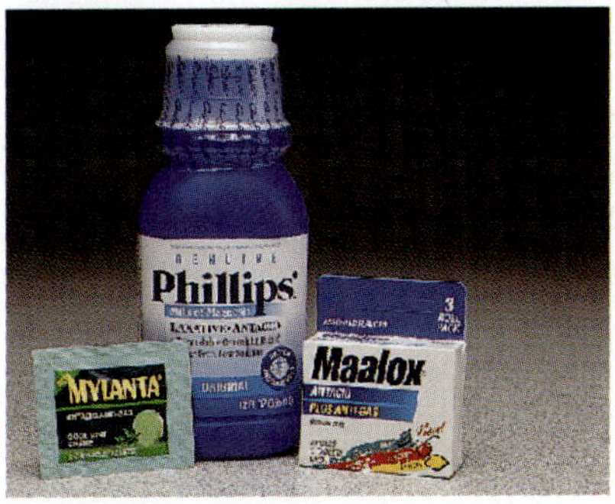

A bottle of milk of magnesia. Water-insoluble hydroxides, such as $Mg(OH)_2$ (milk of magnesia) and $Al(OH)_3$, are used as antacids. One should *never* use a water-soluble ionic hydroxide, because in high concentrations $OH^-(aq)$ is strongly basic. It causes severe burning and scarring of tissue.

constipation. Also, calcium carbonate apparently can cause *increased* acid secretion after a few hours. Temporary relief may be achieved at the expense of a worse problem later. Tums is simply flavored calcium carbonate. Alka 2 and Di-Gel liquid suspension also contain calcium carbonate as the antacid ingredient. The carbonate ion neutralizes acid.

You can make your own aspirin-free "Alka-Seltzer." Simply place half a teaspoon of baking soda in a glass of orange juice. (What is the acid and what is the base in this reaction?)

$$CO_3^{2-}(aq) \quad + \quad 2\,H_3O^+(aq) \longrightarrow CO_2(g) \quad + \quad 3\,H_2O$$

Aluminum hydroxide [$Al(OH)_3$] is another popular antacid ingredient. Like calcium carbonate, it can cause constipation in large doses. The hydroxide ion neutralizes acid.

$$OH^- \quad + \quad H_3O^+ \longrightarrow 2\,H_2O$$

There is concern that antacids containing aluminum ions deplete the body of essential phosphate ions. The aluminum phosphate formed is insoluble and is excreted from the body.

$$Al^{3+}(aq) \quad + \quad PO_4^{3-}(aq) \longrightarrow AlPO_4(s)$$

Aluminum hydroxide is the only antacid in Amphojel. It occurs in combination in many popular products.

Magnesium compounds constitute the fourth category of antacids. These include magnesium carbonate ($MgCO_3$) and magnesium hydroxide [$Mg(OH)_2$]. Milk of magnesia is a suspension of magnesium hydroxide in water. It is sold under a variety of brand names, the best known of which is probably Phillips'. In small doses, magnesium compounds act as antacids. In large doses, they act as laxatives. Magnesium ions are poorly absorbed in the digestive tract. Rather, these small, dipositive ions draw water into the colon (large intestine), causing the laxative effect.

A variety of popular antacids combine aluminum hydroxide, with its tendency to cause constipation, and a magnesium compound, which acts as a laxative. These tend to counteract one another. Maalox and Mylanta are familiar brands.

All antacids are basic compounds. They act by neutralizing hydronium ions in stomach acid.

Rolaids contains the complex substance aluminum sodium dihydroxy carbonate [$AlNa(OH)_2CO_3$]. Both the hydroxide ion and the carbonate ion consume acid. Sodium-free Rolaids contains calcium carbonate and magnesium hydroxide.

Claims of "fast action" are almost meaningless. All acid–base reactions are almost instantaneous. Some tablets may dissolve a little more slowly than others. You can speed their action by chewing them.

Antacids interact with other medications. Anyone taking any type of medication should consult a physician before taking an antacid. Generally, antacids are safe and effective for occasional use in small amounts. All antacids are basic compounds. If you are otherwise in good health, you can choose a product on the basis of price. Anyone with severe or repeated attacks of indigestion should consult a physician. Self-medication in such cases might be dangerous.

10.9 Acids, Bases, and Human Health

Strong acids and bases cause damage on contact with living cells. Their action is nonspecific—all cells, regardless of type, are damaged. These corrosive poisons produce what are known as chemical burns. Once the offending agent is neutralized or removed, the injuries are similar to burns from heat, and they are often treated the same way.

Sulfuric acid (H_2SO_4) is by far the leading chemical product of U.S. industry. More than 40 billion kilograms are produced annually. Most of this acid is used by industry. A major use is in the conversion of phosphate rock to soluble compounds for use as fertilizer. Only small quantities are used in or around the home. Automobile batteries contain sulfuric acid. It is also the major ingredient in one type of drain cleaner.

Sulfuric acid is a powerful dehydrating agent. It takes up water from cellular fluid, rapidly killing the cell. The sulfuric acid molecules dehydrate by reacting with water in the cells to form hydronium ions and hydrogen sulfate ions.

$$H_2SO_4 + H_2O \longrightarrow H_3O^+ + HSO_4^-$$

Hydration of these ions and other secondary reactions may also be involved in the dehydration process.

In Section 10.7 we saw how the burning of sulfur-containing coal produces aerosol mists of sulfuric acid. When this airborne pollutant comes into contact with the alveoli of the lungs, the cells are broken down. The alveoli lose their resilience, which makes it difficult for them to remove carbon dioxide. Such lung damage may contribute to pulmonary emphysema, a condition characterized by increasing shortness of breath. Emphysema is the fastest-growing cause of death in the United States. Most likely the principal factor in the rise of emphysema is cigarette smoking. Air pollution is also known to be a factor, however. For example, the incidence of the disease among smokers is three times as great in St. Louis, where air pollution is rather heavy, as in Winnipeg, Manitoba, where air pollution is rather mild.

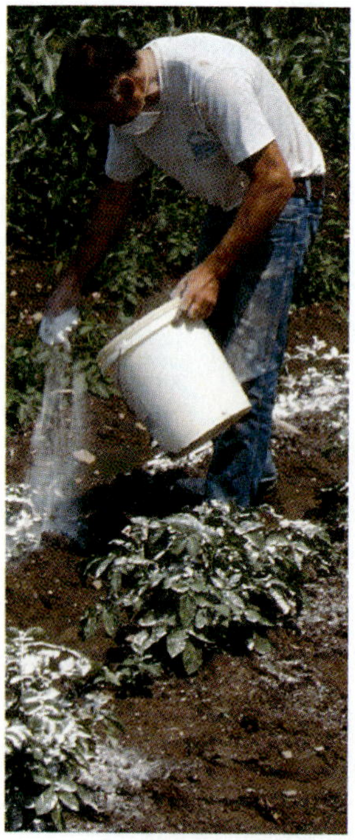

Treating acidic soil with lime makes it "sweeter" (less acidic).

Hydrochloric acid (also called muriatic acid) is used in homes to clean calcium carbonate deposits from toilet bowls. It is used in building construction to remove excess mortar from bricks. In industry, it is used to remove scale or rust from metals. Concentrated solutions (about 38% HCl) cause severe burns, but dilute solutions are considered safe enough for use around the home if handled carefully. The gastric juice in your stomach is a solution containing around 0.5% hydrochloric acid.

Ingestion of sulfuric, hydrochloric, or any other strong acid causes corrosive damage to the digestive tract. As little as 10 milliliters of concentrated (98%) H_2SO_4, taken internally, can be fatal.

Lime (CaO) is the cheapest and most widely used commercial base. It is made by heating limestone ($CaCO_3$) to drive off CO_2. It is the fifth most produced industrial chemical in the United States with an annual production of about 19 billion kilograms. It is used to make mortar and cement and also to "sweeten" acidic soil.

Sodium hydroxide (commonly known as lye) is the strong base most often used in the home. It is used as an oven cleaner in products such as Easy Off, and it is used to open clogged drains with products such as Drano. It destroys tissue rapidly, causing severe chemical burns. Several detergent additives, such as carbonates, silicates, and borates, form strongly basic solutions when dissolved in water. These, too, can cause corrosive damage to tissues, particularly those of the digestive tract and the eyes.

Both acids and bases, even in dilute solutions, break down the protein molecules in living cells. Generally, the fragments are not able to carry out the functions of the original proteins. In cases of severe exposure, the fragmentation continues until the tissue has been completely destroyed. And, within living cells, proteins function properly only at an optimum acidity. If the acidity changes much in either direction, the proteins can't carry out their usual functions properly (see Chapter 21).

When they are misused, acids and bases can be damaging to human health. But acids and bases affect human health in more subtle—and ultimately more important—ways. A delicate balance must be maintained between acids and bases in the blood and other body fluids. If the acidity of the blood changes very much, the blood loses its capacity to carry oxygen. Fortunately, the body has a complex but efficient mechanism for maintaining proper acid–base balance (see Section 28.10).

Summary

The Arrhenius definition of **acid** is any substance that yields hydrogen ions (H^+) in aqueous solution. Because a hydrogen *atom* is one proton and one electron, a hydrogen *ion* is simply a proton. Therefore a more general definition of **acid** is any substance that donates protons in a chemical reaction.

A proton in aqueous solution is always associated with a water molecule. The species has the formula H_3O^+ and is called a **hydronium ion.**

When an acid is dissolved in water, some or all of the acid molecules break up into hydronium ions plus anions. When all the molecules ionize so that many hydronium ions are formed, the acid is a **strong acid.** When only some of the molecules ionize so that relatively few hydronium ions are formed, the acid is a **weak acid.**

The Arrhenius definition of **base** is any substance that yields hydroxide ions (OH^-) in aqueous solution. A more general definition of **base** is any substance that accepts protons in a chemical reaction.

A **strong base** is one that dissociates completely when dissolved in water and so yields many hydroxide ions. A **weak base** is one that ionizes only slightly in water and so produces relatively few hydroxide ions.

Because many acids are made by the reaction between a nonmetal oxide and water, nonmetal oxides are called **acid anhydrides.** Because many bases are made by the reaction between a metal oxide and water, metal oxides are called **basic anhydrides.**

A **salt** is a compound containing the anion of an acid and the cation of a base. Example: NaCl (sodium chloride, table salt) is the salt formed when hydrochloric acid and sodium hydroxide react:

$$Na^+ + OH^- + H_3O^+ + Cl^- \rightarrow 2\,H_2O + Na^+ + Cl^-$$

Any such reaction between an acid and a base is a **neutralization** because the two species neutralize each other as they react to form water and the (dissolved) salt. Once the water is evaporated, the solid remaining is pure salt.

The reaction between an acid and either a carbonate (CO_3^{2-}) salt or a bicarbonate (HCO_3^-) salt produces the weak acid carbonic acid (H_2CO_3), which immediately decomposes to gaseous carbon dioxide and water. Example:

$$K_2CO_3 + 2\,HCl \longrightarrow H_2CO_3 + 2\,K^+ + 2\,Cl^-$$
$$H_2CO_3 \longrightarrow H_2O + CO_2(g)$$

Key Terms

acid (10.1)
acid anhydride (10.5)
amine (10.4)
antacid (10.8)
base (10.1)
basic anhydride (10.5)
diprotic acid (10.1)
hydronium ion (H_3O^+) (10.1)
monoprotic acid (10.1)
net ionic equation (10.6)
neutralization (10.6)
polyprotic acid (10.1)
proton acceptor (10.1)
proton donor (10.1)
salt (10.1)
strong acid (10.2)
strong base (10.4)
triprotic acid (10.1)
weak acid (10.2)
weak base (10.4)

Review Questions

1. List five general properties of acidic solutions.
2. List four general properties of basic solutions.
3. According to the Arrhenius theory, all acids have one element in common. What is that element? Are all compounds containing that element acids? Explain.
4. List five general properties of acidic aqueous solutions.
5. List four general properties of basic aqueous solutions.
6. What ion is responsible for the properties of acidic aqueous solutions?
7. What ion is responsible for the properties of basic aqueous solutions?
8. Which, if any, of the properties listed in Questions 4 and 5 remain after an acid and a base neutralize each other?
9. Suggest some ways in which you might determine whether a particular water solution contains an acid or a base.
10. Can a substance be a Brønsted-Lowry acid if it does not contain H atoms? Are there any characteristic atoms that must be present in a Brønsted-Lowry base?
11. Give examples of a monoprotic, a diprotic, and a triprotic acid.
12. What are the characteristic features of a *polyprotic* acid? Is CH_4 a polyprotic acid? Explain.
13. What is an acid anhydride? What is the anhydride of H_2SO_4?
14. What is a basic anhydride? What is the anhydride of $Ba(OH)_2$?
15. How does one brand of household ammonia differ from another?
16. Write the equation for the decomposition of carbonic acid.
17. Name some of the active ingredients in antacid tablets.
18. What is aqueous ammonia? Why is it sometimes called ammonium hydroxide?
19. What is the leading chemical product of U.S. industry?
20. What is the effect of strong acids on clothing?
21. What is the effect of strong acids and strong bases on skin?
22. How do corrosive acids and bases destroy living tissue?
23. Should a person who has hypertension be advised to use baking soda or milk of magnesia as an antacid? Why?
24. How does magnesium hydroxide, in large doses, act as a laxative? Would other magnesium compounds have similar effects?

Problems

Strong and Weak Acids and Bases

25. In aqueous solution, which of the following substances are *strong* acids, which are *weak* acids, and which are neither?
    a. CH_3OH b. HBr
    c. HCOOH d. HNO_2
26. In aqueous solution, which of the following substances are *strong* bases, which are *weak* bases, and which are neither?
    a. CH_3NH_2 b. CH_3OH
    c. LiOH d. $Ba(OH)_2$
27. Identify each of the following substances as a strong acid, weak acid, strong base, weak base, or salt.
    a. Na_2SO_4 b. KOH
    c. $CaCl_2$ d. CH_3CH_2COOH
28. Identify each of the following substances as a strong acid, weak acid, strong base, weak base, or salt.
    a. HI b. NH_3 c. NH_4I d. $Ca(OH)_2$
29. Thallium hydroxide (TlOH) is ionic in the solid state and is quite soluble in water. Classify the compound as a strong acid, weak acid, weak base, or strong base.
30. Hydrogen iodide (HI) gas reacts completely with water to form hydronium ions and iodide ions. Classify the compound as a strong acid, weak acid, weak base, or strong base.
31. Hydrogen sulfide (H_2S) gas reacts slightly with water to form relatively few hydronium ions and hydrogen sulfide ions (HS^-). Classify the compound as a strong acid, weak acid, weak base, or strong base.
32. Methylamine (CH_3NH_2) gas reacts slightly with water to form relatively few hydroxide ions and methylammonium ions ($CH_3NH_3^+$). Classify the compound as a strong acid, weak acid, weak base, or strong base.
33. According to the equation, is phenol, C_6H_5OH, an acid or a base? Should it be classified as strong or weak?

$$C_6H_5OH + H_2O \rightleftharpoons C_6H_5O^- + H_3O^+$$

34. According to the equation, is aniline, $C_6H_5NH_2$, an acid or a base? Should it be classified as strong or weak?

$$C_6H_5NH_2 + H_2O \rightleftharpoons C_6H_5NH_3^+ + OH^-$$

Names of Acids and Bases

35. Give the formulas for the following acids and bases.
    a. hydrochloric acid b. sulfuric acid
    c. carbonic acid d. lithium hydroxide
    e. magnesium hydroxide f. potassium hydroxide
36. Give the formulas for the following acids and bases.
    a. nitric acid b. sulfurous acid
    c. phosphoric acid d. hydrosulfuric acid
    e. calcium hydroxide
37. Name the following acids and bases.
    a. NaOH b. H_3PO_4 c. HNO_3
    d. H_2SO_3 e. $Ca(OH)_2$ f. H_2S

38. Name the following acids and bases.
 a. HCl b. H_2SO_4 c. LiOH
 d. H_2CO_3 e. $Mg(OH)_2$
39. What are the names of Br^- and HBr(aq)?
40. Se^{2-} is selenide ion. What is the name of $H_2Se(aq)$?
41. What are the names of NO_2^- and $HNO_2(aq)$?
42. What are the names of PO_3^{3-} and $H_3PO_3(aq)$?
43. $C_2O_4^{2-}$ is oxalate ion. What is the name of $H_2C_2O_4(aq)$?
44. $C_6H_5CO_2^-$ is benzoate ion. What is the name of $C_6H_5CO_2H(aq)$?

Ionization of Acids and Bases

45. Write equations showing the ionization of the following acids.
 a. HI(aq) b. $CH_3CH_2COOH(aq)$
 c. $HNO_2(aq)$ d. $H_2PO_4^-(aq)$
46. Write equations showing the ionization of the following as Brønsted-Lowry weak acids.
 a. HOClO b. CH_3CH_2COOH
 c. HCN d. C_6H_5COOH
47. Write equations showing the ionization of the following acids and bases.
 a. $HNO_3(aq)$ b. KOH(aq)
 c. HCOOH(aq) d. $CH_3NH_2(aq)$
48. Write equations showing the ionization of the following acids and bases.
 a. $HC_2O_4^-(aq)$ b. $Ba(OH)_2(aq)$
 c. $HClO_2(aq)$ d. $C_6H_5NH_2(aq)$

Acidic and Basic Anhydrides

49. Give the formula for the compound formed when sulfur trioxide reacts with water. Is the product an acid or a base?
50. Give the formula for the compound formed when magnesium oxide reacts with water. Is the product an acid or a base?
51. Give the formula for the compound formed when potassium oxide reacts with water. Is the product an acid or a base?
52. Give the formula for the compound formed when carbon dioxide reacts with water. Is the product an acid or a base?

Acid–Base Reactions

53. Use the definitions to identify the first compound in each equation as an acid or a base. (*Hint:* What is *produced* by the reaction?)
 a. $C_5H_5N + H_2O \longrightarrow C_5H_5NH^+ + OH^-$
 b. $C_6H_5OH + H_2O \longrightarrow C_6H_5O^- + H_3O^+$
 c. $CH_3COCOOH + H_2O \longrightarrow CH_3COCOO^- + H_3O^+$
54. Use the definitions to identify the first compound in each equation as an acid or a base.
 a. $C_6H_5SH + H_2O \longrightarrow C_6H_5S^- + H_3O^+$
 b. $CH_3NH_2 + H_2O \longrightarrow CH_3NH_3^+ + OH^-$
 c. $C_6H_5SO_2NH_2 + H_2O \longrightarrow C_6H_5SO_2NH^- + H_3O^+$
55. Write the equation for the reaction of sodium hydroxide with hydrochloric acid.
56. Write the equation for the reaction of lithium hydroxide with nitric acid.
57. Write the equation for the reaction of 1 mol of calcium hydroxide with 2 mol of hydrochloric acid.
58. Write the equation for the reaction of 1 mol of sulfuric acid with 2 mol of potassium hydroxide.
59. Write the equation for the reaction of 1 mol of phosphoric acid with 3 mol of sodium hydroxide.
60. Write the net ionic equation for the reaction of a bicarbonate salt with an acid.
61. Write the net ionic equation for the reaction of a carbonate salt with an acid.
62. Write an equation that describes the effect of acid rain on marble.

Additional Problems

63. According to the Arrhenius theory, are all compounds containing OH groups bases? Explain.
64. Write equations showing how hydrogen phosphate ion, HPO_4^{2-}, can act either as a Brønsted-Lowry acid or as a Brønsted-Lowry base.
65. Slaked lime [$Ca(OH)_2(s)$] can be used to reduce excess acidity in natural waters such as lakes. Write a net ionic equation for the reaction that occurs.
66. With continued use, automatic coffee makers often develop mineral deposits ($CaCO_3$). The manufacturer's instructions generally call for removing the deposit by treatment with vinegar. Write a net ionic equation for the reaction that occurs. (*Hint:* Recall that vinegar contains acetic acid, CH_3COOH.)
67. A paste of sodium hydrogen carbonate (sodium bicarbonate) and water can be used to relieve the pain of an ant bite. The irritant in the ant bite is formic acid (HCOOH). Write a net ionic equation for the reaction that occurs.
68. Which of the following aqueous solutions has the highest concentration of H^+ ion: 0.10 M HCl, 0.10 M H_2SO_4, 0.10 M CH_3COOH, or 0.10 M NH_3?
69. Acids with three or more H atoms can exist in two forms. The *ortho* form has two more H atoms and one more O atom than the *meta* form. For example, H_3PO_4 is sometimes called orthophosphoric acid; metaphosphoric acid is HPO_3. Orthosilicic acid is H_4SiO_4. What is the formula for metasilicic acid? For sodium metasilicate?
70. Boric acid is H_3BO_3. What is the formula for metaboric acid? For sodium metaborate? (See Problem 69.)

11 Acids and Bases II

Laboratory reagent bottles are labeled as "CON" (for concentrated) or "DIL" (for dilute). Most acid and base solutions are made starting with these stock solutions.

Our bodies are largely solutions—quite special solutions, of course, but solutions nonetheless. Delicate balances must be maintained among the many solutes in our blood and other body fluids, but none are more important than the balance between acids and bases. If the acidity of the blood changes very much, its capacity to carry oxygen is diminished. Because many bodily processes produce acids, the control of acidity is literally a matter of life and death.

Our bodies have developed a marvelously complex yet efficient mechanism for maintaining the proper acid–base balance. Before we can talk about this mechanism in a meaningful way, however, we need to develop a few concepts—concepts that are more quantitative in nature than those developed in Chapter 10. It is important to know how exact concentrations of acids and bases are determined and expressed. We must understand the concept of pH, particularly as it relates to the chemistry of the blood and other body fluids. Most important, we must see just how, through substances called buffers, the level of acidity is controlled.

Visualizing the dilution of a solution. When pure water is added to a solution of methanol (CH_3OH) (left), a more dilute solution is produced (right). However, the diluted solution contains the same number of CH_3OH molecules as the original solution.

11.1 Concentrations of Acids and Bases

A common laboratory task is to make a more dilute solution from a more concentrated one. This process, called **dilution,** is a procedure by which we make solutions of a desired concentration. Concentrated solutions, often ones that are commercially available, are kept in stock in a storeroom. For example, concentrated solutions of sulfuric acid (18 M), hydrochloric acid (12 M), and nitric acid (16 M) are readily available. The principle of dilution is that *addition of solvent does not change the amount of solute in a solution but does change its concentration.*

In diluting acid solutions, always add acid to water. If water is added to acid, the heat of reaction can cause the water to boil, spattering the acid and causing burns.

Recall the definition of molarity.

$$\text{Molarity (M)} = \frac{\text{moles of solute}}{\text{liters of solution (V)}}$$

This relationship can be rearranged as

$$\begin{aligned}\text{Moles of solute} &= \text{molarity} \times \text{liters of solution}\\ &= \text{M} \times \text{V}\end{aligned}$$

Because the numbers of moles of solute does not change upon dilution, we can write

$$(\text{Moles of solute})_{\text{conc}} = (\text{moles of solute})_{\text{dil}}$$

From the rearranged relationship we can substitute to obtain

$$\text{M}_{\text{conc}} \times \text{V}_{\text{conc}} = \text{M}_{\text{dil}} \times \text{V}_{\text{dil}}$$

Although the derivation of this expression assumes solution volumes in liters, any unit of volume can be used as long as it is the same on both sides of the equation.

The following examples illustrate the types of calculations involved in dilution problems.

Example 11.1

What volume of an 18.0 M H_2SO_4 stock solution would you use to prepare 0.250 L of 4.00 M H_2SO_4?

Solution

In the dilution formula

$$M_{conc} \times V_{conc} = M_{dil} \times V_{dil}$$

the unknown quantity is the volume of concentrated solution. Let's rearrange the relationship by dividing each side by M_{conc}, thus solving for V_{conc}.

$$V_{conc} = \frac{M_{dil} \times V_{dil}}{M_{conc}} = \frac{4.00\ M \times 0.250\ L}{18.0\ M} = 0.0556\ L$$

To make the solution, then, measure out 0.0556 L (55.6 mL) of 18.0 M H_2SO_4 and add it to enough water to make 250 mL of solution.

Practice Exercise

A stock bottle of aqueous ammonia indicates that the solution is 15.0 M NH_3. What volume of this concentrated solution is needed to make 1.25 L of 2.50 M NH_3 solution?

Example 11.2

What volume of a 15.7 M HNO_3 stock solution would you use to prepare 1.250 L of 2.50 M HNO_3?

Solution

$$M_{conc} \times V_{conc} = M_{dil} \times V_{dil}$$

$$V_{conc} = \frac{M_{dil} \times V_{dil}}{M_{conc}} = \frac{2.50\ M \times 1.250\ L}{15.7\ M} = 0.199\ L\ (\text{or } 199\ mL)$$

Practice Exercise

A stock bottle of aqueous formic acid is 23.5 M HCOOH. What volume of this concentrated solution is needed to make 2.00 L of 1.00 M HCOOH solution?

11.2 Acid–Base Titrations

The concept of molarity gives us a way to obtain a specific number of moles of a solute by measuring out a given volume of solution. Such a volume measurement is often made in a *buret,* a graduated, long glass tube calibrated to deliver precise volumes of solution through a stopcock valve. A buret is the chief instrument used in a **titration,** a procedure in which two reactants in solution react in proportions determined by the chemical equation for the reaction.

Consider, for example, a typical acid–base titration. A measured volume of a solution of an acid of unknown concentration is transferred to a flask. Then, a solution of a base of *known* concentration is added carefully from a buret until the reaction of the acid with the base is just complete. The point at which the acid is just neutralized is called the **equivalence point** of the titration. At that point, the number of moles of OH^- equals the number of moles of H_3O^+ that are in the sample of acid. Figure 11.1 illustrates the titration of an acid solution of unknown concentration with a sodium hydroxide solution of known concentration.

A substance that has the property of changing color when an acid or base is added to it is called an *indicator.*

The equivalence point of a titration reaction often is determined with an **indicator dye,** a substance that changes color as the reaction is completed. Litmus,

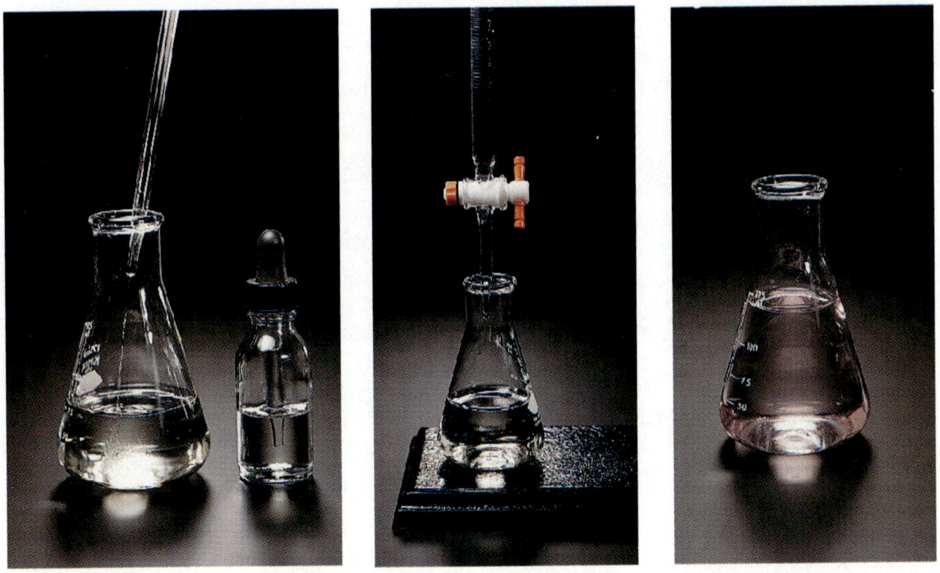

(a) (b) (c)

▲ FIGURE 11.1

The amount of acid or base in a solution is determined by a technique called a titration. In the experiment shown here, (a) a precisely measured volume (for example, 10.00 mL) of a hydrochloric acid solution of unknown concentration is added from a volumetric pipet to a quantity of water in a small flask. A few drops of phenolphthalein indicator solution (in the dropper bottle) is also added to the flask. (b) A sodium hydroxide solution of known concentration (for example, 0.1000 M NaOH) is slowly added from a previously filled buret. (c) As long as the acid is in excess, the solution remains colorless. When the acid has just been neutralized, an additional drop of NaOH(aq) causes the solution to become slightly basic; the phenolphthalein indicator turns a faint pink color. This is the equivalence point of the titration.

mentioned in Section 10.1, is one such indicator. A better indicator for some acid–base titrations is phenolphthalein. In the titration pictured in Figure 11.1, phenolphthalein is colorless in the hydrochloric acid solution and remains so until the equivalence point is reached. With the addition of as little as one drop of NaOH(aq) beyond the equivalence point, however, the reaction mixture becomes basic and the phenolphthalein indicator turns pink. The titration is stopped when the pink color appears.

The following examples illustrate calculations involving acid–base titrations.

Example 11.3

Sodium hydroxide reacts with hydrochloric acid.

$$\text{NaOH(aq)} + \text{HCl(aq)} \longrightarrow \text{NaCl(aq)} + \text{H}_2\text{O}$$

A flask contains 20.00 mL of 0.1030 M HCl. What volume of 0.2010 M NaOH must be added to just neutralize the acid?

Solution

First, convert the quantity of HCl to moles, using the molarity (moles/liter) as a conversion factor.

$$0.02000\ \text{L}\ \cancel{\text{HCl(aq)}} \times \frac{0.1030\ \text{mol HCl}}{1\ \text{L}\ \cancel{\text{HCl(aq)}}} = 0.002060\ \text{mol HCl}$$

Then, use the mole ratio from the chemical equation to relate amounts, in moles, of HCl and NaOH.

$$0.002060\ \cancel{\text{mol HCl}} \times \frac{1\ \text{mol NaOH}}{1\ \cancel{\text{mol HCl}}} = 0.002060\ \text{mol NaOH}$$

Finally, use the molarity of the NaOH(aq) as a conversion factor to convert this amount of NaOH to a volume of NaOH(aq).

$$0.002060\ \cancel{\text{mol NaOH}} \times \frac{1\ \text{L NaOH(aq)}}{0.2010\ \cancel{\text{mol NaOH}}} = 0.01025\ \text{L NaOH(aq)}$$

The titration requires 10.25 mL (0.01025 L) of the NaOH(aq).

Practice Exercise

A flask contains 20.00 mL of KOH(aq) of unknown concentration. Titration of the solution requires 15.62 mL of 0.1104 M H_2SO_4(aq).

$$2\ \text{KOH(aq)} + \text{H}_2\text{SO}_4\text{(aq)} \longrightarrow \text{K}_2\text{SO}_4\text{(aq)} + 2\ \text{H}_2\text{O}$$

Calculate the molarity of the KOH(aq).

Acids, Bacteria, and Ulcers

For many years the conventional wisdom was that ulcers were caused by excessive hydrochloric acid in the stomach and that this excess acidity was caused by stress, spicy foods, alcohol, antiinflammatory drugs such as aspirin, and cigarette smoking. In the 1980s, however, a young Australian physician named Barry J. Marshall showed that most ulcers are caused by bacteria, *Heliocbacter pylori*, that live in the mucus layer that lines the stomach. The bacteria, not the excess acid, erode the stomach lining. Other studies have confirmed Marshall's findings and have also linked these bacteria to stomach cancer. Ulcer treatment has therefore largely changed from diet, antacids, and drugs that block acid secretion, to antibiotics.

The stomach's mucus layer helps to protect the stomach cells from its strong acid contents. The mucus also provides some protection for the bacteria, which ordinarily would be killed by the acid. But the bacteria have their own defense; they produce urease, an enzyme that catalyzes the breakdown of urea, a waste product of the cells' metabolism, to carbon dioxide and ammonia.

$$\underset{\text{urea}}{\text{CO(NH}_2)_2} \xrightarrow{\text{urease}} \text{CO}_2 + \text{NH}_3$$

The ammonia neutralizes the stomach acid in the vicinity of the bacteria. The practice of acid–base chemistry is not limited to humans.

Example 11.4

Sodium hydroxide reacts with sulfuric acid.

$$2\,NaOH(aq) + H_2SO_4(aq) \longrightarrow Na_2SO_4(aq) + 2\,H_2O$$

A flask contains 25.00 mL of $H_2SO_4(aq)$ of unknown concentration. Titration of the sample requires 25.20 mL of 0.1000 M NaOH(aq) for neutralization. What is the molarity of the sulfuric acid solution?

Solution

First, convert the quantity of NaOH to moles, using the molarity (moles/liter) as a conversion factor.

$$0.02520\,\cancel{\text{L NaOH(aq)}} \times \frac{0.1000\text{ mol NaOH}}{1\,\cancel{\text{L NaOH(aq)}}} = 0.002520\text{ mol NaOH}$$

Then, use the mole ratio from the chemical equation to relate moles of H_2SO_4 to moles of NaOH.

$$0.002520\,\cancel{\text{mol NaOH}} \times \frac{1\text{ mol }H_2SO_4}{2\,\cancel{\text{mol NaOH}}} = 0.001260\text{ mol }H_2SO_4$$

Finally, use the definition of molarity to write

$$\frac{0.001260\text{ mol }H_2SO_4}{0.02500\text{ L}} = 0.05040\text{ M }H_2SO_4$$

Practice Exercise

What volume of 0.1000 M HCl is required to just neutralize 40.00 mL of 0.5020 M $Ca(OH)_2$ solution?

$$2\,HCl(aq) + Ca(OH)_2(aq) \longrightarrow CaCl_2(aq) + 2\,H_2O$$

11.3 The pH Scale

When we think of water, we think of H_2O molecules. But even the purest water isn't all H_2O. About 1 molecule in 500 million transfers a proton to another water molecule, yielding a hydronium ion and a hydroxide ion.

$$H_2O + H_2O \rightleftharpoons H_3O^+ + OH^-$$

In other words, water is in equilibrium with hydronium ion and hydroxide ion, although the equilibrium point lies *far to the left*, as suggested by the unequal arrow lengths in the above equation. The experimentally determined equilibrium concentrations in pure water at 25 °C are

$$[H_3O^+] = [OH^-] = 1.0 \times 10^{-7}\text{ M}$$

Keep in mind that the bracketed symbols are shorthand for the concentration of the enclosed ions in moles per liter. Thus $[H_3O^+]$ is read as the molarity of hydronium ion. The product of these ion concentrations at 25 °C is

$$[H_3O^+][OH^-] = (1.0 \times 10^{-7})(1.0 \times 10^{-7}) = 1.0 \times 10^{-14}$$

This product is called the **ion product of water (K_W)**. That is, at 25 °C,

$$K_w = [H_3O^+][OH^-] = (1.0 \times 10^{-7})(1.0 \times 10^{-7}) = 1.0 \times 10^{-14}$$

This ion product relationship is of considerable importance. It doesn't just describe self-ionization in pure water, but it applies to *all aqueous solutions*—that is, to solutions of acids, bases, salts, or even molecular substances. As an illustration, consider Example 11.5.

Example 11.5

What is the $[H_3O^+]$ and the $[OH^-]$ in a solution that is 0.00015 M HCl?

Solution

Because HCl is a strong acid, its ionization is complete.

$$HCl + H_2O \longrightarrow H_3O^+ + Cl^-$$

and because the $[H_3O^+]$ produced by the HCl is so much greater than that found in pure water, we can state with assurance that in this solution

$$[H_3O^+] = 0.00015\text{ M} = 1.5 \times 10^{-4}\text{ M}$$

With the K_w expression, we can now calculate $[OH^-]$ in the solution.

$$[OH^-] = \frac{K_w}{[H_3O^+]} = \frac{1.0 \times 10^{-14}}{1.5 \times 10^{-4}} = 6.7 \times 10^{-11}\text{ M}$$

Practice Exercise

What is the $[H_3O^+]$ and the $[OH^-]$ in a solution that is 0.025 M NaOH?

The *H* in pH stands for "hydrogen" and the *p* refers to "power or potential."

Exponential notation is a convenient way to express a small quantity such as 1.5×10^{-4}, but there's an even more convenient way. In 1909 the Danish biochemist Søren P. L. Sørenson proposed that only the number in the exponent be used to express acidity. His convention, which came to be known as the **pH scale,** is still used today. pH is defined as the negative of the logarithm of $[H^+]$. The reaction of most people on first encountering this definition of pH is "You call this more convenient?" Well, it really is. To determine the pH of a solution that has $[H_3O^+] = 1.0 \times 10^{-n}$, you need only take the exponent of the hydronium ion concentration and reverse its sign. The pH of a solution whose hydronium ion concentration is 1×10^{-4} M is 4. The point is this: it is easier to say, "The pH is 4" than to say, "The hydronium ion concentration is 1×10^{-4} M." They mean the same thing.

Steps for Calculating pH on a Calculator

Step 1: Enter the value for the $[H_3O^+]$.
Step 2: Press the LOG key.
Step 3: Press the +/− (change-of-sign) key.

In Example 11.6, where $[H_3O^+] = 1.0 \times 10^{-5}$ M, following these steps gives a pH value of 5.00. *Step 1:* Enter 1.0, press EXP, enter 5, and press +/−. *Step 2:* Press the LOG key. *Step 3:* Press +/−. You should have a value of 5.00. (Calculator brands differ somewhat; be sure you practice enough on yours to be confident of your answers.)

Example 11.6

What is the pH of a solution that has $[H_3O^+] = 1.0 \times 10^{-5}$?

Solution

$$[H_3O^+] = 1.0 \times 10^{-5}\ M,$$
$$pH = -\log[H_3O^+] = -\log(1.0 \times 10^{-5}) = -(-5.00) = 5.00$$

Practice Exercise

What is the pH of a solution that has $[H_3O^+] = 1.0 \times 10^{-9}$?

Because we can only take logarithms of dimensionless numbers, we use just the numerical value of the molarity of H_3O^+ and not the unit *"M."*

Example 11.7

What is the pH of a solution that has $[H_3O^+] = 4.5 \times 10^{-3}$?

Solution

$$[H_3O^+] = 4.5 \times 10^{-3}\ M$$
$$pH = -\log[H_3O^+] = -\log(4.5 \times 10^{-3}) = -(-2.35) = 2.35$$

Practice Exercise

What is the pH of a solution that has $[H_3O^+] = 2.7 \times 10^{-9}$?

To determine the $[H_3O^+]$ corresponding to a given pH, we do an inverse calculation.

Example 11.8

What is the $[H_3O^+]$ in a solution with pH = 2.19?

Solution

$$-\log[H_3O^+] = pH = 2.19$$
$$\log[H_3O^+] = -2.19$$
$$[H_3O^+] = \text{antilog}(-2.19) = 10^{-2.19} = 6.5 \times 10^{-3}$$

(On a calculator, enter 2.19 and press the +/− key, the INV (or 2nd) key, and then the LOG key. The "inverse logarithm" is commonly called the *antilog*.)

Practice Exercise

What is the $[H_3O^+]$ and $[OH^-]$ in a solution that has a pH of 10.79?

We can also define **pOH** as

$$pOH = -\log[OH^-]$$

A 2.5×10^{-3} M NaOH solution has $[OH^-] = 2.5 \times 10^{-3}$ M and

$$pOH = -\log(2.5 \times 10^{-3}) = -(-2.60) = 2.60$$

The relationship between pH and pOH at 25 °C is

$$pH + pOH = 14.00$$

Thus, the pH of 2.5×10^{-3} M NaOH is

$$pH = 14.00 - pOH = 14.00 - 2.60 = 11.40$$

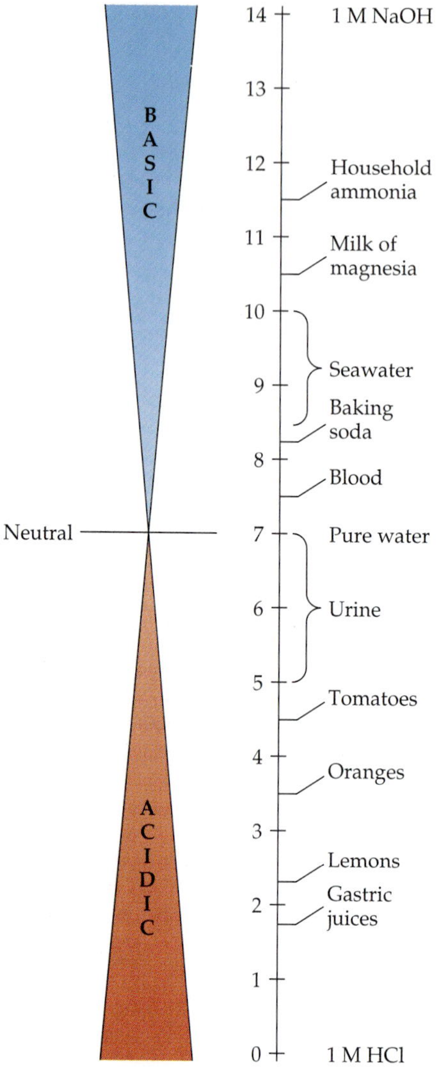

▲ **FIGURE 11.2**

The pH scale. The pH values of common substances range from about 0 to 14.

▲ **FIGURE 11.3**

pH papers change colors over a wide range of pH values.

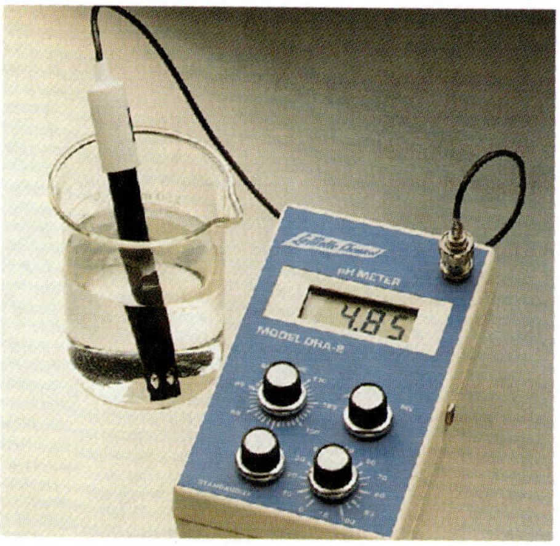

▲ **FIGURE 11.4**

A pH meter uses electrochemical reactions (Chapter 12) to measure the pH of a solution in comparison with a standard.

In pure water, where $[H_3O^+] = [OH^-] = 1.0 \times 10^{-7}$ M, pH and pOH are both 7.00. Pure water and all aqueous solutions with pH = 7.00 are *neutral*. If the pH is less than 7.00, the solution is *acidic;* if the pH is above 7.00, the solution is *basic* or *alkaline*. As a solution becomes more acidic, $[H_3O^+]$ increases and pH decreases. As a solution becomes more basic, $[H_3O^+]$ decreases and pH increases.

Figure 11.2 gives the pH values of a number of familiar materials. You may want to use this as a frame of reference from time to time. Another point to keep in mind is that, because pH is on a logarithmic scale, every unit change in pH represents a *tenfold* change in $[H_3O^+]$. Thus, lemon juice (pH $\approx$ 2.3) is somewhat more than ten times as acidic as orange juice (pH $\approx$ 3.5) and somewhat more than 100 times as acidic as tomato juice (pH $\approx$ 4.5).

The cells in our bodies are bathed in solutions of fairly constant pH. Indeed, even small changes in pH may be fatal. Many chemical reactions, especially those in living cells, are quite sensitive to pH. The control of pH, by compounds called buffers, is described in Sections 11.5 and 11.6.

Various dyes, some synthetic and some from plant and animal sources, have the property of changing color over a range of pH values. We noted in Chapter 10 that litmus, a vegetable dye, turns red in acidic solutions and blue in basic solutions. Actually the change occurs over a range of pH from 4.5 to 8.3. Certain combinations of indicator dyes will exhibit a whole range of colors as the pH changes from strongly basic to strongly acidic (Figure 11.3). By selecting the proper indicator, we can determine the pH of almost any clear, colorless aqueous solution.

More accurate measurements of pH can be made electrically with pH meters. Generally, these instruments can measure pH to a precision of about 0.01 pH unit. Also, colors and turbidity generally do not interfere with electrical measurement of pH as they do with indicator color changes. Thus, pH meters can be used on blood, urine, and other complex mixtures without prior treatment of the specimen. A typical pH meter is illustrated in Figure 11.4.

11.4 Salts in Water: Acidic, Basic, or Neutral?

When an acid reacts with a base, the products are a salt and water. The process is called neutralization, but is the solution neutral? What if we simply take a salt and dissolve it in water? Would the solution be acidic, basic, or neutral? Although we might well expect a salt solution to be neutral, not all of them are. There are many neutral salts, including the familiar sodium chloride and others such as potassium sulfate and sodium nitrate. Some, such as ammonium nitrate and aluminum chloride, are acidic. Others, such as sodium acetate and potassium cyanide, are basic.

How do we tell if a salt is acidic, basic, or neutral? Simple. Just dissolve some of the salt in water and test the solution with an indicator (Figure 11.5) or with a pH meter. That is the experimental way, and who can argue with it? There is another way, though. We can *predict* whether a solution of a salt will be acidic, basic, or neutral by considering the relative strengths of the acid and base from which the salt was made. Just think how much more convenient it is to

◀ **FIGURE 11.5**

This sodium carbonate solution contains a few drops of thymolphthalein indicator. Thymolphthalein is blue when the pH of the solution is greater than 10.6 and colorless when the pH is less than 9.4. The blue color of the indicator shows that the pH of this solution is greater than 10.6. The solution is rather strongly basic as a result of the reaction of CO_3^{2-} as a base with water.

$$CO_3^{2-}(aq) + H_2O \rightleftharpoons HCO_3^{-}(aq) + OH^{-}(aq)$$

The green color of the bromthymol blue indicator indicates that this NaCl(aq) is neutral. Bromthymol blue is yellow below pH 7, green at pH 7, and blue above pH 7.

The yellow color of the bromthymol blue indicator indicates that this $NH_4Cl(aq)$ is acidic.

apply a rule than to go into a laboratory and do an experiment. The rules are as follows:

1. The salt of a strong acid and a strong base forms a neutral solution.
2. The salt of a strong acid and a weak base forms an acidic solution.
3. The salt of a weak acid and a strong base forms a basic solution.
4. The salt of a weak acid and a weak base may form an acidic, basic, or (by chance) neutral solution.

To apply these rules, you must be able to recognize strong acids, strong bases, weak acids, and weak bases (recall Sections 10.2 and 10.4). The following examples will illustrate the process.

Example 11.9

Does NaCl form an acidic, basic, or neutral solution?

Solution

In any aqueous solution, there will always be H^+ (actually, H_3O^+) and OH^-. NaCl dissociates in water to yield $Na^+ + Cl^-$. Therefore, in an aqueous solution of NaCl, there will be the following four ions:

$$Na^+,\ Cl^-,\ H^+,\ OH^-$$

Combine the positive ion of the salt with OH^-, and the negative ion of the salt with H^+.

$$Na^+ + OH^- = NaOH$$
$$Cl^- + H^+ = HCl$$

Then examine the base and the acid that are formed. NaOH is a strong base, and HCl is a strong acid. Therefore, NaCl is a salt of a strong acid and a strong base, and (according to rule 1) it forms a neutral solution.

Practice Exercise

Does KNO_3 form an acidic, basic, or neutral solution?

Example 11.10

Is a solution of NH_4Cl acidic, basic, or neutral?

Solution

The four ions in solution are

$$NH_4^+,\ Cl^-,\ H^+,\ OH^-$$

Combining the appropriate ions, we have

$$NH_4^+ + OH^- = NH_4OH$$
$$H^+ + Cl^- = HCl$$

NH_4OH (really aqueous NH_3) is a weak base, and HCl is a strong acid. Therefore, NH_4Cl is a salt of a weak base and a strong acid, and (according to rule 2) it forms an acidic solution.

Practice Exercise

Is a solution of $(NH_4)_2SO_4$ acidic, basic, or neutral?

Example 11.11

Is a solution of CH_3COONa acidic, basic, or neutral?

Solution

The four ions in solution are

$$Na^+,\ CH_3COO^-,\ H^+,\ OH^-$$

Combining the appropriate ions, we have

$$Na^+ + OH^- = NaOH$$
$$CH_3COO^- + H^+ = CH_3COOH$$

NaOH is a strong base, and CH_3COOH (acetic acid) is a weak acid. Therefore, CH_3COONa is a salt of a weak acid and a strong base, and (according to rule 3) it forms a basic solution.

Practice Exercise

Is a solution of Li_2CO_3 acidic, basic, or neutral?

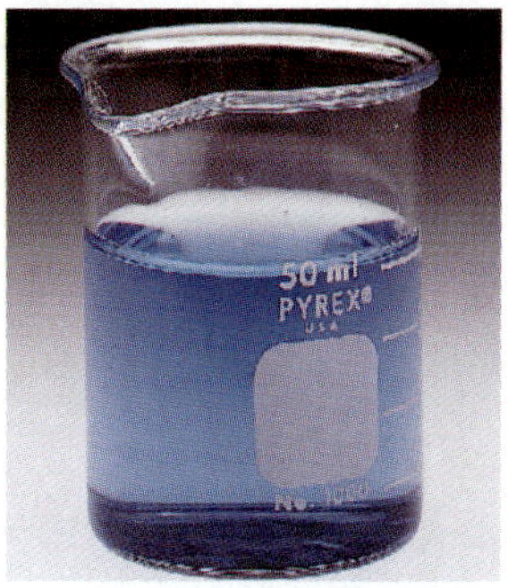

The blue color of the bromthymol blue indicator indicates that this $CH_3COONa(aq)$ is basic.

Example 11.12

Is a solution of CH_3COONH_4 acidic, basic, or neutral?

Solution

The base, NH_4OH (aqueous NH_3), is a weak one. The acid, CH_3COOH, is also weak. There is no way to tell from the rules whether the solution is acidic, basic, or neutral (rule 4).

11.5 Buffers: Control of pH

Who cares about the pH of a salt solution anyway? You do, that's who, for some of these salts play a vital role in the control of the pH of body fluids. If they should fail to function, so will you. Our bodies are acid factories. Our stomachs produce hydrochloric acid. Our muscles produce lactic acid. Starches and sugars produce pyruvic acid when metabolized. Carbon dioxide from respiration produces carbonic acid in the blood. Our bodies must eliminate or neutralize these acids, because excess acidity in the wrong place would kill us rather quickly.

A **buffer solution** is one in which the pH remains nearly constant even if acid or base is added. Chemically, a buffer solution is one that contains a weak acid and one of its salts (or a weak base and one of its salts), usually in approximately equal concentrations. For example, 1 L of a solution that contains 0.1 mol of acetic acid (CH_3COOH) and 0.1 mol of sodium acetate (CH_3COONa) acts as a buffer at pH 4.74, an acidic value.[1] This buffer can absorb significant amounts of additional acid or base without appreciable change in pH (Figure 11.6).

How does a buffer work? It may seem strange that a solution can absorb acid or base without the pH changing appreciably. The explanation, however, is fairly

[1] This may be a bit confusing because we have just spent some time considering the fact that a solution of sodium acetate is slightly basic (i.e., has a pH greater than 7). However, it is a solution containing *only* the salt that is basic. The buffer solution consists not only of a salt of acetic acid but also of some acetic acid itself. It is the presence of this acid that makes the buffer solution acidic.

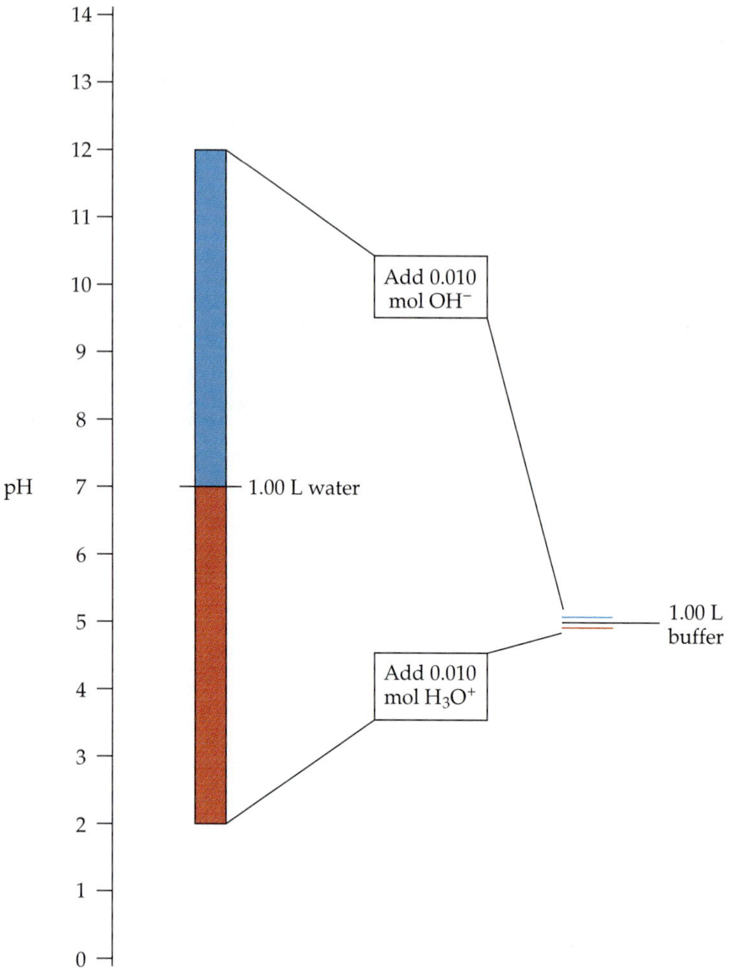

FIGURE 11.6 ▶

Representing buffer action. The addition of 0.010 mol of [H_3O^+] or of [OH^-] to 1.00 L produces a huge change in the pH of pure water and practically no change in the pH of a solution that is 1.00 M in both CH_3COOH and CH_3COONa. The acetic acid–sodium acetate solution is a buffer solution, and pure water has no buffering ability at all.

simple. It follows Le Châtelier's principle (Section 5.13). The buffer solution has a large reservoir of both acid molecules and the anions from the salt. If a strong acid is added, the hydronium ions from the added acid will donate protons to the anions of the buffer to form the weak acid and water.

$$H_3O^+ + CH_3COO^- \rightleftharpoons CH_3COOH + H_2O$$

Although the reaction is reversible to a slight extent, most of the protons are removed from the solution as they are added, and the pH changes hardly at all.

Table 11.1 Some Important Buffers

Buffer Components	Buffer System Names	pH[a]
$CH_3CHOHCOOH/CH_3CHOHCOO^-$	Lactic acid/lactate ion	3.86
CH_3COOH/CH_3COO^-	Acetic acid/acetate ion	4.74
$H_2PO_4^-/HPO_4^{2-}$	Dihydrogen phosphate ion/monohydrogen phosphate ion	7.20
H_2CO_3/HCO_3^-	(Carbon dioxide) carbonic acid/bicarbonate ion	6.46[b]
NH_4^+/NH_3	Ammonium ion/ammonia	9.25

[a] The values listed are for solutions that are 0.1 M in each compound at 25 °C.

[b] This value includes dissolved CO_2 molecules as undissociated H_2CO_3. The value for H_2CO_3 alone is about 3.8.

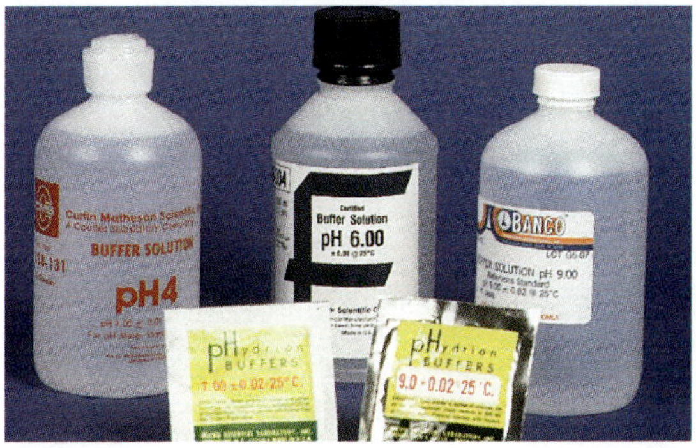

◀ **FIGURE 11.7**
Buffers are available as prepackaged solutions and as powdered ingredients for preparing buffer solutions of specific pH values.

When a strong base is added, the hydroxide ions will react with the hydronium ions formed in the solution by the acetic acid of the buffer.

$$OH^- + H_3O^+ \longrightarrow 2\,H_2O$$

The added hydroxide ions are tied up, and the hydronium ions removed from the solution are immediately replaced by further ionization of the acetic acid in the buffer.

$$CH_3COOH + H_2O \rightleftharpoons CH_3COO^- + H_3O^+$$

The concentration of hydronium ions returns to approximately the original value, and the pH is only slightly changed.

There are many important buffer solutions. Most biochemical reactions, whether they occur in a laboratory or in our bodies, are carried out in buffered solutions (Figure 11.7). The buffers that control the pH of our blood will be discussed in the next section. Table 11.1 lists some buffers of interest and the pH ranges in which they operate.

11.6 Buffers in Blood

The pH of the blood of higher animals is held remarkably constant. In humans, blood plasma normally varies from 7.35 to 7.45 in pH. Should the pH rise above 7.8 or fall below 6.8, due to starvation or disease, the person may suffer irreversible damage to the brain or even die. Fortunately, human blood has at least three buffering systems. Of these, the bicarbonate/carbonic acid (HCO_3^-/H_2CO_3) buffering system is the most important.

If acids are put into the blood, hydronium ions are taken up by the bicarbonate ions to form undissociated carbonic acid and water.

$$HCO_3^- + H_3O^+ \longrightarrow H_2CO_3 + H_2O$$

As long as there is sufficient bicarbonate to take up the added acid, the pH will change little.

The carbonic acid is a weak acid and exists in equilibrium with hydronium and bicarbonate ions.

$$H_2CO_3 + H_2O \rightleftharpoons H_3O^+ + HCO_3^-$$

If bases come into the bloodstream, reacting with hydronium ions to form water, more carbonic acid molecules will ionize to replace the removed hydronium ions. Further, as the carbonic acid molecules are used up, more carbonic acid can be formed from the large reservoir of dissolved carbon dioxide in the blood.

$$CO_2 + H_2O \rightleftharpoons H_2CO_3$$

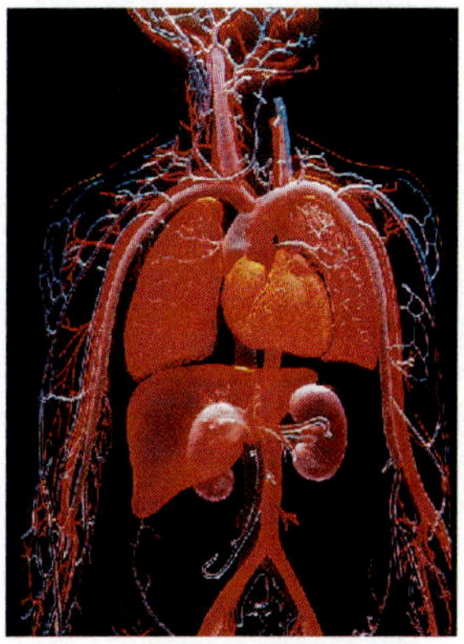

Blood carried by the human circulatory system is maintained at a pH of about 7.4 in a highly buffered system.

Thus, bicarbonate/carbonic acid buffers the blood against either added base or added acid.

Another blood buffer is the dihydrogen phosphate/monohydrogen phosphate ($H_2PO_4^-/HPO_4^{2-}$) system. Any acid reacts with monohydrogen phosphate to form dihydrogen phosphate.

$$HPO_4^{2-} + H_3O^+ \longrightarrow H_2PO_4^- + H_2O$$

The dihydrogen phosphate is a weak acid and exists in equilibrium with hydronium ions and monohydrogen phosphate.

$$H_2PO_4^- + H_2O \rightleftharpoons H_3O^+ + HPO_4^{2-}$$

Any base that comes into the blood will react with hydronium ions to form water. However, more dihydrogen phosphate will ionize to replace these hydronium ions, leaving the pH essentially unchanged.

Proteins act as a third type of blood buffer. These complex molecules (Chapter 21) contain —COO^- groups, which, like acetate ions (CH_3COO^-), can act as proton acceptors. Proteins also contain —NH_3^+ groups, which, like ammonium ions (NH_4^+), can donate protons. If acid comes into the blood, hydronium ions can be neutralized by the —COO^- groups. (The wavy lines represent parts of protein molecules.)

$$\underset{\sim}{COO^-} + H_3O^+ \longrightarrow \underset{\sim}{COOH} + H_2O$$

If base is added, it can be neutralized by the —NH_3^+ groups.

$$\underset{\sim}{NH_3^+} + OH^- \longrightarrow \underset{\sim}{NH_2} + H_2O$$

These three buffers (and perhaps others) act to keep the pH of the blood constant. Buffers can be overridden by large amounts of acid or base; their capacity is not infinite. The blood buffers can be overwhelmed if the body's metabolism goes badly amiss.

11.7 Acidosis and Alkalosis

Have your muscles ever hurt after prolonged physical activity? If so, you have had your blood buffers somewhat overloaded. Muscle contraction produces lactic acid. This acid ionizes somewhat more strongly than carbonic acid and thus tends to lower the pH of the blood (it tends to release more hydronium ions into the blood). Moderate amounts of lactic acid can be handled by the blood buffers. For bicarbonate, the reaction is

$$\underset{\text{Lactic acid}}{CH_3CH(OH)COOH} + HCO_3^- \longrightarrow \underset{\text{Lactate ion}}{CH_3CH(OH)COO^-} + H_2CO_3$$

Excessive amounts of lactic acid overload the buffers, however, and the pH is lowered. Nerve cells respond to the increased acidity by sending a message of pain to the brain.

If the pH of the blood falls below 7.35, the condition is called **acidosis.** If the pH of the blood rises above 7.45, **alkalosis** sets in. These pathological conditions can be caused by faulty respiration or by metabolic problems. In severe cases of starvation, the body gets its energy by the oxidation of stored fats. The products of fat metabolism are acidic, and prolonged starvation leads to acidosis (Section 26.6). Fad diets, such as those that severely limit the intake of carbohydrates, can also lead to acidosis.

The body's excretory system tries to compensate for acidosis or alkalosis by selectively excreting certain compounds. Conversely, the conditions can be brought on by kidney failure or other excretory problems. We discuss these pathological problems further in later chapters.

Summary

Acid and base concentrations are usually expressed in molarity (moles/liter). When a concentrated solution of acid or base is diluted, the concentration of the dilute solution can be calculated from the relationship

$$V_{\text{conc}} \times M_{\text{conc}} = V_{\text{dil}} \times M_{\text{dil}}$$

Both sides in this relationship represent number of moles (liters × moles/liter = moles), and so the equation says that the number of moles in the dilute solution is the same as the number of moles in the concentrated solution.

Titration is a procedure for determining the concentration of an acidic or basic solution. To determine the concentration of an acidic solution, a buret is filled with a basic solution of known concentration, and an exactly measured volume of the acidic solution is put into a receiving flask. An **indicator,** which is a substance that is one color in acidic solutions and another color in basic solutions, is added to the flask, and the basic solution from the buret is added slowly. At the *equivalence point* of the titration, the number of moles of base added to the flask is just equal to the number of moles of acid contained in the flask. Knowing the original volume of the acidic solution, you can calculate its molarity. To determine the concentration of a basic solution, the procedure is reversed: a known volume of the basic solution is put in the receiving flask, along with an indicator, and an acidic solution of known concentration is added from the buret.

Some of the H_2O molecules in any sample of water dissociate into H^+ and OH^- ions. Pure water at 25 °C contains 1×10^{-7} mol of H^+ ions (as H_3O^+) and 1×10^{-7} mol of OH^- ions. The product of these two concentrations is the **ion product of water:**

$$K_w = [H_3O^+]\,[OH^-] = (1 \times 10^{-7})\,(1 \times 10^{-7}) = 1 \times 10^{-14}$$

The ion product is a constant at a given temperature, and so if you add H^+ to water, thereby raising the value of $[H_3O^+]$, the value of $[OH^-]$ must decrease. This decrease occurs as some of the OH^- ions formed when H_2O molecules dissociated recombine with H^+ to re-form H_2O.

The **pH scale** is a way of expressing H^+ concentrations. The pH of any solution is the exponent (ignoring the minus sign) of the H^+ concentration when that concentration is expressed in the form 1×10^{-x}. The **pOH** of a solution is the exponent of the OH^- concentration when that concentration is expressed in the form 1×10^{-x}. Because $[H^+][OH^-]$ is always 1×10^{-14}, it is also true that pH + pOH = 14.

A neutral solution has a pH of 7, an acidic solution a pH below 7, and a basic solution a pH above 7.

When a salt is dissolved in water, the resulting solution may be neutral, acidic, or basic. It is neutral when the salt is made up of the anion of a strong acid and the cation of a strong base (NaCl), acidic when the salt is made up of the anion of a strong acid and the cation of a weak base (NH_4Cl), and basic when the salt is made up of the anion of a weak acid and the cation of a strong base (KNO_2). When the salt anion is from a weak acid and the cation from a weak base, the resulting solution may be neutral, acidic, or basic, depending on the relative strengths of the acid and base.

A **buffer solution** is one that maintains a constant pH even when small amounts of acids or bases are added. Such a solution is made either from a weak acid and one of its salts or from a weak base and one of its salts. Different acid/salt or base/salt combinations produce buffers that are effective in different pH ranges. A large amount of added acid or added base can overwhelm a buffer, allowing the pH to change.

The bicarbonate/carbonic acid buffer is the principal one maintaining the pH of blood. Several pathological conditions in humans can result in blood buffers being overwhelmed. A blood pH above 7.45 is the condition known as **alkalosis,** and a blood pH below 7.35 is **acidosis.**

Key Terms

acidosis (11.7)
alkalosis (11.7)
buffer solution (11.5)
dilution (11.1)
equivalence point (11.2)
indicator dye (11.2)
ion product of water (K_W) (11.3)
pH scale (11.3)
pOH (11.3)
titration (11.2)

Review Questions

1. Define or illustrate
   a. equivalence point
   b. K_w
   c. pH
   d. pOH
   e. titration
   f. indicator
   g. pH meter
   h. buffer
   i. acidosis
   j. alkalosis
2. Indicate whether each pH value represents an acidic, basic, or neutral solution.
   a. 11.2 b. 4.6 c. 7.0 d. 3.4
3. Use acetic acid and acetate ion to explain how a buffer controls pH.
4. Use ammonia and ammonium chloride to explain how a buffer controls pH.
5. Name three buffer systems operating in the blood.
6. What groups in proteins react with added acid and base?
7. If acid is added to an unbuffered solution, will the pH increase or decrease?
8. If someone is suffering from alkalosis, is the blood pH too high or too low?
9. Would 1 mol of NaOH neutralize **(a)** 3.00 mol of H_3PO_4? **(b)** 0.5 mol of H_2SO_4?
10. Would 0.50 mol of $Ca(OH)_2$ neutralize all of the acid in **(a)** 0.50 mol of H_3PO_4? **(b)** 1.00 mol of HCl?
11. A weak acid is titrated with a strong base. Is the solution at the equivalence point acidic, basic, or neutral? Explain.
12. A weak base is titrated with a strong acid. Is the solution at the equivalence point acidic, basic, or neutral? Explain.
13. What special meaning is conveyed when we refer to a *concentrated* acid or base?
14. Is the concentration of a solution changed by dilution? Is the number of moles of solute changed by dilution?
15. Describe how you would determine the concentration of a solution of an acid by using a solution of a base of known concentration.
16. Which of the following aqueous solutions would you expect to have a pH greater than 7, and which would you expect to have a pH lower than 7? **(a)** 0.0025 M HCl; **(b)** 0.0037 M NH_3; **(c)** 1.0 M CH_3COOH; **(d)** 0.050 M CH_3COONa
17. Which of the following 0.05 M aqueous solutions would you expect to have the *lowest* pH? **(a)** K_2SO_4; **(b)** K_2CO_3; **(c)** $KHSO_4$; **(d)** KOH
18. Arrange the following 0.10 M aqueous solutions in order of *increasing* pH. **(a)** HCl; **(b)** KOH; **(c)** CH_3COOH; **(d)** CH_3COONa

Problems

Dilution of Solutions

19. What volume of 12.0 M HCl is required to make 2.00 L of 1.00 M HCl?
20. What volume of 8.89 M HBr is required to make 2.00 L of 1.00 M HBr?
21. What volume of 1.04 M Na_2CO_3 is required to make 0.500 L of 1.00 M Na_2CO_3?
22. What volume of 19.1 M NaOH is required to make 2.00 L of 6.00 M NaOH?
23. What volume of concentrated (18.0 M) sulfuric acid would be required for the following to be made?
    a. 1.25 L of 6.00 M solution
    b. 575 mL of 0.100 M solution
24. You have a stock solution of 12.0 M HCl. How would you prepare the following solutions?
    a. 125 mL of 1.25 M HCl solution
    b. 5.00 L of 6.00 M HCl solution
    c. 1.50 L of 1.00 M HCl solution

Acid–Base Titrations

25. Calculate the molarity of an HCl solution if 20.0 mL of it requires 33.2 mL of 0.150 M NaOH for neutralization.

$$HCl(aq) + NaOH(aq) \longrightarrow NaCl(aq) + H_2O$$

26. Calculate the molarity of an HNO_3 solution if 30.0 mL of it requires 18.3 mL of 0.104 M KOH for neutralization.

$$HNO_3(aq) + KOH(aq) \longrightarrow KNO_3(aq) + H_2O$$

27. Calculate the molarity of a $Ca(OH)_2$ solution if 18.5 mL of it requires 28.2 mL of 0.0302 M HCl for neutralization.

$$Ca(OH)_2(aq) + 2\,HCl(aq) \longrightarrow CaCl_2(aq) + 2\,H_2O$$

28. Calculate the molarity of a $H_2C_2O_4$ solution if 12.5 mL of it requires 25.7 mL of 0.0995 M NaOH for neutralization.

$$H_2C_2O_4(aq) + 2\,NaOH(aq) \longrightarrow Na_2C_2O_4(aq) + 2\,H_2O$$

29. A 20-mL sample of gastric fluid is neutralized by 25 mL of 0.10 M NaOH. What is the molarity of HCl in the fluid? Assume that all the acidity of the gastric fluid is due to HCl.
30. When the stomach isn't being stimulated by food to make more, it produces 0.0023 mol of HCl and 30 to 60 mL of total juices per hour. What range of concentrations of HCl in the stomach does this represent?
31. How many milliliters of 0.100 M H_2SO_4 are required to react with 10.3 mL of 0.404 M $NaHCO_3$?

$$H_2SO_4(aq) + 2\,NaHCO_3(aq) \longrightarrow Na_2SO_4(aq) + 2\,H_2O + 2\,CO_2(g)$$

32. How many milliliters of 0.110 M H_2SO_4 are required to react with 30.0 mL of 0.0887 M $Ba(OH)_2$?

$$H_2SO_4(aq) + Ba(OH)_2(aq) \longrightarrow BaSO_4(s) + 2\,H_2O$$

33. How many milliliters of 0.0195 M HCl are required to titrate (**a**) 25.00 mL of 0.0365 M KOH(aq), (**b**) 10.00 mL of 0.0116 M $Ca(OH)_2$(aq), and (**c**) 20.00 mL of 0.0225 M NH_3(aq)?
34. How many milliliters of 0.0108 M $Ba(OH)_2$(aq) are required to titrate (**a**) 20.00 mL of 0.0265 M H_2SO_4(aq), (**b**) 25.00 mL of 0.0213 M HCl(aq), and (**c**) 10.00 mL of 0.0868 M CH_3COOH(aq)?

pH and pOH

35. What is the pH of each of the following solutions?
    a. 1.0×10^{-2} M HCl **b.** 1.0×10^{-4} M HNO_3
36. What is the pH of each of the following solutions?
    a. 0.00010 M HCl **b.** 0.00010 M HNO_3
    c. 0.10 M HBr
37. What is the pOH of each of the following solutions?
    a. 1.0×10^{-2} M NaOH **b.** 1.0×10^{-3} M KOH
38. What is the pOH of each of the following solutions?
    a. 0.0010 M NaOH **b.** 0.010 M KOH
39. What is the pOH of each of the solutions in Problem 36?
40. What is the pH of each of the solutions in Problem 38?

Note: The following problems require the use of a table of logarithms or a scientific calculator with a base 10 logarithm function.

41. Calculate the pH values of solutions with the following molar hydronium ion concentrations.
    a. 3.3×10^{-3} **b.** 5.7×10^{-5} **c.** 8.1×10^{-4}
42. Calculate the pH values of (**a**) a 3.6×10^{-2} M HCl solution and (**b**) an 8.8×10^{-4} M HNO_3 solution.
43. Calculate the pH of a blood solution that has a hydronium ion concentration of 4.6×10^{-8} M.
44. Calculate the pH of a urine sample that has a hydronium ion concentration of 2.3×10^{-6} M.
45. Calculate the pH of an ammonia solution that has a hydronium ion concentration of 2.0×10^{-12} M.
46. Calculate the pH of a sample of gastric juice that has a hydronium ion concentration of 0.12 M.
47. What is the hydronium ion concentration of a urine sample that has a pH of 5.10?
48. What is the hydronium ion concentration of a lemon juice sample that has a pH of 2.31?

Salt Solutions: Acidic, Basic, or Neutral?

49. Write an equation for the equilibrium established when CH_3COO^- is placed in water.
50. Write an equation for the equilibrium established when NH_4^+ is placed in water.
51. Classify the aqueous solution of each of these salts as acidic, basic, or neutral.
    a. KCl **b.** NaCN **c.** NH_4CN
52. Classify the aqueous solution of each of these salts as acidic, basic, or neutral.
    a. CH_3COOK **b.** $(NH_4)_2SO_4$ **c.** Na_2SO_4

Additional Problems

53. Vinegar is an aqueous solution of acetic acid, CH_3COOH. A 10.00-mL sample of a particular vinegar requires 31.45 mL of 0.2560 M KOH for its titration. What is the molarity of acetic acid in the vinegar? (*Hint:* Write a net ionic equation for the titration reaction.)
54. Most window cleaners are aqueous solutions of ammonia. A 10.00-mL sample of a particular window cleaner requires 39.95 mL of 1.008 M HCl for its titration. What is the molarity of ammonia in the window cleaner? (*Hint:* Write a net ionic equation for the titration reaction.)
55. What volume of CO_2(g), measured at STP, is produced by the reaction of excess H_2SO_4(aq) with 148 g of Na_2CO_3(s)?

$$Na_2CO_3(s) + H_2SO_4(aq) \longrightarrow Na_2SO_4(aq) + H_2O + CO_2(g)$$

56. What volume of SO_2(g), measured at 764 mm Hg and 26 °C, is produced by the reaction of excess HCl(aq) with 212 g of Na_2SO_3(s)?

$$Na_2SO_3(s) + 2\,HCl(aq) \longrightarrow 2\,NaCl(aq) + H_2O + SO_2(g)$$

57. Which of the following 0.010 M solutions has the highest $[H_3O^+]$: CH_3CH_2COOH(aq), HI(aq), NH_3(aq), H_2SO_4(aq), or $Ba(OH)_2$? Explain.
58. A railroad tank car carrying 1.5×10^3 kg of concentrated sulfuric acid derails and spills its load. The acid is 93.2% H_2SO_4 and has a density of 1.84 g/mL. What mass, in kilograms, of sodium carbonate (soda ash) is needed to neutralize the acid? (*Hint:* What is the neutralization reaction?)
59. Household ammonia, used as a window cleaner and for other cleaning purposes, is NH_3(aq). A 31.08-mL portion of 0.9928 M HCl(aq) is required to neutralize the NH_3 present in a 5.00-mL sample of NH_3(aq) in a titration. The reaction that occurs is

$$NH_3(aq) + HCl(aq) \longrightarrow NH_4Cl(aq)$$

What is the molarity of NH_3 in the NH_3(aq) sample?

Special Topic C
Equilibrium Calculations

In Chapters 10 and 11, we described the equilibria established when a weak acid, a weak base, a salt of either, or some combination of these is dissolved in water. Our discussion was entirely qualitative. It is possible to treat these equilibria in a more quantitative fashion. We will now do that, in this special topic, for those who are interested or whose work might require it.

C.1 Equilibrium Constant Expressions

The proportions of reactants and products at equilibrium are determined by a simple relationship. Let's write a generalized reaction,

$$a\,A(g) + b\,B(g) \rightleftharpoons c\,C(g) + d\,D(g)$$

where A and B are reactants and C and D are products. The small letters are the coefficients for the substances. The relationship at a given temperature is given by the expression

$$K = \frac{[C]^c \times [D]^d}{[A]^a \times [B]^b}$$

The quantities in brackets stand for molar concentration at equilibrium. The quantity K is called the **equilibrium constant.** The entire expression is called the **equilibrium constant expression.** The coefficients in the generalized reaction become exponents in the equilibrium constant expression.

Let's consider a specific reaction.

$$H_2(g) + Cl_2(g) \rightleftharpoons 2\,HCl(g)$$

The equilibrium constant expression for this reaction is

$$K = \frac{[HCl]^2}{[H_2] \times [Cl_2]}$$

For the reaction

$$N_2(g) + 3\,H_2(g) \rightleftharpoons 2\,NH_3(g)$$

the equilibrium expression is

$$K = \frac{[NH_3]^2}{[N_2] \times [H_2]^3}$$

Example C.1

Write the equilibrium constant expression for each of the following reactions.

a. $H_2(g) + F_2(g) \rightleftharpoons 2\,HF(g)$
b. $2\,NO(g) + O_2(g) \rightleftharpoons 2\,NO_2(g)$
c. $CO(g) + H_2O(g) \rightleftharpoons CO_2(g) + H_2(g)$

Solution

a. $K = \dfrac{[HF]^2}{[H_2] \times [F_2]}$ **b.** $K = \dfrac{[NO_2]^2}{[NO]^2 \times [O_2]}$

c. $K = \dfrac{[CO_2] \times [H_2]}{[CO] \times [H_2O]}$

Practice Exercise

Write the equilibrium constant expression for each of the following reactions.

a. $2\,HI(g) \rightleftharpoons H_2(g) + I_2(g)$
b. $3\,O_2(g) \rightleftharpoons 2\,O_3(g)$
c. $Xe(g) + 2\,F_2(g) \rightleftharpoons XeF_4(g)$

We can demonstrate that acetic acid is a weak acid by measuring the pH of vinegar. Vinegar is approximately 1 M acetic acid. A strong acid of the same molarity would have a much lower pH, about pH 1.

Table C.1 Ionization Constants of Some Weak Acids in Water at 25 °C

Acid	Simplified Ionization Equilibrium	Ionization Constant, K_a
Inorganic Acids		
Nitrous acid	$HNO_2 \rightleftharpoons H^+ + NO_2^-$	7.2×10^{-4}
Hydrofluoric acid	$HF \rightleftharpoons H^+ + F^-$	6.6×10^{-4}
Hypochlorous acid	$HOCl \rightleftharpoons H^+ + OCl^-$	2.9×10^{-8}
Hypobromous acid	$HOBr \rightleftharpoons H^+ + OBr^-$	2.5×10^{-9}
Hydrocyanic acid	$HCN \rightleftharpoons H^+ + CN^-$	6.2×10^{-10}
Carboxylic Acids		
Chloroacetic acid	$CH_2ClCOOH \rightleftharpoons H^+ + CH_2ClCOO^-$	1.4×10^{-3}
Formic acid	$HCOOH \rightleftharpoons H^+ + HCOO^-$	1.8×10^{-4}
Benzoic acid	$C_6H_5COOH \rightleftharpoons H^+ + C_6H_5COO^-$	6.3×10^{-5}
Acetic acid	$CH_3COOH \rightleftharpoons H^+ + CH_3COO^-$	1.8×10^{-5}

All of the examples on the previous page involve gaseous reactants and products. The same treatment is applied to substances in aqueous solution in the next section. Special cases involving a solid reactant or product are considered in Section 12.8.

C.2 Ionization of Weak Acids

Now let's take another look at the ionization of acetic acid. The reaction is

$$CH_3COOH(aq) + H_2O \rightleftharpoons H_3O^+(aq) + CH_3COO^-(aq)$$

At equilibrium, the equilibrium constant expression is

$$K = \frac{[H_3O^+] \times [CH_3COO^-]}{[CH_3COOH] \times [H_2O]}$$

For dilute solutions, the molar concentration of water, $[H_2O]$, stays nearly constant at 55 M[1]. We can substitute this value into the equation:

$$K = \frac{[H_3O^+] \times [CH_3COO^-]}{[CH_3COOH] \times 55}$$

Rearranging, we get

$$K \times 55 = \frac{[H_3O^+] \times [CH_3COO^-]}{[CH_3COOH]} = K_a$$

[1] In 1.0 L of dilute aqueous solution there is approximately 1.0 L of water, which weighs 1000 g and contains

$$\frac{1000\ \cancel{g}}{18\ \cancel{g}/\text{mol}} = 55 \text{ mol of water}$$

Thus, the concentration of water in 1.0 L of dilute solution is approximately 55 M.

The product of the two constants, $K \times 55$, is a constant. This new constant, K_a, is called the **acid ionization constant** for the weak acid.

Sometimes the involvement of water is omitted, and the ionization of acetic acid is written

$$CH_3COOH(aq) \rightleftharpoons H^+(aq) + CH_3COO^-(aq)$$

The ionization constant becomes

$$K_a = \frac{[H^+] \times [CH_3COO^-]}{[CH_3COOH]}$$

If you recognize that the hydronium ion is simply a hydrated proton, you will see that the two K_a expressions are the same.

The strength of an acid is related to the degree of ionization. The greater the degree of ionization, the stronger the acid. (Strong acids, such as HCl, are 100% ionized in dilute solution.) Also, the greater the degree of ionization, the larger the K_a. Table C.1 lists the K_a values for several common acids. Don't forget: The larger the K_a (the less negative the exponent), the stronger the acid.

Ionization constants can be calculated from measurements of the hydrogen ion concentration, as shown in the following example.

Example C.2

Find the K_a for an organic acid (symbolized by HOrg), if the hydrogen ion concentration of a 0.0200 M solution is 3.0×10^{-4} M. The simplified ionization reaction is

$$HOrg(aq) \rightleftharpoons H^+(aq) + Org^-(aq)$$

and the equilibrium constant expression is

$$K_a = \frac{[H^+][Org^-]}{[HOrg]}$$

Solution

Ionization of the acid produces equal concentrations of H^+ and Org^-.

$$[H^+] = [Org^-] = 3.0 \times 10^{-4}\ M = 0.00030\ M$$

At equilibrium, the concentration of nonionized acid equals its original concentration minus the amount that has ionized to form H^+ and Org^-.

$$[HOrg] = 0.0200 - 0.00030 = 0.0197\ M$$

$$= 1.97 \times 10^{-2}\ M$$

$$K_a = \frac{(3.0 \times 10^{-4})(3.0 \times 10^{-4})}{1.97 \times 10^{-2}} = 4.6 \times 10^{-6}$$

Practice Exercise

What is the K_a for the hypothetical acid HB for which the $[H^+]$ of a 0.010 M solution is 1.6×10^{-4}?

Usually, however, we calculate hydrogen ion concentrations from K_a values found in tables. This approach is illustrated in the following examples.

Example C.3

Calculate the $[H^+]$ in a 0.10 M solution of acetic acid. The equation is
$CH_3COOH(aq) \rightleftharpoons H^+(aq) + CH_3COO^-(aq)$.

Solution

First, write the equilibrium constant expression. (The K_a of acetic acid is given in Table C.1.)

$$K_a = \frac{[H^+][CH_3COO^-]}{[CH_3COOH]} = 1.8 \times 10^{-5}$$

For each CH_3COOH that ionizes, one H^+ and one CH_3COO^- are formed. We can therefore let $x = [H^+] = [CH_3COO^-]$.

Since each H^+ formed represents one CH_3COOH ionized, the equilibrium concentration of CH_3COOH will be $0.10 - x$. Substitution yields

$$\frac{(x)(x)}{0.10 - x} = 1.8 \times 10^{-5}$$

The amount of CH_3COOH ionized is very small compared with the total amount of CH_3COOH that we started with, so let's assume that

$$0.10 - x \cong 0.10 = 1.0 \times 10^{-1}$$

We then have

$$\frac{x^2}{1.0 \times 10^{-1}} = 1.8 \times 10^{-5}$$

$$x^2 = (1.8 \times 10^{-5})(1.0 \times 10^{-1})$$

$$= 1.8 \times 10^{-6}$$

$$x = 1.3 \times 10^{-3}\ M = [H^+]$$

Now let's check our assumption. Is 0.10 changed significantly by subtracting 0.0013?

$$0.10 - 0.0013 \stackrel{?}{=} 0.10$$

Certainly no significant error was introduced.

Practice Exercise

Calculate the $[H^+]$ in a 0.033 M solution of hypobromous acid.

$$HOBr(aq) \rightleftharpoons H^+(aq) + OBr^-(aq)$$

Example C.4

Calculate the $[H^+]$ in a 0.0010 M solution of HCN.

Solution

$$HCN(aq) \rightleftharpoons H^+(aq) + CN^-(aq)$$

$$K_a = \frac{[H^+][CN^-]}{[HCN]} = 6.2 \times 10^{-10}$$

Let $x = [H^+] = [CN^-]$. Then

$$[HCN] = 0.0010 - x \cong 0.0010$$

Substituting, we have

$$\frac{(x)(x)}{1.0 \times 10^{-3}} = 6.2 \times 10^{-10}$$

$$x^2 = 6.2 \times 10^{-13} = 62 \times 10^{-14}$$

$$x = 7.9 \times 10^{-7} = [H^+]$$

Is x small compared with 0.0010? Yes!

Practice Exercise

Calculate the $[H^+]$ in a 0.15 M solution of hypochlorous acid.

$$HOCl(aq) \rightleftharpoons H^+(aq) + OCl^-(aq)$$

Approximations such as those made in the preceding examples are good only if the degree of ionization is small. Such an assumption should always be checked.

Table C.2 Ionization Constants of Some Weak Bases in Water at 25 °C

Base	Ionization Equilibrium	Ionization Constant, K_b
Inorganic Bases		
Ammonia	$NH_3 + H_2O \rightleftharpoons NH_4^+ + OH^-$	1.8×10^{-5}
Hydrazine	$H_2NNH_2 + H_2O \rightleftharpoons H_2NNH_3^+ + OH^-$	8.5×10^{-7}
Amines		
Dimethylamine	$(CH_3)_2NH + H_2O \rightleftharpoons (CH_3)_2NH_2^+ + OH^-$	5.9×10^{-4}
Ethylamine	$CH_3CH_2NH_2 + H_2O \rightleftharpoons CH_3CH_2NH_3^+ + OH^-$	4.3×10^{-4}
Methylamine	$CH_3NH_2 + H_2O \rightleftharpoons CH_3NH_3^+ + OH^-$	4.2×10^{-4}
Hydroxylamine	$HONH_2 + H_2O \rightleftharpoons HONH_3^+ + OH^-$	9.1×10^{-9}
Pyridine	$C_5H_5N + H_2O \rightleftharpoons C_5H_5NH^+ + OH^-$	1.5×10^{-9}
Aniline	$C_6H_5NH_2 + H_2O \rightleftharpoons C_6H_5NH_3^+ + OH^-$	7.4×10^{-10}

C.3 Equilibria Involving Weak Bases

Equilibrium constant expressions can also be written for weak bases. For ammonia, the reaction is

$$NH_3(aq) + H_2O \rightleftharpoons NH_4^+(aq) + OH^-(aq)$$

and the equilibrium constant expression is

$$K = \frac{[NH_4^+][OH^-]}{[NH_3][H_2O]}$$

Since the concentration of water is constant for dilute solutions, we can include $[H_2O]$ with K to give a new constant, K_b.

$$K_b = \frac{[NH_4^+][OH^-]}{[NH_3]}$$

Base ionization constants (K_b) for several weak bases are given in Table C.2.

Example C.5

Calculate the hydroxide ion concentration in a 0.010 M solution of aniline.

$$C_6H_5NH_2(aq) + H_2O \rightleftharpoons C_6H_5NH_3^+(aq) + OH^-(aq)$$

Solution

$$K_b = \frac{[C_6H_5NH_3^+][OH^-]}{[C_6H_5NH_2]} = 4.2 \times 10^{-10}$$

Let $x = [OH^-] = [C_6H_5NH_3^+]$.

$$[C_6H_5NH_2] = 0.010 - x \cong 0.010 = 1.0 \times 10^{-2}$$

Substituting, we have

$$\frac{(x)(x)}{1.0 \times 10^{-2}} = 4.2 \times 10^{-10}$$

$$x^2 = 4.2 \times 10^{-12}$$

$$x = 2.0 \times 10^{-6}\ M = [OH^-]$$

Again, we see that x is small compared with 0.010.

Practice Exercise

Calculate the $[OH^-]$ in a 0.15 M solution of hydroxylamine.

$$HONH_2(aq) + H_2O \rightleftharpoons HONH_3^+(aq) + OH^-(aq)$$

C.4 Calculations Involving Buffers

Let's take another look at solutions that contain both a weak acid and a salt of that acid. We saw in Section 11.5 that such a solution acts as a buffer to resist a change in pH. Now we can calculate the hydrogen ion concentration in such a system.

If sodium acetate is added to a solution of acetic acid, ionization of the acid is considerably decreased. In effect, we are increasing the concentration of acetate ion in the system.

$$CH_3COOH(aq) \rightleftharpoons H^+(aq) + CH_3COO^-(aq)$$

The acetate ions from sodium acetate react with the H^+, forming more un-ionized acetic acid molecules. A greater proportion of the total acetic acid is in the molecular form; ionization is decreased. This is an

The common ion effect. Both solutions contain bromophenol blue indicator. The yellow color indicates that the pH is less than 3.0 in 1.00 M CH_3COOH, and the blue-violet color indicates that the pH is greater than 6.0 in the solution that is 1.00 M in both CH_3COOH and CH_3COONa. The presence of CH_3COO^-, a common ion, raises the pH by about two units and reduces $[H_3O^+]$ about 100-fold in the solution that has both acetic acid and sodium acetate as solutes.

bromophenol blue indicator:	pH < 3.0 yellow	pH > 4.6 blue-violet

example of the **common ion effect.** If an ion common to those in the equilibrium is added, the degree of ionization is decreased. A new equilibrium is established with more acetate ions but fewer H^+. The value of K_a remains unchanged.

Example C.6

What is the $[H^+]$ in a solution that is 0.01 M acetic acid and 0.10 M sodium acetate?

Solution

Use the K_a expression for acetic acid.

$$K_a = \frac{[H^+][CH_3COO^-]}{[CH_3COOH]} = 1.8 \times 10^{-5}$$

Let $x = [H^+]$.

Sodium acetate is completely ionized; thus the concentration of acetate ion equals the sum of $[CH_3COO^-]$ from the salt and $[CH_3COO^-]$ from the ionization of acetic acid.

$$[CH_3COO^-] = 0.10 + x \cong 0.10 = 1.0 \times 10^{-1}$$

The concentration of acetic acid at equilibrium is the original concentration minus the amount ionized.

$$[CH_3COOH] = 0.10 - x \cong 0.10 = 1.0 \times 10^{-1}$$

$$\frac{(x)(1.0 \times 10^{-1})}{1.0 \times 10^{-1}} = 1.8 \times 10^{-5}$$

$$x = 1.8 \times 10^{-5}\ M = [H^+]$$

Was our assumption valid that x was small enough to be ignored when added to or subtracted from 0.10? Yes, 1.8×10^{-5} is negligibly small compared with 0.10—that is, with 1.0×10^{-1}. (This buffer solution tends to keep the hydrogen ion concentration constant at a value of 1.8×10^{-5} mol/L.)

Practice Exercise

Calculate the $[H^+]$ in a solution that is 0.033 M in HOBr and 0.033 M in NaOBr.

The ionization of a weak base is also decreased by the addition of a common ion. Adding ammonium chloride to a solution of ammonia decreases the amount of hydroxide ion in the system.

$$NH_3(aq) + H_2O \rightleftharpoons NH_4^+(aq) + OH^-(aq)$$

Example C.7

Calculate the $[OH^-]$ and $[H^+]$ of a solution that is 0.50 M NH_3 and 0.10 M NH_4^+.

Solution

Use the K_b expression for NH_3.

$$K_b = \frac{[NH_4^+][OH^-]}{[NH_3]} = 1.8 \times 10^{-5}$$

Let $y = [OH^-]$.

$$[NH_4^+] \cong 0.10 = 1.0 \times 10^{-1}$$
$$[NH_3] \cong 0.50 = 5.0 \times 10^{-1}$$

$$\frac{(1.0 \times 10^{-1})(y)}{5.0 \times 10^{-1}} = 1.8 \times 10^{-5}$$

$$y = 9.0 \times 10^{-5} = [OH^-]$$

Now use the relationship

$$K_w = [H^+][OH^-] = 1.0 \times 10^{-14}$$

to calculate $[H^+]$.

$$[H^+] = \frac{1.0 \times 10^{-14}}{9.0 \times 10^{-5}} = 1.1 \times 10^{-10}$$

Buffered solutions are widely used in analytical chemistry, medicine, and biochemistry and in many industrial applications such as leatherworking and dyeing. A rearrangement of the equilibrium constant expression, called the **Henderson-Hasselbalch equation,** is used to calculate the concentration of a buffer. For a weak acid HA, the equation is

$$pH = pK_a + \log \frac{[A^-]}{[HA]}$$

where pK_a is a logarithmic term similar to pH.

$$pK_a = -\log K_a$$

If large quantities of acid or base are added, the capacity of a buffer is exceeded. In general, the more concentrated a solution is in its buffer components, the more added acid or base it is capable of neutralizing. And, as a rule, a buffer is most effective against both added H_3O^+ and added OH^- if $[HA] = [A^-]$. This pH value is called the *optimum pH* of the buffer and it corresponds to a pH value equal to pK_a. That is,

$$pH = pK_a + \log \frac{[A^-]}{[HA]}$$

$$pH = pK_a + \log 1$$

$$pH = pK_a$$

The pH range over which a buffer solution is effective generally is about one pH unit on either side of $pH = pK_a$. For example, the pK_a for acetic acid is

$$pK_a = -\log K_a = -\log (1.8 \times 10^{-5}) = 4.74$$

so the effective pH range for an acetic acid–acetate ion buffer is $3.74 < pH < 5.74$.

Example C.8

What is the pH of a solution that is 0.20 M H_2S and 0.20 M HS^-? The K_a for H_2S is 1×10^{-7}. (Ignore the second ionization of H_2S.)

$$H_2S(aq) \rightleftharpoons H^+(aq) + HS^-(aq)$$

Solution

Since $[H_2S] = [HS^-]$, the Henderson-Hasselbalch equation reduces to

$$pH = pK_a$$
$$pK_a = -\log K_a = -\log (1 \times 10^{-7}) = 7$$
$$pH = pK_a = 7$$

Practice Exercise

What is the pH of a solution that is 0.50 M HF and 0.50 M F^-?

The concentration of HA and A^- are not always equal. Example C.9 illustrates how the Henderson-Hasselbalch equation is used to calculate the pH values for such buffer solutions.

Example C.9

What is the pH of a solution that is 0.10 M HCN and 0.50 M CN^-? The K_a for HCN is 6.2×10^{-10}.

Solution

$$\begin{aligned} pH &= pK_a + \log \frac{[CN^-]}{[HCN]} \\ &= -\log (6.2 \times 10^{-10}) + \log \frac{0.50}{0.10} \\ &= -\log 6.2 - \log 10^{-10} + \log 5.0 \\ &= -0.79 - (-10.00) + 0.70 \\ &= 9.91 \end{aligned}$$

Practice Exercise

What is the pH of a solution that is 0.40 M HNO_2 and 0.25 M NO_2^-? The K_a for HNO_2 is 7.2×10^{-4}.

Key Terms

acid ionization constant (K_a) (C.2)
base ionization constant K_b) (C.3)
common ion effect (C.4)
equilibrium constant (C.1)
equilibrium constant expression (C.1)
Henderson-Hasselbalch equation (C.4)

Review Questions

1. Explain what is meant by the *common ion effect.* How is this effect involved in the functioning of buffer solutions?
2. Which of the following will *suppress* the ionization of formic acid, HCOOH:
   (**a**) NaCl; (**b**) KOH; (**c**) $(HCOO)_2Ca$; or (**d**) Na_2CO_3? Explain.
3. Write the equation for the ionization of each of the following acids.
   a. HBO_2
   b. $HClO_2$
   c. $HC_9H_7O_4$
   d. H_2Se (first ionization only)
4. Write equations for the ionization of each of the following bases (i.e., for the reaction of the base with water to form ions).
   a. $C_4H_9NH_2$
   b. $C_{11}H_{21}O_4N$
   c. C_3H_5N

Problems

Note: You may refer to Table C.1 for K_a values and to Table C.2 for K_b values as necessary.

Equilibrium Constant Expressions

5. Write an equilibrium constant expression for each of the following reactions.
   a. $HOCl(aq) \rightleftharpoons H^+(aq) + OCl^-(aq)$
   b. $HC_6H_7O_6(aq) \rightleftharpoons H^+(aq) + C_6H_7O_6^-(aq)$
   c. $HCOOH(aq) \rightleftharpoons H^+(aq) + HCOO^-(aq)$
6. Write an equilibrium constant expression for each of the following reactions.
   a. $C_5H_5N(aq) + H_2O \rightleftharpoons C_5H_5NH^+(aq) + OH^-(aq)$
   b. $C_2H_5NH_2(aq) + H_2O \rightleftharpoons C_2H_5NH_3^+(aq) + OH^-(aq)$
   c. $HPO_4^{2-}(aq) + H_2O \rightleftharpoons H_2PO_4^-(aq) + OH^-(aq)$

Calculating K_a Values

7. Find the K_a for the acid HZ if the $[H^+]$ of a 0.100 M solution of HZ is 0.0002 M.
8. Find K_b for the base QOH if the $[OH^-]$ of a 0.0200 M solution of QOH is 0.000400 M. The ionization reaction is

$$QOH(aq) \rightleftharpoons Q^+(aq) + OH^-(aq)$$

Equilibria in Solutions of Weak Acids and Weak Bases

9. Calculate the $[H^+]$ in each of these solutions.
   a. 0.010 M CH_3COOH b. 0.20 M C_6H_5COOH
   c. 0.50 M HCN
10. Calculate the $[H^+]$ in each of these solutions.
   a. 0.10 M HCOOH b. 0.15 M HF
   c. 0.050 M HNO_2
11. What is $[OH^-]$ in each of the solutions in Problem 9?
12. What is $[OH^-]$ in each of the solutions in Problem 10?
13. Calculate $[OH^-]$ in each of these solutions.
   a. 0.025 M NH_3 b. 0.10 M CH_3NH_2
   c. 0.10 M $C_6H_5NH_2$
14. Calculate $[OH^-]$ in each of these solutions.
   a. 0.010 M $(CH_3)_2NH$ b. 0.15 M H_2NNH_2
   c. 0.030 M $HONH_2$
15. What is $[H^+]$ in each of the solutions in Problem 13?
16. What is $[H^+]$ in each of the solutions in Problem 14?

Buffer Solutions

17. What is $[H^+]$ in each of the following buffer solutions?
   a. 0.25 M HCN and 0.25 M KCN
   b. 0.50 M HF and 0.20 M NaF
   c. 0.033 M C_6H_5COOH and 0.045 M $C_6H_5COO^-$
18. What is $[H^+]$ in each of the following buffer solutions?
   a. 0.20 M HCN and 0.20 M KCN
   b. 0.50 M HF and 0.20 M NaF
   c. 0.40 M C_6H_5COOH and 0.20 M $C_6H_5COO^-$
19. Calculate $[OH^-]$ in a solution that is 0.40 M NH_3 and 0.040 M NH_4^+. What is $[H^+]$ in the solution?
20. Calculate $[OH^-]$ in a solution that is 0.040 M NH_3 and 0.020 M NH_4^+. What is $[H^+]$ in the solution?

The Henderson-Hasselbalch Equation

21. Calculate the pH of a solution that is 0.040 M HCN and 0.040 M CN^-.
22. Calculate the pH of a solution that is 0.20 M HCOOH and 0.20 M $HCOO^-$.
23. Calculate the pH of a solution that is 0.15 M benzoic acid and 0.15 M benzoate ion.
24. Calculate the pH of a solution that is 0.20 M HF and 0.50 M F^-.
25. Calculate the pH of a solution that is 0.15 M HCOOH and 0.60 M $HCOO^-$.
26. Calculate the pH of a solution that is 0.11 M benzoic acid and 0.96 M benzoate ion.
27. Calculate the pH of a solution that is 0.350 M CH_3CH_2COOH and 0.0786 M CH_3CH_2COOK.

$$CH_3CH_2COOH(aq) + H_2O \rightleftharpoons H_3O^+(aq) + CH_3CH_2COO^-(aq) \quad K_a = 1.3 \times 10^{-5}$$

28. Calculate the pH of a solution that is 0.132 M diethylamine, $(CH_3CH_2)_2NH$, and 0.145 M diethylammonium chloride, $(CH_3CH_2)_2NH_2Cl$.

$$(CH_3CH_2)_2NH(aq) + H_2O \rightleftharpoons (CH_3CH_2)_2NH_2^+(aq) + OH^-(aq) \quad K_b = 6.9 \times 10^{-4}$$

29. What is the pH of a buffer solution that is 0.10 M HOC_6H_5 and 0.10 M $NaOC_6H_5$? The K_a for HOC_6H_5 is 1.0×10^{-10}.
30. What is the pH of a buffer solution that is 0.050 M $C_5H_{11}COOH$ and 0.050 M $C_5H_{11}COO^-$? The K_a for $C_5H_{11}COOH$ is 1.0×10^{-8}.

Additional Problems

31. A buffer solution contains a Brønsted-Lowry acid as one component and a Brønsted-Lowry base as the other. **(a)** Can the two components be HCl and NaOH? Explain. **(b)** Acetic acid contains both CH_3COOH and CH_3COO^-. Can an acetic acid solution alone be considered a buffer? Explain.
32. Calculate the pH of 1.50 M formic acid, HCOOH.

$$HCOOH(aq) + H_2O \rightleftharpoons H_3O^+(aq) + HCOO^-(aq) \quad K_a = 1.8 \times 10^{-4}$$

33. Calculate the pH of a solution of pyridine that has 1.25 g in 125 mL of water solution. (*Hint:* First calculate the molarity of pyridine.)

$$C_5H_5N(aq) + H_2O \rightleftharpoons C_5H_5N^+(aq) + OH^-(aq)$$
$$K_b = 1.5 \times 10^{-9}$$

34. What molarity of hydrazoic acid, HN_3, is required to produce an aqueous solution with a pH of 3.10?

$$HN_3(aq) + H_2O \rightleftharpoons H_3O^-(aq) + N_3^-(aq)$$
$$K_a = 1.9 \times 10^{-5}$$

12 Electrolytes

Chemical reactions in solution are the source of the electric current produced by a battery.

We first encountered ions in Chapter 4. Since then we have discussed ionic bonds, ionic solids, and solutions of ions. The nineteenth-century scientists whose experiments laid the foundation for the ionic theory were working mainly for the joy of discovery and the recognition gained with success. We can only speculate as to whether they imagined the importance of ions in living systems.

The fluids in our bodies are like the salt water of the oceans. In our fetal development, we have gill-like organs, hands that look like fins, and a shape much like that of a fish. Until we are born and breathe the air, we float in the watery darkness of the womb. Even after birth, fluids bathe our tissues and transport the materials that keep them alive. It is now well established that messages are sent to and from the brain in the form of electric signals. These messages are often carried by ions through cellular and intercellular fluids. Certain ions are essential to the proper functioning of all living organisms. They must be present in proper concentrations, however. Too few or too many can be dangerous. We will discuss some of these ions and their properties in this chapter.

12.1 Early Electrochemistry

In 1800, an Italian physicist, Alessandro Volta, invented a battery that produced an electric current. Electrical phenomena had already been subjected to considerable study before this time. (In 1752, Benjamin Franklin determined that lightning was a form of electricity by flying a kite in a thunderstorm.) The importance of Volta's invention lay in the fact that the battery provided a convenient source of electricity, one that could be used by other scientists who wished to study the interaction of matter and electricity.

Within six weeks of its invention, a battery was used by two English chemists, William Nicholson and Anthony Carlisle, to decompose water into hydrogen gas and oxygen gas. They accomplished the decomposition by passing an electric current through a water sample. Humphry Davy, another English chemist, used an electric current to liberate potassium metal from potassium hydroxide (KOH), sodium metal from sodium hydroxide (NaOH), and metallic magnesium, strontium, barium, and calcium from their respective compounds.

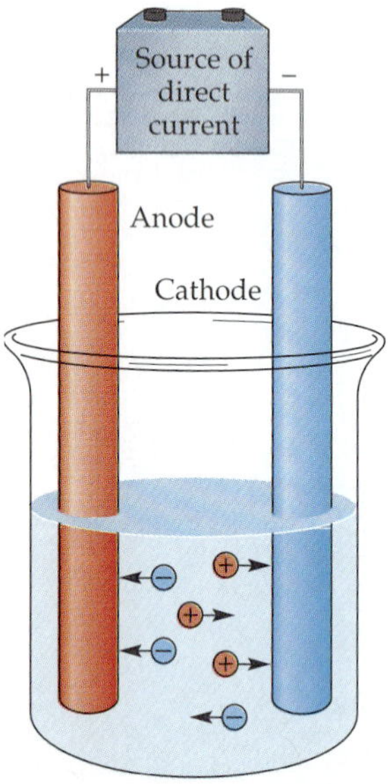

▲ **FIGURE 12.1**

When electric current is passed through an electrolyte, positive ions (cations) move to the cathode and negative ions (anions) move to the anode. Reduction occurs at the cathode, and oxidation takes place at the anode.

Davy's protégé, Michael Faraday, named the process of splitting compounds by means of electricity **electrolysis,** a word of Greek origin that literally means "releasing by electricity." It was Faraday who was responsible for many of the terms used in electrochemistry today. Some of these terms were introduced in Chapter 2, but we offer a brief review here. The carbon rods or metal strips that are connected to a source of electricity (the battery) and inserted into solutions under study are called **electrodes** (Figure 12.1). Electricity, as we now know, is a flow of electrons. In systems used by electrochemists, electrons flow from one electrode to the other. The electrode that has lost electrons and is therefore positively charged is called the **anode.** The electrode that has gained electrons and is negatively charged is called the **cathode.** Electrolysis involves oxidation and reduction reactions. *Oxidation* occurs at the *anode,* where substances give up electrons to the positively charged electrode. *Reduction* occurs at the *cathode,* where substances can pick up electrons from the electrode, which has an excess of them.

In Faraday's studies, electrodes were placed in certain solutions and melts (Section 12.2). When this was done, the electric "circuit" was completed. The battery provided the driving force (the electric potential or voltage), and electric current flowed from one electrode to the other through the solution and then back to the first electrode—that is, around the circuit. Faraday hypothesized that the electric current was carried through the solution (or melt) by atoms that had electric charges. He called these charged atoms *ions,* a term we have used extensively

The work of Faraday and other electrical pioneers generated much interest among scientists who were concerned with the fundamental structure of matter. The excitement of the general public was captured in large part by a purely fictional experiment that related electricity to life. The experiment that excited the imagination of nineteenth- and twentieth-century readers was described by 21-year-old Mary Wollstonecraft Shelley in her 1818 novel, *Frankenstein.* This photograph shows some of the apparatus that Dr. Frankenstein used to generate the electricity that brought his monster to life.

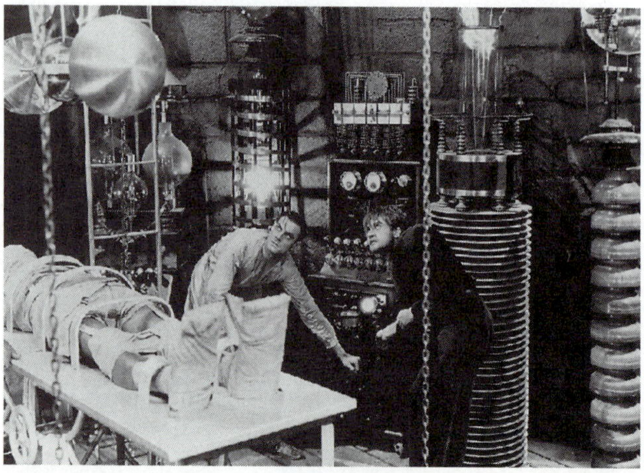

in preceding chapters. You will recall that positively charged ions, which are attracted to the negatively charged electrode (the cathode), are called **cations,** and negatively charged ions, which are attracted to the positively charged anode, are called **anions.** (It is useful to remember that in electrolysis *anions* migrate to the *anode,* and *cations* migrate to the *cathode.*)

12.2 Electrical Conductivity

The electricity with which we are most familiar usually flows in metallic wires. The outer electrons of the metal atoms flow through the wire, while the nucleus and inner-level electrons of the atom remain nearly fixed in place. Most nonmetals are nonconductors. The outer-level electrons of these elements are tightly bound or are shared with neighboring atoms in covalent compounds; they are not free to roam around. Solid ionic compounds do not conduct electricity. Although the ions have electric charges, they occupy fixed positions in a crystal lattice and are unable to move very much in an electric field. When the solid is melted, the lattice is broken down and the ions are free to move about. A substance that conducts electricity when melted or in solution is called an **electrolyte.**

How do we tell when a substance is an electrolyte? One way is to dissolve some of the compound in water and test it with a conductivity apparatus. Figure 12.2, which compares the electrical conductivities of three solutions of equal molarity, allows us to establish three categories based on the following observations.

(a) 1 M $NaCl(aq)$
Strong electrolyte

(b) 1 M $CH_3CH_2OH(aq)$
Nonelectrolyte

(c) 1 M $CH_3COOH(aq)$
Weak electrolyte

▲ **FIGURE 12.2**

Electrolytic properties of aqueous solutions. For electric current to flow, cations and anions must be present and free to flow between the graphite electrodes.

Table 12.1 A Selection of Strong Electrolytes, Weak Electrolytes, and Nonelectrolytes

Compound Name	Formula	Kind of Compound	Electrical Conductivity
Hydrochloric acid	HCl	Strong acid	Strong
Sulfuric acid	H_2SO_4	Strong acid	Strong
Sodium hydroxide	$NaOH$	Strong base	Strong
Sodium chloride	$NaCl$	Salt	Strong
Calcium nitrate	$Ca(NO_3)_2$	Salt	Strong
Acetic acid	CH_3COOH	Weak acid	Weak
Ammonia	NH_3	Weak base	Weak
Methylamine	CH_3NH_2	Weak base	Weak
Sugar (sucrose)	$C_{12}H_{22}O_{11}$	Molecular solid	None
Ethyl alcohol	C_2H_5OH	Molecular liquid	None

- *The light bulb lights up brightly.* The electrical conductivity is *high,* indicating that the solution contains a significant concentration of ions. The substance in solution—the solute—is called a **strong electrolyte.** (We observe this even for a dilute solution of a strong electrolyte.)
- *The bulb doesn't light.* There are few, if any, ions present. The solute is present in *molecular* form and is called a **nonelectrolyte.**
- *The bulb lights but glows only dimly.* The electrical conductivity is *low,* corresponding to a low concentration of ions. The solute is present only partly in ionic form and is called a **weak electrolyte.** (We observe this even for a concentrated solution of a weak electrolyte.)

If we were to test a large number of water-soluble substances by the method suggested in Figure 12.2, we would arrive at the following generalizations.

- Water-soluble ionic compounds and a few molecular compounds (a few acids; the strong acids listed in Table 10.1) are *strong electrolytes.*
- Most molecular compounds are either *nonelectrolytes* or *weak electrolytes.*

The weak acids (Section 10.2) and weak bases (Section 10.4) are also weak electrolytes. Among water-soluble organic compounds, most are nonelectrolytes (for example, alcohols and sugars), but the carboxylic acids (weak acids) and amines (weak bases) are weak electrolytes. Table 12.1 lists some familiar compounds and classifies them according to the behavior of their aqueous solutions toward electric current.

12.3 The Theory of Electrolytes: Ionization and Dissociation

In 1887, Svante Arrhenius proposed a general theory to explain the properties of electrolytes. We encountered a part of that theory in Chapter 10 in our study of acids and bases. Further, Arrhenius's theory has been modified somewhat through the years to account for new data. The modernized theory is summarized here.

1. An electrolyte, when dissolved in water, separates into ions. For salts and strong bases, which are already ionic in the solid state, we can summarize the process by equations such as

$$NaCl(s) \xrightarrow{H_2O} Na^+(aq) + Cl^-(aq)$$
$$Na_2SO_4(s) \xrightarrow{H_2O} 2\,Na^+(aq) + SO_4^{2-}(aq)$$
$$KOH(s) \xrightarrow{H_2O} K^+(aq) + OH^-(aq)$$

These processes are called **dissociation,** the separation of existing ions. For acids and some bases, the process is actually one of *ion formation* or **ionization.**

$$HCl(g) + H_2O \longrightarrow H_3O^+(aq) + Cl^-(aq)$$
$$NH_3(g) + H_2O \longrightarrow NH_4^+(aq) + OH^-(aq)$$

2. When an ionic solid dissociates or when a polar molecule ionizes, the algebraic sum of all positive charges and all negative charges is 0.

$$K_3PO_4(s) \longrightarrow \underset{3(+1)}{3\,K^+(aq)} \underset{+}{+} \underset{(-3)\,=\,0}{PO_4^{3-}(aq)}$$

Solutions as a whole are therefore electrically neutral.

3. Each ion, regardless of size, charge, or shape, has the same effect on boiling-point elevation, freezing-point depression, and osmotic pressure (Chapter 9) as an undissociated molecule would have. This assumption holds quite well for dilute solutions but is not strictly true for concentrated ones. In concentrated solutions, each ion is rather closely surrounded by others of opposite charge. This hinders the movement of the ions; they are not completely free. This decreases the expected activity somewhat. For example, the freezing-point depression of a solution of 1 mol of NaCl in 1 kg of water is 1.8 times (not quite two times) that of a nonelectrolyte at the same concentration.
4. Weak electrolytes react with water to a limited extent and hence provide only a limited number of ions in solution.
5. Nonelectrolytes exist in molecular form in solution or are insoluble salts; they produce no appreciable concentration of ions.

12.4 Electrolysis: Chemical Change Caused by Electricity

If sodium chloride is heated until it melts, the molten salt conducts electricity, as one can see by testing the melt in the conductivity apparatus shown in Figure 12.2. In addition to causing the light bulb to glow, the electric current, as it passes through the melt, causes observable chemical changes. Yellow-green chlorine gas forms at the anode. At the cathode, silvery metallic sodium is formed and is rapidly vaporized by the hot melt. The molten, ionic sodium chloride is decomposed by electrical energy into elemental sodium and chlorine.

$$2\,NaCl + \text{energy} \longrightarrow 2\,Na + Cl_2$$

In crystalline form, sodium chloride does not conduct electricity. The ions occupy relatively fixed positions in the lattice and do not move very much, even under the influence of an electric potential. When sodium chloride melts, the ions are freed to move around. When a battery is connected to the melt through a pair

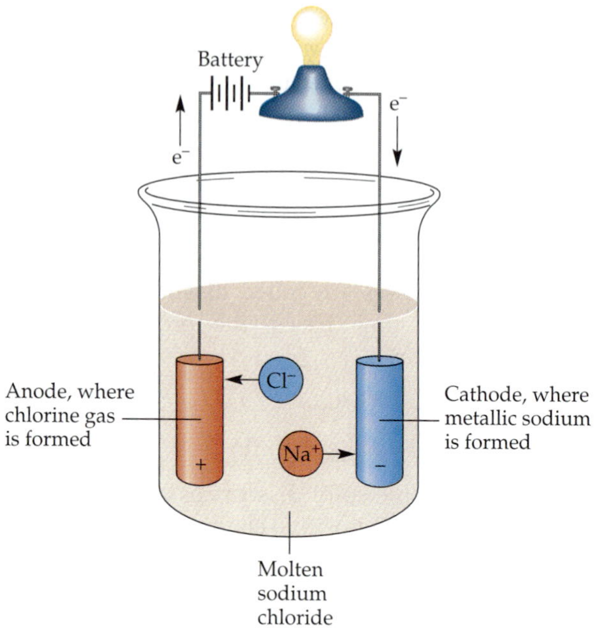

FIGURE 12.3 ▶
The electrolysis of molten sodium chloride.

of electrodes, the sodium ions are attracted to the electron-rich cathode, where they pick up electrons (Figure 12.3).

$$Na^+ + e^- \longrightarrow Na \text{ (at the cathode)}$$

The chloride ions migrate to the electron-poor anode, where they give up electrons.

$$2\,Cl^- \longrightarrow Cl_2 + 2\,e^- \text{ (at the anode)}$$

The battery in the circuit is responsible for seeing to it that this exchange of electrons does not eventually lead to neutralized electrodes. Electrons picked up by the anode from chloride ions are immediately shunted, under the influence of the battery, to the cathode, which has been losing electrons to sodium ions. As long as sodium ions and chloride ions are present in the melt, current will flow. When all of the salt has been converted to elemental sodium and chlorine, the circuit will be broken and current will cease.

This electrolytic reaction is an oxidation–reduction process. Oxidation occurs at the anode, where chloride ions lose electrons. Reduction occurs at the cathode, where sodium ions gain electrons.

Electrolytic processes are used in the production of Al, Li, K, Na, and Mg and in the refining of Cu.

Electrolysis, then, is a process of using electricity to bring about chemical change. The process is useful for the preparation and purification (refining) of many metals. Electrolysis is also used for coating one metal with another, an operation called *electroplating*. Usually the object to be electroplated, such as a spoon, is cast of a cheaper metal. It is then coated with a thin layer of a more attractive and more corrosion-resistant metal, such as gold or silver. The cost of the finished product is far less than that of a corresponding item made entirely of silver or gold. A cell for the electroplating of silver is shown in Figure 12.4. The silver is made the anode, and the spoon is made the cathode. A solution of silver nitrate is used as the electrolyte. Under the influence of the battery in the system (or any

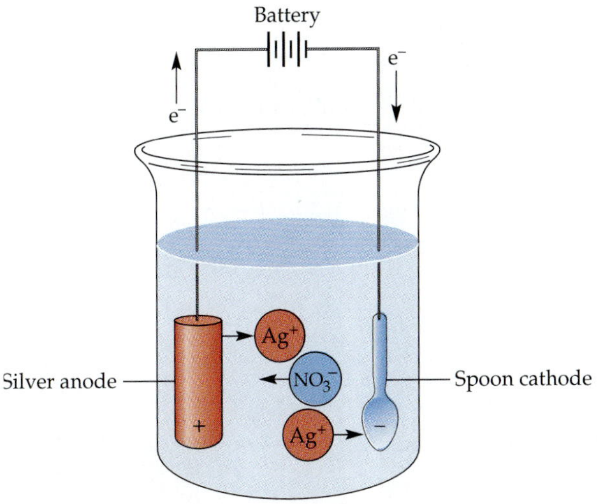

◀ **FIGURE 12.4**
An electrochemical cell for the plating of silver.

voltage source), the silver ions (Ag^+) are attracted to the cathode (spoon), where they pick up electrons and are deposited as silver atoms.

$$Ag^+ + e^- \longrightarrow Ag$$

At the anode, electrons are removed from the silver bar. Some of the silver atoms lose electrons to become silver ions.

$$Ag \longrightarrow Ag^+ + e^-$$

The net process is one in which the silver from the bar is transferred to the spoon. The thickness of the deposit can be controlled by accurate measurement of the amount of current flow and of the duration of the process.

Electrolysis finds some biological applications, including the removal of unwanted hair. In this process, a tiny wire needle is used to supply a mild electric current to the hair root. The chemical changes engendered there kill the living follicle. This is perhaps the only permanent method of hair removal. In the hands of a skilled technician, the method can be clean and safe. It is tedious, however, for each hair root must be treated individually. Similar procedures are sometimes used for the removal of warts or other growths.

12.5 Electrochemical Cells: Batteries

Electricity can cause chemical change. Conversely, chemical change can produce electricity. Dry cells, storage batteries, and fuel cells all convert chemical energy to electrical energy.

When a strip of zinc metal is placed in a solution of copper(II) sulfate, the following reaction takes place.

$$Zn + Cu^{2+} \longrightarrow Zn^{2+} + Cu$$

The zinc atoms give up their outer electrons to the copper(II) ions. (The sulfate ions are not changed in the reaction and can be omitted from the equation.) The zinc metal dissolves, going into solution as Zn^{2+} ions. The copper ions precipitate from the solution as copper metal (Figure 12.5). The Zn is oxidized; the Cu^{2+} ions

FIGURE 12.5 ▶

The photograph on the left shows a blue solution of Cu^{2+}(aq) ions and a sample of zinc metal. When the Zn is added to the Cu^{2+}(aq) solution, the more active Zn displaces the less active copper from solution. The products of the displacement reaction (right) are a reddish-brown precipitate of copper metal and a colorless solution of Zn^{2+}(aq) ions. The equation for the reaction is

$$Cu^{2+}(aq) + Zn(s) \longrightarrow Cu(s) + Zn^{2+}(aq)$$

are reduced. Electrons are transferred directly from zinc atoms to copper(II) ions. To emphasize this transfer of electrons, the reaction can be split into two half-reactions (the oxidation portion and the reduction portion).

$$Zn \longrightarrow Zn^{2+} + 2\,e^- \quad (\textit{oxidation half-reaction})$$
$$Cu^{2+} + 2\,e^- \longrightarrow Cu \quad (\textit{reduction half-reaction})$$

If we place copper ions in one compartment and zinc metal in another, physically separating them (Figure 12.6), electrons have to pass through a wire to get from the zinc metal to the copper ions. This flow of electrons constitutes an **electric current,** and it can be used to run a motor or light a bulb.

In the cell pictured in Figure 12.6, there are two separate compartments. One contains zinc metal in a solution of zinc sulfate, and the other contains copper metal in a solution of copper(II) sulfate (which is blue). Zinc atoms give up electrons much more readily than copper atoms, so electrons flow away from the zinc and toward the copper. The zinc metal slowly dissolves as zinc atoms give up electrons to form zinc ions. The electrons flow through the wire to the copper, where copper ions pick up the electrons to become copper atoms. As time goes by, the zinc bar will slowly disappear, while the copper bar will get bigger, and the blue solution will gradually lose its color. Meanwhile something else also happens. Sulfate ions move from the blue copper sulfate solution to the zinc sulfate solution. Since more and more positively charged zinc ions are being added to the compartment at the left, while fewer and fewer copper ions remain in the compartment at the right, some of the negative sulfate ions must move from the right to the left in order to keep the two solutions electrically neutral. Notice that the porous partition in the cell allows the sulfate ions to move through it. (If the sulfate ions were unable to move through the barrier, the cell would not work.) Each time a zinc atom gives up two electrons, a copper ion picks up two electrons, and one sulfate ion moves from the right compartment to the left.

Why bother with the porous plate? Why not just put both solutions and both electrodes in a single compartment so the sulfate ions don't have to move through pores? If we were to do that, the electrons wouldn't have to make the trip through the wire; they could just be passed from zinc atoms to copper ions, which would now be in contact with one another. And unless the electrons went through the

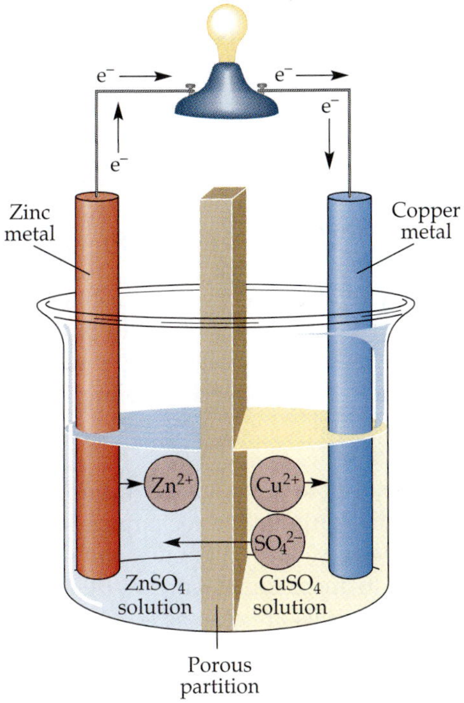

◀ **FIGURE 12.6**
A simple electrochemical cell.

wire, the light bulb wouldn't glow (nor would battery-operated portable radios, hand calculators, or cardiac pacemakers work).

If 1.0 M solutions of $ZnSO_4$ and $CuSO_4$ are used in the cell pictured in Figure 12.6, the system will produce about 1.1 volts at 25 °C. [The *volt* is a measure of electrical potential, or of the tendency of electrons to flow (see Section 12.6)].

The types of batteries used in most flashlights and portable electronic devices are *primary* batteries, so called because the cell reactions are irreversible and the batteries cannot be recharged. During use, reactants are converted to products, and when the reactants are used up, the battery is "dead" and we have to buy a new one. A typical example is the one diagrammed in Figure 12.7, the "dry" cell.

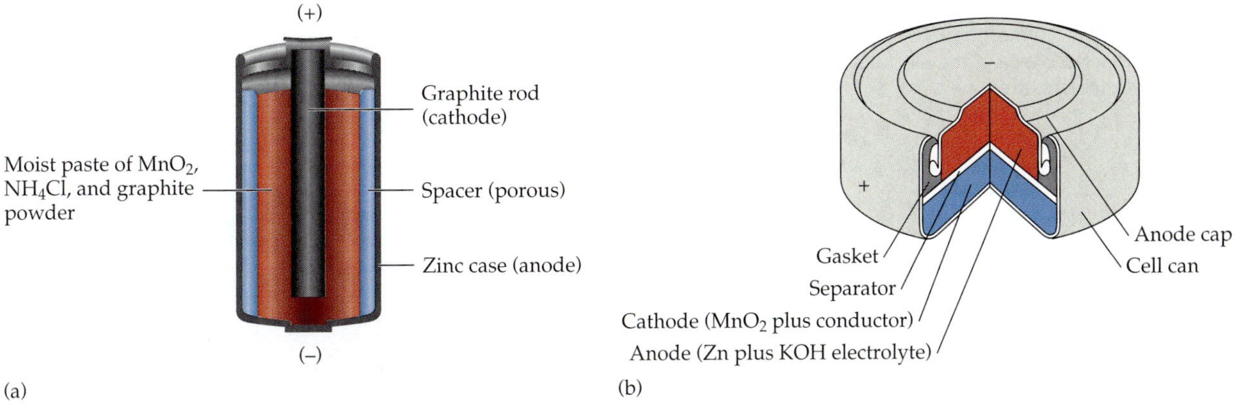

▲ **FIGURE 12.7**
(a) Cross section of a zinc–carbon dry cell. (b) Cutaway view of a miniature alkaline Zn–MnO_2 dry cell.

A zinc container is the anode and an inert carbon (graphite) rod is the cathode. The electrolyte is a moist paste of MnO_2, NH_4Cl, $ZnCl_2$, and carbon black. There is no free-flowing liquid in the cell, and it is for this reason that we call it a "dry" cell. Zinc metal is oxidized to Zn^{2+} at the anode. A simplified form of the overall reaction is

$$Zn(s) + 2\,MnO_2(s) + H_2O \longrightarrow Zn^{2+}(aq) + Mn_2O_3(s) + 2\,OH^-(aq)$$

Alkaline cells have KOH or NaOH as an electrolyte. They contain up to 50% more energy than ordinary dry cells.

In everyday life we call a device that stores chemical energy for later release of electricity a **battery.** A flashlight "battery" consists of a single cell with two electrodes in contact with one or more electrolytes. An automobile battery is a true battery in the sense that it consists of several simple **electrochemical cells** connected to one another.

The lead–acid storage battery used in automobiles is a *secondary* battery. The cell reaction can be reversed and the battery restored to its original condition. The battery can be used through repeated cycles of discharging and recharging. Figure 12.8 shows a portion of a cell in the lead–acid battery. Several anodes and several cathodes are connected together in each cell to increase its current-delivering capacity. Each cell has a voltage of 2.05 V, and six cells are connected together in "series" fashion, + to −, to form a 12-volt battery.

The anodes in the lead–acid storage cell are of a lead alloy, and the cathodes are of a lead alloy impregnated with lead dioxide. The electrolyte is dilute sulfuric acid. The net reaction on discharge is

$$Pb(s) + PbO_2(s) + 4\,H^+(aq) + 2\,SO_4^{2-}(aq) \longrightarrow 2\,PbSO_4(s) + 2\,H_2O$$

As the cell reaction proceeds, $PbSO_4(s)$ precipitates and partially coats both electrodes; the water formed dilutes the $H_2SO_4(aq)$. The cell is *discharged.* By connecting the cell to an external electric energy source (of greater than 12 V), we can force electrons to flow in the opposite direction. The net cell reaction is reversed and the battery is recharged. These lead storage batteries are durable, but they are heavy and contain corrosive sulfuric acid.

FIGURE 12.8 ▶

A lead–acid (storage) cell. The composition of the electrodes, the cell reaction, and the cell voltage are described in the text. Shown here are two anode plates and two cathode plates in "parallel" connections. This type of connection increases the surface area of the electrodes and the capacity of the cell to deliver current.

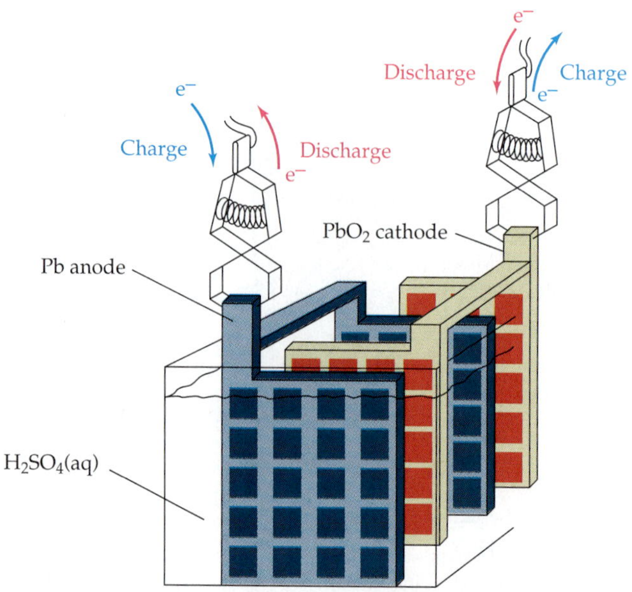

Fuel Cells

Modern civilization runs mainly on fossil fuels, but combustion of fuels and conversion of the evolved heat to electricity is limited to about 30 to 40% efficiency. An electrochemical cell, on the other hand, is able to convert chemical energy to electricity with an efficiency of 90% or more. Because combustion reactions and electrochemical cell reactions both involve oxidation and reduction, why not design an electrochemical cell that has a combustion reaction as its cell reaction? It's already been done. The first such device, now called a **fuel cell,** was constructed over 150 years ago. Its cell reaction is

$$\textit{Fuel} + \textit{oxygen} \longrightarrow \textit{oxidation (combustion) products}$$

In the hydrogen–oxygen cell, hydrogen and oxygen gases flow over separate inert electrodes in contact with an electrolyte such as KOH(aq) at about 200 °C (see Figure 12.9). The reactions are

$$\textit{oxidation:}\quad 2\{H_2(g) + 2\,OH^-(aq) \longrightarrow 2\,H_2O + 2\,e^-\}$$

$$\textit{reduction:}\quad O_2(g) + 2\,H_2O + 4e^- \longrightarrow 4\,OH^-(aq)$$

$$\textit{Net:}\quad 2\,H_2(g) + O_2(g) \longrightarrow 2\,H_2O$$

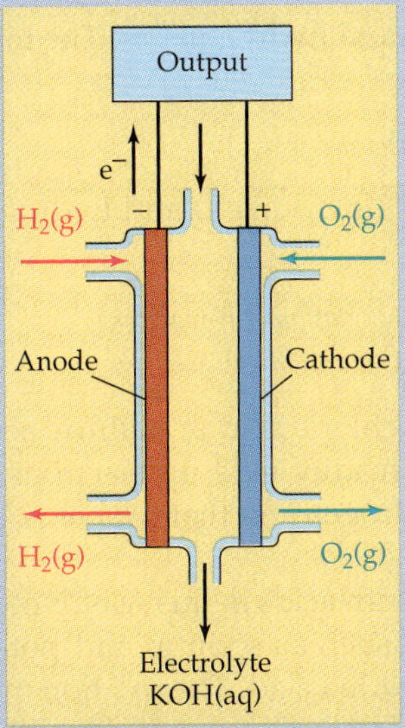

▲ **FIGURE 12.9**
Cross section of a hydrogen-oxygen fuel cell. The electrodes are porous to allow easy access of the gaseous reactants to the electrolyte. Also, the electrode material catalyzes the electrode reactions.

This type of fuel cell has been widely used in space vehicles. In addition to the electricity, the water formed is also a valuable product.

In principle, a fuel cell can run directly on a hydrocarbon fuel, such as natural gas (mostly methane).

$$CH_4(g) + 2\,O_2(g) \longrightarrow CO_2(g) + 2\,H_2O$$

Fuel cells based on the direct conversion of a hydrocarbon fuel to oxidation products are beset with technical problems. Steam re-forming of the fuel to produce $H_2(g)$, for example,

$$CH_4(g) + H_2O(g) \longrightarrow CO(g) + 3\,H_2(g)$$

followed by the use of $H_2(g)$ in a hydrogen–oxygen fuel cell, is more feasible. Even this arrangement has its drawbacks, however, because of the high temperature required in the steam re-forming.

Continuing new developments in fuel cell technology may one day lead to economically viable fuel cells for widespread use in automobiles. Automobiles that use hydrogen–oxygen cells produce no undesirable products (beyond humid air) and are classified as zero-emission vehicles.

Somewhat like a fuel cell is the aluminum–air battery. Aluminum metal serves as a solid fuel that reacts with oxygen from the air.

$$4\,Al(s) + 3\,O_2(g) + 6\,H_2O \longrightarrow 4\,Al(OH)_3(s)$$

Oxygen is reduced at a carbon cathode and aluminum is oxidized at the anode. The electrolyte is NaOH(aq). The battery is recharged by adding aluminum and water. Light in weight compared with a lead–acid battery, the aluminum–air battery may someday find use in electric cars.

Other important commercial batteries include: mercury cells (zinc anode/HgO cathode), which are used in small electronic devices such as calculators and hearing aids; silver oxide cells (Zn anode/Ag_2O cathode), which are used in many watches and cameras; nickel–cadmium or "Ni–Cad" cells (Cd anode/NiO cathode), which are rechargeable and popular for portable radios and cordless appliances; and lithium cells (Li anode/MnO_2 cathode), which are very light in weight and used in devices such as pacemakers. All of these cells have a potassium hydroxide paste between the electrodes.

12.6 The Activity Series

In Section 12.5, we saw that the zinc–copper electrochemical cell produced a voltage of about 1.1 V. This value does not depend on the size of the cell or on the size of the electrodes. The voltage measures the force with which electrons are moved around the circuit, and therefore it measures the tendency of this reaction to occur. Thus, electrochemical cells give a quantitative measure of the relative tendencies of various oxidation–reduction reactions to occur.

If zinc metal is treated with hydrochloric acid, a reaction occurs to produce hydrogen gas and zinc chloride (see Figure 6.5). The net reaction is

$$\mathrm{Zn(s)} \quad + \quad 2\,\mathrm{HCl(aq)} \quad \longrightarrow \quad \mathrm{ZnCl_2(aq)} \quad + \quad \mathrm{H_2(g)}$$

In fact, zinc will react with any acid to yield zinc ions and hydrogen gas. The reaction of zinc with sulfuric acid is

$$\mathrm{Zn(s)} \quad + \quad \mathrm{H_2SO_4(aq)} \quad \longrightarrow \quad \mathrm{ZnSO_4(aq)} \quad + \quad \mathrm{H_2(g)}$$

In these reactions with acids, zinc gives up electrons (is oxidized) and hydrogen ions gain electrons (are reduced).

$$\mathrm{Zn(s)} \quad \longrightarrow \quad \mathrm{Zn^{2+}(aq)} \quad + \quad 2\,\mathrm{e^-} \quad \textit{(oxidation half-reaction)}$$
$$2\,\mathrm{H^+(aq)} \quad + \quad 2\,\mathrm{e^-} \quad \longrightarrow \quad \mathrm{H_2(g)} \quad \textit{(reduction half-reaction)}$$

In contrast, if copper metal is treated with hydrochloric acid or sulfuric acid, no reaction occurs. Why? Because copper does not readily give up electrons to hydrogen ions. We say that zinc is *more active* than hydrogen and that copper is *less active* than hydrogen.

Zinc is one of a number of *active metals* that react with acids in this way. Not all these metals react at the same rate, however. Some, such as sodium and potassium, react violently with plain water—without any added acid. The reaction may produce enough heat to ignite the hydrogen gas, causing an explosion. With sodium, the by-product is a solution of sodium hydroxide.

$$2\,\mathrm{Na(s)} \quad + \quad 2\,\mathrm{H_2O} \quad \longrightarrow \quad 2\,\mathrm{NaOH(aq)} \quad + \quad \mathrm{H_2(g)}$$

A similar reaction occurs with potassium.

$$2\,\mathrm{K(s)} \quad + \quad 2\,\mathrm{H_2O} \quad \longrightarrow \quad 2\,\mathrm{KOH(aq)} \quad + \quad \mathrm{H_2(g)}$$

All the alkali metals (those in Group 1A) react in this manner. Knowing the reaction for sodium, we can write the equation for the reaction of cesium with water.

$$2\,\mathrm{Cs(s)} \quad + \quad 2\,\mathrm{H_2O} \quad \longrightarrow \quad 2\,\mathrm{CsOH(aq)} \quad + \quad \mathrm{H_2(g)}$$

We could have substituted lithium or rubidium for cesium, and we would still have been correct. Similar reactions involving members of the same chemical family are a part of the framework that helps make chemistry more understandable. We discuss the chemistry of some of the families of elements in Special Topic D.

However, a word of caution is in order. Even members of a family differ. They react at different rates. Sometimes even the *kind* of reaction is different. But these exceptions will not concern us very much here.

In contrast to the alkali metals, metals such as zinc and iron do not react appreciably when placed in water. However, if the temperature is raised and zinc is brought into contact with steam, then hydrogen gas is produced. Tin will not react with steam, but it will release hydrogen on contact with acids. Gold won't react even with acids.

It is possible to arrange metals in an **activity series** with the most reactive metals (such as potassium) at the top and the least reactive (such as gold) at the bottom (Table 12.2). The position of a metal in the table reflects its tendency to give up electrons to form ions. All of those listed above hydrogen in the series (the *active metals*) will react with acids to produce hydrogen gas. These metals will give their electrons to the hydrogen ions produced by the acids, thus becoming ions. Metals below hydrogen in the series will not give up electrons to hydrogen ions and hence will not liberate hydrogen gas from acids.

A lot of chemical information is summarized in Table 12.2. Remember that the arrangement of the table indicates how readily these metallic elements give up their electrons. When one of the metals is brought into contact with the ions of another metal, one of two things can happen. Either the metal can transfer electrons to the ions (a reaction occurs) or the ions will not accept the electrons (no reaction occurs). We can use the table to predict what will happen. A metal can transfer electrons to the ions of any metal that appears lower in the table. For

Table 12.2 An Activity Series of Metals

	Reduced Form (metal atom)		Oxidized Form (metal ion)		
	Metal atom	$\xrightarrow{\text{oxidation}}$	Metal ion $+ n\,e^-$		
	Metal atom	$\xleftarrow{\text{reduction}}$	Metal ion $+ n\,e^-$		
	Li	\longrightarrow	Li^+	$+ e^-$	React with
	K	\longrightarrow	K^+	$+ e^-$	cold water,
	Ca	\longrightarrow	Ca^{2+}	$+ 2\,e^-$	steam, or acids,
	Na	\longrightarrow	Na^+	$+ e^-$	releasing hydrogen gas
	Mg	\longrightarrow	Mg^{2+}	$+ 2\,e^-$	React with
	Al	\longrightarrow	Al^{3+}	$+ 3\,e^-$	steam or
	Zn	\longrightarrow	Zn^{2+}	$+ 2\,e^-$	acids,
	Cr	\longrightarrow	Cr^{3+}	$+ 3\,e^-$	releasing
	Fe	\longrightarrow	Fe^{2+}	$+ 2\,e^-$	hydrogen gas
	Cd	\longrightarrow	Cd^{2+}	$+ 2\,e^-$	React with
	Ni	\longrightarrow	Ni^{2+}	$+ 2\,e^-$	acids,
	Sn	\longrightarrow	Sn^{2+}	$+ 2\,e^-$	releasing
	Pb	\longrightarrow	Pb^{2+}	$+ 2\,e^-$	hydrogen gas
	H_2	\longrightarrow	$2\,H^+$	$+ 2\,e^-$	
	Cu	\longrightarrow	Cu^{2+}	$+ 2\,e^-$	Do not
	Ag	\longrightarrow	Ag^+	$+ e^-$	react with
	Hg	\longrightarrow	Hg^{2+}	$+ 2\,e^-$	acids to release
	Au	\longrightarrow	Au^{3+}	$+ 3\,e^-$	hydrogen gas

Relative ease of oxidation

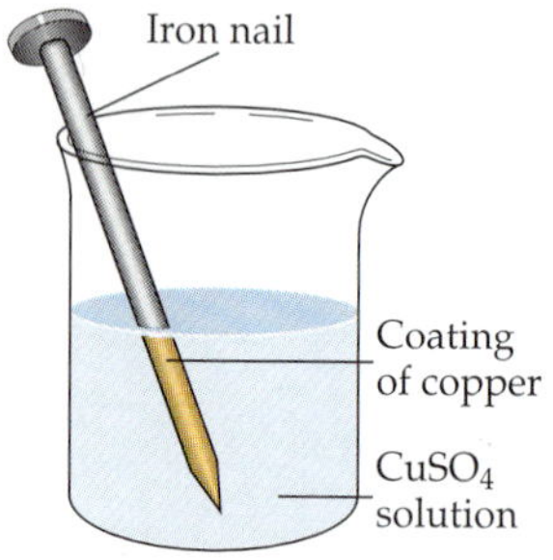

▲ **FIGURE 12.10**

An iron nail placed in a solution of $CuSO_4(aq)$ becomes coated with copper. The equation for the reaction is

$$Fe(s) + Cu^{2+}(aq) \longrightarrow Cu(s) + Fe^{2+}(aq)$$

example, an iron nail placed in a dilute solution of copper(II) sulfate ($CuSO_4$) gradually becomes coated with copper metal (Figure 12.10). Since iron atoms have a greater tendency than copper atoms to give up electrons, the iron atoms transfer electrons to the copper ions.

$$Fe(s) + Cu^{2+}(aq) \longrightarrow Fe^{2+}(aq) + Cu(s)$$

What we observe is the copper metal plating out of solution as it is formed. Some of the iron dissolves as iron(II) ions, but this is not as obvious to the observer.

If we use another system, it is possible to see both processes as they occur. A strip of copper metal placed in a solution of silver nitrate ($AgNO_3$) becomes coated with crystals of metallic silver (Figure 12.11). The silver ions grab electrons from the more reactive copper atoms.

$$2\,Ag^{+}(aq) + Cu(s) \longrightarrow 2\,Ag(s) + Cu^{2+}(aq)$$

There is also visual evidence that the copper is dissolving. A solution of copper ions is blue, whereas a solution of silver ions is colorless. Thus, as the silver metal crystallizes out of solution, the copper ions form and slowly turn the solution blue.

Note that in one of the examples copper loses electrons and in the other it gains them. This is because there are elements both above and below copper in the activity series, and what happens depends only on the *relative tendencies of the elements involved to gain or lose electrons.*

▲ **FIGURE 12.11**

The displacement of $Ag^{+}(aq)$ by Cu(s). (a) A coil of copper wire is immersed in $AgNO_3(aq)$. (b) The shiny deposit that forms on the copper wire is metallic silver. The blue color of the solution indicates the presence of $Cu^{2+}(aq)$

Corrosion

An oxidation–reduction reaction of particular economic importance is the corrosion of metals. Perhaps 20% of all the iron and steel production in the United States each year goes to replace corroded items. Let's look first at the corrosion of iron.

In moist air, iron is oxidized, particularly at a nick or scratch.

$$Fe \longrightarrow Fe^{2+} + 2\,e^-$$

In order for iron to be oxidized, oxygen must be reduced.

$$O_2 + 2\,H_2O + 4\,e^- \longrightarrow 4\,OH^-$$

The net result, initially, is the formation of insoluble iron(II) hydroxide.

$$2\,Fe + O_2 + 2\,H_2O \longrightarrow 2\,Fe(OH)_2$$

This product is usually further oxidized to iron(III) hydroxide.

$$4\,Fe(OH)_2 + O_2 + 2\,H_2O \longrightarrow 4\,Fe(OH)_3$$

Corrosion of iron is an electrochemical reaction of great economic importance. Overall, corrosion costs about $100 billion dollars a year in the United States.

The latter, sometimes written as $Fe_2O_3 \cdot 3H_2O$ is the familiar iron rust.

Oxidation and reduction often occur at separate points on the metal surface. Electrons are transferred through the iron metal. The circuit is completed by an electrolyte in aqueous solution. In the snowbelt, this solution is often the slush from road salt and melting snow. The metal is pitted in an anodic area, where iron is oxidized to Fe^{2+}. These ions migrate to the cathodic area, where they react with the hydroxide ions formed by reduction of oxygen.

$$Fe^{2+}(aq) + 2\,OH^-(aq) \longrightarrow Fe(OH)_2(s)$$

As indicated above, this iron(II) hydroxide is then oxidized to $Fe(OH)_3$, or rust. This process is diagrammed in Figure 12.12. Notice that the anodic area is protected from oxygen by the water film, whereas the cathodic area is exposed to air.

Aluminum is more reactive than iron (see Table 12.2). Therefore, we might expect it to corrode more rapidly, but billions of beer cans testify to the fact that it doesn't. How can this be? It so happens that freshly prepared aluminum quickly forms a thin, hard film of aluminum oxide on its surface. This film is impervious to air and protects the underlying metal from further oxidation.

We should add, however, that corrosion can sometimes be a problem with aluminum. Certain substances, such as salt, can interfere with the protective oxide coating on aluminum, allowing the metal to oxidize. Mag wheels on automobiles have cracked, and planes with aluminum landing gear have had their wheels sheared off because of this problem.

The tarnish on silver is due to the oxidation of the silver surface by hydrogen sulfide (H_2S) in the air. It produces a film of black silver sulfide (Ag_2S) on the metal surface. You can use a silver polish to remove the tarnish, but in doing so, you also lose part of the silver. An alternative method involves the use of aluminum metal to reduce the silver ions back to silver metal.

$$3\,Ag^+(aq) + Al(s) \longrightarrow 3\,Ag(s) + Al^{3+}(aq)$$

This reaction also requires an electrolyte. Sodium bicarbonate ($NaHCO_3$) is usually used. The tarnished silver is placed in contact with aluminum foil and covered with a solution of sodium bicarbonate. A precious metal is conserved at the expense of a cheaper one.

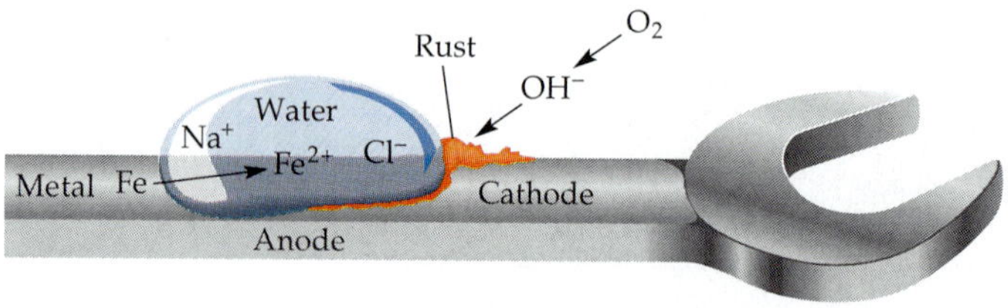

FIGURE 12.12 ▶
The corrosion of iron requires water, $O_2(g)$ from the air, and an electrolyte.

12.7 Precipitation: The Solubility Product Relationship

We have thus far focused our attention on strong electrolytes, those compounds that supply high concentrations of ions in solution. There are also ionic compounds whose solubilities in water are quite low. For example, the solubility of barium sulfate ($BaSO_4$) is 0.0002 g per 100 g of water at 25 °C. This makes barium sulfate a very weak electrolyte in aqueous solution. The fact that a salt is a weak electrolyte or a nonelectrolyte does not necessarily mean that it is unimportant in the chemistry of living systems. On the contrary, for some physiological processes, the relative insolubility of certain salts is of critical importance. To understand these processes, it will be necessary to consider first a new way of describing solubility, or, more correctly, of determining when a salt will start precipitating from solution because its solubility has been exceeded.

When solid barium sulfate is added to water, an equilibrium is established between the undissolved solute and the ions.

$$BaSO_4(s) \rightleftharpoons Ba^{2+}(aq) + SO_4^{2-}(aq)$$

An extremely small amount goes into solution. Even though barium salts are quite toxic, large amounts of barium sulfate can be swallowed or given by enema because very little will dissolve in the solutions of the body. Since barium sulfate is opaque to X-rays, technicians can use it to outline the stomach or intestines for X-ray photographs. The undissolved barium salt scattered throughout the intestines blocks the X-rays. The X-ray film is unexposed in these areas, which therefore appear white in the developed negative. The barium sulfate is later voided from the body unchanged.

Even though we say that barium sulfate is insoluble, the small amount that does dissolve can be measured (as we indicated above). The concentration of barium ions (Ba^{2+}) and sulfate ions (SO_4^{2-}) in a saturated solution of barium sulfate is 0.00001 M (or 1×10^{-5} M) each at 18 °C.

The product of the concentrations of the ions in a saturated solution

$$[Ba^{2+}][SO_4^{2-}]$$

is a constant. At 18 °C, the product is

$$(1 \times 10^{-5})(1 \times 10^{-5}) = 1 \times 10^{-10}$$

This value is called the **solubility product constant** (K_{sp}) for barium sulfate. It is this constant that we will find useful in predicting whether precipitation will occur if a solution containing barium ions is mixed with one containing sulfate ions. Precipitation will occur if the product $[Ba^{2+}][SO_4^{2-}]$ is greater than 1×10^{-10}. Barium sulfate will continue to dissociate until the product is just equal to 1×10^{-10}. At that value, the solution will be saturated. If $[Ba^{2+}][SO_4^{2-}]$ is less than

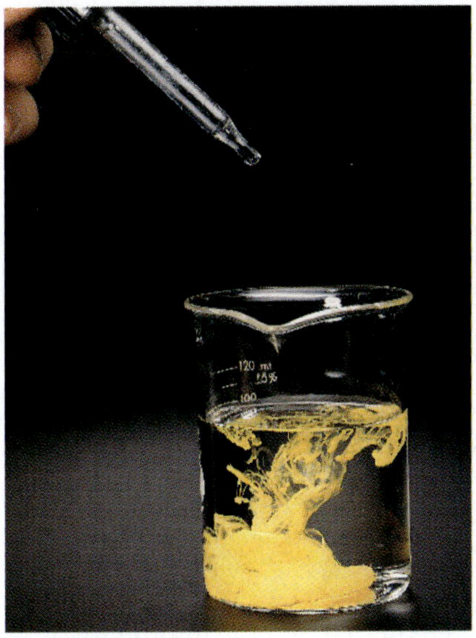

Addition of KI(aq) to $Pb(NO_3)_2(aq)$ yields a yellow precipitate of lead(II) iodide [$PbI_2(s)$].

1×10^{-10}, no barium sulfate precipitate will be formed. The solution will be unsaturated in barium sulfate.

As in the case of the ion product of water, K_w (Section 11.3), it is the *product* of the ion concentrations that is important. If barium sulfate is simply placed in water, the concentration of barium ion equals the concentration of sulfate ion. It is, however, possible to prepare a solution in which the two concentrations are not equal (Example 12.1). The barium ion concentration could be higher than the sulfate ion concentration, or vice versa. Precipitation will always occur when the *product* of the two concentrations exceeds 1×10^{-10}, but precipitation will not necessarily coincide with concentrations of 1×10^{-5} for each ion.

Example 12.1

If 0.001 mol of $BaCl_2$ and 0.0001 mol of Na_2SO_4 are added to 1 L of water, will $BaSO_4$ precipitate? (Assume no volume change.)

Solution

$BaCl_2$ and Na_2SO_4 are soluble salts; each one will completely dissociate.

$$\underset{0.001\text{ mol}}{BaCl_2} \longrightarrow \underset{0.001\text{ mol}}{Ba^{2+}} + \underset{0.002\text{ mol}}{2\,Cl^-}$$

$$\underset{0.0001\text{ mol}}{Na_2SO_4} \longrightarrow \underset{0.0002\text{ mol}}{2\,Na^+} + \underset{0.0001\text{ mol}}{SO_4^{2-}}$$

The 0.001 mol of $BaCl_2$ will yield 0.001 M Ba^{2+} (as well as 0.002 M Cl^-). The 0.0001 mol of Na_2SO_4 will yield 0.0001 M SO_4^{2-} (as well as 0.0002 M Na^+). The product of the barium and sulfate ion concentrations is

$$K_{sp} = [Ba^{2+}][SO_4^{2-}]$$
$$= (1 \times 10^{-3})(1 \times 10^{-4}) = 1 \times 10^{-7}$$

Remember that the smaller negative exponent corresponds to the larger number. For example, $10^{-1} = 0.1$ and $10^{-2} = 0.01$.

Since 1×10^{-7} is greater than 1×10^{-10}, $BaSO_4$ will precipitate until the product of the $[Ba^{2+}]$ and $[SO_4^{2-}]$ is just equal to 1×10^{-10}. The sodium and chloride ions in solution have no tendency to precipitate as sodium chloride salt, because sodium chloride is quite soluble in water.

Practice Exercise

If 2.74×10^{-5} mol of Na_2CrO_4 is added to 225 mL of 1.5×10^{-4} M $AgNO_3(aq)$, will a precipitate form?

$$Ag_2CrO_4(s) \rightleftharpoons 2\,Ag^+(aq) + CrO_4^{2-}(aq)$$

$$K_{sp} = 2.4 \times 10^{-12}$$

The solubility product principle can be used in the preparation of certain compounds. If silver nitrate solution is mixed with sodium chloride solution, slightly soluble silver chloride will precipitate.

$$\underbrace{Ag^+ + NO_3^-}_{\text{Soluble}} + \underbrace{Na^+ + Cl^-}_{\text{Soluble}} \longrightarrow \underbrace{AgCl(s)}_{\text{Insoluble}} + \underbrace{Na^+ + NO_3^-}_{\text{Soluble}}$$

The silver chloride can be removed by filtration—that is, by pouring the mixture through a porous paper that will trap the solid silver chloride but permit passage of the solution. Further, if equimolar quantities of silver nitrate and sodium chloride are used, the filtrate (the solution going through the filter paper) can be evaporated, and the other ions can be obtained as sodium nitrate in crystalline form. The solubility product (K_{sp}) of silver chloride, $[Ag^+][Cl^-]$, is 1.6×10^{-10} at 25 °C. This means that a very small number of silver ions and chloride ions in solution will pass through the filter paper. But the vast majority of these ions will be trapped on the paper as the solid salt.

The solubility products for many salts are known. A few values are given in Table 12.3.

Precipitation is important in the formation of many minerals in nature. Geologists are not the only ones concerned with the formation of mineral deposits, however. Our teeth and bones are largely calcium phosphate salts. One such salt is $Ca_3(PO_4)_2$, sometimes called tricalcium phosphate. Teeth and bones are formed by the precipitation of calcium phosphate salts from solution. In order for this precipitation to occur, the concentrations of ions must exceed the solubility product in the immediate area of deposition. If we assume that the precipitation reaction is

$$3\,Ca^{2+}(aq) + 2\,PO_4^{3-}(aq) \rightleftharpoons Ca_3(PO_4)_2(s)$$

Table 12.3 Selected Solubility Product Constants at 25 °C

Compound	Formula	K_{sp}
Barium carbonate	$BaCO_3$	2.0×10^{-9}
Barium sulfate	$BaSO_4$	1.1×10^{-10}
Calcium carbonate	$CaCO_3$	4.8×10^{-9}
Copper(II) sulfide	CuS	8.7×10^{-36}
Lead chromate	$PbCrO_4$	2.0×10^{-14}
Magnesium carbonate	$MgCO_3$	2.0×10^{-8}
Mercury(II) sulfide	HgS	3.0×10^{-53}
Silver acetate	$AgC_2H_3O_2$	2.0×10^{-3}
Silver bromide	$AgBr$	5.0×10^{-13}
Silver chloride	$AgCl$	1.6×10^{-10}

the solubility product is

$$[Ca^{2+}][Ca^{2+}][Ca^{2+}][PO_4^{3-}][PO_4^{3-}]$$

Each calcium phosphate unit provides three calcium ions (Ca^{2+}) and two phosphate ions (PO_4^{3-}). More simply, we can write the expression as

$$K_{sp} = [Ca^{2+}]^3[PO_4^{3-}]^2$$

At a temperature of 37 °C (normal body temperature), the solubility product constant for calcium phosphate is about 4×10^{-27}. In the blood, the concentration of free calcium ions is about 0.0012 M (1.2×10^{-3} M), and the concentration of phosphate ions is about 1.6×10^{-8} M. Plugging these values into the solubility product relationship, we get

$$(1.2 \times 10^{-3})^3 \times (1.6 \times 10^{-8})^2 = (1.7 \times 10^{-9}) \times (2.6 \times 10^{-16}) = 4.4 \times 10^{-25}$$

(A brief review on the manipulation of exponential numbers is offered in Appendix II.) Since this value is *larger* than the solubility product (4×10^{-27}), we expect precipitation of calcium phosphate and the subsequent growth of bone and teeth.

Our diets contain an ample supply of phosphate ions and therefore calcium ions become the limiting factor in the development of *osteoporosis*, the brittle bone disease. After about the age of 35, people tend to consume less calcium and/or it is absorbed from foods less efficiently. Nutritionists recommend that we eat foods (dairy products) that are rich in calcium, or take calcium supplements.

In growing children, the foregoing description may closely represent the actual mechanism. In adults, however, teeth and bones are no longer growing. What keeps them from getting ever larger? The pH at the area of growth is a little below 7.4 owing to metabolic processes in the cells. The hydrogen ions that are present tie up phosphate ions as hydrogen phosphate ions.

$$PO_4^{3-}(aq) + H^+(aq) \rightleftharpoons HPO_4^{2-}(aq)$$

The phosphate ion concentration is reduced to a value that just satisfies the solubility product relationship at the constant value of 4.0×10^{-27}. Normal bone and tooth maintenance takes place, with neither growth nor diminution.

Two pathological conditions of the mouth and teeth are worthy of mention. If the salivary glands are removed or destroyed, the teeth are no longer bathed by saliva. At the usual pH values, saliva provides just the right concentration of calcium ions and phosphate ions to prevent dissolution of the teeth. With the salivary glands gone, the concentration of the ions rapidly falls below the solubility product constant. When this happens, more calcium phosphate dissolves, and the teeth, if not removed, erode away.

A similar situation occurs when children suffer from chronic acidosis. The blood pH may be as low as 7.1, and the pH in the immediate areas of bone formation is probably even lower. The additional hydrogen ions tie up phosphate ions, lowering the concentration of the latter. Bone growth is greatly hindered, and the child's skeleton is badly formed.

Tooth decay (caries) can also be related to solubility. A combination carbohydrate–protein called *mucin* forms a film, called plaque, on teeth. If it is not removed by brushing and flossing, buildup of plaque continues. Food and bacteria trapped in the plaque metabolize carbohydrates, producing lactic acid (see Chapter 24). Saliva does not penetrate plaque and hence cannot buffer against the buildup of the acid. The pH at the surface of the tooth may go as low as 4.5, and the concentration of phosphate ions in solution is rapidly depleted as hydrogen phosphate ions are formed. The calcium phosphate of the tooth dissolves to replenish the phosphate ion, leaving a cavity in the tooth.

Teeth are also eroded in people suffering from *bulimia,* a condition characterized by binge eating followed by vomiting. Hydrochloric acid vomited from the stomach acts in the mouth to tie up phosphate ions. The pH may drop as low as 1.5, and erosion can be much more rapid than in caries.

Knowledge of solubility product relationships can also be used to bring slightly soluble salts into solution. If a relatively concentrated solution of calcium chloride ($CaCl_2$) is added to a solution with a fair concentration of sodium oxalate ($Na_2C_2O_4$), a precipitate of calcium oxalate (CaC_2O_4) is formed (some kidney stones are primarily calcium oxalate).

$$\underbrace{Ca^{2+} + 2\,Cl^-}_{\text{Soluble}} + \underbrace{2\,Na^+ + C_2O_4^{2-}}_{\text{Soluble}} \longrightarrow \underbrace{CaC_2O_4(s)}_{\text{Insoluble}} + \underbrace{2\,Na^+ + 2\,Cl^-}_{\text{Soluble}}$$

If hydrochloric acid is now added to the precipitate, the precipitate disappears. The solid dissolves because the oxalate ions ($C_2O_4^{2-}$) are tied up by the hydronium ions (from the HCl) to form soluble, slightly ionized oxalic acid.

$$C_2O_4^{2-}(aq) + 2\,H_3O^+(aq) \rightleftharpoons H_2C_2O_4(aq) + 2\,H_2O$$

This decreases the concentration of oxalate ions so that the solubility product constant for calcium oxalate is no longer exceeded. The precipitate dissolves.

Unfortunately, strong hydrochloric acid solutions cannot be used to dissolve kidney stones while they are still in the kidneys. The acid would be much too corrosive to cells. There are chemicals, however, that have shown modest success in dissolving kidney stones. If the stones are calcium oxalate, removing foods containing oxalates (e.g., chocolate, spinach, black tea) from the diet might be of some help.

12.8 The Salts of Life: Minerals

Various inorganic compounds are necessary for the proper growth and repair of our tissues. It is estimated that such minerals represent about 4% of human body weight. Some of these, such as the chlorides (Cl^-), phosphates (PO_4^{3-}), bicarbonates (HCO_3^-), and sulfates (SO_4^{2-}), occur in the blood and other body fluids. Others, such as iron (as Fe^{2+}) in hemoglobin and phosphorus in the nucleic acids (DNA and RNA), are constituents of complex organic compounds.

Minerals essential to one or more living organisms include the elements sodium (Na), magnesium (Mg), potassium (K), phosphorus (P), sulfur (S), chlorine (Cl), calcium (Ca), manganese (Mn), iron (Fe), copper (Cu), cobalt (Co), zinc (Zn), iodine (I), fluorine (F), boron (B), silicon (Si), cadmium (Cd), lithium (Li), vanadium (V), chromium (Cr), bromine (Br), selenium (Se), molybdenum (Mo), nickel (Ni), and arsenic (As). These, along with the structural elements carbon, hydrogen, nitrogen, and oxygen, make up the 29 chemical elements of life. Other elements are sometimes found in body fluids and tissues but are not known to be essential. These include barium (Ba), strontium (Sr), and aluminum (Al). It might yet be discovered that one or more of these is essential.

These minerals serve a variety of functions. Perhaps the most dramatic is that of iodine. A small amount of iodine is necessary for the proper functioning of the thyroid gland. A deficiency of iodine has serious effects, of which *goiter* is perhaps the best known. Iodine is available in seafood. To guard against iodine deficiency, a small amount (0.02% by mass) of potassium iodide (KI) is added to table salt (NaCl). Iodized salt has greatly reduced the incidence of goiter.

Iron(II) ions (Fe^{2+}) are necessary for the proper functioning of the oxygen-transporting compound hemoglobin. Without sufficient iron, there will be a shortage of oxygen supplied to the body tissues. The resulting weakened condition is called *anemia*. Foods especially rich in iron compounds include red meats and liver.

As we have seen, calcium and phosphorus are necessary for the proper development of bones and teeth. Growing children need about 1.5 g of each per day. These elements are available in plentiful quantities in milk. The need of adults for these elements is less widely known but very real. For example, calcium ions are necessary for the coagulation of blood (to stop bleeding) and for maintaining the rhythm of the heartbeat. Phosphorus is necessary for the body to metabolize carbohydrates (Chapter 24). Without phosphorus compounds, we couldn't get *any* energy from those "quick energy" foods. Compounds containing phosphorus play many other essential roles as well. We will encounter a number of these compounds in subsequent chapters.

Sodium chloride in moderate amounts is essential to life. It is important in the exchange of fluids between cells and plasma, for example. The presence of salt increases water retention. A high volume of retained fluids can cause swelling (edema) and high blood pressure (*hypertension*). Over 120 million prescriptions are written each year in the United States for diuretics, drugs that induce urination in an attempt to reduce the volume of retained fluids. Another 30 million prescriptions are written for potassium (K^+) supplements to replace the potassium that is washed out in the excess urine. An estimated 36 million people in the United States suffer from hypertension, and most physicians agree that our diets generally contain too much salt. The American Heart Association recommends that adults limit their salt intake to no more than 3 g (3000 mg) per day. This recommendation is at least a 50% reduction from our present levels of salt intake.

Iron, copper, zinc, cobalt, manganese, molybdenum, calcium, and magnesium are essential to the proper functioning of certain enzymes. These metalloenzymes are necessary to life. The functions of some other minerals are quite complex. Some things are known about how they operate, but a great deal remains to be learned about the role of inorganic chemicals in our bodies. Bioinorganic chemistry is a flourishing area of research.

Summary

Electrolysis is the process in which an electric current passed through either a solution containing ions or a molten ionic substance causes changes in those ions.

An electrolytic cell consists of a source of electricity (usually a battery) and two solid **electrodes:** a positively charged **anode** connected to the positive terminal of the source and a negatively charged **cathode** connected to the negative terminal.

Electrolysis involves oxidation/reduction. The oxidation takes place at the anode as the substance being oxidized gives up electrons to the anode, and the reduction takes place at the cathode as the substance being reduced picks up electrons from the cathode.

Substances that conduct electricity when either dissolved in water or melted are **electrolytes.** Electrolytes can be weak (poor conductors in solution) or strong (good conductors in solution). In general, the more ions a substance produces when mixed with water, the stronger an electrolyte it is. **Nonelectrolytes** are substances that do not conduct electricity when dissolved in water or melted.

Dry cells, storage batteries, and fuel cells are devices that convert chemical energy to electrical energy. The devices resemble an electrolytic cell except that now there is no external source of electricity and the species being reduced and oxidized are separated by a physical barrier. Each electrode is a metal sitting in a solution of some salt of that metal. Because of the barrier, the electrons produced in the oxidation are forced to travel through a metal wire to get to the species being reduced. This movement of electric charge through the wire is an electric current that can be used to do work.

Whether or not a metal is oxidized in an electrolysis depends on the relative tendency of the two metals involved to give up electrons. The ranking of this tendency is called the **activity series** for metals. Any member of the series gives up its electrons to ions of any metal listed below it. All metals above hydrogen in the list are called *active metals*.

Any ionic solid that is only sparingly soluble in water is characterized by a number called its **solubility product constant** (K_{sp}), defined as the product of the concentrations of all the ions the solid produces in a saturated solution. As long as the product of the ion concentrations in a solution of a sparingly soluble solid stays below K_{sp}, more of the ions can be added from an outside source and the solid will not precipitate out of solution. As soon as the addition of ions causes the ion product to exceed K_{sp}, the solid precipitates.

Key Terms

activity series (12.6)
anion (12.1)
anode (12.1)
battery (12.5)
cathode (12.1)
cation (12.1)
dissociation (12.3)
electric current (12.5)
electrochemical cell (12.5)
electrode (12.1)
electrolysis (12.1)
electrolyte (12.2)
fuel cell (12.5)
ionization (12.3)
nonelectrolyte (12.2)
solubility product constant (12.7)
strong electrolyte (12.2)
weak electrolyte (12.2)

Review Questions

1. Define or illustrate the following terms.
   a. cathode **b.** anode
   c. cation **d.** anion
   e. electrolysis **f.** strong electrolyte
   g. weak electrolyte **h.** nonelectrolyte
   i. ionization **j.** ion dissociation
   k. activity series **l.** solubility product constant
2. Why does a solution of 1 mol of NaCl in 1 kg of water freeze at a lower temperature than a solution of 1 mol of sugar in 1 kg of water? Why is the freezing-point depression for the salt solution not quite twice that for the sugar solution?
3. Hydrogen chloride is a covalent molecule. Why does a solution of the gas in water conduct electricity?
4. Write an equation that shows why hydrogen iodide (HI), a gas, forms an aqueous solution that conducts electricity.
5. Lead chromate ($PbCrO_4$) is an ionic compound, yet a saturated aqueous solution of lead chromate does not conduct electricity. Explain.
6. State the main ideas of the modernized version of Arrhenius's theory of electrolytes.
7. Why are water, oxygen, and an electrolyte all required for the corrosion of iron?
8. Why does aluminum corrode more slowly than iron, even though aluminum is more reactive than iron?
9. How does silver tarnish? How can the tarnish be removed without the loss of silver?
10. How does an electrolytic cell differ from an electrochemical cell?
11. Solid sodium chloride does not conduct electricity, but the molten salt does. Explain.
12. Formaldehyde (CH_2O) is a nonelectrolyte. What type of bonding exists in formaldehyde?
13. Is an object to be electroplated made the anode or the cathode in an electrolytic cell? Why?
14. What is goiter? How is it prevented?
15. What is anemia? How is it prevented?
16. Describe the roles of calcium and phosphorus in the proper development of bones and teeth.
17. What is hypertension?
18. What is a diuretic?
19. What causes tooth decay?
20. What is bulimia? How is it related to erosion of the teeth?
21. What happens to bone formation in children who suffer from chronic acidosis? Why?
22. What happens to the teeth when the salivary glands are removed or destroyed? Why?

Problems

Strong Electrolytes, Weak Electrolytes, and Nonelectrolytes

23. Classify the following as strong electrolytes, weak electrolytes, or nonelectrolytes in solution.
    a. KCl **b.** HNO_3 **c.** H_2CO_3
    d. Na_2SO_4 **e.** KOH
24. Classify the following as strong electrolytes, weak electrolytes, or nonelectrolytes in solution.
    a. CH_3OH **b.** $Ca(NO_3)_2$ **c.** CCl_4
    d. $CaCl_2$ **e.** NH_3

Activity Series

25. Complete the following by writing formulas for the expected products, and then balance the equations.
 a. $Ca(s) + HCl(aq) \longrightarrow$
 b. $Ni(s) + HCl(aq) \longrightarrow$
 c. $Mg(s) + HNO_3(aq) \longrightarrow$
26. Complete the following by writing formulas for the expected products, and then balance the equations.
 a. $Zn(s) + H_2SO_4(aq) \longrightarrow$
 b. $Al(s) + H_2SO_4(aq) \longrightarrow$
 c. $Pb(s) + HNO_3(aq) \longrightarrow$
27. Complete the following by writing formulas for the expected products, and then balance the equations.
 a. $Na(s) + H_2O \longrightarrow$ b. $Ba(s) + H_2O \longrightarrow$
28. Complete the following by writing formulas for the expected products, and then balance the equations.
 a. $Ca(s) + H_2O \longrightarrow$ b. $K(s) + H_2O \longrightarrow$
29. Complete the following equations for those reactions that can occur. You may refer to Table 12.2.
 a. $Mg(s) + Cu^{2+}(aq) \longrightarrow$
 b. $Ag(s) + Pb^{2+}(aq) \longrightarrow$
 c. $Fe(s) + Zn^{2+}(aq) \longrightarrow$
 d. $Al(s) + Ni^{2+}(aq) \longrightarrow$
30. Complete the following equations for those reactions that can occur. You may refer to Table 12.2
 a. $Cr(s) + Na^+(aq) \longrightarrow$
 b. $Au(s) + Ag^+(aq) \longrightarrow$
 c. $Sn(s) + K^+(aq) \longrightarrow$
 d. $Ca(s) + Al^{3+}(aq) \longrightarrow$

Solubility Product and Precipitation Criteria

Note: You may refer to Table 12.3 for the K_{sp} values.

31. Will a precipitate form if 0.001 mol of $MgCl_2$ and 0.001 mol of Na_2CO_3 are added to 1.0 L of water? The net ionic reaction is

$$Mg^{2+}(aq) + CO_3^{2-}(aq) \rightleftharpoons MgCO_3(s)$$

32. Will a precipitate form if 0.001 mol of $AgNO_3$ and 0.0001 mol of NaCl are added to 1.0 L of water? The net ionic reaction is

$$Ag^+(aq) + Cl^-(aq) \rightleftharpoons AgCl(s)$$

33. Will a precipitate occur if 1×10^{-6} mol of $Pb(NO_3)_2$ and 1×10^{-5} mol of Na_2CrO_4 are added to 1.0 L of water? The net ionic reaction is

$$Pb^{2+}(aq) + CrO_4^{2-}(aq) \rightleftharpoons PbCrO_4(s)$$

34. If a concentrated solution of sodium acetate is mixed with a solution of silver nitrate, a precipitate of silver acetate is formed. The precipitate dissolves readily when nitric acid is added. Write equations to explain what happens.

Additional Problems

35. What new substances are formed when electricity is passed through the following?
 a. molten KBr b. molten LiCl
 c. molten Al_2O_3
36. Boiler scale ($CaCO_3$) is insoluble in water, yet it readily dissolves in hydrochloric acid. Write the equation that explains what happens.
37. Classify each of the following electrolytes as an acid, a base, or a salt. Write equations to show what happens when each is dissolved in water.
 a. $Ca(OH)_2$ b. HBr c. $Sr(NO_3)_2$
38. Both $Ba(OH)_2$ and H_2SO_4 are strong electrolytes. When equimolar solutions of the two are mixed, the resulting solution is essentially nonconducting. Write the equation for the reaction and explain the observation.
39. Nineteen centuries ago, the Romans added calcium sulfate to wine. It clarifies the wine but also removes any dissolved lead. If 1.0 L of wine is saturated with calcium sulfate, the concentration of SO_4^{2-} is 0.014 M. What concentration of Pb^{2+} would remain in solution? The K_{sp} for lead sulfate is 1.1×10^{-8}, and the net ionic equation is

$$Pb^{2+}(aq) + SO_4^{2-}(aq) \rightleftharpoons PbSO_4(s)$$

40. Phosphates can be removed from water by precipitation using iron(III) sulfate.

$$2\ PO_4^{3-}(aq) + Fe_2(SO_4)_3(aq) \longrightarrow 2\ FePO_4(s) + 3\ SO_4^{2-}(aq)$$

An estimated 7.0 metric tons per day of phosphates enters the East Anglian (United Kingdom) water system. How much iron(III) sulfate would be required to remove this phosphate?
41. Hard water is about 2×10^{-4} M in Ca^{2+}. Water is fluoridated with 1 g of F^- in 10^3 L of water. Will CaF_2 ($K_{sp} = 2 \times 10^{-10}$) precipitate upon fluoridation of hard water?

Special Topic D
Inorganic Chemistry

In this special topic, we take a look at some inorganic chemistry that is important to living systems. It almost seems a contradiction; *inorganic* means nonliving and not derived from life, and *organic* means living, having the characteristics of life, or derived from life. At least, these are the everyday definitions of the terms. So how could inorganic chemicals be important to living organisms?

In the old days, chemists used the terms *organic* and *inorganic* in much the same way as everyone else. They believed that, while they could make many different inorganic chemicals, they could not hope to make organic compounds in the laboratory. They believed, with everyone else, that only living organisms, within their cells, could make organic compounds. Some mysterious *vital force,* they thought, was necessary for the synthesis of organic substances.

A series of experiments in the early 1800s led to the discarding of the vital force theory. Perhaps the most important single step was made in 1828 by Friedrich Wöhler while he was a medical student at the University of Heidelberg. He attempted to prepare ammonium cyanate by heating a mixture of two inorganic substances, lead cyanate and ammonium hydroxide. To his surprise, instead of ammonium cyanate, he obtained crystals of the well-known organic compound urea.

$$Pb(OCN)_2 \ + \ NH_4OH \ \longrightarrow$$

Lead Ammonium
cyanate hydroxide

$$[NH_4OCN] \xrightarrow[\text{heating}]{\substack{\text{rearranges} \\ \text{upon}}} H_2N-\overset{\displaystyle O}{\overset{\displaystyle \|}{C}}-NH_2$$

Ammonium Urea
cyanate

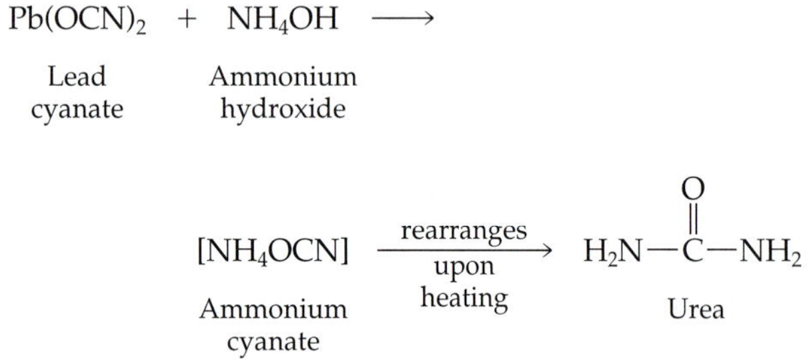

Urea had been isolated from urine in 1780. (Urea is synthesized in the liver, transported to the kidneys, and excreted in the urine.) Wöhler correctly concluded that ammonium cyanate is formed but then rearranges under the influence of heat to yield urea. Urea contains the same number of each kind of atom as ammonium cyanate, but these atoms are arranged differently.

Although a few die-hard vitalists held out for several decades, the vital force theory was practically dead by the middle of the nineteenth century. **Organic chemistry** is now defined as the chemistry of the compounds of carbon. **Inorganic chemistry** is the chemistry of all the other elements. Several later chapters are devoted to organic chemistry. In this section we examine some of the properties of the various groups of elements and relate those properties to electron structure. We start with the noble gases. Then we contrast the reactivity of the elements in other groups with the inactivity of the noble gases.

The more metallic the element is, the easier it is to remove the outermost electrons. Metallic character *decreases* from left to right in the periodic table. It *increases* from top to bottom with increasing atomic size. The alkali metals are all quite metallic, and the halogens are all distinctly nonmetallic. In the middle of the table we see intermediate behavior. Group 4A has carbon, a nonmetal, at the top, and two metals, tin and lead, at the bottom. In between are the *metalloids*—metal-like elements—silicon and germanium. The following brief description of the groups of elements will emphasize periodic trends.

D.1 Group 8A: The Noble Gases

In the last decade of the nineteenth century, a series of elements were discovered that made up an entirely new family, one completely unexpected by Mendeleev (Section 2.6) and his contemporaries. Nonetheless, this new group, called the noble gases, fit neatly between the highly reactive alkali metals (Group 1A) and the active nonmetals of Group 7A. In the usual form of the periodic table, the noble gases are placed to the far right in the column immediately following Group 7A.

The six noble gases are helium, neon, argon, krypton, xenon, and radon. All are components of the atmosphere. Argon is rather abundant, making up nearly 1% by volume of the atmosphere. Xenon, on the other hand, is quite rare, making up only 91

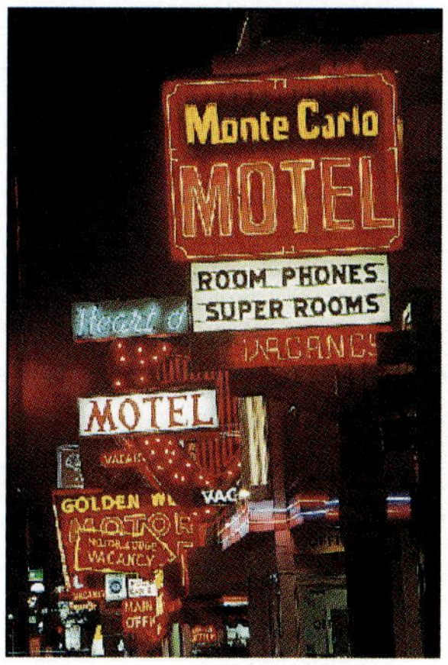

The orange-red color of neon signs results from changes in electronic energy levels in neon atoms.

parts per billion (by volume) of the atmosphere. Radioactive radon seeps from the ground. It makes up only a tiny portion of the atmosphere. It may be a problem, however, when it concentrates in a building with little ventilation.

The noble gases are exceptionally resistant to chemical reaction. This lack of reactivity is a reflection of their electron structure. Helium has the configuration $1s^2$, a filled first energy level. Neon has the structure $1s^2 2s^2 2p^6$, a filled second energy level. The others all have the valence configuration ns^2np^6, an especially stable configuration. Other elements undergo reactions in which they achieve the more stable electron configurations of the noble gases. All the noble gases exist in elemental form as monatomic species.

Because of their lack of reactivity, the noble gases were once called "inert gases." But in 1962 it was discovered that a few compounds of krypton and xenon could be formed, necessitating the change from "inert" to "noble." As yet, despite many attempts, no compounds have been made of the lighter elements helium, neon, and argon. Although this family of elements is no longer called inert, its nobility is still unquestioned. More than any other family, the noble gases disdain interactions with the masses of other elements.

Helium is found in natural gas deposits, where it was formed from elements within the Earth. Helium is used to fill balloons and blimps. Its lifting power is more than 90% that of hydrogen, the lightest of all the gases, and helium has the advantage of being nonflammable. Helium is also used to provide an inert atmosphere for the welding of metals that otherwise might be attacked by oxygen in the air. Liquid helium is used to achieve extremely low temperatures; it boils at only 4.2 K.

Neon is used in lighted signs for advertising. A tube with electrodes is shaped into letters or symbols and is filled with neon at low pressure. An electric current is passed through the tube, causing the atoms to emit their characteristic orange glow.

Argon is the most plentiful of the noble gases. It is used to fill incandescent light bulbs. Unlike nitrogen and oxygen, it does not react with the tungsten filament. It also decreases the tendency of the filament to vaporize, thus extending the filament's life. Fluorescent lights are filled with a mixture of argon and mercury vapor.

Krypton and xenon are too expensive to have many important commercial applications, although krypton has found some use in light bulbs. Radon, although exceedingly rare in the atmosphere, can be collected from the radioactive decay of radium. Sealed in small vials, radon can be used for radiation therapy of certain malignancies. Radon accumulates in well-insulated buildings; it is thought to be a minor cause of lung cancer. (Cigarette smoking is the main cause.)

D.2 Group 1A: The Alkali Metals

The Group 1A elements, the alkali metals, all have the outer electron configuration ns^1. There are six metals in the group: lithium, sodium, potassium, rubidium, cesium, and francium.[1]

In the elemental form, the alkali metals are soft solids with low melting points. Indeed, on a hot day cesium would be a liquid, for it melts at 29 °C. These metals, when freshly cut, are bright and shiny, but they tarnish readily as they become oxidized by the atmosphere. They are the most reactive of the metals: for example, they all react vigorously with water to evolve hydrogen gas. The atoms all have a great tendency to give up one electron each and form +1 ions. Using symbols, we write the process for sodium as

$$Na(s) \longrightarrow Na^+(aq) + e^-$$

Sodium, a soft metal, can be cut with a knife. The freshly cut surface is silvery, but this active metal is soon covered with a thick oxide coating.

[1] Although it is a member of Group 1A, hydrogen is NOT an alkali metal; it isn't even a metal. It exists as H_2 molecules and is a gas at room temperature.

All alkali metals form oxides of the general formula M_2O. Nearly all compounds of the alkali metals are soluble in water.

Lithium salts are found in certain naturally occurring brines. Lithium carbonate (Li_2CO_3) is used in medicine to level out the dangerous "manic" highs that occur in manic-depressive psychoses. Some practitioners also recommend lithium carbonate for the depression stage of the cycle. It appears to act by affecting the transport of chemical substances across cell membranes in the brain.

Sodium salts are quite common. Ordinary table salt (NaCl) supplies the body with chloride ions, necessary for the production of hydrochloric acid by our stomachs. Living tissues also require a balance of sodium ions and potassium ions.

Potassium ion is an essential nutrient for plants. It is generally abundant and is readily available to plants except in soil depleted by high-yield agriculture. The usual form of potassium in commercial fertilizers is potassium chloride (KCl). In animals, potassium ions are the principal positive ions inside cells, and sodium ions are the principal positive ions in the extracellular fluid (see Chapter 28).

D.3 Group 2A: The Alkaline Earth Elements

The six alkaline earth metals are beryllium, magnesium, calcium, strontium, barium, and radium. All have the outer electron configuration ns^2. In the elemental form these metals are fairly soft and reactive. The atoms show a tendency to give up two electrons each and form +2 ions. For example,

$$Mg(s) \longrightarrow Mg^{2+}(aq) + 2e^-$$

All alkaline earth metals form oxides of the general types MO. (Mendeleev based his periodic table to a large degree on the fact that elements within a group formed oxides and hydrides with the same general formula.)

Beryllium is something of an oddball member of the family. Unlike the others, it does not react with water. The metal itself is rather hard, rigid, and strong. Its lightness makes it quite valuable in structural alloys. Beryllium is poisonous in all its forms.

Magnesium ions are essential to both plants and animals. In plants, magnesium ions are incorporated in chlorophyll molecules; Mg^{2+} is therefore essential to photosynthesis. Both calcium and magnesium are essential for proper functioning of the nerves that control muscles.

Calcium ions are necessary for the proper development of bones and teeth. For this reason growing children are usually encouraged to drink milk, a rich source of calcium. Adults also require calcium because it is necessary for clotting of blood and maintenance of a regular heartbeat.

Calcium carbonate (limestone) and other rocks containing calcium ions (Ca^{2+}), magnesium ions (Mg^{2+}), or iron ions (Fe^{2+} or Fe^{3+}) are widely distributed in nature. Water containing calcium, magnesium, or iron ions is known as *hard water.* The ions react with soaps to form curdy precipitates sometimes called bathtub ring (see Section 20.5).

D.4 Group 3A: Boron and Aluminum

Group 3A consists of the metalloid boron and the metals aluminum, gallium, indium, and thallium. All have the electron configuration ns^2np^1. Boric acid (H_3BO_3) is a familiar ingredient of mild antiseptic eye rinses. The other elements in Group 3A are typical metals and tend to form +3 ions.

$$Al(s) \longrightarrow Al^{3+}(aq) + 3\,e^-$$

Aluminum is the most abundant metal in the Earth's crust, but it is tightly bound in compounds in nature. Much energy, mainly electricity, is required to extract aluminum from its principal ore, bauxite [aluminum oxide (Al_2O_3)].

Aluminum is light and strong. A piece of aluminum weighs only one-third as much as a piece of steel the same size. Although it is considerably more active than iron, aluminum corrodes much more slowly. Freshly prepared aluminum metal reacts with oxygen to form a hard, transparent film of aluminum oxide (Al_2O_3) over its surface. The thin film protects the metal from further oxidation. Iron, on the other hand, forms an oxide coating that is porous and flaky. Instead of protecting the metal, this coating encourages further oxidation. Iron rusts; aluminum does not.

D.5 Group 4A: Some Compounds of Carbon

Group 4A is made up of the nonmetal carbon, the metalloids silicon and germanium, and the metals tin and lead. All have the electron configuration ns^2np^2. Of these, carbon is easily the most important. Carbon forms thousands—perhaps millions—of compounds with hydrogen. These compounds, called hydrocarbons, are discussed in detail in

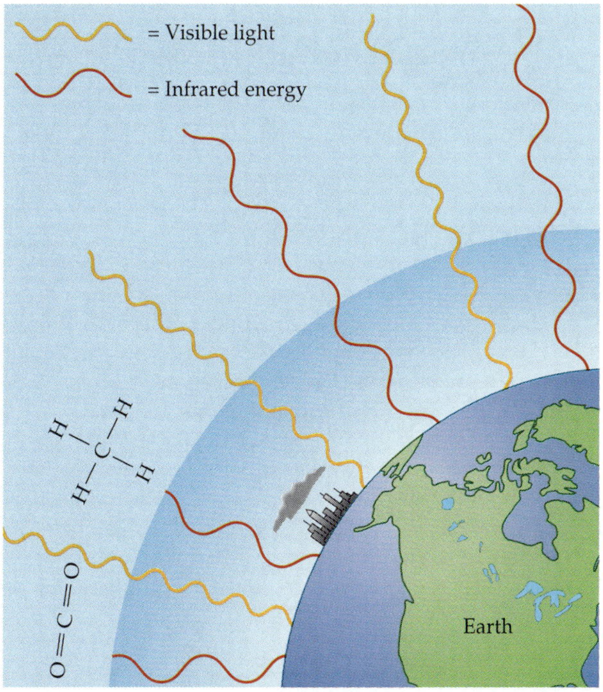

The greenhouse effect. Sunlight passing through the atmosphere is absorbed, warming Earth's surface. The warm surface emits infrared radiation. Some of this radiation is absorbed by CO_2, H_2O, and other gases and is retained in the atmosphere as thermal energy.

Chapter 13. Hydrocarbons and their derivatives are called organic compounds, and their study is called organic chemistry. There are a few simple compounds of carbon, however, that are more like inorganic than organic compounds. Among these are carbon monoxide (CO), carbon dioxide (CO_2), and such minerals as limestone and marble (calcium carbonate, $CaCO_3$).

Carbon exists in two main allotropic forms. **Allotropes** are modifications of an element that can exist in more than one form in the same physical state. The solid element can be found in the form of graphite (the "lead" of pencils) and diamond (the precious jewel). Coal is composed of varying amounts of elemental carbon, from about 6% in peat up to 88% or more in anthracite. When burned, the carbon in coal combines with oxygen to form carbon dioxide.

$$C + O_2 \longrightarrow CO_2$$

Hydrocarbons also are burned as fuels. Methane (CH_4), the simplest hydrocarbon, burns with a hot flame. If sufficient oxygen is present, the main products are carbon dioxide and water.

$$CH_4 + 2\,O_2 \longrightarrow CO_2 + 2\,H_2O$$

The Greenhouse Effect

Not only is carbon dioxide a product of combustion, it is also produced in respiration. Generally, it is regarded as innocuous. Certainly any immediate effect on us is slight. But what about long-term effects?

No matter how cleanly engines and factories operate, they produce $CO_2(g)$ when they burn fossil fuels. The concentration of CO_2 in the atmosphere has increased 18% in this century, an increase largely attributed to the burning of fossil fuels. The concentration continues to increase, and at an accelerating rate, because of the increased burning of carbon fuels. The burning of forests adds CO_2 to the atmosphere, and the clearing of forests eliminates trees and other plants that would otherwise remove CO_2 by photosynthesis. The quantity of carbon dioxide in the atmosphere could double by the year 2030 if we continue our present practices.

The environmental impact of CO_2 and certain other gases is often called the **greenhouse effect.** These gases are transparent to visible light; they let the sun's rays through to warm the surface of Earth. When Earth tries to reradiate energy as infrared radiation into outer space, some of the energy is trapped by molecules of CO_2 and other greenhouse gases.

Human activities add 25 billion metric tons of carbon dioxide to the atmosphere each year, with 22 billion metric tons coming from the burning of fossil fuels. About 15 billion metric tons are removed by plants, the soil, and the oceans, leaving a net addition of 10 billion metric tons per year. The concentration of carbon dioxide is therefore increasing at a rate of 1 ppm per year.

Methane, chlorofluorocarbons, and other trace gases also contribute to global warming. Methane is 20 to 30 times and chlorofluorocarbons 20,000 times as effective as carbon dioxide at holding heat in Earth's atmosphere. Water vapor also acts as a greenhouse gas. When released into the atmosphere, however, water soon falls back to Earth as rain. It therefore affects the climate mostly at the local level.

Many scientists predict a warming trend, but they often differ in their estimates of its magnitude and effect. For example, in 1995 two apparently conflicting reports were released. A study conducted by the U.S. Environmental Protection Agency estimated that the planet would warm just 1 degree Celsius by the year 2050 and only 2 degrees by 2100. This was about half the previous estimates for warming in the next century.

That was the good news. The bad news came in a report from a United Nations panel of scientists projecting growing deserts, dying forests, and flooded coastal areas as a result of global warming.

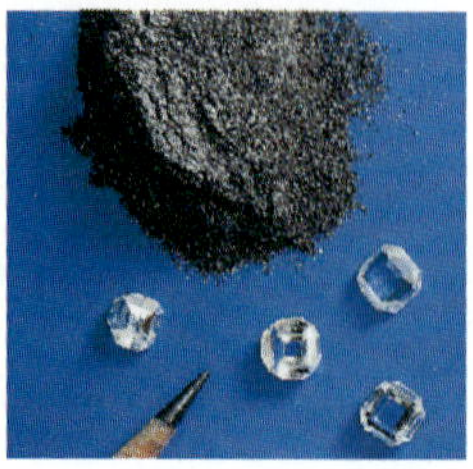

Diamond and graphite are allotropic forms of carbon.

With less oxygen available, poisonous carbon monoxide forms. Carbon monoxide is an invisible, odorless, tasteless gas. It exerts its insidious effect by tying up the hemoglobin in the blood. The normal function of hemoglobin is to transport oxygen. Carbon monoxide binds tenaciously to hemoglobin—once on, it refuses to get off. The hemoglobin is thus prevented from transporting oxygen (see Section 28.9).

D.6 Group 5A: Some Nitrogen Compounds

Group 5A includes the nonmetals nitrogen and phosphorus, the metalloids arsenic and antimony, and the metal bismuth. Their atoms have the electron configuration ns^2np^3. We limit our discussion to the chemistry of nitrogen.

Although nitrogen makes up 78% of the atmosphere, the molecules of nitrogen gas (N_2) cannot be used directly by higher plants or by animals. They first have to be "fixed"—that is, converted to a more readily used form. Certain types of bacteria convert atmospheric nitrogen to nitrates. Lightning also serves to "fix" nitrogen by causing it to combine with oxygen. Nitric oxide (NO) and nitrogen dioxide (NO_2) are formed. The equations are

$$N_2 + O_2 \xrightarrow{\text{lightning}} 2\,NO$$

$$2\,NO + O_2 \longrightarrow 2\,NO_2$$

The nitrogen dioxide reacts with water to form nitric acid (HNO_3).

$$3\,NO_2 + H_2O \longrightarrow 2\,HNO_3 + NO$$

The nitric acid falls in rainwater, adding to the supply of available nitrates in the oceans and the soil. Humans have undertaken substantial intervention in the nitrogen cycle by industrial fixation—the manufacture of nitrogen fertilizers. This intervention

Photochemical Smog

Automobiles also fix nitrogen, as does any high-temperature combustion in air.

$$N_2 + O_2 \longrightarrow 2\,NO$$

This reaction takes place in power plants that burn fossil fuels and in incinerators. The greatest source of such NO(g), however, is in the exhaust fumes of high-compression, high-temperature internal combustion automobile engines. NO(g) is considered an air pollutant because of various reactions in which it participates that yield other pollutants. It reacts with oxygen in the air to form nitrogen dioxide (NO_2), an irritant to the eyes and respiratory system. Red-brown NO_2(g) is often seen in polluted air in major urban centers, but, as with NO(g), NO_2(g) is objectionable mostly for chemical reactions stemming from it. In the presence of sunlight, NO_2 decomposes.

$$NO_2 + \text{sunlight} \longrightarrow NO + O$$

Oxygen *atoms* produced by the photochemical (light-induced) decomposition of NO_2 are highly reactive. These O atoms can react with many substances that are generally available in polluted air. For example, they may react with O_2 molecules to form ozone.

$$O + O_2 \longrightarrow O_3$$

A considerably higher-than-normal concentration of ozone (O_3) in air is one of the hallmarks of **photochemical smog,** air that is polluted with oxides of nitrogen and unburned hydrocarbons, together with ozone and several other components produced by the action of sunlight.

Ozone in the stratosphere protects us from harmful ultraviolet radiation, but ground-level ozone is the main cause of breathing difficulties that some people experience during smog episodes. Another effect of ozone is that it causes rubber to crack and deteriorate. The components of photochemical smog can cause heavy damage to crops, and photochemical smog also reduces visibility. In addition to their role in smog formation, NO(g) and NO_2(g) contribute to the fading and discoloration of fabrics. By forming nitric acid, they contribute to the acidity of rainwater (Section 10.7), which accelerates the corrosion of metals and building materials. They also produce crop damage, although the specific effects of these gases are difficult to separate from those of other pollutants.

Photochemical smog is characterized by an amber haze, like that seen in this view across the city of Buenos Aires, Argentina.

has greatly increased our food supply, because the availability of fixed nitrogen is often the limiting factor in the production of food. In 1912, American farmers produced an average of 26 bushels of corn per acre. Today the yield per acre is almost 100 bushels. This fourfold increase is due in large part to the increased use of nitrogen fertilizers. Not all the consequences of this intervention have been favorable, however; excessive runoff of nitrogen fertilizer has led to serious water-pollution problems in some areas.

D.7 Group 6A: Compounds of Oxygen and Sulfur

Group 6A includes the nonmetals oxygen, sulfur, and selenium, the metalloid tellurium, and the radioactive metal polonium. Their atoms have the electron configuration ns^2np^4. We limit our discussion to the chemistry of oxygen and sulfur.

The element oxygen occurs in the atmosphere mainly as the diatomic molecule O_2. Oxygen is involved in oxidation processes such as combustion (rapid burning) and rusting and other forms of corrosion, and in respiration. Oxygen reacts rapidly with more active metals. Magnesium, for example, burns with a brilliant white flame when ignited in air.

$$2\,Mg + O_2 \longrightarrow 2\,MgO + \text{heat} + \text{light}$$

At room temperature, a reaction occurs on the surface of freshly prepared magnesium metal. This reaction forms a thin, transparent coating of magnesium oxide that is impervious to air, preventing further oxidation. Such oxide coatings are common on metals, making it possible for us to use those otherwise quite reactive metals in utensils and machines. Magnesium, aluminum, and titanium are familiar examples of highly reactive metals that can be used as structural materials because of their ability to form protective oxide coatings.

Generally, oxygen reacts with metals by acquiring electrons and forming oxide ions.

$$O_2 + 4\,e^- \longrightarrow 2\,O^{2-}$$

The oxide ions react with water to form hydroxide ions, the ions that make solutions basic (Section 10.5).

$$O^{2-} + H_2O \longrightarrow 2\,OH^-$$

Oxides of metals, then, are generally basic oxides.

Oxygen also reacts with many nonmetals. Sulfur, for example, burns in air to form sulfur dioxide, a choking, acrid gas.

$$S(s) + O_2(g) \longrightarrow SO_2(g)$$

When dissolved in water, sulfur dioxide reacts to form sulfurous acid.

$$SO_2(g) + H_2O \longrightarrow H_2SO_3(aq)$$

Generally, the oxides of nonmetals form acidic solutions (Section 10.5); thus, we call them acidic oxides.

Small amounts of oxygen occur as *ozone* (O_3), an allotropic form of oxygen. Ozone is quite unstable. At room temperature, it breaks down slowly to ordinary oxygen.

$$2\,O_3(g) \longrightarrow 3\,O_2(g) + \text{heat}$$

Ozone is formed by electrical discharges through oxygen and by ultraviolet lamps. The pungent odor around electrical equipment is due to ozone. Ozone also is formed in photochemical smog. The ozone shield in the upper atmosphere protects us from harmful ultraviolet radiation (see page 331).

Industrial Smog

Polluted air associated with industrial activities is often called **industrial smog.** The presence of SO_2 and SO_3, collectively referred to as SO_x, is an important characteristic of this type of smog. Coal, especially soft coal from the eastern United States, has a relatively high sulfur content. When this coal is burned, sulfur compounds in it also burn, forming sulfur dioxide. Sulfur dioxide is readily absorbed in the respiratory system. It is a powerful irritant and is known to aggravate the symptoms of people who suffer from asthma, bronchitis, emphysema, and other lung diseases.

And things get worse. Although the actual reactions are more complicated than those we show here, the net results are the same. Some of the sulfur dioxide reacts further with oxygen in air to form sulfur trioxide.

$$2\ SO_2(g) + O_2(g) \longrightarrow 2\ SO_3(g)$$

Sulfur trioxide then reacts with water to form sulfuric acid.

$$SO_3(g) + H_2O(l) \longrightarrow H_2SO_4(aq)$$

Fine droplets of this acid form an aerosol mist that is even more irritating to the respiratory tract than sulfur dioxide.

Usually, industrial smog is also characterized by high levels of *particulate matter*, solid and liquid particles of greater than molecular size. The largest particles often are visible in air as dust and smoke. Particulate matter consists mainly of *soot* (unburned carbon) and the mineral matter that occurs in coal. These minerals do not burn. In the roaring fire of a huge boiler in a factory or power plant, some of this solid mineral matter is left behind as *bottom ash*, but much is carried aloft in the tremendous draft created by the fire. This *fly ash* settles over the surrounding area, covering everything with dust. It is also inhaled, contributing to respiratory problems in animals and humans.

The harmful effects of sulfur dioxide and particulate matter may be considerably magnified by their interaction. A certain level of sulfur dioxide, without the presence of particulate matter, might be reasonably safe. A certain level of particulate matter, without sulfur dioxide around, might be fairly harmless. But take these same levels of the two together, and the effect might well be deadly. *Synergistic effects* such as this are quite common whenever certain chemicals are brought together. For example, some forms of asbestos are carcinogenic, and about 35 or 40 of the chemicals in cigarette smoke are carcinogens. Asbestos workers who smoke develop cancer at a much greater rate than do people who are exposed to one carcinogen but not the other.

When the pollutants in industrial smog come into contact with the alveoli of the lungs, the cells are broken down. The alveoli lose their resilience, and it becomes difficult for them to expel carbon dioxide. Such lung damage leads to—or at least contributes to—pulmonary emphysema, a condition characterized by an increasing shortness of breath. (See again Section 10.9.)

The oxides of sulfur and the aerosol mists of sulfuric acid are damaging to plants. Leaves become bleached and splotchy when exposed to sulfur oxides. The yield and quality of farm crops can be severely affected. These compounds are also major ingredients in the production of acid rain.

Sulfur occurs in nature in both the combined and elemental forms. Free sulfur occurs as S_8, a ring of eight atoms. For simplicity, however, sulfur is often represented in equations only by the letter S, just as if it were monatomic.

Sulfur atoms can accept two electrons to form sulfide ions (S^{2-}).

$$S + 2\ e^- \longrightarrow S^{2-}$$

Sulfur gains electrons from the more active metals to form sulfides.

$$Ca(s) + S(s) \longrightarrow CaS(s)$$

D.8 Group 7A: The Halogens

Fluorine, chlorine, bromine, iodine, and astatine make up Group 7A, the halogen family. All halogens are nonmetals, and their atoms have the electron configuration ns^2np^5. The Group 1A metal sodium reacts with chlorine to form sodium chloride, the familiar table salt. Indeed, it is characteristic of the entire group that they react with metals to form salts. The word *halogen* is derived from Greek words meaning "salt former." We will limit our discussion to the first four halogens, because astatine is rare and highly radioactive.

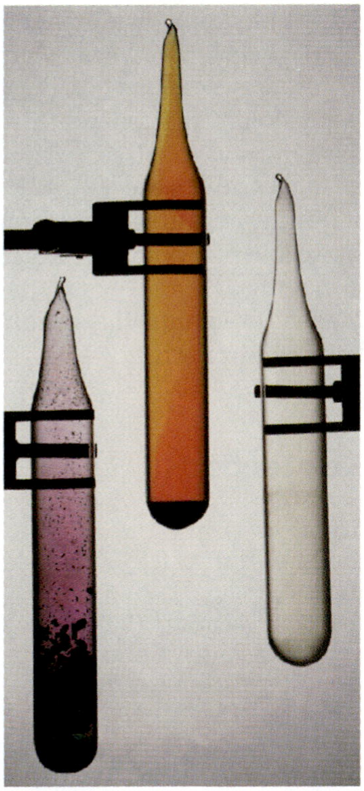

At room temperature, chlorine (Cl_2, right) is a greenish-yellow gas; bromine (Br_2, middle) is a reddish-brown liquid that readily vaporizes; and iodine (I_2, left) is a dark solid that sublimes to form a violet vapor.

In the elemental form, all the halogens exist as diatomic molecules. The atoms can achieve a noble gas configuration by gaining an electron to form a negative ion.

$$F_2 + 2\,e^- \longrightarrow 2\,F^-$$
$$Cl_2 + 2\,e^- \longrightarrow 2\,Cl^-$$

This tendency, combined with that of many metals to give up electrons readily, is responsible for the ability of the halogens to form so many salts.

Fluoride salts, in moderate to high concentrations, are acute poisons. Small amounts of fluoride ion, however, are essential for our well-being. Concentrations of 0.7 to 1.0 part per million, by mass, of sodium fluoride have been added to the drinking water of many communities. Evidence indicates that such fluoridation results in a reduction in the incidence of dental caries (cavities).

Chlorine in the elemental form (Cl_2) is used to kill bacteria in water-treatment plants. Sodium hypochlorite (NaOCl) is an ingredient of common household bleaching solutions. Chlorine is present in our bodies as chloride ion (Cl^-), an ion essential to life.

Several compounds of bromine are of importance. Silver bromide (AgBr) is sensitive to light and is used in photographic film. Sodium bromide (NaBr) and potassium bromide (KBr) have been used medicinally as sedatives. Bromide ions depress the central nervous system. Unfortunately, prolonged intake can cause mental deterioration and other problems. Bromides have been largely replaced by other, presumably safer, pain relievers.

Iodine is another essential nutrient. Compounds of iodine are necessary for the proper action of the thyroid gland. Iodine is also used as a topical antiseptic. Tincture of iodine is a solution of elemental iodine (I_2) in a mixture of alcohol and water. Iodine-releasing compounds, called iodophors, are often used as antiseptics in hospitals.

The halogens also react with hydrogen to form hydrogen halides.

$$H_2(g) + Cl_2(g) \longrightarrow 2\,HCl(g)$$
$$H_2(g) + Br_2(g) \longrightarrow 2\,HBr(g)$$

In water solutions, these compounds form acids. Hydrochloric acid is a familiar example. This acid is present in the stomach and is involved in the digestive process.

D.9 The B Groups: Some Typical Transition Elements

Most of our attention so far has been focused on the A groups of the periodic table. The chemistry of the B groups is also important, if perhaps more complicated. The B groups collectively are often called the *transition elements*. All are metals in the elemental form. They conduct electricity and have a characteristic metallic luster.

The transition elements, in general, are those in which inner electron energy levels are being filled. (Group 1B and 2B elements are exceptions, and yet they share many properties with the other B groups and are often included as transition elements.) Recall that the third period ends with argon, which has eight electrons in its outermost energy level. The fourth period begins with potassium, which has one electron in its fourth energy level. The next element, calcium, has two electrons in its fourth energy level. Recall, however, that the $2n^2$ rule predicts a maximum of 18, not 8, electrons for the third energy level ($2n^2 = 2 \times 3^2 = 2 \times 9 = 18$). The fourth energy level has begun to fill before the third one is full. Things return to "normal" with the next element, scandium. This element skips back and continues to fill the third energy level. The first transition series (from scandium to zinc) corresponds to a filling of the $3d$ sublevel. With gallium ($Z = 31$), we return to an A group element. All inner energy levels are filled, and the next electron enters the outer energy level.

Physical properties vary widely in the transition series. For example, mercury (Hg) is a liquid at room temperature (its melting point is −38 °C). Tungsten (W), however, melts at 3410 °C. Chemically, the transition elements also exhibit a variety of properties. Most form more than one kind of simple ion. Iron is a familiar example. It readily forms both iron(II) and iron(III) ions (Fe^{2+} and Fe^{3+}, respectively).

Many compounds of the transition metals are colored, often brilliantly so. For a given element, the color is different for different ions. Aqueous solutions of Fe^{2+} are often pale green; those of Fe^{3+} are often yellow. With polyatomic ions, color variation often is greater still. An aqueous solution containing manganese(II) ions (Mn^{2+}) is a faint pink. Solutions of MnO_4^{2-} are an intense green, and those of MnO_4^- are deep purple.

Transition metals known to be essential to life are iron, copper, zinc, cobalt, manganese, vanadium, chromium, nickel, tungsten, and molybdenum. Others are sometimes found in body tissues but have not been shown to be essential. Still others, most notably cadmium and mercury, are dangerous poisons.

Within the past few years, fitness enthusiasts have been taking chromium supplements to lose fat and to increase muscle mass. Chromium plays a role in stabilizing glucose levels in the blood and in metabolizing carbohydrates and lipids. To date, there is no evidence to substantiate the claims that chromium supplements are actually beneficial.

Cobalt is a component of vitamin B_{12}, a deficiency of which leads to pernicious anemia. Vitamin B_{12} is found only in animal products. One hazard of a strict vegetarian diet is vitamin B_{12} deficiency.

Iron is essential for the proper functioning of hemoglobin, the red protein molecule involved in oxygen transport (see Section 28.9). This iron must be in the form of the +2 ion. If it is changed to the +3 form, the resulting compound (methemoglobin) is incapable of carrying oxygen. The oxygen-deficiency disease that results is called methemoglobinemia. In infants, this condition is called the blue-baby syndrome.

Solutions of transition metal salts are brightly colored. The colors are due to the following ions (left to right): Mn^{2+}, Fe^{2+}, Co^{2+}, Ni^{2+}, Cu^{2+}, and Zn^{2+}.

Key Terms

allotropes (D.5)
greenhouse effect (D.5)
industrial smog (D.7)
inorganic chemistry (Introduction)
organic chemistry (Introduction)
photochemical smog (D.7)

Review Questions

1. Define each term.
   a. inorganic chemistry b. noble gas
   c. halogen d. alkali metal
   e. alkaline earth metal f. transition element
2. What is the outstanding chemical property of the noble gases? What structural feature accounts for this property?
3. Why is helium preferred to hydrogen for filling blimps, even though hydrogen has greater lifting power?
4. What function does argon serve in an electric light bulb?
5. How does a neon sign work?
6. What similarities in properties are shared by the alkali metals? What structural feature do atoms of these elements share?
7. What similarities in properties are shared by the alkaline earth metals? What structural feature do atoms of these elements share?
8. Which alkaline earth metal is most different from the others? What are some of those differences?
9. Name one use for each element or compound.
   a. chlorine b. iodine c. NaF
   d. AgBr e. NaOCl f. KCl
10. What is the effect on tooth enamel of small amounts of fluoride?
11. What are basic oxides? Use calcium oxide to illustrate your answer.
12. What are acidic oxides? Use sulfur trioxide to illustrate your answer.

13. What alkali metal compound is used to treat manic-depressive psychoses?
14. What is the most abundant metal in the Earth's crust?
15. What is industrial smog? How is it formed?
16. What are typical weather conditions during episodes of (**a**) photochemical smog and (**b**) industrial smog?
17. What is the greenhouse effect? How do greenhouse gases contribute to global warming?
18. What was the vital force theory? How was it overthrown?
19. What is the origin of helium in natural gas wells?
20. Why are the noble gases no longer called the "inert gases"?
21. What are the halogens? Why are they so called?
22. What is nitrogen fixation? Why is it important?
23. What condition(s) lead(s) to formation of carbon monoxide during combustion?
24. How does carbon monoxide exert its poisonous effect?
25. What two alkali metal ions play major roles in maintaining fluid balance in the body?
26. What important molecule in plants incorporates magnesium?
27. Name two functions of calcium ions in the body.
28. What is hard water? How does it affect the action of soaps?
29. List three distinguishing characteristics of transition metals.
30. List four transition metals essential to life. Indicate their functions in the body.
31. Why is aluminum replacing steel wherever possible in automobiles?
32. What is photochemical smog? What chemical compound starts the formation of photochemical smog by absorbing sunlight?
33. What is synergism? Indicate one specific example of a synergistic effect concerning air pollution.
34. What are allotropes? Name two sets of allotropes.

Problems

Lewis Symbols

35. Give the Lewis symbol for each of the following elements.
    a. neon **b.** oxygen **c.** fluorine
36. Give the Lewis symbol for each of the following elements.
    a. potassium **b.** barium **c.** nitrogen
37. Give the Lewis symbol for each of the following ions.
    a. fluoride ion **b.** iodide ion **c.** oxide ion
38. Give the Lewis symbol for each of the following ions.
    a. sulfide ion **b.** potassium ion
    c. strontium ion

Chemical Equations

39. Complete and balance the following equations.
    a. $Li(s) + O_2(g) \longrightarrow$
    b. $CaO(s) + H_2O \longrightarrow$
    c. $S(s) + O_2(g) \longrightarrow$
40. Complete and balance the following equations.
    a. $Ca(s) + O_2(g) \longrightarrow$
    b. $SO_2(g) + H_2O \longrightarrow$
    c. $Ca(s) + S(s) \longrightarrow$

Additional Problems

41. Give Lewis symbols for gallium (Ga) and gallium ion.
42. How do oxygen atoms, oxygen molecules, and ozone differ in structure and properties?
43. How does the burning of sulfur-containing coal lead to acid rain? Give the pertinent chemical equations.
44. Write a series of equations representing the natural fixation of atmospheric nitrogen in an electrical storm.
45. What are the health effects associated with ozone in the stratosphere and at ground level? Why are they not the same?
46. Write equations representing (**a**) the production of nitrogen monoxide (nitric oxide) in an automobile engine and (**b**) the action of sunlight on nitrogen dioxide.

13 Hydrocarbons

Offshore drilling for natural gas and crude oil. Petroleum is the chief raw material for the production of plastics, pesticides, drugs, detergents, and many other important commodities.

Scientists of the eighteenth and nineteenth centuries studied compounds isolated from rocks and ores, from the atmosphere and oceans, and labeled them *inorganic* because they were obtained from nonliving systems. Compounds obtained from plants and animals were called *organic* because they were isolated from organized (living) systems. The early chemists believed that only living organisms could synthesize organic compounds. Although by the middle of the nineteenth century a number of "organic" compounds had been prepared using ordinary laboratory techniques, the labels *organic* and *inorganic* remained.

Today, **organic chemistry** is defined simply as the chemistry of the compounds of carbon.[1] It may seem strange that we divide chemistry into two branches, one that considers only one element and one that covers the 100 plus

[1] This definition is not adhered to strictly; the following compounds often are considered inorganic:

Carbon monoxide (CO)	Carbonates (e.g., Na_2CO_3)	Thiocyanates (e.g., NaSCN)
Carbon dioxide (CO_2)	Bicarbonates (e.g., $NaHCO_3$)	Cyanates (e.g., KOCN)
Carbon disulfide (CS_2)	Cyanides (e.g., KCN)	Carbides (e.g., CaC_2)

The word *organic* has different meanings. Organic fertilizer is organic in the original sense—it is derived from living organisms. Organic foods generally are foods grown without synthetic pesticides or fertilizers. Organic chemistry is the chemistry of compounds of carbon.

remaining elements. However, this division seems more reasonable when one discovers that, of the 15 million compounds that have been characterized, the overwhelming majority contain carbon.

What is so special about carbon that differentiates it from all of the other elements in the periodic table? Carbon has a unique ability to form stable, covalent bonds with itself and with other elements in infinite variations. The molecules thus produced may contain only one or over a million carbon atoms. So complex is the chemistry of carbon that we shall approach its study by dividing its millions of compounds into families. We'll study one family at a time and begin by concentrating on the simpler members of each family. Eventually we shall consider those molecules that deserve to be called organic in the old sense—complex, carbon-containing molecules that determine the forms and functions of living systems. This is the field of biochemistry.

In January 1990 the Chemical Abstract Service of the American Chemical Society recorded the 10 millionth known chemical compound. Over 600,000 new compounds are added to the list each year. About 95% of all the compounds are organic substances.

We might pause to ponder how far science has come since Wöhler's synthesis of urea in 1828 (see page 306). Before that year, scientists believed they could not synthesize even the simplest organic molecule. In 1980, the U.S. Supreme Court ruled that new life forms created in a laboratory could be patented. The patent under consideration described a new organism designed to consume spilled oil and made by changing the genetic heritage of an existing life form. Was this "creation of life in a test tube"? Not quite, but each new experiment in genetic engineering (see Section 23.9) brings that achievement closer. We use thousands of carbon compounds every day without even realizing it because they are silently carrying out important chemical reactions within our bodies. Many of these carbon compounds are so vital that we literally could not live without them.

13.1 Organic Versus Inorganic

Organic compounds, like inorganic ones, obey all the natural laws. Often there is no clear distinction in chemical or physical properties between organic and inorganic molecules. Nevertheless, it is useful to compare typical members of each class, as in Table 13.1. (It must be understood, however, that there are exceptions

Table 13.1 Some Contrasting Properties of Organic and Inorganic Compounds

Organic	Inorganic
1. Low melting points	1. High melting points
2. Low boiling points	2. High boiling points
3. Low solubility in water; high solubility in nonpolar solvents	3. High solubility in water; low solubility in nonpolar solvents
4. Flammable	4. Nonflammable
5. Solutions do not conduct electricity	5. Solutions conduct electricity
6. Chemical reactions are usually slow	6. Chemical reactions are rapid
7. Exhibit isomerism	7. Isomers are limited to a few exceptions (e.g., the transition elements)
8. Exhibit covalent bonding	8. Exhibit ionic bonding

Table 13.2 Comparison of an Inorganic and an Organic Compound

	Benzene	Sodium Chloride
Formula	C_6H_6	NaCl
Solubility in H_2O	Insoluble	Soluble
Solubility in gasoline	Soluble	Insoluble
Flammable?	Yes	No
Melting point	5.5 °C	801 °C
Boiling point	80 °C	1413 °C
Density	0.88 g/cm^3	2.7 g/cm^3
Bonding	Covalent	Ionic

to every entry in this table.) To further illustrate typical organic–inorganic differences, Table 13.2 lists properties of the inorganic compound sodium chloride (NaCl, common table salt) and the organic compound benzene (C_6H_6, a solvent once widely used to strip furniture for refinishing).

▲ **FIGURE 13.1**
Ball-and-stick models of methane and ethane.

13.2 Alkanes: Saturated Hydrocarbons

Before you can understand the large, complex organic molecules on which life is based, you need to learn something about simpler ones. We therefore start with organic compounds containing only two elements, carbon and hydrogen—the **hydrocarbons.** In Sections 4.10 and B.2, we encountered compounds called methane (CH_4) and ethane (C_2H_6), shown in Figure 13.1. These are the first two members of a series of related compounds called **alkanes** or **saturated hydrocarbons**. *Saturated,* in this case, means that each carbon atom is bonded to four other atoms; there are no double or triple bonds in the molecules. Structurally, methane and ethane are generally represented as follows.

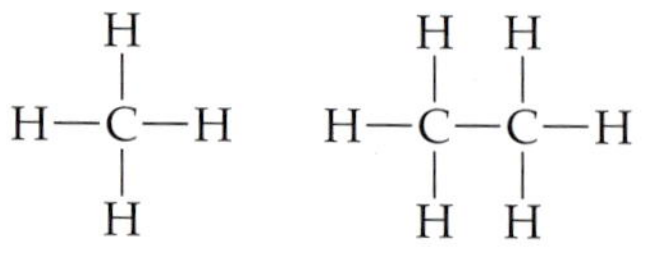

Methane Ethane

These representations make no attempt to accurately portray bond angles in particular or molecular geometry in general (these points are covered in Section B.2).

Alkanes often are called *saturated hydrocarbons.* They are hydrocarbons in which there are only single bonds.

The three-carbon hydrocarbon (C_3H_8) is called propane. A ball-and-stick model is shown in Figure 13.2. The two-dimensional structure is generally written

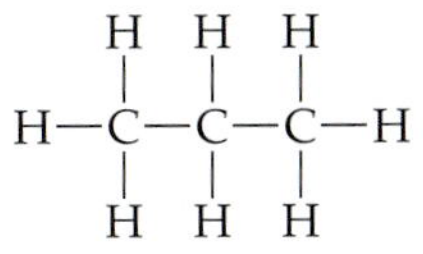

$$\begin{array}{c} \text{H}\quad\text{H}\quad\text{H} \\ |\quad\ \ |\quad\ \ | \\ \text{H}-\text{C}-\text{C}-\text{C}-\text{H} \\ |\quad\ \ |\quad\ \ | \\ \text{H}\quad\text{H}\quad\text{H} \end{array}$$

Propane

Notice that methane, ethane, and propane form a series in which any two adjacent members differ by one carbon atom and two hydrogen atoms—that is, by a CH_2 unit (Figure 13.3). Compounds related in this way are said to make up a **homologous series.** The members in any such series have properties that vary in a regular and predictable manner. The principle called *homology* gives organization to organic chemistry in much the same way that the periodic table gives organization to inorganic chemistry. Instead of a bewildering array of individual carbon compounds, we can study a few members of a homologous series (called *homologs*), and from them deduce the properties of other compounds in the series.

Continuing the homologous series that begins with methane, we add another carbon atom and a pair of hydrogens to get C_4H_{10}. It is rather easy to write a structure corresponding to this formula. We merely string four carbon atoms in a row,

$$-\text{C}-\text{C}-\text{C}-\text{C}-$$

and then we add enough hydrogen atoms to give each carbon atom four bonds:

$$\begin{array}{c} \text{H}\quad\text{H}\quad\text{H}\quad\text{H} \\ |\quad\ \ |\quad\ \ |\quad\ \ | \\ \text{H}-\text{C}-\text{C}-\text{C}-\text{C}-\text{H} \\ |\quad\ \ |\quad\ \ |\quad\ \ | \\ \text{H}\quad\text{H}\quad\text{H}\quad\text{H} \end{array}$$

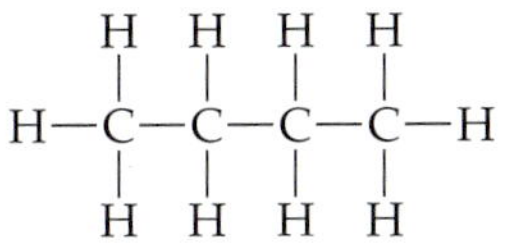

There is a compound called butane that has this structure. But there is another way to put four carbons and 10 hydrogens together. String out three of the carbon atoms and then branch the fourth one off the middle carbon:

$$\begin{array}{c} -\text{C}-\text{C}-\text{C}- \\ | \\ \text{C} \end{array}$$

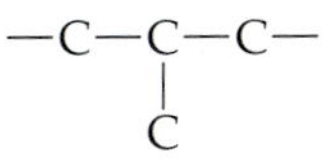

Now we add enough hydrogen atoms to give each carbon four bonds.[2]

$$\begin{array}{c} \text{H}\quad\text{H}\quad\text{H} \\ |\quad\ \ |\quad\ \ | \\ \text{H}-\text{C}-\text{C}-\text{C}-\text{H} \\ |\quad\ \ |\quad\ \ | \\ \text{H}\quad\ |\quad\text{H} \\ \ \ \text{H}-\text{C}-\text{H} \\ \ \ \ \ \ | \\ \ \ \ \ \ \text{H} \end{array}$$

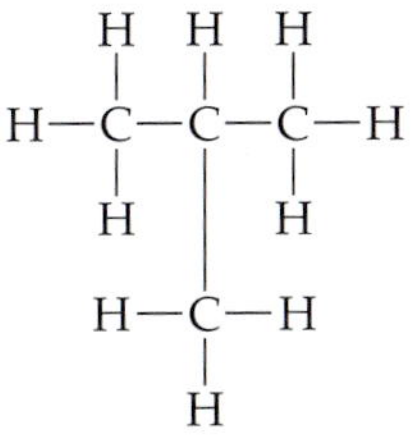

Fortunately for the sake of our structural theory, there is a hydrocarbon that corresponds to this structure. There are two compounds, then, that have the molecular

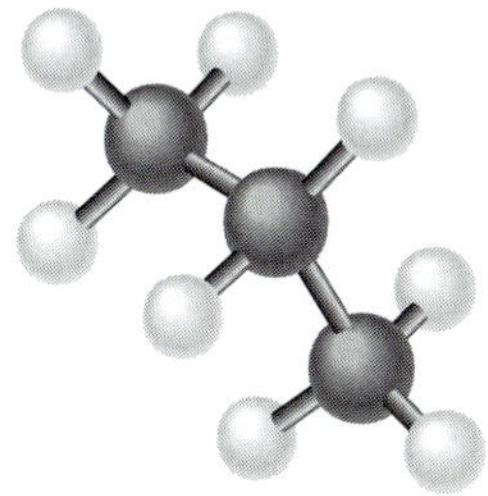

▲ **FIGURE 13.2**

Ball-and-stick model of propane.

$$\begin{array}{cc} \text{H} & \quad\text{H}\quad\text{H} \\ | & \quad|\quad\ \ | \\ \text{H}-\text{C}-\text{H} & \text{H}-\text{C}-\text{C}-\text{H} \\ | & \quad|\quad\ \ | \\ \text{H} & \quad\text{H}\quad\text{H} \\ \text{Methane} & \text{Ethane} \\ (CH_4) & (C_2H_6) \end{array}$$

$$\begin{array}{c} \text{H}\quad\text{H}\quad\text{H} \\ |\quad\ \ |\quad\ \ | \\ \text{H}-\text{C}-\text{C}-\text{C}-\text{H} \\ |\quad\ \ |\quad\ \ | \\ \text{H}\quad\text{H}\quad\text{H} \end{array}$$

Propane
(C_3H_8)

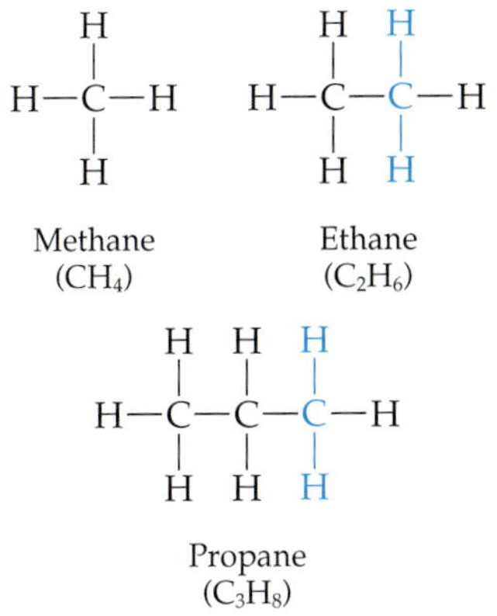

▲ **FIGURE 13.3**

Members of a homologous series. Each succeeding formula incorporates one carbon atom and two hydrogen atoms more than the previous formula.

[2] The bond to the fourth carbon has been drawn longer than the others to avoid congestion. In the molecule, all carbon–carbon bonds are the same length.

formula C_4H_{10}. One boils at −1 °C; the other at −12 °C. Different compounds having the same molecular formula are called **isomers.** To give the two butanes unique names, we call the one with the continuous carbon chain *butane* and the one with the branched chain *isobutane* (Figure 13.4).

The three smallest homologs of the alkane series do not exist in isomeric forms because there is only one way to arrange the atoms in each formula so that each carbon atom has four bonds. It is important to realize that bending a chain does *not* mean an isomer is formed. With butane, for example, there are two (and only two) ways to arrange the carbon and hydrogen atoms. Butane is a continuous four-carbon chain. This chain may be bent as in Figure 13.4, but it is still continuous. The formula of isobutane shows a continuous chain of three carbon atoms only, with the fourth attached as a branch off the middle carbon of the continuous chain. You should make models of these molecules to convince yourself. (Marshmallows and toothpicks work well.)

A continuous chain of carbon atoms is also referred to as a *straight* chain.

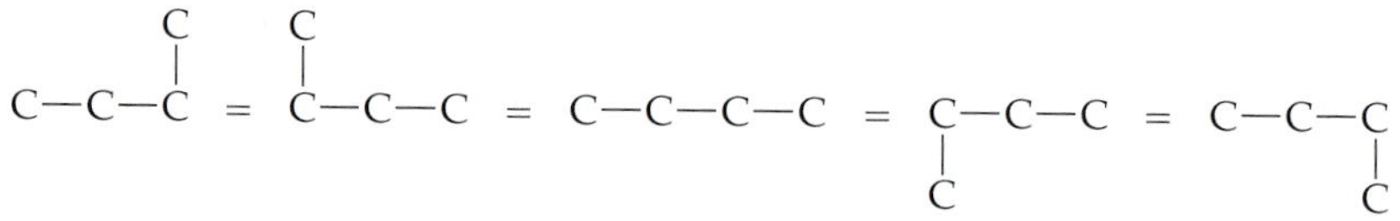

One more step up the homologous alkane series gets us to C_5H_{12}. There are three compounds with this molecular formula. Collectively, these compounds are called *pentanes*. We can name compound I pentane because it has all five carbons in a continuous chain. Compound II can be called isopentane because, like isobutane, it has a single carbon atom branched off the second carbon of the continuous chain. But what shall we call compound III? Let's name it the way the chemists did when it was discovered in 1870. Since the other two pentanes were characterized first, this one was called neopentane (from the Greek *neos*, new).

The alkanes can be represented by the general formula C_nH_{2n+2}, in which n is the number of carbon atoms.

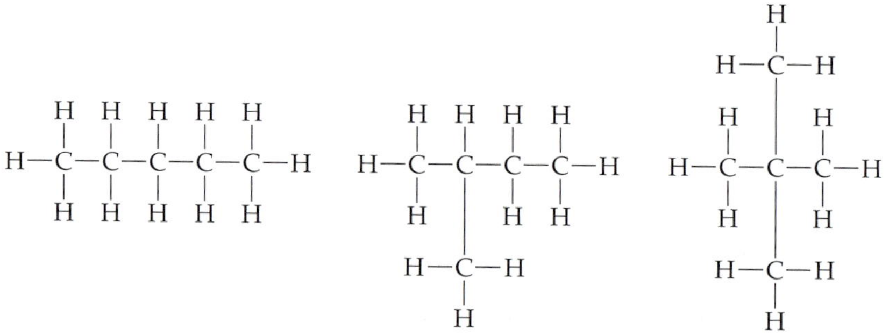

I. Pentane II. Isopentane III. Neopentane

Naming hydrocarbons seems complex, but there is some system to the process (as we shall see in Section 13.5). The first part of the name indicates the

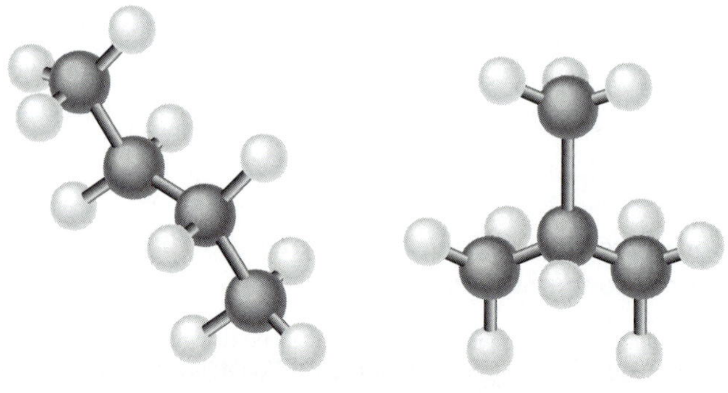

Butane Isobutane

FIGURE 13.4 ▶ Ball-and-stick models of butane and isobutane. These two compounds are isomers; both have the molecular formula C_4H_{10}.

number of carbon atoms in the molecule (Table 13.3). For five or more carbon atoms, the stems are derived from Greek or Latin names for the numbers.

The ending *-ane* in each name indicates an alkane. The name heptane, then, means an alkane having seven carbon atoms. To draw the structure of heptane, we write out a string of seven carbon atoms,

$$-C-C-C-C-C-C-C-$$

and then attach enough hydrogens to give each carbon a valence of 4. Doing so requires three hydrogens on each end carbon and two on each of the others.

$$
\begin{array}{ccccccc}
 & H & H & H & H & H & H & H \\
 & | & | & | & | & | & | & | \\
H- & C- & C- & C- & C- & C- & C- & C-H \\
 & | & | & | & | & | & | & | \\
 & H & H & H & H & H & H & H \\
\end{array}
$$

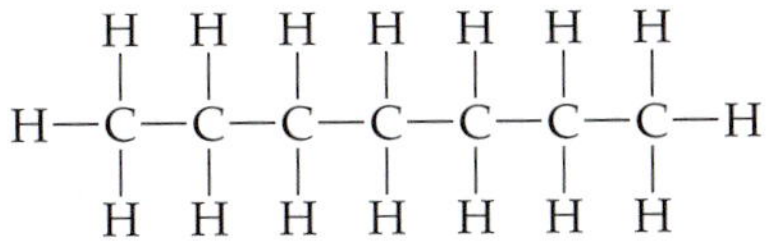

The first 10 straight-chain alkanes are shown in Table 13.4. Notice that the number of isomers increases rapidly with increasing carbon number. There are five hexanes, nine heptanes, and 18 octanes. Even more striking are the 366,319 possible isomers of the alkane whose molecular formula is $C_{20}H_{42}$ and the 62,491,178,805,831 possible isomers of the alkane whose molecular formula is $C_{40}H_{82}$. Not all of these have been isolated or characterized.

Table 13.3 Stems That Indicate the Number of Carbon Atoms in Organic Molecules	
Stem	**Number**
Meth-	1
Eth-	2
Prop-	3
But-	4
Pent-	5
Hex-	6
Hept-	7
Oct-	8
Non-	9
Dec-	10

Table 13.4 Straight-Chain Alkanes			
Name	**Molecular Formula**	**Condensed Structural Formula**	**Number of Possible Isomers**
Methane	CH_4	CH_4	1
Ethane	C_2H_6	CH_3CH_3	1
Propane	C_3H_8	$CH_3CH_2CH_3$	1
Butane	C_4H_{10}	$CH_3(CH_2)_2CH_3$	2
Pentane	C_5H_{12}	$CH_3(CH_2)_3CH_3$	3
Hexane	C_6H_{14}	$CH_3(CH_2)_4CH_3$	5
Heptane	C_7H_{16}	$CH_3(CH_2)_5CH_3$	9
Octane	C_8H_{18}	$CH_3(CH_2)_6CH_3$	18
Nonane	C_9H_{20}	$CH_3(CH_2)_7CH_3$	35
Decane	$C_{10}H_{22}$	$CH_3(CH_2)_8CH_3$	75

13.3 Condensed Structural Formulas

The line formulas such as we have used so far, showing all the carbon and hydrogen atoms and how they are attached to one another, are called **structural formulas.** They convey much more information than simple chemical formulas. For example, the molecular formula C_4H_{10} expresses only the number of atoms in the molecule. It doesn't tell us whether we are dealing with butane or with isobutane. The structural formulas

$$
\begin{array}{cccc}
H & H & H & H \\
| & | & | & | \\
H-C-C-C-C-H \\
| & | & | & | \\
H & H & H & H \\
\end{array}
\qquad \text{and} \qquad
\begin{array}{ccc}
H & H & H \\
| & | & | \\
H-C-C-C-H \\
| & | & | \\
H & | & H \\
 & H-C-H \\
 & | \\
 & H \\
\end{array}
$$

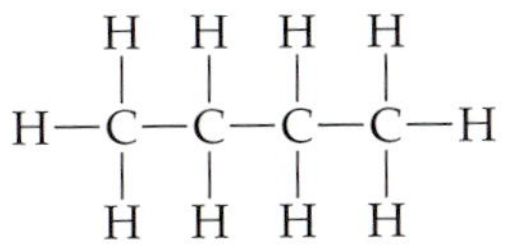

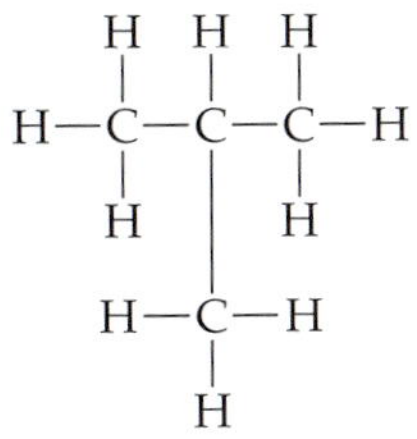

identify the specific isomers by showing the order of attachment of the various atoms.

Unfortunately, structural formulas are difficult to type and take up a lot of space. Chemists often use **condensed structural formulas** to alleviate these problems. The condensed formulas show the hydrogen atoms right next to the carbon atoms to which they are attached. For example, the two isomeric butanes become

$$CH_3{-}CH_2{-}CH_2{-}CH_3 \qquad \text{and} \qquad \begin{array}{c} CH_3{-}\underset{\displaystyle CH_3}{\underset{|}{CH}}{-}CH_3 \end{array}$$

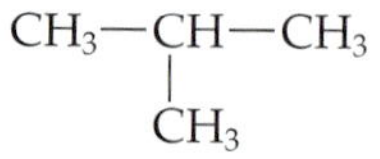

The parentheses in condensed structural formulas tell us that the enclosed grouping of atoms attaches to the adjacent carbon.

Sometimes condensed formulas are further simplified by omitting even more of the bond lines:

$$CH_3CH_2CH_2CH_3 \quad \text{and} \quad CH_3\underset{\displaystyle CH_3}{\underset{|}{C}}HCH_3 \quad [\text{or} \quad (CH_3)_3CH]$$

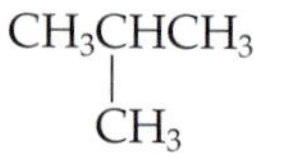

13.4 Alkyl Groups

The group of atoms that results when one hydrogen atom is removed from an alkane is called an **alkyl group.** Thus, the general formula for an alkyl group is C_nH_{2n+1}. The group is named by replacing the *-ane* suffix of the parent hydrocarbon with *-yl* (Table 13.5). It is important to note that alkyl groups are not independent molecules. Rather, they exist as parts of molecules.

Notice that two alkyl groups can be derived from propane. These are called propyl and isopropyl, respectively. Remember that there is only one alkane named propane, but a chain of three carbon atoms can be attached to a longer chain in two ways. One way has the attachment at an end carbon of the three-carbon chain, and the other way has the attachment at the middle carbon. There are many other alkyl groups. The ones that we are most likely to encounter are listed in Table 13.5.

Table 13.5 Common Alkyl Groups

Parent Hydrocarbon		Alkyl Group		Condensed Formula
Methane	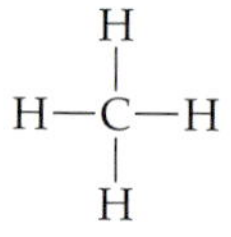	Methyl	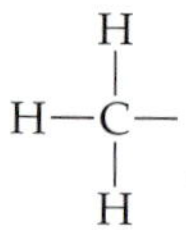	CH_3-
Ethane	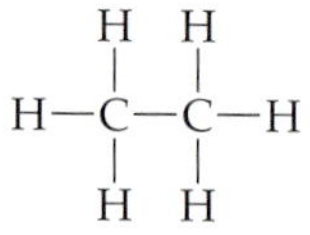	Ethyl	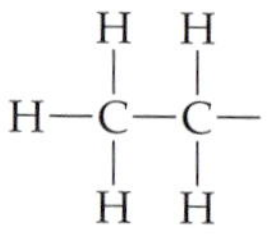	CH_3CH_2- or C_2H_5-
Propane	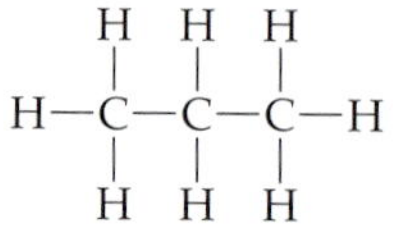	Propyl	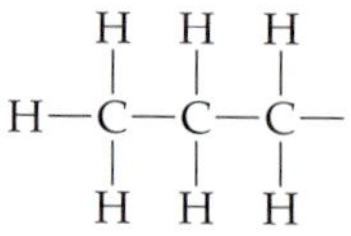	$CH_3CH_2CH_2-$ or C_3H_7-
		Isopropyl	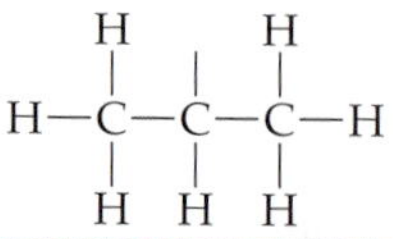	 $CH_3\overset{|}{C}HCH_3$ or $(CH_3)_2CH-$

13.5 The Universal Language: IUPAC Nomenclature

To bring order to the chaotic naming of newly discovered compounds, an international meeting of chemists was held at Geneva, Switzerland, in 1892. The meeting resulted in a simple, unequivocal system for naming organic compounds. The system was modified by the International Union of Chemists and is kept up to date by its successor, the International Union of Pure and Applied Chemistry (IUPAC). What has evolved is a set of rules known as the **IUPAC System of Nomenclature.** The rules for the alkanes are as follows.

1. Saturated hydrocarbons are named according to the longest continuous chain of carbons in the molecule (rather than the total number of carbon atoms). This longest chain is the parent compound.
2. The suffix *-ane* indicates that the molecule is a saturated hydrocarbon.
3. The name of the parent hydrocarbon is modified by noting what alkyl groups are attached to the chain.
4. The chain is numbered, and the position of each substituent alkyl group is indicated by the number of the carbon atom to which it is attached. The chain is numbered in such a way that the substituents occur on the carbon atoms with the lowest numbers.
5. Names of the substituent groups are placed in alphabetical order before the name of the parent compound.
6. If the same alkyl group appears more than once, the numbers of all the carbons to which it is attached are expressed. If the same group appears more than once on the same carbon, the number of that carbon is repeated as many times as the group appears.
7. Hyphens are used to separate numbers from names of substituents; numbers are separated from each other by commas. The number of identical groups is indicated by the Greek prefixes *di-, tri-, tetra-,* etc. These prefixes are NOT considered in determining the alphabetical order of the substituents (e.g., ethyl precedes dimethyl).
8. The last alkyl group named is prefixed to the name of the parent alkane to form one word.

> The term *substituent* refers to any atom or group of atoms that substitutes for a hydrogen atom in an organic molecule.

Table 13.6 contains some examples of the IUPAC system for naming organic compounds. The best way to learn how to name alkanes is by working out examples, not just by memorizing rules. It's easier than it sounds. Try the following.

Example 13.1

Name the compound

$$CH_3-CH_2-\underset{\underset{CH_3}{|}}{CH}-\underset{\underset{CH_3}{|}}{CH}-CH_3$$

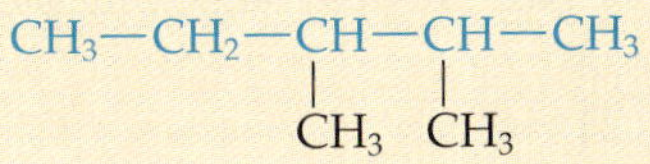

Solution

The longest continuous chain has five carbon atoms, and so the parent compound is pentane. There are methyl groups attached to the second and third carbon atoms of the pentane chain (not the third and fourth; see rule 4, second sentence). The correct name is 2,3-dimethylpentane.

Table 13.6 Examples of IUPAC Nomenclature

Condensed Structural Formula	Rewritten and Numbered	IUPAC Name
$CH_3CH_2CH(CH_3)CH_2CH(CH_3)_2$	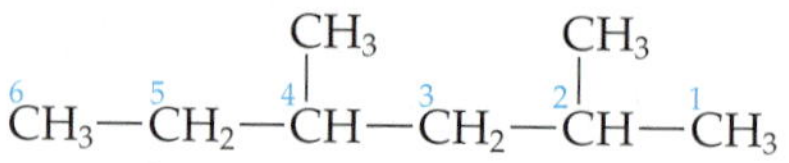	2,4-Dimethylhexane NOT 3,5-Dimethylhexane (use lowest numbers)
$(C_2H_5)_2CHCH(CH_3)CH_2CH_2CH_3$	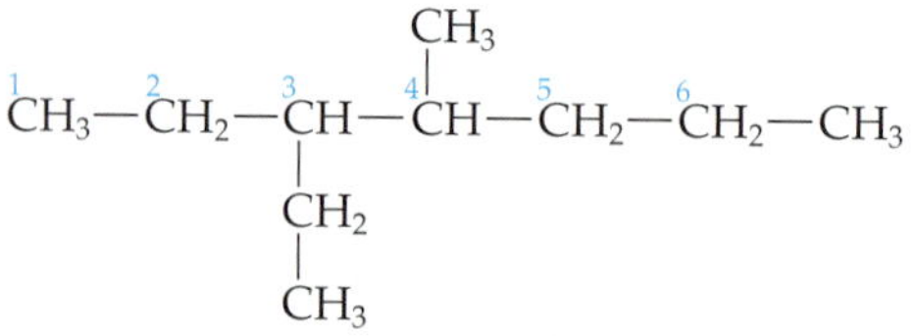	3-Ethyl-4-methylheptane NOT 4-Methyl-5-ethylheptane (use lowest numbers)
$(CH_3)_2CHCH(C_3H_7)_2$	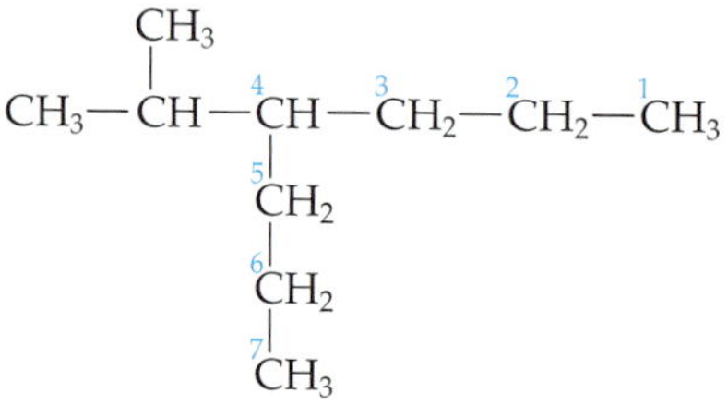	4-Isopropylheptane NOT 2-Methyl-3-propylhexane (use longest chain as parent)
$(CH_3)_2CHCH_2CH_2C(CH_3)_3$	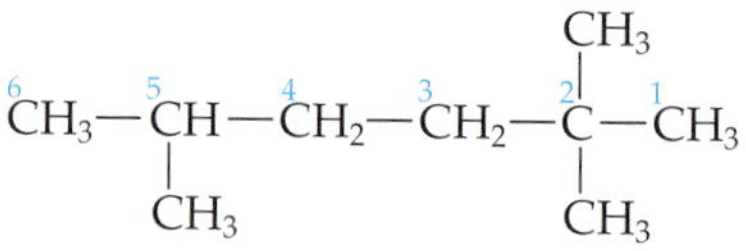	2,2,5-Trimethylhexane NOT 2,5-Trimethylhexane (each substituent must be numbered)
$(C_2H_5)_2CHC(CH_3)(C_2H_5)_2$	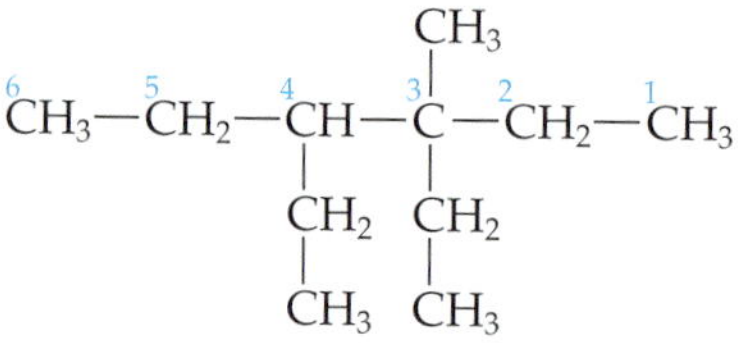	3,4-Diethyl-3-methylhexane NOT 3,4-Ethyl-3-methylhexane (use prefixes when same substituent occurs more than once)

Example 13.2

Name the compound

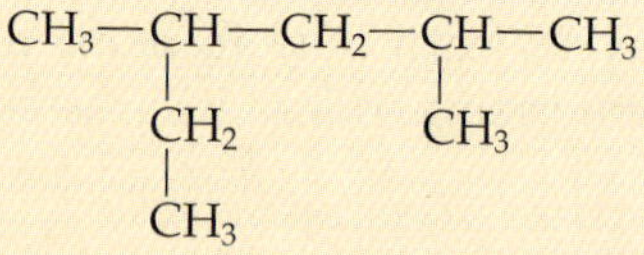

Solution

The correct name is 2,4-dimethylhexane, NOT 2-ethyl-4-methylpentane. This is a fooler. The parent compound is the longest continuous chain, which is not necessarily the chain drawn straight across the page. In this compound, the longest chain is bent:

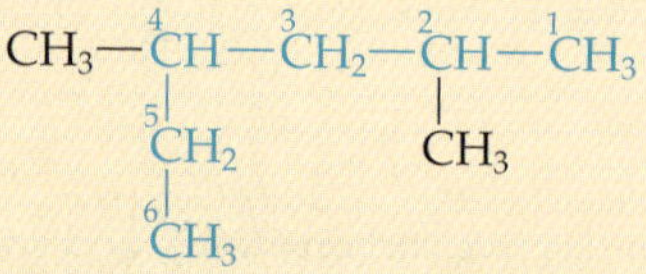

Note: Structural formulas are usually presented with the longest chain written horizontally. We gave this example to emphasize the point of searching out the longest chain.

Example 13.3

Name the compound

$$CH_3CH_2CH_2CH_2 \overset{\overset{\displaystyle CH_3}{\underset{\displaystyle |}{H-\overset{|}{C}-CH_3}}}{\underset{\underset{\displaystyle CH_3}{|}}{\overset{|}{-}\overset{|}{C}-}} CH_2CH_2CH_3$$

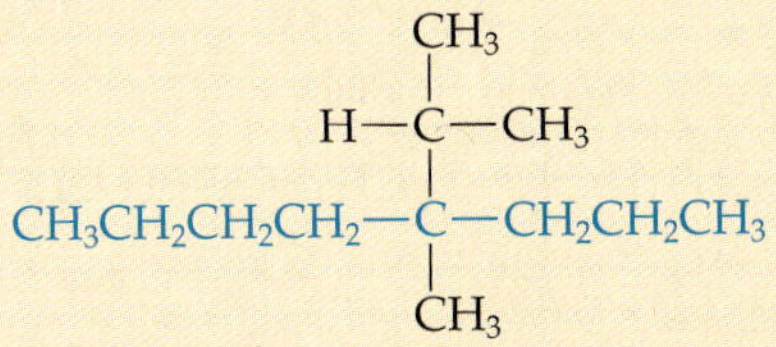

Solution

The correct name is 4-isopropyl-4-methyloctane.

Practice Exercise

Name the following compounds:

a. $CH_3CH_2CHCH_2CH_2CH_3$
 |
 CH_3

b. $CH_3CHCH_2CHCH_3$
 | |
 CH_3 CH_3

c. $CH_3CH_2CHCH_2CH_2CH_3$
 |
 CH_2CH_3

d. $CH_3CH_2CH_2CHCH_2CH_2CH_3$
 |
 $H-C-CH_3$
 |
 CH_3

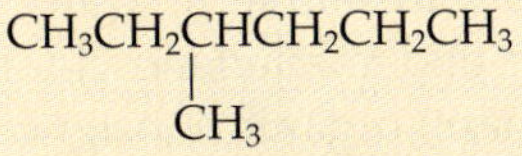

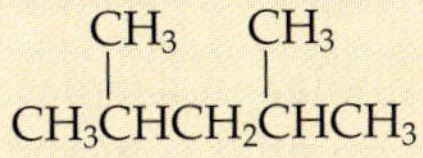

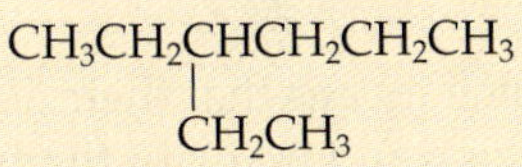

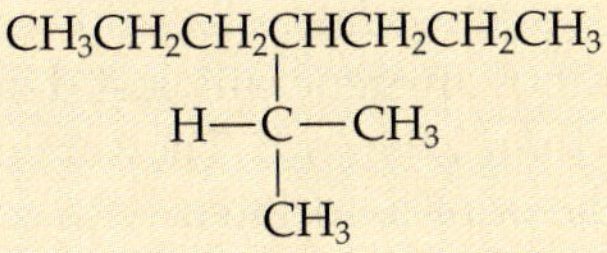

Example 13.4

Draw the structural formula for 4-isopropyl-2-methylheptane.

Solution

In drawing structural formulas, always start with the parent chain, heptane in this case:

$$-C-C-C-C-C-C-C-$$

Then add the groups at their proper positions. You can number the parent chain from either direction as long as you are consistent (don't change directions in the middle of the problem). First place the additional carbons:

$$\underset{1 \quad 2 \quad 3 \quad 4 \quad 5 \quad 6 \quad 7}{C-\overset{\overset{\displaystyle C}{|}}{C}-C-\overset{\overset{\displaystyle C-C-C}{|}}{C}-C-C-C}$$

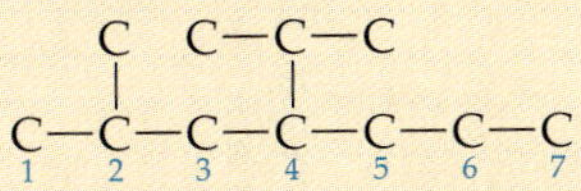

Finally, fill in all the hydrogens (each carbon atom must have four bonds):

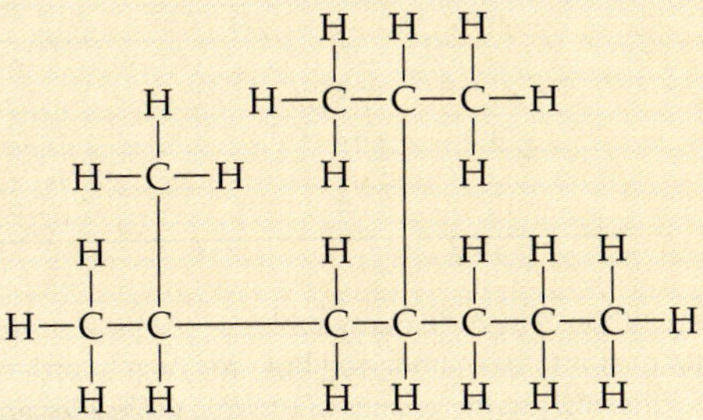

You can condense this formula by writing the hydrogens right next to the carbons to which they are attached:

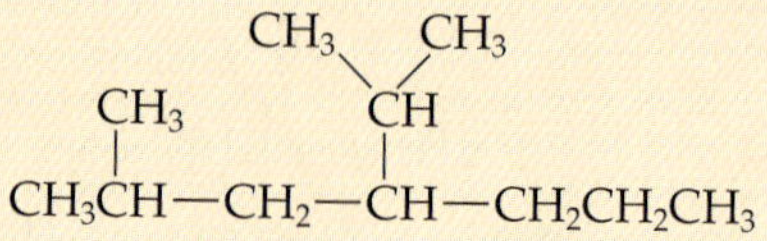

Practice Exercise

Draw the structural formulas for the following compounds:

a. 4-propylheptane

b. 2,2-dimethylbutane

c. 3-ethyl-2-methylpentane

d. 3-isopropyl-3-methyloctane

13.6 Physical Properties of Alkanes

The alkanes are nonpolar molecules. They are therefore insoluble in water, but they are soluble in organic solvents, such as toluene and methylene chloride. Alkanes themselves are good solvents; they dissolve many organic substances of low polarity, such as fats, oils, and waxes.

The alkanes are generally less dense than water, which means that their densities are less than 1.0 g/mL. They are colorless and tasteless, and many of them are odorless. The odor associated with natural gas is not from methane or any of the other alkanes, but rather from an odorant (usually CH_3SH) purposely added to the gas in sufficient quantities to allow for the detection of leaks. Table 13.7 lists some properties of the first 10 straight-chain alkanes.[3] The physical states of substances at room temperature are obtained directly from a knowledge of their melting points and boiling points. Room temperature is considered to be approximately 68 °F or about 20 °C. If the melting point of a substance is above 20 °C, the substance exists as a solid at room temperature. (For example, the first member of the alkane series that is a naturally occurring solid is octadecane, $C_{18}H_{38}$, mp 28 °C.) If the boiling point of a substance is above 20 °C, the substance exists as a

Table 13.7 Physical Properties of Some Alkanes

Name	Molecular Formula	Melting Point (°C)	Boiling Point (°C)	Density (g/mL)	Normal State
Methane	CH_4	−182	−164	—	Gas
Ethane	C_2H_6	−183	−89	—	Gas
Propane	C_3H_8	−190	−42	—	Gas
Butane	C_4H_{10}	−138	−1	—	Gas
Pentane	C_5H_{12}	−130	36	0.63	Liquid
Hexane	C_6H_{14}	−95	69	0.66	Liquid
Heptane	C_7H_{16}	−91	98	0.68	Liquid
Octane	C_8H_{18}	−57	125	0.70	Liquid
Nonane	C_9H_{20}	−51	151	0.72	Liquid
Decane	$C_{10}H_{22}$	−30	174	0.73	Liquid

[3] All organic chemistry textbooks include many tables of physical properties. The figures in these tables are not meant to be memorized but are given in order to present a numerical description of the physical characteristics of organic molecules and to serve as standards of purity.

liquid at room temperature. (Note that pentane, bp 36 °C, is the first liquid alkane.) Similarly, knowledge of the density of a substance indicates whether the substance is lighter or heavier than water. We observe that oil and grease do not mix with water but rather float on the surface. They do not mix with water because they are *insoluble* in water; they float on top of the water because they are *less dense* than water. Densities are also useful in calculations that call for the conversion of a certain mass of a liquid into the corresponding volume of that liquid, or vice versa. (Recall that density equals the mass of a substance divided by its volume.)

This oil slick produced by the Exxon *Valdez* oil spill in Alaska in 1989 provides a reminder that hydrocarbons and water don't mix.

As indicated in Table 13.7, the first four members of the alkane series are gases. Natural gas is composed chiefly of methane, which has a density of about 0.65 g/L. The density of air is about 1.29 g/L. Natural gas, then, is much lighter than air, and it rises in a room in which there is a natural-gas leak. Once the leak is detected and eliminated, the gas can be removed from the room by opening an upper window. On the other hand, the three constituents of bottled gas are much heavier than air. Propane has a density of 1.6 g/L, and the butane isomers have densities of about 2.0 g/L. If bottled gas escapes into a room, it collects near the floor. This presents a much more serious fire hazard than a natural-gas leak because it is more difficult to rid the room of the heavier gas.

As shown in Table 13.7, the boiling points of the straight-chain alkanes increase with increasing molar mass. We shall see that this general rule holds true within all the families of organic compounds whenever the straight-chain homologs of the family are considered. Larger molecules are able to wrap around one another and interact, and thus more energy is required to separate them. In general, a straight-chain isomer has a higher boiling point than its branched-chain isomers. The straight-chain compounds can be likened to strands of spaghetti. The molecules can be very closely packed together, resulting in relatively strong intermolecular dispersion forces of attraction (Section 8.5). Dispersion forces depend on the total area of contact available between two molecules. The greater this area of contact, the greater the attractive force, and therefore the greater amount of heat needed to separate the molecules and reach the boiling point. Branched-chain hydrocarbons are more compact than their straight-chain isomers. Consequently, there is less surface area to interact. The dispersion forces between molecules are weaker, and the molecules can more easily escape from the liquid. Table 13.8 lists the melting points and boiling points of the isomers of butane and pentane.

Notice, from Table 13.8, that there is no trend that enables us to predict melting points. In general more symmetrical isomers tend to have higher melting points than less symmetrical isomers (because symmetrical molecules can pack closer together in the crystal lattice). Contrast 2,2-dimethylpropane with the less symmetrical pentanes in Table 13.8. Contrast, also, the very symmetrical octane

Table 13.8 Physical Properties of the Isomers of Butane and Pentane

IUPAC Name	Condensed Structural Formula	Melting Point (°C)	Boiling Point (°C)
Butane	$CH_3(CH_2)_2CH_3$	−138	−1
2-Methylpropane	$CH_3CH(CH_3)_2$	−159	−12
Pentane	$CH_3(CH_2)_3CH_3$	−130	36
2-Methylbutane	$CH_3CH_2CH(CH_3)_2$	−160	30
2,2-Dimethylpropane	$(CH_3)_4C$	−17	9

$CH_3CH_2CH_2CH_2CH_2CH_2CH_2CH_2CH_2CH_2CH_2CH_2CH_2CH_2CH_2$

$CH_3CH_2CH_2CH_2CH_2CH_2CH_2CH_2CH_2CH_2CH_2CH_2CH_2CH_2CH_2$

$CH_3CH_2CH_2CH_2CH_2CH_2CH_2CH_2CH_2CH_2CH_2CH_2CH_2CH_2CH_2$

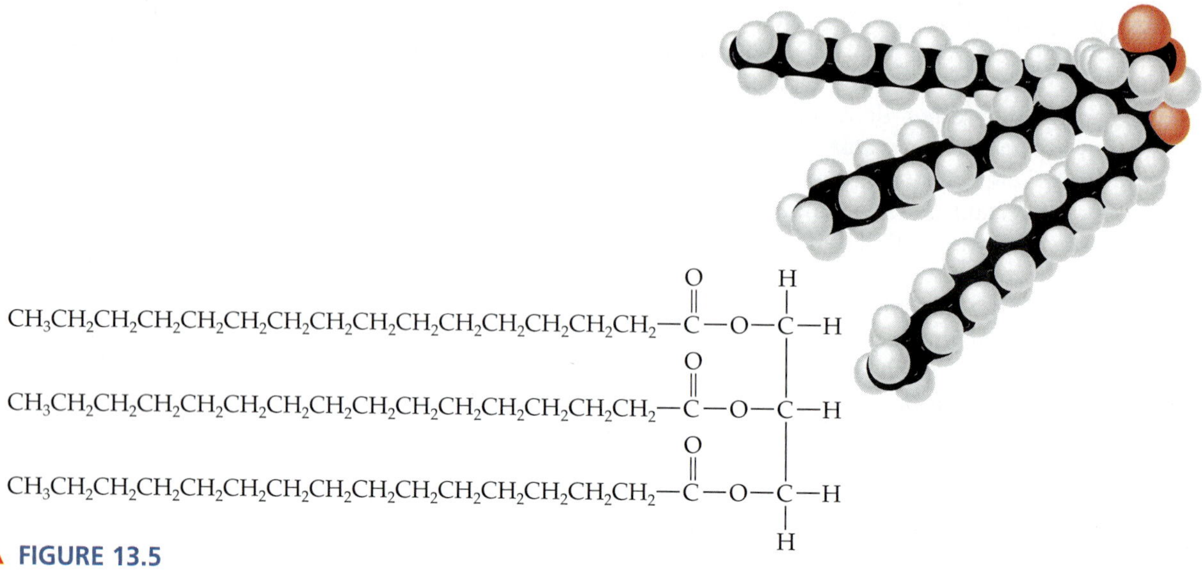

▲ **FIGURE 13.5**

Tripalmitin, a typical fat molecule, has long hydrocarbon chains typical of most lipids.

isomer $(CH_3)_3CC(CH_3)_3$ with the straight-chain octane: the former is a solid and melts at 101 °C; the latter is a liquid whose melting point is −57 °C.

An extensive review of the physical properties of the alkanes has been given here not because these compounds are so important in and of themselves but rather because of their contributions to the properties of other families of organic and biological compounds. For instance, a knowledge of alkane properties is vital to an understanding of the functions of lipids because large portions of their structures consist of segments of alkane groups (Figure 13.5). One of the major functions of phospholipids and sphingolipids (Chapter 20) is to serve as structural components of living tissues. Their biological importance here is dependent upon the presence of both polar and nonpolar groups, which enable them to bridge the gap between water-soluble and water-insoluble phases. This is a requisite in maintaining the selective permeability of cell membranes.

13.7 Physiological Properties of Alkanes

The physiological properties of alkanes vary in a regular way as we proceed through the homologous series. Methane appears to be totally physiologically inert. We could breathe a mixture of 80% methane and 20% oxygen without ill effect. Such a mixture would be flammable, however, and no fire or spark of any kind could be permitted in such an atmosphere. Breathing an atmosphere of pure methane (the "gas" of a gas stove) can lead to death, not because of the presence of methane but because of the absence of oxygen (asphyxia). In high concentrations, the other gaseous alkanes (and vapors of volatile liquid alkanes) act as anesthetics. They can also produce asphyxiation by excluding oxygen.

The lighter liquid alkanes (C_5 to about C_{12}) have varied effects on the body, depending on the part exposed. On the skin, alkanes dissolve body oils. Repeated contact may cause dermatitis. Swallowed, alkanes do little harm while in the stomach. In the lungs, however, they cause "chemical pneumonia" by dissolving fatlike molecules from cell membranes in the alveoli. The cells become less flexible, and the alveoli are no longer able to expel fluids. The buildup of fluids is similar to that which occurs in bacterial or viral pneumonia. People who swallow gasoline, petroleum distillates, or other liquid alkane mixtures should not be made to vomit, as this would increase the chance of their getting the alkanes into the lungs.

Heavier liquid alkanes (those above C_{17}), when applied to the skin, act as emollients (skin softeners). Such alkane mixtures as mineral oil can be used to replace natural skin oils washed away by frequent bathing or swimming. Petroleum jelly (Vaseline is one brand) is a semisolid mixture of hydrocarbons that can be applied as an emollient or simply as a protective film. Water and aqueous solutions (e.g., urine) will not dissolve such a film, which explains why petroleum jelly protects a baby's tender skin from diaper rash.

13.8 Chemical Properties of Alkanes: Little Affinity

The alkanes are the least reactive of all organic compounds. They generally do not react with strong acids, strong bases, most oxidizing agents, and most reducing agents. In fact, the alkanes undergo so few reactions that they are sometimes called **paraffins** (from the Latin *parum affinis,* little affinity).

The alkanes do undergo a few important reactions, most notably combustion (treated here) and halogenation (treated in Section 13.9). When mixed with oxygen at room temperature, alkanes give no apparent reaction. However, when a flame or spark supplies sufficient energy to get things started (the energy of activation), an exothermic (heat-producing) reaction proceeds vigorously. For methane, this reaction, which is called **combustion,** is

$$CH_4 + 2\,O_2 \longrightarrow CO_2 + 2\,H_2O + \text{heat}$$

If the reactants are adequately mixed and there is sufficient oxygen, the only products are carbon dioxide, water, and the all-important heat (for cooking foods, heating homes, and drying clothes). Conditions are rarely ideal, however, and other products are frequently formed. When the oxygen supply is limited, carbon monoxide is a by-product:

$$2\,CH_4 + 3\,O_2 \longrightarrow 2\,CO + 4\,H_2O$$

This reaction is responsible for dozens of deaths each year from unventilated or improperly adjusted gas heaters. (Similar reactions with similar results occur with kerosene heaters.) At the high temperatures achieved in some combustion reactions, some molecular atmospheric nitrogen (which makes up 80% of the atmosphere) is converted to oxides. Further, petroleum usually contains some sulfur compounds, and combustion reactions involving sulfur yield sulfur dioxide. Thus, the same combustion reactions that heat our homes, power our industries, and propel our automobiles also produce most of our air pollution.

13.9 Halogenated Hydrocarbons

Alkanes react with chlorine and bromine in the presence of ultraviolet light or at a high temperature to yield organochlorine and organobromine compounds. Fluorine combines explosively with most hydrocarbons, whereas iodine is relatively unreactive. In general, **halogenated hydrocarbons** are those in which one or more hydrogens have been replaced by halogen atoms:

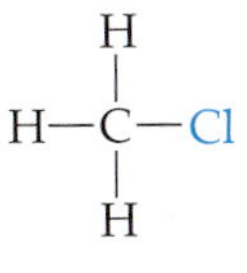

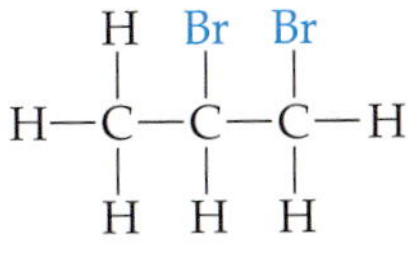

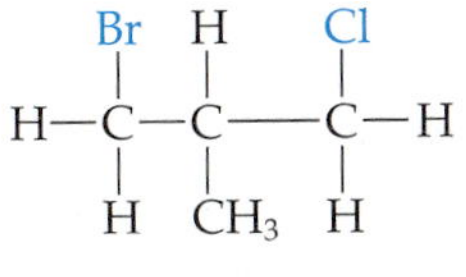

Replacement of only one hydrogen atom gives an **alkyl halide.** These compounds are given *common names* (which differ from official IUPAC names) that consist of two parts. The first is the name of the alkyl group; the second is the stem of the name of the halogen, with the ending *-ide.*

Example 13.5

Give the common name for CH_3CH_2Br.

Solution

The alkyl group (CH_3CH_2-) is ethyl. The halogen is bromine. The compound is therefore ethyl bromide.

Example 13.6

Give the common name for $(CH_3)_2CHCl$.

Solution

The alkyl group is isopropyl. The halogen is chlorine. The compound is therefore isopropyl chloride.

In the IUPAC system, halogen substituents are indicated by the prefixes fluoro-, chloro-, bromo-, and iodo-. The prefix is used with the name of the parent alkane, with numbers to indicate the position of the halogen if necessary.

Example 13.7

Give the IUPAC name for

$$\overset{1}{C}H_3\overset{2}{C}H\overset{3}{C}H_2\overset{4}{C}H_2\overset{5}{C}H_3$$
$$|$$
$$Cl$$

Solution

The parent alkane is pentane. The name of the compound is 2-chloropentane.

Example 13.8

Give the IUPAC name for

$$\overset{1}{C}H_3\overset{2}{C}H\overset{3}{C}H_2\overset{4}{C}H\overset{5}{C}H_2\overset{6}{C}H_3$$
$$\quad\;\; |\qquad\;\; |$$
$$\quad CH_3\quad Br$$

Solution

The parent alkane is hexane. The compound is named 4-bromo-2-methylhexane. (As usual, the substituents are given in alphabetical order.)

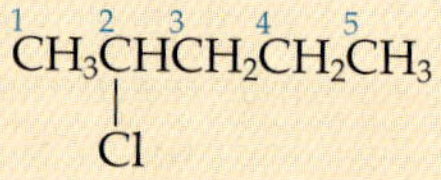

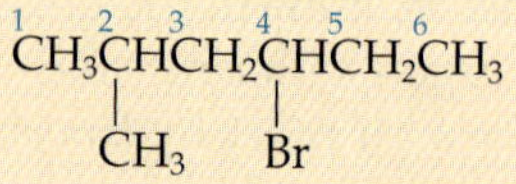

When more than one hydrogen on a hydrocarbon molecule is replaced by halogen atoms, the result is a wide variety of interesting and often useful compounds. Consider methane. One hydrogen can be replaced by chlorine, yielding methyl chloride (CH_3Cl). Methyl chloride is a refrigerant. It is also used in the manufacture of silicones and synthetic rubber, and as a general methylating agent. A second hydrogen can be substituted to yield dichloromethane, or methylene chloride (CH_2Cl_2), a common solvent used, for example, to extract most of the caffeine from coffee to make the decaffeinated brands. It boils at 40 °C; hence it can easily be removed by distillation if one wishes to recover the solute.

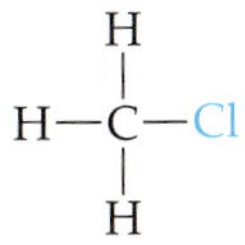

Methyl chloride

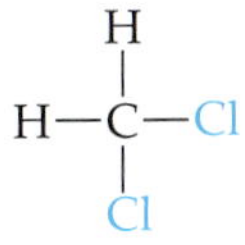

Methylene chloride

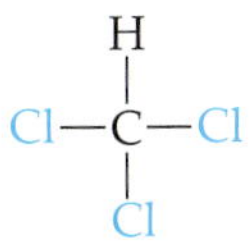

Chloroform

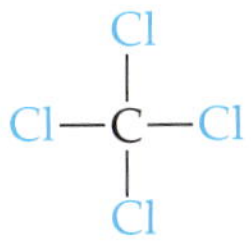

Carbon tetrachloride

A third methane hydrogen can be replaced by chlorine, giving chloroform—trichloromethane ($CHCl_3$), one of the first anesthetics. It has been largely replaced by safer, less toxic chemicals for that use, but it is still an important commercial and industrial solvent.

Replacing all four of methane's hydrogens with chlorine gives carbon tetrachloride (CCl_4), also called tetrachloromethane. Carbon tetrachloride has been used as a dry-cleaning solvent and in fire extinguishers. It is no longer recommended for either use. Exposure to carbon tetrachloride (or to most of the other chlorinated hydrocarbons, for that matter) can cause severe liver damage. Even when breathed in small amounts, the vapor can cause serious illness if the exposure is prolonged. Use of a carbon tetrachloride fire extinguisher in conjunction with water to put out a fire can be deadly. Carbon tetrachloride reacts with water at high temperatures to form phosgene ($COCl_2$), an extremely poisonous gas used during World War I.

With ethane, replacement of only one hydrogen gives ethyl chloride, which is used as an external local anesthetic. When sprayed on the skin, it begins to evaporate, and this cools the area enough to make it insensitive to pain. Replacement of two hydrogens can give either of two isomers. 1,2-Dichloroethane is commonly used as a solvent, particularly for rubber.

Ethyl chloride (chloroethane) is used as a spray-on local anesthetic to treat athletic injuries.

Halogenated Hydrocarbons and Ozone Depletion

Alkanes substituted with fluorine and chlorine atoms have been used as the dispersing gases in aerosol cans, as foaming agents for plastics, and as refrigerants. Properly called **chlorofluorocarbons (CFCs),** they are also known as *freons,* from their DuPont trade name. Three typical formulas are

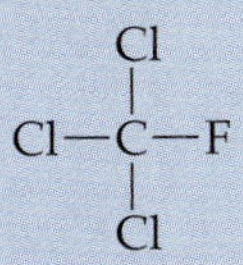

Freon 11

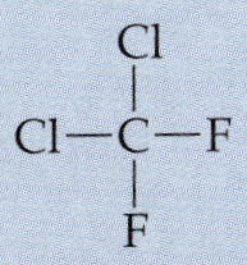

Freon 12

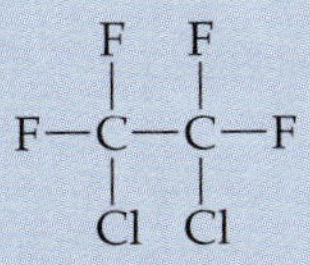

Freon 114

At room temperature, the chlorofluorocarbons are either gases or low-boiling-point liquids. They are essentially insoluble in water and inert toward most other substances. These properties make them ideal propellants in aerosol cans, but unfortunately, their inertness allows these compounds to persist in the environment.

Chlorofluorocarbons diffuse into the upper atmosphere, where they are broken down by ultraviolet radiation. Chlorine atoms formed in this process break down the ozone (O_3) that protects the Earth from harmful ultraviolet radiation:

$$O_3 + \text{ultraviolet light} \longrightarrow O_2 + O$$
$$CF_2Cl_2 + \text{ultraviolet light} \longrightarrow CF_2Cl\cdot + Cl\cdot$$
$$Cl\cdot + O_3 \longrightarrow ClO\cdot + O_2$$
$$\cdot ClO + O \longrightarrow Cl\cdot + O_2$$

Note that the last step yields another chlorine atom that can break down another molecule of ozone. The third and fourth steps are repeated many times; thus, the decomposition of one molecule of chlorofluorocarbon can result in the destruction of many molecules of ozone.

In 1995, Mario Molina, Sherwood Rowland, and Paul Crutzen shared the Nobel Prize in chemistry for sounding the alarm about the depletion of the Earth's protective ozone layer. The research of the three scientists predicting an ozone "hole" laid the groundwork for its discovery in 1985 over the South Pole.

The U.S. National Research Council predicts a 2–5% increase in skin cancer for each 1% depletion of the ozone layer. In 1984, the NRC predicted a 2–4% depletion by late in the twenty-first century. Worries escalated in 1987 with the discovery that ozone concentrations over Antarctica had plummeted to a record low. This thinning or "hole" in the ozone layer has been getting worse since 1979. A similar but less dramatic thinning has been detected in the Arctic.

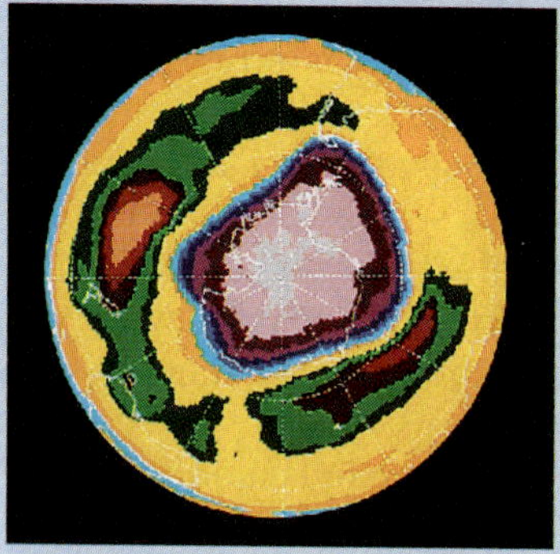

The ozone hole over Antarctica. Ozone in the upper atmosphere shields Earth's surface from ultraviolet radiation from the sun, which can cause skin cancer in humans and is also harmful to other animals and to some plants. Ozone "holes" in the upper atmosphere (the gray, pink, and purple areas at the center) are large areas of substantial ozone depletion. They occur mainly over Antarctica from late August through early October and fill in about mid-November. Ozone depletion has also been noted over the Arctic regions.

Winter and early spring stratospheric ozone depletion at middle to high latitudes is estimated to be 2–10% over the last 25 years. Satellite data show that the ozone layer has been thinning at a rate of 0.5% per year since 1978. In the fall of 1995, scientists reported that the ozone hole over the Antarctic was growing faster than ever and was roughly the size of Europe. Every 1% drop in ozone means about 1.5 percent more ultraviolet radiation reaching the surface of the earth.

In the United States, chlorofluorocarbons have been banned from most aerosol preparations and are being phased out as refrigerants. They escape into the atmosphere from refrigerators and air conditioners through leaks and when the appliances are carelessly discarded.

Worldwide action to reduce use of CFCs and related compounds began in 1987, and an international agreement calling for a total ban on the release of CFCs by 1996 was reached in 1992. The concentration of CFCs in the stratosphere has peaked and will decline over the next century.

Other such compounds as nitrogen oxides from agricultural activities and from aircraft that fly in the stratosphere also deplete the ozone shield.

The current strategy for dealing with the ozone depletion problem is to replace the offensive CFCs and other chlorine- or bromine-containing compounds that might diffuse into the stratosphere with more benign substances. Fluorocarbons (HFCs), which have no Cl or Br to form radicals, are one alternative. Another is to use hydrochlorofluorocarbons (HCFCs); these molecules break down more readily in the troposphere, and fewer ozone-destroying molecules reach the stratosphere.

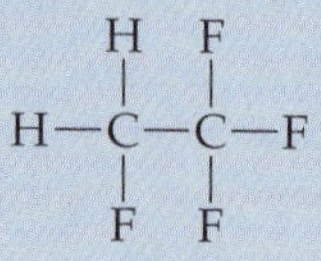

An HFC

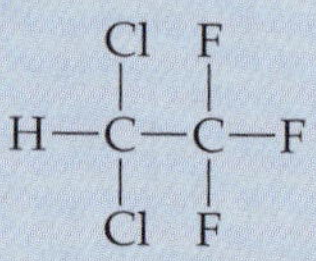

An HCFC

Replacement of three ethane hydrogen atoms with chlorine atoms also gives two isomers, both used as commercial and industrial solvents. 1,1,1-Trichloro-ethane has been used for cleaning molds that are used in the fabrication of plastic items.

Replacement of four ethane hydrogens with chlorine also leads to isomers. Only one is of importance, however—1,1,2,2-tetrachloroethane. This compound is widely used but highly toxic. It dissolves rubber, cellulose acetate, and other complex organic materials. It is also used to sterilize soil and as a component of weed killers and insecticide preparations.

13.10 Cycloalkanes

The hydrocarbons we have encountered so far have been open-ended chains. Carbon and hydrogen atoms can also hook up in other arrangements, some of them quite interesting. There is a synthetic hydrocarbon that has the formula C_3H_6, with the three carbons joined in a *ring* or *cycle*. The compound is called cyclopropane (Figure 13.6). In addition to having an interesting structure, cyclo-propane has some intriguing properties. It is a potent, quick-acting anesthetic with few undesirable side effects. It is seldom used in surgery, however, because it forms explosive mixtures with air at nearly all concentrations.

Names of cycloalkanes are formed by addition of the prefix *cyclo-* in front of the name of the open-chain compound having the same number of carbon atoms as there are in the ring. Thus, the name for the cyclic compound C_4H_8 is cyclobu-tane. Names and structures of several cycloalkanes are given in Figure 13.7. The carbon atoms in cyclic compounds form regular geometric figures, and in the very abbreviated structures shown in Figure 13.7, each corner of the geometric figure represents a carbon plus as many hydrogens as needed to give the carbon four bonds. For example, the triangle stands for C_3H_6,

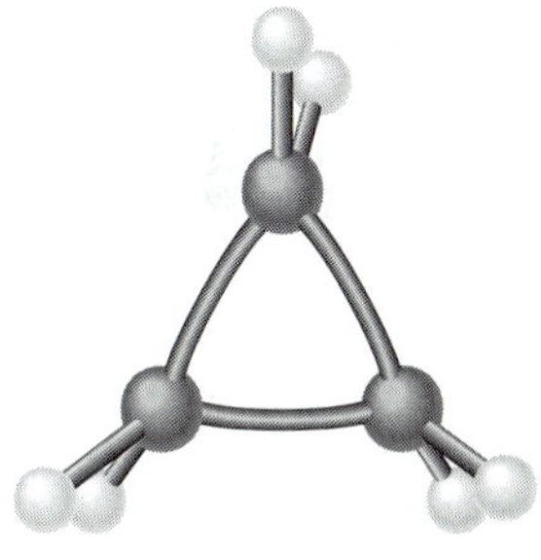

▲ **FIGURE 13.6**

Ball-and-stick model of cyclopropane.

$$
\begin{array}{c}
H_2 \\
C \\
\diagup\;\diagdown \\
H_2C\!-\!CH_2
\end{array}
$$

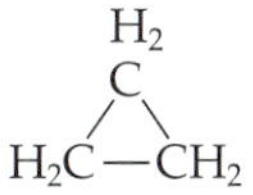

and the hexagon for C_6H_{12}.

$$
\begin{array}{c}
H_2 \\
C \\
H_2C \qquad CH_2 \\
| \qquad\qquad | \\
H_2C \qquad CH_2 \\
CH_2
\end{array}
$$

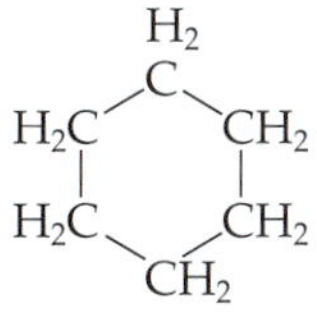

Cyclic compounds may contain attached substituent groups. If there is only one substituent on the ring, the substituent does not have to be numbered because

◀ **FIGURE 13.7**

Some cycloalkanes.

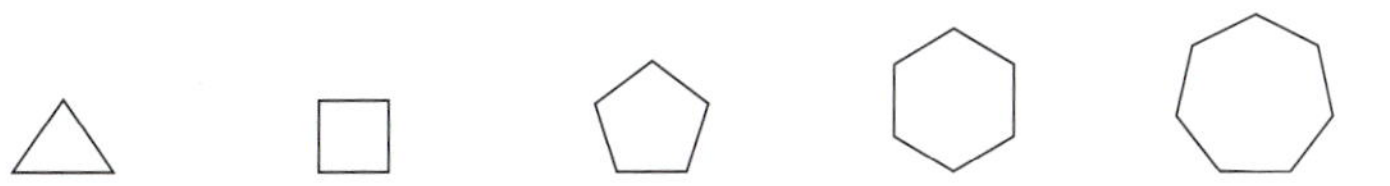

Cyclopropane Cyclobutane Cyclopentane Cyclohexane Cycloheptane

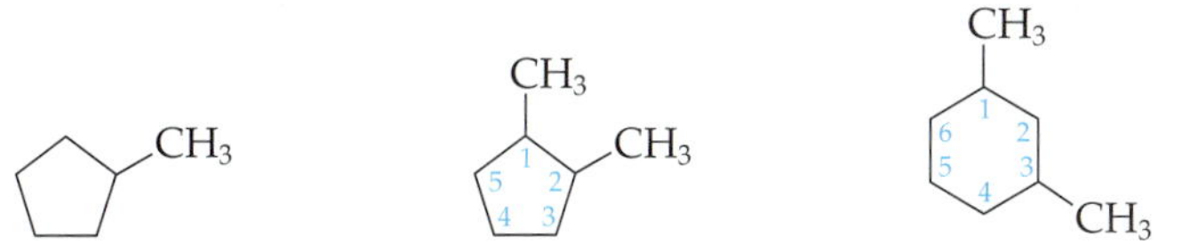

Methylcyclopentane 1,2-Dimethylcyclopentane 1,3-Dimethylcyclohexane

all positions on the ring are equivalent. When there is more than one substituent, numbers are required. The ring carbons are numbered so that the carbons bearing the substituents have the lowest numbers.

The major distinction between the cyclic and noncyclic alkanes arises from differences in geometric configuration. There is free rotation about all carbon–carbon bonds of open–chain alkanes. This is not the case with the cyclic compounds because the ring structure holds the carbons rigidly in place. Free rotation is impossible without disruption of the ring structure. For cyclic compounds, the position of any substituent relative to the ring becomes extremely important. (The substituent is situated either above or below the ring.) A new form of isomerism is possible, which we discuss in Section 18.4. For now, let it suffice to say that the compounds at the left are not the same, but are isomers.

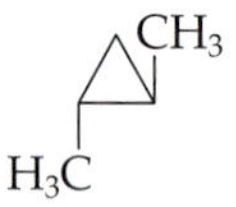

Example 13.9

Draw structures for (a) cyclooctane, (b) ethylcyclohexane, (c) 1,1,2-trimethylcyclobutane, and (d) 1-ethyl-1,2,5,5-tetramethylcycloheptane.

Solution

a.
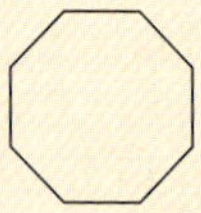

b.
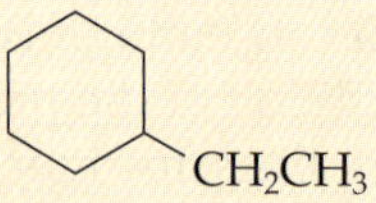

c.
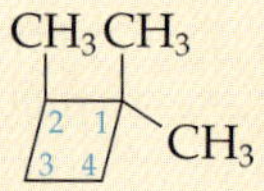

d.
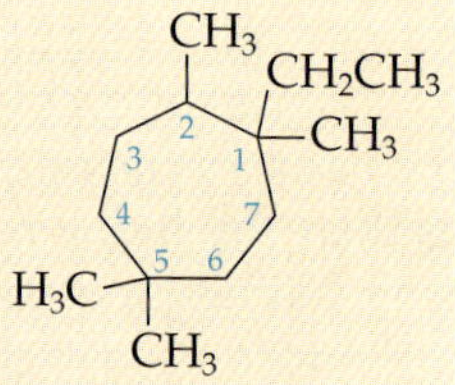

Practice Exercise

Draw structures for (a) cyclopropane and (b) 1-ethyl-2-methylcyclopentane.

The physical, chemical, and physiological properties of cyclic hydrocarbons are generally quite similar to those of the corresponding open-chain compounds. Cycloalkanes (with the exception of cyclopropane, which has a highly strained ring) act very much like noncyclic alkanes. Cyclic hydrocarbons with five- and six-membered rings occur in petroleum from certain areas; California crude, for instance, is particularly rich in these compounds. Like all other hydrocarbons, cyclic hydrocarbons burn.

13.11 Alkenes: Structure and Nomenclature

Not all hydrocarbons are as resistant to reaction as are the alkanes. In fact, the next family we consider is quite reactive. This family, called **alkenes** (note the *-ene* ending), is characterized by the presence of a carbon–carbon double bond. Names, structures, and physical properties of a few representative alkenes are given in Table 13.9. Like all other hydrocarbons, the alkenes are insoluble in water but soluble in organic solvents.

We have used only condensed structural formulas in this table. Thus, $CH_2{=}CH_2$ stands for

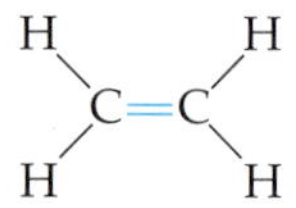

Table 13.9 Physical Properties of Some Selected Alkenes

IUPAC Name	Molecular Formula	Condensed Structure	Melting Point (°C)	Boiling Point (°C)
Ethene	C_2H_4	$CH_2{=}CH_2$	−169	−104
Propene	C_3H_6	$CH_3CH{=}CH_2$	−185	−47
1-Butene	C_4H_8	$CH_3CH_2CH{=}CH_2$	−185	−6
1-Pentene	C_5H_{10}	$CH_3CH_2CH_2CH{=}CH_2$	−138	30
1-Hexene	C_6H_{12}	$CH_3(CH_2)_3CH{=}CH_2$	−140	63
1-Heptene	C_7H_{14}	$CH_3(CH_2)_4CH{=}CH_2$	−119	94
1-Octene	C_8H_{16}	$CH_3(CH_2)_5CH{=}CH_2$	−102	121

The double bond is shared by the two carbon atoms and does not involve the hydrogens, although the condensed formula does not make this point obvious.

The physical properties of alkenes are quite similar to those of the corresponding alkanes. Like alkanes, alkenes are insoluble in water, and they float on water.

Compare the molecular formulas of the alkenes in Table 13.9 with those of the alkanes in Table 13.4. The molecular formula for ethane is C_2H_6, but the formula for the two-carbon alkene is C_2H_4. The alkene has two fewer hydrogens than the alkane because each alkene carbon atom involved in the double bond gives up one hydrogen atom. Because the double-bond carbons do not have the maximum number of attached atoms (four), alkenes are called **unsaturated hydrocarbons.** Alkenes with one double bond have the general formula C_nH_{2n}.

The first two alkenes in Table 13.9, ethene and propene, are most often called by their common names, ethylene and propylene, respectively. Ethylene ($CH_2{=}CH_2$) is a major commercial chemical. The U.S. chemical industry produces about 18 billion kg of ethylene annually, making it the most important of all synthetic organic chemicals. More than half of this ethylene goes into the manufacture of polyethylene, one of the most familiar plastics (Section 13.14). Another one-sixth is converted to ethylene glycol (Section 14.7), the principal component of most brands of antifreeze for automobile radiators.

Propylene ($CH_3CH{=}CH_2$) is also an important industrial chemical. It is converted to plastics, isopropyl alcohol (Chapter 14), and a variety of other products.

Naming Alkenes

Although there is only one alkene with the formula C_2H_4 (ethene) and only one with the formula C_3H_6 (propene), there are several alkenes with the formula C_4H_8. Common names are hardly helpful in naming the many isomers of the higher alkenes. For the most part, the IUPAC system is used for these compounds. Some of the IUPAC rules for alkenes are as follows.

1. All have names ending in *-ene.*
2. The longest chain of atoms *containing the double bond* is the parent compound. The name has the same stem as the alkane having the same number of carbon atoms. Thus, the compound $CH_3CH{=}CH_2$, with three carbon atoms, is named *propene.*
3. When it is necessary to indicate the position of the double bond, the first carbon of the two that are doubly bonded is given the lowest possible number. The compound $CH_3CH{=}CHCH_2CH_3$, for example, has the double bond between the second and third carbon atoms. Its name is 2-pentene.

4. Substituent groups are named as with alkanes. Their position is indicated by a number. Thus,

$$CH_3CHCH_2CH{=}CHCH_3$$
$$| $$
$$CH_3$$

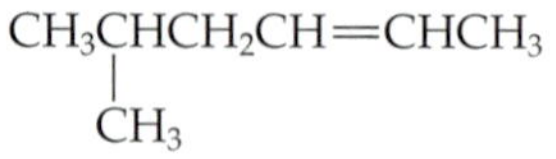

is 5-methyl-2-hexene. Note that the numbering of the parent chain is always done in such a way as to give the double bond the lowest number, even if that forces a substituent to have a higher number. We say that the double bond has priority in numbering.

Table 13.10 shows the IUPAC names of some representative alkenes and cycloalkenes. Study the table and then try the following examples.

Example 13.10

Name the compound

$$CH_3CH{=}CHCH_2CH{-}CHCH_3$$
$$| \quad |$$
$$CH_3 \quad CH_3$$

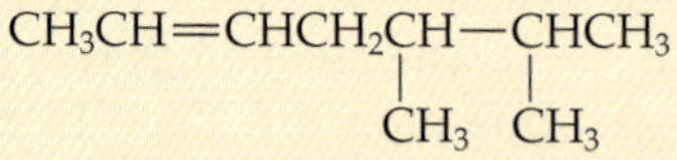

Solution

The longest chain containing the double bond has seven carbon atoms. To give the first carbon of the double bond the lowest number, we start numbering from the left:

$$\overset{1}{C}H_3\overset{2}{C}H{=}\overset{3}{C}H\overset{4}{C}H_2\overset{5}{C}H{-}\overset{6}{C}H\overset{7}{C}H_3$$
$$| \quad |$$
$$CH_3 \quad CH_3$$

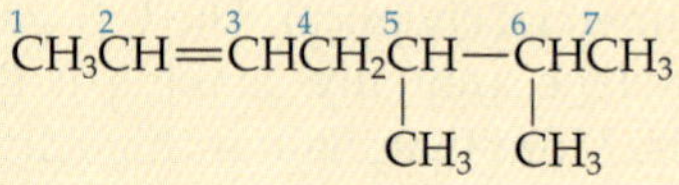

The name of the compound is 5,6-dimethyl-2-heptene.

Example 13.11

Name the compound

$$CH_2{=}C{-}CH_2CH_3$$
$$|$$
$$CH_2CH_3$$

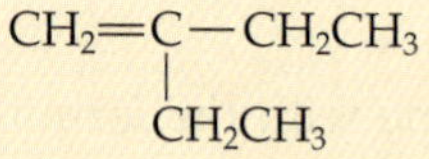

Solution

The name is 2-ethyl-1-butene. The longest chain contains five carbon atoms, but this chain does not contain the double bond and so the parent name is not pent-. The longest continuous chain *containing the double bond* incorporates only four carbon atoms, and this four-carbon chain serves as the parent compound.

Practice Exercise

Name the following compounds.

a. $CH_3C{=}CHCHCH_2CH_3$
 $\quad |\qquad\quad|$
 $\quad CH_3 \quad CH_2CH_3$

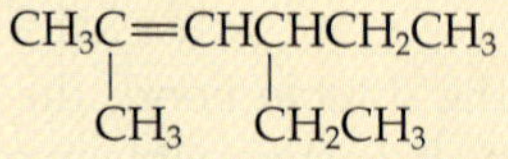

b.

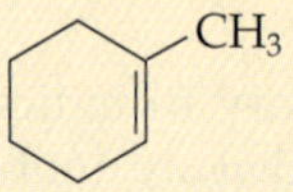

Table 13.10 IUPAC Names of Representative Alkenes and Cycloalkenes

Condensed Structural Formula	Rewritten and Numbered	IUPAC Name
$CH_3(CH_2)_2CH{=}CH_2$	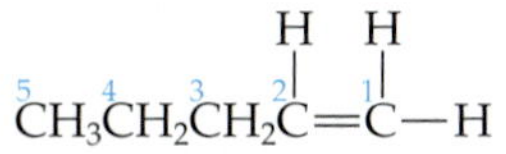	1-Pentene
$CH_3CH{=}CHCH_2CH_3$	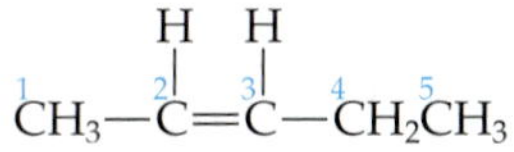	2-Pentene
$CH_3CH_2CH{=}C(CH_3)_2$	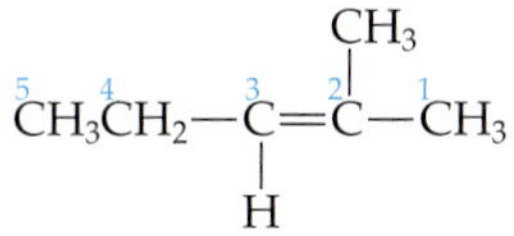	2-Methyl-2-pentene
$(CH_3)_2CHC(CH_3){=}CHCH_3$	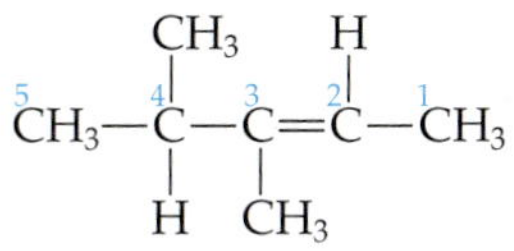	3,4-Dimethyl-2-pentene
$CH_3CH_2CH(CH_3)C(C_3H_7){=}CH_2$	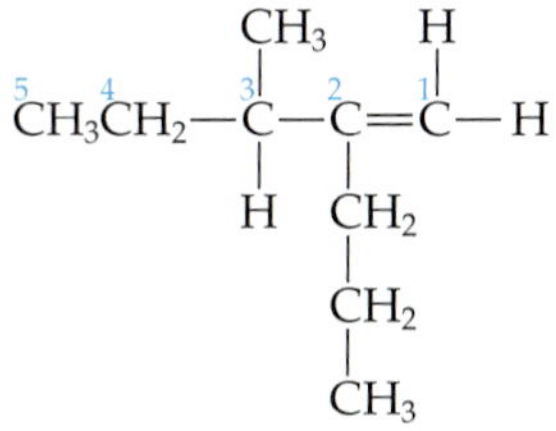	3-Methyl-2-propyl-1-pentene
		Cyclopentene
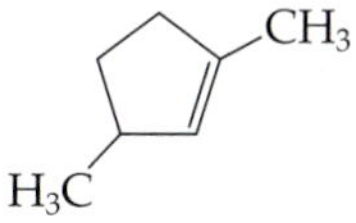	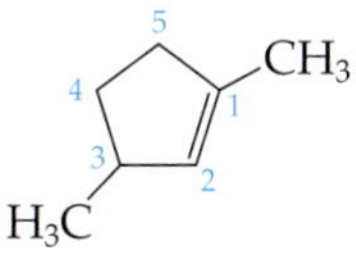	1,3-Dimethylcyclopentene
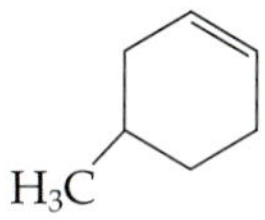	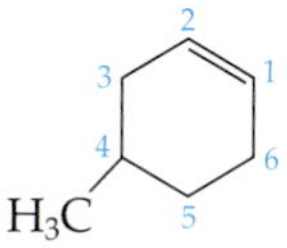	4-Methylcyclohexene
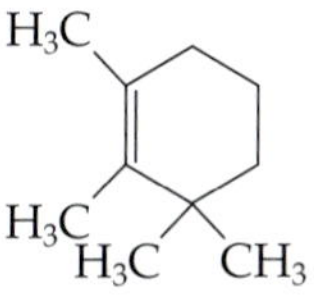	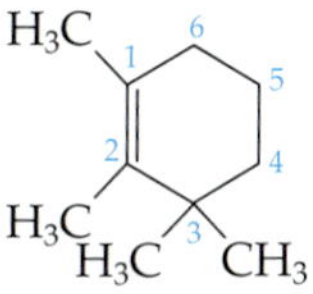	1,2,3,3-Tetramethylcyclohexene
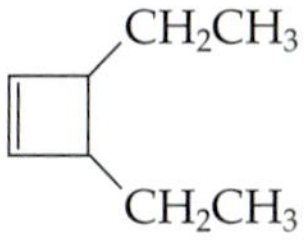	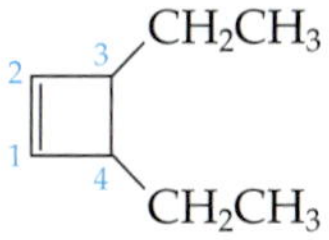	3,4-Diethylcyclobutene

Example 13.12

Draw the structural formula for 3,4-dimethyl-2-pentene.

Solution

First write the parent chain of five carbons:

$$C—C—C—C—C$$

Then add the double bond between the second and third carbons:

$$\overset{1}{C}—\overset{2}{C}=\overset{3}{C}—\overset{4}{C}—\overset{5}{C}$$

Now add the groups at their proper positions:

$$\left(\text{or} \quad CH_3—CH=\overset{\overset{\displaystyle CH_3}{|}}{C}——\overset{\overset{\displaystyle CH_3}{|}}{CH}—CH_3 \right)$$

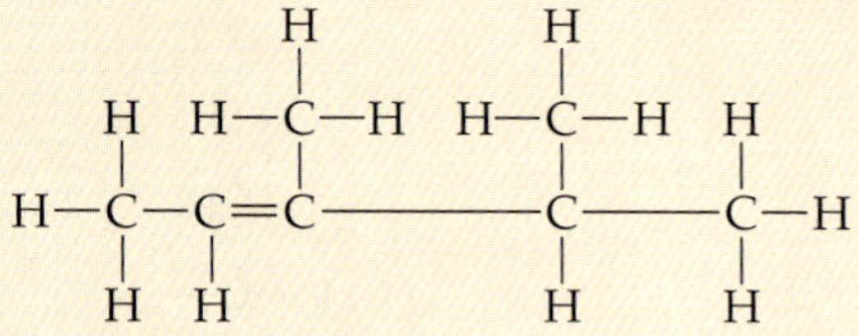

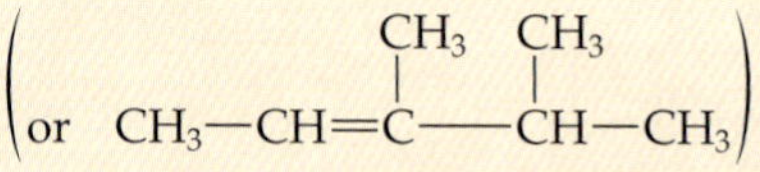

Practice Exercise

Draw structural formulas for the following compounds.

a. 3-ethyl-2-methyl-1-hexene b. 3-isopropylcyclopentene

The double bond of alkene molecules, like the ring structure of cycloalkanes, imposes geometric restrictions. Recall that, according to VSEPR theory (Section B.1), each doubly bonded carbon atom lies in the center of an equilateral triangle (to minimize the repulsive forces among the three regions of electron density). The carbon atoms of a double bond and the two atoms bonded to each carbon all lie in a single plane (Figure 13.8). Free rotation about doubly bonded carbon atoms is *not* possible without rupturing the bond. Therefore, the relative positions of substituent groups located above or below the double bond become very significant. The nomenclature employed in situations such as this, as well as the consequences of restricted rotation, are discussed in Chapter 18.

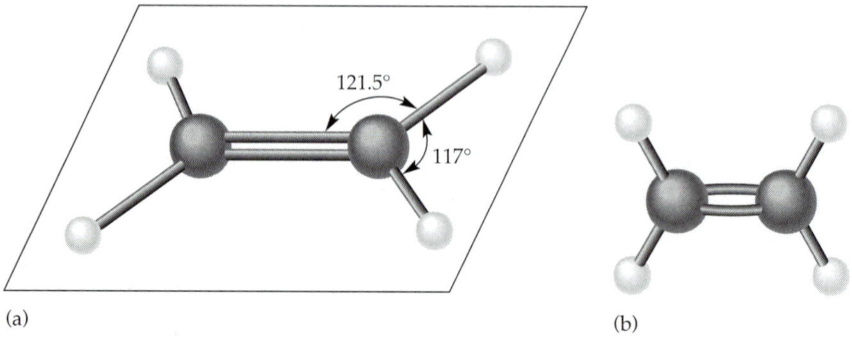

FIGURE 13.8 ▶

(a) Planar configuration of the carbon-to-carbon double bond.
(b) Ball-and-spring model of ethylene (ethene).

121.5°

117°

(a)

(b)

13.12 Alkenes and Living Things

The physiological properties of the alkenes are similar to those of the alkanes. Ethylene has found some use as an inhalation anesthetic. Like the gaseous alkanes, ethylene can cause unconsciousness and even death by asphyxiation. Large amounts of liquid and solid (or mixtures of liquid and solid) alkenes are seldom encountered. They would probably act on or in our bodies much as the alkanes do.

Alkenes occur widely in nature. Ripening fruits and vegetables give off ethylene, which triggers further ripening. Fruit processors artificially introduce ethylene to hasten the normal ripening process; 1 kg of tomatoes can be ripened by exposure to as little as 0.1 mg of ethylene for 24 hours. Unfortunately, the tomatoes don't taste much like those that ripen on the vine.

Other alkenes that occur in nature include 1-octene, a constituent of lemon oil, and octadecene ($C_{18}H_{36}$), found in fish liver. Dienes (which have two double bonds) and polyenes (three or more double bonds) are also common. Butadiene ($CH_2{=}CH{-}CH{=}CH_2$) is found in coffee. A hexadecadiene ($C_{16}H_{30}$) occurs in olive oil. Lycopene and the carotenes are isomeric polyenes ($C_{40}H_{56}$) that give the attractive red, orange, and yellow colors to watermelons, tomatoes, carrots, and other vegetables and fruits. Vitamin A, essential to good vision, is derived from a carotene (see Special Topic H). The world would be a much less colorful place without alkenes.

13.13 Chemical Properties of Alkenes

Like the alkanes—and all other hydrocarbons—the alkenes burn. While this is a hazard you should remember when working with alkenes, these compounds are not commercially important as fuels. We can write an equation for the combustion of ethene that is similar to the one for alkane combustion:

$$C_2H_4 \; + \; 3\,O_2 \; \longrightarrow \; 2\,CO_2 \; + \; 2\,H_2O \; + \; \text{heat}$$

The typical reactions of the alkenes are **addition reactions.** One of the bonds in the double bond is broken, permitting each of the involved carbon atoms to bond to an additional atom or group. The originally doubly bonded carbons are still attached by the remaining single bond. Perhaps the simplest addition reaction is that which occurs with hydrogen in the presence of a nickel (Ni), platinum (Pt), or palladium (Pd) catalyst:

Alkenes typically undergo addition reactions in which all the atoms of the reactants are incorporated into a single product.

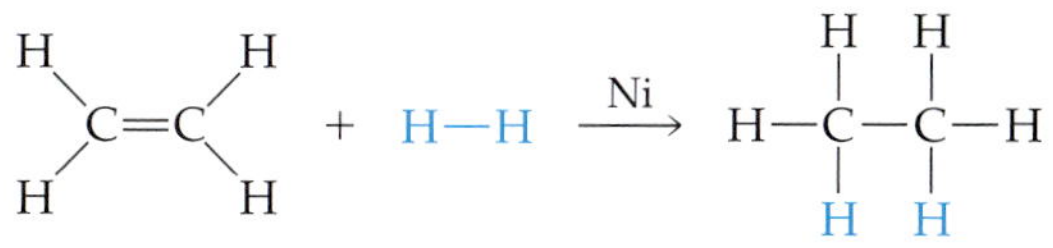

The product of this reaction is an alkane having the same carbon skeleton as the alkene. This addition of hydrogen to an unsaturated molecule is called **hydrogenation** and was once widely used in industry to convert unsaturated vegetable oils to saturated fats. The difference between the liquid oil and the solid fat is due to a difference in the number of double bonds present; there are more in the unsaturated oil and fewer in the saturated fat. Vegetable oils are now more readily accepted and, in fact, are frequently preferred by the consumer, but some oils are still hydrogenated so that the vegetable product will resemble an animal product. Margarine, for example, resembles butter by virtue of hydrogenation (see Chapter 20).

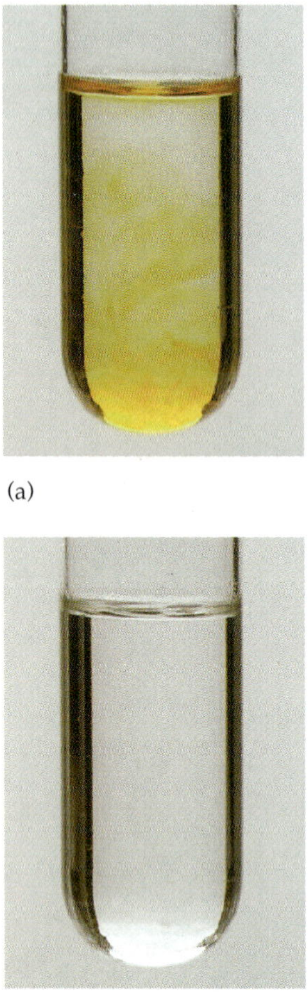

(a)

(b)

Alkenes readily add halogen molecules. Indeed, the reaction with bromine is often used to test for alkenes (Figure 13.9). Solutions of bromine are brownish red. When an alkene is added to such a solution, the color disappears because the alkene reacts with the bromine:

$$\underset{\substack{\text{Ethene}}}{\underset{\text{H}}{\overset{\text{H}}{\diagup}}\text{C}=\text{C}\underset{\text{H}}{\overset{\text{H}}{\diagdown}}} + \underset{\substack{\text{Bromine}\\\text{(brownish red)}}}{\text{Br}-\text{Br}} \longrightarrow \underset{\substack{\text{1,2-Dibromoethane}\\\text{(colorless)}}}{\text{H}-\overset{\overset{\text{H}}{|}}{\underset{\underset{\text{Br}}{|}}{\text{C}}}-\overset{\overset{\text{H}}{|}}{\underset{\underset{\text{Br}}{|}}{\text{C}}}-\text{H}}$$

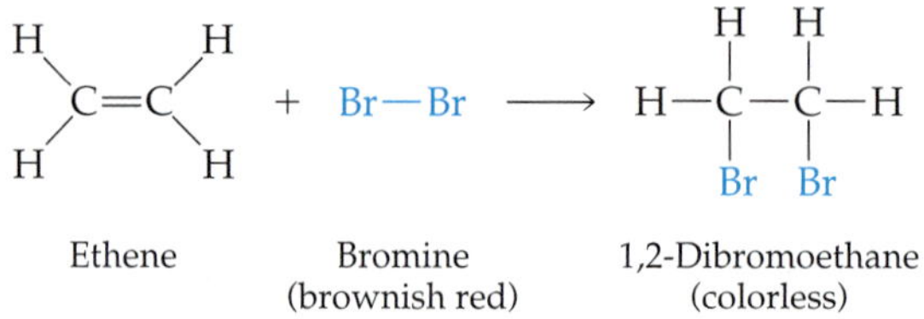

Another important addition reaction is that between alkenes and water. This reaction, called **hydration,** requires the presence of a mineral acid, such as sulfuric acid, as a catalyst:

$$\underset{\substack{\text{Ethene}}}{\underset{\text{H}}{\overset{\text{H}}{\diagup}}\text{C}=\text{C}\underset{\text{H}}{\overset{\text{H}}{\diagdown}}} + \underset{\substack{\text{Water}}}{\text{H}-\text{OH}} \xrightarrow{\text{H}_2\text{SO}_4} \underset{\substack{\text{Ethyl alcohol}}}{\text{H}-\overset{\overset{\text{H}}{|}}{\underset{\underset{\text{H}}{|}}{\text{C}}}-\overset{\overset{\text{H}}{|}}{\underset{\underset{\text{OH}}{|}}{\text{C}}}-\text{H}}$$

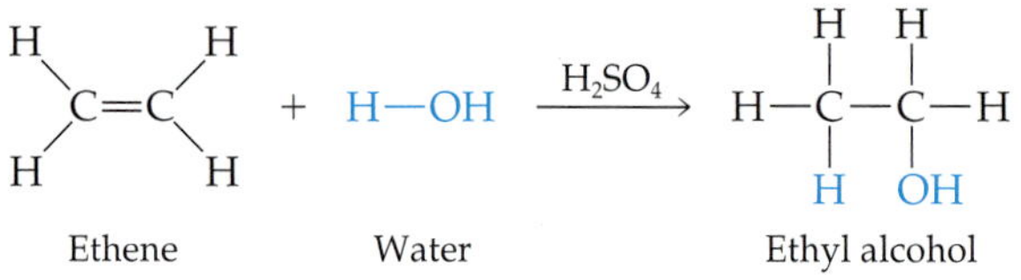

Vast quantities of ethyl alcohol, for use as an industrial solvent, are made from ethylene. This alcohol is structurally identical to that used in alcoholic beverages. However, federal law requires that all drinking alcohol be produced by the natural process called fermentation. (See Section 14.4 for a more extensive discussion of the hydration reaction.)

FIGURE 13.9 ▲

(a) When bromine is added to a saturated hydrocarbon such as cyclohexane, the brownish-red color remains because the bromine dissolves in the cyclohexane but does not react with it. (b) When bromine is added to an unsaturated hydrocarbon such as cyclohexene, the brownish-red color disappears because the bromine adds to the double bond of the cyclohexene. This reaction serves as a convenient test for the presence of a carbon-to-carbon double bond in a molecule.

Example 13.13

Write equations for the reaction between $CH_3CH=CHCH_3$ and (a) H_2 (Ni catalyst), (b) Br_2, and (c) H_2O (H_2SO_4 catalyst).

Solution

In each reaction, the reagent adds across the double bond:

a. $CH_3CH=CHCH_3 + H_2 \xrightarrow{\text{Ni}} CH_3CH-CHCH_3$ or $CH_3CH_2CH_2CH_3$
 $\qquad\qquad\qquad\qquad\qquad\qquad\qquad\quad |\quad\; |$
 $\qquad\qquad\qquad\qquad\qquad\qquad\qquad\; H\;\; H$

b. $CH_3CH=CHCH_3 + Br_2 \longrightarrow CH_3CH-CHCH_3$
 $\qquad\qquad\qquad\qquad\qquad\qquad\qquad\qquad\;\; |\quad\;\; |$
 $\qquad\qquad\qquad\qquad\qquad\qquad\qquad\;\; Br\;\; Br$

c. $CH_3CH=CHCH_3 + H_2O \xrightarrow{\text{H}_2\text{SO}_4} CH_3CH-CHCH_3$ or $CH_3CH_2CHCH_3$
 $\qquad\qquad\qquad\qquad\qquad\qquad\qquad\qquad\quad |\quad\;\; |\qquad\qquad\qquad\quad |$
 $\qquad\qquad\qquad\qquad\qquad\qquad\qquad\qquad\; H\;\; OH\qquad\qquad\qquad OH$

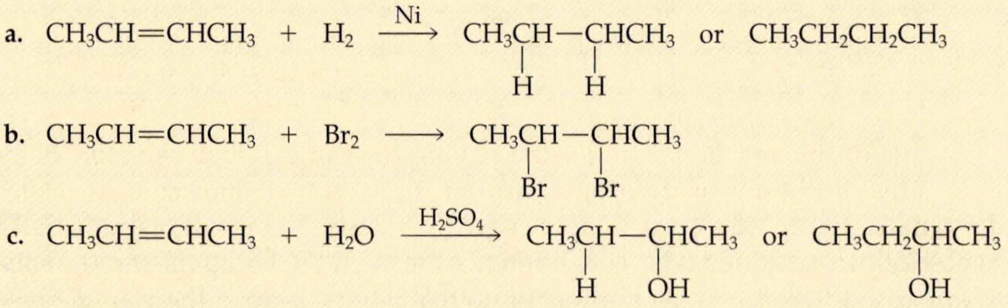

Practice Exercise

Write equations for the reaction of ⬠ with each of the following.
a. H_2 (Ni catalyst) **b.** Cl_2 **c.** H_2O (H_2SO_4 catalyst)

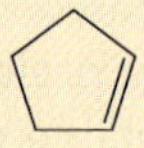

13.14 Polymerization

The most important commercial reaction of alkenes is **polymerization.** Chemists call giant molecules assembled from much smaller ones **polymers** (from the Greek *poly,* many, and *meros,* parts). In a laboratory (and in living systems, for that matter), the preparation of polymers involves the hooking together of many smaller molecules. The small molecules, the building blocks, are called **monomers** (from the Greek *monos,* one and *meros,* parts). A polymer is as different from its monomer as a long strand of spaghetti is from a tiny speck of flour. For example, polyethylene, the familiar waxy material used to make plastic bags, is made from the monomer ethylene, which is a gas.

There are two general types of polymerization reactions: addition polymerization and condensation polymerization. In **addition polymerization,** the monomers add to one another in such a way that the polymer contains all the atoms of the starting monomers. Under high pressure and temperature and in the presence of a catalyst, ethylene molecules are made to join together in long chains (Figure 13.10). The polymerization can be represented by the reaction of a few monomer units:

A giant bubble of tough, transparent plastic film emerges from a die of an extruding machine. The film is used in packaging, consumer products, and food services.

$$\cdots + \underset{H}{\overset{H}{C}}{=}\underset{H}{\overset{H}{C}} \; + \; \underset{H}{\overset{H}{C}}{=}\underset{H}{\overset{H}{C}} \; + \; \underset{H}{\overset{H}{C}}{=}\underset{H}{\overset{H}{C}} \; + \; \cdots \longrightarrow \cdots \underset{H}{\overset{H}{C}}{-}\underset{H}{\overset{H}{C}}{-}\underset{H}{\overset{H}{C}}{-}\underset{H}{\overset{H}{C}}{-}\underset{H}{\overset{H}{C}}{-}\underset{H}{\overset{H}{C}} \cdots$$

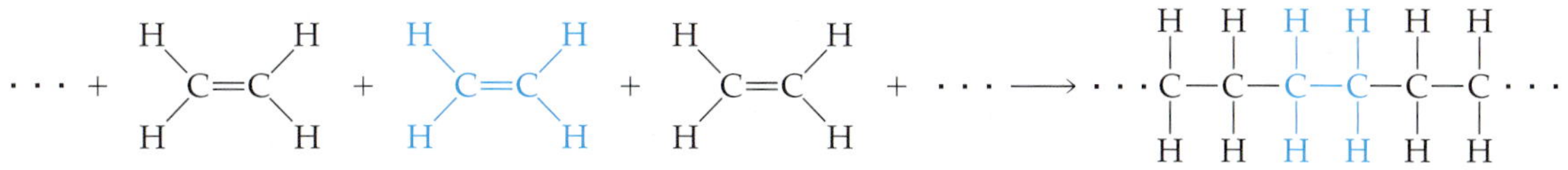

The dotted lines in the formula of the product are like etc.'s: they indicate that the structure extends for many units in each direction. Notice that the two carbon atoms and four hydrogen atoms of each monomer molecule are incorporated into the polymer structure.

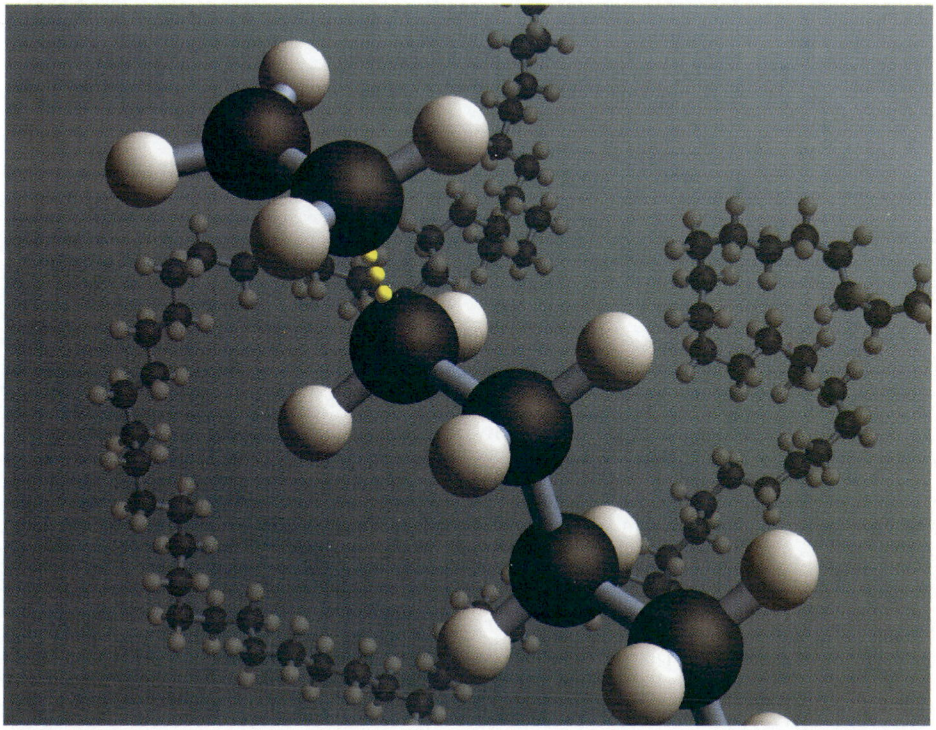

◄ **FIGURE 13.10**

The formation of poly-ethylene. In the synthesis of polyethylene, many monomer units join together to form huge polymer molecules. In this computer-generated representation, the yellow dots indicate a new bond forming as an ethylene molecule (upper left) is added to the growing polymer chain.

The Many Uses of Polyethylene

Polyethylene was invented shortly before the start of World War II. Before long, it was used for insulating cables in a top-secret invention—radar—which helped British pilots spot enemy aircraft before the aircraft became visible to the naked eye. Polyethylene proved to be tough and flexible and an excellent electrical insulator. It could withstand both high and low temperatures. Without polyethylene, the British could not have had effective radar, and without radar the Battle of Britain might have been lost. The invention of this simple plastic may have changed the course of history.

Today, there are two principal kinds of polyethylene produced by the use of different catalysts and different reaction conditions. *High-density polyethylenes (HDPE)* have largely linear molecules that pack closely together. These linear molecules can assume a fairly ordered, crystalline structure; this structure gives high-density polyethylenes greater rigidity and higher tensile strength than other ethylene plastics. Linear polyethylenes are used for such things as threaded bottle caps, toys, bottles, and gallon milk jugs.

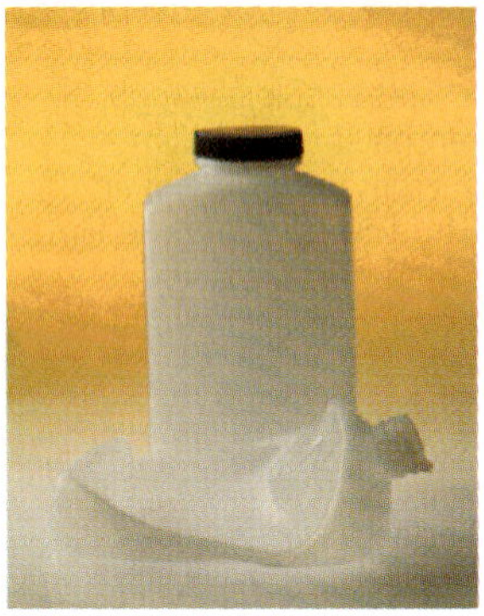

These polyethylene bottles were heated in the same oven for the same length of time. The one that melted has branched polyethylene molecules; the other has unbranched molecules.

Low-density polyethylenes (LDPE), on the other hand, have a lot of side chains branching off the polymer molecules. The branches prevent the molecules from packing closely together and assuming a crystalline structure. Low-density polyethylenes are waxy, bendable plastics that have lower melting points than high-density polyethylenes. Objects made of HDPE hold their shape in boiling water, while those made of LDPE become severely deformed at this temperature. Low-density polyethylenes are used to make plastic bags, plastic film, squeeze bottles, electric wire insulation, and many common household products.

A third type of polyethylene, linear low-density polyethylene (LLDPE), is becoming increasingly important. It is a copolymer of ethylene and a higher alkene, such as 1-hexene. The structure of a segment of LLDPE would be something like

$$\begin{array}{rcl} \sim CH_2CH_2-CH_2 & CH & -CH_2CH_2\sim \\ & | & \\ & CH_2 & \\ & | & \\ & CH_2 & \\ & | & \\ & CH_2 & \\ & | & \\ & CH_3 & \end{array}$$

where the horizontal portion shows an ethylene residue at each end, and the middle portion depicts a segment from 1-hexene. More than 70% of the LLDPE produced is used to make film.

Natural materials such as proteins (Chapter 21), natural rubber, cellulose, starch (Chapter 19), and complex silicate minerals are polymers. Artificial materials such as fibers, films, plastics, semisolid resins, and synthetic rubbers are also polymers. Synthetic polymers represent more than half of the compounds produced by the chemical industry (Table 13.11). They are widely used today, but there is no doubt that there will be new kinds of polymers and even wider use in the future. We already have new polymers that conduct electricity, amazing new adhesives, and synthetic materials stronger than steel but much lighter in weight. In many applications, substitution of plastics for natural materials saves energy. Their use in place of metals in automobiles and airplanes saves weight and therefore saves fuel. Although plastics present some environmental problems, they have become such an important part of our daily lives that we would find it difficult to live without them.

Table 13.11 Some Addition Polymers

Monomer	Polymer	Polymer Name	Some Uses
$H_2C{=}CH_2$ Ethylene	$\left[\begin{array}{c} H\ \ H \\ -C-C- \\ H\ \ H \end{array}\right]_n$	Polyethylene	Plastic bags, bottles, toys, electrical insulation
$H_2C{=}CH{-}CH_3$ Propylene	$\left[\begin{array}{c} H\ \ H \\ -C-C- \\ H\ \ CH_3 \end{array}\right]_n$	Polypropylene	Indoor-outdoor carpeting, bottles, luggage
$H_2C{=}CH_2$⬡ Styrene	$\left[\begin{array}{c} H\ \ H \\ -C-C- \\ H\ \ \bigcirc \end{array}\right]_n$	Polystyrene	Simulated wood furniture, styrofoam insulation, cups, toys, packing materials
$H_2C{=}CH{-}Cl$ Vinyl chloride	$\left[\begin{array}{c} H\ \ H \\ -C-C- \\ H\ \ Cl \end{array}\right]_n$	Poly(vinyl chloride), PVC	Plastic wrap, simulated leather (Naugahyde), phonograph records, garden hoses, rainwear, floor covering
$H_2C{=}CCl_2$ 1,1-Dichloroethene (Vinylidene chloride)	$\left[\begin{array}{c} H\ \ Cl \\ -C-C- \\ H\ \ Cl \end{array}\right]_n$	Poly(vinylidene chloride), Saran	Food wrap, seatcovers
$F_2C{=}CF_2$ Tetrafluoroethylene	$\left[\begin{array}{c} F\ \ F \\ -C-C- \\ F\ \ F \end{array}\right]_n$	Polytetrafluoro-ethylene, Teflon	Nonstick coating for cooking utensils, electrical insulation, lubricant, bearings
$H_2C{=}CH{-}CN$ Cyanoethylene (Acrylonitrile)	$\left[\begin{array}{c} H\ \ H \\ -C-C- \\ H\ \ CN \end{array}\right]_n$	Polyacrylonitrile, Orlon, Acrilan, Creslan, Dynel	Yarns, wigs, paints
$H_2C{=}CH{-}O{-}\overset{\displaystyle O}{\overset{\|}{C}}{-}CH_3$ Vinyl acetate	$\left[\begin{array}{c} H\ \ H \\ -C-C- \\ H\ \ O-C-CH_3 \\ \ \ \ \ \ \ \| \\ \ \ \ \ \ \ O \end{array}\right]_n$	Poly(vinyl acetate), PVA	Adhesives, textile coatings, chewing gum resin, paints
$H_2C{=}\overset{\displaystyle CH_3}{\underset{\displaystyle \underset{O}{\|}}{C}}{-}C{-}O{-}CH_3$ Methyl methacrylate	$\left[\begin{array}{c} H\ \ CH_3 \\ -C-C- \\ H\ \ C-O-CH_3 \\ \ \ \ \ \| \\ \ \ \ \ O \end{array}\right]_n$	Poly(methyl metha-crylate), Lucite, Plexiglas	Glass substitute, bowling balls

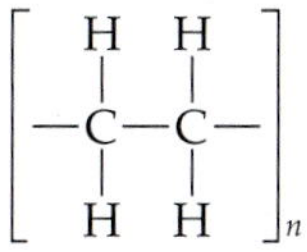

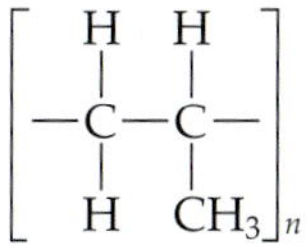

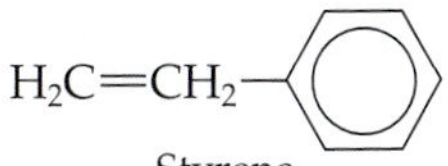

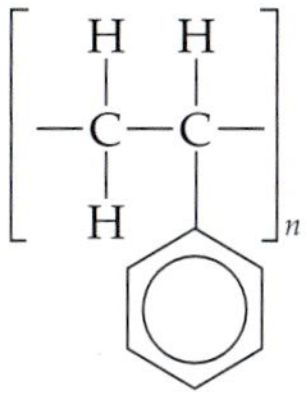

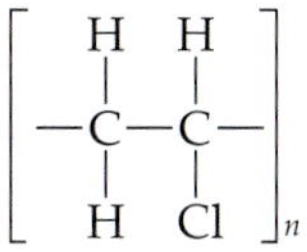

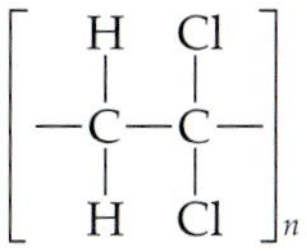

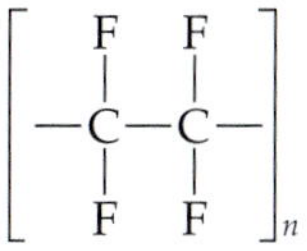

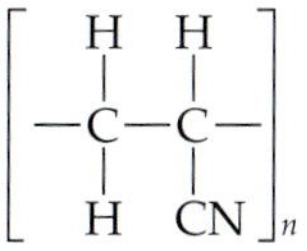

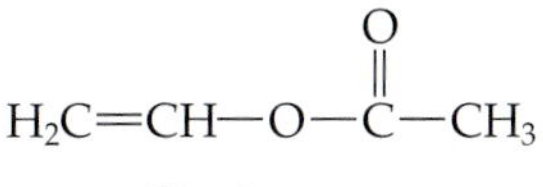

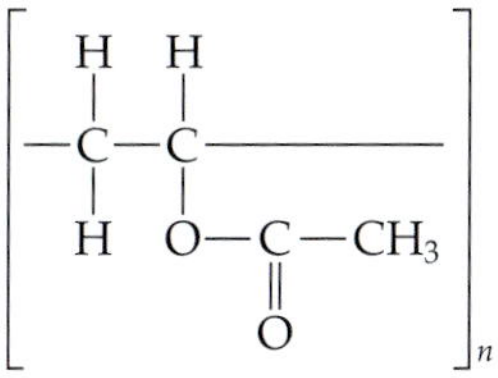

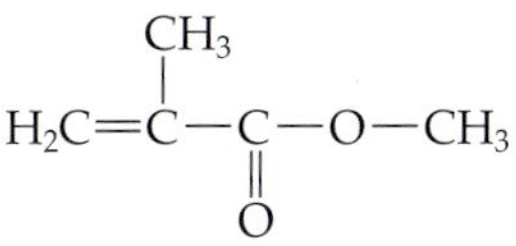

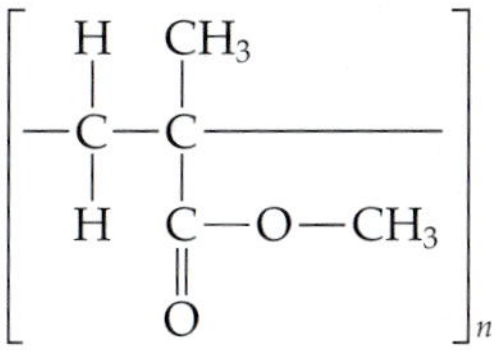

The numbers in the recycling symbols indicate the compositions of these plastic containers.

Disposal of Plastics

By mass, plastics make up about 8% of all solid waste in the United States, but by volume they make up about 21%. This large volume has created a serious problem because most of our solid waste goes into landfills, and we are rapidly running out of available landfill space. The very features that make plastics so useful—durability and resistance to many things in the environment—become disadvantages when we dispose of them. Once they are dumped, they do not go away. You see them littering parks, sidewalks, and highways, and if you go out to the middle of the ocean you see them there, too. Small fish have been found dead with their digestive tracts clogged by bits of plastic foam ingested with their food.

About half of our waste plastic is from packaging. One approach to the plastics disposal problem is to make plastic packages that are biodegradable or photodegradable (broken down in the presence of bacteria or light). This approach has been tried with plastic bags, six-pack ring connectors, and food cartons. Most biodegradable plastics are starch-based synthetic polymers. Photodegradable plastics usually contain a light-sensitive additive. Of course, it is important that the package remain intact and not start decomposing while it is still being used. So far, many people seem reluctant to pay extra for garbage bags that are designed to fall apart.

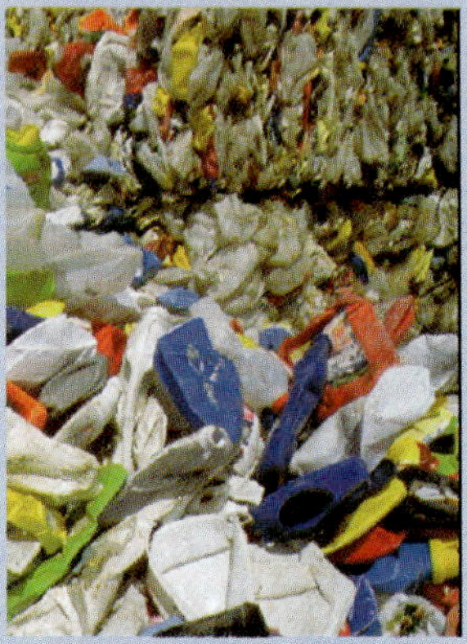

Plastics can be recycled. Separated, individual polymers can be melted and formed into new objects. Polyester bottles are melted and spun into fibers for carpets and insulating fabrics.

Perhaps the best way to deal with discarded plastic is to recycle it. To do that, the plastic must be collected, sorted, chopped, melted, and then remolded. Collection may be the hardest step in the process. It works best when there is strong community cooperation.

The separation step has been simplified by having code numbers (Table 13.12) stamped on most plastic bottles and other objects. These codes make it possible to sort the containers by their compositions. In general, the lower the number, the greater the ease with which the material can be recycled.

At present, the only plastic items being recycled on a large scale are HDPE milk jugs, which are being remolded into detergent bottles, and PET soda bottles, which are being turned into fiber, mainly for carpets. Small companies are recycling such items as polystyrene containers and polyethylene bags. One recycler is turning polycarbonate water bottles into automobile bumpers, and others are converting milk jugs into artificial lumber.

Only about 2% of waste plastic is currently being recycled. However, there are many possible uses for recycled plastic, and the shortage of landfill space is becoming critical. Recycling of plastics is a field with great potential for growth.

Table 13.12 Code for Separating Plastics for Recycling

Plastic	Abbreviation	Code
Polyethylene terephthalate	PET	1
High-density polyethylene	HDPE	2
Polyvinyl chloride	PVC	3
Low-density polyethylene	LDPE	4
Polypropylene	PP	5
Polystyrene	PS	6
All others		7

13.15 Alkynes

In alkenes, carbon atoms in the double bond share two pairs of electrons. Carbon atoms can also share three pairs of electrons, forming triple bonds. Compounds containing such bonds are called **alkynes.** The common name of the simplest alkyne is acetylene (C_2H_2; see Figure 13.11). Its structure is

$$H{-}C{\equiv}C{-}H$$

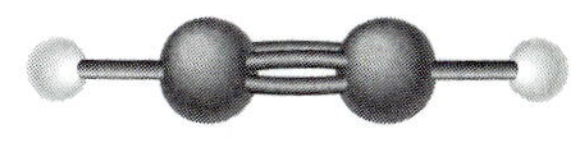

▲ **FIGURE 13.11**
Ball-and-spring model of acetylene.

About 10% of all acetylene produced is used in oxyacetylene torches for cutting and welding metals. The flame from such a torch can be very hot. Most acetylene, however, is converted to chemical intermediates that are in turn used to make vinyl and acrylic plastics, fibers, and resins and a variety of other chemical products.

The common name, acetylene, sounds very much like ethylene or propylene. Remember, however, that acetylene is an alkyne, whereas ethylene and propylene are alkenes.

The alkynes are similar to the alkenes in both physical and chemical properties. For example, they undergo many of the typical addition reactions of alkenes. Like ethene, acetylene has been used as an anesthetic for surgery. At higher concentrations, it causes narcosis and asphyxia. The IUPAC nomenclature for alkynes parallels that of the alkenes, except that the family ending is *-yne* rather than *-ene.* The official name for acetylene is ethyne.

13.16 Benzene

In 1825, Michael Faraday isolated a hydrocarbon from a sample of illuminating gas made from whale oil. He determined that the compound had the simplest formula CH_2 (based upon the then-current belief that the atomic mass of carbon was 6), and he named it "bicarburet of hydrogen." Later the compound was given the name benzene because it could be obtained by distilling benzoic acid with calcium oxide. During the ensuing decades, the concept of atomic mass became more clearly defined, and vapor-density measurements established the molecular formula of benzene to be C_6H_6.

One of the many possible structural formulas corresponding to this molecular formula is 1,3,5-cyclohexatriene. The compound would be a slightly lopsided hexagon having three double bonds in a six-membered ring (Figure 13.12a). The high degree of unsaturation would imply a very high reactivity. Chemists soon discovered, however, that benzene is unreactive and behaves more like an alkane than an alkene. It does not react readily with bromine, which, as mentioned in Section 13.13, is a test for detecting alkenes. In fact, all the evidence from laboratory work pointed to a molecule in which all six carbon atoms were equivalent.

In 1865, the German chemist Friedrich August Kekulé proposed a structure that could account for all of the known chemical properties of benzene. His theory was that the benzene molecule consists of a cyclic, hexagonal, planar structure of six carbon atoms with alternate single and double bonds. Each carbon atom is

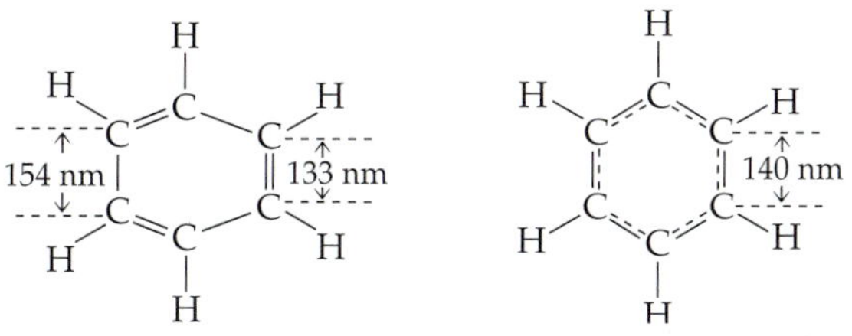

(a) (b)

◀ **FIGURE 13.12**
(a) The hypothetical 1,3,5-cyclohexatriene molecule would be a lopsided hexagon because double bonds are shorter than single bonds. (b) The actual benzene molecule is a regular hexagon having all sides 140 nm long.

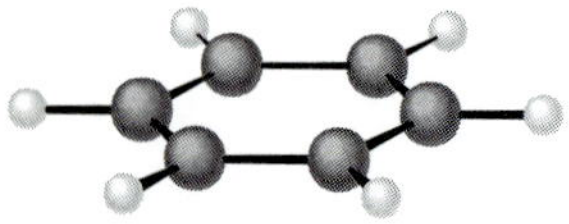

A computer-generated structure of benzene

bonded to only one hydrogen atom. He accounted for the equivalence of all six carbon atoms by suggesting that the double bonds are not static but rather are mobile and oscillate from one position to another:

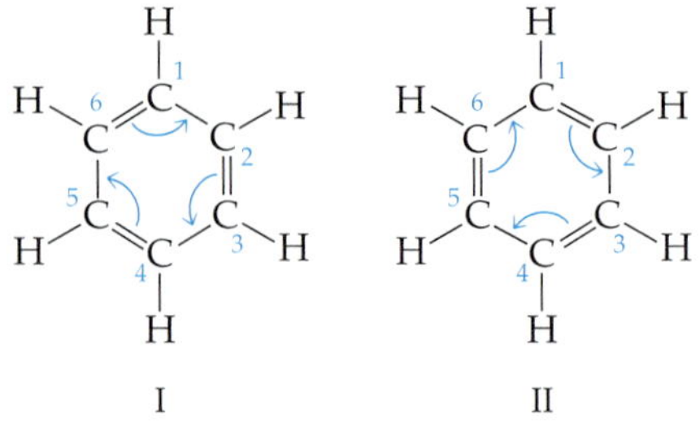

Structures I and II differ from each other only in the positions of the double bonds, and although they satisfy the requirements for equivalent hydrogens, they do not explain why benzene does not behave like an unsaturated hydrocarbon. Furthermore, X-ray-diffraction measurements indicate that all the carbon–carbon bonds in benzene are the same length, 140 nm (Figure 13.12b). This value falls between the length of a carbon-carbon double bond (133 nm) and that of a carbon–carbon single bond (154 nm).

Although the term *resonance* is much used in organic chemistry, it is not unique to organic compounds. Many inorganic compounds (e.g., O_3, SO_2, and NO_3^-) also exhibit resonance.

To accommodate all these findings, chemists have postulated that benzene exhibits resonance. **Resonance** is a word used to describe the phenomenon in which no single classical Lewis structure adequately accounts for the experimentally observed properties of a molecule (such as bond energies and bond distances). According to the resonance theory, the benzene molecule is a **resonance hybrid** of structures I and II. That is, neither of the two structures actually exists, but taken together they represent the true structure of the molecule. We depict this situation with a double-headed arrow ($\longleftrightarrow$) that connects the contributing structures (Figure 13.13a). This resonance arrow should be clearly distinguished from the pair of half-headed arrows ($\rightleftharpoons$) that indicate an equilibrium condition.

Some chemists combine the two contributing forms into the single structure depicted in Figure 13.13b. The inner circle indicates that the valence electrons are shared equally by all six carbon atoms (that is, the electrons are *delocalized*, or spread out, over all the carbon atoms). This method is a valid shortcut device for writing benzene, but it does not adequately account for all the electrons in the molecule. The representation of benzene that contains alternating single and double bonds (*resonance* or *Kekulé forms*) is the best model for keeping track of electrons. The circle within a hexagon better describes the molecule by indicating the equal sharing of the electrons and the identical bond lengths within the molecule. In this text, we shall use the hexagon with inscribed circle. Remember that when either representation is used, it is understood that each corner of the hexagon is occupied by one carbon atom. Attached to each carbon atom is one hydrogen atom, and all the atoms lie in the same plane. Any other atom, or group of atoms, substituted for a hydrogen atom must be shown to be bonded to a particular corner of the hexagon.

(a) 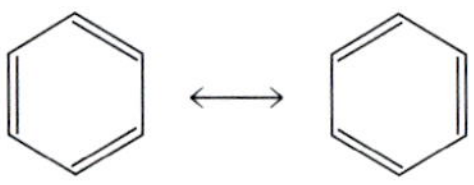

(b) 

▲ **FIGURE 13.13**

Two methods of representing the structure of benzene. (a) Kekulé structures; (b) hexagon with inscribed circle.

One final note: taken literally, the term "resonance" is a misnomer. It indicates oscillation from one structure to another or oscillation of an electron pair from one bond to another. *There is no oscillation.* A mule, the offspring of a male donkey and a mare, can be considered a hybrid of a donkey and a horse. However, it is not a horse at one instant and a donkey at another. It is always a mule, a distinctive animal that has some of the characteristics of a horse and some of a donkey. Likewise, any molecule that exhibits resonance has *one* real structure that is never

Kekulé's Dream

There is a story told that one evening Kekulé fell asleep while sitting in front of a fire. He dreamed about chains of atoms having the forms of twisting snakes. Suddenly one of the snakes caught hold of its own tail, forming a whirling ring. Kekulé awoke, freshly inspired, and spent the remainder of the night working on his now-famous hypothesis. He is said to have written, "Let us learn to dream, gentlemen, and then perhaps we shall learn the truth."

any of the rather unsophisticated Lewis structures used to describe it. The difficulty arises because we cannot adequately portray the molecule, not because of the molecule itself.

13.17 Aromatic Hydrocarbons: Structure and Nomenclature

Benzene and similar compounds are referred to as **aromatic hydrocarbons.** This is because quite a few of the first benzene-like substances to be discovered had strong aromas. Even though many benzene derivatives have turned out to be odorless, the name stuck. Today the term "aromatic" in organic chemistry is applied to any compound that contains a benzene ring or has certain properties similar to those of benzene. (All the nonaromatic hydrocarbons that we have considered—the alkanes, alkenes, and alkynes—are referred to collectively as **aliphatic compounds** to distinguish them from aromatic compounds. "Aliphatic" originally meant that the source of the compound was a fat. Today, however, it simply means "not aromatic.")

Benzene is a liquid that smells like gasoline, boils at 80 °C, and freezes at 5.5 °C.

Aromatic hydrocarbons have both common names and systematic names. Some aromatic compounds are referred to exclusively as derivatives of benzene, whereas others are more frequently denoted by their common names. Note that in the following structures, it is immaterial whether the substituent is written at the top, side, or bottom of the ring: a hexagon is symmetrical, and therefore all positions are equivalent.

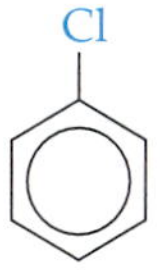

Chlorobenzene

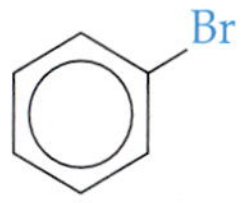

Bromobenzene

Nitrobenzene

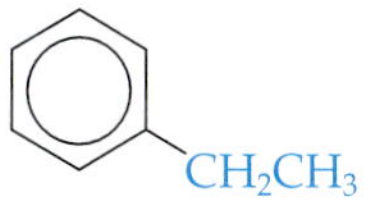

Ethylbenzene

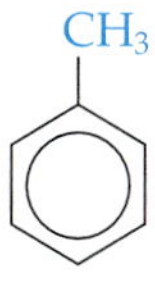

Toluene
(Methylbenzene)

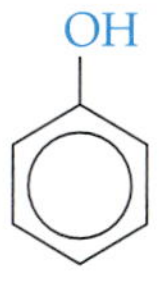

Phenol
(Hydroxybenzene)

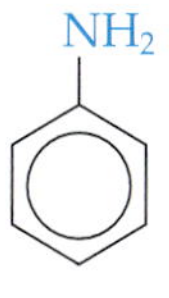

Aniline
(Aminobenzene)

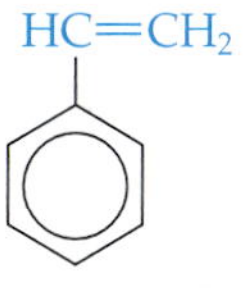

Styrene
(Vinylbenzene)

A complication arises when there is more than one substituent because now all the positions on the hexagon are no longer equivalent and relative positions must be designated. In the case of a disubstituted benzene, one nomenclature system uses the prefixes *ortho (o-)*, *meta (m-)*, and *para (p-)*. Ortho designates

1,2-disubstitution, meta designates 1,3-disubstitution, and para designates 1,4-disubstitution:

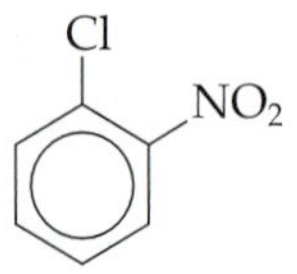

o-Chloronitrobenzene

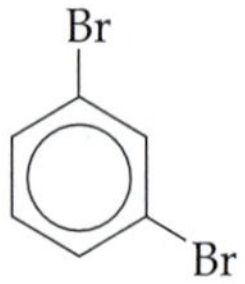

m-Dibromobenzene

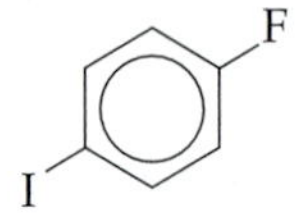

p-Fluoroiodobenzene

Alternatively, the ring is numbered and the substituent names are listed in alphabetical order. The first substituent is given the lowest number. When a common name is used, the carbon atom that bears the group responsible for the name is carbon 1:

Xylene is the common name for dimethylbenzene. Hence, there are three xylene isomers: *o*-, *m*-, and *p*-dimethylbenzene.

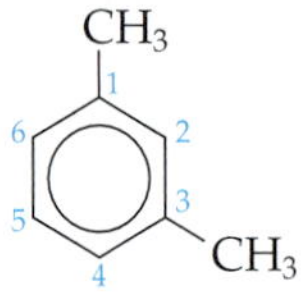

m-Xylene
(1,3-Dimethylbenzene)

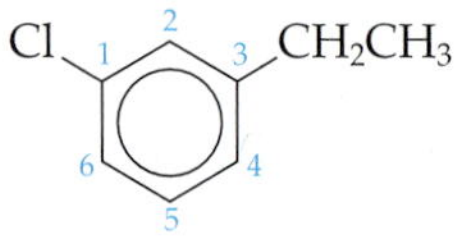

m-Chloroethylbenzene
(1-Chloro-3-ethylbenzene)

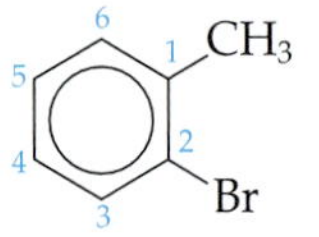

o-Bromotoluene
(2-Bromotoluene)
(1-Bromo-2-methylbenzene)

If there are more than two substituents on a benzene ring, the numbering is determined by the requirement that the numbers give the smallest possible sum (substituents are listed in alphabetical order).

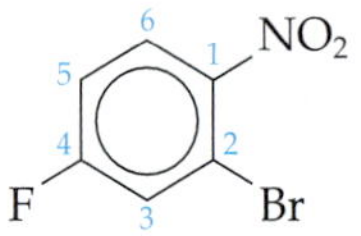

2-Bromo-4-fluoro-1-nitrobenzene

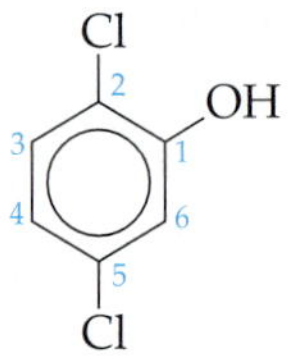

2,5-Dichlorophenol
(1,4-Dichloro-2-hydroxybenzene)

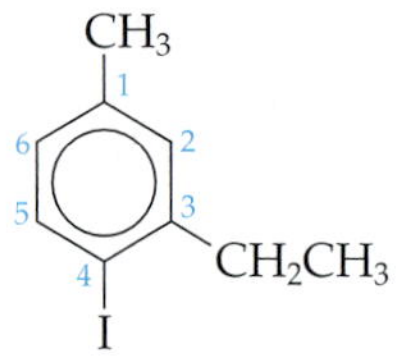

3-Ethyl-4-iodotoluene
(3-Ethyl-4-iodo-1-methylbenzene)

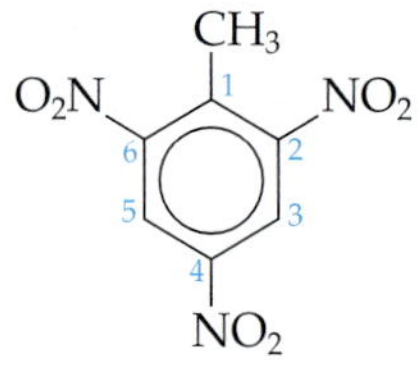

2,4,6-Trinitrotoluene
(TNT—an explosive)

Occasionally, an aromatic group is a substituent that is bonded to an aliphatic compound or to another aromatic ring. The group of atoms remaining when a hydrogen atom is removed from an aromatic compound is called an **aryl group.** The most common aryl group is the one derived from benzene (C_6H_5—). It is called *phenyl*, a name derived from *pheno*, an old name for benzene.[4]

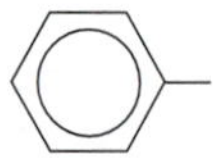
Phenyl group

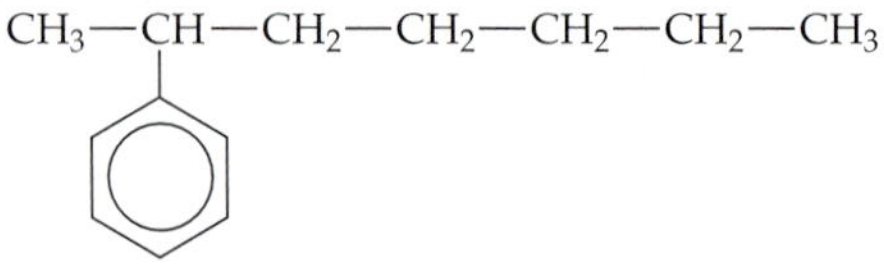

2-Phenylheptane

[4] This terminology is confusing because it would seem that the group derived from benzene should be called benzyl. The problem is compounded by the fact that another group is called benzyl. Replacement of one of the methyl hydrogens of toluene gives the *benzyl* group, $C_6H_5CH_2$—.

Benzyl group

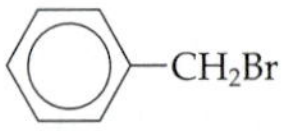

Benzyl bromide

Mention should be made of some common aromatic hydrocarbons that are not substituted benzenes but are fused benzene rings.

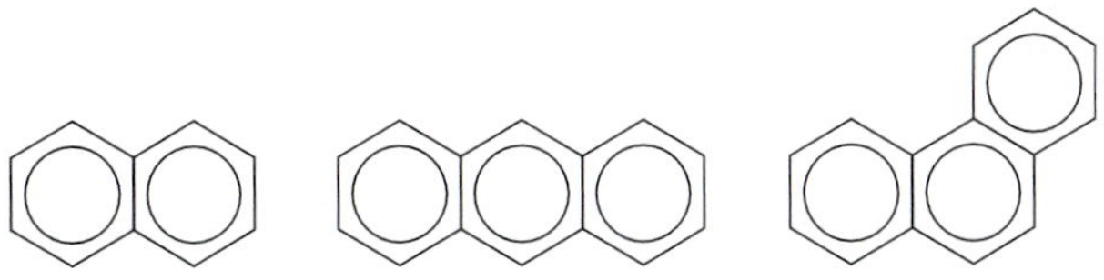

Naphthalene	Anthracene	Phenanthrene
mp 80 °C	mp 218 °C	mp 101 °C
bp 218 °C	bp 342 °C	bp 340 °C

These three substances are colorless, crystalline solids obtained from coal tar. Naphthalene has a pungent odor and is commonly used in mothballs. Anthracene is an important starting material in the manufacture of certain dyes. A large group of naturally occurring substances, the steroids, contain the phenanthrene structure (see Section 20.9).

Polycyclic aromatic compounds do not exist in unprocessed coal itself but are formed by the intense heating needed to distill coal tar. For many years, it has been known that workers in coal-tar refineries are susceptible to a type of skin cancer known as tar cancer. Investigation has shown that a number of polycyclic aromatic hydrocarbons can cause cancer when applied to the skin. Such compounds are called **carcinogens** (cancer producers). One of the most active carcinogenic compounds, benzpyrene, occurs in coal tar and has been isolated from cigarette smoke, automobile exhaust gases, and charcoal-broiled steaks. It is estimated that more than 1000 tons of benzpyrene are emitted into the air over the United States each year. Only a few milligrams of benzpyrene is required to induce cancer in experimental animals.

The mechanism by which carcinogens cause cancer has not yet been elucidated. One hypothesis is that carcinogens are not polycyclic hydrocarbons themselves but one or more of their metabolites. (As we shall learn in Chapter 24, metabolites are the products of chemical transformations in living cells.) Figure 13.14 indicates the conversion of benzpyrene via a multistep oxidation sequence to yield a highly carcinogenic diolepoxide metabolite. To a certain extent, fused polycyclic hydrocarbons are formed whenever organic molecules are heated to high temperatures. It is the current belief that lung cancer is caused by carcinogenic compounds formed when the tobacco in cigarettes is burned.

Aromatic hydrocarbons containing only one benzene ring are generally liquids, and polyring aromatics are generally solids. All are insoluble in water but soluble in organic solvents. Aromatic hydrocarbons are readily combustible. Unlike aliphatic hydrocarbons, which burn with a relatively clean flame, aromatic compounds burn with a very sooty flame.

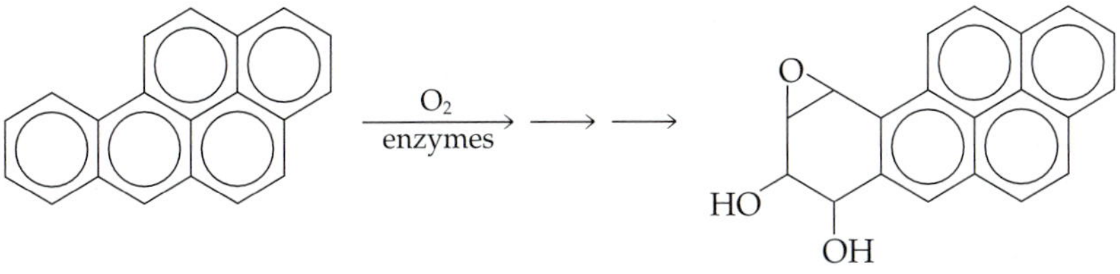

Benzpyrene A diolepoxide

Figure 13.14
Benzpyrene is metabolized in the body to produce an active carcinogen.

13.18 Uses of Benzene and Benzene Derivatives

Most of the benzene used commercially comes from petroleum. Benzene is employed industrially as a starting material for the production of such products as detergents, drugs, dyes, insecticides, and plastics. Benzene was once widely used as an organic solvent, but we now know it to be a poisonous substance with both short- and long-term effects. Inhalation of large concentrations of benzene can cause nausea and even death due to respiratory or heart failure. Repeated exposure leads to a progressive disease in which the ability of the bone marrow to make new blood cells is eventually destroyed. This results in a condition called *aplastic anemia*, in which there is a decrease in the numbers of both the red and white blood cells.

Unleaded gasolines contain about 2% benzene. For most people, exposure to benzene comes from inhaling gasoline fumes, or being around a cigarette smoker. For smokers, cigarettes overwhelm all other sources of benzene.

Because of these hazards, many chemical laboratories have replaced benzene with toluene as a general solvent.[5] Toluene is used in the production of dyes, drugs, and explosives and as a solvent in lacquers. It is commonly used as a preservative for urine specimens, and is added to fuels to improve their octane

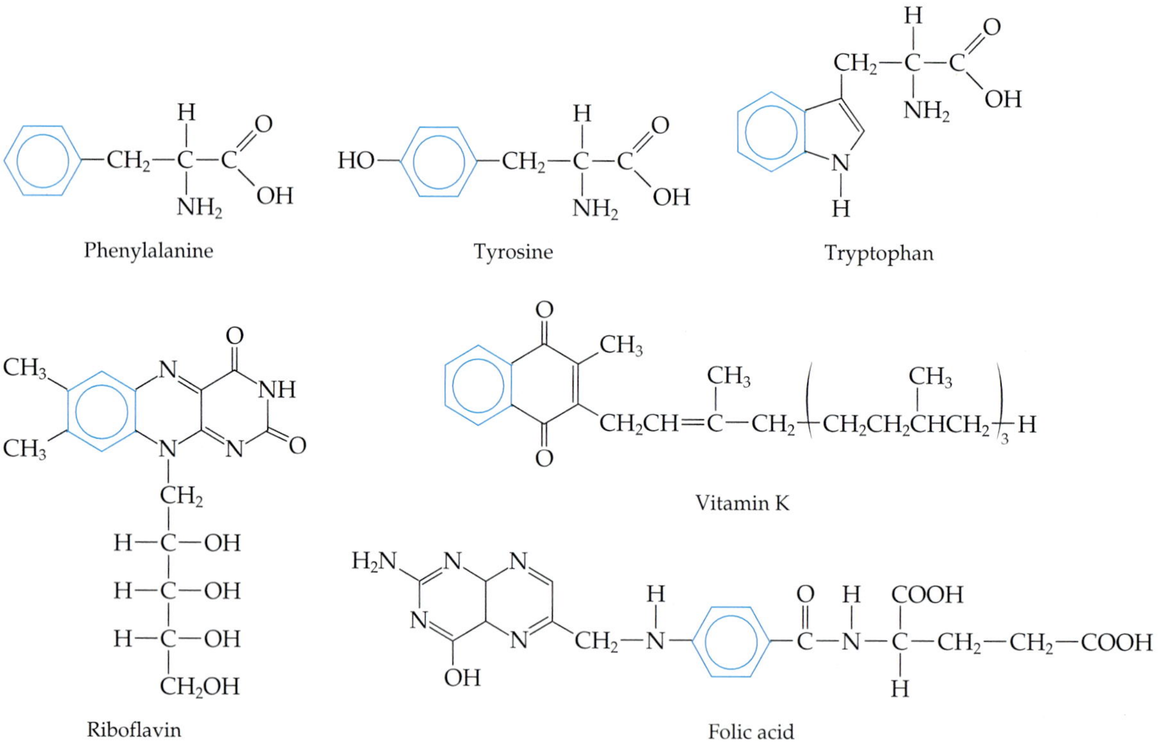

▲ Figure 13.15
Some biologically important compounds that cannot be synthesized by animals. Each contains a benzene ring that must be supplied by some substance in the diet.

[5] The maximum allowed concentration for an 8-hour exposure to toluene is 100 ppm, whereas the allowable exposure to benzene for the same period of time is 1 ppm. There are no toxic symptoms attributable to toluene until concentrations reach 200 ppm. Care must be taken in using toluene because most commercial toluene contains benzene as an impurity. If the concentration of benzene in toluene is less than 1%, the toluene is safe to use. *However,* toluene has been shown to cause birth defects, so pregnant women should avoid breathing toluene vapors. Prolonged inhalation of toluene should be avoided by everyone.

numbers (see Section E.2). Trinitrotoluene (TNT), unlike nitroglycerin (see Section 14.7), is not sensitive to shock on jarring and must be exploded by a detonator.

Nitrobenzene is used extensively in the manufacture of aniline, the parent compound of many dyes and drugs. Phenol containing a small amount of water is a liquid and in this form is referred to as carbolic acid. It is a good antiseptic and germicide, but its use is limited because of its toxicity (see Section 14.8).

The xylenes are good solvents for grease and oil and are used for cleaning microscope slides and optical lenses. The xylenes are also used to raise the octane numbers of unleaded gasolines.

Substances containing the benzene ring are commonly found in both the animal and plant kingdoms, although they are more abundant in the latter. Plants have the ability to synthesize the benzene ring from carbon dioxide, water, and inorganic materials. Animals are incapable of this synthesis but are dependent on benzene compounds for their survival. Therefore, animals must obtain these compounds from the food that they ingest. Included among the aromatic compounds necessary for animal metabolism are the amino acids phenylalanine, tyrosine, and tryptophan and certain vitamins such as vitamin K, riboflavin, and folic acid (Figure 13.15). In addition, a great majority of drugs contain the benzene ring (see Special Topics F and G).

Summary

Organic chemistry is the chemistry of carbon compounds. Carbon is able to form stable covalent bonds both with other carbon atoms and with atoms of other elements, and this property allows carbon to form the millions of carbon compounds we call *organic compounds.*

Carbon has four electrons in its outermost shell (the second shell), and this shell is complete with eight electrons. Thus, carbon can form up to four covalent bonds. Any carbon forming four single bonds is said to be a *saturated carbon.* (Joined to the maximum number of other atoms, the carbon is *saturated* with atoms.)

The simplest organic compounds are **hydrocarbons,** those containing only hydrogen and carbon. Hydrocarbons in which each carbon is bonded to four other atoms are called **alkanes** or **saturated hydrocarbons.** They have the general formula C_nH_{2n+2}, which means that any given alkane differs from the alkane containing one fewer carbon by one $-CH_2-$ unit. Any family of compounds in which adjacent members differ from each other in this way is called a **homologous series,** and the individual members are *homologs.*

Two or more compounds having the same molecular formula but different structural formulas are called **isomers** of each other:

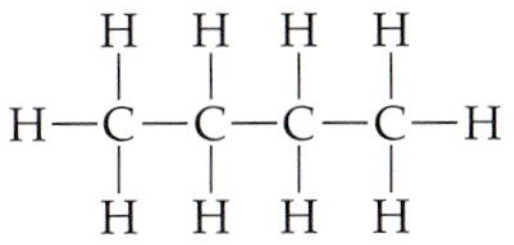

C_4H_{10}

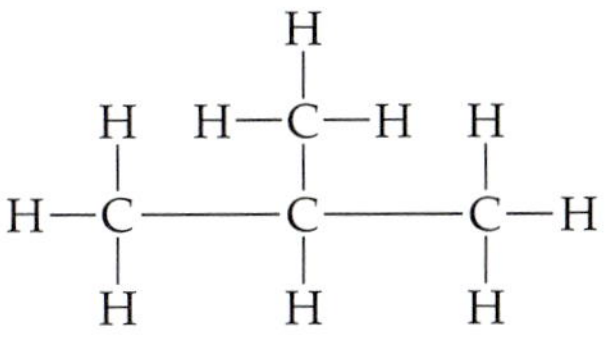

C_4H_{10}

There are no isomeric forms for the three smallest alkanes, but beginning with butane (C_4H_{10}), all other alkanes have isomeric forms.

The unit resulting when one hydrogen is removed from an alkane is an **alkyl group:** $CH_3CH_2CH_2-$.

Alkanes are chemically inert for the most part and do not react with acids, bases, oxidizing agents, or reducing agents. They do burn, however, as in the combustion reaction: $CH_4 + 2\ O_2 \longrightarrow CO_2 + 2\ H_2O$ + heat and can accept one or more halogen atoms (F, Cl, Br, I) as a replacement for hydrogen; the result is a **halogenated hydrocarbon:**

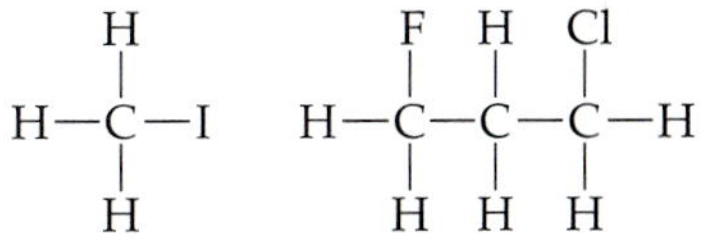

Cycloalkanes are hydrocarbons that exist as closed rings rather than as straight-chain or branched-chain molecules:

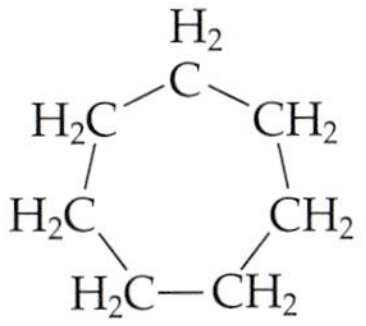

Cycloheptane C_7H_{14}

Any hydrocarbon containing either a double bond or a triple bond is said to be an **unsaturated hydrocarbon:**

Alkenes are hydrocarbons containing a carbon-to-carbon double bond: $\diagdown C = C \diagup$ The general formula for alkenes containing only one double bond is C_nH_{2n}. Alkenes can be straight-chain, branched-chain, or cyclic.

More reactive than alkanes, alkenes burn and also add across the double bond:

combustion:

$$C_2H_4 + 3\,O_2 \longrightarrow 2\,CO_2 + 2\,H_2O + \text{heat}$$

addition of hydrogen (**hydrogenation**)

$$H_2C{=}CH_2 + H_2 \longrightarrow H_3C{-}CH_3$$

addition of halogen (**halogenation**)

$$H_3C{-}CH{=}CH_2 + I_2 \longrightarrow H_3C{-}\underset{\underset{\displaystyle I}{|}}{\overset{\overset{\displaystyle H}{|}}{C}}{-}\underset{\underset{\displaystyle I}{|}}{\overset{\overset{\displaystyle H}{|}}{C}}{-}H$$

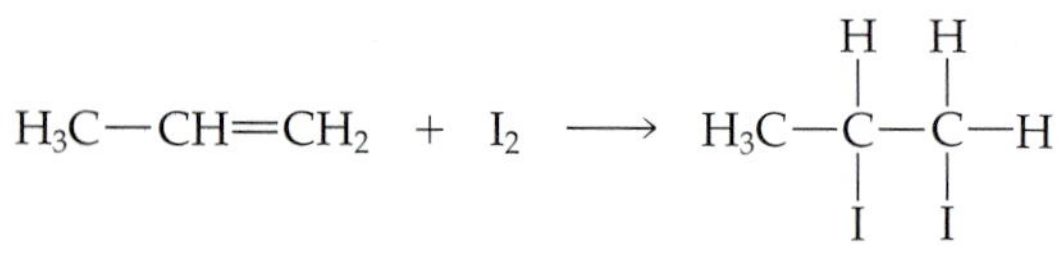

addition of water (**hydration**)

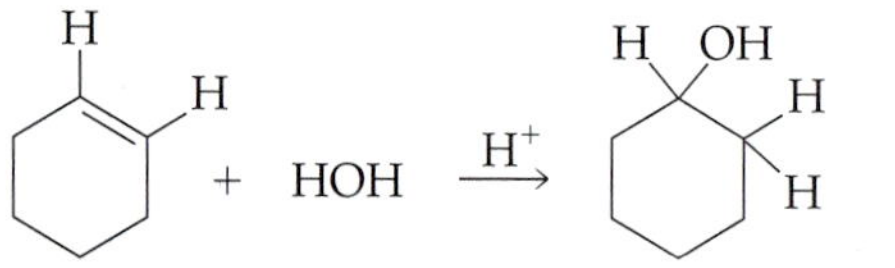

A third reaction of alkenes is that they condense to form long molecules in the process called **polymerization:**

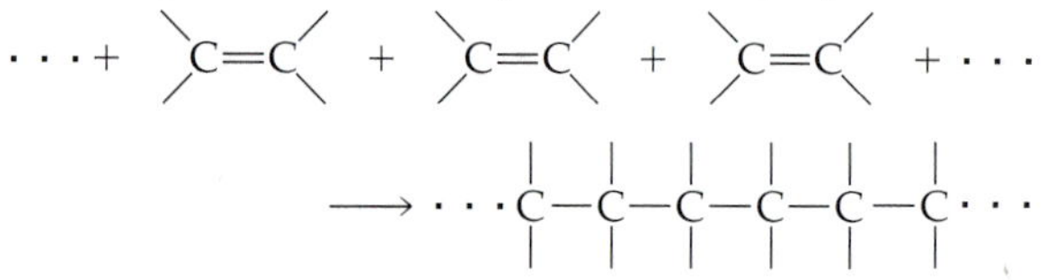

The reactant units are **monomers,** and the product is a **polymer.**

Unsaturated hydrocarbons that contain carbon-to-carbon triple bonds, $-C{\equiv}C-$, are called **alkynes.**

The cyclic unsaturated hydrocarbon *benzene*, C_6H_6, is an extremely stable ring, undergoing none of the reactions expected of alkenes. Because of this stability, it is believed that the ring is not one of alternating single and double bonds but rather a **resonance hybrid** in which the "extra" electrons are *delocalized* over all six carbons:

Compounds containing one or more benzene rings (or other similarly stable resonance-hybrid units) are called *aromatic compounds*. (Any hydrocarbon not containing an aromatic unit is an **aliphatic compound.**)

One or more of the hydrogens on a benzene ring can be replaced by other atoms. When two hydrogens are replaced, the product name is based on the relative position of the replacement atoms (or atom groups):

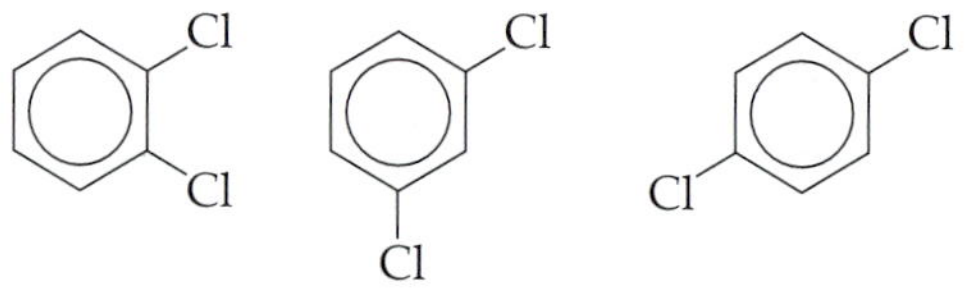

Ortho Meta Para

Key Terms

addition polymerization (13.14)
addition reaction (13.13)
aliphatic compound (13.17)
alkane (13.2)
alkene (13.11)
alkyl group (13.4)
alkyl halide (13.9)
alkyne (13.15)
aromatic hydrocarbon (13.17)
aryl group (13.17)
carcinogen (13.17)

chlorofluorocarbon (CFC) (13.9)
combustion (13.8)
condensed structural formula (13.3)
halogenated hydrocarbon (13.9)
homologous series (13.2)
hydration (13.13)
hydrocarbon (13.2)
hydrogenation (13.13)
isomer (13.2)
IUPAC System of Nomenclature (13.5)

monomer (13.14)
organic chemistry (Introduction)
paraffin (13.8)
polymer (13.14)
polymerization (13.14)
resonance (13.16)
resonance hybrid (13.16)
saturated hydrocarbon (13.2)
structural formula (13.3)
unsaturated hydrocarbon (13.11)

Review Questions

1. List three ways in which a typical organic compound differs from a typical inorganic one.

2. Define, illustrate, or give an example for each of the following terms.

a. hydrocarbon
b. alkane
c. paraffin
d. saturated hydrocarbon
e. unsaturated hydrocarbon
f. substituent
g. alkene
h. alkyne
i. alkyl group
j. isomer
k. polymer
l. aromatic compound
m. aliphatic compound
n. phenyl group
o. alkyl halide
p. chlorofluorocarbon
q. homologous series
r. carcinogen

3. How many carbon atoms are there in (a) ethane, (b) heptane, (c) butane, and (d) nonane?
4. What compounds contain fewer carbons than propane and are homologs of propane?
5. Using appropriate examples, distinguish between:
   a. alkyl group and alkane
   b. alkyl group and aryl group
   c. common name and IUPAC name
   d. propyl group and isopropyl group
   e. straight-chain alkane and branched-chain alkane
6. Distinguish between:
   a. natural gas and bottled gas
   b. physical properties and chemical properties
   c. exothermic reaction and endothermic reaction
7. Classify each of the following as saturated or unsaturated.
   a. $CH_3C(CH_3){=}CH_2$ (structure: $CH_3C{=}CH_2$ with CH_3 on the central C)
   b. $CH_3C{\equiv}CCH_3$
   c. $CH_3{-}C(CH_3)_2{-}CH_3$ (central C bearing CH_3 above and below)
   d. 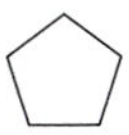

8. Write the molecular formula for each of the following.
   a. 
   b. 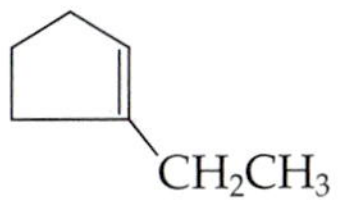
   c. 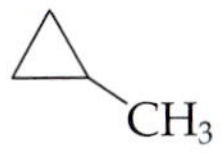
   d. 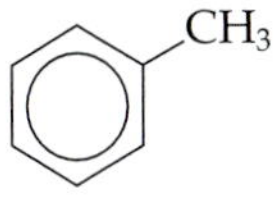
9. Indicate whether each compound is aromatic or aliphatic.
   a. 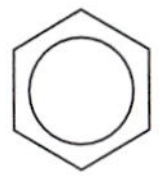
   b. 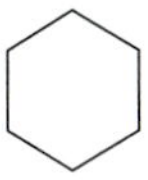
   c. 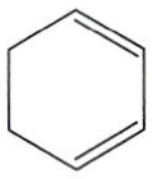
   d. 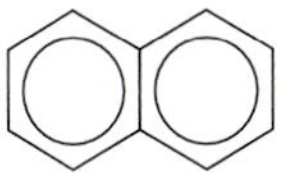
10. Identify each substitution pattern as meta, ortho, or para.
   a. 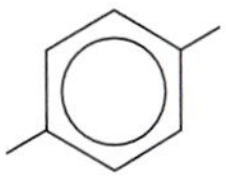
   b. 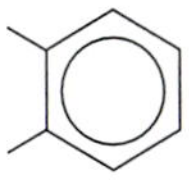
   c. 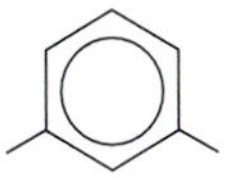
11. What is the danger in swallowing a liquid alkane?
12. Distinguish between lighter and heavier liquid alkanes in terms of their effects on the skin.
13. Describe a physiological effect of some polycyclic aromatic hydrocarbons.
14. What are some of the hazards associated with the use of benzene?
15. What physiological effect is shared by ethene and cyclopropane?
16. List several uses of polyethylene.
17. What is addition polymerization?
18. What structural feature usually characterizes molecules used as monomers in addition polymerization?

Problems

Organic Versus Inorganic

19. Classify each compound as organic or inorganic.
   a. C_6H_{10} b. $CoCl_2$ c. $C_{12}H_{22}O_{11}$
20. Classify each compound as organic or inorganic.
   a. CH_3NH_2 b. $NaNH_2$ c. $Cu(NH_3)_6Cl_2$
21. Which member of each pair has a higher melting point?
   a. CH_3OH and $NaOH$ b. CH_3Cl and KCl
22. Which member of each pair has a higher melting point?
   a. $C_{20}H_{42}$ and $C_{40}H_{82}$ b. CH_4 and LiH

Isomers

23. Indicate whether the structures in each set represent the same compound or isomers.
   a. CH_3CH_3 and $CH_3{-}CH_3$ (drawn vertically)
   b. $CH_3CH_2{-}CH_3$ (CH_3 drawn above the CH_2) and $CH_3CH_2CH_3$
   c. $CH_3CH_2CH(CH_3){-}CH_2(CH_3)$ (CH_3 above CH and above CH_2) and $CH_2(CH_3)CH_2CH(CH_3)CH_3$ (CH_3 above the first CH_2 and above CH)

24. Indicate whether the structures in each set represent the same compound or isomers.

a. 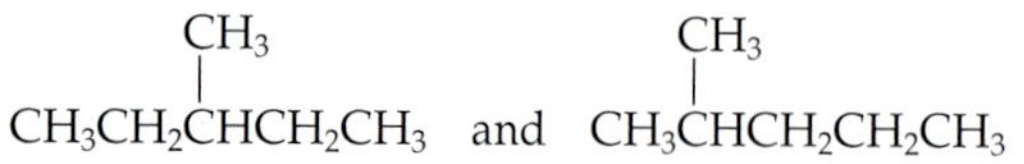

b. 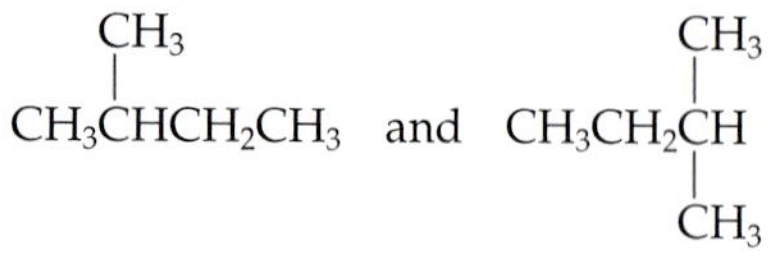

25. Indicate whether the structures in each set represent the same compound or isomers.

a. $CH_3CH=CHCH_3$ and $CH_3CH_2CH=CH_2$

b. 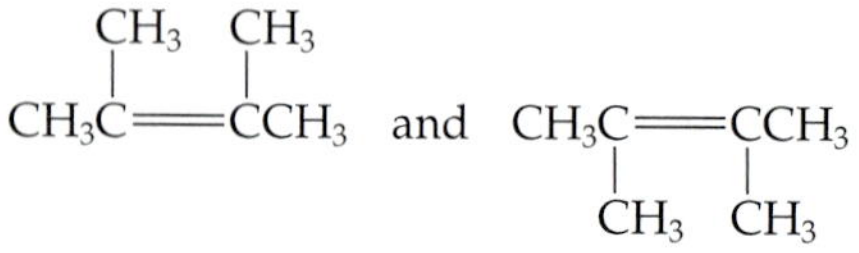

c. 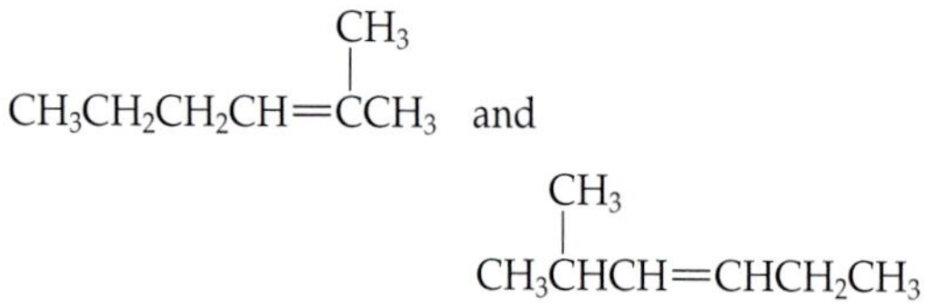

26. Indicate whether the structures in each set represent the same compound or isomers.

a. 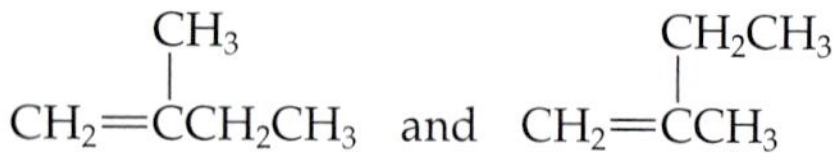

b. 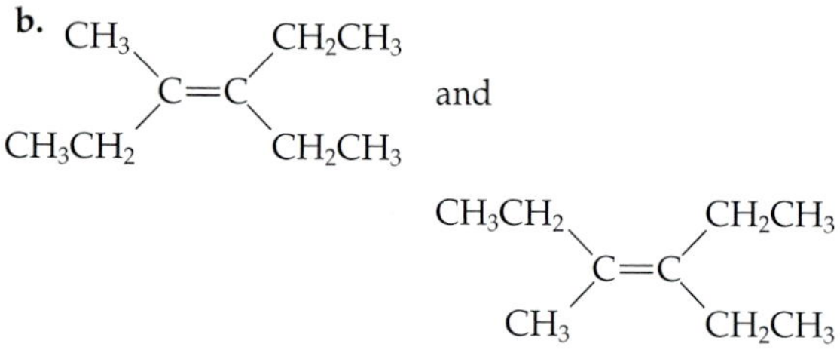

Alkanes: Structures and Names

27. Draw the structural formula for each compound.
a. heptane b. 3-methylpentane
c. 2,2,5-trimethylhexane d. 4-ethyl-3-methyloctane

28. Draw the structural formula for each compound.
a. 2-methylpentane
b. 4-ethyl-2-methylhexane
c. 2,2,3,3-tetramethylbutane
d. 4-ethyl-3-isopropyloctane

29. Name the following compounds by the IUPAC system.
a. $CH_3CH_2CH(CH_3)CH_2CH_3$
b. $(CH_3)_2CHCH(CH_3)_2$

30. Name the following compounds by the IUPAC system.
a. $CH_3CH(CH_2CH_3)_2$
b. $(CH_3)_3CCH_2C(CH_3)_2CH_2CH_3$

31. Draw the structural formulas for the following alkyl groups.
a. ethyl b. isopropyl

32. Draw the structural formulas for the following alkyl groups.
a. methyl b. propyl

33. Draw the structural formulas for both four-carbon alkanes (C_4H_{10}). Identify butane and isobutane, and give the IUPAC name for the latter.

34. Draw the structural formulas for the five isomeric hexanes (C_6H_{14}). Name each by the IUPAC system.

Cyclic Hydrocarbons

35. Name the following compounds by the IUPAC system.

a. 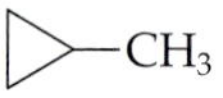

b. 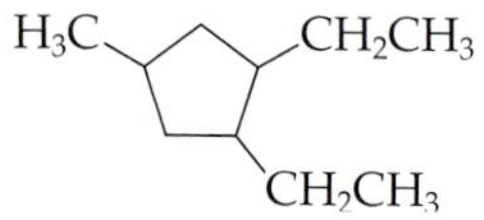

c. 

d. 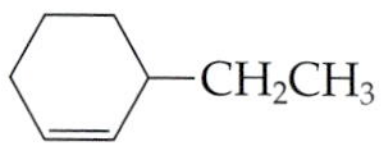

36. Name the following compounds by the IUPAC system.

a. 

b. 

c. 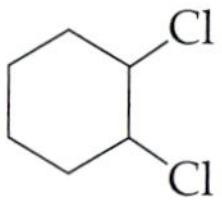

d. 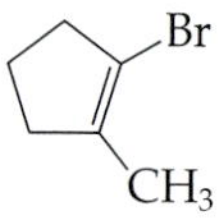

e. 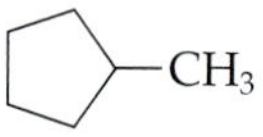

37. Draw the structural formula for each compound.
a. ethylcyclobutane
b. 3-chlorocyclopentene

38. Draw the structural formula for each compound.
a. 1,3-diethylcyclopentane
b. 4-methylcyclohexene

Halogenated Hydrocarbons

39. Draw the structural formula for each compound.
a. methyl chloride b. chloroform

40. Draw the structural formula for:
a. ethyl bromide b. carbon tetrachloride

41. Draw the structural formulas for the two isomers that have the molecular formula C_3H_7Br. Give the common name and the IUPAC name of each.

42. Draw the structural formulas for the four isomers that have the molecular formula C_4H_9Br. Give the IUPAC name of each.

Alkenes and Alkynes

43. Draw the structural formula for each compound.
a. acetylene b. cyclohexene
c. 3-isopropyl-1-hexyne d. 2,3-dimethyl-2-butene

44. Draw the structural formula for each compound.
    a. 1,2-dimethylcyclobutene
    b. 3-ethyl-2-pentene
    c. cyclooctyne
    d. 4-methyl-2-hexene
45. Draw the structural formula for each compound.
    a. 2-methyl-2-pentene b. 5-methyl-1-hexene
46. Draw the structural formula for each compound.
    a. 2-ethyl-1-butene
    b. 2,4,6,6-tetramethyl-2-heptene
47. Name the following compounds by the IUPAC system.
    a. $CH_2{=}C(CH_3)CH_2CH_2CH_3$
    b. $CH_3CH_2CH{=}C(CH_3)_2$
    c. $(CH_3)_2C{=}CHCH_2CH(CH_3)_2$
48. Name the following compounds by the IUPAC system.
    a. $(CH_3)_2C{=}CHCH_3$
    b. $CH_3CHClCH_2CH{=}CHCH_2CHCl_2$
    c. $(CH_3)_3CCH{=}C(CH_3)CH_2CH_3$

Aromatic Compounds

49. Draw the structural formula for each compound.
    a. toluene
    b. *m*-diethylbenzene
    c. 2,4-dinitrotoluene
50. Draw the structural formula for each compound.
    a. *p*-dichlorobenzene
    b. naphthalene
    c. 1,2,4-trimethylbenzene
51. Name the following compounds by the IUPAC system.

a. 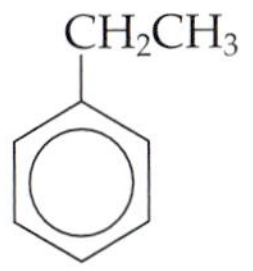

b. 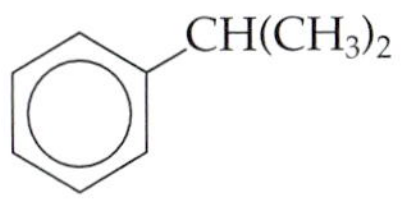

c. 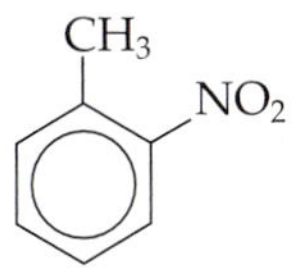

d. 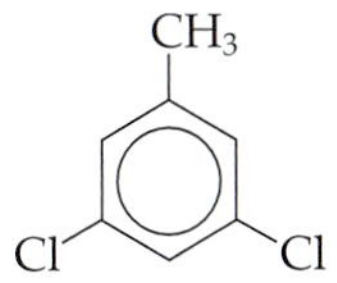

52. Name the following compounds by the IUPAC system.

a. 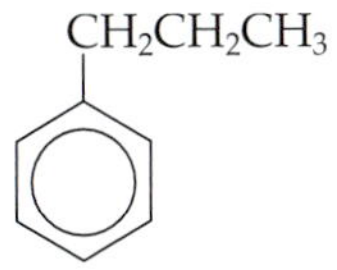

b. 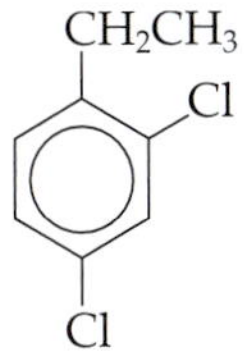

c. 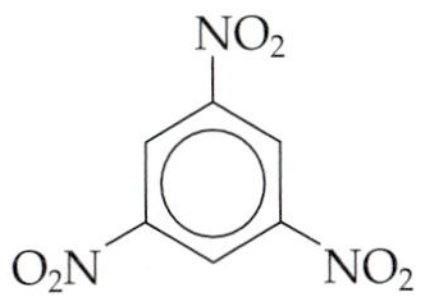

d. 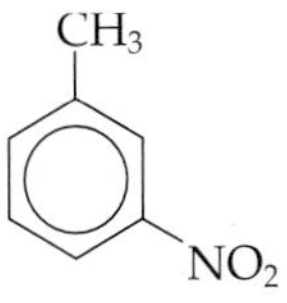

Physical Properties

53. Which member of each pair has the higher boiling point?
    a. pentane or butane
    b. $(CH_3)_2CHCH(CH_3)_2$ or $CH_3(CH_2)_4CH_3$
    c. cyclopentane or cyclohexane
    d. $CH_3(CH_2)_5CH_3$ or $CH_3(CH_2)_7CH_3$
54. Which member of each pair has the higher melting point?
    a. pentane or butane
    b. neopentane or pentane
    c. cyclopentane or cyclohexane
    d. $CH_3(CH_2)_5CH_3$ or $CH_3(CH_2)_7CH_3$

Chemical Reactions

55. Complete the following equations.
    a. $(CH_3)_2C{=}CH_2 + Br_2 \longrightarrow$
    b. $CH_2{=}C(CH_3)CH_2CH_3 + H_2 \xrightarrow{Ni}$
    c. 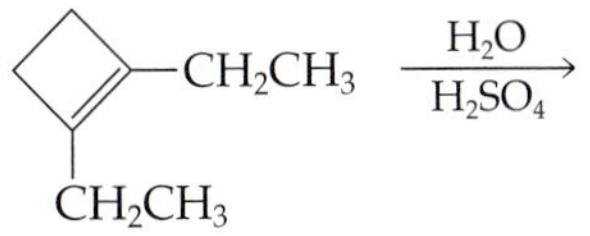

56. Complete the following equations.
    a. $CH_2{=}CHCH{=}CH_2 + 2\,H_2 \xrightarrow{Ni}$
    b. $(CH_3)_2C{=}C(CH_3)_2 \xrightarrow[H_2SO_4]{H_2O}$
    c. 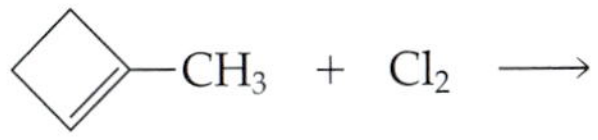

57. Give the reagents required for the following transformations.
    a. $H_2C{=}CHCH_3 \longrightarrow CH_3CH_2CH_3$
    b. 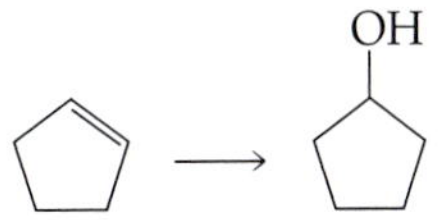

58. Give the reagents required for the following transformations.
    a. $CH_3CH{=}CHCH_3 \longrightarrow CH_3CH(OH)CH_2CH_3$
    b. $CH_2{=}CHCH_3 \longrightarrow CH_2(Cl)CH(Cl)CH_3$
59. List the starting materials required to complete the following transformations.
    a. ? 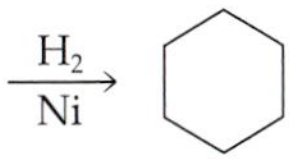

    b. ? 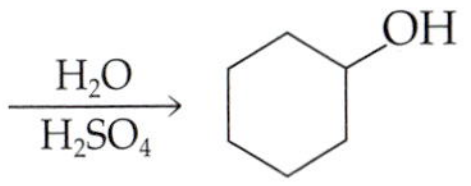

60. List the starting materials required to complete the following transformations.

 a. ? $\xrightarrow{Cl_2}$

$$H-\overset{\displaystyle Cl}{\underset{\displaystyle H}{\overset{\displaystyle |}{\underset{\displaystyle |}{C}}}}-\overset{\displaystyle Cl}{\underset{\displaystyle H}{\overset{\displaystyle |}{\underset{\displaystyle |}{C}}}}-H$$

 b. ? $\xrightarrow[H_2SO_4]{H_2O}$ $CH_3\overset{\displaystyle OH}{\overset{\displaystyle |}{C}}HCH_3$

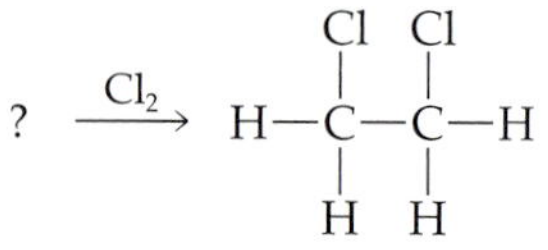

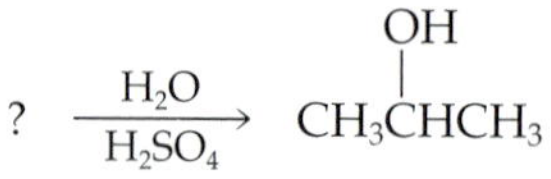

Additional Problems

61. Three isomeric pentenes, X, Y, and Z, can be hydrogenated to 2-methylbutane. Addition of chlorine to Y gives 1,2-dichloro-3-methylbutane and addition of chlorine to Z gives 1,2-dichloro-2-methylbutane. Write the structural formulas for the three isomers.

62. Pentane and 1-pentene are both colorless, low-boiling liquids. Give a simple test that distinguishes the two compounds. Indicate what you would observe.

63. What is wrong with each of the following names? Draw the structural formula and give the correct name for each compound.
 a. 2-dimethylpropane
 b. 2,3,3-trimethylbutane
 c. 2,4-diethylpentane
 d. 3,4-dimethyl-5-propylhexane

64. What is wrong with each of the following names? Draw the structural formula and give the correct name for each compound.
 a. 2-methyl-4-heptene
 b. 2-ethyl-3-hexene
 c. 2,2-dimethyl-3-pentene
 d. 4-bromocyclobutene

65. You find an unlabeled jar containing a solid that melts at 48 °C. It ignites readily and burns cleanly. The substance is insoluble in water and floats on the surface of the water. Is the substance likely to be organic or inorganic?

66. Write equations for the complete combustion of each of the following.
 a. natural gas (methane)
 b. a typical petroleum hydrocarbon (such as octane)

67. The complete combustion of benzene forms carbon dioxide and water:

$$C_6H_6 + O_2 \longrightarrow CO_2 + H_2O$$

 Balance the equation. What mass of carbon dioxide is formed by the complete combustion of 39.0 g of benzene?

68. The density of a gasoline sample is 0.690 g/mL. On the basis of the complete combustion of octane, calculate the amount of carbon dioxide and water formed per gallon (3.78 L) of the gasoline when used in an automobile.

Special Topic E
Petroleum

Compounds of carbon are essential to all life processes, a subject that occupies much of the remainder of this book. But carbon compounds, especially the hydrocarbons (Chapter 13), constitute much of our energy supply. Carbon compounds are the basis of many structural building materials and of most of our medicines. In this special topic, we look at petroleum and at some of the products that we obtain from it.

E.1 Natural Gas and Gasoline

Some organic compounds are still obtained from plants and animals, but most come ultimately from the fossilized carbon materials coal and petroleum. **Petroleum** is a complex mixture of hydrocarbons (alkanes) produced by the decomposition of animal and vegetable matter that has been entrapped in the Earth's crust for a long time. Most of the **petrochemicals** so vital to our modern economy are derived from these alkanes. A portion of the remaining petrochemicals, the aromatics, are derived from coal.

Petroleum, as it comes from the ground, is of limited use. So that it will better suit our needs, we separate it into fractions by boiling it in a distillation column (Figure E.1). The lighter molecules, those containing one to four carbon atoms each, come off the top of the column. The next fraction contains, for the most part, molecules having from five to 12 carbon atoms. These and other fractions are listed in Table E.1.

Crude oil and natural gas are the liquid and gaseous components of petroleum. Natural gas is about 80% methane and 10% ethane, and the remaining 10% is a mixture of the higher alkanes. Methane is a product of the bacterial decay of plant and marine organisms that have become buried beneath the Earth's surface. It is also produced by

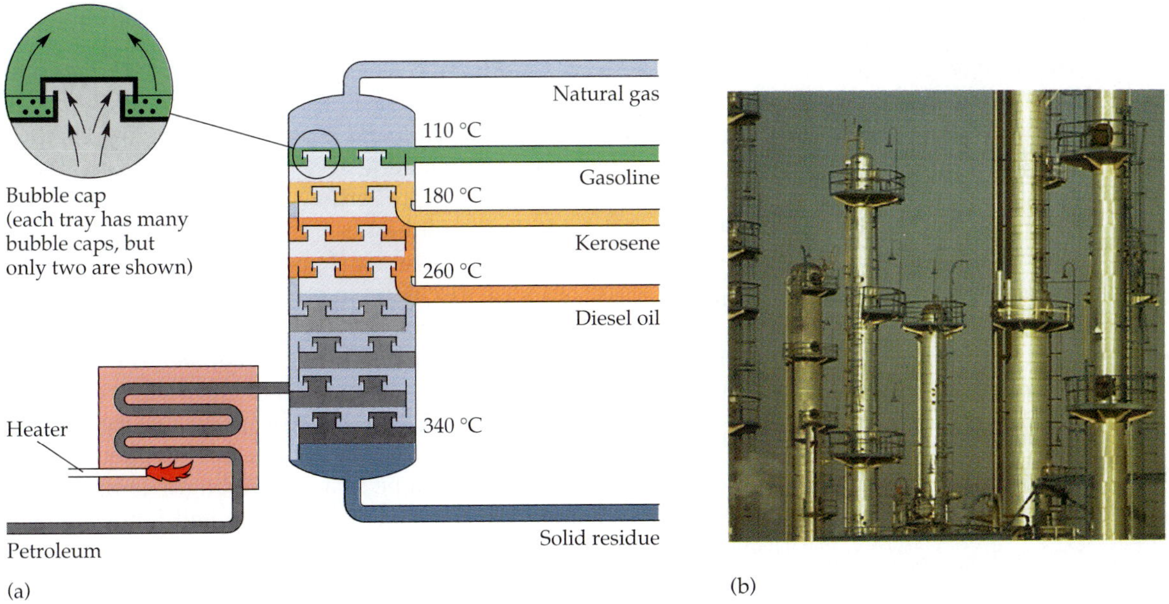

▲ **FIGURE E.1**

(a) A schematic of an oil refinery distillation column. Petroleum is vaporized by heating with very hot steam at the bottom of a tall column. The components with higher boiling points condense quite low in the column, and those with lower boiling points move farther up. Fractions with different compositions are removed from the column at different heights. (b) Petroleum distillation columns.

Table E.1 Typical Petroleum Fractions

Fraction	Typical Range of Hydrocarbons	Approximate Range of Boiling Point (°C)	Typical Uses
Natural gas	CH_4 to C_4H_{10}	Below 40	Fuel, starting materials for plastics
Gasoline	C_5H_{12} to $C_{12}H_{26}$	40–200	Fuel, solvents
Kerosene	$C_{12}H_{26}$ to $C_{16}H_{34}$	175–275	Diesel fuel, jet fuel, home heating; cracking to gasoline
Heating oil	$C_{15}H_{32}$ to $C_{18}H_{38}$	250–400	Industrial heating, cracking to gasoline
Lubricating oil	$C_{17}H_{36}$ and up	Above 300	Lubricants
Residue	$C_{20}H_{42}$ and up	Above 350 (some decomposition)	Paraffin, asphalt, road tar, roofing tar

the microbial decomposition of organic matter in sewage treatment plants. Because methane was first isolated in marshes, it was given the name "marsh gas." Methane is used primarily as a cooking fuel (and in bunsen burners). Natural gas is the cleanest of the fossil fuels because it contains the least amount of sulfur compounds; little sulfur dioxide is produced when natural gas is burned.

Propane and the butanes are familiar fuels. They are usually supplied under pressure in tanks. Although they are gases at ordinary temperatures and under normal atmospheric pressure, they are liquefied under pressure and are sold as liquefied petroleum (LP) gas. Butane, liquefied under pressure, can be seen in disposable butane cigarette lighters. When the release lever is pressed on these lighters, the butane comes under atmospheric pressure, and some of it vaporizes and is ignited by a spark.

	$C_{13}H_{28}$	$C_{13}H_{26}$	$C_{13}H_{24}$
	$C_{12}H_{26}$	$C_{12}H_{24}$	$C_{12}H_{22}$
	$C_{11}H_{24}$	$C_{11}H_{22}$	$C_{11}H_{20}$
$C_{14}H_{30} \xrightarrow{heat}$	$+$	$+$	$+$ etc.
	$C_{10}H_{22}$	$C_{10}H_{20}$	$C_{10}H_{18}$
	C_9H_{20}	C_9H_{18}	C_9H_{16}
	CH_4	C_2H_4	C_2H_2

Smaller saturated hydrocarbons (having from 1 to 13 carbon atoms)

Unsaturated hydrocarbons (having from 2 to 13 carbon atoms)

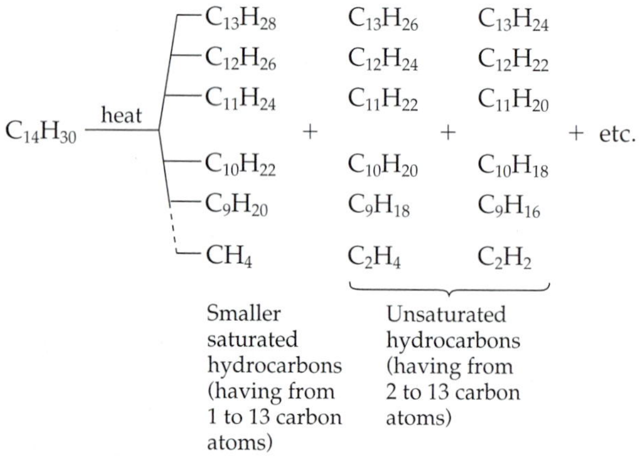

▲ **FIGURE E.2**

Formulas of some of the products formed when $C_{14}H_{30}$, a typical molecule in kerosene, is cracked. Note the great variety of hydrocarbons with fewer carbon atoms that are formed.

Since gasoline is generally the petroleum fraction most in demand, the fractions with higher boiling points are often in excess supply. These are converted to gasoline when they are heated in the absence of air. This process, called *cracking*, breaks the big molecules apart. The process is illustrated in Figure E.2, where $C_{14}H_{30}$ is used as an example. Not only does cracking convert some of the molecules to those in the gasoline range, but it also results in a variety of useful by-products. The unsaturated hydrocarbons are starting materials for the manufacture of many plastics, detergents, and drugs. Future shortages of petroleum will mean a great deal more than just scarce, high-priced gasoline.[1]

The cracking process described here is a crude, yet illustrative, example of how chemists modify nature's materials to meet our needs and desires. Starting with coal tar or petroleum, the chemist can create a dazzling array of substances having a wide variety of properties. These include plastics, pain killers, antibiotics, stimulants, depressants, and detergents, to name just a few. Many of these materials are discussed in other chapters.

Gasoline, like the petroleum from which it is derived, is a mixture of hydrocarbons. The typical alkanes in gasoline have formulas ranging from C_5H_{12} to $C_{12}H_{26}$. Because there are many isomeric forms, particularly for the higher members of the group, gasoline is an exceedingly complex mixture of alkanes. There are also small amounts of other

[1] In today's energy-conscious world, the unit of commerce on the oil market is the *barrel*. One barrel of crude oil equals 42 gallons. Of this, approximately 45% is converted to gasoline, 30% to heating oil, and 10% to jet fuel. Only about 5% of this precious liquid is used in the petrochemical industry to manufacture detergents, dyes, fertilizers, pesticides, plastics, etc. The remainder is sold as aviation gasoline, lubricating oils, greases, and asphalt.

hydrocarbons present and even some sulfur- and nitrogen-containing compounds. The gasoline fraction of petroleum as it comes from the distillation column is called **straight-run gasoline.** It doesn't burn very well in modern, high-compression automobile engines. Chemists have learned how to modify it in a variety of ways to make it burn more smoothly.

E.2 The Octane Rating of Gasoline

Early in the development of the automobile engine, scientists learned that some types of hydrocarbons burned more evenly and were less likely to ignite prematurely than others. Ignition before the piston was in proper position led to a knocking in the engine. Scientists soon were able to correlate good performance with branched-chain hydrocarbons. An arbitrary performance standard, called the **octane rating,** was established in 1927. The best performer in a laboratory test engine was a compound that became known as isooctane (it was one of 18 isomeric octanes tested). Isooctane was assigned a value of 100 octane. The unbranched-chain compound, heptane, caused a very bad knock and so was given an octane rating of 0. A gasoline rated 90 octane was one that performed the same as a mixture that was 90% isooctane and 10% heptane.

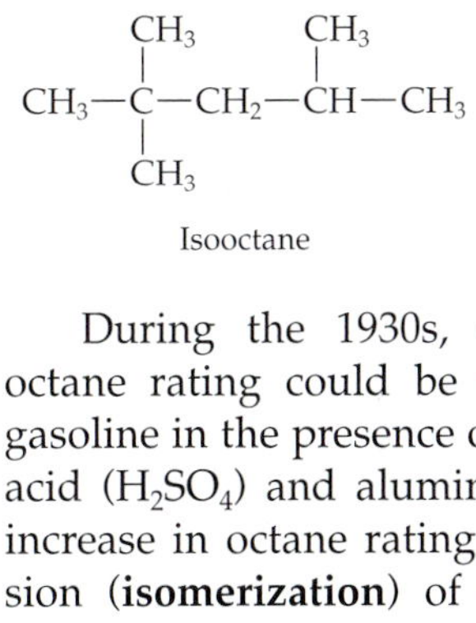

$$
\begin{array}{cc}
\underset{\substack{|\\ CH_3}}{\overset{\substack{CH_3\\ |}}{CH_3-C-CH_2}}-\underset{\substack{|\\ CH_3}}{CH}-CH_3 & CH_3CH_2CH_2CH_2CH_2CH_2CH_3
\end{array}
$$

Isooctane Heptane

During the 1930s, chemists discovered that octane rating could be improved by heating the gasoline in the presence of catalysts such as sulfuric acid (H_2SO_4) and aluminum chloride ($AlCl_3$). This increase in octane rating was attributed to conversion (**isomerization**) of unbranched structures to highly branched ones. For example, heptane molecules could be isomerized to a branched structure:

$$
CH_3CH_2CH_2CH_2CH_2CH_2CH_3 \xrightarrow[\text{heat}]{H_2SO_4}
$$

$$
CH_3CH_2-\underset{\substack{|\\ CH_3}}{CH}-\overset{\substack{CH_3\\ |}}{CH}-CH_3
$$

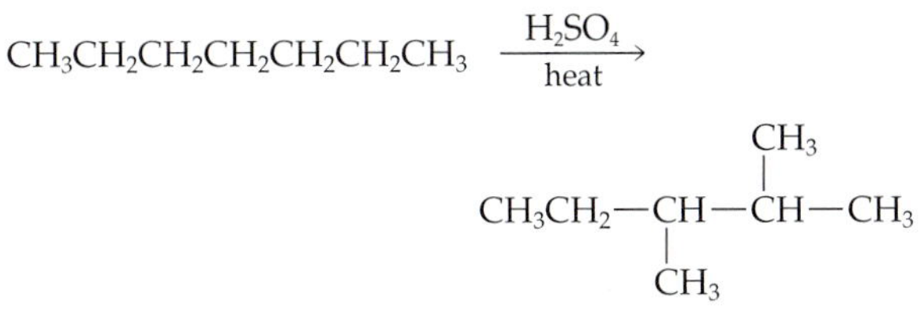

Certain chemical substances were discovered that, when added in small amounts, substantially improved the antiknock quality of gasoline. Chief among these additives was tetraethyllead. This compound, when added in amounts as small as 1 mL per liter of gasoline (about 1 part per thousand), would increase the octane rating by 10 or more.

$$
C_2H_5-\underset{\substack{|\\ C_2H_5}}{\overset{\substack{C_2H_5\\ |}}{Pb}}-C_2H_5
$$

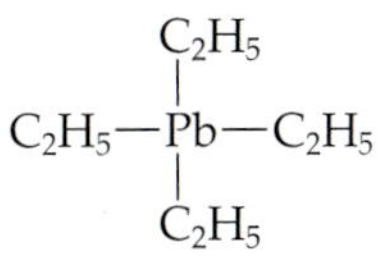

Tetraethyllead

Lead is toxic, however.[2] Large amounts of it have entered the environment through the combustion of leaded gasoline in automobiles. Further, lead fouls the catalytic converters used in modern automobiles. For these reasons, unleaded fuels have been developed. Since 1994, almost no leaded fuel has been available in the United States. The only exceptions are fuels for some farm and off-highway vehicles. Europe has been much slower in its phaseout of leaded motor fuel.

Scientists have developed a number of ways to get high octane ratings in unleaded fuels. For example, petroleum refineries use **catalytic reforming** to convert low-octane alkanes to high-octane aromatic compounds. For example, hexane (octane rating of 25) is converted to benzene (octane rating of 106):

$$
CH_3CH_2CH_2CH_2CH_2CH_3 \xrightarrow[\text{heat}]{\text{catalyst}} \bigcirc + 4\,H_2
$$

(C_6H_{14}) (C_6H_6)

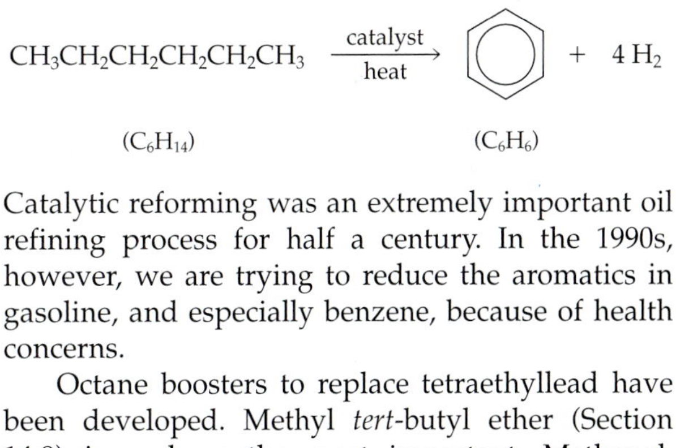

Catalytic reforming was an extremely important oil refining process for half a century. In the 1990s, however, we are trying to reduce the aromatics in gasoline, and especially benzene, because of health concerns.

Octane boosters to replace tetraethyllead have been developed. Methyl *tert*-butyl ether (Section 14.9) is perhaps the most important. Methanol, ethanol, and *tert*-butyl alcohol also are used. None of these is nearly as effective as tetraethyllead in boosting the octane rating. They must therefore be used in fairly large quantities. The amount that can be used in gasoline is limited by solubility problems. Methanol in excess of 5% and ethanol in excess of 10% tend to separate from the gasoline, especially if moisture gets into the fuel.

[2] Lead is especially toxic to the brain. Even small amounts can cause learning disabilities in children.

Unlike gasoline, which is made up of hydrocarbons, methyl *tert*-butyl ether and the various alcohols and their derivatives all contain oxygen. They are sometimes referred to as "oxygenates." These oxygenates are added to gasoline not only for their octane improvement, but also, and perhaps more important, for their ability to decrease carbon monoxide in the auto exhaust gas.

Key Terms

catalytic reforming (E.2)
isomerization (E.2)
octane rating (E.2)
petrochemical (E.1)
petroleum (E.1)
straight-run gasoline (E.1)

Review Questions

1. What kinds of compounds make up the bulk of petroleum?
2. What is straight-run gasoline?
3. List three ways to increase the octane rating of gasoline.
4. What is the main component of natural gas? What other compounds are present?
5. What chemical change occurs during catalytic reforming?
6. List five products made from petroleum.
7. How is natural gas formed in nature?
8. What are some advantages of natural gas as a fuel?

14 Alcohols, Phenols, and Ethers

Ethanol, the beverage alcohol, is made by fermentation. Grapes are the principal source of the ethanol found in wine.

The three families of organic compounds discussed in this chapter occur widely in nature. The human race has been quick to adapt these materials to its own use. The earliest written histories record that primitive peoples used the compound we know as alcohol. According to Genesis, Noah planted a vineyard after the Flood, drank wine from its grapes, and became drunk.

Human ingenuity may have reached some sort of peak in finding sources of *aqua vitae,* the water of life. Alcohol has been obtained from the fermentation of fruits, grains, potatoes, rice, and even cacti. It was prescribed as medicine in the twelfth century but has been most frequently used without such justification. What we know as alcohol is actually only one member of a family of organic compounds known by the same name. The family also includes such familiar substances as cholesterol and the carbohydrates.

Anyone who has ever been in a hospital would recognize the pungent, antiseptic odor of phenol, the simplest member of the second family of compounds to be introduced in this chapter. And the name of the third family considered in this chapter, the ethers, has become almost synonymous with anesthesia.

$$R\text{—}OH \;=\; CH_3OH \;\text{ or }\; CH_3\overset{\displaystyle OH}{\underset{\displaystyle H}{C}}CH_3 \;\text{ or }\; CH_3CH_2CH_2CH_2OH \;\text{ or } ...$$

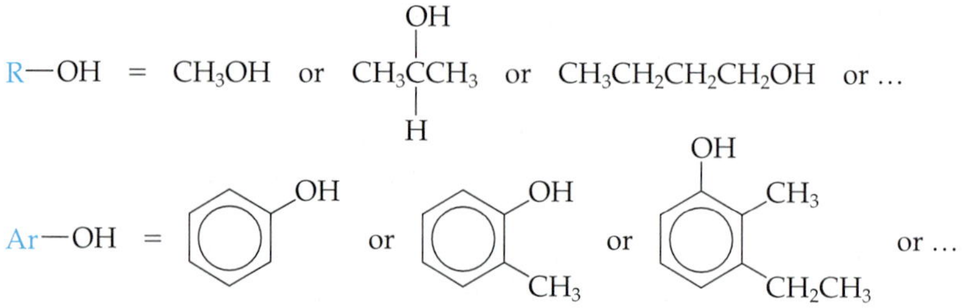

FIGURE 14.1 ▶

In the notation of organic chemistry, R represents any alkyl group and Ar represents any aromatic group.

$$Ar\text{—}OH \;=\; \text{(ring)}OH \;\text{ or }\; \text{(ring)}\underset{CH_3}{OH} \;\text{ or }\; \overset{OH}{\text{(ring)}}\underset{CH_2CH_3}{CH_3} \;\text{ or } ...$$

Why are we taking up three families in one chapter? We do so because the members of these families have something in common: they can all be considered organic derivatives of water.

Consider the water molecule:

$$\underset{H\qquad H}{O}$$

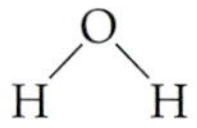

This molecule is a bent molecule containing a central oxygen atom attached to two hydrogen atoms. If one hydrogen is replaced with an alkyl group, for which we use the symbol R, as shown in Figure 14.1, we have

$$\underset{R\qquad H}{O}$$

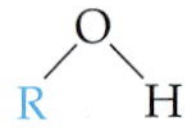

This is the general formula for the alcohol family. The alkyl group may be methyl, ethyl, isopropyl, or an aliphatic group too complicated to have a simple name. As long as the carbon attached to the hydroxyl group (—OH) is aliphatic, the compound is an alcohol.

If the hydroxyl group is attached directly to an aromatic ring, a different family of compounds is produced. Compounds in which an aryl group (Figure 14.1) is attached to a hydroxyl group are called *phenols* (see Section 14.8).

$$\underset{Ar\qquad H}{O}$$

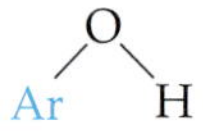

Replacement of both hydrogen atoms of water by alkyl or aryl groups gives rise to an *ether* (see Section 14.9).

$$\underset{R\qquad R}{O} \qquad \underset{R\qquad Ar}{O} \qquad \underset{Ar\qquad Ar}{O}$$

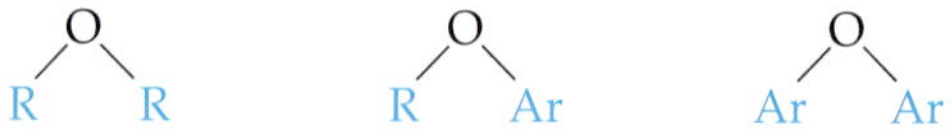

14.1 The Functional Group

The chemistry of the phenols is sufficiently different from that of the alcohols to justify treating the two classes of compounds as separate, if closely related, families. Nonetheless, for both families, the chemistry is largely determined by the hydroxyl group. Nearly all the characteristic reactions of alcohols and many of those of the phenols take place at the hydroxyl group. Even the physical properties are determined to a large extent by the presence of the hydroxyl group. Such a group of atoms, which confers characteristic chemical and physical properties on a family of organic compounds, is called a **functional group.**

The hydroxyl group is one functional group. We have already encountered others. The carbon–carbon double bond (C=C) in alkenes and the carbon–carbon triple bond (C≡C) in alkynes are functional groups. In both instances, these structural features confer on the members of the families a particular chemical reactivity; for example, alkenes and alkynes tend to undergo addition reactions. In addition, the halogens in halogenated hydrocarbons are functional groups, although we did not consider in detail the particular reactions associated with these groups.

The alkanes are characterized by their *lack* of a distinct functional group. Other functional groups will serve as unifying concepts for the next three chapters. Some of the more important functional groups are listed in Table 14.1. For ready reference, this table is also reproduced on the inside back cover.

Table 14.1 Selected Organic Functional Groups

Name of Family	Functional Group	General Formula(s) of Family
Alkane	None	R—H
Alkene	—C(—)=C(—)—	R—C(R)=C(R)—R
Alkyne	—C≡C—	R—C≡C—R
Alcohol	—C(—)(—)—O—H	R—O—H
Phenol	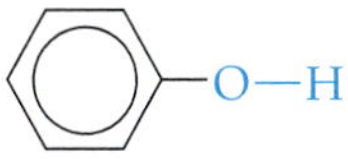	Ar—O—H
Ether	—C(—)(—)—O—C(—)(—)—	R—O—R
Aldehyde	—C(=O)—H	R—C(=O)—H
Ketone	—C(=O)—	R—C(=O)—R
Amine	—C(—)(—)—N(—)—	R—N(H)—H R—N(H)—R R—N(R)—R
Carboxylic acid	—C(=O)—O—H	R—C(=O)—O—H
Ester	—C(=O)—O—C(—)(—)—	R—C(=O)—O—R
Amide	—C(=O)—N(—)—	R—C(=O)—N(H)—H R—C(=O)—N(H)—R R—C(=O)—N(R)—R

14.2 Classification and Nomenclature of Alcohols

The properties of alcohols depend on the arrangement of the carbon atoms in the molecule. Alcohols can be grouped into three classes that serve to distinguish these different structural arrangements from one another. The classes are known as primary, secondary, and tertiary, and an alcohol is classified according to the type of carbon atom to which the hydroxyl group is attached.

1. A **primary (1°) carbon atom** is one attached directly either to no other carbons or to one other carbon:

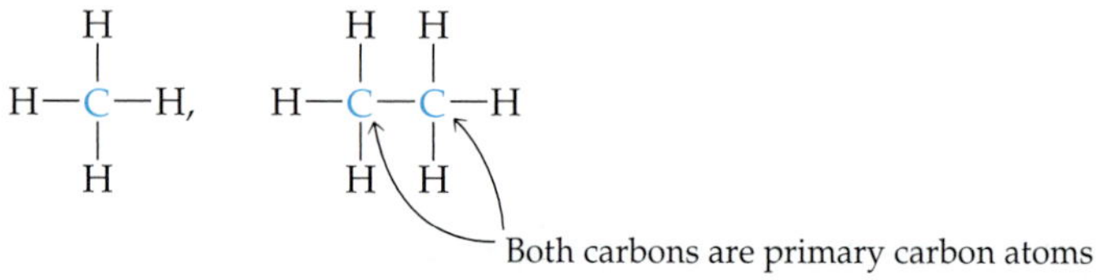

2. A **secondary (2°) carbon atom** is one attached to two other carbon atoms:

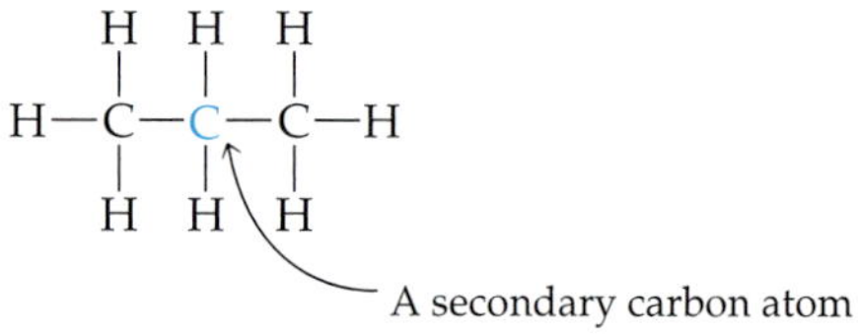

3. **A tertiary (3°) carbon atom** is one attached to three other carbon atoms:

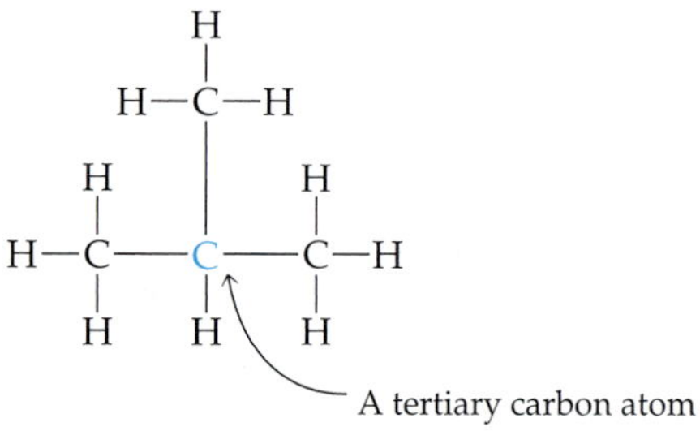

Therefore, a **primary alcohol** is one in which the hydroxyl group replaces one of the hydrogens on a primary carbon, a **secondary alcohol** is one whose hydroxyl group replaces one of the hydrogens on a secondary carbon atom, and a **tertiary alcohol** has its hydroxyl group replacing the hydrogen on a tertiary carbon. Table 14.2 presents the nomenclature and classification of some of the simpler alcohols.

As shown in the table, the common names of the lower members of the alcohol family are formed in a manner similar to that used for alkyl halides. The name of the alkyl group is followed by the word *alcohol* to indicate the presence of the hydroxyl group. The IUPAC system is used in naming the higher homologs, and here the designations primary, secondary, and tertiary are unnecessary. Alcohols are named under the IUPAC system as follows:

1. The longest continuous chain of carbons containing the —OH group is taken as the parent compound.
2. The chain is numbered from the end closer to the hydroxyl group.
3. The number that indicates the position of the hydroxyl group is prefixed to the name of the parent hydrocarbon, and the *-e* ending of the parent alkane is

Table 14.2 Classification and Nomenclature of Some Alcohols

Structural Formula	Class of Alcohol	Common Name	IUPAC Name
CH_3OH	Primary	Methyl alcohol	Methanol
CH_3CH_2OH	Primary	Ethyl alcohol	Ethanol
$CH_3CH_2CH_2OH$	Primary	Propyl alcohol	1-Propanol
CH_3CHCH_3 (OH)	Secondary	Isopropyl alcohol	2-Propanol
$CH_3CH_2CH_2CH_2OH$	Primary	Butyl alcohol	1-Butanol
$CH_3CH_2CHCH_3$ (OH)	Secondary	*sec*-Butyl alcohol	2-Butanol
CH_3CHCH_2OH (CH_3)	Primary	Isobutyl alcohol	2-Methyl-1-propanol
$CH_3-\overset{OH}{\underset{CH_3}{C}}-CH_3$	Tertiary	*tert*-Butyl alcohol	2-Methyl-2-propanol
⬡—OH	Secondary	Cyclohexyl alcohol	Cyclohexanol
⬡—CH_2OH	Primary	Benzyl alcohol	Phenylmethanol

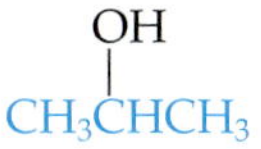

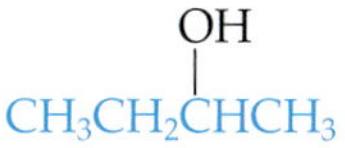

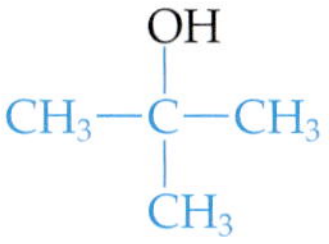

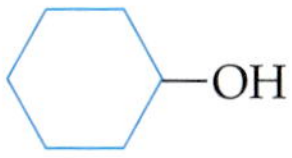

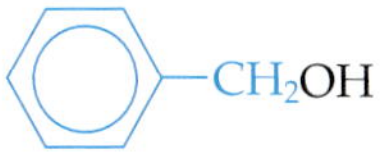

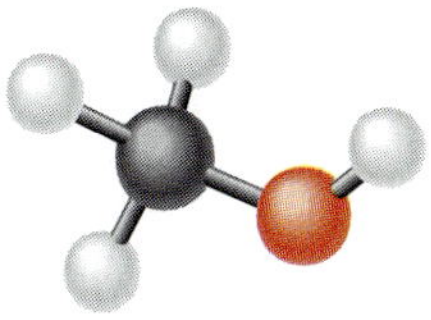

Methanol

Ball-and-stick model of methanol.

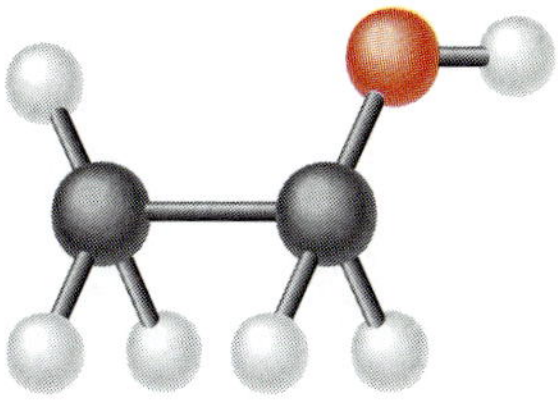

Ethanol

Ball-and-stick model of ethanol.

replaced by the suffix *-ol*. (When the hydroxyl group is on carbon 1 of a ring, the "1" is not needed in the name.)

4. If more than one hydroxyl group appears in the same molecule (polyhydroxy alcohols), the suffixes *-diol*, *-triol*, etc., are used. In these cases, the *-e* ending of the parent alkane is retained.

$$\overset{OH}{\underset{CH_3}{CH_3CH_2CCH_3}}$$

2-Methyl-2-butanol

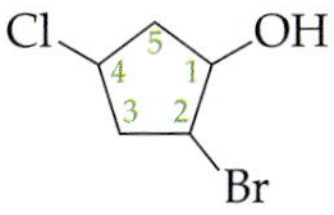

2-Bromo-4-chlorocyclopentanol

$$H-\overset{OH}{\underset{H}{C}}-\overset{OH}{\underset{H}{C}}-H$$

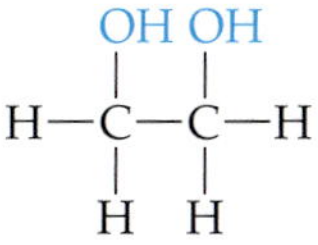

1,2-Ethanediol
(Ethylene glycol)

$$H-\overset{H}{\underset{OH}{C}}-\overset{H}{\underset{OH}{C}}-\overset{H}{\underset{OH}{C}}-H$$

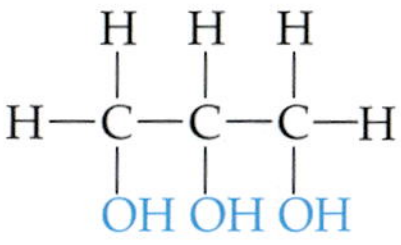

1,2,3-Propanetriol
(Glycerol)

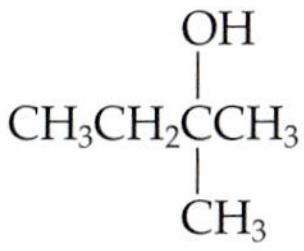

$$\underset{6\quad5\quad4\quad3\,|\,2\quad1}{CH_3CHCH_2CCH_2CH_3}$$ with CH_3 and OH

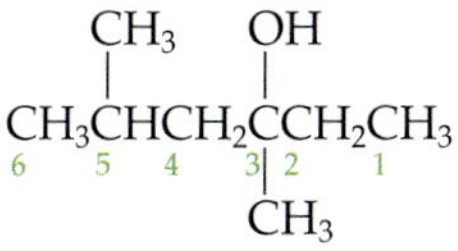

3,5-Dimethyl-3-hexanol

$$\underset{7\quad6\quad5\quad4\quad3\quad2\quad1}{CH_3CHCH_2CH_2CHCH_2CH_3}$$ with CH_3 and OH

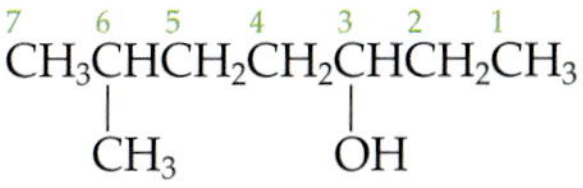

6-Methyl-3-heptanol

Example 14.1

Give the IUPAC name for

$$\underset{10\ 9\quad8\quad7\quad6\quad5\quad4\quad3\quad2\ 1}{CH_3CH_2CHCH_2CHCH_2CH_2CHCH_2CH_3}$$
with CH_3 CH_3 OH

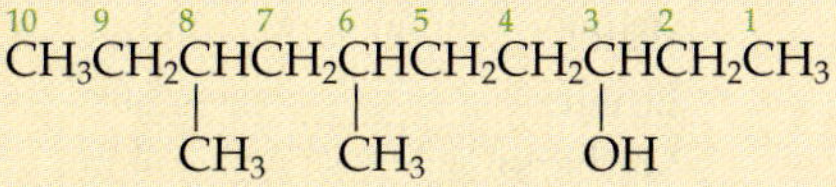

Solution

The carbons are numbered from the end closer to the —OH group. The name is 6,8-dimethyl-3-decanol (not 3,5-dimethyl-8-decanol).

Practice Exercise

Give the IUPAC name for

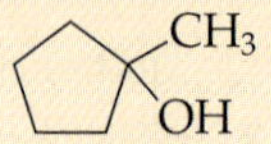

14.3 Physical Properties of Alcohols

It was stated earlier that alcohols are considered derivatives of water. This relationship becomes particularly apparent, especially for the lower homologs, when we discuss the physical and chemical properties of the alcohols. Remember that the alkanes methane, ethane, and propane are gases and insoluble in water. In contrast, the alcohols methanol, ethanol, and propanol are liquids and are completely soluble in water. Replacement of a single hydrogen atom with a hydroxyl group greatly changes solubility and physical state. These differences result from the hydrogen–bonding capabilities of the alcohols. Hydrogen bonding (Section 8.4) is responsible for the intermolecular attractions between alcohol molecules. Even the lightest homolog of the series is liquid at room temperature. Alcohols can form hydrogen bonds with water molecules as well, and so the lower homologs of the series are water–soluble.

Table 14.3 lists the molar masses and the boiling points of some common compounds. The table shows that substances that have similar molar masses do not always have similar boiling points. The relatively high boiling points of alcohols are a direct result of strong intermolecular attractions. Recall that boiling point is a rough measure of the amount of energy necessary to separate a liquid molecule from its nearest neighbors. If the nearest neighbors are attached to that molecule by means of hydrogen bonds, a considerable amount of energy must be supplied to break those bonds. Only then can the molecule escape from the liquid into the gaseous state.

Figure 14.2 illustrates hydrogen bonding in water and in the alcohols. This drawing reveals why water boils at a higher temperature than methyl alcohol, even though water is a lighter molecule. The oxygen atom and both hydrogen atoms of the water molecule participate in hydrogen bonding to three, or even four, adjacent water molecules. Because the alkyl group of an alcohol molecule does not participate in hydrogen bonding, the molecule is associated to only two other alcohol molecules. More energy is required to disrupt three or four inter-

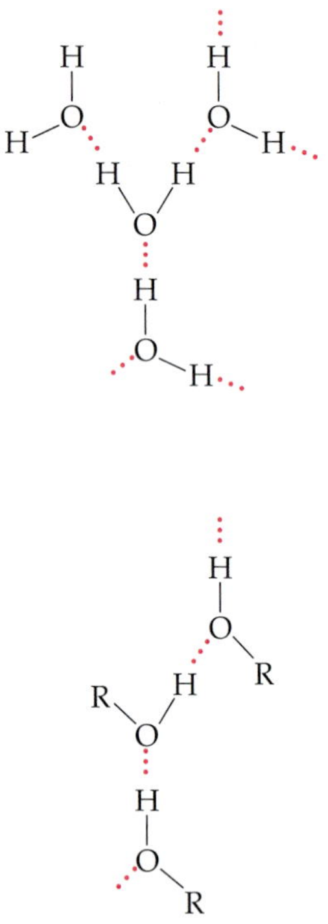

▲ **FIGURE 14.2**

Intermolecular hydrogen bonding (top) in water and (bottom) in an alcohol.

Table 14.3 Comparison of Boiling Points and Molar Masses

Formula	Name	Molar Mass	Boiling Point (°C)
CH_4	Methane	16	−164
HOH	Water	18	100
C_2H_6	Ethane	30	−89
CH_3OH	Methanol	32	65
C_3H_8	Propane	44	−42
CH_3CH_2OH	Ethanol	46	78
C_4H_{10}	Butane	58	−1
$CH_3CH_2CH_2OH$	1-Propanol	60	97

molecular bonds than two, and thus greater energy is needed to vaporize the water. (The energy required to break a hydrogen bond is about 5 kcal/mol. Although this is distinctly less than the energy required to break any of the intramolecular bonds in water or alcohol, it is still an appreciable amount of energy. Its significance is evidenced by the boiling point differences.)

Polarity and hydrogen bonding are significant factors in the water solubility of alcohols. A common expression among chemists is "like dissolves like," implying that polar solvents dissolve polar solutes, and nonpolar solvents dissolve nonpolar solutes. Care must be taken, however, not to apply this generalization to all cases. All alcohol molecules are polar, yet not all alcohols are water-soluble. On the other hand, all alcohols are soluble in most common nonpolar solvents (toluene, ether, hexane). Only the lower homologs of the alcohol family have an appreciable solubility in water. As the length of the carbon chain increases, water solubility decreases.

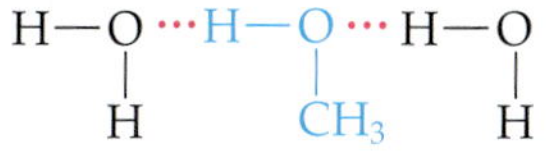

The differences in water solubility can be explained in the following manner. The hydroxyl group confers polarity and water solubility upon the alcohol molecule. The alkyl group confers nonpolarity and water insolubility. Whenever the hydroxyl group represents a substantial portion of a molecule, the molecule is water–soluble. (The hydroxyl group can be thought of as dragging the remainder of the molecule into the water structure.) As the size of the alkyl group increases, the alcohols become more like alkanes (they become more insoluble in water) and less like water. Decyl alcohol ($CH_3CH_2CH_2CH_2CH_2CH_2CH_2CH_2CH_2CH_2OH$) is insoluble in water. The hydroxyl group's ability to form hydrogen bonds is almost totally overshadowed by the lack of attraction between water molecules and the long alkane chain. (Figure 14.3).

Consider Table 14.4, which lists the solubilities of the butyl alcohols in water. The large differences in the water solubilities of these isomeric alcohols cannot be attributed to differences in molar mass. Instead, they are the result of the different geometric shapes. The very compact *tert*-butyl alcohol molecules experience weaker intermolecular attractions and therefore are more easily surrounded by water molecules. Hence, *tert*-butyl alcohol has a lower boiling point (83 °C) than any of its isomers (all of which boil above 100 °C) and a higher solubility in water.

In summary, solubility depends on the balance of polar and nonpolar groups within a molecule, as well as molecular shape. The more polar a molecule and the more compact its shape, the greater its water solubility. Molecules that can effectively form hydrogen bonds to water dissolve in water. Each functional group, such as the hydroxyl, that can form hydrogen bonds to water can carry into solution an alkyl group of up to four or five carbon atoms. Thus, we frequently find that the borderline of water solubility in a family of organic compounds occurs at four or five carbon atoms.

Table 14.4 Solubilities of the Butyl Alcohols in Water

Alcohol	Formula	Solubility ($g/100\ g\ H_2O$)
Butyl	$CH_3CH_2CH_2CH_2OH$	8
Isobutyl	$(CH_3)_2CHCH_2OH$	11
sec-Butyl	$CH_3CH_2CH(OH)CH_3$	12.5
tert-Butyl	$(CH_3)_3COH$	Completely soluble

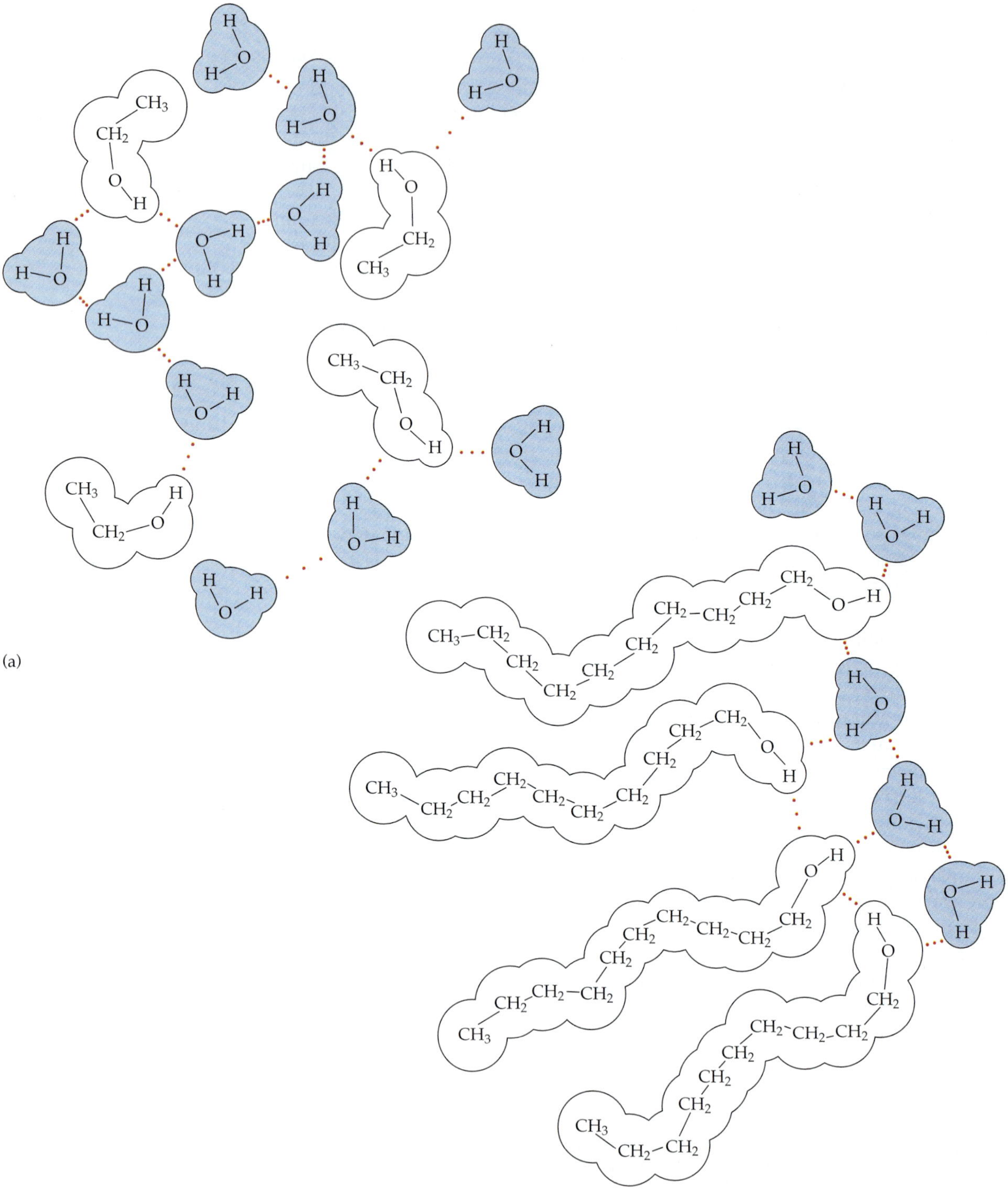

(a)

(b)

▲ FIGURE 14.3

(a) Hydrogen bonding between ethanol molecules and water molecules accounts for the solubility of ethanol in water. (b) Water molecules interact with 1-decanol molecules only near the hydroxyl end. The water molecules are unable to surround the 1-decanol molecules, and therefore 1-decanol is insoluble in water.

14.4 Preparation of Alcohols

Most simple alcohols are made by the hydration of alkenes (the addition of water to the double bond—Section 13.13). Since alkenes are made by the cracking of petroleum, our industrial supply of most alcohols (especially ethyl and isopropyl) depends to a large extent on the availability of oil.

Ethanol is made by the hydration of ethylene in the presence of sulfuric acid:

$$\underset{H}{\overset{H}{\diagdown}}C=C\underset{H}{\overset{H}{\diagup}} \;+\; H-OH \;\xrightarrow{H^+}\; H-\overset{\overset{\textstyle H}{|}}{\underset{\underset{\textstyle H}{|}}{C}}-\overset{\overset{\textstyle H}{|}}{\underset{\underset{\textstyle OH}{|}}{C}}-H$$

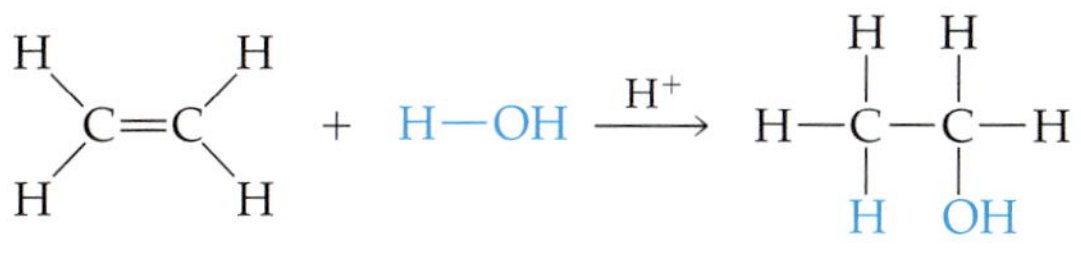

In a similar manner, isopropyl alcohol is produced by the addition of water to propylene:

$$H-\overset{\overset{\textstyle H}{|}}{\underset{\underset{\textstyle H}{|}}{C}}-\overset{\overset{\textstyle H}{|}}{\underset{\underset{\textstyle H}{|}}{C}}=C\underset{H}{\overset{H}{\diagup}} \;+\; H-OH \;\xrightarrow{H^+}\; H-\overset{\overset{\textstyle H}{|}}{\underset{\underset{\textstyle H}{|}}{C}}-\overset{\overset{\textstyle H}{|}}{\underset{\underset{\textstyle OH}{|}}{C}}-\overset{\overset{\textstyle H}{|}}{\underset{\underset{\textstyle H}{|}}{C}}-H$$

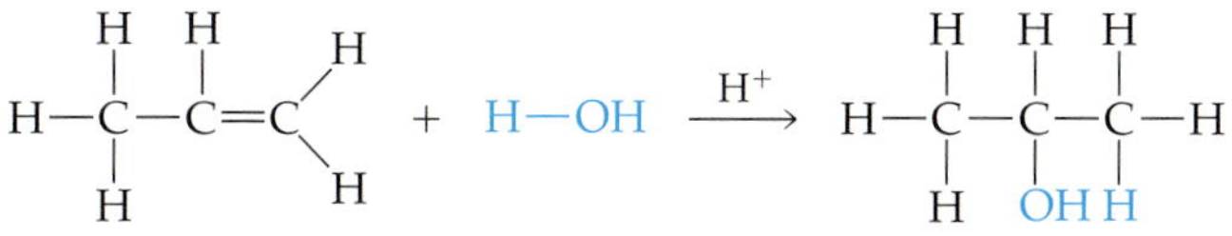

With 2-methylpropene, the product is 2-methyl-2-propanol:

$$H-\overset{\overset{\textstyle H}{|}}{C}=\overset{\overset{\textstyle CH_3}{|}}{C}-CH_3 \;+\; H-OH \;\xrightarrow{H^+}\; H-\overset{\overset{\textstyle H}{|}}{\underset{\underset{\textstyle H}{|}}{C}}-\overset{\overset{\textstyle CH_3}{|}}{\underset{\underset{\textstyle OH}{|}}{C}}-CH_3$$

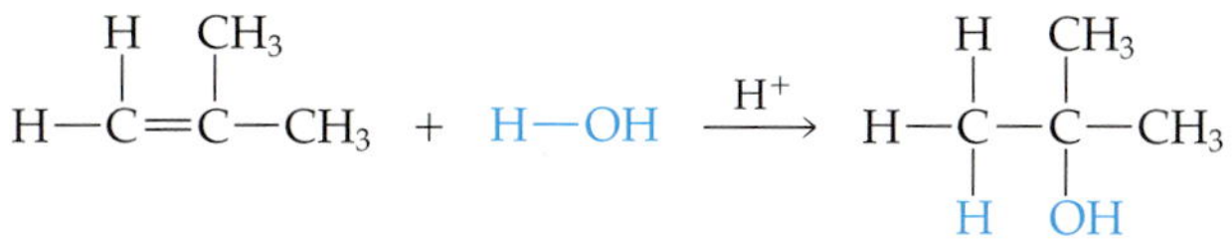

Note that in the last two reactions the hydrogen goes on the carbon atom (of the two involved in the double bond) that has the most hydrogens already bonded to it. The hydroxyl group goes on the carbon with fewer hydrogens. Thus, addition of water to propylene always gives isopropyl alcohol, never propyl alcohol:

$$CH_3CH=CH_2 \;+\; H-OH \;\xrightarrow{H^+}\; \begin{cases} \xrightarrow{\text{always}} \; \overset{\overset{\textstyle OH}{|}}{CH_3CHCH_3} \\[1.5em] \xrightarrow{\text{never}} \; CH_3CH_2CH_2OH \end{cases}$$

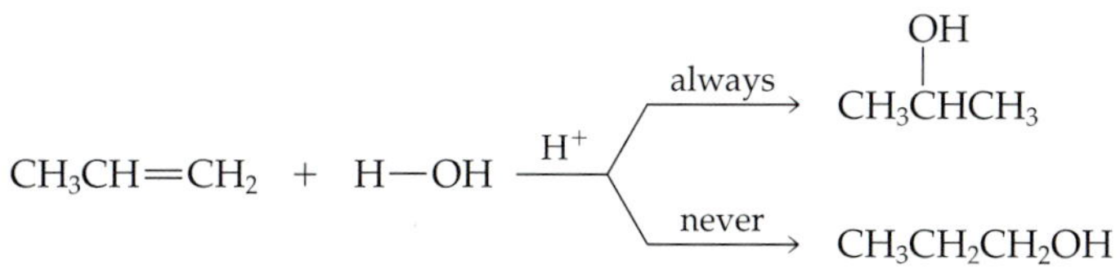

The above rule, in a more general form, was first formulated in 1870 by Vladimir V. Markovnikov, a Russian chemist. It is widely known as **Markovnikov's rule.** Sometimes the rule is stated (somewhat facetiously) as "the rich get richer" or "them that has, gets."

Example 14.2

Which alcohols are formed by the hydration of (**a**) 2-methyl-1-pentene and (**b**) 1-methylcyclopentene?

Solution

First write out the structural formulas and count the number of hydrogens directly bonded to each double–bond carbon:

a.

$$CH_3CH_2CH_2-\overset{\overset{\displaystyle CH_3}{|}}{C}=\overset{\overset{\displaystyle H}{|}}{C}-H$$

This C has no hydrogen. This C has 2 hydrogens.

b.

This C has 1 hydrogen. This C has no hydrogen.

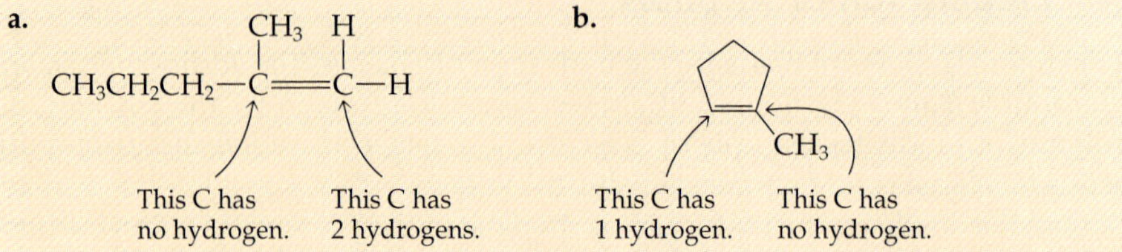

According to Markovnikov's rule, the hydrogen goes to the carbon that has the most hydrogens. The hydroxyl group goes to the other carbon:

a.

$$CH_3CH_2CH_2\overset{\overset{\displaystyle CH_3}{|}}{C}=\overset{\overset{\displaystyle H}{|}}{C}-H \;+\; HOH \;\xrightarrow{H^+}\; CH_3CH_2CH_2\overset{\overset{\displaystyle CH_3}{|}}{\underset{\underset{\displaystyle OH}{|}}{C}}\;\;\;\;\overset{\overset{\displaystyle H}{|}}{\underset{\underset{\displaystyle H}{|}}{C}}-H$$

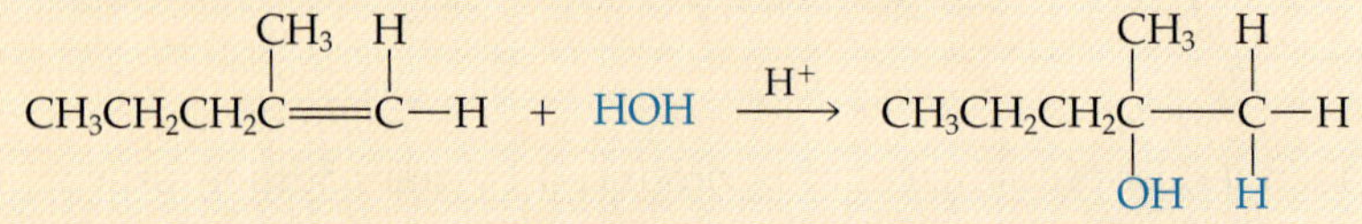

2-Methyl-1-pentene 2-Methyl-2-pentanol

b.

$$+ \; HOH \;\xrightarrow{H^+}\;$$

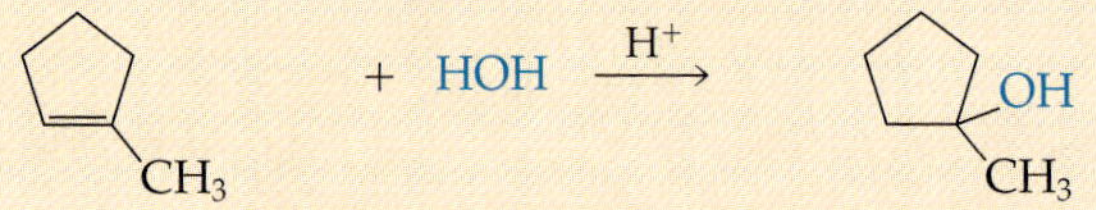

1-Methylcyclopentene 1-Methylcyclopentanol

Practice Exercise

Which alcohols are formed by the hydration of (a) 2-methyl-2-pentene and (b) 1-ethyl-cyclobutene?

Although alkene hydration is an important industrial source of alcohols, it is seldom used in the laboratory because other methods are more convenient. Alkene hydration is very common in biochemistry, however, and many hydroxy compounds in living systems are formed in this manner. The following reaction, for example, occurs in the Krebs cycle (see Figure 25.1).

$$HOOC-\overset{\overset{\displaystyle H}{|}}{C}=\overset{\overset{\displaystyle }{}}{C}-COOH \;+\; HOH \;\underset{}{\overset{enzyme}{\rightleftharpoons}}\; HOOC-\overset{\overset{\displaystyle H}{|}}{\underset{\underset{\displaystyle H}{|}}{C}}-\overset{\overset{\displaystyle OH}{|}}{\underset{\underset{\displaystyle H}{|}}{C}}-COOH$$

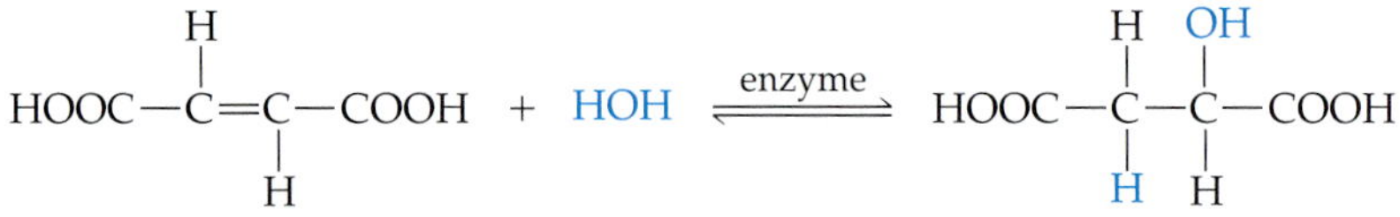

Fumaric acid Malic acid

Prior to 1923, methanol was prepared by the destructive distillation of wood—hence its common name, *wood alcohol*. When wood is heated to 450 °C in the absence of air, it decomposes to charcoal and a volatile fraction (Figure 14.4). Two to three percent of this fraction is methanol and can be separated from the other components (acetic acid and acetone) by fractional distillation. On the average, 1 ton of wood produces about 35 lb of methanol. Today, methanol is prepared more economically by combining hydrogen and carbon monoxide at high temperature and pressure in the presence of a zinc oxide–chromium oxide catalyst:

About 1.7 billion gallons of methanol are produced each year in the U.S. by catalytic reduction of carbon monoxide with hydrogen gas.

$$2\,H_2 \;+\; CO \;\xrightarrow[ZnO,\,Cr_2O_3]{200\ atm,\ 350\ °C}\; CH_3OH$$

The production of alcoholic spirits is one of the oldest known chemical reactions. Even in biblical times, ethanol was prepared by the fermentation of sugars or starch from various sources (potatoes, corn, wheat, rice, etc.). Biochemical

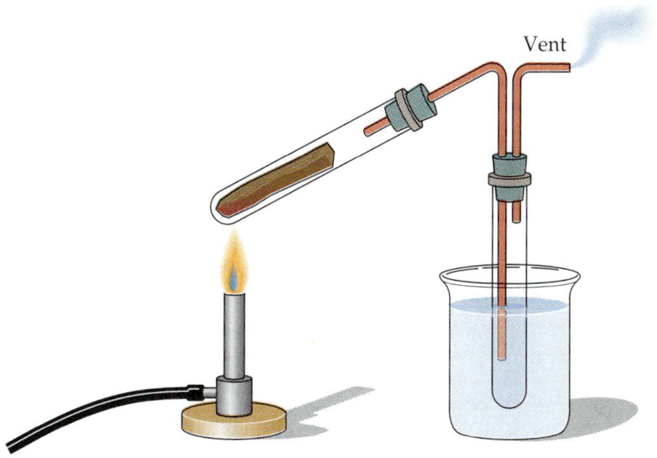

FIGURE 14.4

An apparatus for the destructive distillation of wood. The wood is heated in an enclosed tube, and the methanol is condensed in the second tube by the cold water in the beaker. Gases formed in the process are burned as they exit the vent tube.

investigations have shown that fermentation is catalyzed by enzymes found in yeast and proceeds by an elaborate multistep mechanism considered in Chapter 24. The equation for the overall process can be written

$$\underset{\text{Starch}}{(C_6H_{10}O_5)_x} \xrightarrow{\text{enzymes}} \underset{\text{Glucose}}{C_6H_{12}O_6} \xrightarrow{\text{enzymes}} \underset{\text{Ethanol}}{2\,C_2H_5OH} + 2\,CO_2$$

On an industrial scale, either molasses from sugar cane or starches from various grains are fermented by yeast to ethanol. It is ethanol that most people are referring to when they say "alcohol," meaning liquor to be drunk. The greatest use of ethanol is as a beverage. Wines contain about 12% ethanol by volume, champagnes 14 to 20%, beers and ciders 4%, and whiskey, gin, and brandy 40 to 50%. The alcoholic content of a beverage is indicated by a measure known as *proof spirit*.[1] The proof value is twice the alcoholic content by volume, and thus whiskey

Ethanol can be made by fermentation of nearly any type of sugary or starchy material: *(left to right)* wine from rice, vodka from potatoes, aperitif from artichoke, raki from raisins, and wine from grapes.

[1] The term has its origin in a seventeenth-century English method for testing whiskey. Dealers were perhaps too often tempted to increase profits by adding water to the booze. The whiskey to be tested was poured on gunpowder and the mixture was ignited. Since an ethanol–water solution will ignite when the alcohol concentration is about 50%, this solution scored 100 in the test, "proof" of the spirit content in the whiskey. If the powder did not burn, it was due to the presence of too much water in the whiskey.

that is 50% alcohol is said to be 100 proof. Fermented alcohol can be concentrated to as much as 95% (190 proof) by distillation. Such grain alcohol is frequently used as a solvent for drugs meant for internal consumption.

14.5 Physiological Properties of Alcohols

People often want to know whether or not something is a *poison*. The question is difficult to answer. Toxicity depends on the nature of the substance, the amount, and the route by which it is taken into the body.

The simple alcohols are all poisonous to humans to some degree. In an attempt to quantify the degree of toxicity, scientists use the term **LD$_{50}$** to indicate the *lethal dose* of a chemical to 50% of a population of test animals. Like humans, individual animals respond differently to various poisons. Some are killed by amounts much smaller than the LD$_{50}$; others survive considerably larger amounts. The LD$_{50}$ term, then, is only approximate for animals. Extrapolation to human toxicities can introduce even larger errors. The dosage usually is expressed as the amount of tested substance per kilogram of body weight of the test animal. The smaller the LD$_{50}$ value, the smaller the quantity of the substance required to kill the animal and therefore the more toxic the substance. Table 14.5 lists LD$_{50}$ values for alcohols administered orally to rats.

Methanol

Ingestion of as little as 15 mL of methanol can cause blindness; 30 mL (1 fluid ounce) can cause death.

Note that no LD$_{50}$ is given for methanol. Although its short-term toxicity is not terribly high, methanol can cause permanent blindness or death, even in small concentrations. Each year many accidents are attributed to this alcohol, which is frequently mistaken for its less harmful relative ethanol. Methanol should never be applied to the body, nor should its vapors be inhaled, because it is readily absorbed through the skin and respiratory tract.

The reason methanol is so dangerous[2] is that humans and other primates have liver enzymes that oxidize primary alcohols to compounds called aldehydes (Section 14.6). Ethanol, for example, is oxidized to acetaldehyde:

$$CH_3CH_2OH \xrightarrow{\text{liver enzymes}} \begin{array}{c} H \\ CH_3 \end{array}\!\!\!C{=}O$$

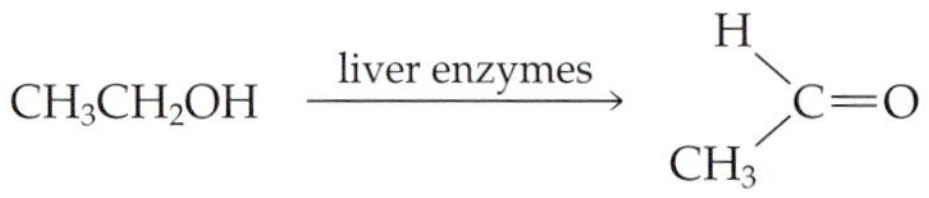

Ethanol Acetaldehyde

Table 14.5 Lethal Oral Doses (in Rats) for Some Alcohols

Alcohol	Structure	Boiling Point (°C)	LD$_{50}$ (g/kg body weight)	Uses
Methyl alcohol	CH_3OH	64	—	Solvent, fuel additive
Ethyl alcohol	CH_3CH_2OH	78	7.06	Solvent, beverages
Propyl alcohol	$CH_3CH_2CH_2OH$	97	1.87	Solvent
Isopropyl alcohol	$CH_3CHOHCH_3$	82	5.8	Solvent, body rubs
Butyl alcohol	$CH_3CH_2CH_2CH_2OH$	118	4.36	Solvent
Hexyl alcohol	$CH_3(CH_2)_4CH_2OH$	156	4.59	—
Ethylene glycol	$HOCH_2CH_2OH$	198	8.54	Antifreeze
Glycerol	$HOCH_2CHOHCH_2OH$	290 (decomposes)	>25	Moisturizer

[2] Methanol is not particularly toxic to horses, rats, and some other animals. These animals are deficient in the enzymes that oxidize alcohols to aldehydes. Toxicity studies in other animals cannot always be extrapolated to humans. In many cases, however, trends in toxicities can be judged from animal studies.

The acetaldehyde is in turn oxidized to acetic acid, a normal constituent of cells. The acetic acid can then be oxidized to carbon dioxide and water.

Similarly, methanol is oxidized to formaldehyde:

$$CH_3OH \xrightarrow{\text{liver enzymes}} \underset{H}{\overset{H}{>}}C=O$$

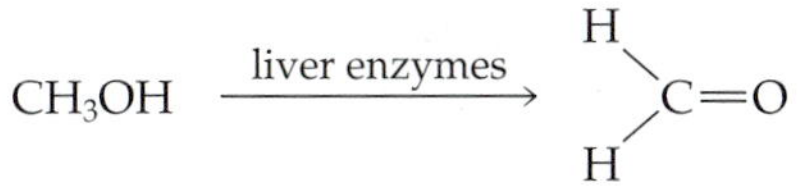

Methanol Formaldehyde

Formaldehyde reacts rapidly with the components of cells. It causes proteins to be coagulated, in much the same way that an egg is coagulated by cooking. It is this property of formaldehyde that accounts for the great toxicity of methanol. The LD_{50} for formaldehyde administered orally to rats is 0.070 g per kilogram of body weight. For acetaldehyde under the same conditions, LD_{50} is 1.9 g per kilogram of body weight. Thus, formaldehyde is about 27 times as toxic to rats as acetaldehyde. Indeed, the antidote for methanol poisoning has long been ethanol, administered intravenously. The ethanol preferentially loads up the liver enzymes in humans and other primates. If the enzymes are tied up oxidizing ethanol to acetaldehyde, they cannot catalyze the oxidation of the methanol to the dangerously toxic formaldehyde. Thus, the unoxidized methanol is gradually excreted from the body.

Despite its toxicity, methanol is a valuable industrial solvent. Its largest use is as the starting material for the commercial synthesis of formaldehyde. It is also used in windshield washer fluids and as a solvent for paint, gum, and shellac. Methanol, along with ethanol (see following), is mixed with gasoline and sold as motor fuel.

Ethanol

Ethyl alcohol is potentially toxic to humans. Rapid ingestion of 1 pint (about 500 mL) of pure ethanol would kill most people. Ethanol freely crosses into the brain, where it depresses the respiratory control center, resulting in failure of the respiratory muscles in the lungs and hence suffocation. Ethanol is believed to act on the nerve cell membranes, causing a diminution in speech, thought, cognition, and judgment. Excessive ingestion over a long period of time leads to deterioration of the liver (cirrhosis) and loss of memory and may lead to strong physiological addiction. Addiction to ethanol (alcoholism) is the most serious drug problem in the United States. It has been estimated that there are about 40 times as many alcoholics (about 10 million) as there are heroin addicts in the United States. If ethanol is diluted (as in alcoholic beverages) and consumed in small quantities, it is relatively safe. The body possesses enzymes that have the capacity to metabolize it to carbon dioxide and water (see Chapter 24).

Ethanol not intended for beverage purposes is commercially prepared by the hydration of ethylene, a by-product of the petroleum industry. The ethanol so produced is 95% ethanol and 5% water. The water that remains in this mixture cannot be removed by ordinary distillation because 95% ethanol is a constant-boiling mixture (an **azeotrope**). This 95% ethanol is used in chemical laboratories as a solvent. If 100% ethanol is needed, special procedures must be employed to prepare it. One method is to dry the ethanol over calcium oxide for several hours and then distill the remaining ethanol, known as **absolute** (100%) **alcohol.**

Ethanol is used as a solvent for perfumes, medicinal formulations (tinctures), lacquers, varnishes, and shellacs. It denatures enzymes in bacteria (see Section 21.13) and for this reason is widely used as an antiseptic in mouthwashes and

An azeotropic mixture is a constant-boiling mixture of two components that are present in a fixed ratio. A mixture of 95% ethanol and 5% water is an azeotropic mixture that boils at 78 °C.

Alcohol causes blood vessels to dilate. The resulting increased flow of blood through the capillaries beneath the skin imparts a feeling of warmth and a reddish hue to the skin.

Alcohol in the Blood

Contrary to popular belief, alcohol is a depressant of the central nervous system, not a stimulant. The illusionary stimulation comes from its effect of depressing brain areas responsible for judgment. The resulting lack of inhibitions and restraints may cause one to feel "stimulated." People under the influence of alcohol suffer from diminished control of their judgment and actions and hence may endanger themselves and/or others (especially if they are driving). In most states, a blood alcohol concentration (BAC) of 0.1% (100 mg of alcohol in 100 mL of blood) is legal evidence of intoxication. A BAC of 0.5-1% leads to coma and death (Table 14.6).

Table 14.6 Approximate Relationship Between Drinks Consumed, Blood Alcohol Concentration, and Effect for a 70-kg (154-lb) Moderate Drinker

Number of Drinks[a]	Blood Alcohol Concentration (% by volume)	Effect[b]
2	0.05	Mild sedation; tranquillity
4	0.10	Lack of coordination
6	0.15	Obvious intoxication
10	0.30	Unconsciousness
20	0.50	Possible death

[a] Rapidly consumed 30-mL (1-oz) "shots" of 90 proof whiskey, 360-mL (12-oz) bottles of beer, or 150-mL (5-oz) glasses of wine.
[b] An inexperienced drinker would be affected more strongly, or more quickly, than one who is ordinarily a moderate drinker. Conversely, an experienced *heavy* drinker would be affected less.

Heavy drinking alters brain cell function, causes nerve damage, and shortens the life span by contributing to diseases of the liver, cardiovascular system, and virtually every other organ of the body.

aerosol disinfectants. Ethanol is also employed in the synthesis of other organic compounds. When used for such industrial purposes, it is not subject to a federal tax (more than $20/gallon in most states). To ensure the legitimate use of tax-free ethanol, the government treats it with certain additives that make it unfit to drink. Such ethanol is known as **denatured alcohol.** Common denaturants are methanol and 2-propanol. These compounds are toxic but do not interfere with the solvent properties of the ethanol.

Because the U.S. has a large corn surplus, there is a generous government subsidy for gasohol producers who use alcohol from corn.

The use of ethanol (and methanol) as a blend component of gasoline (the blend is called "gasohol") began in the United States in 1979. Favorable results have been achieved with mixtures containing 80 to 90% unleaded gasoline and 10 to 20% alcohol. Since ethanol can be obtained easily from grain and methanol from coal, wood chips, or municipal refuse, this represents one method of augmenting our dwindling fuel supplies. Brazil, which has an abundance of sugar cane (and no oil), is producing cars designed to burn only ethanol. In addition to its use as a fuel extender, ethanol and methanol also increase the octane rating of the gasoline with which they are blended.

2-Propanol

A 70% solution of isopropyl alcohol is commonly referred to as rubbing alcohol. It has a high vapor pressure, and its rapid evaporation from the skin produces a cooling effect. Rubbing alcohol, like ethanol, denatures bacterial enzymes and is used as an antiseptic to cleanse the skin before a blood sample is taken or an injection is given. Isopropyl alcohol is toxic when ingested but, compared to methanol, is less readily absorbed through the skin. Though more toxic than ethanol, it less often causes fatalities. Instead, it can induce vomiting; it doesn't stay down long enough to kill you. Much of the isopropyl alcohol produced industrially is for the manufacture of acetone (see Section 15.5) and to introduce the isopropyl group into organic molecules.

14.6 Chemical Properties of Alcohols

Chemical reactions in alcohols occur mainly at the functional group. They may, however, involve hydrogen atoms attached to the hydroxyl-bearing carbon or even those on an adjacent carbon. We discuss three major kinds of reactions of the alcohols. Dehydration and oxidation are considered here. Esterification is covered in Section 16.8.

Dehydration of Alcohols

Dehydration (removal of water) is usually accomplished by adding concentrated sulfuric acid to the alcohol and heating the resulting mixture. The hydroxyl group is removed from the alcohol carbon, and a hydrogen atom is removed from an adjacent carbon, giving an alkene:

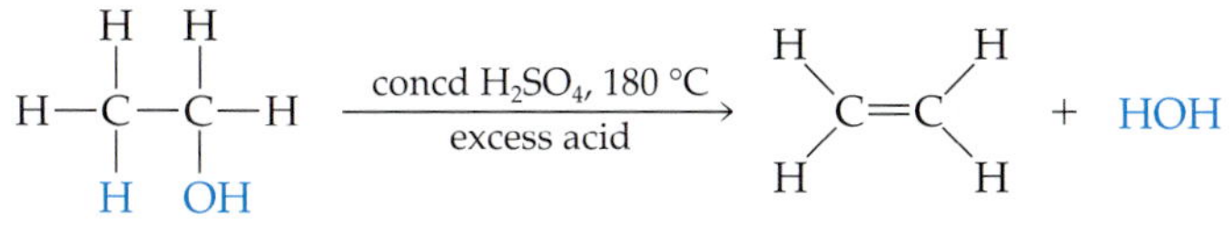

Ethanol Ethylene

Under the proper conditions, it is possible to perform a dehydration involving two molecules of alcohol. The hydroxyl group of one alcohol and only the hydrogen of the hydroxyl group of the second alcohol molecule are removed. The two organic groups remaining combine to form an ether molecule (see Section 14.9):

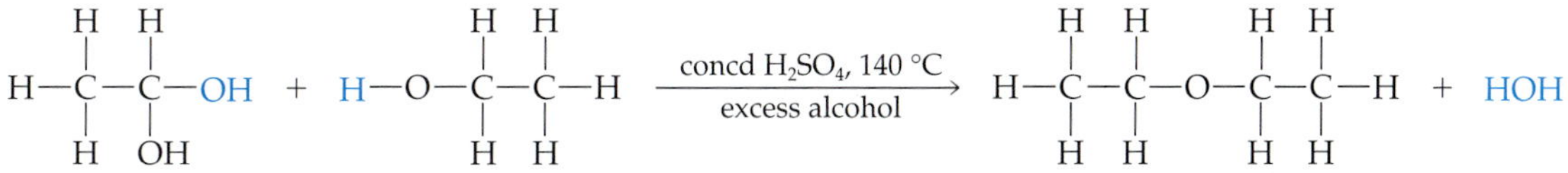

Two molecules of ethanol Diethyl ether

Thus, depending on conditions, one can prepare either alkenes or ethers by dehydration of alcohols. At 180 °C and with an excess of H_2SO_4, dehydration of ethanol gives ethylene as the main product. At 140 °C and with an excess of ethanol, the main product of dehydration of ethanol is diethyl ether.

Dehydration (and its reverse, hydration) reactions occur continuously in cellular metabolism. In these biochemical dehydrations, enzymes serve as catalysts instead of acids, and the reaction temperature is 37 °C instead of the elevated temperatures required in the laboratory. The following reaction occurs in the Embden–Meyerhof pathway (see Figure 24.12):

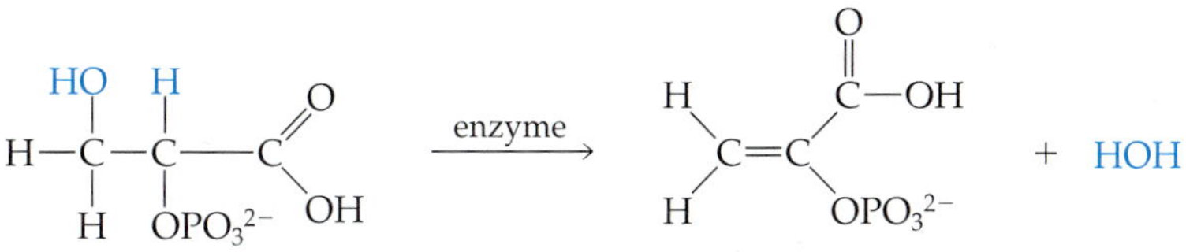

2-Phosphoglyceric acid Phosphoenolpyruvic acid

Look carefully at this equation. The compounds involved are more complex than the ethanol and ethylene we used in our previous example, but the reaction is not. Ignore all other functional groups that are in the molecule. In this reaction, these other functional groups remain unchanged in the product. The only thing that has happened is that a hydrogen and hydroxyl group have been eliminated from the starting material, and the product contains a double bond. The point is that if you know the chemistry of a particular functional group, you know the chemistry of a thousand or a hundred thousand different compounds. Alcohols have a potential for undergoing dehydration, and you will find that big ones, little ones, and ones that incorporate other functional groups all dehydrate if conditions are right.

Dehydration to form simple ethers in biological systems is perhaps less common than the reaction to form an alkene. However, many important reactions, such as the formation of glycosides from sugars (Chapter 19), are at least technically dehydrations leading to etherlike compounds.

Oxidation of Alcohols

The Breathalyzer test to detect drunk drivers is based on the color change associated with the reduction of dichromate ion. If a suspect's breath causes the color to change to green (Cr^{3+}), his/her blood contains more than the legal level of alcohol. Further tests are then carried out at the police station to confirm the suspicion.

Primary and secondary alcohols are readily oxidized. We saw earlier how methanol and ethanol are oxidized by liver enzymes to form aldehydes. Such reactions can also be carried out in the laboratory with chemical oxidizing agents. For example, in acid solution, potassium dichromate oxidizes ethyl alcohol to acetaldehyde:

$$8\,H^+ + \underset{\substack{\text{Dichromate ion}\\\text{(orange)}}}{Cr_2O_7^{2-}} + 3\,C_2H_5OH \longrightarrow \underset{\substack{\text{Chromium (III)}\\\text{ion (green)}}}{2\,Cr^{3+}} + 3\,C_2H_4O + 7\,H_2O$$

Similarly, propyl alcohol is oxidized to propionaldehyde. The balanced equation for this reaction is quite complicated, even if we write only the net ionic equation:

$$3\,CH_3CH_2CH_2OH + 8\,H^+ + Cr_2O_7^{2-} \longrightarrow 3\,CH_3CH_2CHO + 2\,Cr^{3+} + 7\,H_2O$$

In situations like this, organic chemists have a tendency to simplify everything until only the change involving the organic molecules is shown. Thus, the above reaction would be simplified to

$$\underset{\text{Propyl alcohol}}{CH_3CH_2CH_2OH} \xrightarrow[H^+]{K_2Cr_2O_7} \underset{\text{Propionaldehyde}}{CH_3CH_2\overset{\overset{\displaystyle O}{\|}}{C}{-}H}$$

The required inorganic reagents are written either above or below the arrow. The inorganic by-products are ignored (they're still formed, but we just ignore them in this form of the equation). In this way, all attention is focused on the organic starting material and product, and less time is spent balancing the frequently complicated equations.

The abbreviated form of this particular equation indicates that a primary alcohol is oxidized to an aldehyde. We shall see, in Chapter 15, that aldehydes are even more easily oxidized than alcohols and yield carboxylic acids. If one wishes to isolate the aldehyde initially formed in the oxidation of the alcohol, it is necessary to remove it from contact with the oxidizing agent. This can be done by distilling the aldehyde from the reaction mixture as it forms.

Secondary alcohols are oxidized to compounds called *ketones* (see Chapter 15). Oxidation of isopropyl alcohol by dichromate gives acetone:

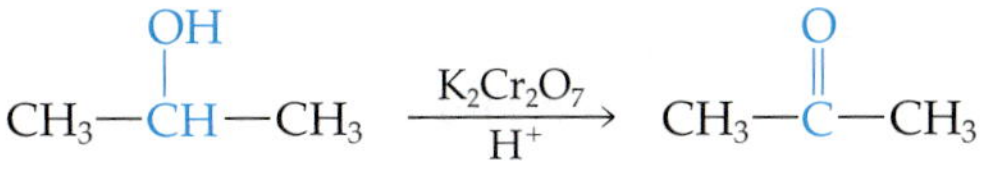

$$\underset{\substack{\text{Isopropyl alcohol}\\\text{(a secondary alcohol)}}}{CH_3-\overset{\displaystyle OH}{\overset{|}{CH}}-CH_3} \quad\xrightarrow[\text{H}^+]{\text{K}_2\text{Cr}_2\text{O}_7}\quad \underset{\substack{\text{Acetone}\\\text{(a ketone)}}}{CH_3-\overset{\displaystyle O}{\overset{\|}{C}}-CH_3}$$

Unlike aldehydes, ketones are relatively resistant to further oxidation and special precautions to isolate the product of this reaction are not necessary.

As we saw in the preceding section, alcohol oxidation is important in living organisms. Indeed, enzyme-controlled oxidation reactions provide the energy cells need in order to do useful work. One step in the metabolism of carbohydrates (Chapter 25) involves the oxidation of the secondary alcohol group in isocitric acid to a ketone group:

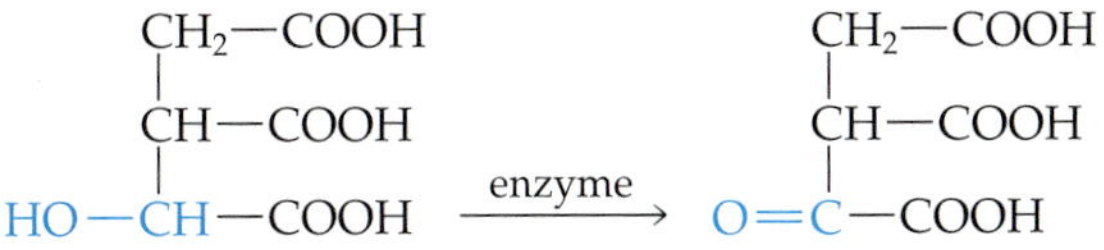

$$\underset{\text{Isocitric acid}}{\begin{array}{l} CH_2-COOH \\ | \\ CH-COOH \\ | \\ HO-CH-COOH \end{array}} \quad\xrightarrow{\text{enzyme}}\quad \underset{\text{Oxalosuccinic acid}}{\begin{array}{l} CH_2-COOH \\ | \\ CH-COOH \\ | \\ O{=}C-COOH \end{array}}$$

Note that the overall reaction is identical to the conversion of isopropyl alcohol to acetone. The complications of structure that distinguish isocitric acid in no way interfere with the characteristic reaction of its secondary alcohol group.

Tertiary alcohols are resistant to oxidation, because the carbon atom bonded to a hydroxyl group is not also bonded to a hydrogen. The oxidation reactions we have described involve the formation of a carbon–oxygen double bond. Thus, the hydroxyl carbon in the alcohol must be able to release one of the atoms attached to it so that it can form the double bond with oxygen. The carbon–hydrogen bond is easily broken under oxidative conditions, but the carbon–carbon bond is not. Therefore tertiary alcohols are not easily oxidized.

The following equations summarize the differences in the oxidation (indicated by the symbol [O]) of the various classes of alcohols:

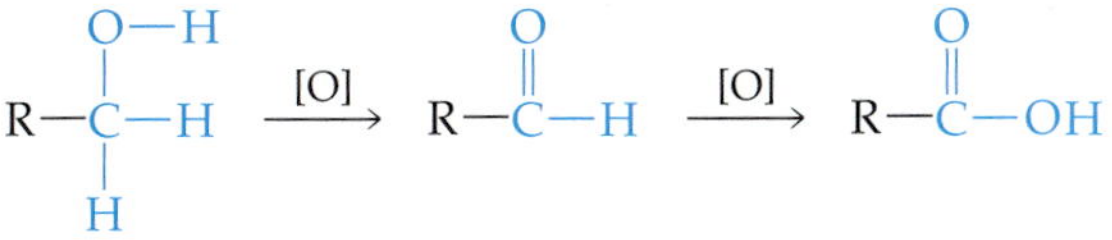

$$\underset{\text{Primary alcohol}}{R-\overset{\displaystyle O-H}{\overset{|}{\underset{\underset{\displaystyle H}{|}}{C}}}-H} \quad\xrightarrow{[O]}\quad \underset{\text{Aldehyde}}{R-\overset{\displaystyle O}{\overset{\|}{C}}-H} \quad\xrightarrow{[O]}\quad \underset{\text{Carboxylic acid}}{R-\overset{\displaystyle O}{\overset{\|}{C}}-OH}$$

$$\underset{\text{Secondary alcohol}}{R-\overset{\displaystyle OH}{\overset{|}{\underset{\underset{\displaystyle H}{|}}{C}}}-R} \quad\xrightarrow{[O]}\quad \underset{\text{Ketone}}{R-\overset{\displaystyle O}{\overset{\|}{C}}-R}$$

$$\underset{\text{Tertiary alcohol}}{R-\overset{\displaystyle OH}{\overset{|}{\underset{\underset{\displaystyle R}{|}}{C}}}-R} \quad\xrightarrow{[O]}\quad \text{No reaction}$$

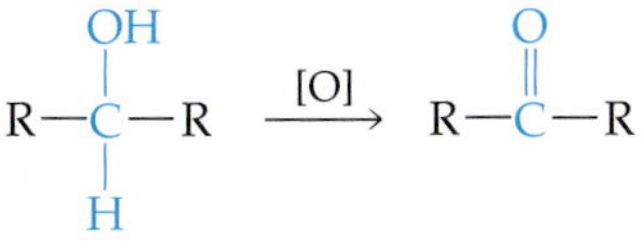

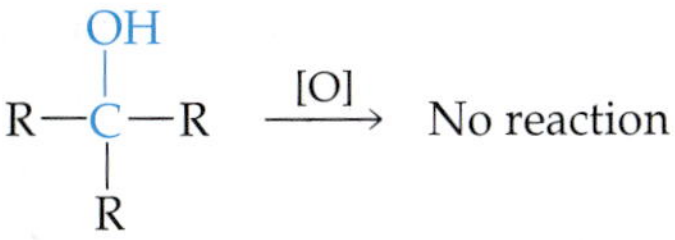

Example 14.3

Write equations for the reactions of the following alcohols with $K_2Cr_2O_7$ and H_2SO_4:

a. 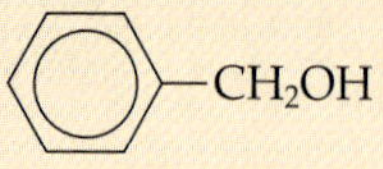 **b.** 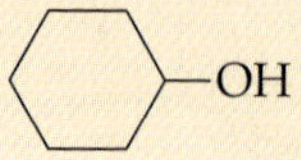 **c.** 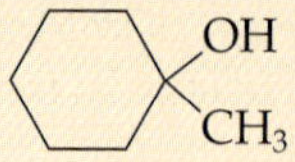

Solution

First recognize the class of each alcohol: (**a**) primary, (**b**) secondary, and (**c**) tertiary. Therefore,

a. 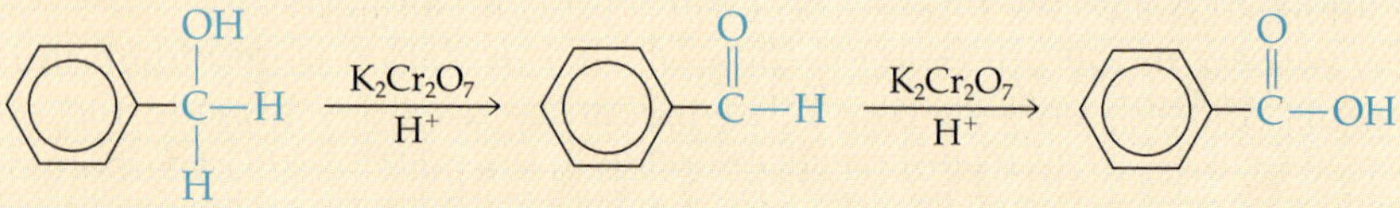

b. 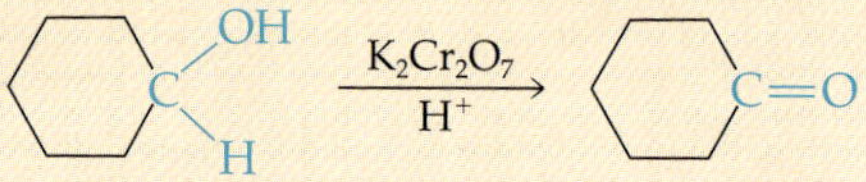

c. 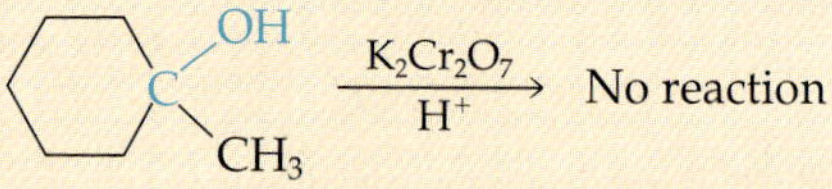

Practice Exercise

Write equations for the reactions of the following alcohols with $K_2Cr_2O_7$ and H_2SO_4:

a. $CH_3CH_2CH_2CH_2CH_2OH$ **b.** $CH_3CH_2CH_2CH(OH)CH_3$ **c.** $CH_3CH_2C(OH)(CH_3)CH_3$

14.7 Multifunctional Alcohols: Glycols and Glycerol

The simple alcohols we have met so far contain only one hydroxyl group each. They are called **monohydric** alcohols. Several important alcohols contain more than one hydroxyl group per molecule. They are called **polyhydric alcohols.** Those with two hydroxyl groups are said to be **dihydric alcohols,** and those with three hydroxyl groups are called **trihydric alcohols.**

Dihydric alcohols are often called **glycols.** The most important of these is ethylene glycol. This compound is the main ingredient in permanent antifreeze mixtures for automobile radiators. Ethylene glycol is a sweet, colorless, somewhat viscous liquid. Because of the two hydroxyl groups, there is extensive intermolecular hydrogen bonding. Thus, ethylene glycol has a high boiling point (198 °C) and does not boil away when used as antifreeze. It is also completely miscible with water. A solution of 60% ethylene glycol in water does not freeze until the temperature falls to −49 °C (−56 °F). (The color of most commercial antifreezes is due to additives.) Ethylene glycol is also used in the manufacture of polyester fiber (Dacron) and the magnetic film (Mylar) used in tapes for recorders and computers.

Ethylene glycol is quite toxic. As with methanol, its toxicity is due to a metabolite. Liver enzymes oxidize the ethylene glycol to oxalic acid:

$$\underset{\text{Ethylene glycol}}{HOCH_2{-}CH_2OH} \xrightarrow{\text{liver enzymes}} \underset{\text{Oxalic acid}}{HO{-}\overset{O}{\overset{\|}{C}}{-}\overset{O}{\overset{\|}{C}}{-}OH}$$

This compound crystallizes as its calcium salt, calcium oxalate (CaC_2O_4), in the kidneys, leading to renal damage, which can lead to kidney failure and death. As with methanol poisoning, the usual treatment for ethylene glycol poisoning is ethanol, administered to load up and thus block the liver enzymes from catalyzing the conversion of ethylene glycol to oxalic acid.

Another common dihydric alcohol is propylene glycol. The physical properties of this compound are quite similar to those of ethylene glycol. Its physiological properties, however, are quite different. Propylene glycol is essentially nontoxic, and it can be used as a solvent for drugs. It is also used as a moisturizing agent for foods. Like other alcohols, propylene glycol can be oxidized by liver enzymes:

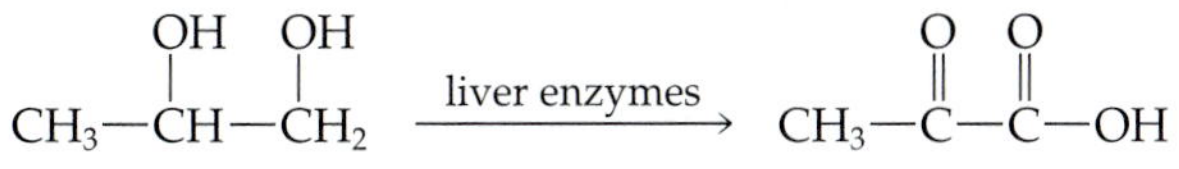

$$\underset{\text{Propylene glycol}}{CH_3-\overset{\overset{\displaystyle OH}{|}}{CH}-\overset{\overset{\displaystyle OH}{|}}{CH_2}} \quad \xrightarrow{\text{liver enzymes}} \quad \underset{\text{Pyruvic acid}}{CH_3-\overset{\overset{\displaystyle O}{\|}}{C}-\overset{\overset{\displaystyle O}{\|}}{C}-OH}$$

In this case, however, the product is pyruvic acid, a normal intermediate in carbohydrate metabolism (Chapter 24).

Glycerol (also called glycerin) is the most important trihydric alcohol. It is a sweet, syrupy liquid. Essentially nontoxic, it is a product of the hydrolysis of fats and oils. Glycerol has widespread industrial use, including:

1. Preparation of hand lotions and cosmetics
2. Additive in inks, tobacco products, and plastic clays to prevent dehydration (glycerol is hygroscopic)
3. Constituent of glycerol suppositories
4. Sweetening agent and solvent for medicines
5. Lubricant
6. Starting material in the production of plastics, surface coatings, and synthetic fibers
7. Source of nitroglycerin

The equation for the preparation of nitroglycerin shows that three molecules of nitric acid are required for every molecule of glycerin. The glycerin must be very pure to ensure stability of the product:

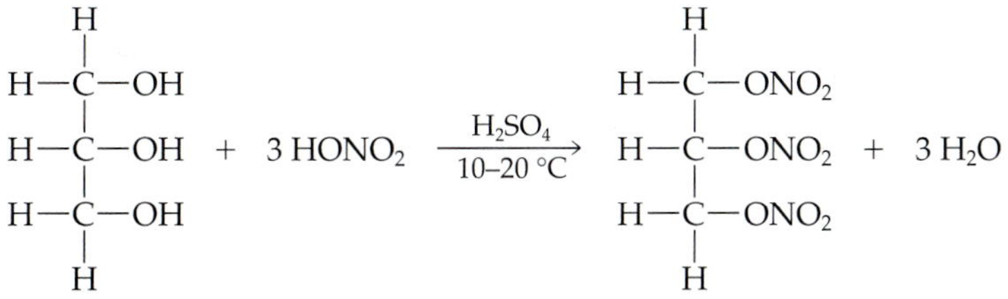

$$\underset{\substack{\text{Glycerol} \\ \text{(Glycerin)}}}{\begin{matrix} H \\ | \\ H-C-OH \\ | \\ H-C-OH \\ | \\ H-C-OH \\ | \\ H \end{matrix}} + \; 3\,HONO_2 \;\; \xrightarrow[\text{10-20 °C}]{H_2SO_4} \;\; \underset{\substack{\text{Glycerol trinitrate} \\ \text{(Nitroglycerin)}}}{\begin{matrix} H \\ | \\ H-C-ONO_2 \\ | \\ H-C-ONO_2 \\ | \\ H-C-ONO_2 \\ | \\ H \end{matrix}} + \; 3\,H_2O$$

Nitroglycerin is a pale yellow, oily liquid that detonates upon slight impact. The explosive power arises from the extremely rapid conversion of a small volume of liquid into a large volume of hot, expanding gases.

$$4\,C_3H_5(ONO_2)_3(l) \; \longrightarrow \; 6\,N_2(g) \; + \; 12\,CO_2(g) \; + \; 10\,H_2O(g) \; + \; O_2(g)$$

The reaction produces temperatures above 3000 °C and pressures above 2000 atm. The explosion wave caused by such temperatures and pressures is enormous, accounting for the damaging effect of the detonation. The shock wave can travel at

Nitroglycerin

Nitroglycerin was first prepared in 1846 by the Italian chemist Sobrero, who was lucky that he lived to tell of his discovery. Sobrero mixed nitric acid and glycerin, and the ensuing explosion nearly killed him. It was not until 15 years later that the famed Swedish chemist and inventor Alfred Nobel discovered a method to prepare and transport the compound safely. He found that a type of diatomaceous earth, a clay-like material, was capable of absorbing the nitroglycerin, thus rendering it insensitive to shock. The stabilized mixture was referred to as *dynamite*. Unless exploded by means of a percussion cap or a detonator containing lead azide [$Pb(N_3)_2$], it was quite stable.

The production of dynamite was a major breakthrough. With dynamite, the construction of canals, dams, highways, mines, and railroads became much easier. Its use as a weapon in warfare greatly disturbed the Nobel family, however, and by Alfred Nobel's will a trust fund was established to provide an annual award for an outstanding contribution toward peace. (A trust fund was also set up to offer annual awards for contributions in the fields of chemistry, physics, literature, and medicine or physiology.)

It is surprising that a compound so sensitive to shock is also used as a drug to relieve angina pectoris—sharp chest pains caused by an insufficient supply of oxygen to the heart muscle. Nitroglycerin is administered in tablet form (mixed with nonactive ingredients), as an alcoholic solution (spirit of glyceryl trinitrate), or in the form of a patch from which the drug is absorbed through the skin. Nitroglycerin functions as a vasodilator. It relaxes cardiac muscle and smooth muscle in the smaller blood vessels, thus increasing the supply of blood (and hence oxygen) to the heart.

Alfred Nobel (1833–1896), inventor of dynamite and founder of the famed Nobel Prizes.

speeds up to 9000 m/s (about 20,000 mi/h), causing the explosion to occur at a rate far faster than that of other chemical reactions.

14.8 Phenols

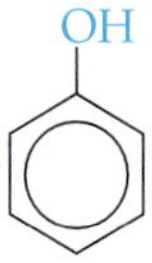

Phenol
(Carbolic acid)

As we learned in the opening of this chapter, compounds in which a hydroxyl group is attached directly to an aromatic ring are called **phenols.** The parent compound, C_6H_5OH, is itself called phenol. It is a white crystalline compound that has a distinctive ("hospital smell") odor. Other compounds may be named as derivatives of phenol, but most of those of interest to us are best known by special, nonsystematic names.

The phenols generally are either low-melting-point solids or oily liquids. Most are only sparingly soluble in water. They have found wide use as antiseptics (substances that kill microorganisms on living tissue) and as disinfectants (substances intended to kill microorganisms on furniture, fixtures, floors, and around the house in general).

The first widely used antiseptic was phenol, which was also called carbolic acid. Joseph Lister used it for antiseptic surgery in 1867. Unfortunately, phenol doesn't kill only undesirable microorganisms. It kills all types of cells. Applied to the skin, it can cause severe burns. In the bloodstream, it is a systemic poison—that is, one that is carried to and affects all parts of the body. Its severe side effects led to searches for safer antiseptics, a number of which have been found.

One of the most active phenolic antiseptics is 4-hexylresorcinol. It is much more powerful than phenol as a germicide and has fewer undesirable side effects. Indeed, it is safe enough to be used as the active ingredient in some mouthwashes and in antiseptic throat lozenges such as Sucrets.

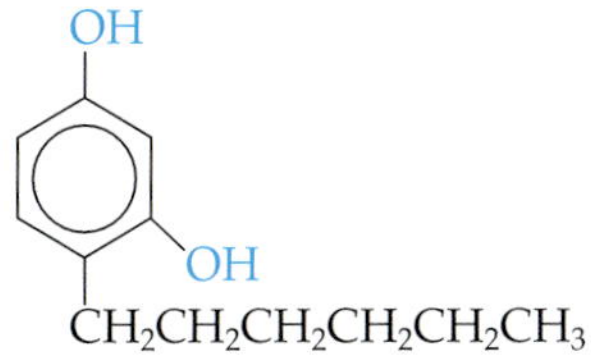

The methyl derivatives of phenols are called *cresols*. They are important ingredients in the wood preservative creosote.

OH
$CH_2CH_2CH_2CH_2CH_2CH_3$
4-Hexylresorcinol

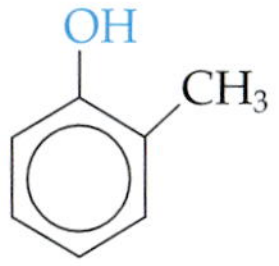

o-Cresol *m*-Cresol *p*-Cresol

Unlike the three cresol isomers, the three dihydroxybenzenes have individual names. They all have commercial significance, and two of them are important components of biochemical molecules. Hydroquinone occurs in a coenzyme (Section 22.2), and catechol forms part of the structure of certain neurotransmitters termed catecholamines (see Sections G.2 and G.3).

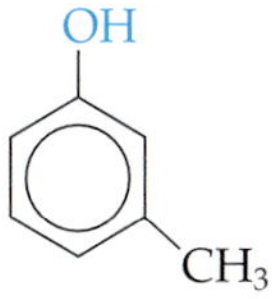

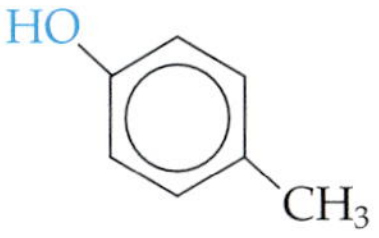

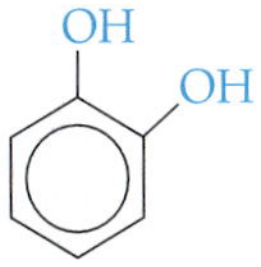

Catechol Resorcinol Hydroquinone

Hexachlorophene was once widely used in germicidal cleaning solutions (pHisohex) and as an ingredient in deodorant soaps and other cosmetics. In the United States, products contained at most 3% hexachlorophene. The compound was generally considered a safe and effective antibacterial agent. In 1972, however, the picture changed rapidly. An outbreak of neurological disease among infants in northeastern France was traced to a baby powder called Bébé that contained over 20% hexachlorophene. More than 30 infants died. The U.S. Food and Drug Administration acted quickly. Hexachlorophene was banned from all products intended for over-the-counter sales. It is still available for prescription use and for use in hospitals—in concentrations not to exceed 3%.

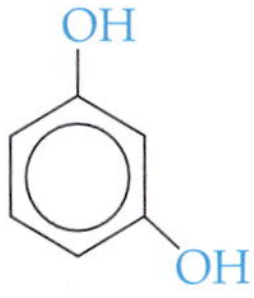

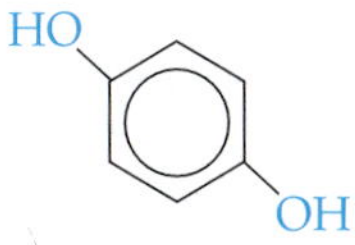

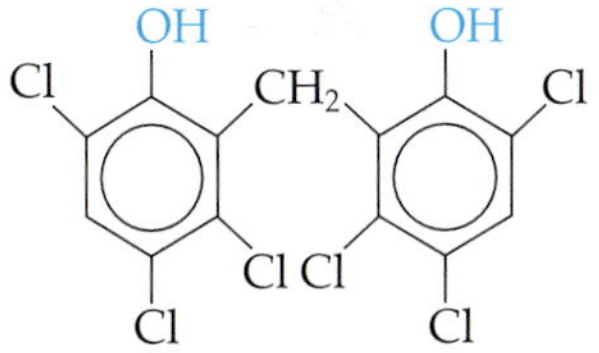

Hexachlorophene

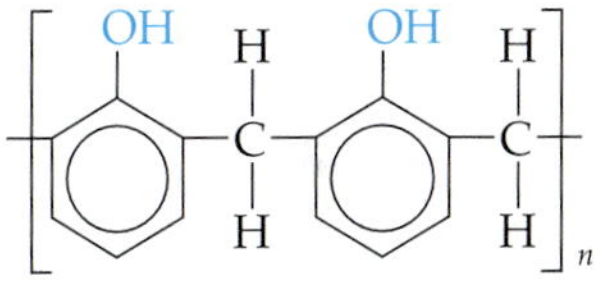

Bakelite

The most important commercial reaction of phenols is the condensation with formaldehyde to yield phenolic polymers (Bakelite). Bakelite was used initially as an electrical insulator and later to form plastic parts for the automotive and radio industries. Phenol is also used in the production of phenolphthalein, an acid–base indicator.

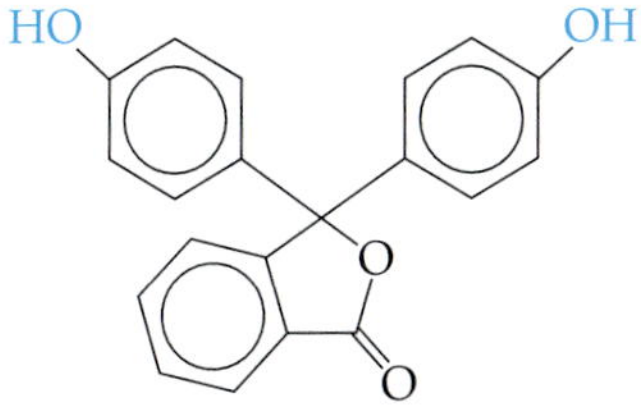

One of the major distinguishing characteristics between phenols and alcohols is that phenols are slightly acidic ($K_a \cong 10^{-10}$), whereas alcohols are neutral. Phenols can be neutralized by strong bases, but they are too weakly acidic to react with weak bases such as aqueous sodium bicarbonate. The latter reaction serves to distinguish the phenols from the carboxylic acids (Chapter 16), which do react with $NaHCO_3$:

Phenolphthalein

$$CH_3CH_2OH + NaOH(aq) \longrightarrow \text{No reaction}$$

$$\text{—OH} + NaOH(aq) \longrightarrow \text{—O}^- \ Na^+(aq) + HOH$$

$$\text{—OH} + NaHCO_3(aq) \longrightarrow \text{No reaction}$$

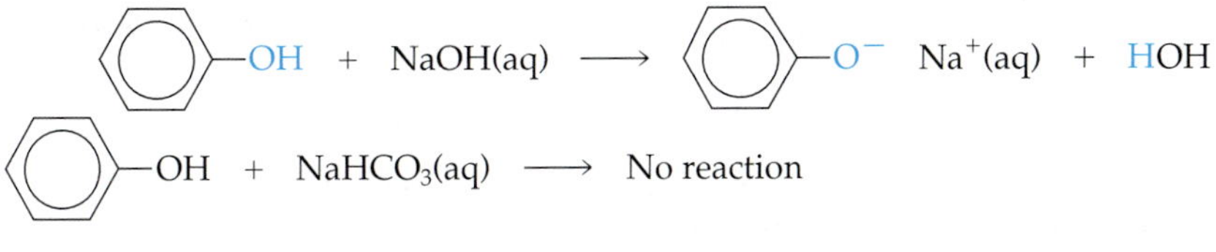

14.9 Ethers

As mentioned at the beginning of this chapter, **ethers** may be considered to be derivatives of water in which both hydrogen atoms have been replaced by alkyl or aryl groups. They may also be considered derivatives of an alcohol in which the hydroxyl hydrogen has been replaced by an organic group:

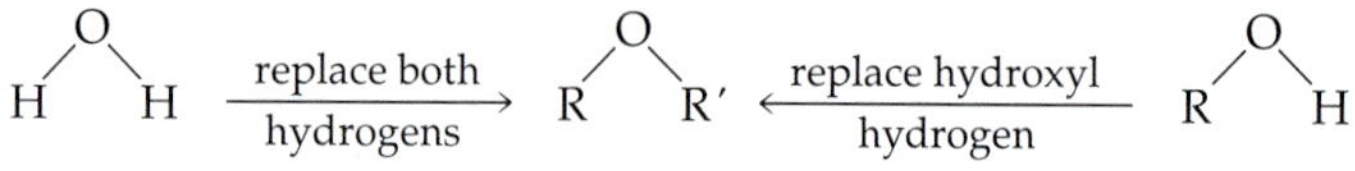

$$\underset{\text{Water}}{\overset{\text{O}}{\underset{\text{H}\qquad\text{H}}{}}} \xrightarrow[\text{hydrogens}]{\text{replace both}} \underset{\text{Ether}}{\overset{\text{O}}{\underset{\text{R}\qquad\text{R}'}{}}} \xleftarrow[\text{hydrogen}]{\text{replace hydroxyl}} \underset{\text{Alcohol}}{\overset{\text{O}}{\underset{\text{R}\qquad\text{H}}{}}}$$

The general formula for the ethers is R—O—R′. When both R-groups are the same, the compound is a *symmetrical* ether. When R and R′ are different, the ether is *unsymmetrical*.

Symmetrical Ether		*Unsymmetrical Ether*	
R—O—R	H_3C—O—CH_3	R—O—R′	H_3C—O—CH_2CH_3

CH_3—O—CH_3

Dimethyl ether

CH_3—O—CH_2CH_3

Ethyl methyl ether

CH_3CH_2—O—CH_2CH_3

Diethyl ether

R—O⋯⋯H—O
R H

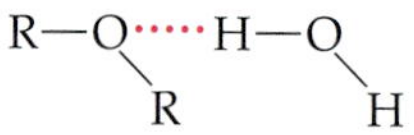

Simple ethers are simply named. Just name the groups attached to oxygen and then add the generic name *ether*. For symmetrical ethers, the group name should be preceded by the prefix *di-*, although the prefix is sometimes dropped in common usage. The names methyl ether and dimethyl ether refer to the same compound, but the latter is preferred.

Ether molecules have no hydrogen atom on oxygen. Therefore the molecules in a pure liquid ether are incapable of intermolecular hydrogen bonding. Given their molar mass, then, the ethers have quite low boiling points. Indeed, ethers have boiling points about the same as those of alkanes of comparable molar mass and much lower than those of the corresponding alcohols (Table 14.7).

Ether molecules do have an oxygen atom, however, and so can hydrogen bond with water molecules. Consequently, the ethers have about the same water solubilities as their isomeric alcohols. (For example, dimethyl ether and ethanol are completely soluble in water, whereas diethyl ether and 1-butanol are soluble to the extent of 8 g/100 mL of water.)

Chemically, the ethers are quite inert. Like the alkanes, they do not react with the usual oxidizing agents, reducing agents, or bases. Their inertness makes ethers excellent solvents for organic materials. Often called simply "ether," diethyl ether is often used in the extraction of organic compounds from plant and animal materials or from mixtures of organic and inorganic substances. The volatile ether is then easily removed by evaporation, and the desired organic components are left

Table 14.7 Comparison of Boiling Points of Alkanes, Alcohols, and Ethers

Formula	Name	Molar Mass	Boiling Point (°C)
$CH_3CH_2CH_3$	Propane	44	−42
CH_3OCH_3	Dimethyl ether	46	−25
CH_3CH_2OH	Ethyl alcohol	46	78
$CH_3CH_2CH_2CH_2CH_3$	Pentane	72	36
$CH_3CH_2OCH_2CH_3$	Diethyl ether	74	35
$CH_3CH_2CH_2CH_2OH$	Butyl alcohol	74	117

Anesthesia

A **general anesthetic** acts on the brain to produce unconsciousness as well as insensitivity to pain. [A local anesthetic (see Section G.5) renders one part of the body insensitive to pain yet leaves the patient conscious.]

$$CH_3CH_2—O—CH_2CH_3$$

Diethyl ether

Diethyl ether was the first general anesthetic. It was introduced into surgical practice in 1846 by a Boston dentist, William Morton. Inhalation of ether vapor produces unconsciousness by depressing the activity of the central nervous system. Ether is relatively safe because there is a fairly wide gap between the effective level of anesthesia and the lethal dose. The disadvantages are its high flammability and postanesthetic nausea and vomiting.

$$N{=}N—O$$

Nitrous oxide
(Dinitrogen monoxide)

Nitrous oxide, or laughing gas (N_2O), was tried by Morton without success before he tried ether. It was discovered by Joseph Priestley in 1772. Its narcotic effect was noted, and it soon came to be used widely at laughing gas parties among the nobility. Nitrous oxide, mixed with oxygen, finds some use in modern anesthesia. It is quick-acting but not very potent. Concentrations of 50% or greater must be used to be effective. When nitrous oxide is mixed with ordinary air instead of oxygen, not enough oxygen gets into the patient's blood, and permanent brain damage can result.

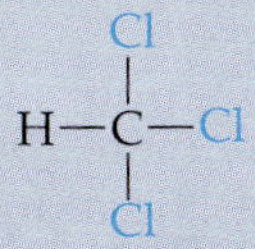

Chloroform
(Trichloromethane)

Chloroform ($CHCl_3$) was introduced as a general anesthetic in 1847. Its use quickly became popular after Queen Victoria gave birth to her eighth child while anesthetized by chloroform in 1853. Chloroform was used widely for years. It is nonflammable and produces effective anesthesia, but it has a number of serious drawbacks. For one, it has a narrow safety margin; the effective dose is close to the lethal dose. It also causes liver damage, and it must be protected from oxygen during storage to prevent the formation of deadly phosgene gas.

Modern anesthetics include fluorine-containing compounds such as halothane, enflurane, and methoxyflurane (Figure 14.5). These compounds are nonflammable and relatively safe for the patient. Their safety, particularly that of halothane, for operating-room personnel, however, has been questioned. For example, female operating room workers suffer a higher rate of miscarriages than the general population.

Modern surgical practice has moved away from the use of a single anesthetic. Generally, a patient is given an intravenous anesthetic such as thiopental (Section G.6) to produce unconsciousness. The gaseous anesthetic then is administered to provide insensitivity to pain and to keep the patient unconscious. A relaxant, such as curare, also may be employed. Curare and related compounds produce profound relaxation thus, only light anesthesia is required. This practice avoids the hazards of deep anesthesia.

The potency of an anesthetic is related to its solubility in olive oil: the more soluble in olive oil, the more potent as an anesthetic. This unusual observation has led many scientists to believe that anesthetics act by dissolving in the fatty membranes (Section 20.9) surrounding nerve cells. The resultant changes in the fluidity and shape of the membranes apparently decrease the ability of sodium ions to pass into the nerve cells, thereby blocking the firing of nerve impulses.

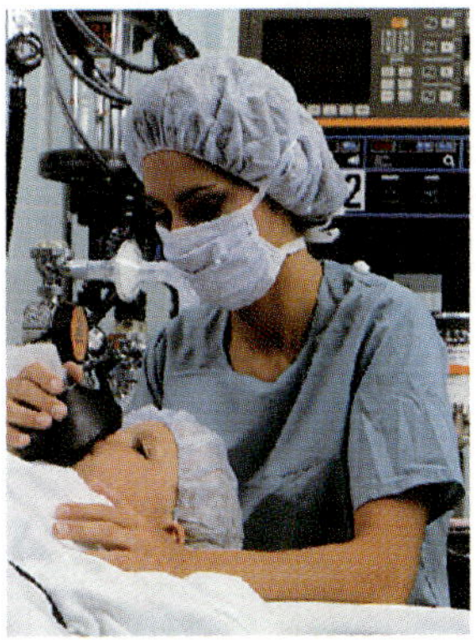

Inhalant anesthetic agents are used to render patients unconscious and insensitive to pain before surgery.

The ideal anesthetic should quickly make the patient unconscious but allow a quick return to consciousness, have few side effects, and be safe to handle.

Curare is the arrow poison used by South American Indian tribes. Large doses of curare kill by causing a complete relaxation of all muscles. Death occurs because of respiratory failure.

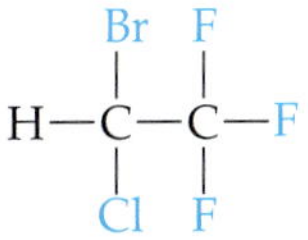

Halothane

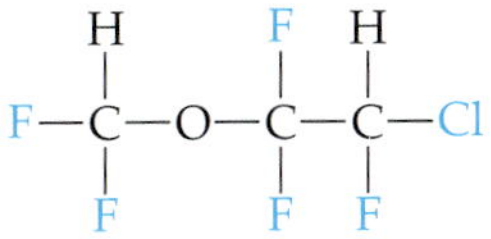

Enflurane

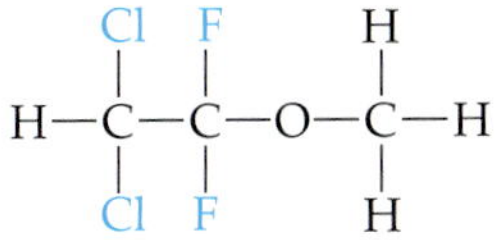

Methoxyflurane

▲ **FIGURE 14.5**

Three modern general anesthetics.

behind.[3] Ether is extremely hygroscopic. A freshly opened can of ether will immediately pick up about 1 to 2% water from the moisture in the air. Special techniques (dry box, nitrogen atmosphere) are employed in handling the compound when reaction conditions call for anhydrous ether.

Use of diethyl ether in the laboratory produces unusual hazards. The compound is quite volatile and extremely flammable. The vapors form an explosive mixture with air. They are also heavier than air and can travel long distances along a tabletop or the floor to reach a flame or spark and set off an explosion. Hence, open flames are not permitted in a laboratory in which ether is being used. Ether fires cannot be extinguished with water because the ether is less dense than water and floats on top of it. The use of carbon dioxide fire extinguishers is recommended.

Diethyl ether should not be stored in an ordinary refrigerator. Even at low temperatures, it has sufficient vapor pressure to form an explosive mixture with air. A spark can ignite the vapors. Special explosion-proof refrigerators, with sealed electrical equipment to prevent contact between spark and flammable vapors, are available for safe storage of volatile, flammable liquids.

Still another hazard with ethers is that, upon standing, they react with oxygen from the air to form peroxides:

$$\underset{\text{Diethyl ether}}{CH_3CH_2{-}O{-}CH_2CH_3} + O_2 \longrightarrow \underset{\text{A peroxide}}{CH_3\underset{\underset{O-O-H}{|}}{C}H{-}O{-}CH_2CH_3}$$

Peroxides are less volatile than ether and are concentrated in the residue left behind during a distillation or evaporation. The concentrated peroxides are highly explosive and sensitive to both shock and heat. Hospitals avoid these problems by buying only those amounts of ether sufficient for immediate use and by keeping containers tightly closed and away from strong light, which catalyzes peroxide formation. Ether suspected of containing peroxides should be treated with a reducing agent, such as alkaline ferrous sulfate solution, before being used.

Summary

A **functional group** is any atom or atom group that confers characteristic properties to a family of compounds. The **hydroxyl group,** —OH, is one example, and carbon–carbon double and triple bonds are two others.

A **primary carbon atom** is one attached to one other carbon, a **secondary carbon atom** is one attached to two other carbons, and a **tertiary carbon atom** is one attached to three other carbons:

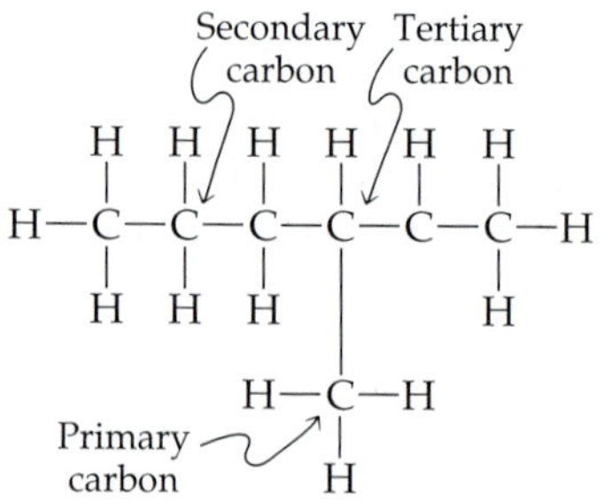

[3] The extracting power of ether has made it the solvent of choice among cocaine users. "Freebasing" first involves separating the cocaine from other substances by extracting it into ether. Enormous quantities of ether are used in extracting cocaine from the coca plant. Drug enforcement officials in the United States have been able to apprehend some cocaine manufacturers by keeping track of the shipments of large quantities of ether.

Alcohols are aliphatic compounds in which one or more hydrogen atoms have been replaced by a hydroxyl group. A **primary alcohol** is an aliphatic compound in which the hydroxyl group replaces a hydrogen on a primary carbon; a **secondary alcohol** is one in which the hydroxyl group replaces a hydrogen on a secondary carbon atom; and a **tertiary alcohol** is one in which the hydroxyl group replace a hydrogen on a tertiary carbon atom:

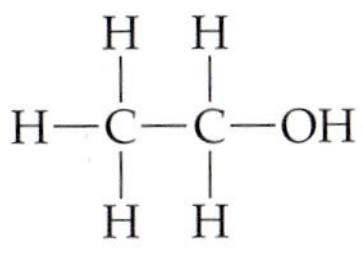

Primary alcohol

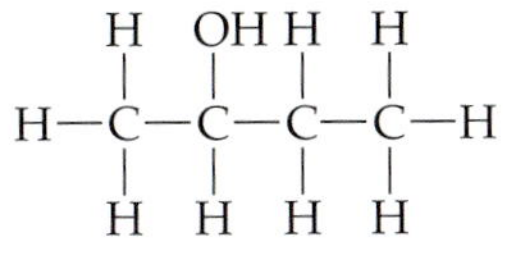

Secondary alcohol

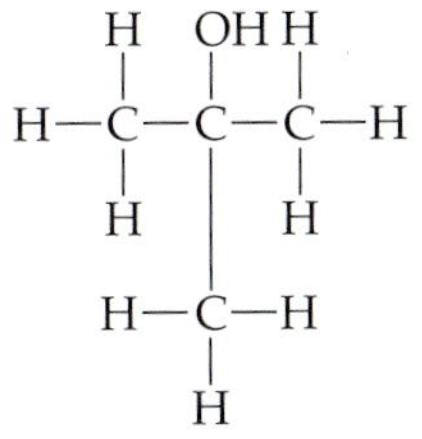

Tertiary alcohol

Alcohols are synthesized by alkene hydration:

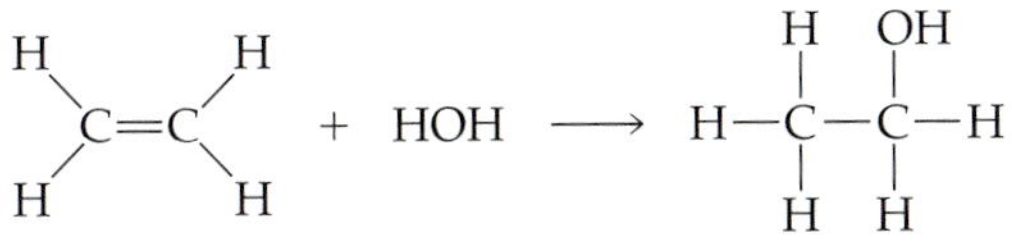

Markovnikov's rule states that, in alkene hydration, the H of HOH goes to the alkene carbon bonded to the higher number of hydrogen atoms:

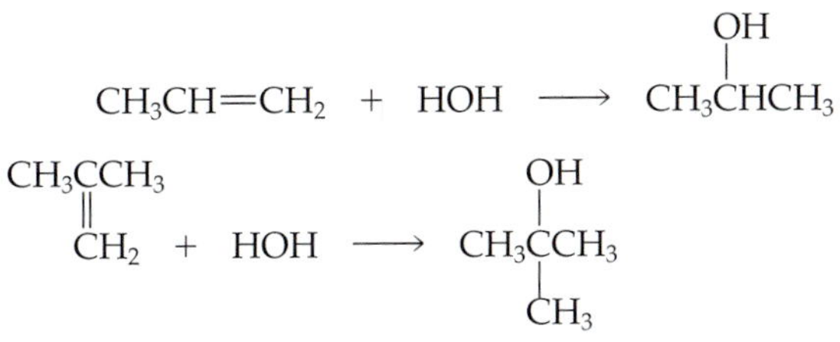

When a molecule of water is removed from an alcohol in a dehydration step, the result is either an alkene or an ether:

$$CH_3CH_2OH \xrightarrow{[H^+]} H_2C{=}CH_2 + HOH$$

$$2\,CH_3CH_2OH \xrightarrow{[H^+]} CH_3CH_2OCH_2CH_3 + HOH$$

Primary alcohols are oxidized to aldehydes, and secondary alcohols are oxidized to ketones. Tertiary alcohols are not easily oxidized:

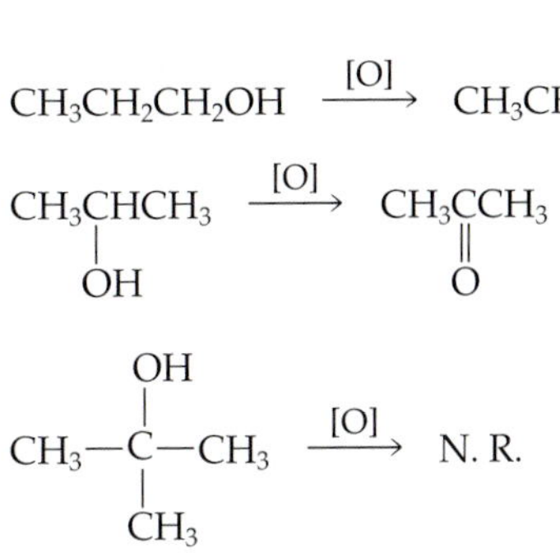

Alcohols containing one hydroxyl group are **monohydric alcohols,** those containing two are **dihydric alcohols,** and those containing three are **trihydric alcohols.** Dihydric alcohols are usually called **glycols.** The commercially most important glycols are ethylene glycol, CH_2OHCH_2OH, the main component in antifreeze, and propylene glycol, $CH_3CHOHCH_2OH$, a solvent for drugs and moistening agent in foods.

Phenols are compounds having the hydroxyl group attached to an aromatic ring:

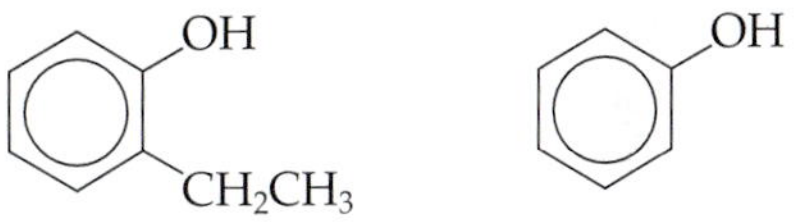

Ethers are compounds in which an oxygen atom is joined to two organic groups:

$H_3C{-}O{-}CH_3$

Symmetrical ether

$H_3C{-}O{-}CH_2CH_2CH_3$

Unsymmetrical ether

Key Terms

absolute alcohol (14.5)
alcohol (Introduction)
azeotrope (14.5)
denatured alcohol (14.5)
dihydric alcohol (14.7)
ether (14.9)
functional group (14.1)
general anesthetic (14.9)
glycol (14.7)
hydroxyl group (Introduction)
LD_{50} (14.5)
Markovnikov's rule (14.4)
monohydric alcohol (14.7)
phenol (14.8)
primary alcohol (14.2)
primary (1°) carbon atom (14.2)
secondary alcohol (14.2)
secondary (2°) carbon atom (14.2)
tertiary alcohol (14.2)
tertiary (3°) carbon atom (14.2)
trihydric alcohol (14.7)

Review Questions

1. What is a functional group?
2. Give the structure of and name the functional group in:
   a. alkenes b. alcohols c. ethers
3. Define the following terms.
   a. absolute alcohol
   b. general anesthetic
   c. antiseptic
   d. disinfectant
   e. 86 proof
   f. LD_{50}
   g. azeotrope
   h. gasohol
4. What is denatured alcohol? Why is some alcohol denatured?
5. Why is methanol so much more toxic to humans than ethanol?
6. Why is ethylene glycol so much more toxic to humans than propylene glycol?
7. What chemical compound is used in the treatment of acute methanol or ethylene glycol poisoning? How does it work?
8. Why is ethyl alcohol the only primary alcohol that can be prepared by the hydration of an alkene?
9. State Markovnikov's rule.
10. In the preparation of diethyl ether from ethanol, why is it so critical to maintain the reaction temperature between 130 and 150 °C?
11. Methanol is not particularly toxic to rats. If methanol were newly discovered and tested for toxicity in laboratory animals, what would you conclude about its safety for human consumption?
12. Tetrahydrocannabinol (THC) is the principal active ingredient in marijuana. What functional groups are present in the THC molecule?

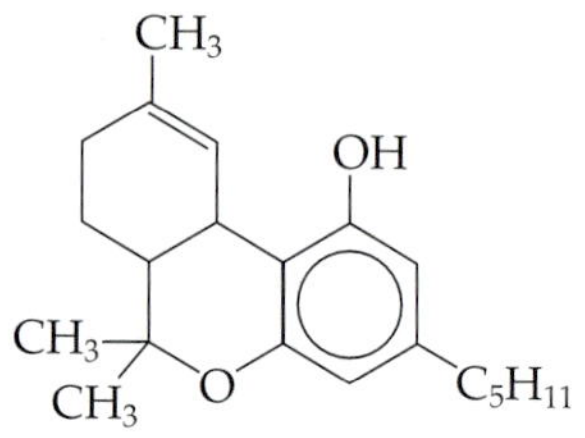

Tetrahydrocannabinol
(THC)

13. What is a polyhydric alcohol?
14. What is a glycol?
15. What precautions must be taken when using diethyl ether as a solvent in a laboratory experiment?
16. Ethyl alcohol, like rubbing alcohol, is often used for sponge baths. What property of alcohols makes them useful for this purpose?

Problems

Alcohols: Names and Structural Formulas

17. Name:
   a. $CH_3CH_2CH_2CH_2CH_2CH_2OH$
   b. $CH_3CH_2CH_2CH_2CHOHCH_3$
18. Name:
   a. $CH_3CH_2CHOHC(CH_3)_3$
   b. $(CH_3)_2CHCH_2OH$
19. Name:
   a. $CH_3CHOHCH_2CHCl_2$
   b. $(CH_3)_2COHCBr_2CH_3$
20. Name:
   a. 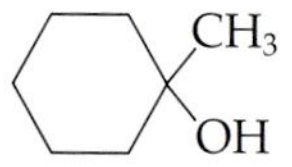
   b. $CH_3CH_2CH_2COH(CH_2CH_3)_2$
21. Give structural formulas for:
   a. 3-hexanol
   b. 3,3-dimethyl-2-butanol
22. Give structural formulas for:
   a. cyclopentanol
   b. 4-methyl-2-hexanol
23. Give structural formulas for:
   a. 4,5-dimethyl-3-heptanol
   b. 2-ethyl-1-phenyl-1-butanol
   c. propylene glycol
24. Give structural formulas for:
   a. 2-bromo-2-chlorocyclobutanol
   b. 3-phenylcyclopentanol
   c. glycerol

Ethers: Names and Structural Formulas

25. Name:
   a. $CH_3CH_2CH_2OCH_2CH_2CH_3$
   b. 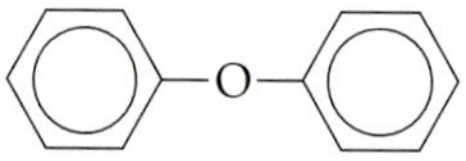
26. Name:
   a. $CH_3CH_2OCH(CH_3)_2$
   b. 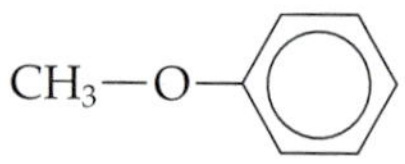
27. Give structural formulas for:
   a. ethyl methyl ether
   b. phenyl benzyl ether
28. Give structural formulas for:
   a. diisopropyl ether
   b. cyclopropyl propyl ether

Phenols: Names and Structural Formulas

29. Name:

a. 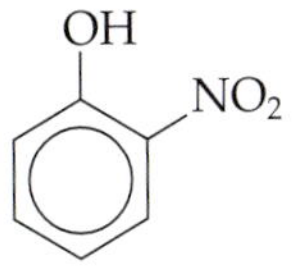

b. 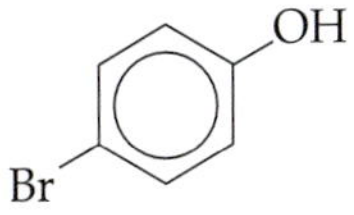

30. Name:

a. 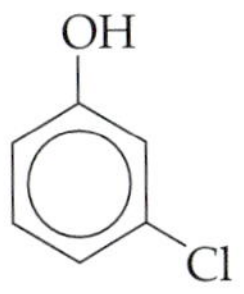

b. 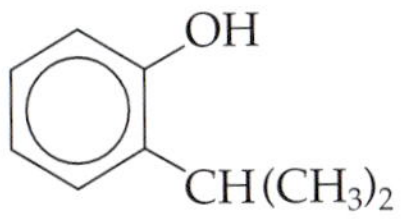

31. Give structural formulas for:
  a. *m*-iodophenol
  b. *p*-methylphenol (*p*-cresol)

32. Give structural formulas for:
  a. 2,4,6-trinitrophenol (picric acid)
  b. 3,5-diethylphenol

Chemical Reactions

33. Classify each conversion as oxidation, dehydration, or hydration (only the organic starting material and product are shown):

a. $CH_3OH \longrightarrow H{-}\overset{\overset{H}{|}}{C}{=}O$

b. $CH_3\overset{\overset{OH}{|}}{C}HCH_3 \longrightarrow CH_3CH{=}CH_2$

c. $CH_2{=}CHCH_2CH_3 \longrightarrow CH_3CHOHCH_2CH_3$

34. Classify each conversion as oxidation, dehydration, or hydration (only the organic starting material and product are shown.):

a. $CH_3\overset{\overset{OH}{|}}{C}HCH_3 \longrightarrow CH_3\overset{\overset{O}{\|}}{C}CH_3$

b. $HOOCCH{=}CHCOOH \longrightarrow HOOCCH_2\overset{\overset{OH}{|}}{C}HCOOH$

c. $2\ CH_3OH \longrightarrow CH_3OCH_3$

35. Each of the four isomeric butyl alcohols is treated with potassium dichromate in acid. Draw the product (if any) expected from each reaction.

36. Write an equation for the dehydration of 2-propanol to yield (**a**) an alkene and (**b**) an ether.

37. Draw the structural formula of the ether formed by the *intra*molecular dehydration of:

$$HOCH_2CH_2CH_2CH_2CH_2OH$$

38. Draw the alkene formed by the dehydration of cyclohexanol.

39. Give the structural formula of the product:

a. $CH_2{=}CHCH_2CH_3 \xrightarrow{H^+,\ H_2O}$

b. 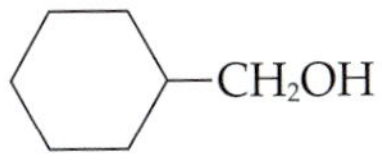 $\xrightarrow[H^+]{K_2Cr_2O_7}$

40. Give the structural formula of the product:

a. $CH_3CHOHCH_3 \xrightarrow[H^+]{KMnO_4}$

b. 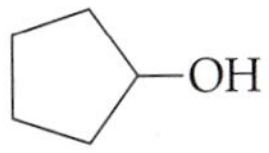 $\xrightarrow[180\ °C]{\text{concd } H_2SO_4}$

41. What reagents are necessary to carry out the following conversions?

a. $CH_3CH{=}CH_2 \xrightarrow{?} CH_3\overset{\overset{OH}{|}}{C}HCH_3$

b. $CH_3\underset{\underset{OH}{|}}{C}HCH_3 \xrightarrow{?} CH_3\underset{\underset{O}{\|}}{C}CH_3$

c. $2\ CH_3CH_2OH \xrightarrow{?} CH_3CH_2OCH_2CH_3$

42. What reagents are necessary to carry out the following conversions?

a. $CH_2{=}\underset{\underset{CH_3}{|}}{C}{-}CH_3 \xrightarrow{?} CH_3\underset{\underset{CH_3}{|}}{\overset{\overset{OH}{|}}{C}}{-}CH_3$

b. $CH_3CH_2OH \xrightarrow{?} CH_2{=}CH_2$

c. $CH_3CH_2CH_2OH \xrightarrow{?} CH_3CH_2\overset{\overset{O}{\|}}{C}{-}H$

43. Give the structure of the alkene from which each of the following alcohols is made by reaction with water in acidic solution:

a. $CH_3\underset{\underset{OH}{|}}{C}HCH_3$

b. 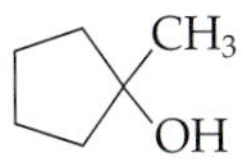

c. $CH_3\underset{\underset{CH_3}{|}}{\overset{\overset{CH_3}{|}}{C}}{-}OH$

44. Give the structure of the alkene from which each of the following alcohols is made by reaction with water in acidic solution:

a. CH_3CH_2OH

b. —OH

c. $CH_3CHCH_2CH_3$
 |
 OH

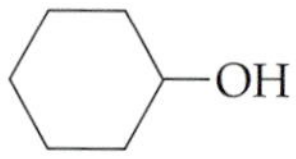

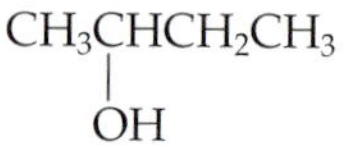

45. Write an equation for the reaction (if any) of phenol with aqueous (**a**) NaOH and (**b**) $NaHCO_3$.

46. Write an equation for the ionization of phenol in water.

Physical Properties

(Answer Problems 47–50 without consulting tables.)

47. Arrange in order of increasing boiling point: ethanol, 1-propanol, methanol.

48. Arrange in order of increasing boiling point: butane, ethylene glycol, 1-propanol.

49. Arrange in order of increasing solubility in water: methanol, 1-butanol, 1-octanol.

50. Arrange in order of increasing solubility in water: pentane, propylene glycol, diethyl ether.

Additional Problems

51. Give the IUPAC names for the compounds commonly known as:
 a. grain alcohol **b.** wood alcohol
 c. rubbing alcohol **d.** carbolic acid

52. Without consulting tables, arrange in order of increasing boiling point: diethyl ether, propylene glycol, 1-butanol.

53. In addition to ethanol, the fermentation of grain produces other organic compounds collectively called fusel oils (FO). The four principal FO components are 1-propanol, isobutyl alcohol, 3-methyl-1-butanol, and 2-methyl-1-butanol. Draw a structural formula for each. (FO is quite toxic and accounts in part for hangovers.)

54. Give the name and one use for:
 a. CH_3OH
 b. CH_3CH_2OH
 c. $CH_3CHOHCH_3$
 d. CH_2OHCH_2OH
 e. $CH_2OHCHOHCH_2OH$
 f. —OH

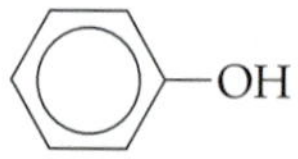

55. Give structural formulas for the eight isomeric pentyl alcohols ($C_5H_{12}O$). (*Hint:* Three are derived from pentane, four from isopentane, and one from neopentane.) Name the alcohols by the IUPAC system.

56. Classify the alcohols in Problem 55 as primary, secondary, or tertiary.

57. Draw and name the isomeric ethers that have the formula $C_5H_{12}O$.

58. a. Menthol is an ingredient in mentholated cough drops and nasal sprays. It produces a cooling, refreshing sensation when rubbed on the skin and so is used in shaving lotions and cosmetics. What is its IUPAC name?

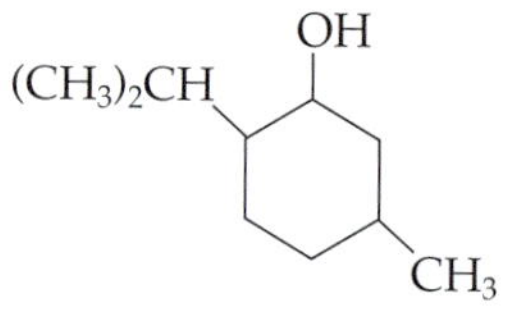

Menthol Thymol

 b. The aromatic equivalent of menthol is thymol, the flavoring constituent of thyme. Give two names for thymol.

59. Benzyl alcohol and *p*-cresol are isomers. Write the formula for each. Compare their solubilities in (**a**) water and (**b**) an aqueous solution of NaOH.

60. Write the equation for the production of ethanol by the addition of water to ethylene. How much ethanol can be made from 14.0 kg of ethylene?

61. The label on a bottle of light wine indicates that 100 mL of the wine furnishes 70 Cal (food calories), 0.2 g of protein, 5.77 g of carbohydrates, and 0.0 g of fat. Assuming that carbohydrates and proteins furnish 4 Cal/g each, that alcohol furnishes 7 Cal/g, and that no other caloric nutrients are present, (**a**) how many Calories are provided by the alcohol in a 100-mL serving of the wine? What percentage of the total Calories is provided by alcohol? (**b**) How many grams of alcohol are there in each 100-mL serving? (**c**) The density of alcohol is 0.789 g/mL. How many milliliters of alcohol are there in each 100-mL serving? What is the percent alcohol by volume?

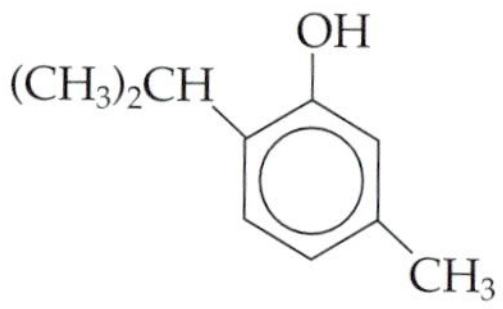

15 Aldehydes and Ketones

The odors and flavors of cinnamon and many other spices and foods are due to aldehydes.

What do certain hormones, vanilla flavor, a biological tissue preservative, and fresh cucumbers have in common? The answer: a carbonyl functional group. The carbonyl group is characteristic of aldehydes and ketones, the families we consider in this chapter. As the preceding list indicates, this functional group and these two families of compounds are found in a most diverse company of products. Both the tempting aromas associated with cinnamon, vanilla, and fresh baked goods and the sickeningly sweet smell of some rancid foods are associated with the carbonyl group.

The aldehydes and ketones offer us an opportunity to study the carbonyl group in its simplest surroundings. In Chapter 16, we consider more complicated functional groups incorporating the carbonyl group. And we ultimately find ourselves running into this ubiquitous grouping of atoms in carbohydrates, fats, proteins, nucleic acids, hormones, vitamins, and the host of organic compounds critical to living systems. But first things first. Let's begin by focusing on the carbonyl group in aldehydes and ketones.

15.1 The Carbonyl Group: A Carbon–Oxygen Double Bond

We first encountered a functional group containing a double bond in the alkenes. In the members of that family, two carbon atoms share four electrons (two pairs) to form a carbon–carbon double bond. In the alcohols, we saw a functional group

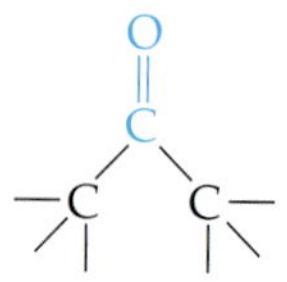

A ketone

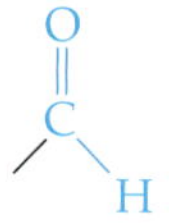

An aldehyde

in which an oxygen atom is attached to a carbon atom. The **carbonyl group** incorporates a feature of each of these functional groups. It has a carbon bonded to oxygen and a double bond—a carbon–oxygen double bond:

$$\text{>C=O}$$

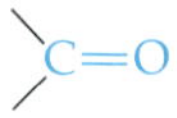

The carbonyl double bond, like the alkene double bond, tends to undergo addition reactions. Unlike the alkene double bond, however, the carbonyl bond involves an oxygen atom and so is highly polar. That polarity confers certain special properties on aldehydes and ketones.

What is the difference between a ketone and an aldehyde? It appears to be rather trivial at first sight. In **ketones,** two carbon groups are attached to the carbonyl carbon. These general formulas all represent ketones:

$$
\begin{array}{ccc}
\overset{\text{O}}{\underset{R\;\;\;\;R}{\overset{\|}{C}}} &
\overset{\text{O}}{\underset{R\;\;\;\;Ar}{\overset{\|}{C}}} &
\overset{\text{O}}{\underset{Ar\;\;\;\;Ar}{\overset{\|}{C}}}
\end{array}
$$

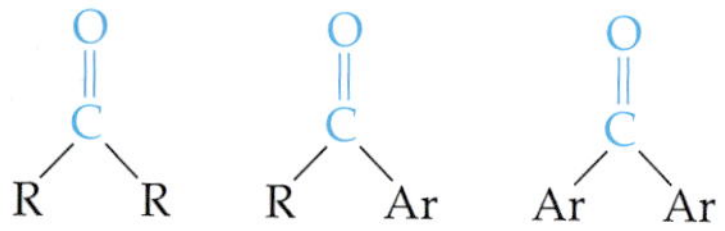

In **aldehydes,** at least one of the attached groups must be hydrogen. These compounds are all aldehydes:

$$
\begin{array}{ccc}
\overset{\text{O}}{\underset{H\;\;\;\;H}{\overset{\|}{C}}} &
\overset{\text{O}}{\underset{R\;\;\;\;H}{\overset{\|}{C}}} &
\overset{\text{O}}{\underset{Ar\;\;\;\;H}{\overset{\|}{C}}}
\end{array}
$$

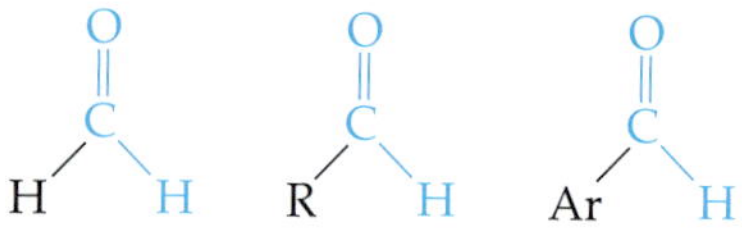

Because they contain the same functional group, aldehydes and ketones share many common properties. They are different from one another in other respects, however—different enough to warrant their classification into two families.

15.2 How to Name the Common Aldehydes

Aldehydes can be obtained by the removal of hydrogen from an alcohol, and the name is derived from the two words *alcohol dehyd*rogenation. Both common and IUPAC names are frequently used for aldehydes, with common names predominating for the lower homologs. The common names are taken from the names of the acids into which the aldehydes can be converted by oxidation (represented by [O]):

$$
H-C\overset{\text{O}}{\underset{\text{H}}{\diagup\!\!\!\backslash}} \quad \xrightarrow{[O]} \quad H-C\overset{\text{O}}{\underset{\text{OH}}{\diagup\!\!\!\backslash}}
$$

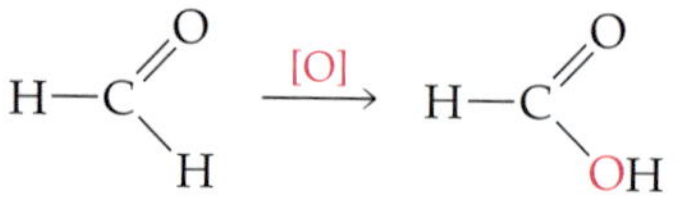

Formaldehyde Formic acid

$$
CH_3-C\overset{\text{O}}{\underset{\text{H}}{\diagup\!\!\!\backslash}} \quad \xrightarrow{[O]} \quad CH_3-C\overset{\text{O}}{\underset{\text{OH}}{\diagup\!\!\!\backslash}}
$$

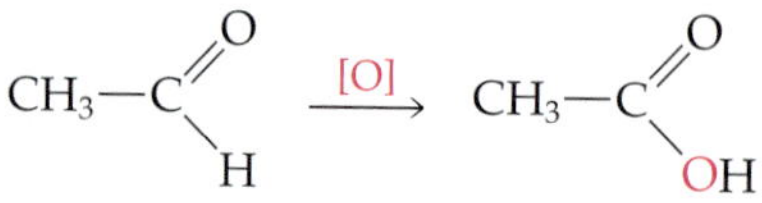

Acetaldehyde Acetic acid

Ball-and-stick model of formaldehyde (methanal).

Ball-and-stick model of acetaldehyde (ethanal).

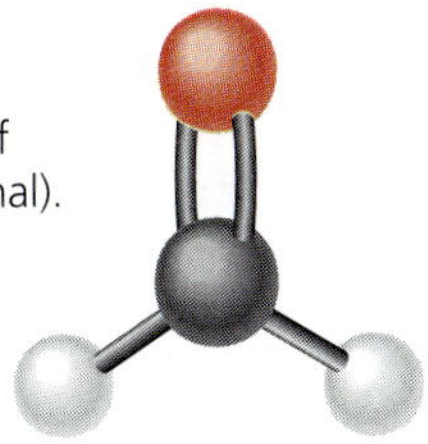

Formaldehyde Acetaldehyde

The IUPAC names of aldehydes are derived from those of the corresponding alkanes. Select the longest continuous chain of carbon atoms that contains the functional group. Take the name of the alkane having that number of carbon atoms, drop the *-e*, and add the ending *-al*.[1] The aldehyde functional group takes precedence over all the groups discussed in previous chapters. Thus, the carbonyl carbon is always considered to be carbon 1, and it is unnecessary to designate this group by number. Examples of aldehyde nomenclature are provided in Table 15.1.

$$\overset{5}{C}-\overset{4}{C}-\overset{3}{C}-\overset{2}{C}-\overset{1}{C}\overset{\displaystyle O}{\underset{H}{\Big\langle}}$$

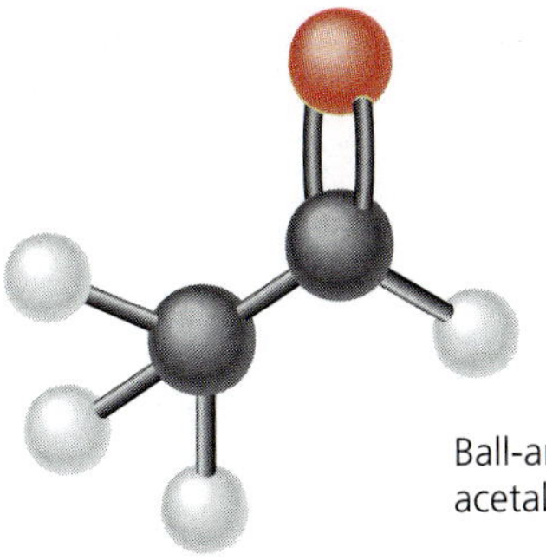

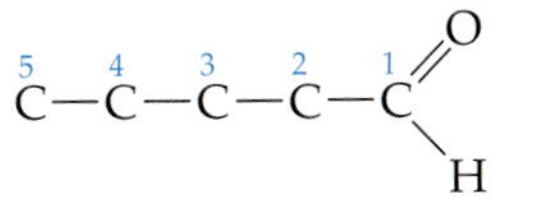

Example 15.1

Give the IUPAC name for

$$CH_3CH_2CH_2CH(CH_3)C\overset{\displaystyle O}{\underset{H}{\Big\langle}}$$

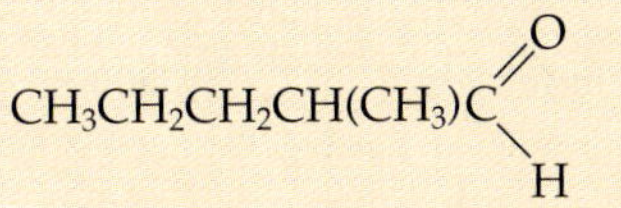

Solution

There are five carbon atoms in the longest continuous chain and a methyl group on the second carbon:

$$\overset{5}{C}H_3\overset{4}{C}H_2\overset{3}{C}H_2\overset{2}{C}H-\overset{1}{C}\underset{\underset{CH_3}{|}}{\overset{\displaystyle O}{\Big\langle}}{}_H$$

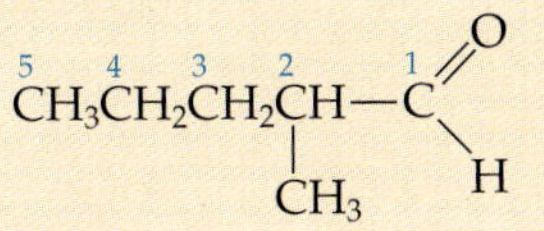

The name is 2-methylpentanal.

Practice Exercise

Give the IUPAC name for $C(CH_3)_3CH_2C\overset{\displaystyle O}{\underset{H}{\Big\langle}}$.

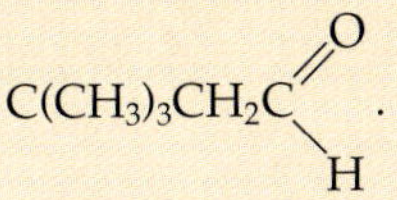

[1] Because the IUPAC ending for alcohols is *-ol*, there is occasionally confusion unless care is exercised in writing and pronouncing the IUPAC names of these two families. The one-carbon alcohol is methanol, with the ending pronounced like the *ol* in old. The one-carbon aldehyde is methanal, with the ending pronounced like the man's name *Al*.

Table 15.1 Nomenclature of Aldehydes

Molecular Formula	Condensed Structural Formula	Common Name	IUPAC Name
CH_2O	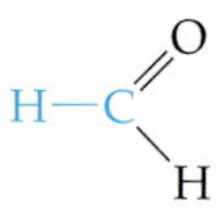	Formaldehyde	Methanal
C_2H_4O	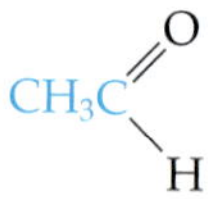	Acetaldehyde	Ethanal
C_3H_6O	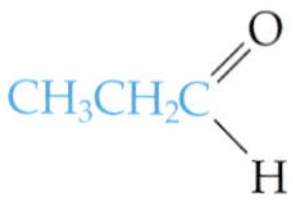	Propionaldehyde	Propanal
C_4H_8O		Butyraldehyde	Butanal
C_4H_8O	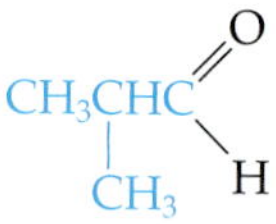	Isobutyraldehyde	2-Methylpropanal
$C_3H_6O_3$	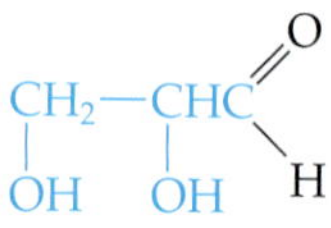	Glyceraldehyde	2,3-Dihydroxypropanal
$C_5H_{10}O$		Valeraldehyde	Pentanal
$C_5H_{10}O$	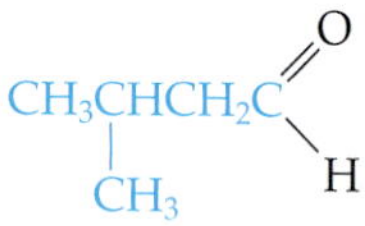	Isovaleraldehyde	3-Methylbutanal (NOT 2-Methylbutanal)
C_7H_6O	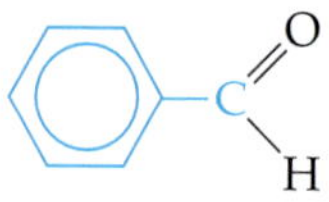	Benzaldehyde	Benzaldehyde (Phenylmethanal)
C_8H_8O	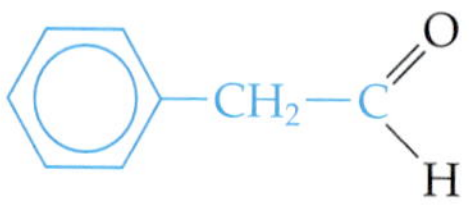	Phenylacetaldehyde	Phenylethanal
$C_8H_8O_3$	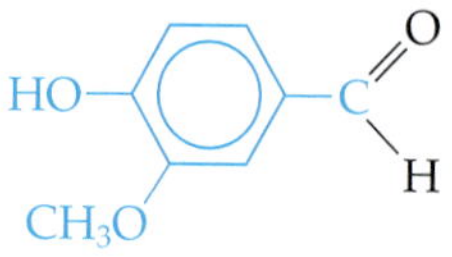	Vanillin (odor of vanilla)	4-Hydroxy-3-methoxybenzaldehyde
C_9H_8O	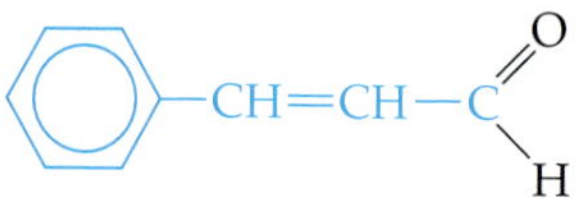	Cinnamaldehyde (odor of cinnamon)	3-Phenyl-2-propenal

Example 15.2

Write the structural formula for 7-chlorooctanal.

Solution

From the "octan-", we know that there are eight carbon atoms in the longest continuous chain. There is a chlorine atom on the seventh carbon atom, numbering from the carbonyl group and counting the carbonyl carbon as carbon 1:

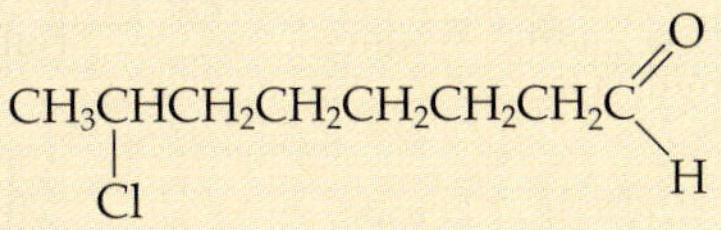

$$CH_3CHCH_2CH_2CH_2CH_2CH_2C\begin{smallmatrix}O\\ \\H\end{smallmatrix}$$
$$\hspace{1.2cm}|$$
$$\hspace{1.2cm}Cl$$

Practice Exercise

Write the structural formula for 5-bromo–3-iodoheptanal.

15.3 Naming the Common Ketones

The carbonyl group in a ketone must be attached to two carbon groups. Therefore, the simplest ketone has three carbon atoms. It is known far and wide by the name *acetone*. (It was first prepared from acetic acid.) The name is unique and does not correspond to the first in a series of similar common names. Generally, ketone common names consist of the names of the groups attached to the carbonyl group, followed by the word *ketone*. (Note the similarity to the naming of ethers.) Another name for acetone, then, is *dimethyl ketone*. With four carbon atoms, we have ethyl methyl ketone. If names for the groups attached to the carbonyl group are known, this common naming system can be applied.

In the IUPAC system, the longest continuous chain containing the carbonyl carbon is the parent chain. The *-e* ending of the corresponding alkane name is dropped and replaced with *-one*. The IUPAC name for acetone thus is propanone, and that for ethyl methyl ketone is butanone. In higher ketones, a number indicates the position of the carbonyl carbon. The chain is numbered so that the carbonyl carbon has the lowest possible number. In cyclic ketones, it is understood that the carbonyl carbon is number 1. Table 15.2 illustrates the nomenclature for some of the ketones.

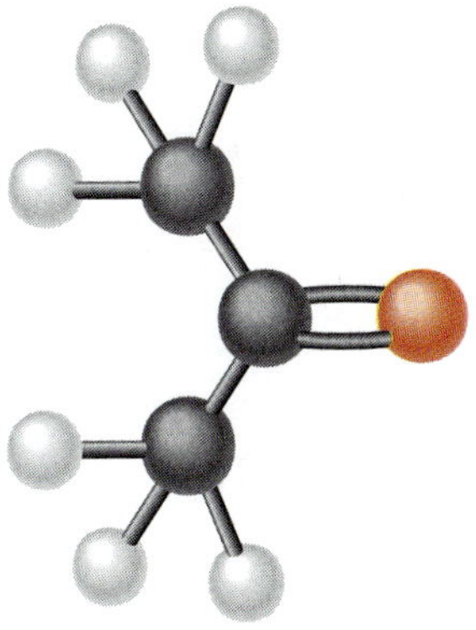

Acetone

Ball-and-stick model of acetone (propanone).

Example 15.3

Write the structural formula for 4-methyl–3-hexanone.

Solution

The "hexan-" tells us that the longest chain has six carbon atoms. The "3" means that the carbonyl carbon is carbon 3 in this chain, and the "4" tells us that there is a methyl group at the fourth carbon:

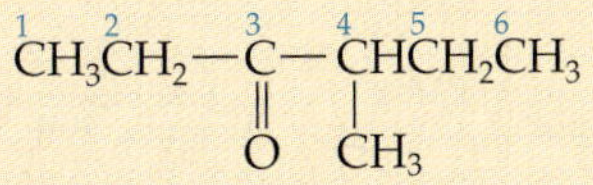

$$\overset{1}{C}H_3\overset{2}{C}H_2-\overset{3}{C}-\overset{4}{C}H\overset{5}{C}H_2\overset{6}{C}H_3$$
$$\hspace{2.3cm}\|\hspace{0.4cm}|$$
$$\hspace{2.3cm}O\hspace{0.3cm}CH_3$$

Practice Exercise

Write the structural formula for 1,5-dibromo–4-ethyl–2-heptanone.

Table 15.2 Nomenclature of Ketones

Molecular Formula	Condensed Structural Formula	Common Name	IUPAC Name
C_3H_6O	$CH_3\overset{O}{\overset{\|}{C}}CH_3$	Acetone (dimethyl ketone)	Propanone
C_4H_8O	$CH_3\overset{O}{\overset{\|}{C}}CH_2CH_3$	Ethyl methyl ketone	Butanone
C_4H_6O	$CH_2{=}CH\overset{O}{\overset{\|}{C}}CH_3$	Methyl vinyl ketone	3-Buten-2-one (NOT 1-Buten-3-one)
$C_5H_{10}O$	$CH_3CH_2\overset{O}{\overset{\|}{C}}CH_2CH_3$	Diethyl ketone	3-Pentanone
$C_5H_{10}O$	$CH_3CH_2CH_2\overset{O}{\overset{\|}{C}}CH_3$	Methyl propyl ketone	2-Pentanone
$C_5H_{10}O$	$CH_3\underset{CH_3}{\underset{\|}{C}}H\overset{O}{\overset{\|}{C}}CH_3$	Isopropyl methyl ketone	3-Methyl-2-butanone (NOT 2-Methyl-3-butanone)
$C_6H_{10}O$	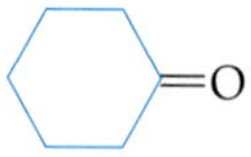	Cyclohexanone	Cyclohexanone
C_8H_8O	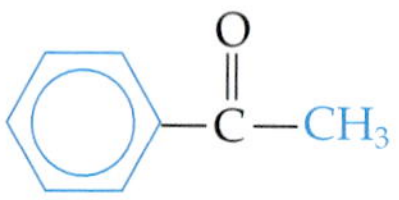	Acetophenone (Methyl phenyl ketone)	Phenylethanone
$C_{13}H_{10}O$	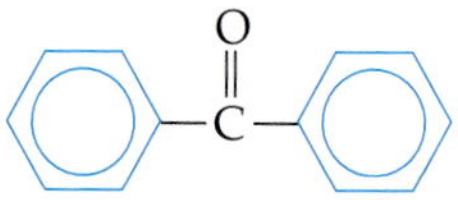	Benzophenone (Diphenyl ketone)	Diphenylmethanone

Example 15.4

Give the IUPAC name for

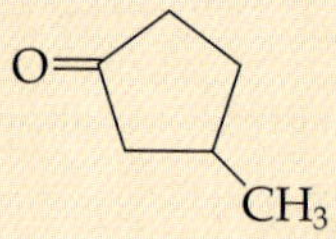

Solution

There are five carbon atoms in the ring, and the carbonyl carbon is always carbon 1. Therefore, the methyl group is on the third carbon:

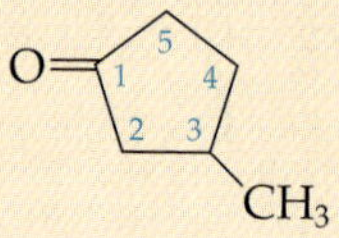

The name is 3-methylcyclopentanone.

Practice Exercise

Give the IUPAC name for

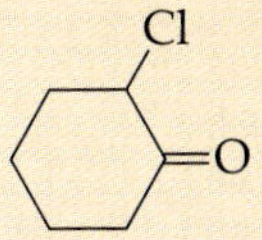

Example 15.5

Give the IUPAC name for

Solution

$$\overset{1}{C}H_3\overset{2}{C}H(CH_3)-\overset{3}{C}(=O)-\overset{4}{C}H(CH_3)\overset{5}{C}H_3$$

The name is 2,4-dimethyl-3-pentanone.

15.4 Physical Properties of Aldehydes and Ketones

The carbon and oxygen of the carbonyl group share two pairs of electrons, but they do not share them equally (Section 4.9). The electronegative oxygen has a much greater attraction for the bonding electron pairs. Thus, the electron density is greater at the oxygen end of the bond and less at the carbon end. The carbon is left with a partial positive charge and the oxygen with a partial negative charge:

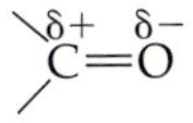

The polarity of the carbon–oxygen double bond is greater than that of the carbon–oxygen single bond. As we learned in Section 8.3, charge separation in a molecule leads to dipole interactions. Indeed, double-bond polarity is great enough to affect the boiling points of aldehydes and ketones, whereas the polar single bonds in ethers have little effect on boiling points (Table 15.3). The dipolar

Table 15.3 Boiling Points of Compounds Having Similar Molar Masses but Different Types of Intermolecular Forces

Compound	Family	Molar Mass	Type of Intermolecular Forces	Boiling Point (°C)
$CH_3CH_2CH_2CH_3$	Alkane	58	Dispersion only	−1
$CH_3OCH_2CH_3$	Ether	60	Weak dipole	6
$CH_3CH_2CH(=O)$	Aldehyde	58	Strong dipole	49
$CH_3CH_2CH_2OH$	Alcohol	60	Hydrogen bonding	97

Table 15.4 Physical Properties of Selected Aldehydes and Ketones

Compound	Formula	Boiling Point (°C)	Solubility in Water (g/100 g H_2O)
Formaldehyde	HCHO	−21	Miscible
Acetaldehyde	CH_3CHO	20	Miscible
Propionaldehyde	CH_3CH_2CHO	49	16
Butyraldehyde	$CH_3CH_2CH_2CHO$	76	7
Valeraldehyde	$CH_3CH_2CH_2CH_2CHO$	103	Slightly soluble
Benzaldehyde	C_6H_5CHO	178	0.3
Acetone	CH_3COCH_3	56	Miscible
Ethyl methyl ketone	$CH_3COCH_2CH_3$	80	26
Methyl propyl ketone	$CH_3COCH_2CH_2CH_3$	102	6.3
Diethyl ketone	$CH_3CH_2COCH_2CH_3$	101	5

forces in aldehydes and ketones however, are not comparable to the hydrogen bonding between molecules of an alcohol.

With the exception of the gaseous formaldehyde, the majority of aldehydes are liquids. (The physical state of acetaldehyde, bp 20 °C, depends on the temperature of the laboratory; in warm rooms acetaldehyde exists as a gas.) Although the lower members of the series have pungent odors, many higher aldehydes have pleasant odors and are used in perfumes and artificial flavorings. The hydrogens of water molecules can form hydrogen bonds with the carbonyl oxygen; thus the solubility of aldehydes is about the same as that of alcohols and ethers. Formaldehyde and acetaldehyde are soluble in water; as the carbon chain increases, water solubility decreases. The borderline of solubility occurs at about four carbon atoms per oxygen atom. All aldehydes are soluble in organic solvents and, in general, are less dense than water.

The physical properties of the ketones are almost identical to those of the corresponding aldehydes. Acetone has a pleasant odor, and it is the only ketone that is completely soluble in water. The higher homologs are colorless liquids, are slightly soluble in water, and, unlike the aldehydes, have rather bland odors. Table 15.4 lists some physical constants for several aldehydes and ketones.

15.5 Preparation of Aldehydes and Ketones

In Section 14.6 we learned that primary and secondary alcohols are oxidized to aldehydes and ketones, respectively. However, in aqueous solutions the product aldehyde forms a hydrate that is further oxidized to a carboxylic acid. Therefore, organic solvents are used whenever aldehydes are prepared from alcohols. The reagent of choice is chromic oxide in combination with pyridine, methylene chloride, and HCl. In organic solvents, chromium(VI) compounds are mild oxidizing agents that can oxidize primary alcohols to aldehydes without oxidizing the aldehydes to acids. We shall see in Chapters 24 and 25 that the enzyme-catalyzed oxidation of alcohols to aldehydes and ketones is of great significance in biological systems.

$$R-CH_2OH \xrightarrow{[O]} R-C{\overset{O}{\underset{H}{}}}$$

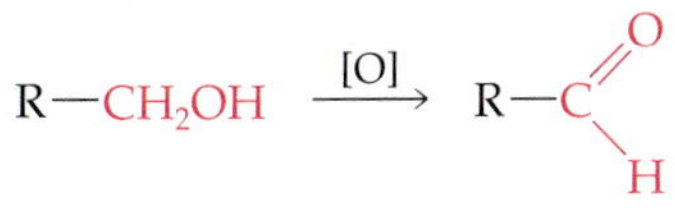

A primary alcohol An aldehyde

$$3\ CH_3CH_2CH_2CH_2OH\ +\ CrO_3\ +\ 3\ HCl\ \xrightarrow[CH_2Cl_2]{pyridine}\ 3\ CH_3CH_2CH_2C{\overset{O}{\underset{H}{}}}\ +\ CrCl_3\ +\ 3\ H_2O$$

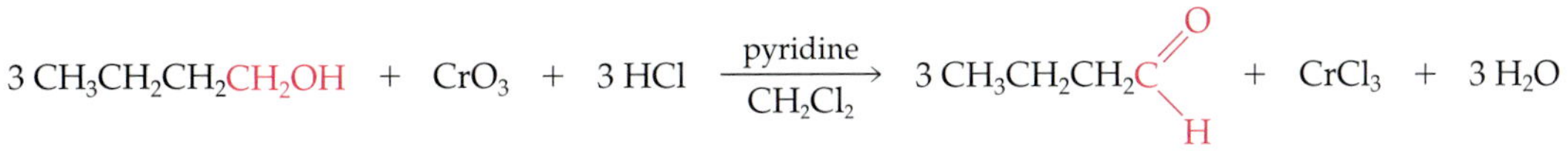

1-Butanol (orange) Butanal (green)

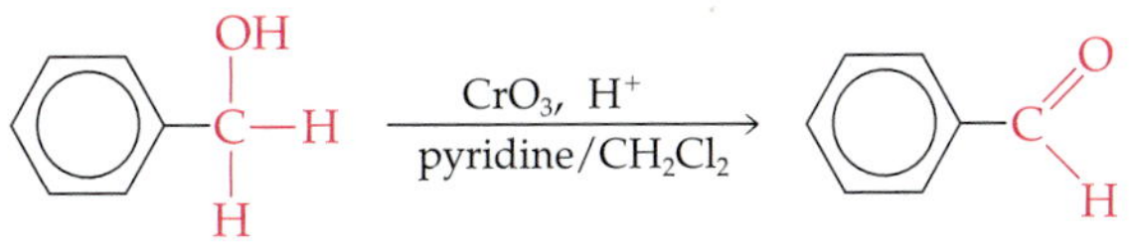

Benzyl alcohol $\xrightarrow[\text{pyridine/CH}_2\text{Cl}_2]{\text{CrO}_3,\ \text{H}^+}$ Benzaldehyde

Like the aldehydes, ketones are obtained by alcohol oxidation. However, the alcohol must be a secondary alcohol, and no special reagents are needed because the ketone is not susceptible to further oxidation. Although many oxidants are used, the most commonly employed are chromium(VI) compounds and sulfuric acid:

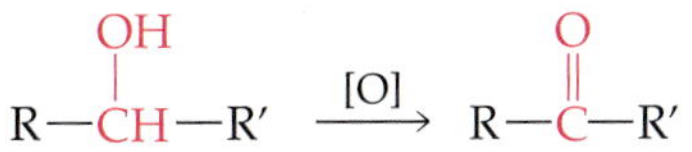

$$R-\underset{\underset{\text{A secondary}}{\overset{\overset{OH}{|}}{CH}}}-R'\ \xrightarrow{[O]}\ R-\overset{\overset{O}{||}}{C}-R'$$

A secondary A ketone
alcohol

$$CH_3\underset{\overset{|}{OH}}{CH}CH_3\ \xrightarrow[\text{H}_2\text{SO}_4]{\text{K}_2\text{Cr}_2\text{O}_7}\ CH_3\overset{\overset{O}{||}}{C}CH_3$$

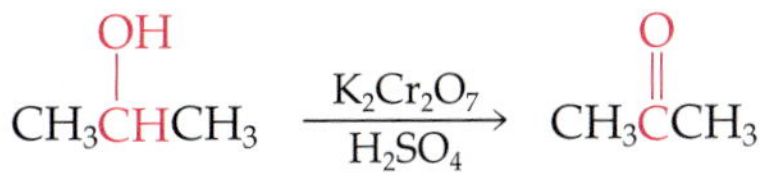

Isopropyl alcohol Acetone

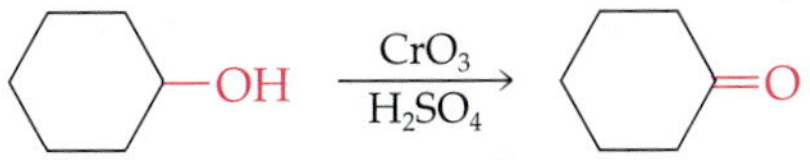

Cyclohexanol $\xrightarrow[\text{H}_2\text{SO}_4]{\text{CrO}_3}$ Cyclohexanone

As we shall see in Chapters 24 and 25, the electrons released when alcohols are oxidized to carbonyl compounds are converted to a form of energy that can be utilized by living cells. These biochemical oxidation reactions are carried out at body temperature (~37 °C) and are catalyzed by enzymes. Notice that in the following reaction the enzyme selectively catalyzes the oxidation of the secondary alcohol group to a ketone, but it does not oxidize the primary alcohol group to an aldehyde. We discuss enzyme specificity in Section 22.4:

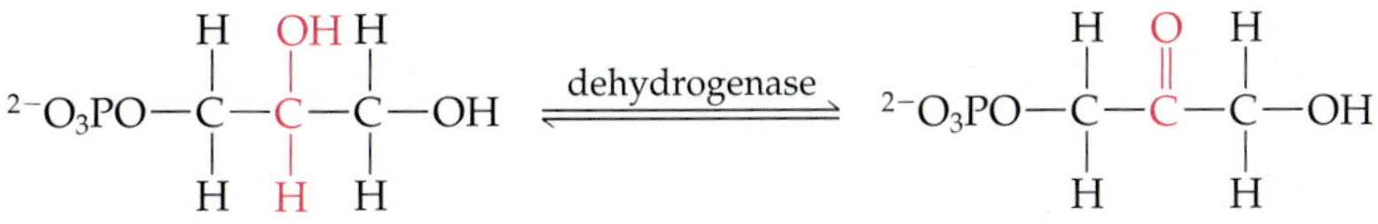

Glycerol 3-phosphate $\xrightarrow{\text{dehydrogenase}}$ Dihydroxyacetone phosphate

15.6 Chemical Properties of Aldehydes and Ketones

Oxidation

Aldehydes are readily oxidized to carboxylic acids, whereas ketones resist oxidation:

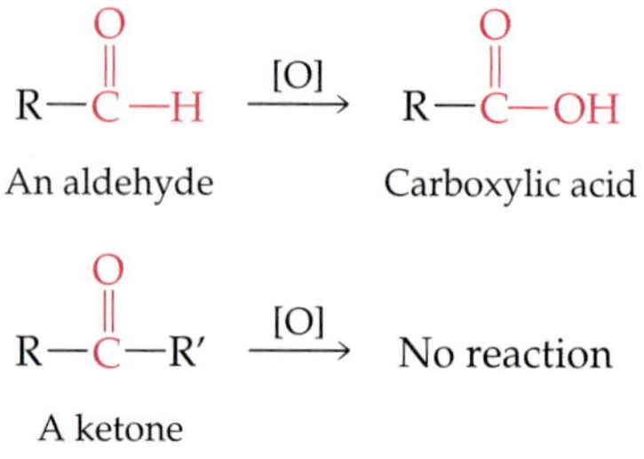

The aldehydes are, in fact, among the most easily oxidized of organic compounds, and this fact helps chemists identify them. Through the use of oxidizing agents, aldehydes can be distinguished not only from ketones but also from alcohols if the reagent is gentle enough. One such test reagent was invented by Bernhard Tollens (1841–1918) at the University of Göttingen in Germany. *Tollens's reagent* employs silver ion as the mild oxidizing agent. In order for the silver ion to be kept in solution, it must be complexed by two ammonia molecules:

$$H_3N{-}Ag^+{-}NH_3$$

When Tollens's reagent oxidizes an aldehyde, the silver ion is reduced to free silver:

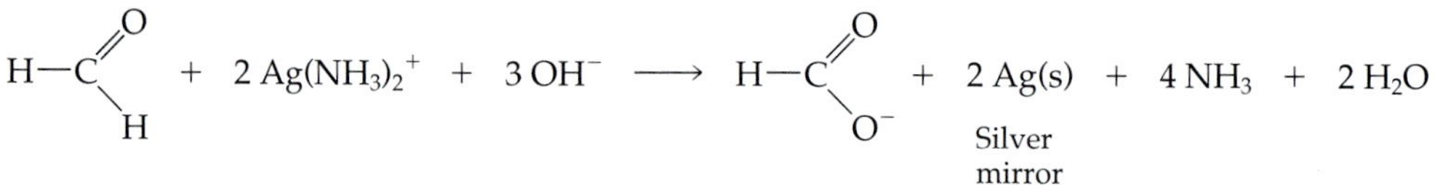

The silver, when deposited on a clean glass surface, produces a beautiful mirror. Indeed, mirrors are often silvered by means of the Tollens reaction. The reducing agent of choice is often the sugar glucose (which contains an aldehyde functional group) rather than a simple aldehyde. Ordinary ketones do not react with Tollens's reagent.

Two other test reagents, Benedict's and Fehling's, use alkaline solutions of copper(II) ion (Cu^{2+}). The source of the ion is copper(II) sulfate. Because Cu^{2+} forms an insoluble hydroxide in basic solution, another reagent must be added to the solution to keep the copper ion from precipitating out as the hydroxide. In *Benedict's solution,* sodium citrate serves this purpose; copper remains in solution as the copper(II) citrate ion. The additional reagent in *Fehling's solution* is sodium potassium tartrate (Rochelle salt), with which copper forms the water-soluble copper(II) tartrate ion. The blue color of these solutions is due to the presence of copper(II) ion complexes. A positive test for the aldehyde group is evidenced by a color change to brick red, indicating the presence of the copper(I) oxide:

The copper(II) ion is the oxidizing agent and therefore must be the substance that is reduced, in this case to copper(I) oxide.

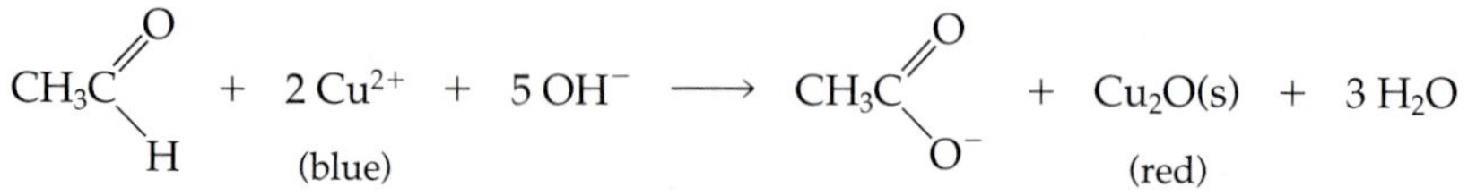

Although ketones resist oxidation by ordinary laboratory oxidizing agents, it is possible to force their oxidation. In particular, both aldehydes and ketones undergo the extreme form of oxidation we call combustion; in other words, they burn. Acetone is a common organic solvent. Neither it nor any other volatile, flammable organic solvent should be used around open flames, heating elements, or other possible sources of ignition.

Reduction

Aldehydes and ketones are readily reduced to the corresponding primary and secondary alcohols, respectively. A wide variety of reducing agents may be used.

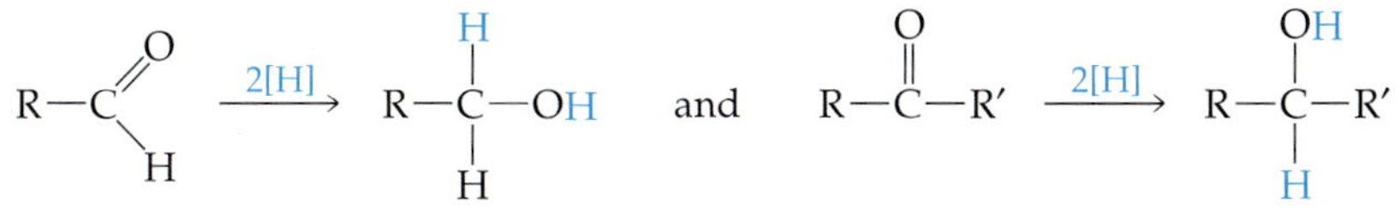

Carbonyl compounds can also be reduced to alcohols by hydrogen gas in the presence of a metal catalyst (catalytic hydrogenation). However, this method suffers from the disadvantages that many of the catalysts (Pt, Pd, Ru) are expensive and other functional groups are also reduced.

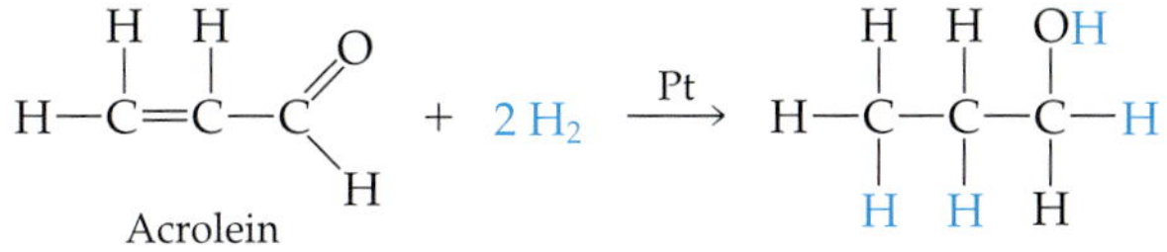

Acrolein (2-Propenal) 1-Propanol

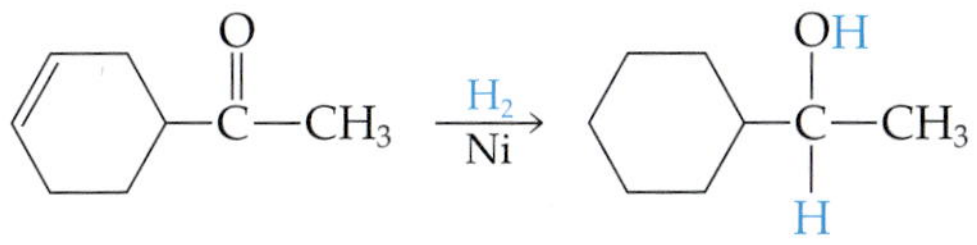

Two extremely important biochemical carbonyl reduction reactions are discussed in Chapter 24. They are the reduction of acetaldehyde to ethyl alcohol and the reduction of pyruvic acid to lactic acid. In each case, the enzyme that catalyzes the reaction contains the coenzyme NADH, which is the reducing agent. (See Table J.1 for the structure of the coenzyme.)

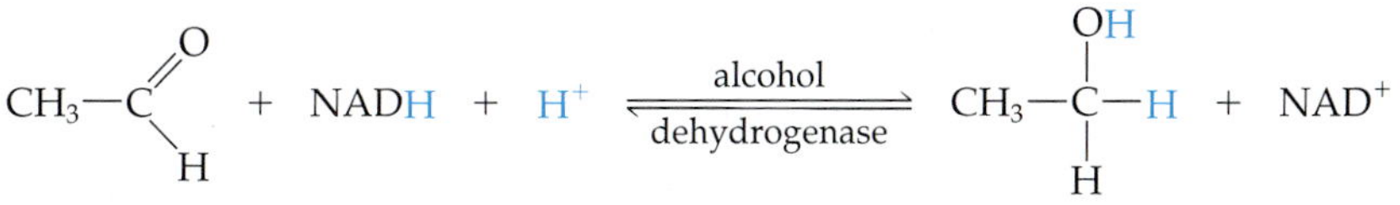

Acetaldehyde Ethyl alcohol

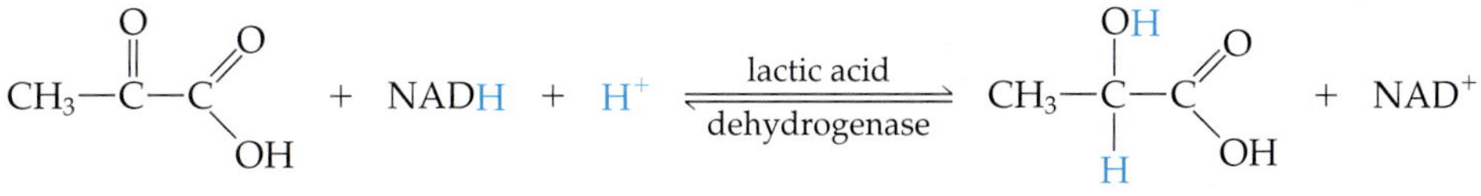

Pyruvic acid Lactic acid

Hydration of Carbonyl Compounds

Formaldehyde dissolves readily in water and even *reacts* with it:

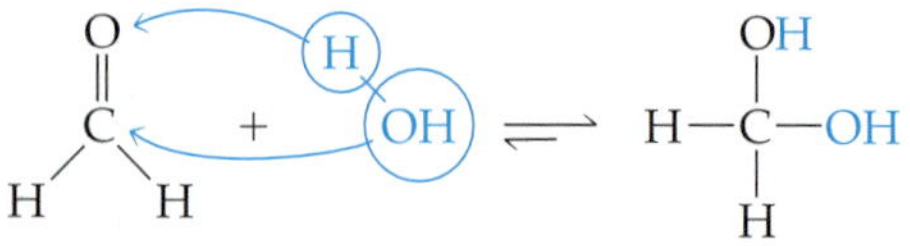

The process is an addition reaction, analogous to the hydration of the carbon–carbon double bond of an alkene. The net result is that a hydrogen from water is added to the carbonyl oxygen and a hydroxyl group from water becomes attached to the carbonyl carbon. The product is called a **hydrate.** It readily breaks down to re-form formaldehyde and water. At equilibrium at 20 °C, the hydrate predominates. Indeed, only 1 molecule in 10,000 exists as free formaldehyde. The other 9999 are in the form of the hydrate.

Acetaldehyde is also hydrated in aqueous solution, but to a lesser extent than formaldehyde:

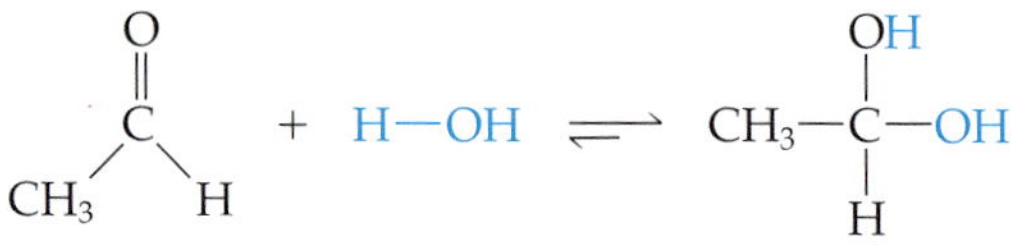

Out of 10,000 molecules, about 4200 are in the form of the free aldehyde at equilibrium. Still, that leaves 5800 in the hydrated form. Generally, higher aldehydes and ketones are even less hydrated, existing primarily in the free aldehyde (or ketone) form at equilibrium in water.

In most cases, it is impossible to isolate the hydrates from solution. Attempts to do so result in loss of water and regeneration of the carbonyl compound. An exception is the hydrate of trichloroacetaldehyde (chloral):

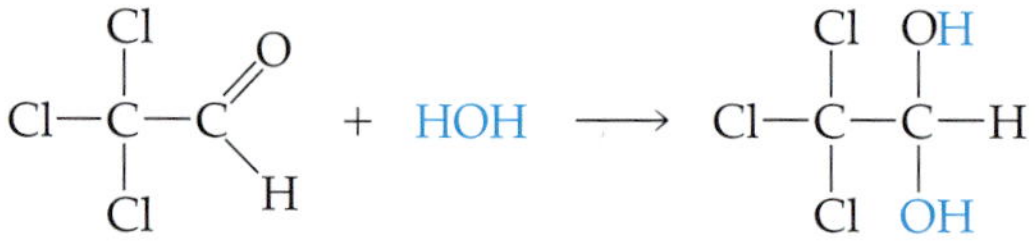

Chloral Chloral hydrate

Chloral hydrate is a stable solid, soluble in water, and one of the very few organic compounds that possess two hydroxyl groups on the same carbon atom. It is a powerful sedative and soporific (sleep-inducing drug). Chloral hydrate has had wide use in medicine. It is perhaps even better known in fictional mystery stories. Slipped into someone's drink, the mixture is called a "Mickey Finn" or "knockout drops." Such combinations of alcohol and chloral hydrate—two "downers"—are exceedingly dangerous. A little too much, and the unfortunate victim may be put to sleep permanently.

Addition of Alcohols to Carbonyl Groups

Alcohols add to the carbonyl group of aldehydes and ketones in much the same way as water does. The addition of 1 mol of an alcohol to 1 mol of an aldehyde or ketone yields a **hemiacetal** or **hemiketal,** respectively. In the presence of an anhydrous acid catalyst, equilibrium is rapidly established, and the equilibrium favors the carbonyl compounds (the reactants). As with the hydrates, simple hemiacetals and hemiketals are generally not stable enough to be isolated:

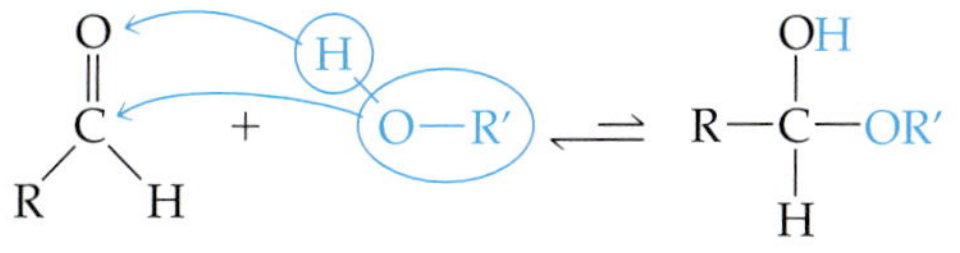

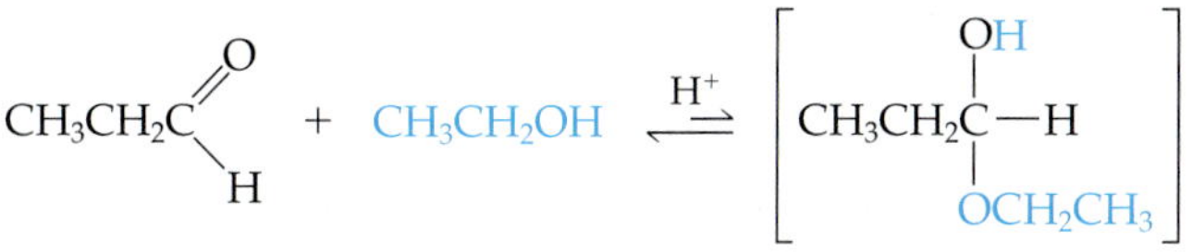

Unstable hemiacetal

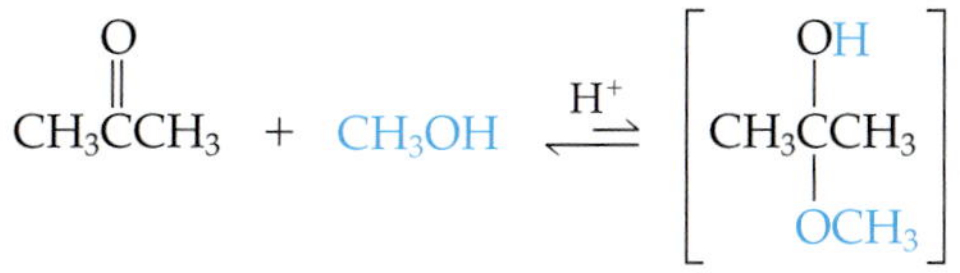

Unstable hemiketal

When the alcohol and carbonyl groups occur within the same molecule, however, the equilibrium favors the formation of cyclic hemiacetals and hemiketals. These cyclic compounds result from *intramolecular* interaction between the —OH and C=O groups. (The cyclization reactions are of particular significance in our discussion of the structures of monosaccharides; see Section 19.5):

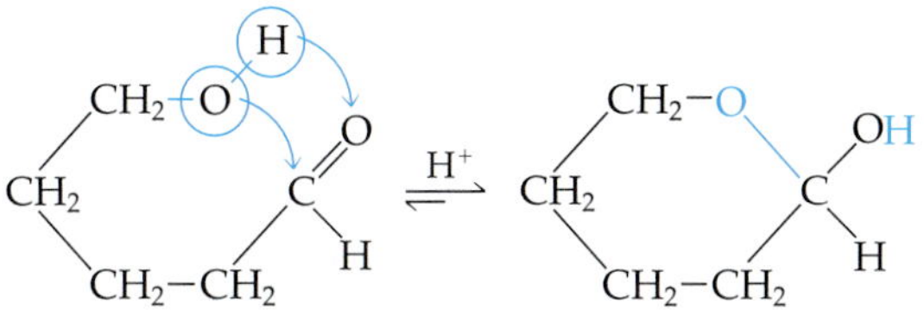

5-Hydroxypentanal   Cyclic hemiacetal

Hemiacetals can be made to react further with alcohols. If dry hydrogen chloride gas is bubbled into a solution of aldehyde in excess alcohol, an **acetal** is formed. Unlike hemiacetals and hydrates, acetals are stable and can be isolated. The reaction for acetaldehyde and methanol is

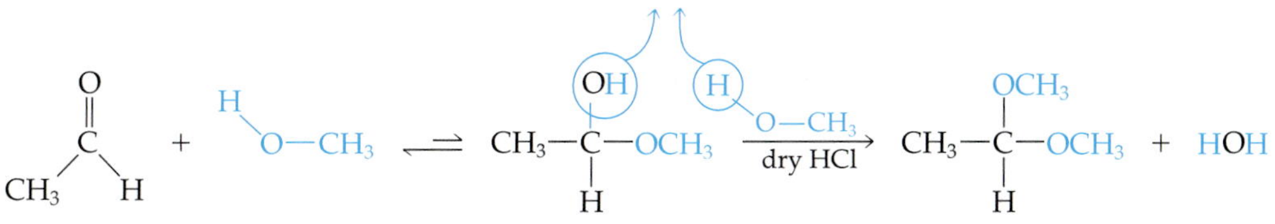

A hemiacetal (unstable)   An acetal (stable)

First an alcohol molecule adds to the double bond of the aldehyde to form the hemiacetal. Then the hydroxyl group of the hemiacetal and the hydrogen from the hydroxyl group of a second alcohol molecule are eliminated, and the two remaining pieces combine to form the acetal. A **ketal** is formed when the starting material is a ketone rather than an aldehyde.

Acetals and ketals are resistant to oxidation. For this reason, acetal or ketal formation is often used to protect the functional group of aldehydes or ketones while other chemical operations are performed on the molecules. An aldehyde is converted to an acetal, an oxidation reaction is then carried out on another part of the molecule, and finally the aldehyde is regenerated from the acetal. The carbonyl group is easily regenerated by aqueous acid:

$$CH_3-\overset{OCH_3}{\underset{H}{C}}-OCH_3 \ + \ H_2O \ \xrightarrow{H^+} \ CH_3-C\overset{O}{\underset{H}{\diagup}} \ + \ 2\,CH_3OH$$

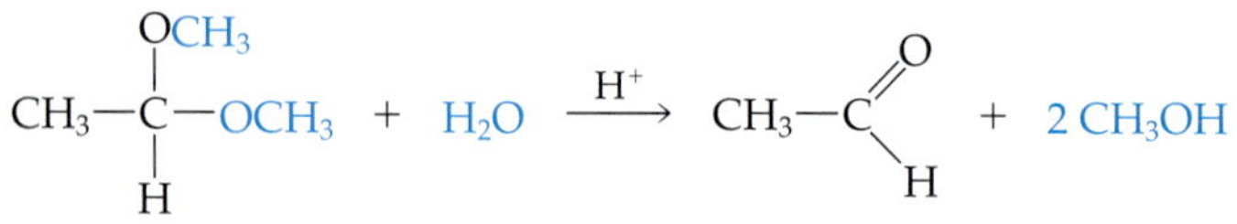

In an interesting application, the antibiotic chloramphenicol is treated with acetone, and a protective cyclic ketal is formed that masks the bitter taste of the drug. Both of the alcohol hydroxyl groups required to form the ketal are attached to a single molecule in this product:

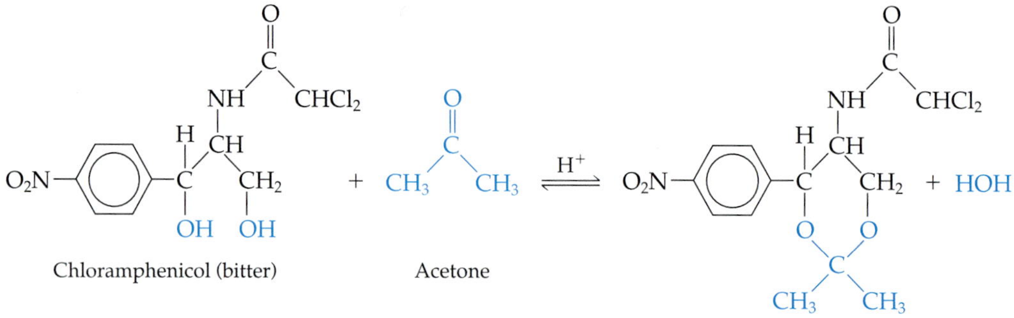

Chloramphenicol (bitter) Acetone A cyclic ketal (not bitter)

Acids in the digestive tract convert the cyclic ketal back to chloramphenicol. Chloramphenicol is a powerful, but hazardous, antibiotic. It is used only when other, less dangerous drugs are ineffective. In about 1 person in 20,000 to 40,000 (depending on dosage), chloramphenicol causes fatal aplastic anemia.

Example 15.6

Complete the following equations:

a. $CH_3CH_2C\overset{O}{\underset{H}{\diagup}} \ + \ 2\,CH_3CH_2OH \ \xrightarrow{H^+}$

b. $CH_3\overset{O}{\overset{\|}{C}}CH_3 \ + \ 2\,CH_3OH \ \xrightarrow{H^+}$

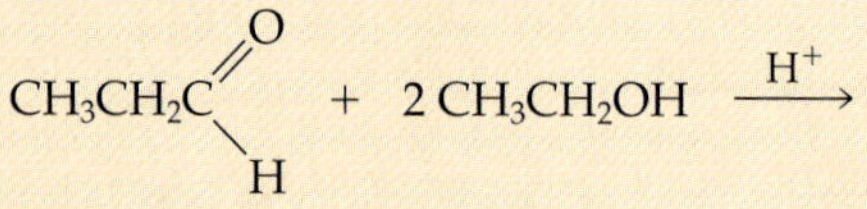

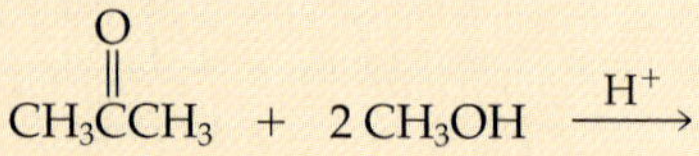

Solution

Realize that each reaction involves two steps. First the addition of 1 mol of alcohol to 1 mol of the carbonyl compound:

a. $CH_3CH_2C\overset{O}{\underset{H}{\diagup}} \ + \ CH_3CH_2OH \ \underset{}{\overset{H^+}{\rightleftharpoons}} \ CH_3CH_2\overset{OH}{\underset{OCH_2CH_3}{C}}-H$

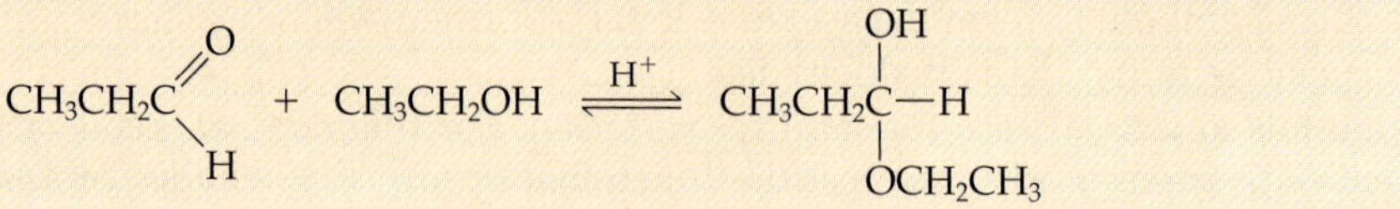

b. $CH_3\overset{\displaystyle O}{\overset{\|}{C}}CH_3$ + CH_3OH $\xrightleftharpoons{H^+}$ $CH_3\overset{\displaystyle OH}{\underset{\displaystyle OCH_3}{\overset{|}{\underset{|}{C}}}}CH_3$

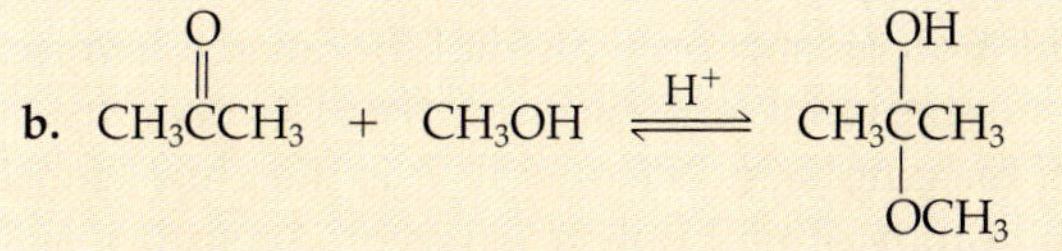

This is followed by the interaction of the hemiacetal and hemiketal with a second mole of the alcohol:

a. $CH_3CH_2\overset{\displaystyle OH}{\underset{\displaystyle OCH_2CH_3}{\overset{|}{\underset{|}{C}}}}-H$ + CH_3CH_2OH $\xrightarrow{H^+}$ $CH_3CH_2\overset{\displaystyle OCH_2CH_3}{\underset{\displaystyle OCH_2CH_3}{\overset{|}{\underset{|}{C}}}}-H$ + HOH

 A hemiacetal An acetal

b. $CH_3\overset{\displaystyle OH}{\underset{\displaystyle OCH_3}{\overset{|}{\underset{|}{C}}}}CH_3$ + CH_3OH $\xrightarrow{H^+}$ $CH_3\overset{\displaystyle OCH_3}{\underset{\displaystyle OCH_3}{\overset{|}{\underset{|}{C}}}}CH_3$ + HOH

 A hemiketal A ketal

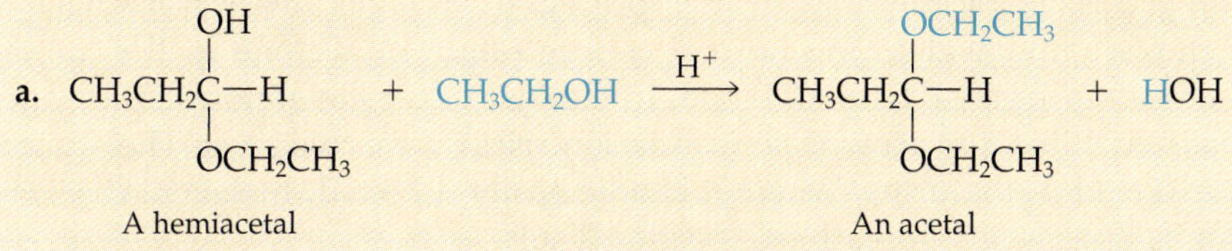

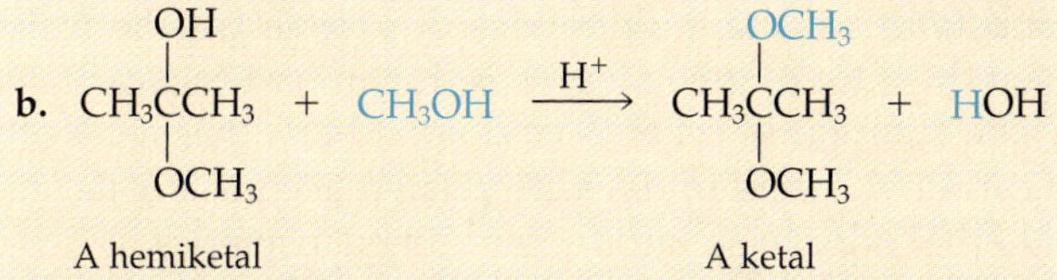

Practice Exercise

Complete the equation

+ $2\ CH_3CH_2CH_2OH$ $\xrightarrow{H^+}$

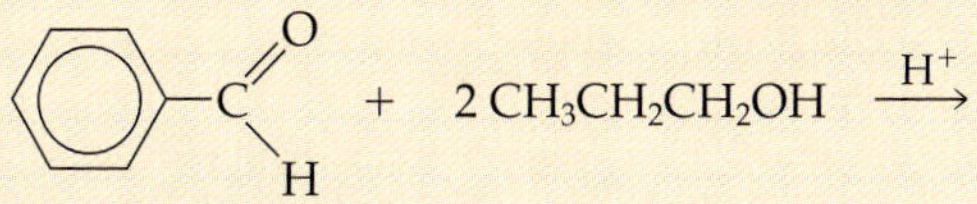

15.7 Some Common Carbonyl Compounds

Formaldehyde is the simplest and industrially the most important member of the aldehyde family. It is manufactured from methanol and oxygen in the air by passing methanol vapor over a copper or silver catalyst at temperatures above 300 °C. Formaldehyde is a colorless gas that has an extremely irritating odor. Because of its reactivity, it cannot be handled easily in the gaseous state and is therefore dissolved in water and sold as a 37 to 40% aqueous solution (such a solution is called *formalin*).

The largest use of formaldehyde is as a reagent for the preparation of many other organic compounds and for the manufacture of polymers such as Bakelite, Formica, and Melmac. Formaldehyde can denature proteins (see Section 21.14), rendering them insoluble in water and resistant to decay bacteria. For this reason it is used in embalming solutions and in the preservation of biological specimens. Formalin is also used as a general antiseptic in hospitals to sterilize gloves and surgical instruments. However, its use as an antiseptic, preservative, and embalming fluid has declined in recent years because formaldehyde is suspected of being carcinogenic.

Acetaldehyde is an extremely volatile, colorless liquid. It is prepared by the catalytic (Ag) oxidation of ethyl alcohol or the catalytic (PdCl$_2$) oxidation of ethylene. It is a starting material for the preparation of many other organic compounds, such as acetic acid, ethyl acetate, and chloral. Acetaldehyde is formed as a metabolite in the fermentation of sugars and in the detoxification of alcohol in the liver (see Chapter 24).

Formaldehyde is present in wood smoke. Because it kills bacteria, it is one of the compounds responsible for the preservative effect in smoked foods.

Acetone is the simplest and most important ketone. It is produced in large quantities by the catalytic (Ag) oxidation of isopropyl alcohol. Because it is miscible with water as well as with most organic solvents, acetone finds its chief use as an industrial solvent (for example, for paints and lacquers). It is the chief (sometimes the only) ingredient in some brands of nail polish remover. Acetone is also an important intermediate in the preparation of chloroform, iodoform, dyes, methacrylates, and many other complex organic compounds.

Acetone is formed in the human body as a by-product of lipid metabolism. Normally it does not accumulate to an appreciable extent because it is oxidized to carbon dioxide and water. The normal concentration of acetone in the human body is less than 1 mg/100 mL of blood. In the case of certain abnormalities, such as diabetes mellitus, the acetone concentration rises above this level. The acetone is then excreted in the urine, where it can be easily detected. In severe cases, its odor can be noted on the breath (see Section 26.6).

Many other familiar substances contain aldehydes or ketones as the active principles (Figure 15.1). Even the odor of green leaves is due in part to a carbonyl compound: *cis*–3-hexenal. The compound *trans*–2-*cis*–6-nonadienal has a cucumber odor. These and other carbonyl compounds (with related acetals, ketals, and alcohols) impart a "green" herbal odor to shampoos and other cosmetics.

Several steroid hormones (Special Topic I) have the carbonyl functional group as an integral part of their structure. Progesterone is a hormone secreted by the ovaries. It stimulates the growth of cells in the uterus wall, preparing it for attachment of a fertilized egg. Testosterone is the main male sex hormone. These (and other) sex hormones affect our development and our lives in most fundamental ways.

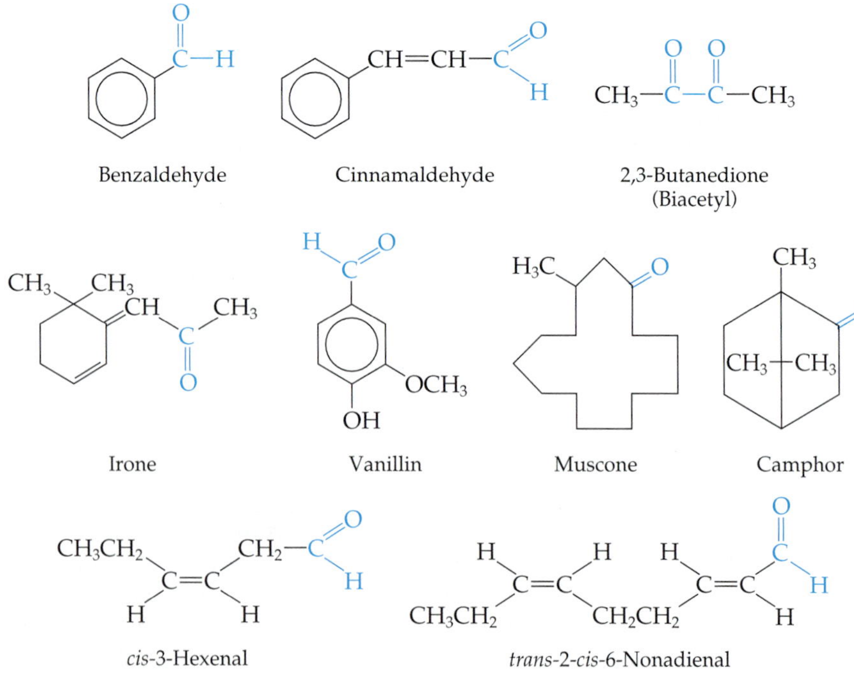

FIGURE 15.1 ▶

Some interesting aldehydes and ketones. Benzaldehyde is an oil found in almonds. Cinnamaldehyde is oil of cinnamon. 2,3-Butanedione is a butter flavoring, and irone is responsible for the odor of violets. Vanillin gives vanilla its flavor. Muscone is musk oil, an ingredient in perfumes, and camphor is used in some insect repellents. *cis*–3-Hexenal provides an herbal odor, and *trans*–2-*cis*–6-nonadienal gives a cucumber odor.

Summary

The **carbonyl group,** as shown below, is the defining feature of **aldehydes,** carbonyl compounds in which at least one bond on the carbonyl group is a carbon–hydrogen bond,

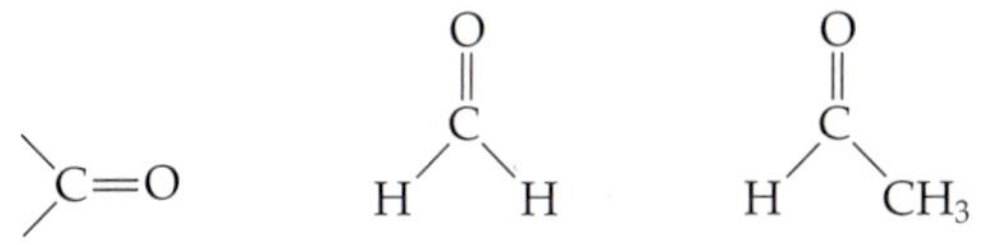

Carbonyl Aldehyde Aldehyde

and of **ketones,** carbonyl compounds in which both available bonds on the carbonyl carbon are carbon–carbon bonds,

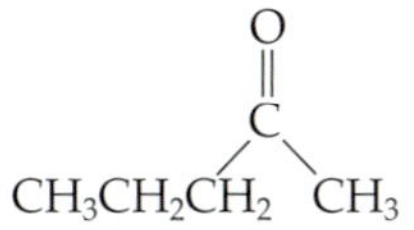

Ketone

Aldehydes are synthesized via oxidation of primary alcohols. The aldehyde can be further oxidized to a carboxylic acid, and for this reason aldehydes are distilled from the reaction mixture as they form.

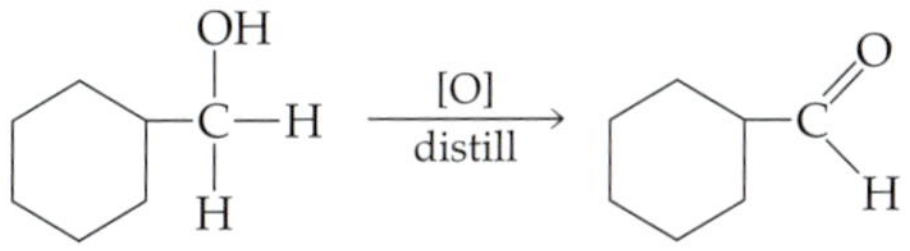

Primary alcohol Aldehyde

Ketones are prepared via oxidation of secondary alcohols:

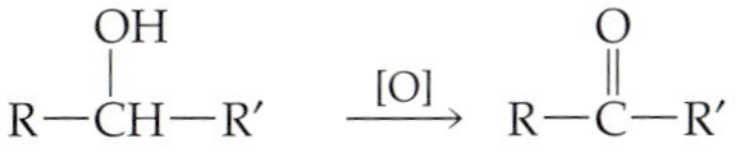

Secondary alcohol Ketone

In the presence of even mild oxidizing agents, aldehydes are oxidized to carboxylic acids, but ketones are not oxidized:

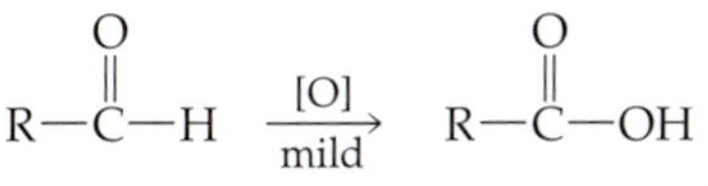

Aldehyde Carboxylic acid

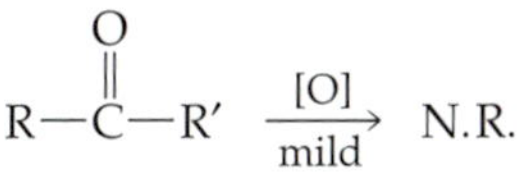

Ketone

Both aldehydes and ketones undergo combustion, and both can be reduced to alcohols:

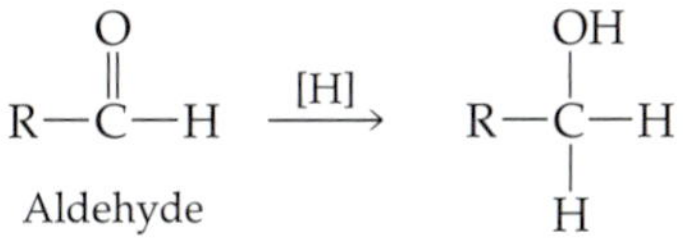

Aldehyde Primary alcohol

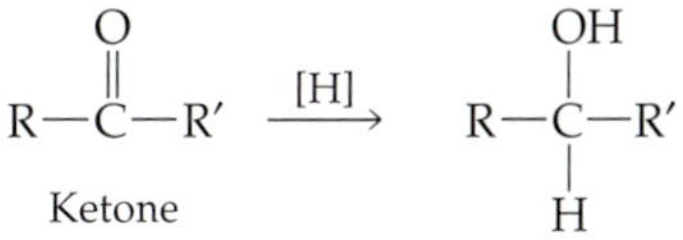

Ketone Secondary alcohol

Aldehydes add water across the carbonyl double bond to form molecules known as **hydrates:**

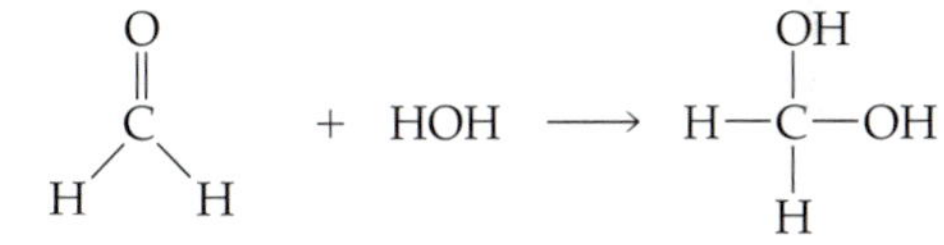

Formaldehyde Hydrate

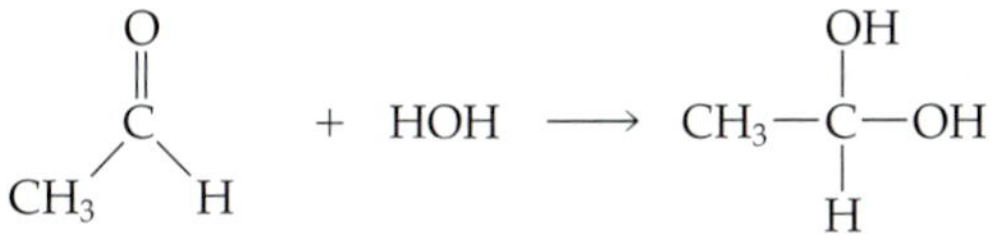

Acetaldehyde Hydrate

When aldehydes add an alcohol across the carbonyl group, the result is a **hemiacetal.** When this alcohol addition takes place with a ketone, the product is a **hemiketal:**

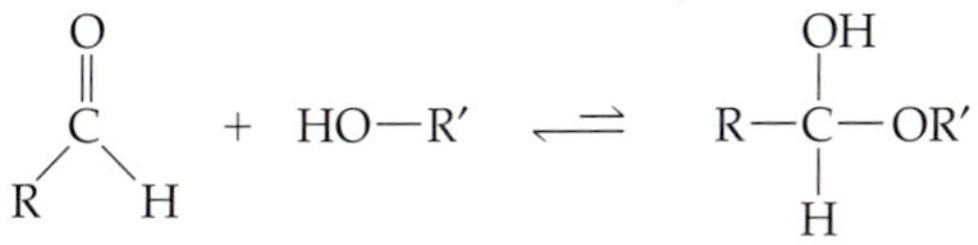

Aldehyde Alcohol Hemiacetal (unstable)

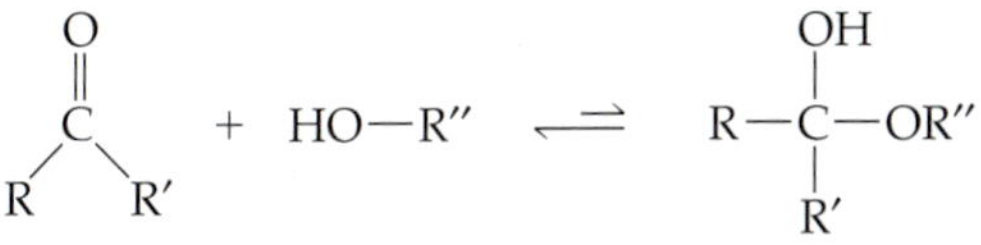

Ketone Alcohol Hemiketal (unstable)

In the presence of excess alcohol and dry HCl, the hemiacetal and hemiketal react further to form an **acetal** or a **ketal:**

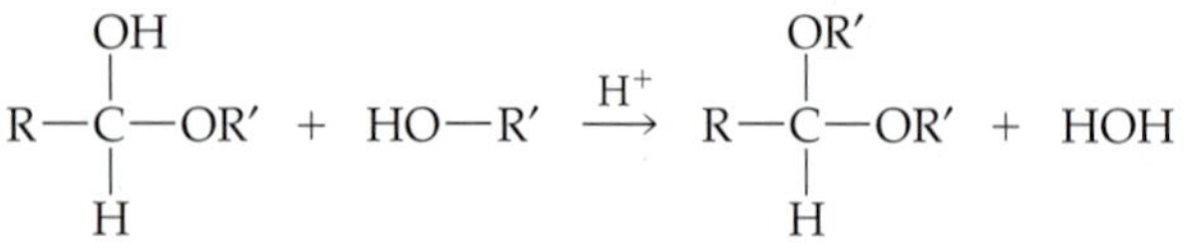

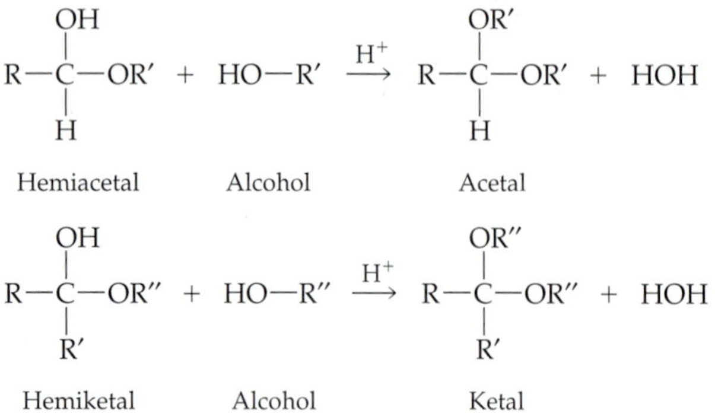

$$\underset{\text{Hemiacetal}}{\overset{\displaystyle \text{OH}}{\underset{\displaystyle \text{H}}{R-C-OR'}}} + \underset{\text{Alcohol}}{HO-R'} \xrightarrow{H^+} \underset{\text{Acetal}}{\overset{\displaystyle \text{OR'}}{\underset{\displaystyle \text{H}}{R-C-OR'}}} + HOH$$

$$\underset{\text{Hemiketal}}{\overset{\displaystyle \text{OH}}{\underset{\displaystyle \text{R'}}{R-C-OR''}}} + \underset{\text{Alcohol}}{HO-R''} \xrightarrow{H^+} \underset{\text{Ketal}}{\overset{\displaystyle \text{OR''}}{\underset{\displaystyle \text{R'}}{R-C-OR''}}} + HOH$$

Alcohol addition can also take place intramolecularly in molecules that contain both a hydroxyl group and a carbonyl group. In this case the product is a *cyclic hemiacetal* or *cyclic hemiketal*.

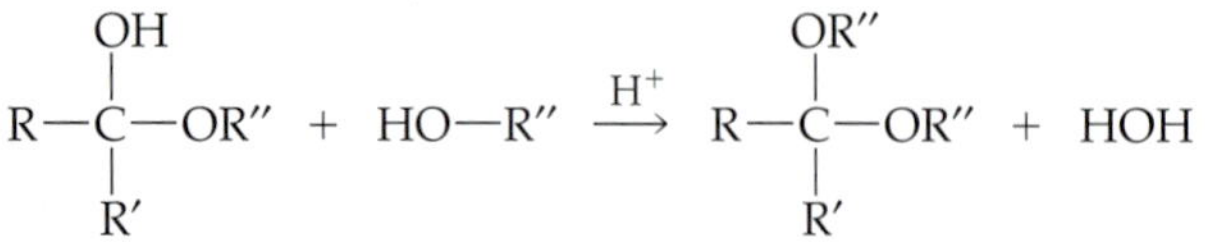

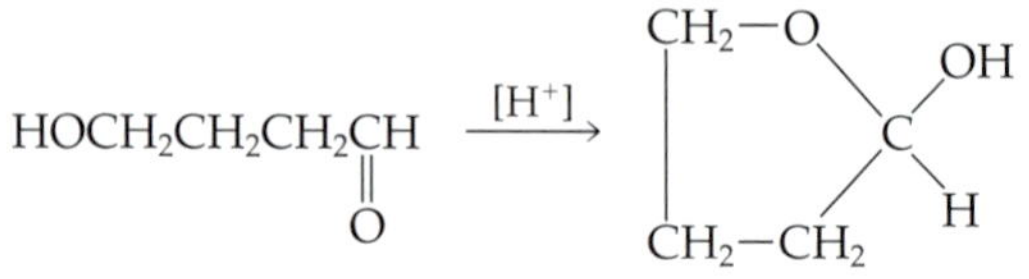

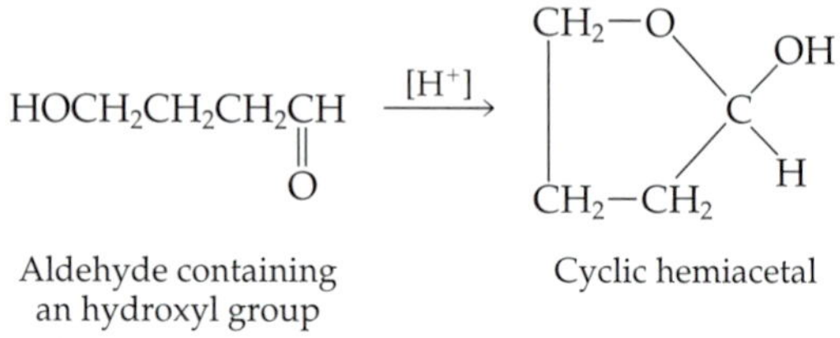

$$\underset{\substack{\text{Aldehyde containing} \\ \text{an hydroxyl group}}}{HOCH_2CH_2CH_2CH{\overset{\displaystyle \parallel}{\underset{\displaystyle O}{}}}} \xrightarrow{[H^+]} \underset{\text{Cyclic hemiacetal}}{\text{(structure)}}$$

Key Terms

acetal (15.6)
aldehyde (15.1)
carbonyl group (15.1)

hemiacetal (15.6)
hemiketal (15.6)
hydrate (15.6)

ketal (15.6)
ketone (15.1)

Review Questions

1. Name three aldehydes or ketones that serve as active principles in flavors or aromas.
2. Account for the fact that the yield when primary alcohols are oxidized to aldehydes is usually lower than the yield when secondary alcohols are oxidized to ketones.
3. Name the three functional groups on the vanillin molecule (Figure 15.1).

4. Name the three functional groups on the testosterone molecule.

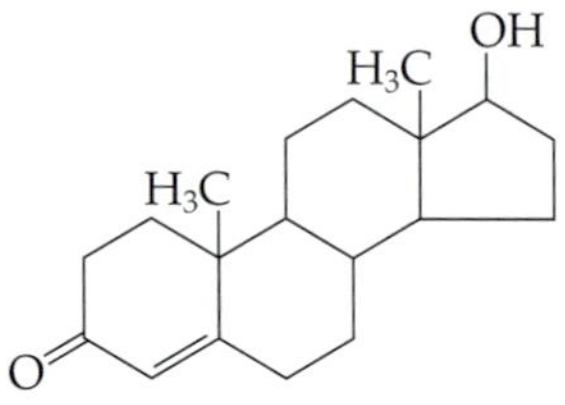

Testosterone

Problems

Names and Structural Formulas

5. Name:

 a. (benzaldehyde structure)

 b. $CH_2OHCH_2C\overset{\displaystyle O}{\underset{\displaystyle H}{}}$

 c. $(CH_3)_3CCH_2CH_2C\overset{\displaystyle O}{\underset{\displaystyle H}{}}$

 d. (2-chlorobenzaldehyde structure)

 c. $(CH_3CH_2)_2CHC\overset{\displaystyle O}{\underset{\displaystyle H}{}}$

 d. (benzene)$-CH_2C(CH_3)_2C\overset{\displaystyle O}{\underset{\displaystyle H}{}}$

6. Name:

 a. $CH_3CH_2CH_2C\overset{\displaystyle O}{\underset{\displaystyle H}{}}$

 b. $CH_3CHClCCl_2C\overset{\displaystyle O}{\underset{\displaystyle H}{}}$

7. Name:

 a. $CH_3CH_2\overset{\displaystyle O}{\overset{\displaystyle \parallel}{C}}CH_2CH(CH_3)_2$ b. (cyclopentanone structure) =O

 c. $CH_3\overset{\displaystyle O}{\overset{\displaystyle \parallel}{C}}CH_2CH_2CH_3$ d. $(CH_3)_3C\overset{\displaystyle O}{\overset{\displaystyle \parallel}{C}}CHBrCH_3$

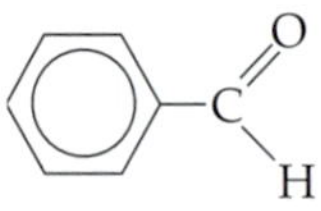

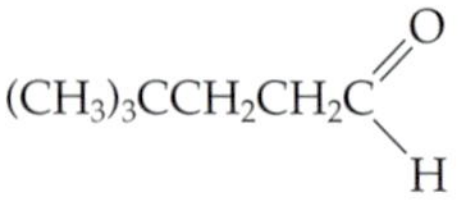

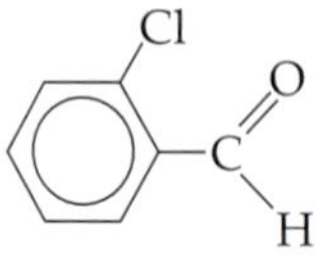

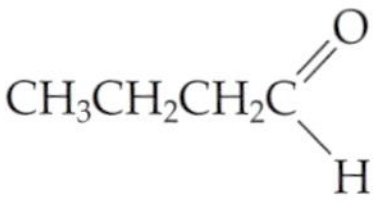

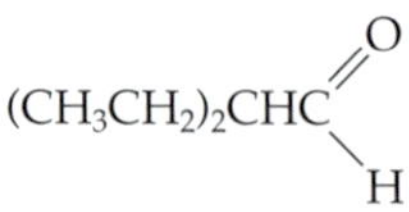

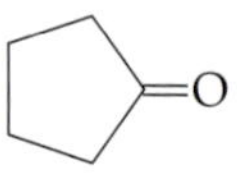

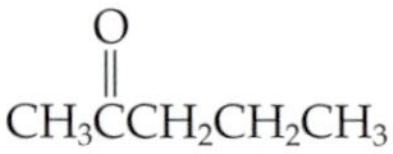

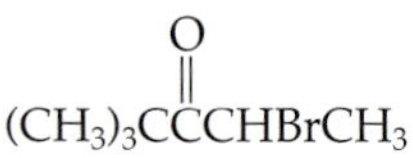

8. Name:

a. $(CH_3)_2CHCCHCl_2$ (with C=O above second C)

b. $CH_3CH_2CH(CH_3)CCH_3$ (with C=O above)

c. (benzene ring)—$CH(CH_3)CCH_3$ (with C=O above)

d. CH_3—(cyclohexane ring with I at top and =O at right)

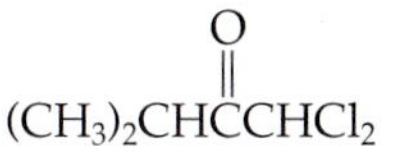

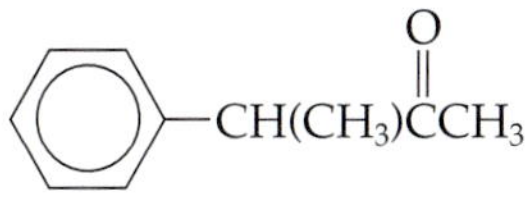

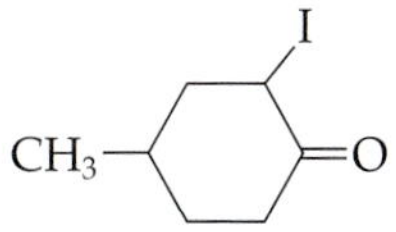

9. Give structural formulas for:
 a. butyraldehyde **b.** 3-methylheptanal
 c. *p*-nitrobenzaldehyde

10. Give structural formulas for:
 a. 5-ethyloctanal **b.** 2-chloropropanal
 c. 2,5-dimethylhexanal

11. Give structural formulas for:
 a. 2-hexanone **b.** 3-bromo–2-heptanone
 c. 4-methylcyclohexanone

12. Give structural formulas for:
 a. 1-phenyl–2-butanone
 b. 2-iodo–2-methyl–4-octanone
 c. 2-hydroxy–3-pentanone

Physical Properties

13. Which compound has the higher boiling point: acetone or 2-propanol?

14. Which compound has the higher boiling point: butanal or 1-butanol?

15. Which compound has the higher boiling point: dimethyl ether or acetaldehyde?

16. Which compound has the higher boiling point: acetone or isobutane?

Preparation of Aldehydes and Ketones

17. Give the structures of the alcohols that could be oxidized to:
 a. 4-methylcyclohexanone
 b. 2,2-dimethylpropanal
 c. 3-bromopentanal

18. Give the structures of the alcohols that could be oxidized to:
 a. 2-pentanone
 b. phenylethanal
 c. *o*-methylbenzaldehyde

Chemical Reactions

19. Write the equations for the reactions of acetaldehyde with:
 a. 1 mol of CH_3OH

 b. 2 mol of CH_3OH, with dry HCl present
 c. 1 mol of $HOCH_2CH_2OH$, with dry HCl present

20. Write the equations for the reactions of acetaldehyde with:
 a. Cu^{2+}
 b. $K_2Cr_2O_7$
 c. hydrogen gas with a nickel catalyst

21. Write the equations for the reactions, if any, of acetone with the reagents in Problem 19.

22. Write the equations for the reactions, if any, of acetone with the reagents in Problem 20.

23. Indicate whether Tollens's reagent could be used to distinguish between the compounds in each set and explain your reasoning.
 a. 1-pentanol and pentanal
 b. 2-pentanol and 2-pentanone
 c. pentanal and 2-pentanone
 d. pentanal and pentane
 e. 2-pentanone and pentane

24. Assume that a *stronger* oxidizing agent, such as $K_2Cr_2O_7$, could be used as a test for distinguishing among compounds. For each set in Problem 23, indicate whether this reagent would distinguish between the two compounds. Explain your reasoning.

25. What reagent would you use to distinguish between 2-pentanone and 2-pentanol? What would you observe when the reagent was added?

26. What reagent would you use to distinguish between 2-pentanone and pentanal? What would you observe when the reagent was added?

27. List the reagents necessary to carry out the following conversions.

a. $CH_3CH_2C\overset{O}{\underset{H}{\big<}}$ $\xrightarrow{?}$ $CH_3CH_2C\overset{O}{\underset{O^-}{\big<}}$ + $Ag(s)$

b $CH_3CH_2CH_2CH_2OH$ $\xrightarrow{?}$ $CH_3CH_2CH_2C\overset{O}{\underset{H}{\big<}}$

c. CH_3—$C\overset{O}{\underset{CH_3}{\big<}}$ $\xrightarrow{?}$ CH_3—$\overset{OCH_3}{\underset{CH_3}{\overset{|}{\underset{|}{C}}}}$—$OCH_3$

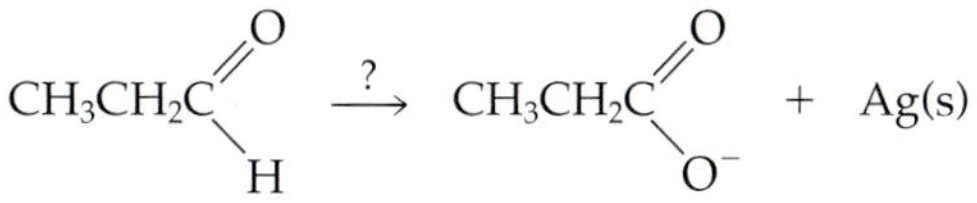

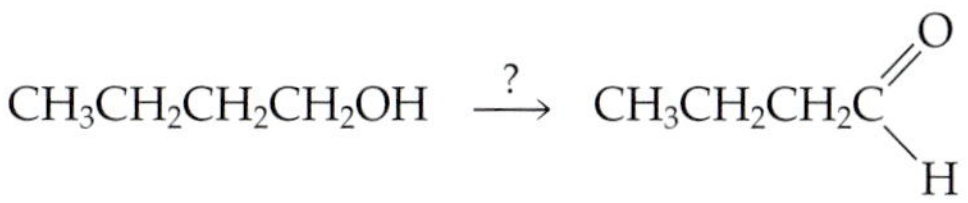

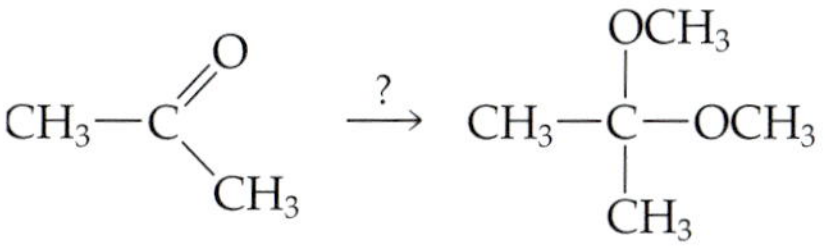

28. List the reagents necessary to carry out the following conversions.

a. $(CH_3)_2CH\overset{OH}{\overset{|}{C}}HCH_3$ $\xrightarrow{?}$ $(CH_3)_2CH\overset{O}{\overset{\|}{C}}CH_3$

b CCl_3—$C\overset{O}{\underset{H}{\big<}}$ $\xrightarrow{?}$ CCl_3—$\overset{OH}{\underset{H}{\overset{|}{\underset{|}{C}}}}$—$OH$

c. (cyclopentanone, ring with =O) $\xrightarrow{?}$ (cyclopentane ring with —OH)

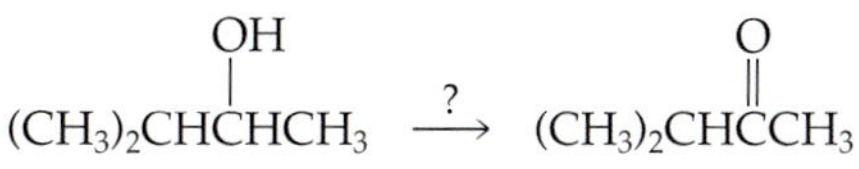

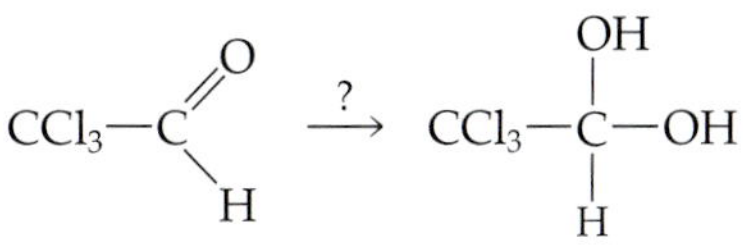

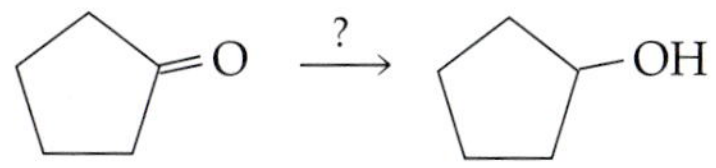

Hydrates, Hemiacetals, and Acetals

29. Which of the following compounds are hemiacetals?

a. CH₃CH₂CHOCH₃
\quad |
\quad OH

b. CH₃CH₂CHOCH₃
\quad |
\quad OCH₃

c. CH₃CH₂CHOH
\quad |
\quad OH

d. ⬡—CHOCH₂CH₃
$\quad\quad\quad$ |
$\quad\quad\quad$ OH

e. ⬡—CH₂CH₂CHOH
$\quad\quad\quad\quad\quad$ |
$\quad\quad\quad\quad\quad$ OH

f. ⬡—CHOCH₂CH₃
$\quad\quad\quad$ |
$\quad\quad\quad$ OCH₂CH₃

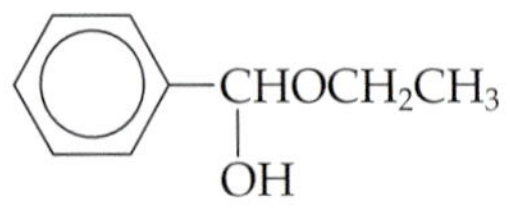

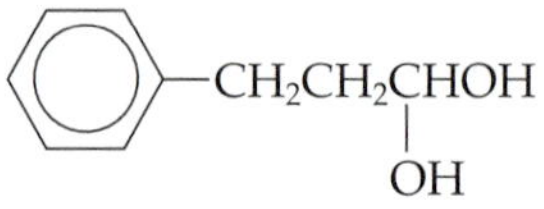

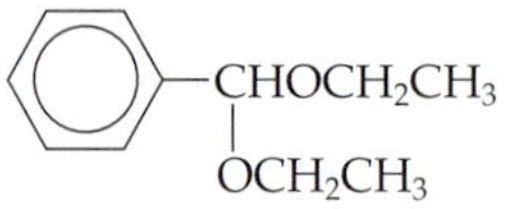

30. Which of the substances in Problem 29 are acetals?
31. Which of the substances in Problem 29 are hydrates?
32. Draw the hemiacetal formed from the intramolecular reaction of

$$HOCH_2CH_2CH_2CH_2C \overset{\displaystyle O}{\underset{\displaystyle H}{\diagdown}}$$

Additional Problems

33. Draw structures, and give common and IUPAC names, for the four isomeric aldehydes having the formula $C_5H_{10}O$.
34. Draw structures, and give common and IUPAC names, for the three isomeric ketones having the formula $C_5H_{10}O$.
35. As we shall see in Chapter 19, 2,3-dihydroxypropanal and 1,3-dihydroxyacetone are important carbohydrates. Draw their structural formulas.
36. Glutaraldehyde (pentanedial) is a germicide that is replacing formaldehyde as a sterilizing agent. It is less irritating to the eyes, nose, and skin. Draw the structural formula of glutaraldehyde.
37. Chloral (CCl_3CHO), which forms a stable hydrate, also forms a stable hemiacetal. Give the structure of the hemiacetal formed by the reaction of chloral with methanol.
38. What is the effective chemical reagent in each of the following?
a. Benedict's solution \quad b. Fehling's solution
c. Tollens's reagent
39. Which of the compounds in Figure 15.1 would give a positive Benedict's test?

16 Carboxylic Acids and Derivatives

The perfume industry is quite dependent on the fragrance of esters.

Organic acids were known long before the inorganic acids were isolated. We studied some inorganic acids (HCl and H_2SO_4) first; however, primitive tribes were more familiar with organic acids, such as the acetic acid they obtained when their fermentation reactions went awry and produced vinegar instead of alcohol. Naturalists of the seventeenth century knew that the sting of a red ant's bite was due to an organic acid which that pest injected into the wound. And it was long recognized that the crisp, tart flavor of citrus fruits was produced by an organic compound appropriately called citric acid. The acetic acid of vinegar, the formic acid of red ants, and the citric acid of fruits all belong to the same family of compounds, the carboxylic acids.

A number of derivatives of carboxylic acids are also important. The amides, of which proteins (Chapter 21) are perhaps the most spectacular example, and the esters, which include fats (Chapter 20), are two classes of acid derivatives we shall consider most carefully. Two synthetic fibers are also classed within these two families of derivatives. Nylon, like silk and wool, is a polyamide. Dacron is a polyester.

In this chapter, we look at simple carboxylic acids, esters, and amides. The more complex worlds of lipids and proteins we shall save for later chapters.

16.1 Carboxylic Acids and Their Derivatives: The Functional Groups

We spoke of the carbonyl group in Chapter 15, and there we noted that it is this functional group that determines the chemistry of the aldehydes and ketones. The carbonyl group is also incorporated in carboxylic acids and their derivatives. However, in these compounds, the carbonyl group is only one part of the functional group that characterizes these families.

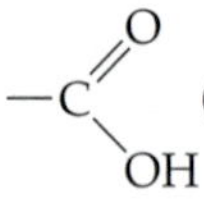
(or —COOH)

The carboxyl group

The functional group of the **carboxylic acids** is the **carboxyl group.** This group can be considered a combination of the *carb*onyl group (>C=O) and the hydr*oxyl* group (—OH), but it has characteristic properties of its own.

The **amide** functional group has nitrogen attached to the carbonyl group. The properties of the amide functional group are different from those of the simple carbonyl group and different from those of simple nitrogen-containing compounds, called amines (Chapter 17).

The functional group of the **esters** looks a little like that of an ether and a little like that of a carboxylic acid. As you should now suspect, compounds in this group react neither like carboxylic acids nor like ethers, but rather make up a distinctive family.

We keep talking about the *derivatives* of carboxylic acids. All of the families we discuss in this chapter, excluding the carboxylic acids themselves, are regarded as derived from the acids. In each case, the hydroxyl group of the acid's functional group is replaced with some other group in the derivative. Table 16.1 gathers all of these functional groups in one location to permit you to compare and contrast them. The table also offers an example (with common and IUPAC names) for each type of compound. We shall consider nomenclature in more detail as we take up each family separately.

Table 16.1 Carboxylic Acid Derivatives

Family	Functional Group	Example	Common Name	IUPAC Name
Carboxylic acid	—C(=O)—OH	$CH_3C(=O)OH$	Acetic acid	Ethanoic acid
Amide	—C(=O)—N<	$CH_3C(=O)NH_2$	Acetamide	Ethanamide
Ester	—C(=O)—O—C<	$CH_3C(=O)OCH_3$	Methyl acetate	Methyl ethanoate

16.2 Some Common Carboxylic Acids: Structures and Names

Most of the organic acids we consider are derived from natural sources. The acids most frequently encountered are known by their common names. Many of these are based upon Latin and Greek names and are related to the source of the acid.

The simplest carboxylic acid is formic acid. It was first obtained by the distillation of ants (Latin *formica*, ant). The bite of an ant smarts because the ant injects formic acid as it bites. The stings of wasps and bees also contain formic acid (as well as other poisonous materials).

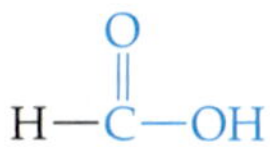

Acetic acid can be made by the aerobic fermentation of a mixture of cider and honey. This produces a solution (vinegar) that contains 4 to 10% acetic acid, plus a number of other compounds that give vinegar its flavor. Acetic acid is probably the most familiar *weak* acid used in educational and industrial chemistry laboratories.

The third member of the homologous series of acids, propionic acid, is seldom encountered in everyday life. The fourth member is more familiar, at least by its odor. If you've ever smelled rancid butter, you probably wish you hadn't—but you know what butyric acid smells like. It is one of the most foul-smelling substances imaginable. Butyric acid can be isolated from butterfat or synthesized in the laboratory. It is one of the ingredients of body odor. Extremely small amounts of this and other chemicals enable bloodhounds to track fugitives.

The acid with the carboxyl group attached directly to a benzene ring is called benzoic acid. In general, carboxylic acids can be represented by the formula RCOOH, where R can be either an alkyl or an aryl group.

Table 16.2 lists several members of the carboxylic acid family and the derivations of their common names. When common names are used, substituted acids are named by locating the position of the substituent group by means of the Greek letters α, β, γ, δ, etc., rather than numbers. These letters refer to the position of the carbon atom in relation to the carboxyl carbon:

$$H-\overset{\overset{\displaystyle O}{\|}}{C}-OH$$

Formic acid

$$CH_3-\overset{\overset{\displaystyle O}{\|}}{C}-OH$$

Acetic acid

$$CH_3CH_2-\overset{\overset{\displaystyle O}{\|}}{C}-OH$$

Propionic acid

$$CH_3CH_2CH_2-\overset{\overset{\displaystyle O}{\|}}{C}-OH$$

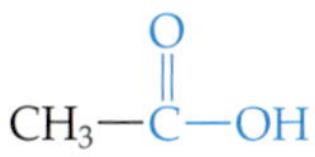

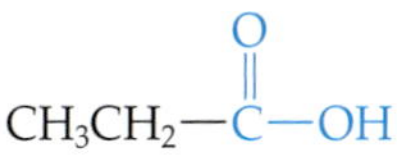

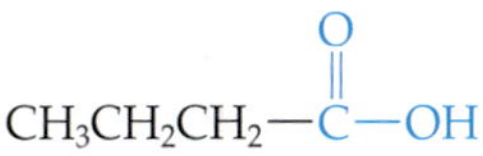

Butyric acid

Benzoic acid

Table 16.2 Some Common Aliphatic Carboxylic Acids

Condensed Formula	IUPAC Name	Common Name	Derivation of Common Name
HCOOH	Methanoic acid	Formic acid	Latin *formica*, ant
CH₃COOH	Ethanoic acid	Acetic acid	Latin *acetum*, vinegar
CH₃CH₂COOH	Propanoic acid	Propionic acid	Greek *protos*, first, and *pion*, fat
CH₃CH₂CH₂COOH	Butanoic acid	Butyric acid	Latin *butyrum*, butter
CH₃(CH₂)₃COOH	Pentanoic acid	Valeric acid	Latin *valere*, powerful
CH₃(CH₂)₄COOH	Hexanoic acid	Caproic acid	
CH₃(CH₂)₆COOH	Octanoic acid	Caprylic acid	Latin *caper*, goat
CH₃(CH₂)₈COOH	Decanoic acid	Capric acid	
CH₃(CH₂)₁₀COOH	Dodecanoic acid	Lauric acid	Laurel tree
CH₃(CH₂)₁₂COOH	Tetradecanoic acid	Myristic acid	*Myristica fragrans* (nutmeg)
CH₃(CH₂)₁₄COOH	Hexadecanoic acid	Palmitic acid	Palm tree
CH₃(CH₂)₁₆COOH	Octadecanoic acid	Stearic acid	Greek *stear*, tallow

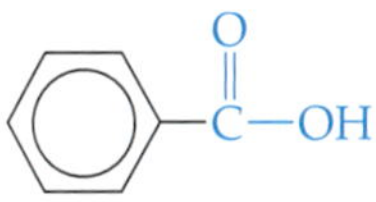

Acetic acid is a familiar laboratory weak acid. It is also the active ingredient in vinegar.

$$C-C-C-C-C\overset{O}{\underset{\delta\quad\gamma\quad\beta\quad\alpha}{\diagdown}}OH$$

$$CH_3-\overset{CH_3}{\underset{H}{\overset{|}{C}}}-C\overset{O}{\diagdown}OH \qquad CH_3-\overset{H}{\underset{OH}{\overset{|}{C}}}-CH_2-C\overset{O}{\diagdown}OH$$

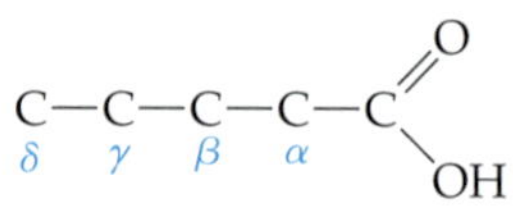

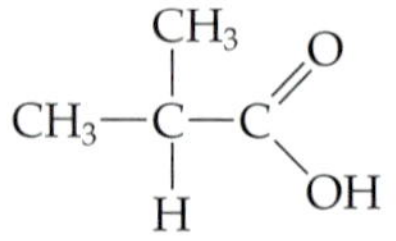

α-Methylpropionic acid β-Hydroxybutyric acid

In the IUPAC system, the parent hydrocarbon is taken to be the one that corresponds to the longest continuous chain containing the carboxyl group. The *-e* ending of the parent alkane is replaced by the suffix *-oic,* and the word *acid* follows. As with aldehydes, the carboxyl carbon is understood to be carbon 1. If substituents are attached to the parent chain of the acid, numbers are used to indicate the substituted carbon. Remember: Greek letters with common names, numbers with IUPAC names.

Example 16.1

Give the common and IUPAC names for

$$CH_3CH_2\underset{\underset{CH_3}{|}}{CH}COOH$$

Solution

The longest continuous chain contains four carbon atoms; the compound is therefore named as a substituted butyric (or butanoic) acid. The methyl substituent is at the alpha carbon in the common system, or at the number 2 carbon in the IUPAC system. The compound is α-methylbutyric acid or 2-methylbutanoic acid.

Practice Exercise

Give the IUPAC name for $CH(CH_3)_2CH_2COOH$.

Example 16.2

Draw the structural formula for α,β-dichloropropionic acid.

Solution

Propionic acid contains three carbons:

$$\overset{\beta}{C}-\overset{\alpha}{C}-C\overset{O}{\diagdown}OH$$

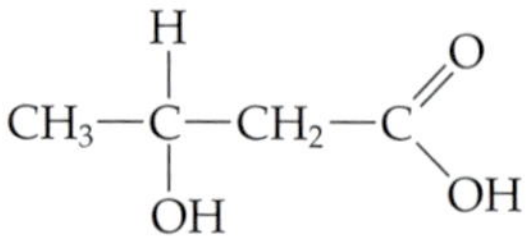

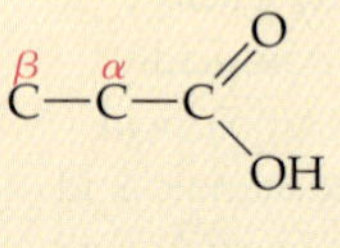

Two chlorine atoms must be attached to the parent chain, one at the alpha carbon and one at the beta carbon:

$$\underset{\overset{\displaystyle |}{H}}{\overset{\overset{\displaystyle Cl}{|}}{H}}-\underset{\overset{\displaystyle |}{H}}{\overset{\overset{\displaystyle Cl}{|}}{C}}-C\overset{\displaystyle O}{\underset{\displaystyle OH}{\diagup}}$$

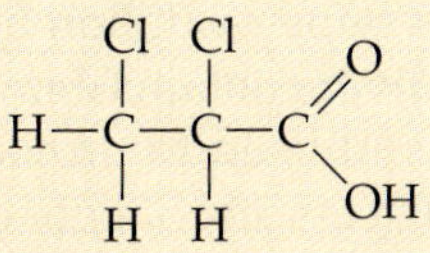

Practice Exercise

Draw a structural formula for 4-bromo–5-methylhexanoic acid.

Derivatives of the aliphatic dicarboxylic acids are very important in biological systems. These acids are almost always referred to by their common names; a mnemonic for remembering these names is given in Table 16.3.

Table 16.3 Aliphatic Dicarboxylic Acids		
Formula	**Common Name**	**Mnemonic**
$\overset{\displaystyle O}{\underset{\displaystyle HO}{\diagdown}}C-C\overset{\displaystyle O}{\underset{\displaystyle OH}{\diagup}}$	Oxalic acid	Oh
$HOOC-CH_2-COOH$	Malonic acid	My
$HOOC-(CH_2)_2-COOH$	Succinic acid	Such
$HOOC-(CH_2)_3-COOH$	Glutaric acid	Good
$HOOC-(CH_2)_4-COOH$	Adipic acid	Apple
$HOOC-(CH_2)_5-COOH$	Pimelic acid	Pie

16.3 Preparation of Carboxylic Acids

Few carboxylic acids occur free in nature. Many aliphatic acids, particularly those containing even numbers of carbon atoms, are found combined with glycerol in fats (Chapter 20). Some of the acids are thus available from the hydrolysis of fats.

Carboxylic acids are the final oxidation products of aldehydes and/or primary alcohols. The acid produced contains the same number of carbon atoms as did the precursor aldehyde or alcohol.

General Equation

$$RCH_2OH \xrightarrow{[O]} R-C\overset{\displaystyle O}{\underset{\displaystyle OH}{\diagup}}$$

$$R-C\overset{\displaystyle O}{\underset{\displaystyle H}{\diagdown}} \xrightarrow{[O]} R-C\overset{\displaystyle O}{\underset{\displaystyle OH}{\diagup}}$$

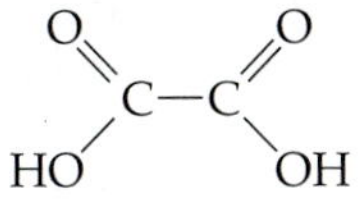

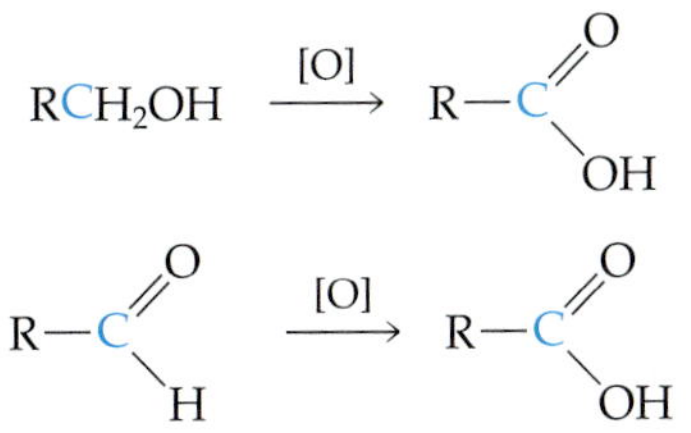

Specific Equation

$$CH_3CH_2OH \xrightarrow[H_2SO_4]{K_2Cr_2O_7} CH_3COOH$$

Ethanol Acetic acid

Our bodies accomplish this same oxidation of alcohols to acids whenever we drink alcoholic beverages. The liver contains enzymes that convert the ethyl alcohol to acetic acid. Acetic acid is utilized to provide energy (Section 25.1), or it is converted to fat (Section 26.3). Excess alcohol not oxidized in the liver continues to circulate in the blood and eventually causes intoxication.

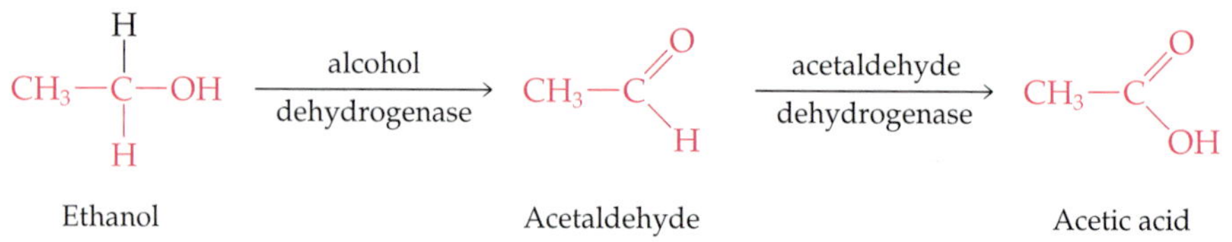
alcohol
dehydrogenase
acetaldehyde
dehydrogenase
Ethanol
Acetaldehyde
Acetic acid

16.4 Physical Properties of Carboxylic Acids

The first nine members of the carboxylic acid series are colorless liquids that have very disagreeable odors. The odor of vinegar is that of acetic acid; the odor of rancid butter is primarily that of butyric acid. Caproic acid is present in the hair and the secretions of goats. The acids from C_5 to C_{10} all have "goaty" odors (odor of Limburger cheese). These acids are produced by the action of skin bacteria on perspiration oils; hence the odor of unaired locker rooms ("essence of old gym sneakers"). The acids above C_{10} are waxlike solids, and because of their low volatility they are practically odorless.

Carboxylic acid molecules are highly polar and form strong intermolecular hydrogen bonds. Consequently, these compounds have higher boiling points than even the alcohols of comparable molar masses. Ethyl alcohol (46 g/mol) boils at 78 °C, while formic acid (same molar mass) boils at 100 °C. Similarly, propyl alcohol (molar mass 60 g/mol) boils at 97 °C, while acetic acid (same molar mass) boils at 118 °C.

There is good evidence that, even in the vapor phase, some of the hydrogen bonds between acid molecules are not broken. The structure of the carboxyl group permits two molecules to hydrogen-bond very strongly to one another:

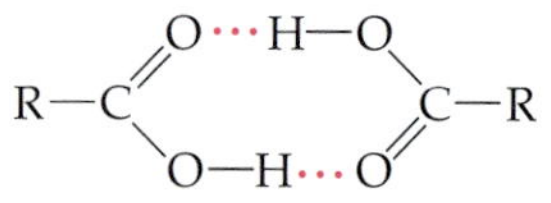

In many situations, the interaction is so strong that the **dimer** (two-molecule unit) acts as a single particle. Osmotic pressures, freezing-point depressions, and other properties are frequently less than one would expect when carboxylic acids are the solute. That is because we are counting as two separate molecules a combination that really is behaving as one molecule.

The carboxyl group readily hydrogen-bonds to water molecules. The acids having one to four carbon atoms are colorless liquids that are completely miscible with water. Solubility decreases with increasing number of carbon atoms: hexanoic acid ($C_6H_{12}O_2$) is soluble only to the extent of 1.0 g per 100 g of water; palmitic acid ($C_{16}H_{32}O_2$) is essentially insoluble. The aromatic carboxylic acids are odorless solids that are sparingly soluble in water. All the acids are soluble in such organic solvents as alcohol, toluene, methylene chloride, and ether.

Table 16.4 lists physical constants for the first 10 members of the aliphatic carboxylic acid family. Notice that the melting points show no regular increase

Table 16.4 Physical Constants of Carboxylic Acids

Formula	Name of Acid	Melting Point (°C)	Boiling Point (°C)	Solubility (g/100 g H_2O)	K_a (25 °C)
HCOOH	Formic	8	100	Miscible	1.8×10^{-4}
CH_3COOH	Acetic	17	118	Miscible	1.8×10^{-5}
CH_3CH_2COOH	Propionic	−22	141	Miscible	
$CH_3(CH_2)_2COOH$	Butyric	−5	163	Miscible	
$CH_3(CH_2)_3COOH$	Valeric	−35	187	5	
$CH_3(CH_2)_4COOH$	Caproic	−3	205	1	
$CH_3(CH_2)_5COOH$	Enanthic	−8	224	0.24	1.5×10^{-5}
$CH_3(CH_2)_6COOH$	Caprylic	16	238	0.07	
$CH_3(CH_2)_7COOH$	Pelargonic	14	254	0.03	
$CH_3(CH_2)_8COOH$	Capric	31	268	0.02	

with increasing molar mass. Pure acetic acid freezes at 16.6 °C. Because this is only slightly below normal room temperature (about 20 °C), acetic acid solidifies when cooled only slightly. In the poorly heated laboratories of a century or so ago in northern North America and Europe, acetic acid often froze on the reagent shelf. For that reason, pure acetic acid (sometimes referred to as concentrated acetic acid) came to be known as *glacial acetic acid*, a name that survives to this day.

16.5 Chemical Properties of Carboxylic Acids: Neutralization

In Chapter 10, we defined an acid as a compound that (1) turns blue litmus red, (2) neutralizes bases, (3) reacts with active metals to give off hydrogen, and (4) tastes sour. Of these four properties, you are probably most familiar with the last. Vinegar is sour because it contains acetic acid. Grapefruits and lemons are sour because they contain citric acid. Sour milk contains lactic acid. The acids we eat are, for the most part, organic acids. Historically, the first organic acids came from plant or animal matter—that is, from organisms. Now many organic acids are synthesized in the laboratory from petroleum products.

Those carboxylic acids that are water-soluble form moderately acidic solutions. They will change litmus from blue to red. All carboxylic acids, whether water-soluble or not, react with aqueous solutions of sodium hydroxide, sodium carbonate, and sodium bicarbonate to form salts:

$$RCOOH + NaOH(aq) \longrightarrow RCOO^- Na^+(aq) + H_2O$$
$$2\,RCOOH + Na_2CO_3(aq) \longrightarrow 2\,RCOO^- Na^+(aq) + H_2O + CO_2$$
$$RCOOH + NaHCO_3(aq) \longrightarrow RCOO^- Na^+(aq) + H_2O + CO_2$$

In these reactions, the carboxylic acids act just the way inorganic acids act: they neutralize basic compounds. With solutions of carbonate and bicarbonate ions, they also form carbon dioxide gas.

The carboxylic acids are weak acids, and therefore tend to ionize only slightly in aqueous solution:

$$RCOOH + H_2O \rightleftharpoons RCOO^- + H_3O^+$$

In order to distinguish between degrees of weakness, we can order some organic compounds (and some inorganic ones) according to their relative acidities:

Strongest acid Weakest acid

$$H_2SO_4, HNO_3, HCl \ > \ RCOOH \ > \ H_2CO_3 \ > \ ArOH \ > \ H_2O \ > \ ROH \ > \ RH$$

| Mineral acids | Carboxylic acids | Carbonic acid | Phenols | Water | Alcohols | Alkanes |

Water is frequently used as a dividing line for acidity and basicity. Because we live in a water world, we quite naturally regard water as neutral. In aqueous solutions, if something is more acidic than water, it is treated as an acid (as are phenols and carboxylic acids). If a compound is less acidic than water (as are alcohols and alkanes), it is not regarded as an acid. Neither alcohols nor alkanes affect the pH of aqueous solutions.

Because of the difference in relative acidities among organic compounds, solubility behavior is often an identifying feature of carboxylic acids. Carboxylic acids that are insoluble in water dissolve in aqueous hydroxide, carbonate, or bicarbonate because the insoluble acids react to form ionic salts that are water-soluble. Solution in aqueous sodium bicarbonate, with the formation of carbon dioxide bubbles, is characteristic of carboxylic acids.

Example 16.3

Write equations for the reactions of decanoic acid with (a) NaOH and (b) $NaHCO_3$.

Solution

a. $CH_3(CH_2)_8C$(=O)(OH) $+ \ NaOH \ \longrightarrow \ CH_3(CH_2)_8C$(=O)($O^- Na^+$) $+ \ HOH$

b. $CH_3(CH_2)_8C$(=O)(OH) $+ \ NaHCO_3 \ \longrightarrow \ CH_3(CH_2)_8C$(=O)($O^- Na^+$) $+ \ HOH \ + \ CO_2$

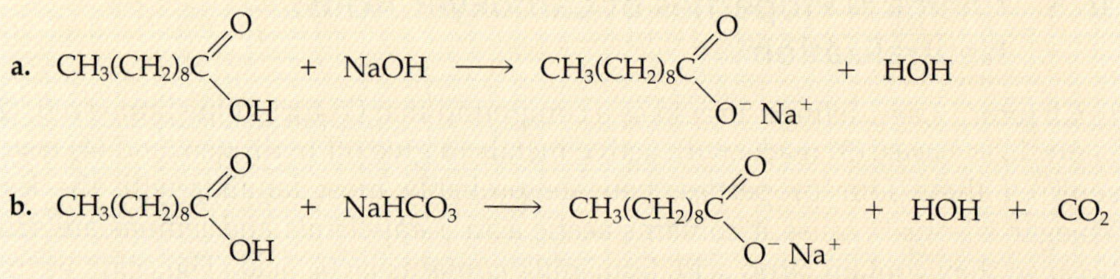

Practice Exercise

Write equations for the reactions of benzoic acid with (a) NaOH and (b) $NaHCO_3$.

Organic salts are named in the same manner as the inorganic salts: name of cation followed by name of organic anion. The name of the anion is obtained by dropping the -ic ending of the acid name and replacing it with the suffix -ate. This rule applies whether we are using common names or IUPAC names:

CH_3C(=O)($O^- Li^+$)

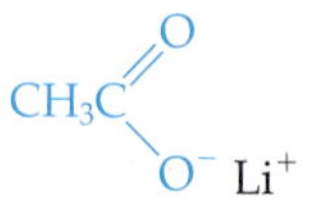

Lithium acetate
(Lithium ethanoate)

$CH_3CH_2CH_2C$(=O)($O^- K^+$)

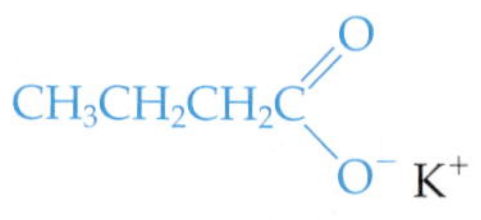

Potassium butyrate
(Potassium butanoate)

C_6H_5—C(=O)($O^- Na^+$)

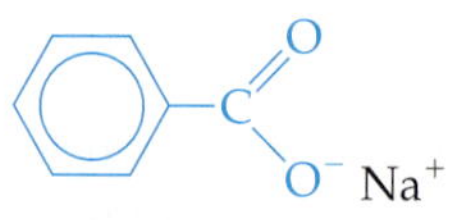

Sodium benzoate

The sodium or potassium salts of long-chain carboxylic acids are called soaps. We discuss the chemistry of soaps in considerable detail in Section 20.5.

$$CH_3CH_2CH_2CH_2CH_2CH_2CH_2CH_2CH_2CH_2CH_2CH_2CH_2CH_2CH_2CH_2CH_2 \!-\! \overset{\displaystyle O}{\underset{\displaystyle O^- \; Na^+}{C}}$$

Sodium stearate (a soap)

Other organic salts that have commercial significance are those used as preservatives. They act to prevent spoilage by inhibiting the growth of bacteria and fungi. Calcium and sodium propionate are added to processed cheese and bakery goods; sodium benzoate is used as a preservative in cider, jellies, pickles, and syrups; and sodium and potassium sorbate are added to fruit juices, sauerkraut, soft drinks, and wine. (Look for these salts on ingredient labels the next time you are shopping.)

$$\left(CH_3CH_2\overset{\displaystyle O}{\underset{\displaystyle O^-}{C}} \right)_2 Ca^{2+} \qquad\qquad CH_3CH{=}CHCH{=}CH\overset{\displaystyle O}{\underset{\displaystyle O^- \; K^+}{C}}$$

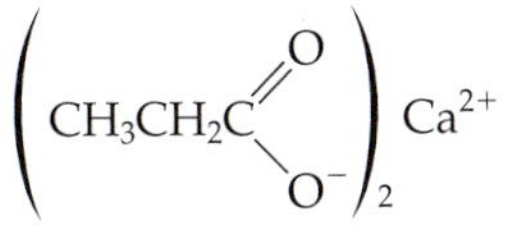

Calcium propionate Potassium sorbate

16.6 An Ester by Any Other Name . . .

Esters are perhaps the most important derivatives of the carboxylic acids. The general formula for an ester is RCOOR′, where R may be a hydrogen, an alkyl group, or an aryl group and R′ may be alkyl or aryl but *not* hydrogen.

Esters are widely distributed in nature. Their occurrence is particularly important in fats and vegetable oils, which are esters of long-chain fatty acids and glycerol (Chapter 20). Esters of phosphoric acid are of the utmost importance to life, and they are enumerated in Section 16.10.

Esters are named in a manner similar to that used for naming organic salts. The group name of the alkyl or aryl portion is given first and is followed by the name of the acid portion. In both common and IUPAC nomenclature, the *-ic* ending of the parent acid is replaced by the suffix *-ate* (Table 16.5).

Example 16.4
Give the common and IUPAC names for

$$CH_3C\overset{\displaystyle O}{\underset{\displaystyle OCH_3}{\big\langle}}$$

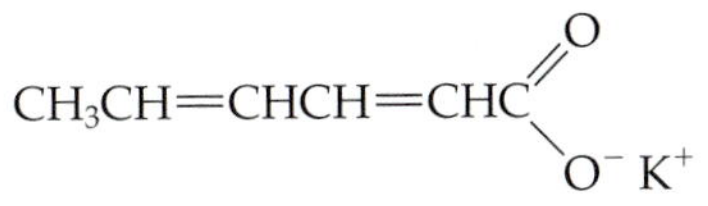

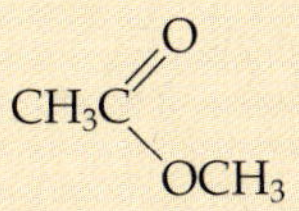

Solution
The alkyl group attached to oxygen is methyl. The

$$CH_3C\overset{\displaystyle O}{\underset{\displaystyle O-}{\big\langle}}$$

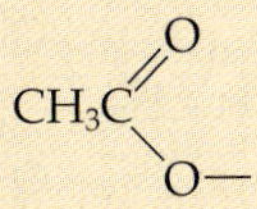

is derived from acetic acid (which has two carbons), and so its name is acetate. The compound is methyl acetate. The IUPAC name for the two-carbon acid unit is ethanoate, and so the IUPAC name is methyl ethanoate.

Table 16.5 Nomenclature of Esters		
Formula	**Common Name**	**IUPAC Name**
H—C(=O)O—CH$_3$	Methyl formate	Methyl methanoate
CH$_3$—C(=O)O—CH$_3$	Methyl acetate	Methyl ethanoate
CH$_3$—C(=O)O—CH$_2$CH$_3$	Ethyl acetate	Ethyl ethanoate
CH$_3$CH$_2$C(=O)O—CH$_2$CH$_3$	Ethyl propionate	Ethyl propanoate
CH$_3$CH$_2$CH$_2$C(=O)O—CH$_2$CH$_2$CH$_3$	Propyl butyrate	Propyl butanoate
CH$_3$CH$_2$CH$_2$C(=O)O—CHCH$_3$ (CH$_3$)	Isopropyl butyrate	Isopropyl butanoate
C$_6$H$_5$—C(=O)O—CH$_2$CH$_3$	Ethyl benzoate	Ethyl benzoate
CH$_3$C(=O)O—C$_6$H$_5$	Phenyl acetate	Phenyl ethanoate

Example 16.5

Give the common and IUPAC names for

$$CH_3CH_2C(=O)OCH_2CH_2CH_3$$

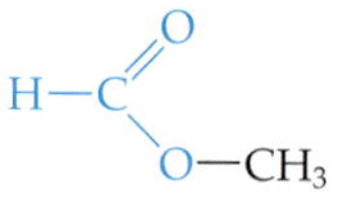

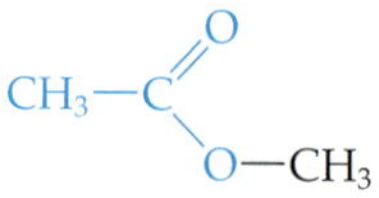

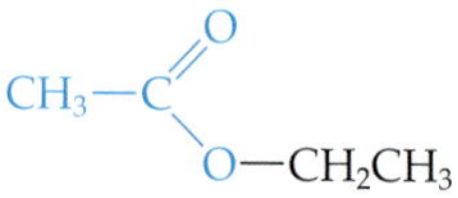

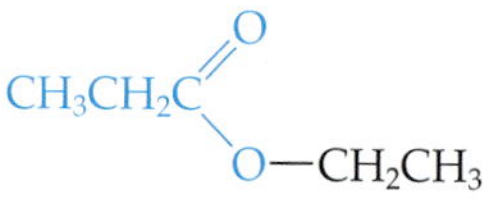

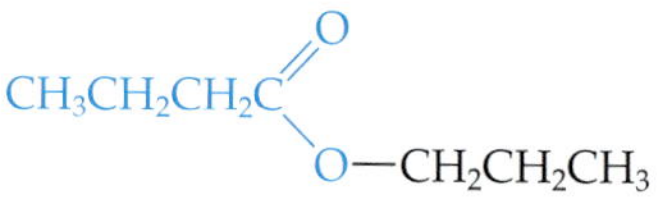

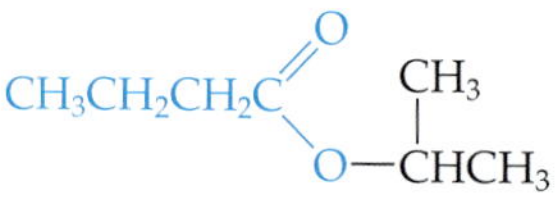

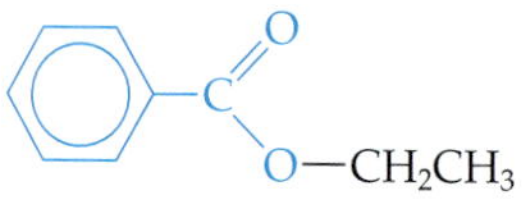

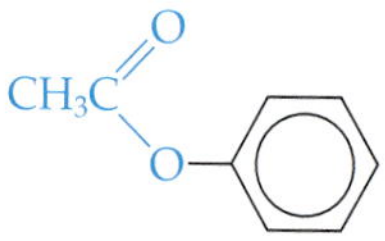

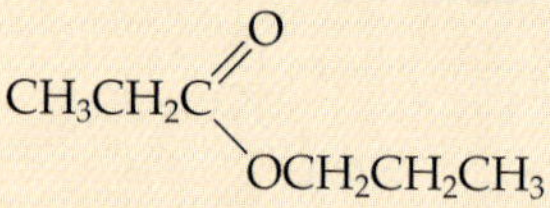

Solution

The alkyl group attached directly to oxygen is propyl. The part of the molecule derived from the acid,

$$CH_3CH_2C(=O)O—$$

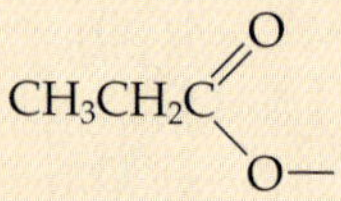

has three carbon atoms. It is thus called propionate or, by IUPAC terminology, propanoate. The common name of the ester is therefore propyl propionate, and the IUPAC name is propyl propanoate.

Practice Exercise

Give the IUPAC name for

CH$_3$O–C(=O)–C$_6$H$_5$

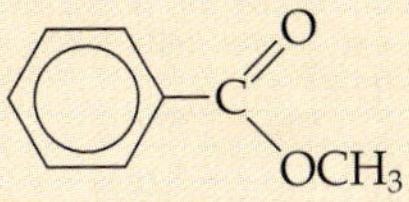

Example 16.6

What is the name of this ester?

C$_6$H$_5$–C(=O)–O–C$_6$H$_5$

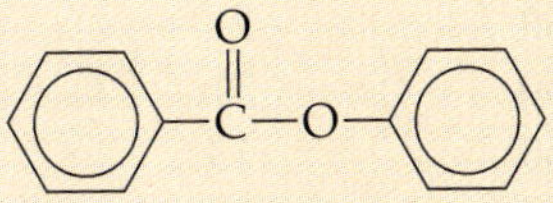

Solution

The group attached to oxygen is phenyl. The acid portion corresponds to the benzoate group (from benzoic acid). Therefore, the compound is phenyl benzoate.

Example 16.7

Draw the structure for ethyl pentanoate.

Solution

It is easier to start with the acid portion. Draw the pentanoate (five-carbon) group first:

$$CH_3CH_2CH_2CH_2C(=O)O-$$

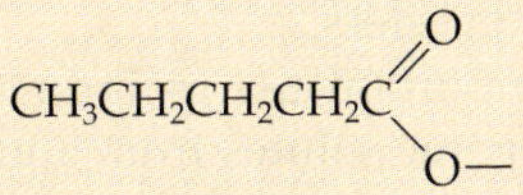

Then simply attach the ethyl group to the bond that ordinarily holds the hydrogen in the carboxylic acid:

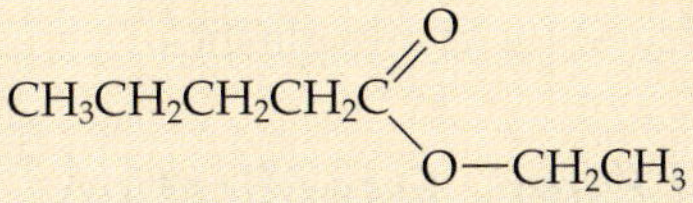

$$CH_3CH_2CH_2CH_2C(=O)O-CH_2CH_3$$

Practice Exercise

Draw the structure for phenyl butanoate.

16.7 Physical Properties of Esters

Unlike the carboxylic acids from which they are derived, the esters generally have pleasant odors and are often responsible for the characteristic fragrances of fruits and flowers. Once a flower or fruit has been chemically analyzed, flavor chemists can attempt to duplicate the natural odor or taste. They are seldom completely successful, but they often get close enough for practical purposes. Esters are used in the manufacture of perfumes and as flavoring agents in the confectionary and

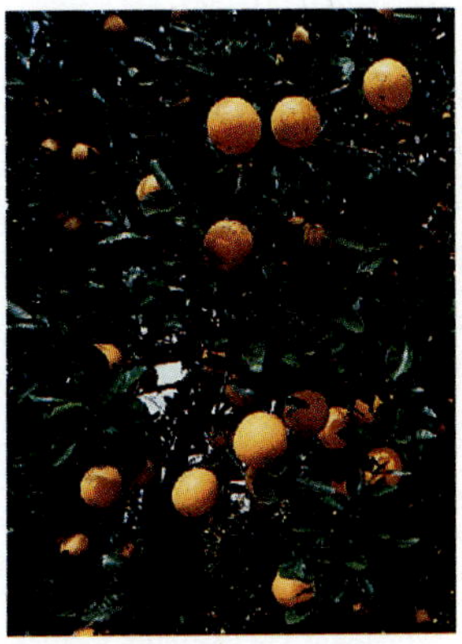

The distinctive aroma and flavor of oranges are due in part to octyl acetate, an ester formed from 1-octanol (octyl alcohol) and acetic acid.

Table 16.6 Physical Properties of Esters

Formula	Name	Molar Mass	Melting Point (°C)	Boiling Point (°C)	Aroma
$HCOOCH_3$	Methyl formate	60	−99	32	
$HCOOCH_2CH_3$	Ethyl formate	74	−80	54	Rum
CH_3COOCH_3	Methyl acetate	74	−98	57	
$CH_3COOCH_2CH_3$	Ethyl acetate	88	−84	77	
$CH_3CH_2COOCH_3$	Methyl propionate	88	−88	80	
$CH_3CH_2COOCH_2CH_3$	Ethyl propionate	102	−74	99	
$CH_3CH_2CH_2COOCH_3$	Methyl butyrate	102	−85	102	Apple
$CH_3CH_2CH_2COOCH_2CH_3$	Ethyl butyrate	116	−101	121	Pineapple
$CH_3COO(CH_2)_4CH_3$	Amyl acetate	130	−71	148	Banana
$CH_3COOCH_2CH_2CH(CH_3)_2$	Isoamyl acetate	130	−79	142	Pear
$CH_3COOCH_2C_6H_5$	Benzyl acetate	150	−51	215	Jasmine
$CH_3CH_2CH_2COO(CH_2)_4CH_3$	Amyl butyrate	158	−73	185	Apricot
$CH_3COO(CH_2)_7CH_3$	Octyl acetate	172	−39	210	Orange

soft drink industries. (A mixture of nine esters is used to produce an artificial raspberry flavor.)

Ester molecules are polar but incapable of forming intermolecular hydrogen bonds with one another. Esters thus have considerably lower boiling points than the isomeric carboxylic acids. As one might expect, the boiling points of esters are about intermediate between those of ketones and ethers of comparable molar mass. Because ester molecules can form hydrogen bonds with water molecules, esters of low molar mass are somewhat water-soluble. Borderline solubility occurs in those molecules that have three to five carbon atoms. Table 16.6 lists the physical properties of some common esters.

Esters find their most important use as industrial solvents. Ethyl acetate is commonly used to extract organic solutes from aqueous solutions. It is also used as a nail polish remover and is a major constituent of some paint removers. Cellulose nitrate is dissolved in ethyl acetate and butyl acetate to form lacquers. The solvent evaporates as the lacquer "dries," leaving a thin protective film on the surface to which the lacquer was applied. Esters having high boiling points are used as softeners (plasticizers) for brittle plastics.

16.8 Preparation of Esters: Esterification

Some esters are prepared by direct esterification of the carboxylic acid. This conversion is accomplished by heating a carboxylic acid with an alcohol in the presence of a mineral acid catalyst:

$$R-\overset{\displaystyle O}{\underset{\displaystyle OH}{C}} \ + \ R'-OH \ \overset{H^+}{\rightleftharpoons} \ R-\overset{\displaystyle O}{\underset{\displaystyle OR'}{C}} \ + \ H_2O$$

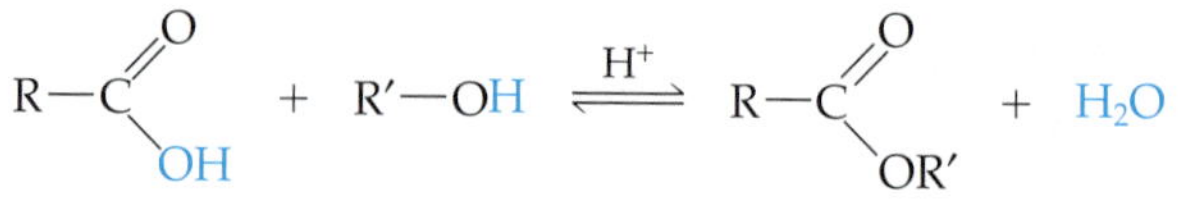

An alcohol molecule condenses with an acid molecule, splitting out water to form an ester. The reaction is reversible, and it soon comes to equilibrium. Experimental data reveal that, for any esterification reaction, the composition of the equilibrium mixture is governed by the value of the equilibrium constant. Consideration of the law of chemical equilibrium and Le Châtelier's principle (Section 5.13) allows predetermination of reaction conditions that produce a maximum yield of the desired ester.

If the reaction involves an inexpensive alcohol, such as methanol, excess alcohol can be used to drive the reaction toward completion. In the preparation of methyl benzoate, for example, 10 mol of CH_3OH may be used for each mole of benzoic acid:

$$\text{(benzoic acid)} \; + \; CH_3OH \; \underset{}{\overset{H^+}{\rightleftharpoons}} \; \text{(methyl benzoate)} \; + \; H_2O$$

(large excess)

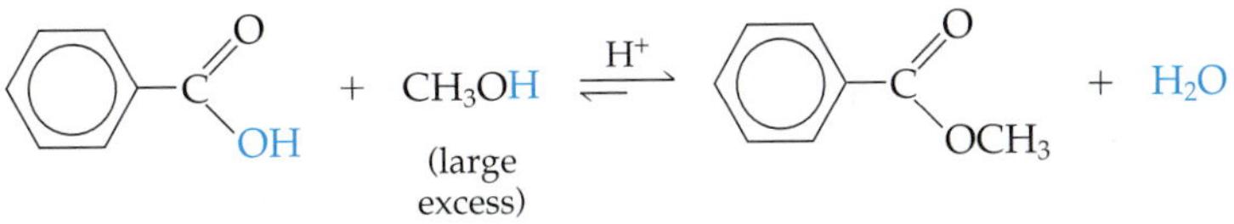

Forcing the reaction in this way can give a 75% yield of methyl benzoate.

Similarly, if the acid is cheap, it can be employed in excess. In the preparation of butyl acetate, acetic acid is used in a molar ratio of 2:1 (or greater):

$$CH_3C\!\!\begin{array}{c}O\\OH\end{array} \; + \; CH_3CH_2CH_2CH_2OH \; \overset{H^+}{\rightleftharpoons} \; CH_3C\!\!\begin{array}{c}O\\OCH_2CH_2CH_2CH_3\end{array} \; + \; H_2O$$

(excess)

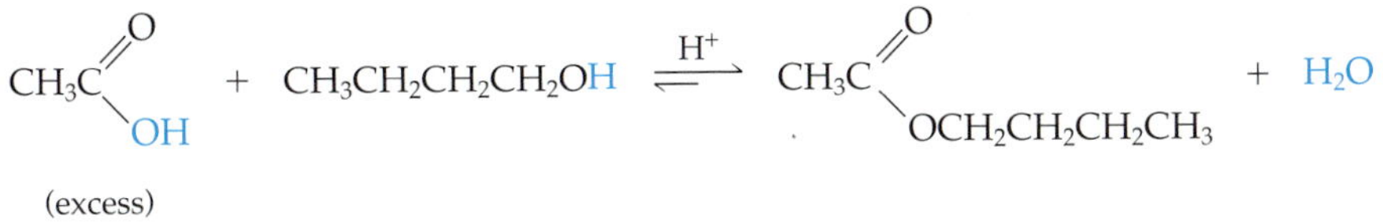

A third method of driving esterification toward completion involves removal of the product water as it is formed. This is easily accomplished if the acid, the alcohol, and the ester all boil at temperatures well above 100 °C:

$$CH_3CH_2CH_2C\!\!\begin{array}{c}O\\OH\end{array} \; + \; CH_3CH_2CH_2CH_2OH \; \overset{H^+}{\rightleftharpoons} \; CH_3CH_2CH_2C\!\!\begin{array}{c}O\\OCH_2CH_2CH_2CH_3\end{array} \; + \; H_2O$$

| Butyric acid (bp 164 °C) | Butyl alcohol (bp 118 °C) | Butyl butyrate (bp 165 °C) | Distilled from mixture (bp 100 °C) |

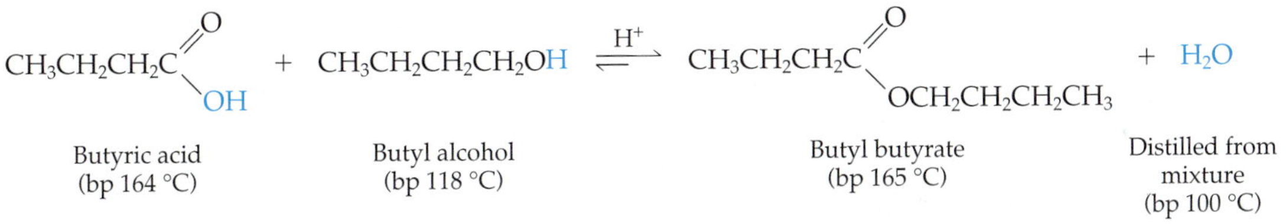

In general, the reaction between an acid and an alcohol is extremely slow, and a catalyst must be employed to speed it up. Sulfuric acid is commonly used because it is both an acid and a good dehydrating agent. Thus it serves both to increase the rate of reaction (by acting as a catalyst), and to shift the equilibrium to the right (by reacting with water to form hydronium ions).

A commercially important esterification reaction occurs between a dicarboxylic acid and a dialcohol. Such a reaction yields an ester that contains a free carboxyl group at one end and a free alcohol group at the other end. Hence further condensation reactions can occur to produce **polyester** polymers. The most significant polyester, polyethylene terephthalate (PET), is made from terephthalic acid and ethylene glycol monomers.

$$n \; HO-CH_2CH_2-OH \; + \; n \; HO-\overset{O}{\underset{}{C}}-\text{(benzene)}-\overset{O}{\underset{}{C}}-OH \; \longrightarrow \; \left[\!O-CH_2CH_2-O-\overset{O}{\underset{}{C}}-\text{(benzene)}-\overset{O}{\underset{}{C}}\right]_n \; + \; 2n \; H_2O$$

| Ethylene glycol | Terephthalic acid | Polyethylene terephthalate (a polyester) |

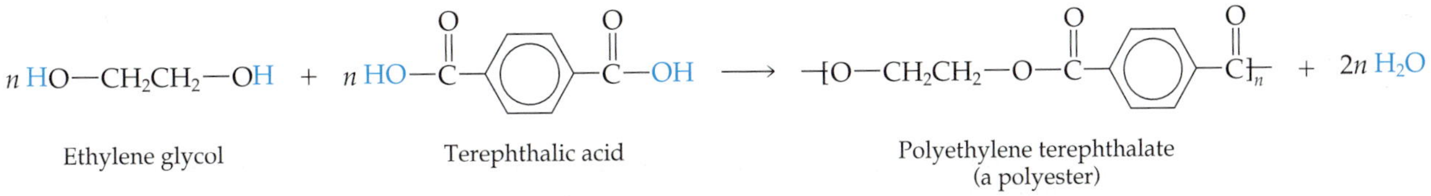

The polyester molecules make excellent fibers (Figure 16.1) that are spun into thread or yarn and marketed under the trade names Dacron or Fortrel, depending

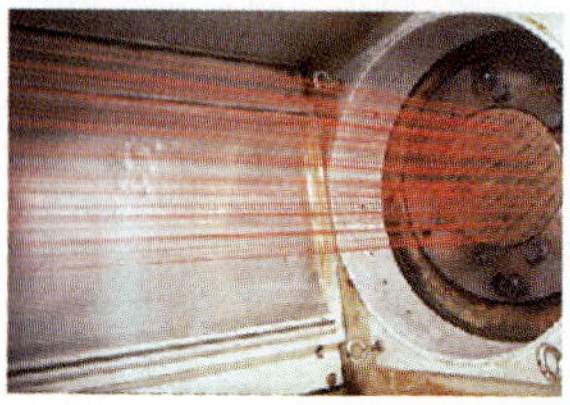

on the manufacturer. Dacron polyester is found in permanent press garments, carpets, tires, and many other products. Dacron is inert and is used in surgery (in the form of a mesh) to repair or replace diseased sections of blood vessels. The same polyester, when formed as a film rather than fiber, is called Mylar. When magnetically coated, Mylar tape is used in audio and video cassettes.

16.9 Chemical Properties of Esters: Hydrolysis

Esters are neutral compounds, unlike the acids from which they are formed. Esters typically undergo chemical reactions in which the alkoxy ($-OR'$) group is replaced by another group. One such reaction is **hydrolysis,** or splitting with water. Hydrolysis is catalyzed by either acid or base. Acidic hydrolysis is simply the reverse of esterification. The ester is refluxed with a large excess of water containing a strong acid catalyst. However, the equilibrium is unfavorable for ester hydrolysis, and so the reaction never goes to completion.

General Equation

$$R-\overset{\displaystyle O}{\underset{\displaystyle OR'}{C}} \ + \ HOH \ \underset{\Delta}{\overset{H^+}{\rightleftharpoons}} \ R-\overset{\displaystyle O}{\underset{\displaystyle OH}{C}} \ + \ R'OH$$

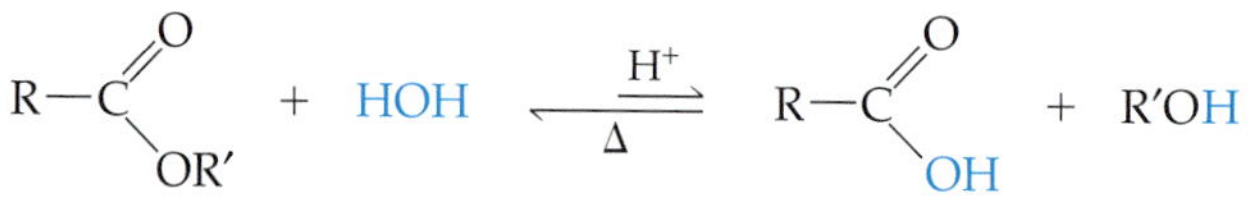

Specific Equation

$$\text{C}_6\text{H}_5-\overset{\displaystyle O}{\underset{\displaystyle OCH_3}{C}} \ + \ HOH \ \underset{\Delta}{\overset{H^+}{\rightleftharpoons}} \ \text{C}_6\text{H}_5-\overset{\displaystyle O}{\underset{\displaystyle OH}{C}} \ + \ CH_3OH$$

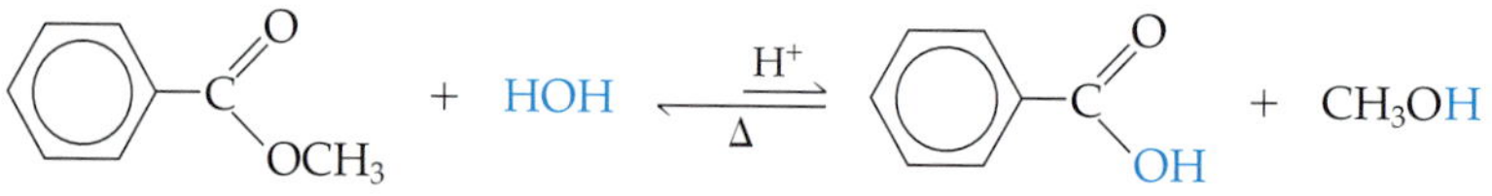

Methyl benzoate Benzoic acid Methanol

Example 16.8

Write an equation for the acid hydrolysis of ethyl acetate.

Solution

Remember that in acid hydrolysis, water splits the ester bond. The H of water joins to the oxygen in the $-OR$ part of the original ester, and the OH joins to the carbonyl carbon:

$$CH_3\overset{\displaystyle O}{\underset{\displaystyle OCH_2CH_3}{C}} \ + \ HOH \ \overset{H^+}{\rightleftharpoons} \ CH_3\overset{\displaystyle O}{\underset{\displaystyle OH}{C}} \ + \ CH_3CH_2OH$$

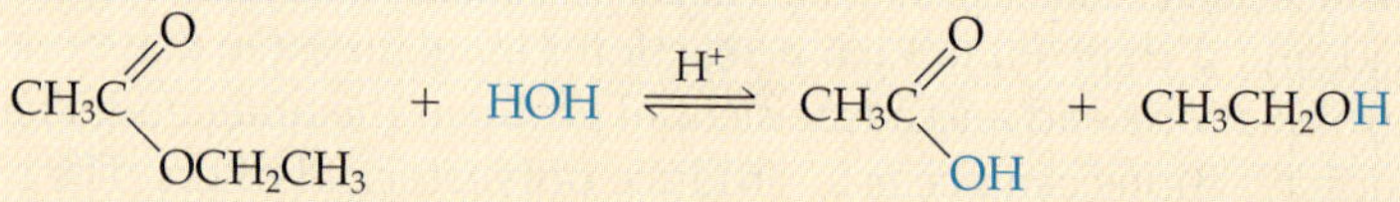

Ethyl acetate Acetic acid Ethanol

Practice Exercise

Write an equation for the acid hydrolysis of isopropyl propanoate.

When a base (such as sodium hydroxide or potassium hydroxide) is used to hydrolyze an ester, the reaction goes to completion because the carboxylic acid is removed from the equilibrium by its conversion to a salt. Organic salts do not react with alcohols, so the reaction is essentially irreversible. Accordingly, ester hydrolysis is usually carried out in basic solution. Because soaps are prepared by the alkaline hydrolysis of fats and oils, the term **saponification** (Latin *sapon*, soap; *facere*, to make) is used to describe the alkaline hydrolysis of all esters (see also Section 20.5). Note that in the saponification reaction, the base is a reactant and thus is not a catalyst:

General Equation

$$R-C\underset{OR'}{\overset{O}{\vphantom{x}}} \quad + \quad NaOH \quad \xrightarrow{\Delta} \quad R-C\underset{O^-\,Na^+}{\overset{O}{\vphantom{x}}} \quad + \quad R'OH$$

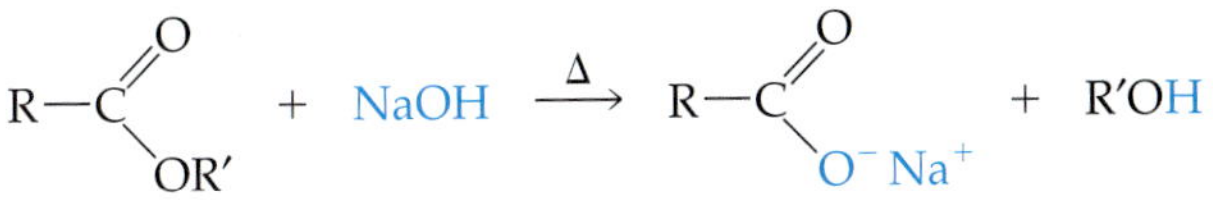

Specific Equation

$$CH_3-C\underset{OCH_2CH_3}{\overset{O}{\vphantom{x}}} \quad + \quad NaOH \quad \xrightarrow{\Delta} \quad CH_3-C\underset{O^-\,Na^+}{\overset{O}{\vphantom{x}}} \quad + \quad CH_3CH_2OH$$

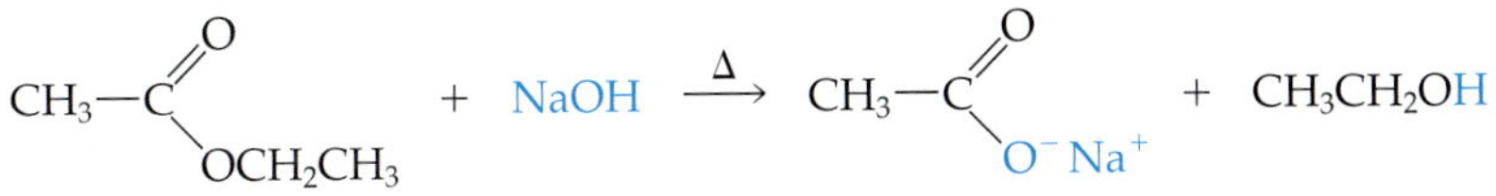

 Ethyl acetate Sodium acetate Ethanol

Example 16.9

Write an equation for the hydrolysis of methyl benzoate in a potassium hydroxide solution.

Solution

In basic hydrolysis, the molecule of base splits the ester linkage. The hydrogen of the base joins the oxygen of the split-off group, and the oxy cation of the base joins the carbonyl carbon:

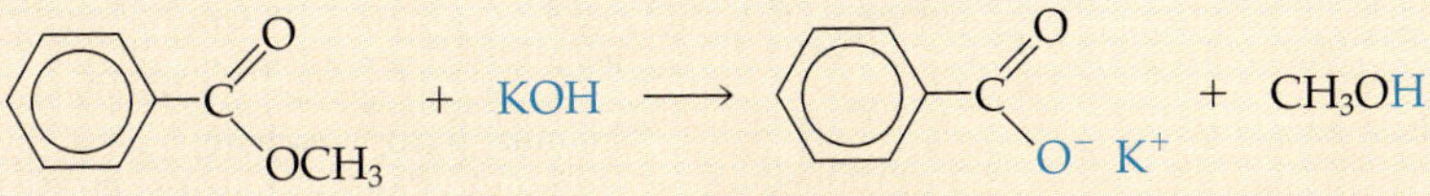

Practice Exercise

Write an equation for the hydrolysis of phenyl methanoate in a sodium hydroxide solution.

16.10 Esters of Phosphoric Acid

Esters can also be prepared by reacting alcohols with inorganic acids (e.g., HNO_3, H_2SO_4, H_3PO_4). Esters formed in this way, called *inorganic esters*, do not contain the C=O group; instead they contain an N=O, S=O, or P=O group. In Section 14.7 we mentioned one such ester—nitroglycerin (glycerol trinitrate), which is the ester formed from glycerol and nitric acid. Perhaps the most important inorganic esters in biochemistry are those of phosphoric acid and two of its anhydrides, pyrophosphoric acid[1] and triphosphoric acid. Phosphate or pyrophosphate esters are present in every plant and animal cell. They are biochemical intermediates in the transformation of food into usable energy (in the form of ATP). Phosphate esters

[1] Pyrophosphoric acid is an anhydride that can be considered as being formed when one molecule of water is removed from two molecules of phosphoric acid:

$$HO-\underset{\underset{OH}{|}}{\overset{\overset{O}{\|}}{P}}-OH \quad + \quad HO-\underset{\underset{OH}{|}}{\overset{\overset{O}{\|}}{P}}-OH \quad \xrightarrow{heat} \quad HO-\underset{\underset{OH}{|}}{\overset{\overset{O}{\|}}{P}}-O-\underset{\underset{OH}{|}}{\overset{\overset{O}{\|}}{P}}-OH \quad + \quad HOH$$

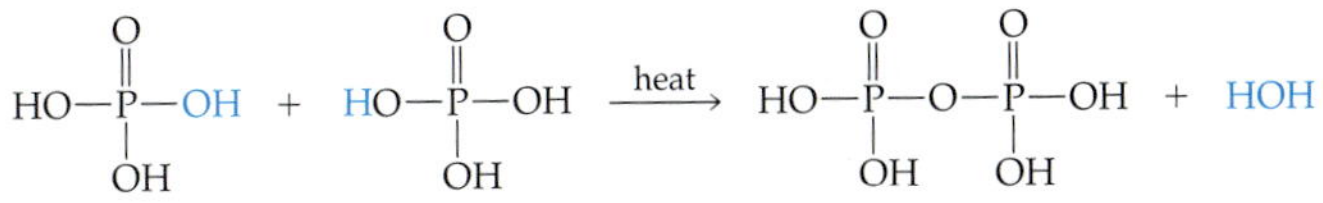

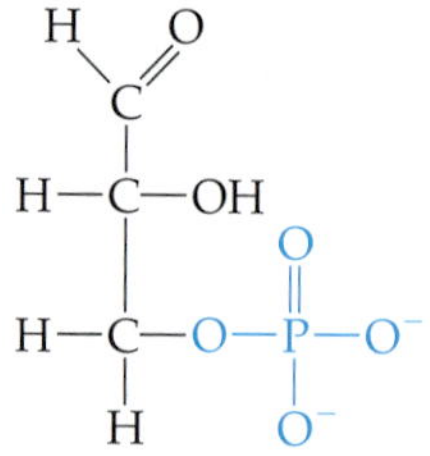

Glyceraldehyde 3-phosphate

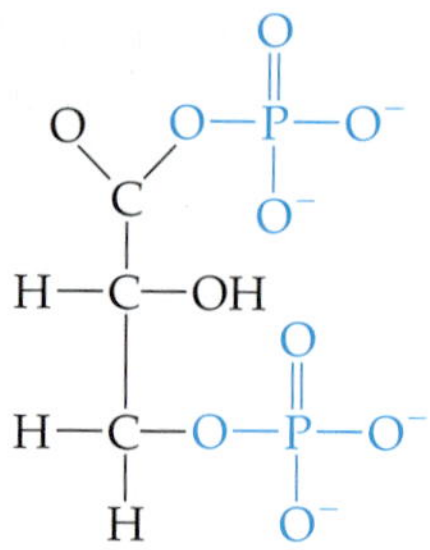

1,3-Diphosphoglyceric acid

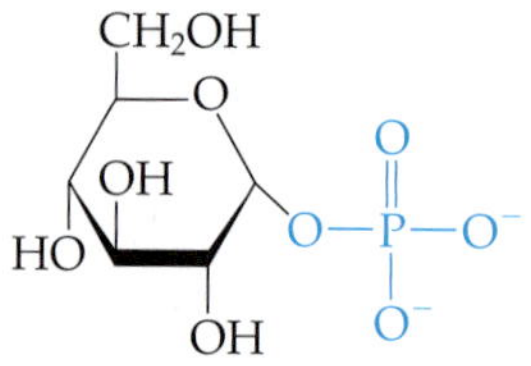

Glucose 1-phosphate

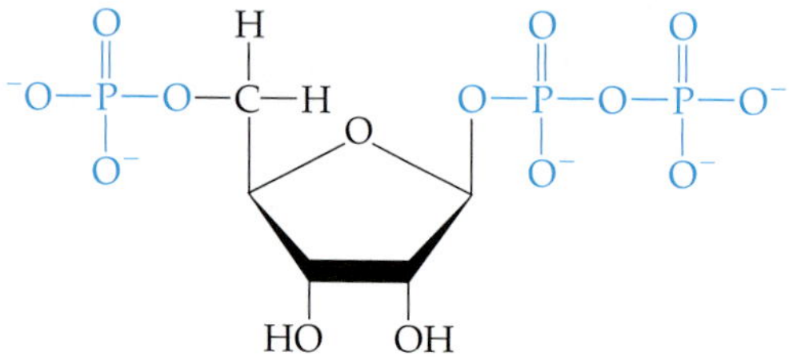

Phosphoribosyl pyrophosphate

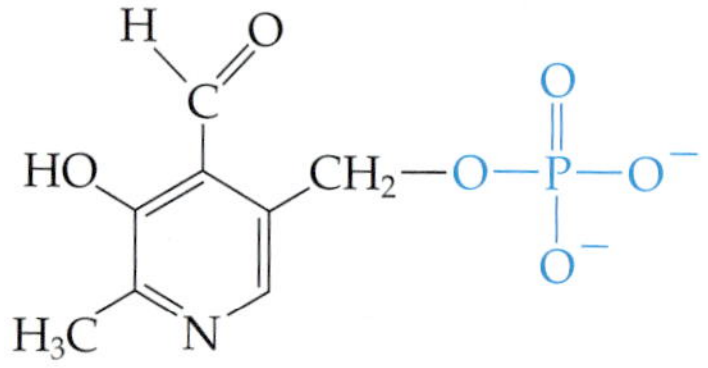

Pyridoxal phosphate

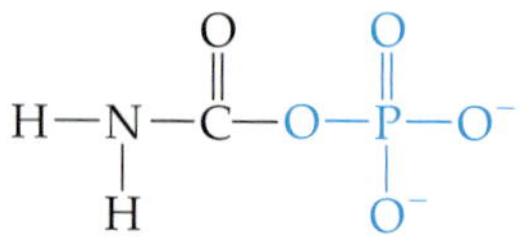

Carbamyl phosphate

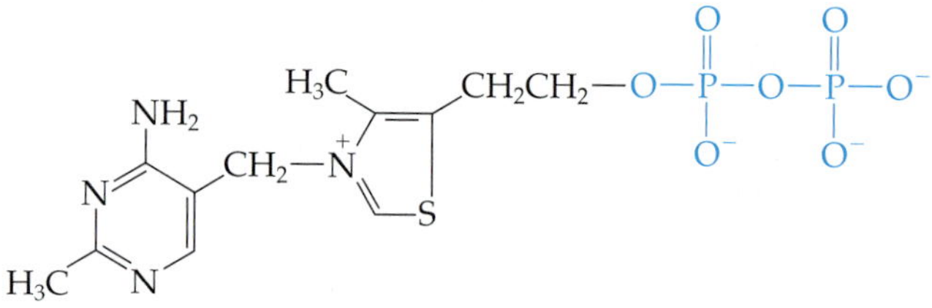

Thiamine pyrophosphate

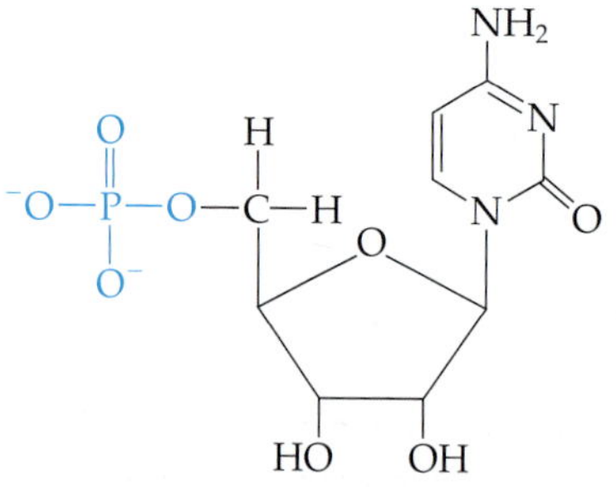

Cytidine nucleotide

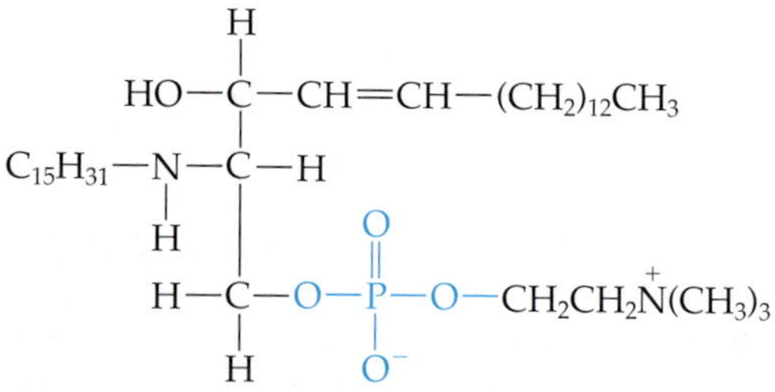

Sphingomyelin

FIGURE 16.2 ▶ Some phosphate compounds of biological importance.

are also important structural constituents of phospholipids (Section 20.7), nucleic acids (Chapter 23), coenzymes (Table J.1), and insecticides.

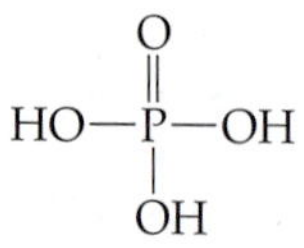

Phosphoric acid

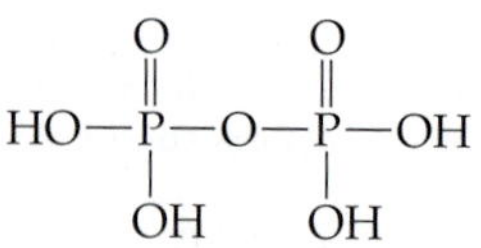

Pyrophosphoric acid

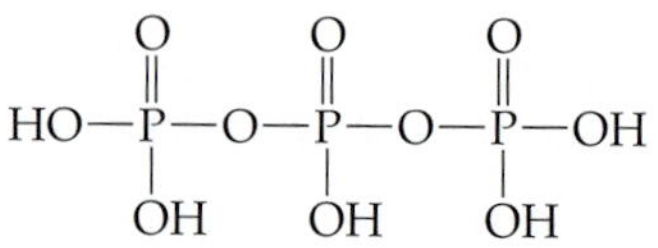

Triphosphoric acid

Phosphoric acid is triprotic and can form monoalkyl, dialkyl, and trialkyl esters as one or more hydrogen atoms are replaced by alkyl groups:

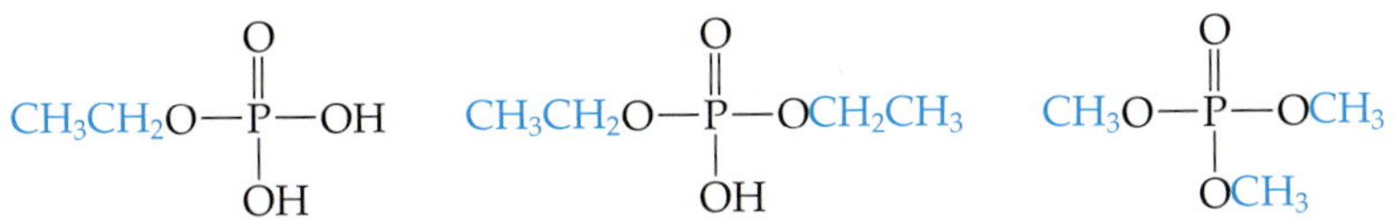

Ethyl dihydrogen phosphate — Diethyl hydrogen phosphate — Trimethyl phosphate

A wide variety of phosphate esters occur naturally and are compounds of central importance in metabolism. The majority of the substances obtained from our food must be converted to phosphate esters before they can be used by the cells. Many of them are formed by phosphorylation reactions—the interaction of alcohols with anhydrides of phosphoric acid:

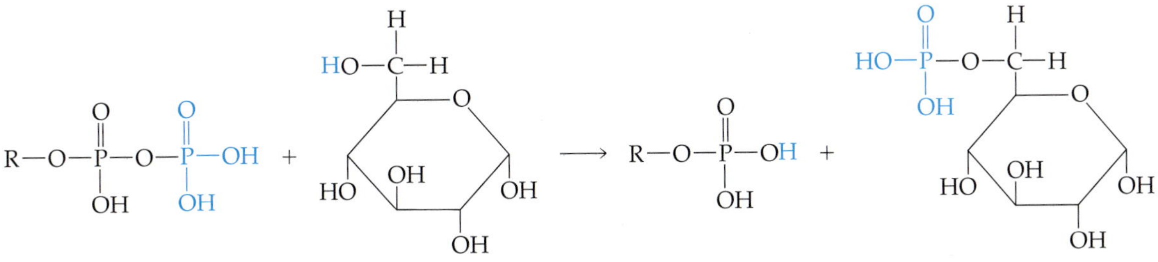

A polyphosphate — Glucose — Glucose 6-phosphate

Figure 16.2 presents only a few of the phosphate esters found in living cells. At physiological pH values (≈7), the phosphate groups are ionized.

16.11 Amides: Structures and Names

In the amide functional group, a nitrogen is attached to a carbonyl group. If the two remaining bonds to nitrogen are attached to hydrogen atoms, the compound is called a simple amide. If one or both of the two remaining bonds to nitrogen are attached to alkyl or aryl groups, the compound is called a substituted amide. The carbonyl carbon–nitrogen bond is referred to as the **amide linkage.** This bond is very stable and is found in the repeating units of protein molecules (Chapter 21), in nylon, and in many other industrial polymers.

The amide group — A simple amide — A substituted amide — Formamide (Methanamide)

Amides are named as derivatives of carboxylic acids. The *-ic* ending of the common name or the *-oic* ending of the IUPAC name is replaced with the suffix *-amide.*

Example 16.10

Name the compound

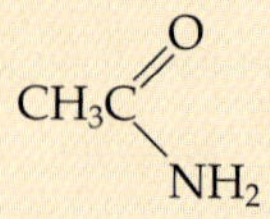

Solution

This amide is derived from acetic acid. Drop the *-ic* suffix, attach the ending *-amide,* and you have the name: acetamide (or ethanamide in the IUPAC system).

Example 16.11

Name the compound

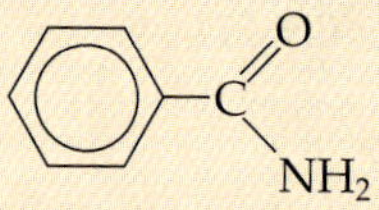

Solution

This amide is derived from benzoic acid. Drop the *-oic,* add *-amide,* and you have it: benzamide.

Practice Exercise

Draw a structural formula for pentanamide.

In substituted amides, alkyl groups attached to the nitrogen atom are named as substituents. Instead of using a Greek letter or a number to specify location, chemists indicate the group's attachment to nitrogen by a capital letter N. If the substituent on nitrogen is phenyl, the compound is named as an anilide. The -ic or -oic ending of the acid name is replaced with -anilide instead of -amide:

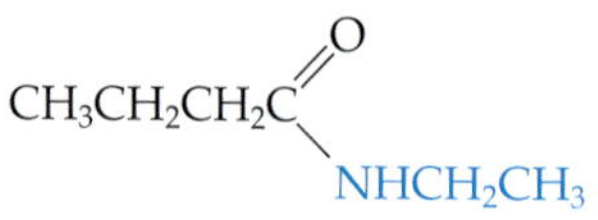

N-Ethylbutyramide

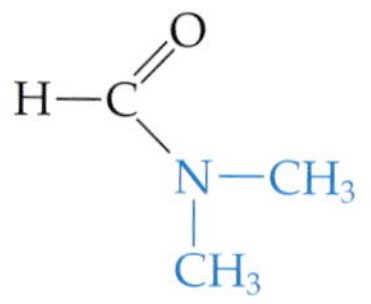

N,N-Dimethylformamide

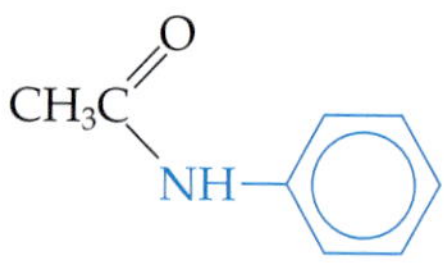

Acetanilide

Example 16.12

Name the compound

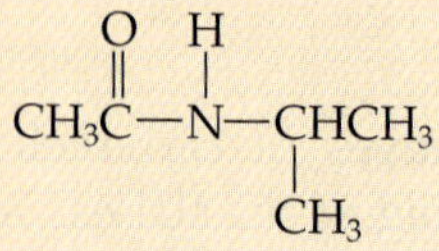

Solution

The acid portion of the molecule (the portion incorporating the carbonyl group and to one side of the nitrogen) contains two carbon atoms and is derived from acetic acid. This compound is therefore named as a substituted acetamide. The substituent attached directly to the nitrogen is an isopropyl group. The name of the compound is *N*-isopropylacetamide. (The IUPAC name is *N*-isopropylethanamide.)

Practice Exercise

Draw a structural formula for *N,N*-dimethylpropanamide.

16.12 Physical Properties of Amides

With the exception of formamide, which is a liquid, all unsubstituted amides are solids (Table 16.7). Most amides are colorless and odorless. The lower members of the series are soluble in water, with borderline solubility occurring in those that have five or six carbon atoms.

Table 16.7 Physical Constants of Some Unsubstituted Amides

Formula	Name	Melting Point (°C)	Boiling Point (°C)	
$HCONH_2$	Formamide	2	193	Soluble in water
CH_3CONH_2	Acetamide	82	222	
$CH_3CH_2CONH_2$	Propionamide	81	213	
$CH_3CH_2CH_2CONH_2$	Butyramide	115	216	
$C_6H_5CONH_2$	Benzamide	132	290	Insoluble

The amides have high boiling points and melting points. This phenomenon, as well as the water solubility of the amides, is a result of the polar nature of the amide group and the formation of hydrogen bonds (Figure 16.3). (Similar hydrogen bonding plays a critical role in determining the structure and properties of

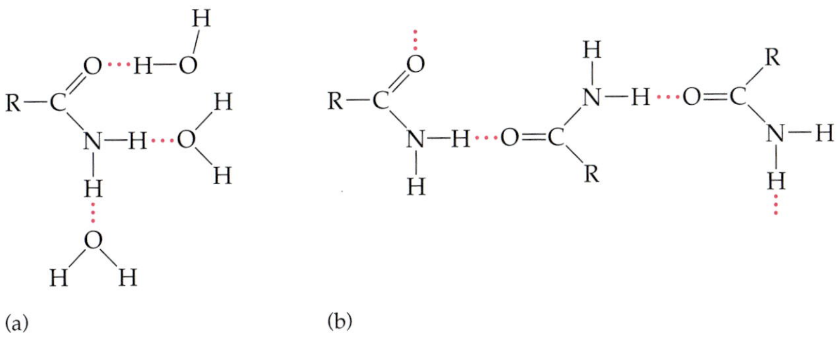

FIGURE 16.3
(a) Hydrogen bonding of amides with water molecules. (b) Intermolecular hydrogen bonding in amides.

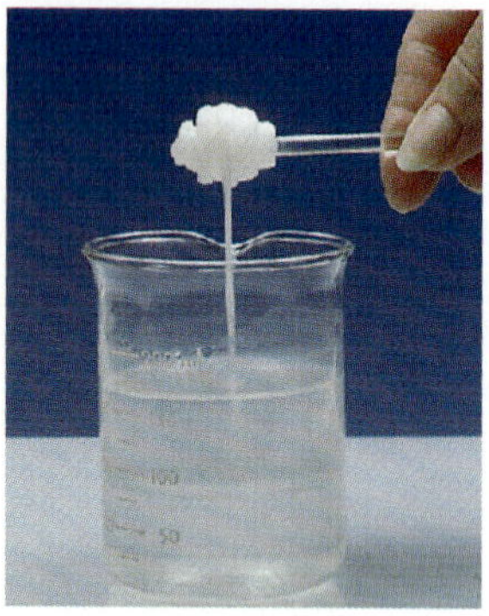

In 1934, Wallace Carothers and coworkers at E. I. du Pont Company produced nylon, the first synthetic fiber. Today, students carry out a variation of this polymerization reaction as a laboratory experiment.

proteins, DNA, RNA, and other giant molecules so important to life processes.) Electrostatic forces and hydrogen bonding combine to account for the very strong intermolecular attractions found in the amides. Note, however, that disubstituted amides have no hydrogens bonded to nitrogen and thus are incapable of hydrogen bonding. *N,N*-Dimethylacetamide has a melting point of $-20\,°C$, which is about $100\,°C$ lower than the melting point of acetamide.

16.13 Synthesis of Amides

The addition of ammonia to a carboxylic acid results in the formation of an amide, but the reaction is very slow at room temperature. The ammonium salt of the acid is formed first; then water can be split out if the reaction temperature is maintained above $100\,°C$. The second step is reversible; the equilibrium favors salt formation. Continuous removal of the water shifts the equilibrium to the right:

$$CH_3C\!\!\begin{array}{c}O\\\\OH\end{array} \;+\; NH_3 \;\longrightarrow\; CH_3C\!\!\begin{array}{c}O\\\\O^-NH_4^+\end{array} \;\rightleftharpoons^{\Delta}\; CH_3C\!\!\begin{array}{c}O\\\\NH_2\end{array} \;+\; HOH$$

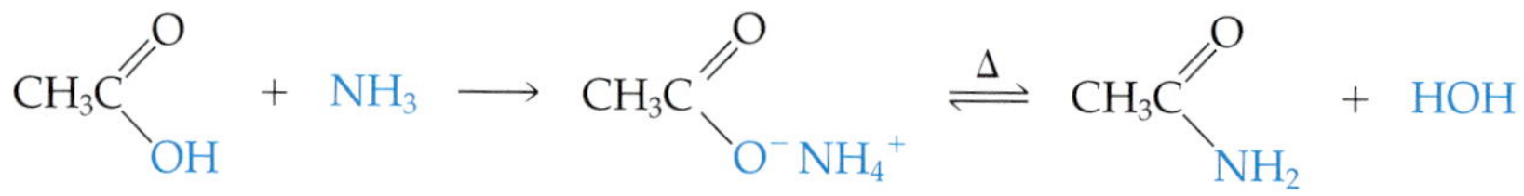

Acetic acid Ammonium acetate

In Section 16.8, we showed the condensation polymerization reaction that forms polyesters. An even more important reaction is one that yields polyamides. Again, two difunctional monomers are employed, usually adipic acid and 1,6-hexanediamine. The monomers condense by splitting out water to form a new product, which is still difunctional and thus can react continuously to yield a **polyamide** polymer:

$$n\,H\!-\!NCH_2CH_2CH_2CH_2CH_2CH_2N\!-\!H \;+\; n\,HO\!-\!CCH_2CH_2CH_2CH_2C\!-\!OH \xrightarrow[\text{10 atm}]{270\,°C}$$

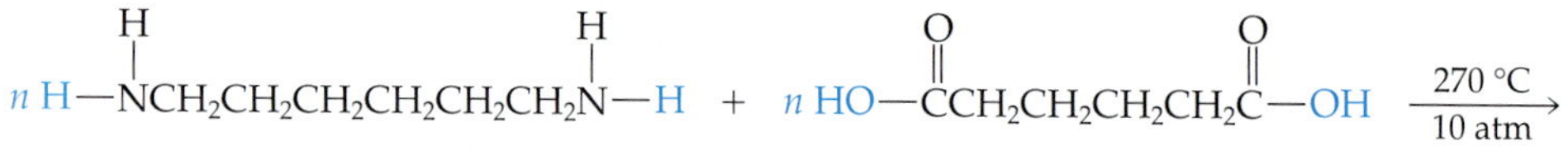

1,6-Hexanediamine Adipic acid

$$-N\!\!\begin{array}{c}H\end{array}\!\!\left[\begin{array}{c}O\\\\C\end{array}\!-\!(CH_2)_4\!-\!\begin{array}{c}O\\\\C\end{array}\!-\!N\!\!\begin{array}{c}H\end{array}\!-\!(CH_2)_6\!-\!N\!\!\begin{array}{c}H\end{array}\right]_n\!\!\begin{array}{c}O\\\\C\end{array}\!- \;+\; 2n\,HOH$$

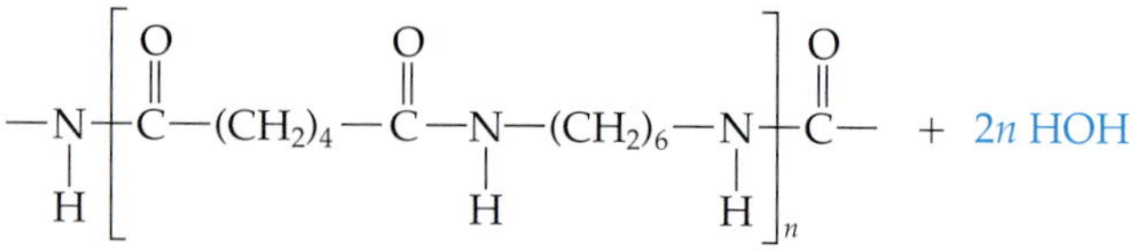

Nylon 66
(a polyamide)

Nylon is the collective name of several different synthetic polyamide fibers. Nylon 66 is the most common; the 66 designation refers to the number of carbon atoms in each monomer. It is possible to make other nylons by varying the number of carbons in either the dicarboxylic acid or the diamine. Nylon is a remarkable polymer. It is stable in dilute acids or bases, has a high melting point ($260\,°C$), and is very strong. Nylon is one of the most widely used synthetic fibers; for example, it is used in rope, sails, carpets, clothing, tires, brushes, and parachutes. It also can be molded into blocks for use in electrical equipment, gears, bearings, and valves.

16.14 Chemical Properties of Amides: Hydrolysis

Generally, the amides are neutral compounds. Furthermore, they resist hydrolysis in plain water, even upon prolonged heating. In the presence of added acid or base, however, hydrolysis proceeds at a moderate rate. Acidic hydrolysis of a simple amide gives a carboxylic acid and an ammonium salt:

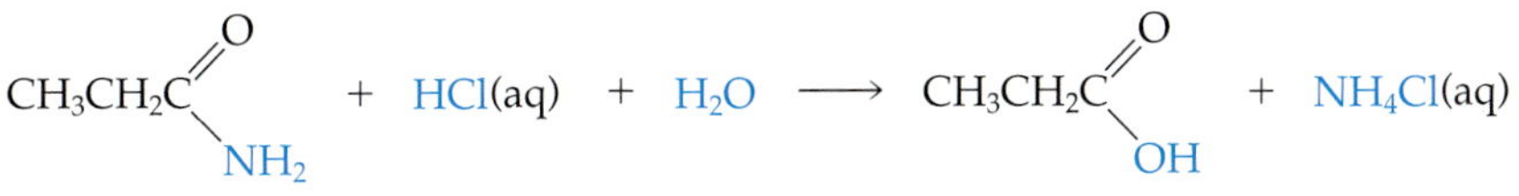
CH3CH2C(=O)NH2 + HCl(aq) + H2O ⟶ CH3CH2C(=O)OH + NH4Cl(aq)

Basic hydrolysis gives a salt of the carboxylic acid and ammonia:

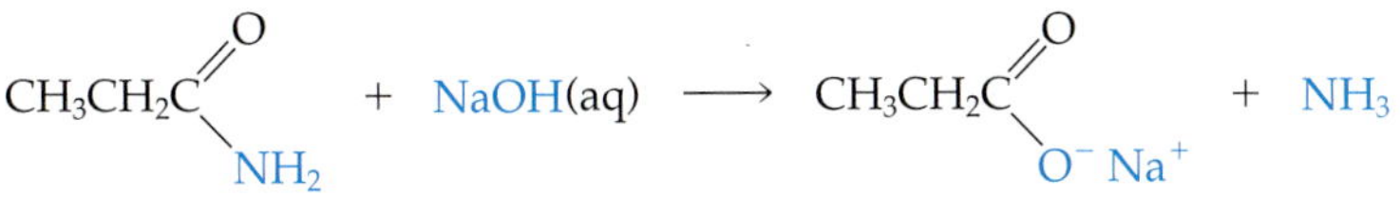
CH3CH2C(=O)NH2 + NaOH(aq) ⟶ CH3CH2C(=O)O⁻ Na⁺ + NH3

It may be easier to see why the products of the two reactions differ if we consider the hydrolysis products that would form if the reaction could be carried out in the absence of added acid or base:

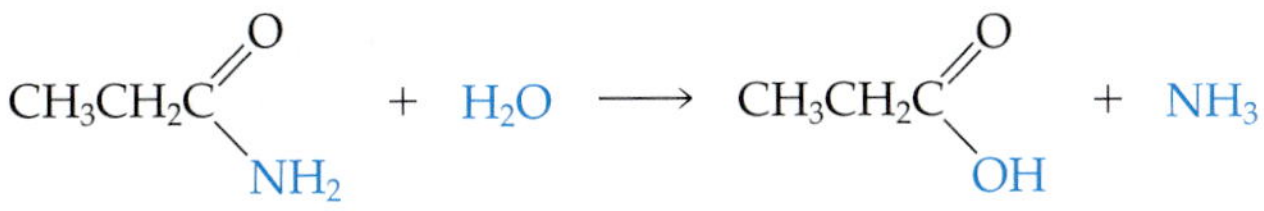
CH3CH2C(=O)NH2 + H2O ⟶ CH3CH2C(=O)OH + NH3

Note: This is not a possible reaction.

The products of this reaction are an acid (the carboxylic acid) and a base (ammonia). If we carry out the hydrolysis in the presence of hydrochloric acid, some of the hydrochloric acid reacts with the ammonia to form ammonium chloride. If, instead, we use sodium hydroxide to speed the reaction, some of this base will react with the carboxylic acid to form the sodium salt of the carboxylic acid.

There are two points to be made here. One is that the several hydrolysis reactions discussed in this chapter are closely related and should be considered as variations on a theme rather than as separate reactions. The second point is that chemical principles are not confined within chapters. If acids react with bases to form salts in Chapter 10, they also react that way in this chapter. If a reaction under consideration happens to produce an acidic product, that product always exhibits any and all properties we have previously ascribed to acids.

Example 16.13

Write an equation for the hydrolysis of butyramide in the presence of HCl.

Solution

Remember that in the acid hydrolysis of amides, the products are the organic acid and an ammonium salt:

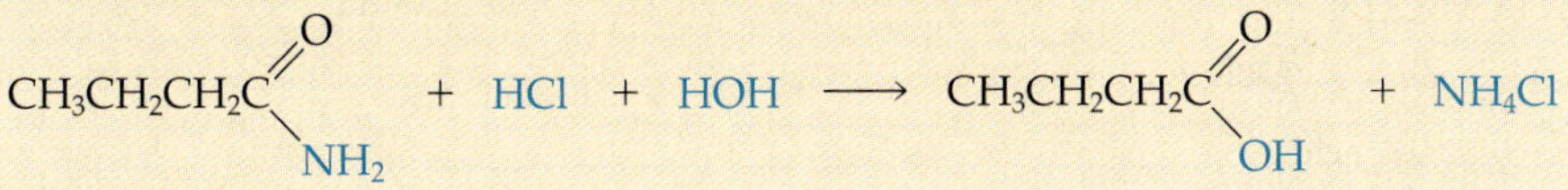
CH3CH2CH2C(=O)NH2 + HCl + HOH ⟶ CH3CH2CH2C(=O)OH + NH4Cl

Practice Exercise

Write an equation for the hydrolysis of benzamide in the presence of HCl.

Example 16.14

Write an equation for the hydrolysis of butyramide in the presence of KOH.

Solution

In basic hydrolysis of amides, the products are the salt of the carboxylic acid and ammonia:

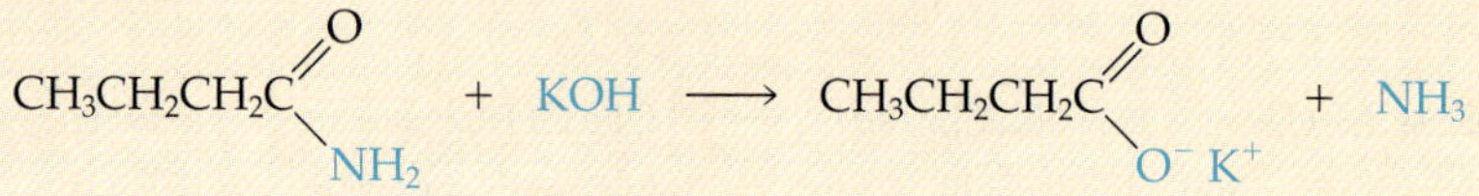

$$CH_3CH_2CH_2C\overset{O}{\underset{NH_2}{\big\|}} \quad + \quad KOH \quad \longrightarrow \quad CH_3CH_2CH_2C\overset{O}{\underset{O^- K^+}{\big\|}} \quad + \quad NH_3$$

Practice Exercise

Write an equation for the hydrolysis of benzamide in the presence of NaOH.

The hydrolysis of amides is of more than theoretical interest. Digestion of proteins (Section 27.1) involves the hydrolysis of amide bonds. Clothes made of nylon have been known to disintegrate in air polluted with sulfuric acid mist. The acid catalyzes the hydrolysis of the amide bonds that hold the long chains of the nylon molecules together. Perhaps it is worth some time to consider what the same polluted air does to the proteins of our lungs.

Summary

Carboxylic Acids

Carboxylic acids contain the functional group —COOH, called the **carboxyl group.** The fourth bond on the carboxyl carbon may be either with an alkyl group, as in acetic acid (CH_3COOH), or an aryl group, as in benzoic acid (C_6H_5COOH).

A given carboxylic acid is formed by oxidation of the aldehyde containing the same number of carbon atoms:

$$CH_3CH_2CH_2C\overset{O}{\underset{H}{\big\|}} \quad \longrightarrow \quad CH_3CH_2CH_2C\overset{O}{\underset{OH}{\big\|}}$$

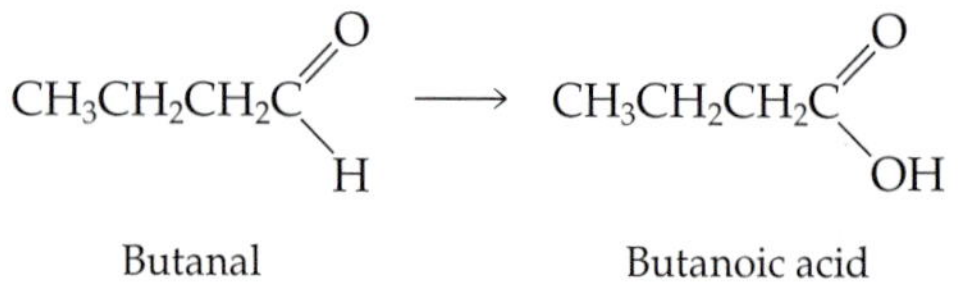

Butanal Butanoic acid

Because aldehydes are formed from primary alco-

hols, these alcohols are also a starting material for carboxylic acids:

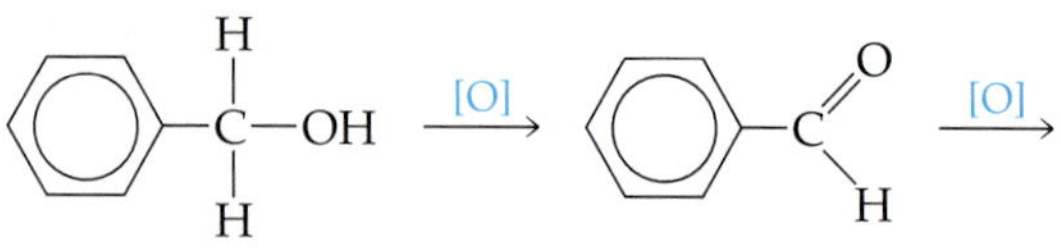

Benzyl alcohol Benzaldehyde

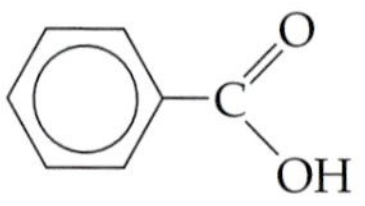

Benzoic acid

Carboxylic acids have strong, often disagreeable odors. They are highly polar molecules, and readily form intermolecular hydrogen bonds. For this reason, they have comparatively high boiling points. The hydrogen bonding in carboxylic acids is so strong that in many cases the acid exists not as a collection of individual molecules but rather as a collection of two-molecule units called **dimers.** Such dimers cause the boiling point of carboxylic acids to be unusually high.

Carboxylic acids are weak acids. They react with bases to form salts and with carbonate and bicarbonate salts to form carbon dioxide gas.

Esters

The compound formed when the hydroxyl hydrogen of a carboxylic acid is replaced either by an alkyl or an aryl group is called an **ester:**

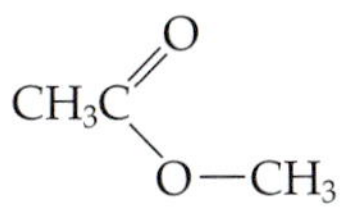

Methyl acetate

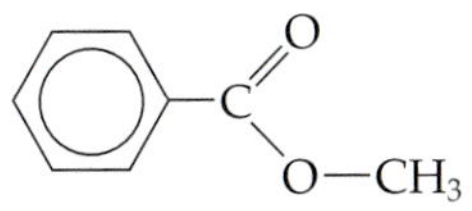

Methyl benzoate

Esters are pleasant-smelling compounds that are responsible for the fragrances of flowers and fruits. They have lower-than-expected boiling points because, even though ester molecules are somewhat polar, they cannot form intermolecular hydrogen bonds. They can hydrogen-bond with water, however, and consequently the lighter esters are water-soluble.

One way to synthesize esters is to combine a carboxylic acid and an alcohol under acidic conditions; the reaction is called an *esterification:*

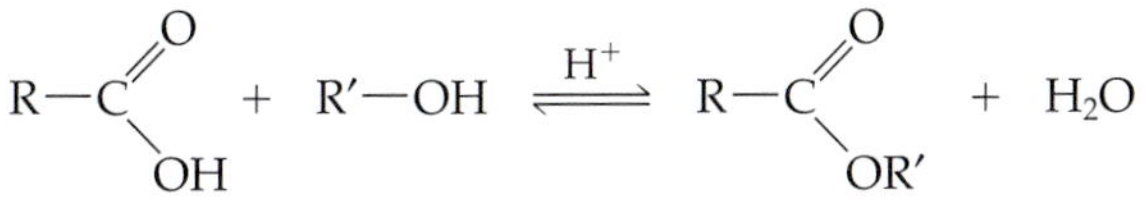

Esters are neutral compounds, and one of their most common reactions is **hydrolysis,** a reaction with water. When ester hydrolysis is carried out under acidic conditions, the reaction is essentially the reverse of esterification:

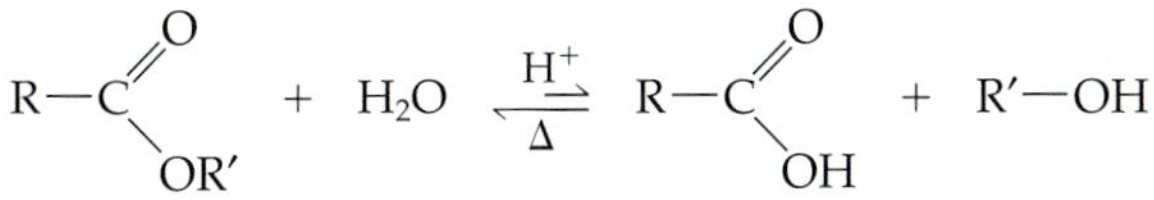

When ester hydrolysis is carried out under basic conditions, the process is called **saponification:**

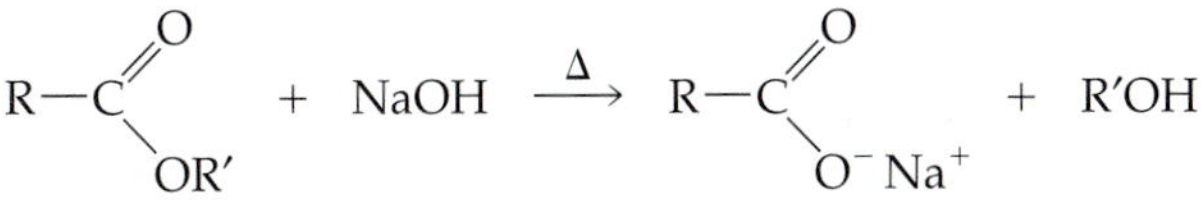

Inorganic acids also react with alcohols to form esters. Some of the most important esters in biochemistry are those formed from phosphoric acid.

Amides

Organic compounds containing a carbonyl group bonded to a nitrogen atom are **amides,** and the carbon–nitrogen bond is an **amide linkage.**

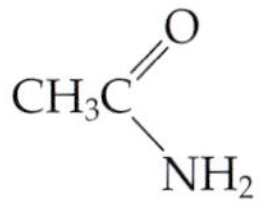

Acetamide
(a simple amide)

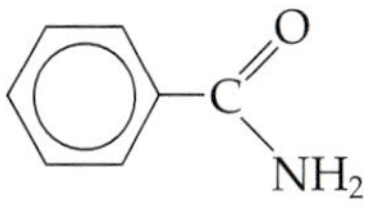

Benzamide
(a simple amide)

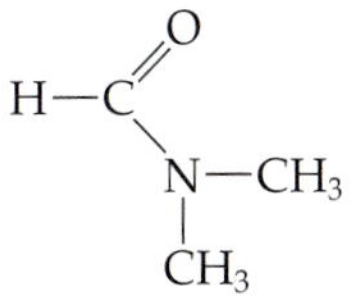

N,N-Dimethylformamide
(a substituted amide)

Most amides are colorless and odorless, and the lighter ones are soluble in water. Because they are polar molecules, amides have comparatively high boiling and melting points.

Amide synthesis involves the addition of ammonia to a carboxylic acid:

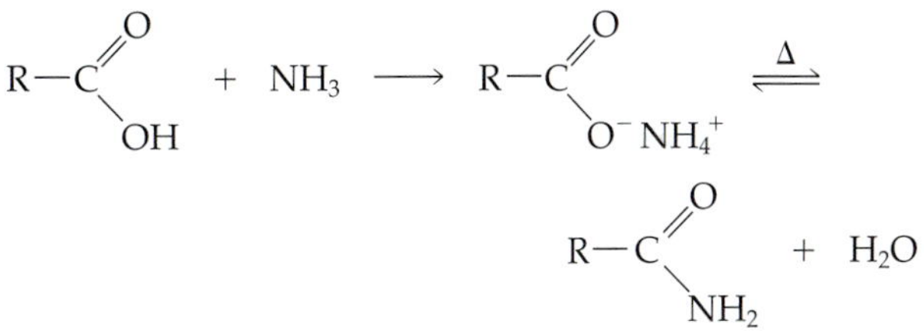

Amides are neutral compounds. They resist hydrolysis in water, but addition of either an acid or a base allows the reaction to proceed:

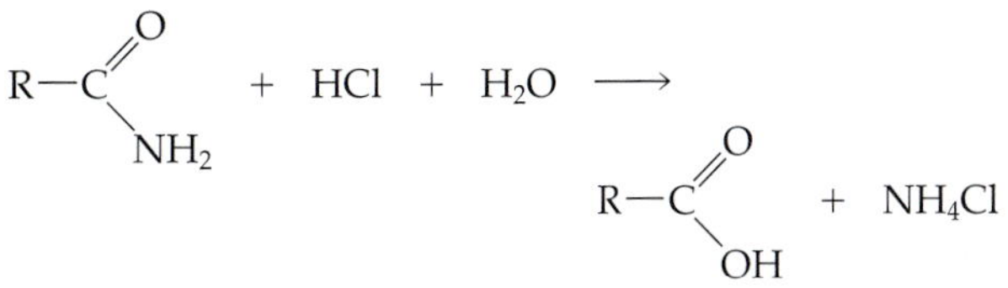

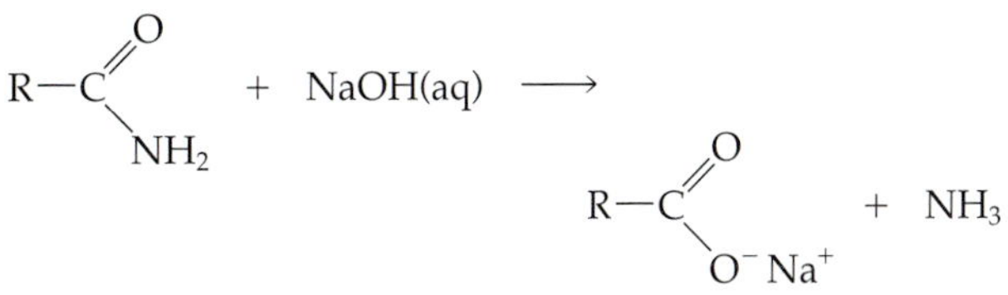

Key Terms

amide (16.1)
amide linkage (16.11)
carboxyl group (16.1)
carboxylic acid (16.1)
dimer (16.4)
ester (16.1)
hydrolysis (16.9)
polyamide (16.13)
polyester (16.8)
saponification (16.9)

Review Questions

1. Draw the functional group in:
   a. aldehydes **b.** ketones
   c. carboxylic acids **d.** esters
   e. ethers **f.** amides
2. Give the common name for the straight-chain carboxylic acids containing the given number of carbon atoms.
   a. 1 **b.** 2 **c.** 3 **d.** 4
3. Of the families of compounds discussed in this chapter, which is known for its characteristically unpleasant odors? Which for its characteristically pleasant aromas?
4. Define and illustrate:
   a. neutralization **b.** esterification
   c. acid hydrolysis **d.** saponification

Problems

Carboxylic Acids: Names and Structural Formulas

5. Draw structural formulas for:
   a. heptanoic acid
   b. 3-methylbutanoic acid
   c. 2,3-dibromobenzoic acid
   d. *m*-isopropylbenzoic acid
6. Draw structural formulas for:
   a. *o*-nitrobenzoic acid
   b. *p*-chlorobenzoic acid
   c. 3-chloropentanoic acid
   d. 3-phenylpropanoic acid
7. Draw structural formulas for:
   a. oxalic acid
   b. β-hydroxybutyric acid
8. Draw structural formulas for:
   a. α-chloropropionic acid
   b. phenylacetic acid
9. Name:
   a. $(CH_3)_2CHCH_2COOH$
   b. $(CH_3)_3CCH(CH_3)CH_2COOH$
   c. $CH_2OHCH_2CH_2COOH$
   d. $(CH_3)_2CHCH_2CH(CH_3)COOH$
10. Name:
   a. $CH_3(CH_2)_8COOH$
   b. $(CH_3)_2CHCCl_2CH_2CH_2COOH$
   c. $CH_3CHOHCH(CH_2CH_3)CHICOOH$
   d. 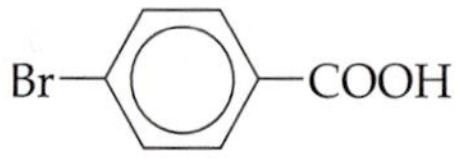

Salts: Names and Structural Formulas

11. Draw structural formulas for:
   a. potassium acetate
   b. calcium propanoate
12. Name:
   a. 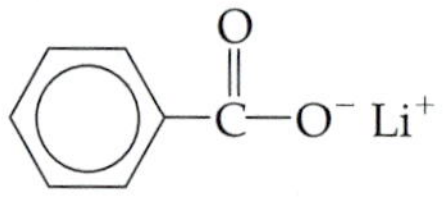

   b. $CH_3CH_2CH_2\overset{O}{\overset{\|}{C}}—O^- \ NH_4^+$

Esters: Names and Structural Formulas

13. Draw structural formulas for:
   a. methyl acetate
   b. phenyl acetate
14. Draw structural formulas for:
   a. ethyl pentanoate
   b. ethyl 3-methylhexanoate
15. Draw structural formulas for:
   a. ethyl benzoate
   b. phenyl benzoate
16. Draw structural formulas for:
   a. ethyl butyrate
   b. isopropyl propionate

17. Name:

a. [benzene ring]—$\overset{\overset{\displaystyle O}{\|}}{C}$—O—$CH_3$

b. CH_3—O—$\overset{\overset{\displaystyle O}{\|}}{C}$—H

c. $CH_3CH_2\overset{\overset{\displaystyle O}{\|}}{C}$—O—$CH_2CH_3$

18. Name:

a. $CH_3CH_2CH_2O$—$\overset{\overset{\displaystyle O}{\|}}{C}$—$CH_3$

b. $CH_3CH_2\overset{\overset{\displaystyle O}{\|}}{C}$—$OCH_2CH_2CH_3$

c. $CH_3CH_2CH_2\overset{\overset{\displaystyle O}{\|}}{C}$—O—[benzene ring]

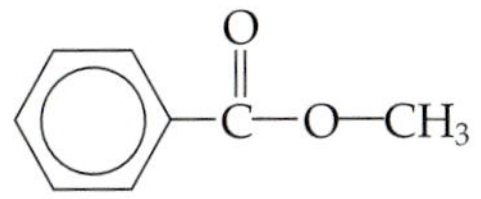

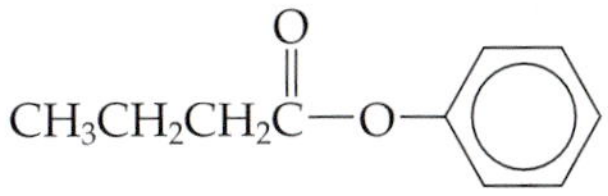

Amides: Names and Structural Formulas

19. Draw structural formulas for:
 a. butanamide
 b. hexanamide
 c. *N*-methylacetamide
20. Draw structural formulas for:
 a. formamide
 b. propionamide
 c. *N,N*-dimethylbenzamide
21. Name:

a. [benzene ring]—$\overset{\overset{\displaystyle O}{\|}}{C}$—$NH_2$

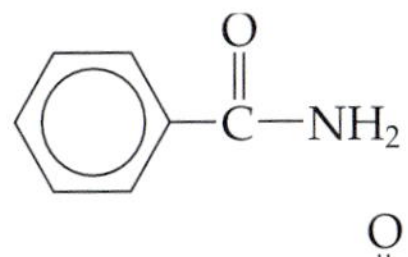

b. $CH_3CH_2CH(CH_3)\overset{\overset{\displaystyle O}{\|}}{C}$—$NH_2$

c. $CH_3\overset{\overset{\displaystyle O}{\|}}{C}$—$NH_2$

22. Name:

a. $CH_3CH_2\overset{\overset{\displaystyle O}{\|}}{C}$—$\overset{\overset{\displaystyle }{N}}{\underset{\underset{\displaystyle CH_3}{|}}{}}$—$CH_3$

b. Cl—[benzene ring]—$\overset{\overset{\displaystyle O}{\|}}{C}$—$NH_2$

c. $CH_3CH_2CH_2\overset{\overset{\displaystyle O}{\|}}{C}$—NH—[benzene ring]

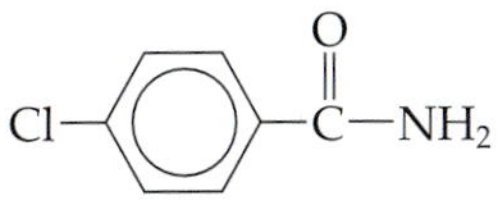

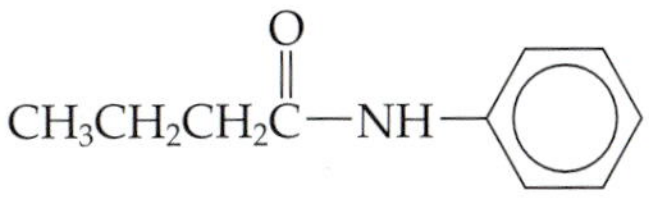

Physical Properties

23. Which compound has the higher boiling point? Explain.

$CH_3CH_2CH_2$—O—CH_2CH_3 $CH_3CH_2CH_2\overset{\overset{\displaystyle O}{\|}}{C}$—OH

 I II

24. Which compound has the higher boiling point? Explain.

$CH_3CH_2CH_2CH_2CH_2OH$ $CH_3CH_2CH_2\overset{\overset{\displaystyle O}{\|}}{C}$—OH

 I II

25. Which compound has the higher boiling point? Explain.

$CH_3CH_2CH_2\overset{\overset{\displaystyle O}{\|}}{C}$—$NH_2$ $CH_3\overset{\overset{\displaystyle O}{\|}}{C}$—O—$CH_2CH_3$

 I II

26. Which compound has the higher boiling point? Explain.

$CH_3CH_2CH_2\overset{\overset{\displaystyle O}{\|}}{C}$—OH $CH_3CH_2\overset{\overset{\displaystyle O}{\|}}{C}$—O—$CH_3$

 I II

27. Which compound is more soluble in water? Explain.

$CH_3\overset{\overset{\displaystyle O}{\|}}{C}$—OH $CH_3CH_2CH_2CH_3$

 I II

28. Which compound is more soluble in water? Explain.

CH_3CH=$CHCH_3$ $CH_3\overset{\overset{\displaystyle O}{\|}}{C}$—$NH_2$

 I II

29. Which compound is more soluble in water? Explain.

$CH_3\overset{\overset{\displaystyle O}{\|}}{C}OCH_3$ $CH_3CH_2CH_2CH_2\overset{\overset{\displaystyle O}{\|}}{C}OCH_2CH_3$

 I II

30. Which compound is more soluble in water? Explain.

[benzene ring]—$\overset{\overset{\displaystyle O}{\|}}{C}$—OH [benzene ring]—$\overset{\overset{\displaystyle O}{\|}}{C}$—$O^-$ Na^+

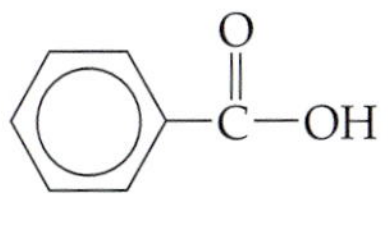

 I II

Chemical Reactions

31. Write equations for the reactions of butyric acid with (**a**) aqueous NaOH and (**b**) aqueous $NaHCO_3$.
32. Write equations for the reactions of benzoic acid with (**a**) aqueous NaOH and (**b**) aqueous $NaHCO_3$.

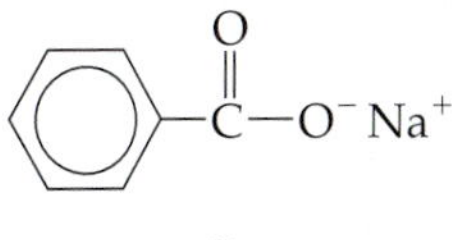

33. Write an equation for the acid-catalyzed hydrolysis of ethyl acetate.

34. Write an equation for the base-catalyzed hydrolysis of ethyl acetate.

35. Write an equation for the acid-catalyzed hydrolysis of benzamide.

36. Write an equation for the base-catalyzed hydrolysis of benzamide.

37. Complete the following equations.

a. $CH_3CH_2\overset{\overset{\displaystyle O}{\|}}{C}-OH \xrightarrow{NaOH}$

b. [benzene ring with two COOH groups] $\xrightarrow{\text{excess NaHCO}_3}$

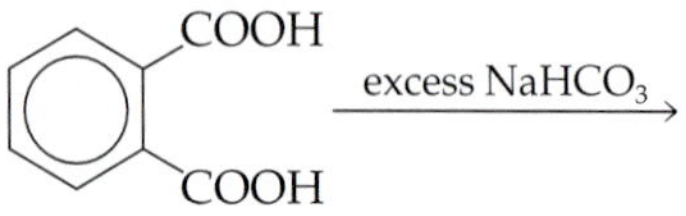

38. Complete the following equations.

a. $HOOC-COOH \xrightarrow{\text{excess NaOH}}$

b. [benzene ring]$-COOH \xrightarrow{KOH}$

39. Complete the following equations.

a. [benzene ring]$-\overset{\overset{\displaystyle O}{\|}}{C}-OCH_2CH_2CH_3 + NaOH \longrightarrow$

b. [cyclohexane ring]$-OH + CH_3\overset{\overset{\displaystyle O}{\|}}{C}OH \underset{\longleftarrow}{\overset{H^+}{\longrightarrow}}$

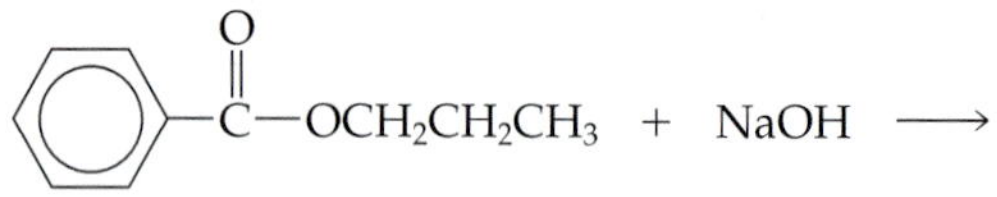

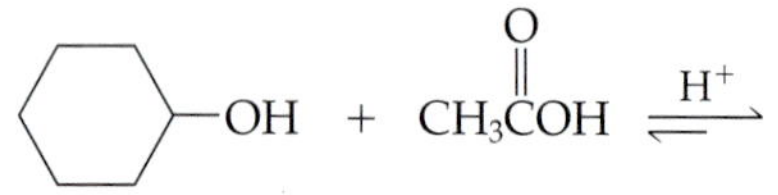

40. Complete the following equations.

a. [benzene ring]$-CH_2OH + CH_3\overset{\overset{\displaystyle O}{\|}}{C}OH \underset{\longleftarrow}{\overset{H^+}{\longrightarrow}}$

b. $CH_3\overset{\overset{\displaystyle O}{\|}}{C}-OCH(CH_3)_2 + KOH(aq) \longrightarrow$

41. Complete the following equations.

a. $CH_3\overset{\overset{\displaystyle O}{\|}}{C}-OH + CH_3CH_2CH_2OH \underset{\longleftarrow}{\overset{H^+}{\longrightarrow}}$

b. $HO-\overset{\overset{\displaystyle O}{\|}}{C}CH_2\overset{\overset{\displaystyle O}{\|}}{C}-OH + 2\,CH_3OH \underset{\longleftarrow}{\overset{H^+}{\longrightarrow}}$

42. Complete the following equations.

a. $CH_3CH_2CH_2O-\overset{\overset{\displaystyle O}{\|}}{C}-$[benzene ring] $+ H_2O \underset{\longleftarrow}{\overset{H^+}{\longrightarrow}}$

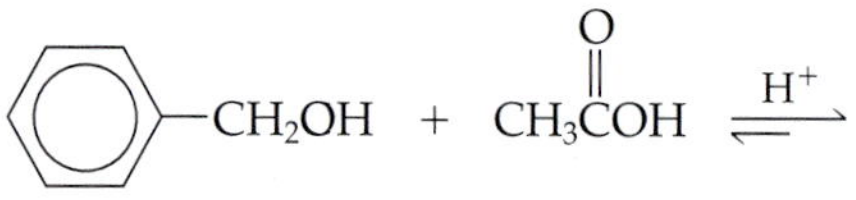

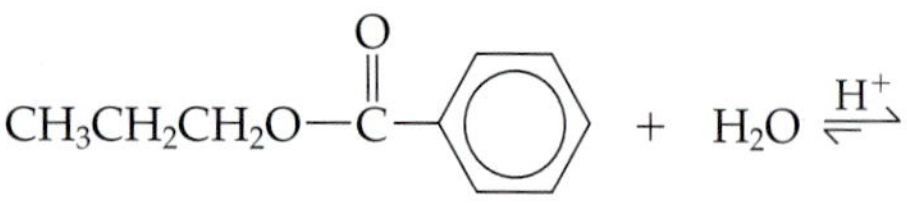

b. $(CH_3)_2CH-\overset{\overset{\displaystyle O}{\|}}{C}-O-CH_2CH_3 + H_2O \underset{\longleftarrow}{\overset{H^+}{\longrightarrow}}$

43. Complete the following equations.

a. $CH_3\overset{\overset{\displaystyle O}{\|}}{C}-NH_2 + HCl + H_2O \longrightarrow$

b. [benzene ring]$-\overset{\overset{\displaystyle O}{\|}}{C}-\underset{\underset{\displaystyle CH_3}{|}}{N}-CH_3 + NaOH(aq) \longrightarrow$

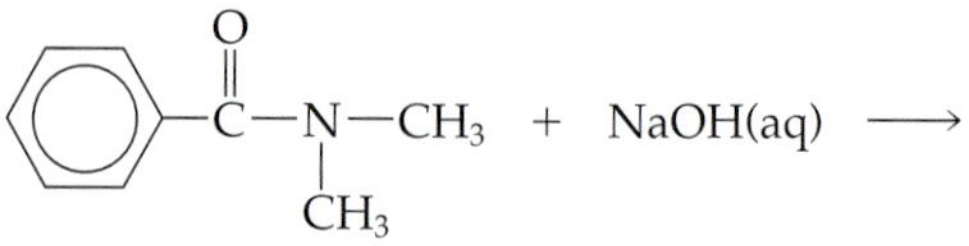

44. Complete the following equations.

a. $CH_3CH_2\overset{\overset{\displaystyle O}{\|}}{C}-NH_2 + KOH(aq) \longrightarrow$

b. [benzene ring]$-\overset{\overset{\displaystyle O}{\|}}{C}-NHCH_3 + HCl + H_2O \longrightarrow$

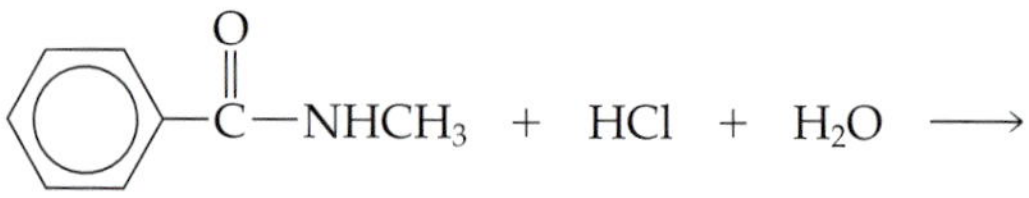

45. List the reagents necessary to carry out the following conversions.

a. $CH_3CH_2\underset{\underset{\displaystyle CH_3}{|}}{C}HCH_2OH \xrightarrow{?} CH_3CH_2\underset{\underset{\displaystyle CH_3}{|}}{C}H\overset{\overset{\displaystyle O}{\|}}{C}-OH$

b. $CH_3\overset{\overset{\displaystyle O}{\|}}{C}-H \xrightarrow{?} CH_3\overset{\overset{\displaystyle O}{\|}}{C}OH$

c. [benzene ring]$-\overset{\overset{\displaystyle O}{\|}}{C}-OH \xrightarrow{?}$ [benzene ring]$-\overset{\overset{\displaystyle O}{\|}}{C}-O^-Na^+$

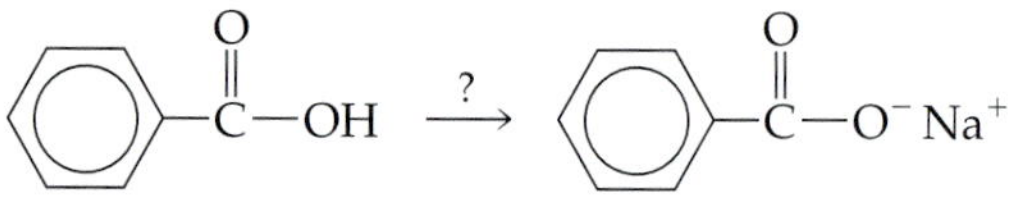

46. List the reagents necessary to carry out the following conversions.

a. [benzene ring]$-\overset{\overset{\displaystyle O}{\|}}{C}-OH \xrightarrow{?}$ [benzene ring]$-\overset{\overset{\displaystyle O}{\|}}{C}-O^-NH_4$

b. [benzene ring]$-\overset{\overset{\displaystyle O}{\|}}{C}-OH \xrightarrow{?}$ [benzene ring]$-\overset{\overset{\displaystyle O}{\|}}{C}-OCH_3$

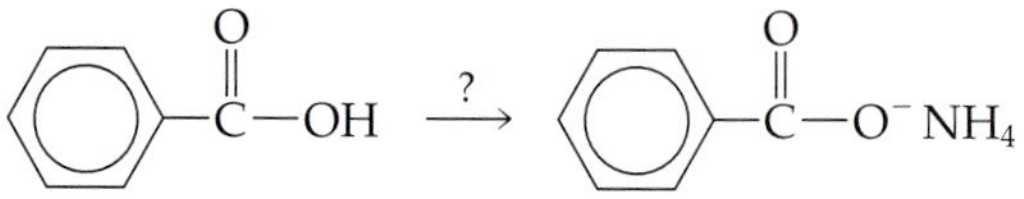

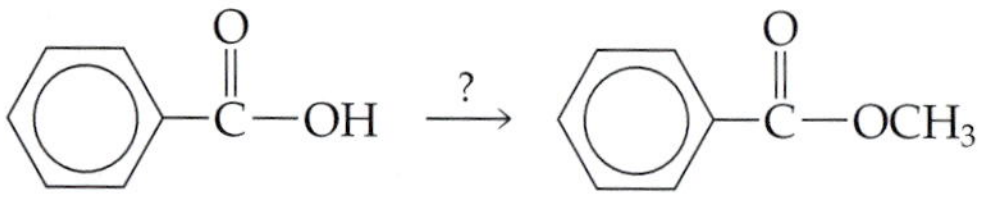

47. List the reagents necessary to carry out the following conversions.

a. $CH_3CH_2CH_2CH_2OH \xrightarrow{?} CH_3CH_2CH_2CH_2O\overset{\overset{\displaystyle O}{\|}}{C}CH_3$

b. $CH_3CH_2CH_2\overset{\overset{\displaystyle O}{\|}}{C}-OCH_3 \xrightarrow{?}$

$CH_3CH_2CH_2\overset{\overset{\displaystyle O}{\diagup\!\!\diagdown}}{C}\!\!\diagdown_{O^-Li^+} + CH_3OH$

48. List the reagents necessary to carry out the following conversions.

a. [benzene ring]$-\overset{\overset{\displaystyle O}{\|}}{C}-NH_2 \xrightarrow{?}$ [benzene ring]$-\overset{\overset{\displaystyle O}{\|}}{C}-OH + NH_4Br$

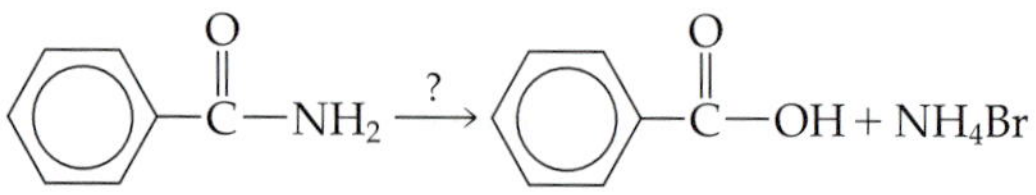

b. 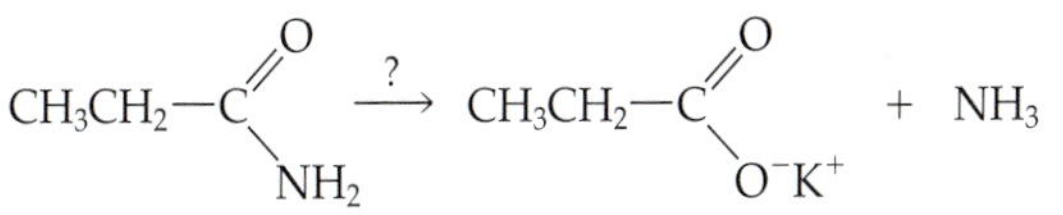

Phosphorus Compounds

49. Draw the structural formulas for:
   a. diethyl hydrogen phosphate
   b. methyl dihydrogen phosphate
   c. triphosphoric acid

50. Name:

a. 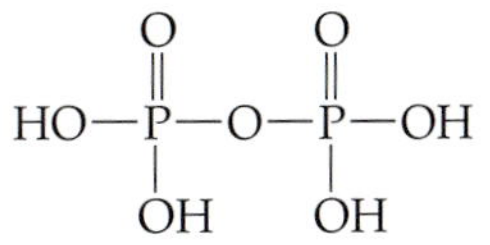

b. 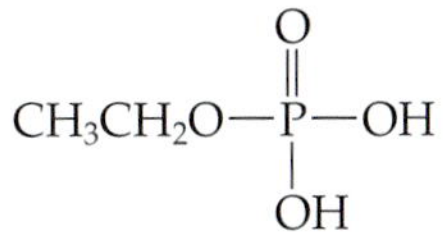

Additional Problems

51. All the following compounds are isomers. Circle and name the functional groups in each.

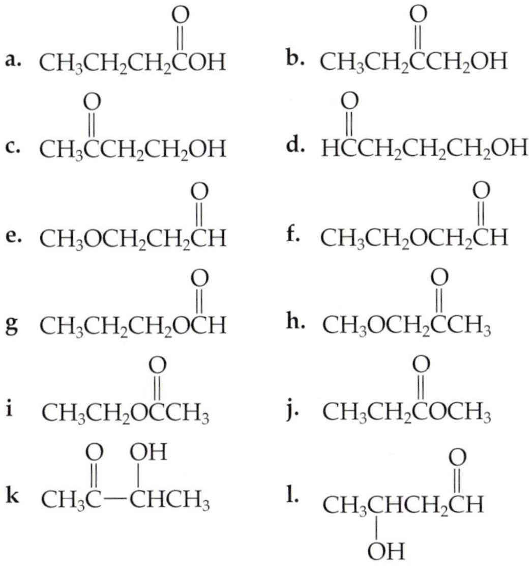

52. Arrange in order of increasing acidity, with the least-acidic compound first: (**a**) benzoic acid, (**b**) benzyl alcohol ($C_6H_5CH_2OH$), (**c**) phenol, (d) toluene.
53. Without consulting tables, arrange the following compounds in order of increasing boiling point: (**a**) butyl alcohol, (**b**) methyl acetate, (**c**) pentane, (**d**) propionic acid.
54. Name and draw structural formulas for all the isomeric amides that have the molecular formula C_4H_9NO.
55. Benzoic acid is insoluble in water. If the reactions described in Problem 32 were carried out in test tubes, what would you observe?
56. Offer explanations for the following facts.
   a. Even though it has a higher molar mass, ethyl acetate has a lower boiling point (57 °C) than either methyl alcohol (65 °C) or formic acid (100 °C).
   b. Sodium benzoate is soluble in water, whereas benzoic acid is insoluble.
   c. The alkaline hydrolysis of esters is irreversible, whereas the acidic hydrolysis of esters is reversible.
   d. Both acidic hydrolysis and alkaline hydrolysis of amides are irreversible.
57. From which alcohol might each acid be prepared via oxidation with acidic dichromate?
   a. CH_3CH_2COOH
   b. HOOCCOOH
   c. HCOOH
   d. $(CH_3)_2CHCH_2COOH$
58. A lactone is a cyclic ester. What product is formed in each reaction?

a. 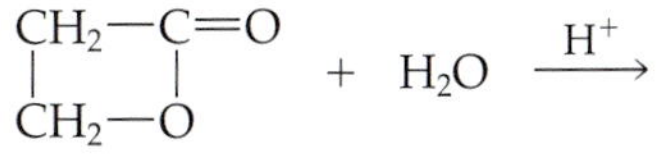

b. 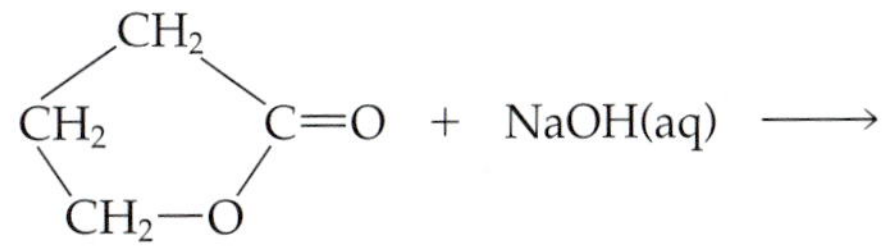

59. A lactam is a cyclic amide. What product is formed in each reaction?

a. 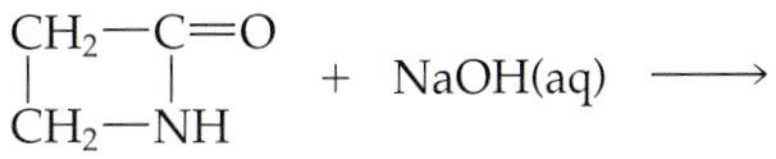

b. 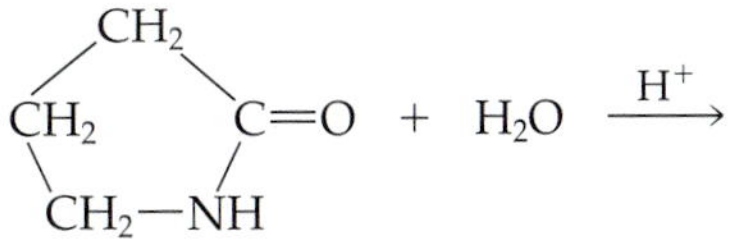

60. An ester with the molecular formula $C_6H_{12}O_2$ was hydrolyzed in aqueous acid to yield an acid (Y) and an alcohol (Z). Oxidation of the alcohol with potassium permanganate gave the identical acid (Y). What is the structural formula of the ester?
61. The neutralization of 125 mL of a 0.400 M NaOH solution requires 5.10 g of a monocarboxylic acid. Write all possible structural formulas for the acid.
62. If 3.00 g of acetic acid reacts with excess methanol, how many grams of methyl acetate are formed?
63. How many milliliters of a 0.100 M barium hydroxide solution are required to neutralize 0.500 g of dichloroacetic acid?

Special Topic F
Drugs: Some Carboxylic Acids, Esters, and Amides

People have long sought relief from pain and discomfort. Alcohol, opium, cocaine, and marijuana have been used as medicines for centuries. Often they have been used for their pleasurable effects, not just for relief of pain. According to the broadest definition, a **drug** is any chemical substance that affects an individual in such a way as to bring about physiological, emotional, or behavioral change. In this special topic, we discuss several important drugs that are carboxylic acids, esters, or amides.

F.1 Aspirin and Other Salicylates

Soon after acetic anhydride[1] became available, in the nineteenth century, chemists began to acetylate a variety of physiologically active compounds. Such structural modifications often change the properties of drugs to enhance their effectiveness or to minimize undesirable side effects. Two such cases, aspirin and heroin, are described here.

The first successful synthetic pain relievers were derivatives of salicylic acid (Figure F.1). Salicylic acid was first isolated from willow bark in 1860, although an English clergyman named Edward Stone had reported to the Royal Society as early as 1763 that an extract of willow bark was useful in

[1] Acetic anhydride can be considered as being formed when one molecule of water is removed from two molecules of acetic acid.

$$CH_3\overset{O}{\overset{\|}{C}}-OH \ + \ HO-\overset{O}{\overset{\|}{C}}CH_3 \ \longrightarrow \ CH_3\overset{O}{\overset{\|}{C}}-O-\overset{O}{\overset{\|}{C}}CH_3 \ + \ HOH$$

reducing fever. Salicylic acid is itself a good **analgesic** (pain reliever) and **antipyretic** (fever reducer), but it is sour and irritating when taken orally. Chemists sought to modify the structure of the molecule to remove this undesirable property while retaining (or even improving) the desirable properties.

The first modification was simple neutralization of the acid. The salt sodium salicylate was first used as a pain reliever in 1875. It was less unpleasant to swallow, but it proved to be highly irritating to the lining of the stomach. Phenyl salicylate (salol) was introduced in 1886. It passed unchanged through the stomach. In the small intestine, it was hydrolyzed to the desired salicylic acid, but phenol, which is rather toxic, also was formed.

Methyl salicylate is an oil found in numerous plants and has a fragrance associated with wintergreen. Commercially, it is used in perfumes and for flavoring candy. It finds widespread use as the pain–relieving ingredient in liniments such as Ben–Gay. When rubbed on the skin, this ester has the unusual ability of penetrating the surface. Hydrolysis then occurs, liberating salicylic acid, which relieves the soreness. Methyl salicylate also causes a mild, burning sensation when applied to the skin, thus serving as a counterirritant for sore muscles.

Acetylsalicylic acid, to which the German Bayer Company assigned the trade name Aspirin, was first introduced in 1899. It soon became the best-selling drug in the world. Over 55 billion aspirin tablets (under many trade names) are now consumed annu-

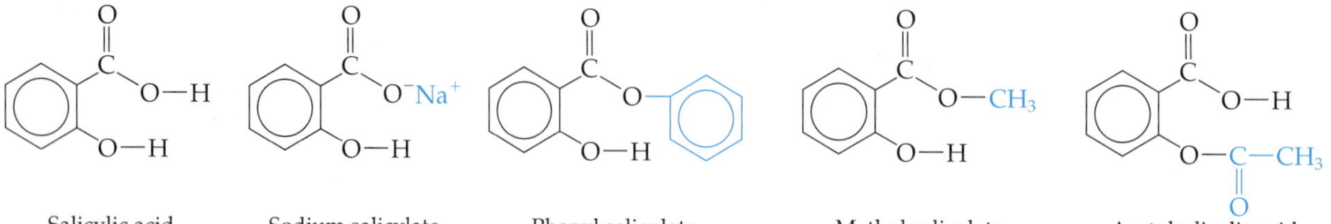

Salicylic acid	Sodium salicylate	Phenyl salicylate (Salol)	Methyl salicylate	Acetylsalicylic acid

▲ **FIGURE F.1**

Salicylic acid and some of its derivatives.

ally in North America. Acetylsalicylic acid is made by treating salicylic acid with acetic anhydride. In this reaction, the hydroxyl group of the phenol reacts like that of an alcohol.

Since aspirin is the most widely used drug in the world, let's take a close look at it. Aspirin is a chemical compound. Like other compounds, its properties are invariant. Each aspirin *tablet* usually contains 325 mg of the compound, held together with an inert binder (usually starch). Aspirin has been tested extensively. The conclusions of impartial studies are invariably the same: the only significant difference between brands is price. In fact, nearly all of the 14 million kg of aspirin produced annually in the United States is made by two companies, Dow Chemical and Monsanto. Only two drug companies, Sterling Drug (which makes Bayer) and Norwich–Eaton, make their own aspirin.

"Buffered" aspirin contains antacids but is not truly buffered. A *buffer* reacts with either acid or base and keeps the pH of a solution essentially constant. The antacids neutralize acid only; they do not buffer. Some people experience mild stomach irritation when they take aspirin on an empty stomach. Eating a little food first or drinking a full glass of water with aspirin is just as effective as taking "buffered" tablets.

"Extra–strength" formulations simply contain 500 mg of aspirin rather than the usual 325 mg. They have no other active ingredients. Simple arithmetic tells us that three plain aspirin tablets are equal to two extra–strength tablets in dosage, and they are usually much lower in price.

Aspirin relieves minor aches and pains and suppresses inflammation. Aspirin acetylates, and thus inhibits, an enzyme (cyclooxygenase) necessary for the synthesis of prostaglandins (see Section I.6). Among their many functions, prostaglandins are involved in inflammation, increased blood pressure, and the contraction of smooth muscle. Elevated concentrations of prostaglandins appear to activate pain receptors in the tissues, making the tissues more sensitive to any pain stimulus. Generally speaking, prostaglandins enhance inflammatory effects, whereas aspirin diminishes them. Aspirin does not cure whatever is causing the pain; it merely kills the messenger. Aspirin often is the initial drug of choice for treatment of arthritis, a disease characterized by the inflammation of joints and connective tissues.

Aspirin also acts as an **anticoagulant**—it inhibits the clotting of blood. Aspirin should not be used by people facing surgery, childbirth, or other hazards involving the possible loss of blood (nonuse should start a week before the hazard). On the other hand, small daily doses seem to lower the risk of coronary heart attack and stroke, presumably by the same anticoagulant action that causes bleeding in the stomach. For people with heart problems, many doctors now prescribe one aspirin tablet to be taken every other day. (Some doctors prescribe daily doses of one baby aspirin, 81 mg.)

Fevers are induced by substances called **pyrogens,** compounds produced by and released from leukocytes and other circulating cells. Pyrogens usually use prostaglandins as secondary mediators. Fevers therefore can be reduced by aspirin and other prostaglandin inhibitors.[2]

Our bodies try to fight off infections by elevating body temperature. Mild fevers (those below 39 °C, or 102 °F) usually are therefore best left untreated. High fevers, however, can cause brain damage and require immediate treatment.

Aspirin is not effective for severe pain—for example, pain from a migraine headache. Prolonged use of aspirin, as for arthritic pain, can lead to gastrointestinal disorders. Like all other drugs, aspirin is somewhat toxic. It is the drug most often involved in the accidental poisoning of children. The toxicities of aspirin and other drugs are listed in Table F.1.

Use of aspirin to treat children feverish with flu or chicken pox is associated with **Reye's syndrome.** This syndrome is characterized by vomiting, lethargy, confusion, and irritability. Fatty degeneration of the liver and other organs can lead to death unless treatment is begun promptly. Just how aspirin use enhances the onset of Reye's syndrome is not known, but the correlation is quite strong. Aspirin products bear a warning not to use aspirin to treat children with fevers.

Still another hazard associated with aspirin use is allergic reaction. In some people, an allergy to aspirin can cause skin rashes, asthmatic attacks, and even loss of consciousness. Some doctors claim that the allergic reaction may be delayed 3 to 5 hours, so the victim may not associate the reaction with aspirin. Susceptible individuals must be careful to avoid aspirin by itself or in any combination with other drugs.

[2] Some pyrogens do not work through prostaglandins. Aspirin doesn't affect those pyrogens.

Table F.1 Acute Toxicities of Chemicals Presently or Formerly Used in Over-the-Counter Drugs

Chemical Compound	LD_{50}[a]
Acetaminophen	338[b]
Acetanilide	800
Aspirin	1500
Caffeine	355 (246)
Diphenhydramine	500
Ibuprofen	(1050)
Methyl salicylate	887
Naproxen	534
Phenacetin	1650
Piroxicam	360[b]

[a] LD_{50} values are for oral administration of the drug in rats in milligrams per kilogram of body weight (unless otherwise noted). Values in parentheses are for male rats.
[b] Orally in mice.

Source: Susan Budavari (Ed.), *The Merck Index,* 11th ed., Rathway, NJ: Merck and Co., 1989.

F.2 Aspirin Substitutes and Combination Pain Relievers

People allergic to aspirin may safely take substitute medicines (Figure F.2). The most common is acetaminophen. This compound gives pain relief and fever reduction comparable to the action of aspirin. Unlike aspirin, however, it is not effective against inflammation, so it is of limited use to people with arthritis. Neither does it induce bleeding, so it is often used to relieve the pain that follows minor surgery. The fact that acetaminophen does not promote bleeding probably accounts for the fact that it is used in hospitals more frequently than aspirin. Acetaminophen usually costs more than aspirin, especially in the form of highly advertised brands such as Tylenol. Regular acetaminophen tablets are 325 mg; "extra–strength" forms are 500 mg. Overuse of acetaminophen is linked to liver and kidney damage, especially in those who drink a lot of alcohol.

Three antiinflammatory drugs—ibuprofen, naproxen, and ketoprofen (see again Figure F.2)—are also used as aspirin substitutes. These drugs may be a bit more effective than aspirin in treating inflammation, and they relieve mild pain and reduce fevers. They are usually much more expensive than aspirin.

Much of what we spend on analgesics is spent on combinations of aspirin and other drugs. Is our money well spent? What are those other drugs? Do they really add anything to the effectiveness of the medication? These are questions that can be answered rather simply. Let's look first at the chemical compounds in some of the more familiar "combination pain relievers."

For many years, the most familiar combination was aspirin, phenacetin, and caffeine (APC). This combination was available under a variety of trade names, or it could be purchased as APC tablets USP, usually at a lower price than the proprietary medications. Phenacetin has about the same effectiveness as aspirin in reducing fever and relieving minor aches and pains. It has been implicated in damage to the

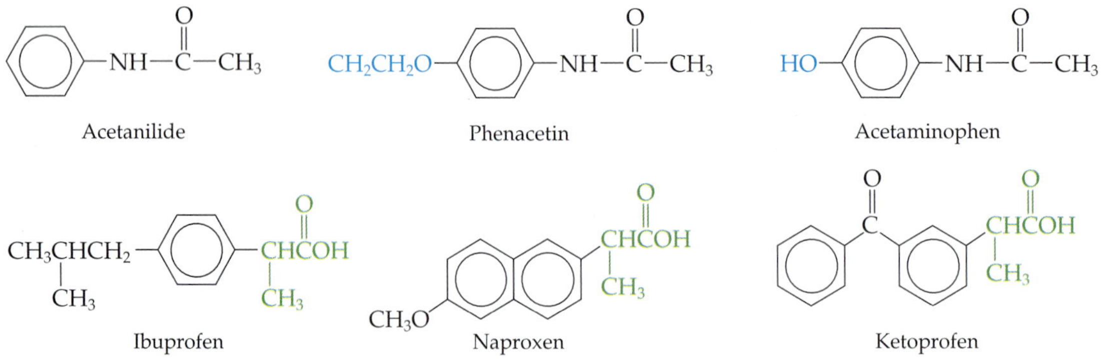

▲ FIGURE F.2

Some aspirin substitutes. Phenacetin and acetaminophen are derivatives of acetanilide. Ibuprofen (Advil, Nuprin), naproxen (Aleve), and ketoprofen (Orudis KT) have long been available by prescription but are now available over-the-counter; all three are derivatives of propionic acid. Phenacetin and acetanilide are no longer used.

The opium poppy flower (a) and seed pod (b).

kidneys, in blood abnormalities, and as a likely carcinogen, however. The U.S. FDA banned further use of phenacetin in 1983.

Anacin, which, along with Empirin and Excedrin, was once an APC formulation, now contains only aspirin and caffeine. Caffeine is a mild stimulant found in coffee, tea, and cola syrup. There is no reliable evidence that caffeine significantly enhances the effect of aspirin. In fact, evidence indicates that for fever reduction, caffeine *counteracts* the action of aspirin. Combinations containing caffeine are therefore *less effective* than plain aspirin for this use. Excedrin also contains caffeine, along with aspirin and acetaminophen.

Many other combination products are available. Brands and formulations change frequently. Extensive studies show repeatedly that, for most people, plain aspirin is the cheapest, safest, and most effective product.

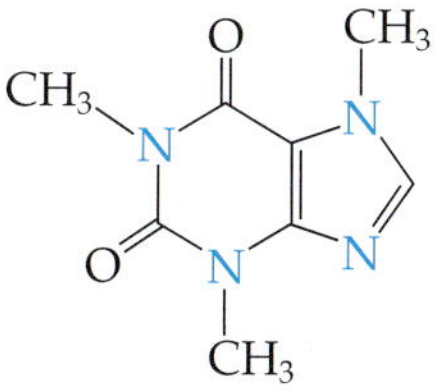

Caffeine

F.3 Opium Alkaloids

Many plants produce **alkaloids,** nitrogen-containing compounds that have physiological activity. Nicotine from tobacco and caffeine from coffee are familiar alkaloids. The opium poppy, *Papaver somniferum,* produces a number of alkaloids, including morphine and codeine. These and related compounds are called opiates. The opiates act as analgesics (pain relievers) and as **narcotics** (they induce *narcosis,* a state of profound stupor).

Morphine, first isolated in 1805, is still the most important narcotic analgesic in medicine. It remains the standard against which new analgesics are measured. Although morphine can be synthesized in the laboratory, it is still obtained from opium. When pure, morphine is an odorless, white, crystalline solid having a bitter taste, and it is insoluble in water.

Commercial opium is the dried juice from the unripened seed pod of the poppy plant. It is a complex mixture containing more than 20 compounds. The principal alkaloid, morphine, makes up about 10% of the weight of raw opium. Raw opium was used in many of the patent medicines of the nineteenth century. Ayer's Cherry Pectoral, Jayne's Expectorant, Pierce's Golden Medical Discovery, and Mrs. Winslow's Soothing Syrup were but a few.

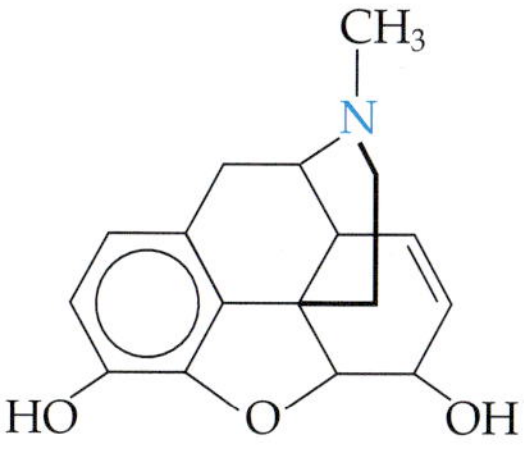

Morphine

Morphine and other narcotics were placed under the federal government's control by the Harrison Act of 1914. Morphine is still used by

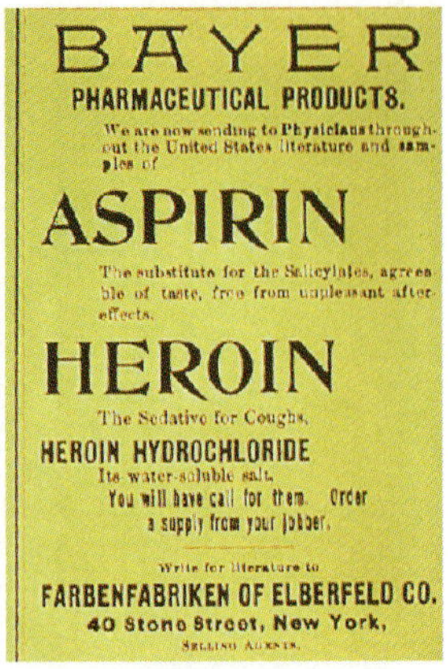

Heroin was regarded as a safe medicine in 1900; it was widely used as a sedative for coughs.

prescription for relief of severe pain. It also induces lethargy, drowsiness, confusion, euphoria, chronic constipation, and depression of the respiratory system. It is strongly addictive if administered in amounts greater than the prescribed doses or for a period longer than the prescribed time.

Slight changes in the molecular architecture of morphine produce altered physiological properties. Replacement of one of the $-OH$ groups by an $-OCH_3$ group produces codeine. Actually, codeine occurs in opium to an extent of about 0.5%. It is usually synthesized, however, by methylating the more abundant morphine molecules. About 55,000 kg of codeine is produced each year in the United States, enough for 16 doses of 15 mg for every person in the country.

Codeine is similar to morphine in its physiological action, except that codeine is less potent and has less tendency to induce sleep. It is also thought to be less addictive. In amounts of less than 2.2 mg/mL, codeine is exempt from stringent narcotics regulations and is used in a few "controlled substance" cough syrups.

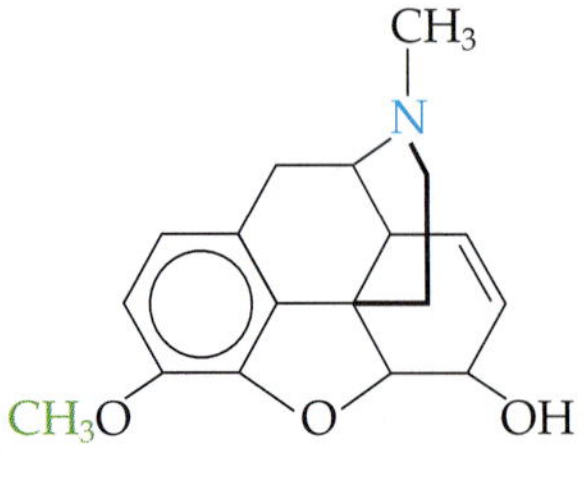

Codeine
(Methylmorphine)

In the laboratory, reaction of morphine with acetic anhydride produces heroin. This morphine derivative was first prepared by chemists at the Bayer Company of Germany in 1874. It received little attention until 1890, when it was proposed as an antidote for morphine addiction. Shortly thereafter, Bayer widely advertised heroin as a sedative for coughs, often in the same ads as aspirin. It soon was found, however, that heroin induced addiction more quickly than morphine and that heroin addiction was harder to cure.

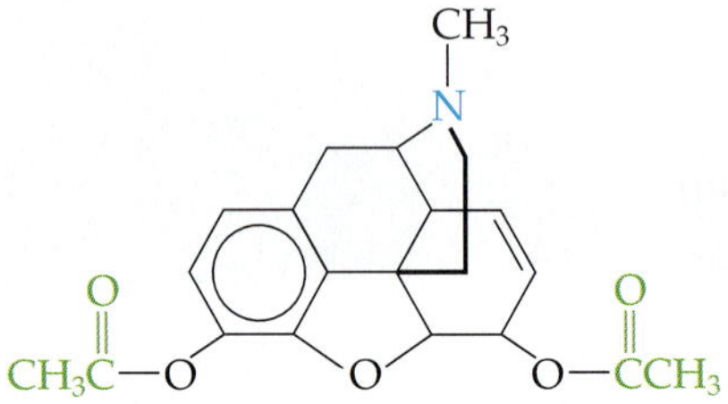

Heroin
(Diacetylmorphine)

The physiological action of heroin is similar to that of morphine, except that heroin seems to produce a stronger feeling of euphoria for a longer period of time. Heroin is not legal in the United States, even by prescription. It has, however, been advocated for use in pain relief in terminal cancer patients and has been so used in Britain.

Deaths from heroin usually are attributed to overdose, most frequently caused by the user not being aware of the amount of heroin in a given amount of powder. As an illustration, the office of the chief medical examiner for New York City analyzed 132 samples of drugs, supposedly heroin, that had been confiscated on the streets. Twelve contained no heroin at all. The remaining 120 varied from 1 to 77% heroin. A user could think he or she was getting a dose of 1 unit, when he or she was actually getting 77 times as much—a catastrophic overdose.

F.4 Synthetic Narcotics: Analgesia and Addiction

Much research has gone into developing a drug that is as effective as morphine in relieving pain but is not addictive. Perhaps the best known of the synthetic narcotics is meperidine (Demerol). Meperidine is somewhat less effective than morphine, but it has the advantage that it does not cause nausea. Repeated use, unfortunately, does lead to addiction.

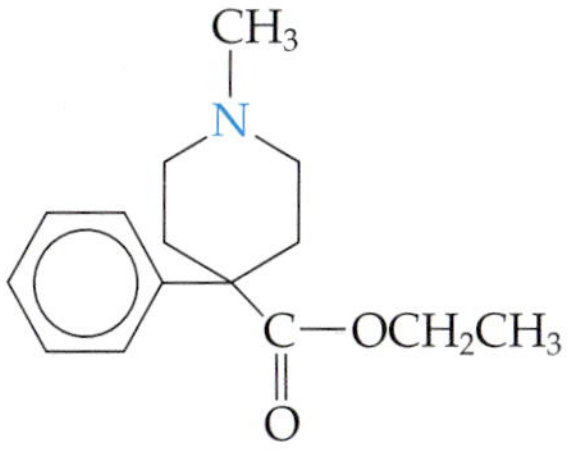

Meperidine
(Demerol)

Another synthetic narcotic is methadone. This drug has been widely used to treat heroin addiction. Like heroin, methadone is highly addictive. However, when taken orally, it does not induce the sleepy stupor characteristic of heroin intoxication. Unlike a heroin addict, a person on methadone maintenance usually is able to hold a productive job. Methadone is available free in clinics. If an addict who has been taking methadone reverts to heroin, the methadone in his or her system effectively blocks the euphoric rush normally given by heroin and so reduces the addict's temptation to use heroin.

Methadone maintenance is not a perfect answer. Perhaps it is not even a good one. When injected into the body, methadone produces an effect similar to that of heroin, and methadone has been diverted for illegal use in this manner. And an addict on methadone is still an addict.

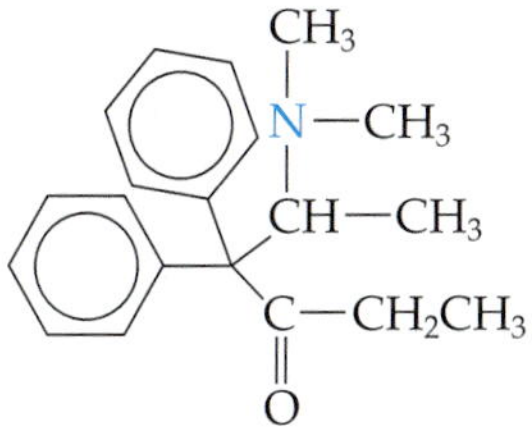

Methadone

Chemists have synthesized thousands of morphine analogs. Only a few have shown significant analgesic activity. Most are addictive. Morphine acts by binding to receptors in the brain. Molecules that mimic the action of a drug are called **agonists.** An **antagonist** blocks the action of a drug. Morphine antagonists are molecules that block the action of morphine, most likely by blocking the receptors. Some molecules have both agonist and antagonist effects. These substances show great promise as analgesics. An example is pentazocine (Talwin). It is less addictive than morphine and yet effective for relief of pain. There is some hope that an effective,

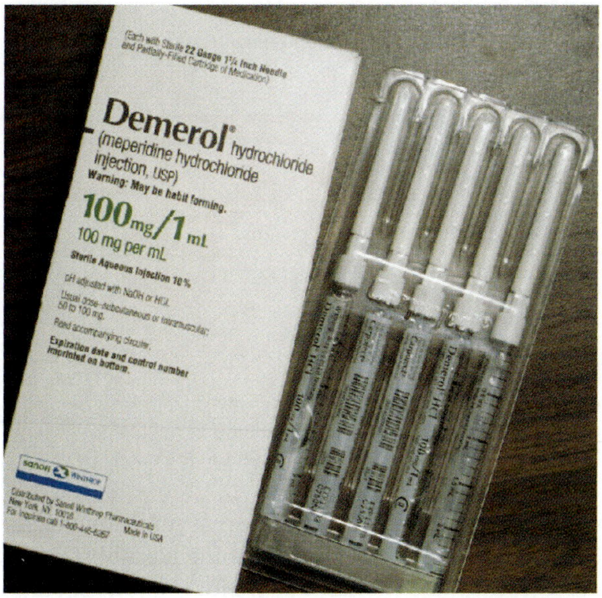

Dosage forms of meperidine (Demerol), a synthetic narcotic.

nonaddictive analgesic will be developed, but to date the two effects seem inseparable.

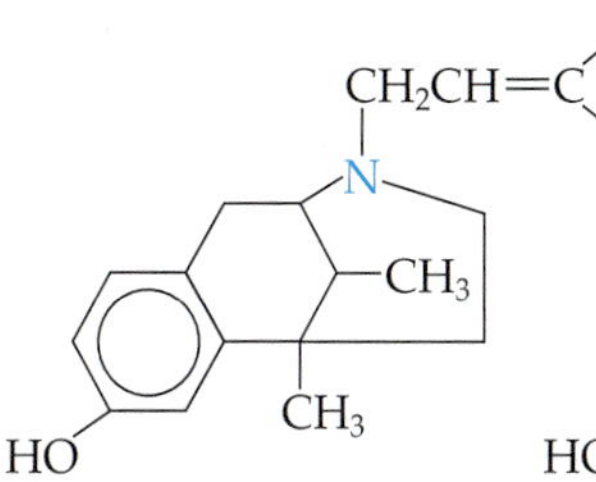

Pentazocine
(Talwin)

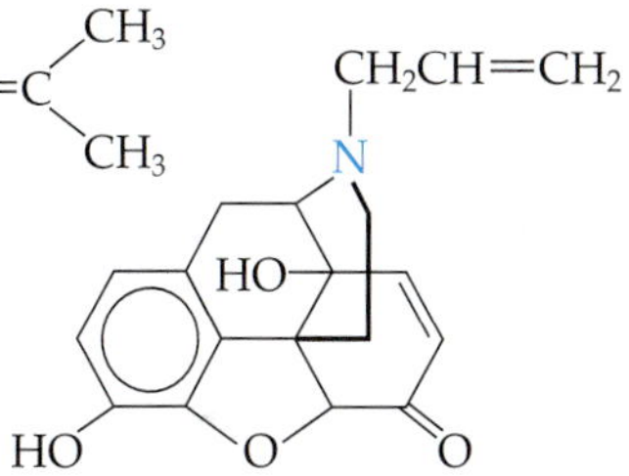

Naloxone

Pure antagonists such as naloxone are of value in treating opiate addicts. Overdosed addicts can be brought back from death's door by an injection of naloxone. Long–acting antagonists can block the action of heroin for as much as a month, thus aiding an addict in overcoming his or her addiction.

F.5 A Natural High: The Brain's Own Opiates

Morphine acts by fitting specific receptor sites on neurons in the brain (Figure F.3). The existence of these opiate receptors was first demonstrated in 1973 by Solomon Snyder and Candace Pert at Johns Hopkins University School of Medicine.

Why should the human brain have receptors for a plant-derived drug such as morphine? There

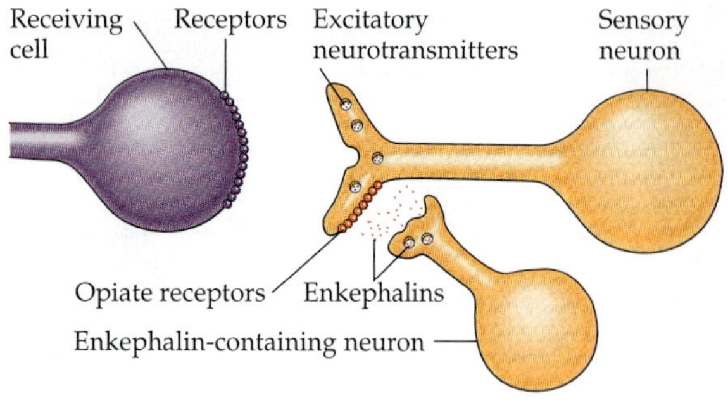

▲ **FIGURE F.3**

A proposed mechanism for opiate activity. Enkephalins bind to the opiate receptors and block the release of neurotransmitters that would convey the pain message to the brain.

seemed to be no good reason, so several investigators started a search for morphinelike substances produced by the human body. Not one but several such substances, called **endorphins** ("endogenous morphines"), were soon found. Each was a short peptide chain composed of amino acid units (Chapter 21). Endorphins containing five amino acid units are called **enkephalins.** There are two enkephalins, and they differ only in the amino acid at the end of the chain. *Leu*-enkephalin has the sequence Tyr-Gly-Gly-Phe-Leu, and *Met*-enkephalin is Tyr-Gly-Gly-Phe-Met.[3] Other endorphins having chains of 30 amino acids also were found.

The enkephalins have been synthesized in the laboratory and shown to be potent pain relievers. Their use in medicine is quite limited, however, because, after being injected, they are rapidly broken down by the enzymes that hydrolyze proteins. It is hoped, though, that enkephalin analogs more resistant to hydrolysis can be employed as morphine substitutes for the relief of pain. Unfortunately, both the natural enkephalins and the synthetic analogs seem to be addictive.

It appears that endorphins are released as a response to pain deep in the body. Bruce Pomeranz of the University of Toronto has collected evidence that acupuncture anesthetizes somewhat by stimulating the release of the brain "opiates." The long needles stimulate deep sensory nerves that cause the release of the peptides that then block the pain signals. Endorphin release also has been used to explain other phenomena once thought to be largely psychological. A soldier wounded in battle feels no pain until the skirmish is over, because the body has secreted its own painkiller. The production of these compounds during strenuous athletic activity also may explain the "high" reported by distance runners.

[3] The three–letter symbols are abbreviations for amino acids; see Table 21.1.

F.6 LSD: A Hallucinogenic Drug

One of the most interesting amides of all is the *N,N*–diethylamide of lysergic acid, better known as LSD (from the German *lysergsaure diethylamid*) or "acid." The physiological properties of this compound were discovered quite accidentally by Albert Hofmann in 1943. Hofmann, a chemist at the Sandoz Laboratories in Switzerland, unintentionally ingested some LSD. He later took 250 μg, which he considered a small dose, to verify that LSD had caused the symptoms he had experienced. Hofmann had a very rough time for the next few hours, exhibiting such symptoms as visual disturbance and schizophrenic behavior.

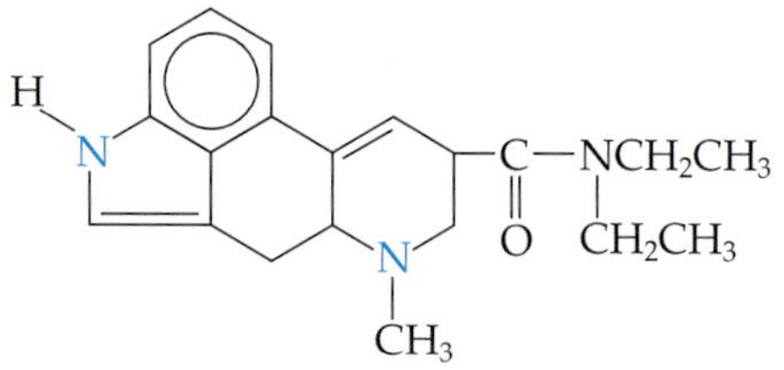

Lysergic acid diethylamide
(LSD)

Lysergic acid is obtained from ergot, a fungus that grows on rye.[4] It is converted to the diethylamide by treatment with thionyl chloride ($SOCl_2$) followed by diethylamine. Note that a part of the LSD structure resembles that of serotonin (see Section G.2). LSD seems to act as a serotonin agonist.

LSD is a potent drug, as indicated by the small amount required for a person to experience its fantastic effects. The usual dose is probably about 10 to 100 μg. No wonder Hofmann had a bad time with 250 μg! To give you an idea of how small 10 μg is, let's compare that amount of LSD with the amount of aspirin in one tablet—one aspirin tablet contains 325,000 μg of aspirin.

[4] Several useful drugs are obtained from the ergot fungus. Ergotamine shrinks blood vessels in the brain; it is used to treat migraine headaches. Ergonovine causes small blood vessels to contract. It is used to induce uterine contractions and thus reduce bleeding after childbirth. Both compounds, like LSD, are amides of lysergic acid.

Is LSD a dangerous drug? A few facts are known, but most are disputed. In 1967, Maimon Cohen of the State University of New York at Buffalo reported that LSD damages chromosomes, especially those of the leukocytes (white blood cells). The report received wide publicity, but additional studies produced mixed results. Fear of damage to germ cells—and the subsequent birth of deformed babies—caused a decline in the use of LSD. The question still has not been resolved.

The great concern over the problem is due, in part, to the thalidomide tragedy of the late 1950s and early 1960s. Thalidomide was a completely legal, amidelike drug used as a tranquilizer.

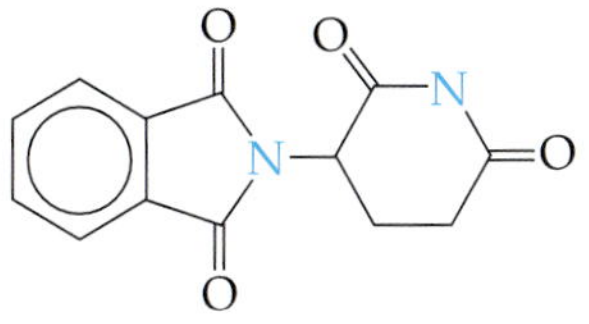

Thalidomide

It was considered so safe, based on laboratory studies, that it often was prescribed for pregnant women. In Germany it was available without a prescription. It took several years for the human population to provide evidence that laboratory animals had not. The drug had a disastrous effect on developing human embryos. Women who had taken the drug during the first 12 weeks of pregnancy had babies that suffered from phocomelia, a condition characterized by shortened or absent arms and legs and other physical defects. The drug was used widely in Germany and Great Britain, and these two countries bore the brunt of the tragedy. The United States escaped relatively unscathed because an official of the FDA had believed that there was evidence to doubt the drug's safety and therefore had not approved it for use in the United States.

F.7 Marijuana: Some Chemistry of *Cannabis*

Although marijuana is neither an ester nor an amide, we discuss it here because of its significance. (In the United States, marijuana is second only to alcohol as an intoxicant.) Even though many books have been written about marijuana, all we know for certain about the drug fills only a few pages. Let's look at some chemistry of the weed and at some of the ways chemists are involved with the marijuana problem.

The weed *Cannabis sativa* has long been useful. The stems yield tough fibers for making rope, and the plant has been used as a drug in tribal religious rituals. The term **marijuana** refers to a preparation made from the leaves, flowers, seeds, and small stems of the plant. These are generally dried and smoked and referred to as "pot." They contain a variety of chemical substances, many still unidentified. The principal active ingredient, however, is tetrahydrocannabinol (THC). Actually, there are several active cannabinoids in marijuana; only one is shown here.

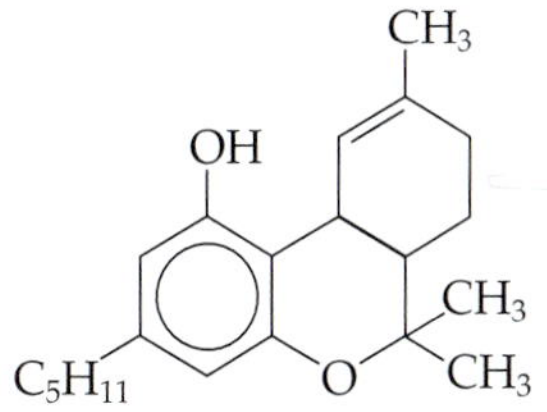

Tetrahydrocannabinol
(THC)

Marijuana plants vary considerably in THC content. Most of the marijuana sold in North America has a THC content of about 1%. Plants native to the United States have a low THC content, usually about 0.1%. Potency depends on the genetic variety of plant, not to any significant extent on the climate or the soil where it is grown.

More potent preparations sometimes are made from marijuana. By selecting only the flowering tops and tender top leaves, you get a stronger product called *ganja*. Jamaican ganja has a THC content of between 4 and 8%. Indian ganja is generally somewhat less potent. By collecting only the resinous secretions of the flowering parts, you get a product called hashish, or "hash" (THC between 5 and 12%). Liquid hash and hash oil are probably solvent extracts of marijuana.

The physiological effects of marijuana are difficult to measure, partly because of variable THC content. A marijuana variety of standard potency is now grown and supplied for controlled clinical studies. With this standard product, some effects can be measured in reproducible experiments. Smoking *Cannabis* increases the pulse rate, distorts the sense of time, and impairs some complex motor functions. These effects can be measured easily. Other results also have been noted widely, if less quantitatively. Marijuana smoking sometimes induces a euphoric sense of lightness—a floating sensation. Sometimes it causes a feeling of anxiety. Often, the user has an impression of brilliance, although studies have shown no mind-expanding effects. Users sometimes experience hallucinations, although these are much less frequent than with LSD. Marijuana seems to heighten one's enjoyment of food, with users relishing beans as much as they normally would enjoy steak.

The long-term effects of marijuana use are more difficult to evaluate. Some people claim that smoking marijuana leads to the use of harder drugs. There is little objective evidence of this. More heroin addicts start drug use with alcohol than with marijuana.

There is some evidence—both direct and indirect—that marijuana causes brain damage. Studies in rats indicate that marijuana causes brain lesions. Heavy users often are lazy, passive, and mentally sluggish. It is difficult, however, to prove that these are effects of marijuana use. There are millions of users, and even if the observed brain damage is due to marijuana, it is less extensive than that caused by alcohol.[5]

Some people claim that excessive use of marijuana leads to psychoses. The drug has long been known to induce short-term psychotic episodes in those already predisposed and in others who take excessive amounts. Long-term psychoses, however, occur at the same rate among regular marijuana users as among the general population.

One of the more interesting reports has come from two surgeons at the Harvard Medical School. Menelaos Aliapoulios and John Harman claim to have treated 13 young men for gynecomastia (enlarged breasts). All were heavy marijuana users. Their breasts also discharged a white, milky liquid. The painful swelling receded in three of the men after they stopped smoking *Cannabis,* but three others needed surgery. The two doctors are convinced that marijuana contains a "feminizing ingredient." There is a slight structural similarity between THC and the female hormones (Special Topic I). Some studies indicate that THC binds weakly to estrogen receptors.

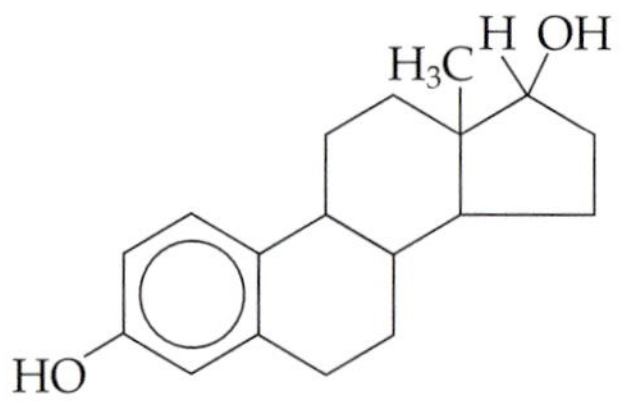

Estradiol, a female hormone

Despite its apparent feminizing properties, THC in both high and low doses causes an initial rise in testosterone levels in men. With high doses of THC, however, the slight increase is followed by a rapid fall to below-normal testosterone levels.

Chemists have isolated the active components in marijuana and synthesized them. They can monitor the THC content of marijuana as well as monitor THC levels in the bloodstream and identify the breakdown products. They cannot, however, tell how it changes the body chemistry or what its long-term effects are.

Perhaps the most significant harm from marijuana comes through its impairment of complex motor functions—such as those used in driving an automobile. Incidentally, chemists have developed a THC detector similar to the one used to test blood alcohol, but it is too expensive for widespread use.

Unlike alcohol, THC persists in the bloodstream for several days because it is soluble in fats. The products of its breakdown remain in the blood for as long as 8 days. The persistence of these chemicals inside the body indicates that some of a given dose may still be active in the body at the time another dose is taken. This might account for the fact that an experienced pot smoker can get high on a dose that doesn't affect a novice.

Marijuana has some legitimate medical uses. It reduces pressure in the eyes of people who have glaucoma. If not treated, the buildup of pressure eventually causes blindness. Also, marijuana relieves the nausea that afflicts cancer patients undergoing radiation treatment and chemotherapy.

[5] The gene for a THC receptor was cloned in 1990 (see Section 23.9). The receptors are found in movement-control centers, thus explaining the loss of coordination in those intoxicated with the drug. The memory and cognition areas of the brain also are rich in THC receptors. This explains why marijuana users often do poorly on tests. There are few receptors in the brainstem, where breathing and heartbeat are controlled. This correlates with the fact that it is difficult to get a lethal dose of marijuana.

Key Terms

agonist (F.4)
alkaloid (F.3)
analgesic (F.1)
antagonist (F.4)
anticoagulant (F.1)
antipyretic (F.1)

drug (Introduction)
endorphin (F.5)
enkephalin (F.5)
marijuana (F.7)
narcotic (F.3)
pyrogen (F.1)
Reye's syndrome (F.1)

Review Questions

1. Define, explain, or give an example of each of the following terms.
   a. drug
   b. analgesic
   c. antipyretic
   d. pyrogen
   e. Reye's syndrome
   f. narcotic
   g. opium
   h. agonist
   i. antagonist
   j. endorphin
   k. enkephalin
2. How does heroin differ in structure from morphine? How does it differ in physiological properties?
3. a. What are the structure and chemical name for aspirin?
   b. In what ways may one brand of aspirin differ from another?
   c. In what ways must brands of aspirin be the same?
4. What is an alkaloid? Name several alkaloids.
5. What are the effects of a hallucinogenic drug?
6. How does methadone work as a treatment for heroin addiction?
7. How are endorphins related to (**a**) the anesthetic effect of acupuncture and (**b**) the absence of pain in a badly wounded soldier?
8. What are some of the problems in the clinical evaluation of LSD?
9. What are some of the problems in the clinical evaluation of marijuana?
10. What is the role of the chemist in the marijuana controversy?
11. How might marijuana have a feminizing effect on males?
12. Why is tetrahydrocannabinol retained in the body for several days?
13. In what areas of the brain are THC receptors found? How does the location of these receptors explain some of the effects of THC as a drug?
14. A "maximum-strength" aspirin has 1000 mg of aspirin per two-tablet dose. How many "regular" 325-mg aspirin tablets would you take to get about the same amount of aspirin?

Problems

15. To what family of organic compounds does ibuprofen belong?
16. What alkyl group is attached to the *para* position of the benzene ring of the ibuprofen molecule? Can you see how the generic name ibuprofen was derived?
17. List the reagents necessary to carry out the following conversions.

a.

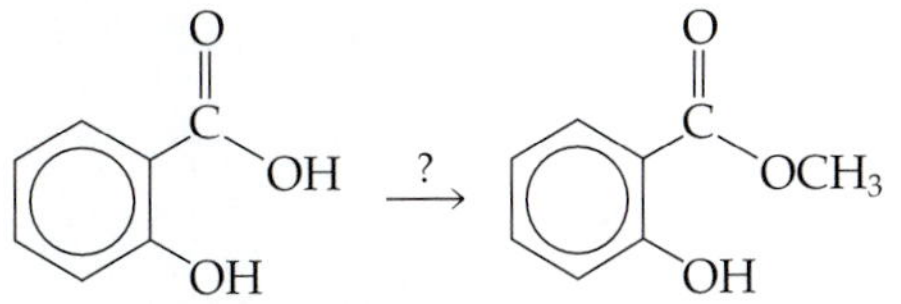

b.

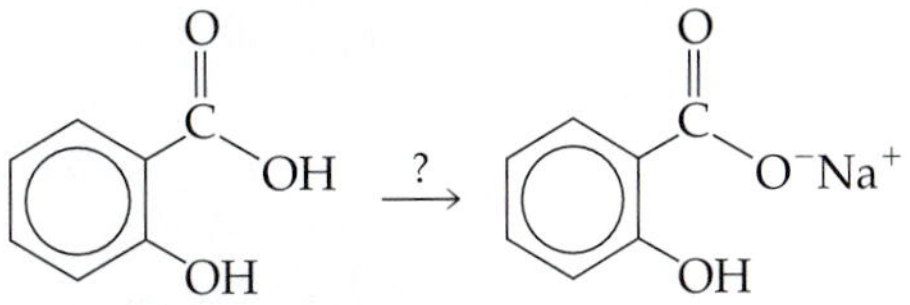

18. List the reagents necessary to carry out the following conversions.

a.

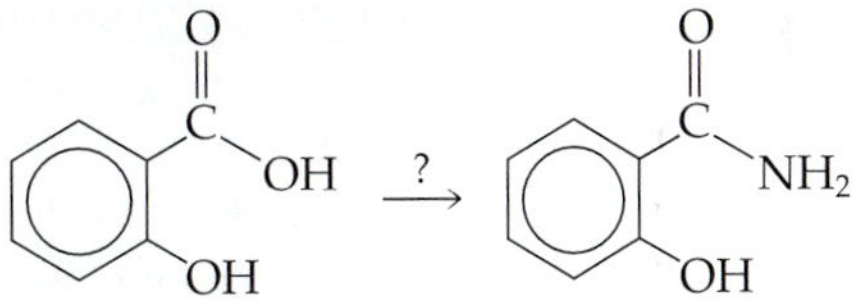

b.

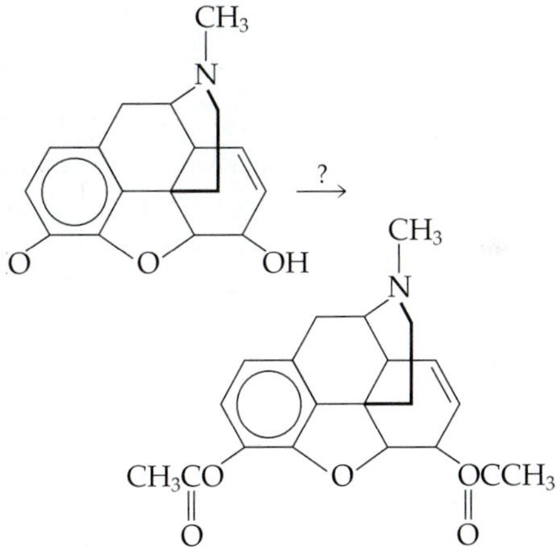

Projects

19. Examine the labels of at least five "combination" pain relievers (e.g., Excedrin, Empirin, Anacin). Make a list of the ingredients in each. Look up the properties (medical use, dosage, side effects, toxicity) in a reference work such as *The Merck Index*.
20. Do a cost analysis on at least five brands of plain aspirin, calculating the cost per gram.

17 Amines and Derivatives

A nerve cell in the human brain. Some amines are vital to communication between cells; others are used to block these signals.

If carbon compounds are the basis of life, nitrogen compounds are the bases of life (pun intended). Recall from Chapter 10 that ammonia is a nitrogen-containing weak base. Amines are alkyl (and aryl) derivatives of ammonia, and they too are weak bases. These organic bases occur in living (and especially in once-living but now-decaying) organisms.

Nitrogen is an essential constituent of many physiologically active compounds. All enzymes—indeed, all proteins—contain nitrogen. Many vitamins and hormones contain nitrogen, as do most drugs. And nitrogenous bases are part of the complex structure of the compounds that carry our genetic heritage, the nucleic acids DNA and RNA (bases in acids!). In this chapter, we discuss the amines generally, and in Special Topic G, we discuss a number of related nitrogen-containing compounds that exhibit interesting physiological effects. The discussion of proteins and nucleic acids we shall save for later chapters.

We shall also save for subsequent chapters a consideration of the implications of the following facts. Plants can take inorganic nitrogen, usually in the form of nitrate or ammonium salts, and combine it with carbon compounds from photosynthesis to make all the organic nitrogen compounds they require. But animals are not quite so clever. They require in their diet some preformed organic nitrogen compounds that are essential to their health but that they themselves cannot synthesize.

17.1 Structure and Classification of Amines

Chapter 14 shows that alcohols and ethers can be considered derivatives of water. In a similar manner, amines are derived from ammonia. Amines are classified according to the number of carbon atoms bonded directly to the nitrogen atom. A **primary (1°) amine** has one alkyl (or aryl) group on the nitrogen, a **secondary (2°) amine** has two, and a **tertiary (3°) amine** has three (Figure 17.1).

This use of the terms *primary, secondary,* and *tertiary* must be distinguished from their previous use in connection with alcohols (Section 14.2). For example, consider structures I and II in Figure 17.2. Compound I is a primary amine because only one of the nitrogen's bonds is attached to a carbon atom, but compound II is a *secondary* alcohol. When determining whether an alcohol is primary, secondary, or tertiary, one counts the number of carbons bonded to the carbon attached to the oxygen, not to the oxygen itself. Another difference can be seen in structures III and IV. Compound III is a secondary amine, but compound IV is an ether (*not* an alcohol, secondary or otherwise). When there is only one carbon group attached to oxygen, the compound is an alcohol; when there are two, the compound is an ether. In contrast, whether there are one or two (or three) alkyl or aryl groups attached to nitrogen, the compounds are all classified as amines.

Let's look at the structure of ammonia and the amines a little more closely. In ammonia, three hydrogens are bonded to nitrogen. The nitrogen also has an unshared pair of electrons:

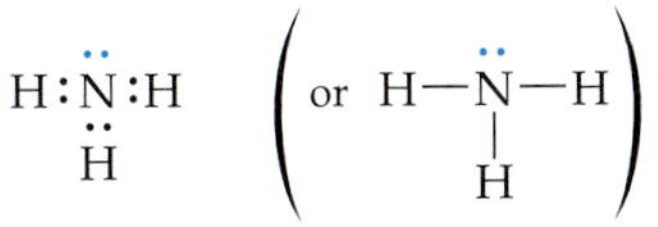

Ammonia can undergo a reaction in which it shares its normally nonbonding electrons. It can accept a proton and form the ammonium ion:

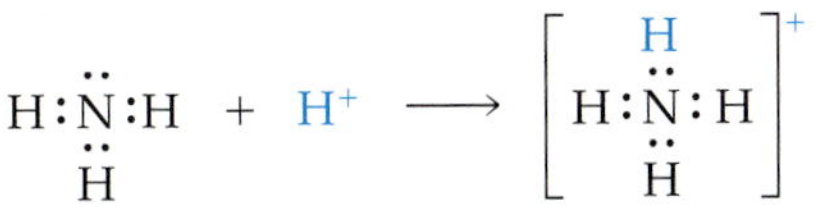

This, of course, is how ammonia reacts as a base. An amine also has an unshared pair of electrons, and it too can act as a base. This reaction we consider in detail in Section 17.4. For the moment, let's concentrate on the fact that in the ammonium ion, nitrogen is bonded to four hydrogens. Just as amines are derived from ammonia by replacement of one or several of the hydrogens with alkyl or aryl groups, so too can substituted ammonium ions be derived from the simple ammonium ion. Any or all of the hydrogens on the ammonium ion can be replaced by alkyl (or aryl) groups:

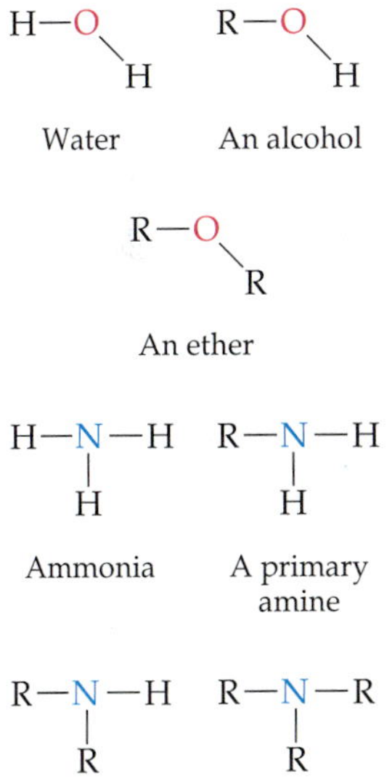

▲ **FIGURE 17.1**
Amines are derived from ammonia in a manner similar to that in which alcohols and ethers are derived from water.

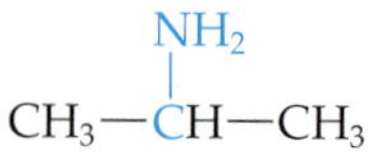

$$NH_2$$
$$CH_3-CH-CH_3$$

I

Primary amine

$$OH$$
$$CH_3-CH-CH_3$$

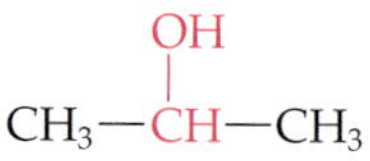

II

Secondary alcohol

$$CH_3-N-CH_3$$
$$\quad\quad|$$
$$\quad\quad H$$

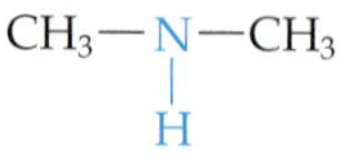

III

Secondary amine

$$CH_3-O-CH_3$$

IV

Ether

▲ **FIGURE 17.2**

Organic nitrogen-containing compounds are classified in a different manner from organic oxygen-containing compounds.

$$\left[\begin{array}{c} H \\ | \\ R-N-H \\ | \\ H \end{array}\right]^+ \quad \left[\begin{array}{c} R \\ | \\ R-N-H \\ | \\ H \end{array}\right]^+ \quad \left[\begin{array}{c} R \\ | \\ R-N-R \\ | \\ H \end{array}\right]^+ \quad \left[\begin{array}{c} R \\ | \\ R-N-R \\ | \\ R \end{array}\right]^+$$

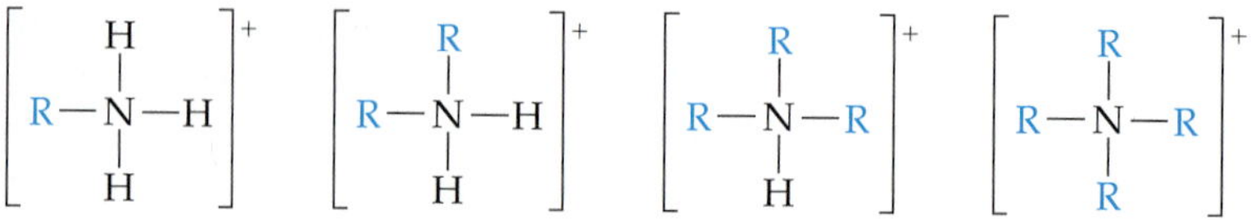

The ion in which all four hydrogens are replaced by alkyl groups is a *quaternary* ammonium ion. Compounds that incorporate this type of ion are called **quaternary (4°) ammonium salts.**

17.2 Naming Amines

To name simple aliphatic amines, we merely specify the alkyl groups attached to nitrogen and add the suffix *-amine*.

Example 17.1

Name and classify $(CH_3)_2CHNH_2$.

Solution

The alkyl group attached to nitrogen is isopropyl; thus the name is isopropylamine. There is only one alkyl group attached to the nitrogen, so the amine is primary.

Example 17.2

Name and classify $CH_3CH_2NHCH_2CH_3$.

Solution

There are two ethyl groups attached to the nitrogen. The compound is diethylamine, a secondary amine.

Example 17.3

Name and classify $CH_3NHCH_2CH_2CH_3$.

Solution

There are a methyl group and a propyl group on the nitrogen. The compound is methylpropylamine, a secondary amine.

Example 17.4

Name and classify

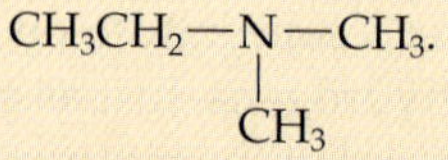

$$CH_3CH_2-N-CH_3.$$
$$\quad\quad\quad\quad\;|$$
$$\quad\quad\quad\quad CH_3$$

Solution

There are two methyl groups and one ethyl group on the nitrogen. The compound is ethyldimethylamine, a tertiary amine.

Practice Exercise

Name and classify:

a. 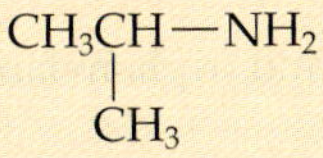   **b.** $(CH_3CH_2)_2NCH_3$ **c.** 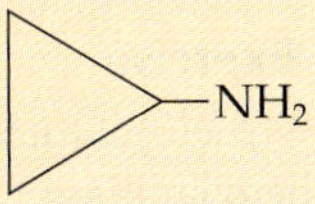

The primary amine in which the nitrogen is attached directly to a benzene ring is called *aniline*. Aryl amines are named as derivatives of this parent compound. Compounds in which the nitrogen is attached to both a benzene ring and an alkyl group are also named as derivatives of aniline. The alkyl groups are named first, and their position of attachment (i.e., at the nitrogen atom) is indicated by the capital letter *N*.

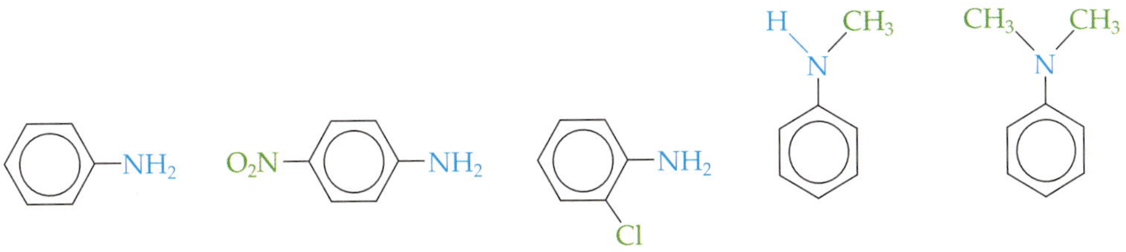

Aniline   *p*-Nitroaniline   *o*-Chloroaniline   *N*-Methylaniline   *N,N*-Dimethylaniline

Example 17.5

Name the compound 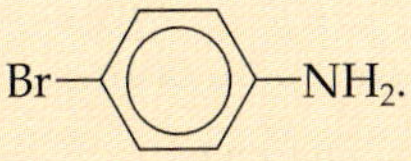.

Solution

The compound is named as a derivative of aniline. It is *p*-bromoaniline.

Practice Exercise

Name the compound 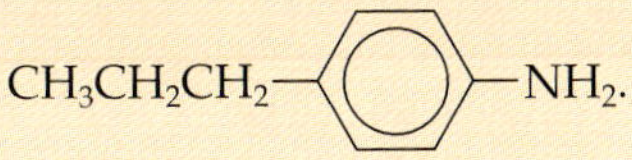.

Example 17.6

Draw the structural formulas for *p*-ethylaniline and *N*-ethylaniline.

Solution

Both compounds are derivatives of aniline. The first compound is a primary amine having an ethyl group located *para* to the *amino* ($-NH_2$) group:

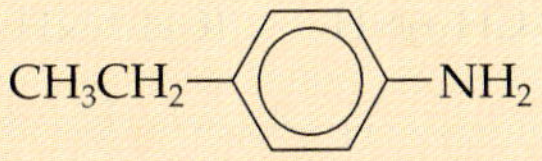

The second compound is a secondary amine in which the ethyl group is attached at the nitrogen.

$H_2NCH_2CH_2CH_2CH_2NH_2$

1,4-Diaminobutane
(Putrescine)

$H_2NCH_2CH_2CH_2CH_2CH_2NH_2$

1,5-Diaminopentane
(Cadaverine)

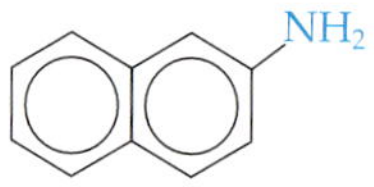

β-Naphthylamine

▲ **FIGURE 17.4**

Some amines of interest. Putrescine and cadaverine have odors indicated by their names. β-Naphthylamine is a carcinogen.

Aromatic amines generally are quite toxic. They are readily absorbed through the skin, and you must exercise caution when working with these compounds. Several aromatic amines, including β-naphthylamine (Figure 17.4), are potent carcinogens.

17.4 Amines as Bases

As we noted in Section 17.1, ammonia is basic; it can accept a proton from water to form ammonium ions and hydroxide ions:

$$:NH_3 + H_2O \rightleftharpoons NH_4^+ + OH^-$$

Ammonia is a weak base; this equilibrium strongly favors the un-ionized form. Similarly, amines have an unshared electron pair on nitrogen and can accept a proton:

$$CH_3\ddot{N}H_2 + H_2O \rightleftharpoons CH_3NH_3^+ + OH^-$$

Simple aliphatic amines are somewhat more basic than ammonia, although still much less basic than compounds such as sodium hydroxide. Aromatic amines, such as aniline, are much weaker bases than ammonia.

Nearly all amines, including those that are not very soluble in water, will react with strong acids to form water-soluble salts:

$$
\begin{array}{ccc}
& CH_3 & \\
& | & \\
CH_3-N: & + \ HNO_3(aq) \ \longrightarrow & \\
& | & \\
& CH_3 &
\end{array}
\qquad
\begin{array}{c}
CH_3 \\
| \\
CH_3-NH^+NO_3^-(aq) \\
| \\
CH_3
\end{array}
$$

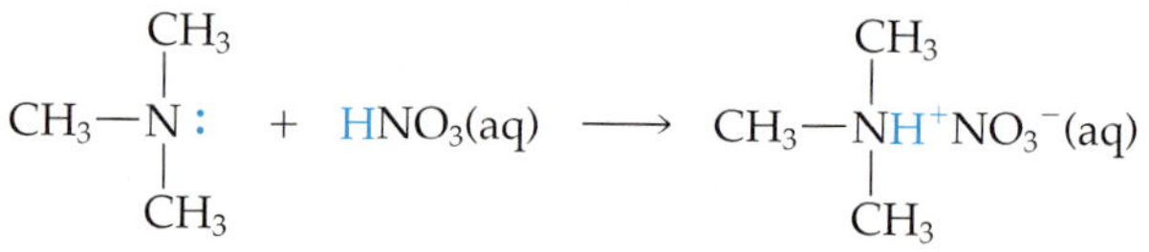

Trimethylamine

Trimethylammonium nitrate

Amine salts are named like other salts: the name of the cation is followed by that of the anion. Remember that the ions formed from aliphatic amines are named as substituted ammonium ions (Example 17.8).

Example 17.9

Name the salt

$$[CH_3NH_2CH_2CH_3]^+ \ CH_3COO^-$$

Solution

The cation has a methyl and an ethyl group attached to nitrogen and is therefore the ethylmethylammonium ion. The anion is the acetate ion. The salt is therefore ethylmethylammonium acetate.

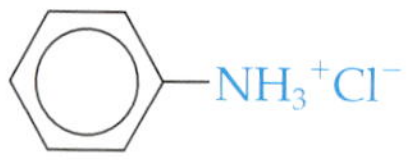

Anilinium chloride

Salts of aniline are named as anilinium compounds. An older system, still in use for naming drugs, calls the salt of aniline and hydrochloric acid "aniline hydrochloride." By this older system, the formula of the compound is frequently drawn to correspond to the name. Keep in mind that these compounds are really ionic—they are salts—even though the name and formula seem to indicate a loose

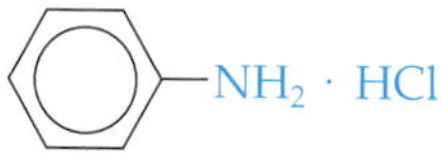

"Aniline hydrochloride"

$H_2NCH_2CH_2CH_2CH_2NH_2$

1,4-Diaminobutane
(Putrescine)

$H_2NCH_2CH_2CH_2CH_2CH_2NH_2$

1,5-Diaminopentane
(Cadaverine)

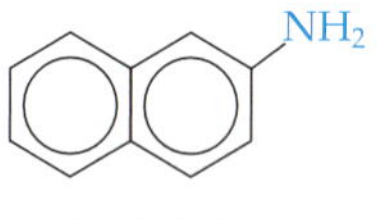

β-Naphthylamine

▲ **FIGURE 17.4**

Some amines of interest. Putrescine and cadaverine have odors indicated by their names. β-Naphthylamine is a carcinogen.

Aromatic amines generally are quite toxic. They are readily absorbed through the skin, and you must exercise caution when working with these compounds. Several aromatic amines, including β-naphthylamine (Figure 17.4), are potent carcinogens.

17.4 Amines as Bases

As we noted in Section 17.1, ammonia is basic; it can accept a proton from water to form ammonium ions and hydroxide ions:

$$:NH_3 + H_2O \rightleftharpoons NH_4^+ + OH^-$$

Ammonia is a weak base; this equilibrium strongly favors the un-ionized form. Similarly, amines have an unshared electron pair on nitrogen and can accept a proton:

$$CH_3\overset{..}{N}H_2 + H_2O \rightleftharpoons CH_3NH_3^+ + OH^-$$

Simple aliphatic amines are somewhat more basic than ammonia, although still much less basic than compounds such as sodium hydroxide. Aromatic amines, such as aniline, are much weaker bases than ammonia.

Nearly all amines, including those that are not very soluble in water, will react with strong acids to form water-soluble salts:

$$
\begin{array}{ccc}
\overset{\displaystyle CH_3}{\underset{\displaystyle CH_3}{CH_3-N:}} + HNO_3(aq) & \longrightarrow & \overset{\displaystyle CH_3}{\underset{\displaystyle CH_3}{CH_3-NH^+NO_3^-(aq)}}
\end{array}
$$

Trimethylamine Trimethylammonium nitrate

Amine salts are named like other salts: the name of the cation is followed by that of the anion. Remember that the ions formed from aliphatic amines are named as substituted ammonium ions (Example 17.8).

Example 17.9

Name the salt

$$[CH_3NH_2CH_2CH_3]^+ \; CH_3COO^-$$

Solution

The cation has a methyl and an ethyl group attached to nitrogen and is therefore the ethylmethylammonium ion. The anion is the acetate ion. The salt is therefore ethylmethylammonium acetate.

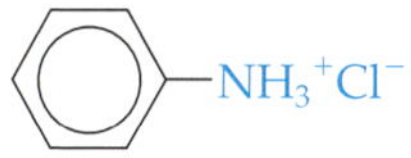

Anilinium chloride

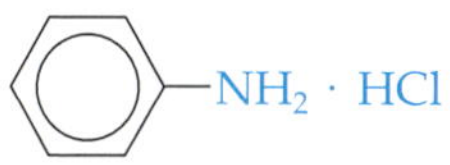

"Aniline hydrochloride"

Salts of aniline are named as anilinium compounds. An older system, still in use for naming drugs, calls the salt of aniline and hydrochloric acid "aniline hydrochloride." By this older system, the formula of the compound is frequently drawn to correspond to the name. Keep in mind that these compounds are really ionic—they are salts—even though the name and formula seem to indicate a loose

Practice Exercise

Name $CH_3CH_2\overset{+}{N}H_3$ and $(CH_3CH_2)_4\overset{+}{N}$.

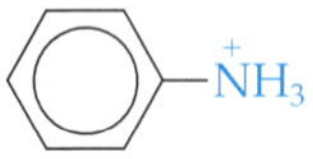

Anilinium ion

Ions in which one hydrogen of the ammonium ion is replaced by a benzene ring are named as anilinium ions instead of ammonium ions. Such ions are prepared from the corresponding aniline, and the name reflects this fact.

17.3 Physical Properties of Amines

Primary and secondary amines have hydrogens bonded to nitrogen; thus, they are capable of intermolecular hydrogen bonding (Figure 17.3a). These forces are not as strong as those between alcohol molecules (which have hydrogens bonded to oxygen, a more electronegative element than nitrogen). Amines boil at higher temperatures than alkanes but at lower temperatures than alcohols of comparable molar mass. For example, compare the boiling points of CH_3NH_2 (−6 °C) and CH_3OH (65 °C). Tertiary amines have no hydrogen bonded to nitrogen and therefore cannot form intermolecular hydrogen bonds. They have boiling points comparable to those of the ethers (Table 17.1).

All three classes of amines can hydrogen-bond to water (Figure 17.3b). Amines of low molar mass are quite soluble in water, the borderline of water solubility coming at five or six carbon atoms.

Amines have interesting (!) odors. The simple ones smell very much like ammonia. Higher aliphatic amines smell like decaying fish. Or perhaps we should put it the other way around: decaying fish give off odorous amines. The stench of rotting fish is due in part to putrescine and cadaverine (Figure 17.4), two compounds that are diamines. They arise from the decarboxylation of ornithine and lysine, respectively, amino acids found in animal cells (see Section 27.5).

Table 17.1 Physical Properties of Some Amines and Comparable Oxygen-Containing Compounds

Compound	Class	MM	Boiling Point (°C)
Butylamine	1°	73	78
Diethylamine	2°	73	55
Butyl alcohol	—	74	118
Propylamine	1°	59	49
Trimethylamine	3°	59	3
Ethyl methyl ether	—	60	6

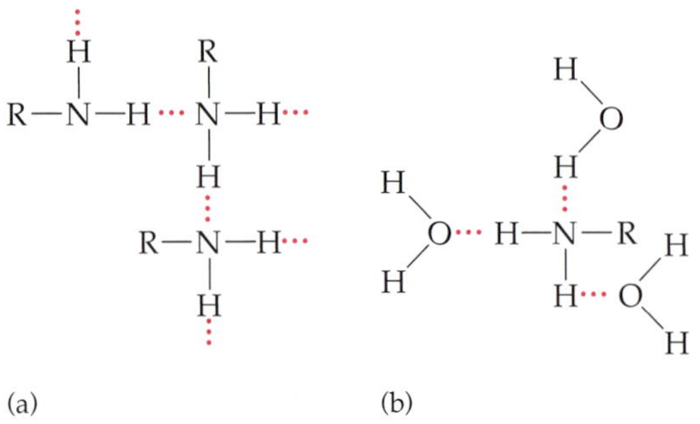

FIGURE 17.3
(a) Intermolecular hydrogen bonding in amines. (b) Hydrogen bonding between amine and water molecules.

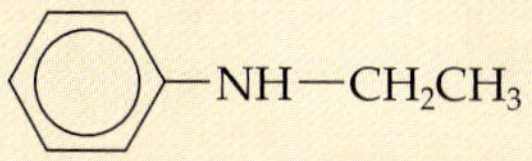

Practice Exercise

Draw the structural formulas for diphenylamine (*N*-phenylaniline) and triphenylamine (*N,N*-diphenylaniline).

In naming amines that incorporate other functional groups and amines in which the alkyl groups cannot be simply named, the amino group is named as a substituent:

$H_2N{-}CH_2CH_2{-}OH$

2-Aminoethanol
(Ethanolamine)

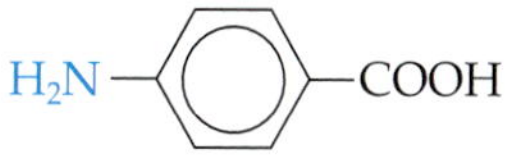

p-Aminobenzoic acid

$$\begin{array}{c} CH_3CH_2\underset{\displaystyle NH_2}{\underset{|}{C}}HCH_2CH_2CH_3 \end{array}$$

3-Aminohexane

Example 17.7

Draw the structural formula for 2-amino–3-methylpentane.

Solution

Always start with the parent compound. First draw the five-carbon pentane chain. Then attach a methyl group at the third carbon atom and an amino group at the second:

$$CH_3\overset{\displaystyle NH_2}{\overset{|}{C}}H{-}\overset{\displaystyle CH_3}{\overset{|}{C}}H{-}CH_2CH_3$$

Practice Exercise

Draw the structural formula for 2-amino-3-ethyl-1-phenylheptane.

Ammonium ions in which one or more hydrogens are replaced with alkyl groups are named in a manner analogous to that used for simple amines. The alkyl groups are named as substituents, and the parent species is regarded as the ammonium ion.

Example 17.8

Name the following ions:

$$CH_3\overset{+}{N}H_3 \qquad (CH_3)_2\overset{+}{N}H_2 \qquad (CH_3)_3\overset{+}{N}H \qquad (CH_3)_4\overset{+}{N}$$

Solution

The ions are, in order, methylammonium, dimethylammonium, trimethylammonium, and tetramethylammonium ions.

Practice Exercise

Name and classify:

a. CH_3CH-NH_2
 |
 CH_3

b. $(CH_3CH_2)_2NCH_3$

c. [triangle]—NH_2

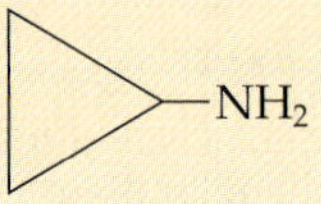

The primary amine in which the nitrogen is attached directly to a benzene ring is called *aniline*. Aryl amines are named as derivatives of this parent compound. Compounds in which the nitrogen is attached to both a benzene ring and an alkyl group are also named as derivatives of aniline. The alkyl groups are named first, and their position of attachment (i.e., at the nitrogen atom) is indicated by the capital letter *N*.

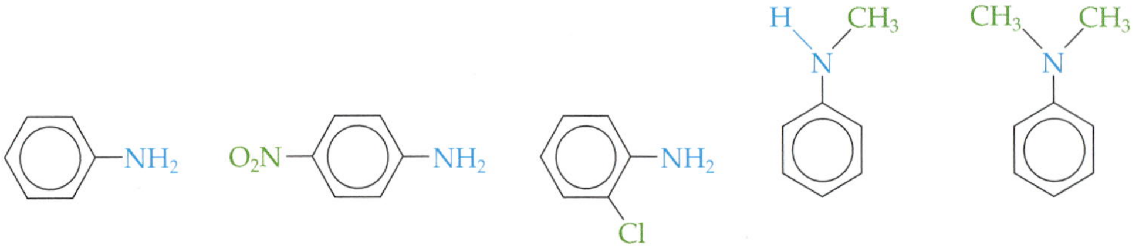

Aniline *p*-Nitroaniline *o*-Chloroaniline *N*-Methylaniline *N,N*-Dimethylaniline

Example 17.5

Name the compound Br—[benzene ring]—NH_2.

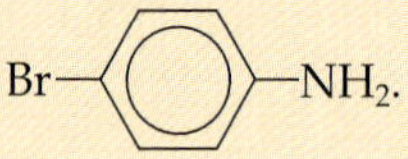

Solution

The compound is named as a derivative of aniline. It is *p*-bromoaniline.

Practice Exercise

Name the compound $CH_3CH_2CH_2$—[benzene ring]—NH_2.

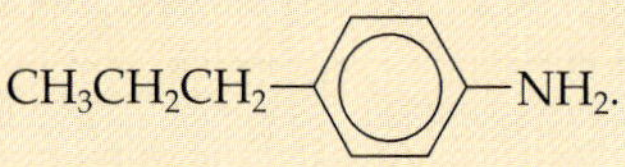

Example 17.6

Draw the structural formulas for *p*-ethylaniline and *N*-ethylaniline.

Solution

Both compounds are derivatives of aniline. The first compound is a primary amine having an ethyl group located *para* to the *amino* ($-NH_2$) group:

$$CH_3CH_2-\text{[benzene ring]}-NH_2$$

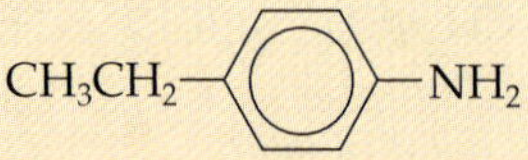

The second compound is a secondary amine in which the ethyl group is attached at the nitrogen.

association of molecules. The properties of the compounds (solubility, for example) are those characteristic of salts.

To facilitate injection in aqueous solution and/or for ease of transport through the blood, pharmaceutical companies often convert an insoluble amine into the more soluble amine salt. For instance, procaine is soluble only to the extent of 0.5 g in 100 g of water. The hydrochloride is soluble to the remarkable degree of 100 g in 100 g of water. Procaine hydrochloride, perhaps better known by the trade name Novocaine, is widely used as a local anesthetic.

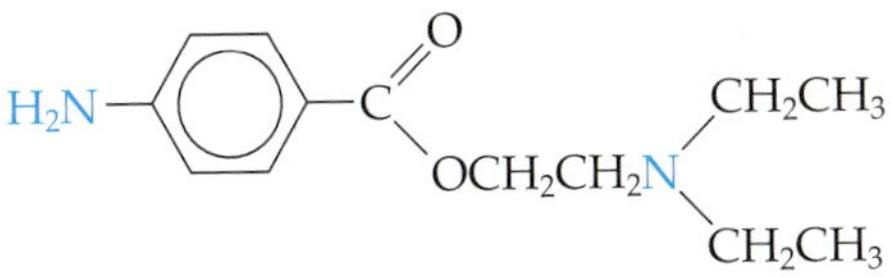

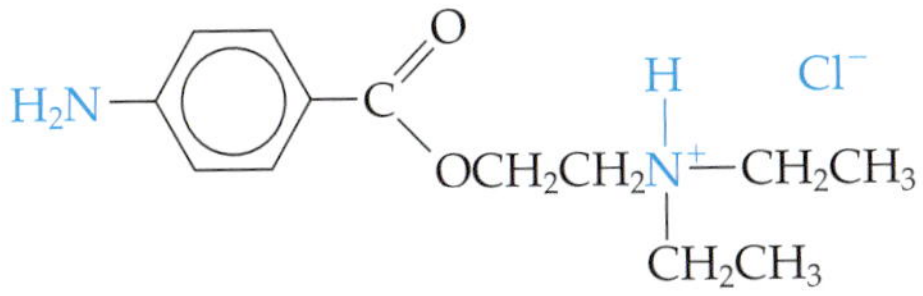

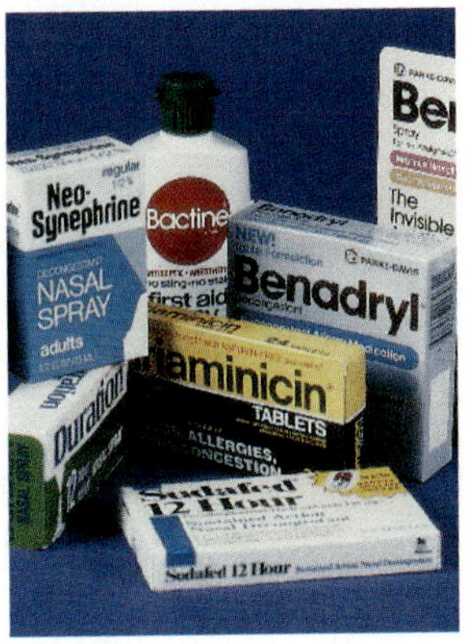

Amine hydrochloride salts are the main active ingredients in a variety of over-the-counter drugs.

Procaine

Procaine hydrochloride (Novocaine)

We also use the chemistry of amines when we put lemon juice on fish. The unpleasant fishy odor is due to amines. The citric acid in the juice converts the amines to nonvolatile salts, thus reducing the odor.

You know that a weak acid and its salt (e.g., acetic acid and sodium acetate) can be used in the preparation of buffer solutions (Section 11.5). Similarly, amines and their salts can also be used in the preparation of buffers. Whereas the acid/acid salt combination yields a solution buffered at acidic pH values, the amine/amine salt buffer stabilizes the pH in a basic range. An important example is tris(hydroxymethyl)aminomethane, often called simply tris. This compound and its salt, tris hydrochloride, buffer in the range of pH 7-9. Amine buffers find wide use in the cosmetics and textile industries, in cleaning compounds, and in biochemical research. Tris is also used in the treatment of metabolic acidosis (see Section 28.10).

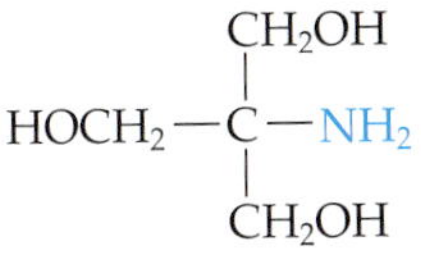

$$HOCH_2 - \underset{\underset{CH_2OH}{|}}{\overset{\overset{CH_2OH}{|}}{C}} - NH_2$$

"Tris"

$$HOCH_2 - \underset{\underset{CH_2OH}{|}}{\overset{\overset{CH_2OH}{|}}{C}} - \overset{+}{N}H_3Cl^-$$

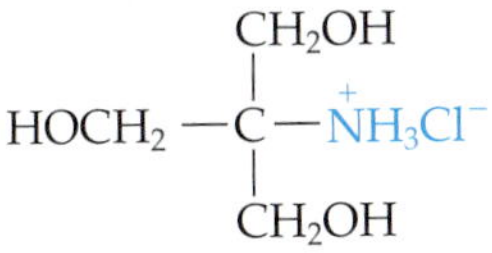

Tris hydrochloride

17.5 Other Chemical Properties of Amines

In the preceding section, we saw that water-soluble amines give basic solutions and that amines generally react with mineral acids to form salts. In Section 16.13, we saw that carboxylic acids react with ammonia when heated to produce unsubstituted amides:

$$R - \overset{\overset{O}{\|}}{C} - OH \ + \ H - \underset{\underset{}{\overset{H}{|}}}{N} - H \ \xrightarrow{\text{heat}} \ R - \overset{\overset{O}{\|}}{C} - NH_2 \ + \ HOH$$

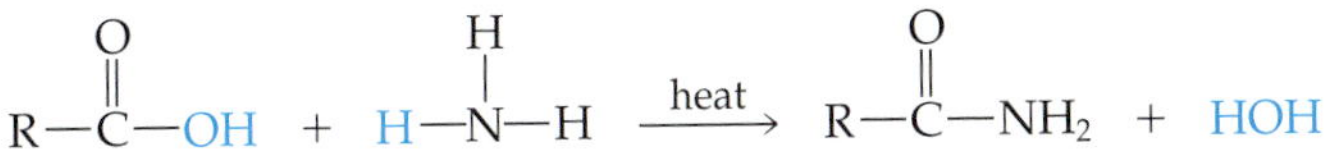

Amines can react with carboxylic acids to produce substituted amides:

$$R - \overset{\overset{O}{\|}}{C} - OH \ + \ H - \underset{\underset{}{\overset{H}{|}}}{N} - R' \ \longrightarrow \ R - \overset{\overset{O}{\|}}{C} - \underset{\underset{}{\overset{H}{|}}}{N} - R' \ + \ HOH$$

In chemistry laboratories, this reaction is carried out using special reactive derivatives of the carboxylic acid instead of the acid itself. We shall see in Chapter 21 that, in living organisms, the formation of the amide bond is under the control of enzymes. Proteins are substituted amides formed through the interaction of amino acids. The amino group of one amino acid reacts with the carboxyl group of another amino acid.

$$H-\underset{\underset{R}{|}}{\overset{\overset{H}{|}}{N}}-\underset{\underset{R}{|}}{\overset{\overset{H}{|}}{C}}-\overset{\overset{O}{\|}}{C}-OH \; + \; H-\underset{\underset{R'}{|}}{\overset{\overset{H}{|}}{N}}-\underset{}{\overset{\overset{H}{|}}{C}}-\overset{\overset{O}{\|}}{C}-OH \longrightarrow H-\overset{\overset{H}{|}}{N}-\overset{\overset{H}{|}}{\underset{\underset{R}{|}}{C}}-\overset{\overset{O}{\|}}{C}-\overset{\overset{H}{|}}{N}-\overset{\overset{H}{|}}{\underset{\underset{R'}{|}}{C}}-\overset{\overset{O}{\|}}{C}-OH \; + \; HOH$$

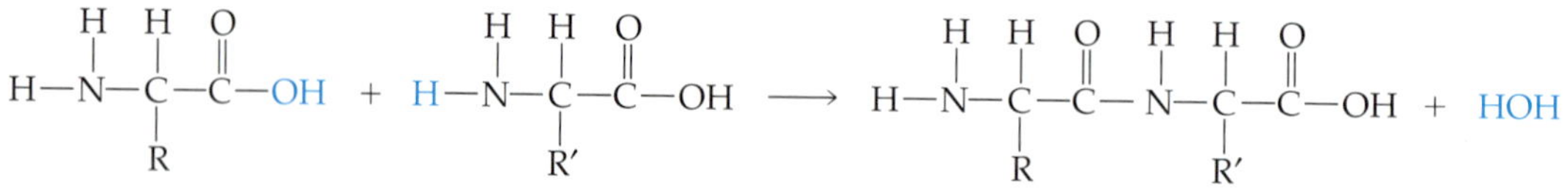

Two additional reactions of amines are of considerable importance to those studying biological processes. First, amines react with nitrous acid. The nature of the products depends on the class of the amine. Primary amines give a quantitative yield of nitrogen gas:

$$R-NH_2 \, (1°) \; + \; HNO_2 \longrightarrow N_2(g) \; + \; \text{other products}$$

Because nitrous acid is unstable, it is usually made *in situ* (right in the reaction vessel) by addition of hydrochloric acid to sodium nitrite:

$$NaNO_2(aq) \; + \; HCl(aq) \longrightarrow HNO_2(aq) \; + \; NaCl(aq)$$

When a primary amine is added to this solution, bubbles of nitrogen gas can be seen escaping. These bubbles are an indication that the amine is a primary one, because secondary and tertiary amines do not release nitrogen gas. Therefore, this reaction serves as a qualitative test for primary amines. (If the amount of the original amine is carefully measured and the nitrogen is collected and its volume measured and corrected to conditions of standard temperature and pressure, the reaction can be used for the quantitative determination of primary amino groups.) One molecule of nitrogen (N_2) is liberated for each free amino group. The procedure is referred to as the Van Slyke method and is used especially with amino acids and proteins (see page 570).

Secondary amines react with nitrous acid to form oily *N*-nitroso compounds:

$$R-\underset{\underset{R}{|}}{\overset{}{N}}-H \, (2°) \; + \; HO-N{=}O \longrightarrow R-\underset{\underset{R}{|}}{\overset{}{N}}-N{=}O \; + \; HOH$$

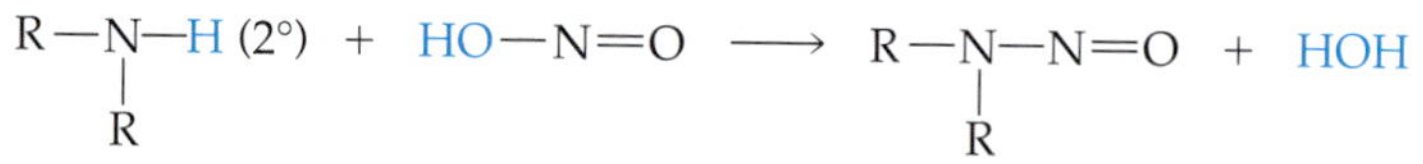

The appearance of an oil when an amine is added to a solution of nitrous acid indicates that the amine is probably secondary. However, the test is no longer recommended because most *N*-nitroso compounds are potent carcinogens. [It may be that, as with the polycyclic aromatic compounds (Section 13.17), the carcinogens are not the nitrosoamines themselves but rather some metabolite of the nitrosoamines.] Because our diet contains sodium nitrite and our stomachs contain hydrochloric acid, nitrosoamines are formed from secondary amines in the breakdown products of the food we eat. It has been postulated that the high rates of stomach cancer in countries in which prepared meats are common in the diet are due to the nitrites found in these products.

In the 1980s, the Department of Agriculture required meat firms to reduce the sodium nitrite added to preserve bacon from 200 to 120 ppm. The amount of nitrite added to ham and frankfurters was also reduced, and other preservatives are being used. The meat industry points out, however, that there is no evidence to link the ingestion of nitrites and the incidence of stomach and intestinal cancer.

Sodium nitrite is used as a food additive in cured meats such as frankfurters, ham, and cold cuts. The NO_2^- serves two functions. It inhibits growth of bacteria, especially *Clostridium botulinum*, which produces the potentially fatal food poisoning known as *botulism*. It also preserves the red color of the meat and thereby its appetizing appearance.

They say that the amount of nitrite that occurs naturally in foods far surpasses the amount that is used as an additive. Nitrosoamines have also been found in alcoholic beverages, cosmetics, and pesticides.

Tertiary amines also react with nitrous acid. Generally, the only product is a salt. Sometimes, however, the tertiary amine is cleaved to a secondary one, which then can form a nitroso derivative.

A second reaction used to detect the presence of amines is the reaction with ninhydrin. Some amines react with ninhydrin to form a purple to blue anion:

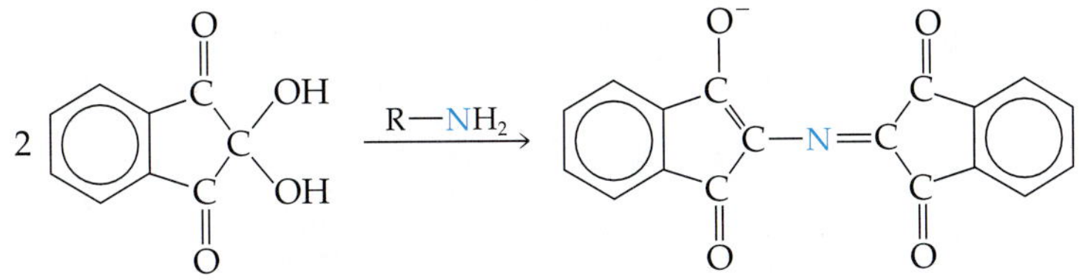

The ninhydrin test is especially suited to the detection of amino acids. Mixtures of amino acids, such as those that result from the hydrolysis of proteins, can be separated by a process called paper chromatography. At the conclusion of such a separation, the individual amino acids are scattered at different locations on the paper chromatogram. The amino acids cannot be seen, however, until the chromatogram is sprayed with ninhydrin and the colored ions become visible. Then the position of the constituents of the original mixture can be compared with those of known amino acids, and an identification can be made on this basis.

17.6 Heterocyclic Amines

In Chapter 13, a variety of cyclic hydrocarbons were introduced. All the atoms in the rings of these compounds are carbon atoms. There are other cyclic compounds in which nitrogen, oxygen, sulfur, or some other atom is incorporated in the ring. These are called **heterocyclic compounds** (Greek *heteros*, other), and a few are shown in Figure 17.5.

The compounds pyrrole and pyrrolidine each have four carbon atoms and one nitrogen atom in a ring. Pyrrole is an aromatic compound and has properties similar to those of benzene. Pyrrolidine, which contains four more hydrogen atoms than does pyrrole, behaves like an aliphatic amine. Imidazole also has a five-membered ring, but it contains two nitrogen atoms and only three carbon atoms. Like pyrrole, imidazole has aromatic properties.

Pyridine and piperidine each have five carbon atoms and one nitrogen atom. Pyridine is aromatic; piperidine is aliphatic. Another six-membered heterocycle is pyrimidine, which has two nitrogen atoms and four carbon atoms. Pyrimidine is another aromatic compound.

Other heterocyclic compounds have two rings that share a common side (a situation we encountered with naphthalene). Indole has a benzene ring fused with a pyrrole ring. Purine has a pyrimidine ring sharing a side with an imidazole structure. Bases related to purine and pyrimidine make up a part of the structure of the nucleic acids, compounds that compose the genetic material of cells and that direct protein synthesis. Nucleic acids are discussed in Chapter 23, in which we encounter one of the truly outstanding examples of the critical importance of the shapes of molecules and of molecular structure in general.

Many heterocyclic amines occur naturally in plants. Like most other amines, these compounds are basic. They are called **alkaloids** (Section F.3), a name that

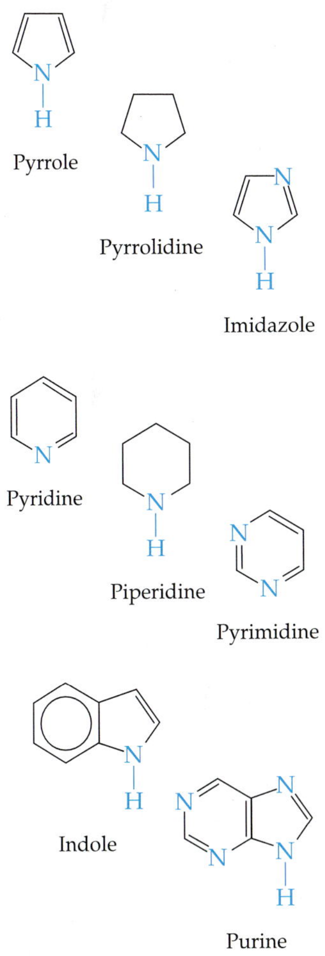

▲ FIGURE 17.5
Some heterocyclic amines.

Conium maculatum (poison hemlock).

Jacques Louis David's painting *The Death of Socrates* (1787) shows Socrates about to drink the cup of hemlock to carry out the death sentence decreed by the rulers of Athens.

means "like alkalis." Knowledge of many of these, at least in their crude forms, dates back to antiquity. Opium, which contains about 10% morphine as the principal alkaloid, has been used for thousands of years, although morphine was not isolated until 1805.

When the Greek philosopher Socrates was accused of corrupting the youth of Athens in 399 B.C.E., he was given the choice of exile or death. He chose the latter and implemented his decision by drinking a cup of hemlock. His hemlock was probably prepared from the fully grown but unripened fruit of *Conium maculatum*, or poison hemlock. The fruit would probably have been carefully dried and then brewed into a tea. Hemlock contains several alkaloids, but the principal one is coniine. Coniine causes nausea, weakness, paralysis, and ultimately—as in the case of Socrates—death.

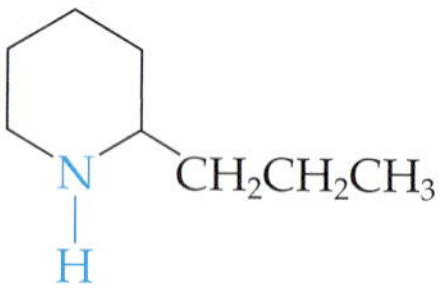

Coniine

Hemlock tea, anyone?

Summary

Amines are nitrogen-containing organic molecules derived from the inorganic ammonia molecule. A **primary amine** has one organic group bonded to the nitrogen, a **secondary amine** has two organic groups bonded to the nitrogen, and a **tertiary amine** has three:

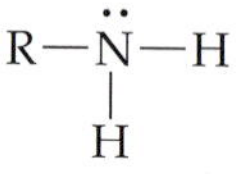

Primary amine

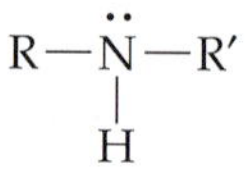

Secondary amine

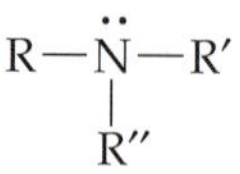

Tertiary amine

When a fourth organic group is bonded to the nitrogen via the nitrogen's unshared electron pair, the result is a *quaternary ammonium ion:*

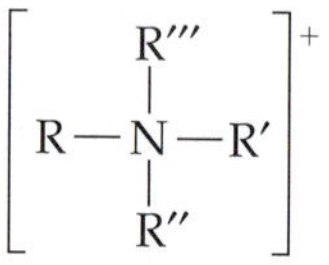

Quaternary ammonium ion

Amines are basic compounds, aliphatic amines being more basic than ammonia and aromatic amines being less basic than ammonia. They react with strong acids to produce an ammonium salt:

$$\underset{\text{Amine}}{R-\overset{\overset{R'}{|}}{\underset{\underset{H}{|}}{N}}:} + \underset{\text{Strong acid}}{HNO_3} \longrightarrow \underset{\text{Ammonium salt}}{R-\overset{\overset{R'}{|}}{\underset{\underset{H}{|}}{N}}H^+NO_3^-}$$

and with carboxylic acids, upon heating, to yield amides:

$$\underset{\text{Carboxylic acid}}{R-\overset{\overset{O}{\|}}{C}-OH} + \underset{\text{Amine}}{H-\overset{\overset{H}{|}}{N}-R'} \xrightarrow{\text{heat}} \underset{\text{Amide}}{R-\overset{\overset{O}{\|}}{C}-\overset{\overset{H}{|}}{N}-R'} + HOH$$

Primary, secondary, and tertiary amines can be distinguished from each other according to how they react with nitrous acid. Primary amines yield gaseous nitrogen in a reaction known as the *Van Slyke method*. Secondary ones yield *N*-nitroso products that cause an oily film on the solution surface, and tertiary ones yield only salts:

$$\underset{\text{Primary amine}}{RNH_2} + \underset{\text{Nitrous acid}}{HNO_2} \longrightarrow \underset{\text{Gas}}{N_2(g)} + \text{other products}$$

$$\underset{\text{Secondary amine}}{R-\underset{\underset{R'}{|}}{N}-H} + \underset{\text{Nitrous acid}}{HO-N=O} \longrightarrow \underset{\substack{N\text{-nitroso compound}\\ \text{(oily film)}}}{R-\underset{\underset{R'}{|}}{N}-N=O} + HOH$$

$$\underset{\text{Tertiary amine}}{R-\overset{\overset{R''}{|}}{\underset{\underset{R'}{|}}{N}}} + \underset{\text{Nitrous acid}}{HNO_2} \longrightarrow \underset{\text{Salt}}{R-\overset{\overset{R''}{|}}{\underset{\underset{R'}{|}}{N}}H^+NO_2^-}$$

A cyclic compound in which the ring contains one or more noncarbon atoms is called a **heterocyclic compound.** There are many heterocyclic amines, including the physiologically important purine and pyrimidine:

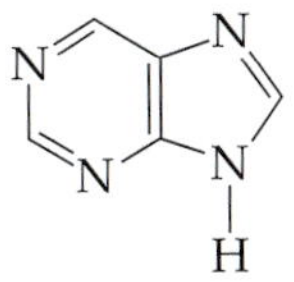

Purine

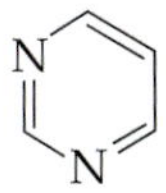

Pyrimidine

Alkaloids are heterocyclic amines found in many plants. Morphine in the opium plant and coniine in hemlock are two examples.

Key Terms

alkaloid (17.6)
heterocyclic compound (17.6)
primary (1°) amine (17.1)
quaternary (4°) ammonium salt (17.1)
secondary (2°) amine (17.1)
tertiary (3°) amine (17.1)

Review Questions

1. Contrast the physical properties of amines with those of amides.
2. What chemical reaction occurs when lemon juice is added to fish? Why is this desirable?
3. Tell whether each of the following compounds forms an acidic, basic, or neutral solution in water.
   a. $CH_3CH_2NH_2$ **b.** CH_3CH_2OH
   c. $CH_3\overset{\overset{O}{\|}}{C}OH$ **d.** $CH_3\overset{\overset{O}{\|}}{C}NH_2$
4. Describe the odors of amines.
5. Why are nitrite food additives considered by some people to be health hazards?
6. What is meant by the term *heterocyclic compound?*

Problems

Classification of Compounds

7. Classify as an amine, an amide, both, or neither:

a. 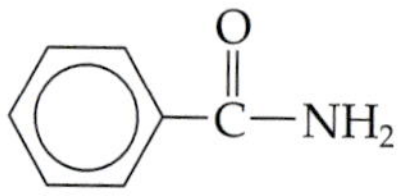

b. 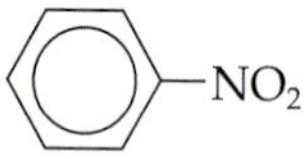

c. $H_2NCH_2CH_2CH_2\overset{\overset{\large O}{\|}}{C}NH_2$

8. Classify as an amine, an amide, both, or neither:

a. $(CH_3)_4N^+I^-$

b. $CH_3CH_2NH\overset{\overset{\large O}{\|}}{C}CH_2CH_3$

c. $CH_3CH_2NHCH_2CH_3$

9. Identify the following compounds as amines, alcohols, phenols, or ethers. Classify any amines and alcohols as primary (1°), secondary (2°), or tertiary (3°).

a. $CH_3CH_2CH_2OH$

b. $CH_3CH_2CH_2NH_2$

c. $CH_3\overset{\overset{\large OH}{|}}{C}HCH_3$

d. $CH_3\overset{\overset{\large NH_2}{|}}{C}HCH_3$

e. 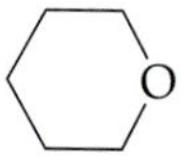

f. 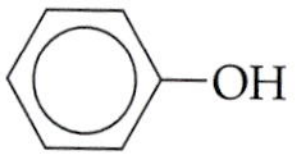

10. Identify the following compounds as amines, alcohols, phenols, or ethers. Classify any amines and alcohols as primary, secondary, or tertiary.

a. $CH_3CH_2NHCH_2CH_3$

b. $CH_3CH_2OCH_2CH_3$

c. $CH_3-\underset{\underset{\large CH_3}{|}}{N}-CH_3$

d. $CH_3-\underset{\underset{\large CH_3}{|}}{\overset{\overset{\large OH}{|}}{C}}-CH_3$

e. 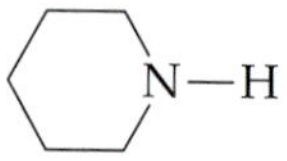

f. 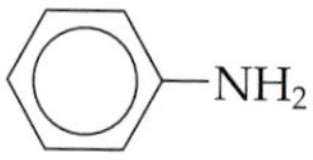

Amines: Structures and Names

11. Draw structural formulas for:
    a. dimethylamine
    b. diethylmethylamine
    c. 2-amino-1-cyclobutanol
    d. 2-aminoethanol

12. Draw structural formulas for:
    a. 3-aminopentane
    b. 1,6-diaminohexane
    c. cyclohexylamine
    d. ethylphenylamine

13. Draw structural formulas for:
    a. aniline
    b. *m*-bromoaniline
    c. pyrimidine
    d. *N*-ethylaniline

14. Draw structural formulas for:
    a. pyridine
    b. purine
    c. *N,N*-dimethylaniline
    d. 3,5-dichloroaniline

15. Name the following compounds:
    a. $CH_3CH_2CH_2NH_2$
    b. $(CH_3)_2CHNHCH_3$
    c. $(CH_3CH_2)_3N$
    d. $CH_3CH(NH_2)CH_2CH_2CH_3$

16. Name the following compounds:

a. 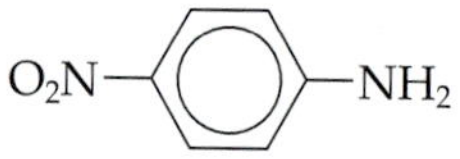

b. 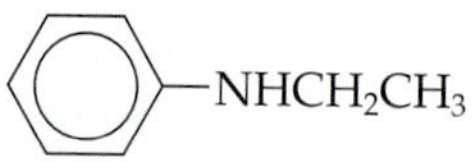

Names and Formulas of Amine Salts

17. Draw structural formulas for:
    a. anilinium bromide
    b. tetramethylammonium chloride

18. Draw structural formulas for:
    a. ethylmethylammonium chloride
    b. anilinium nitrate

19. Name the following compounds:
    a. $[CH_3CH_2NH_2CH_2CH_3]^+ Br^-$
    b. $(CH_3CH_2)_4N^+ I^-$

20. Name the following compounds:

a. (benzene ring)—$NH_3^+Cl^-$

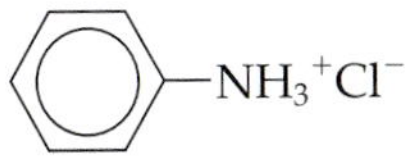

b. $(CH_3)_4N^+NO_3^-$

Physical Properties

21. Which compound has the higher boiling point, butylamine or pentane? Explain.

22. Which compound has the higher boiling point, butylamine or butyl alcohol? Explain.

23. Which compound has the higher boiling point, trimethylamine or propylamine? Explain.

24. Which compound has the higher boiling point, CH_3NH_2 or $CH_3CH_2CH_2CH_2CH_2NH_2$? Explain.

25. Which compound is more soluble in water, $CH_3CH_2CH_3$ or $CH_3CH_2NH_2$? Explain.

26. Which compound is more soluble in water:
$CH_3CH_2CH_2NH_2$ or
$CH_3CH_2CH_2CH_2CH_2CH_2CH_2NH_2$? Explain.

27. Which compound is more soluble in water? Explain.

$$\underset{CH_2CH_2CHCH_2CHCH_3}{\overset{NH_2 \quad CH_3 \quad CH_3}{| \quad | \quad |}} \quad or \quad \underset{CH_2CH_2CHCH_2CHCH_3}{\overset{NH_2 \quad NH_2 \quad NH_2}{| \quad | \quad |}}$$

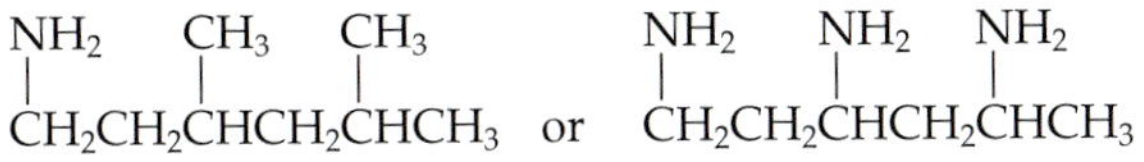

28. Which compound is more soluble in water? Explain.

Cl—(benzene ring)—NH_2 or (benzene ring)—$NH_3^+Cl^-$

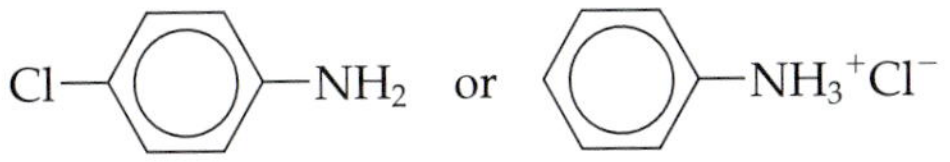

Chemical Reactions

29. Draw the structural formula of the salt formed in the reaction

$$CH_3N + HBr \longrightarrow$$

30. Draw the structural formula of the salt formed in the reaction

(benzene ring)—$NHCH_3 + HNO_3 \longrightarrow$

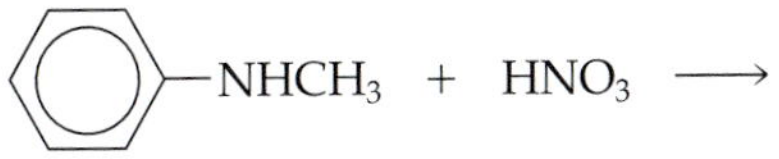

31. Draw the structural formula of the salt formed in the reaction

$$\underset{CH_3}{\overset{}{CH_3-N-CH_3}} + H_2SO_4 \longrightarrow$$

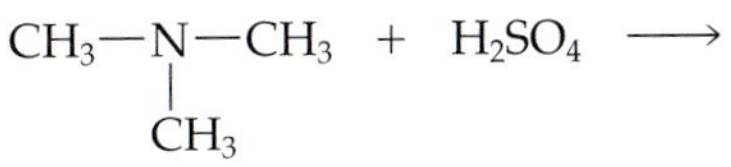

32. Draw the structural formula of the salt formed in the reaction

(piperidine ring with N–H) $+ HCl \longrightarrow$

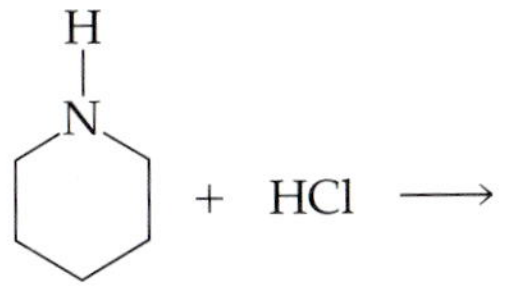

33. Draw the amide, if any, derived from hexanoic acid and butylamine.

34. Draw the amide, if any, derived from

$$CH_3CH_2\overset{\overset{\displaystyle O}{\|}}{C}-OH \quad and \quad CH_3-\overset{\overset{\displaystyle H}{|}}{N}-CH_3$$

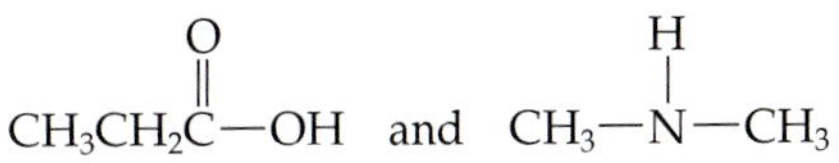

35. Draw the amide, if any, derived from benzoic acid and aniline.

36. Draw the amide, if any, derived from

$$CH_3\overset{\overset{\displaystyle O}{\|}}{C}-OH \quad and \quad CH_3-\underset{\underset{\displaystyle CH_3}{|}}{\overset{}{N}}-CH_3$$

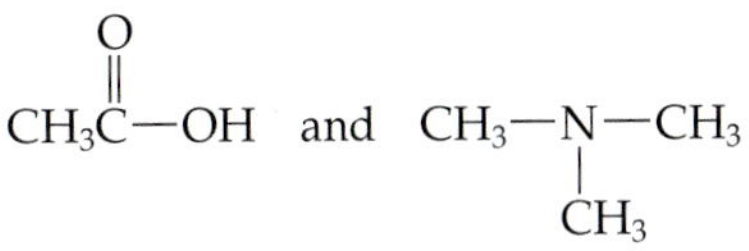

37. Draw the carboxylic acid and amine from which the following amide was formed.

$$CH_3CH_2\underset{\underset{\displaystyle CH_3}{|}}{N}-\overset{\overset{\displaystyle O}{\|}}{C}CH_2CH_3$$

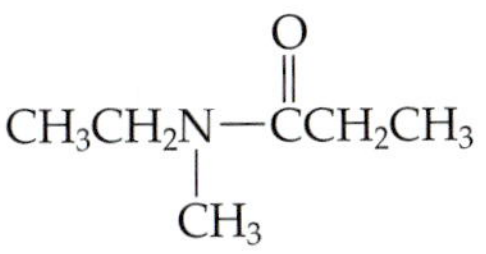

38. Draw the carboxylic acid and amine from which the following amide was formed.

(benzene ring)—$\overset{\overset{\displaystyle O}{\|}}{C}-\underset{\underset{\displaystyle CH_3}{|}}{N}-CH_3$

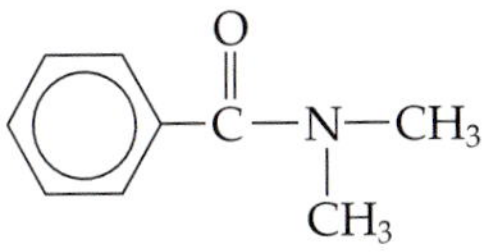

39. Draw the carboxylic acid and amine from which the following amide was formed.

(benzene ring)—$NH\overset{\overset{\displaystyle O}{\|}}{C}CH_2CH_3$

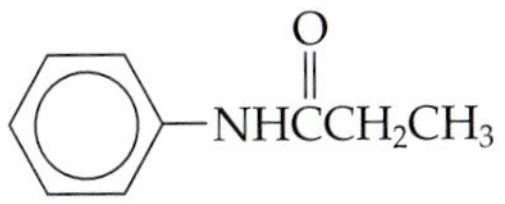

40. Draw the carboxylic acid and amine from which the following amide was formed.

(piperidine ring)$N-\overset{\overset{\displaystyle O}{\|}}{C}-CH_3$

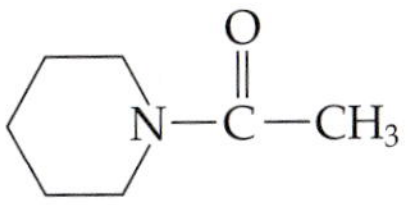

41. Draw the structural formula for the principal organic product formed in the reaction

(benzene ring)—$\overset{\overset{\displaystyle O}{\diagup\!\!\diagdown}}{C}\underset{O^-NH_4^+}{}$ $\xrightarrow{\text{heat}}$

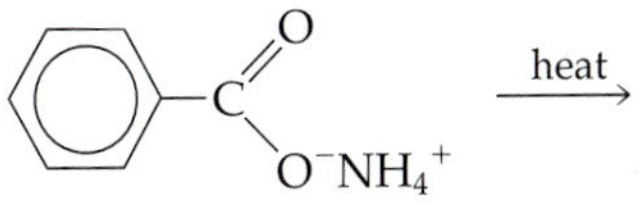

42. Draw the structural formula for the principal organic product formed in the reaction

$$CH_3NHCH_2CH_3 + HNO_2 \longrightarrow$$

43. List the reagents needed to carry out the following conversions.

a.

$$CH_3-\underset{\underset{CH_3}{|}}{\overset{\overset{CH_3}{|}}{C}}-NH_2 \xrightarrow{?} CH_3-\underset{\underset{CH_3}{|}}{\overset{\overset{CH_3}{|}}{C}}-NH_3{}^+Cl^-$$

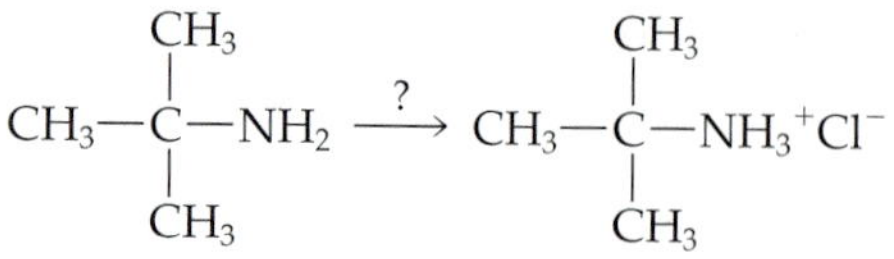

b.

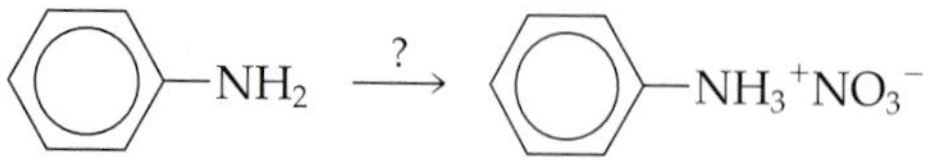

44. List the reagents needed to carry out the following conversions.

a.

$$CH_3NHCH_3 \xrightarrow{?} \overset{\overset{N=O}{|}}{CH_3NCH_3}$$

b.

$$HOCH_2-\underset{\underset{CH_2OH}{|}}{\overset{\overset{CH_2OH}{|}}{C}}-NH_2 \xrightarrow{?}$$

$$HOCH_2-\underset{\underset{CH_2OH}{|}}{\overset{\overset{CH_2OH}{|}}{C}}-NH_3{}^+Cl^-$$

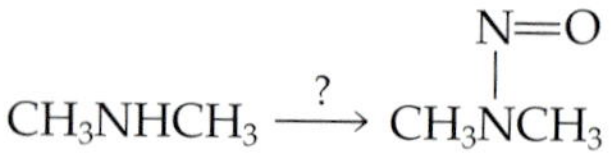

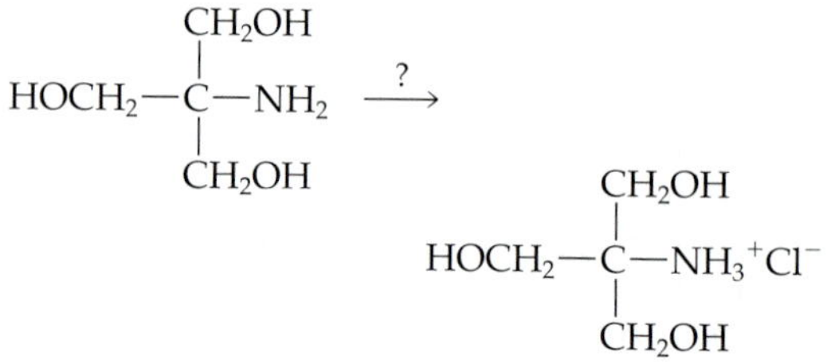

Additional Problems

45. Amine X is insoluble in water but dissolves readily in aqueous hydrochloric acid. Explain.

46. Draw structural formulas for the eight isomeric amines that have the molecular formula $C_4H_{11}N$. Give each a common name, and classify it as primary, secondary, or tertiary.

47. Draw structural formulas for the five isomeric amines that have the molecular formula C_7H_9N and contain a benzene ring. Name each compound, and classify it as primary, secondary, or tertiary.

48. A carboxylic acid group and an amino group combined to form the amide functional group in the following compound. Draw the two starting materials for the reaction.

$$? + ? \longrightarrow H_2N-CH_2\overset{\overset{O}{||}}{C}-NHCH_2\overset{\overset{O}{||}}{C}-OH$$

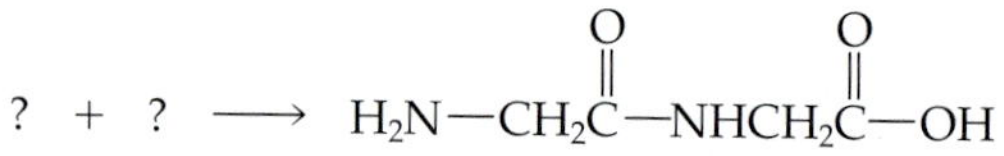

49. Write equations for the reaction of anthranilic acid (*o*-aminobenzoic acid) with each of the following.
 a. NaOH b. HCl
 c. H_2SO_4 d. CH_3OH, H^+

50. How many milliliters of 0.150 M hydrochloric acid are required to neutralize 0.250 g of diethylamine?

Special Topic G
Brain Amines and Related Drugs

Some drugs are rather simple amines. Others, including some we discuss in other chapters, are alkaloids (Section F.3). In this special topic, we look at some amines and related compounds that affect our mental state, render us insensitive to pain, put us to sleep, and calm our anxieties. First, however, let's take a look at nerve cells and how they work.

G.1 Some Chemistry of the Nervous System

The nervous system is made up of billions of interconnected **neurons** (nerve cells). The brain operates with a power output of about 25 W and has capacity for about 10 trillion bits of information. Neurons vary a great deal in shape and size. One type is shown in Figure G.1. The essential parts are the cell body, the axon, and the dendrites. We discuss here only those neurons that make up the involuntary (autonomic) nervous system. These neurons carry messages between the organs and glands that act involuntarily (such as the heart, the digestive organs, and the lungs) and the brain and spinal column (which make up the **central nervous system**).

Although the axons on a given neuron may be up to 60 cm long, there is no continuous pathway from an organ to the central nervous system. Messages must be transmitted across tiny, fluid-filled gaps called **synapses** (Figure G.2). When an electric signal from the brain reaches the end of an axon, specific chemicals (called **neurotransmitters**) that carry the signal across the synapse to the next neuron are liberated. There are perhaps a few dozen neurotransmitters. (We have previously mentioned the endorphins and the enkephalins in Section F.5.) Each has a specific function. Messages are carried to other neurons, to muscles, and to the endocrine glands (such as the adrenal glands). Each neurotransmitter fits a specific receptor site on the receptor cell (Figure G.3). Many drugs (and some poisons) act by mimicking the action of the neurotransmitter. Others act by blocking the receptor and preventing

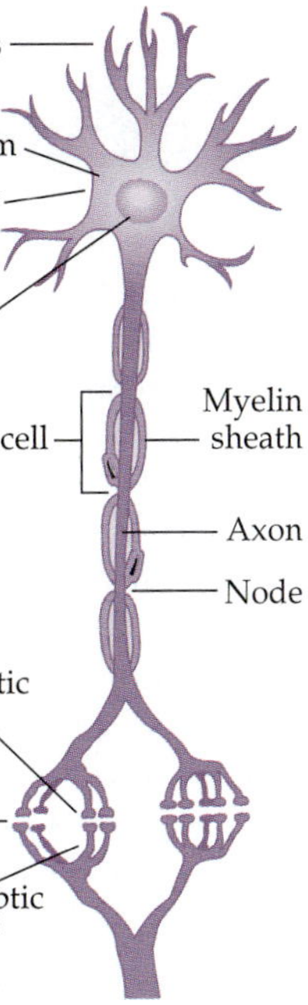

FIGURE G.1 ▲

Diagram of a human nerve cell.

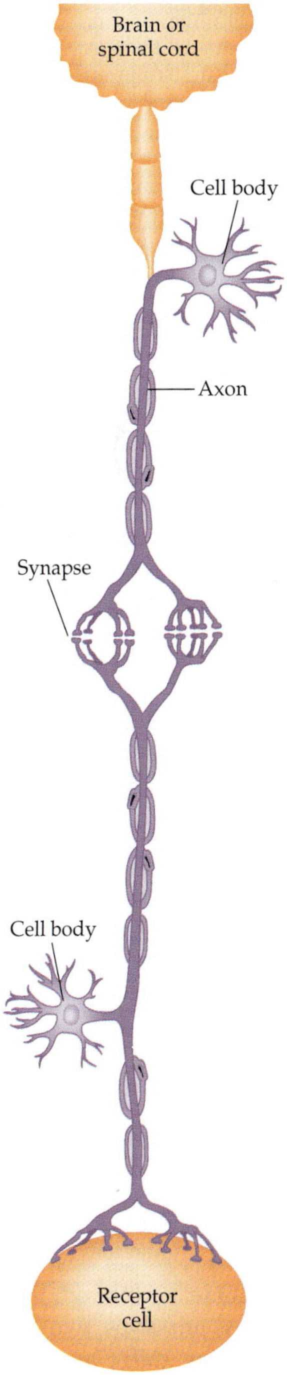

FIGURE G.2 ▶

Diagram of the pathway by which messages are transmitted to (and from) a receptor cell in a gland or an organ from (and to) the central nervous system.

FIGURE G.3 ▶

Diagram of a synapse. When an electric signal reaches the presynaptic nerve ending, neurotransmitter molecules are released from the vesicles. They migrate across the synapse to the receptor cell where they fit specific receptor sites.

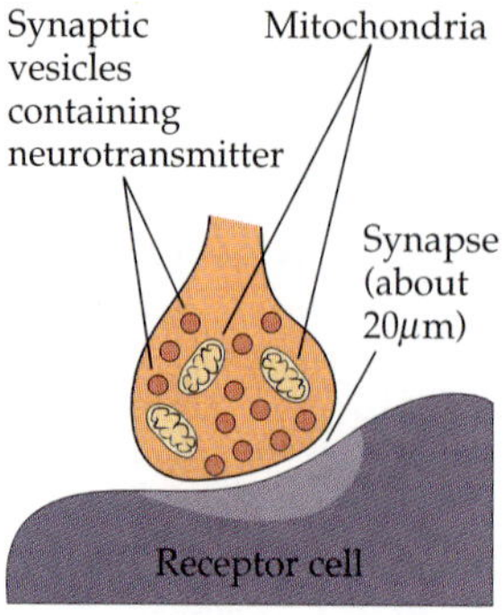

the neurotransmitter from acting on it. Several of the neurotransmitters are amines, as are some of the drugs that affect the chemistry of the human brain.

G.2 Brain Amines

We all have our ups and downs in life. These moods probably result from multiple causes, but it appears likely that a variety of chemical compounds formed in the brain are involved. Before we consider these ups and downs, however, let's take a look at epinephrine, an amine formed in the adrenal glands.

Commonly called *adrenaline,* epinephrine is secreted by the adrenal glands. A tiny amount of epinephrine causes a great increase in blood pressure. When a person is under stress or frightened, the flow of adrenaline prepares the body for fight or flight. Because culturally imposed inhibitions prevent fighting or fleeing in most modern situations, the adrenaline-induced supercharge is not used. This sort of frustration has been implicated in some forms of mental illness.

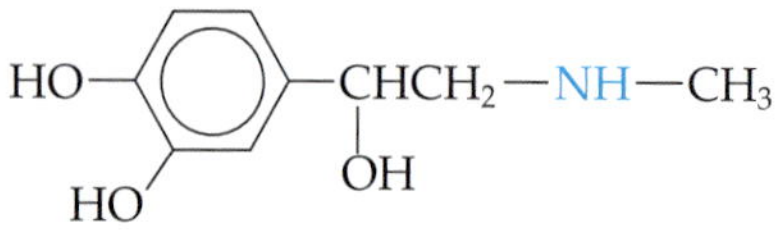

Epinephrine
(Adrenaline)

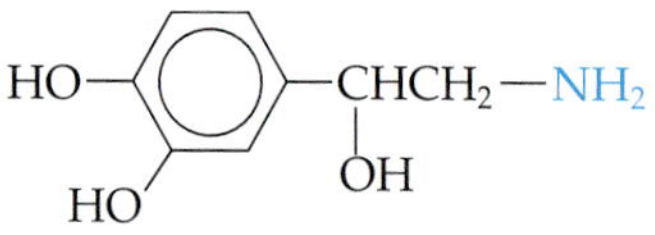

Norepinephrine

One simplified biochemical theory of mental illness involves two brain amines. One is norepinephrine (NE), a relative of epinephrine. NE is a neurotransmitter formed in the brain. When formed in excess, NE causes the person to be elated—perhaps even hyperactive. In large excess, NE induces a manic state. A deficiency of NE, on the other hand, could cause depression.

Our cells have at least six different receptors that are activated by NE and related compounds. NE agonists (drugs that enhance or mimic its action) are stimulants. NE antagonists (drugs that block the action of NE) slow down various processes. Drugs called beta blockers reduce the stimulant action of epinephrine and NE on various kinds of cells. Propranolol (Inderal) is used to treat cardiac arrhythmias, angina, and hypertension by lessening slightly the force of the heart beat. Unfortunately, it also causes lethargy and depression. Metoprolol (Lopressor) acts selectively on cells in the heart. It can be used by hypertensive patients who have asthma because it does not act on receptors in the bronchi.

Norepinephrine is synthesized in the body from the amino acid tyrosine. The synthesis is complex and proceeds through several intermediates (Figure G.4). Each step is catalyzed by one or more enzymes.

Toxic Gases and the Learning Process

Gases such as carbon monoxide (CO), nitrogen monoxide (NO, commonly called nitric oxide), and hydrogen sulfide (H_2S) have long been known for their toxic effects. Imagine the surprise when scientists discovered that brain cells make NO and use it for communication.

A living cell constantly senses its environment. It changes in response to information flowing into and out of the cell. Most chemical messengers are complex substances such as norepinephrine and serotonin. But NO and possibly CO and H_2S are also chemical messengers. NO is formed in cells from arginine, a nitrogen-containing amino acid (Chapter 21) in an enzyme-catalyzed reaction. NO kills invading microorganisms, probably by deactivating iron-containing enzymes in much the same way that CO destroys the oxygen-carrying capacity of hemoglobin. NO also helps to regulate blood pressure and is involved in the formation of long-term memories.

CO is formed in cells by the enzyme-catalyzed oxidation of heme. Like NO, it seems to be involved in the long-term potentiation of learning. H_2S has been found in the brain cells of rats, but there is as yet no proof that the cells make it. There is evidence that H_2S stimulates certain receptors that strengthen the connections between brain cells, an indication of long-term learning. Learning can be a gas!

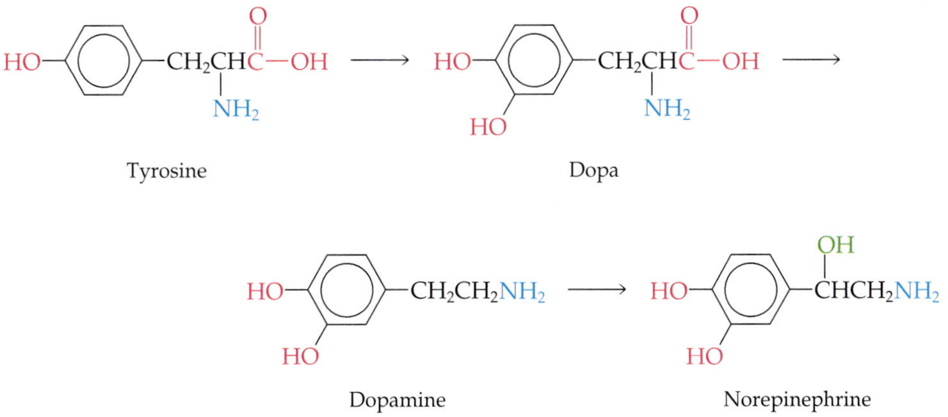

◀ FIGURE G.4

The biosynthesis of norepinephrine from tyrosine.

The intermediate compounds also have physiological activity: dopa has been used successfully in the treatment of Parkinson's disease, and dopamine has been employed to treat low blood pressure (see Section 27.5). Because tyrosine is a component of our diets, it may well be that our mental state depends to a fair degree on what we eat.

Another brain amine is serotonin, also a neurotransmitter. It is produced in the body from the amino acid tryptophan.

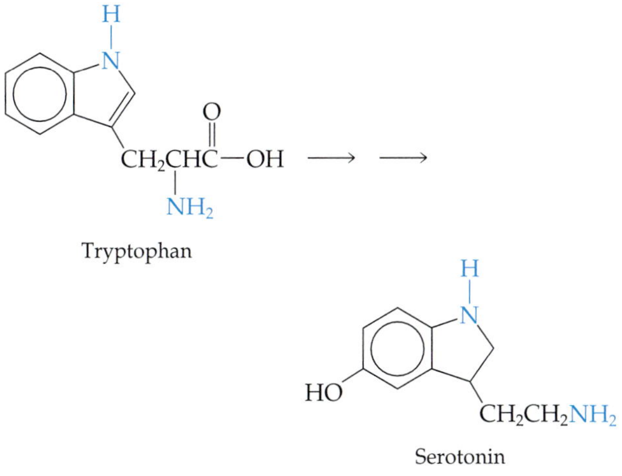

Serotonin is involved in sleep, sensory perception, and the regulation of body temperature. Its exact role in mental illness is not clear. A metabolite of serotonin, 5-hydroxyindoleacetic acid (5-HIAA) is found in unusually *low* levels in the spinal fluid of violent suicide victims.[1] This indicates that abnormal serotonin metabolism may play a role in depression. Serotonin agonists are used to treat depression and anxiety, and are being used experimentally to treat obsessive-compulsive disorder. Serotonin antagonists are used to treat migraine headaches and to relieve the nausea caused by cancer chemotherapy.

Nearly one out of every 10 people in the United States suffers from mental illness. Over half the patients in hospitals are there because of mental problems. When the biochemistry of the brain is more fully understood, mental illness may be cured (or at least alleviated) by administration of drugs. In subsequent sections, we see just how far we have already come in learning to control our moods with drugs. As is true for so many things, the potential for good that such compounds represent is matched by a potential for abuse.

G.3 Stimulant Drugs: Amphetamines

Among the more widely known stimulant drugs are a variety of synthetic amines related to phenylethylamine (Figure G.5). Note the similarity between these molecules and epinephrine and norepinephrine; all are derived from the basic phenylethylamine structure. The amphetamines probably act as stimulants by mimicking the natural brain amines.

Amphetamine[2] and methamphetamine have been widely abused. Amphetamine has been extensively used for weight reduction. It has also been

[1] Levels of 5-HIAA also are low in murderers and other violent offenders. Levels of this serotonin metabolite are higher than normal in persons with obsessive-compulsive disorders, sociopaths, and people who have guilt complexes.

[2] Amphetamine is not a single compound but a mixture of two stereoisomers (see Section 18.5) marketed under the trade name Benzedrine.

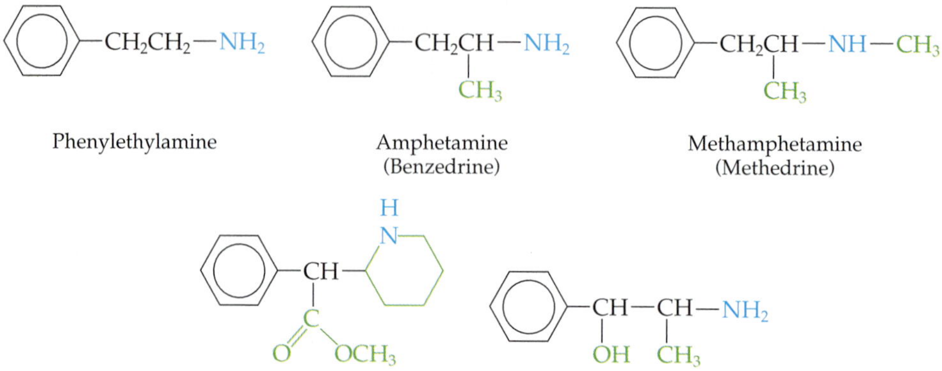

FIGURE G.5 ▶

Phenylethylamine and related compounds.

employed for treating mild depression and narcolepsy, a rare form of sleeping sickness. Amphetamine induces excitability, restlessness, tremors, insomnia, dilated pupils, increased pulse rate and blood pressure, hallucinations, and psychoses. It is no longer recommended for weight reduction. It was found that, generally, any weight loss was only temporary. The greatest problem, however, was the diversion of vast quantities of amphetamines into the illegal drug market.

Table G.1 Toxicities of Various Drugs[a]

Drug	LD_{50} (mg/kg body weight)	Method of Administration	Experimental Animal
Local anesthetics			
Lidocaine	292	Oral	Mice
Procaine	45	Intravenous	Mice
Cocaine	17.5	Intravenous	Rats
Barbiturates			
Barbital	600	Oral	Mice
Pentobarbital	118	Oral	Rats
Phenobarbital	162	Oral	Rats
Amobarbital	212	Subcutaneous	Mice
Thiopental	149	Intraperitoneal	Mice
Narcotics			
Morphine	500	Subcutaneous	Mice
Heroin	21.8	Intravenous	Mice
Meperidine	170	Oral	Rats
Stimulants			
Caffeine	355	Oral	Rats
Nicotine	230	Oral	Mice
	0.3	Intravenous	Mice
Amphetamine	180	Subcutaneous	Rats
Methamphetamine	70	Intraperitoneal	Mice
Mescaline	370	Intraperitoneal	Rats

[a] Comparisons of toxicities in different animals—and extrapolation to humans—are at best crude approximations. The method of administration can have a profound effect on the observed toxicity. *Source:* Susan Budavari (Ed.), *The Merck Index*, 11th ed. Rahway, NJ: Merck and Co., 1989.

Amphetamines are inexpensive. Armed forces personnel, truck drivers, and college students have been among the heavy users.

Methamphetamine has a more pronounced psychological effect than amphetamine. Generally, the "speed" that abusers inject into their veins is methamphetamine.[3] Such injections, at least initially, are said to give the abuser a euphoric rush. Shooting methamphetamine is quite dangerous, however, because the drug is relatively toxic (see Table G.1).

Another amphetamine derivative, phenylpropanolamine, is widely used as an over-the-counter appetite suppressant. Like its relatives, this compound is a stimulant. Studies show that it is at best marginally effective as a diet aid, and it poses a threat to people with hypertension. Nevertheless, sales of phenylpropanolamine are 1 billion tablets, or $150 million, each year.

One controversial use of amphetamines is their employment in the treatment of attention deficit disorder (ADD) in children. The drug of choice is often methylphenidate (Ritalin). The drug acts as a stimulant for adults but seems to have the opposite effect on children. It calms them and helps them filter out extraneous stimuli, enabling them to focus their attention on the task at hand. This use has been criticized as "leading to drug abuse" and as "solving the teacher's problem, not the child's."

G.4 Caffeine, Nicotine, and Cocaine

Coffee, tea and some soft drinks[4] contain the mild stimulant caffeine. The effective dose of caffeine is about 200 mg, corresponding to about two cups of strong coffee or tea. Caffeine is also available in tablet form as a stay-awake or keep-alert drug. The best-known brands are probably No-Doz and Vivarin. No-Doz contains about 100 mg of caffeine per tablet; each Vivarin tablet has 200 mg.

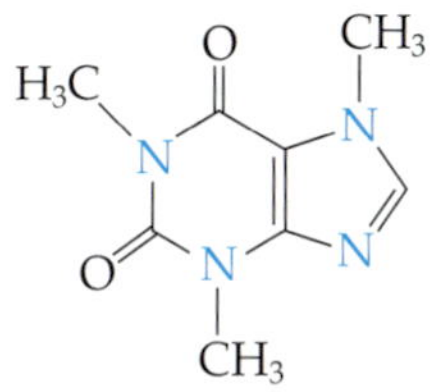

Caffeine

Is caffeine addictive? The "morning grouch" syndrome indicates that it may be mildly so.

Another common stimulant is nicotine. This drug is taken by smoking or chewing tobacco. Nicotine is highly toxic to animals (Table G.1). It is especially deadly when injected; the lethal dose for a human is estimated to be about 50 mg. Nicotine has been used in agriculture as a contact insecticide. In humans, it seems to have a transient stimulant effect. This initial response is followed by depression.

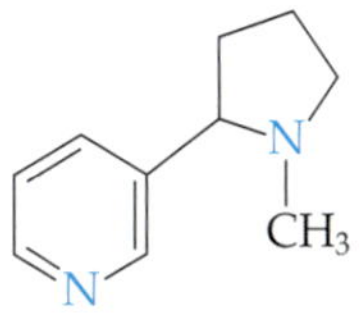

Nicotine

Is nicotine addictive? Casual observation of a person trying to quit smoking seems to indicate that it is.[5] Consider the 1972 memorandum by a Philip Morris scientist who noted that "no one has ever become a cigarette smoker by smoking cigarettes without nicotine." He suggested that the company "think of the cigarette as a dispenser for a dose unit of nicotine."

Cocaine, first used as a local anesthetic (Section G.5), is also a powerful stimulant. The drug is obtained from the leaves of a shrub that grows almost exclusively on the eastern slopes of the Andes Mountains. Many of the Indians living in and around the area of cultivation chew coca leaves—mixed with lime and ashes—for their stimulant effect. Cocaine used to arrive in the United States as the salt cocaine hydrochloride. Now much of it

[3] Like other amine drugs, the amphetamines normally are used in the form of hydrochloride salts (Section 17.4). A freebase form of methamphetamine is used like crack cocaine (Section G.4) for smoking. This form is called "ice" because it is a clear crystalline solid that resembles the solid form of water.

[4] Each year a million kilograms of caffeine are added to food in the United States. Most of it goes into soft drinks.

[5] Clonidine, a drug used to treat high blood pressure, reduces nicotine withdrawal symptoms.

comes in the form of broken lumps of the free base. This form is called *crack cocaine*.[6]

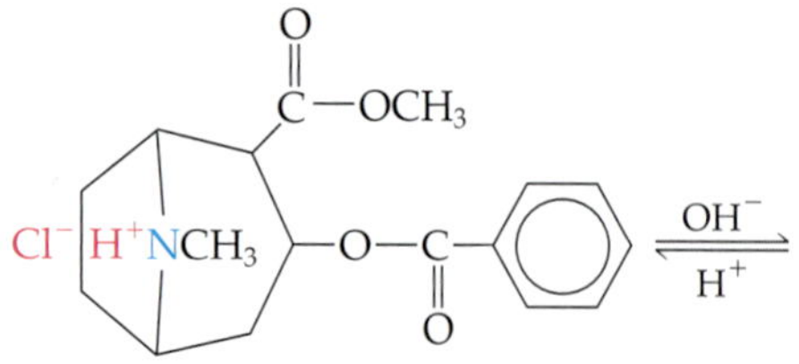

Cocaine hydrochloride

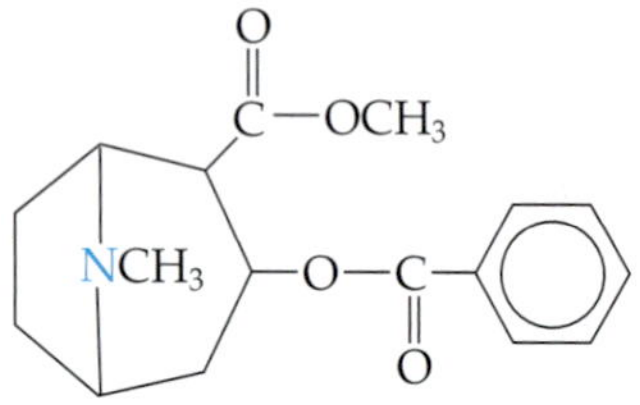

Cocaine (crack)

[6] Crack is the same chemical as the free base that occasionally made the news several years ago. When cocaine hydrochloride was the main street form of the drug, some people made the free base by reacting the salt with a strong base. The person then used a solvent to extract the free base (pure cocaine) from solution. Some free basers were badly burned when the solvent (such as diethyl ether) was accidentally ignited.

Because it is soluble in water, cocaine hydrochloride is snorted and is readily absorbed through the watery mucous membranes of the nose. Crack is more volatile than the hydrochloride. It vaporizes at the temperature of a burning cigarette. When smoked, cocaine reaches the brain in 15 seconds. It acts by preventing the neurotransmitter dopamine from being taken back up after it is released by nerve cells. High levels of dopamine are therefore available to stimulate the pleasure centers of the brain. After the binge, dopamine is depleted in less than an hour. This leaves the user in a pleasureless state and (often) craving more cocaine.

Use of cocaine increases stamina and reduces fatigue, but the effect is short-lived. Stimulation is followed by depression. Once quite expensive and limited to use mainly by the wealthy, cocaine is now available in cheap and potent forms. Hundreds, including several well-known athletes, have died from cocaine overdose.

G.5 Local Anesthetics

An **anesthetic** is any substance that causes either unconsciousness or insensitivity to pain. **Local anesthetics** are drugs that block transmission of nerve signals when applied to nerve tissue. They act on all

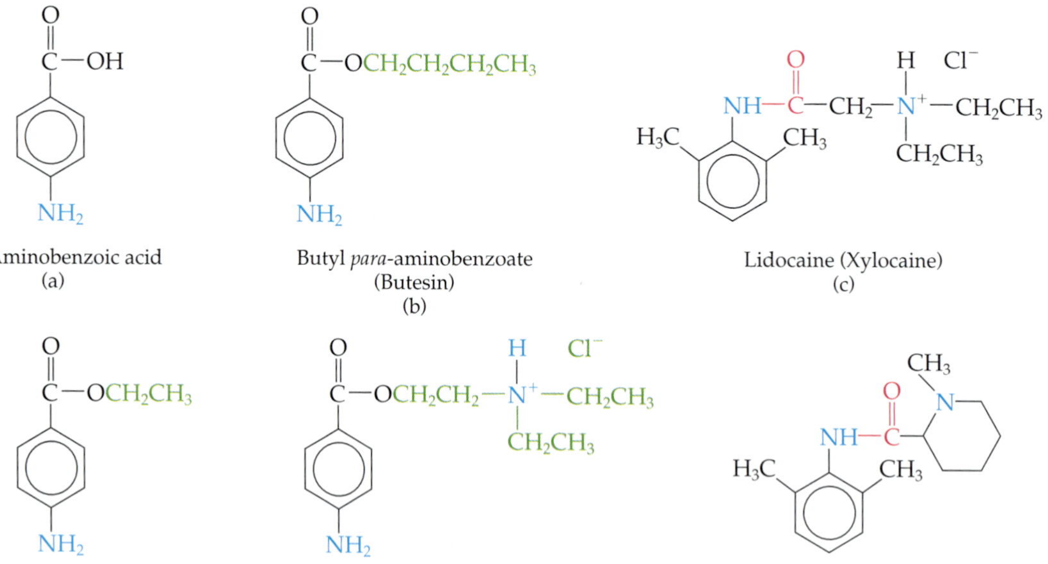

para-Aminobenzoic acid (a)

Butyl *para*-aminobenzoate (Butesin) (b)

Lidocaine (Xylocaine) (c)

Ethyl *para*-aminobenzoate (Benzocaine) (b)

Procaine (Novocaine) (b)

Mepivicaine (c)

▲ **FIGURE G.6**

Some local anesthetics. Three (b) are derived from *p*-aminobenzoic acid (a). The other two (c) contain amide functional groups. All usually are used in the form of the hydrochloride salt, which is more soluble in water than the free base.

kinds of nerve cells and on all parts of the nervous system.

For dental work and minor surgery, it is usually desirable to deaden the pain in one part of the body only. The first local anesthetic to be used successfully was cocaine. Its structure was determined by Richard Willstätter in 1898. Even before Willstätter's work, there were attempts to develop synthetic compounds having similar properties.

Certain esters of *p*-aminobenzoic acid (PABA) act as local anesthetics (Figure G.6). The ethyl and butyl esters are used to relieve the pain of burns and open wounds. These are applied as ointments, usually in the form of salts.

More powerful in their anesthetic action are a series of PABA derivatives that have a nitrogen atom in the alkyl group of the ester. Perhaps the best known of these is procaine (Novocaine), first synthesized by Alfred Einhorn in 1905. Procaine can be injected as a local anesthetic, or it can be injected into the spinal column to deaden the entire lower portion of the body. Local anesthetics work by blocking nerve impulses to the brain. When the block involves the spinal cord, messages of pain from the lower parts of the body are prevented from reaching the brain.

The local anesthetic of choice nowadays is often lidocaine or mepivicaine. Each compound is highly effective and yet has a fairly low toxicity (Table G.1). Note that lidocaine and mepivicaine are not derivatives of *p*-aminobenzoic acid, but they do share some structural features with the compounds that are.

G.6 Barbiturates: Sedation, Sleep, and Synergism

As a family of compounds, the barbiturates display a wide variety of properties. They can be employed to produce mild sedation, deep sleep, and even death.

Barbituric acid was first synthesized in 1864 by Adolph von Baeyer, a young student of August Kekulé (Chapter 13). The term **barbiturates,** according to Willstätter, came about because, at the time of the discovery, von Baeyer was infatuated with a girl named Barbara. The word comes from *Barbara* and *urea.* Curiously, in the United States, the names of the barbiturates end in *-al* even though they are ketones rather than aldehydes (Figure G.7). The British spelling uses the suffix *-one.*

The medicinal value of the barbiturates was discovered in 1903 by Joseph von Mering. Several thousand barbiturates have been synthesized through the years, but only a few have found widespread use in medicine. Pentobarbital (Nembutal) is employed as a short-acting hypnotic drug. Before the discovery of the modern tranquilizers, pentobarbital was used widely to calm anxiety and other disorders of psychic origin.

Phenobarbital (Luminal) is a long-acting drug. It, too, is a hypnotic and can be used as a sedative. It is employed widely as an anticonvulsant for epileptics and brain-damaged people. Thiopental (Pentothal)[7] is used widely in anesthesia.

The barbiturates were once used in small doses as sedatives. The dose for sedation was generally a few milligrams. In larger doses (about 100 mg), barbiturates induce sleep. They were once the sleeping pills so widely used—and abused—by middle-class, often middle-aged, people. The lethal dose is in the vicinity of 1500 mg. Barbiturates are the drugs of choice for many suicides—news reports list the cause of death as "an overdose of sleeping pills." There is also potential for accidental overdose. After a couple of tablets, the person becomes groggy. If

[7] Thiopental has been investigated as a possible "truth drug." It does seem to help psychiatric patients recall traumatic experiences. It also helps uncommunicative individuals talk more freely. It does not, however, prevent one from withholding the truth or even from lying. No true truth drug exists.

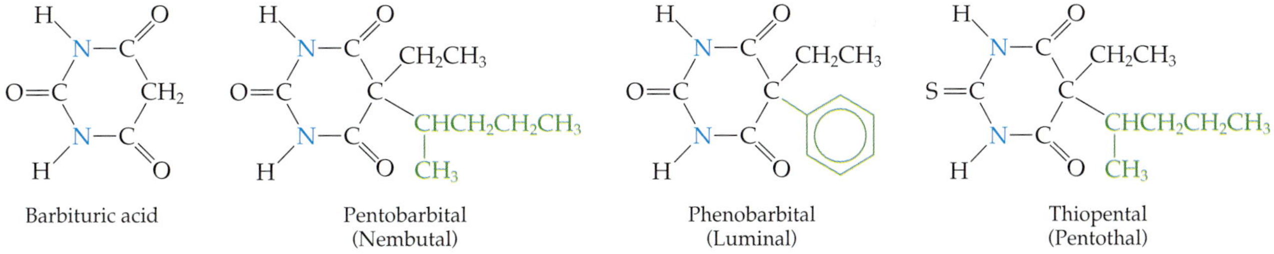

▲ **FIGURE G.7**

Barbituric acid and some barbiturate drugs.

unable to remember taking the sleeping pills, he or she may take more pills.

The barbiturates are especially dangerous when ingested along with ethyl alcohol. This combination produces an effect much more drastic than just the sum of the effects of two depressants. The effect of a barbiturate is enhanced by as much as 200-fold when taken with an alcoholic beverage. This effect of one drug enhancing the action of another is called **synergism.**

The barbiturates are strongly addictive. Habitual use leads to tolerance, which means that ever-larger doses are required to get the same degree of intoxication. Barbiturates are legally available by prescription only, but they are a part of the illegal drug scene also. They are known as "downers" because of their depressant, sleep-inducing effects.

The side effects of barbiturates are similar to those of alcohol. Abuse leads to hangovers, drowsiness, dizziness, and headaches. Withdrawal symptoms are often severe, accompanied by convulsions and delirium. In fact, some medical authorities now say that withdrawal from barbiturates is more dangerous—that is, more likely to cause death—than withdrawal from heroin.

Barbiturates are cyclic amides. Notice, however, that the barbiturate ring resembles that of thymine, one of the bases found in nucleic acids. Evidence indicates that barbiturates may act by substituting for thymine (or cytosine or uracil) in nucleic acids, thus interfering with protein synthesis (see Chapter 23).

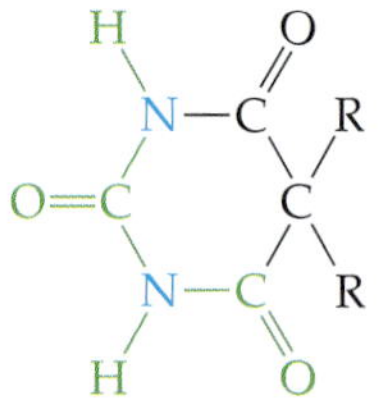

A barbiturate

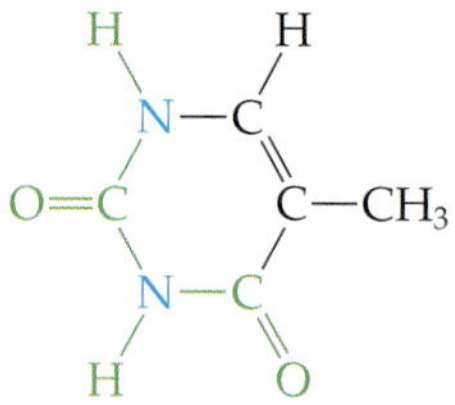

Thymine

G.7 Dissociative Anesthetics: Ketamine and PCP

Ketamine, like thiopental an intravenous anesthetic, is called a **dissociative anesthetic**—it induces hallucinations similar to those reported by people who have had near-death experiences. They seem to remember observing their rescuers from a vantage point above it all, or moving through a dark tunnel toward a bright light. Unlike thiopental, ketamine seems to affect associative pathways before it hits the brain stem.

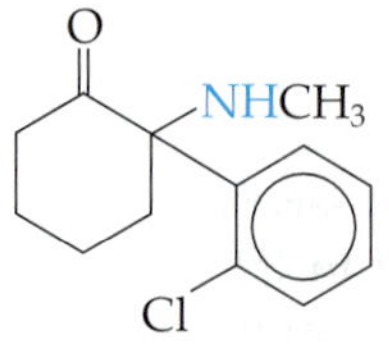

Ketamine

Little is known of the action of ketamine at the molecular level. If it acts by fitting receptors in the body, we can assume that our bodies produce their own chemicals that fit those receptors. These compounds may be synthesized or released only in extreme circumstances—such as in near-death experiences.

Closely related to ketamine is phencyclidine (PCP), known on the street as "angel dust" or "crystal." PCP is soluble in fat and has no appreciable water solubility. It is stored in fatty tissue and released when the fat is metabolized; this accounts for the "flashbacks" commonly experienced by users.

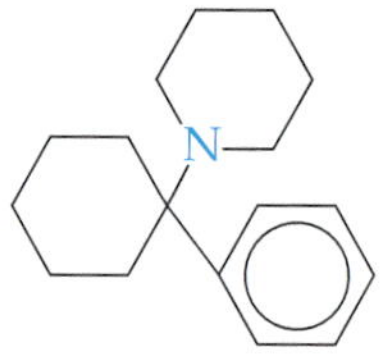

Phencyclidine
(PCP)

PCP is an important part of the illegal drug scene. It is cheap and easily prepared. It was tested and found to be too dangerous for human use, but it has been used as an animal tranquilizer and marketed for this purpose under the trade name Sernylan. Many users experience bad "trips" with PCP. About 1 in 1000 develops a severe form of schizophrenia. Laboratory tests show that PCP depresses the immune system. This could lead to increased risk of infection. Despite these well-known problems, every few years a new crop of young people appears on the scene to be victimized by this hog tranquilizer.

From a pharmacological viewpoint, the compound is of interest because it acts in various ways. It can both stimulate and depress the central nervous system; it is a hallucinogen and an analgesic. There are few drugs that seem to induce so wide a range of effects. The immediate danger is that

PCP is an extremely toxic substance, particularly when mixed with alcohol (it triggers violent, psychotic behavior). PCP is considered more dangerous than heroin because medically nothing is known about its effects, nor are there drugs available to counteract it. It is very easy to overdose on PCP to the point of convulsions.

G.8 Antianxiety Agents

The hectic pace of life in the modern world has driven people to seek rest and relaxation in chemicals. Ethyl alcohol is undoubtedly the most widely used tranquilizer. The drink before dinner—to "unwind" from the tensions of the day—is very much a part of the American way of life. Many people, however, seek their relief in other chemical forms.

Several over-the-counter drugs—Cope, Vanquish, and Compoz among others—claim to be able to help us cope with or vanquish our problems or at least to compose ourselves in the face of minor adversity. Such products usually contain a little aspirin plus an antihistamine. The latter has a side effect of making one drowsy. These products have come under attack by consumer groups as being worthless at best and perhaps even dangerous.

Another class of widely used antianxiety drugs is the benzodiazepines, compounds that feature a seven-member heterocyclic ring (Figure G.8). Of these, perhaps the best known are diazepam (Valium), lorazepam (Ativan), alprazolam (Xanax), and chlordiazepoxide (Librium). These drugs are among the most widely prescribed medications in the United States.[8]

The benzodiazepine derivatives are sometimes called "minor tranquilizers." They often are used to treat anxiety. Antianxiety agents make people feel better simply by making them feel dull and insensitive. They do not solve any of the underlying problems that cause anxiety.

Like most other mind-altering drugs, the benzodiazepines act by fitting specific receptors. Presumably our bodies produce compounds that fit

[8] Certain benzodiazepines, such as flurazepam (Dalmane) and triazolam (Halcion), are used to treat insomnia. They do help a person fall asleep, but the kind of sleep achieved is not restful. These pills may be useful to help someone get through tough times, but they do not cure insomnia or correct the conditions that cause it.

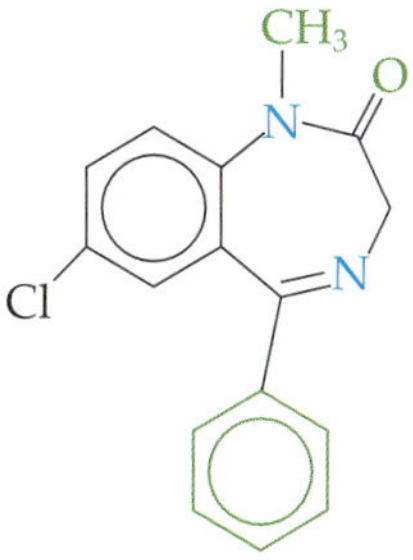

Diazepam
(Valium)

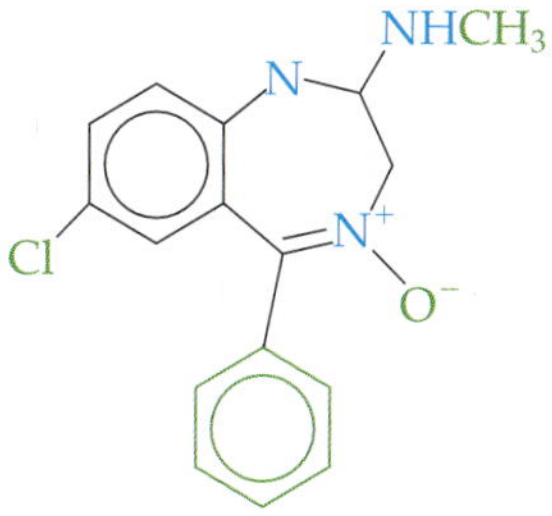

Chlordiazepoxide
(Librium)

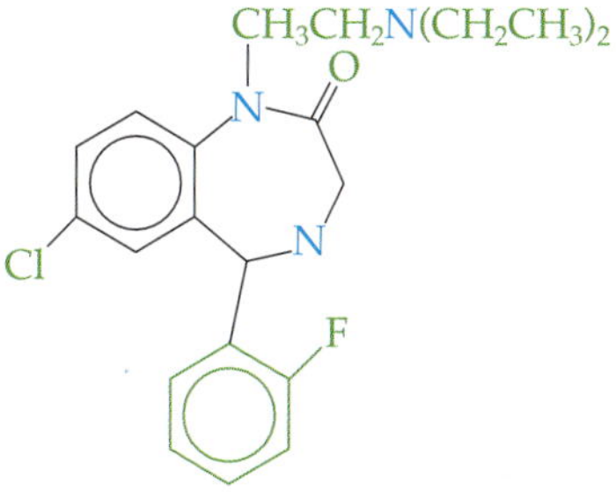

Flurazepam
(Dalmane)

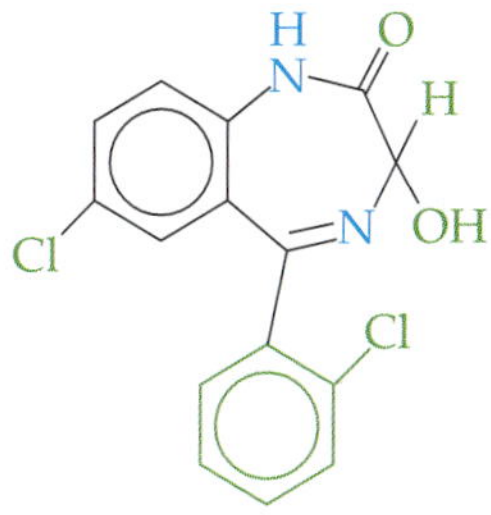

Lorazepam
(Ativan)

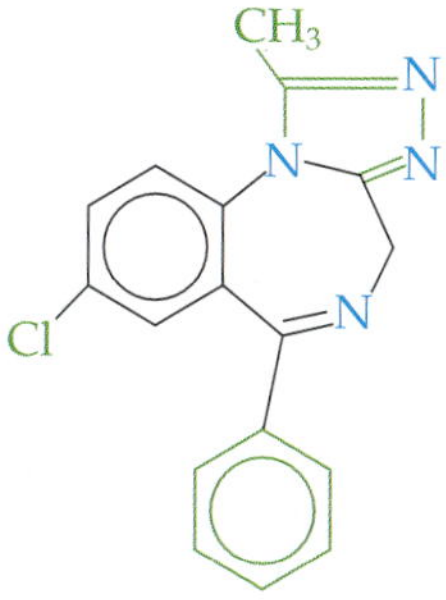

Alprazolam
(Xanax)

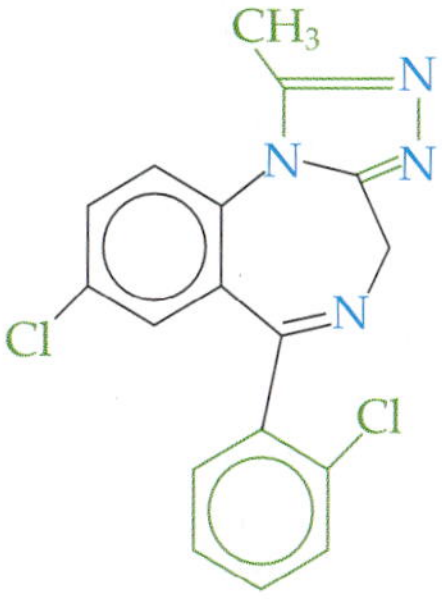

Triazolam
(Halcion)

◀ **FIGURE G.8**

Some benzodiazepine drugs.

these receptors. To date, no such compound has been found. Rather, scientists have found compounds, called β-carbolines, that act on the brain's anxiety receptors to produce *terror.* There is yet so much to learn about the chemistry of the brain.

Anyway, what price tranquility? After 20 years of use, benzodiazepines were found to be addictive. People trying to get off the drugs after prolonged use go into painful withdrawal.

G.9 Antipsychotic Agents

For centuries, the people of India used the snakeroot plant, *Rauwolfia serpentina,* to treat a variety of ailments including fever, snakebite and other poisonings, and—most important—maniacal forms of mental illness. Western scientists became interested in the plant near the middle of the twentieth century—after disdaining such remedies as quackery for many generations.

In 1952, rauwolfia was introduced into American medical practice as a hypertensive (blood-pressure-reducing) agent by Robert Wilkins of Massachusetts General Hospital. In the same year, Emil Schlittler of Switzerland isolated an active alkaloid, which he named reserpine, that has an impressive (intimidating?) structure.

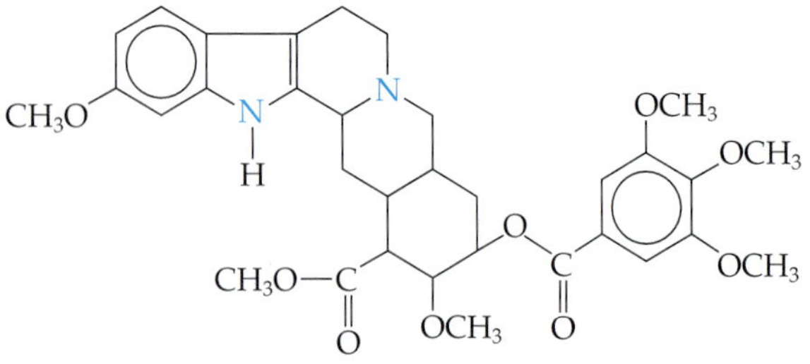

Reserpine

Reserpine was found not only to reduce blood pressure but also to bring about sedation. The latter finding attracted the interest of psychiatrists, who found reserpine so effective that by 1953 it had replaced electroshock therapy for 90% of psychotic patients.

Also in 1952, chlorpromazine (Thorazine) was tried as a tranquilizer on psychotic patients in the United States. The drug had been tested in France as an antihistamine. Medical workers there noted that it calmed mentally ill people who were being treated for allergies. Chlorpromazine was found to be quite effective in controlling the symptoms of schizophrenia. It truly revolutionized mental illness therapy.

Chlorpromazine is one of a group of related compounds called phenothiazines. Several of these compounds are used in medicine. Promazine itself is a tranquilizer, but it is much less potent than chlorpromazine.

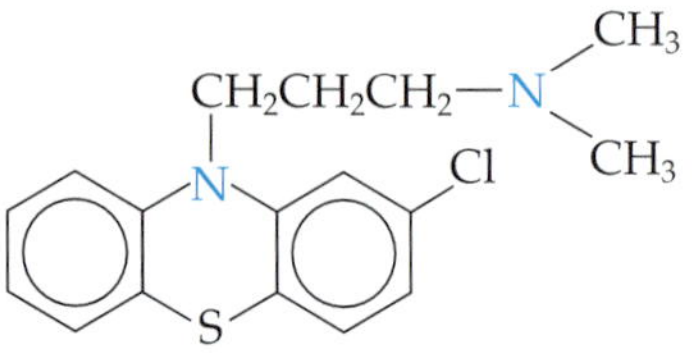

Chlorpromazine
(Thorazine)

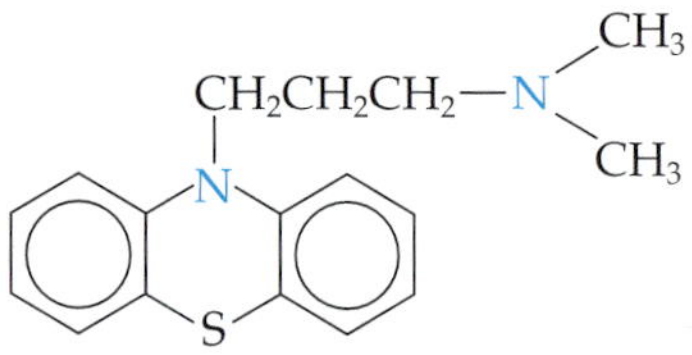

Promazine

The phenothiazines are dopamine antagonists. Dopamine (see Figure G.4) is important in the control of detailed motion (such as grasping small objects), in memory and emotions, and in exciting the cells of the brain. Some researchers think that schizophrenic patients produce too much dopamine; others believe that they have too many dopamine receptors. In either case, blocking the action of dopamine relieves the symptoms of schizophrenia.

The antipsychotic drugs (so-called "major tranquilizers") have been one of the real triumphs of chemical research. They have served to greatly reduce the number of patients confined to mental hospitals by controlling the symptoms of schizophrenia to the extent that 95% of all schizophrenics no longer need hospitalization.

G.10 Antidepressants

It is interesting to note that slight changes in structure can result in profound changes in properties. Replacing the sulfur atom of promazine with a $-CH_2CH_2-$ group produces imipramine (Tofranil), a compound that is not a tranquilizer at all. Rather, it is an antidepressant. Another common tricyclic (three-ring) antidepressant drug is amitriptylene (Elavil), in which the ring nitrogen atom is replaced by a carbon atom.

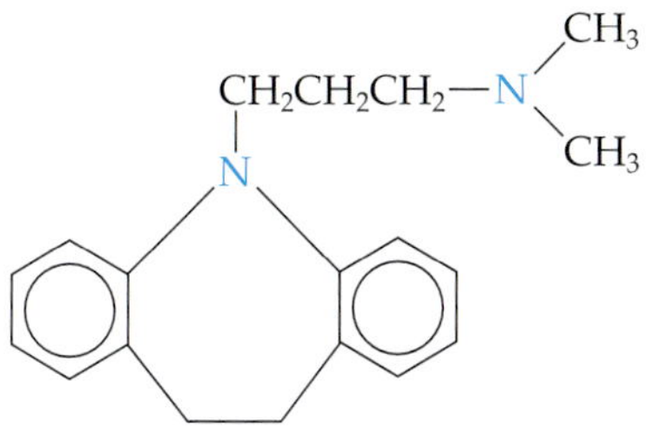

Imipramine
(Tofranil)

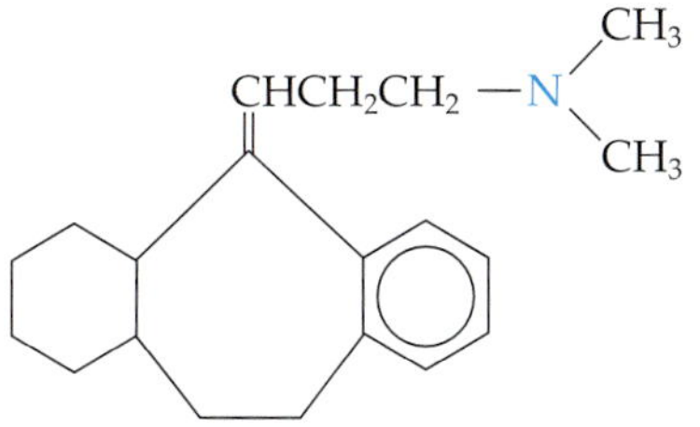

Amitriptylene
(Elavil)

Although the tricyclic antidepressants have been around since the 1950s, they have been only mildly successful. In the first place, there is a very narrow range in which the dose is both safe and effective. Low doses have very little effect, and higher doses quickly become toxic. There are also undesirable side effects, such as nausea, headache, dizziness, loss of appetite, or grogginess, as well as more serious problems such as jaundice or high blood pressure.

Since 1988, several new antidepressants have been introduced, the most popular by far being fluoxetine (Prozac). In 1993, antidepressant sales exceeded $3 billion, and more than 40% of the sales were for Prozac. Doctors write prescriptions for Prozac at the rate of about a million per month. They prescribe it to help people cope with gambling problems, obesity, fear of public speaking, or premenstrual syndrome (PMS). Sometimes they prescribe it for healthy people just to help them loosen up a little or to have a more cheerful disposition. The drug works by enhancing the effect of serotonin, blocking its reabsorption by the cells. It is safer than the older antidepressants and more easily tolerated.

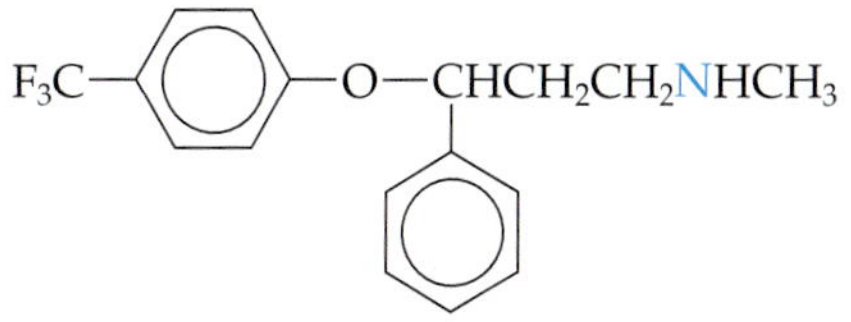

Fluoxetine
(Prozac)

Key Terms

anesthetic (G.5)
barbiturate (G.6)
central nervous system (G.1)
dissociative anesthetic (G.7)
local anesthetic (G.5)
neuron (G.1)
neurotransmitter (G.1)
synapse (G.1)
synergism (G.6)

Review Questions

1. Define or identify each of the following terms.
   a. neuron
   b. synapse
   c. neurotransmitter
2. Which two naturally occurring amines are presently considered to play major roles in the biochemistry of mental health? What are their proposed roles?
3. a. Which amino acids serve as precursors for the amines of Problem 2?
   b. How may our mental state in part be related to our diet?
4. How do amphetamines exert a stimulant effect?
5. How does cocaine exert a stimulant effect?
6. What is a local anesthetic? How does a local anesthetic work?
7. What is a general anesthetic? How does a general anesthetic work?
8. Name two dissociative anesthetics. How do they work?
9. a. What do we mean when we say that amphetamines are "uppers"?
   b. Why are barbiturates called "downers"?
10. What is synergism?
11. Drugs such as lithium carbonate and reserpine block the release of norepinephrine. How might these drugs be useful for treating manic patients?
12. Electroconvulsive therapy (shock treatment) induces the release of norepinephrine. What sort of mental problems are treated with this therapy?

Problems

Structural Formulas and Functional Groups

13. a. What is the basic structure common to all barbiturate molecules?
    b. How is the basic structure modified to change the properties of individual barbiturate drugs?
14. Examine the structure of the reserpine molecule (page 470) and identify the following.
    a. five ether functional groups
    b. two amine functional groups
    c. two ester functional groups
15. Acebutolol (Sectral) is used as a drug for the treatment of heart disease (angina and arrhythmias) and hypertension. There are five functional groups in the compound. Name the five families of organic compounds to which acebutolol could be assigned.

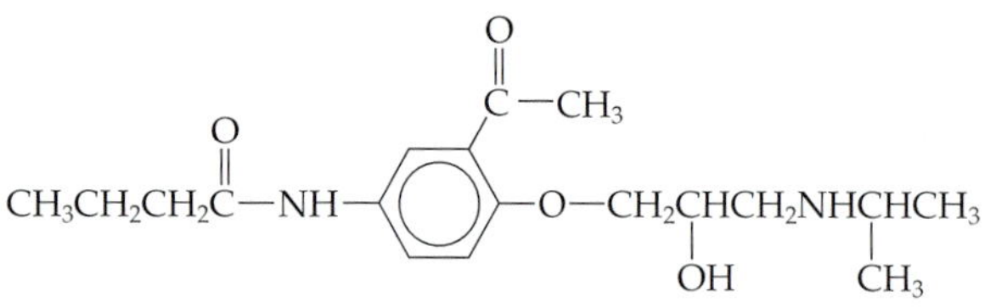

16. Labetalol (Labelol) is used as a drug for the treatment of angina and hypertension. Circle the four functional groups in the molecule, and name the families of organic compounds that incorporate these functional groups.

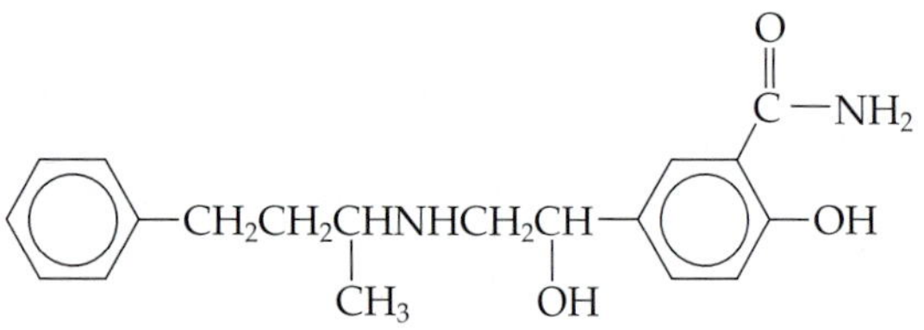

Toxicities

17. When administered intravenously to rats, procaine and cocaine have LD_{50} values of 50 mg/kg and 17.5 mg/kg, respectively. Which drug is more toxic?
18. If the minimum lethal dose (MLD) of amphetamine is 5 mg per kilogram of body weight, what would be the MLD for a 70-kg person? Can toxicity studies on animals always be extrapolated to humans?

Additional Problems

19. Cocaine is usually used in the form of the salt cocaine hydrochloride and sniffed up the nose, where it is readily absorbed through the watery mucous membranes. Some prefer to take their cocaine by smoking it (mixed with tobacco, for example). Before smoking, the cocaine hydrochloride must be converted back to the free base (that is, to the molecular form). Explain the choice of dosage form for each route of administration.
20. Haloperidol (Haldol) is one of the most widely prescribed antipsychotic drugs. What five functional groups are present in the molecule?

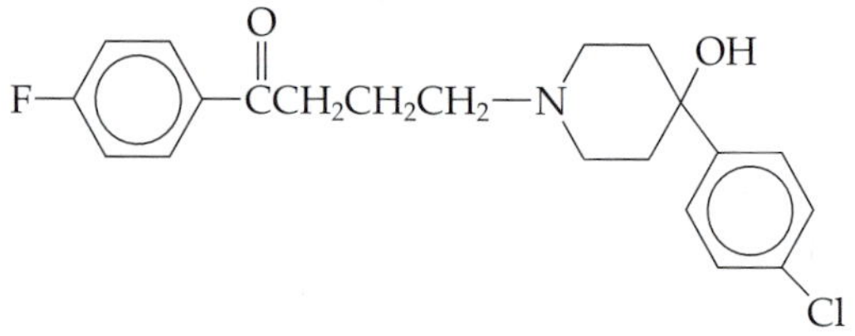

18 Stereoisomerism

A space-filling model and its mirror image. Three-dimensional shapes of molecules such as these are vital to our understanding of life processes.

Isomerism has been defined as the phenomenon whereby two or more *different* compounds are represented by *identical* molecular formulas. There are two main categories of isomers: structural isomers and stereoisomers.

As the term implies, **structural isomers** are compounds that have different structural formulas. Two types of structural isomers have been mentioned—positional isomers and functional-group isomers. *Positional isomers* result from the presence of an atom or a group of atoms at different positions on the carbon chain. Examples of such isomers were discussed in the sections on alkanes, alkyl halides, and alcohols (Table 18.1). Two molecules that have the same molecular formula but contain different functional groups are *functional-group isomers*. Examples of functional-group isomers of alcohols and ethers, aldehydes and ketones, and acids and esters are given in Table 18.2.

The second major category of isomerism is *stereoisomerism*, or space isomerism. Unlike structural isomers, stereoisomers have the identical order of

Table 18.1 Examples of Positional Isomers

ALKANES	
$CH_3CH_2CH_2CH_3$	$CH_3CH(CH_3)_2$
Butane	Isobutane
ALKYL HALIDES	
CH_3CHCl_2	CH_2ClCH_2Cl
1,1-Dichloroethane	1,2-Dichloroethane
ALCOHOLS	
$CH_3CH_2CH_2OH$	$CH_3CHOHCH_3$
1-Propanol	2-Propanol

Table 18.2 Examples of Functional Group Isomers

ALCOHOLS AND ETHERS	
CH_3CH_2OH	CH_3OCH_3
Ethanol	Dimethyl ether
ALDEHYDES AND KETONES	

Propanal	Acetone
ACIDS AND ESTERS	
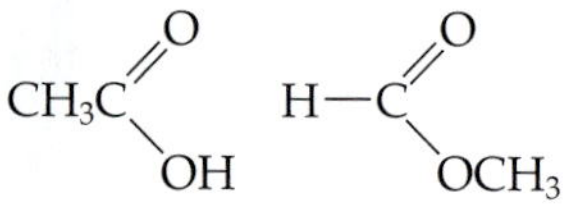	
Acetic acid	Methyl formate

atoms and identical functional groups. They differ only with respect to the spatial arrangement of atoms or groups of atoms within the molecule. Therefore, **stereoisomers** are isomers that have the same structural formulas but differ in the arrangement of atoms in three-dimensional space. Much of what we know about stereoisomers comes from their effect on plane-polarized light. Let's take a look at the nature of this light before we examine the molecules that act upon it.

18.1 Polarized Light and Optical Activity

First of all, let's establish a convention for drawing waves. A wave is something that goes up and down or increases and decreases or varies regularly in some such way. Here's a wave coming toward you while moving to the right and left across the page.

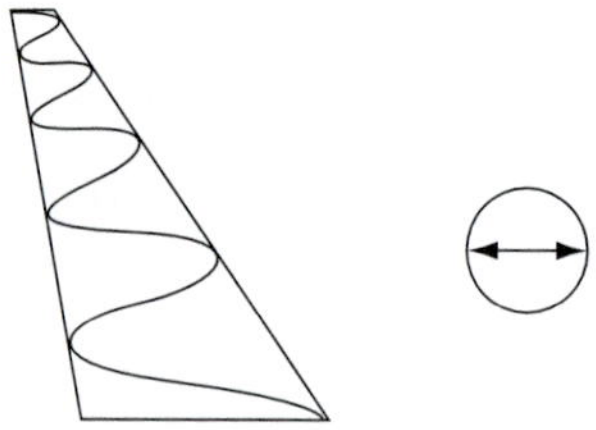

The arrow in the circle is our convention for indicating the same thing. When you see such an arrow, you are supposed to imagine a wave moving out of the page toward you while vibrating back and forth in the direction indicated by the arrow.

Here's another example.

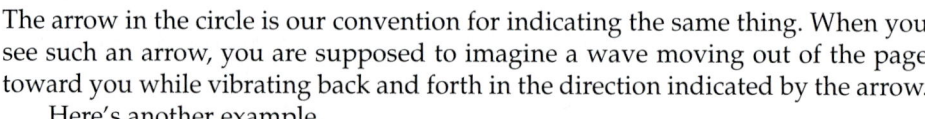

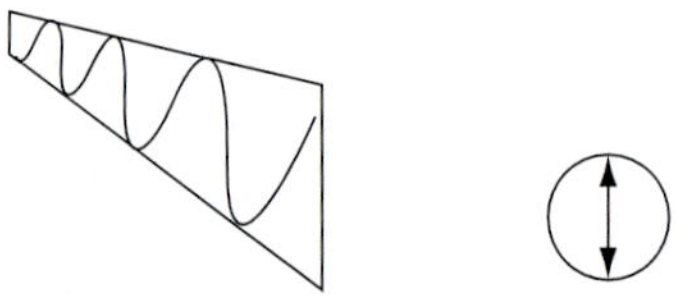

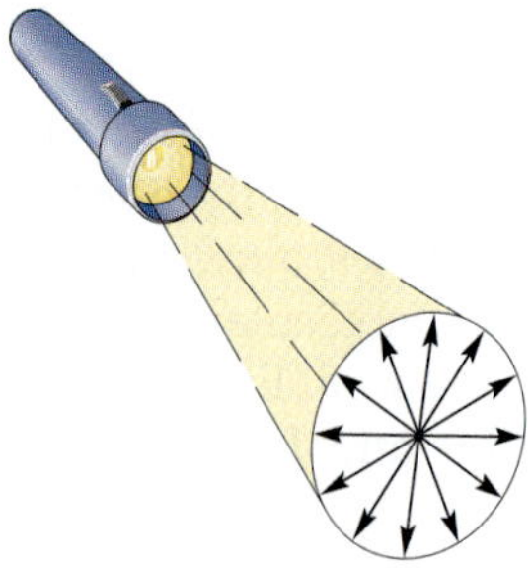

▲ **FIGURE 18.1**
The beam of light in this illustration is not polarized.

This wave is coming toward you in an up-and-down motion. The vertical double-headed arrow stands for the same motion.

Ordinary light may be described as a wave (i.e., it has characteristics that are associated with wavelike motion). A beam of ordinary light can be pictured as a bundle of waves, some of which move up and down, some sideways, and others at all conceivable angles (Figure 18.1). While such light can be described as ordinary, a more scientific term is **nonpolarized light,** which brings us to what we wanted to talk about all along—polarized light.

The waves of **polarized light** vibrate in a single plane. Both of the beams of light shown in Figure 18.2 are polarized. The two polarized beams differ only in the angle of the plane of polarization (represented in our drawings by the orientation of the double-headed arrows).

Sunlight, in general, is not polarized, nor is the light from an ordinary light bulb, nor the beam of light from an ordinary flashlight. One way to get polarized light is to pass ordinary light through Polaroid sheets, such as those used for the lenses of some sunglasses. These lenses are made by carefully orienting organic compounds in plastic to produce a material that permits only light vibrating in a single plane to pass through (Figure 18.3). To the eye, polarized light doesn't "look" any different from nonpolarized light. We can detect polarized light, however, by using a second sheet of polarizing material (Figure 18.4).

Certain substances act on polarized light by rotating the plane of vibration. Such substances are said to be **optically active.** The extent of optical activity is measured in an instrument called a **polarimeter** (Figure 18.5). The device consists of two polarizing lenses, one called the *polarizer* and the other called the *analyzer.*

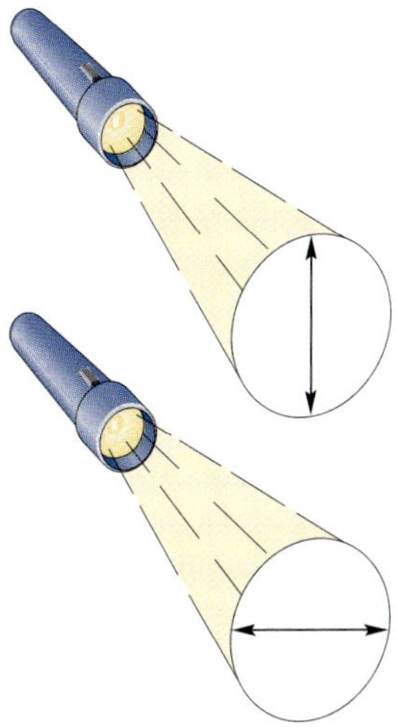

▲ **FIGURE 18.2**

Both of the beams of light in this illustration are polarized. The planes of polarization differ.

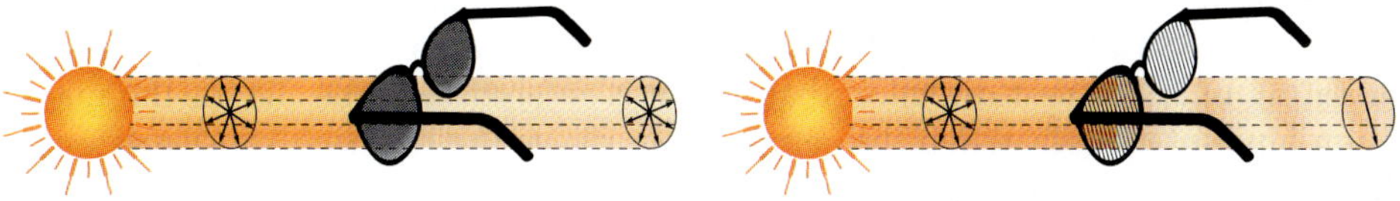

(a) (b)

▲ **FIGURE 18.3**

Ordinary sunglasses (a) dim light by preventing some of it from passing through the lenses. They do not discriminate among light waves vibrating at different angles; rather they filter all the waves to some extent. Light reaching the eyes through these glasses is nonpolarized. Sunglasses with Polaroid lenses (b) selectively pass light waves vibrating in a single plane. Light waves vibrating in other planes do not pass through. Light reaching the eyes through Polaroid sunglasses is plane-polarized.

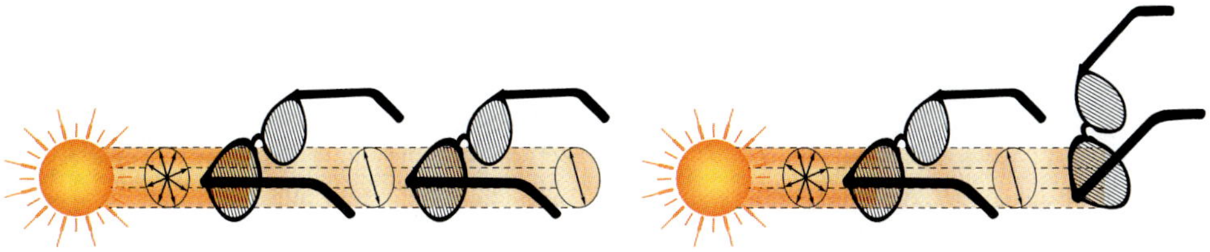

(a) (b)

▲ **FIGURE 18.4**

(a) The light that passes through the first Polaroid lens is polarized. The second pair of glasses is oriented like the first; its Polaroid lens therefore passes the polarized light. (b) The second pair of glasses is oriented 90° to the first. The plane of polarization of the light that made it through the first Polaroid lens is oriented incorrectly for the second lens. No light gets through the second lens.

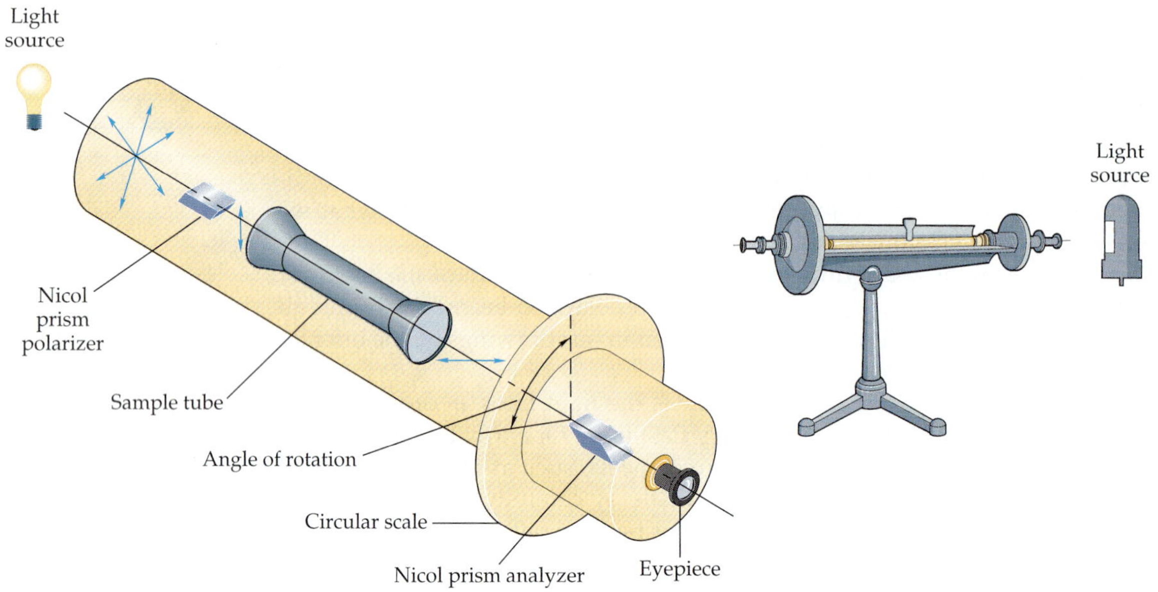

▲ **FIGURE 18.5**
A polarimeter.

With the sample tube empty or containing distilled water, maximum light reaches the observer's eye when the polarizer and the analyzer are aligned so that both pass light vibrating in the same plane. When an optically active substance[1] is placed in the sample tube, that substance rotates the plane of polarization of the light passing through. The polarized light emerging from the sample tube is vibrating in a different direction than when it entered the tube. To see the maximum amount of light when the sample is in place, the observer must rotate the analyzing lens to accommodate this change in the plane of polarization. The angle of rotation, indicated by a pointer on the analyzing lens, corresponds to the change in the plane of polarization caused by the sample.

The size of the rotation angle depends not only on the structure of the optically active material but also on the length of the sample tube, the concentration of the solution, and even the color of light used and its temperature. However, scientists have agreed to certain standard conditions for reporting the angle of rotation. When a rotation is calculated and reported with these conditions taken into account, the value is referred to as the **specific rotation** (symbol $[\alpha]$) and is a physical constant as characteristic of the material as are its melting point, boiling point,

FIGURE 18.6 ▶
Direction of rotation of analyzer.

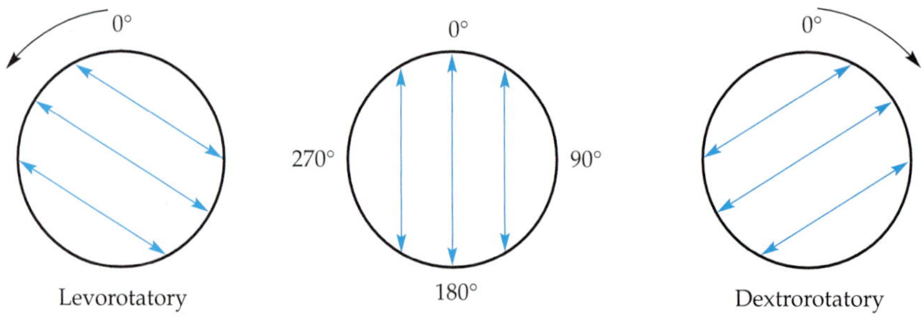

[1] The sample may be a pure gas, a pure liquid, a pure crystalline solid, or a solute dissolved in an appropriate solvent.

density, and solubility. For example, the specific rotation of an aqueous solution of sucrose is +66.5.

Some optically active substances rotate the plane of polarized light to the right (clockwise from the observer's point of view). These compounds are said to be **dextrorotatory** (Latin *dexter,* right); substances that rotate light to the left (counterclockwise) are **levorotatory** (Latin *laevus,* left) (Figure 18.6). To denote the direction of rotation, a positive sign (+) is given to dextrorotatory substances and a negative sign (−) to levorotatory substances. Sucrose is said to be dextrorotatory because it rotates plane-polarized light 66.5° in the clockwise direction, and it is designated as (+)-sucrose.

18.2 Chiral Centers

So far, optically active compounds have been described, and certain terms have been defined for use in dealing with them. Certain fundamental questions have yet to be answered:

1. Why are some compounds optically active?
2. What are the spatial arrangements of the atoms in these compounds?
3. How do these compounds differ from one another and from compounds that are not optically active?

The story of the discovery and explanation of optically active isomers is one of the most interesting sagas in the history of chemistry. You may wish to consult a more comprehensive organic chemistry textbook or, for full enlightenment, the original papers of Pasteur (1848), of Wislicenus (1873), and of van't Hoff and Le Bel (both 1874). The key to their explanation of optical activity in organic compounds was the *tetrahedral carbon atom.*

Recall that the configuration of the methane molecule is tetrahedral. If any or all hydrogen atoms are replaced by other atoms or groups of atoms, the tetrahedral arrangement about the central carbon atom is retained (Figure 18.7). If

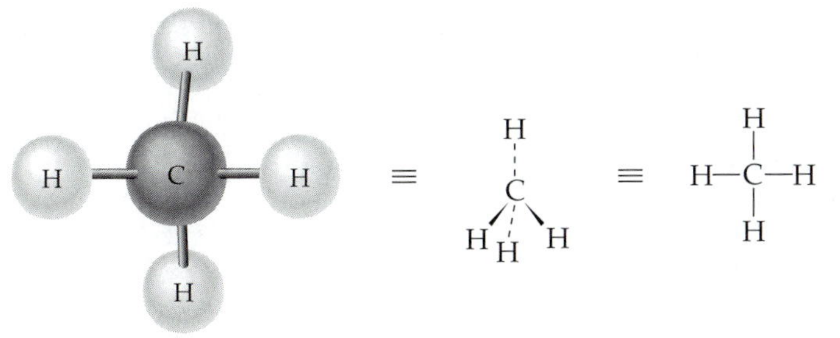

(a)

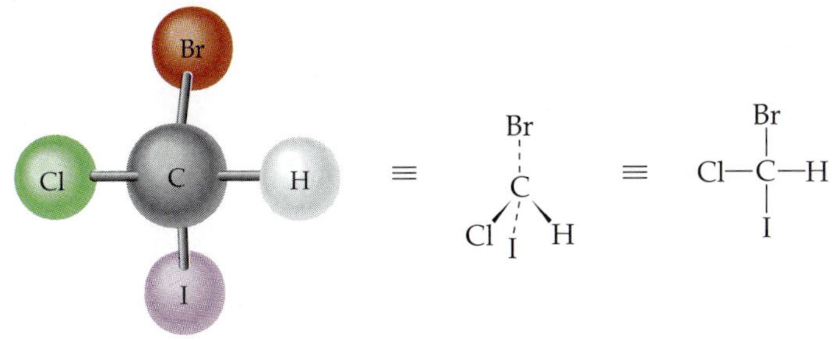

(b)

FIGURE 18.7

Tetrahedral configurations of (a) methane and (b) a trisubstituted derivative of methane.

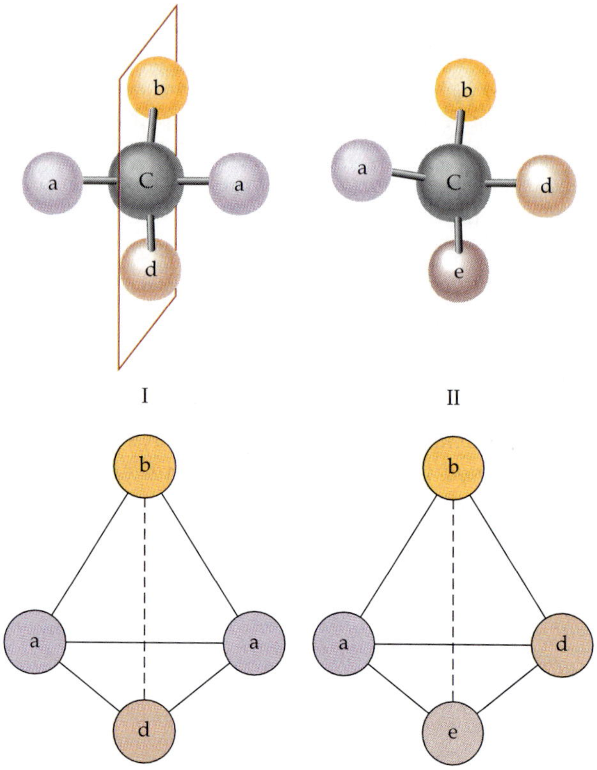

FIGURE 18.8 ▶
Compound I has two similar and two dissimilar substituents. Compound II has four dissimilar substituents.

additional carbon atoms are added and if they all are joined by single bonds, the configuration about each of the carbon atoms is tetrahedral.

Consider the two generalized molecules shown in Figure 18.8. Compound I contains two identical substituents and two different substituents bonded to the central carbon atom, and compound II contains four dissimilar substituents. Compound I is a symmetrical molecule; that is, a plane of symmetry passes through b, d, and the central carbon atom. Compound II is a nonsymmetrical molecule; you can't draw a plane of symmetry anywhere through the molecule.

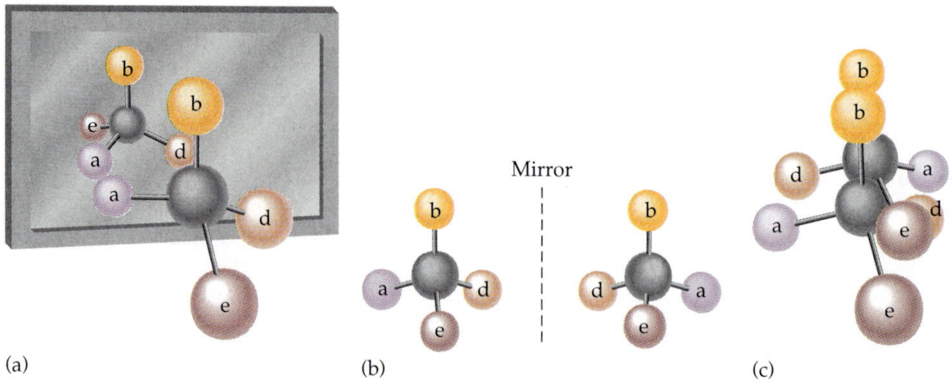

▲ **FIGURE 18.9**
(a) and (b) Mirror images of a nonsymmetrical molecule. (c) A nonsymmetrical molecule cannot be superimposed on its mirror image.

To understand better why compound I is symmetrical and compound II is nonsymmetrical, we can place both compounds in front of a mirror and attempt to impose the mirror images upon the original molecules (Figure 18.9). A symmetrical molecule is *superimposable* on (is identical to) its mirror image, whereas a nonsymmetrical molecule cannot be superimposed upon its mirror image. In doing the superimposing, you can twist and turn the bonds but none of them can be broken.[2]

A useful analogy can be drawn between *nonsuperimposable* (nonmatching) mirror-image compounds and your right and left hands (Figure 18.10a). Regardless of how you twist and turn them, you cannot superimpose your right hand upon your left, or vice versa. This is because, although your right and left hands are nearly perfect mirror images of each other, they are *not* identical. This difference becomes immediately apparent if you try to place your right hand in a left-hand glove. The general property of "handedness" is called **chirality** (Greek *cheir,* hand). An object that is not superimposable upon its mirror image is said to be *chiral*. An object that is superimposable upon (that is, identical to) its mirror image is **achiral** (Figure 18.10b).

The word *chiral* (pronounced KYE-ral) is now widely accepted and has displaced the earlier terms dissymmetric and asymmetric.

A **chiral center** is an atom in a molecule or ion that has four different groups attached. A carbon atom with four different groups attached is sometimes called a

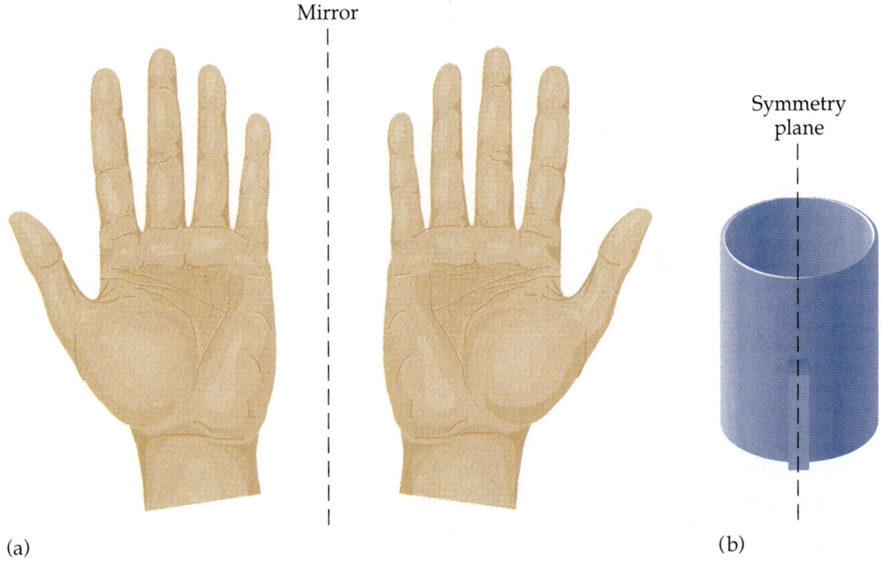

▲ FIGURE 18.10

(a) The left and right hands are nonsuperimposable mirror images; they are chiral. (b) This coffee mug is achiral; a plane of symmetry passes through it. The cup is superimposable on its mirror image.

[2] These molecules, like any others, can be turned upside down or spun about or tipped forward or backward, just as you can stand on your feet or on your head or lie on your back, etc. If you had a mole on your right arm, however, no matter what position you assumed, the mole would still be on your right arm. Your mirror image would always have a left-arm mole. No amount of spinning or turning would cause the mole to change from your right to your left arm. The molecules pictured in Figure 18.9 also have such a distinguishing arrangement. If, for these molecules, we call atom d the *head* and the side to which atom e is attached the *front,* then one compound always has atom d on its right side and the other always has d on its left side.

chiral carbon atom. A chiral molecule is one that is *not* identical with (that is, *not* superimposable on) its mirror image. A molecule containing only one chiral center is always chiral. Some molecules that contain more than one chiral center may be achiral, as we shall see in Section 18.3.

Molecules that are nonsuperimposable (nonidentical) mirror images of each other are called **enantiomers** (Greek *enantios,* opposite). Enantiomers have identical physical and chemical properties, and they rotate plane-polarized light the same number of degrees—*but in opposite directions.*

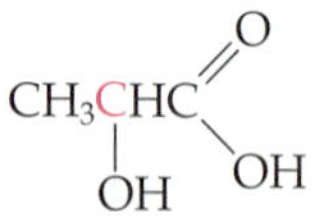
Lactic acid

Let's look at some real molecules that have chiral carbons. A usual representation of lactic acid is given in the margin. Examination of the structure reveals that the second carbon atom has four different groups attached—a hydrogen atom, a carboxyl group, a hydroxyl group, and a methyl group. This carbon, then, is a chiral center. There should be *two* lactic acids. Are there? Yes! (Why, it's enough to give you faith in chemical theory!) One lactic acid is found in sour milk. It is levorotatory and can be designated as (−)lactic acid. Another lactic acid is found in muscle tissue, particularly after exercise. It is dextrorotatory and is called (+) lactic acid. How are the two related? They are enantiomers—one is the mirror image of the other. Using perspective formulas, the enantiomers can be represented as

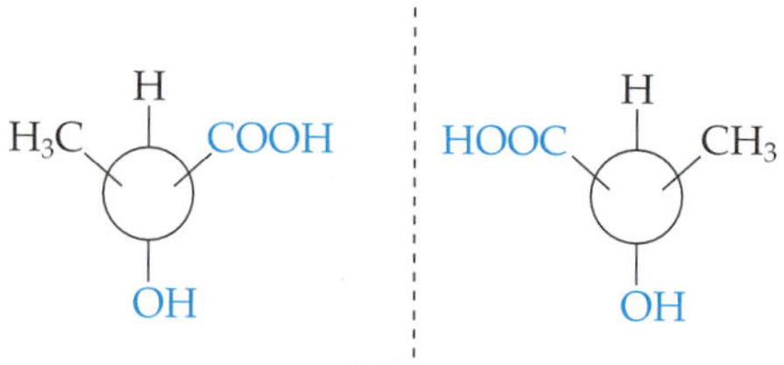

Enantiomers can be drawn using "flat" formulas, as in

but such drawings require more of *you.* You have to visualize that the horizontal bonds project out of the plane of the page toward you and the vertical bonds project below the page. Because they are easier to draw, we shall use these "flat" formulas (often called **Fischer projections**) to represent stereoisomers.

In what ways do the actual lactic acid enantiomers differ from each other? In many respects, they seem more alike than different. All of their physical properties are identical save one, the direction in which they rotate the plane of polarized light. One has a specific rotation of +2.6; the other, −2.6. Only the sign is different. Simple chemical properties are also the same. Both form acidic solutions. Both neutralize bases. Both form esters. Indeed, when reacting with *achiral* molecules, the two enantiomers exhibit identical chemical properties. (Such common reagents as water, hydroxide ion, and ethyl alcohol do not contain chiral centers and are achiral.) It is only when the lactic acid isomers react with other chiral molecules that they behave differently. They may react at different rates and to different extents. The products formed will have different properties. Are these differences really important? In living cells, reactions are controlled by enzymes, and enzymes are chiral. This means that enantiomers may behave quite differently from one another in living cells. Enantiomers may have different tastes and smells. One may be an effective drug and the other worthless. One may be essential to health and the other toxic. Quite literally, we may be talking about differences between life and death.

When lactic acid is made from pyruvic acid in the laboratory, it shows *no* optical activity:

$$CH_3-C\underset{O}{\overset{O}{\|}}-C\overset{O}{\underset{OH}{\|}} \xrightarrow[Ni]{H_2} CH_3-\underset{H}{\overset{OH}{\underset{|}{C}}}-C\overset{O}{\underset{OH}{\|}} + CH_3-\underset{OH}{\overset{H}{\underset{|}{C}}}-C\overset{O}{\underset{OH}{\|}}$$

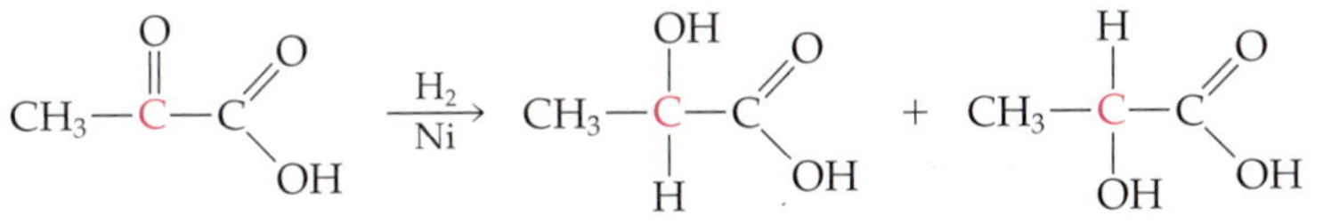

Pyruvic acid Racemic lactic acid
[50%(+)-lactic acid and 50%(−)-lactic acid]

How can this be? The lactic acid has a chiral center, so why isn't it optically active? The answer: in syntheses of this sort, the (+) and (−) forms are formed in exactly equal amounts. Such a mixture of enantiomers is called a **racemic mixture**. It shows no optical activity because it contains equal amounts of molecules with equal but opposite rotatory power. Everything cancels out. Racemic lactic acid is designated (±). A racemic mixture may exhibit physical properties different from those of the pure enantiomers (Table 18.3).

In summary, for compounds that contain only one chiral carbon atom, there always exist two isomers that are nonsuperimposable mirror images (enantiomers), a dextrorotatory form and a levorotatory form. A convenient method of drawing the enantiomer of an optically active compound is to maintain the positions of two of the substituents (usually the larger ones) about the chiral center and invert the positions of the other two.

Table 18.3 Properties of Lactic Acids

Form	Melting Point (°C)	Specific Rotation	pK_a
(+)	53	+2.6	3.8
(−)	53	−2.6	3.8
(±)	16.8	0	3.8

Example 18.1

2-Methyl-1-butanol exists in two optically active forms. The specific rotation of one is +5.756, and for the other it is −5.756. Draw structural formulas for these enantiomers.

Solution

Write the structural formula for one enantiomer and identify the chiral center. (To allow for valid comparisons, the convention is to draw the carbon chain vertically.) Then generate the other enantiomer by interchanging the positions of two of the groups about the chiral center while maintaining the positions of the other two groups.

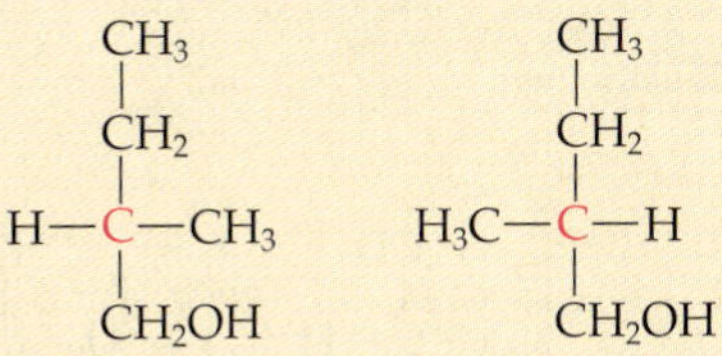

$$\begin{array}{c} CH_3 \\ | \\ CH_2 \\ | \\ H-C-CH_3 \\ | \\ CH_2OH \\ \\ I \end{array} \qquad \begin{array}{c} CH_3 \\ | \\ CH_2 \\ | \\ H_3C-C-H \\ | \\ CH_2OH \\ \\ II \end{array}$$

It is not possible to tell which isomer is dextrorotatory and which is levorotatory merely by inspection of the structural formulas. The distinction can be made only by measuring the optical rotation of each compound in a polarimeter.

Practice Exercise

Draw structural formulas for the enantiomers of 2-butanol.

"Mirror, mirror on the wall, who is the enantiomerest of them all?"

Enantiomers are nonsuperimposable mirror images.

$$CH_3CH-CH-CH_2CH_3$$
$$\ \ \ \ \ \ | \ \ \ \ |$$
$$\ \ \ \ OH \ \ OH$$

2,3-Pentanediol

18.3 Multiple Chiral Centers

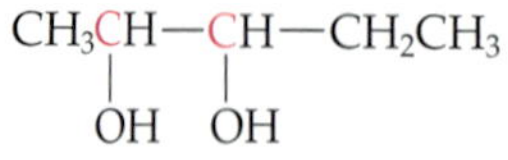

A molecule may have more than one chiral center. Indeed, simple carbohydrate molecules generally have several each. Giant molecules, such as starch, cellulose, and the proteins, may have several hundred or even several thousand chiral centers. Let us look first, however, at molecules with just two.

Consider 2,3-pentanediol. There are *four ways* in which the groups can be arranged about the two chiral centers (Figure 18.11). Note that structures I and II are enantiomers; they are nonsuperimposable mirror images of one another. Compounds III and IV make up another set of enantiomers. What is the relationship, though, between structures II and III? They are stereoisomers because they differ only in their spatial arrangement; they are not enantiomers, however, because they are not mirror images. Such pairs of isomers are called **diastereomers.** Note that II and IV are also diastereomers, as are I and III and I and IV. Diastereomers generally have *different* physical properties (such as boiling point and solubility, as well as specific rotation). Unlike enantiomers, they can be separated by distillation or fractional crystallization.

The first chemist to postulate the existence of multiple stereoisomeric forms was Jacobus van't Hoff (first Nobel prize in chemistry, 1901). He formulated a statement (**van't Hoff's rule**) that makes it possible to predict the total number of possible stereoisomers for a molecule containing more than one chiral center. *The maximum number of different configurations is 2^n, where n is the number of chiral carbon atoms.* The rule is best illustrated with an example.

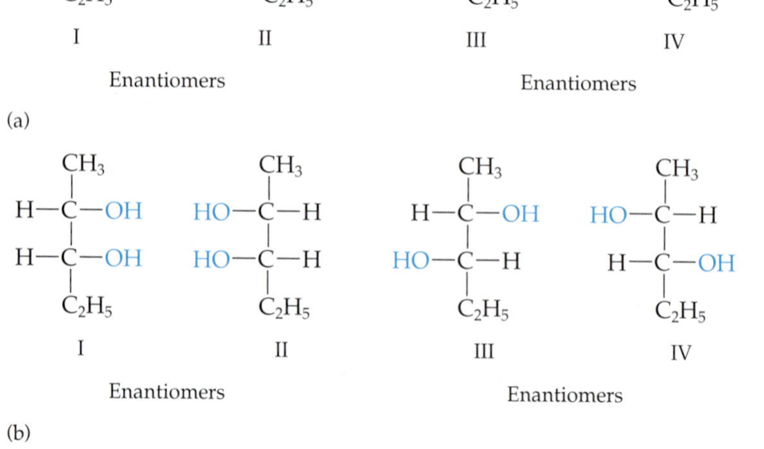

FIGURE 18.11 ▶

The four stereoisomeric 2,3-pentanediols. (a) Perspective drawings. (b) Flat projection formulas. Note that in Fischer projections of molecules with more than one chiral center, the carbon chain is arranged vertically.

Example 18.2

Draw all the possible stereoisomers of 2-methyl-1,3-butanediol.

Solution

1. First draw the structural formula and note the number of chiral carbon atoms. Using the formula 2^n, calculate the number of possible stereoisomers.

$$CH_3-\underset{\underset{H}{|}}{\overset{\overset{OH}{|}}{C}}-\underset{\underset{H}{|}}{\overset{\overset{CH_3}{|}}{C}}-CH_2OH$$

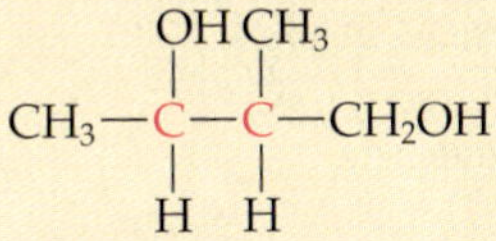

 There are two chiral carbons, and therefore there are 2^2, or 4, possible stereoisomers.

2. Since two configurations are possible for each chiral carbon, there are two sets of mirror images:

Mirror

$$\begin{array}{c} CH_3 \\ H-\overset{|}{C}-OH \\ H-\overset{|}{C}-CH_3 \\ CH_2OH \end{array} \qquad \begin{array}{c} CH_3 \\ HO-\overset{|}{C}-H \\ H_3C-\overset{|}{C}-H \\ CH_2OH \end{array}$$

Mirror

$$\begin{array}{c} CH_3 \\ H-\overset{|}{C}-OH \\ H_3C-\overset{|}{C}-H \\ CH_2OH \end{array} \qquad \begin{array}{c} CH_3 \\ HO-\overset{|}{C}-H \\ H-\overset{|}{C}-CH_3 \\ CH_2OH \end{array}$$

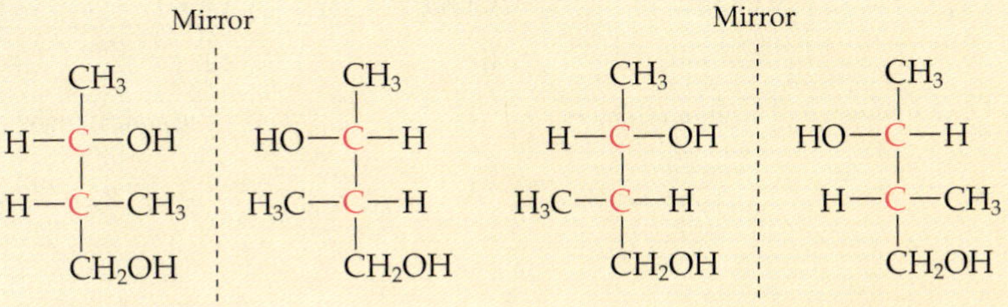

Practice Exercise

Draw all the possible stereoisomers of 2,3-dibromobutanal.

Louis Pasteur (1822–1895), a French chemist who invented the process, now called pasteurization, of heating milk to kill bacteria and retard spoilage. Pasteur's work also led to the germ theory of disease and to immunization procedures. He also discovered the stereoisomerism associated with enantiomers.

Let us consider one last set of compounds before we move on to the next section. These compounds, the tartaric acids, were involved in the earliest studies relating structure and optical activity. The investigator was Louis Pasteur, and the compounds he studied included a new type of stereoisomer.

Tartaric acid, like 2,3-pentanediol, contains two chiral carbon atoms. There is, however, one notable difference between tartaric acid and 2,3-pentanediol. In tartaric acid, each chiral carbon is attached to the same four groups: $-COOH$, $-OH$, $-H$, and $-CH(OH)COOH$. In 2,3-pentanediol, one chiral carbon is attached to a methyl group, whereas the other is attached to an ethyl group. This difference is significant, as we shall see.

Writing out the perspective formulas of tartaric acid (Figure 18.12), we see a pair of enantiomers (I and II). The other apparent pair (III and IV), however, are not enantiomers; they are not even isomers. They are, in fact, the same compound. The structures can be superimposed by rotating one of them 180° in the plane of the page (Figure 18.13). It is important to stress once more that enantiomers are not simply mirror images of one another; instead, they are *nonsuperimposable* mirror images. Every molecule has a mirror image. An achiral molecule is superimposable upon (identical with) its mirror image; a chiral molecule is not. Structures III and IV in Figure 18.12 are mirror images, but they are superimposable and therefore identical. The corresponding structures for 2,3-pentanediol in Figure 18.11 are not superimposable because the CH_3 group and C_2H_5 group are different.

$$HOOC-\underset{\underset{OH}{|}}{CH}-\underset{\underset{OH}{|}}{CH}-COOH$$

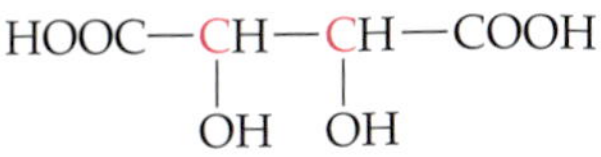

Tartaric acid

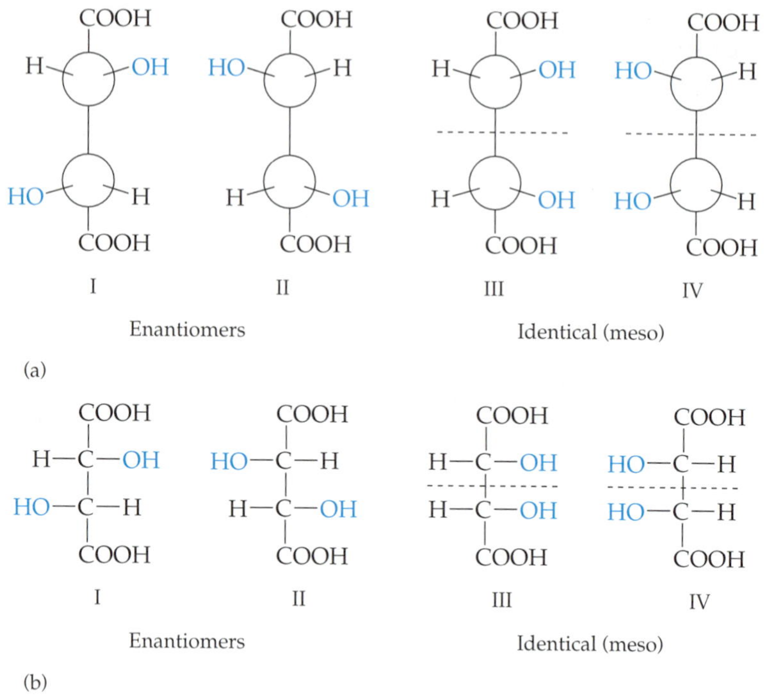

FIGURE 18.12 ▶

The three stereoisomeric tartaric acids.
(a) Perspective drawings. (b) Flat projection formulas. Note that the meso form has an internal plane of symmetry.

The *single* compound represented by structures III and IV in Figure 18.12 is termed a **meso compound.** It is a diastereomer of compound I and compound II. All meso compounds are optically inactive.[3] They contain at least two chiral centers but have an internal symmetry plane; in Figure 18.12 we have indicated this mirror plane in structures III and IV by a dashed line. The meso molecule as a whole is *not* chiral. Table 18.4 lists all the forms of tartaric acid.

Let us summarize here the various conditions that result in a lack of optical activity. A compound (such as ethanol, CH_3CH_2OH) may contain no chiral center. It is not chiral (i.e., it is achiral) and therefore not optically active. A meso compound contains chiral centers, but it is not chiral because it also contains an internal mirror plane. It, too, is optically inactive. A racemic mixture contains chiral molecules but is not optically active because the dextrorotatory molecules cancel out the effects of the levorotatory ones.

[3] A meso compound is superimposable on its mirror image even though it contains chiral carbon atoms. For this reason, it is incorrect to say that all molecules that contain chiral carbon atoms are chiral and thus optically active. Another example of a meso compound is ribitol, whose structural formula is

$$
\begin{array}{c}
CH_2OH \\
| \\
H-C-OH \\
| \\
\text{- - -}\,H-C-OH\,\text{- - - Plane of symmetry} \\
| \\
H-C-OH \\
| \\
CH_2OH
\end{array}
$$

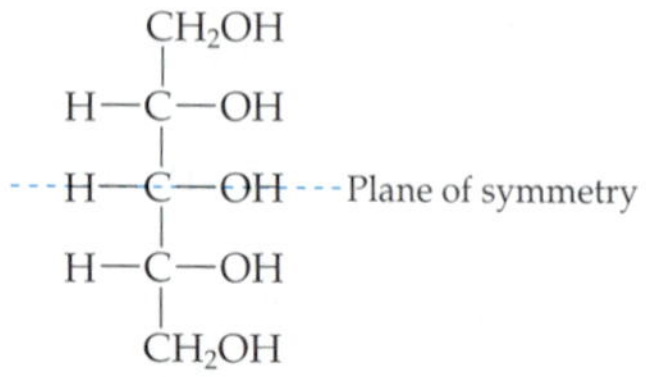

Structures that have meso forms have fewer stereoisomers than predicted by van't Hoff's rule. Tartaric acid, for example, has two chiral carbon atoms but exists in only three stereoisomeric forms, including one that is meso.

(a)

(b)

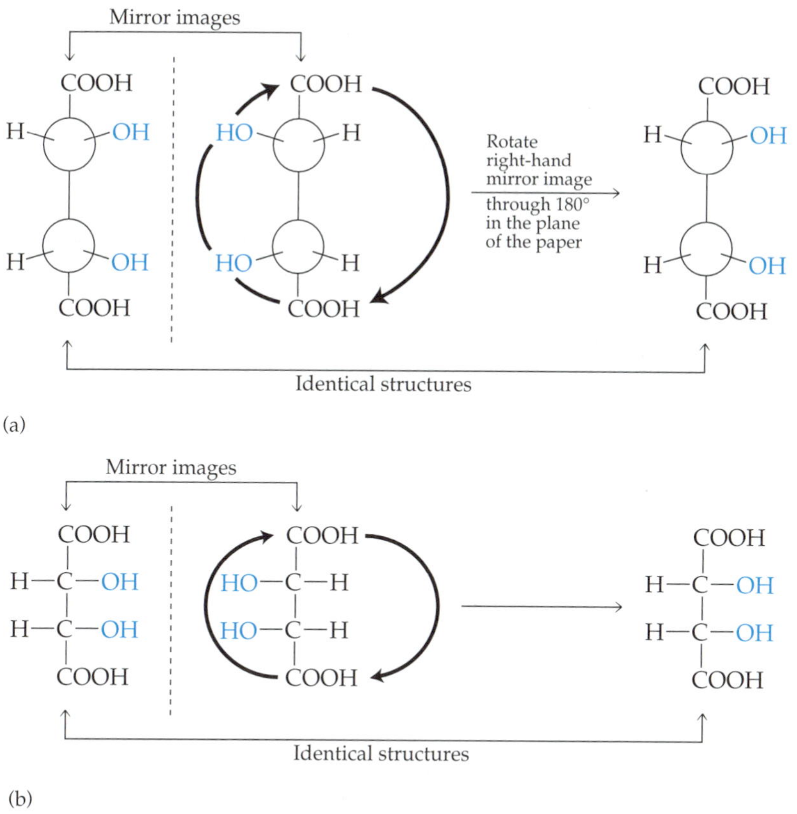

◀ **FIGURE 18.13**

meso-Tartaric acid is superimposable on its mirror image. It exists *not* as a pair of enantiomers but only as a single compound. (a) Perspective drawings. (b) Flat projection formulas.

Table 18.4 Properties of the Tartaric Acids

Form	Melting Point (°C)	Specific Rotation	pK_a	Solubility (g/100 g H_2O)
(+)	168–170	+12.0	2.93	133
(−)	168–170	−12.0	2.93	133
(±)	206	0	2.96	21
Meso	140	0	3.11	125

18.4 Geometric Isomerism (Cis–Trans Isomerism)

It was mentioned in Section 13.11 that cyclic molecules and those containing a carbon–carbon double bond have certain restrictions placed upon them. As we see in this section, these restrictions lead to another kind of stereoisomerism.

In Section 13.11, three isomers of butene, C_4H_8, were identified:

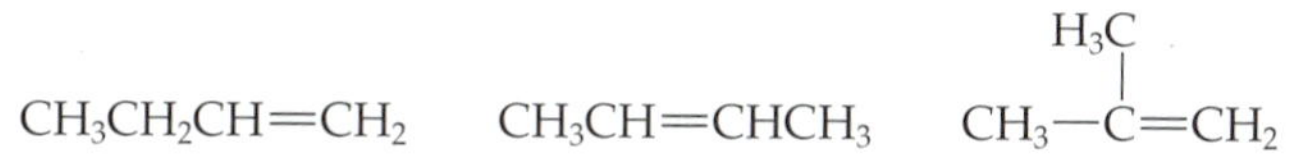

$$CH_3CH_2CH=CH_2 \qquad CH_3CH=CHCH_3 \qquad CH_3-\overset{\overset{H_3C}{|}}{C}=CH_2$$

1-Butene 2-Butene 2-Methylpropene
I II III

Experimental evidence has shown, however, that there are not three but rather four different butene molecules, all having distinctly different physical properties. The fourth isomer also has structure II. Although these two isomeric 2-butenes are structural isomers of I and III, they are not structural isomers of one another. Our knowledge of stereoisomerism leads us to the conjecture that these two molecules might differ in the spatial configuration of their atoms. Because they have different physical properties, however, they cannot be enantiomers. Therefore, a different type of configurational explanation must be sought to account for this phenomenon. The explanation is based upon the geometric arrangement of the carbon–carbon double bond.

Recall that the two carbon atoms of a C=C double bond and the four atoms attached to them are all in the same plane and that rotation around the double bond is prevented. (This is in sharp contrast to the free rotation enjoyed by carbon atoms linked to one another by single bonds.) Ball-and-stick models indicate two ways to arrange the atoms of 2-butene that are in keeping with its structural formula (Figure 18.14). These three-dimensional models are more simply represented as

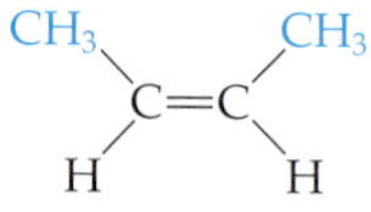

$$
\underset{\substack{\text{IIa}\\[2pt]\textit{cis-2-Butene}\\ \text{mp } -139\,^\circ\text{C}\\ \text{bp } 4\,^\circ\text{C}}}{\underset{H}{\overset{CH_3}{\diagdown}}C\!=\!C\underset{H}{\overset{CH_3}{\diagup}}}
\qquad
\underset{\substack{\text{IIb}\\[2pt]\textit{trans-2-Butene}\\ \text{mp } -106\,^\circ\text{C}\\ \text{bp } 1\,^\circ\text{C}}}{\underset{H}{\overset{CH_3}{\diagdown}}C\!=\!C\underset{CH_3}{\overset{H}{\diagup}}}
$$

In structure IIa, the two methyl groups lie on the same side of the molecule, and this compound is the **cis** isomer (Latin *cis*, on this side). The methyl groups of structure IIb are on opposite sides of the molecule; it is the **trans** isomer (Latin *trans*, across). Because of the restriction on free rotation about the double bond, the structures are clearly nonsuperimposable and hence not identical. *cis*-2-Butene and *trans*-2-butene are geometric isomers of each other.

Geometric isomers are compounds that have different configurations because of the presence of a rigid structure in the molecule. Geometric isomers are diastereomers because they are stereoisomers that are not enantiomers. For alkenes, there are *only two* geometric isomers that correspond to each double bond (cis and trans).

FIGURE 18.14 ▶

Cis-2-Butene and *trans*-2-butene.

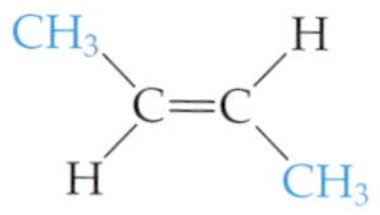

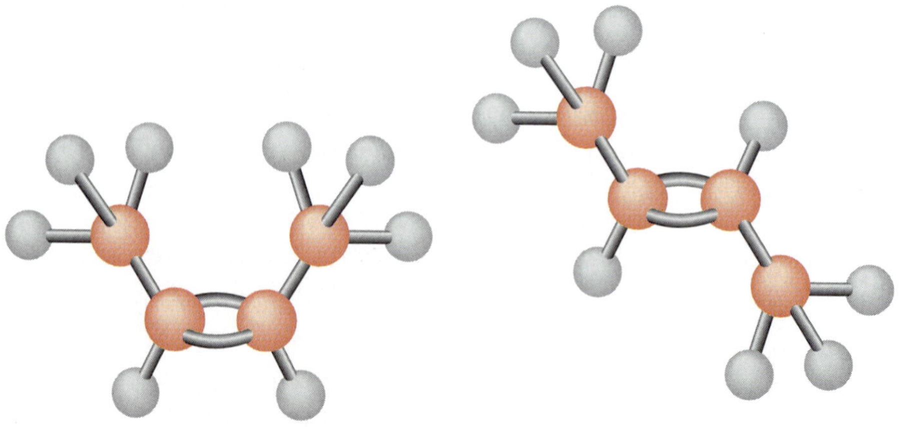

cis-2-Butene trans-2-Butene

We can draw two *seemingly* different propenes:

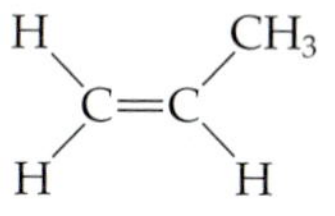
IV

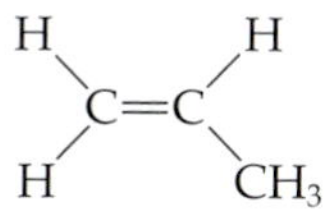
V

However, these two structures are not really different from each other. If you could pick either molecule up from the page and flip it over top to bottom, you would see that the two formulas are identical:

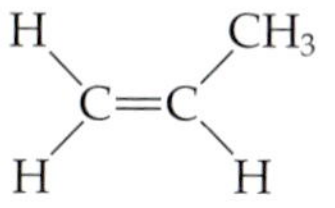
IV

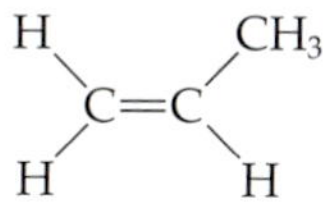
V (flipped over)

The same thing *cannot* be done with cis and trans isomers. If we start with drawings of the two 2-butenes (VI and VII),

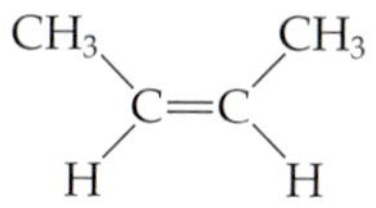
VI

cis-2-Butene

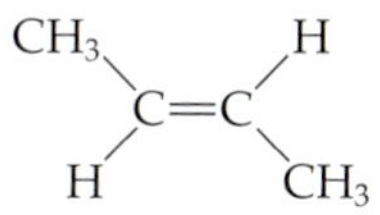
VII

trans-2-Butene

and then flip the trans isomer top to bottom, the resulting structure is still clearly different from the cis isomer:

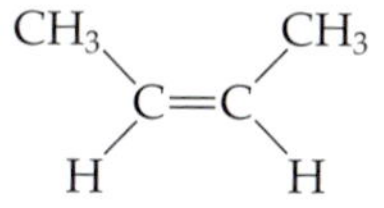
VI

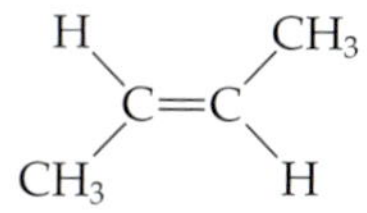
VII (flipped over)

As propene proves, the mere presence of a double bond is not the only criterion (nor is it a necessary one) for geometric isomerism. The requirements for such isomerism are (1) that rotation be restricted in the molecule and (2) that there be two nonidentical groups on *each* doubly bonded carbon. In structures VI and VII, the doubly bonded carbon on the left has a hydrogen group and a methyl group (two different groups), and the doubly bonded carbon on the right has a hydrogen group and a methyl group (two different groups). Thus, 2-butene exists as cis and trans isomers. Propene (structures IV and V) has a doubly bonded carbon with two hydrogens (two identical groups) attached. The second requirement for geometric isomerism is not fulfilled, therefore, and this compound does *not* exist as cis and trans isomers. One of the doubly bonded carbons in propene does have two different groups attached, but the rules require that *both* carbons have two different groups.

In general, when two identical (or nearly identical) substituents are on the same side of the double bond, the compound is the cis isomer. The trans isomer is the one in which similar groups are on opposite sides of the double bond:

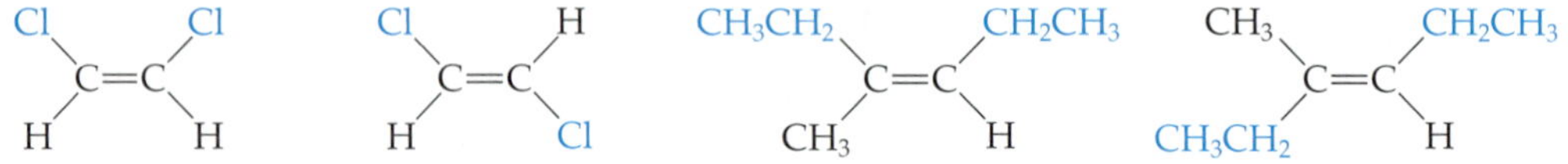

cis-1,2-Dichloroethene
mp −80 °C
bp 60 °C

trans-1,2-Dichloroethene
mp −50 °C
bp 48 °C

cis-3-Methyl-3-hexene

trans-3-Methyl-3-hexene

Example 18.3

Draw all alkenes with the formula C_5H_{10} and indicate which ones exist as cis and trans isomers. Give the IUPAC name for each isomer.

Solution

First we draw the various possible carbon skeletons incorporating a double bond (no bond angles are implied):

$CH_2{=}CHCH_2CH_2CH_3$ — 1-Pentene, VIII

$CH_3CH{=}CHCH_2CH_3$ — 2-Pentene, IX

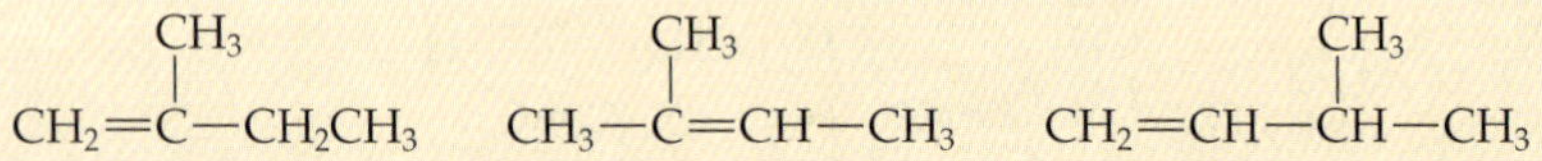

2-Methyl-1-butene
X

2-Methyl-2-butene
XI

3-Methyl-1-butene
XII

Of these, only IX exists as cis and trans isomers:

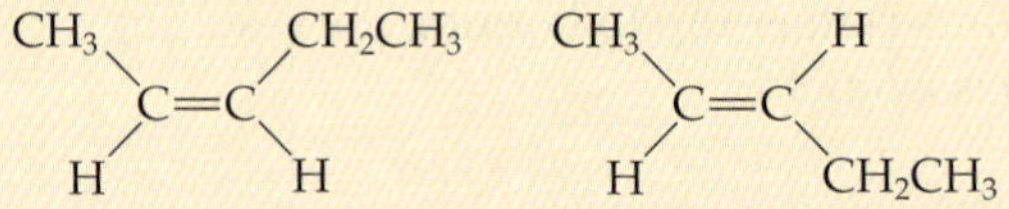

cis-2-Pentene

trans-2-Pentene

Structures VIII, X, and XII each have two hydrogens on one of their doubly bonded carbon atoms, and structure XI has two methyl groups on one of its doubly bonded carbons.

Practice Exercise

Draw and name all alkenes having the formula $C_3H_4Br_2$ and indicate which ones exist as cis and trans isomers.

Maleic and fumaric acids are classic examples of geometric isomers that have widely different chemical and physical properties (Figure 18.15). Because of the proximity of its carboxyl groups, maleic acid readily loses water to form an anhydride upon gentle heating:

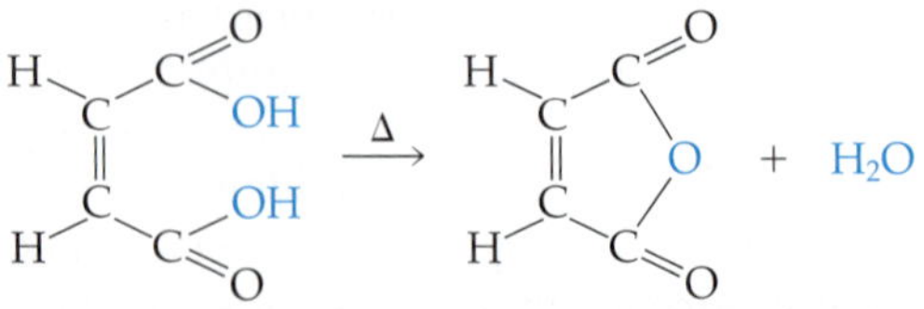

Maleic acid

Maleic anhydride

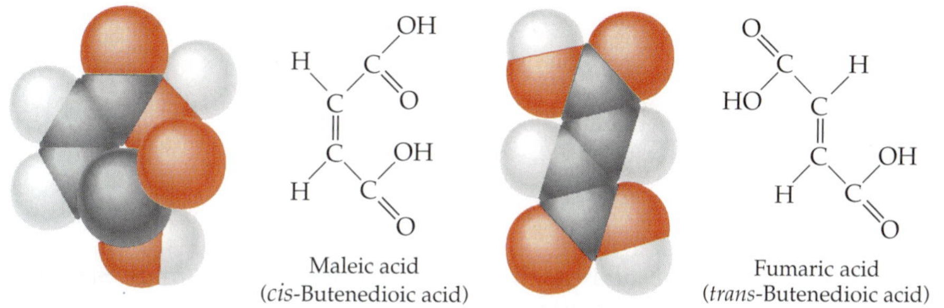

Maleic acid
(*cis*-Butenedioic acid)
mp 130°
Density 1.59 g/cm^3
Solubility in H_2O 78.8 g/100 mL

Fumaric acid
(*trans*-Butenedioic acid)
mp 287°
Density 1.64 g/cm^3
Solubility in H_2O 0.7 g/100 mL

FIGURE 18.15 Space-filling models, structural formulas, and properties of maleic acid and fumaric acid.

Fumaric acid is incapable of anhydride formation under the same reaction conditions. If it is heated to high temperatures (~300 °C), it rearranges to form maleic acid, which then loses water to form the anhydride. We shall see in Section 25.1 that fumaric acid is the isomer produced and utilized by enzymes in living cells.

Recall from Section 13.10 that the bonding in cycloalkanes also imposes geometric constraints on the groups bonded to the ring carbon atoms. Common to all ring structures is the inability of groups to rotate about any of the ring carbon–carbon bonds. Therefore, groups can be either on the same side of the ring (cis) or on opposite sides of the ring (trans). For our purposes here, we represent all cycloalkanes as planar structures, but we clearly indicate the positions of the groups, either above or below the plane of the ring:

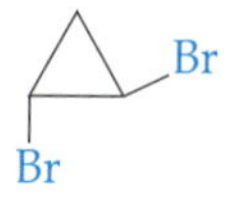

trans-1,2-Dibromocyclopropane

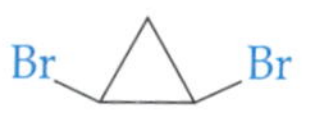

cis-1,2-Dibromocyclopropane

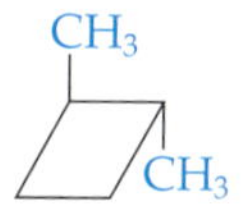

trans-1,2-Dimethylcyclobutane

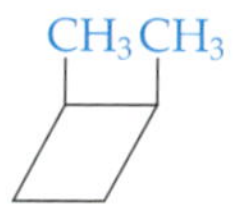

cis-1,2-Dimethylcyclobutane

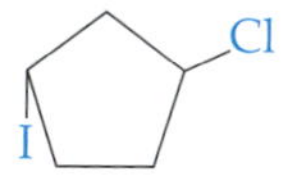

trans-1-Chloro-3-iodocyclopentane

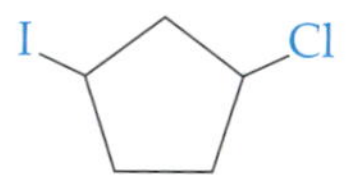

cis-1-Chloro-3-iodocyclopentane

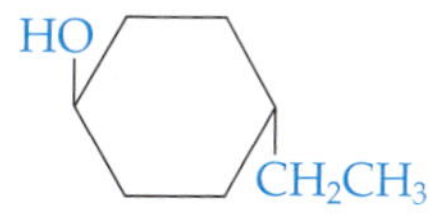

trans-4-Ethylcyclohexanol

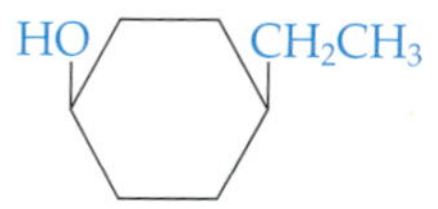

cis-4-Ethylcyclohexanol

18.5 Biochemical Significance

Molecular configurations are of the utmost importance in biochemistry. The example of the two enantiomers of lactic acid has already been cited in Section 18.2. The dextrorotatory form is isolated from muscle tissue, and the levorotatory isomer is found in yeast and some bacteria. Laboratory synthesis of lactic acid from either acetaldehyde or pyruvic acid produces a racemic mixture of (±)-lactic acid; it is impossible to synthesize chemically only one of the two chiral forms using achiral reagents.

The obvious question, then, is how do muscle cells synthesize only (+)-lactic acid, and yeast only the (−)-isomer? The explanation arises many times during the study of biochemistry. Enzymatic control is the answer.

MSG was once used heavily in Chinese foods. For several years MSG was thought to be responsible for a set of symptoms (including headaches and a feeling of weakness) known as Chinese-restaurant syndrome. Experts who reviewed 230 studies concluded that MSG was not the cause of the syndrome. This "not guilty" finding has been confirmed by further studies.

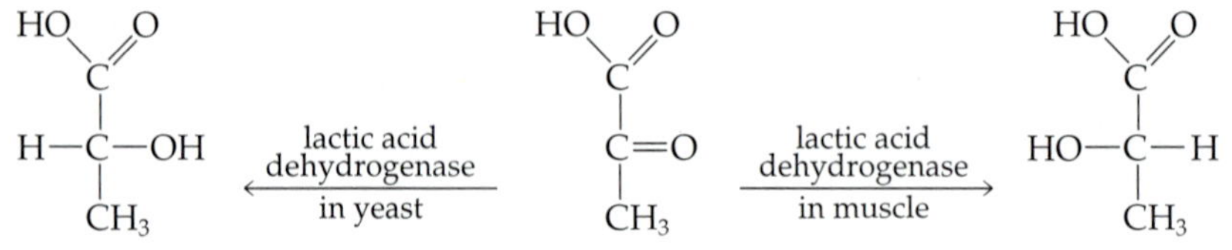

(−)-Lactic acid   Pyruvic acid   (+)-Lactic acid

Enzymes are biological catalysts that are chiral organic compounds. Similarly, almost every organic compound that occurs in living organisms is one enantiomer of a pair. Foods and medicines must have the proper molecular configurations if they are to be beneficial to the organism. For example, the popular flavoring agent Accent is levorotatory monosodium glutamate.[4] The dextrorotatory form of this salt would not enhance the flavor of meat because our taste buds could not recognize it (Figure 18.16). Similarly, the natural form of epinephrine is levorotatory. It has a physiological activity about 15 to 20 times greater than that of its dextrorotatory enantiomer.

In Section G.3 we mentioned that the stimulant drug amphetamine is not a pure compound. Rather, it is a mixture of two enantiomers:

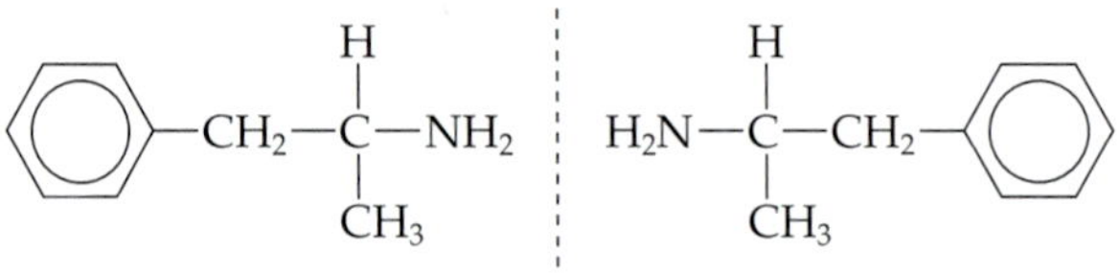

The dextrorotatory form is a stronger stimulant than its levorotatory isomer. Dexedrine is the trade name for the pure dextro isomer. Benzedrine is the trade name for a mixture of the two isomers in equal amounts. Dexedrine is two to four times as active as Benzedrine.

Because of the dangers of pesticides, scientists have been searching for alternative methods to control insects. One method involves the use of chemicals

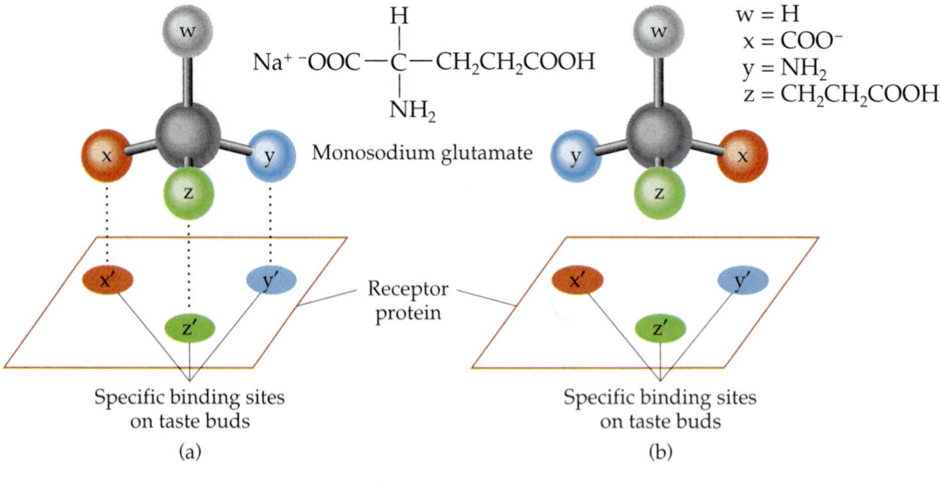

FIGURE 18.16 ▶
(a) The levorotatory stereoisomer of monosodium glutamate fits precisely at bonding sites on a receptor protein of our taste buds. (b) The dextrorotatory isomer does not fit and therefore cannot bind to the receptor sites.

[4] L-Monosodium glutamate (MSG) does not itself impart any taste, but it enhances the flavor of foods to which it is added. Although glutamates are found naturally in proteins, there is evidence that huge excesses can be harmful. MSG can numb portions of the brains of laboratory animals. It also may cause birth defects when administered in large amounts.

known as pheromones. **Pheromones** are chemicals used for communication between members of the same insect species. Insects emit pheromones for a variety of purposes, such as sending an alarm, social regulation, attracting a mate, trail marking, and territorial marking.

In a few insect species, including the boll weevil, the male emits the pheromone.

The most important pheromone for insect control is the sex attractant. The females of many insect species depend upon an attractant to lure males for mating. These chemicals are remarkably powerful; a few drops can attract males within a range of 2 miles. Traps baited with the sex attractant, and also containing an insecticide, can be used to lure all the males of that species to their deaths. One such compound is trimedlure, which has been found to be strongly attractive to the male Mediterranean fruit fly. Trimedlure has eight possible stereoisomers, which vary considerably in their ability to attract male flies. The fly is most strongly drawn to the isomer in which the methyl and ester groups are trans to each other:

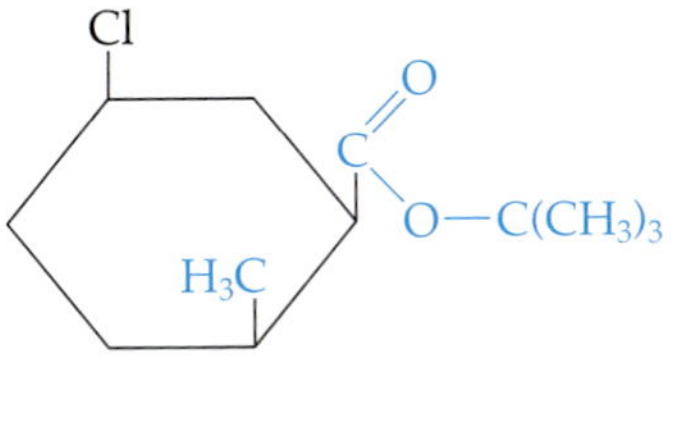

Methyl and ester groups are cis.

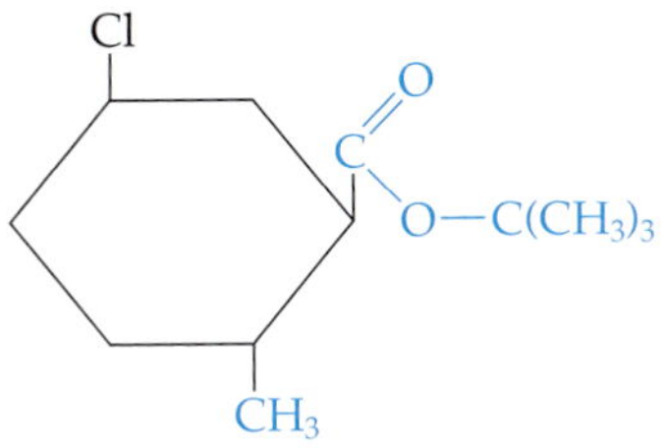

Methyl and ester groups are trans.

Sex attractants of more than 30 insects have been identified. In most cases, just one of the possible stereoisomers is physiologically active (Figure 18.17).

The subtle differences in structural configurations of organic molecules are of primary importance to life. We shall deal with the vitally important stereoselectivity of enzymes only after examining the compositions of the three major classes of biochemical compounds—carbohydrates, lipids, and proteins. It is necessary to observe strictly the proper configurational formulas of these compounds. If enzymes can recognize such subtle differences of shape and structure, so must we.

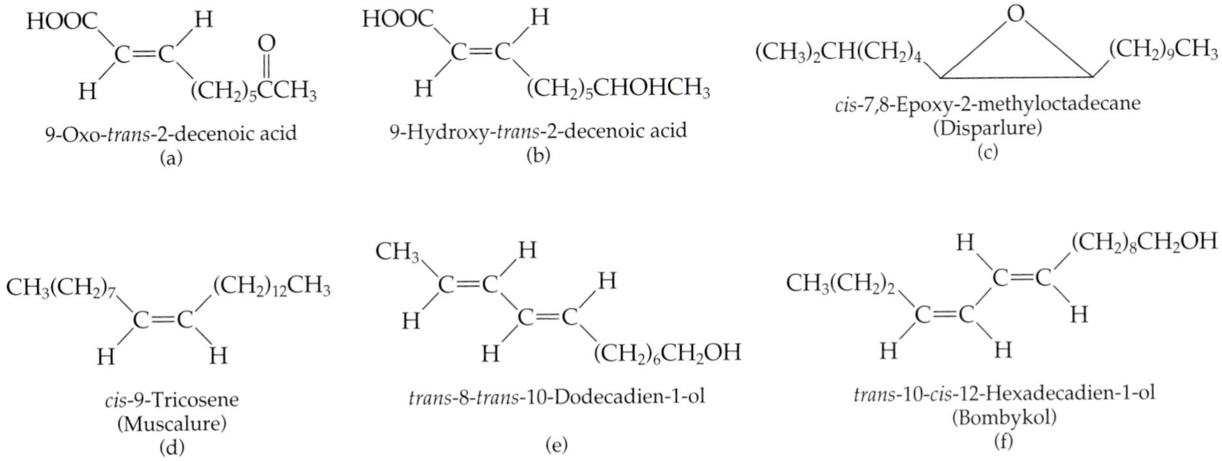

▲ **FIGURE 18.17**

Sex attractants of some female insects. (a) and (b) Queen honeybee, (c) gypsy moth, (d) common housefly, (e) coddling moth, (f) silkworm moth.

Summary

In **structural isomers,** the isomers differ from one another either in the position in which a certain bond is found, as in

$$CH_3CH_2CHCH_3 \qquad CH_2CH_2CH_2CH_3$$
$$\quad\ |\qquad\qquad\qquad |$$
$$\quad Br\qquad\qquad\qquad Br$$

$$C_4H_9Br \qquad\qquad C_4H_9Br$$

or in how the same atoms are combined into different functional groups, as in

$$CH_3CH_2OH \qquad CH_3OCH_3$$

$$C_2H_6O \qquad\qquad C_2H_6O$$

$$\text{Alcohol} \qquad\qquad \text{Ether}$$

In **stereoisomers,** both the order in which atoms are joined and the functional groups are identical; all that differs is the three-dimensional spatial arrangement of the atoms.

Ordinary light has waves vibrating in all directions and for this reason is described as being **nonpolarized light.** When most waves are filtered out of a light beam so that all the remaining ones vibrate in the same plane, that light is **polarized light.** Substances that rotate the plane of vibration of polarized light are **optically active.** Those that rotate the plane to the right are **dextrorotatory** (+) compounds, those that rotate it leftward are **levorotatory** (−) compounds.

A **chiral center** is one bonded to four different groups, and one chiral center in a molecule makes the whole molecule chiral. A chiral molecule and its mirror image are *nonsuperimposable,* and two molecules that are nonsuperimposable mirror images of each other are a special type of *stereoisomer* called **enantiomers:**

$$
\begin{array}{ccc}
A & & A \\
| & & | \\
B-C-D & & D-C-B \\
| & & | \\
E & \text{Mirror} & E
\end{array}
$$

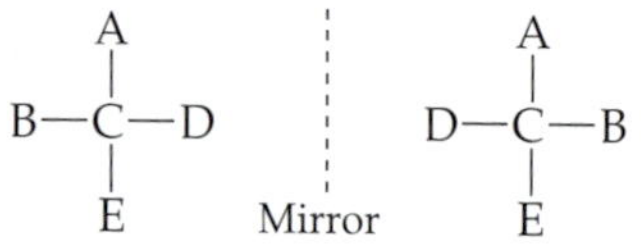

Chiral molecule Mirror image

$$
\begin{array}{cc}
Br & Br \\
| & | \\
CH_3CH_2-C-CH_3 & CH_3-C-CH_2CH_3 \\
| & | \\
H & H
\end{array}
$$

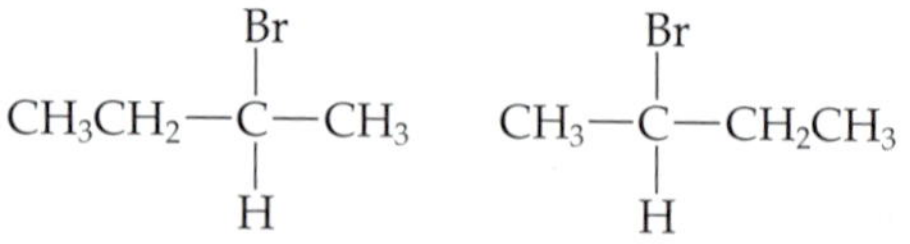

Chiral molecule I Chiral molecule II

enantiomers

Physical properties of the members of an enantiomeric pair are identical save one: one member rotates polarized light in one direction, and the other rotates polarized light in the opposite direction. Chemical properties of the members of an enantiomeric pair are the same when they react with **achiral** substances, but differ when they react with other *chiral* compounds.

A **racemic mixture** is a mixture containing equal amounts of both compounds of an enantiomeric pair; a solution of this mixture has no optical activity because the rotation caused by one enantiomer is exactly canceled by the rotation caused by the other.

Having more than one chiral center in a molecule may result in **diastereomers,** defined as two or more molecules that are stereoisomers of each other but not enantiomers.

$$
\begin{array}{cc}
CH_3 & CH_3 \\
| & | \\
H-C-OH & HO-C-H \\
| & | \\
H-C-OH & HO-C-H \\
| & | \\
Br & Br
\end{array}
$$

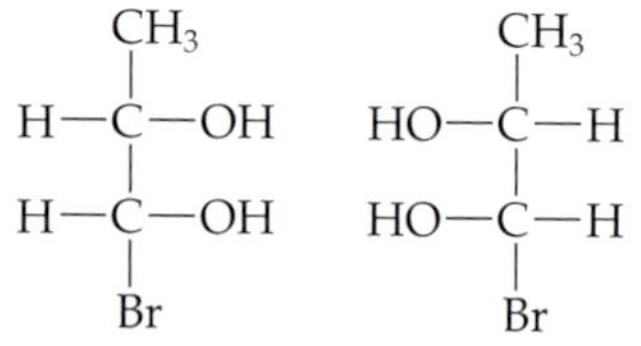

$$C_3H_7O_2Br$$

Enantiomers
(nonsuperimposable mirror images)

$$
\begin{array}{cc}
CH_3 & CH_3 \\
| & | \\
HO-C-H & H-C-OH \\
| & | \\
HO-C-H & HO-C-H \\
| & | \\
Br & Br
\end{array}
$$

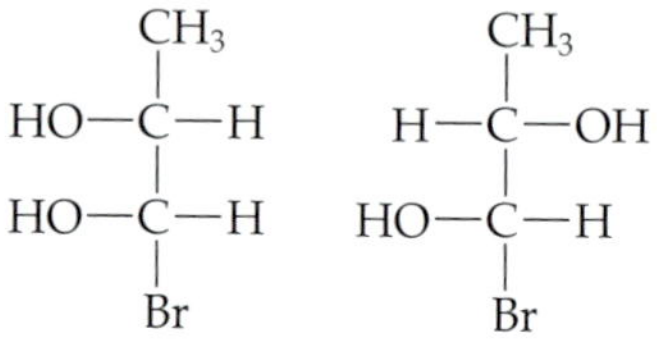

$$C_3H_7O_2Br$$

Diastereomers

The maximum number of possible stereoisomers in a chiral molecule is given by **van't Hoff's rule** as 2^n, with n being the number of chiral carbons.

$$
\begin{array}{c}
CH_2OH \\
| \\
H-C-OH \\
| \\
---H-C-OH--- \text{Plane of symmetry} \\
| \\
H-C-OH \\
| \\
CH_2OH
\end{array}
$$

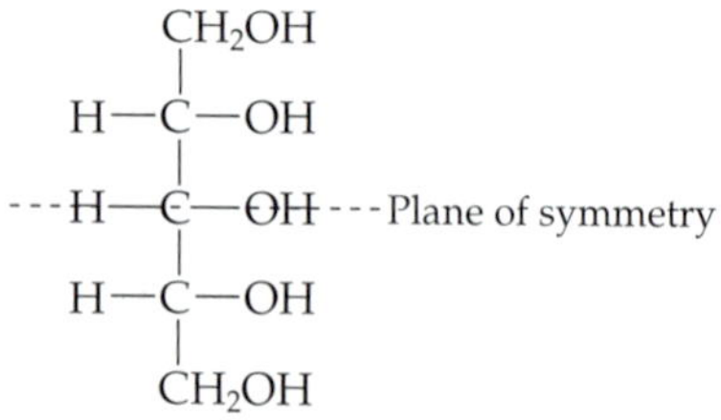

Meso

Meso compounds are *not* optically active because of their internal symmetry plane. The rotation caused by one part of the molecule is canceled by the rotation caused by the comparable part of the molecule on the other side of the symmetry plane.

Geometric isomers are stereoisomers in which molecules differ from each other only in their configuration around a rigid part of the molecule, the rigidity frequently being caused by a carbon–carbon double bond or by the presence of a ring structure. The molecule having two identical atoms or groups on the same side of the molecule is the **cis** isomer; the one having the two groups on opposite sides of the molecule is the **trans** isomer:

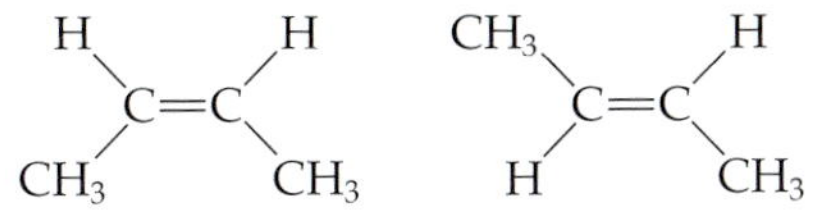

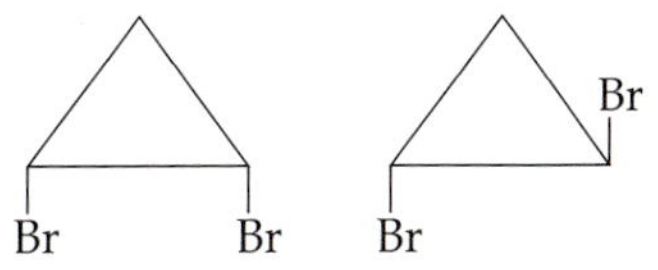

cis trans

The requirements for geometric isomerism are

- restricted rotation in some part of the molecule
- dissimilar groups bonded to each side in the restricted part of the molecule

Geometric isomers are diastereomers but not enantiomers.

Key Terms

achiral (18.2)
chiral center (18.2)
chirality (18.2)
cis (18.4)
dextrorotatory (18.1)
diastereomers (18.3)
enantiomers (18.2)
Fischer projection (18.2)

geometric isomer (18.4)
levorotatory (18.1)
meso compound (18.3)
nonpolarized light (18.1)
optically active (18.1)
pheromone (18.5)
polarimeter (18.1)

polarized light (18.1)
racemic mixture (18.2)
specific rotation (18.1)
stereoisomer (Introduction)
structural isomer (Introduction)
trans (18.4)
van't Hoff's rule (18.3)

Review Questions

1. Define:
 a. chiral center
 b. enantiomer
 c. polarimeter
 d. optically active
 e. specific rotation
 f. geometric isomer
 g. polarized light
 h. diastereomers
 i. stereoisomer
 j. meso compound
 k. racemic mixture
 l. levorotatory
 m. dextrorotatory
 n. pheromone

2. Are these structures mirror images of each other? Are they superimposable?

$$\begin{array}{ccc} & CH_3 & & & CH_3 \\ H-&C&-OH & HO-&C&-H \\ & CH_3 & & & CH_3 \end{array}$$

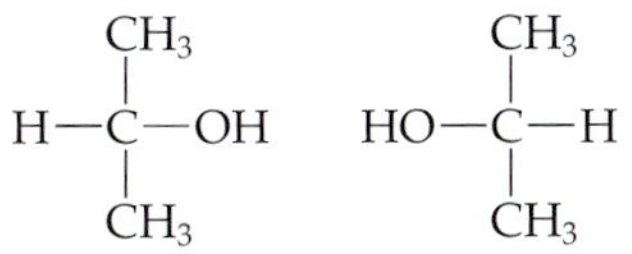

3. Are these structures mirror images of each other? Are they superimposable?

$$\begin{array}{ccc} & H & & H \\ CH_3CH_2-&C&-CH_3 & CH_3-&C&-CH_2CH_3 \\ & OH & & OH \end{array}$$

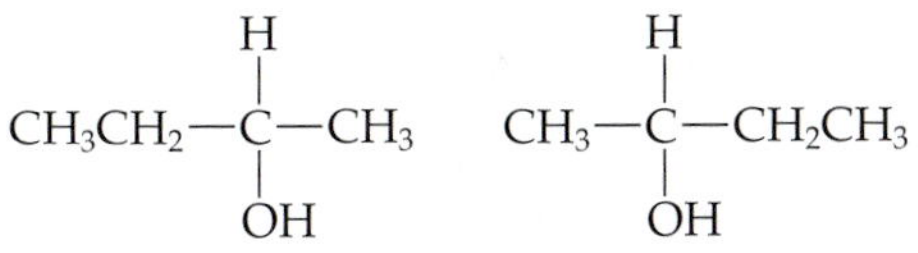

4. Compare (+)lactic acid and (−)lactic acid with respect to

 a. boiling point
 b. melting point
 c. specific rotation
 d. solubility in H_2O
 e. reaction with ethanol
 f. reaction with (+)-*sec*-butylamine

5. (−)Menthol melts at 43 °C, boils at 212 °C, and has a density of 0.890 g/cm^3 and a specific rotation of −50. List the corresponding properties of (+)menthol.

6. Would (+)nicotine and (−)nicotine react the same way with (**a**) HCl? (**b**) (+)lactic acid? (**c**) (−)tartaric acid?

7. 1-Butene reacts with HCl to form 2-chlorobutane. Does the reactant have a chiral center? Does the product have a chiral center? Would 2-chlorobutane formed in this manner show optical activity? Why or why not?

8. Write formulas for pairs of isomers that represent the following.
 a. positional isomers
 b. functional group isomers
 c. enantiomers
 d. diastereomers
 e. noncyclic geometric isomers
 f. cyclic geometric isomers

9. In what ways do enantiomers resemble each other? How do they differ from each other?

10. How does a meso compound differ from a racemic mixture?

Problems

Chirality

11. Use an asterisk (*) to indicate which of the carbon atoms shown in color are chiral centers.

 a. ⬡—CH$_2$CH—NH$_2$
 　　　　　　｜
 　　　　　　CH$_3$

 b.
 　　　　　CH$_2$CH$_2$CH$_3$
 　　　　　｜
 　　O　　｜　　O
 　　‖　　｜　　‖
 H$_2$NCOCH$_2$CCH$_2$OCNH$_2$
 　　　　　｜
 　　　　　CH$_3$

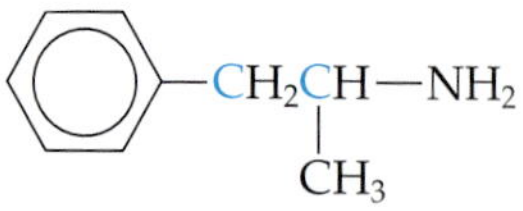

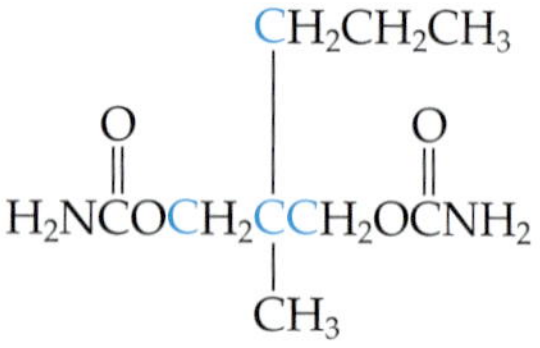

12. Circle each chiral carbon atom in:

 a. HO—⬡—CHCH$_2$—NH—CH$_3$
 　　　｜　　　　　｜
 　　HO　　　　　OH

 Epinephrine

 b.
 　　O
 　　‖
 　　C—CH$_2$—CH$_2$—CH—C
 HO　　　　　　　　　　｜　　‖
 　　　　　　　　　　NH$_2$　O$^-$ Na$^+$

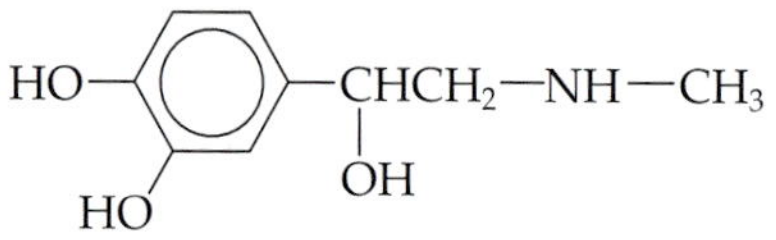

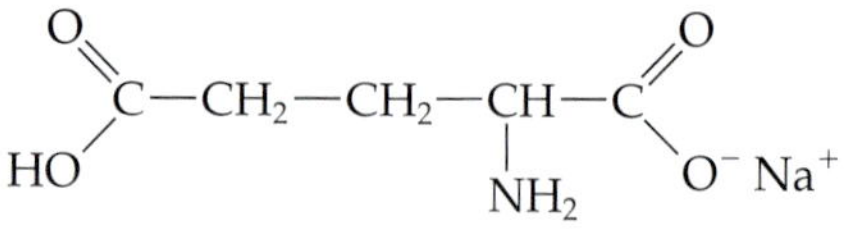

 MSG

13. Circle each chiral carbon atom in:

 a. CH$_3$CHCH$_2$OH
 　　　｜
 　　　OH

 b. CH$_3$CHCOOH
 　　　｜
 　　　NH$_2$

 c. C$_6$H$_5$CH$_2$CHCH$_3$
 　　　　　　｜
 　　　　　　NH$_2$

 d. CH$_3$CHCH$_2$CH$_3$
 　　　｜
 　　　Br

 e. CH$_3$CHCHO
 　　　｜
 　　　OH

 f. CH$_3$CH—CHCH$_3$
 　　　｜　　｜
 　　　OH　OH

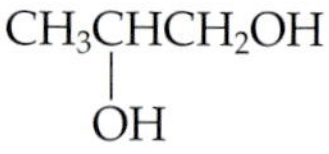

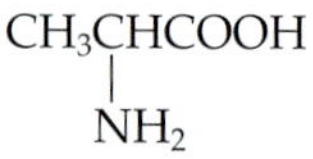

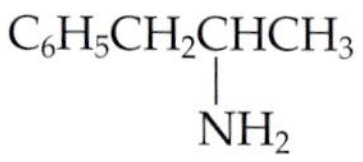

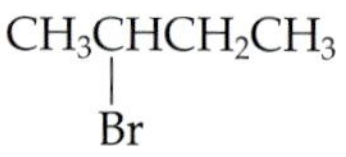

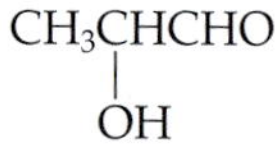

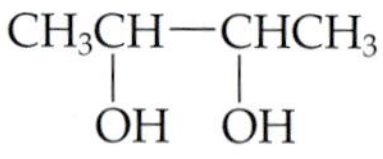

14. Circle each chiral carbon atom in:

 a.
 　　　　　　　　O　　　　　　　O
 　　　　　　　　‖　　　　　　　‖
 H$_2$N—CH—C—NH—CH—C—OCH$_3$
 　　　　｜　　　　　　　｜
 　　　　CH$_2$　　　　　　CH$_2$
 　　　　｜　　　　　　　｜
 　　　　COOH　　　　　　⬡

 Aspartame

15. Indicate whether the compound is chiral.

 a. CH$_3$CHCH$_2$CH$_2$CH$_3$
 　　　｜
 　　　NH$_2$

 b. H$_2$N—⬡—COCH$_2$CH
 　　　　　　　‖　　　　｜
 　　　　　　　O　　　　CH$_3$
 　　　　　　　　　　　　｜
 　　　　　　　　　　　　CH$_3$

 c.
 　　　　　　　　O
 　　　　　　　　‖
 　　　　　　　　C
 HO—C　　　　O
 　　‖
 　　C—C　　OH
 　｜　｜　｜
 OH　H　C—CH$_2$OH
 　　　　｜
 　　　　H

 d. CH$_3$CHCH$_2$CCHCH$_3$
 　　　｜　　　‖　｜
 　　　CH$_3$　　O　CH$_3$

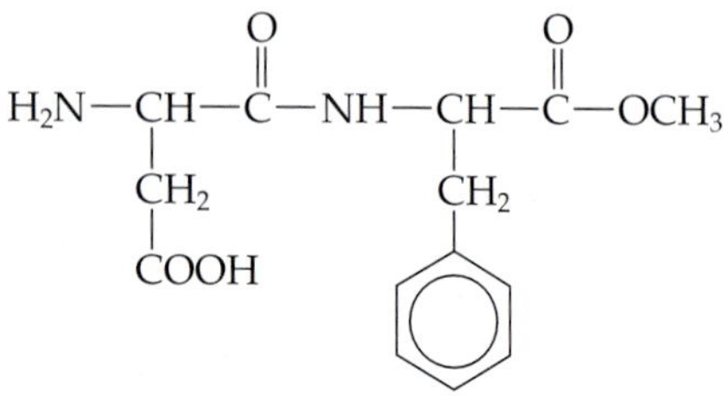

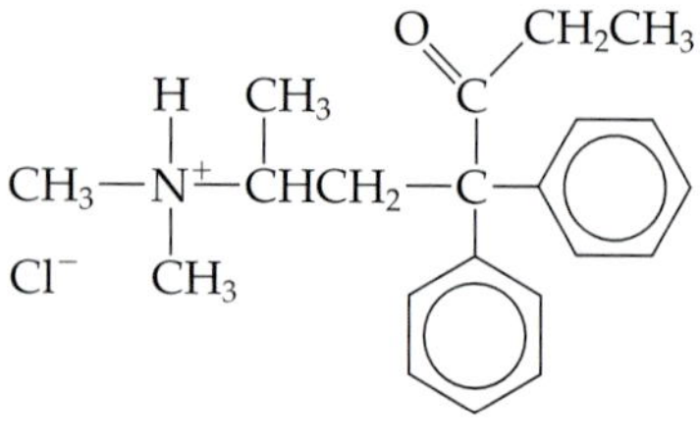

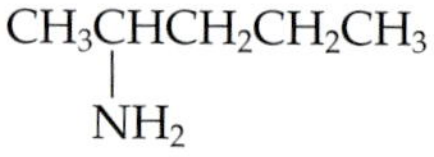

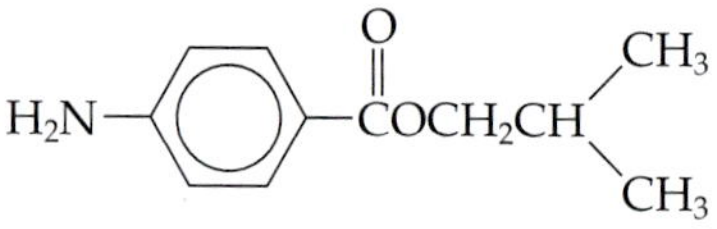

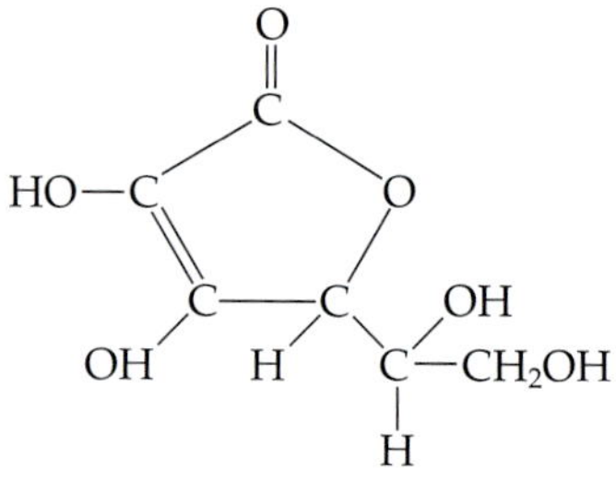

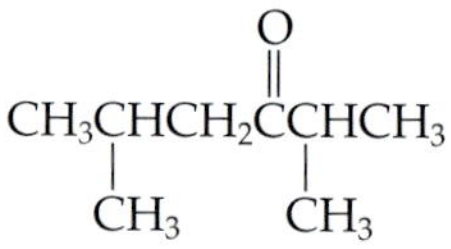

16. Indicate whether the molecule as drawn is chiral.

 a.
 　　　　CH$_3$
 　　　　｜
 　H—C—OH
 　　　　｜
 　　　　CH$_3$

 b.
 　　　H　H
 　　　｜　｜
 CH$_3$—C—C—CH$_3$
 　　　｜　｜
 　　　Br　Br

 c.
 　　　H　Br
 　　　｜　｜
 CH$_3$—C—C—CH$_3$
 　　　｜　｜
 　　　Br　H

 d.
 　　　H　H
 　　　｜　｜
 CH$_3$—C—C—CH$_3$
 　　　｜　｜
 　　　H　Cl

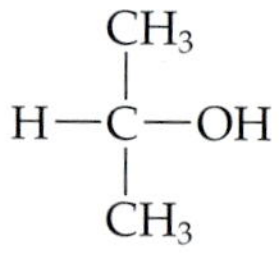

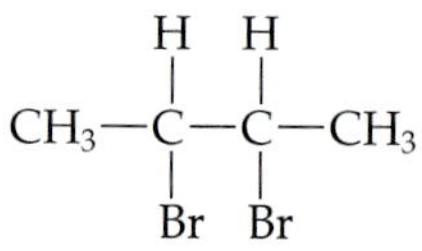

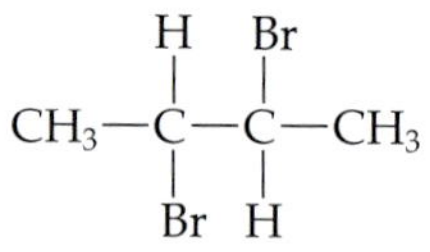

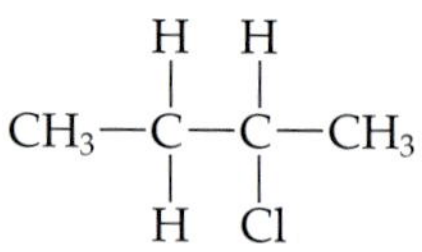

Enantiomers

17. Draw the enantiomers of:
 a. 2-butanol
 b. 2-methylpropanoic acid

18. Draw the enantiomers of:
 a. 2,3-dihydroxypropanal
 b. 2-bromobutanoic acid

19. Draw Fischer projections (flat) for 2,3-dichlorobutanal. Label pairs of enantiomers.

20. Draw Fischer projections (flat) for the stereoisomers of bromochlorofluoromethane.

b.
　　　　　　　　　O　　CH$_2$CH$_3$
　　　　　　　　　‖　／
　　　　　　　　　C
　　H　CH$_3$
　　｜　｜
CH$_3$—N$^{\pm}$—CHCH$_2$—C—⬡
　　｜　｜
Cl$^-$　CH$_3$
　　　　　　　　　⬡

Methadone

Multiple Chiral Centers

21. Which of the following can exist in the meso form?

a. CH₂OH b. CHO c. COOH d. CH₃
 | | | |
 CHOH CHOH CHOH CHCl
 | | | |
 CHOH CHOH CHOH CHCl
 | | | |
 CH₂OH CH₂OH CH₂OH CH₃

22. Which of the following can exist in the meso form?

a. CH₃CH—CHCH₃
 | |
 OH OH

b. CH₃CH—CHCH₂CH₃
 | |
 Br Br

c. HOOCCH—CHCOOH
 | |
 OH OH

Van't Hoff's Rule

23. How many stereoisomers are there for each of the following?

a. CH₃CHCH₂CHCH₂CH₃
 | |
 OH OH

b. CH₃CH—CH—CHCH₂CH₃
 | | |
 Br Br Br

24. How many stereoisomers are there for the following?

CH₂—CH—CH—CH—C—CH₂
 | | | | ‖ |
 OH OH OH OH O OH

Geometric (Cis–Trans) Isomers

25. Write the formulas of the geometric isomers for the following compounds. Label them cis and trans. If there are no geometric isomers, write None.
 a. 2-bromo-2-pentene
 b. 3-hexene
 c. 4-methyl-2-pentyne
 d. 1,1-dibromo-1-butene
 e. 2-butenoic acid (CH₃CH=CHCOOH)
 f. 4-methyl-2-pentene

26. Write the formulas of the geometric isomers for the following compounds. Label them cis and trans. If there are no geometric isomers, write None.
 a. 2,3-dimethyl-2-pentene
 b. 1,1-dibromo-2-ethylcyclopropane
 c. 1,2-dibromocyclohexene
 d. 2-chlorocyclohexanol
 e. 1-bromo-3-chlorocyclobutane
 f. 1,2,3-trimethylcyclopropane

Additional Problems

27. Draw Fischer projections (flat) for 2,3-dibromo-1-butanol. Label pairs of enantiomers.

28. Benzedrine is a racemic mixture of (+)- and (−)-amphetamine. The pure dextrorotatory enantiomer, Dexedrine, has a much greater physiological activity than either the racemic mixture or the pure levorotatory form. Draw the structural formulas for the enantiomers of amphetamine (see Figure G.5).

29. Draw Fischer projection formulas for (+), (−), and meso forms of 2,3-butanediol. Which is meso? Can you tell which is (+) and which is (−)?

30. In Chapter 14, Problem 55, you are asked to draw structures for the eight isomeric pentyl alcohols. Which of these could exist in enantiomeric forms? (That is, which molecules have chiral centers?) Draw Fischer projection formulas for each pair of enantiomers.

31. One isomer of 3-phenyl-2-butanol is shown at the right. Draw the Fischer projection formulas for the remaining three stereoisomers. Predict boiling point and specific rotation for any of the other three isomers that you can. Include a statement explaining why you were not able to predict some of the properties.

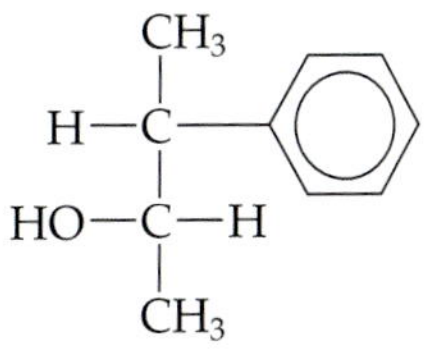

 CH₃
 |
 H—C—⬡
 |
 HO—C—H
 |
 CH₃

bp (25 mm Hg) 118 °C
[α] +30.9

32. 1,2-Dimethylcyclobutane exists as cis and trans isomers. One isomer is chiral, the other is achiral. Draw mirror images for both sets of geometric isomers, and identify which isomers are identical and which are enantiomers. Are *cis*- and *trans*-1,2-dimethylcyclobutane diastereomers?

33. For 1,2-dibromocyclopropane, there are three stereoisomers. Draw their structures.

34. Draw the other geometric isomers for all the pheromones shown in Figure 18.17.

35. Urushiol is an unsaturated phenolic compound that is the active agent in poison ivy and poison sumac. Draw all its geometric isomers.

$$\text{HO} \quad \text{OH}$$

(benzene ring)—$(CH_2)_7CH{=}CHCH_2CH{=}CH(CH_2)_2CH_3$

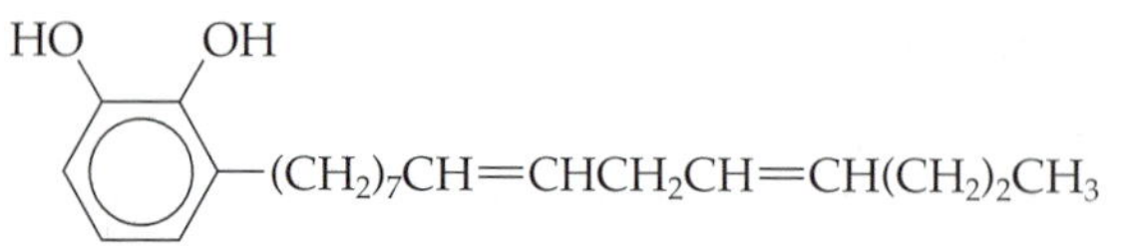

36. If the four bonds of carbon were directed toward the corners of a square, how many isomers of CH_2BrCl would exist? Draw them.
37. Draw and name all the geometric isomers (both noncyclic and cyclic) corresponding to the molecular formula C_5H_{10}.
38. Only one aldehyde isomer of pentanal is optically active. What is its formula? Draw the two enantiomers of this compound.
39. Give the formula for the smallest noncyclic alkane that could be optically active.

40. An alcohol has the molecular formula $C_4H_{10}O$.
 a. Write all the possible structural formulas for the alcohol.
 b. If the alcohol can be separated into two optically active forms, which of the formulas in part (a) is correct?
41. β-Hydroxybutyric acid occurs in the urine of diabetics in amounts up to 30 g/day. Draw the structure of both enantiomers. (*Hint:* Place the carbon atoms in a vertical column with the carboxyl group at the top.)
42. Lysergic acid diethylamide (LSD) contains two chiral carbon atoms (see Section F.6). Identify which carbon atoms are chiral and then draw structural formulas of the four stereoisomers of LSD. [Only one of these four isomers, (+)-lysergic acid diethylamide, is physiologically active.]

Project

43. Examine the labels of some common household products (detergents, foods, drugs, sprays, cosmetics).

Write structural formulas for the chemical compounds contained in these products.

Special Topic H
Molecules to See and to Smell

We perceive the world around us through our five senses: sight, sound, smell, taste, and touch. We discussed the association of taste with acid-base chemistry in Chapter 10. In this special topic, we look at some of the chemistry of seeing and smelling.

H.1 Vision: Cis–Trans Isomerism

The retina of the human eye contains two kinds of receptor cells, rods and cones. The compound 11-*cis*-retinal forms a complex with various proteins, and these complexes are the photosensitive chemicals found in the rods and cones. Rhodopsin, a complex of 11-*cis*-retinal and a protein called opsin, is the visual pigment found in the cones. When light strikes the retina, a complex series of reactions is initiated by a cis–trans isomerization. This reaction is termed a **photochemical isomerization** (photoisomerization) because the energy of light causes the geometric change. The only function of light in vision is to alter the shape of the absorbing molecule, 11-*cis*-retinal, to the trans configuration (Figures H.1 and H.2). This primary event in vision takes only a few picoseconds (1 ps = 10^{-12} s).

Proteins in the eye are altered by this single photochemical act, and some energy is released, triggering a nerve impulse. The impulse is transmitted via the optic nerve to the brain. Then an enzyme converts *trans*-retinal back to *cis*-retinal so it can bind to opsin to await the next exposure to light. Notice from Figure H.3 that *trans*-retinal is analogous to vitamin A in all respects except that the former contains a carbonyl group whereas the latter has a primary alcohol group. Vitamin A is the reduced form of *trans*-retinal (or, to say the same thing another way, *trans*-retinal is the oxidized form of vitamin A), and the oxidation–reduction interconversion is essential to the chemical events of vision.

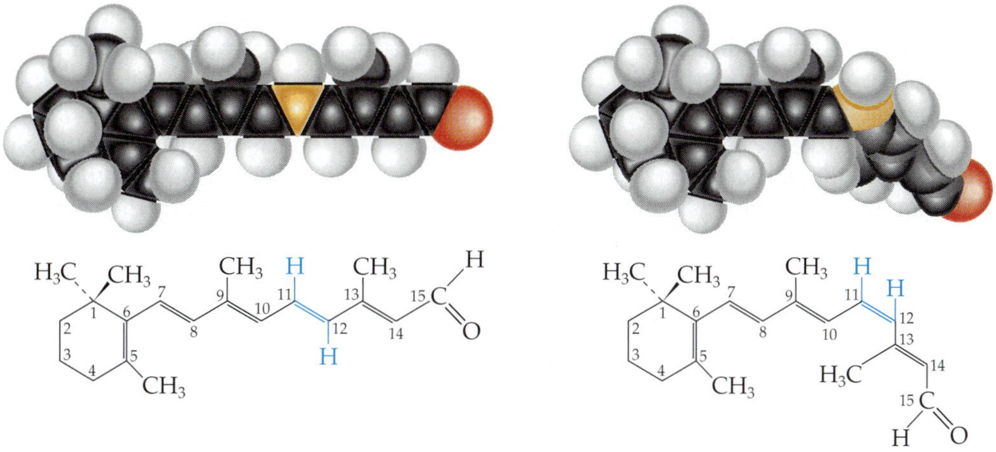

All–trans retinal

11–*cis*–Retinal

▲ **FIGURE H.1**

The crucial molecule of vision is retinal, $C_{15}H_{28}O$. Retinal combines with proteins called opsins to form visual pigments. Because the nine-carbon chain has alternating double and single bonds, it can assume a variety of bent forms. Two isomers of retinal are shown here. In the space-filling models, carbon atoms are dark except for C-11 which is colored. Hydrogen atoms are light, and the red atom on C-15 is oxygen. When tightly bound to opsin, retinal is in the bent and twisted form known as 11-*cis*-retinal. When struck by light, it straightens out into the all-trans configuration. This simple photochemical reaction is the basis for vision.

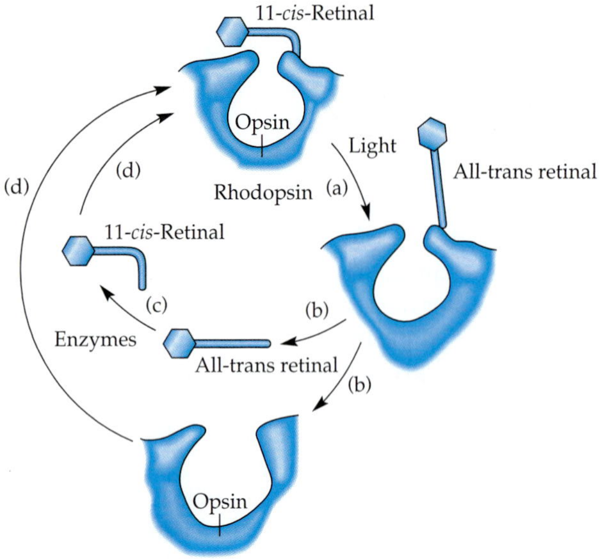

▲ **FIGURE H.2**

(a) When light strikes rhodopsin, 11-*cis*-retinal is isomerized to the all-trans isomer. This change in structure is accompanied by a change in electrical potential. The photoisomerization of the retinal triggers a nerve impulse. (b) The rhodopsin complex then splits into opsin and all-trans retinal. (c) The all-trans retinal is converted back to 11-*cis*-retinal in an enzyme-catalyzed reaction. The 11-*cis*-retinal again complexes with opsin (d), completing the visual cycle.

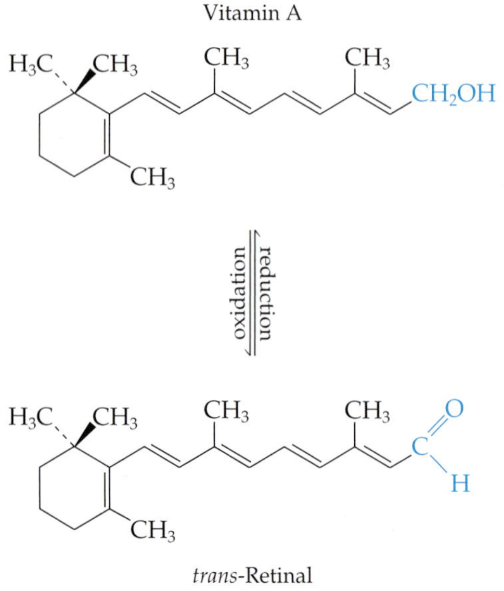

▲ **FIGURE H.3**

Vitamin A and all-trans retinal are interconverted by oxidation–reduction reactions.

It is interesting to note that, of all the tissues in the body, the retina has the highest respiration rate. You may think of seeing as a rather passive activity, but vision is metabolically demanding.

The nerve impulses triggered during the retinal *cis*-to-*trans* conversions are interpreted by the brain as an image—in other words, we see. Some retinal is lost during the regeneration of rhodopsin. It must be replaced by vitamin A from the bloodstream.

H.2 Odors: The Shapes of Things We Smell

Whether or not a substance has an odor depends on whether it can excite the olfactory nerve endings in the nose. The current theory of olfaction was first postulated by R. W. Moncrieff. He proposed that the olfactory system is composed of a limited number of receptor cells, each one associated with a distinct primary odor. Furthermore, all odorous molecules produce their effects by fitting closely on the receptor sites of these cells. (This hypothesis is essentially the same as the lock-and-key theory that we discuss

FIGURE H.4 ▶

The structural formulas of chemicals having camphorlike odors do not look much alike, but their sizes and shapes are similar. They all fit the bowl-shaped receptor for camphoraceous molecules.

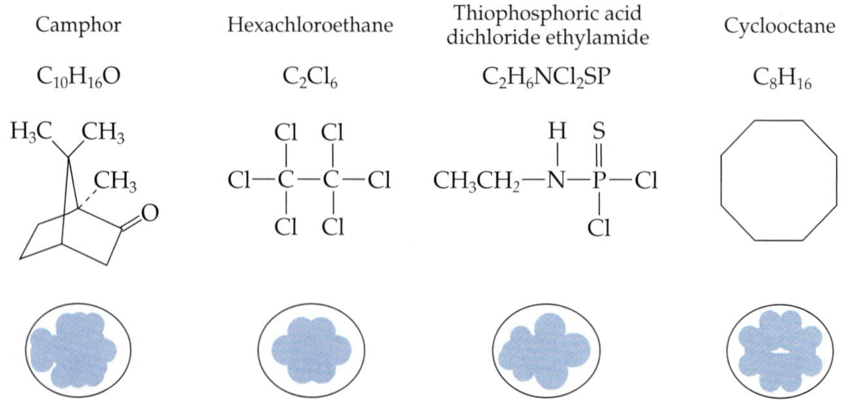

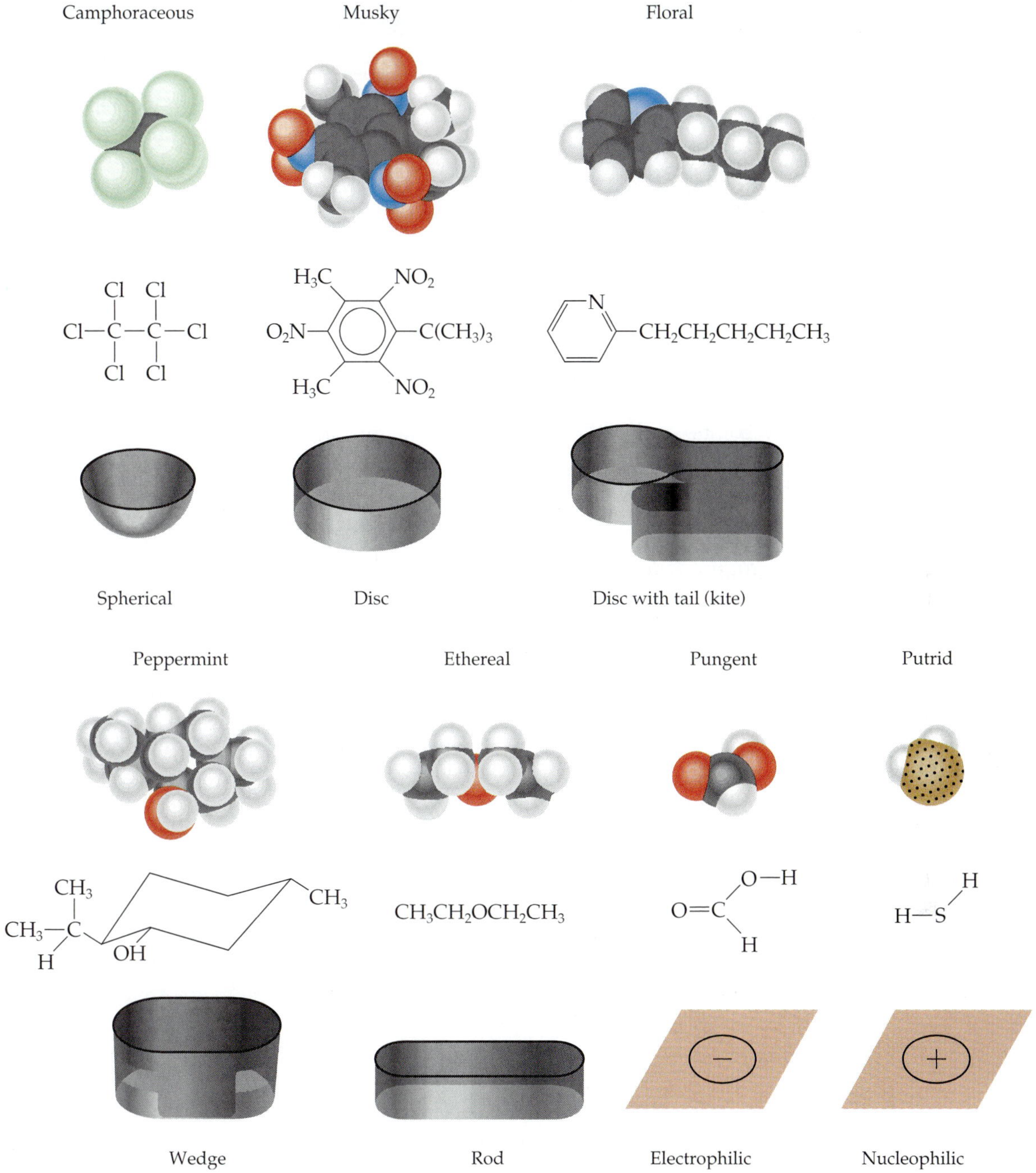

▲ **FIGURE H.5**

The shapes of olfactory receptor sites are shown for each primary odor along with a molecule representative of the odor. The first five, respectively, are hexachloroethane, xylene musk, α-amylpyridine, $(-)$-menthol, and diethyl ether. Pungent (formic acid) and putrid (hydrogen sulfide) molecules fit on the receptor sites because of polarity rather than shape.

in connection with enzymes in Section 22.3.) Many additional kinds of odor receptors have been identified in recent years, but in essence the major factor in determining the odor of a substance is the three-dimensional geometric shape of its molecules (Figures H.4 and H.5).

In 1952, John Amoore identified seven primary odors (Table H.1). Every known odor can be made

Table H.1 Primary Odors

Primary Odor	Chemical Example	Familiar Substance
Camphoraceous	Camphor	Moth repellent
Musky	Pentadecanolactone	Angelica root oil
Floral	3-Methyl-1-phenyl-3-pentanol	Roses
Peppermint	Menthone	Mint candy
Ethereal	Ethylene dichloride	Dry-cleaning fluid
Pungent	Acetic acid	Vinegar
Putrid	Butyl mercaptan	Rotten eggs

from these seven by mixing them in various proportions (in the same way that all the colors can be made from the three primary colors—red, yellow, and blue). In addition, Amoore described the sizes, shapes, and chemical affinities of the seven kinds of receptor sites that recognize each primary odor. These receptor sites must necessarily have a distinctive shape, size, and chemical affinity so as to accept only those molecules having the correct geometric configuration or polarity.

If the molecules of a substance can fit into more than one receptor site, the brain will receive a more complex signal, and the odor will be perceived as a combination of those associated with the primary sites occupied by the molecules. The olfactory receptors are similar to those of taste (the taste buds). These two senses function together, and they aid in our food selection, because many foods are smelled at the same time as they are tasted.

Key Term

photochemical isomerization (H.1)

Review Questions

1. What kind of chemical transformation is the key to the visual process?
2. What is the relationship of vitamin A to the visual process?
3. What is the chemical basis of olfaction?

19 Carbohydrates

These desserts are rich in carbohydrates and appeal greatly to our sense of taste. Most are also rich in fats—and in calories.

Almost everyone knows what carbohydrates are: they're what you eat or don't eat depending on whose diet book you fancy. In fact, as any dietitian or nutritionist will tell you, carbohydrates must be included in any well-balanced diet. They are the body's primary source of energy. Energy is stored in the complex molecular structure of the carbohydrates when these compounds are synthesized by green plants from carbon dioxide and water. The entire biosphere, and that includes us, is dependent on the endothermic reaction in which simple compounds plus energy yield complex compounds. When we metabolize the complex compounds, the atoms rearrange themselves back into simple compounds and, in the process, release their stored energy for our use. We'll discuss the details of these metabolic processes in Chapters 24 and 25. Here, however, let us look at the basic chemistry of carbohydrates as we begin the third and final section of this book: **biochemistry,** the chemical reactions that take place in living organisms.

The photosynthetic reactions in green plants produce glucose. This simple sugar can be used by the plant for energy, or it can be converted to a more complex form—starch—to serve as a source of energy for later use, perhaps as nourishment for the plant's seeds. Some of the glucose is converted to cellulose, the structural material of plants. We can gather and eat the parts of plants that store energy—seeds, roots, tubers, fruits—and use some of that energy for ourselves. (We can eat the cellulose, too, but we can't get any energy from it, for reasons we'll study in Section 19.8.)

In contrast, most animal tissue contains a comparatively small amount of carbohydrate (less than 1% in humans). Plants are able to synthesize their own carbohydrates from the carbon dioxide of the air and water taken from the soil. Animals are incapable of this synthesis, and therefore they are dependent upon the plant kingdom as a source of these vital compounds. We humans use carbohydrates not only for our food (about 60 to 65% by mass of the average diet) but also for our clothing (cotton, linen, rayon), shelter (wood), fuel (wood), and paper (wood).

19.1 Carbohydrates: Definitions and Classifications

A formal definition of carbohydrates is difficult. Chemically, *carbohydrates* are polyhydroxy aldehydes or ketones or compounds that can be hydrolyzed to form such compounds.

Carbohydrates are compounds containing carbon, hydrogen, and oxygen. They include the starches, the sweet-tasting compounds called sugars, and structural materials such as cellulose. The term *carbohydrate* has its origin in a misinterpretation of the molecular formulas of many of these substances. For example, the formula for glucose is $C_6H_{12}O_6$, but we could also represent this molecule as a "carbon hydrate" $(C{\cdot}H_2O)_6$.

Carbohydrates are not hydrates of carbon, however. They are alcohols—they all contain the hydroxyl (—OH) functional group. Most also contain either a real or a latent carbonyl (C=O) group. [By a *latent* carbonyl group, we mean a func-

The principal sugars in our diets are sucrose (cane or beet sugar), glucose (corn syrup), and fructose (fruit sugar, often in the form of high-fructose corn syrup).

tional group such as a hemiacetal or an acetal (Section 15.6) that can be more or less readily converted to a carbonyl group.] Consequently, some carbohydrates give reactions as aldehydes or ketones even though the "carbonyl" may exist primarily in a hemiacetal form. Complex carbohydrates containing the acetal function can be hydrolyzed like other acetals to simpler compounds that give reactions typical of aldehydes or ketones.

Simple carbohydrates, those that cannot be further hydrolyzed, are called **monosaccharides.** Carbohydrates than can be hydrolyzed to two monosaccharide units are called **disaccharides,** and carbohydrates that can be hydrolyzed to many monosaccharide units are called **polysaccharides.**

Because carbohydrate molecules contain several —OH groups, the molecules can form an extensive network of intermolecular hydrogen bonds. Because of this extensive bonding, monosaccharides and disaccharides are crystalline solids at room temperature. They have relatively high melting points and often char before melting. Carbohydrate molecules also can form hydrogen bonds to water molecules, and so the simpler ones are readily soluble in water. For example, 100 g of glucose dissolves in 100 mL of water at 25 °C.

19.2 Monosaccharides: General Terminology

The general names for the monosaccharides are obtained in a manner analogous to the naming of organic compounds by the IUPAC system. The number of carbon atoms in the molecule is denoted by the appropriate stem, and *-ose* is the generic ending for any sugar. For example, the terms triose, tetrose, pentose, and hexose signify three-, four-, five-, and six-carbon monosaccharides, respectively. In addition, those monosaccharides that contain an aldehyde group are called **aldoses;** those containing a ketone group are **ketoses.** By combining these terms, both the type of carbonyl group and the number of carbon atoms in the molecule are easily expressed. Thus, monosaccharides are generally referred to as aldotetroses, aldopentoses, ketopentoses, ketoheptoses, etc. Glucose and fructose are specific examples of an aldose and a ketose, respectively.

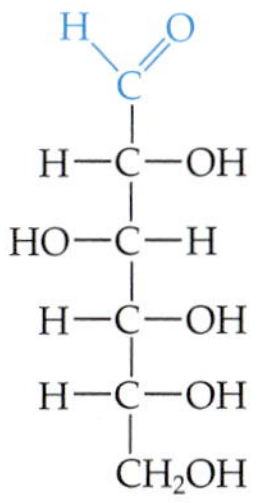

Glucose
(an aldohexose)

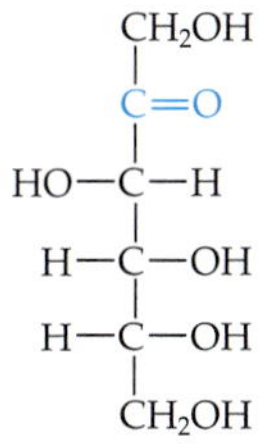

Fructose
(a ketohexose)

19.3 Stereochemistry

The simplest sugars are the trioses. Two trioses derived by oxidation of the triol glycerol are important intermediates in metabolism. Dihydroxyacetone is a ketotriose, and glyceraldehyde is an aldotriose. Dihydroxyacetone does not contain a chiral center, but glyceraldehyde does and thus exists in two optically active forms. Except for the direction in which they rotate plane-polarized light, the two isomers have identical physical properties. One form has a specific rotation of +8.7, the other a specific rotation of −8.7. The great German chemist Emil Fischer initiated the convention of projecting the formulas onto a two-dimensional plane so that the aldehyde group is written at the top, with the hydrogen and hydroxyl written to the right and left. (Formulas of chiral molecules represented in this manner are referred to as Fischer projections, Fischer models, or Fischer configurations.) Arbitrarily, Fischer then decided that the formula of glyceraldehyde in which the hydroxyl group is positioned to the right of the chiral carbon atom represents the dextrorotatory isomer. He assigned the letter D as its prefix. The levorotatory isomer, in which the —OH group is

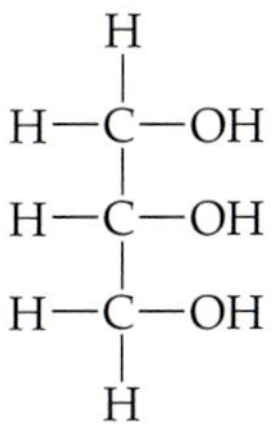

Glycerol

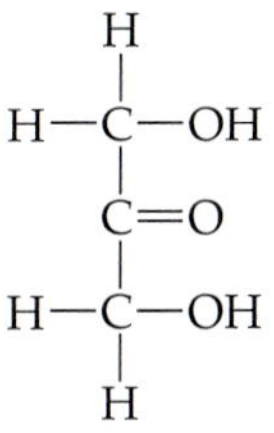

Dihydroxyacetone

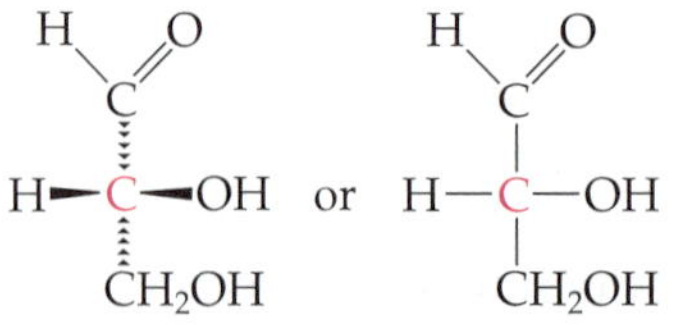

D-Glyceraldehyde

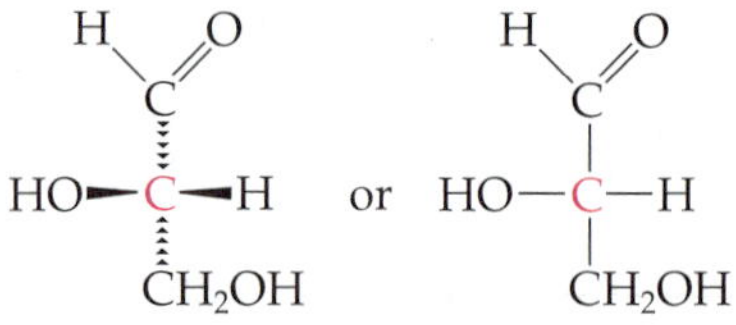

L-Glyceraldehyde

positioned to the left of the chiral carbon atom, was accordingly assigned the letter L as its prefix.[1]

The two forms of glyceraldehyde are especially important because the more complex sugars can be considered to be derived from them. They therefore serve as a reference point for designating and drawing all other monosaccharides. Sugars whose Fischer projections terminate in the same configuration as D-glyceraldehyde are designated as **D sugars;** those derived from L-glyceraldehyde are designated as **L sugars.** Humans cannot obtain energy from L carbohydrates (although beings on some other planet, with enzymes that are mirror images of ours, might be able to use only L forms).

The penultimate (next to last) carbon has been chosen, by convention, to be the reference carbon atom. It is the chiral carbon atom farthest from the aldehyde or ketone group.

The letters D and L often mislead the beginning student. It must be emphasized that these prefixes serve only to signify the *absolute configuration* of a molecule. A D sugar is one that has the same configuration about the *penultimate* carbon atom as D-glyceraldehyde has (H on the left, OH on the right). The letters do not in any way refer to the optical rotation of the molecule. D-Glyceraldehyde just happens to be dextrorotatory. It is not to be expected that all compounds derived from D-glyceraldehyde will have the same optical rotation, for such compounds will have additional chiral carbon atoms.

For example, there are two aldopentoses of interest to us. One is D-ribose, the sugar unit that occurs in ribonucleic acids (RNA). Related to D-ribose but not an isomer of it (its molecular formula is different) is D-2-deoxyribose, the sugar unit that occurs in deoxyribonucleic acids (DNA). As the 2-*deoxy-* implies, this sugar is "missing" an oxygen on the second carbon atom. Both of these D sugars are levorotatory.

The direction in which it rotates plane-polarized light is a specific property of each optically active molecule. It is not at all dependent upon the configuration about the penultimate carbon. The symbols (+) and (−) are used to denote the optical rotation of the monosaccharides: (+) indicates a clockwise, or dextrorotatory, rotation, and (−) indicates a counterclockwise, or levorotatory, rotation.

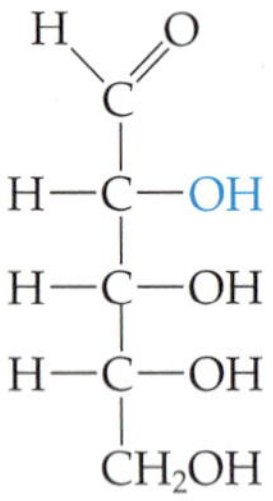

D-(−)-Ribose

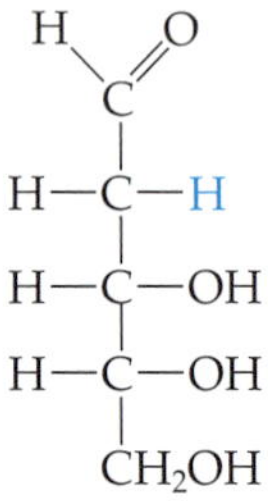

(−)-2-Deoxy-D-ribose

Aldohexoses contain four chiral carbon atoms, and thus there are eight enantiomeric pairs, or 16 isomers. Fortunately for students of biochemistry, only three of the 16 isomers are commonly found in nature: D-(+)-glucose, D-(+)-mannose, and D-(+)-galactose. (All 16 isomers have been prepared synthetically.) There are quite a few ketohexoses, too, but we shall deal with only a single ketohexose here: D-(−)-fructose. All of the sugars we discuss in the remainder of this chapter belong to the D family. If no family designation is given, you can assume that the compound is a D sugar.

[1] Fischer's arbitrary assignment proved to be correct. In 1951, chemists, with the aid of X-ray crystallography, determined the absolute configurations of the glyceraldehyde enantiomers and found that the D isomer is indeed dextrorotatory.

19.4 Hexoses

Glucose

D-Glucose is the most abundant sugar found in nature. It is commonly found in fruits, especially in ripe grapes, and for this reason it is often referred to as *grape sugar.* It is also known as *dextrose,* a name that derives from the fact that the predominant natural form of the sugar is dextrorotatory.

Most of the carbohydrates taken in by the body are eventually converted to glucose in a series of metabolic pathways that produce energy for our cells. Since glucose requires no digestion, it can be given intravenously (as a 5% mass-volume solution) to patients who are unable to take food orally. Glucose is the circulating carbohydrate of animals; hence the name *blood sugar.* The blood contains about 180 mg/dL glucose, and normal urine may contain anywhere from a trace to 200 mg/dL glucose.

Commercially, glucose is made by the hydrolysis of starch. In the United States, cornstarch is used in the process; in Europe, the starch is obtained from potatoes. Glucose is only 74% as sweet as table sugar (sucrose), but it has the same caloric value per gram.

The structure of D-glucose is illustrated in Figure 19.1. This formula follows our convention of writing the aldehyde group at the top and the primary alcohol group at the bottom. Glucose is a D sugar because the hydroxyl group at the fifth carbon (the chiral center farthest from the carbonyl group) is on the right. In fact, all the hydroxyl groups except the one at the third carbon are to the right. You should learn to draw this formula for glucose, as well as the analogous formulas for the other three monosaccharides in Figure 19.1.

Mannose

D-Mannose is a component of the polysaccharide mannan, found in some berries. A particularly good source is "vegetable ivory," the endosperm (white fleshy part) of palm nuts. Buttons were once widely made of this material, with the waste being hydrolyzed to mannose. A structure for mannose is also presented in Figure 19.1. Note that the configuration differs from that of glucose only at the second carbon atom.

Galactose

D-Galactose is formed by the hydrolysis of lactose, a disaccharide composed of a glucose unit and a galactose unit. Galactose does not occur in nature in the uncombined state. The galactose needed by the human body for the synthesis of

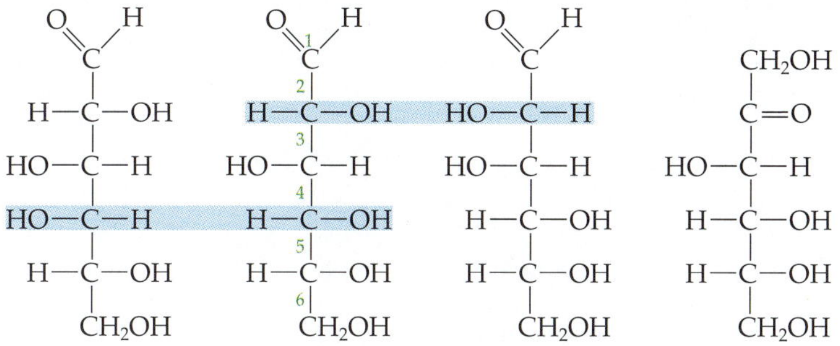

D-(+)-Galactose D-(+)-Glucose D-(+)-Mannose D-(−)-Fructose

FIGURE 19.1 Structures of four important hexoses. The first three are aldoses; fructose is a ketose.

Aspartame

"Sweet" is one of the four primary taste sensations that can be distinguished by the taste buds on the surface of the tongue. The others are sour, salty, and bitter. Although sweetness is commonly associated with most mono- and disaccharides, it is not a specific property of carbohydrates. Many sugars are sweet to varying degrees, but several organic compounds have been synthesized that are far superior as sweetening agents. These synthetic compounds have no caloric value, and therefore they are useful for those persons (e.g., diabetics) who must minimize their carbohydrate intake.

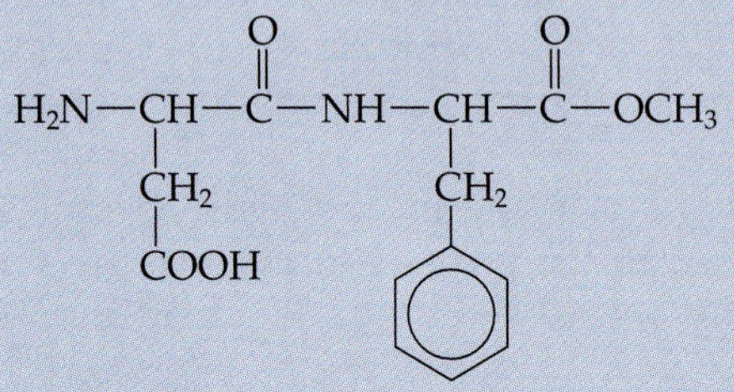

Aspartame

The popular commercial products NutraSweet and Equal are both aspartame.

In 1981, after eight years of extensive testing (and controversy regarding its safety), the FDA approved the use of the low-calorie sweetener *aspartame* (L-aspartyl-L-phenylalanine methyl ester). This white crystalline compound is about 160 times sweeter than sucrose and does not leave the bitter aftertaste often associated with saccharin. The FDA approved aspartame for use in more than 70 products, including soft drinks (NutraSweet), cereals, gelatins, and chewing gum, and as tablets to be used as sugar substitutes. It is interesting that the two constituent amino acids are not sweet. L-Aspartic acid has a flat taste, and L-phenylalanine is bitter.

Aspartame is used as a sweetener for a wide variety of foods because it can blend well with other food flavors. In the body, aspartame is hydrolyzed to aspartic acid, phenylalanine, and methanol. The small amount of methanol does not seem to be a problem, but the release of phenylalanine is a matter of concern to those on low-phenylalanine diets (see Section 23.8). A report was released in the early 1980s that pregnant women who consume aspartame may have babies with permanent brain damage, and there was a later report of similar damage to infants who ingested it during the six months following birth. In 1985, the American Medical Association completed its investigations and concluded that aspartame is safe, and only people who are sensitive to phenylalanine need regulate their intake. (It should be noted that the shelf life of aspartame is limited because at high temperatures the molecule breaks down and loses its sweetness.)

Acesulfame K (Sunette) also is now approved for use as an artificial sweetener in the United States. It can survive the high temperatures of cooking processes, whereas aspartame is broken down by heat.

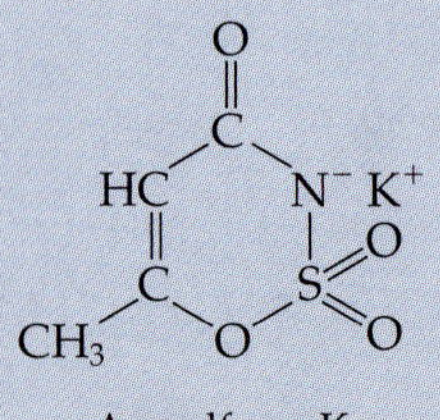

Acesulfame K

lactose (in the mammary glands) is obtained by the conversion of D-glucose to D-galactose. In addition, galactose is an important constituent of the glycolipids (see Section 20.8) that occur in the brain and in the myelin sheath of nerve cells. The genetic disease galactosemia results from the absence of an enzyme that converts galactose to glucose (see Section 23.8). Figure 19.1 also shows the structure of galactose. Notice that the configuration differs from that of glucose only at the fourth carbon atom.

Fructose

D-Fructose, whose structure is shown in Figure 19.1, is the only naturally occurring ketohexose. It occurs, along with glucose and sucrose, in honey (which is 40% fructose) and sweet fruits. Fructose (Latin *fructus*, fruit) is also referred to as *levulose* because it has a specific rotation that is strongly levorotatory (−92.4). It is the sweetest sugar (1.7 times sweeter than sucrose). Many nonsugars, however, are several hundred or several thousand times as sweet (Table 19.1). Fructose is the only sugar found in the semen of bulls and men. It is the major energy source for spermatozoa and is formed in the prostate gland. Structurally, fructose is a 2-ketohexose. From the third through the sixth carbon atoms, its structure is the same as that of glucose.

Table 19.1 Relative Sweetness of Some Compounds (Sucrose = 100)

Compound	Relative Sweetness
Acesulfame K	20,000
Glucose	74
Fructose	173
Lactose	16
Sucrose	100
Maltose	33
Saccharin	50,000
Aspartame	16,000

High-fructose corn syrup is made by using enzymes to convert much of the glucose in the syrup to fructose.

19.5 Monosaccharides: Cyclic Structures

So far we have represented the monosaccharides as hydroxy aldehydes and ketones. These representations are useful in studying many of the properties (Section 19.6) of these simple sugars. However, in Section 15.6 we mentioned that aldehydes and ketones react with alcohols to form hemiacetals and hemiketals:

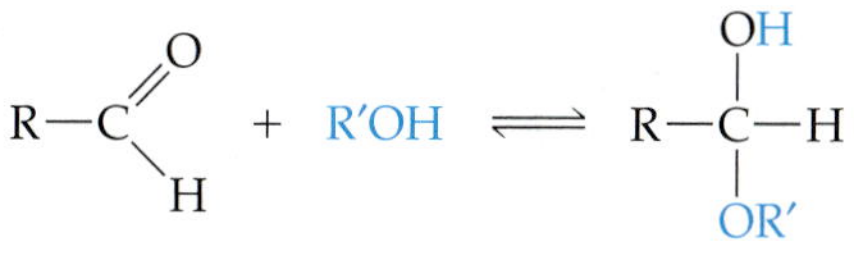

Aldehyde Alcohol Hemiacetal

You should not be surprised, then, to find that hydroxyl groups and carbonyl groups conveniently located on the same molecule react with one another and that consequently monosaccharides (larger than tetroses) exist mainly as cyclic hemiacetals:

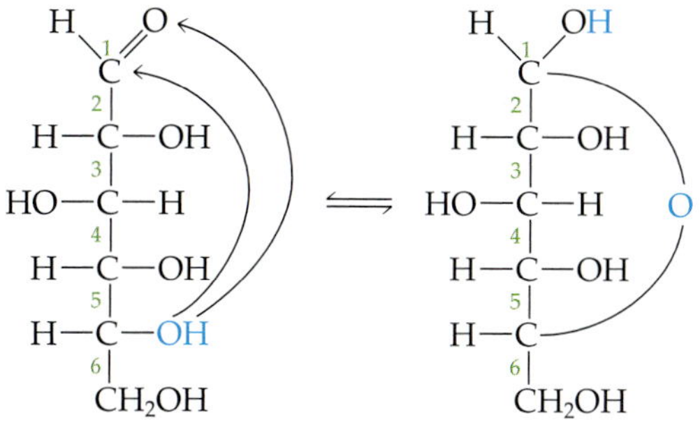

Hydroxy aldehyde Cyclic hemiacetal

"Now wait a minute," you say. "Things are bad enough with all these complications, and now you ignore a hydroxyl group located right next to the carbonyl in

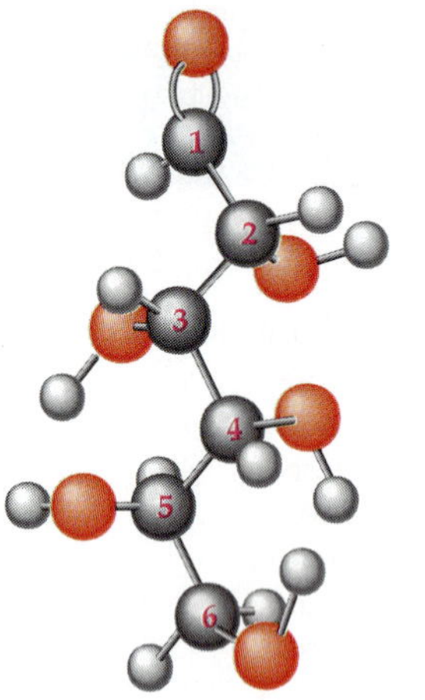

Extended

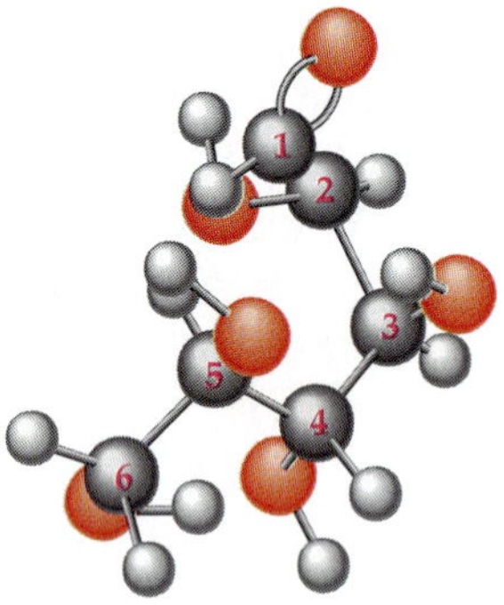

Folded

▲ **FIGURE 19.2**

Models of two arrangements of the free-aldehyde form of glucose. Note that the O atom on C-5 in the folded model is much nearer the carbonyl carbon than is the O atom on C-5 in the extended model.

order to play around with the hydroxyl at the fifth carbon. And you had to write a long, silly-looking bond to do it!" A reasonable objection—which is why we must introduce a different type of formula. In this new formula, we'll take into account approximately correct bond angles. Figure 19.2 shows two models of the free-aldehyde form of glucose. Notice how the molecule can fold around on itself. The center structure in Figure 19.3 is drawn to resemble the folded model. Note that the hydroxyl on the fifth carbon is quite near the carbonyl carbon, and so it's not surprising that this hydroxyl group reacts with the carbonyl. When the reaction occurs, the carbonyl oxygen may be pushed either up or down, giving rise to two cyclic hemiacetal forms. The structure on the left, with the hydroxyl on the first carbon projected downward, represents what is called the *alpha (α) form.* That on the right, with the hydroxyl on the first carbon pointed upward, is the *beta (β) form.*

Crystalline glucose may exist in either the alpha or the beta form. The two forms have different properties. The alpha form melts at 146 °C, and the beta form melts at 150 °C. In solution, an equilibrium mixture is formed. You can start out with either pure crystalline hemiacetal form, but as soon as it is dissolved in water, the unstable hemiacetal group opens to form the free carbonyl and then closes to either the alpha or beta hemiacetal, reopening and reclosing in succession. This interconversion is referred to as **mutarotation** (Latin *mutare,* to change).[2] At equilibrium, the mixture is about 36% alpha and 64% beta. There is less than 0.02% of the open-chain aldehyde form. Nevertheless, that is enough to give most of the characteristic reactions of aldehydes. As the small amount of free aldehyde is used up in a reaction, the hemiacetal forms open up to yield more free aldehyde. Thus, *all* the molecules may eventually react as aldehyde entities, even though very little free aldehyde is present at any given time.

In this book, we use the convention first suggested by an English chemist, W. N. Haworth, for representing the formulas of the cyclic forms of sugars. The molecules are drawn as planar hexagonal slabs with darkened edges toward the viewer. Ring carbon atoms and the hydrogen atoms directly attached to them are not shown. The disposition of the hydroxyl groups, positioned either above or below the plane of the ring, is sufficient to define the correct configuration of the molecule. Any group of atoms written to the right in the Fischer projection

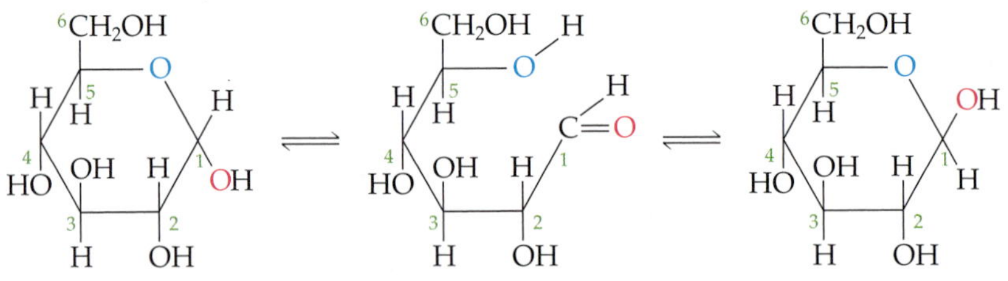

α-Glucose   Open-chain form   β-Glucose

▲ **FIGURE 19.3**

In aqueous solution, glucose exists as an equilibrium mixture of these three forms.

---

[2] The two forms of glucose also differ in the way they affect plane-polarized light. α-D-Glucose has a specific rotation of +112, and β-D-glucose has a specific rotation of +18.7. If either is placed in solution, the observed rotation slowly changes (the substance undergoes mutarotation) to an equilibrium value of +52.7.

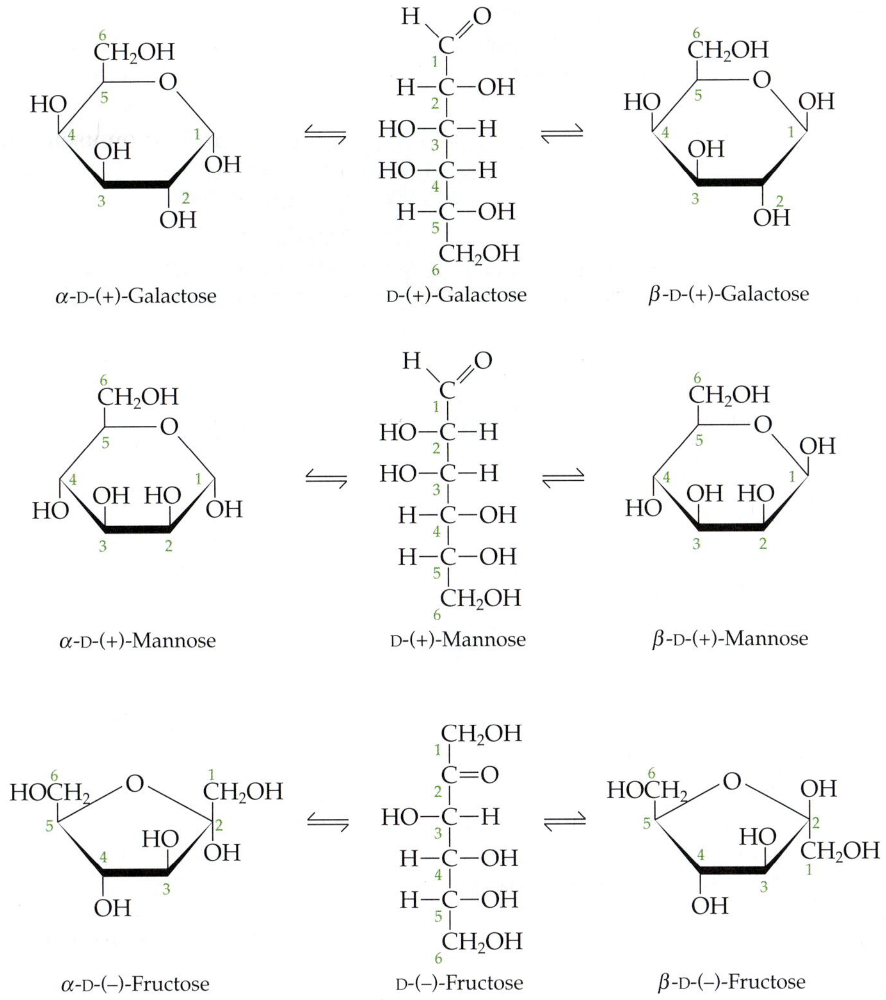

FIGURE 19.4 Mutarotation of D-galactose, D-mannose, and D-fructose.

appears below the plane of the ring, and any group written to the left appears above the plane in the cyclic forms.

Intramolecular hemiacetal formation is not unique to glucose. It occurs in galactose, mannose, and the naturally occurring aldopentoses and aldoheptoses. Fructose and other ketoses form intramolecular hemiketals. Figure 19.4 illustrates the equilibrium among the three forms of D-galactose, D-mannose, and D-fructose. Notice that galactose and mannose, like glucose, form a six-member cyclic structure. Fructose can also exist in this form, but it is most commonly found in nature as the five-member ring shown here.

The difference between the alpha and beta forms of the sugars may seem trivial, but keep in mind that such differences are often crucial in biochemical reactions. We shall encounter some examples of this principle later in this chapter.

19.6 Properties of Some Monosaccharides

Glucose, mannose, galactose, and fructose are crystalline solids at room temperature. With five hydroxyl groups per molecule, these sugars are quite soluble in water (Figure 19.5).

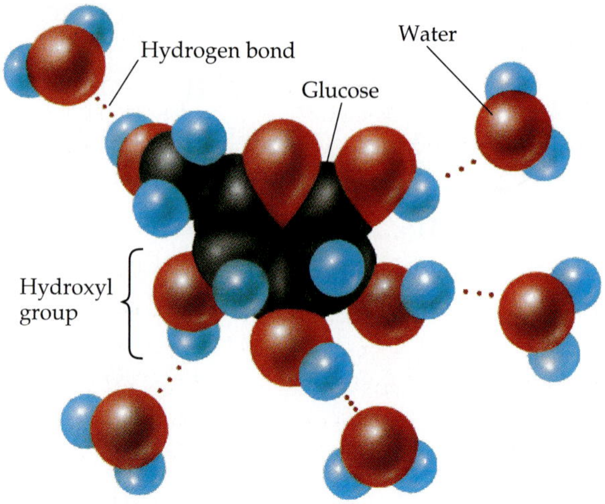

FIGURE 19.5 ▶

Extensive hydrogen bonding between sugar molecules and water molecules leads to the considerable water solubility of most sugars.

Chemically, these monosaccharides undergo the reactions to be expected from their functional groups. The hydroxyl groups react to form esters and ethers. These reactions, though, are more important commercially for the polysaccharide cellulose (Section 19.10) than for the simpler sugars.

One important reaction is the oxidation of the aldehyde group, which is one of the most easily oxidized organic functional groups. This can be accomplished by any mild oxidizing agent. Tollens's, Benedict's, and Fehling's reagents (Section 15.6) are frequently used. The Tollens's test is based on the reduction of silver ions, and both Benedict's and Fehling's tests involve the reduction of copper complexes. Any carbohydrate capable of this reduction without first undergoing hydrolysis is said to be a **reducing sugar**.[3]

$$\text{CHO} + \text{Ag(NH}_3)_2^+ \longrightarrow \text{COO}^- + \text{Ag(s)}$$

| An aldose | Tollens's reagent (clear solution) | | Carboxylate anion | Silver mirror |

$$\text{CHO} + \text{Cu(citrate)}_2^{2-} \longrightarrow \text{COO}^- + \text{Cu}_2\text{O(s)}$$

Benedict's reagent (blue solution) Brick-red precipitate

$$\text{CHO} + \text{Cu(tartrate)}_2^{2-} \longrightarrow \text{COO}^- + \text{Cu}_2\text{O(s)}$$

Fehling's reagent (blue solution) Brick-red precipitate

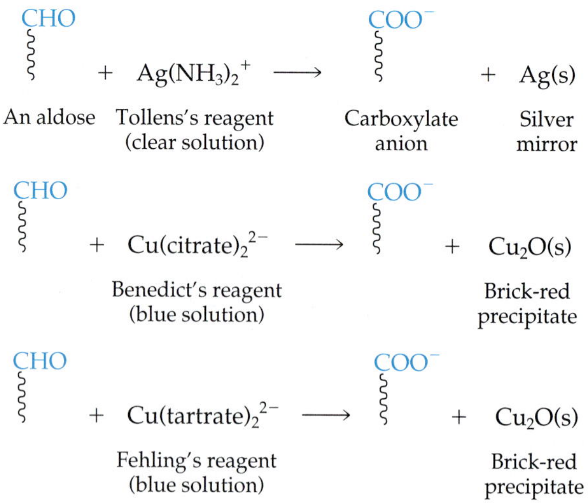

[3] It should not be surprising that aldoses are reducing sugars, but ketoses also give a positive test. The explanation is based on a reaction we have not discussed (tautomerism— see an organic chemistry text). In alkaline solution, an equilibrium exists between the ketoses and the aldoses. Since the oxidizing reagents commonly used to detect reducing sugars are prepared in a basic solution, all monosaccharides act as reducing sugars.

These reactions have been used as simple and rapid diagnostic tests for the presence of glucose in blood or urine. For example, Clinitest tablets, which are used in clinical laboratories to test for sugar in the urine, contain cupric ions and are based on Benedict's test. A green color indicates very little sugar, whereas a brick-red color indicates sugar in excess of 2 g/100 mL of urine.

19.7 Disaccharides

Disaccharides ($C_{12}H_{22}O_{11}$) are composed of two monosaccharide units joined by an acetal linkage (Section 15.6). They differ from one another in what the constituent monosaccharides are and in the type of acetal linkage connecting them. There are three common disaccharides—maltose, lactose, and sucrose. Hydrolysis of 1 mol of disaccharide yields 2 mol of monosaccharide. Using word equations, we can write

$$\begin{array}{lclcl} \text{Maltose} & + & H_2O & \longrightarrow & 2\ \text{Glucose} \\ \text{Lactose} & + & H_2O & \longrightarrow & \text{Glucose} + \text{Galactose} \\ \text{Sucrose} & + & H_2O & \longrightarrow & \text{Glucose} + \text{Fructose} \end{array}$$

All three disaccharides are white crystalline solids at room temperature. Sucrose is quite soluble in water (200 g in 100 mL), and lactose is moderately soluble (20 g in 100 mL). All three molecules are too large to pass through cell membranes. Now let's look at each of these sugars, in turn, in more detail.

Sucrose

Sucrose is known as beet sugar, cane sugar, table sugar, or simply sugar. It is probably the largest-selling pure organic compound in the world. As its names imply, sucrose is obtained from sugar canes and sugar beets (whose juices contain 14 to 20% of the sugar) by evaporation of the water and recrystallization. The dark brown liquid that remains after crystallization of the sugar is sold as molasses.

Maple syrup is the concentrated sap of the sugar maple tree. It is a solution of sugars—about 65% sucrose with small amounts of glucose and fructose.

A molecule of sucrose may be envisioned as resulting from the combination of one molecule of α-D-glucose and one molecule of β-D-fructose; a molecule of water is eliminated in the process:

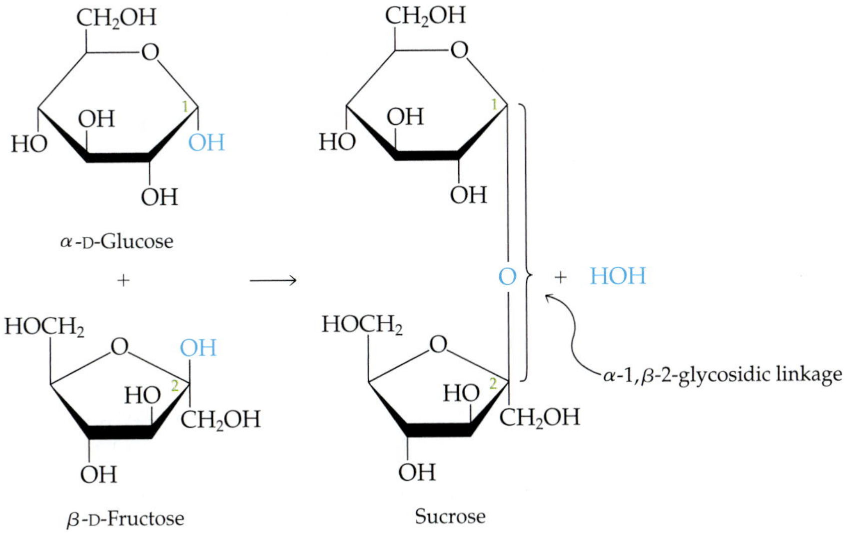

Recall from Section 15.6 that the reaction of an alcohol with a hemiacetal yields an acetal. When two monosaccharides combine, the carbon–oxygen–carbon linkage that joins the components of the acetal is called a **glycosidic linkage.**

The unique feature that characterizes the sucrose molecule is its acetal linkage. This linkage involves the hydroxyl group on the carbon in position 1 of α-D-glucose and the hydroxyl group on C-2 of β-D-fructose. By convention, sugars are read from left to right (or top to bottom). This connecting linkage is therefore an α-1, β-2-glycosidic linkage. This bonding bestows certain properties upon sucrose that are quite different from those of the other disaccharides.

For one thing, sucrose, unlike other disaccharides, is incapable of mutarotation. Thus, it exists in only one form both in the solid state and in solution. The presence of the α-1, β-2-glycosidic linkage makes it impossible for sucrose to exist in the alpha or beta configuration or in the open-chain form. This is a direct result of the fact that the acetal carbon of the glucose ring and the ketal carbon of the fructose ring have both been tied up in the formation of the α-1, β-2 (head-to-head) linkage. As long as the sucrose molecule remains intact, it cannot "uncyclize" to form the open-chain structure. Sucrose, therefore, does not undergo reactions that are typical of aldehydes and ketones, and we have now encountered our first nonreducing sugar.

The human body is unable to utilize sucrose or any other disaccharide directly because the molecules are too large to pass through cell membranes. Therefore, the disaccharide must first be broken down by hydrolysis into its two constituent monosaccharide units. In the body, this hydrolysis reaction is catalyzed by enzymes. The same hydrolysis reaction can be carried out in a test tube with dilute acid as a catalyst, but the reaction rate is much slower. The equation for the hydrolysis reaction is the reverse of the one given for the formation of sucrose. The product is an equimolar mixture of glucose and fructose. This 1:1 mixture is called **invert sugar:**[4]

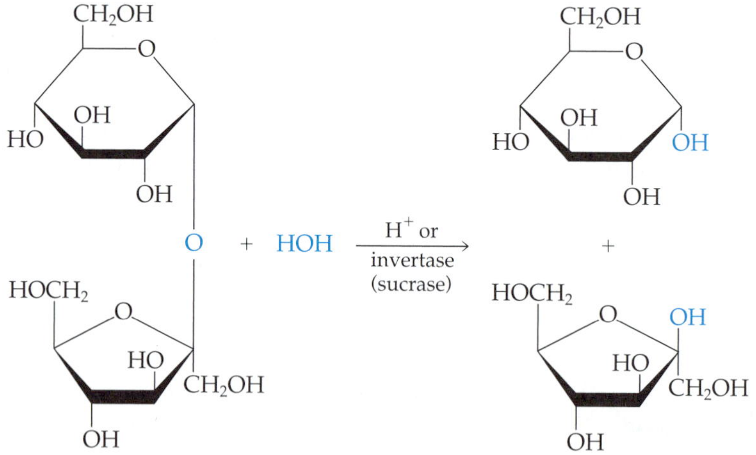

This hydrolysis reaction has several practical applications. Because sucrose can exist in only one molecular configuration, it is one of the most readily crystallizable sugars. Invert sugar has a much greater tendency to remain in solution. In the manufacture of jelly and candy and in the canning of fruit, crystallization of the sugar is undesirable. Therefore, conditions leading to the hydrolysis of sucrose

[4] Sucrose is dextrorotatory with a specific rotation of 66.5. During hydrolysis, the observed rotation drops off and eventually becomes negative. This is because fructose has a high negative specific rotation (−92.4 at equilibrium) that more than balances the positive rotation of glucose (+52.7 at equilibrium). Because the sign of rotation changes during the reaction, the process is known as *inversion* and the products as *invert sugar.*

Chocolate-Covered Cherries

If you are a devotee of chocolate-covered cherries, you may wonder how the manufacturer makes the liquid-filled cherry center liquid without making a hole in the chocolate covering. The secret of liquefying the center is a chemical reaction that takes place after the candy is made. The process uses an enzyme, invertase, to convert sucrose into invert sugar.

The liquid interior of a chocolate-covered cherry is a result of the hydrolysis of sucrose.

Before being dipped in chocolate, the cherries are coated with a sugary paste containing invertase. Once the paste hardens, the cherries are dipped in chocolate and then stored for one to two weeks. During this storage period, the invertase converts sucrose to a 1:1 ratio of glucose and fructose (invert sugar). In effect, the outer part of the cherry liquefies in its own syrup. It also pleases the sweet tooth of the consumer because one of the sugars formed, fructose, is sweeter than sucrose.

The preparation of the chocolate for the coating is also important. It must be of an optimum particle size to please the palate. If the particles are too small, the chocolate will feel slimy when eaten; if too large, it will feel gritty.

are employed in these processes. Since fructose is sweeter than sucrose, the hydrolysis adds to the sweetening effect. Bees carry out this reaction when they make honey.

The average American consumes more than 100 pounds of sucrose every year. About two-thirds of this amount is ingested in soft drinks, presweetened cereals, and other highly processed foods. The widespread use of sucrose has generated much adverse publicity. Various health magazines have reported that excess sugar causes cancer, heart disease, migraine headaches, hyperkinetic children, obesity, and tooth decay. Only the latter two claims have been substantiated. As we shall see in Chapter 26, carbohydrates are converted to fat when the caloric intake exceeds the body's requirements. Sucrose does cause tooth decay by serving as part of the plaque that sticks to a tooth. The bacteria contained within the plaque use sucrose both as an adhesive and as a food source. We shall see in Chapter 24 that bacteria can decompose sugar to lactic acid, and this acid corrodes the teeth and leads to gum destruction. The best way to remove plaque is by daily flossing, and the amount of plaque formed can be decreased by reducing your intake of sucrose.

Some food faddists claim that raw sugar is much better for you than refined sugar. Raw sugar does contain a few trace minerals, but hardly enough to make it a lot more desirable than refined sugar. People in the United States probably consume too much sugar—whether raw or refined.

Maltose

Maltose occurs to a limited extent in sprouting grain. Its major source, however, is the partial hydrolysis of starch. In the manufacture of beer, maltose is liberated by the action of malt (germinating barley) on starch, and for this reason it is often referred to as *malt sugar*. Maltose is about 30% as sweet as sucrose.

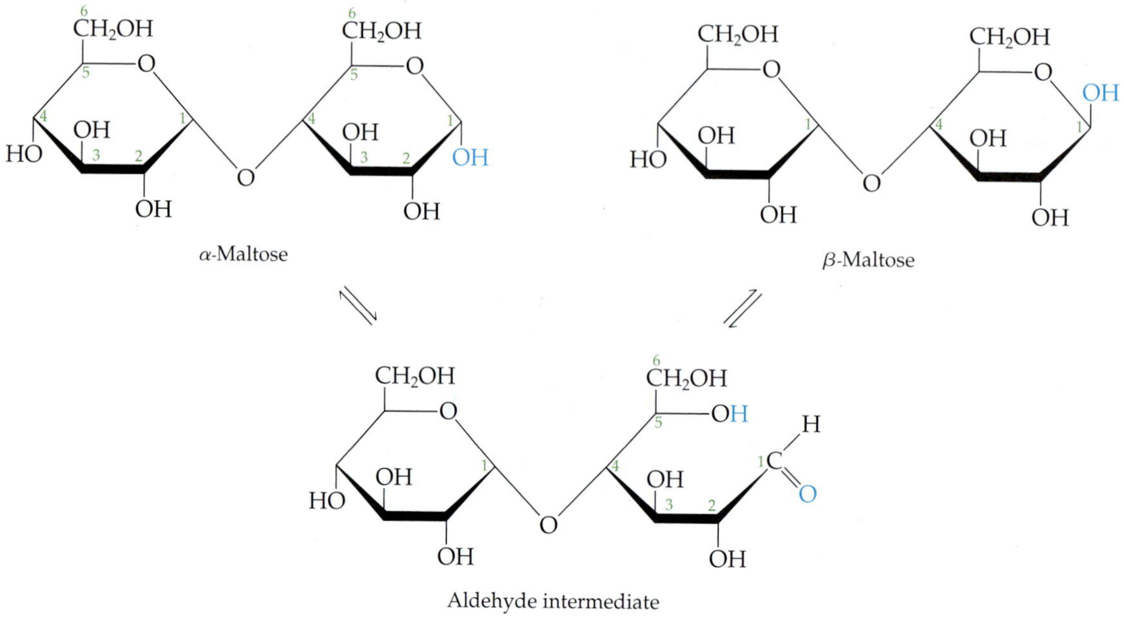

▲ **FIGURE 19.6**
Equilibrium mixture of maltose isomers.

Maltose is a reducing sugar, and it exhibits mutarotation. An equilibrium mixture of the maltose isomers is highly dextrorotatory, $[\alpha] = +136$ (Figure 19.6). When maltose is hydrolyzed, either enzymatically or by means of an acid catalyst, two molecules of D-glucose are produced. The formula of maltose must therefore incorporate two glucose molecules in such a way that one free hemiacetal hydroxyl group exists. The glucose units in maltose are joined in a *head-to-tail* fashion through an alpha linkage from C-1 of one glucose molecule to C-4 of the second glucose molecule (that is, an α-1,4-glycosidic linkage):

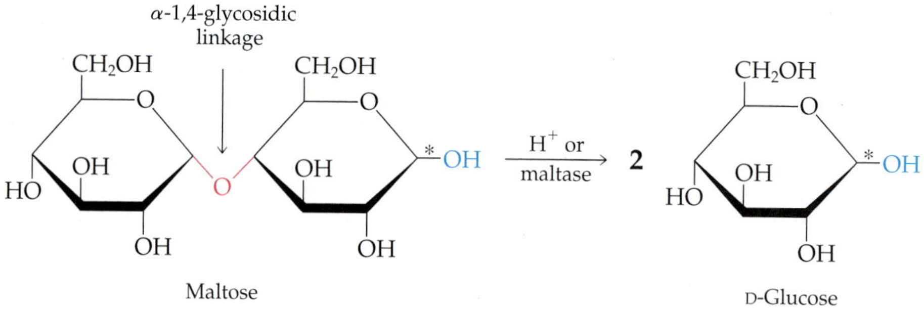

(*We use this convention for writing the hydroxyl group on the hemiacetal carbon when we do not wish to specify either the α or the β isomer.)

Lactose

Lactose is known as *milk sugar* because it occurs in the milk of humans, cows, and other mammals. Human milk contains about 7.5% lactose, whereas cow's milk, which is not as sweet, contains about 4.5% lactose. Unlike most other carbohydrates, which are plant products, lactose is one of the few carbohydrates associated exclusively with the animal kingdom. (The biosynthesis of lactose is confined to the mammary tissue.) It is produced commercially from whey, which is obtained as a by-product in the manufacture of cheese. Lactose is one of the lowest ranking sugars in terms of sweetness (about one-sixth as sweet as sucrose). It is a reducing sugar and exhibits mutarotation. An equilibrium mixture of lactose has a specific rotation of +54. The alpha form of the sugar is of commercial importance as an infant food and in the production of penicillin. Drug dealers use lactose to "cut" their heroin and thus increase their profits.

Lactose is composed of one molecule of D-galactose and one molecule of D-glucose joined by a β-1,4-glycosidic bond. The two monosaccharides are obtained from lactose by acid hydrolysis or by catalytic action of the enzyme lactase:

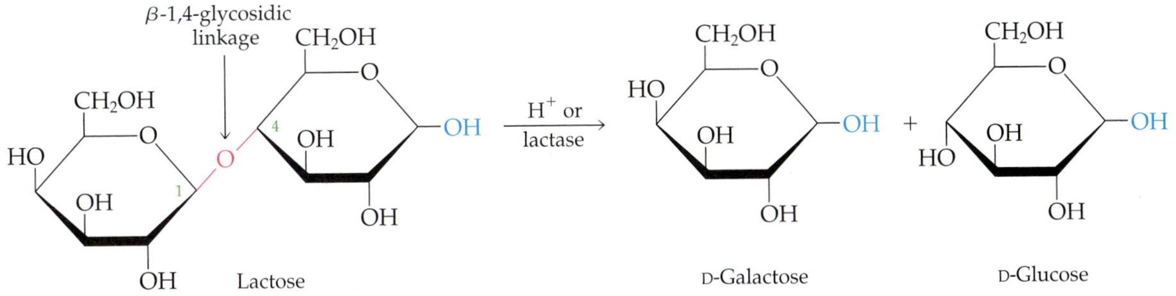

Lactose

Yeasts can metabolize sucrose and maltose but not lactose because the yeasts do not contain lactase. Certain bacteria can metabolize lactose, and they form lactic acid as one of the products. As we shall see (Section 21.13), this reaction is responsible for the "souring" of milk.

Lactose Intolerance

Lactose makes up about 40% of an infant's diet during the first year of life. Infants and small children have one form of lactase in their small intestines and therefore can easily digest lactose. However, adults have a less active form of the enzyme, and about 70% of the world's adult population (especially Africans and Asians) have some lactase deficiency. People who suffer from **lactose intolerance** are unable to digest the sugar found in milk. For some people, the inability to synthesize sufficient enzyme increases with age. Up to 20% of the U.S. population suffers some degree of lactose intolerance.

Some of the unhydrolyzed lactose passes into the colon, and its presence tends to draw water from the interstitial fluid into the intestinal lumen by osmosis. At the same time, intestinal bacteria may act on the lactose to produce organic acids and gases. The intake of water, plus the bacterial decay products, leads to the abdominal distention, cramps, and diarrhea that are symptoms of the condition.

Symptoms disappear completely if milk is excluded from the diet. (Many food stores now carry special brands of milk that have been pretreated with lactase to hydrolyze the lactose.) When milk is cooked or fermented, the lactose is at least partially hydrolyzed. For this reason, people with lactose intolerance may still be able to enjoy cheese, yogurt, or cooked foods containing milk with little or no problem. The most common treatment for lactose intolerance is lactase preparations (e.g., Lactaid), which are available in liquid and tablet form. These are taken orally with dairy foods to assist in their digestion.

The enzyme that acts on lactose is now readily available in drug stores.

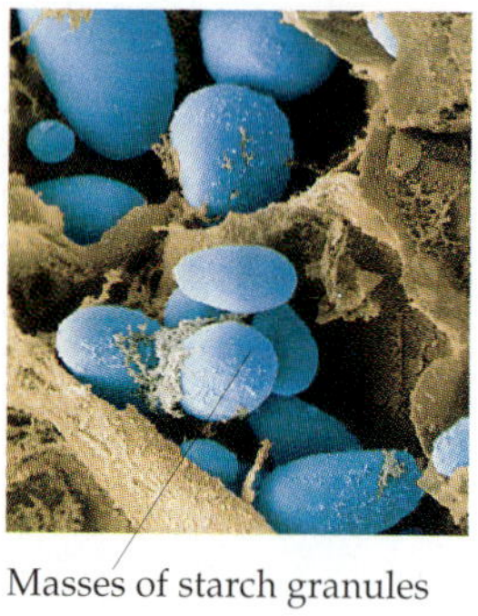

▲ **FIGURE 19.7**

Starch forms water-insoluble granules, such as those that make up the bulk of the cells in a potato.

19.8 Polysaccharides

The polysaccharides are the most abundant carbohydrates in nature. They serve as reserve food substances and as structural components of plant cells. Polysaccharides are high-molecular weight (25,000 to 15,000,000 u) polymers of monosaccharides joined together by glycosidic linkages. Biochemically, the three most significant polysaccharides are starch, glycogen, and cellulose. These three are also referred to as *homopolymers* because each yields only one type of monosaccharide (D-glucose) upon complete hydrolysis. [Inulin, another homopolymer, is found in the tubers of the Jerusalem artichoke and is composed entirely of fructose units. *Heteropolymers* may contain sugar acids, amino sugars, or noncarbohydrate substances; they are very common in nature (gums, pectins, hyaluronic acid) but are not discussed in this text.] The polysaccharides are nonreducing carbohydrates, are not sweet-tasting (probably because of their limited solubility), and do not undergo mutarotation.

Starch

Starch is the most important source of carbohydrates in the human diet. (It accounts for more than 50% of our carbohydrate intake.) Starch occurs in plants in the form of granules (Figure 19.7). Starch granules are particularly abundant in seeds (especially the cereal grains) and tubers, where they serve as a storage form of carbohydrate. The breakdown of starch to glucose nourishes the plant during periods of reduced photosynthetic activity. We often think of potatoes as a "starchy" food, and yet other plants contain a much greater percentage of starch (potatoes 15%, wheat 55%, corn 65%, and rice 75%).

Starch is a mixture of two polymers, **amylose** and **amylopectin,** which can be separated from each other by physical and/or chemical methods. Natural starches consist of about 25% amylose and 75% amylopectin. Amylose is a straight-chain polysaccharide composed entirely of D-glucose units joined by α-1,4-glycosidic linkages, as in maltose. Thus, amylose might be thought of as either polymaltose or polyglucose. There may be 60 to 300 glucose units per chain.

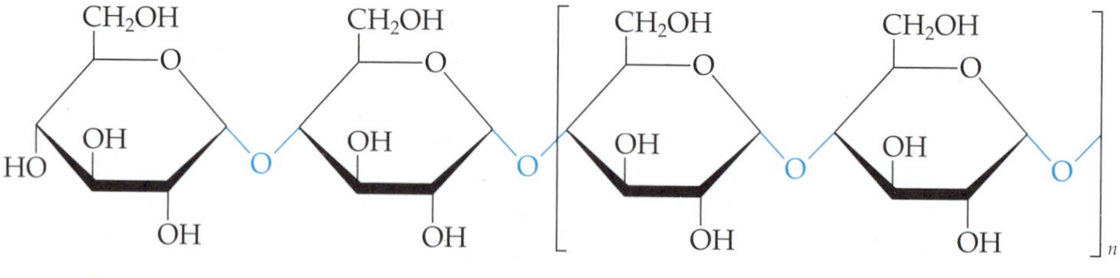

Amylose Repeating unit: n = 30–150

Experimental evidence indicates that amylose is not a straight chain of glucose units. Rather, it is coiled like a spring with six glucose monomers per turn (Figure 19.8a). When coiled in this fashion, amylose has just enough room in its core to accommodate an iodine molecule. The characteristic blue-violet color that starch gives when treated with iodine is due to the formation of the amylose-iodine complex. The test is sensitive enough to detect even minute amounts of starch in solution.

Amylopectin is a branched-chain polysaccharide composed of glucose units linked primarily by α-1,4-glycosidic bonds but with occasional α-1,6-glycosidic linkages, which are responsible for the branching. It has been estimated that there may be 300 to 6000 glucose units in amylopectin and that branching occurs about

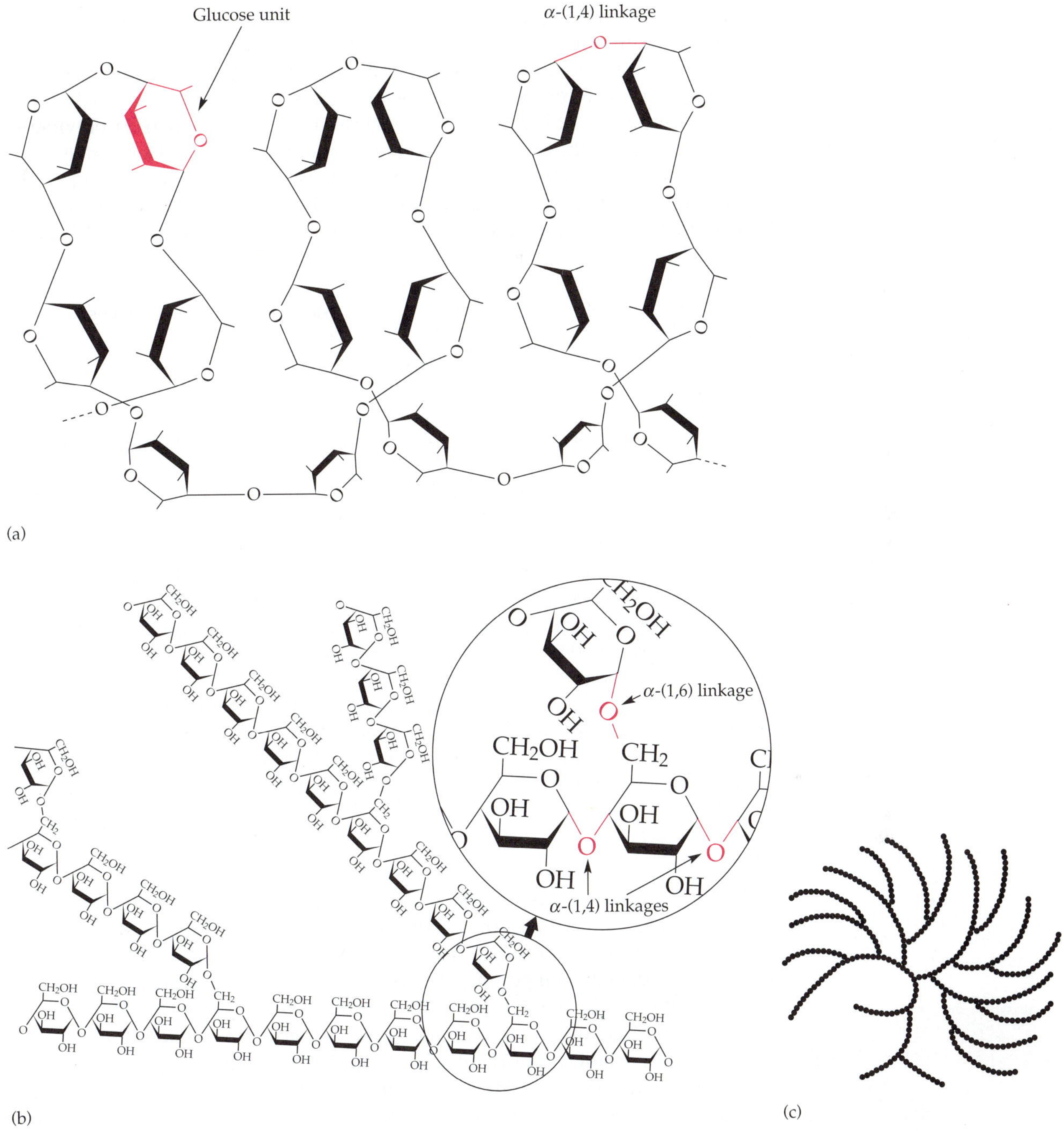

▲ FIGURE 19.8

(a) The conformation of the amylose chain. (b) Branch points of amylopectin. (c) Branched array of glucose units in amylopectin or glycogen.

once every 25 to 30 units (Figure 19.8b). The helical structure of amylopectin is disrupted by the branching of the chain, so instead of the deep blue-violet color amylose gives with iodine, amylopectin produces a less intense reddish brown.

Commercial starch is a white powder. The complete hydrolysis of starch (amylose and amylopectin) yields, in three successive stages, dextrins, maltose, and glucose. Dextrins are glucose polysaccharides of intermediate size. The shine and stiffness imparted to clothing by starch are due to the presence of dextrins formed when the clothing is ironed. Because of their characteristic stickiness upon wetting, dextrins are used as adhesives on stamps, envelopes, and labels and as pastes and mucilages. Since dextrins are more easily digested than starch, they are extensively used in the commercial preparation of infant foods (Dextrimaltose). A dried mixture of dextrins, maltose, and milk is used in the preparation of malted milk. Starch can be hydrolyzed by heating in the presence of dilute acid. In the human body, it is degraded sequentially by several enzymes known collectively as amylase:

The symbol Δ is often used to indicate that a reaction requires heat.

$$\text{Starch} \xrightarrow[\text{amylase}]{\text{H}^+,\, \Delta \text{ or}} \text{Dextrins} \xrightarrow[\text{amylase}]{\text{H}^+,\, \Delta \text{ or}} \text{Maltose} \xrightarrow[\text{maltase}]{\text{H}^+,\, \Delta \text{ or}} \text{Glucose}$$

Glycogen

Glycogen, often called animal starch, is the reserve carbohydrate of animals. Practically all mammalian cells contain some stored carbohydrate in the form of glycogen. However, it is especially abundant in the liver (4 to 8% per weight of tissue) and in skeletal muscle cells (0.5 to 1.0%). Glycogen in liver and muscle tissue is arranged in granules (Figure 19.9), clusters of small particles. When fasting or during periods of starvation, animals draw upon these glycogen reserves to obtain the glucose needed to maintain metabolic balance.

About 70% of the total glycogen in the body is stored in muscle cells. Although the percent of glycogen is higher in the liver, the much greater mass of skeletal muscle stores a greater total amount of glycogen.

In terms of structure, glycogen is quite similar to amylopectin, but glycogen is more highly branched and its branches are shorter (8 to 12 glucose units in length) than those in amylopectin. When treated with iodine, glycogen gives a reddish brown color. Glycogen can be broken down into its D-glucose subunits by acid-hydrolysis or by the same enzymes that attack starch. In animals, the enzyme phosphorylase catalyzes the breakdown of glycogen to phosphate esters of glucose (see Section 24.5).

Cellulose

Cellulose is a fibrous carbohydrate found in all plants; it is the structural component of plant cell walls. Since the Earth is covered with vegetation, cellulose is the most abundant of all carbohydrates, accounting for over 50% of all the carbon found in the vegetable kingdom. Cotton fibrils and filter paper are almost entirely cellulose (about 95%), wood is about 50% cellulose, and the dry weight of leaves is about 10 to 20% cellulose. From an industrial and economic standpoint, cellulose is the most important carbohydrate. The largest use of cellulose is in the manufacture of paper and paper products. Although there is increasing use of synthetic fibers, rayon (made from cellulose) and cotton still account for over 70% of textile production.

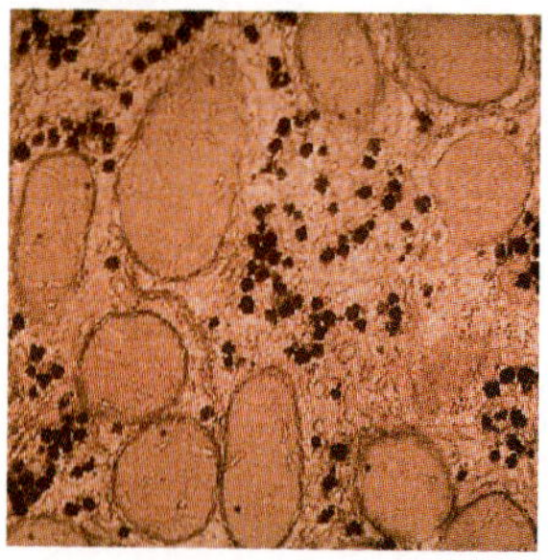

▲ FIGURE 19.9

Electron micrograph of glycogen granules in a rat liver cell.

Like amylose, cellulose is a linear polymer of glucose. It differs, however, in that the glucose units (about 2000 to 3000) are joined by β-1,4-glycosidic linkages. The linear nature of the cellulose chains allows a great deal of hydrogen bonding between hydroxyl groups on adjacent chains. As a result, the chains are closely packed into fibers (Figure 19.10), and there is little interaction with water or with any other solvent. Cotton and wood, for example, are completely insoluble in water and have considerable mechanical strength. Since there is no helical structure, cellulose does not bind to iodine to give a colored product.

◀ **FIGURE 19.10**

Electron micrograph of the cell wall of an alga. The wall consists of successive layers of cellulose fibers in parallel arrangement.

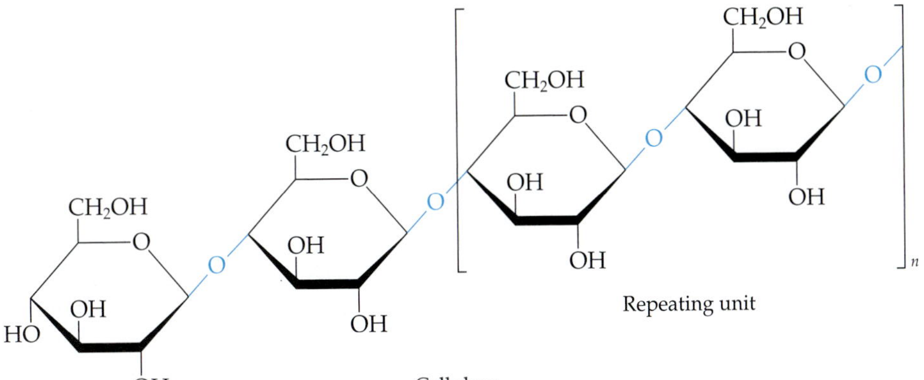

Cellulose

Cellulose yields D-glucose upon complete acid hydrolysis, and yet humans (and all other vertebrates) cannot utilize cellulose as a source of glucose. We can eat potatoes, but we can't eat grass. Our digestive juices lack enzymes that can hydrolyze the β-glycosidic linkages found in cellulose. Many microorganisms and herbivorous animals (cows, horses, sheep) can digest cellulose. These higher animals can do so only because they have in their digestive tracts microorganisms whose enzyme (cellulase) catalyzes cellulose hydrolysis. Termites also contain cellulase-secreting microorganisms and thus can subsist on a wood diet. This once again demonstrates the extreme stereospecificity of biochemical processes.

Although it has no nutritive value, cellulose makes up the greater part of dietary fiber. The fibrous portions of plants (stems, peels, and seeds) are rich in fiber. Bran, celery, beans, apples, raspberries, and figs are good sources of dietary fiber.

Since so much of the carbon in the biosphere exists as cellulose, considerable research has gone into converting it to glucose or some other form of food for humans. There are enzymes that will degrade cellulose, but to do so efficiently requires some pretreatment of the fibers. Perhaps some day, through the ingenuity of research chemists, we will be able to eat grass and straw—after proper conversion, of course. For the moment we must continue to depend on reactions such as

$$\underset{\text{(grass)}}{\text{Cellulose}} + \text{Cow} \longrightarrow \underset{\text{(milk, meat, butter)}}{\text{Carbohydrates, Proteins, and Fats}}$$

Dietary Fiber

The importance of fiber in the human diet was often ignored or overlooked until the mid-1970s. Since then, several studies have been published that have the public excited about fiber. One group of findings suggests a relationship between a low-fiber diet and colon cancer. First, people in developed countries are much more likely to get colon cancer than people in underdeveloped nations. People in developed countries eat diets rich in highly processed, low-fiber foods, while those in more "primitive" areas eat high-fiber diets, which lead to frequent and robust bowel movements. Low-fiber diets therefore result in less frequent bowel action, with high retention times for feces in the colon. Bacteria act upon the materials in the colon. (Indeed, bacteria, both living and dead, make up about one-third of the dry weight of feces.) With a high-fiber diet, the materials seldom remain in the colon for more than 1 day. With a low-fiber diet, the retention time can be as long as 3 days, allowing for prolonged bacterial activity that produces a high level of mutagenic chemicals. Chemicals that are mutagenic often are also carcinogenic.

Some scientists accept this line of reasoning and so believe that a low-fiber diet (less than 20 g/day) is associated with a higher incidence of colon cancer (as well as higher incidences of such other problems as diverticulosis, coronary heart disease, atherosclerosis, gallstones, and hemorrhoids). Other scientists dispute some of these claims, believing that the culprits are not lack of fiber but rather meat and saturated fats (Chapter 20). They point out that a high-fiber diet (about 100 g/day) is a largely vegetarian diet, and thus individuals consume relatively little meat and saturated fat.

Another study suggests that a high-fiber diet is useful in the treatment of diabetes (see Section 24.4). It has long been known that diabetics ought to avoid ingestion of rapidly digested sugars such as glucose and sucrose and eat only carbohydrates that are slowly digested (e.g., starches high in amylopectin). The presence of fiber in the diet reduces the rate of absorption of glucose, and therefore the peak blood sugar concentration is lowered.

What we now call dietary fiber was once called roughage *or referred to as* bulk in the diet.

Summary

Carbohydrates, compounds containing carbon, hydrogen, and oxygen, include sugars, starch, glycogen, and cellulose. All carbohydrates contain alcohol functional groups, and most also contain either an aldehyde or a ketone group (often in the form of an acetal or ketal):

$$
\begin{array}{cc}
\begin{array}{c}
H\quad O \\
\diagdown \diagup \\
C \\
| \\
(CHOH)_n \\
| \\
CH_2OH
\end{array}
&
\begin{array}{c}
CH_2OH \\
| \\
C{=}O \\
| \\
(CHOH)_n \\
| \\
CH_2OH
\end{array}
\end{array}
$$

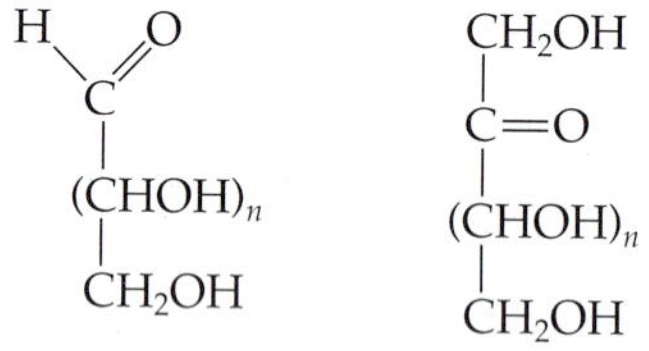

The simplest carbohydrates are **monosaccharides.** Those that can be hydrolyzed to two monosaccharide units are **disaccharides,** and those that can be hydrolyzed to many monosaccharide units are **polysaccharides.** Most sugars are either monosaccharides or disaccharides. Cellulose, glycogen, and starch are polysaccharides.

A sugar is designated as being D **sugar** or L **sugar** according to how, in a Fischer projection of the molecule,

the hydrogen and hydroxyl group are attached to the *penultimate* carbon, which means the carbon immediately before the terminal alcohol carbon. If the structure at this carbon is the same as that of D-glyceraldehyde (—OH to the right), the sugar is a D sugar; if the configuration at the penultimate carbon is that of L-glyceraldehyde (—OH to the left), the sugar is an L sugar. The D/L notation in no way relates to whether the sugar is dextrorotatory or levorotatory. An L sugar can rotate the plane of polarized light to the right; a D sugar can rotate it to the left. The direction of optical activity is indicated by a plus or a minus sign following the D/L part of the name. Thus D-(−)-ribose has the —OH to the right on the penultimate carbon of a Fischer projection but rotates polarized light to the left.

The carbon chain in glucose can bend into a ring shape, with the oxygen of the hydroxyl group on carbon 5 forming a hemiacetal bond with the carbonyl carbon and the hydrogen of that hydroxyl group combining with the carbonyl oxygen to form a new hydroxyl group on carbon 1:

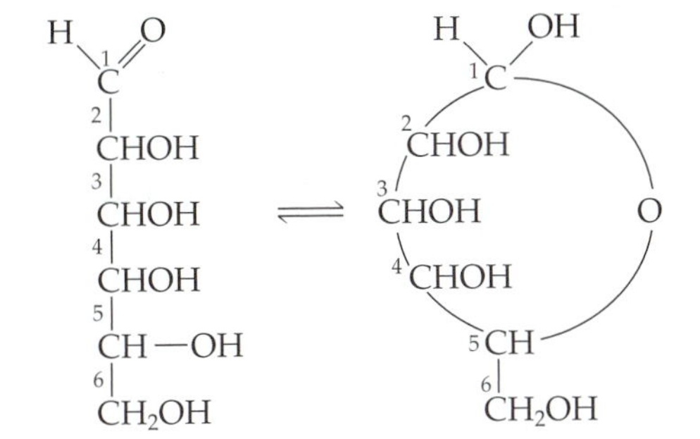

Glucose in solution exists as an equilibrium mixture of three forms: α-, β-, and open-chain. The *alpha* form is drawn with the hydroxyl group on the carbonyl carbon pointing downward, the *beta* form, with it pointing upward:

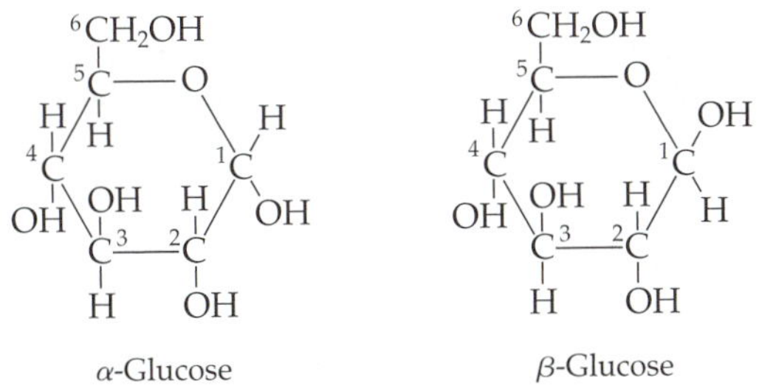

Any solid sugar can be all alpha or all beta. Once the sample is dissolved in water, however, the hemiacetal ring opens up into the straight-chain form and then closes to the other hemiacetal form. These interconversions occur back and forth until an equilibrium mixture is achieved, and the process is called **mutarotation.**

Monosaccharides form esters and ethers at the hydroxyl groups, and the carbonyl group is easily oxidized by Tollens's, Benedict's, or Fehling's reagents. Any mono- or disaccharide containing a latent carbonyl group is a **reducing sugar.**

The disaccharide *sucrose* (table sugar) consists of a glucose unit and a fructose unit joined by an acetal linkage. The linkage is designated an α-1, β-2-glycosidic linkage because it involes the carbon-1 hydroxyl group of α-D-glucose and the carbon-2 hydroxyl group of β-D-fructose. Sucrose is not a reducing sugar because there is no latent carbonyl group, and it cannot undergo mutarotation because of the restrictions imposed by this linkage.

The disaccharide *maltose* contains two glucose units joined in an α-1,4-glycosidic linkage. Maltose is a reducing sugar because one of its glucose units exists as a hemiacetal, so that the carbonyl group is available.

The disaccharide *lactose* comprises a D-galactose unit and a D-glucose unit joined via a β-1,4-glycosidic linkage. It is a reducing sugar; it undergoes mutarotation.

Starch, the principal carbohydrate of plants, is made up of the polysaccharides **amylose** (20–30%) and **amylopectin** (80–20%). Ingested by humans, starch is hydrolyzed to glucose and used as the body's energy source. *Glycogen* is the polysaccharide animals use to store excess ingested carbohydrates. Similar in structure to amylopectin, glycogen is hydrolyzed to glucose whenever the animal needs energy for some metabolic process. The polysaccharide *cellulose* is the structural component of plant cells. It is a linear polymer of glucose units joined by β-1,4-glycosidic linkages. It is indigestible in the human body, but digestible by many microorganisms.

Key Terms

aldose (19.2)
amylopectin (19.8)
amylose (19.8)
biochemistry (Introduction)
carbohydrate (19.1)
D sugar (19.3)
disaccharide (19.1)
glycosidic linkage (19.7)
invert sugar (19.7)
ketose (19.2)
L sugar (19.3)
lactose intolerance (19.8)
monosaccharide (19.1)
mutarotation (19.5)
polysaccharide (19.1)
reducing sugar (19.6)

Review Questions

1. Define:
   a. triose
   b. aldose
   c. hexose
   d. disaccharide
   e. polysaccharide
   f. aldopentose
   g. ketotetrose
   h. invert sugar
   i. mutarotation
   j. glycosidic linkage
2. Draw formulas for D-glyceraldehyde and L-glyceraldehyde. What do the prefixes mean?
3. What is meant by a latent carbonyl group?
4. What is a reducing sugar?
5. Are all monosaccharides and disaccharides soluble in water? Explain.
6. Why are (+)-glucose and (−)-fructose both classified as D sugars?
7. How can it be shown that a solution of α-D-glucose exhibits mutarotation?

8. How does ribose differ from deoxyribose?
9. What purposes do starch and cellulose serve in plants?
10. What purpose does glycogen serve in animals?
11. Identify these sugars by their proper names:
 a. blood sugar b. milk sugar c. dextrose
 d. levulose e. table sugar f. malt sugar
12. How do amylose and amylopectin differ from each other? How are they similar?
13. How do amylose and cellulose differ from each other? How are they similar?
14. How do amylopectin and glycogen differ from each other? How are they similar?

Problems

Classification

15. Specify whether each of the following is a D sugar or an L sugar:

a.

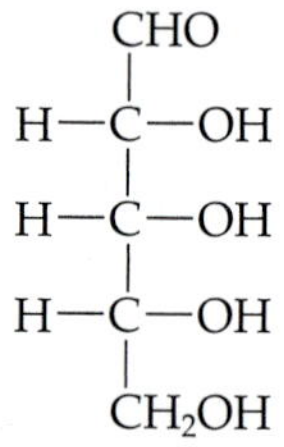

b.

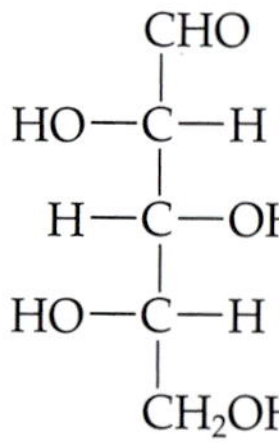

16. Specify whether each of the following is a D sugar or an L sugar:

a.

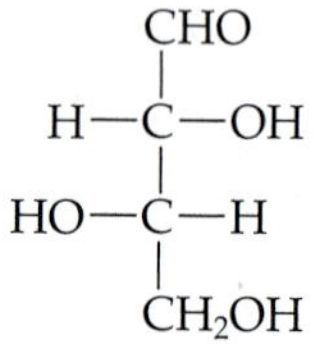

b.

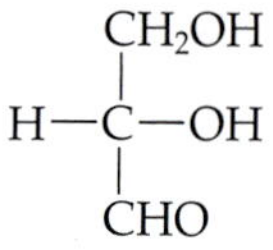

17. Identify as an aldose or a ketose:
 a. D-glyceraldehyde b. D-ribose
 c. D-deoxyribose d. D-fructose
18. Identify as an aldose or a ketose:
 a. D-glucose b. L-fructose
 c. D-galactose d. L-mannose
19. Identify as a triose, tetrose, pentose, or hexose:
 a. L-glucose
 b. D-deoxyribose
20. Identify as a triose, tetrose, pentose, or hexose:
 a. D-fructose
 b. L-glyceraldehyde

Structural Formulas

21. Draw a ketotetrose.
22. Draw an aldoheptose.
23. From memory, draw formulas for the open-chain forms of D-glucose and D-mannose.
24. From memory, draw formulas for the open-chain forms of D-galactose and D-fructose.
25. Draw the cyclic structure for α-D-glucose.
26. Draw the cyclic structure for β-D-fructose.
27. Draw the structure for β-D-glucose.
28. By reference to Problem 27, draw the structure for β-D-galactose.

Glycosides

29. For each of these abbreviated sugar formulas, indicate whether the glycosidic linkage is alpha or beta.

a.

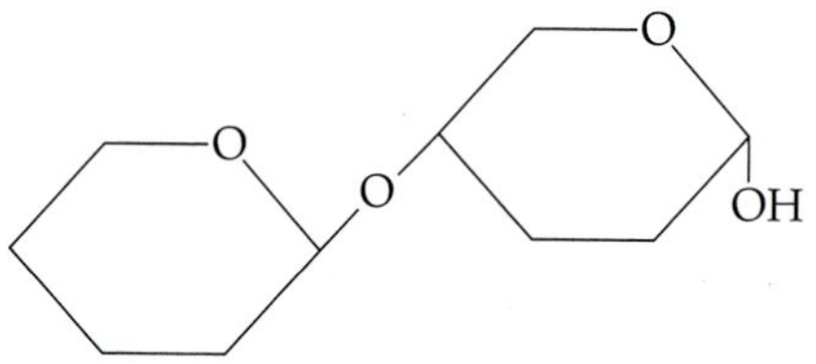

b.

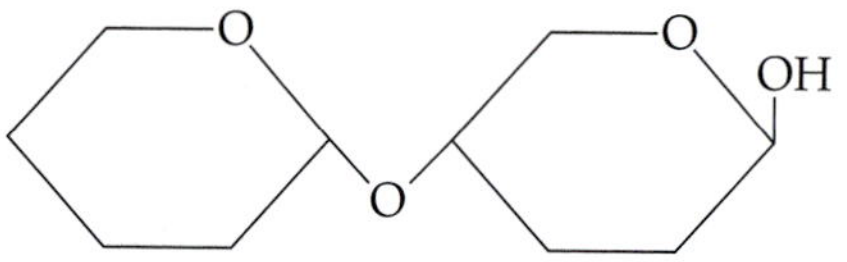

30. For each of these abbreviated sugar formulas, indicate whether the glycosidic linkage is alpha or beta.

a.

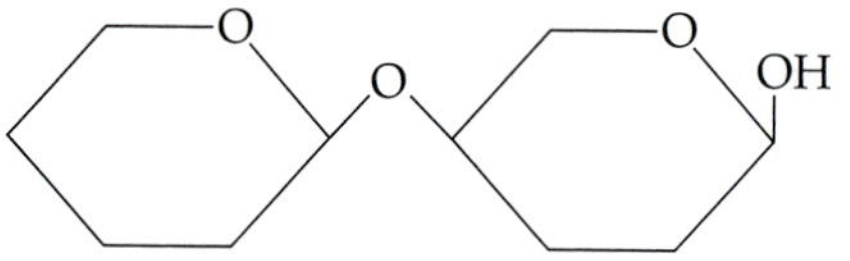

b.

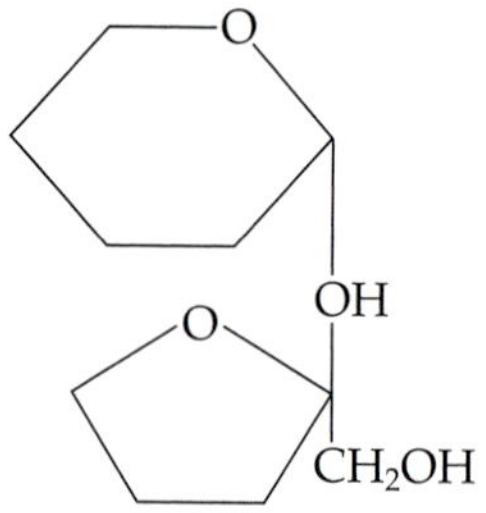

31. What is the orientation of the hydroxyl group at the hemiacetal carbon of structures (**a**) and (**b**) in Problem 29? Which structure (if any) shown in Problem 29 is *not* a reducing sugar?
32. What is the orientation of the hydroxyl group at the hemiacetal carbon of structure (**a**) in Problem 30? Which structure (if any) shown in Problem 30 is not a reducing sugar?

33. The structure of a methyl glycoside of glucose is

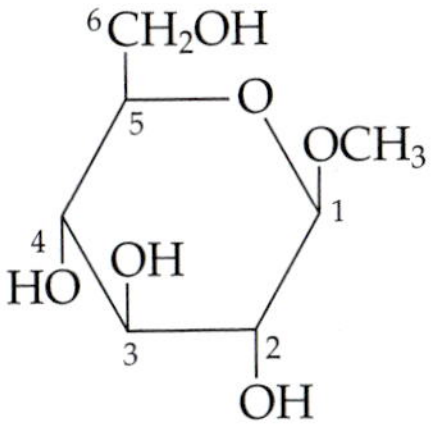

a. Is C-1 in the alpha or the beta arrangement?
b. Is the compound a reducing sugar?
c. Will it give a positive test with Benedict's reagent?

34. Melibiose is a disaccharide that occurs in some plant juices. Its structure is

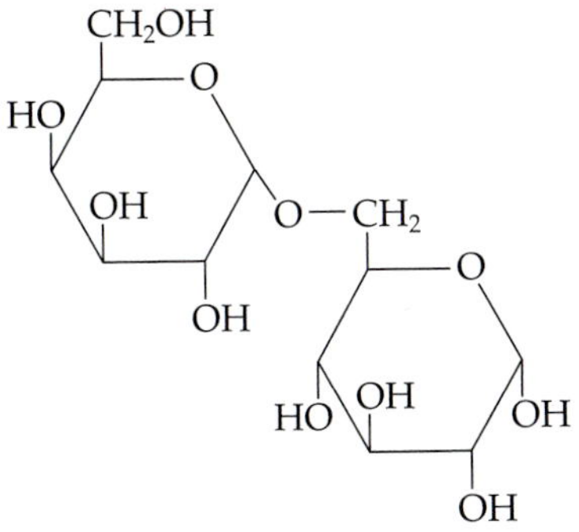

What monosaccharide units are incorporated in melibiose? What type of linkage (α or β) joins the two rings of melibiose? Is melibiose a reducing sugar? If so, circle the hemiacetal carbon and indicate whether the hydroxyl group is α or β.

35. Gentiobiose is a disaccharide composed of two glucose units joined by a β-1,6-glycosidic linkage. Draw the structure of gentiobiose.

36. The disaccharide cellobiose has two D-glucose units joined by a β-1,4 linkage. Draw the alpha form of cellobiose.

Chemical Reactions

37. Which of the following gives a positive Benedict's test?
a. L-galactose b. levulose c. D-mannose
d. maltose e. cellulose f. glycogen

38. Which of the following gives a positive Benedict's test?
a. lactose b. ribose c. inulin
d. starch e. sucrose f. invert sugar

39. What monosaccharide(s) is (are) obtained from the hydrolysis of each of the following?
a. starch b. cellulose c. maltose

40. What monosaccharide(s) is (are) obtained from the hydrolysis of each of the following?
a. sucrose b. glycogen c. lactose

41. List the reagents necessary for the following conversions:

a.

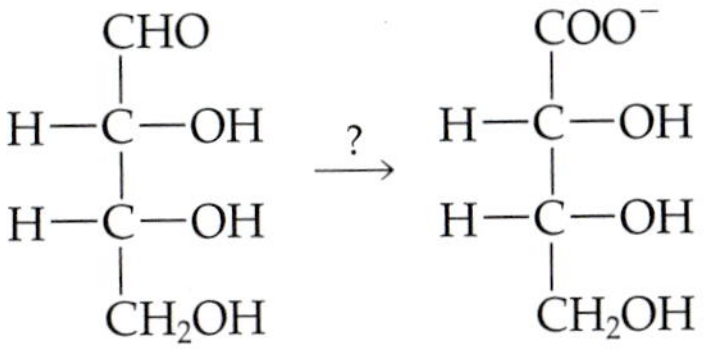

b.

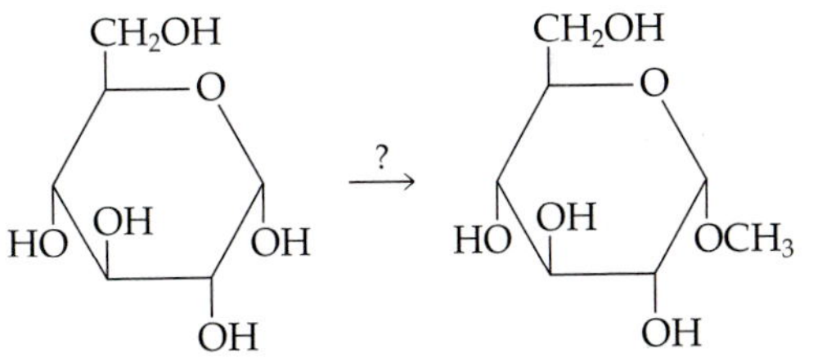

42. List the reagents necessary for the following conversion:

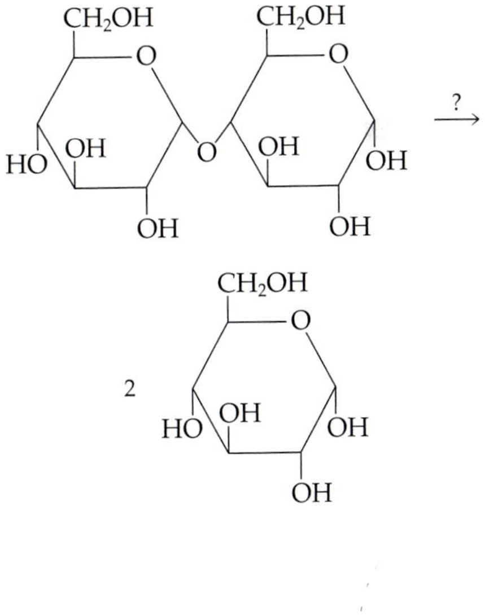

Additional Problems

43. Knowing that mannose differs from glucose only in the configuration at the second carbon, draw the cyclic structure for α-D-mannose.

44. What structural characteristics are necessary if a disaccharide is to be a reducing sugar? Draw the structure of a hypothetical nonreducing disaccharide composed of two aldohexoses.

45. Raffinose is a trisaccharide (found in beans and sugar beets) containing D-galactose, D-glucose, and D-fructose. The enzyme α-galactase catalyzes the hydrolysis

of raffinose to galactose and sucrose. Draw the structure of raffinose. (The linkage from galactose to the glucose unit is α-1,6.)

46. In the schematic below, Glc represents glucose. What substance is indicated?

\cdots Glc-Glc

\cdots Glc-Glc-Glc-Glc

\cdots Glc-Glc-Glc-Glc-Glc-Glc-Glc-Glc-Glc

\cdots Glc-Glc-Glc-Glc-Glc-Glc-Glc-Glc-Glc-Glc-Glc-Glc-Glc

\cdots Glc-Glc-Glc-Glc-Glc-Glc-Glc-Glc

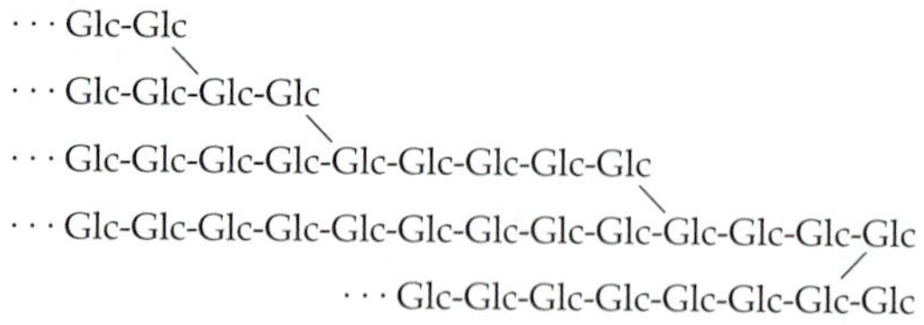

47. Xylulose, found in the urine of humans suffering from a condition called pentosuria, has the structure

$$
\begin{array}{c}
CH_2OH \\
| \\
C=O \\
| \\
H-C-OH \\
| \\
HO-C-H \\
| \\
CH_2OH
\end{array}
$$

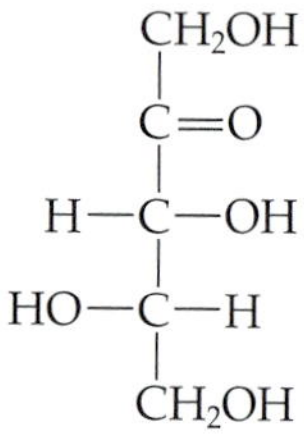

Classify xylulose as fully as possible.

48. Erythrulose, which can be prepared from D-fructose, has the structure

$$
\begin{array}{c}
CH_2OH \\
| \\
C=O \\
| \\
HO-C-H \\
| \\
CH_2OH
\end{array}
$$

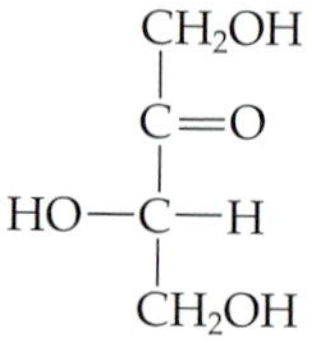

Classify erythrulose as fully as possible.

49. D-Glucose can be oxidized at C-1 to form D-gluconic acid, at C-6 to yield D-glucuronic acid, and at both C-1 and C-6 to yield D-glucaric acid. Draw structures of these three oxidation products.

50. What monosaccharide units make up the disaccharide lactulose (below)?

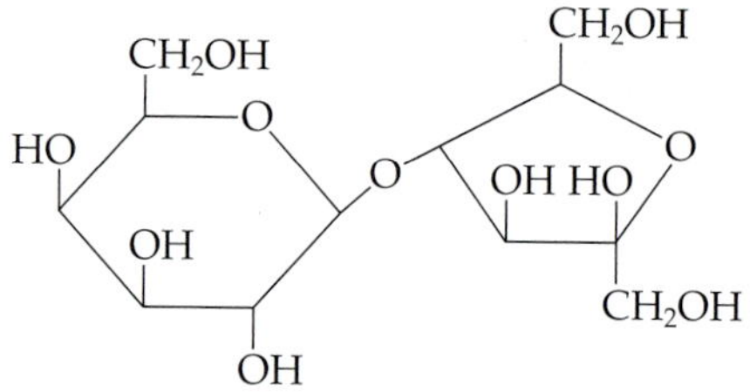

20 Lipids

Soap film. Soaps are made in great quantities from fats and oils, two kinds of lipids.

The food we eat is divided into three primary groups: carbohydrates (Chapter 19), proteins (Chapter 21), and lipids (which we discuss in this chapter). The best-known lipids are fats, which, gram for gram, pack about twice the caloric content of carbohydrates. Although this may be bad news for the dieter, it says something about the efficiency of nature's designs. The body has a limited capacity for storing carbohydrates. It can tuck away a bit of glycogen in the liver or in muscle tissue, but carbohydrates, primarily in the form of glucose, are meant to serve the body's *immediate* energy needs. If we intend to store energy reserves, then the more energy we can pack into a given space, the better off we are. The oxidation of fats supplies about 9 kcal/g, whereas the oxidation of carbohydrates supplies only 4 kcal/g (see Chapter 26). The body, an efficient organism, is geared to store fats, and its capacity for doing so is astounding. Most of us store enough energy in the form of fats to last a month or so, but there is a recorded instance of a man weighing 486 kg. If all that energy were stored as carbohydrate, he would have weighed a 1000 kg or more.

The body's ability to store fats may elicit from you feelings of disgust or despair rather than awe. However, a quick summary of the other functions of fats and the functions of other kinds of lipids in the body may provide a more positive picture of these essential compounds. Lipids are important components of brain and nervous tissue, and serve as protective padding and insulation for vital organs. Without lipids in our diets, we'd be deficient in the fat-soluble vitamins, A, D, E, and K. Most important, lipids make up the major part of the membranes of each of the 10 trillion cells in our bodies.

20.1 What Is a Lipid?

Of the three types of foodstuffs, two are classified by functional groups. As Chapter 19 states, carbohydrates are polyhydroxy aldehydes or ketones. The proteins, as we shall soon see, are polyamides. But lipids are not poly anything in particular. They tend to be either esters or compounds that can form esters, but that takes in a lot of territory and doesn't even hint at the wide variation found among the lipids.

Because they are defined by solubility, a physical property, rather than by chemical structure, it is not surprising that there are a great many different kinds of lipids.

What makes a lipid a lipid is its solubility. Considering that water is the major solvent in living systems and that reactions of physiological importance tend to take place in aqueous solutions, it is not surprising that insolubility in water should be considered a noteworthy feature when found in important body constituents. Lipids are soluble in relatively nonpolar organic solvents such as carbon tetrachloride, hexane, and diethyl ether (the so-called fat solvents), but they are generally insoluble in water.

Compounds isolated from body tissues are classified as **lipids** if they are more soluble in organic solvents than in water. Included in this category are esters of glycerol and the fatty acids (or phosphoric acid), compounds that incorporate sugar units or a complicated amino alcohol called sphingosine, and steroids such as cholesterol. Because of this broad variation in structure, we can't present a general formula for lipids. We shall, instead, consider one subclass at a time and try to point out similarities and differences in structure as we go along. Figure 20.1 indicates a scheme for classifying the lipids.

Cream, butter, margarine, cooking oils, and foods fried in fats and oils are rich in lipids.

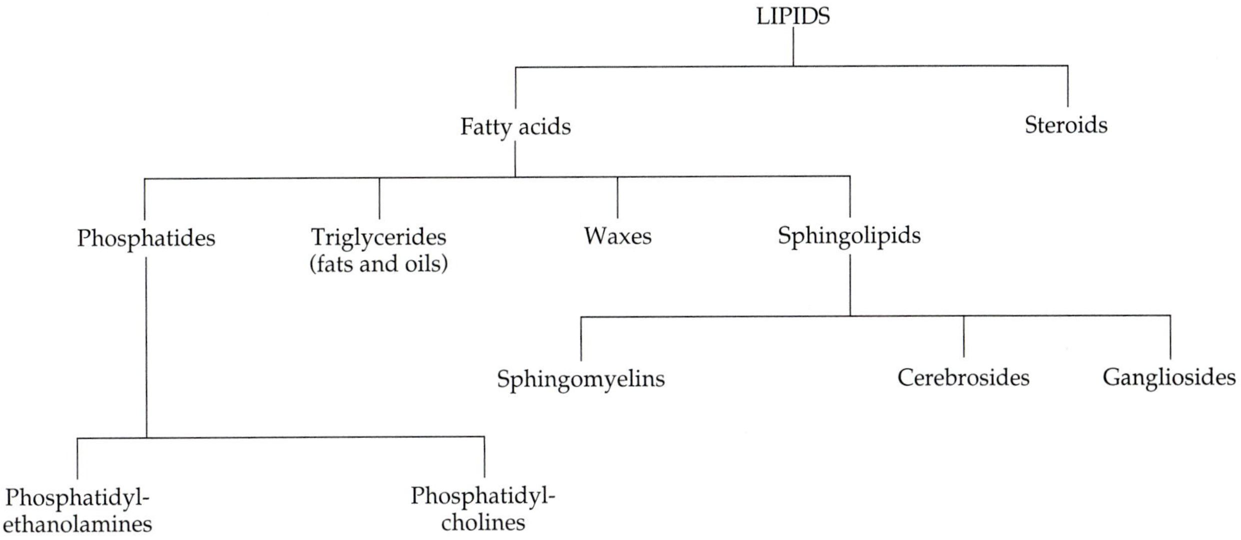

▲ **FIGURE 20.1**
A scheme that organizes the major types of lipids on the basis of structural relationships. Fatty acids are the simplest in structure. Several other types of lipids contain or are derived from fatty acids. All steroids are based on a structure with four fused rings.

20.2 Fatty Acids

Fatty acids are so named because they are structural components of fats and oils. Chemically, **fatty acids** are generally long-chain carboxylic acids. More than 70 have been identified in nature. Nearly all contain an even number of carbon atoms. Few are branched. Some, the unsaturated fatty acids, contain one or more carbon-to-carbon double bonds. Free fatty acids are rare, occurring in nature in only small amounts. Fats, oils, and other lipids, however, provide a reservoir from which fatty acids can be obtained. Table 20.1 lists some common fatty acids and one important source for each.

Table 20.1 Some Fatty Acids in Natural Fats

Abbreviated Formula[a]	Condensed Structure	Melting Point (°C)	Name	Source
C_3H_7COOH	$CH_3CH_2CH_2COOH$	−8	Butyric acid	Butter
$C_5H_{11}COOH$	$CH_3(CH_2)_4COOH$	−3	Caproic acid	Butter
$C_7H_{15}COOH$	$CH_3(CH_2)_6COOH$	−17	Caprylic acid	Coconut oil
$C_9H_{19}COOH$	$CH_3(CH_2)_8COOH$	31	Capric acid	Coconut oil
$C_{11}H_{23}COOH$	$CH_3(CH_2)_{10}COOH$	44	Lauric acid	Palm kernel oil
$C_{13}H_{27}COOH$	$CH_3(CH_2)_{12}COOH$	54	Myristic acid	Oil of nutmeg
$C_{15}H_{31}COOH$	$CH_3(CH_2)_{14}COOH$	63	Palmitic acid	Palm oil
$C_{17}H_{35}COOH$	$CH_3(CH_2)_{16}COOH$	70	Stearic acid	Beef tallow
$C_{17}H_{33}COOH$	$CH_3(CH_2)_7CH{=}CH(CH_2)_7COOH$	13	Oleic acid	Olive oil
$C_{17}H_{31}COOH$	$CH_3(CH_2)_3(CH_2CH{=}CH)_2(CH_2)_7COOH$	−5	Linoleic acid	Soybean oil
$C_{17}H_{29}COOH$	$CH_3(CH_2CH{=}CH)_3(CH_2)_7COOH$	−11	Linolenic acid	Fish oil
$C_{19}H_{31}COOH$	$CH_3(CH_2)_4(CH{=}CHCH_2)_4CH_2CH_2COOH$	−50	Arachidonic acid	Liver

[a] Saturated fatty acids have the general formula $C_nH_{2n+1}COOH$; unsaturated fatty acids are of the form $C_nH_{2n-1}COOH$, $C_nH_{2n-3}COOH$, $C_nH_{2n-5}COOH$, and so on.

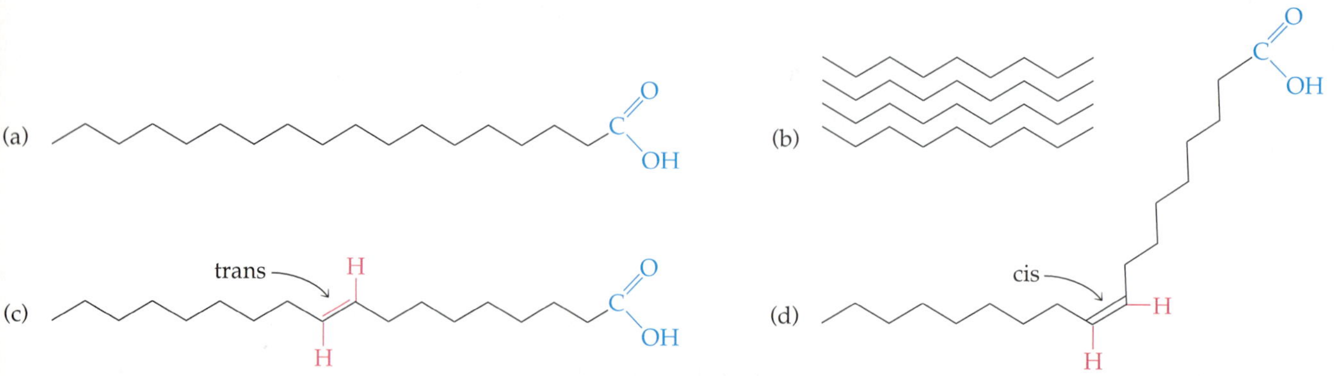

▲ **FIGURE 20.2**

(a) A schematic representation of a stearic acid molecule. (b) These saturated acid molecules stack nicely in a crystal lattice. (c) Elaidic acid has a trans double bond and also can stack neatly. (d) Oleic acid, with its cis double bond, will not fit neatly into a crystalline arrangement.

The tetrahedral bond angles of carbon require that the chain of saturated fatty acid molecules assume a zigzag configuration (Figure 20.2a), but the molecule viewed as a whole is relatively straight. (Each angle in these zigzag formulas represents one carbon atom in the fatty acid chain.) Such molecules fit nicely into a crystal lattice (Figure 20.2b), a capability that gives fatty acids and the fats derived from them relatively high melting points. A few unsaturated fatty acids exist in the *trans* configuration (Figure 20.2c), but the great majority occur in the *cis* configuration. This results in a severe bend in the molecules (Figure 20.2d). These molecules don't stack neatly, and so the attractions between molecules are smaller. Consequently, the unsaturated fatty acids (and unsaturated fats) have lower melting points. Most are liquids at room temperature.

20.3 Fats and Oils

Do not confuse the term *oil,* used here to refer to a particular group of lipids, with the hydrocarbon petroleum oils.

Fats and oils are the most abundant lipids found in nature. Both types of compounds are called **triglycerides** because they consist of *esters* composed of *three fatty acids* joined to *glycerol*, a *trihydroxy alcohol:*

The systematic name for these esters is *triacylglcerols,* but they are commonly called by the old name, triglycerides.

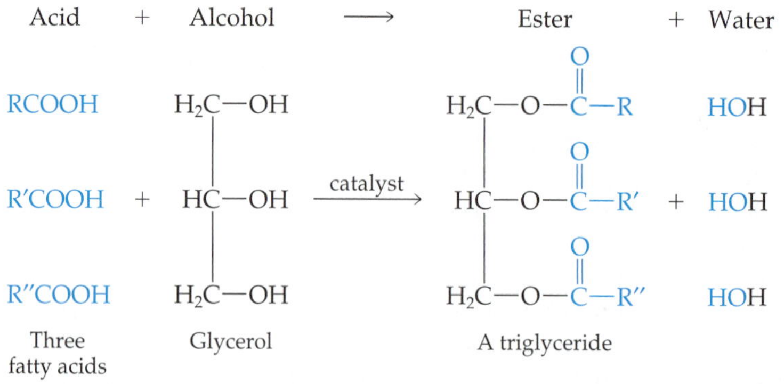

If all three hydroxyl groups of the glycerol molecule are esterified with the same fatty acid, the resulting ester is called a **simple triglyceride.** Although some simple triglycerides have been synthesized in the laboratory, they rarely occur in nature. All of the triglycerides obtained from naturally occurring fats and oils contain two or three different fatty acid components and are thus termed **mixed triglycerides.**

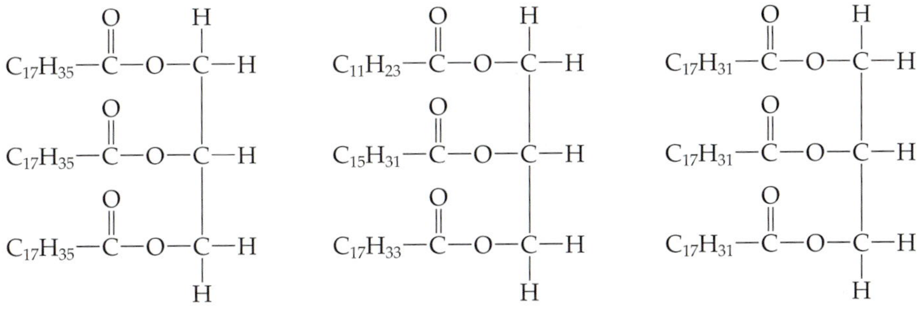

Glyceryl stearate (Tristearin) (a simple triglyceride) mp 71 °C

Glyceryl lauropalmitooleate (a mixed triglyceride)

Glyceryl linoleate (Trilinolein) (a simple triglyceride) mp 9 °C

No single formula can be written to represent the naturally occurring fats and oils because they are highly complex mixtures of molecules in which many different fatty acids are represented. Table 20.2 shows the fatty acid compositions of some common fats and oils. Notice that there is a fairly wide range of values. The range is wide because the composition of any given fat or oil varies depending on the plant or animal species involved as well as dietetic and climatic factors. To cite just two examples, lard from corn-fed hogs is more highly saturated than lard from peanut-fed hogs, and linseed oil from cold climates is more unsaturated than linseed oil from warm climates. Palmitic acid is the most abundant of the saturated fatty acids, and oleic acid is the most abundant unsaturated fatty acid. Unsaturated fatty acids predominate over saturated ones in most plants and animals.

Tallow is a fat obtained from cattle and sheep. Lard is a fat obtained from pigs.

Table 20.2 Fatty Acid Components of Some Common Fats and Oils

	Component Fatty Acids (%)[a]						
	Lauric (C_{12})	Myristic (C_{14})	Palmitic (C_{16})	Stearic (C_{18})	Oleic (C_{18})	Linoleic (C_{18})	Linolenic (C_{18})
Fats							
Butter	1–4	8–13	25–32	8–13	22–29	2–4	
Tallow		2–3	24–32	20–25	37–43	2–3	
Lard		1–2	25–30	12–16	40–50	3–8	
Edible Oils							
Coconut oil[b]	44–50	13–18	7–10	1–4	5–8	1–3	
Palm oil[c]		1–6	32–47	1–6	40–52	2–11	
Olive oil	0–1	0–2	7–20	2–3	53–86	4–22	
Peanut oil		0–1	6–11	3–6	40–65	17–38	
Cottonseed oil		0–3	17–23	1–3	23–44	34–55	
Corn oil		1–2	8–12	2–5	29–49	34–56	
Soybean oil		0–1	6–10	2–5	20–30	50–60	2–10
Safflower oil			6–7	2–3	12–14	75–80	0–2
Nonedible Oil							
Linseed oil		0–1	5–9	4–7	9–29	8–29	45–67

[a] Totals less than 100% indicate the presence of lower or higher acids in small amounts.
[b] Coconut oil is highly saturated. It contains an unusually high percentage (53 to 70%) of the low-melting C_8, C_{10}, and C_{12} saturated fatty acids. Coconut oil is a liquid in the warmer, tropical climates, but at room temperature in the temperate zone it is a solid.
[c] Palm oil is highly saturated because of the large percentage of palmitic acid.

One classification of triglycerides is made on the basis of their physical states at room temperature. A lipid is called a **fat** if it is a solid at 25 °C and an **oil** if it is a liquid at the same temperature. (These differences in melting points reflect differences in the degree of unsaturation of the constituent fatty acids.) Furthermore, lipids obtained from animal sources are usually solids, whereas oils are generally of plant origin. Therefore we commonly speak of animal fats and vegetable oils. Coconut and palm oils, which are highly saturated, and fish oils, which are relatively unsaturated, are notable exceptions to the general rule.

Fats and oils are often classified according to the degree of unsaturation of the fatty acids they incorporate. *Saturated* fatty acids contain no double bonds, *monounsaturated* fatty acids contain one double bond per molecule, and *polyunsaturated* fatty acids are those that have two or more double bonds. Saturated fats contain a high proportion of saturated fatty acids, and polyunsaturated oils incorporate mainly unsaturated fatty acids.

Saturated fats have been implicated, along with cholesterol, a steroid (Section 20.9), in one type of *arteriosclerosis* (hardening of the arteries). There is a strong correlation between diets rich in saturated fats and incidence of the disease. This correlation has led to a concern over the relative amounts of saturated and unsaturated fats in our diets. Most nutritionists are advising us to use olive oil and canola oil (rape seed oil) as our major source of dietary lipids. Olive oil and canola oil contain a high percentage of monounsaturated fatty acids, which have been shown to lower LDL cholesterol (see Section 20.10).

20.4 Properties of Fats and Oils

Butter is prepared from milk by churning its cream, a process that causes the fat globules to coalesce; the liquid that remains is called buttermilk.

As previously mentioned, the triglycerides may be either liquids (oils) or noncrystalline solids (fats) at room temperature. Contrary to popular belief, *pure* fats and oils are colorless, odorless, and tasteless. The characteristic colors, odors, and flavors associated with these lipids are imparted to them by foreign substances that have been absorbed by the lipids and are soluble in them. For example, the yellow color of butter is due to the presence of the pigment carotene; the taste of butter is a result of two compounds, diacetyl ($CH_3COCOCH_3$) and 3-hydroxy–2-butanone ($CH_3COCHOHCH_3$), produced by bacteria in the ripening of the cream. Fats and oils are lighter than water, having densities of about 0.8 g/cm^3. They are poor conductors of heat and electricity and therefore serve as excellent insulators for the body.

Fats and oils undergo a variety of chemical reactions; the most important is hydrolysis. Triglycerides are esters. They can be hydrolyzed in either acidic or basic media. Acid hydrolysis is of little importance, however, because it is difficult to dissolve fats in acidic media. Basic hydrolysis is of considerable importance in the making of soap and is discussed in Section 20.5. When we eat fats, they are hydrolyzed by enzymes (lipases) in our bodies. This process is discussed in Chapter 26.

The degree of unsaturation of a fat or an oil is usually measured in terms of the **iodine number.** Recall that chlorine and bromine add readily to carbon–carbon double bonds (Section 13.13). Iodine also adds, but less readily:

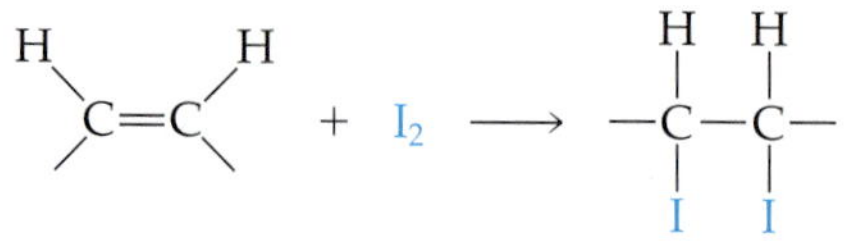

The iodine number of a fat or oil is the number of grams of iodine that react with 100 g of fat or oil. The more double bonds in a lipid, the more iodine is required for the addition reaction; thus, a high iodine number means a high degree of unsaturation. Representative iodine numbers are given in Table 20.3. Notice the generally lower values for the animal fats (butter, tallow, lard) compared with those for the vegetable oils.

Animal fats, as mentioned earlier, have been implicated in raising cholesterol and clogging arteries, leading to cardiovascular disease. For this reason, food processors often want vegetable oils available in solid form. A large-scale commercial industry has been developed for the purpose of transforming vegetable oils into edible fats. The chemistry of this conversion process is essentially identical to the catalytic hydrogenation reaction described for alkenes in Section 13.13:

$$ \underset{H}{\overset{H}{\diagup}}C=C\underset{\diagdown}{\overset{H}{\diagup}} \quad + \quad H_2 \quad \xrightarrow{\text{Ni}} \quad -\overset{\overset{\displaystyle H}{|}}{\underset{\underset{\displaystyle H}{|}}{C}}-\overset{\overset{\displaystyle H}{|}}{\underset{\underset{\displaystyle H}{|}}{C}}- $$

Margarine, a butter substitute, and vegetable shortening, a lard substitute, consist of vegetable oils that have been partially hydrogenated. (If all the bonds were hydrogenated, the product would become hard and brittle like tallow.) If reaction conditions are properly controlled, it is possible to prepare a fat with a desirable physical consistency (soft and pliable). In this manner, inexpensive and abundant vegetable oils (cottonseed, corn, soybean) are converted into oleomargarine and cooking fats (Crisco, for example). The consumer would get much greater unsaturation by using the oils directly, but most people would rather spread margarine than pour oil on their toast. Table 20.4 lists iodine numbers for butter and various kinds of margarines.

In the preparation of margarine, the partially hydrogenated oils are mixed with water, salt, and nonfat dry milk. Flavoring agents, coloring agents, and vitamins A and D are added to approximate butter. (Preservatives and antioxidants are also added.) The peanut oil in peanut butter has been partially hydrogenated to prevent the oil from separating out. Today, because of the possible connection between saturated fats and arterial disease (see Section 20.10), many people are cooking with the vegetable oils (especially olive oil) rather than with the hydrogenated products.

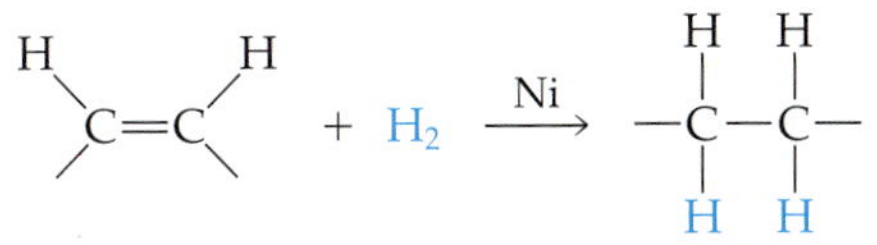

Table 20.3 Typical Iodine Numbers for Some Fats and Oils[a]

Fat or Oil	Iodine Number
Coconut oil	8–10
Butter	25–40
Beef tallow	30–45
Palm oil	37–54
Lard	45–70
Olive oil	75–95
Peanut oil	85–100
Cottonseed oil	100–117
Corn oil	115–130
Fish oils	120–180
Soybean oil	125–140
Safflower oil	130–140
Sunflower oil	130–145
Linseed oil	170–205

[a]Most oils are from plant sources. Three fats and one oil (in color) come from animals.

During the hydrogenation of vegetable oils, an isomerization reaction produces some *trans* fatty acids (recall Figure 20.2c). Recent studies showed that these trans acids, like saturated fatty acids, raise cholesterol levels and increase the incidence of coronary heart disease. As a result, consumers are now being advised to use soft or liquid margarine and to reduce their total fat consumption.

Table 20.4 Iodine Numbers and Comparative Unsaturation Ratings of Butter and Various Margarines

Food Product	Iodine Number	Comparative Unsaturation[a]
Butter	27	100
Margarines		
Hard type A	68	252
Hard type B	72	267
Hard type C	77	285
Soft type D	84	311
Soft type E	88	326
Liquid type F	90	333
Liquid type G	93	344

[a] Calculated by dividing the iodine number of the substance by the iodine number of butter and multiplying the result by 100.

Hydrogenation of an oil. A solid vegetable fat like Crisco is made by bubbling hydrogen through a vegetable oil in the presence of a catalyst such as nickel metal.

On standing at room temperature in contact with moist air, fats and oils soon turn rancid. This rancidity, characterized by a disagreeable odor, results from two reactions. Hydrolysis of the ester bonds releases volatile fatty acids. Butter, for example, yields foul-smelling butyric, caprylic, and capric acids. Microorganisms present in the air furnish the lipases that catalyze the process. Hydrolytic rancidity can easily be prevented by storing a fat or oil covered in a refrigerator.

Oxidation of the unsaturated fatty acid components by oxygen in the air also produces a variety of volatile, odorous compounds. The structural unit

$$\sim CH{=}CH{-}CH_2{-}CH{=}CH\sim$$

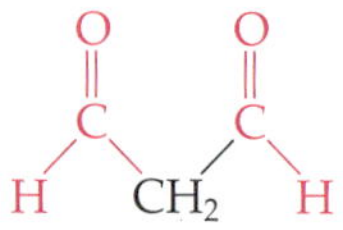

Malonaldehyde

in linoleic and linolenic acids is readily oxidized. One particularly offensive product, formed by the cleavage of both double bonds, is a compound called malonaldehyde. The stale, sweaty odor of the unwashed skin results when fats and oils excreted by the body are oxidized to C_5–C_{10} carboxylic acids (recall Section 16.4).

Rancidity is a major concern of the food industry, and food chemists are always seeking new and better substances to act as **antioxidants.** Such compounds are added in very small amounts (0.001 to 0.01%) to suppress rancidity. They have a greater affinity for oxygen than the lipids to which they are added and thus function by preferentially depleting the supply of absorbed oxygen. We shall see in Special Topic J that two of the vitamins (C and E) have antioxidant properties.

20.5 Soaps

Animal fats are available in large quantities as a by-product of the meat-packing industry. The fatty acids and many other long-chain organic compounds are derived from these fats. The most important derivatives are soaps. The reaction that converts animal fats to soaps is called, logically, **saponification,** which we discussed originally as a reaction of esters (Section 16.9). Soap-making is one of the oldest organic syntheses known, second only to the production of ethyl alcohol (fermentation). Even though cave people, the Phoenicians (600 B.C.E.), and the Romans made soap from animal fat and wood ash, the widespread production of soap did not begin until the 1700s.

The old method of soap production consisted of treating molten tallow with a slight excess of alkali in large open vats. The mixture was heated, and steam was bubbled through it. After saponification was completed, the soap was precipitated by the addition of sodium chloride, then filtered and washed several times with water. It was then dissolved in water and reprecipitated by the addition of more sodium chloride. The glycerol was recovered from the aqueous wash solutions.

Today most soaps are prepared by a continuous process wherein triglycerides (frequently tallow and/or coconut oil) are hydrolyzed by water under high pressures and temperatures (700 lb/in.^{2} and 200 °C). Sodium carbonate is used to neutralize the fatty acids:

$$\begin{array}{l} CH_2OOC(CH_2)_nCH_3 \\ | \\ CHOOC(CH_2)_nCH_3 \\ | \\ CH_2OOC(CH_2)_nCH_3 \end{array} \xrightarrow[\text{heat, pressure}]{H_2O} \text{Glycerol} + \underset{\text{Fatty acid}}{3\,CH_3(CH_2)_nCOOH} \xrightarrow{Na_2CO_3} \underset{\substack{\text{Sodium salt of a fatty acid} \\ \text{(a soap)}}}{3\,CH_3(CH_2)_nCOO^-Na^+}$$

The crude soap is used as industrial soap without further processing. Pumice or sand may be added to produce scouring soap. Other ingredients, such as dyes and perfumes are added to produce colored soaps and fragrant soaps, respectively. If air is blown through molten soap, a floating soap is produced. Such a soap is not necessarily purer than other soaps; it merely contains more air. Ordinary soap is a mixture of the sodium salts of various fatty acids. Potassium soaps (soft soap) are more expensive but produce a finer lather and are more soluble. They are used in liquid soaps, shampoos, and shaving creams. Tincture of green soap is an alcoholic solution of a potassium soap that is commonly used in hospitals.

Many soaps claim to have deodorant action, but few have any active deodorant other than the soap itself.

Dirt and grime usually adhere to skin, clothing, and other surfaces because they are combined with body oils, cooking fats, lubricating greases, and a variety of similar substances that act like glues. Because oils are not miscible with water, washing with water alone does little good.

Soap molecules have a dual nature. One end is ionic and dissolves in water; the other end is like a hydrocarbon and dissolves in oils. Often, the ionic end is referred to as **hydrophilic** (water-soluble) and the nonpolar end as **hydrophobic** (repelled by water). We can illustrate the cleansing action of soap schematically (Figure 20.3). The hydrocarbon "tails" dissolve in the oil; the ionic "heads" remain in the aqueous phase. In this manner, the oil is broken into tiny droplets called *micelles* (Section 20.8) and dispersed throughout the solution. The droplets don't coalesce because of the repulsion of the charged groups (the carboxyl anions) on their surfaces. The oil and water thus form an emulsion, with soap acting as the emulsifying agent. With the oil no longer "gluing" it to the surface (skin, cloth, dish), the dirt can be easily removed by the water.

The suffix *phil* means to love, and the suffix *phob* means to fear or hate. Hydrophobic substances are sometimes said to be *lipophilic.*

Soaps work by enabling grease and oil to mix with water and thus to be removed when the soap solution is rinsed away.

For cleaning clothes and for many other purposes, soap has been largely replaced by synthetic detergents because soaps have two rather serious shortcomings. One is that in acidic solutions, soaps are converted to free fatty acids:

$$\underset{\text{A soap}}{CH_3(CH_2)_{16}COO^-Na^+} + H^+ \longrightarrow \underset{\text{A fatty acid}}{CH_3(CH_2)_{16}COOH} + Na^+$$

The fatty acids, unlike soap, don't have an ionic end. Lacking the necessary dual nature, they can't emulsify the oil and dirt; that is, they do not exhibit any cleaning action. What is more, these fatty acids are insoluble in water and separate out as a greasy scum. To counteract this lack of cleaning action in acidic solution, various alkaline substances are added to laundry soap formulations to keep the pH high. These basic compounds include carbonates and silicates.

(a)

(b)

FIGURE 20.3 ▶

Cleaning action of soap visualized. (a) Sodium palmitate, $CH_3(CH_2)_{14}COO^-Na^+$, a typical soap. (b) A microscopic view of soap action. In an oil droplet suspended in water, the hydrocarbon "tails" of the soap molecules are immersed in the oil and the ionic ends extend into the water. The attractive forces between these ionic "heads" and the water molecules cause the oil droplet to be suspended in the water.

The second and more serious disadvantage of soap is that it doesn't work well in hard water. Hard water is water that contains certain metal ions, particularly magnesium, calcium, and iron ions. The soap anions react with these metal ions to form greasy, insoluble curds:

$$\underset{\text{Soap (soluble)}}{2\,CH_3(CH_2)_{16}COO^-Na^+} + Ca^{2+} \longrightarrow \underset{\text{Bathtub ring (insoluble)}}{(CH_3(CH_2)_{16}COO^-)_2\,Ca^{2+}} + 2\,Na^+$$

Emulsions

When violently shaken with water, fats are broken into tiny (submicroscopic) particles and dispersed throughout the water. Such a mixture is called an **emulsion.** Unless a third substance has been added, the emulsion breaks down rapidly. The droplets then recombine and float to the surface of the water. Soap, certain types of gum, and protein can stabilize the emulsion by forming protective coatings on the fat droplets that prevent them from coming together.

Many foods are emulsions. Milk is an emulsion of butterfat in water. The stabilizing agent in milk is a protein called casein. Mayonnaise is an emulsion of salad oil in water, stabilized by egg yolk.

These deposits make up the familiar "bathtub ring." They leave freshly washed hair sticky and are responsible for the "telltale gray" of the family wash.

The term **detergent** is a general one meaning any cleansing agent. Even though soaps fall under this broad definition, the popular use of the word generally refers to *synthetic detergents*, also called *syndets*. Syndets have the desirable property of not forming precipitates with the ions of hard water. There are close to a thousand synthetic detergents commercially available in the United States, and worldwide production exceeds 25 million tons. Although most synthetic detergents are now made from petroleum, early ones were made from fats. Some, such as sodium lauryl sulfate (sodium dodecyl sulfate), still are. Sodium lauryl sulfate is widely used in specialty products such as toothpastes and shampoos. Sodium dodecylbenzenesulfonate is found in many laundry detergents.

$$CH_3(CH_2)_{11}O-\overset{\overset{\displaystyle O}{\|}}{\underset{\underset{\displaystyle O}{\|}}{S}}-O^-Na^+$$

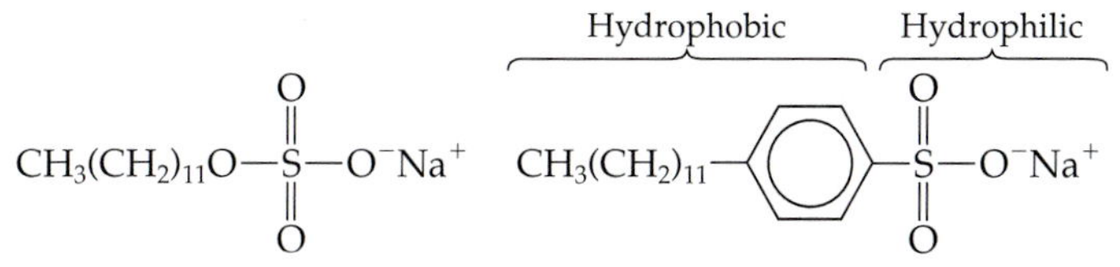

Sodium lauryl sulfate
(in shampoos)

Hydrophobic Hydrophilic

$$CH_3(CH_2)_{11}-\bigcirc-\overset{\overset{\displaystyle O}{\|}}{\underset{\underset{\displaystyle O}{\|}}{S}}-O^-Na^+$$

Sodium dodecylbenzenesulfonate
(in laundry detergents)

20.6 Waxes

Waxes are esters formed from long-chain fatty acids and long-chain monohydroxy alcohols. (Household paraffin wax, which is a mixture of high-molar-mass hydrocarbons, has waxlike properties but is not a wax according to our definition.) The general formula for a wax, then, is the same as that of a simple ester. For a wax, however, R and R′ are limited to alkyl groups containing large numbers of carbon atoms.

$$R-\overset{\overset{\displaystyle O}{\|}}{C}-O-R'$$

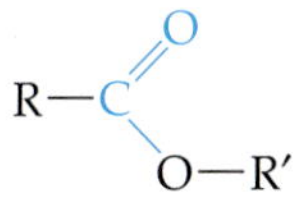

A wax
(a simple ester)

Most natural waxes are mixtures of such esters. Many also contain free alcohols, hydrocarbons, and esters of diprotic acids, hydroxy acids, and diols. All have similar properties; they feel "waxy," are insoluble in water, and melt at temperatures above body temperature (37 °C) and below the boiling point of water (100 °C).

Waxes are not as easily hydrolyzed as triglycerides and therefore are useful as protective coatings. Plant waxes on the surfaces of leaves, stems, flowers, and fruits protect the plant from dehydration and from invasion by harmful microorganisms. Carnauba wax, largely myricyl cerotate ($C_{25}H_{51}COOC_{30}H_{61}$), is obtained from the leaves of certain Brazilian palm trees and is used extensively in floor waxes, automobile waxes, and furniture polish.

Animal waxes also serve as protective coatings. They are found on the surfaces of feathers, skin, and hair and help to keep these surfaces pliable and water-repellent. Earwax, for example, protects the delicate lining of the inner ear. The waxy coating on the feathers of water birds (ducks, gulls) helps them to stay afloat. If this wax is dissolved as a result of the bird swimming in an oil slick, the feathers become wet and heavy; the bird cannot maintain its buoyancy and will drown.

Beeswax is the material from which bees construct honeycombs. Upon saponification, beeswax yields alcohols and fatty acid salts. Figure 20.4 shows some of the typical molecules found in beeswax. It is used in such household products as candles and shoe polish.

$$CH_3(CH_2)_{34}\overset{O}{\underset{O-(CH_2)_{35}CH_3}{C}} \qquad CH_3(CH_2)_{24}\overset{O}{\underset{O-(CH_2)_{29}CH_3}{C}}$$

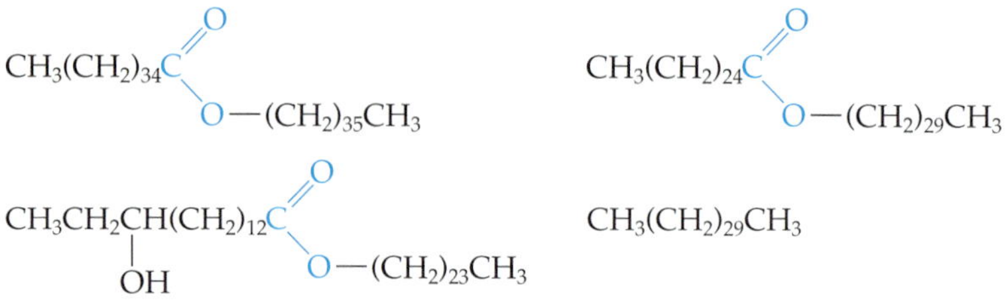

FIGURE 20.4 ▶

Three esters and an alkane found in beeswax.

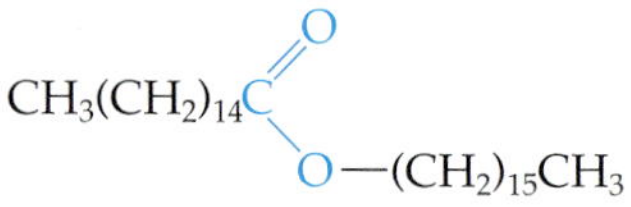

$$CH_3CH_2CH(CH_2)_{12}\overset{O}{\underset{\underset{OH}{|}}{C}}{}_{O-(CH_2)_{23}CH_3} \qquad CH_3(CH_2)_{29}CH_3$$

Commercial operations that harvest beeswax can always leave behind enough honey for the bees to live on. Obtaining the wax called spermaceti, however, is a little harder on the creature that produces it. Spermaceti (mp 42 to 50 °C) crystallizes when oil from the head of the sperm whale is cooled, and so whales must be killed before the product can be obtained. The principal constituent of spermaceti is cetyl palmitate. Esters of lauric, myristic, and palmitic acids are also present, as are esters of higher alcohols. Whales and other marine species store these waxes as metabolic fuels. Once widely used in ointments, cosmetics, soaps, and candles, spermaceti is now in short supply because the sperm whale has been hunted almost to extinction.

$$CH_3(CH_2)_{14}\overset{O}{\underset{O-(CH_2)_{15}CH_3}{C}}$$

Cetyl palmitate

Lanolin is a wax from sheep's wool. It is a mixture of esters and polyesters of 33 alcohols and 36 fatty acids. Some of the alcohols are steroids (similar to cholesterol). Lanolin is used as a base for ointments and cosmetic lotions.

Many natural waxes have been replaced by synthetic materials, mainly polymers. By careful control of the molecular weights of these polymers, the properties of natural waxes can be closely duplicated. For example, Carbowax, a polymer of ethylene glycol ($HOCH_2CH_2OH$), is available in a wide range of average molecular weights. Synthetic waxes are used in cosmetics and ointments and in certain industrial processes.

20.7 Phospholipids and Glycolipids

Phospholipids are phosphorus-containing lipids and are found in all living organisms. They are particularly abundant in the membranes surrounding individual cells and in the membranes surrounding certain organelles within the cell. Phospholipids are the most polar of lipids, and they contain both hydrophilic and hydrophobic groups. This characteristic property is important in the structure of membranes, as we shall see in Section 20.8.

Glycolipids are sugar-containing lipids that are confined to the outer surface of the cell membrane. Glycolipids provide cells with distinguishing surface markers that may serve in cellular recognition and cell-to-cell communication.

Phosphatides

Recall that glycerol is a trihydroxy alcohol and so combines with acids to form esters. When all three acids are fatty acids, the esters formed are triglycerides. But glycerol can also form esters with inorganic acids such as phosphoric acid. The **phosphatides,** a most important class of phospholipids, are esters of glycerol in which there are two fatty acid groups and one phosphoric acid unit. Further, the phosphoric acid is also esterified with another alcohol molecule, usually an amino alcohol (Figure 20.5). Since these compounds are quite complicated structurally, it

$$CH_2O-\overset{\overset{\displaystyle O}{\|}}{C}-R$$

$$CH-O-\overset{\overset{\displaystyle O}{\|}}{C}-R'$$ Fatty acid units

Glycerol unit

$$CH_2O-\overset{\overset{\displaystyle O}{\|}}{\underset{\underset{\displaystyle O^-}{|}}{P}}-O-CH_2CH_2-\overset{+}{N}R_3$$

Amino alcohol unit

Phosphoric
acid unit

(a)

(b)
$$CH_2O-\text{Fatty acid 1}$$
$$CHO-\text{Fatty acid 2}$$
$$CH_2O-\text{Phosphate}-\text{Amino alcohol}$$

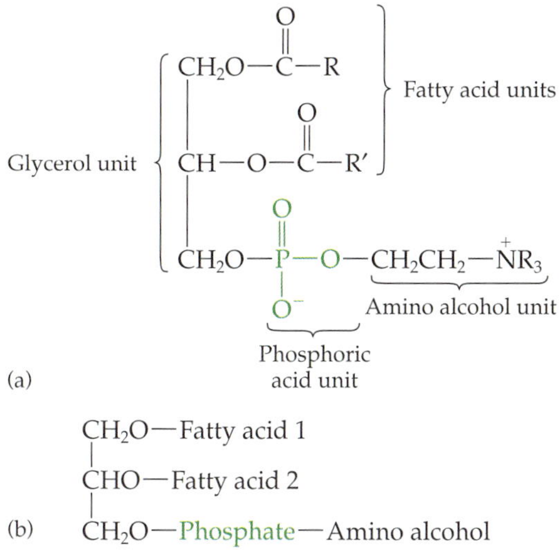

◀ **FIGURE 20.5**

(a) Structural formula and (b) schematic representation of a phosphatide.

is useful to relate their structure to those of other lipids we have studied. Notice that the phosphatide molecule is identical to a triglyceride up to the phosphoric acid part. Let's simplify things by representing those parts schematically (as in Figure 20.5b). We will then write out the structure of the amino alcohol to emphasize the parts that are different.

Two common phosphatides are shown in Figure 20.6. When the phosphatide contains the *ethanolamine* structural group (we've shown the amine group ionized), the compounds are called **cephalins.** The cephalins are found in brain tissue and nerves. They are also involved in blood clotting.

When choline is the amino alcohol unit, the compounds are called **lecithins.** Lecithins occur in all living organisms. They, too, are important constituents of nerve and brain tissue. Egg yolks are especially rich in lecithins. Commercial-grade lecithins are available from soybeans. These lecithins are widely used in foods as emulsifying agents. Many candy bars list lecithin among their ingredients.

$$HOCH_2CH_2\overset{+}{N}H_3$$

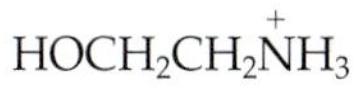

Ethanolamine
(2-Aminoethanol)

$$HOCH_2CH_2\overset{\overset{\displaystyle CH_3}{|}}{\underset{\underset{\displaystyle CH_3}{|}}{\overset{+}{N}}}-CH_3$$

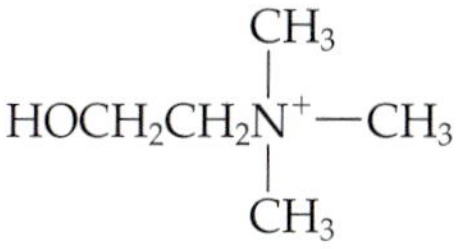

Choline

Sphingolipids

Like the phosphatides, **sphingolipids** contain a phosphoric acid unit. They are classified separately, however, because they are based on the unsaturated amino alcohol sphingosine rather than glycerol. **Sphingomyelins** are the "simplest"

$$CH_2O-\text{Fatty acid 1}$$
$$CHO-\text{Fatty acid 2}$$
$$CH_2O-\text{Phosphate}-OCH_2CH_2\overset{+}{N}H_3$$

Phosphatidylethanolamine
(a cephalin)

$$CH_2O-\text{Fatty acid 1}$$
$$CHO-\text{Fatty acid 2}$$
$$CH_2O-\text{Phosphate}-OCH_2CH_2-\overset{\overset{\displaystyle CH_3}{|}}{\underset{\underset{\displaystyle CH_3}{|}}{\overset{+}{N}}}-CH_3$$

Phosphatidylcholine
(a lecithin)

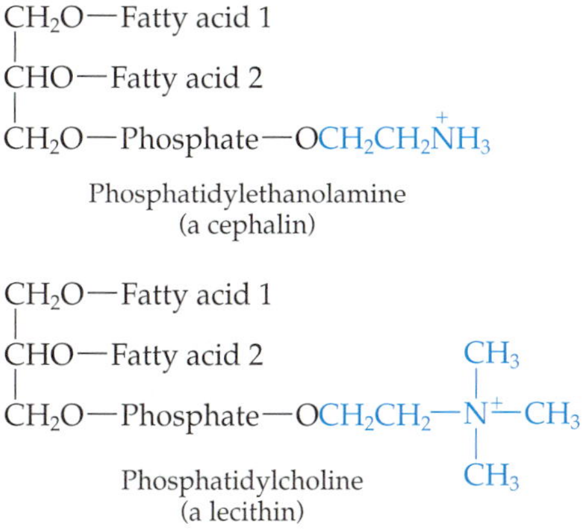

◀ **FIGURE 20.6**

Two common phosphatides.

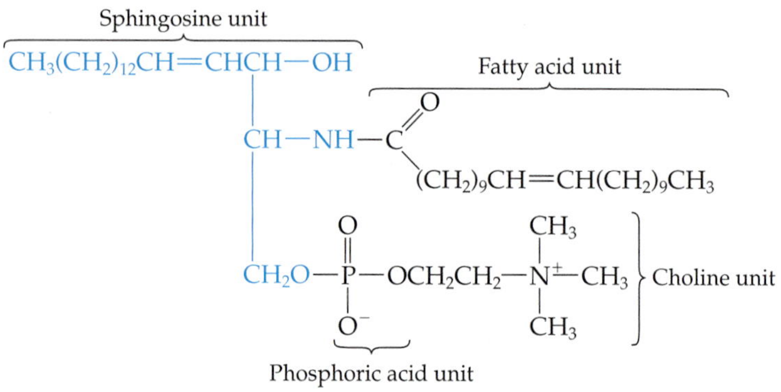

FIGURE 20.7 ▶

A sphingomyelin.

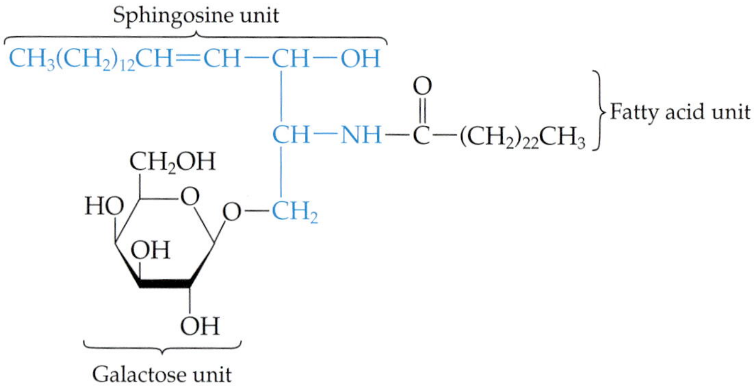

FIGURE 20.8 ▶

Cerebrosides are sphingolipids that contain a sugar unit.

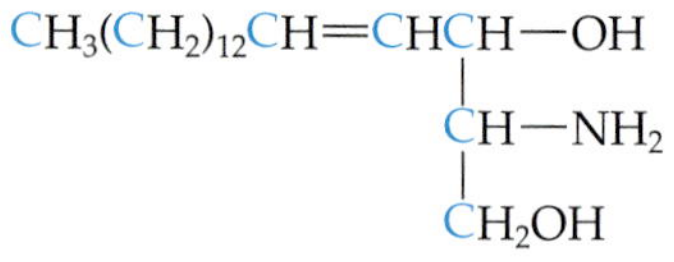

Sphingosine

Note: Both cerebrosides and gangliosides contain sugar groups in place of a phosphate group. They therefore are glycolipids.

sphingolipids, each containing fatty acid, phosphoric acid, sphingosine, and choline units (Figure 20.7). Sphingomyelins are important constituents of the myelin sheath that surrounds the axon of a nerve cell. Multiple sclerosis is one of several diseases related to a fault in the myelin sheath.

Most animal cells contain a group of sphingolipids called **cerebrosides** (Figure 20.8). Cerebrosides have a galactose unit, a fatty acid unit, and a sphingosine unit. The structure resembles that of a sphingomyelin, except that a sugar unit is connected where the sphingomyelin has a choline phosphate group. Cerebrosides are important constituents of the membranes of nerve and brain cells. They are believed to play a principal role in the membrane wrapping process that is unique to myelination. Gaucher's disease, a hereditary affliction, results from the substitution of glucose for galactose in the cerebroside. Large amounts of these abnormal cerebrosides accumulate, causing enlargement of the liver and the spleen.

Related compounds, called **gangliosides,** are also found in cell membranes. The structure of these compounds is even more complex than that of the cerebrosides. Gangliosides usually contain a branched chain of three to eight monosaccharides and/or substituted sugars. They are most prevalent in the outer membranes of nerve cells, although they also occur in smaller quantities in the outer membranes of most other cells. There is considerable variation in their sugar components, and about 130 varieties of gangliosides have been identified. It is the sequence of sugars that most often determines cell-to-cell recognition and communication (e.g., blood group antigens).

20.8 Cell Membranes

The components of a living cell are enclosed within a membrane. Both plant cells (Figure 20.9) and animal cells (Figure 20.10) have cell nuclei that contain nucleic acids (Chapter 23). Everything between the cell membrane and the nuclear membrane—including the fluids and a variety of subcellular components such as the mitochondria and ribosomes—is called the **cytoplasm.** We will encounter several components of the cytoplasm in following chapters. Note that plant cells have rigid *cell walls* that surround and protect the cell, but animal cells have only the cell membrane. Membranes of all cells have a similar structure, but membrane function varies tremendously from one organism to another and from one cell to another within a single organism. This diversity arises mainly from the different proteins and phospholipids in the membrane.

When polar lipids such as phospholipids and glycolipids are placed in water, they disperse and form clusters of molecules in any one of three arrangements: *micelles, monolayers,* and *bilayers.* **Micelles** are aggregations of molecules that contain both polar and nonpolar groups. The hydrocarbon "tails" of these lipids, being hydrophobic, are directed inward away from the water; the hydrophilic "heads" are directed outward into the water. Each micelle may contain thousands of lipid molecules (Figure 20.11a). Polar lipids also form *monolayers,* which are layers one molecule thick on the surface of water (Figure 20.11b). The polar heads stick into the water, and the nonpolar tails stick up into the air. *Bilayers* are layers that have hydrophobic tails sandwiched between hydrophilic heads sticking outward into the water (Figure 20.11c). *It is such layers that make up every cell membrane.*

The three major classes of lipid molecules in the membrane bilayer are phospholipids, glycolipids (about 5%) and cholesterol (see Section 20.9). Roughly equal

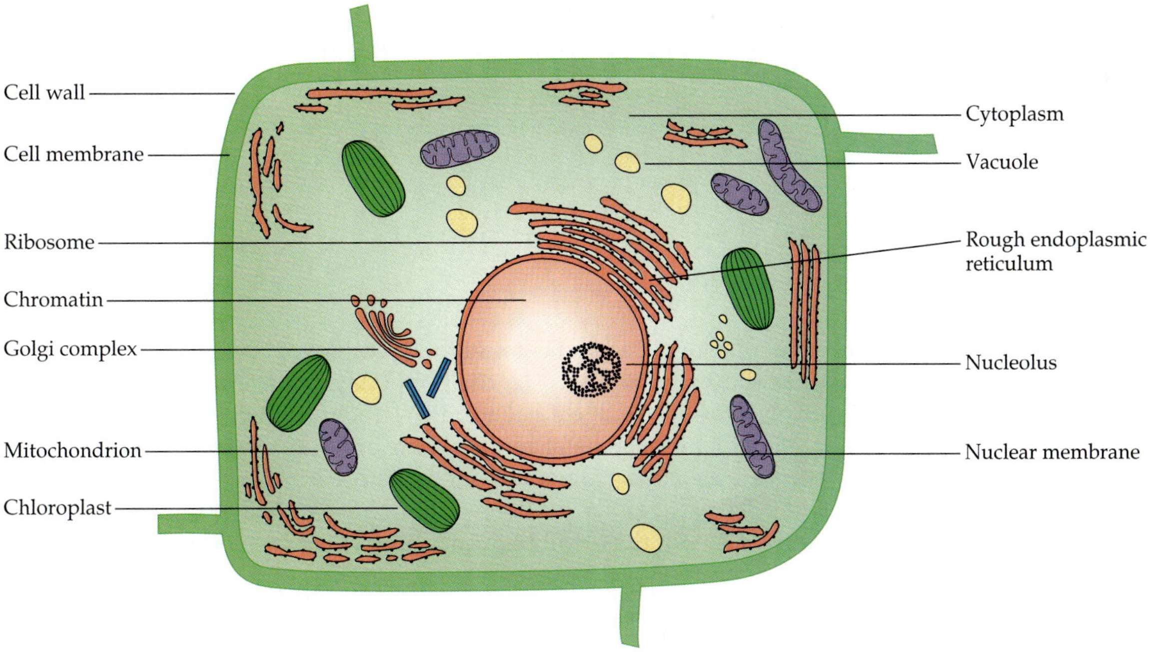

▲ **FIGURE 20.9**
An idealized plant cell. Not all the structures shown here occur in every type of plant cell.

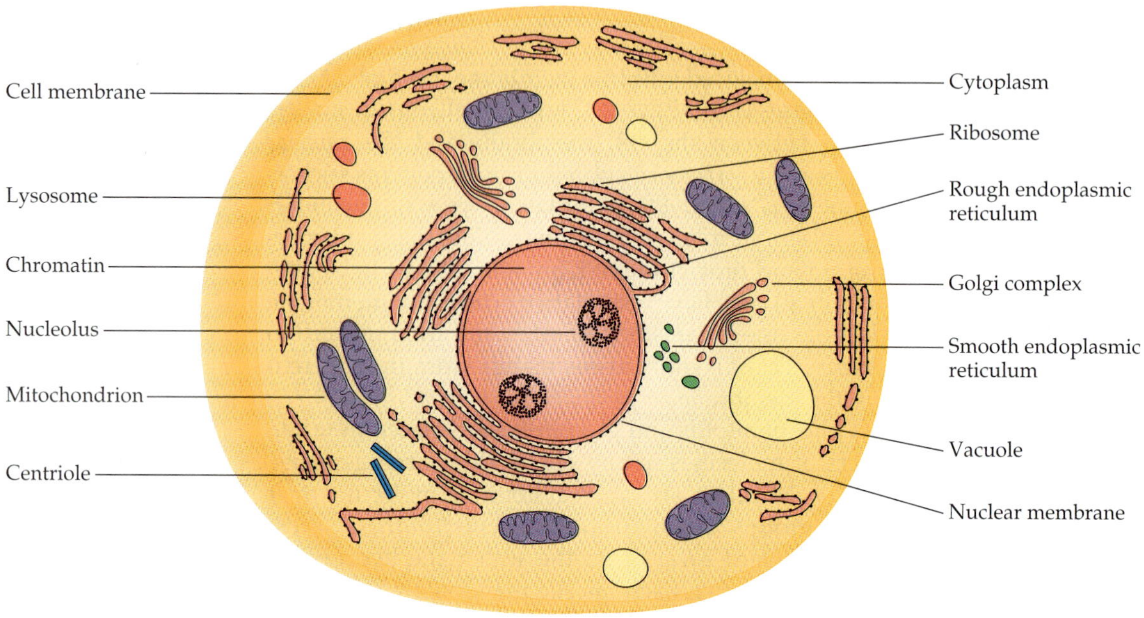

▲ **FIGURE 20.10**

An idealized animal cell. The entire range of structures shown here seldom occurs in a single animal cell. Each kind of tissue has cells specific to the function of that tissue. Muscle cells are different from neurons and neurons differ from red blood cells, and so on.

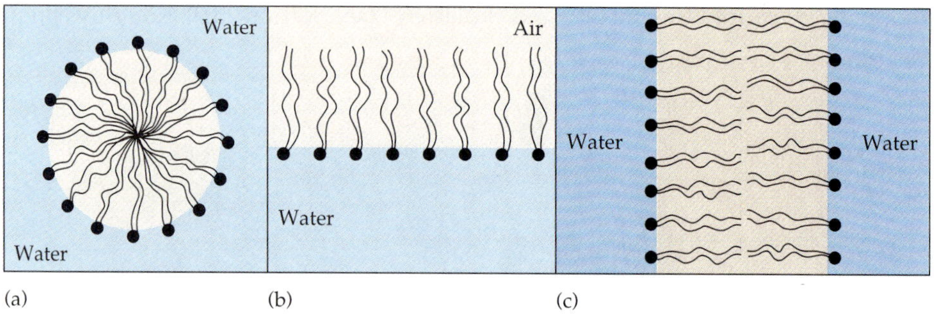

FIGURE 20.11 ►

(a) Micelle, (b) monolayer, and (c) bilayer formed when polar lipids are added to water.

numbers of phospholipid and cholesterol molecules are present in the membranes of cells. As shown in Figure 20.12, the lipid bilayer consists of two rows of phospholipid molecules arranged tail to tail. The hydrophobic tails (the fatty acid portions) interact by means of hydrophobic and dispersion forces.[1] At the same time, they are isolated from the aqueous environment that exists within and outside the cells. The polar portions of the phospholipids project from the inner and outer surfaces of the membrane and interact with water molecules.

[1] This interaction is relatively weak because of the presence of unsaturated (cis-branching) fatty acids. As a result, the lipid portion is not rigid but is quite fluid and allows movement within the membrane. Cholesterol is also a key moderator of membrane fluidity. It breaks up the hydrophobic and dispersion forces of the fatty acid chains by fitting between these chains.

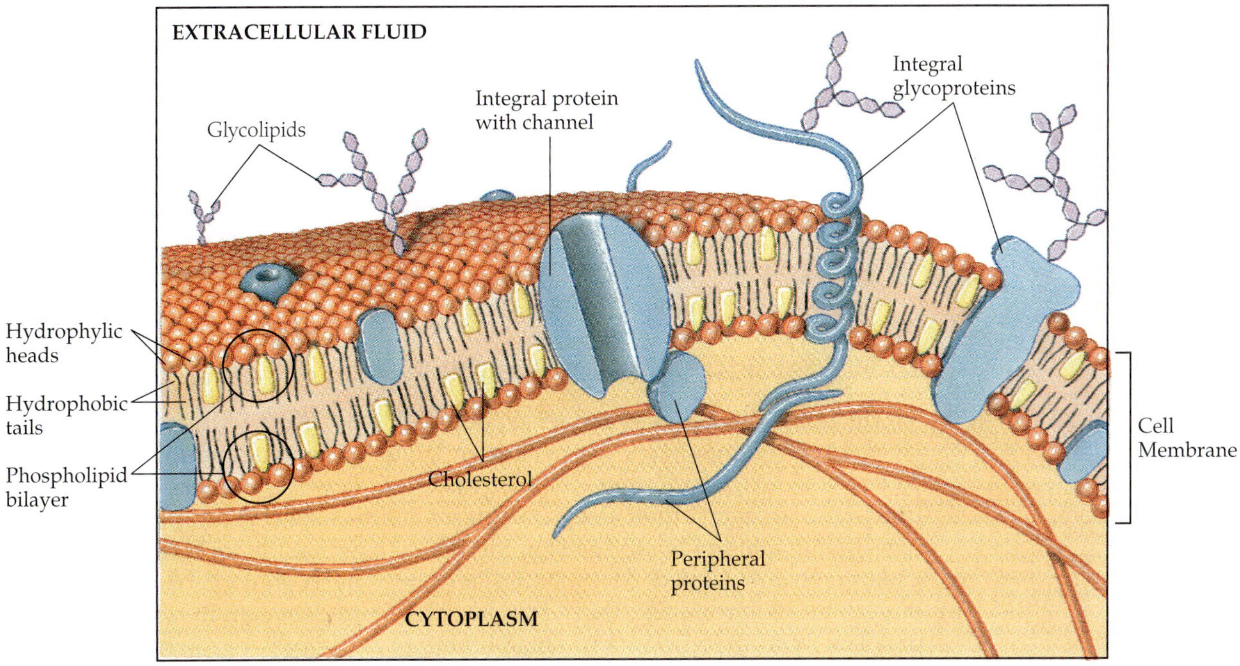

▲ **FIGURE 20.12**

Schematic diagram of a cell membrane. The membrane is a phospholipid bilayer with embedded cholesterol and protein molecules. Short polysaccharide chains are attached to the outer surface.

Biological membranes were once viewed as inert barriers that just served to contain the cell and the organelles (nucleus, mitochondrion, Golgi apparatus, etc.) within the cell. Now, however, we recognize membranes as receptors of external stimuli and as active participants in the process by which materials are imported into and exported from cells. Most cell membranes are composed of about 50% (by mass) lipid and 50% protein, although large variations from these percentages can exist in certain cells.[2] The membranes are referred to as *semipermeable* because only certain materials are allowed to pass from one side to the other. For example, lipid-soluble molecules can cross the membrane by diffusing through the phospholipid bilayer; large molecules that are not lipid-soluble cannot diffuse through the membrane.

If membranes were composed only of lipids, they would act as barriers to the passage of ions or polar molecules. Instead, the passage of polar species across the membrane is facilitated by proteins that move about in the "sea" of lipids. There

Cell membranes become quite rigid at low temperatures as their constituent phospholipids tend to solidify. Warm-blooded animals have no problem with this; they maintain membrane fluidity simply by maintaining body temperature. Membrane rigidity does present a problem, however, to cold-blooded animals that stay active in cold weather. To adapt to the cold, some of these cold-blooded animals are able to make their cell membranes more fluid by increasing the degree of unsaturation of membrane phospholipids, thus lowering the solidification temperature.

[2] The percentage of each component of the membrane is related to the function of the particular tissue. For example, the myelin sheath of nerve cells can contain up to 80% lipid because the major function of the membrane is insulation and protection. The inner mitochondrial membrane, on the other hand, is unique in having a large protein component (75 to 80%). The proteins (enzymes) within the membrane play an integral role in the energy-conversion function of the mitochondria (see Section 25.2). It is important to emphasize that these are percents by *mass*. Lipid molecules are much smaller than proteins; hence there are always more lipid than protein molecules. In an average membrane there are 50 lipid molecules for one protein molecule.

are two classes of proteins in the cell membrane. **Integral proteins** span the hydrophobic interior of the bilayer, whereas **peripheral proteins** are more loosely associated with the membrane surface. Peripheral proteins appear to be attached to integral proteins by hydrogen bonds and electrostatic forces.

Small ions and water-soluble molecules enter and leave the cell by way of channels through the integral proteins. It appears that there are special carrier proteins to facilitate the passage of certain molecules (e.g., hormones and neurotransmitters). A specific interaction occurs between the carrier protein and the molecule being transported.

20.9 Steroids: Cholesterol and Bile Salts

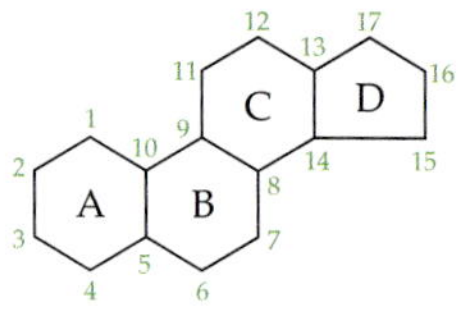

Perhydrocyclopentanophenanthrene (the steroid skeleton)

All the lipids discussed so far are saponifiable. They react with aqueous alkali to yield simpler components such as glycerol, fatty acids, amino alcohols, or sugars. Any lipids extracted from cellular material, however, contain a small but important fraction that does not react with alkali. The most important nonsaponifiable lipids are the steroids. These compounds include the bile salts, cholesterol and related compounds, hormones such as cortisone, and the all-important sex hormones. Those steroids with hormonal activity are discussed in detail in Special Topic I. Very small amounts of steroids or slight variations in structure or in the nature of substituent groups effect profound changes in biological activity.

More than 40 steroidal compounds have been found in nature. They occur in plant and animal tissues, yeasts, and molds, but not in bacteria, and may exist either free or combined with fatty acids or carbohydrates. All steroids have a perhydrocyclopentanophenanthrene ring system, which consists of a completely saturated phenanthrene moiety fused to a cyclopentane ring. The rings are designated by capital letters, and the carbon atoms are numbered as shown at the left.

The term *moiety* means part or portion. In chemistry it is used to denote a particular group within a large molecule.

Cholesterol does not occur in plants, but it is the best known and most abundant (about 240 g) steroid in the human body. About one-half of the total body cholesterol is present in cell membranes interspersed among the phospholipid molecules (recall Figure 20.12). Much of the cholesterol in the body is converted to cholic acid, which is used in the formation of bile salts. Cholesterol is also an important precursor in the biosynthesis of the sex hormones, adrenal hormones, and vitamin D. Excess cholesterol not utilized by the body is released from the liver and transported by the blood to the gallbladder. Normally, it stays in solution and is secreted into the intestine (in the bile) to be eliminated. Sometimes the cholesterol precipitates in the gallbladder, producing gallstones. Its name is derived from this source (Greek *chole,* bile; *stereos,* solid).

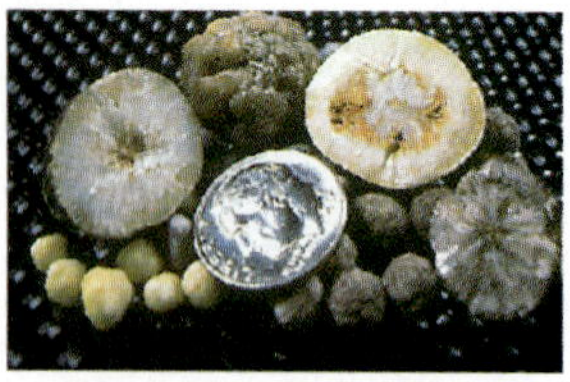

An assortment of human gallstones, which are nearly pure cholesterol. (The dime gives an indication of relative sizes.)

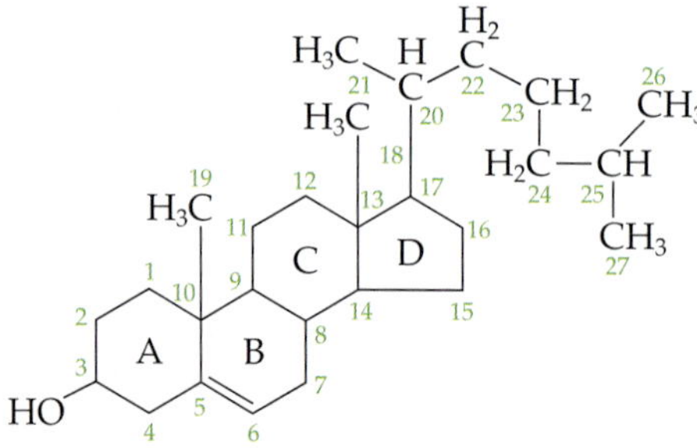

Cholesterol

Most meats and foods derived from animal products such as eggs, butter, cheese, and cream are particularly rich in cholesterol. The average American consumes about 600 mg of cholesterol each day. The liver is the major site of cholesterol biosynthesis, although other tissues (intestines, adrenals, gonads) are also involved. The human body synthesizes about 1 g of cholesterol each day; all 27 carbon atoms are derived from acetyl CoA molecules (see Section 26.4). The plasma cholesterol level controls the synthesis of cholesterol by the liver. When the cholesterol level in the blood exceeds 150 mg/100 mL, the rate of cholesterol biosynthesis is halved. Hence, if cholesterol is present in the diet, a feedback mechanism suppresses its biosynthesis in the liver. However, this is not a 1:1 ratio. The reduction in biosynthesis does not equal the amount of cholesterol ingested. Fasting also inhibits cholesterol biosynthesis because of the limited availability of acetyl CoA. Conversely, diets high in carbohydrate or fat tend to accelerate cholesterol biosynthesis because they increase the amount of acetyl CoA in the liver. The lipids of fish and poultry contain relatively more unsaturated fatty acids than the lipids of beef, lamb, and pork. For this reason, fish (fish oils in particular) and poultry are recommended for people who wish to lower their blood cholesterol levels because of the suspected correlation between the cholesterol level in the blood and certain types of heart disease (see Section 20.10).

One large egg contains 213 mg of cholesterol. Most health authorities recommend a maximum of 300 mg/day of cholesterol in the diet.

The cholesterol content of blood varies considerably with age, diet, and sex. Young adults average about 170 mg of cholesterol per 100 mL of blood, whereas males at age 55 may have 250 mg/100mL or higher (because the rate of cholesterol metabolism decreases with age). Females tend to have lower blood cholesterol levels than males.

Bile is a yellowish green liquid having a pH of 7.8 to 8.6; its composition is shown in Table 20.5. About 500 mL of bile is secreted each day by the gallbladder. Bile serves as a route for the excretion of drugs, of end products from hemoglobin breakdown (see Section 28.9), and of heavy metal ions.

Table 20.5 Composition of Bile

Component	Percent
Water	97
Bile salts	0.7
Inorganic salts	0.7
Bile pigments	0.2
Fatty acids	0.15
Lecithin	0.1
Fat	0.1
Cholesterol	0.06

Bile salts are the most important constituents of bile, and their major function is to aid in the digestion of dietary lipids. The bile salts are sodium salts of amide-like combinations of bile acids and glycine or the rare amino acid taurine ($H_2NCH_2CH_2SO_3^-$). They are synthesized from cholesterol in the liver, stored in the gallbladder, and then secreted in bile into the small intestine. Bile salts are highly effective detergents because they contain both hydrophobic and hydrophilic groups. Thus, they act as emulsifying agents—they break down large fat globules into smaller ones and keep these smaller globules suspended in the aqueous digestive environment (see Section 26.2). Enzymes can then hydrolyze the fat molecules more effectively. Bile salts also aid in the absorption of fatty acids, cholesterol, and the fat-soluble vitamins by forming complexes (micelles) that can diffuse into cells lining the intestines.

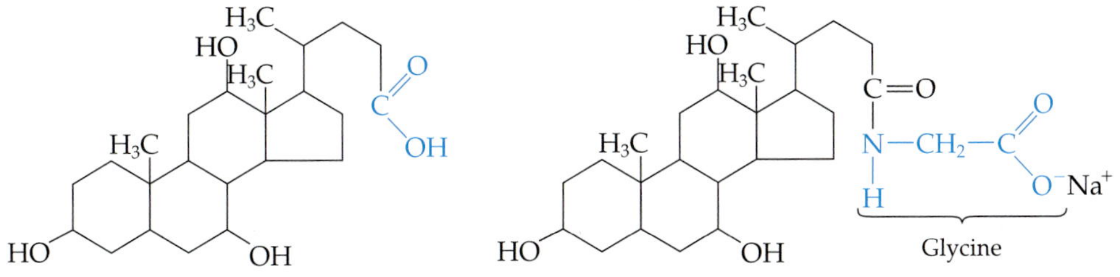

Cholic acid (a bile acid)

Sodium glycocholate (a bile salt)

20.10 Cholesterol and Cardiovascular Disease

In the past 30 years, few subjects in the nutrition field have attracted as much public attention as cholesterol. Today, everything from margarine and vegetable oils to egg substitutes and meat analogs is advertised on the basis that it contains little or no cholesterol. Cholesterol is believed to be a primary factor in the development of atherosclerosis, coronary heart disease, and stroke. As the leading cause of death in the United States, heart attack and stroke together take about 660,000 lives a year.

We get half of our total fats, three-fourths of all our saturated fats, and all of our cholesterol from animal products such as meat, milk, cheese, and eggs. Advertising claims that a vegetable oil contains no cholesterol are simply silly; *no* vegetable product contains cholesterol.

Scientists generally agree that elevated cholesterol levels in the blood, as well as high blood pressure and cigarette smoking, are associated in humans with an increased risk of heart attack. A long-term investigation by NIH showed that, among men aged 30 to 49, the incidence of coronary heart disease was five times greater for those whose cholesterol levels were above 260 mg/100 mL of serum than for those with cholesterol levels of 200 mg/100 mL or less.

Several alternative theories have been proposed to explain the cause of atherosclerosis. The most recent one suggests that defects in the lipid-transporting system are responsible for a buildup of lipids in the blood, which eventually triggers plaque formation. Because lipids such as cholesterol are not soluble in water, they cannot be transported in the blood (an aqueous medium) unless they are complexed with water-soluble proteins as lipoproteins. Lipoproteins generally are classified according to their density and composition. There are four broad categories: *chylomicrons* (density of less than 0.94 g/mL and made in the intestine), *very-low-density lipoproteins* (density of 0.94 to 1.006 and made mainly in the liver), *low-density lipoproteins* (1.006 to 1.063), and *high-density lipoproteins* (1.063 to 1.21). The density of lipoproteins is determined by the relative contents of protein and lipid. Since lipids are less dense than proteins, lipoproteins containing a greater amount of lipid are less dense than those containing a greater proportion of protein. The chylomicrons contain up to 99% by weight of lipids, and the very-low-density lipoproteins (VLDLs) contain up to about 90% lipids. The low-density lipoproteins (LDLs) contain about 80% lipids, and the high-density lipoproteins (HDLs) only about 50%. The protein component makes up the remainder of the lipoprotein molecule (e.g., HDLs are composed of 50% lipid and 50% protein—Table 20.6).

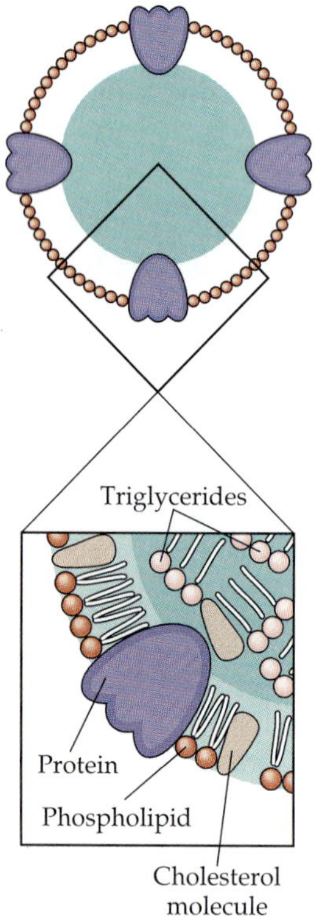

Plasma lipoproteins are roughly spherical. The surface consists largely of phospholipids, free cholesterol, and proteins. The core contains mostly triglycerides and cholesterol esters.

In research on cholesterol and its role in heart disease, the types of lipoproteins that have received the greatest attention in recent years have been the LDLs and HDLs. The reason is that they almost always contain a higher percentage of cholesterol than do the chylomicrons or the VLDLs, which serve to transport the triglycerides. One of the most fascinating discoveries in this field is that choles-

Table 20.6 Compositions of Lipoproteins Isolated from Normal Subjects

Lipoprotein Class	Density Range (g/mL)	Composition (wt %) Protein	Triglyceride	Cholesterol Free	Cholesterol Ester	Phospholipid
Chylomicrons	<0.94	1–2	85–95	1–3	2–4	3–6
VLDL	0.94–1.006	6–10	50–65	4–8	16–22	15–20
LDL	1.006–1.063	18–22	4–8	6–8	45–50	18–24
HDL	1.063–1.21	45–55	2–7	3–5	15–20	26–32

terol bound to HDLs reduces a person's risk of developing coronary heart disease. On the other hand, cholesterol bound to LDLs increases that risk. Notice from Table 20.6 that LDLs are the major cholesterol-carrying lipoproteins, whereas chylomicrons and VLDLs are the major carriers of triglycerides in the plasma.

Research evidence indicated that atherosclerosis and coronary heart disease are associated with elevated levels of serum LDLs, rather than with serum lipoproteins in general. It was also reported that the level of serum LDLs was a better predictor of coronary heart disease risk than the level of serum cholesterol. (The normal concentration of LDLs in humans is about 120 mg/100 mL of serum). Persons who, because of hereditary factors, have high levels of LDLs in their blood have a higher incidence of coronary heart disease.

Most serum cholesterol is transported as an LDL–cholesterol complex, which delivers the cholesterol directly to cells that need it. Low-density lipoproteins contain about 55% cholesterol, whereas high-density lipoproteins contain only about 25% cholesterol. LDLs are believed to promote coronary heart disease by first penetrating the coronary artery wall, where they are broken down enzymatically to cholesterol, cholesterol esters, and protein. The cholesterol and cholesterol esters are then oxidized in the artery wall, becoming major parts of the atherosclerotic plaque.

How do HDLs reduce the risk of developing coronary heart disease? No one knows for sure, but two theories have been suggested:

1. HDLs competitively inhibit the uptake of LDLs by cells by occupying, and thus blocking, the LDL receptor sites on the cell membranes.[3]
2. One role of HDLs is to transport excess cholesterol from the cells to the liver. Therefore HDLs aid in removing cholesterol from blood and the smooth muscle cells of the arterial wall and transporting it to the liver, where it can be metabolized.

Assuming that HDL helps to protect the body against coronary heart disease, what can be done to increase serum levels of this lipoprotein? One way is by sustained exercise (e.g., aerobic training). It was reported that marathon runners had decidedly higher HDL levels (~65 mg/100 mL) than did a group of sedentary men. Another way is to lose weight. It was shown that HDL levels can be increased by losing weight. Finally, we should mention a method that has received quite a bit of publicity. It was reported that HDL levels in the blood can be increased (to about 80 to 100 mg/100 mL) by drinking alcohol *in moderation.* The amount of alcohol used in this study was equivalent to about three 12-oz bottles of beer a day.

There is also statistical evidence that fish oils can prevent heart disease. Researchers at the University of Leiden in the Netherlands have found that Greenlanders, who eat a lot of fish, have a low risk of heart disease despite a diet

[3] LDL receptor proteins on the cell membranes are vital to the uptake of cholesterol from the blood. If these receptors are absent or deficient, the inherited condition called familial hypercholesterolemia results. Cholesterol levels in the blood rise dramatically, and the excess cholesterol that can't enter cells is deposited in certain tissues, particularly in skin and tendons, and in arterial plaques. Most individuals who are homozygous for the disorder have almost no LDL receptors, and they die of coronary artery disease in childhood. Somewhat less than 1% of the U.S. population is heterozygous; they have about half the number of receptors and often develop atherosclerosis by age 50.

Eicosapentaenoic acid and docosahexaenoic acid are referred to as omega-3 fatty acids because the endmost double bond is three carbons from the methyl end of the chain. (Omega, ω, is the last letter in the Greek alphabet, thus the omega position is the one farthest from the carboxyl group.)

high in total fat and cholesterol. The probable effective agents are the polyunsaturated fatty acids such as eicosapentaenoic acidand docosahexaenoic acid (so-called omega-3 fatty acids):

$$CH_3CH_2(CH{=}CHCH_2)_5CH_2CH_2COOH \qquad CH_3CH_2(CH{=}CHCH_2)_6CH_2COOH$$

ω-3-Eicosapentaenoic acid   ω-3-Docosahexaenoic acid

Other studies have shown that diets with added fish oil lead to lower cholesterol and triglyceride levels in the blood.

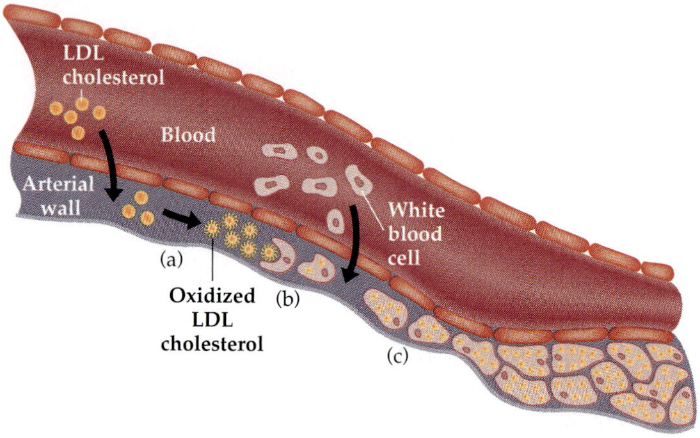

▲ **FIGURE 20.13**

The constriction of arteries. (a) LDL cholesterol accumulates and is oxidized. (b) The oxidation products, which can damage arterial tissue, attract white blood cells. (c) The white blood cells become engorged with the oxidation products and accumulate within the arterial wall, eventually narrowing the artery.

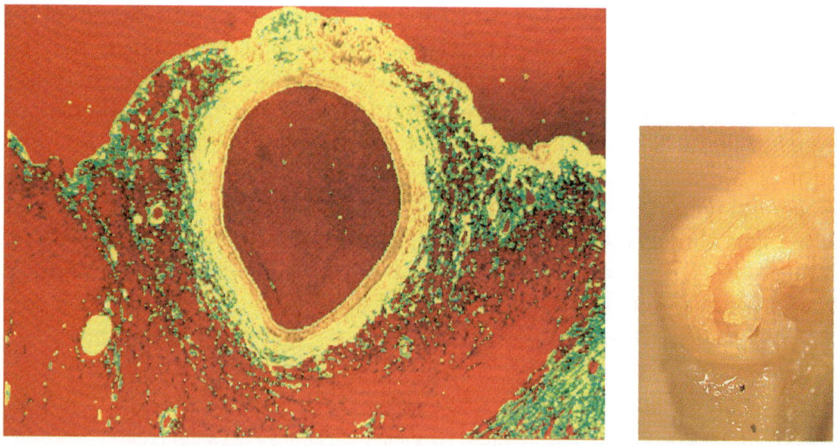

(a) (b)

▲ **FIGURE 20.14**

(a) Normal artery. (b) Photomicrograph of a cross section of an artery plugged with fatty deposits and a clot.

Atherosclerosis

Atherosclerosis causes a loss of elasticity in the large arteries and a thickening of the arterial walls. The thickening results from the deposition of LDL cholesterol in the cells that line the surface of arteries (Figure 20.13). When LDL cholesterol lodges in arterial walls, the cholesterol may become oxidized by free radicals. White blood cells migrate into the arterial cells and attempt to cleanse the cells by consuming the oxidation products. The enlarged white blood cells accumulate within the arterial lining, causing plaques that narrow the arteries. That tissue buildup and the resulting constriction reduce blood flow, diminish oxygen supply, and lead to high blood pressure. (Note that high blood pressure is also aggravated by such other factors as lack of exercise, obesity, heredity, stress, and smoking.)

Occasionally, the plaque ruptures through the lining into the interior of the artery. This rupture stimulates the blood platelets to initiate blood clots (see Section 28.8). These clots further obstruct the vessels, and may completely block the artery (Figure 20.14). Arterial clots are responsible for the most serious consequence of atherosclerosis: heart attack.

Heart attacks occur when one of the coronary arteries (arteries that supply the heart muscle itself) is blocked. If a blood clot suddenly breaks loose, it may be carried to a narrower part of the artery, obstructing blood flow. Deprived of nutrients and oxygen, the heart muscle once served by the blocked artery rapidly and painfully dies. If the area is small, the heart may be able to continue without it and the victim may recover. Death of large areas of heart muscle is almost instantly fatal. Although heart attacks are the major cause of death from atherosclerosis, cholesterol deposits and clots form in arteries throughout the body. A clot or a plaque deposit that obstructs an artery supplying the brain can cause a stroke, with the same results as if the artery had burst.

Heart disease kills more Americans than all other diseases combined. Every year, approximately 1.5 million people suffer heart attacks, and almost half of them die. Of those who survive, many are left with permanent damage, and a significant number have subsequent attacks. A heart attack usually does not happen suddenly. The body has an early warning system. The American Heart Association has compiled a list of early warning signals:

1. One of the first signs is pressure or pain in the middle of the chest. That's where the heart is, not on the left as many believe.
2. The pain can get worse and spread through the whole chest as well as down the left arm.
3. The pain may also spread to both arms, shoulders, neck, or jaw. A sensation of pressure, fullness, or squeezing may occur in the abdomen and is often mistaken for indigestion.
4. Pain may occur in any one or a combination of these areas at the same time. It could go away and return later. The pain is often accompanied by sweating, nausea, vomiting, and/or shortness of breath.

At the first sign of these symptoms, go to the nearest hospital emergency room at once! Get there as fast as possible. Have someone take you or dial 911 (where available) and tell the operator that there is a heart attack victim at your location. In the hospital, insist that it is an emergency—that you have chest pain (or other symptoms) and may be having a heart attack.

Summary

Lipids, found in the body tissue of all organisms, are compounds that are more soluble in organic solvents than in water. A **fatty acid** is a lipid molecule having a long hydrocarbon chain terminating in a carboxylic acid group. The lipids known as **fats** and **oils** are esters composed of three fatty acids joined to the trihydroxy alcohol glycerol:

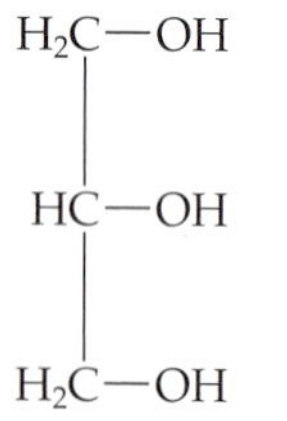

Glycerol Fat or oil

Both fats and oils are classified as triacylglycerols, or, more commonly, as **triglycerides.** Fats are triglycerides that are solid at room temperature, and oils are triglycerides that are liquid at room temperature. Fats are found mainly in animals, and lipids, mainly in plants.

A triglyceride molecule containing three like R-groups is a **simple triglyceride.** When the three R-groups are not identical, the molecule is a **mixed triglyceride.** *Saturated triglycerides* are those containing mainly saturated fatty-acid chains (few C=C bonds); *unsaturated triglycerides* contain mainly unsaturated fatty acid chains. The **iodine number** of a triglyceride is a measure of how unsaturated the molecule is: the higher the iodine number the higher the number of C=C bonds in the molecule.

Saponification is the process whereby a triglyceride is converted to a soap:

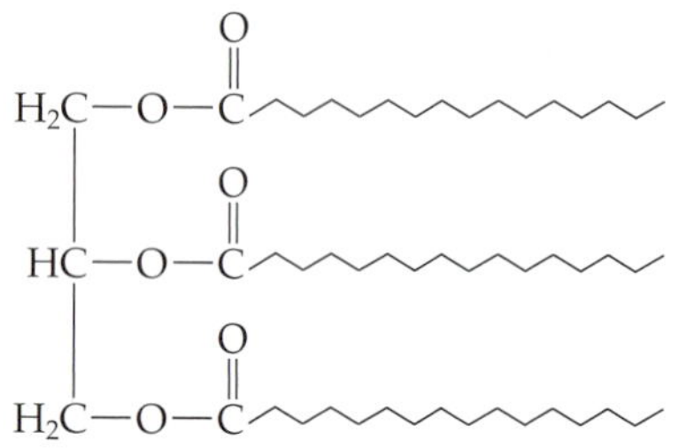

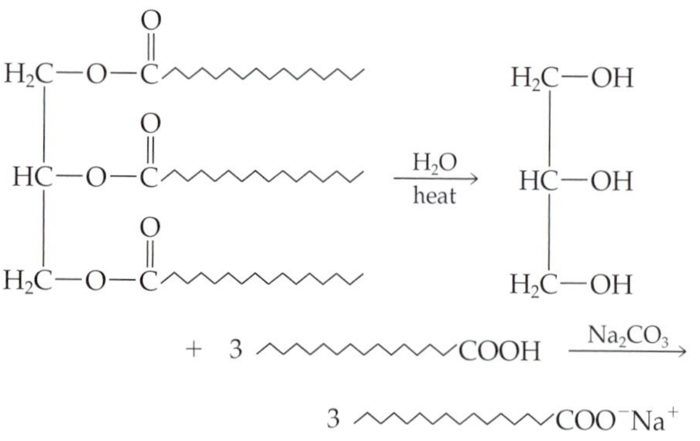

Soap

The —COONa end of the soap molecule is **hydrophilic** (water-loving) and so is soluble in water; the hydrocarbon part of the molecule is **hydrophobic** (water-hating) and thus more soluble in organic solvents than in water. Soap cleans when its hydrophobic end dissolves any lipid attached to dirt particles. Pulled by the hydrophilic end, plus the hydrophobic end, the attached lipid goes into the wash water. The "degreased" dirt particle then leaves the surface to which it is clinging and disperses in the wash water.

A **wax** is an ester in which the precursor carboxylic acid is a long-chain fatty acid and in which the precursor alcohol is a long-chain monohydroxy alcohol.

Phospholipids are lipids in which one of the —OH groups on the trihydroxy alcohol combines with a phosphorus-containing compound instead of a third fatty acid:

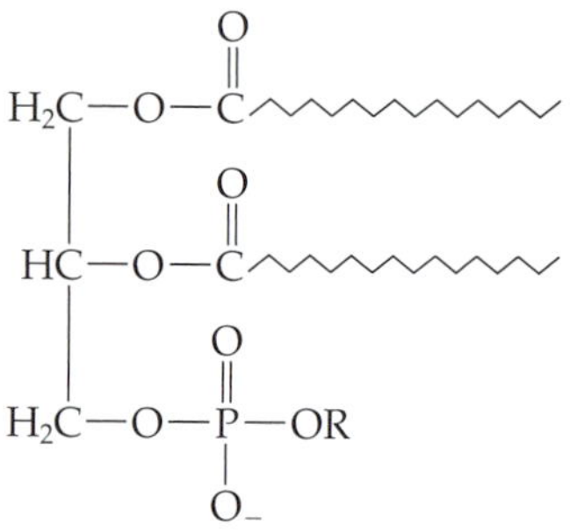

In **phosphatides,** the P is also joined to an amino alcohol unit:

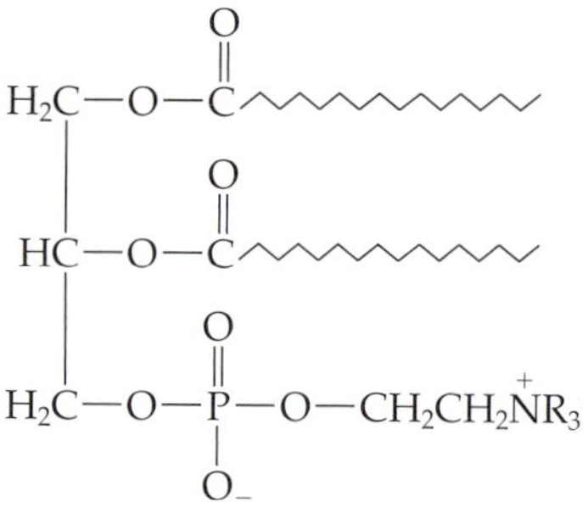

The phosphatides known as **cephalins** contain an ethanolamine group ($HOCH_2CH_2N^+H_3$). The phosphatides known as **lecithins** contain choline [$HOCH_2CH_2N^+(CH_3)_3$] as the amino alcohol group.

Sphingolipids are lipids for which the precursor is not glycerol but rather the amino alcohol sphingosine:

$$CH_3(CH_2)_{12}CH=CHCH-OH$$
$$CH-NH_2$$
$$CH_2-OH$$

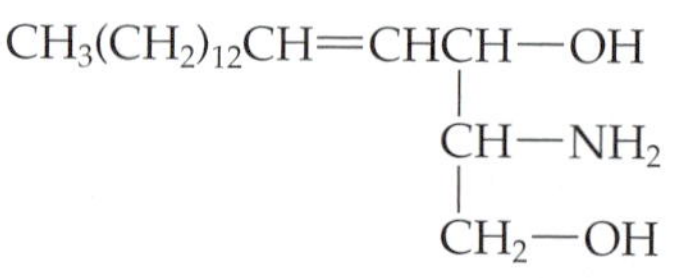

A **glycolipid** can have either glycerol or sphingosine as a precursor, but the defining feature is a sugar substituted at one of the —OH groups:

$$\text{H}_2\text{C}-\text{O}-\overset{\overset{\displaystyle O}{\|}}{\text{C}}\sim\!\sim\!\sim\!\sim\!\sim\!\sim$$

$$\text{HC}-\text{O}-\overset{\overset{\displaystyle O}{\|}}{\text{C}}\sim\!\sim\!\sim\!\sim\!\sim\!\sim$$

$$\text{H}_2\text{C}-\text{sugar}$$

$$\text{CH}_3(\text{CH}_2)_{12}\text{CH}=\text{CHCHOH}$$
$$\text{HC}-\text{NH}-$$
$$\text{H}_2\text{C}-\text{sugar}$$

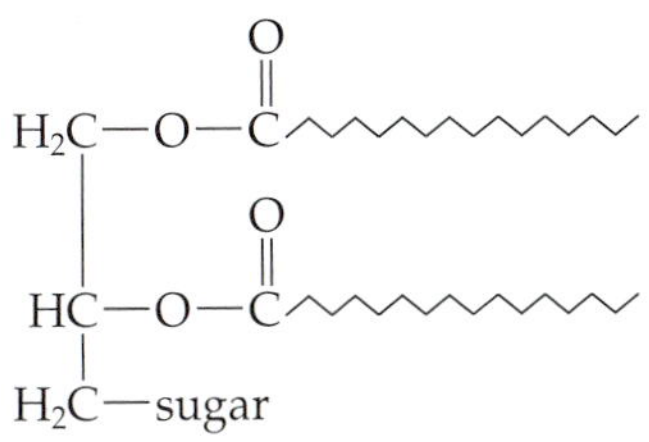

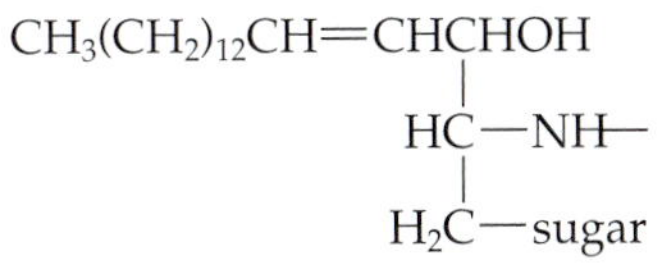

Every living cell is enclosed by a *cell membrane* made up of two layers of polar lipids arranged with their hydrophobic ends facing each other in a *lipid bilayer.* The bilayer contains mainly phospholipids, glycolipids, and the steroid cholesterol. Imbedded in the bilayer are **integral proteins,** and loosely associated with the bilayer along the interior surface of the cell are **peripheral proteins.** Any integral protein having either or both of its hydrophilic ends poking out of the bilayer and thus exposed to the liquid inside the cell or the interstitial fluid (or to both) is called a *transmembrane protein.* The purpose of integral and peripheral proteins is to transport ionic and polar substances back and forth across the cell membrane as the cell either needs these substances for metabolism or needs to get rid of them as waste products.

Almost all lipids can be saponified, but some cannot be. **Steroids** are one type of nonsaponifiable lipid. The steroid *cholesterol* is found in animal cells but never in plant cells. It is a main component of all cell walls and a precursor for hormones, vitamin D, and bile salts.

Key Terms

antioxidant (20.4)
bile (20.9)
cephalin (20.7)
cerebroside (20.7)
cytoplasm (20.8)
detergent (20.5)
emulsion (20.5)
fat (20.3)
fatty acid (20.2)
ganglioside (20.7)

glycolipid (20.7)
hydrophilic (20.5)
hydrophobic (20.5)
integral protein (20.8)
iodine number (20.4)
lecithin (20.7)
lipid (20.1)
micelle (20.8)
mixed triglyceride (20.3)
oil (20.3)

peripheral protein (20.8)
phosphatide (20.7)
phospholipid (20.7)
saponification (20.5)
simple triglyceride (20.3)
sphingolipid (20.7)
sphingomyelin (20.7)
steroid (20.9)
triglyceride (20.3)
wax (20.6)

Review Questions

1. Define or explain:
 a. triglyceride
 b. phosphatide
 c. iodine number
 d. antioxidant
 e. micelle
 f. semipermeable
 g. steroid
2. Distinguish between terms in the following pairs:
 a. fat and oil
 b. fat and wax
 c. saponifiable lipid and nonsaponifiable lipid
 d. cis double bond and trans double bond
 e. simple triglyceride and mixed triglyceride
 f. butter and margarine
 g. oxidative rancidity and hydrolytic rancidity
 h. hydrophobic and hydrophilic
 i. saponification and emulsification
 j. monolayer and bilayer
 k. hard water and soft water
 l. hard soap and soft soap
 m. soaps and syndets
 n. LDLs and HDLs
 o. phosphatides and sphingolipids
 p. lecithin and cephalins
 q. bile and bile salts
3. What functions does fat serve in the body?
4. Contrast the physical properties of fats and fatty acids. In what solvents are fats and oils soluble?
5. Why do unsaturated fatty acids have lower melting points than saturated fatty acids?
6. Which has the higher iodine number—tristearin or triolein?
7. Which would you expect to have a higher iodine number—corn oil or beef tallow? Explain.
8. In the reaction used for the determination of iodine number, what functional group in a fat molecule reacts with the reagent?

9. Which would you expect to have the higher iodine number—hard margarine or liquid margarine? Explain.
10. What triglyceride is formed by the complete hydrogenation of triolein? of trilinolein?
11. What products are formed when a triglyceride is (a) hydrolyzed? (b) saponified?
12. a. What compound with a disagreeable odor is formed when butter becomes rancid?
 b. What leads to its formation?
 c. How can rancidity be prevented?
13. Can waxes be converted to soaps? Explain.
14. Name four waxes obtained from plants and animals, and give one use for each.
15. Briefly describe how soaps clean.
16. a. What structural features are necessary for a compound to be a good detergent?
 b. What advantages do synthetic detergents have over soap?
 c. What are the disadvantages of detergents?
17. Why are phospholipids referred to as polar lipids?
18. What general structural feature do phosphatides share with soaps and detergents?
19. Which of the following can diffuse through the lipid portion of a cell membrane?
 a. glucose
 b. NaCl
 c. $CH_3CH_2OCH_2CH_3$
 d. CH_3CH_2OH
20. Discuss the roles of cholesterol, saturated fats, and fish oils in atherosclerosis.
21. What is an integral protein? What is its function?
22. What is a peripheral protein? Where is it located on the cell membrane?

Problems

Classification

23. Which of these fatty acids are saturated and which are unsaturated? How many carbon atoms are there in each?
 a. caproic acid b. oleic acid
 c. stearic acid
24. Which of these fatty acids are saturated, and which are unsaturated? How many carbon atoms are there in each?
 a. palmitic acid b. linolenic acid
 c. caprylic acid
25. Which compounds are classified as steroids?
 a. tristearin b. cholesterol
 c. lecithin
26. Which of the following compounds are classified as steroids?
 a. prostaglandins b. cephalin
 c. cholic acid
27. Which of the following are derived from glycerol, which from sphingosine, and which from neither?
 a. fats
 b. cerebrosides
 c. waxes
28. Which of the following are derived from glycerol, which from sphingosine, and which from neither?
 a. phosphatides
 b. oils
 c. steroids

Structures and Names

29. Write structural formulas for the following.

a. palmitic acid
b. linolenic acid
c. triolein
30. Write structural formulas for the following.
 a. glyceryl palmitate
 b. cetyl (1-hexadecyl) stearate
 c. a triglyceride likely to be found in cottonseed oil
31. Write structural formulas for the following.
 a. a highly unsaturated oil
 b. sodium oleate
 c. calcium myristate
32. Using a circle to represent the ionic group and zigzag lines to represent the two fatty acid units, draw the arrangement of phosphatide molecules in a micelle and in a bilayer.

Chemical Reactions

33. Draw the structural formulas for the products of this saponification reaction:

$$CH_2-O-\overset{\displaystyle O}{\overset{\|}{C}}-CH_2CH_2CH_2CH_2CH_2CH_2CH_3$$
$$CH-O-\overset{\displaystyle O}{\overset{\|}{C}}-CH_2CH_2CH_2CH_2CH_3 \xrightarrow[\Delta]{NaOH}$$
$$CH_2-O-\overset{\displaystyle O}{\overset{\|}{C}}-CH_2CH_2CH_2CH_2CH_2CH_2CH_2CH_2CH_3$$

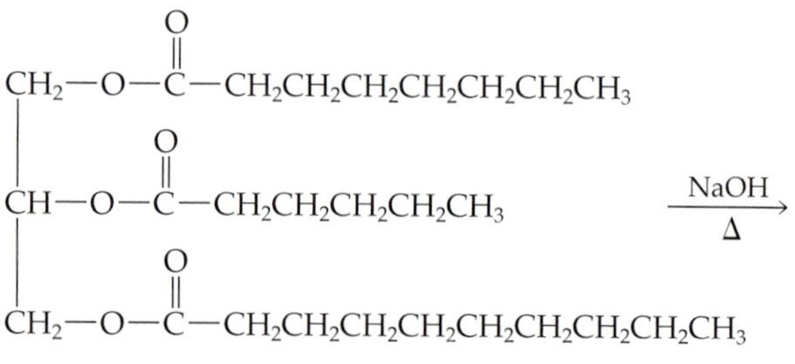

34. Write the equation for the saponification of glyceryl trilaurate.

Additional Problems

35. How would the equation for Problem 34 differ if the compound being saponified were triolein?

36. Start with one molecule each of glycerol, palmitic acid, oleic acid, phosphoric acid, and choline, and construct a typical lecithin molecule. Circle all the ester bonds.

37. In cerebrosides, what type of bond joins the fatty acid to sphingosine? What type of bond joins the sugar unit to sphingosine?

38. Phosphatidylinositol is a phosphatide found in cell membranes. Given the structure of inositol, draw a structural formula for phosphatidylinositol. (Linkage occurs at the colored hydroxyl group.)

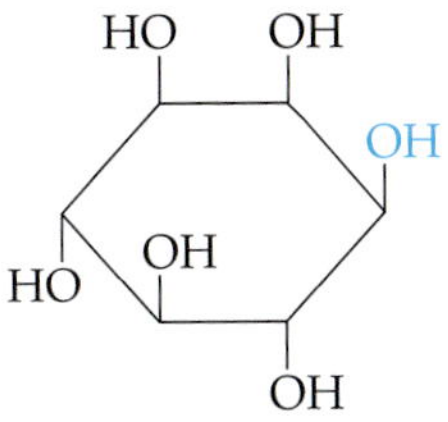

Inositol

39. Draw the basic steroid skeleton.

40. The melting point of elaidic acid (Figure 20.2c) is 52 °C.
 a. How does this compare with the melting points of stearic acid and oleic acid? Explain.
 b. Would you expect the melting point of *trans*-hexadecenoic acid,

 $$CH_3(CH_2)_5CH=CH(CH_2)_7COOH,$$

to be lower or higher than that of elaidic acid? Why?

41. Draw the structure of the cerebroside that has palmitic acid as its fatty acid and glucose as its sugar.

42. How many mixed triglycerides are possible by combining stearic acid and oleic acid with glycerol?

43. A principal wax in spermaceti is an ester of palmitic acid (hexadecanoic acid) and cetyl alcohol (1-hexadecanol). Draw the structure of the ester.

44. Identify each component in the phospholipid:

$$CH_2OC(CH_2)_{16}CH_3$$
$$CHOC(CH_2)_7CH=CH(CH_2)_7CH_3$$
$$CH_2OP-OCH_2CH_2N^+-CH_3$$

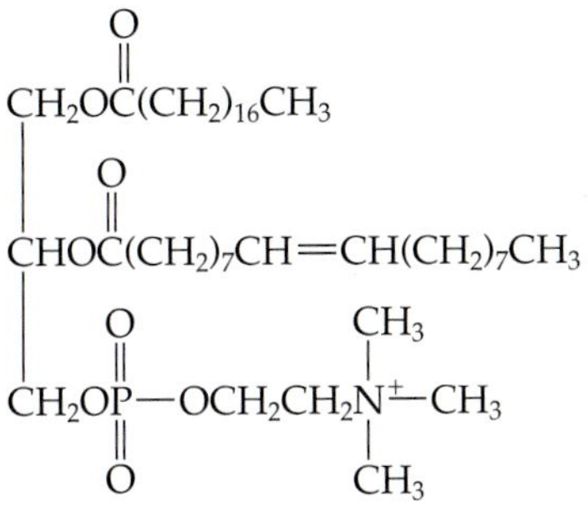

45. Serine is an amino acid that has the structure

$$HOCH_2CHCOOH$$
$$NH_2$$

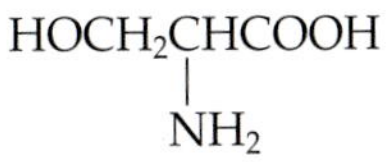

Give the structure for phosphatidylserine.

Projects

46. Examine the labels on margarines and shortenings and list the oils used in the various brands.

47. Examine several detergents, and note the active ingredient in each. What other additives are present?

Special Topic I
Hormones

Like the vitamins (Special Topic J), hormones are organic compounds. Unlike vitamins, hormones can be synthesized in the body. Both vitamins and hormones play critical biochemical roles. The most striking similarity, however, is in their effective amounts. Even in very low concentrations, hormones produce dramatic changes in the body. This special topic, then, deals with a class of compounds that, like the vitamins, illustrate the old adage that good things come in small packages.

I.1 The Endocrine System

The **hormones** are synthesized in the endocrine glands (Figure I.1) and are then discharged directly into the circulatory system. They serve as "chemical messengers." Hormones released in one part of the body signal profound physiological changes in other parts of the body (heart, liver, muscle, kidney, etc.). They cause reactions to speed up or slow down. In this way they control growth, metabolism, reproduction, and many other functions of body and mind. Hormones can affect (1) the permeability of cell membranes, (2) the rate of enzymatic reactions, and (3) the rate of synthesis of certain proteins.

If we consider hormones as messengers, the pituitary gland must be viewed as the central dispatcher or control. Many of the pituitary hormones control the production of hormones by other endocrine glands. The removal of portions of the pituitary gland results in atrophy of other endocrine glands. Shut down the central control, and you ultimately shut down much of the endocrine system. The pituitary itself responds to hormone signals from the hypothalamus. The hypothalamus secretes hormones, called releasing factors, that trigger the production of the pituitary hormones. The hypothalamus is triggered by nerve impulses. The sequence is this: A nerve impulse signals the hypothalamus, the hypothalamus signals the pituitary, the pituitary signals some target endocrine gland, the gland signals some target tissue, and the tissue responds in a specific way (Figure I.2).

Let us take just a moment to consider the extraordinary complexity of this system. For example, in response to neural stimulation, the hypothalamus produces thyrotropin-releasing factor (TRF). The TRF reaches the pituitary and causes that gland to produce thyroid-stimulating hormone (TSH). In the presence of TSH, the thyroid gland releases the hormone thyroxine. Thyroxine signals the cells to increase their metabolic rate. As the level of thyroxine builds up, a feedback mechanism causes the pituitary to slow down its production of thyroid-stimulating hormone. This, in turn, slows the production of thyroxine by the thyroid, which results in a slowing of the metabolic rate, which gets us back to where we started. The cycle includes the

FIGURE I.1 ▶

Approximate locations of endocrine glands in the human body.

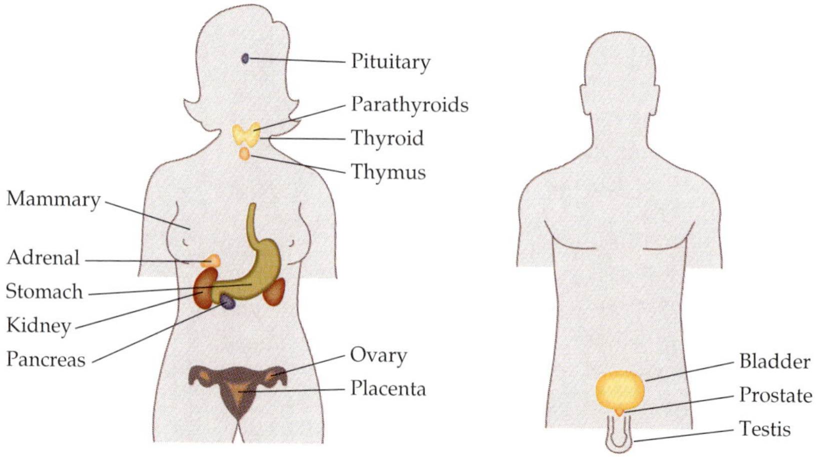

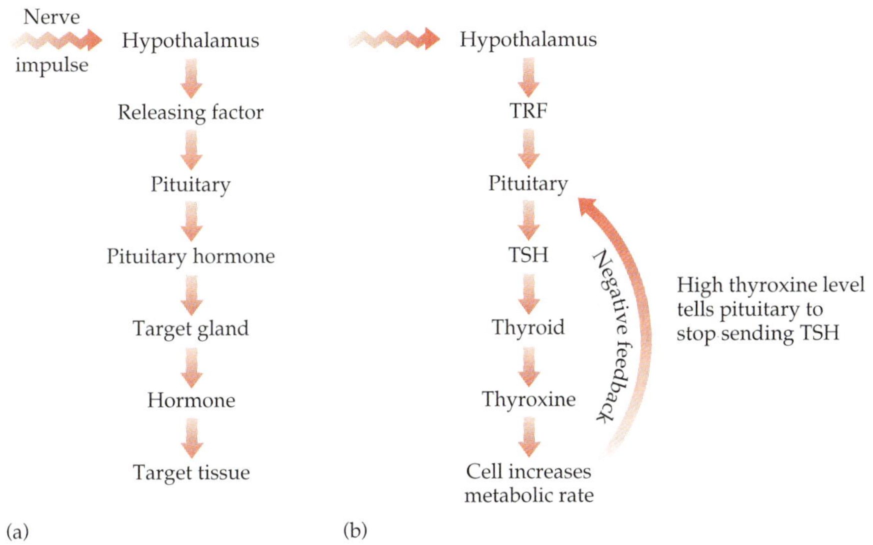

FIGURE I.2

(a) General example and (b) specific example of the sequence of events in the release of hormones.

synthesis of proteins and peptides (through the complex process described in Chapter 23). It includes the multistep synthesis of the amino acid thyroxine and, ultimately, the speeding up of that complicated combination of reactions we lump together under the heading *metabolism*.

In a healthy individual, all of this happens routinely, in perfect balance and with no conscious direction. And that's just one of the myriad interrelated biochemical processes that life requires. That it all works, and works so well, is nothing short of miraculous.

Many important human hormones and their physiological effects are listed in Table I.1 (page 555). We shall discuss the hormones vasopressin and oxytocin in Chapter 21. The pancreatic hormones insulin and glucagon are considered in Chapter 24. In the remaining sections of this special topic, we focus our attention on some steroid hormones and the prostaglandins.

The steroid hormones (e.g., adrenocortical hormones, sex hormones) are lipids. They are soluble in the lipid components of the cell membrane and can easily diffuse into cells. Inside the cell they combine with specific receptor molecules in the cytoplasm. They may influence enzymatic reactions directly, or the steroid–receptor complex can enter the nucleus. In the nucleus, steroids bind to specific sites on DNA, where they increase the rate of synthesis of mRNA (see Section 23.5), thus increasing the rate of biosynthesis of cellular enzymes. Because the primary effect of steroid hormones is regulating protein synthesis, their effects are much slower than those of other hormones (i.e., hours rather than minutes).

I.2 Adrenocortical Hormones

The outer part, or cortex, of the adrenal[1] gland (Figure I.3) uses cholesterol to produce a mixture of many steroids essential to life. These steroids constitute a family of hormones of which aldosterone and cortisol are the major representatives.

Aldosterone is called a *mineralocorticoid*. This name alludes to its function in regulating the exchange of sodium, potassium, and hydrogen ions. Although aldosterone acts on most cells in the body, it is particularly effective in enhancing the rate of reabsorption of sodium ions in the kidney tubule and in increasing the secretion of potassium ions and/or hydrogen ions by the tubule. Since concentration of sodium ions is the major factor in water retention in the tissues, aldosterone also promotes water retention and reduces urine output. It thus supplements the action of vasopressin (Section 21.6).

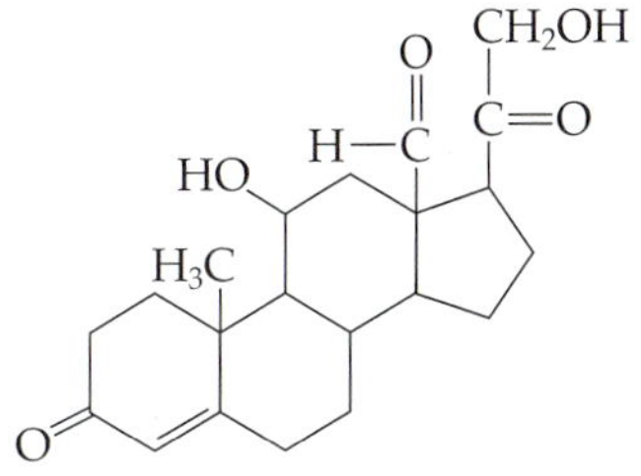

Aldosterone

[1] The term *adrenal* comes from the gland's location in the body, *ad*jacent to the *renal* (kidney).

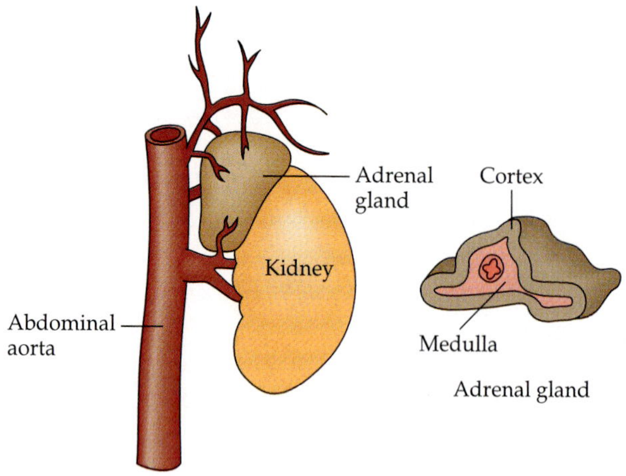

FIGURE I.3

The anatomical location and composition of the adrenal gland.

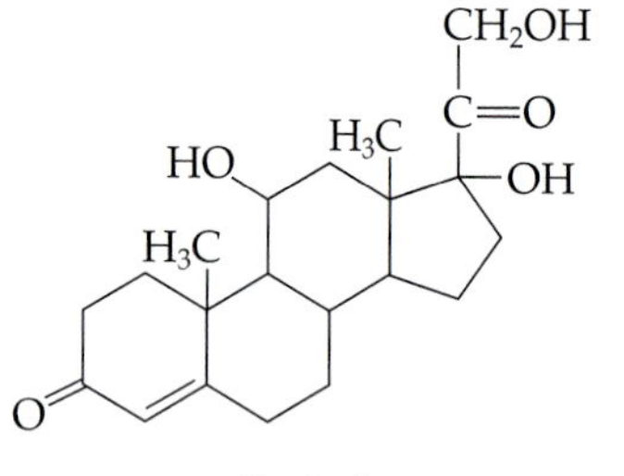

Cortisol
(Hydrocortisone)

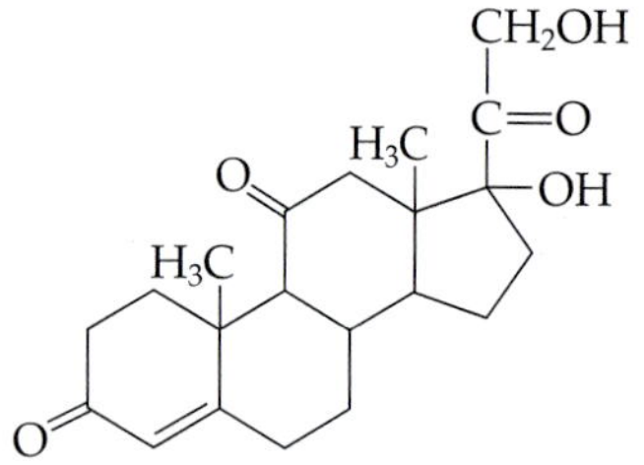

Cortisone

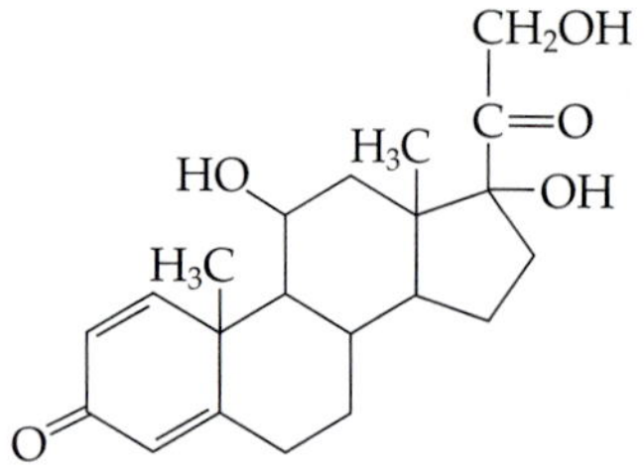

Prednisolone

Cortisol and its keto derivative, cortisone, are called *glucocorticoids*. These hormones regulate a number of key metabolic reactions (e.g., they increase glucose production and mobilize fatty acids and amino acids). They also inhibit the inflammatory response.[2] Glucocorticoids are used as drugs for immunosuppression after transplant operations and in the treatment of autoimmunity, severe skin allergies, and rheumatoid arthritis. The hormones or their analogs are injected, taken orally, or applied directly to the site of inflammation.

Pharmaceutical companies have been able to obtain large quantities of synthetic cortisone by means of an involved process utilizing cholic acid isolated from the bile juices of cattle. Prolonged use of cortisone can have serious side effects, including high blood pressure, wasting of muscles, and resorption of bone. Cortisone has been supplemented with a synthetic analog, prednisolone, which is effective in much smaller doses, thereby greatly reducing the side effects. Prednisolone has been used in combination with an antibiotic in the treatment of autoimmunity and for immunosuppression after kidney and liver transplant operations.

---

[2] **Inflammation** is a tissue response to injury or stress (e.g., chemical irritants, exposure to radiation or to extreme temperature). The blood vessels in the surrounding area become dilated to bring extra blood to the affected area. (This accounts for the redness associated with inflammation.) Dilated blood vessels are more permeable, and there is a tendency for fluid to leave the blood and enter the damaged tissue, causing swelling. Prostaglandins (Section I.6) contribute to the inflammatory response by (1) promoting vasodilation of the blood vessels, (2) increasing the permeability of the capillaries, and (3) stimulating the pain receptors. It may be that the glucocorticoids are effective antiinflammatory agents because they inhibit an enzyme (phospholipase) necessary for the synthesis of arachidonic acid, a prostaglandin precursor.

Table I.1 Some Human Hormones and Their Physiological Effects

Name	Gland and Tissue	Chemical[a] Nature	Effect
Various releasing and inhibitory factors	Hypothalamus	Peptide/ Protein	Trigger or inhibit release of pituitary hormones
Human growth hormone (HGH)	Pituitary, anterior lobe	Protein	Controls the general body; controls bone growth
Thyroid-stimulating hormone (TSH)	Pituitary, anterior lobe	Protein	Stimulates growth of the thyroid gland and production of thyroxine
Adrenocorticotrophic hormone (ACTH)	Pituitary, anterior lobe	Protein	Stimulates growth of the adrenal cortex and production of corticol hormones
Follicle-stimulating hormone (FSH)	Pituitary, anterior lobe	Protein	Stimulates growth of follicles in ovaries of females, sperm cells in testes of males
Luteinizing hormone (LH)	Pituitary, anterior lobe	Protein	Controls production and release of estrogens and progesterone from ovaries, testosterone from testes
Prolactin	Pituitary, anterior lobe	Protein	Maintains the production of estrogens and progesterone, stimulates the formation of milk
Vasopressin	Pituitary, posterior lobe	Protein	Stimulates contractions of smooth muscle; regulates water uptake by the kidneys
Oxytocin	Pituitary, posterior lobe	Protein	Stimulates contraction of the smooth muscle of the uterus; stimulates secretion of milk
Parathyroid	Parathyroid	Protein	Controls the metabolism of phosphorus and calcium
Thyroxine	Thyroid	Amino acid derivative	Increases rate of cellular metabolism
Insulin	Pancreas, beta cells	Protein	Increases cell usage of glucose; increases glycogen storage
Glucagon	Pancreas,	Protein alpha cells	Stimulates conversion of liver glycogen to glucose
Cortisol	Adrenal gland, cortex	Steroid	Stimulates conversion of proteins to carbohydrates
Aldosterone	Adrenal gland, cortex	Steroid	Regulates salt metabolism; stimulates kidneys to retain Na^+ and excrete K^+
Epinephrine (adrenaline)	Adrenal gland, medulla	Amino acid derivative	Stimulates a variety of mechanisms to prepare the body for emergency action, including the conversion of glycogen to glucose
Norepinephrine (noradrenaline)	Adrenal gland, medulla	Amino acid derivative	Stimulates sympathetic nervous system; constricts blood vessels, stimulates other glands
Estradiol	Ovary, follicle	Steroid	Stimulates female sex characteristics; regulates changes during menstrual cycle
Progesterone	Ovary, corpus luteum	Steroid	Regulates menstrual cycle; maintains pregnancy
Testosterone	Testes	Steroid	Stimulates and maintains male sex characteristics

[a] Protein hormones are discussed in Chapter 21 and Chapter 24.

I.3 Sex Hormones and Anabolic Steroids

The sex hormones are a class of steroids secreted by the gonads (ovaries or testes), the placenta, and the adrenal glands. The primary male sex hormones (called **androgens**), *testosterone* and *androstenedione*, are produced in the testes (and in lesser amounts in the adrenal cortex and the ovaries). They control the primary sexual characteristics of males, that is, the development of the male genital organs and the continued production of sperm. Androgens are also responsible for the development of secondary male characteristics, such as facial hair, deep voice, and muscle strength. Men generally have larger muscles than women because men have more testosterone.

Two sex hormones are of particular importance in females. *Progesterone* prepares the uterus for pregnancy and prevents the further release of eggs from the ovaries during pregnancy. The **estrogens** are mainly responsible for the development of female secondary sexual characteristics, such as breast development and increased deposition of adipose tissue in the breast, buttock, and thighs.

Progesterone is the precursor for the synthesis of the androgens, and the estrogens are synthesized from the androgens. Estrone is derived from androstenedione, and estradiol (the major estrogen hormone) is formed from testosterone. Both males and females produce androgens and estrogens. The difference between the sexes is in the amounts of secreted hormones, not the total absence of one group or the other. Notice that the male and female hormones exhibit only very slight structural differences. Their physiological effects, however, differ enormously.

There has been a great deal of controversy generated by the widespread use of **anabolic steroids** by athletes. The drugs in question are synthetic androgens that stimulate protein synthesis (especially in skeletal muscle cells) without affecting the sex glands. We mentioned that testosterone is responsible for an increase in the amount of muscle cells (and more aggressive behavior), as well as its virilizing activity. However, testosterone is not very active if taken orally because it is metabolized in the liver. The incorporation of a methyl group at C–17 prevents this metabolism. Introduction of a second double bond in ring A produces a compound, methandienone (Dianabol), that has anabolic activity (stimulation of protein synthesis) but little of the virilizing effects of testosterone.

There are not good controlled studies that demonstrate the effectiveness of anabolic steroids. They do seem to work—at least for some people—but the side effects are many. In males, side effects include testicular atrophy and loss of function, impotence, acne, liver damage, edema (swelling), elevated cholesterol levels, and growth of breasts. Liver cancer is now showing up at an alarming rate in athletes who began using steroids in the 1960s.

Anabolic steroids act as male hormones (androgens). They make women more masculine. They help women build larger muscles, but they also induce balding, development of extra body hair, deepening of the voice, and menstrual irregularities.

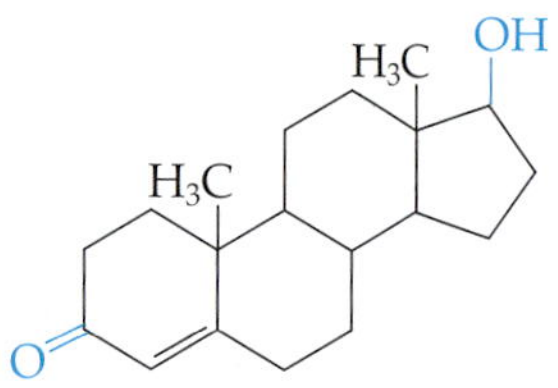

Testosterone

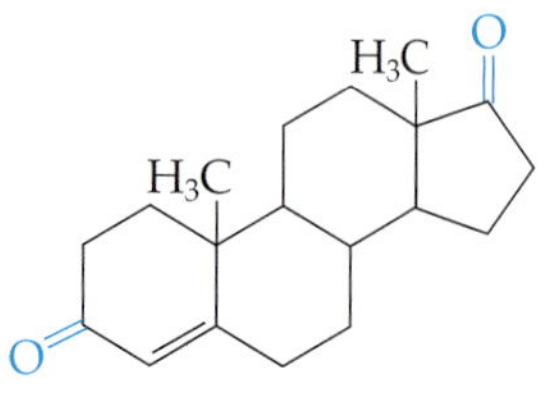

Androstenedione

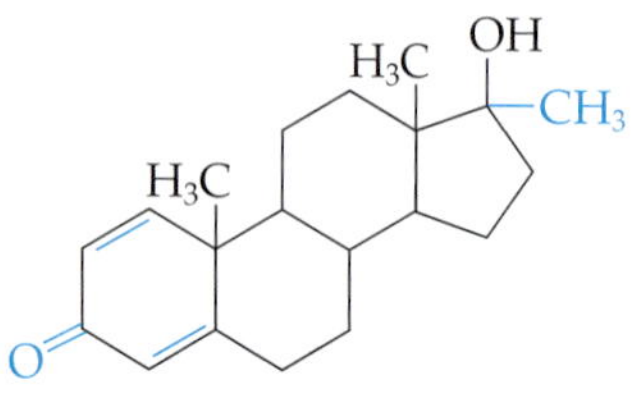

Methandienone
(Dianabol)

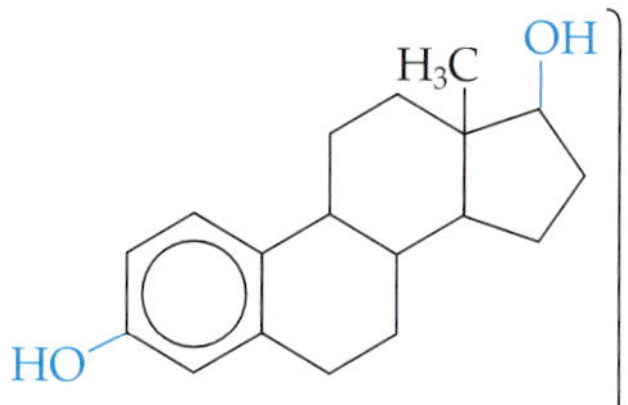

Estradiol

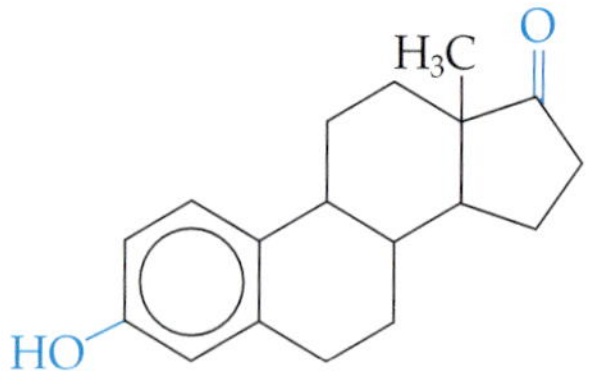

Estrone

Estrogens

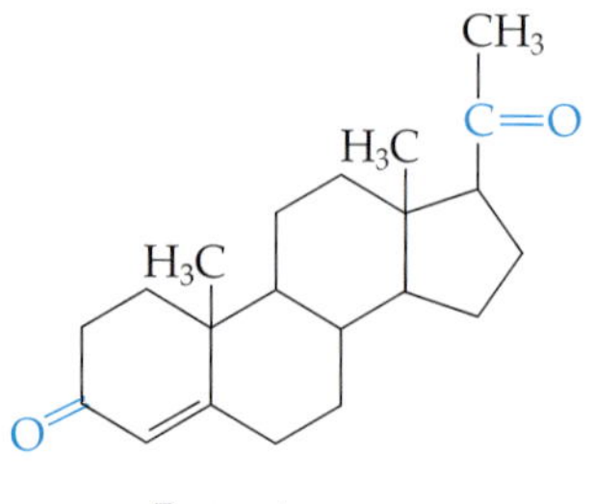

Progesterone

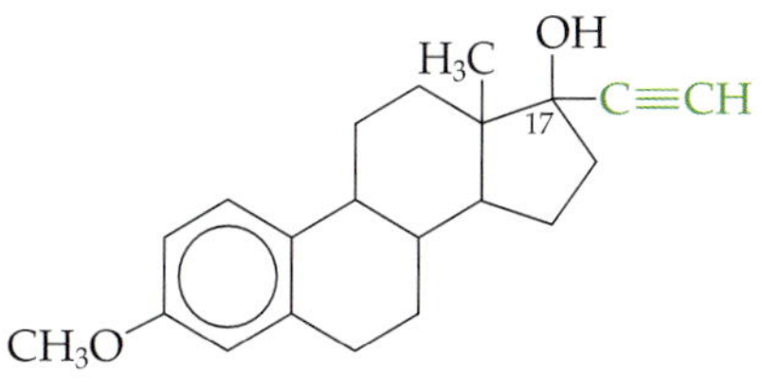

Mestranol

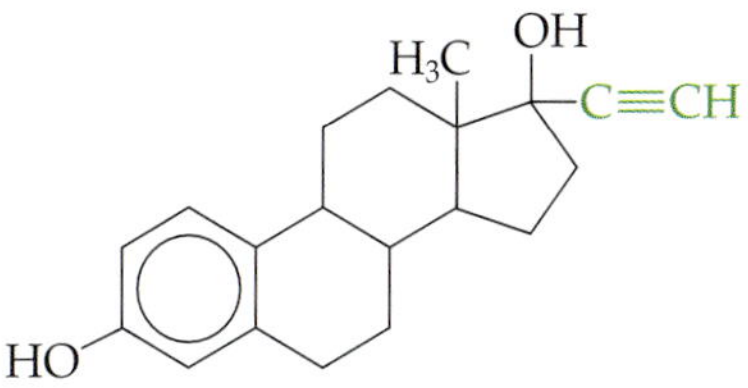

Ethinyl estradiol

(analogs of the estrogens)

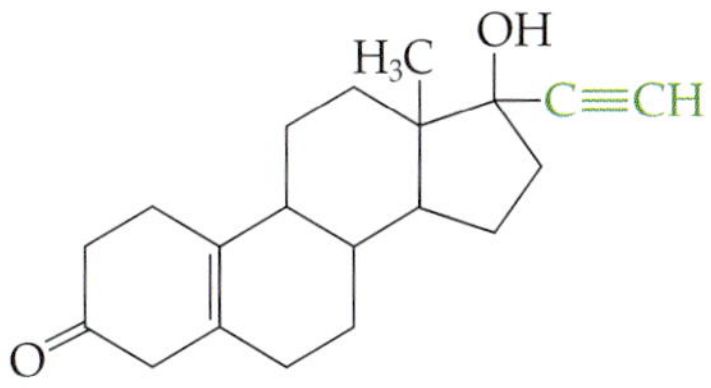

Norethynodrel

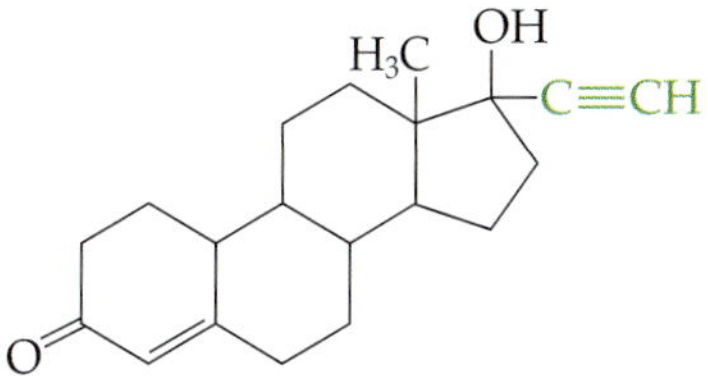

Norethindrone

(analogs of progesterone)

Chemists can detect the presence of synthetic steroids in the body by monitoring the urine and testing for degradation products. Anabolic steroids are marketed for use in the treatment of senile debility, anorexia, and anemia and during convalescence.

Sex hormones—both natural and synthetic—are sometimes used therapeutically. For example, a woman who has had her ovaries removed may be given female hormones to compensate for those no longer produced by the ovaries. Some of the earliest chemical compounds employed in cancer chemotherapy were sex hormones. Testosterone was used to treat carcinoma of the breast in females, and estrogens were given to males to treat carcinoma of the prostate. Sex hormones are also important in sex-change operations. Before corrective surgery, hormones are administered to promote the development of the proper secondary sexual characteristics.

I.4 Conception and Contraceptives

When taken regularly, synthetic derivatives of the female sex hormones prevent ovulation. The oral contraceptives are usually mixtures of analogs of progesterone and the estrogens.[3] For example, Enovid is a combination of norethynodrel and mestranol, whereas Ortho-Novum contains norenthindrone and ethinyl estradiol. Most of the combination pills sold in the United States contain 1 mg of the progesterone analog and less than 0.03 mg of the estrogen analog.

Norethynodrel, norethindrone, and related compounds are called **progestins** because they mimic the action of progesterone. The progestin acts by establishing a state of false pregnancy (Figure I.4). A synthetic estrogen is added to regulate the menstrual cycle. A woman does not ovulate when she is pregnant (or in the state of false pregnancy established by the progestin). Since the woman does not ovulate, she cannot conceive.

Oral contraceptives have been used in the United States since 1960 by millions of women. They appear to be safe in most cases, but some women experience hypertension, acne, or abnormal bleeding. The pills increase the risk of blood clotting in some women, but so does pregnancy. Blood clots can clog arteries and cause death by stroke or heart attack. The death rate associated with birth control pills is about 3 in 100,000, only one-tenth of that associated with childbirth (which is about 30 per 100,000). The FDA advises women over 40, and any women who smoke, to use some other method of contraception. For these women, the risk of using birth control pills is greater than the risk of childbirth.

Progesterone is essential for the maintenance of pregnancy. If its action is blocked, pregnancy could not be established or maintained. Rousel-Uclaf, a French subsidiary of Hoescht, has developed a drug that blocks the action of progesterone. Mifepristone (RU-486), the "morning after" pill, is available in France and China, where it has replaced a substantial number of surgical abortions. A woman who wishes to abort a pregnancy takes three mifepristone tablets, followed in a few days by an injection of a prostaglandin (Section I.6). The lining of the uterine

[3] The prevention of ovulation is also effected by administration of progesterone and estradiol, but these hormones must be injected into the body for maximum results. It is the $C\equiv C$ group in the synthetic analogs that confers upon them the ability to be taken orally and to function in the same manner as the steroids produced by the body.

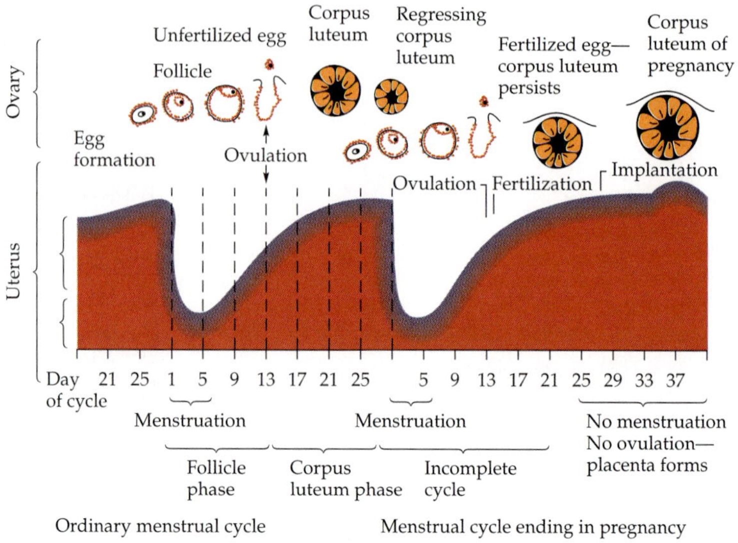

FIGURE I.4 ▶

Changes in the ovary and the uterus during the menstrual cycle. Both pregnancy and the pseudopregnancy caused by the birth control pill prevent ovulation.

wall and the implanted fertilized egg are sloughed off and pregnancy is terminated.

RU-486 has other medical uses, too. It can increase uterine contractions when labor has stalled during childbirth. It appears to trigger lactation in mothers and increases milk production. It seems to slow the growth of certain types of cancer. And it is also being studied as a treatment for Cushing's syndrome, which results from excessive production of cortisone.

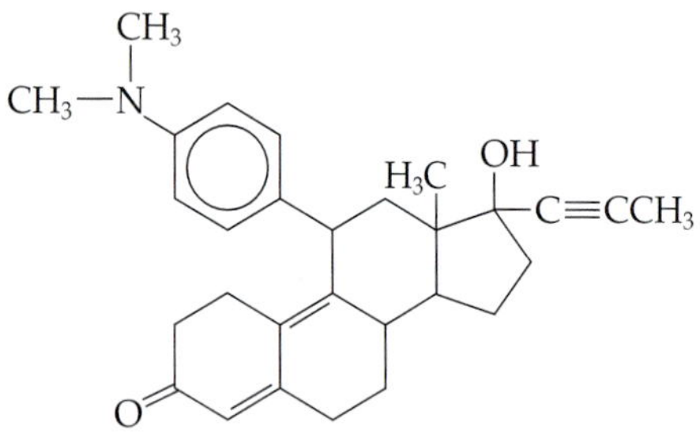

Mifepristone

In 1990 the FDA approved the use of an under-the-skin implant for birth control. Levonorgestrel (Norplant), a synthetic analog of progesterone, is incorporated into six plastic capsules, each about 2.5 cm long and the size of a matchstick. These are then implanted under the skin of a woman's upper arm. Release of the drug over time prevents pregnancy for five years with a failure rate of only 0.2%. It has about the same side effects as the progestin-only minipill. (Levonorgestrel is the progestin used in some ordinary birth control pills.)

Why do females have to bear the responsibility for contraception? Why not a pill for males? Recent studies have shown that many men—including a majority of the younger ones—are willing to share the risks and the responsibility of contraception. Nevertheless, there are biological reasons for females to bear the burden: women get pregnant when contraception fails, and in females, contraception has only to interfere with one monthly event—ovulation. On the other hand, males produce sperm continuously.

Despite these arguments, a good deal of research has gone into male contraceptives, but results so far have not been very successful. Estrogens would work, but they would bring about the development of female characteristics in men, including a complete loss of interest in sexual relations with women.

Gossypol, a pigment found in cottonseed, has been used extensively in China as a contraceptive for males.

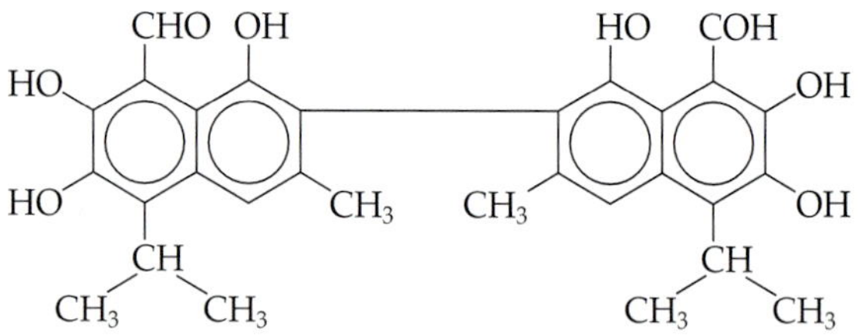

Gossypol

A drug called danazol has also been tested and found safe and effective. Danazol is testosterone enanthate. Administered along with testosterone, it suppresses sperm production, and the effect is reversed when the drug is withdrawn. Like the Pill for women, danazol often causes weight gain in men. The main problem, however, is that the drug is too expensive for widespread use. Perhaps the best bet for a male contraceptive is some sort of drug that doesn't stop sperm formation but does block sperm growth and/or transport.

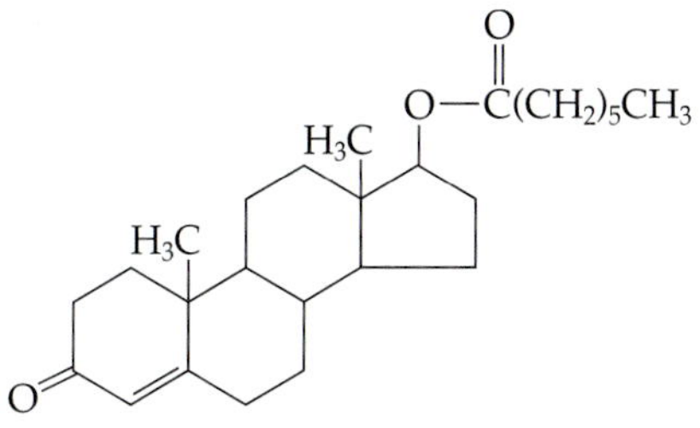

Testosterone enanthate
(Danazol)

The availability of vasectomies—simple surgical procedures that block the emission of sperm—has lessened the demand for a male contraceptive. A major drawback to vasectomy, however, is that it is often irreversible. A safe, cheap, and effective male contraceptive is still a goal of biochemical research.

The control of human reproduction is an issue subject to great moral, political, and legal controversy. Much more research is needed to increase our understanding of the biochemistry and physiology of reproduction. Armed with new knowledge, chemists might be able to design drugs acceptable to people of divergent views.

I.5 Estrogen Replacement Therapy

In most women, menstrual cycles continue into the late forties, at which time they become increasingly irregular and finally cease altogether. This period is called *menopause.* Follicles no longer mature, ovulation does not occur, and the plasma estrogen concentration sharply decreases.

As a result of lower estrogen levels, the female secondary sexual characteristics are modified. In addition, there is no longer the feedback control by estrogens over the secretion of FSH and LH. These hormones are released continuously by the pituitary gland, and they seem to be responsible for some of the unpleasant sensations that menopausal women experience (e.g., hot flashes, irritability, anxiety, and fatigue). These discomforts can be overcome by the administration of daily doses of estrogens or estrogen substitutes. The drugs also reduce the risk of brittle bones (osteoporosis), reduce the risk of heart disease, and may also protect against fatal colon cancer.

Estrogens such as estrone, estradiol, ethinyl estradiol, and mestranol are used in estrogen replacement therapy (ERT), particularly in menopausal and postmenopausal women. Other synthetic compounds that are structurally similar to the estrogens are also used. One such compound is *diethylstilbestrol* (DES). Other medical applications of DES are suppression of lactation, treatment of breast and prostate carcinoma, treatment of acne, and postcoital contraception. It has also been used to stimulate female fertility. However, DES was found to cause mammary tumors in some strains of mice. Additionally, daughters born to women who took DES during pregnancy developed rare forms of vaginal cancer (clear cell adenocarcinoma) when they were in their early twenties. The FDA has ordered that no estrogens may be prescribed during pregnancy. The FDA also removed from the market DES tablets that were used as "morning after pills." Today the use of DES as a postcoital contraceptive is lawful only following rape or incest. (Note that the trans form of DES structurally approximates the estrogen steroids.)

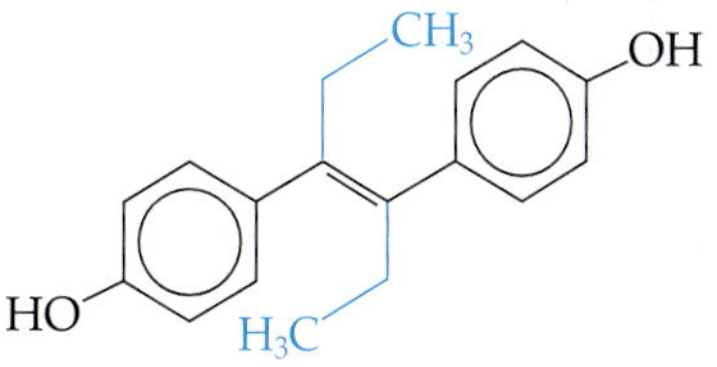

DES

Within the past decade, many investigators have suggested that ERT substantially protects women against heart disease. Heart disease is the leading cause of death in women, killing more women each year than all cancers combined. But many women and their physicians have been wary of estrogen because of some reports linking it to a slightly elevated risk of breast and uterine cancers. Of the more than 43 million American women over 50, fewer than one-third take ERT.

In 1991 the *New England Journal of Medicine* published the results of the largest study to date on ERT and women (48,000 menopausal female nurses over a 10-year period). Researchers from Harvard Medical School had found that women who took estrogen had only half the risk of heart attacks and death from coronary heart disease that nonusers did. The report suggested that even healthy women should consider taking estrogen after menopause.

However, the increased risk of contracting breast cancer is still valid. Usually progesterone is prescribed along with estrogen to counter the increased risk of cancer of the uterine lining that estrogen alone can cause. In considering ERT, women should consult with their doctors to be fully informed of the risks versus the benefits.

Research in this area is ongoing. The Women's Health Initiative, a prospective study of 25,000 women designed by the National Institutes of Health, should provide definitive results on hormones and heart disease, cancer, and osteoporosis over the next decade.

I.6 Prostaglandins

Perhaps no compounds since birth control pills have stimulated as much activity in pharmaceutical companies as have the group of compounds called prostaglandins. More than 6000 papers are published each year on these compounds. **Prostaglandins** are hormonelike substances that were originally isolated from semen found in the prostate gland. In the mature male, the prostate gland secretes about 0.1 mg/day of prostaglandins. However, they are biosynthesized by most mammalian tissue, and they affect almost all organs in the body.

Prostaglandins are a family of unsaturated fatty acids, each containing 20 carbon atoms and having the same basic skeleton as prostanoic acid (Figure I.5a). The major classes are PGA, PGB, PGE, and PGF followed by a subscript that denotes the number of double bonds outside the five-carbon ring. Prostaglandins are not stored as such in cells. Rather, they are synthesized on demand from arachidonic acid, a 20-carbon polyunsaturated fatty acid that is released from phospholipids in cell membranes (by the action of the enzyme phospholipase). Arachidonic acid is then converted to an endoperoxide by an enzyme complex called prostaglandin cyclooxygenase. The endoperoxide intermediate can then be transformed to the

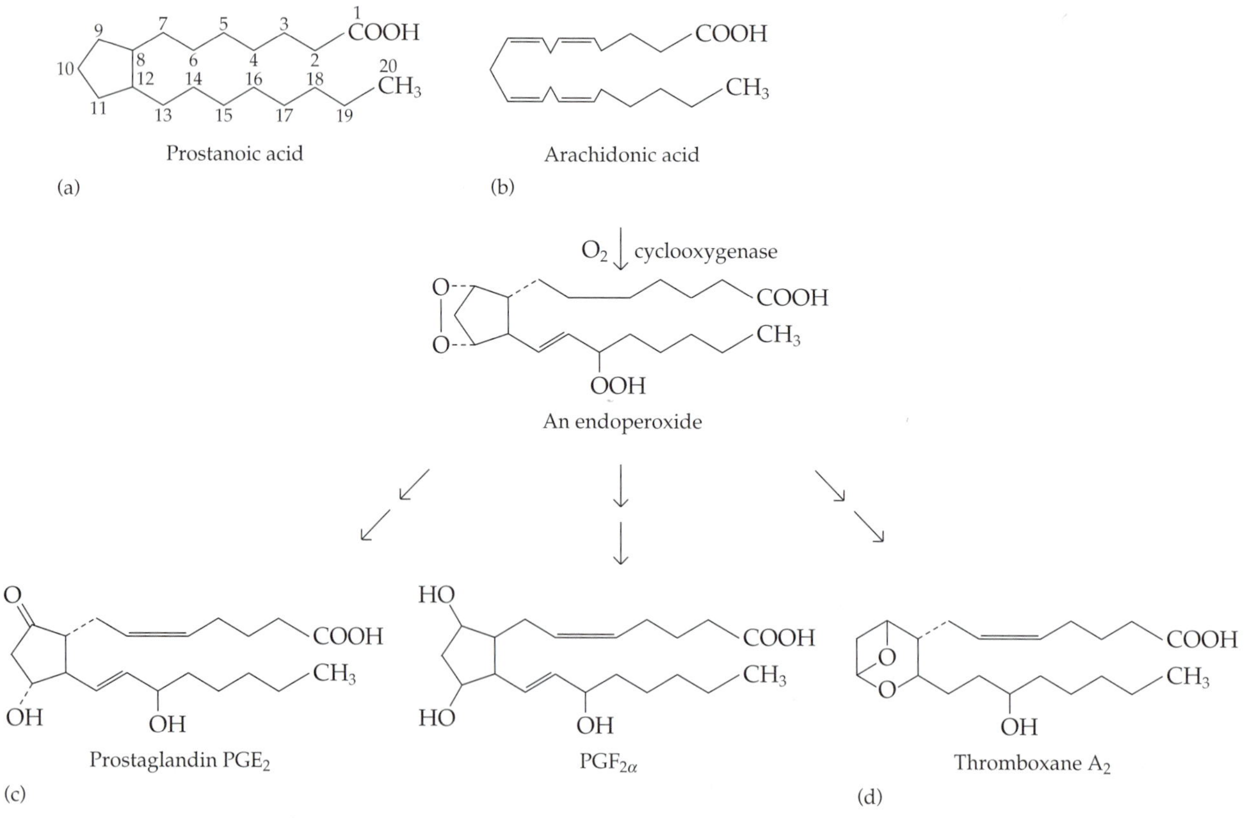

▲ FIGURE I.5

The arachidonic acid cascade. (a) Prostanoic acid is the parent compound of the prostaglandins. (b) Arachidonic acid is released from membrane phospholipids and is the precursor for the synthesis of the prostaglandins (c) and the related thromboxanes (d).

prostaglandins or to a related group of compounds, the thromboxanes. This sequence of reactions is sometimes referred to as the *arachidonic acid cascade* (Figure I.5).

The prostaglandins are among the most potent biological substances known. Slight structural changes are responsible for quite distinct biological effects; however, all prostaglandins exhibit some ability to induce smooth muscle contraction, to lower blood pressure, and to contribute to the inflammatory response (see footnote 2, page 554). We mentioned earlier that certain steroid drugs (cortisol and its synthetic analogs) exert their antiinflammatory action by inhibiting the release of arachidonic acid from membrane phospholipids (i.e., they inhibit *phospholipase*). On the other hand, aspirin and the other nonsteroidal antiinflammatory agents (e.g., indomethacin—Indocin; ibuprofin—Motrin, Advil, Nuprin) obstruct the synthesis of prostaglandins by inhibiting cyclooxygenase, which converts arachidonic acid to endoperoxides.

Their wide range of physiological activity led to the synthesis of hundreds of prostaglandins and their analogs. Two such derivatives are now in use in the United States to induce labor. Others have been employed clinically to lower or increase blood pressure, to inhibit stomach secretions, to relieve nasal congestion, to provide relief from asthma, and to prevent the formation of the blood clots associated with heart attacks and strokes. Thromboxane A_2 (made in blood platelets) induces blood clotting by stimulating blood platelet aggregation. Recall (Section F.1) that one of the side effects of aspirin is a prolonged bleeding time. This is due to the inhibition of platelet aggregation by blocking biosynthesis of thromboxane A_2.

$PGF_{2\alpha}$ is used in cattle breeding. A prize cow is treated with a hormone and with $PGF_{2\alpha}$ to induce the release of many ova. The ova are then fertilized with sperm from a champion bull. The developing embryos are implanted in less valuable cows. This enables a farmer to get several calves a year from one outstanding cow.

The major clinical use of prostaglandins and their analogs is in induction of abortion. Their mechanism is uncertain, but it is different from that of the steroids. The prostaglandins cause regression of the corpus luteum, uterine contractions, and abortion of the embryo. Because they induce abortion, prostaglandins would have to be taken only once a month or only if a menstrual period were missed. Obviously, a great deal of controversy will be generated if and when these compounds become commercially available.[4] Practically every major pharmaceutical company now has active prostaglandin research programs under way to develop new syntheses and to discover new natural sources of prostaglandins.

Key Terms

anabolic steroid (I.3)
androgens (I.3)
estrogens (I.3)
hormone (I.1)
inflammation (I.2)
progestin (I.4)
prostaglandin (I.6)

Review Questions

1. What is the sequence of events that results in the release of hormones from an endocrine gland?
2. Define and give an example of:
   a. hormone **b.** androgen
   c. estrogen **d.** progestin
3. Match each hormone to the gland that produces it:
   a. thyroxine **(1)** pancreas
   b. TSH **(2)** thyroid
   c. insulin **(3)** pituitary
   d. epinephrine **(4)** adrenal
4. What gland produces releasing factors?
5. What gland is the target of releasing factors?
6. Describe the general sequence of events by which a nerve impulse is translated by the endocrine system into a changed physiological state.
7. What is the general structural classification of the compounds incorporated in birth control pills?
8. How do birth control pills work?
9. What structural feature renders a synthetic steroid sex hormone effective orally?
10. What are the differences in biological function between the mineralocorticoids and the glucocorticoids?
11. What fatty acid is the precursor of the prostaglandins?
12. List some potential therapeutic uses of prostaglandins.

[4] As therapeutic agents, prostaglandins must be administered by either intravenous or intrauterine injection. They cannot be taken orally because they are rapidly degraded in the digestive tract. A major goal of researchers is to develop prostaglandin analogs that are orally effective. Several of these drugs will likely be on the market soon, after further testing has demonstrated their relative safety and confirmed their therapeutic value.

21 Proteins

Silkworms and cocoons in an egg box. The caterpillars produce silk, a natural protein used to make clothing.

Carbohydrates, lipids, and proteins—these are the three classes of foods. All are essential to life, but proteins perhaps are closest to the stuff of life itself. No living part of the human body—or any other organism, for that matter—is completely without protein. There is protein in the blood, in the muscles, in the brain, and even in tooth enamel. The smallest cellular organisms, bacteria, contain protein. Viruses, so small that they make bacteria look like giants, are nothing but large molecules of nucleoproteins. (Nucleoproteins are combinations of proteins and nucleic acids—see Chapter 23).

Each type of cell makes its own specific kinds of proteins. Proteins serve as the structural material for animals, much as cellulose does for plants. Muscle tissue is largely protein; so are skin and hair. Proteins are made in different forms in different animals: silk, wool, nails, claws, feathers, horns, and hoofs are all proteins.

Whereas carbohydrates and lipids are used primarily as energy sources, the primary function of proteins is body building and maintenance. Lipids and carbohydrates are stored by the body as energy reserves, but proteins are not stored to any appreciable extent. It is possible for humans to survive for a short period of time on a diet consisting of protein, vitamins, and minerals. We could not survive over the same period of time on a protein-free diet containing lipids, carbohydrates, vitamins, and minerals.

Proteins are polymeric molecules (linear polymers of amino acids) that vary greatly in molecular dimensions. Their molecular weights may range from several thousand to several million daltons. In addition to carbon, hydrogen, and oxygen, all proteins contain nitrogen, and many also contain sulfur, phosphorus, and traces of other elements. The composition of most proteins is remarkably constant at about 51% carbon, 7% hydrogen, 23% oxygen, 16% nitrogen, 1 to 3% sulfur, and less than 1% phosphorus.

The **dalton** is a unit of mass used by biologists. It is equivalent to the atomic mass unit: 1/12 the mass of an atom of carbon-12, 1.66×10^{-24}g. A 30,000-dalton protein has a molecular weight of 30,000 g/u.

In an overcrowded, hungry world, protein is of increasing importance. The rich nations have it—in the form of beefsteak, fish, fowl, soybeans. The poor nations need it but can't afford meat. Many make do with rice or corn and beans. A nation's use of sulfuric acid has long been considered an indication of its industrial development. Perhaps its consumption of protein is a better indication of the quality of life of its people, for without protein no nation can have the healthy, vigorous people vital to progress.

Leather, silk, wool, gelatin, and meat tenderizer are all rich in proteins.

21.1 Amino Acids

Proteins may be defined as compounds of high molecular weight consisting largely or entirely of chains of amino acids. In this respect, proteins may be considered to be polymers analogous to the polysaccharides. However, 20 *different* structural monomeric units are commonly found in proteins, and these are the amino acids. The proteins in all living species, from bacteria to humans, are constructed from the basic set of 20 amino acids. Several other amino acids (e.g., hydroxyproline), which occur to some extent in certain proteins, are all derivatives of the common amino acids and are modified *after* incorporation into the protein chain. With the exception of proline (which contains an alpha secondary nitrogen atom), the amino acids that are the building blocks of proteins are characterized by a primary amino group bonded to the alpha carbon.

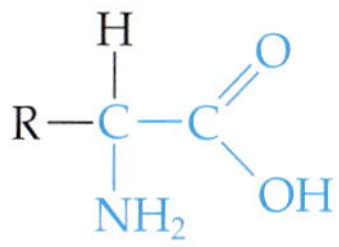

An α-amino acid

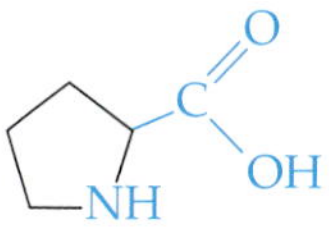

Proline

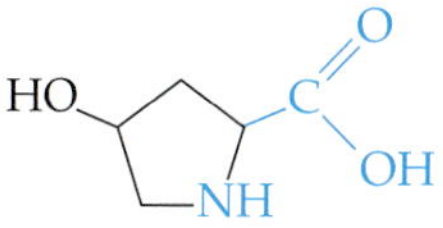

Hydroxyproline

Each amino acid has unique characteristics as a result of the size, shape, solubility, and ionization properties of the different R-groups. (The R-group is referred to as the *amino acid side chain*.) As we shall see in Section 21.8, the side chains of amino acids exert a profound effect on the conformation and the biological activity of proteins.

Amino acids can be classified in several ways. We choose to group them, according to the nature of their side chains, into four classes: (1) nonpolar, (2) polar but neutral, (3) acidic, and (4) basic. The structures of the common amino acids, their three-letter abbreviations, and certain of their distinctive features are given in Table 21.1.

The amino acids are known exclusively by their common names, because the IUPAC names are too cumbersome. Asparagine was the first amino acid to be isolated (1806) and was given its name because it was obtained from protein found in asparagus juice. Glycine, the major amino acid found in gelatin, received its name because of its sweet taste (Greek *glykys*, sweet).

There are more than 150 other amino acids of physiological importance that are not derived from proteins. Most have been isolated from plants, but some are found in animals. These amino acids perform important biological functions (for example, as intermediates in metabolic pathways), either as single molecules or combined in molecules of relatively small size (Figure 21.1).

Table 21.1 Naturally Occurring Amino Acids

Name	Abbrev.	Structural Formula	Molar Mass	Distinctive Features
1. Amino Acids with a Nonpolar R-Group				
Alanine	Ala	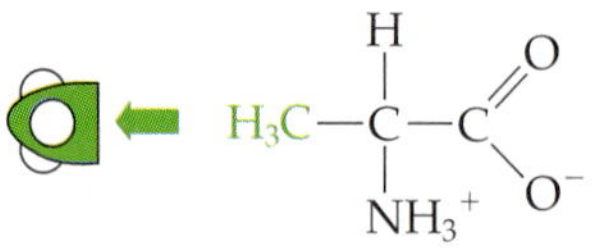	89	The least hydrophobic member of this class because of its small R group (methyl).
Valine*	Val	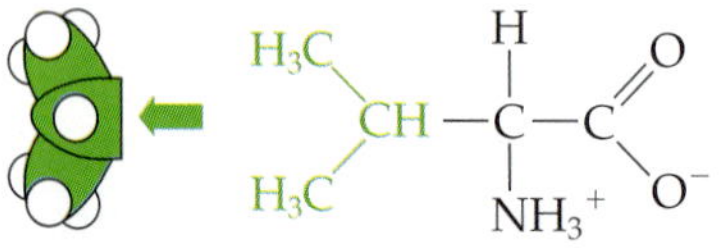	117	
Leucine*	Leu	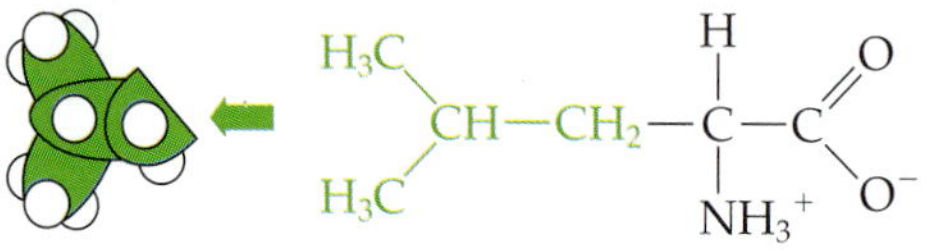	131	Most animals cannot synthesize branched-chain amino acids. They are therefore essential in the diet.
Isoleucine*	Ile	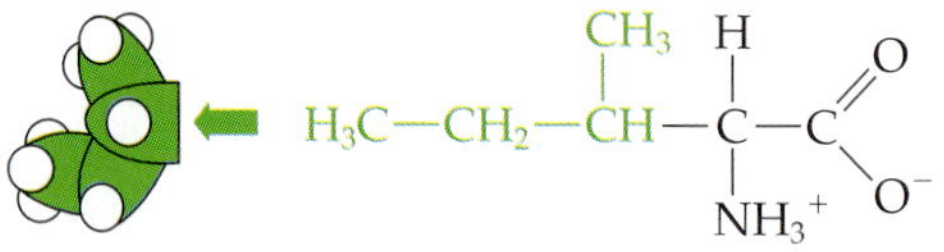	131	
Phenylalanine*	Phe	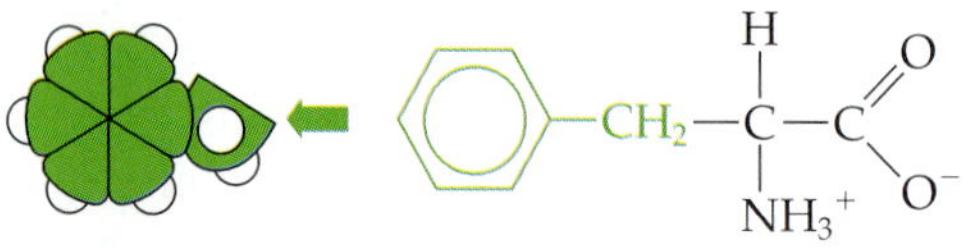	165	
Tryptophan	Trp	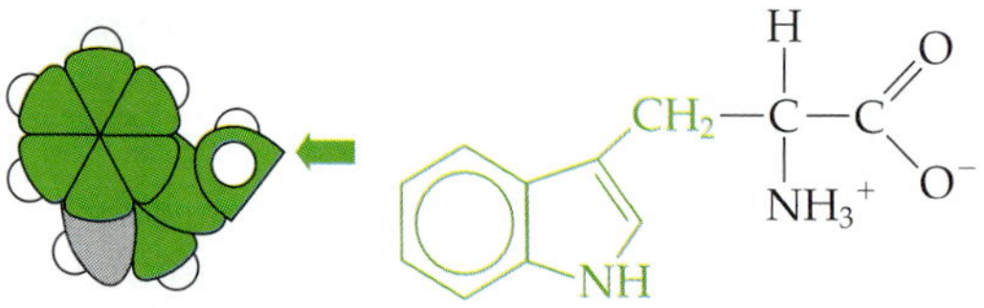	204	A heterocyclic amino acid (a derivative of indole).
Methionine*	Met	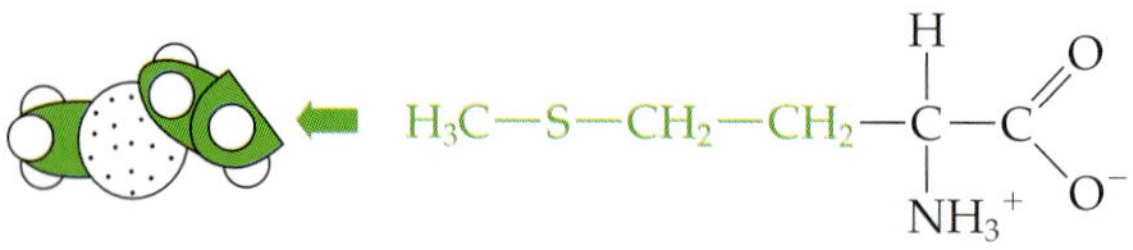	149	Contains a sulfur atom in the nonpolar side chain and is important as a donor of methyl groups.
Proline	Pro	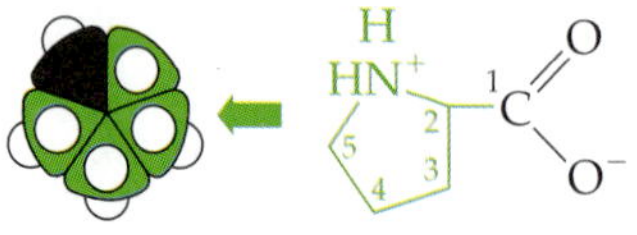	115	Contains a secondary amino group rather than a primary amino group and so is referred to as an *α-imino acid*. A major constituent of the structural protein collagen. Hydroxylation of proline yields 4-hydroxyproline (Hyp), which is also abundant in collagen.

Table 21.1 Cont.

Name	Abbrev.	Structural Formula	Molar Mass	Distinctive Features
2. Amino Acids with a Polar but Neutral R-Group				
Glycine	Gly	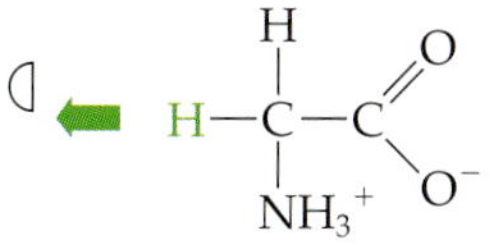	75	The only amino acid lacking a chiral carbon. Sometimes classified as a nonpolar amino acid, but its single hydrogen R-group is too small to influence the polarity of the molecule.
Serine	Ser	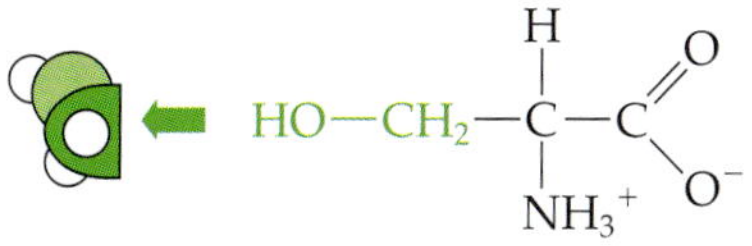	105	Occurs at the active site of many enzymes. The hydroxyl group may take part in the usual alcoholic reactions such as ester formation.
Threonine*	Thr	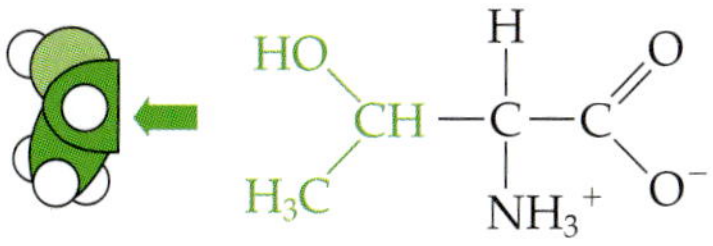	119	Named for its similarity to the sugar threose (contains two chiral carbons).
Cysteine	Cys	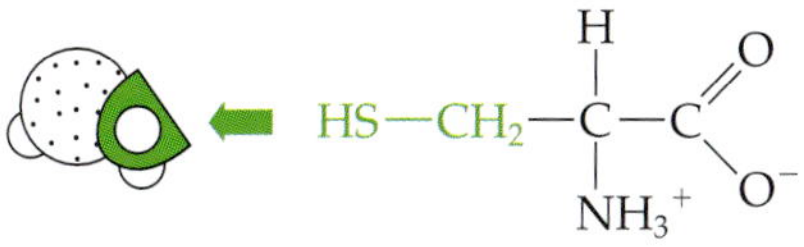	121	Often occurs in proteins in its oxidized form, cystine. The disulfide bond of cystine serves in many proteins as a crosslink between loops of a single chain or between two separate polypeptide chains. (These disulfide bonds are indicated by heavy crosslinks in Figures 21.4 to 21.6).
Tyrosine	Tyr	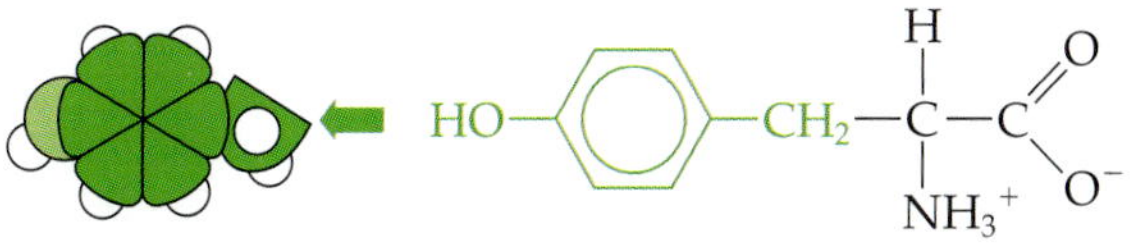	181	The *p*-hydroxy derivative of phenylalanine.
Asparagine	Asn	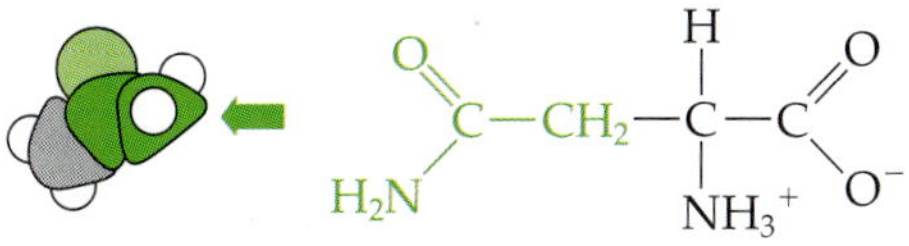	132	The amide of aspartic acid.
Glutamine	Gln	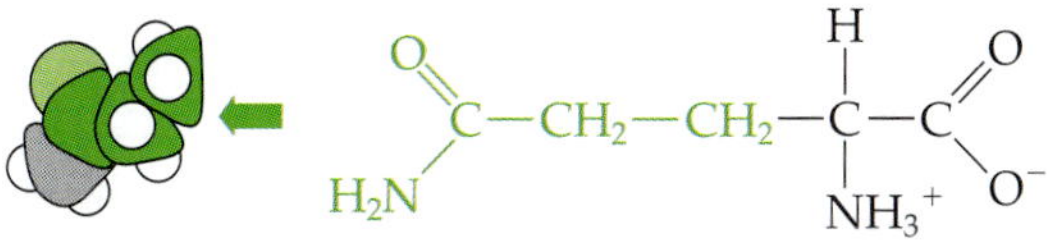	146	The amide of glutamic acid.

Table 21.1 Cont.

Name	Abbrev.	Structural Formula	Molar Mass	Distinctive Features
3. Acidic Amino Acids				
Aspartic acid	Asp	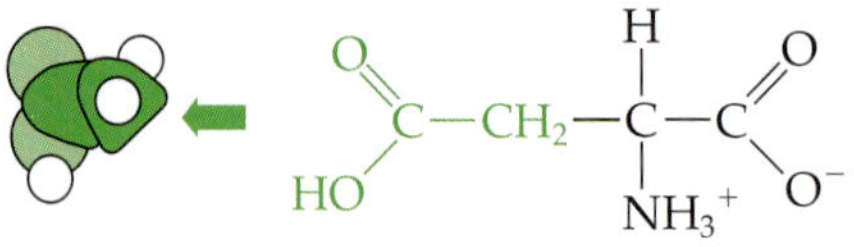	133	The second carboxyl group ionizes. These amino acids are therefore negatively charged at physiological pH.
Glutamic Acid	Glu	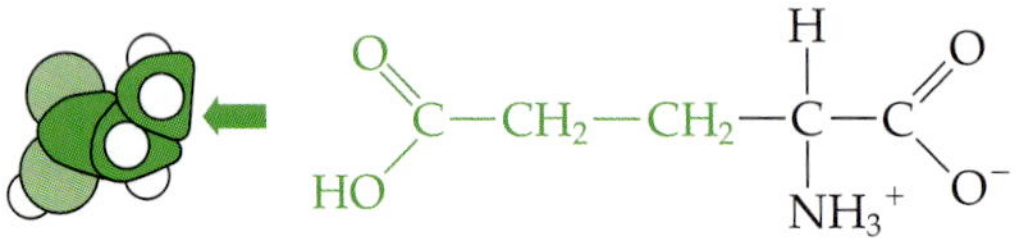	147	
4. Basic Amino Acids				
Lysine*	Lys	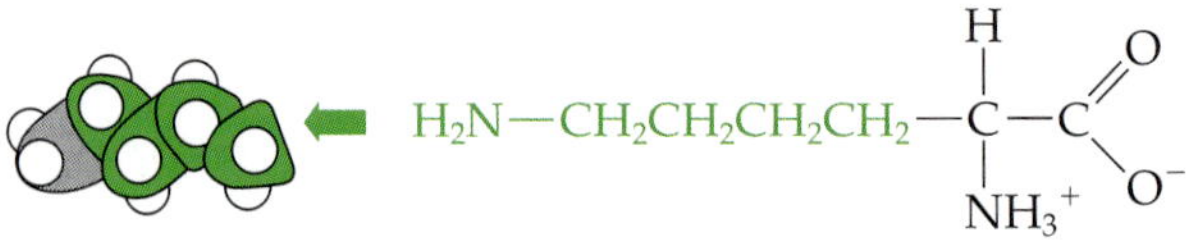	146	The ϵ-amino group of lysine is protonated and thus positively charged at physiological pH.
Arginine*	Arg	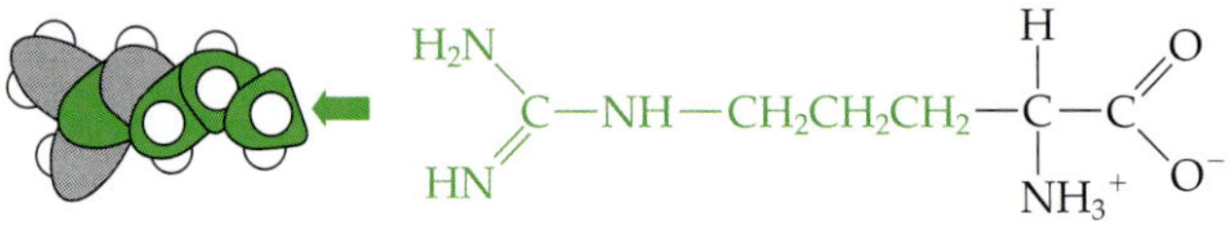	174	Almost as strong a base as NaOH because the guanidyl cation is stabilized by resonance. 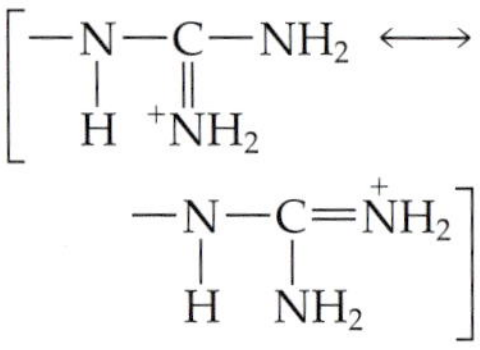
Histidine*	His	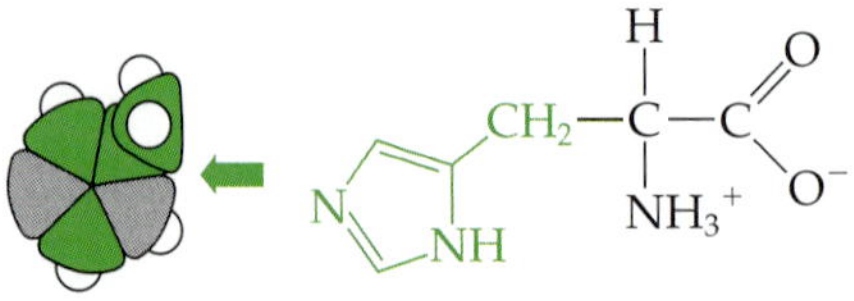	154	The only amino acid whose R group has a pK_a (6.0) near physiological pH and whose isoelectric pH (7.6; see Section 21.3) is near physiological pH. Thus the imidazole ring can be charged (+) or uncharged in the physiological pH range.

*An essential amino acid; see Section 27.3.

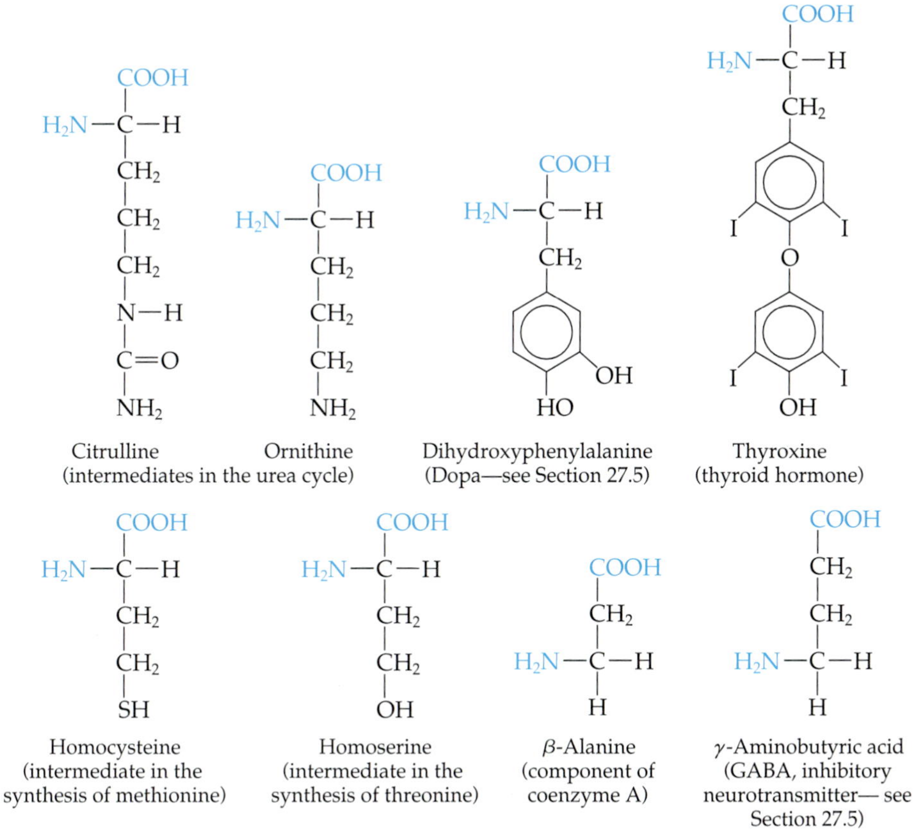

FIGURE 21.1 Some biologically important nonprotein amino acids.

21.2 General Properties of Amino Acids

Configuration

Notice in Table 21.1 that glycine is the only amino acid whose α carbon atom is **not** a chiral center. Therefore, with the exception of glycine, the amino acids are optically active and may exist in either the D or the L enantiomeric form. Once again, the reference compound for the assignment of configuration is glyceraldehyde. (The amino group of the amino acid takes the place of the hydroxyl group of glyceraldehyde.)

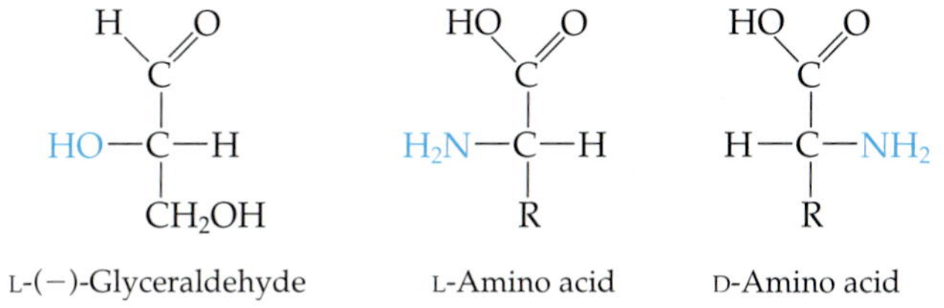

It is interesting to note that the naturally occurring sugars belong to the D series, whereas nearly all known plant and animal proteins are composed entirely of L-amino acids. Recall that the letters D and L refer only to a specific configuration (Section 19.3), not to the direction of optical rotation. For example, in a neutral solution L-alanine is dextrorotatory, whereas L-serine is levorotatory. The optical rotation of any amino acid is very much dependent upon the pH of the solution.

Certain bacteria contain D-amino acids in their cell walls. *Streptococcus faecalis* requires D-alanine, and *Staphylococcus aureus* needs D-glutamic acid. Several antibiotics (e.g., actinomycin D and the gramicidins) contain varying amounts of D-leucine, D-phenylalanine, and D-valine.

Dipolar Ion Structure

The amino acids are colorless, nonvolatile, crystalline solids, melting with decomposition at temperatures above 200 °C. Glycine, alanine, proline, threonine, lysine, and arginine are quite soluble in water. The others are sparingly soluble in varying degrees. All amino acids are insoluble in nonpolar organic solvents. Their properties diverge widely from those of their unsubstituted carboxylic acid analogs. Organic acids of comparable molar mass are liquids or low-melting solids that are soluble in organic solvents but have limited solubility in water. In fact, the properties of the amino acids are more similar to those of inorganic salts than to those of amines or organic acids.

The saltlike character of the amino acids is more readily accounted for if we assign a dipolar ion (also called *inner salt* or **zwitterion**) structure to amino acids in the solid state and in neutral solution (as in Table 21.1). Since amino acids contain both acidic (–COOH) and basic (–NH_2) groups within the same molecule, we may postulate an intramolecular neutralization reaction leading to salt formation:

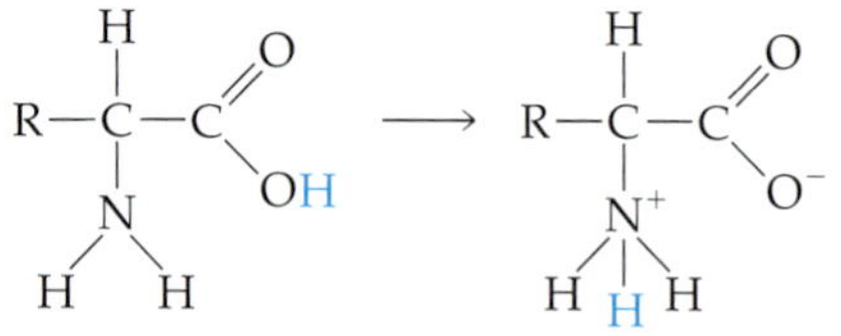

Zwitterion form
of an amino acid
(a dipolar ion)

21.3 Reactions Of Amino Acids

Amino acids can act either as acids or as bases. Indeed, they (and the proteins) act as buffers in living organisms. In the presence of added acid, the carboxylate group of the zwitterion captures protons (note that the product is a positive ion):

$$H_3\overset{+}{N}-\underset{\substack{| \\ R}}{CH}-COO^- + H^+ \longrightarrow H_3\overset{+}{N}-\underset{\substack{| \\ R}}{CH}-COOH$$

If base is added, protons are removed from the amino group of the zwitterion and a negative ion is formed:

$$H_3\overset{+}{N}-\underset{\substack{| \\ R}}{CH}-COO^- + OH^- \longrightarrow H_2N-\underset{\substack{| \\ R}}{CH}-COO^- + HOH$$

In both instances, the amino acid acts to maintain the pH of the system—that is, to tie up added acid and base.

At some intermediate pH value, an amino acid exists almost entirely as the zwitterion. That particular pH at which an amino acid exists in solution as a zwitterion is called the **isoelectric pH.** At the isoelectric pH, the positive and negative charges on an amino acid (or protein) balance, and the molecule as a whole is electrically neutral. At its isoelectric pH, an amino acid (zwitterion) behaves very much like the salt of a weak acid and a weak base. As we noted in Section 11.4, the solution of such a salt can be slightly acidic, slightly basic, or neutral. It depends on which is stronger—the acid or the base from which the salt was formed. Each amino acid has a characteristic isoelectric pH. The neutral amino acids (with

Table 21.2 Isoelectric pH Values of Some Representative Amino Acids

Amino Acid	Type	Isoelectric pH	Typical pK_a of the Side Chain Group
Alanine	Neutral, nonpolar	6.0	—
Valine	Neutral, nonpolar	6.0	—
Serine	Neutral, polar	5.7	—
Threonine	Neutral, polar	6.5	—
Aspartic acid	Acidic	3.0	3.9
Glutamic acid	Acidic	3.1	4.1
Histidine	Basic	7.7	6.0
Lysine	Basic	10.0	10.8
Arginine	Basic	10.8	12.5

unionizable side chains) have isoelectric pH values ranging from 5.0 to 6.5. Basic amino acids (those in which the side chain incorporates a basic group) have relatively high isoelectric pH values. Acidic amino acids have quite low isoelectric pH values (Table 21.2).

By adjusting the pH of a solution, one can vary the net charge on an amino acid, thereby causing amino acids (or proteins) to migrate at different rates in an electric field. This process of separating mixtures of amino acids is called **electrophoresis.** In a typical experiment, a paper strip saturated with a buffer solution at a particular pH is suspended between two reservoirs of the buffer (Figure 21.2). A sample of the solution of amino acids is applied to the center of the paper, and an electric potential is applied between the two buffer solutions. Any amino acids having an isoelectric pH equal to the pH of the buffer are mainly in the zwitterion form and so do not migrate. Any amino acids primarily in the anionic form move toward the positive electrode, and any mainly in the cationic form migrate toward the negative electrode. Amino acids of different sizes move at different rates, even if both have the same electric charge. After a period of time, the various amino

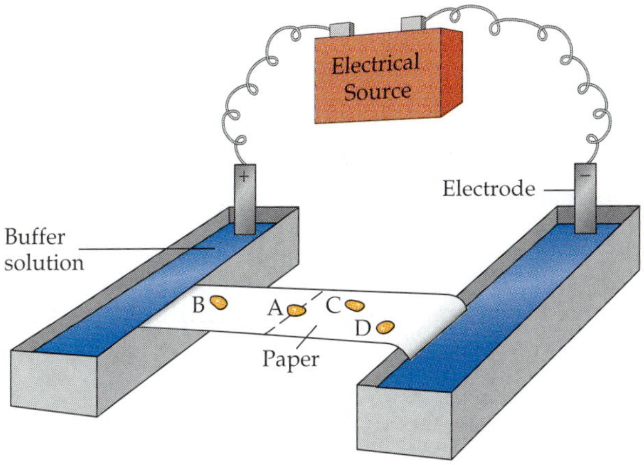

FIGURE 21.2

An electrophoresis apparatus. Amino acid A is in the zwitterion form and has not migrated; B exists as an anion and has therefore moved toward the positive electrode; C and D are in the cationic form and have therefore migrated toward the negative electrode.

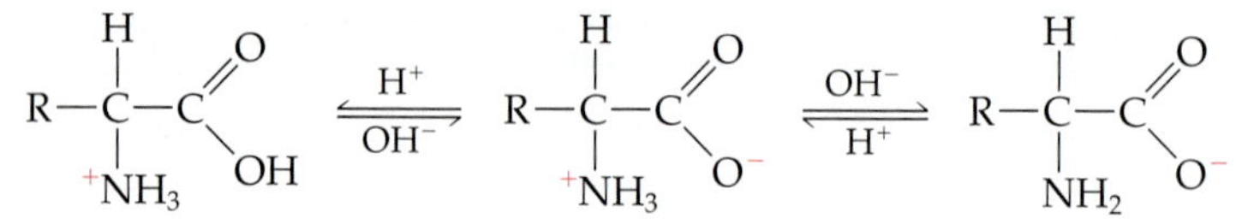

FIGURE 21.3 ▶ Acid–base behavior of neutral amino acids.

acids will have separated into individual spots on the paper. Figure 21.3 illustrates the forms of the neutral amino acids in acidic, neutral, and basic solutions.

Simple chemical tests also can be used to identify amino acids. These compounds undergo reactions characteristic of carboxylic acids and amines. One of possible interest is that with nitrous acid. This reagent reacts with free amino groups, and nitrogen gas is formed (recall page 454):

$$R{-}\underset{\underset{^{+}NH_3}{|}}{CH}{-}COO^- + HNO_2 \longrightarrow R{-}\underset{\underset{OH}{|}}{CH}{-}COOH + N_2(g) + H_2O$$

This reaction is the basis of the Van Slyke method for determining the number of free amino groups in a protein. The ninhydrin test (page 455) gives a purple color when amino acids are present and serves as a qualitative test for amino acids.

The preceding reactions and other similar ones are quite important in the detection and separation of amino acids. The most important reaction of all, however, is the polymerization reaction that forms peptides and proteins (Sections 21.4 and 23.6).

21.4 The Peptide Bond

Proteins are condensation polymers of amino acids. How are the monomer units held together? In Section 17.5 we discussed the reaction between amines and carboxylic acids to form substituted amides. If we heat the salt of an amine and a carboxylic acid, an amide is formed:

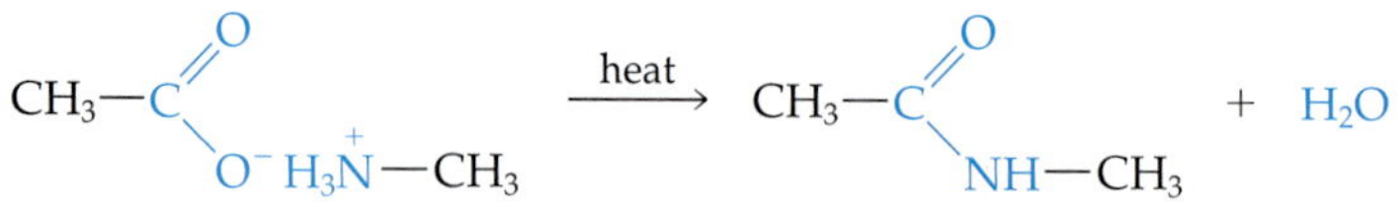

Methylammonium acetate N-Methylacetamide

Similarly, the amino group on one amino acid molecule can react with the carboxyl group on another. A molecule of water is split out and an amide linkage is formed:

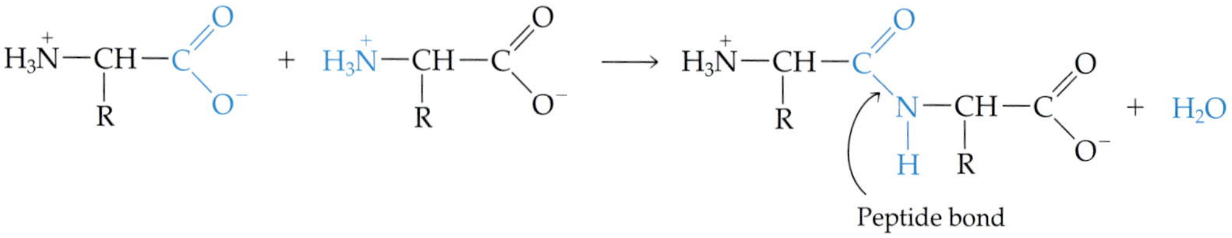

The amide linkage is called a **peptide bond** when it joins two amino acid units. Note that in the product molecule, there is still a reactive amino group on the left and a carboxyl group on the right. Each of these can react further to join more amino acid units. This process can continue until thousands of units have joined to form a gigantic molecule—a polymer called a protein. (Synthesis of proteins in living organisms is quite complex and is discussed in more detail in Chapter 23.)

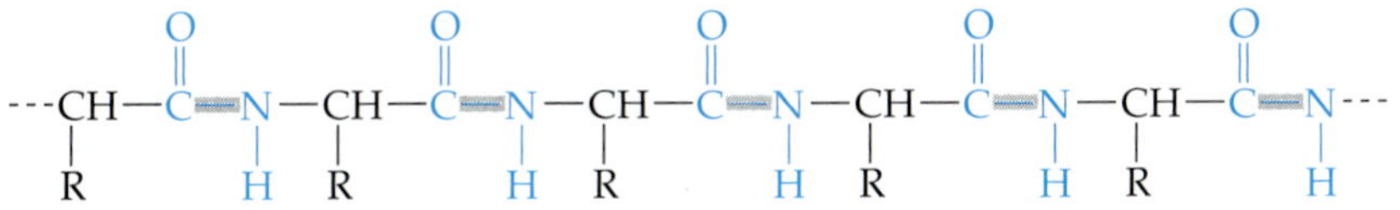

When only two amino acids are joined, the product is called a dipeptide:

Prefixes (di, tri, tetra, etc.) are used to indicate the number of amino acids that are joined together. The general term **peptide** is often used to designate a combination of an unspecified number of amino acids.

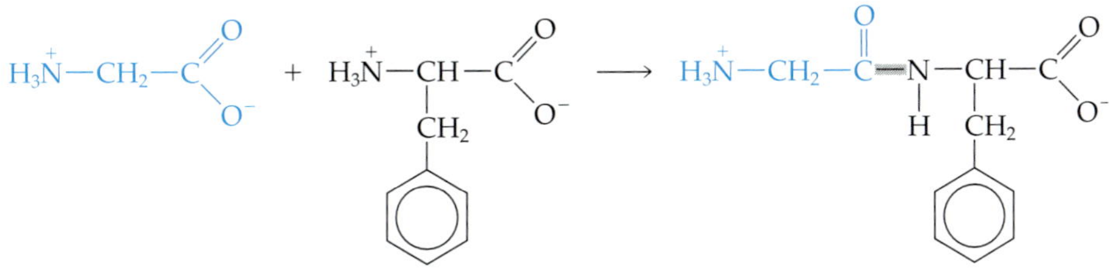

Glycine Phenylalanine Glycylphenylalanine (a dipeptide)

When three amino acids are combined, the substance is a tripeptide:

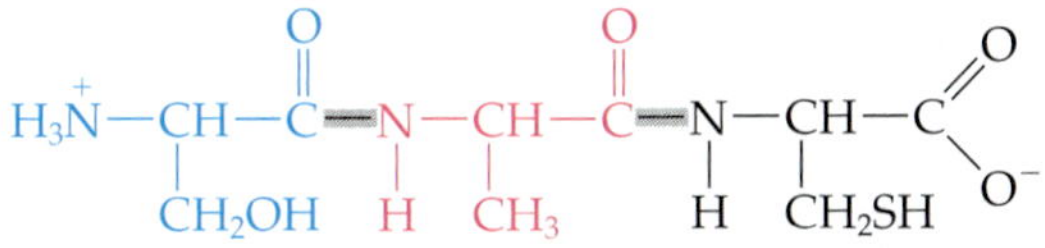

Serylalanylcysteine (a tripeptide)

By convention, we represent the structure of peptides beginning with the amino acid whose amino group is free (the *N-terminal end*). The other end, therefore, contains a free carboxyl group and is referred to as the *C-terminal end.* Each amino acid in the molecule, with the exception of the C-terminal amino acid, is named as an acyl group in which the suffix *-ine* is replaced by *-yl.*

Practice Exercise

Draw structures for the tripeptides

a. valylmethionylthreonine

b. prolylaspartyltryptophan

c. tyrosyllysylglutamine

Combinations with more than 10 amino acid units are often simply called **polypeptides.** When the molecular weight of a polypeptide exceeds 10,000 daltons, it is called a protein. The distinction between polypeptides and proteins is an arbitrary one, and it is not always precisely applied. However, this is not meant to imply that a protein is composed of only one polypeptide chain. The enzymes lysozyme (Figure 21.4) and ribonuclease (Figure 21.5) contain 129 and 124 amino acids, respectively, in only one polypeptide chain, but the hormone insulin (Figure 21.6) contains two polypeptide chains (one chain contains 21 amino acids, the other 30), and the hemoglobin molecule contains four polypeptide chains (see Figure 21.15).

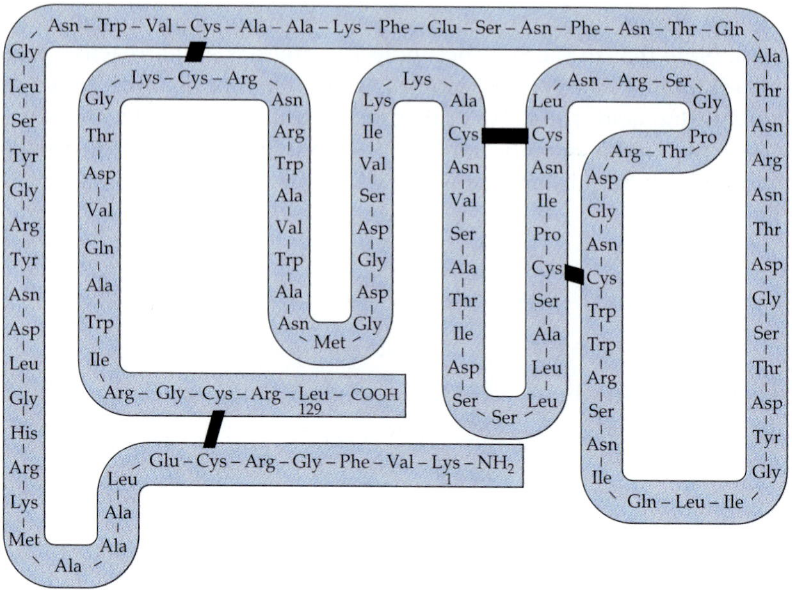

FIGURE 21.4 ▶

The amino acid sequence of lysozyme, a protein enzyme found in humans, plants, and the whites of eggs. It destroys invading bacteria by catalyzing the cleavage of polysaccharide chains that form part of the bacterial cell wall. Without a rigid cell wall, a sudden influx of water bursts the bacterial cell. The thick black lines represent disulfide linkages (see page 582).

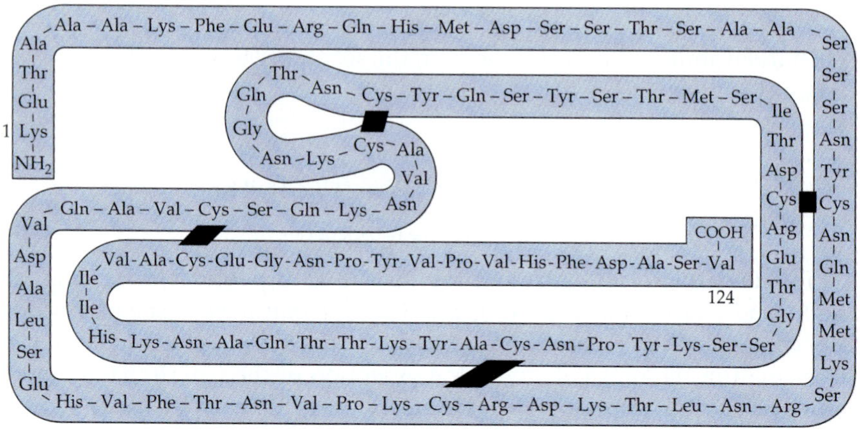

FIGURE 21.5 ▶

The amino acid sequence of bovine ribonuclease, a protein enzyme that catalyzes the hydrolysis of ribonucleic acid (RNA).

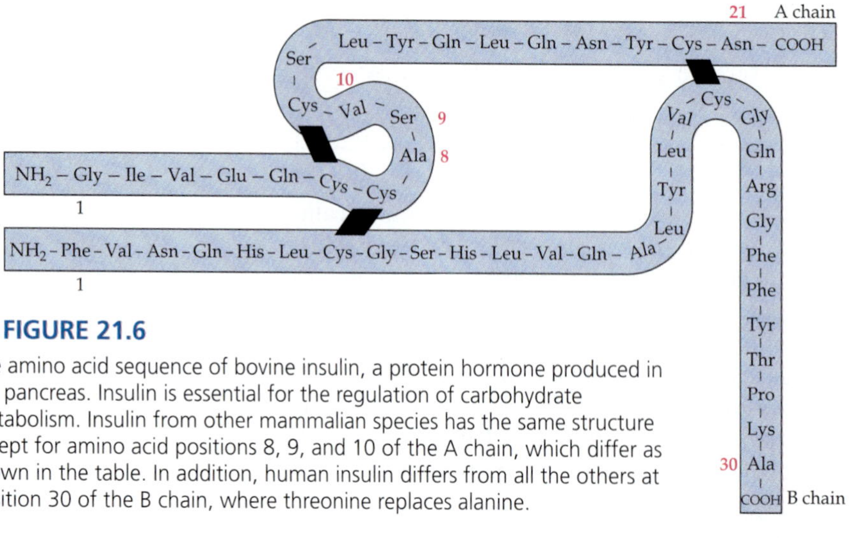

Species	Amino Acid Position #8	#9	#10
Cow	Ala	Ser	Val
Sheep	Ala	Gly	Val
Horse	Thr	Gly	Ile
Human	Thr	Ser	Ile
Pig	Thr	Ser	Ile
Whale	Thr	Ser	Ile
Dog	Thr	Ser	Ile

▲ FIGURE 21.6

The amino acid sequence of bovine insulin, a protein hormone produced in the pancreas. Insulin is essential for the regulation of carbohydrate metabolism. Insulin from other mammalian species has the same structure except for amino acid positions 8, 9, and 10 of the A chain, which differ as shown in the table. In addition, human insulin differs from all the others at position 30 of the B chain, where threonine replaces alanine.

21.5 The Sequence of Amino Acids

For peptides and proteins to be physiologically active, it is not enough that they incorporate certain amounts of specific amino acids. The order, or *sequence*, in which the amino acids are connected is also of critical importance. Glycylalanine is different from alanylglycine.

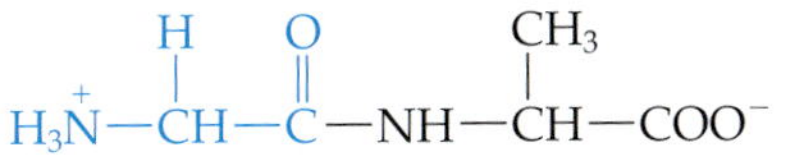

$$
\underset{\text{Glycylalanine (Gly-Ala)}}{H_3\overset{+}{N}-CH-\overset{\overset{\displaystyle O}{\|}}{C}-NH-\overset{\overset{\displaystyle CH_3}{|}}{CH}-COO^-}
\qquad
\underset{\text{Alanylglycine (Ala-Gly)}}{H_3\overset{+}{N}-\overset{\overset{\displaystyle CH_3}{|}}{CH}-\overset{\overset{\displaystyle O}{\|}}{C}-NH-\overset{\overset{\displaystyle H}{|}}{CH}-COO^-}
$$

Although the difference seems minor, the two substances behave differently in the body.

As the length of a peptide chain increases, the possible sequential variations become very large. And this potential for many different arrangements is exactly what one needs in a material that makes up such diverse things as hair and skin and eyeballs and toenails and a thousand different enzymes. To appreciate the enormous complexity of protein molecules, consider that the average protein contains anywhere from 100 to 300 amino acids. It is estimated that a typical animal cell contains about 9000 different proteins, and the human body contains over 100,000 different protein molecules, all of which are characterized by different sequential arrangements of the 20 fundamental building blocks.

Just as we can make millions of different words with our 26-letter English alphabet, we can make millions of different proteins with the 20 or so different amino acids. Just as one can write gibberish with the English alphabet, one can make nonfunctioning proteins by putting together the *wrong sequence* of amino acids. Yet while the correct sequence is ordinarily of utmost importance, it is not always absolutely required. Just as you can sometimes make sense of incorrectly spelled English words, a protein with a small percentage of "incorrect" amino acids may continue to function.[1] It may not function as well, however, as a protein having the correct sequence. And sometimes a seemingly minor change can have a disastrous effect. Some people have hemoglobin with a single incorrect amino acid unit in about 300. That "minor" error is responsible for sickle cell anemia, an inherited condition that ordinarily proves fatal (Section 28.9).

21.6 Some Peptides of Interest

There are several naturally occurring peptides that possess significant biological activity. In Section F.5, we said that the brain produces a variety of peptides, several of which act like morphine to relieve pain. Rather, we should say that morphine acts like these peptides because the brain was making peptides long before people discovered the painkilling effect of the juice of the opium poppy.

[1] An interesting fact about living systems is that different ones contain, in many cases, almost identical protein molecules. Insulin is a striking example of this phenomenon (see Figure 21.6). Not only do a variety of mammalian insulin molecules have similar primary structures, but they also have the same biochemical properties. This is fortunate for some diabetics who develop an allergy to a particular type of insulin—they can often be treated with insulin from another species that does not produce the allergic reaction. In this instance, alteration of some of the amino acids does not cause the protein to have an altered physiological effect. This explanation of the special physiological functions of proteins in terms of their structure will become apparent when we consider the concept of the *active site* in relation to enzymes in Section 22.3. (As we shall see in Section 23.9, human insulin is now being produced by recombinant DNA technology, and this is diminishing the allergy problem.)

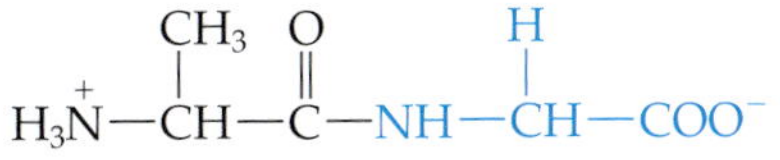

Table 21.3 A Comparison of the Effects of Oxytocin and Vasopressin

Structure or Function Affected	Vasopressin	Oxytocin
Water diuresis	Inhibits	Has no effect on
Blood pressure	Elevates	Slightly lowers
Coronary arteries	Constricts	Slightly dilates
Intestinal contractions	Stimulates	Has questionable effect on
Uterine contractions	Stimulates	Stimulates
Ejection of milk	Slightly stimulates	Stimulates

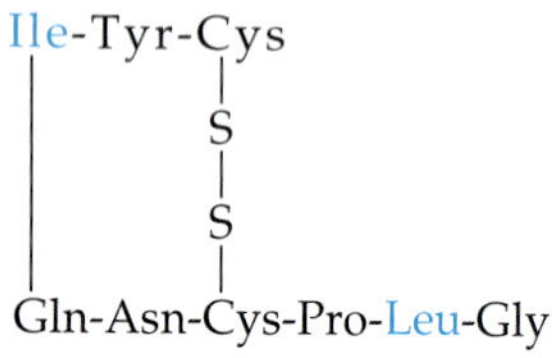

Oxytocin

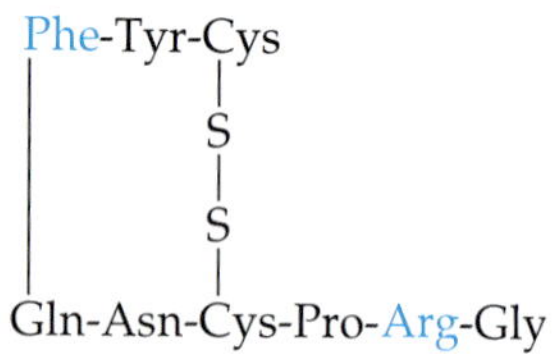

Vasopressin

The hormones *oxytocin* and *vasopressin* are cyclic nonapeptides produced by the pituitary gland. Notice that seven of the nine amino acids are identical in both peptides, and yet their physiological effects are markedly different. Oxytocin stimulates lactation and causes the contraction of smooth muscles in the uterine wall. It is often administered at childbirth to induce labor. Vasopressin is called the *antidiuretic hormone* (ADH) because it acts on the kidneys to reduce the amount of water excreted. (**Diuretics** are substances that increase the volume of urine, and therefore any substance that reduces the volume of urine is an antidiuretic.)[2] Thus, the major function of vasopressin is to increase water reabsorption in the kidney. After drinking alcohol, many people excrete more urine than can be accounted for by the volume of water in their drinks. It is thought that alcohol inhibits the secretion of vasopressin, and therefore the kidney does not reabsorb as much water. A deficiency of vasopressin, or an inability of the kidney to respond to vasopressin, results in diabetes insipidus, in which too much urine is excreted (>10 L/day). This disease (which is treated by administering vasopressin) should not be confused with diabetes mellitus (see Section 24.4). In addition, vasopressin stimulates the contractions of muscles in the walls of blood vessels and thus increases the blood pressure. It has been used to overcome low blood pressure caused by shock following surgery.

Table 21.3 offers a comparison of the effects of vasopressin and oxytocin. Remember the great similarity in the structures of these compounds as you look at the table.

Bradykinin, a nonapeptide produced in the blood by the cleavage of larger protein molecules, has the amino acid sequence

Arg-Pro-Pro-Gly-Phe-Ser-Pro-Phe-Arg

In terms of activity, bradykinin, a peptide, sounds almost like the prostaglandins (Section I.6), which are lipids. It is a potent biochemical that lowers blood pressure, stimulates smooth muscle tissue, increases capillary permeability, and causes pain. The reverse peptide,

Arg-Phe-Pro-Ser-Phe-Gly-Pro-Pro-Arg

has been synthesized. It shows none of the activity of bradykinin.

The octapeptide angiotensin II is produced in the kidneys.

Asp-Arg-Val-Tyr-Ile-His-Pro-Phe

This substance is the most powerful vasoconstrictor known. It acts to maintain blood pressure. Some forms of hypertension probably involve overproduction of

[2] Diuretics are often given to people with high blood pressure (see Section 28.7) to cause the loss of water and sodium ions, both of which contribute to the elevated blood pressure.

Table 21.4 Some Proteins Whose Sequences of Amino Acids Are Known

Protein	Function	Number of Amino Acids
Enzymes		
Ribonuclease	Hydrolyzes RNA	124
Lysozyme	Cleaves bacterial cell walls	129
Papain	Digests proteins	212
Trypsinogen	Digests proteins	229
Chymotrypsinogen	Digests proteins	245
Carbonic anhydrase (human)	Hydrates CO_2	260
Subtilisin (a bacterial protease)	Digests bacterial proteins	274
Carboxypeptidase A (bovine)	Digests proteins	307
Alcohol dehydrogenase (horse)	Oxidizes alcohol	374
Others		
Nisin	Antibiotic	29
Insulin	Variety of metabolic functions	51
Trypsin inhibitor (bovine pancreas)	Inhibits trypsin	58
Cytochrome c (human, horse, pig, rabbit, chicken)	Electron transport	104
Cytochrome c (yeast)	Electron transport	108
Hemoglobin (human)	Oxygen transport	
alpha chain		141
beta chain		146
Calmodulin (bovine)	Calcium transport	148
Myoglobin	Oxygen storage	153
Tobacco mosaic virus protein subunit	Virus protein	158
Myelin (bovine)	Protects nerve cells	170
Myelin (human)	Protects nerve cells	172
Human growth hormone	Necessary for normal growth	191
α-Casein (bovine)	Milk protein	199
Human serum albumin	Blood protein	584
Immunoglobulin, IgG	Antibody	1320

angiotensin II. Drugs that act to suppress its production are important in the control of hypertension.

There are many other physiologically important polypeptides. Some of nature's most potent toxins, such as snake venom and bacterial toxins, are polypeptides. The amino acid sequences are known for over 3000 polypeptides, and partial sequences have been determined for many others (Table 21.4). Some of these polypeptides contain hundreds of amino acid units. Many have been synthesized in laboratories around the world. The first polypeptide hormone synthesized was oxytocin, in 1953. Synthetic polypeptides have the same physiological properties as the corresponding natural ones.

21.7 Classification of Proteins

There are many kinds of proteins. Each has its own characteristic composition, amino acid sequence, and three-dimensional shape. Because of their great complexity, protein molecules cannot possibly be classified systematically in the same way as the carbohydrates and lipids are categorized—that is, on the basis of structural similarities. *One way to classify proteins is based on solubility.* Some proteins, such as those that make up hair, skin, muscles, and connective tissue, are fiberlike. These *fibrous proteins* are insoluble in water. They usually serve structural, connective, and protective functions. Examples of fibrous proteins are keratins, collagens, myosins, and elastins. Hair and the outer layer of skin are composed of *keratin.* Connective tissues contain *collagen. Myosins* are muscle proteins and are involved in contraction and extension of muscles. *Elastins* are found in the elastic tissue of artery walls and in ligaments.

Globular proteins, the other major class, are soluble in aqueous media. The protein chains of globular proteins are folded so that the molecule as a whole is roughly spherical. The mixtures of globular proteins and water are actually colloidal dispersions rather than true solutions. Familiar examples are the *albumins.* Egg albumin is obtained from egg whites. Serum albumin is present in blood, and it plays a major role in maintaining a proper balance of osmotic pressures in the body (Section 28.6). *Globulins* are a second group of globular proteins. Hemoglobin and myoglobin are two examples. Another is the serum globulins, which are a part of our defense against disease.

21.8 Structure of Proteins

The structure of proteins is generally discussed at four organizational levels. The **primary structure** of a protein refers to the number and sequence of the amino acids in its polypeptide chain(s). To specify primary structure, we start (by convention) at the left with the free amino group end, and then write out the sequence of amino acids in the protein. What "holds" the primary structure together are the peptide bonds between the amino acid units.

Protein molecules aren't just arranged at random as tangled threads. The chains are held together in unique configurations. The term **secondary structure** refers to the fixed arrangement of the polypeptide backbone. This arrangement may be a helix (spiral) as in wool, a pleated sheet as in silk, or whatever. The term **tertiary structure** refers to the unique three-dimensional shape that results from

FIGURE 21.7 ▶

The tertiary structure of a protein like myoglobin. This tube or "sausage" model shows the coiled protein backbone and the approximate volume occupied by the protein.

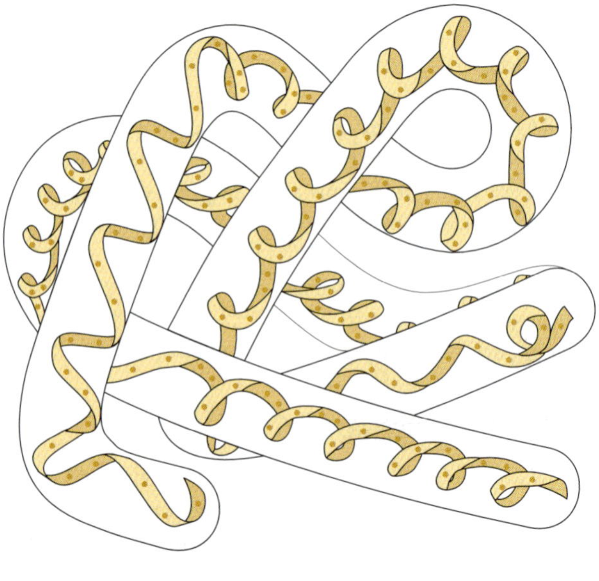

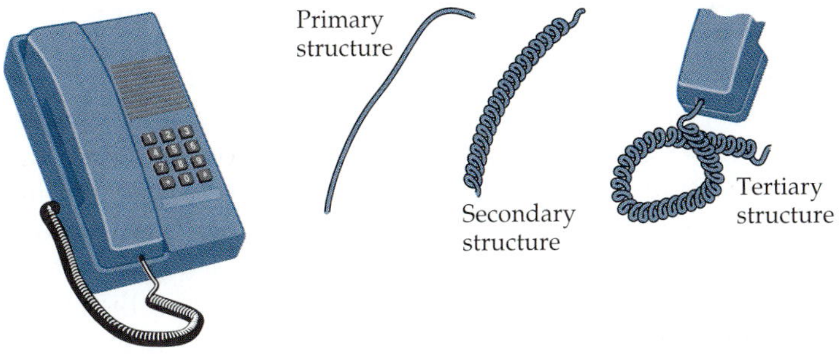

◀ **FIGURE 21.8**
Three levels of structure of a telephone cord.

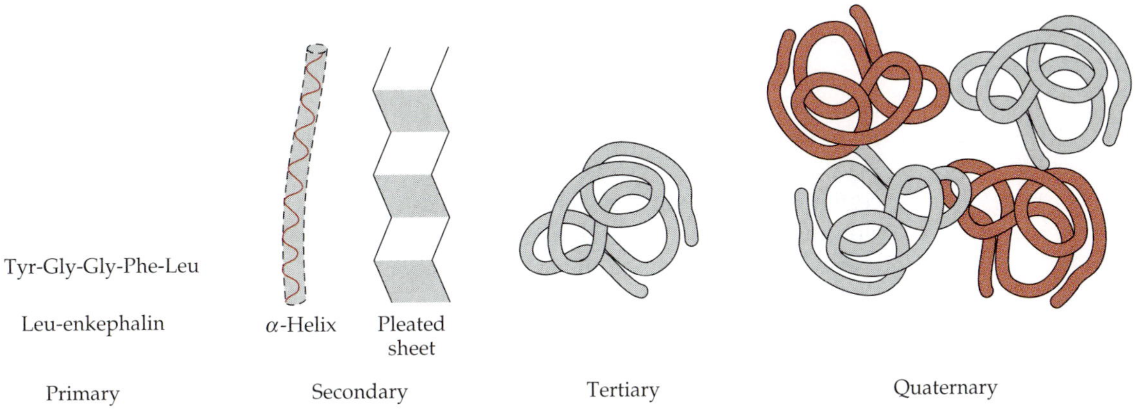

▲ **FIGURE 21.9**
A schematic representation of the four levels of structure in proteins.

the precise folding and bending of the protein backbone. An example is the protein chain in globular proteins, which is folded into a compact spherical shape (Figure 21.7). The tertiary structure of a protein is intimately involved with the proper biochemical functioning of that protein, as we shall see in the next chapter.

We can relate these three levels of organization to a more familiar object. Think of the coiled cord on a telephone receiver. The cord starts out as a long, straight wire (Figure 21.8). We'll call that the primary structure. The wire is coiled into a helical arrangement. That's its secondary structure. When the receiver is hung up, the coiled cord folds into a particular pattern. That would be its tertiary structure.

Some proteins contain more than one polypeptide chain (as subunits), and this multichain arrangement causes the molecule to have another level of structure. The **quaternary structure** of a protein describes the way in which the subunits are packed together in the protein molecule. Hemoglobin is the most familiar example of a protein having quaternary structure. The four polypeptide chains are arranged in a specific pattern (see Figure 21.15). We consider hemoglobin in much greater detail in Chapter 28.

The next sections of this chapter will help you gain some insight into secondary, tertiary, and quaternary organization. A schematic representation of the four levels of protein structure is shown in Figure 21.9.

21.9 Secondary Structure of Proteins

Two major considerations are involved in the secondary structure of proteins. The first involves the manner in which the protein chain is folded and bent; the second involves the nature of the bonds that stabilize this structure.

Based upon X-ray studies, Linus Pauling and Robert Corey postulated that some proteins have a spiral shape (that is, they are shaped like a helix). This shape is best visualized as a spring coiled about an imaginary cylinder (Figure 21.10). The spiral, or helix, is stabilized by hydrogen bond formation between the amide hydrogen of one peptide bond and the carbonyl oxygen above it, which is located on the next turn of the helix. This *intrachain* hydrogen-bonded structure is designated as α-helical. X-ray data indicate that the helix makes one turn for every 3.6 amino acids, and that the side chains of these amino acids project outward from the coiled backbone. The α-keratins, found in hair and wool, are exclusively α-helical in conformation.

Not all proteins assume a helical conformation. Some proteins, such as gamma globulin, chymotrypsin, and cytochrome c, have little or no helical structure. Other proteins, such as hemoglobin and myoglobin, are helical in certain regions of the polypeptide chain; the remaining portions assume random conformations. The polypeptide chains of structural proteins such as silk fibroin and certain enzymes such as carboxypeptidase A and lysozyme are aligned side by side in a sheetlike arrangement. In these proteins, segments of the polypeptide chains lie next to one another and run either parallel or antiparallel, with *interchain* hydrogen bonding connecting the adjacent strands (Figure 21.11). This structural arrangement is designated as the pleated sheet conformation, and it occurs when two extended polypeptide chains (or two separate regions on the same chain) are aligned side by side.

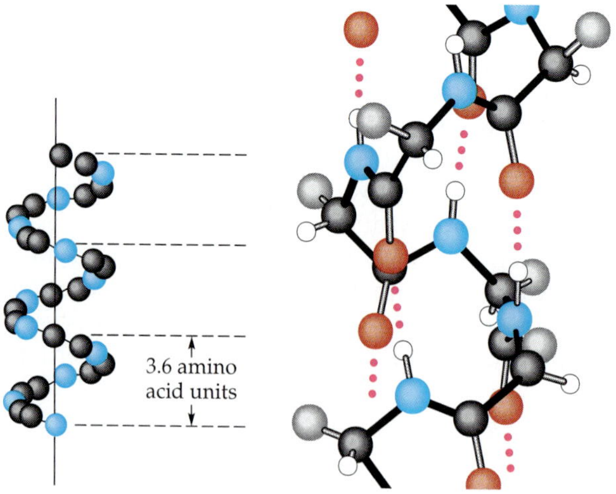

(a) Skeletal representation (b) Ball-and-stick model

(c) Space-filling model

▲ **FIGURE 21.10**

Three representations of the α-helical conformation of a protein chain. (a) The skeletal representation best shows the helix. (b) Intrachain hydrogen bonding between turns of the helix is shown in the ball-and-stick model. (c) The space-filling model shows the shape of a short segment of the chain.

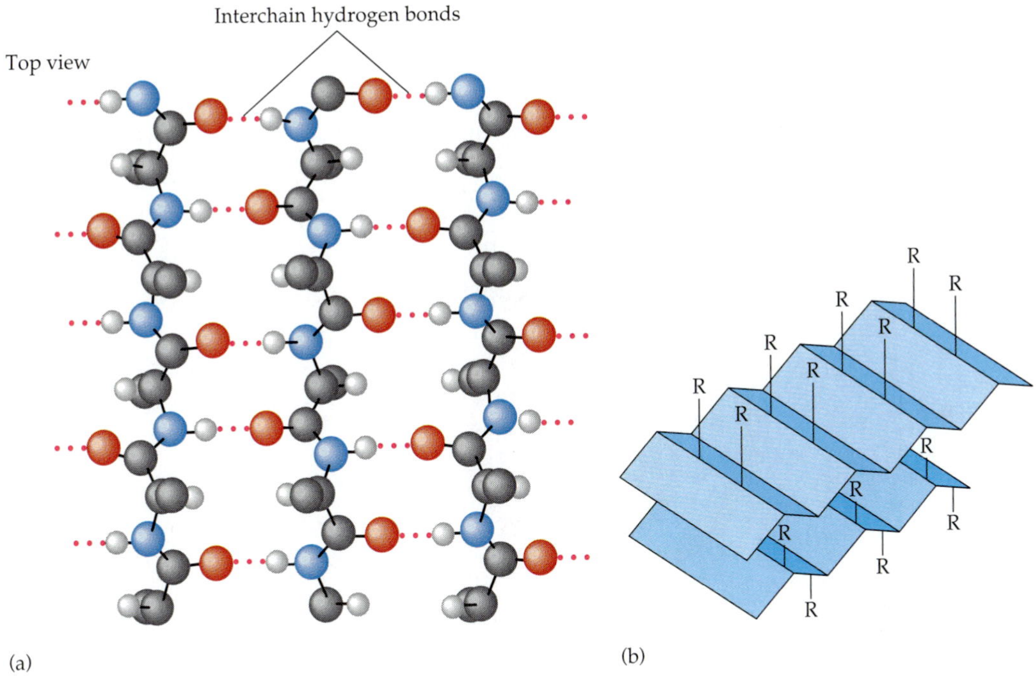

▲ **FIGURE 21.11**

Pleated sheet conformation of protein chains. (a) Ball-and-stick model. (b) Schematic drawing emphasizing the pleats. The peptide bonds lie in the plane of the pleated sheet. The side chains extend above or below the sheet and alternate along the chain. The protein chains are held together by interchain hydrogen bonds.

The physical characteristics of wool and silk are a result of their structural conformations. Wool is very flexible and extensible. It can stretch to twice its normal length without breaking, and the fiber will return to its original state upon release of tension. (Think of how the coiled cord on a telephone can be stretched.) The stretching process involves breaking hydrogen bonds along turns of the α-helix (covalent bonds remain intact). The disulfide bonds between helices, together with re-formed hydrogen bonds, provide the forces that operate to restore the helix when tension is released.

Silk, on the other hand, is already stretched out in the pleated sheet conformation. Silk fibers have the hydrogen-bonded layers arranged one over the other. The properties of silk—strength, flexibility, and resistance to stretching—are a consequence of its structure. Breaking the fibers involves rupturing thousands of hydrogen bonds or breaking covalent bonds, and because the chains are already fully extended, the fibers cannot be stretched easily.

21.10 Tertiary Structure of Proteins

The linkages responsible for the tertiary structure of a protein are a function of the nature of the amino acid side chains within the molecule. Globular proteins are extremely compact, almost spherical in shape. Such proteins have their nonpolar side chains directed toward the interior of the molecule (the hydrophobic or nonaqueous region) and their polar side chains project outward from the surface of the molecule toward the aqueous environment. The resulting picture is very similar to that of a micelle, which was discussed in connection with the properties of cell membranes (Section 20.8). Some of the linkages that contribute to the

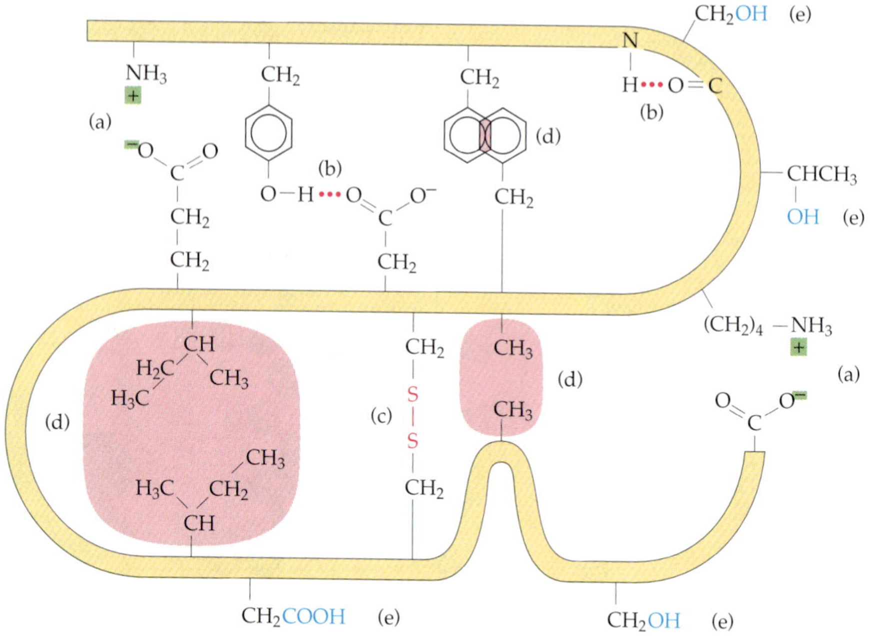

FIGURE 21.12 ▶

Bonds that stabilize the tertiary structure of proteins. (a) Salt linkages, (b) hydrogen bonds, (c) disulfide linkages, (d) hydrophobic interactions, (e) polar groups that interact with water.

tertiary structure of proteins are shown in Figure 21.12. Table 21.5 gives an indication of the relative strengths of interactions involving the noncovalent bonds found in proteins.

Salt Linkages

Salt linkages (ionic bonds) result from electrostatic interactions between positively and negatively charged groups on the side chains of the basic and acidic amino acids. For example, the mutual attraction between an aspartic acid carboxylate ion and a lysine ammonium ion helps to maintain a particular folded area of the protein.

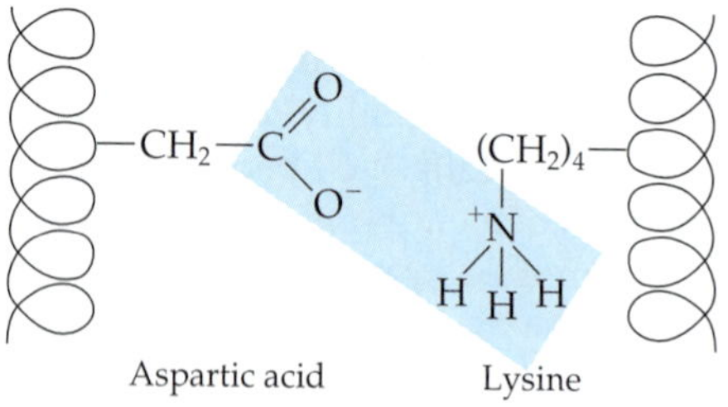

Hydrogen Bonding

Hydrogen bonds are formed principally between the side chains of the polar amino acids and between a carboxyl oxygen and a hydrogen donor group. The hydrogen-bonding capabilities of the terminal amino group of lysine and the terminal carboxyl groups of aspartic acid and glutamic acid are pH-dependent. These groups can serve as both hydrogen-bond acceptors and hydrogen-bond

Table 21.5 Noncovalent Bonds and Interactions in Polypeptides

Example	Type of Bond	Approximate Stabilization Energy (kcal/mol)
	Hydrogen bond between peptides	2–5
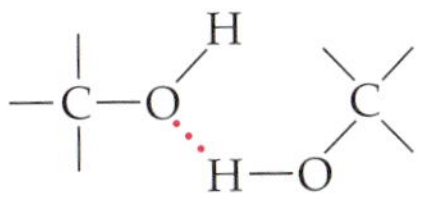	Hydrogen bond between neutral groups	2–5
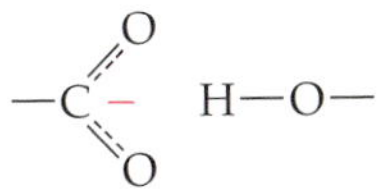	Hydrogen bond between neutral and charged groups	2–5
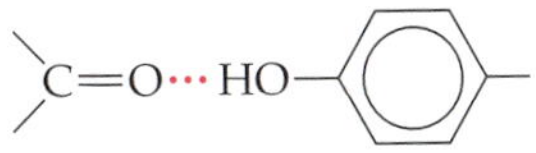	Hydrogen bond between peptide and R group	2–5
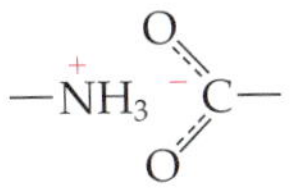	Salt linkage (ionic bond) between charged groups (strongly dependent on distance)	<10
$-CH_3$ CH_3-	Hydrophobic interaction	0.3
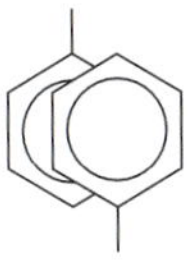	Hydrophobic interaction—stacking of aromatic rings	1.5
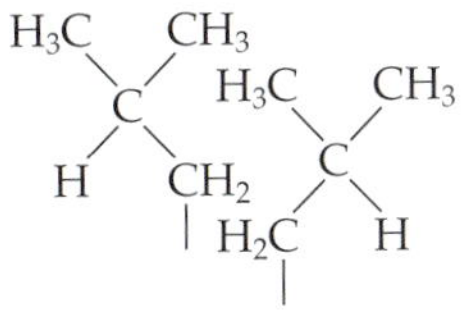	Hydrophobic interaction	1.5
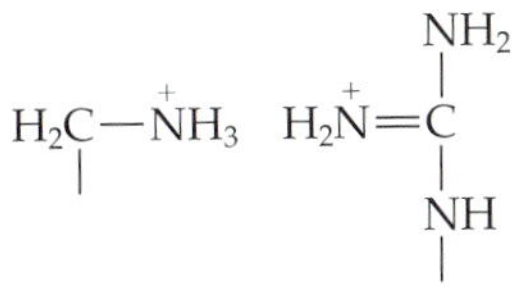	Repulsive interactions between similarly charged groups (strongly dependent on distance)	<–5

donors only over a certain range of pH. Hydrogen bonds (as well as salt linkages) are extremely important in the interaction of proteins with other molecules.

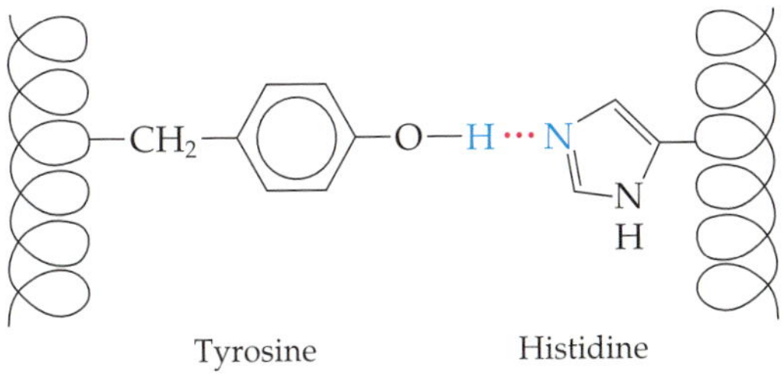

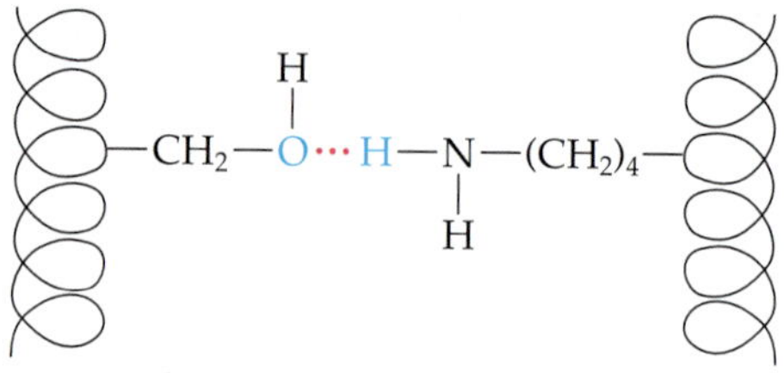

Serine Lysine

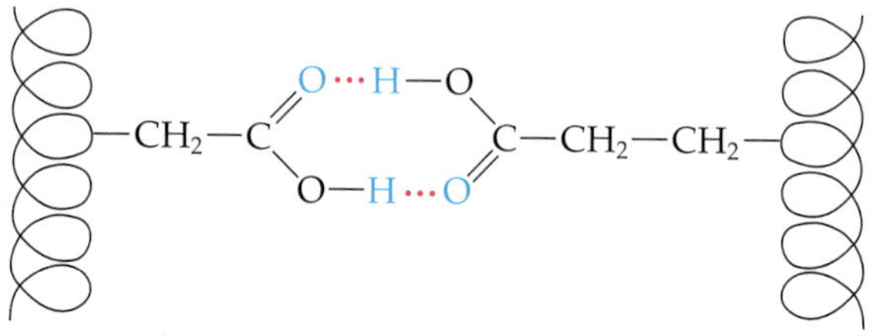

Aspartic acid Glutamic acid

—O—H···O—

Strong H-bond

—O—H···O—

Weak H-bond

A significant feature of hydrogen bonds is that they are highly directional. The strongest hydrogen bond results when the hydrogen donor and the acceptor atom are colinear. If the acceptor atom is at an angle to the covalently bonded hydrogen atom, the hydrogen bond is much weaker.

Disulfide Linkages

Two cysteine residues may come in proximity as the protein molecule folds. The disulfide linkage results from the subsequent oxidation of the highly reactive sulfhydryl (—SH) groups to form cystine.

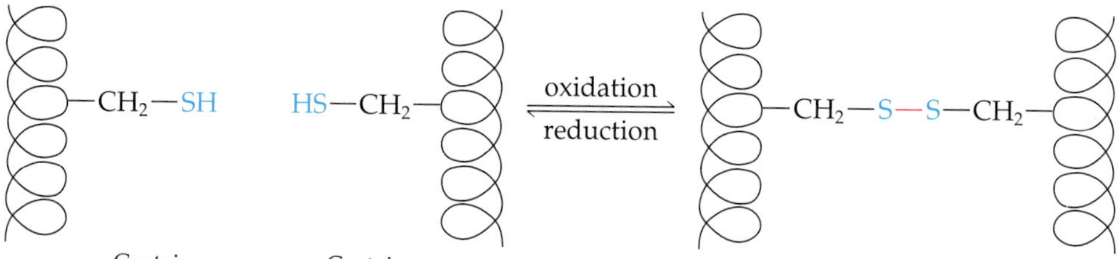

Cysteine Cysteine Cystine

This disulfide bridge is the second most important covalent interaction involved in protein structure. (Recall that the peptide bond is the most important covalent interaction in the structure of proteins.) Intrachain disulfide linkages are frequently found in proteins as a general aid to the stabilization of the tertiary structure. Note however, that one or more of these bonds may join one portion of a polypeptide chain covalently to another, thus interfering with the helical structure. Interchain disulfide bonds are important forces that link two separate polypeptide chains. Such linkages are clearly indicated in the protein structures given in Figures 21.4, 21.5, and 21.6.

Hydrophobic Interactions

Still weaker are the hydrophobic interactions between nonpolar side chains. These links can be important, however, when other types of interactions are either missing or minimized. The hydrophobic interactions are made stronger by the cohesiveness of the water molecules surrounding the protein. Nonpolar side chains minimize their exposure to water by clustering together on the inside folds of the protein in close contact with one another. Hydrophobic interactions become fairly significant in structures such as that of silk, in which a high proportion of amino acids in the protein have nonpolar side chains.

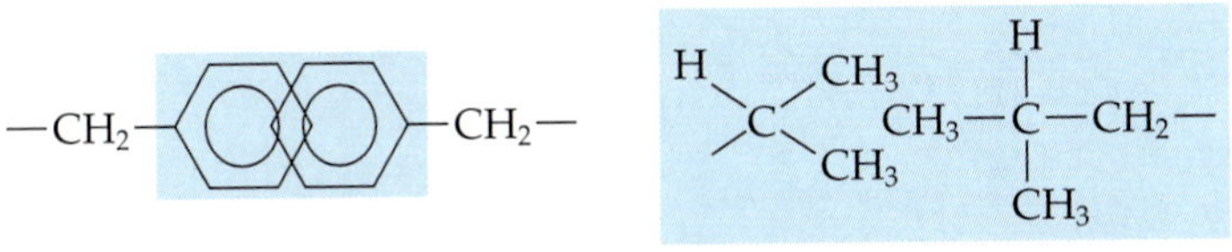

21.11 Quaternary Structure of Proteins

The quaternary structure is stabilized by the same kinds of forces (ionic bonds, hydrogen bonds, interchain disulfide bonds, and hydrophobic bonds) that are involved in maintaining the tertiary structure.

J. C. Kendrew and M. F. Perutz (Nobel Prize winners in 1962) were able, through the use of X-ray diffraction studies, to elucidate completely the primary, secondary, and tertiary structures of myoglobin, a protein consisting of 153 amino acids arranged in a single chain (Figure 21.13). Myoglobin can combine reversibly with molecular oxygen, and it functions to store oxygen in muscle cells. It is particularly abundant in marine animals such as whales, seals, and porpoises, enabling them to remain under water for prolonged periods. (In humans, myoglobin is found mainly in heart muscle.) The secondary structure of myoglobin involves the coiling of this chain into an alpha helix (about 70% of the protein strand has a spiral conformation). The tertiary structure results from the nonuniform folding of the chain to form a stable compact structure. The polar side chains

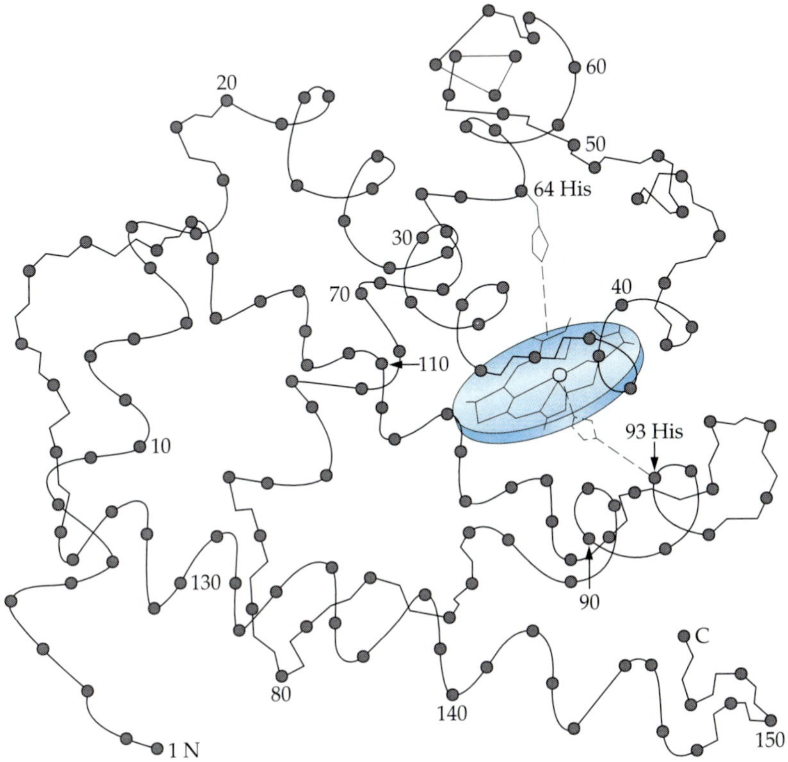

FIGURE 21.13 The conformation of myoglobin deduced from X-ray diffraction studies. The disc shape represents the heme group.

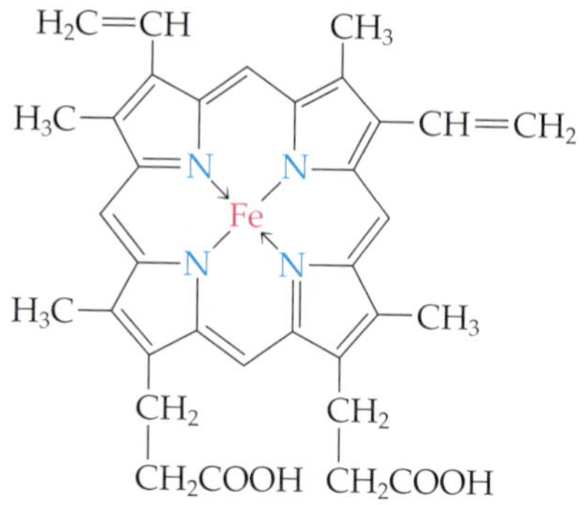

FIGURE 21.14 ▶

Heme is a complex organometallic compound present in both myoglobin and hemoglobin.

are on the outside of the molecule, and almost all of the nonpolar ones are on the inside. The shape of the final molecule includes a hole that nicely accommodates a **heme** unit (an organometallic complex that is the oxygen-binding component of myoglobin; see Figure 21.14). The tertiary structure also brings two amino acid side chains into position to anchor the heme unit to the protein portion of the myoglobin molecule.

The structure of the hemoglobin molecule was also deduced by Kendrew and Perutz. It consists of four polypeptide chains—two identical alpha chains (141 amino acids each) and two identical beta chains (146 amino acids each), as shown in Figure 21.15. Each chain is very similar in structure to the single polypeptide chain of myoglobin. Since each chain contains a heme group, one hemoglobin molecule can bind four molecules of oxygen. The four hemoglobin subunits are held together by noncovalent surface interactions between the polar side chains

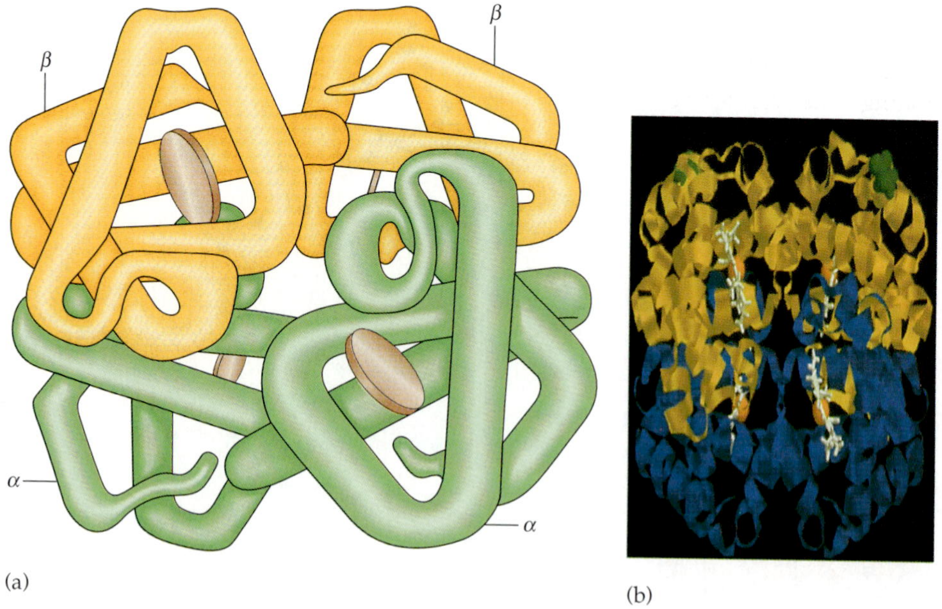

▲ **FIGURE 21.15**

(a) The quaternary structure of hemoglobin. The heme units are shown as discs within the folds of the four polypeptide chains. The four subunits fit together to give hemoglobin an almost spherical shape. (b) Computer graphic representation of the hemoglobin molecule from a human red blood cell.

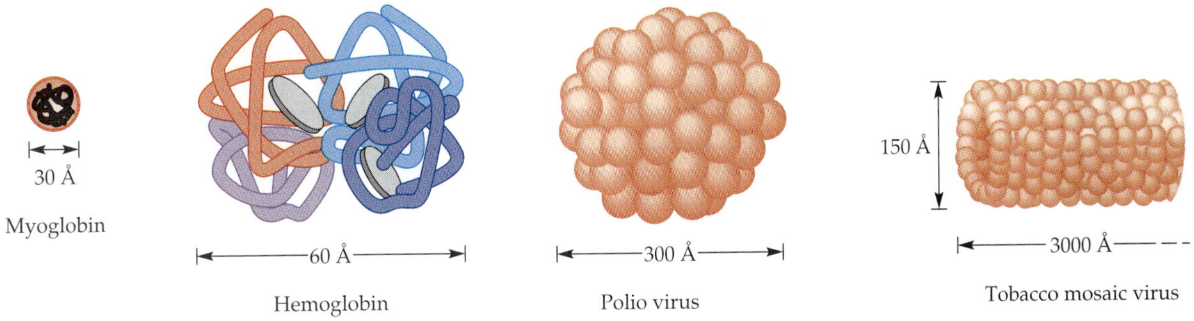

▲ FIGURE 21.16
Schematic diagrams of the structures of several proteins.

(and probably hydrophobic interactions as well). As we shall see in Section 28.9, the chief function of hemoglobin is to transport oxygen within the red blood cells.

Myoglobin has no quaternary structure (since it is composed of a single polypeptide chain), whereas the protein coats of several viruses are composed almost entirely of polypeptide subunits arranged in a highly ordered conformation. The polio virus contains 130 polypeptide chains. The tobacco mosaic virus contains a grand total of about 345,000 amino acids arranged in 2130 individual polypeptide chains that are assembled around a central core of nucleic acids (Figure 21.16). The average molecular weight of each polypeptide chain is about 18,000 daltons.

The principal protein of connective tissues is *collagen* (Figure 21.17). It is the most abundant of all proteins in higher vertebrates. Most of the organic portions of skin, bones, tendons, and teeth are collagen. It also occurs in most other parts of the body as fibrous inclusions. In all, collagen makes up about 25% of all the protein in the human body.

Like other fibrous proteins, collagen is not readily digestible. Treatment with boiling water converts collagen to *gelatin*, which is not only water-soluble but also digestible. The cooking of meat converts part of the tough connective tissue to gelatin, making the meat more tender.

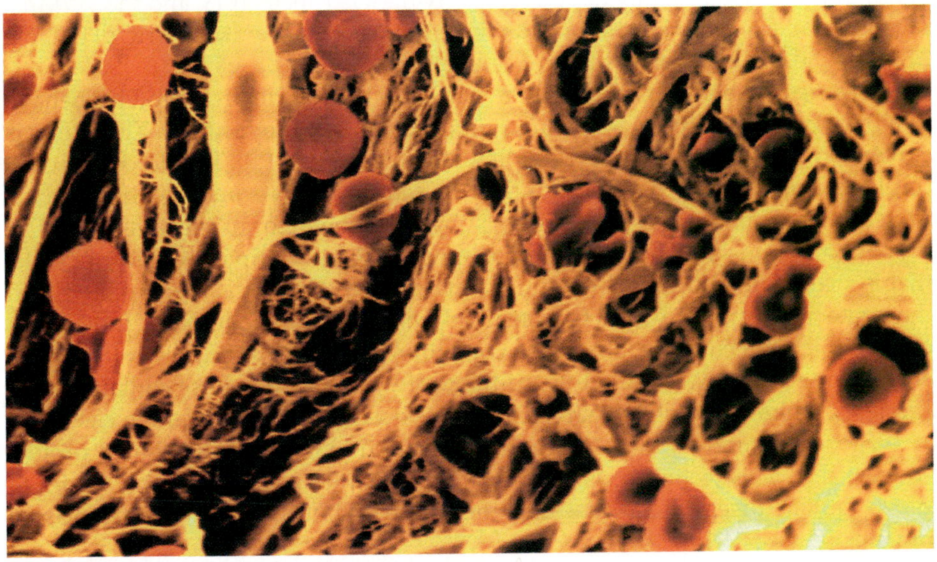

◀ FIGURE 21.17
Electron micrograph of human connective tissue, showing collagen fibers (yellow strands) and red blood cells.

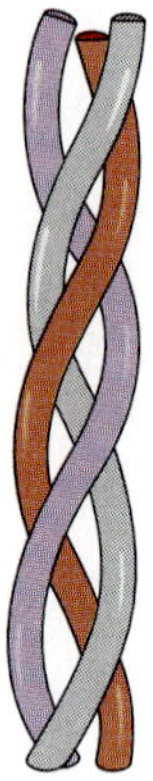

▲ FIGURE 21.18
The protein collagen is a rigid cable made from three left-handed polypeptide chains. Each chain contains about 1000 amino acid units. The chains are held together by interchain hydrogen bonds.

Collagen is about 33% glycine, with another 20 to 25% consisting of proline and hydroxyproline. Collagen also contains acidic and basic amino acids. It does not contain enough of the essential amino acids; nutritionally speaking, the gelatin derived from it is a poor-quality protein (see Section 27.3).

Collagen consists of three protein chains, each wound about its own axis in a left-handed helix (Figure 21.18). The three chains are wrapped around one another like the three strands of a rope and are cross-linked by interchain hydrogen bonding. In tendons, the collagen fibers are arranged in parallel bundles to yield structures that have nearly the tensile strength of steel wire but little or no capacity to stretch. In bones and teeth, the collagen cables form the matrix upon which the network of calcium salts is built.

Collagen chains are also somewhat cross-linked by covalent bonds. As an animal grows older, the extent of cross-linking increases and the meat gets tougher. Collagen is of considerable commercial importance. The process of tanning increases the degree of cross-linking, converting skin to leather. The soluble gelatin derived from collagen is used in food, film emulsions, and glue and in many other ways.

21.12 Electrochemical Properties Of Proteins

When amino acids combine to form the polypeptide chain(s) of protein molecules, the majority of their amino and carboxyl groups are tied up in the peptide bonds. However, the side chains of aspartic and glutamic acids and those of lysine, arginine, and histidine all retain their acidic and basic groups. In proteins, just as in the free amino acids, these groups exist in solution as charged species such as $-COO^-$ and $-NH_3^+$. Accordingly, proteins are also amphoteric substances. Because all proteins contain some of the acidic and basic amino acids, positive and negative charges are found throughout the molecule.

At its isoelectric pH, the protein molecule as a whole is electrically neutral. It may contain many ionized groups, but the positively charged side chains are exactly balanced by negatively charged ones. The isoelectric pH is characteristic of a given protein. It is dependent on the number, kind, and arrangement of the acidic and basic groups within the molecule. Proteins that have a high proportion of basic amino acids usually have a relatively high isoelectric pH, and those with a preponderance of acidic amino acids have a relatively low isoelectric pH. Table 21.6 lists the isoelectric pH values of several proteins.

Table 21.6 Isoelectric pH Values of Various Proteins

Protein	Isoelectric pH
Pepsin	<1.1
Silk fibroin	2.2
Pepsinogen	3.7
Casein	4.6
Egg albumin	4.7
Serum albumin	4.8
Urease	5.0
Insulin	5.3
Fibrinogen	5.5
Catalase	5.6
Hemoglobin	6.8
Myoglobin	7.0
Ribonuclease	9.5
Cytochrome c	10.6
Lysozyme	11.0

Because of the presence of ionized groups in their structures, proteins behave as either cations or anions, depending upon the pH of the solution. This effect is exploited in the electrophoresis of proteins (recall Figure 21.2). The process of electrophoresis is a very powerful tool that is used to separate and identify specific proteins in a mixture of proteins by subjecting them to an electric field. The protein mixture is applied on a solid support, such as a strip of cellulose acetate, that is soaked in a buffer solution at a certain pH. A current is then applied, and the proteins migrate toward the positively charged electrodes. For proteins of comparable molecular weight, those containing the greatest number of negative charges migrate most rapidly toward the positive electrode; those containing the greatest number of positive charges move most rapidly to the negative electrode. A dye (such as ninhydrin or Amido Black) is used to make the separated protein spots visible. This technique is used on blood samples in hospital laboratories to assess certain diseases by detecting the relative concentrations of the plasma proteins (Figure 21.19).

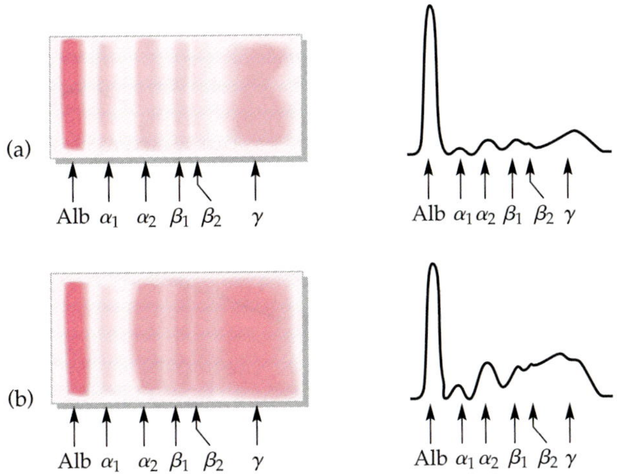

FIGURE 21.19

(a) An electrophoresis pattern of normal blood. (b) A pattern of abnormal blood that has elevated γ-globulin, indicating possible infection, collagen disorder, or liver disease.

The solubility of proteins in water is greatly dependent on pH. As a general rule, a protein is least soluble at its isoelectric pH. The size of many proteins places them in the category of colloids (Section 9.11). At pH values other than the isoelectric pH, the molecules carry a net charge. These charges on the surface of the colloidal proteins repel the other colloidal particles and keep them from coalescing. Thus, they form colloidal dispersions. At the isoelectric pH, however, the colloidal protein molecules are electrically neutral and no longer repel one another. Therefore they come together to form larger aggregates that eventually precipitate from solution.

Casein, for example, is the major protein component of milk, and it precipitates in the form of white curds at its isoelectric pH of 4.6. The souring of milk results from the production of lactic acid by bacteria. The lactic acid lowers the pH of milk from its normal value of about 6.6 to about 4.6. Casein is used in the manufacture of cheese. It can be obtained either by adding acid to milk or by bacterial action.

21.13 Denaturation of Proteins

In many ways, proteins are remarkable compounds. Their highly organized structures are truly masterworks of chemical architecture. But highly organized structures tend to have a certain delicacy, and this is true of many proteins. We define **denaturation** as the process in which a protein is rendered incapable of performing its assigned function. If the protein can't do its job, we say it has been *denatured.* (Sometimes denaturation is equated with the precipitation or coagulation of a protein. Our definition is a bit broader.) The process is sometimes reversible, but usually it is not. You have certainly observed the denaturation of egg albumin. The clear egg "white" turns to an opaque white when the egg is boiled or fried. What you have observed is the denaturation and coagulation of the albumin. No one has yet reversed that process!

The primary structure of proteins is quite sturdy. In general, it takes fairly vigorous conditions for peptide bonds to be hydrolyzed (although chemists have devoted much effort to developing gentler methods, and enzymes manage to hydrolyze proteins with remarkable ease). At the secondary and tertiary levels,

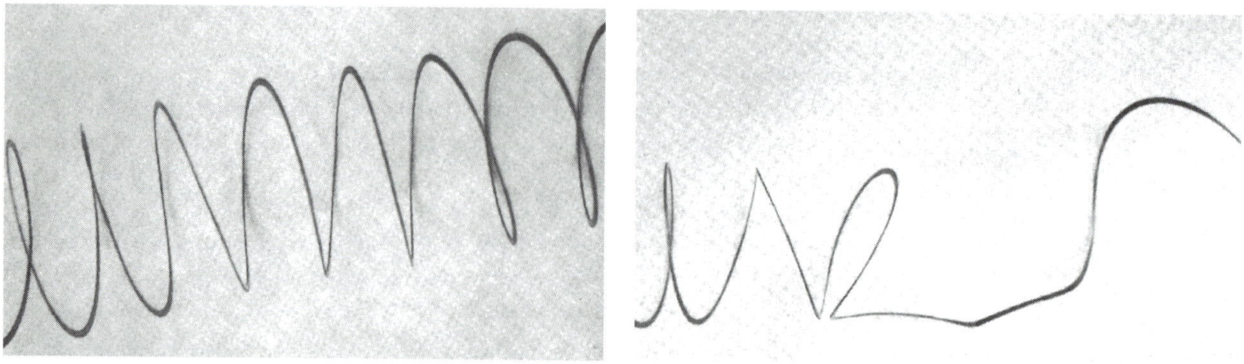

(a)

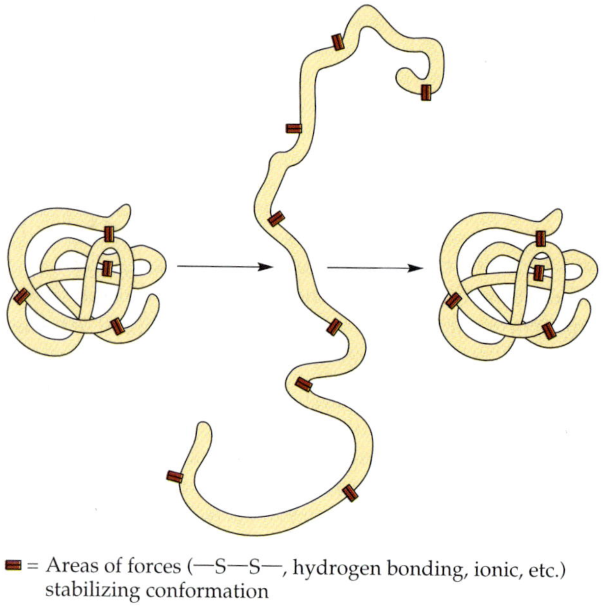

▲ **FIGURE 21.20**

Denaturation of a protein. (a) Irreversible denaturation. The coiled spring represents the helical structure of a protein when the elastic limit of the helix is exceeded, the shape is irreversibly altered. (b) The globular protein is folded into the tertiary conformation necessary for its functioning. The denatured protein can assume various random conformations. It is not active, but under proper conditions, it may refold to the active conformation.

■■ = Areas of forces (—S—S—, hydrogen bonding, ionic, etc.) stabilizing conformation

(b)

however, proteins are quite vulnerable to attack (Figure 21.20). A wide variety of reagents and conditions can cause protein denaturation. Some of them are outlined here.

Heat and Ultraviolet Radiation

Heat and ultraviolet radiation supply kinetic energy to protein molecules, causing their atoms to vibrate more rapidly and thus disrupting the relatively weak hydrogen bonds and hydrophobic bonds. Most proteins are denatured when heated above 50 °C, and this results in coagulation of the protein. Heat and ultraviolet radiation are employed in sterilization techniques because they denature the enzymes in bacteria and, in so doing, destroy the bacteria. Denatured proteins are usually easier to chew and easier for enzymes to digest; hence we cook most of our protein-containing food.

Lead Poisoning

Lead compounds are quite toxic. Metallic lead is converted to Pb^{2+} in the body. We can excrete about 2 mg of lead per day. Our intake from air, food, and water is generally less than that, so normally we do not accumulate toxic levels. If intake exceeds excretion, however, lead builds up in the body and chronic irreversible lead poisoning results.

Lead poisoning is a major problem with children, particularly those in areas containing old, run-down buildings. Some children develop a craving that causes them to eat unusual things. Children with this syndrome (called *pica*) eat chips of peeling, lead-based paints. These children probably also pick up lead compounds from the streets, where they have been deposited by automobile exhausts. They also may get lead from canned milk and other sources. In all, thousands of children suffer from lead poisoning each year. Such poisoning often leads to mental retardation and neurological disorders through damage to the brain and nervous system.

The U.S. Environmental Protection Agency estimates that lead poisoning contributes to 123,000 cases of hypertension and to 680,000 miscarriages annually, and it retards the growth of 7000 children each year. These problems add $635 million to health care costs annually.

Treatment with Organic Compounds

Ethyl alcohol, formaldehyde, urea, and rubbing alcohol are capable of forming intermolecular hydrogen bonds with protein molecules, thus disrupting the intramolecular hydrogen bonding within the molecule. A 70% isopropyl alcohol solution is used as a disinfectant in cleansing the skin before an injection. The alcohol denatures the protein (enzymes in particular) of any bacteria present in the area of the injection. A 70% alcohol solution effectively penetrates the bacterial cell wall, whereas 100% alcohol coagulates proteins at the surface, forming a crust that prevents the alcohol from entering into the cell (Figure 21.21).

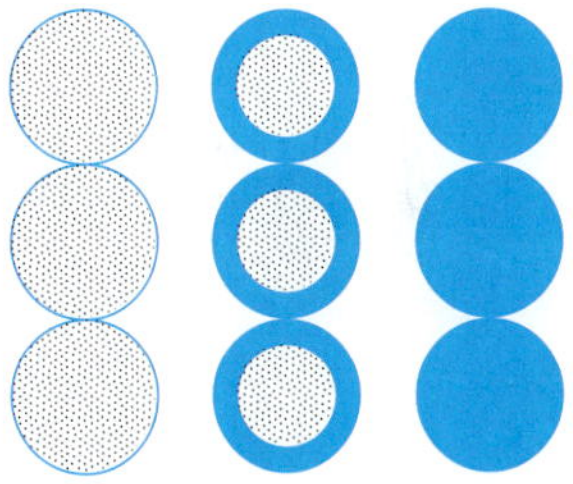
(a) (b) (c)

▲ **FIGURE 21.21**
Effect of isopropyl alcohol on bacteria. Dark areas represent coagulated protein. (a) Bacteria before application of alcohol. (b) After application of 100% alcohol. (c) After application of 70% alcohol, which is more effective than 100% alcohol.

Salts of Heavy Metal Ions

The heavy metal cations Hg^{2+}, Ag^{+}, and Pb^{2+} form very strong bonds with the carboxylate anions of the acidic amino acids and with the sulfhydryl groups of cysteine. Therefore they disrupt salt linkages and disulfide linkages and cause the protein to precipitate out of solution as insoluble metal–protein salts. This property makes some of the heavy metal salts suitable for use as antiseptics. For example, a 1% solution of silver nitrate (also called lunar caustic), which is used to prevent gonorrhea infections in the eyes of newborn infants, and mercuric chloride, another antiseptic, precipitate the proteins in infectious bacteria.

Most heavy metal salts are toxic when taken internally because they precipitate the proteins of all the cells with which they come into contact.[3] Substances high in protein, such as egg whites and milk, are used as antidotes for heavy metal poisoning. If a person who has ingested mercury is fed raw eggs immediately, the mercury reacts with egg protein in the stomach rather than with other, more essential proteins. The stomach contents must then be pumped out or vomited to

Hatter's disease (which probably afflicted the Mad Hatter in *Alice's Adventures in Wonderland*) was a form of chronic mercury poisoning. Mercury compounds were used to convert fur to felt for felt hats.

[3] The danger of lead and/or mercury poisoning has evoked considerable environmental concern. Lead salts are no longer used as pigments in paints, and most of the lead has been removed from gasoline. Mercury compounds still occur in water systems because large quantities of mercury and mercury salts have been dumped by industries into streams and lakes. The mercury is taken in by fish, and then humans eat the fish.

Permanent Waving

The chemistry of curly hair is interesting. Hair is protein, and adjacent protein chains are held together by disulfide linkages. To put a permanent wave in the hair, you use a lotion containing a reducing agent such as thioglycolic acid ($HSCH_2COOH$). This wave lotion ruptures the disulfide linkages (Figure 21.22), allowing the protein chains to be pulled apart as the hair is held in a curled position on rollers. The hair is then neutralized with a mild oxidizing agent such as hydrogen peroxide. Disulfide linkages are formed in new positions to give shape to the hair.

The same chemical process can be used to straighten naturally curly hair. The change in curliness depends only on how you arrange the hair after the disulfide bonds have been reduced and before the linkages have been restored. As with permanent dyes, permanent curls grow out as new hair is formed.

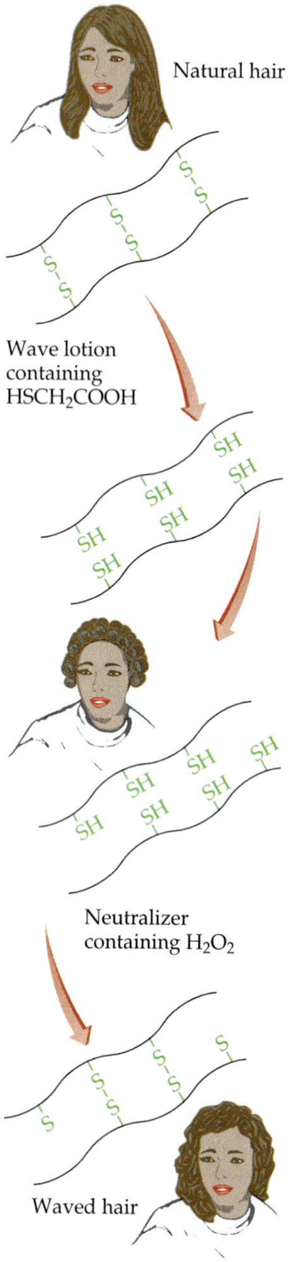

▲ **FIGURE 21.22**
Permanent waving of hair is accomplished by breaking disulfide linkages and then reforming them in new positions.

prevent the ultimate digestion of the egg protein and the consequent release of mercury ions within the body. Quite clearly, the technique works only for acute poisonings and not for the far more common chronic mercury poisoning.

Alkaloid Reagents

Picric acid and tannic acid are called alkaloid reagents because they were originally used to study the structures of the alkaloids (morphine, cocaine, quinine). They function in a manner analogous to the heavy metal cations, but the picrate and tannate anions combine with the positively charged amino groups in proteins to disrupt the salt linkages. In the manufacture of leather, tannic acid is used to precipitate the proteins in animal hides. This is the process called *tanning*. Tannic and picric acids are sometimes used in the treatment of burns. These acids combine with the protein in the exposed areas to form a crust over the wounds that excludes air and stops the loss of body fluids. The loss of water and salts is the most significant cause of shock and the fatalities that result from burns. In an emergency, tea can serve as a source of tannic acid for the treatment of severe burns.

There are many other ways of denaturing proteins that we have not discussed (introduction of radical changes in pH, for example). The point should be clear, however. The very complexity that makes proteins so versatile also makes them vulnerable. There is a considerable range of vulnerability. The delicately folded globular proteins are much more readily denatured than are the tough, fibrous proteins of hair and skin.

On the other hand, there is increasing evidence that a carefully unfolded protein, given the proper conditions and enough time, will refold and may again exhibit biological activity. Such evidence suggests that, for these molecules, primary structure determines secondary and tertiary structure. A given sequence of amino acids seems naturally to adopt its particular three-dimensional arrangement if conditions are right.

We have emphasized structure in this chapter. In other chapters we concentrate on the functions of several kinds of proteins, particularly the enzymes.

Summary

A **protein** is any large polymer molecule made up of amino acid units. Nearly all the proteins in all living species are made from the same 20 amino acids, called the building blocks of life.

There are four classes of amino acids: nonpolar, polar but neutral, acidic, and basic. In the solid state and in neutral solutions, amino acids exist as **zwitterions,** a form that is charged but is electrically neutral.

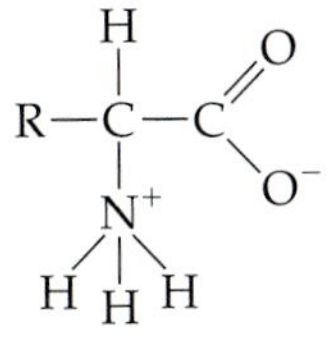

Their existence in the zwitterion form makes amino acids behave much like inorganic salts. They can act, depending on reaction conditions, as either acids or bases, with the result that proteins act as buffers:

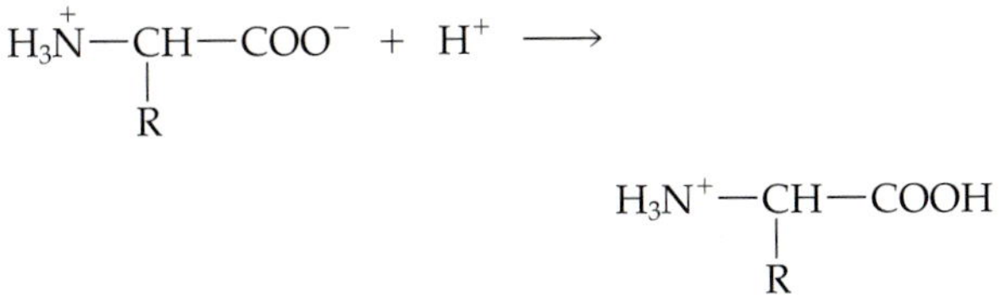

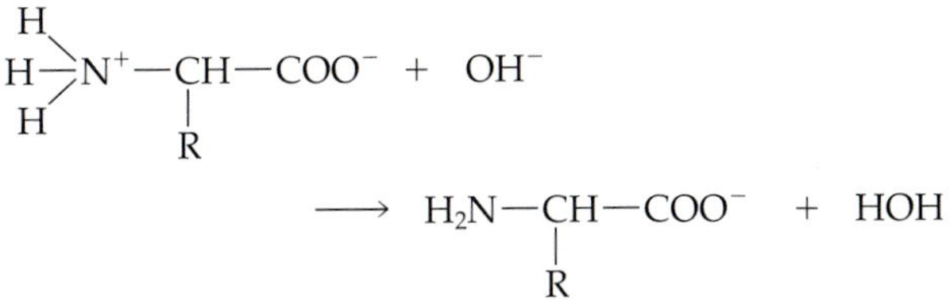

The pH at which an amino acid exists as the zwitterion is called the **isoelectric pH.** The isolectric pH lies somewhere between the pH at which the amino acid acts as an acid and that at which it acts as a base. At its isoelectric pH, an amino acid behaves like the salt of a weak acid and a weak base.

The amino acids in a protein are linked together by **peptide bonds:**

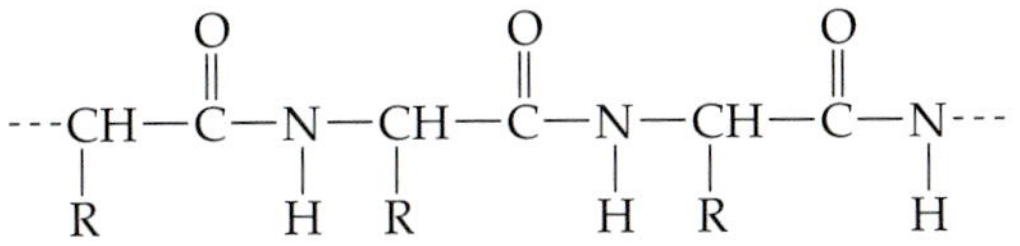

Protein chains containing ten or fewer amino acids are usually referred to as **peptides,** with a prefix—**di-, tri-,** etc., through **deca**—indicating the number of amino acids. Chains containing more than ten amino acid units are **polypeptides,** and when the molecular weight of the chain exceeds about 10,000 daltons, the term *protein* is used. This naming scheme is used loosely, however, and in most contexts, the terms can be used interchangeably.

Proteins are classified as being globular or fibrous, depending on their solubility in water. *Globular proteins* are soluble in water; *fibrous proteins* are not.

Protein molecules can have as many as four levels of structure. The **primary structure** is the sequence of amino acids in the chain. The **secondary structure** is the shape—helical or pleated-sheet—of the chain. The **tertiary structure** is the overall three-dimensional shape of the molecule that results from the way the chain bends and folds in on itself. Proteins that consists of more than one chain also have **quaternary structure,** which refers to the way the multiple chains are packed together.

The intramolecular and intermolecular forces that result in secondary, tertiary, and quaternary structure are of four types:

1. Hydrogen bonding between a carbonyl group on one chain and an amide group either on the same chain or on a neighboring chain:

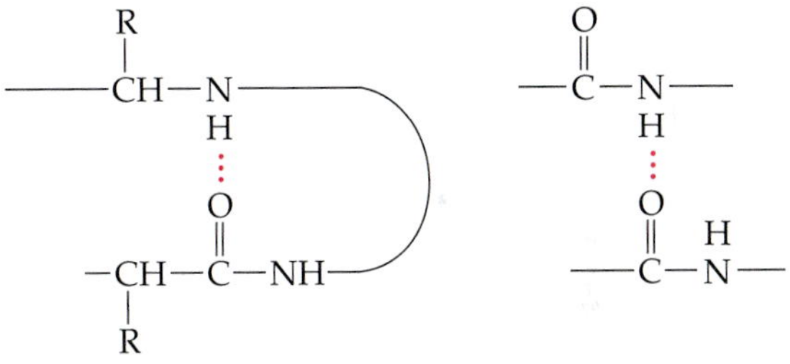

2. Salt linkages between one acidic and one basic side chain:

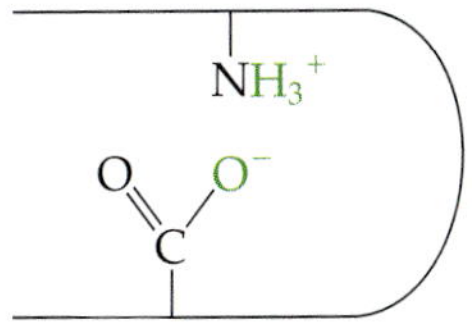

3. Disulfide linkages between cysteine units:

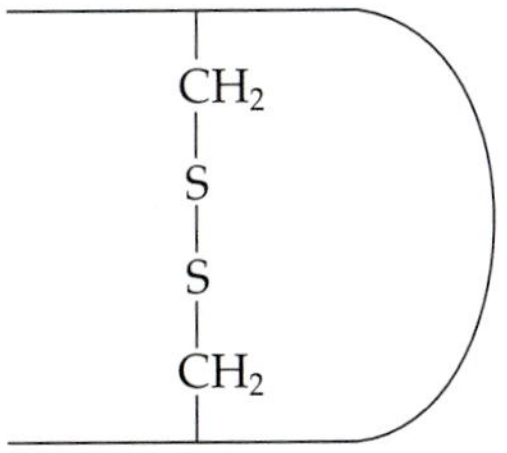

4. Hydrophobic interactions with nonpolar side chains:

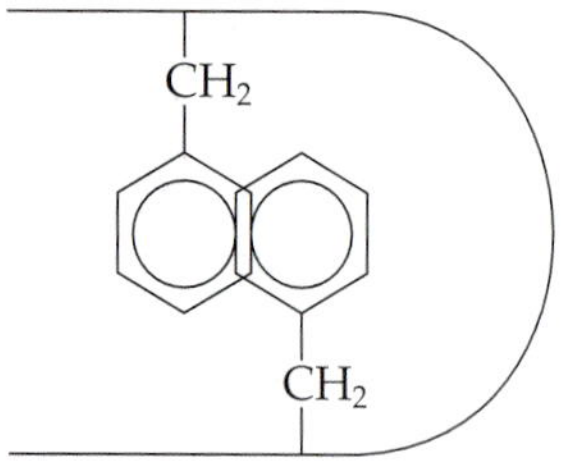

Proteins are amphoteric because they contain acidic and basic side chains that exist as $-COO^-$ and $-NH_3^+$ in solution. At the isoelectric pH of a given protein, these positive and negative charges occur in equal numbers. Proteins containing a preponderance of basic amino acids (lysine, arginine, and histidine) have a high isoelectric pH; those containing large numbers of acidic amino acids (aspartic acid or glutamic acid) have a low isoelectric pH.

Because of their complexity, protein molecules are delicate and can be easily changed by a number of chemical agents. A *denatured* protein is one that has been changed such that it can no longer do its physiological job, a process called **denaturation.** Heat and ultraviolet radiation denature a protein by increasing the atomic vibrations in the molecule and thereby rupturing the hydrogen bonds and hydrophobic bonds giving structure to the protein. Organic compounds, such as alcohols, denature a protein by forming *inter*molecular hydrogen bonds with the protein and thereby causing *intra*molecular hydrogen bonds to rupture. Heavy metal cations tend to bond with carboxylate anions and denature a protein by rupturing salt and disulfide linkages in order to form metal-protein salts. Alkaloid reagents also denature a protein by breaking salt linkages, in this case as the anionic form of the reagent combines with the amino cation parts of the protein molecule.

Key Terms

dalton (introduction)
denaturation (21.13)
disulfide linkage (21.10)
diuretic (21.6)
electrophoresis (21.3)
heme (21.11)
isoelectric pH (21.3)
peptide (21.4)
peptide bond (21.4)
polypeptide (21.4)
primary structure (21.8)
protein (21.1)
quaternary structure (21.8)
secondary structure (21.8)
tertiary structure (21.8)
zwitterion (21.2)

Review Questions

1. Define, describe, or illustrate:
   a. peptide bond
   b. tripeptide
   c. disulfide linkage
   d. salt linkage
   e. hydrophobic interaction
   f. primary structure
   g. secondary structure
   h. tertiary structure
   i. quaternary structure
   j. zwitterion
   k. isoelectric pH
   l. diuretic
   m. globular protein
   n. fibrous protein
   o. electrophoresis
   p. denaturation
2. What is the general structure for an α-amino acid?
3. Identify the amino acid that fits each description:
   a. found in asparagus
   b. most abundant amino acid in gelatin
   c. abundant in collagen but not found in most other proteins
4. To which family of mirror-image isomers do almost all naturally occurring amino acids belong?
5. What is the difference between a polypeptide and a protein?
6. Amino acid units in a protein are connected by peptide bonds. What is another name for the functional group linking the units of proteins?
7. Which class of proteins shows greater solubility in aqueous solution—fibrous or globular?
8. Which class of proteins is more easily denatured—fibrous or globular?
9. Describe some ways of denaturing a protein.
10. What level(s) of structure is(are) ordinarily disrupted in denaturation?
11. Is denaturation of protein usually reversible?

12. Distinguish between the terms in each pair:
 a. N-terminal amino acid and C-terminal amino acid
 b. oxytocin and vasopressin
 c. alpha-helix and pleated sheet
 d. primary structure and secondary structure
 f. tertiary structure and quaternary structure
 g. interchain and intrachain hydrogen bonds
 h. hemoglobin and myoglobin
13. What occurs when milk becomes sour?

14. Give a brief molecular-level description of what occurs when:
 a. an egg is boiled
 b. surgical instruments are sterilized
 c. a dilute solution of silver nitrate is used as a disinfectant in the eyes of newborn infants
15. Why is a 70% alcohol solution more effective as a disinfectant than a 100% alcohol solution?
16. Discuss the use of egg white as an antidote for heavy metal poisoning.

Problems

Amino Acids: Structures and Names

17. Draw the side chains of:
 a. proline b. lysine c. tyrosine
18. Draw the side chains of:
 a. aspartic acid b. cysteine c. tryptophan
19. Write the structural formulas for:
 a. glycine b. alanine c. valine
20. Write the structural formulas for:
 a. serine b. phenylalanine c. isoleucine
21. Write the structural formula for an amino acid that has an acidic side chain, and give the name of the compound.
22. Write a structural formula for an amino acid that has a basic side chain, and give the name of the compound.
23. Identify the amino acid that fits each description:
 a. contains a secondary amino group
 b. contains a heterocyclic ring
 b. contains a benzene ring
 d. not found in proteins
24. Identify the amino acid that fits each description:
 a. most basic of the amino acids
 b. contains a sulfhydryl group
 c. contains a phenolic group
 d. contains a branched chain

Peptides

25. Write structural formulas for:
 a. glycylalanine b. alanylglycine
26. Write the structural formula for phenylalanylglycyl-alanine.
27. Draw the structural formula for Ser-Ala-Gly.
28. Name this peptide by the three-letter abbreviation system.

$$\underset{H_3\overset{+}{N}CH}{\overset{CH_3}{|}}-\overset{O}{\overset{||}{C}}-NH\underset{CH}{\overset{\overset{OH}{|}}{\underset{|}{CH_2}}}-\overset{O}{\overset{||}{C}}-NH\underset{CH}{\overset{\overset{SH}{|}}{\underset{|}{CH_2}}}-\overset{O}{\overset{||}{C}}-NH\underset{CH}{\overset{\overset{C_6H_5}{|}}{\underset{|}{CH_2}}}-\overset{O}{\overset{||}{C}}-O^-$$

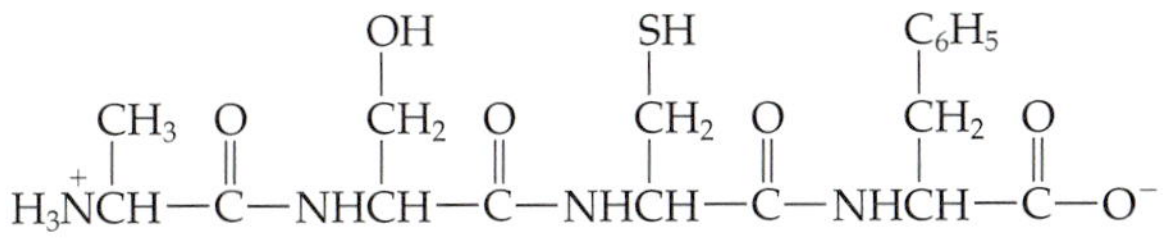

Acid–Base Reactions

29. Write a structural formula for the anion formed when glycine reacts with a base.
30. Write a structural formula for the cation formed when glycine reacts with an acid.
31. Under what conditions does a protein have (a) a net positive charge, (b) a net negative charge, and (c) a net zero charge?
32. How do the properties of a protein at its isoelectric pH differ from its properties in solutions at other pH values?

Structure of Proteins

33. Describe the structure of silk and explain how its properties reflect its structure. What name is given to the secondary structure of silk?
34. Describe the structure of wool and relate this structure to wool's elasticity. What name is given to the secondary structure of wool protein?
35. Name the four types of interactions that maintain the tertiary structure of proteins.
36. The following sets of amino acids are involved in maintaining the tertiary structure of a peptide. In each case, identify the type of interaction involved.
 a. aspartic acid and lysine
 b. phenylalanine and alanine
 c. serine and lysine
 d. two cysteines
37. Classify these proteins as fibrous or globular:
 a. albumin
 b. myosin
 c. collagen
38. Classify these proteins as fibrous or globular:
 a. hemoglobin
 b. keratins
 c. myoglobin

Additional Problems

39. Describe the structure of collagen.

40. Identify two amino acids that contain more than one chiral carbon atom.

41. A direct current was passed through a solution containing alanine, histidine, and aspartic acid at pH 6.0. One amino acid migrated to the cathode, one migrated to the anode, and one remained stationary. Match each behavior with the correct amino acid.

42. Give the structural formulas for the products of the acid hydrolysis of:

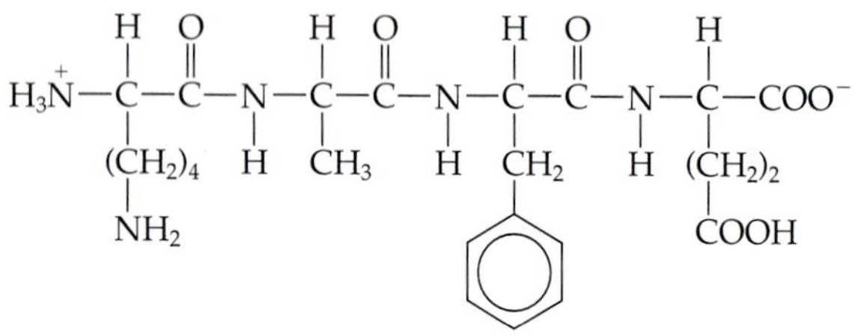

$$H_3\overset{+}{N}-\underset{\underset{NH_2}{\overset{|}{\underset{|}{(CH_2)_4}}}}{\overset{\overset{H}{|}}{C}}-\overset{\overset{O}{\|}}{C}-\underset{\overset{|}{H}}{N}-\underset{\underset{CH_3}{\overset{|}{|}}}{\overset{\overset{H}{|}}{C}}-\overset{\overset{O}{\|}}{C}-\underset{\overset{|}{H}}{N}-\underset{\underset{CH_2}{\overset{|}{|}}}{\overset{\overset{H}{|}}{C}}-\overset{\overset{O}{\|}}{C}-\underset{\overset{|}{H}}{N}-\underset{\underset{COOH}{\overset{|}{\underset{|}{(CH_2)_2}}}}{\overset{\overset{H}{|}}{C}}-COO^-$$

43. Glutathione (γ-glutamylcysteinylglycine) is a tripeptide found in all cells of higher animals. It contains a glutamic acid joined in an unusual peptide linkage involving its γ-carboxyl group. Draw the structure of glutathione.

44. What is the difference between the type of hydrogen bonding that occurs in secondary structures of proteins and the type in tertiary structures?

45. Proteins help to maintain the pH of an organism. How can they perform this function?

46. One of the neurotransmitters involved in the sensing of pain is a polypeptide called substance P. It is an undecapeptide released by nerve terminals in response to pain. Its primary structure is Arg-Pro-Lys-Pro-Gln-Gln-Phe-Phe-Gly-Leu-Met. What is the net electric charge (+ or −) of this polypeptide at pH (**a**) 1.0, (**b**) 6.0, and (**c**) 13.0?

47. Bacteria synthesize D-alanine and use it in the biosynthesis of cell walls. Draw the structure of D-alanine.

48. A deuterated derivative (one in which ordinary hydrogen, 1_1H, has been replaced by deuterium, 2_1H or D), 2-deutero-3-fluoro-D-alanine, acts as an antibiotic by inhibiting cell wall synthesis. Draw the structure of this derivative.

49. Two cysteine units joined through a disulfide linkage are sometimes considered a different amino acid called cystine. Draw the structure of cystine.

50. Draw the structure of γ-aminobutyric acid (GABA). Is GABA found in proteins? What is its role in the body?

51. The isoelectric pH of silk fibroin is 2.2. Which amino acid is likely to be present in large amounts: (**a**) aspartic acid, (**b**) histidine, or (**c**) lysine?

52. Write equations to show how alanine can act as a buffer.

53. What do we mean when we say that hemoglobin shows species variation?

54. Which amino acid is more likely to be in a pleated sheet protein—alanine or phenylalanine?

55. For each of the following amino acids, state whether it is more likely to be on the inside or the outside of a globular protein:

 a. phenylalanine **b.** aspartic acid
 c. serine **d.** lysine
 e. leucine **f.** glutamic acid

56. Carbohydrates are incorporated into *glycoproteins*. How does the incorporation of sugar units affect the solubility of a protein?

57. Aspartame, L-aspartyl-L-phenylalanine methyl ester, is an artificial sweetener that is about 160 times sweeter than sucrose. It is the active ingredient in the sugar substitute NutraSweet. Draw the structure of aspartame (see page 506).

22 Enzymes

Milking a snake to obtain its venom. Snake venoms contain several enzymes that are significant in clinical studies.

The various reactions that occur in biological systems are in many respects identical to those we have already discussed (for example, hydration of unsaturated bonds; interconversions of alcohols, aldehydes, and acids; and hydrolysis of esters and amides). In fact, scientists have been able to duplicate in the test tube (*in vitro* conditions) many of the reactions commonly carried out in living organisms (*in vivo* conditions). There is, however, one significant difference—the rate at which the two types of reactions occur. *In vivo* reactions take place about 100 to 1 million times faster than the corresponding *in vitro* reactions. Some *in vivo* reactions occur in milliseconds (1 ms = 10^{-3} s); others are more rapid, and rates are measured in microseconds (1 μs = 10^{-6} s). For example, it has been estimated that the reactions involved in the transmittal of a nerve impulse take between 1 and 3 millionths of a second (1 to 3 μs).

If *in vitro* reactions are to take place at an appreciable rate, drastic reaction conditions must be employed. Such conditions, which include high temperatures and the use of potent oxidizing or reducing agents and/or strong acids or bases, are all lethal to living systems. *In vivo* reactions are carried out at body temperature (~37 °C) and in the physiological pH range (pH ~7). The agents employed by cells to bring about reactions under these conditions are the highly efficient, highly specific catalysts called enzymes. Life as we know it would be impossible without enzymes because nearly all cell functions depend directly or indirectly on them.

Recall from Section 5.10 that a *catalyst* is any substance that increases the rate of a chemical reaction without being consumed in the reaction. The catalyst does *not* change the position of equilibrium in a reversible reaction; it only increases the rate at which equilibrium is attained. An **enzyme** can be defined as a complex organic catalyst produced by living cells. Most enzymes operate within the cell that produces them and are thus termed *intracellular.* A typical human cell contains about 2000 enzymes, and these enzymes catalyze more than 100 reactions each minute. If the enzyme's usual site of catalytic activity is outside the cell that produces it (as in the case of the digestive enzymes), the enzyme is designated as *extracellular.*

The hydrolysis of sucrose affords a good example by which to distinguish enzyme action from the classic concept of a catalyst. If we were to exclude bacteria and molds from it, a solution of sucrose in water could be kept indefinitely without undergoing hydrolysis to any appreciable extent. If we added hydrochloric acid and heated the reaction mixture, hydrolysis would take place, producing glucose and fructose. If we added the enzyme invertase (sucrase) instead of the acid, the reaction would take place at a greater rate and the solution would not have to be heated at all. Furthermore, hydrochloric acid catalyzes the hydrolysis of lactose and maltose as well as that of sucrose. Invertase is specific for sucrose alone and will not catalyze the hydrolysis of any other disaccharide.

Several thousand enzymes are known. Many are available commercially in varying grades of purity. Some have even been isolated as highly purified crystals.

22.1 Classification and Naming of Enzymes

The first enzymes to be discovered were named according to their source or method of discovery. The enzyme pepsin, which aids in the hydrolysis of proteins, is found in the digestive juices of the stomach (Greek *pepsis,* digestion). Ptyalin, found in saliva (Greek *ptyalon,* spittle), starts the hydrolysis of starches in the mouth.

As more enzymes were discovered, a more systematic nomenclature developed. The substance upon which an enzyme acts is known as its **substrate.** Enzymes are most commonly named by adding the suffix *-ase* to the root of the name of the substrate. Thus, for example, urease is the enzyme that catalyzes the hydrolysis of urea, and sucrase is the enzyme that catalyzes the hydrolysis of sucrose. A lipase is an enzyme that catalyzes the cleavage of lipids. A dipeptidase aids in the splitting of dipeptides into amino acids. Sometimes an enzyme is named after the products that are formed as a result of its catalytic activity, as in the case of invertase, another name for sucrase (Section 19.7).

To avoid the continued haphazard naming of enzymes, the International Union of Biochemistry, in 1961, recommended that enzymes be systematically classified according to the general type of reaction they catalyze. There are six major types:

1. **Oxidoreductases** catalyze all the reactions in which one compound is oxidized and another is reduced. *Oxidases* operate in physiological oxidation

reactions. They are essential in cells that use oxygen to produce energy. The oxidoreductases also include the *dehydrogenases,* enzymes that oxidize by the removal of hydrogen.

2. **Transferases** facilitate the transfer of groups such as methyl, amino, and acetyl from one molecule to another. *Transaminases* catalyze the transfer of an amino group from one molecule to another. This reaction is involved in the removal of the amino group during the metabolism of amino acids. *Kinases* are involved in the transfer of phosphate groups.
3. **Hydrolases** are enzymes that catalyze hydrolysis reactions. These include *lipases,* which act on fats and other lipids; *carbohydrases,* which speed up the hydrolysis of carbohydrates to monosaccharides; and *proteases* and *peptidases,* which catalyze the hydrolysis of proteins and peptides.
4. **Lyases** aid in the removal of certain groups without hydrolysis. Examples are the *decarboxylases,* which catalyze the removal of carboxyl groups.
5. **Isomerases,** as their name implies, catalyze the conversion of a compound to another compound that is isomeric with the first.
6. **Ligases** are involved in the formation of new bonds between carbon and nitrogen, oxygen, sulfur, or another carbon atom.

22.2 Characteristics of Enzymes

Enzymes are globular proteins. Some enzymes, such as pepsin, trypsin, and ribonuclease, are simple proteins—they consist entirely of amino acid chains. Others contain a nonprotein component necessary to the proper functioning of the enzyme. Such a component is called a **cofactor.** There are two types of cofactors: inorganic ions (e.g., Zn^{2+}, Mn^{2+}) and organic molecules. If the cofactor is an organic molecule, it is called a **coenzyme.** Many coenzymes are vitamins or are derived from vitamin molecules. These important chemical substances and their relationships to vitamins are discussed in detail in Special Topic J.

For more than 50 years, one of the central tenets of biochemistry was that *all* enzymes were proteins. In the 1980s, however, Thomas Cech and Sidney Altman identified some ribonucleic acids (RNA; see Chapter 23) that catalyze cellular reactions. In 1989, Cech and Altman were awarded the Nobel prize in chemistry for their discovery of "ribozymes."

The polypeptide segment of any enzyme containing a cofactor is called an **apoenzyme.** Neither the cofactor nor the apoenzyme has enzymatic activity by itself. The catalytically active apoenzyme–cofactor complex is called the **holoenzyme.**

$$\underset{\substack{\text{Nonprotein}\\\text{(inactive)}}}{\text{Cofactor}} + \underset{\substack{\text{Protein}\\\text{(inactive)}}}{\text{Apoenzyme}} \rightleftharpoons \underset{\text{(active)}}{\text{Holoenzyme}}$$

Some enzymes, such as the protein-digesting enzymes (peptidases), are secreted in larger, inactive forms known as *zymogens* or **proenzymes.** This is a protective feature that prevents the active form of these enzymes from digesting the protein in the walls of the digestive tract. Pepsinogen and trypsinogen are two such compounds that have been carefully studied.

Pepsinogen is secreted from the cells of the stomach. The acid in the stomach converts pepsinogen to pepsin, an active protease. The reaction is also an **autocatalytic reaction:** it is catalyzed by the product pepsin. That is, once some pepsin is formed, it speeds up the reaction, which produces more pepsin. Activation involves the removal of 42 amino acid residues, as small peptides, from the pepsinogen molecule:

$$\text{Pepsinogen} + H_3O^+ \xrightarrow{\text{pepsin}} \text{Pepsin} + \text{Small peptides}$$

Trypsinogen is synthesized in the pancreas and then secreted into the small intestine, where it is activated in response to the presence of food in the intestine. A hexapeptide is cleaved from the proenzyme molecule, enabling the new protein to attain the conformation essential to its catalytic activity.

22.3 Mode of Enzyme Action

In 1888, the Swedish chemist Svante Arrhenius proposed a scheme to account for catalytic activity. He suggested that a catalyst combines with a reactant to form an intermediate compound that is more reactive than the initial uncombined species. The formation of an intermediate compound provides a lower-energy pathway than that of the uncatalyzed reaction (Figure 22.1). This, in effect, lowers the activation energy of the reaction, accounting for the increased rate of reaction. Enzymes reduce activation energies more effectively than other catalysts, thus enabling biochemical reactions to proceed at relatively low temperatures. Note that the overall amount of energy absorbed or released in the reaction is not altered by the enzyme.

This scheme applies to all catalytic reactions, whether inorganic, organic, or biochemical. It is generally believed that enzymatic reactions occur in at least two steps. In the first step, a molecule of the enzyme (E) and a molecule of the substrate (S) collide and react to form an intermediate compound called the *enzyme–substrate complex* (E–S). (This step is reversible because the complex can break apart, yielding the original substrate and the free enzyme.) The enzyme–substrate complex may or may not react with additional substances (water, oxidizing or reducing agents, acids or bases) to form products (P), which are then released from the surface of the enzyme:

General Reaction

$$\text{S} + \text{E} \rightleftharpoons \text{E—S}$$

$$\text{E—S} \longrightarrow \text{P} + \text{E}$$

Specific Example

$$\text{Sucrose} + \text{Sucrase} \rightleftharpoons \text{Sucrose–sucrase complex}$$

$$\text{Sucrose-sucrase} + H_2O \longrightarrow \text{Glucose} + \text{Fructose} + \text{Sucrase}$$

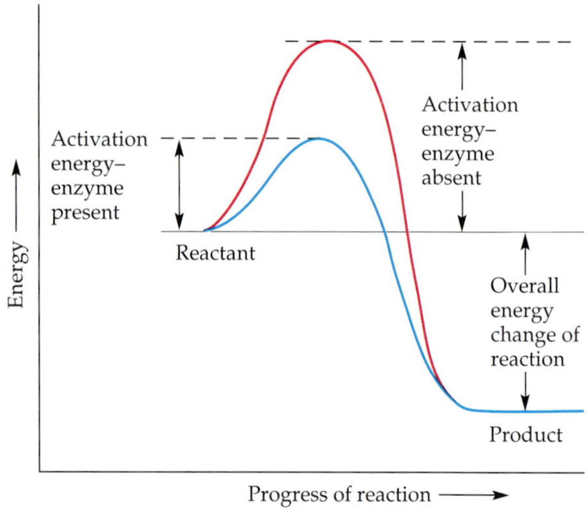

FIGURE 22.1 ▶
Energy diagram for the progress of a chemical reaction, showing the effect of an enzyme.

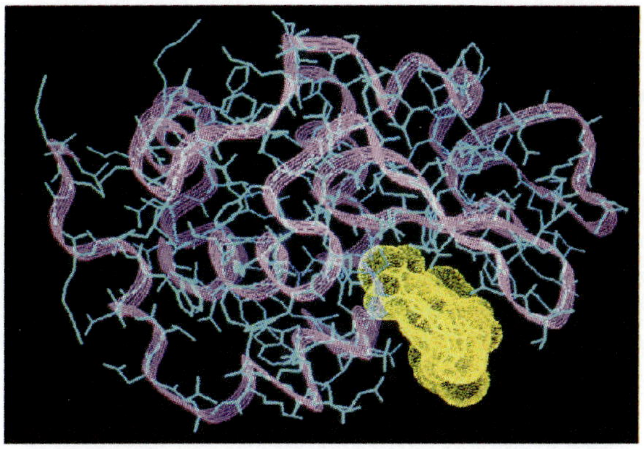

◀ **FIGURE 22.2**

An enzyme is a large, three-dimensional molecule that has a crevice with an active site. According to the lock-and-key model, only a substrate with a shape and structure complementary to those of the active site can fit the enzyme. The active site is shown at the top in this computer-generated structure. The fit of the substrate (blue) is also indicated.

The existence of an enzyme–substrate complex has been verified by spectroscopic and kinetic experiments. The enzyme and substrate are held together by hydrogen bonds and electrostatic interactions between functional groups. In addition, it has been demonstrated that the structural features or functional groups essential to the formation of the enzyme–substrate complex occur at a specific location on the surface of the enzyme. This section of the enzyme, which combines with the substrate and at which the substrate is transformed to products, is called the **active site** of the enzyme. The active site is often a cleft or crevice in the exterior of the molecule (Figure 22.2). It possesses a unique conformation (as well as correctly positioned bonding groups) that is complementary to the structure of the substrate. Thus the two molecules are able to fit together in much the same manner as a key fits into a tumbler lock. This **lock-and-key theory** of enzyme action is illustrated in Figure 22.3. It portrays an enzyme as conformationally rigid

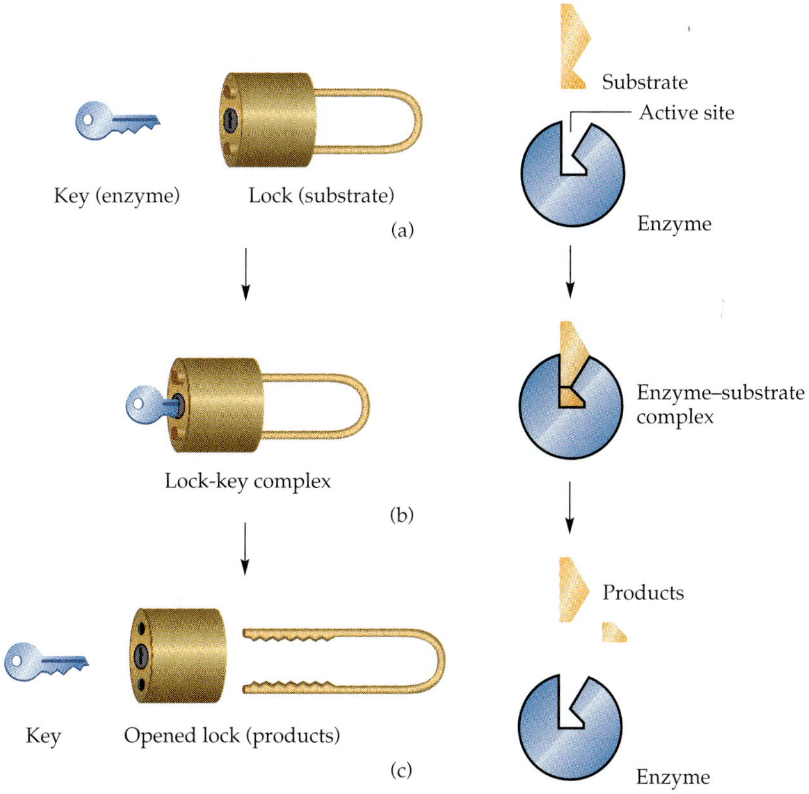

◀ **FIGURE 22.3**

The lock-and-key model of enzyme action. (a) The substrate and the active site of the enzyme have complementary structures and bonding groups; they fit together as a key fits a lock. (b) The catalytic reaction occurs as the two are bonded together in the enzyme–substrate complex. (c) The products of the reaction leave the surface of the enzyme, freeing it to combine with another substrate molecule.

and able to bond only to substrates that are structurally suitable. This restrictiveness explains the high degree of specificity associated with enzymes.

The use of X-ray crystallography in determining the precise three-dimensional structures of enzymes was a major advance in understanding their catalytic activity. It was observed that the binding of some substrates leads to a large conformational change in the enzyme. In 1963, D. E. Koshland Jr. augmented the lock-and-key theory by suggesting that the binding site of an enzyme is not a rigid structure. Instead, some enzymes undergo a change in conformation when they react with a substrate molecule to form an activated complex. The active site has a shape complementary to that of the substrate only *after* the substrate is bound. After catalysis, the enzyme resumes its original structure. In Koshland's words,

> To explain the enzyme's ability to discriminate between similar compounds, a "fit" between the substrate and a portion of the enzyme surface seems essential. However, it appears possible that this fit is not a static one in which a rigid "positive" substrate fits on a rigid "negative" template, but rather, is a dynamic interaction in which the substrate induces a structural change in the enzyme molecule, as a hand changes the shape of a glove.[1]

This **induced-fit theory** is an attractive proposal because it explains several experimental findings incompatible with the lock-and-key model. Koshland cites examples of compounds that bind to enzymes without undergoing further reaction as well as other compounds that are sterically not suited for the active site but nevertheless react catalytically. According to Koshland, the active site of an enzyme consists of two components. One (a *contact group*) is responsible for substrate specificity, and the other (*a catalytic group*) is responsible for catalysis. The active site is a flexible region that can be induced to fit several structurally similar compounds. However, only the proper substrate is capable of correct alignment with the catalytic groups as well (Figures 22.4 and 22.5).

The manner in which an enzyme transforms a substrate into product(s) has been extensively studied. We know that the reactants are brought into proximity as they bind to the enzyme, and this proximity increases the frequency of collisions. Because the enzyme properly aligns each reactant, the effectiveness of each collision is also increased. As yet, however, the detailed mechanism by which enzymes

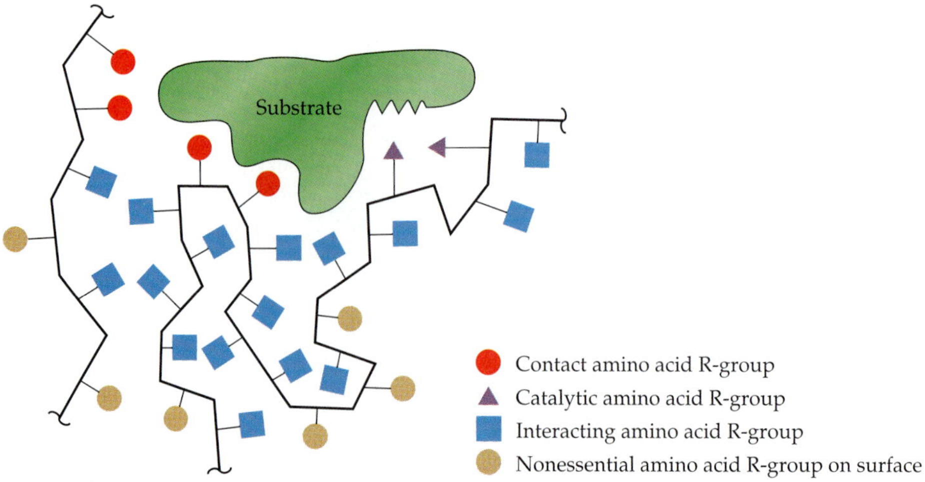

FIGURE 22.4 ▶

Schematic representation of an active site. The fit between contact amino acid R-groups and the substrate determines specificity. Catalytic R-groups act on the substrate bond, which is indicated by the zig-zag line. The R-groups that interact with each other maintain the three-dimensional structure of the enzyme.

[1] D. E. Koshland, Jr., *Science*, **142**, 1533 (1963).

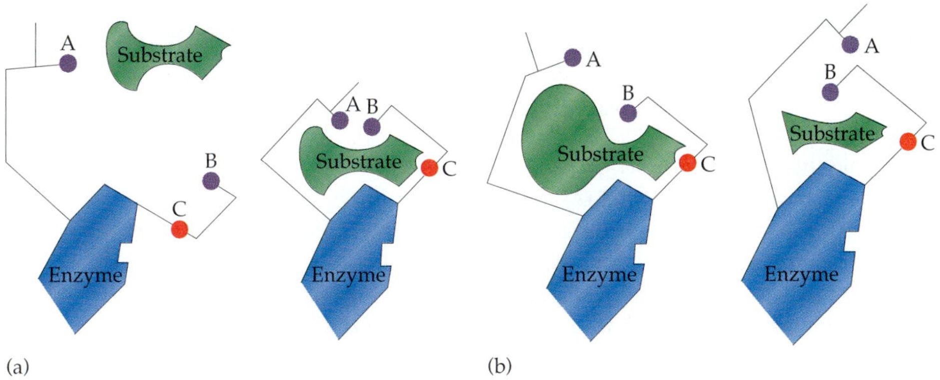

FIGURE 22.5

Schematic representation of a flexible active site. (a) Substrate binding induces the proper alignment of catalytic groups A and B so that reaction takes place. (b) Compounds that are too large or too small are bound but fail to induce proper alignment of catalytic groups; no reaction occurs.

increase the rate of reactions more efficiently than other catalysts is incompletely understood. We still do not know, for example, whether the full catalytic activity of an enzyme resides in its protein structure as a whole or only in a small region associated with the active site. Small peptides have been cleaved from some enzymes (such as ribonuclease) without appreciable loss of catalytic activity.

Specific amino acid side chains (such as the hydroxyl group of serine, the sulfhydryl group of cysteine, and the imidazole group of histidine) are part of the active sites of various enzymes. The active site probably consists of amino acids from several different positions along the protein chain. These amino acids can be brought next to one another as a result of the folding and bending of the polypeptide chain (or chains). The fact that enzymes are inactivated by denaturation points to the importance of the secondary and tertiary structures in maintaining the active site in a precise three-dimensional arrangement. It is known that two amino acids, histidine-57 and serine-195, are involved in the active site of chymotrypsin (Figure 22.6). The polar groups at this site no doubt interact strongly with the peptide

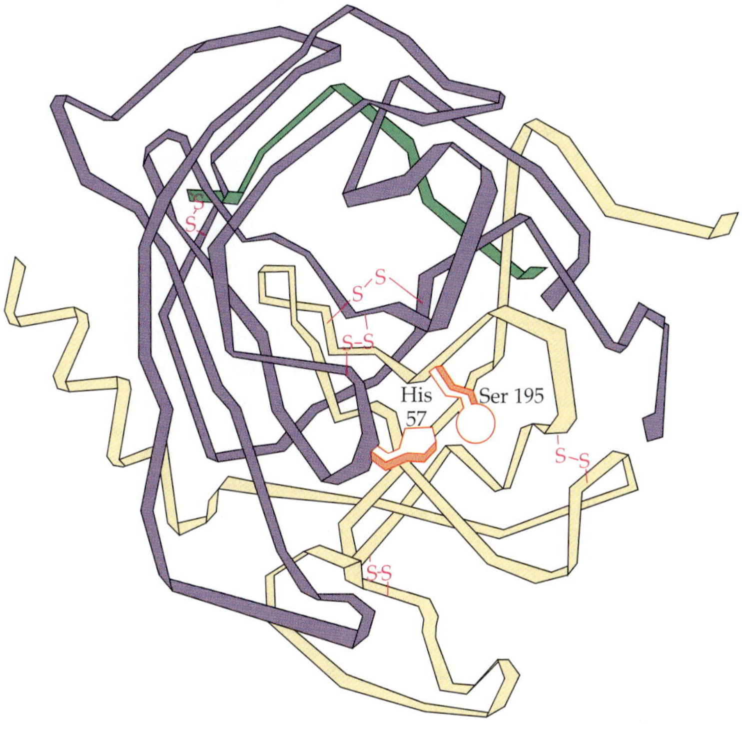

FIGURE 22.6

Model of α-chymotrypsin. The enzyme consists of 245 amino acids in three polypeptide chains, as is indicated by the three colors. The chains are interconnected by disulfide linkages. The active site is in the region of histidine-57 and serine-195. Chymotrypsin is a digestive enzyme that catalyzes the hydrolysis of proteins in the small intestine.

Table 22.1 Some Important Enzymes

Enzyme	Reaction Catalyzed	Turnover Number (per minute)
Carbonic anhydrase	$H_2O + CO_2 \rightleftharpoons H_2CO_3$	36,000,000
Catalase	$2 H_2O_2 \rightleftharpoons 2 H_2O + O_2$	5,600,000
Fumarase	Fumaric acid \rightleftharpoons Malic acid	1,200,000
Lactate dehydrogenase	Pyruvic acid \rightleftharpoons Lactic acid	60,000
Succinate dehydrogenase	Succinic acid \rightleftharpoons Fumaric acid	1,150
DNA polymerase I	Addition of nucleotides to DNA chain	900
Lysozyme	Hydrolysis of specific polysaccharide bonds	30

linkage that is to be hydrolyzed. The remaining amino acids presumably function in maintaining the active site in the correct geometrical configuration to provide for maximum catalytic activity. They also impart specificity to the molecule.

Table 22.1 lists some important enzymes and gives an indication of how rapidly some of them work. The **turnover number** is the number of substrate molecules converted to product in 1 min by one enzyme molecule.

22.4 Specificity of Enzymes

One characteristic that distinguishes an enzyme from all other types of catalysts is its *substrate specificity*. Recall, for example, that hydrogen ions catalyze the hydrolysis of disaccharides, polysaccharides, lipids, and proteins with complete impartiality, whereas different enzymes are required in all four cases. Enzyme specificity is a result of the uniqueness of the active site of each enzyme. This uniqueness is a function of the chemical nature, electric charge, and spatial arrangements of the groups located there. Enzyme specificity is crucial in chemical reactions in the cell. It ensures, for the most part, that the proper reactions occur in the proper place at the proper time. A wide range of enzyme specificities exist, and they are arbitrarily grouped as follows.

Absolute Specificity

Enzymes that have *absolute specificity* catalyze a particular reaction for one particular substrate only and have no catalytic effect on substrates that are closely related. Urease, for example, catalyzes the hydrolysis of urea but not of methylurea, thiourea, or biuret. Absolute specificity is rare among enzymes characterized to date.

$$H_2N-\overset{\overset{\displaystyle O}{\|}}{C}-NH_2 + H_2O \xrightarrow{\text{urease}} CO_2 + 2 NH_3$$

Urea

$$H_2N-\overset{\overset{\displaystyle O}{\|}}{C}-NH-CH_3 \qquad H_2N-\overset{\overset{\displaystyle S}{\|}}{C}-NH_2 \qquad H_2N-\overset{\overset{\displaystyle O}{\|}}{C}-NH-\overset{\overset{\displaystyle O}{\|}}{C}-NH_2$$

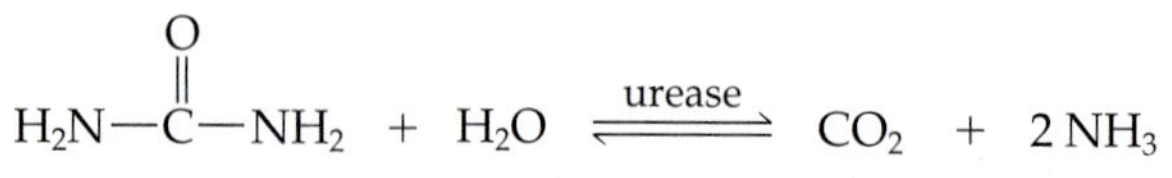

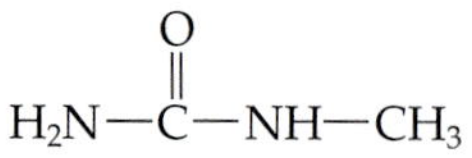

Methylurea Thiourea Biuret

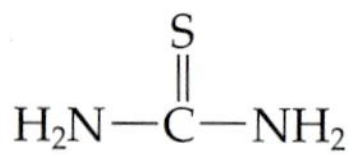

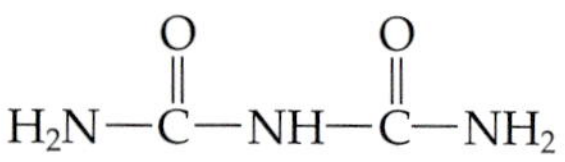

Stereochemical Specificity

Because enzymes are chiral molecules, they show a markedly high degree of *stereochemical specificity*—they have specificity for one stereoisomeric form of the substrate. This is analogous to the binding of monosodium glutamate to the taste buds (see Figure 18.16). L-Lactic acid dehydrogenase catalyzes the oxidation of the L-Lactic acid in muscle cells. D-Lactic acid, found in certain microorganisms, does not bind to the enzyme. Fumarase adds water to fumaric acid but not to its cis isomer, maleic acid.

Group Specificity

Enzymes that have *group specificity* are less selective in that they act upon structurally similar molecules that have the same functional groups. Many of the peptidases fall into this category. Chymotrypsin hydrolyzes peptide bonds involving the carboxyl groups of the aromatic amino acids. Carboxypeptidase attacks peptides from the carboxyl end of the chain, cleaving the amino acids one at a time.

Linkage Specificity

Enzymes that have *linkage specificity* are the least specific of all because they attack a particular kind of chemical bond, irrespective of the structural features in the vicinity of the linkage. The lipases, which catalyze the hydrolysis of ester linkages in lipids, are an example of this type of enzyme.

22.5 Factors That Influence Enzyme Activity

The single most important property of an enzyme is its catalytic activity. Since enzymes are protein catalysts, they are affected both by factors that affect proteins and by factors that affect catalysts in general. The activity of an enzyme can be measured by monitoring the reaction it catalyzes. The reaction rate is determined by observing either the rate at which substrate disappears or the rate at which product forms. In such experiments, the reaction rate is the only variable; all other experimental conditions are held constant.

Concentration of Substrate

The rate of an enzymatic reaction increases as the substrate concentration increases until a limiting rate is reached. At this point, further increase in the substrate concentration produces no significant change in the reaction rate. At excess substrate concentrations, practically all the enzyme molecules are saturated with the substrate at any given instant. Extra substrate molecules must wait until the enzyme–substrate complexes have dissociated to yield products plus free enzymes before they can undergo reaction.

It's as if 10 taxis (enzyme molecules) were waiting at a taxi stand to take people (substrate) on a 10-min trip to a concert hall, one passenger at a time. If only 5 people are present at the stand, the rate of their arrival at the concert hall is 5 people in 10 min. If the number of people at the stand is increased to 10, the rate increases to 10 arrivals in 10 min. With 20 people at the stand, the rate would still be 10 arrivals in 10 min. The taxis have been "saturated." If the taxis could carry 2 or 3 passengers each, the same principle would apply. The rate would simply be higher (20 or 30 people in 10 min) before it leveled off.

This relationship is illustrated in Figure 22.7 and can be summarized in terms of the two equations given in Section 22.3. At low and adequate substrate

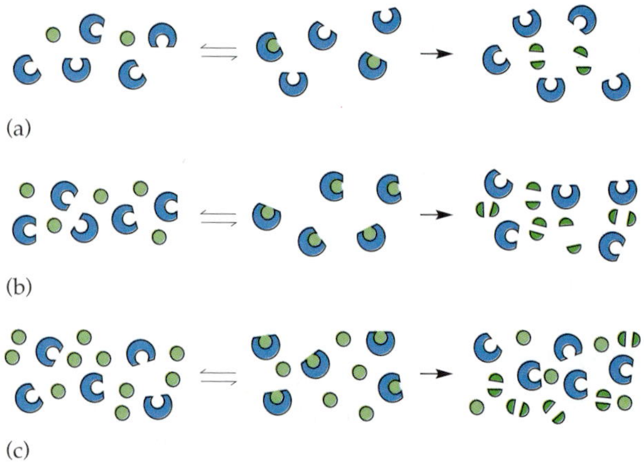

FIGURE 22.7 ▶

Schematic representation of relative concentrations of enzyme and substrate. (a) Low substrate concentration. (b) Adequate substrate concentration. (c) Excess substrate concentration.

concentrations, the formation of the E–S complex is the rate-determining step. At high substrate concentrations, the slowest step is the dissociation of the E–S complex. Figure 22.8 is a characteristic plot of an enzyme-catalyzed reaction, and it is taken as further evidence of the existence of the enzyme–substrate intermediate.

Concentration of Enzyme

Because in essentially all practical cases the concentration of enzyme is much lower than the concentration of substrate (we always have more people than taxis), the rate of an enzyme-catalyzed reaction is directly dependent upon the enzyme concentration (Figure 22.9). This is not a new concept. The reaction rate of any catalytic reaction increases as the concentration of the catalyst is increased (the more taxis, the more people can be transported). At any given time, the cellular concentration of enzyme is determined by its rate of synthesis and its rate of degradation. The cellular concentration can be increased (enzyme induction) or decreased (enzyme suppression) according to the needs of the organism.

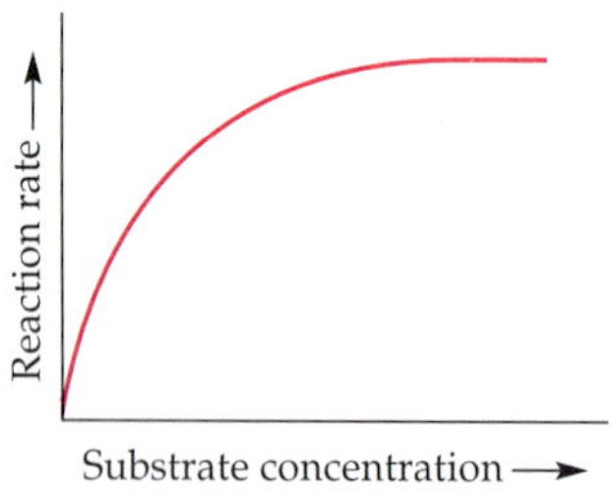

▲ **FIGURE 22.8**

Effect of substrate concentration on the rate of a reaction that is catalyzed by a fixed amount of enzyme.

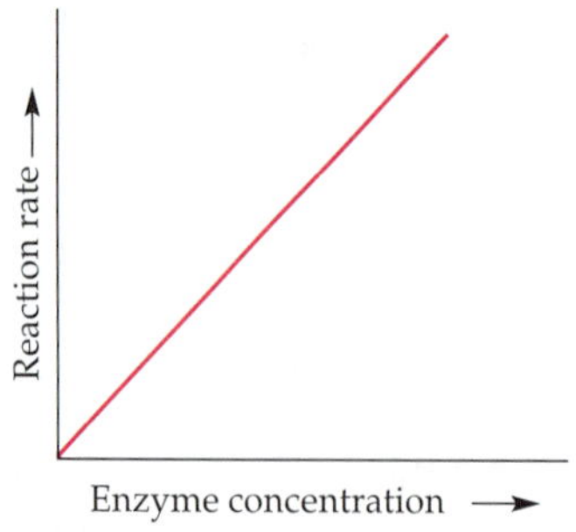

▲ **FIGURE 22.9**

Effect of enzyme concentration on the reaction rate.

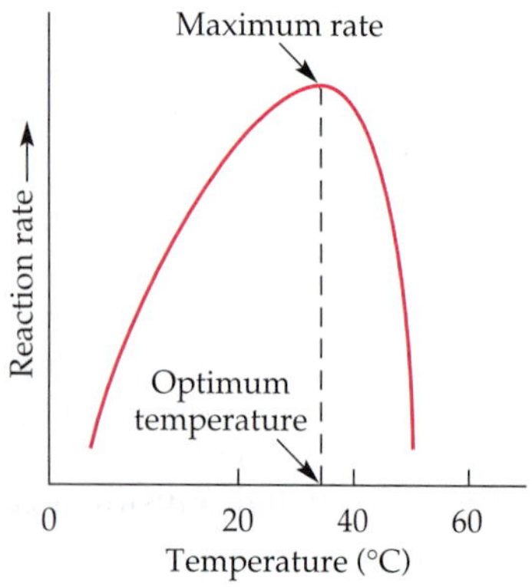

▲ **FIGURE 22.10**

Effect of temperature on the rate of an enzymatic reaction.

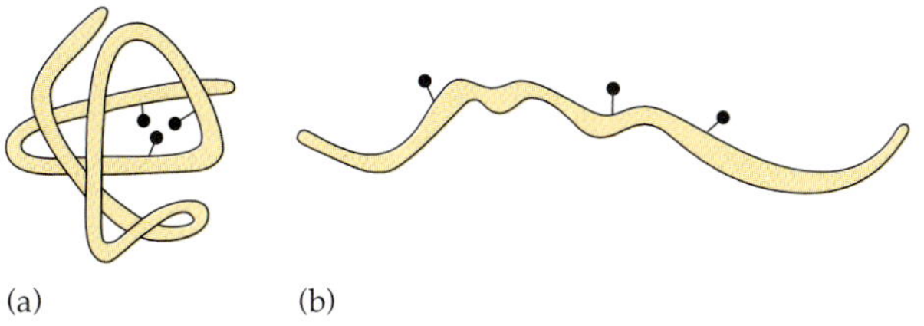

◀ **FIGURE 22.11**

(a) Representation of an active site in an enzyme. (b) Heating denatures the enzyme, and the groups of the active site are no longer close to one another.

Temperature

A rule of thumb for most chemical reactions is that a rise in temperature of 10 °C approximately doubles the reaction rate. (This is due to an increase in the number of molecules that possess sufficient kinetic energy to exceed the activation energy.) To some extent, this rule holds for all enzymatic reactions. After a certain point, however, an increase in temperature causes a decrease in reaction rate, as indicated in Figure 22.10. The temperature that affords maximum activity is known as the **optimum temperature** for the enzyme in question. Most enzymes of warm-blooded animals have optimum temperatures of about 37 °C (98 °F).

The decrease in rate is a direct consequence of the fact that enzymes are proteins and thus denatured by heat. Heating disrupts the secondary and tertiary structures of the enzymes, causing a disorientation of the active site. This disorientation renders the active site inaccessible to the substrate (Figure 22.11).

At 0 °C and 100 °C, the rate of enzyme-catalyzed reactions is nearly zero. This fact has several practical applications. We sterilize objects by placing them in boiling water so as to denature the enzymes of any bacteria that may be in or on the objects. We refrigerate and freeze our food to slow down enzyme activity and preserve the food. Animals go into hibernation because their body temperature decreases in winter, and as a result the rates of their metabolic processes decrease. The food required to maintain this lowered metabolic rate is provided by fat reserves stored in their tissues.

Hydrogen Ion Concentration

Being proteins, enzymes are sensitive to changes in pH. Extreme values of pH (whether high or low) can denature the protein. However, *any* change in pH, even a small one, alters the degree of ionization of acidic and basic groups both on the enzyme and on the substrate. If any ionizable groups are located at the active site and if a certain charge is necessary in order for the enzyme to bind its substrate, an enzyme molecule that has even one of these charges neutralized loses its catalytic activity.

An enzyme will exhibit maximum activity over a narrow pH range in which the molecule exists in its proper charged form. The median value of this pH range is known as the **optimum pH** of the enzyme (Figure 22.12). With the notable exception of gastric juice, most body fluids have pH values between 6 and 8. This is essential because most enzymes exhibit optimal activity in the physiological pH range 7.0 to 7.5. However, there are a few enzymes that have optimum pH values outside the usual physiological range. For example, the optimum pH for pepsin is 2.0 in the stomach, and that for trypsin is 8.0 in the small intestine.

The enzyme lysozyme destroys bacteria by hydrolyzing certain polysaccharide bonds that occur in their cell walls. The optimum pH for lysozyme is 5.0. Activity declines rapidly at both higher and lower pH values. Lysozyme is active only when an aspartic acid side chain is ionized (COO^-) and a glutamic acid side

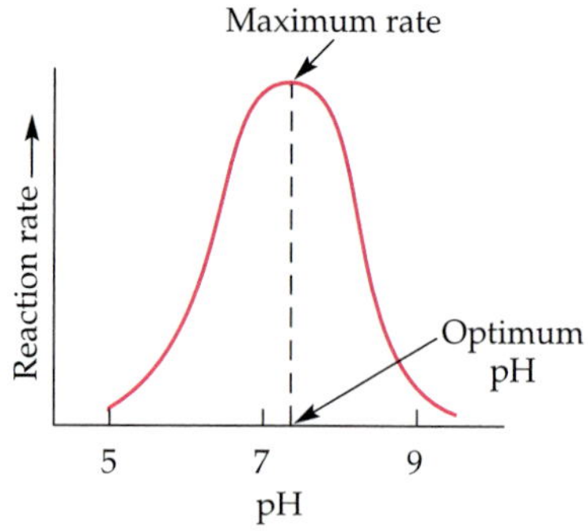

FIGURE 22.12 ▶
Effect of pH on the rate of an enzymatic reaction.

chain is not (COOH). At low pH values, both are protonated. At high pH values, both are ionized. In either case, the enzyme doesn't function.

22.6 Enzyme Inhibition: Poisons

In the preceding section, we noted that enzymes are inactivated by increased temperatures and by changes in pH. In a sense, then, temperature and hydrogen ion concentration can be considered factors that *inhibit* enzyme activity. In fact, any physical change or chemical reagent that denatures protein adversely affects the rate of enzymatic reactions. This type of enzyme inhibition is referred to as *nonspecific* inhibition because it affects all enzymes in the same manner. In contrast, a *specific* inhibitor exerts its effect upon a single enzyme or a group of related enzymes. The inhibition of enzyme activity is one of the most important control mechanisms in living organisms. Many poisons act to inhibit specific enzymes. (Table 22.2)

Competitive Inhibition

A **competitive inhibitor** is any compound that bears a close structural resemblance to a particular substrate and competes with that substrate for binding at the same active site on the enzyme. The inhibitor is not acted upon by the enzyme and so remains bound to the enzyme, preventing the substrate from approaching the active site.

Table 22.2 Poisons as Enzyme Inhibitors

Poison	Formula	Enzyme Inhibited	Action
Cyanide	CN^-	Cytochrome oxidase, catalase	Binds Fe^{3+} cofactor
Fluoride	F^-	Enolase	Binds Mg^{2+} cofactor
Sulfide	S^{2-}	Phenolase	Binds Cu^{2+} cofactor
Arsenate	AsO_4^{3-}	Glyceraldehyde 3-phosphate dehydrogenase	Substitutes for phosphate
Iodoacetate	ICH_2COO^-	Triose phosphate dehydrogenase	Binds to cysteine sulfhydryl group
Nerve gas	$F{-}P(=O)(OCH(CH_3)_2){-}OCH(CH_3)_2$	Acetylcholinesterase	Binds to serine hydroxyl group

The degree of competitive inhibition depends upon the relative concentrations of substrate and inhibitor. If the inhibitor is present in relatively large quantities, it blocks the active sites on all the enzyme molecules, and complete inhibition results. However, formation of the inhibitor–enzyme complex is reversible. Increased substrate concentration permits displacement of the inhibitor from the active site. Competitive inhibition can be completely reversed by addition of large excesses of substrate.

The reversible nature of competitive inhibition has provided much information about the enzyme–substrate complex and about the specific groups involved at the active sites of various enzymes. Pharmaceutical companies have synthesized drugs that can competitively inhibit metabolic processes in bacteria (Section 22.7) and in cancer cells (see Section K.6).

A classic example of competitive inhibition is the effect of malonic acid on the enzyme activity of succinic acid dehydrogenase (Figure 22.13). Malonic acid is a homolog of the enzyme's normal substrate, succinic acid. The malonic acid molecule binds to the active site because the spacing of its carboxyl groups is not greatly different from that of succinic acid. No catalytic reaction occurs, and malonic acid remains bonded to the enzyme. We discuss this reaction again in connection with carbohydrate metabolism (Section 25.1).

Noncompetitive Inhibition

A **noncompetitive inhibitor** is a substance that can combine with either the free enzyme or the enzyme–substrate complex. The noncompetitive inhibitor binds to the enzyme at a position relatively remote from the active site and in so doing alters the three-dimensional conformation of the enzyme. This alteration changes the configuration of the active site, with one of two results. Either the E–S complex does not form at its normal rate or, once formed, it does not decompose at the normal rate to yield products. Since the inhibitor does not structurally resemble

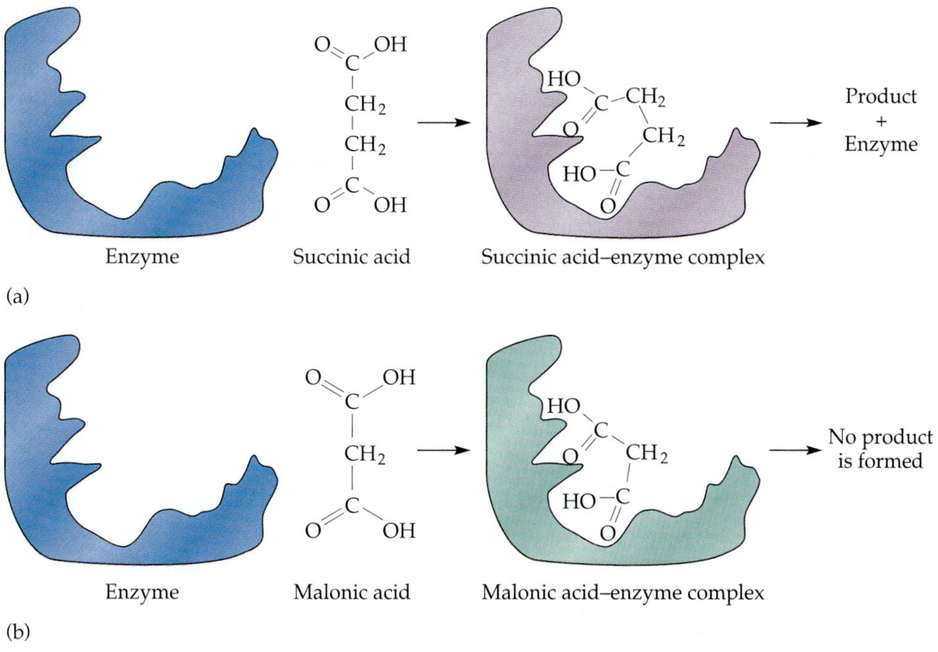

FIGURE 22.13
(a) Succinic acid binds to the enzyme succinic acid dehydrogenase. A dehydrogenation reaction occurs and the product, fumaric acid, is released from the enzyme. (b) Malonic acid also binds to the active site of succinic acid dehydrogenase. In this case, however, no subsequent reaction occurs, and malonic acid remains bound to the enzyme.

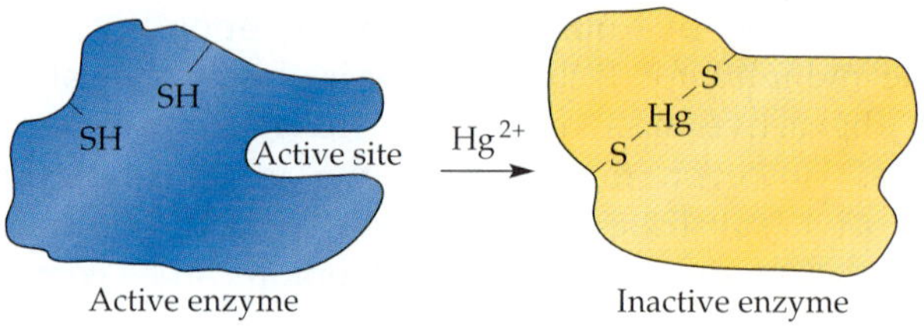

▲ FIGURE 22.14

Mercury poisoning is an example of noncompetitive inhibition of an enzyme. Hg^{2+} ions react with sulfhydryl groups to change the conformation of the enzyme and destroy the active site.

the substrate, the addition of excess substrate does *not* reverse the inhibitory effects.

Many enzymes contain reactive groups, such as $-COO^-$, $-NH_3^+$, $-SH$, or $-OH$, that are essential for maintaining the proper three-dimensional conformation of the enzyme. Any chemical reagent that is capable of combining with one or more of these groups inhibits the enzyme. The heavy metal ions Ag^+, Hg^{2+}, and Pb^{2+} have strong affinities for carboxylate and sulfhydryl groups. We have already discussed their toxic effects with regard to protein denaturation (Section 21.13). Similarly, these ions react with the sulfhydryl groups of enzymes, rendering them inactive. This can occur at a position removed from the active site (Figure 22.14). Poisoning by lead or mercury ions is an example of noncompetitive inhibition.

Irreversible Inhibition

An **irreversible inhibitor** inactivates enzymes by forming strong bonds to a particular group at the active site. The inhibitor does not resemble the substrate, and the inhibitor–enzyme bond is so strong that the inhibition cannot be reversed by addition of excess substrate. Irreversible inhibition was formerly considered to be a type of noncompetitive inhibition, but it is now recognized as a distinct type of inhibition.

The nerve gases, especially diisopropylfluorophosphate (DFP), irreversibly inhibit biological systems by forming an enzyme–inhibitor complex with a specific hydroxyl group (of serine) situated at the active sites of certain enzymes. The peptidases trypsin and chymotrypsin contain such serine groups and so are inhibited by DFP.

$$\boxed{\text{Enzyme}}\!-\!OH + F-\overset{\overset{\displaystyle O}{\|}}{P}-(OR)_2 \longrightarrow \boxed{\text{Enzyme}}\!-\!O-\overset{\overset{\displaystyle O}{\|}}{P}(OR)_2 + HF$$

Metalloenzymes (enzymes that require the presence of a metal ion cofactor) are irreversibly inhibited by substances that form strong complexes with the metal. Traces of hydrogen cyanide inactivate iron-containing enzymes, such as catalase and cytochrome oxidase. Oxalic and citric acids inhibit blood clotting by forming complexes with calcium ions which are necessary for the activation of the enzyme thromboplastin (see Section 28.8).

Inhibition of Nerve Transmission

The action of organophosphorus compounds is another example of irreversible inhibition. These compounds act on an enzyme that is an essential part of the process of nerve transmission. Nerve cells (neurons) interact with other nerve cells, and with muscles and glands, at junctions called *synapses* (refer to Figure G.3). Nerve impulses are transported across synapses by small molecules known as *neurotransmitters*. Neurons use a number of different molecules as neurotransmitters. The two major neurotransmitters are acetylcholine and norepinephrine (see Section G.2).

Acetylcholine is synthesized from acetyl coenzyme A (Section 25.1) and choline and is stored in special vesicles at the axon ends of neurons. The arrival of a nerve impulse leads to the release of acetylcholine into the synapse.[2] The acetylcholine molecules then diffuse across the synapse, where they combine with specific receptor protein molecules embedded in the postsynaptic membrane of the adjacent neuron (or muscle, or gland). Binding of acetylcholine to the receptors causes a change in membrane permeability of the receiving neuron (or muscle, or gland). Sodium ions then move into the cell, potassium ions move out, and this ion flux causes the signal (the action potential) to be sent along the entire neuron until it reaches another synapse.

It is interesting to note that manic-depressive people are overly sensitive to acetylcholine; they seem to have too many receptors. This may account for their wild swings in mood.

Once the impulse has been passed on, the acetylcholine must be immediately deactivated so that the receptor molecules can receive the next stimulus. The deactivation occurs by the hydrolysis of acetylcholine to choline and acetic acid, through the catalytic activity of the enzyme *acetylcholinesterase* (which is located in the synapse):

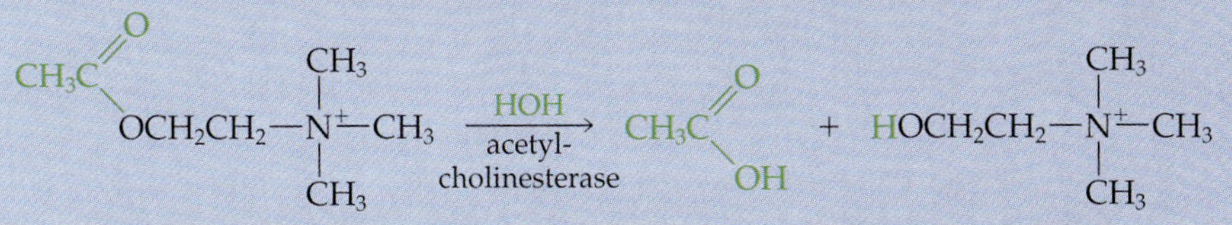

Acetylchloline Acetic acid Choline

This enzyme is characterized by an extremely high turnover number. It is estimated that the enzyme-catalyzed hydrolysis reaction occurs in 40 μs ($40 x 10^{-6}$s). This speed is essential because neurons can transmit 1000 impulses per second as long as each postsynaptic membrane is continually available to receive new acetylcholine molecules.[3]

After the breakdown of acetylcholine, the receptor neuron releases the hydrolysis products and is then ready to receive further impulses. Other enzymes convert the acetic acid and choline back to acetylcholine, completing the cycle.

People with Alzheimer's disease produce too little acetylcholine for proper brain function.

The organophosphorus nerve poisons (Figure 22.15) affect all biological systems in a similar manner. The polar phosphorus–oxygen bond attaches tightly to acetylcholinesterase, preventing the enzyme from performing its normal function (Figure 22.16). If the breakdown of acetylcholine is blocked, this messenger compound builds up, causing the receptor nerves to "fire" repeatedly, to be continuously "on." This overstimulates the muscles, glands, and organs. The heart beats wildly and irregularly. The victim goes into convulsions and dies quickly.

Nerve, muscle, and other cells sensitive to acetylcholine are said to be *cholinergic*. Pharmacologists have developed drugs that affect cholinergic nerves by either enhancing or blocking the action of acetylcholine on the receptor cells.

[2] Botulism toxin, one of the most poisonous substances known, and certain snake venoms act by *inhibiting the release of acetylcholine* from the axon. Thus these toxic substances effectively block nerve transmissions that use acetylcholine as their neurotransmitter. Paralysis sets in and death occurs, usually by respiratory failure.

[3] Curare, used for centuries on the arrows of South American Indians, causes skeletal muscle paralysis. Curare exerts its effects by competing with acetylcholine for the receptor sites on the postsynaptic muscle cell membranes. Thus it *blocks the receptor sites* and prevents the message from being transmitted from nerve to muscle.

Succinylcholine, an inhibitor of acetylcholine, is used to produce muscular relaxation in surgical procedures. On the other hand, an insufficient secretion of acetylcholine by motor neurons results in the condition called myasthenia gravis. The person suffers from muscular weakness and may have trouble contracting the muscles associated with breathing, chewing, eye movements, and speaking. This condition can be treated by administering neostigmine, which inhibits acetylcholinesterase. If the enzyme is inhibited, enough acetylcholine may accumulate in the neuromuscular synapse to stimulate muscle contraction.

$$(CH_3)_3\overset{+}{N}CH_2CH_2O\overset{O}{\overset{\|}{C}}CH_2CH_2\overset{O}{\overset{\|}{C}}OCH_2CH_2\overset{+}{N}(CH_3)_3$$

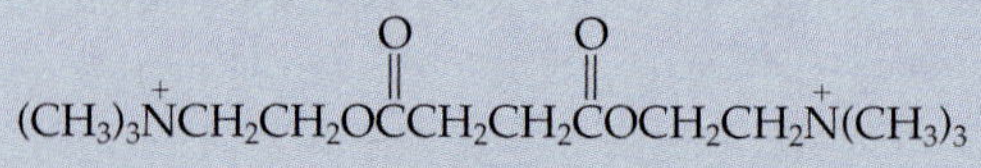

Succinylcholine

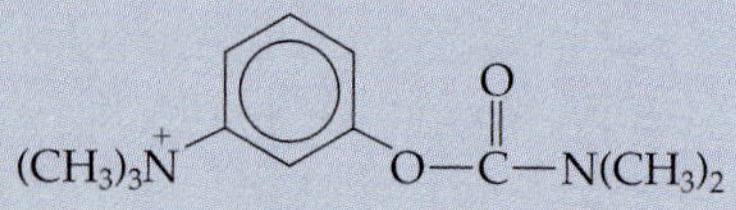

Neostigmine

FIGURE 22.15 ▶

Some organophosphorus compounds. Malathion and parathion are insecticides. Tabun and sarin are nerve poisons designed for use in chemical warfare. Sarin was used in the 1995 terrorist attack on the Tokyo subway that killed 12 people and sickened thousands.

Malathion

Tabun
(agent GA)

Parathion

Sarin
(agent GB)

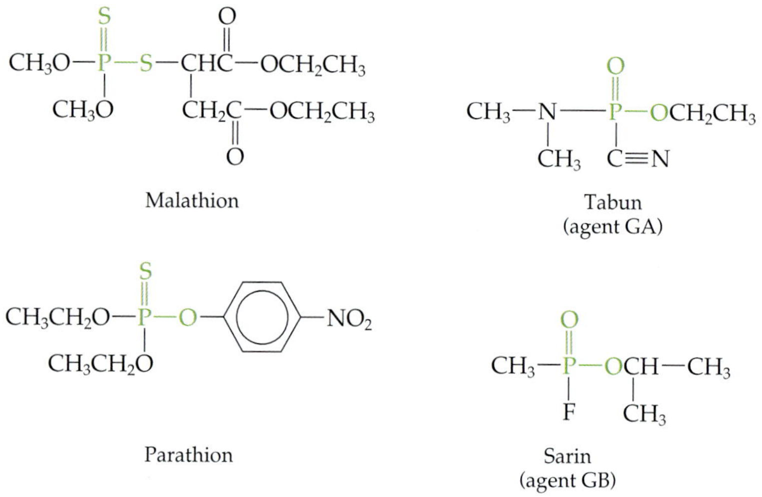

FIGURE 22.16 ▶

(a) Acetylcholinesterase catalyzes the hydrolysis of acetylcholine to acetic acid and choline. (b) An organophosphate ties up acetylcholinesterase, preventing it from breaking down acetylcholine.

(a)

(b)

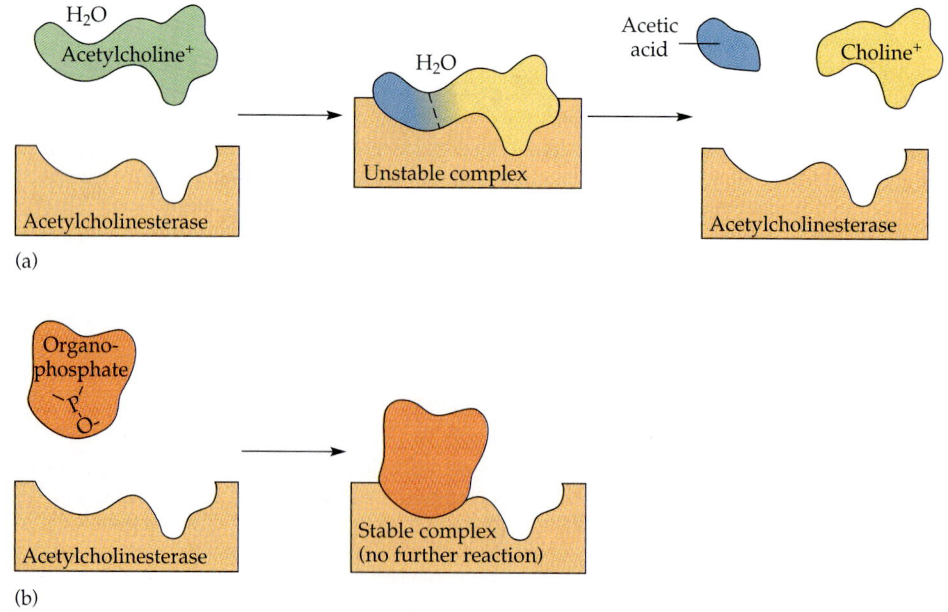

22.7 Chemotherapy

Chemotherapy is the use of chemicals (drugs) to destroy infectious microorganisms and cancer cells without damaging the cells of the host. From bacteria to humans, the metabolic pathways of all living organisms are quite similar, and so the discovery of safe and effective chemotherapeutic agents is a formidable task. It is now well established that many chemotherapeutic drugs function by inhibiting a critical enzyme in the cells of the invading organism.

In 1904, Paul Ehrlich, a German chemist, realized that certain chemicals could be used to control or cure infectious diseases. Ehrlich coined the term *chemotherapy*, a shorter version of the term "chemical therapy."

Chemotherapy is widely used in the treatment of cancer patients. **Antineoplastic drugs** (substances that inhibit the growth of cancer cells), such as 5-fluorouracil and 6-mercaptopurine, interfere with the production of DNA and RNA in tumor cells by substituting for the pyrimidine and purine bases (see Section 23.1).

Antimetabolites

An **antimetabolite** is a substance whose structure closely resembles that of the normal substrate (the *metabolite*) of an enzyme and *competitively* inhibits a significant metabolic reaction. One of the earliest (1935) and best understood antimetabolites is the synthetic antibacterial agent sulfanilamide (Figure 22.17). Its effectiveness rests on its structural similarity to *p*-aminobenzoic acid,[4] a compound vital to the growth of many pathogenic bacteria. The vitamin folic acid (Section J.6) serves as a coenzyme for several important biochemical processes. We humans obtain folic acid from food we eat. Bacteria can synthesize the folic acid they need *only if* they have access to *p*-aminobenzoic acid. When bacteria encounter sulfanilamide, a bacterial enzyme readily incorporates the drug into a pseudofolic acid. This altered folic acid not only cannot function as a proper coenzyme, it also serves as a competitive inhibitor of the enzyme. The bacteria are

In Section K.6, we examine some antimetabolites that are anticancer agents.

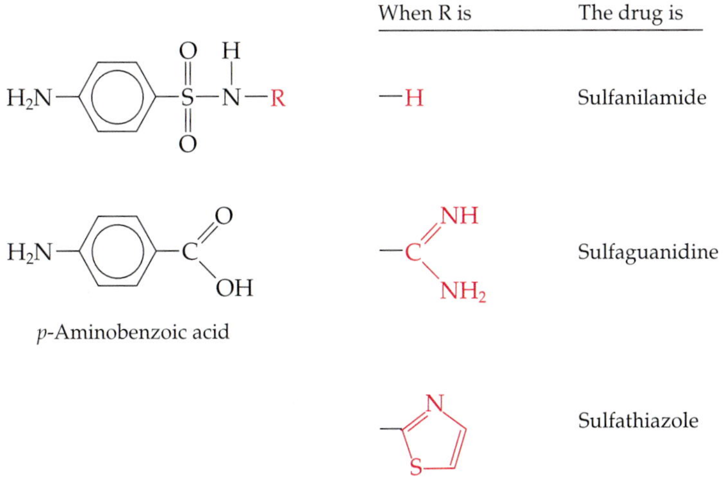

◀ **FIGURE 22.17**
Sulfa drugs interfere with the normal metabolism of *p*-aminobenzoic acid.

[4] It is interesting to note that *p*-aminobenzoic acid (PABA) has become widely known as an ingredient in suntan lotions. PABA acts as a sun filter by absorbing the short-wavelength ultraviolet rays that are responsible for causing sunburn. In most sunscreen products, the concentration of PABA is indicated by skin-protection factor (SPF) ratings. The larger the number, the higher the concentration of PABA and thus the greater the effectiveness against sunburn. The ethyl ester of PABA is Benzocaine, a local anesthetic found in a wide variety of over-the-counter products including first aid and sunburn sprays, foot powders, cough medicines, and appetite control products.

unable to make compounds such as certain amino acids and nucleotides, and so they die.

Sulfanilamide is not harmful to humans (or to other mammals) because we cannot synthesize folic acid but must obtain it, preformed, from our diets. After the drug was recognized as an antibacterial agent, many other sulfanilamide derivatives (sulfa drugs) were synthesized and found to be even more effective in this capacity. Many lives were saved during World War II as a result of these popularly named "wonder drugs." Soldiers carried packages of powdered sulfa drugs to sprinkle on open wounds to prevent infection.

Unfortunately, prolonged use of sulfa drugs causes a number of side effects, particularly kidney damage, and so they have been largely replaced by the penicillins and other antibiotics. However, they are still prescribed for some specific infections against which they are highly effective, such as infections of the bladder and urinary tract. Some newer sulfa drugs are used in the treatment of tuberculosis and leprosy, and they are widely used in veterinary medicine.

Antibiotics

The term *antibiotic* is considered synonymous with *antibacterial*, although many antiviral and anticancer drugs are also antibiotics.

Although some antibiotics are believed to function as antimetabolites, the terms are not synonymous. An **antibiotic** is a compound produced by one microorganism (bacterium, mold, yeast) but toxic to another microorganism. Antibiotics, many of which can now be synthesized in the laboratory, constitute no well-defined class of chemically related substances. Instead, they possess the common property of effectively inhibiting a variety of enzymes essential to bacterial growth.

Penicillin, one of the most widely used antibiotics in the world, was fortuitously discovered by Alexander Fleming in 1928. In 1938, Ernst Chain and Howard Florey isolated it in pure form and proved its effectiveness as an antibiotic. (The three scientists received the Nobel Prize for physiology or medicine in 1945.) Penicillin was first introduced into medical practice in 1941. It functions by interfering with the synthesis of cell walls of reproducing bacteria. Penicillin inhibits an enzyme (transpeptidase) that catalyzes the last step in bacterial cell wall biosynthesis. This step involves the joining of long polysaccharide chains by short peptide chains. The new cell walls are defective, and subsequently the bacterial cells burst. Since human cells have cell membranes and not cell walls, they are not affected.

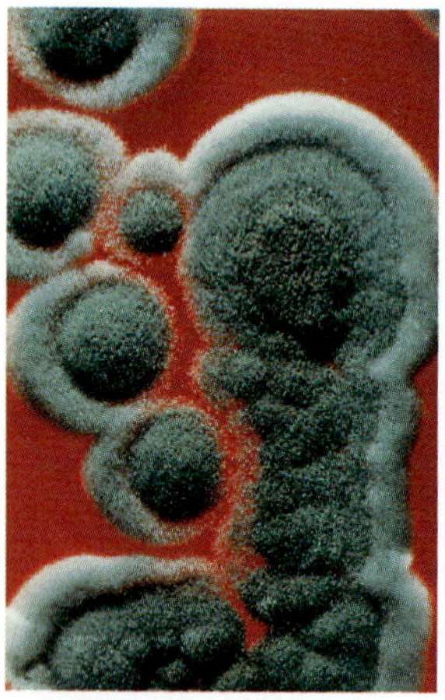

Penicillin molds. These symmetrical colonies of mold are *Penicillium chrysogenum*, a mutant form of which now produces almost all of the world's commercial penicillin.

Several naturally occurring penicillins have been isolated. All have the empirical formula $C_9H_{11}O_4SN_2R$ and contain a four-member ring fused to a five-member ring (Figure 22.18). The various R-groups are obtained by the addition of the appropriate organic compounds to the culture medium.

The penicillins are effective against Gram-positive bacteria (bacteria that are stained by Gram's dye) and a few Gram-negative bacteria (including *E. coli*). They have proved effective in the treatment of diphtheria, gonorrhea, pneumonia, syphilis, many pus infections, and certain types of boils. Penicillin G was the earliest penicillin to be used on a wide scale. However, it cannot be administered orally because it is quite unstable and the acidic pH of the stomach causes a rearrangement to an inactive derivative. Penicillin V, ampicillin, and amoxicillin, on the other hand, are acid-stable, and they are the major oral penicillins.

Some strains of bacteria become resistant to penicillin by a mutation that allows them to synthesize an enzyme, penicillinase, that breaks down the antibiotic (by cleavage of the amide linkage in the four-member ring). To combat these strains, scientists have been able to synthesize penicillin analogs (such as methicillin) that are not inactivated by penicillinase.

When R is		The drug is
$-CH_2-\bigcirc$		Penicillin G
$-CH_2-S-CH_2-CH=CH_2$		Penicillin O
$-CH_2-O-\bigcirc$		Penicillin V
$-\underset{NH_2}{CH}-\bigcirc$		Ampicillin
$-\underset{NH_2}{CH}-\bigcirc-OH$		Amoxicillin
CH_3O \bigcirc CH_3O		Methicillin

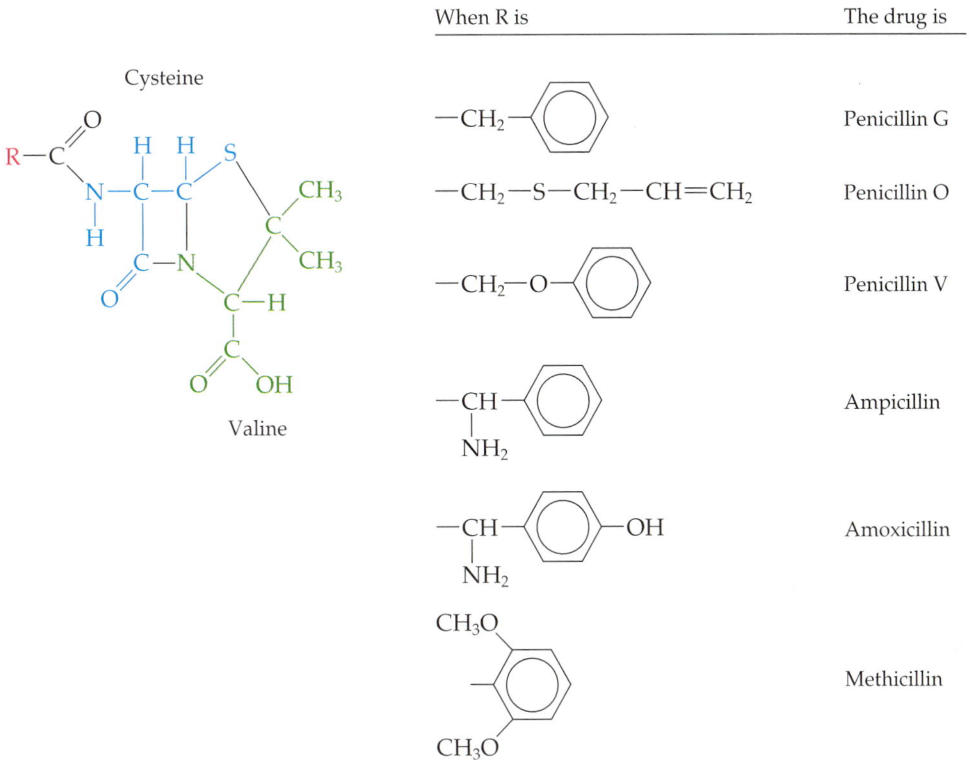

▲ **FIGURE 22.18**

The penicillins differ only in the identity of their R-groups. Note that the amino acids valine and cysteine are incorporated into the penicillin structure.

Some people (perhaps as many as 5% of the population) are allergic to penicillin and therefore must be treated with other antibiotics. The allergic reaction is so severe that a fatal coma may occur if penicillin is inadvertently administered to a sensitive individual. Fortunately, a number of other antibiotics have been discovered (Figure 22.19). Most are the products of microbial synthesis (e.g., aureomycin, streptomycin). Others are made by chemical modifications of antibiotics (e.g., semisynthetic penicillins, tetracyclines), and some are manufactured entirely by chemical synthesis (e.g., chloramphenicol). They have proved to be as effective as penicillin in destroying infectious microorganisms. Many of these antibiotics exert their effects by blocking protein synthesis in microorganisms. (See Table K.1 for a listing of some of the major bacterial diseases.)

During their early history, antibiotics were considered miracle drugs. The number of deaths from blood poisoning, pneumonia, and other infectious diseases was reduced substantially by the use of antibiotics. Only six decades ago, a person with a major infection almost always died. Today, such deaths are rare. Six decades ago, pneumonia was a dread killer of people of all ages. Today, it kills only the very old or those ill from other causes. The antibiotics have indeed worked miracles in our time, but even miracle drugs are not without problems. It wasn't long after the drugs were first used that disease organisms began to develop strains resistant to the drugs. Before erythromycin (an antibiotic obtained from *Streptomyces erythreus*) was used very much, all strains of staphylococci could be handled readily by the drug. After it had been put into extensive use, resistant strains began to appear. Staph infections became a serious problem in

Amoxicillin was the most widely prescribed drug in the United States in 1988. However, by 1992, it had dropped to 30th place because many bacteria had become resistant to it.

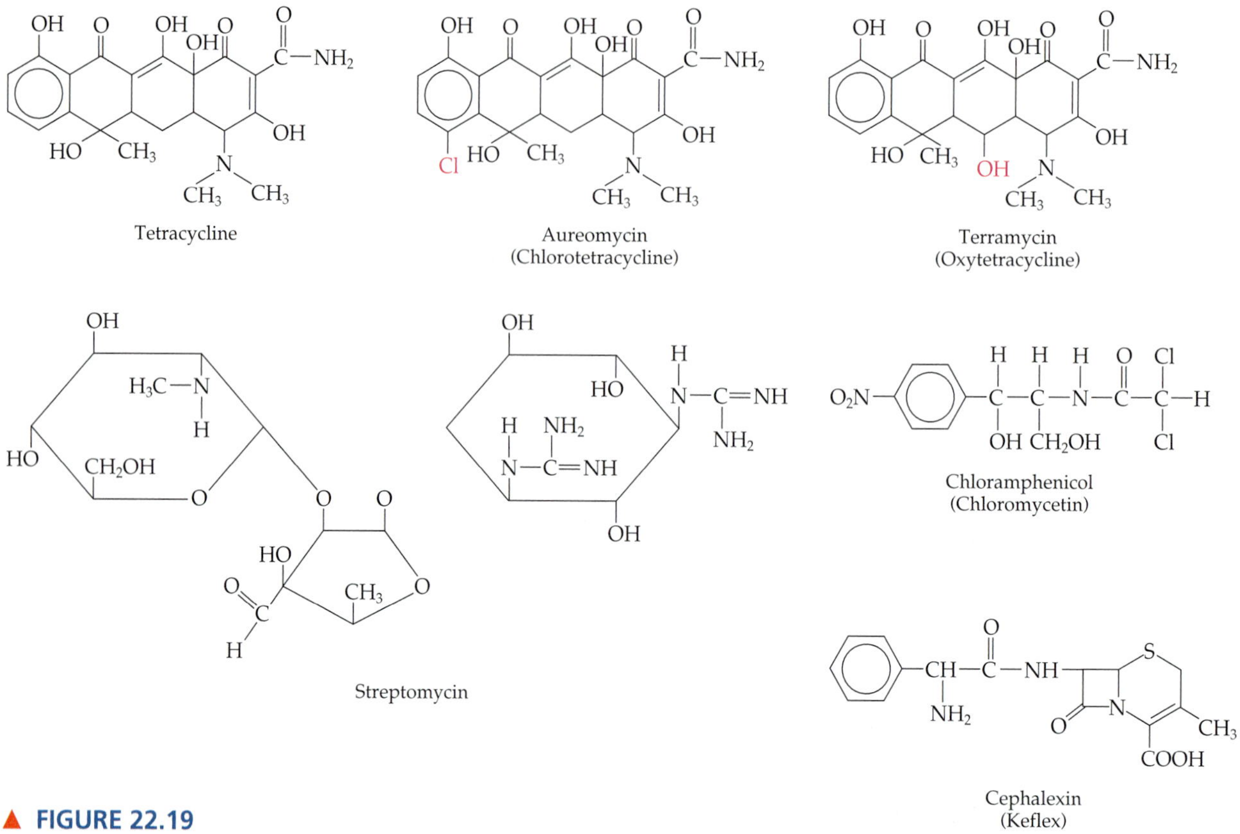

▲ FIGURE 22.19

Structures of some common antibiotics. Note the structural similarities among the tetracyclines. Streptomycin is a glycoside containing an amino derivative of glucose. Chloramphenicol bears a resemblance to epinephrine. Cephalexin is a member of the cephalosporins, antibiotics related to the penicillins.

FIGURE 22.20 ▶

A hospital form for clinical analysis of a blood sample.

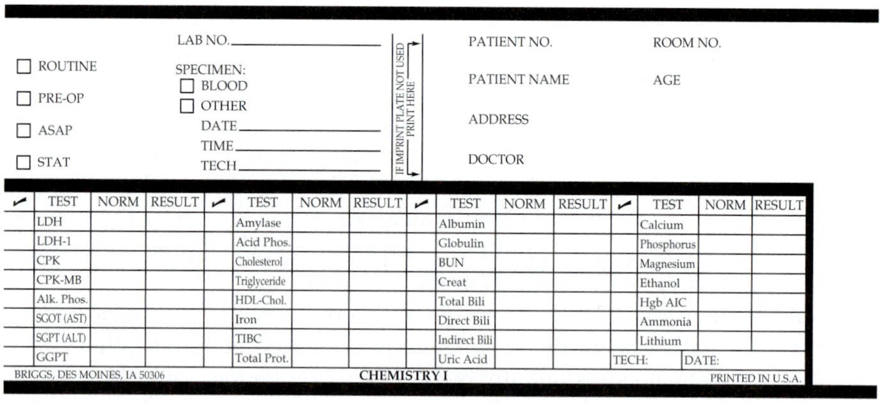

☐ ROUTINE
☐ PRE-OP
☐ ASAP
☐ STAT

LAB NO.______
SPECIMEN:
☐ BLOOD
☐ OTHER
DATE______
TIME______
TECH______

IF IMPRINT PLATE NOT USED PRINT HERE

PATIENT NO. ROOM NO.
PATIENT NAME AGE
ADDRESS
DOCTOR

✓	TEST	NORM	RESULT	✓	TEST	NORM	RESULT	✓	TEST	NORM	RESULT	✓	TEST	NORM	RESULT
	LDH				Amylase				Albumin				Calcium		
	LDH-1				Acid Phos.				Globulin				Phosphorus		
	CPK				Cholesterol				BUN				Magnesium		
	CPK-MB				Triglyceride				Creat				Ethanol		
	Alk. Phos.				HDL-Chol.				Total Bili				Hgb AIC		
	SGOT (AST)				Iron				Direct Bili				Ammonia		
	SGPT (ALT)				TIBC				Indirect Bili				Lithium		
	GGPT				Total Prot.				Uric Acid			TECH:		DATE:	

BRIGGS, DES MOINES, IA 50306 CHEMISTRY I PRINTED IN U.S.A.

Drug-resistant tuberculosis now accounts for one out of every seven new cases. (More than 5% of these resistant cases have already died.)

hospitals. People who had gone to the hospital to be cured got serious bacterial infections instead. Some even died from staph infections picked up at hospitals.

In a race to stay ahead of resistant strains of bacteria, scientists continue to seek new antibiotics. The penicillins now have partially been displaced by related compounds called cephalosporins. Perhaps the most notable of these is cephalexin (Keflex). Unfortunately, some strains of bacteria already show resistance to cephalexin.

Diagnostic Applications of Enzymes

The measurement of enzyme activity in such body fluids as plasma and serum has become a valuable tool in medical diagnosis. Certain enzymes that function in the plasma, such as those involved in blood clotting, are continually secreted into the blood by the liver. Most other enzymes, however, are normally present in plasma in very low concentrations. They are derived from the routine destruction of erythrocytes, leukocytes, and other cells. When cells die, their soluble enzymes leak out of the cells and enter the bloodstream. Since not all cells contain the same complement of enzymes, those that are specific to a particular organ can be important in aiding diagnosis.

An abnormally high level of a particular enzyme in the blood often indicates specific tissue damage, as in hepatitis and myocardial infarction (*myo*, muscle; *cardi*, heart; an *infarct* is an area of dead tissue). For example, elevated blood levels of creatine kinase (CK; see Section 24.8) and glutamic-oxaloacetic transaminase (GOT; see Section 27.5) accompany some forms of severe heart disease. A blood analysis that shows high levels of CK may indicate that the heart muscle has suffered serious damage. On the other hand, many forms of strenuous (and healthful) physical activity also result in elevated CK levels. The enzyme mediates the reaction that serves as one source of energy for muscle contraction. Indeed, it is even possible for the CK level to rise simply because someone who hates needles has tensed up while waiting for the blood sample to be taken. Nonetheless, as Figure 22.20 suggests, analysis for specific enzymes is considered one valuable source of data on which to base a medical diagnosis. Table 22.3 lists the commonly assayed enzymes that are used in clinical diagnoses.

In a *myocardial infarction,* or heart attack, the coronary circulation becomes blocked and the cardiac muscle cells die from lack of oxygen. The affected tissue then degenerates, creating a nonfunctional area known as an *infarct.*

Table 22.3 Some Important Enzymes for Clinical Diagnoses

Enzyme Assayed	Organ or Tissue Affected
α-Amylase	Pancreas
Alkaline phosphatase	Bone, liver
Acid phosphatase	Prostate
Creatine kinase (CK)	Muscle, heart
Glutamic-oxaloacetic transaminase (GOT)	Heart, liver
Glutamic-pyruvic transaminase (GPT)	Liver
Lactic dehydrogenase (LDH)	Heart, liver
Alanine aminotransferase	Liver
Aspartate aminotransferase	Heart, liver

Modern medical practices have automated and computerized the assay procedures for most serum enzymes. It is important to note that the precise patterns of enzyme changes in certain tissue diseases are characteristic. For example, in a myocardial infarction, the GOT/GPT ratio is usually high; the reverse is true in liver disease. A summary of changes in serum enzyme levels following a heart attack is illustrated in Figure 22.21.

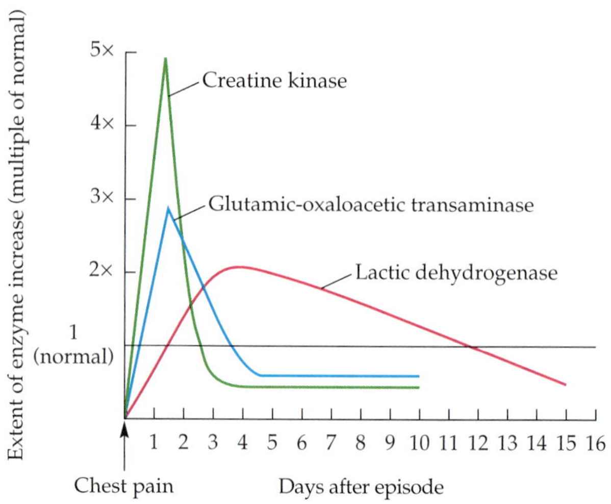

FIGURE 22.21 ▶
Typical changes in serum enzyme levels following a heart attack.

Summary

An **enzyme** is an organic catalyst produced by a living cell. Because enzymes are such powerful catalysts, the reactions they promote occur at body temperature and very quickly. Without the help of enzymes, these reactions would require high temperatures and long reaction times.

The molecule upon which an enzyme acts is the **substrate** for that enzyme. Enzymes are classified into six groups according to the reactions they catalyze: **oxidoreductases, transferases, hydrolases, lyases, isomerases,** and **ligases.**

Enzymes are globular proteins. Simple ones contain only one or more amino acid chains. Complex ones comprise one or more amino acid chains joined to **cofactors** (an inorganic ion or an organic **coenzyme**). The polypeptide part of a complex enzyme is called an **apoenzyme.** The combined enzyme/cofactor entity is a **holoenzyme.**

Enzymes that would harm the cell in which they are synthesized are produced as an inactive form called a *zymogen* or a **proenzyme.**

An enzyme works by lowering the activation energy a reaction needs in order to happen. The enzyme binds to its substrate, and then the product molecule breaks off:

$$S + E \longrightarrow ES \longrightarrow P + E$$

The **lock-and-key theory** of how an enzyme and substrate bond pictures a rigid enzyme of unchanging configuration that binds to the appropriate substrate. In other words, only one site on the enzyme and only one on the substrate are involved. The **induced-fit theory** holds that the enzyme active site is not rigid but rather can change and therefore can in some cases bind to *several different* substrates at different times. In this model, the enzyme active site has a *contact group* part and a *catalytic group* part. The contact group bonds to one part of a substrate, and the catalytic group is situated at that part of the substrate that is going to be the site of catalysis. Within narrow limits, this arrangement allows the enzyme to change shape to accommodate a group of substrates that are shaped almost alike.

The **turnover number** for an enzyme is the number of substrate molecules converted to product by one enzyme molecule in one minute.

Enzymes have varying degrees of specificity:

- *absolute specificity*—the enzyme can catalyze only one particular reaction for only one particular substrate
- *stereochemical specificity*—the enzyme acts on only one stereoisomeric form of a substrate
- *group specificity*—the enzyme acts on structurally similar substrates, all of which have some functional group in common
- *linkage specificity*—the enzyme acts on a particular type of linkage in any number of substrates, regardless of how similar or dissimilar the substrate structures are

Several factors influence enzyme activity:

- Reaction rate increases as *substrate concentration* increases up to a limiting rate.
- Reaction rate is directly dependent on *enzyme concentration.*
- Reaction rate increases with temperature up to the point at which the heat denatures the enzyme. At that point the reaction rate drops off sharply. Every enzyme has an **optimum temperature** at which it operates most effectively.
- Enzyme activity is extremely sensitive to pH. Each enzyme functions only in a narrow range of pH values surrounding its **optimum pH.** For all enzymes except those involved in digestion, the optimum pH lies between 7.0 and 7.5. Optimum pH values for

digestive enzymes are 2.0 for pepsin (stomach) and 8.0 for trypsin (small intestine).

Substances that interfere with enzyme function are inhibitors. A **competitive inhibitor** looks very much like the substrate for a particular enzyme. It binds to the active site on the enzyme and stays there, keeping the enzyme from binding with its substrate. The inhibitor–enzyme bond is relatively weak, so that a high concentration of the true substrate can force the inhibitor to detach from the active site.

A **noncompetitive inhibitor** binds at a site remote from the active site and consequently can bind either to a free enzyme or to an enzyme–substrate complex. The attached inhibitor changes the configuration of the enzyme molecule. As a result, the active site is altered so that, if the enzyme is free, its substrate cannot bond. If the enzyme is bound in an enzyme–substrate complex, the substrate cannot detach from the deformed active site, and so no product forms.

An **irreversible inhibitor** inactivates enzymes by forming strong bonds to a particular group at the active site. Addition of excess substrate cannot force this type of inhibitor to detach.

An **antimetabolite** is a chemotherapeutic substance that competitively inhibits some enzymatic reaction in a pathogen that is causing a human host to be ill. Without the necessary reaction taking place in its body, the pathogen dies. An **antibiotic** is a substance produced by one microorganism and lethal to another.

Key Terms

active site (22.3)
antibiotic (22.7)
antimetabolite (22.7)
antineoplastic drug (22.7)
apoenzyme (22.2)
autocatalytic reaction (22.2)
chemotherapy (22.7)
coenzyme (22.2)
cofactor (22.2)
competitive inhibitor (22.6)
enzyme (Introduction)
holoenzyme (22.2)
hydrolase (22.1)
induced-fit theory (22.3)
irreversible inhibitor (22.6)
isomerase (22.1)
ligase (22.1)
lock-and-key theory (22.3)
lyase (22.1)
noncompetitive inhibitor (22.6)
optimum pH (22.5)
optimum temperature (22.5)
oxidoreductase (22.1)
proenzyme (22.2)
substrate (22.1)
transferase (22.1)
turnover number (22.3)

Review Questions

1. Define and, where appropriate, give an example for:
   a. enzyme **b.** substrate
   c. holoenzyme **d.** proenzyme
   e. active site **f.** chemotherapy
   g. cofactor **h.** enzyme specificity
   i. irreversible inhibition **j.** sulfa drug
2. Distinguish between:
   a. *in vitro* and *in vivo*
   b. intracellular enzyme and extracellular enzyme
   c. sucrose and sucrase
   d. trypsin and trypsinogen
   e. contact group and catalytic group
   f. absolute specificity and stereochemical specificity
   g. group specificity and linkage specificity
   h. optimum temperature and optimum pH
   i. competitive inhibitor and noncompetitive inhibitor
   j. antimetabolite and antibiotic
3. What is the substrate for each of the following?
   a. maltase **b.** cellulase
   c. peptidase **d.** lipase
4. In what ways are enzymes similar to ordinary chemical catalysts? How are they different? What is the effect of an enzyme on an equilibrium system?
5. Why are enzymes more specific than inorganic catalysts?
6. Animals can digest starch but not cellulose. Explain.
7. Identify three intracellular enzymes and three extracellular enzymes.
8. Which enzyme is more specific—urease or carboxypeptidase?
9. What is meant by the *turnover number?*
10. How do mercury ions act as poisons? Are their actions reversible or irreversible?
11. What is acetylcholine? Describe the acetylcholine cycle.
12. How do organophosphorus compounds act as poisons?
13. How do sulfa drugs kill bacteria?
14. How does penicillin kill bacteria?
15. How do resistant bacteria deactivate penicillin?
16. What is an enzyme inhibitor?

17. Why is it that only a relatively few enzyme molecules can catalyze the conversion of many molecules of substrate?
18. Compare the lock-and-key theory with the induced-fit theory of Koshland.
19. How is it possible that two amino acids that are relatively far apart in the primary structure of a protein chain can be in proximity at the active site?
20. Why should enzymes consist of 100 or more amino acid units when only a few amino acid units are involved in the active site?

Problems

Classification

21. To which of the six major types of enzymes does each of the following belong?
    a. decarboxylase
    b. peptidase
    c. transaminase
22. To which of the six major types of enzymes does each of the following belong?
    a. dehydrogenase
    b. kinase
    c. lipase

Enzyme Action

23. Why does the body synthesize trypsin as the proenzyme trypsinogen rather than make the enzyme directly?
24. How does the body activate trypsinogen—that is, how is trypsinogen converted to trypsin?
25. The concentration of substrate X is low. What happens to the rate of the enzyme-catalyzed reaction if the concentration of X is doubled?
26. An enzyme has an optimum pH of 7.4. What happens to the activity of the enzyme (**a**) if the pH drops to 6.8 and (**b**) if the pH rises to 8.0?
27. A bacterial enzyme has an optimum temperature of 35 °C. Will the enzyme be more or less active at normal body temperature? Will it be more or less active if the patient has a fever of 40 °C?
28. Explain why enzymes become inactive above and below (**a**) the optimum temperature and (**b**) the optimum pH.
29. How does increased enzyme concentration affect the rate of a reaction?
30. Which type of R-group interaction (i.e., hydrogen-bonding, hydrophobic, etc.) at the active site of an enzyme would bind to each of the following groups on the substrate?
    a. $-COOH$ b. $-COO^-$
    c. $-NH_2$ d. $-NH_3^+$
    e. $-OH$ f. $-SH$
    g. $-CH(CH_3)_2$ h. $-C_6H_5$ (phenyl)

Classification and Naming

31. Alcohol dehydrogenase catalyzes the conversion of ethanol to acetaldehyde. The active enzyme consists of a protein molecule and a zinc ion. Identify the following components of this reaction system.
    a. substrate b. cofactor c. apoenzyme
32. Can the zinc ion mentioned in Problem 31 be called a coenzyme? Explain.
33. Succinic acid dehydrogenase is active only in combination with a nonprotein organic molecule called flavin adenine dinucleotide (FAD). Is FAD a cofactor? Is it a coenzyme?
34. Which has the longer polypeptide chain—prothrombin or thrombin?
35. What enzyme is involved in the conversion of lactose to galactose and glucose? To what class of enzymes does it belong?
36. In a commercial process, glucose (in corn syrup) is converted enzymatically to fructose (to form high-fructose corn syrup). What type of enzyme is involved? (Recall that both glucose and fructose have the formula $C_6H_{12}O_6$.)
37. Describe two ways to get around the problem of bacterial resistance.

Additional Problems

38. Acetylcholinesterase has aspartic acid, histidine, and serine residues at the active site. In acidic solution, the enzyme is inactive, but activity increases as the pH rises. Explain.
39. Alcohol dehydrogenase is involved in the oxidation of both methanol and ethanol. How might ethanol work as an antidote for methanol poisoning?
40. Experimentally, how could you distinguish a competitive inhibitor from a noncompetitive inhibitor?
41. Is oxaloacetic acid ($HOOCCH_2COCOOH$) an inhibitor of succinic acid dehydrogenase? What type of inhibitor is it?
42. Explain why antimetabolites and antibiotics can both be classified as antiseptics.

Special Topic J
Vitamins

Carbohydrates, fats, and proteins are the three major classes of foods. To remain healthy, we must take in relatively large amounts of these substances. They are not, however, the only nutrients we require. Some of our needs are satisfied only by vitamins and minerals. The minerals, inorganic ions of critical importance to our health, were discussed as a group in Section 12.8. Vitamins are considered in this special topic.

We normally eat three meals a day in order to satisfy our need for carbohydrates, fats, and proteins, but all the necessary vitamins can be packed into a single small pill. So small are the required amounts that not even a vitamin pill is necessary. We can get all we need simply by eating a balanced diet. Nature has thoughtfully incorporated minute but adequate amounts of vitamins into our various foodstuffs.

J.1 What Are Vitamins?

Vitamins are organic compounds that cannot be synthesized by an organism but nevertheless are essential for the maintenance of normal metabolism and therefore must be included in the diet. The absence or shortage of a vitamin results in a vitamin-deficiency disease.

One such disease, called scurvy, had plagued seamen since early times. In 1747, a Scottish navy surgeon, James Lind, showed that scurvy could be prevented by the inclusion of fresh fruit or vegetables in the diet. A convenient fresh fruit to carry on long voyages was the lime. British ships put to sea with barrels of limes aboard. The sailors ate a lime or two every day and remained free of scurvy. British sailors came to be known as "lime eaters" or simply "limeys."

In 1897, the Dutch scientist Christiaan Eijkman showed that polished rice lacked something found in the hulls of whole-grain rice. Lack of that "something" caused the disease beriberi, which was then quite a problem in the Dutch East Indies. Within the next two decades a British scientist, F. G. Hopkins, fed a synthetic diet of carbohydrates, fats, proteins, and minerals to a group of rats. The rats were unable to sustain healthy growth. Again, something was missing. Eijkman and Hopkins shared the 1929 Nobel prize in physiology or medicine for their important discoveries.

In 1912, Casimir Funk, a Polish biochemist, coined the word *vitamine* (from the Latin *vita*, life) for these missing factors. Funk thought that they all contained the amino group. The final *e* was dropped after it was found that not all the factors were amines. The generic term became *vitamin*.

Since organisms differ in their synthetic abilities, a substance that is a vitamin for one species may not be so for another. Over the years, scientists have isolated 13 vitamins needed by humans. Researchers today are in general agreement that no more vitamins are likely to be found. One reason is that about 50 years has elapsed since the last one (vitamin B_{12}) was discovered, despite an intensive search for others. Another reason is that many people have lived for years on intravenous solutions containing only the known vitamins and other nutrients without exhibiting signs of vitamin deficiency.

The vitamins are divided into two broad categories, the **fat-soluble vitamins,** including A, D, E, and K, and the **water-soluble vitamins,** made up of the B complex and vitamin C. All the fat-soluble vitamins incorporate a high proportion of hydrocarbon structural elements. There are one or two oxygen atoms present, but the compounds as a whole are nonpolar. In contrast, a water-soluble vitamin contains a high proportion of the electronegative atoms oxygen and nitrogen, which can form hydrogen bonds to water; therefore, the molecule as a whole is soluble in water.

Most water-soluble vitamins act as coenzymes or are required for the synthesis of coenzymes (see Section J.6). The fat-soluble vitamins have more varied functions. In general, the fat-soluble vitamins are obtained from fish, liver, dairy products, green vegetables, and vegetable oils. Fat-soluble vitamins, if taken in high doses, can accumulate in hazardously large amounts (hypervitaminosis) because they are stored in body fat. People who consume too much vitamin D, for example, can develop bone pain, bonelike deposits in the kidneys, and mental retardation.

The body has a limited capacity to store water-soluble vitamins. It excretes anything over the amount that can be immediately used. Water-soluble vitamins must be taken at frequent intervals, whereas a single dose of a fat-soluble vitamin can be used by the body over several weeks. The vitamin content of some foods can be lost when the food is cooked in water and then drained. The water-soluble vitamins go down the drain with the water.

Minimum daily requirements (MDRs) of the vitamins have been set by examining the levels below which deficiency diseases occur.[1] (General warning signs of vitamin deficiency are slow healing of wounds, tiredness, and frequent illness.) There is no agreement, however, on the optimum levels of dietary vitamins. Vitamins have been the subject of more fads and more misrepresentations than any other group of nutrients. Claims have been made that, among other things, vitamins cure cancer, arthritis, and mental illness; increase sexual potency; prevent colds; and overcome muscular weakness. It is small wonder that the public is baffled by such claims, which most registered dietitians reject.

J.2 Vitamin A

Vitamin A occurs only in the animal kingdom. It was first isolated from halibut oil and is also present in cod liver oil and in butter. However, the plant pigment β-carotene is a precursor substance (a **provitamin**) that can be converted to vitamin A by animals, and thus most green and yellow vegetables (carrots, lettuce, spinach, yams) are good sources of the vitamin (Figure J.1).

Some foods are fortified by the addition of vitamin A obtained by extraction from fish liver oils or made synthetically. The recommended daily allowance of vitamin A is 0.7 mg. If vitamin A is present in excess, it is stored in the liver. Adult livers can store enough vitamin A to last for several months. On the other hand, the livers of infants and children do not store much of the vitamin. Consequently, infants and children are more likely to develop vitamin A deficiencies if their diets are inadequate.

In Special Topic H, we discussed the well-known role of vitamin A in vision. We also know that a deficiency of this vitamin affects most of the body's organs. What we do not know is the detailed biochemistry of vitamin A as it relates to all these other organs. We can, however, describe some of the effects of vitamin A deficiency.

Vitamin A is required for normal growth. Young animals fed a diet lacking in the vitamin simply fail to grow. One of the earliest manifestations of vitamin A deficiency is a loss of night vision. Mucous membranes may harden, dry, and crack. In cases of severe deprivation, victims may exhibit xerophthalmia, a condition characterized by inflammation of the eyes and eyelids, leading ultimately to infection and blindness. Vitamin A also stimulates fluid secretion by the epithelial cells of the eye. Thus, if the dietary supply of vitamin A is inadequate, the cornea of the eye becomes dried, or keratinized. The cornea then becomes extremely vulnerable, and

[1] The RDA (recommended daily allowance) is a government estimate of the amount of vitamins that the average healthy person should eat daily to maintain good nutrition and health. The RDA should not be confused with the MDR. The MDR is the smallest amount of a nutrient that if ingested will prevent a nutritional deficiency. The RDA significantly exceeds the MDR for each nutrient.

FIGURE J.1 ▶

Animals are able to convert β-carotene to vitamin A.

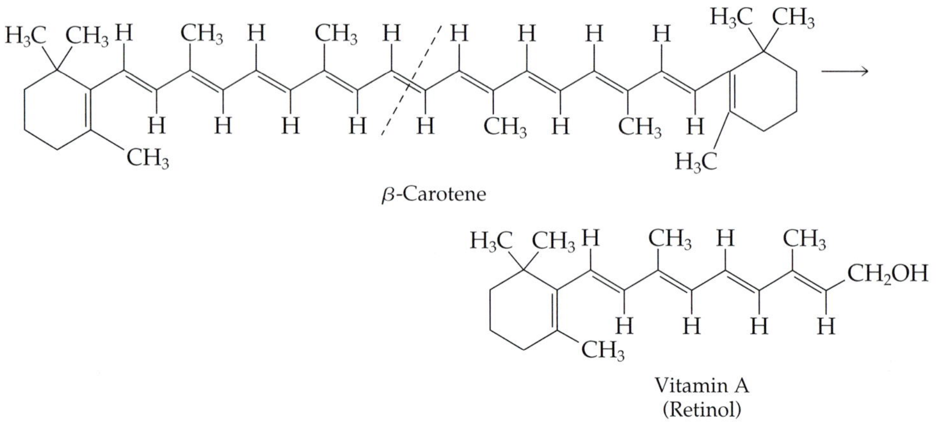

even the slightest nick or scratch may cause it to perforate, which leads to blindness. Blindness in this case is brought about by a vitamin A deficiency disease called keratomalacia. This disease is the major cause of blindness in young children in most of the developing countries, and it is estimated to affect hundreds of thousands of children throughout the world. Health workers in those countries often carry injectible solutions of vitamin A for emergency treatment of such cases.

Vitamin A is important to the growth and maintenance of epithelial tissue. In the past, some physicians used large doses of vitamin A for treating acne. Not only were the doses potentially harmful (painful joints, loss of hair), but they were not really effective. However, it has been found that a synthetic derivative of vitamin A (13-*cis*-retinoic acid) appears promising as a drug for treating severe acne.

There is also evidence that vitamin A may confer resistance against certain kinds of cancer. This may help to explain the anticancer activity that has been noted for cruciferous vegetables, such as broccoli, cauliflower, brussels sprouts, and cabbage. Although they do not contain vitamin A, all these vegetables are rich in beta-carotene, which is a precursor to vitamin A. More research into this phenomenon is currently under way.

Vitamin A is found in high concentration in fish liver oils. Liver, eggs, fish, butter, and cheese are also good sources. It is interesting to note that polar bear liver has so much vitamin A that it is toxic. These large animals eat seals that eat fish. Fat-soluble vitamin A is concentrated in each step of the food chain. Eskimos who kill a polar bear know better than to eat its liver. Large excesses cause irritability, dry skin, and a feeling of pressure inside the head. Massive doses of vitamin A administered to pregnant rats result in malformed offspring.

J.3 Vitamin D

Several chemical compounds have vitamin D activity. Only two commonly occur in foods or are used in drugs and food supplements. Each is formed from a precursor by the action of ultraviolet light (Figure J.2). Vitamin D_2 (ergocalciferol) is synthesized by

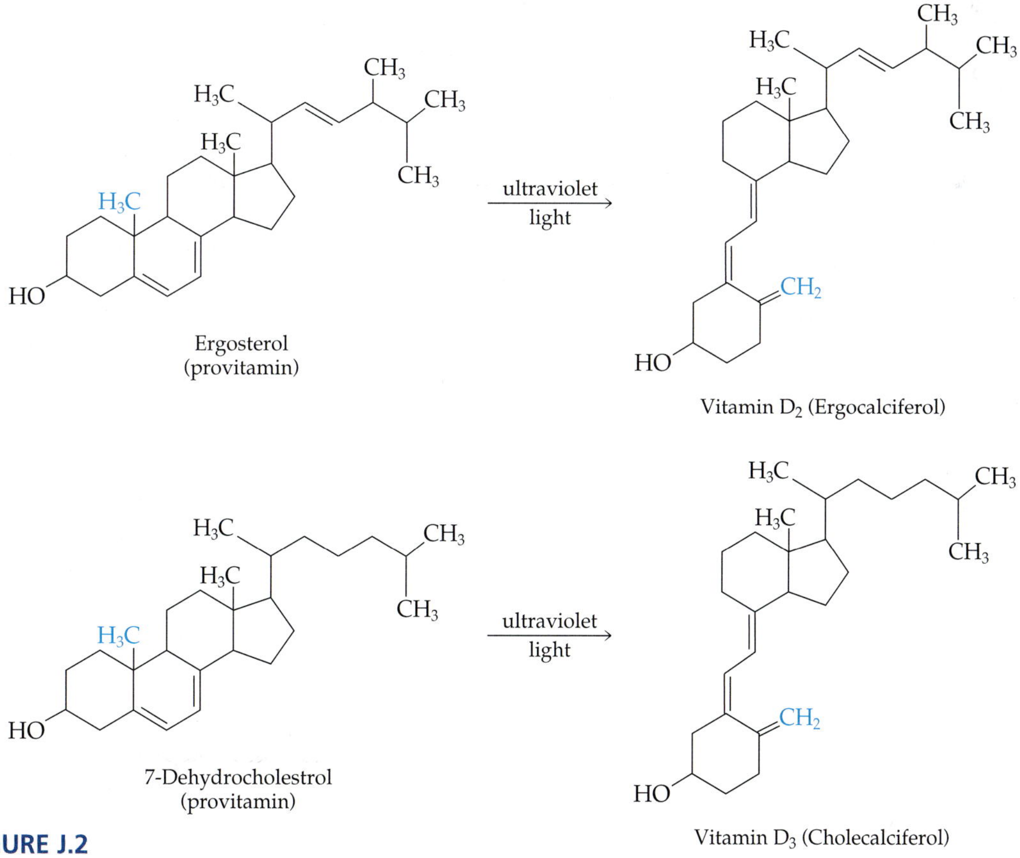

▲ **FIGURE J.2**

Formation of two forms of vitamin D by action of ultraviolet light on provitamins.

irradiation of ergosterol, a compound found in yeast and other molds. Vitamin D_3 (cholecalciferol) is formed in the skin of animals by the action of sunlight on 7-dehydrocholesterol. The two vitamins differ only in the structure of their side chains; D_2 contains an extra carbon and a double bond. (There is no vitamin D_1. The material that was originally given this designation proved to be a mixture of vitamins D_2 and D_3.)

Vitamin D increases the body's ability to absorb calcium and phosphorus. Deficiency in infants and growing children results in abnormal bone formation, a condition known as rickets. The condition is characterized by bowed legs, knobby bone growths where the ribs join the breastbone (called a "rachitic rosary"), pigeon breast, and poor tooth development. In adults, because bone is no longer growing, rickets does not develop. Women who are deficient in vitamin D may develop osteomalacia, a condition characterized by fragile bone structure. This condition is rare but may occur after several pregnancies.

Most foods contain little or no vitamin D. The best natural sources are fish liver oils and egg yolks. Irradiated ergosterol (from yeast) is added to milk (10 μg per quart) and margarine as a supplemental source of vitamin D.

Vitamin D is the "sunshine vitamin." Individuals with a reasonable proportion of their skin exposed to sunlight rarely suffer from a vitamin D deficiency. It is recommended that children receive about 20 μg of vitamin D in their daily diet. For adults, exposure to sunlight for 30 minutes satisfies the daily requirements of the vitamin.

Vitamin D, like vitamin A, is fat-soluble. Amounts taken in excess are stored in body fat. The effects of large overdoses are even more severe than with vitamin A. Too much vitamin D can cause pain in the bones, nausea, diarrhea, and weight loss. Bonelike material may be deposited in kidney tubules, in blood vessels, in heart, stomach, and lung tissue, and in joints.

J.4 Vitamin E

As with vitamin D, there are several compounds that have vitamin E activity. The compounds are called tocopherols; the most potent of these is α-tocopherol:

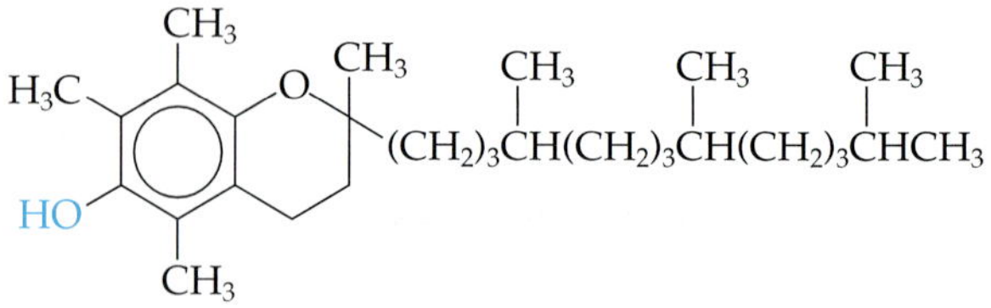

α-Tocopherol

Recent evidence shows that an *oxidized* form of LDL cholesterol is deposited in arteries (Section 20.10). Vitamin E acts to prevent the oxidation of cholesterol; thus it seems to have considerable value in maintaining the cardiovascular system. Vitamin E has been used to treat coronary heart disease, angina, rheumatic heart disease, high blood pressure, arteriosclerosis, varicose veins, and a number of other cardiovascular problems. It is also an anticoagulant that has been useful in preventing blood clots after surgery.

Rats deprived of vitamin E become sterile. Because of this, vitamin E is sometimes called the antisterility vitamin. Low intake of vitamin E can also lead to muscular dystrophy, a disease of the skeletal muscles.

Vitamin E is a potent antioxidant that can inactivate free radicals. It is generally believed that much of the physiological damage from aging is due to the production of free radicals. Indeed, vitamin E has also been referred to as the antiaging vitamin.

A vitamin E deficiency can lead to a deficiency in vitamin A. Vitamin A can be oxidized to an inactive form when vitamin E is no longer present to act as an antioxidant. Vitamin E also collaborates with vitamin C in protecting blood vessels and other tissues against oxidation. Vitamin E is the fat-soluble antioxidant vitamin, and vitamin C is the water-soluble antioxidant vitamin. Oxidation of unsaturated fatty acids in the cell membrane can be prevented or reversed by vitamin E, which is itself oxidized in the process. Vitamin C can then restore vitamin E to its unoxidized form.

Vitamin E is available in wheat germ oil, green vegetables, vegetable oil, egg yolks, and meat. Most nutritionists contend that it would be nearly impossible to eat a diet deficient in vitamin E. Vitamin E is fat-soluble, and some is stored in the body. Large doses may waste money, but they do not seem to have harmful effects. A belief in the efficacy of vitamin E, however, might lead some people to postpone needed medical treatment.

J.5 Vitamin K

Vitamin K has a fused ring system related to the structure of naphthalene (Section 13.17). One of the rings contains two carbonyl groups, an arrangement that has the special name *quinone*. Attached to the quinone ring are alkyl groups. One of these is usually methyl. The other has 20 or more carbon atoms. Many compounds have vitamin K activity; the structure of one of them is

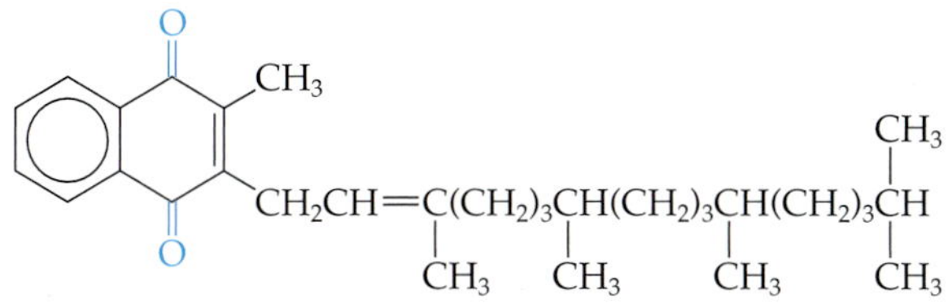

Vitamin K_1 (Phylloquinone)

Vitamin K, like vitamins A, D, and E, is insoluble in water but soluble in fats and fat solvents. It is necessary for the formation of prothrombin (Section 28.8), one of the enzyme precursors involved in blood clotting. (The vitamin got its name from the Danish word *koagulation*.) A vitamin K deficiency increases the time required for blood to clot. Symptoms are bleeding under the skin and in muscles, leading to ugly "bruises" from what would otherwise be minor blows. Infants lacking in vitamin K may die from hemorrhaging in the brain. Increased vitamin K intake by pregnant women has lowered the incidence of this disease in newborn infants. Good sources of vitamin K are spinach and other green leafy vegetables. Synthetic compounds with vitamin K activity are readily available. Vitamin K deficiencies in humans are rare because intestinal bacteria synthesize as much as the body requires. Prolonged treatment with antibiotics has the adverse effect of killing these bacteria, and the body's supply of vitamin K is temporarily reduced.

J.6 The B Complex

In Chapter 22 we discussed how some enzymes require coenzymes in order to function properly. Many coenzymes are vitamin B derivatives.

There is really no vitamin B. What was once called vitamin B has long since been recognized to be a complicated mixture of factors. The term **B complex** is now used to designate a group of water-soluble vitamins found together in many food sources. Table J.1 lists the members of the B complex, their structures and sources, and the coenzymes derived from the vitamins.

There appears to be no toxicity connected with the B vitamins, with the possible exception of vitamin B_6, which apparently can cause neurological damage in some people if taken in extremely large daily doses. As a group, the B vitamins are important for maintaining the skin and the nervous system.

Thiamine (Vitamin B_1)

Thiamine is necessary for the normal metabolism of carbohydrates. It is converted in the body to the pyrophosphate, which is a coenzyme in the decarboxylation of pyruvic acid to acetyl CoA and α-ketoglutaric acid to succinyl CoA (see Section 25.1). The coenzyme is also involved in the synthesis of ribose, used by the body to produce nucleotides and nucleic acids.

A deficiency of thiamine in the diet leads to the *beriberi* syndrome, characterized by deterioration of the cardiovascular and nervous systems. This disease is a serious health problem in the Far East because rice, the major food in the region, has a relatively low thiamine content. Alcoholism is the most common cause of thiamine deficiency in the United States because alcohol is the major caloric contributor to an alcoholic's diet, and therefore there is a low vitamin intake. A severe form of beriberi can also occur in infants of nursing mothers whose diets are deficient in thiamine.

Synthetic vitamin B_1 is added to enrich the vitamin contents of bread and flour. Thiamine is not stored in the body to any significant degree; excesses are excreted in the urine. The vitamin is destroyed in foods that are cooked for prolonged periods at temperatures over 100 °C.

Riboflavin (Vitamin B_2)

Riboflavin is essential for mammalian cells. A lack in the human diet causes well-defined symptoms, among them dermatitis (skin inflammation), glossitis (tongue inflammation), and anemia. Riboflavin is converted in the body to the coenzymes FAD and FMN, and these function in oxidation–reduction reactions in the metabolism of carbohydrates and lipids. Riboflavin is destroyed by light and thus does not have a long stability in food products. It is stable at ordinary cooking temperatures.

Table J.1 Vitamins and Coenzymes

Vitamin and RDA[a]	Coenzyme	Source of Vitamin
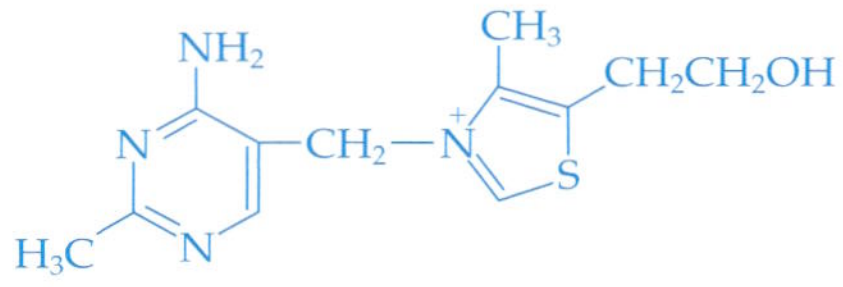 Thiamine (B_1) RDA = 1.5 mg	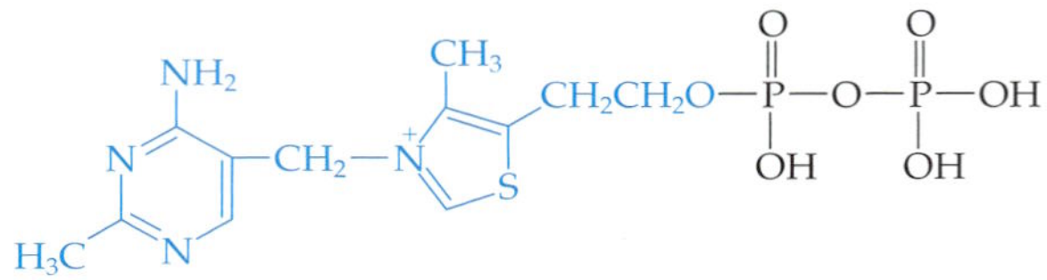 Thiamine pyrophosphate (TPP)	Germ of cereal grains, legumes, nuts, milk, and brewers yeast
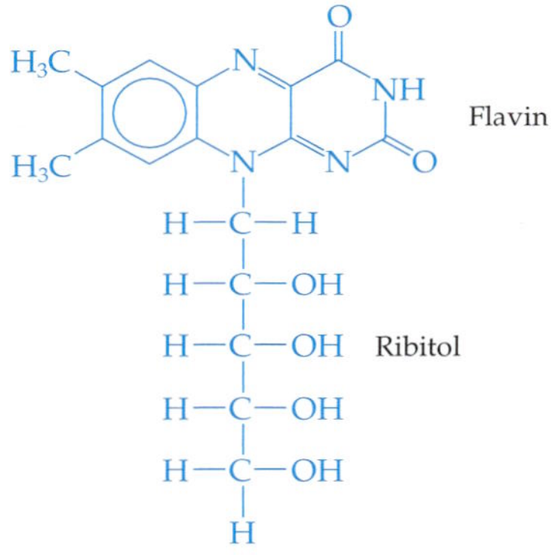 Riboflavin (B_2) RDA = 1.7 mg	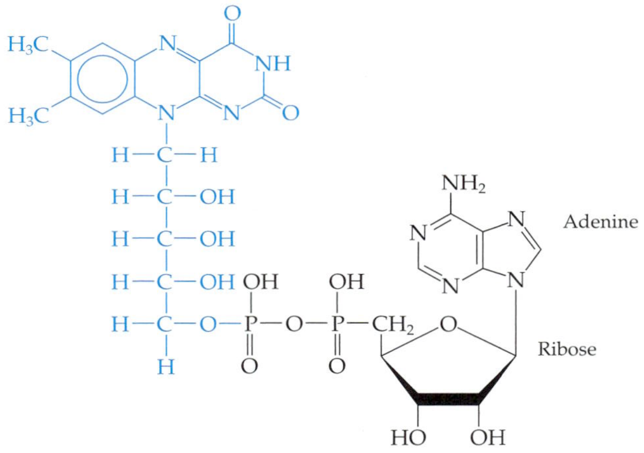 Flavin adenine dinucleotide (FAD)[b]	Milk, red meat, liver, egg white, green vegetables, whole wheat flour (or fortified white flour), and fish
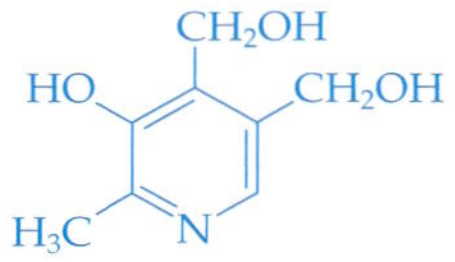 Pyridoxine (B_6) RDA = 2.0 mg	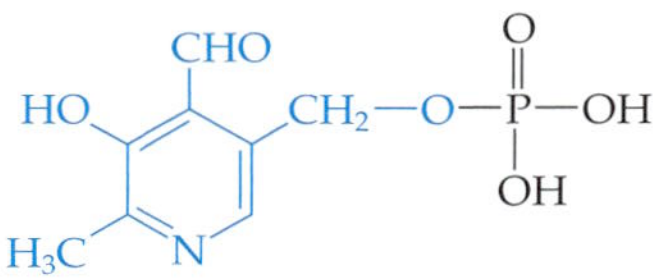 Pyridoxal phosphate	Eggs, liver, yeast, peas, beans, and milk

Table J.1 Cont.

Vitamin and RDA[a]	Coenzyme	Source of Vitamin
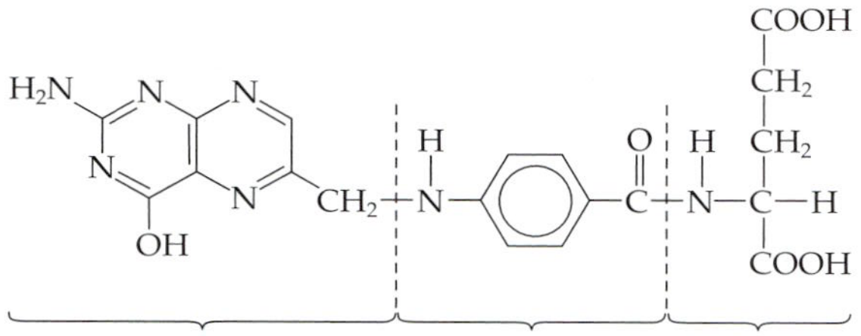 Pterin moiety; *p*-Aminobenzoic acid moiety; Glutamic acid Folic acid (F) RDA = 0.2 mg	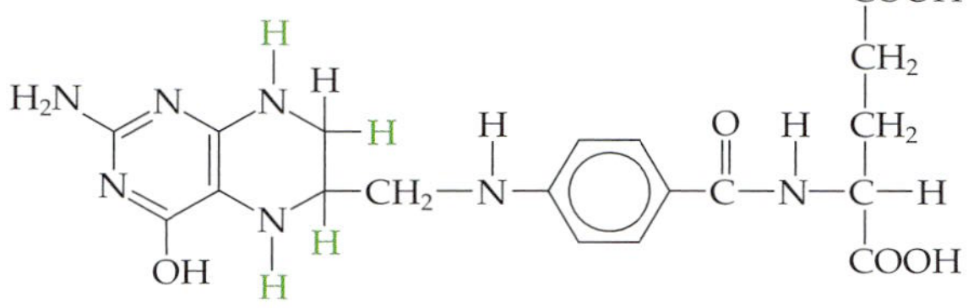 Tetrahydrofolic acid (FH_4)	Liver, kidney, mushrooms, yeast, and green leafy vegetables
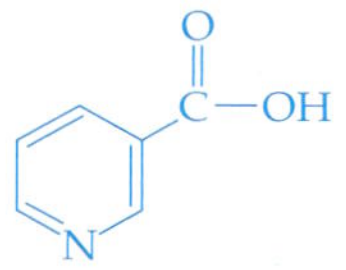 Nicotinic acid (Niacin)	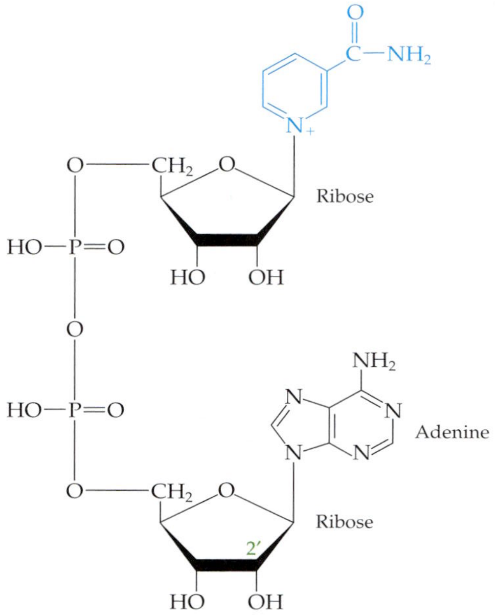 Nicotinamide adenine dinucleotide (NAD^+)[c]	Red meat, liver, collards, turnip greens, yeast, and tomato juice
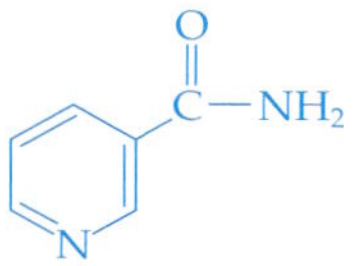 Nicotinamide RDA = 20 mg		
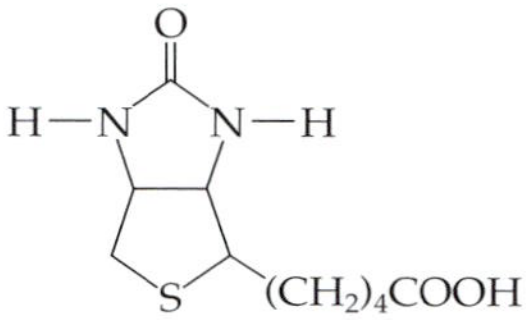 Biotin RDA = 0.3 mg	Biotin is both a vitamin and a coenzyme	Beef liver, yeast, peanuts, chocolate, and eggs

Table J.1 Cont.

Vitamin and RDA[a]	Coenzyme	Source of Vitamin
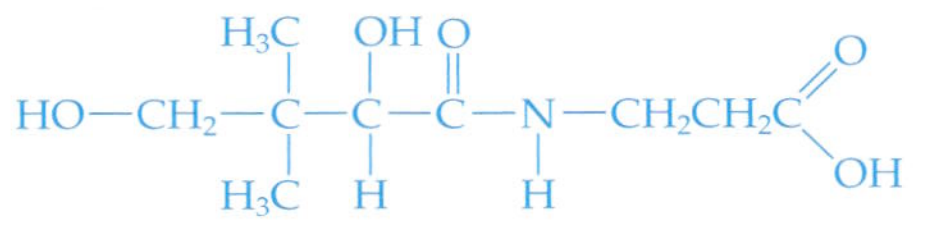 Pantothenic acid RDA = 10 mg	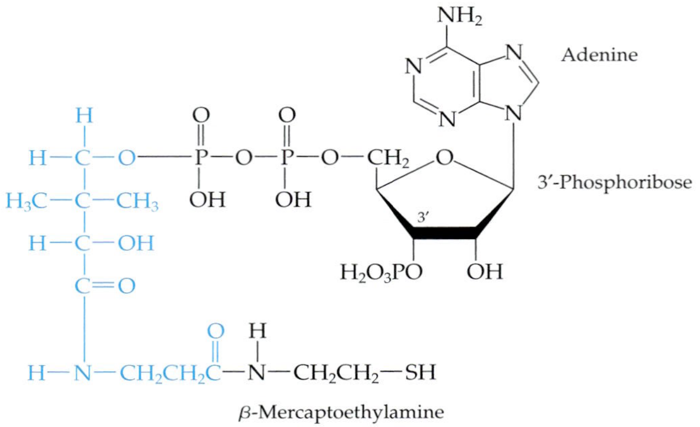 Coenzyme A (CoA-SH)	Liver, eggs, yeast, and milk
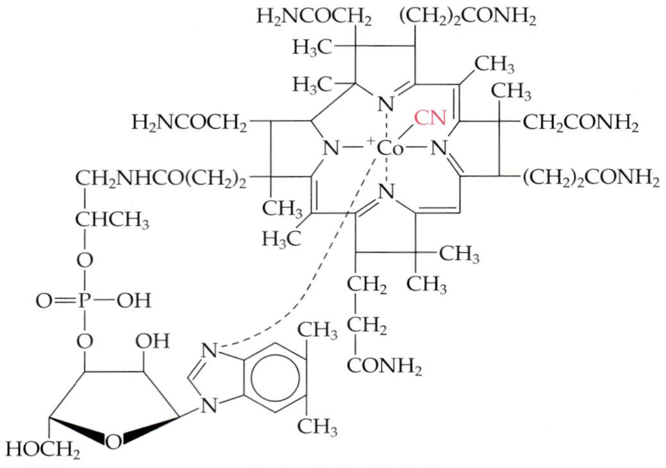 Cyanocobalamin (B_{12}) RDA = 2 μg	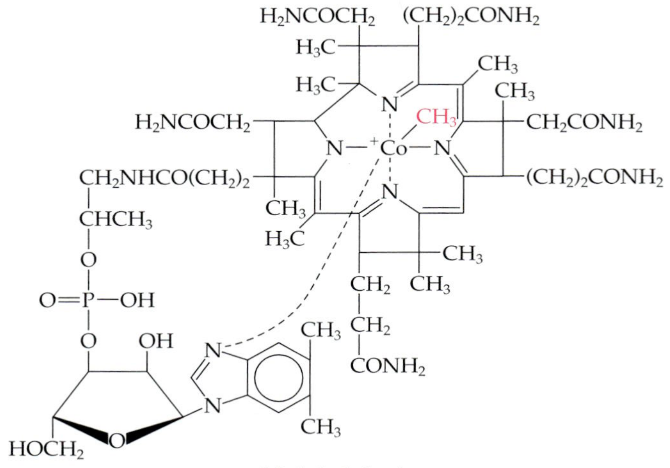 Methylcobalamin	Liver, meat, eggs, and fish (not found in plants)

[a] The Recommended Daily Allowance (RDA) is based on a 70-kg adult consuming 3000 cal/day. The body's requirement for most of the B vitamins increases during pregnancy and lactation. See also footnote 1, Section J.1.

[b] When there is only a phosphate group bonded to the terminal carbon of riboflavin, the coenzyme is named flavin mononucleotide (FMN).

[c] When there is an additional phosphate group on the 2'-hydroxyl group of the ribose moiety, the coenzyme is named nicotinamide adenine dinucleotide phosphate ($NADP^+$).

Pyridoxine (Vitamin B_6)

Pyridoxine occurs in the tissues and body fluids of virtually all living organisms. The coenzyme pyridoxal phosphate is required for a wide variety of metabolic transformations of amino acids (Section 27.5). Vitamin B_6 has been found to be helpful for people suffering from arthritis by shrinking the synovial membranes that line the joints. It also reduces the wrist swelling and pain caused by carpal tunnel syndrome.

Clinical symptoms of vitamin B_6 deficiency include lesions of the skin and mucosa, anemia, irritability, apathy, and neuronal dysfunction including convulsions. Because vitamin B_6 enhances the decarboxylation of L-dopa, the vitamin should be avoided by patients receiving L-dopa to treat Parkinson's disease (see Section 27.5). Vitamin B_6 is stable at normal cooking temperatures but is sensitive to light.

Folic Acid

Folic acid was first discovered in green leafy vegetables, and its name is derived from the Latin word for leaf, *folium*. It is critically important in preventing both spina bifida and anencephaly (birth defects of the neural tube).

Folic acid is reduced to its coenzyme, tetrahydrofolic acid, which acts as a carrier of one-carbon units (e.g., as formyl or methyl groups) in the formation of such compounds as choline, heme, and nucleic acids. Deficiency of folic acid affects purine biosynthesis, and clinical symptoms include anemia and gastrointestinal disturbances.

Folic acid is synthesized by intestinal microorganisms, and it can be absorbed into the general circulation. It is readily destroyed by cooking. As we saw in Section 22.7, the sulfa drugs interfere with the bacterial biosynthesis of folic acid. The anticancer drug methotrexate inhibits the conversion of folic acid to tetrahydrofolic acid. Without the coenzyme, cells cannot grow because they cannot replicate their DNA.

Nicotinic Acid (Niacin) and Nicotinamide

Nicotinic acid and its amide are equally effective in supplying human needs. The vitamin is best known for its ability to prevent *pellagra* in humans. In the early 1900s, pellagra was particularly prevalent in the southern United States and was directly associated with low-grade starchy (corn) diets. Pellagra is characterized by loss of appetite and weakness, followed by diarrhea, dermatitis, mental disorders, and death in severe cases. Niacin offers some relief from arthritis and rheumatism. It also helps in lowering the blood cholesterol level. Nicotinamide serves as a component of coenzymes (NAD^+ and $NADP^+$) for a wide variety of enzymes that catalyze oxidation–reduction reactions. Some nicotinic acid (about 10%) is synthesized from tryptophan. The vitamin is not destroyed by cooking, although some is lost to dissolution in the cooking water.

Biotin

Biotin is widely distributed as a cell constituent of animal and human tissue. It functions as a coenzyme in carboxylation reactions. It is a carbon carrier in both carbohydrate and lipid metabolism. Because biotin is synthesized by intestinal microorganisms in large quantities, biotin deficiency seldom occurs in humans. Deficiency can be produced, however, by antibiotics that inhibit the growth of intestinal bacteria. Also, raw egg white contains a protein, avidin, that binds biotin and prevents its absorption from the intestinal tract. An artificially produced deficiency of biotin in humans causes dermatitis, anorexia, nausea, muscle pains, and depression. Biotin is stable at normal cooking temperatures.

Pantothenic Acid

Pantothenic acid is a precursor for the biosynthesis of coenzyme A, which is important as a carrier of acyl groups. (The name coenzyme A resulted from its involvement in enzymatic acetylation reactions.) Pantothenic (from the Greek word meaning *from everywhere*) acid has a widespread distribution in foods, and deficiency in humans is practically unknown. Symptoms produced by experimental feeding of an antagonist include nausea, fatigue, and burning cramps in the limbs. This vitamin is stable at moderate cooking temperatures but is destroyed at high temperatures.

Cyanocobalamin (Vitamin B_{12})

Vitamin B_{12} is a complex cobalt-containing structure that has similarities to the heme group of hemoglobin. The vitamin is converted to two coenzymes, one of which is methylcobalamin (the $-CN$ group is replaced by $-CH_3$). Methylcobalamin acts as a methyl group donor, and it is essential for cell growth and replication and for the maintenance of neural function (maintaining the myelin sheath). Vitamin B_{12} is formed only by certain bacteria that live in a symbiotic relationship with their hosts. It is stored in various tissues, particularly the liver.

Vitamin B_{12} is associated with the disease *pernicious anemia*. This dietary disease is characterized by the presence of abnormally large, immature, fragile red blood cells. It is accompanied by gastrointestinal disturbances and lesions of the spinal cord with loss of muscular coordination (ataxia).

Pernicious anemia is usually caused by poor absorption of the vitamin from the intestinal tract rather than by any lack of vitamin B_{12}. Cells of the stomach lining synthesize a glycoprotein, called the *intrinsic factor*, that specifically binds vitamin B_{12} and transports the vitamin into intestinal cells for its subsequent transfer to the blood. Pernicious anemia patients lack or have a deficiency of the intrinsic factor and cannot absorb the ingested vitamin B_{12}. Elderly people often have a decreased synthesis of intrinsic factor and must receive vitamin B_{12} by injection directly into the bloodstream in order to avoid anemia. Since plants do not contain vitamin B_{12}, pernicious anemia symptoms are sometimes observed among strict vegetarians. Vitamin B_{12} is stable during most cooking procedures.

J.7 Vitamin C

Vitamin C is ascorbic acid, a white, crystalline solid quite soluble in water. As with vitamin E, the role of vitamin C in nutrition is still a subject of controversy. Whereas vitamin E protects the lipid portion of cells, vitamin C (a highly polar compound) serves as an antioxidant in the aqueous regions. Vitamin C participates in several biological oxidation reactions, such as the hydroxylation of proline and lysine groups in collagen. It reacts with oxygen and/or oxidizing agents to form dehydroascorbic acid:

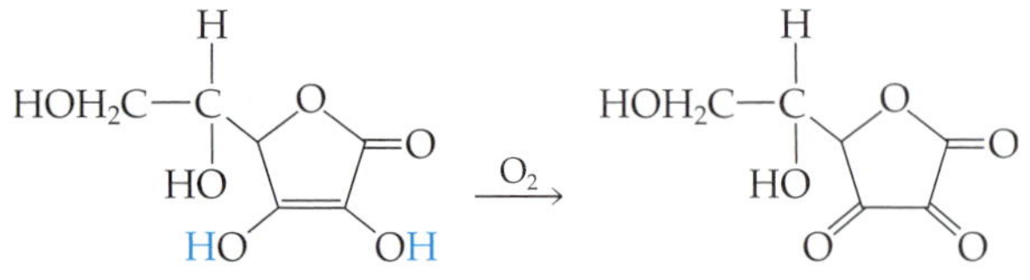

Ascorbic acid (Vitamin C)   Dehydroascorbic acid

As mentioned earlier, James Lind discovered that citrus fruit was effective in treating sailors suffering from scurvy, a weakening of the collagenous tissues. The symptoms are swollen gums; loose teeth; sore joints; thin, porous bones; bleeding under the skin; and slow healing of wounds. It was not until 1932, however, that the vitamin responsible for preventing scurvy was isolated from citrus fruit. Vitamin C was the first dietary component to be recognized as essential for preventing a human disease. It is widely distributed in plants[2] and animals. However, humans, other primates, guinea pigs, and some bats, birds, and fish lack an enzyme necessary for its biosynthesis.

Linus Pauling (1901–1994), winner of two Nobel prizes (for chemistry in 1954 and for peace in 1962), advocated massive doses of vitamin C to prevent the common cold.

The controversy surrounding vitamin C received its impetus in 1970, when Linus Pauling published his best-selling book *Vitamin C and the Common Cold*. He stated that vitamin C in doses ranging from 1 to 5 g a day could prevent colds and that as much as 15 g a day could cure a cold. Scientists investigating Pauling's claims have obtained conflicting results. (The RDA of vitamin C for adults is 60 mg.)

It was Pauling's contention that vitamin C not only helps prevent (or lessen the effects of) the common cold but also has value in preventing and treating influenza, as well as a number of other viral diseases. Vitamin C, he believed, has many functions, including the strengthening of the immune system. It does promote the healing of wounds and burns, and it has demonstrated an ability to heal gastric ulcers. It also seems to play an important role in maintaining the body's collagen supply. Like

[2] As is well known, vitamin C is particularly abundant in vegetables and citrus fruits, but for the vitamin to be useful the food must be reasonably fresh because ascorbic acid is slowly oxidized by air. Vitamin C is one of the least stable vitamins. It can be destroyed by heat, light, and alkali as well as by oxidizing agents. Therefore, cooking vegetables destroys an appreciable amount of vitamin C activity. It deteriorates even when kept for long periods in the refrigerator in well-capped bottles.

vitamin E, it is an antioxidant, and these two vitamins, along with beta-carotene, have been making headlines because of their antioxidant activity. Antioxidants often act as anticarcinogens.

Pauling had long insisted that the power of vitamin C extends to anticancer activity, and many in the medical community are beginning to agree. Indeed, vitamin C does inhibit the formation of nitrosamines (which are known carcinogens) under conditions such as those in the stomach.

There is mounting evidence that vitamin C is essential for the efficient functioning of the immune system. The interferons have been recognized as agents in the immune system. (Interferons are large molecules formed by the action of viruses on their host cells. By producing an interferon, a virus can interfere with the growth of another virus.) It has been demonstrated that increased vitamin C increases the body's production of interferons.

Low levels of vitamin C have been linked to the formation of cataracts and to the development of glaucoma. Low intake of vitamin C can also lead to gingivitis and periodontal disease.

For whatever reasons, whether it is to prevent colds, to protect against cancer, or just to maintain an optimum state of health, there are many people who are now taking vitamin C supplements on a regular basis. Rarely do they take the larger megadoses that Pauling recommended, but they do continue to take the supplements year after year. If vitamin C is not helping all these people to stay healthier, at least they seem to believe that it is.

In a national survey, people were asked to rate their own state of health on a scale from excellent to poor. Those who had high blood levels of vitamin C were most likely to rate their health as "excellent" or "very good." Those with low vitamin C levels tended to rate their health as "fair" or "poor."

Key Terms

B complex (J.6)
fat-soluble vitamin (J.1)
provitamin (J.2)
vitamin (J.1)
water-soluble vitamin (J.1)

Review Questions

1. Define and give an example of
   a. vitamin **b.** provitamin **c.** coenzyme
2. Compare and contrast vitamins and minerals (Section 12.9) with regard to
   a. inorganic or organic
   b. essentiality in the diet
   c. amounts needed by the body
3. Which is likely to be the more dangerous—an excess of a water-soluble vitamin or an excess of a fat-soluble vitamin? Why?
4. Could a one-a-month vitamin pill satisfy all human requirements? Explain.
5. How are vitamins related to coenzymes?
6. If boiled vegetables are served as part of a meal, would the water-soluble or fat-soluble vitamins originally present be lost? Why?
7. What foods, in general, are good sources of the B vitamins?
8. Why is vitamin D called the "sunshine vitamin"?
9. What biochemicals are protected by vitamin E's antioxidant effect?
10. What is meant by the term B *complex?*
11. Which vitamin is a part of the coenzyme NAD^+?
12. Which vitamin is a part of the coenzyme FAD?
13. Which vitamin is a part of coenzyme A?
14. What is the structural difference between water-soluble and fat-soluble vitamins?

Problems

Names and Classification

15. Match each compound with its designation as a vitamin:

Compound	**Designation**
Ascorbic acid	Vitamin A
Ergocalciferol	Vitamin B_{12}
Cyanocobalamin	Vitamin C
Retinol	Vitamin D
Tocopherol	Vitamin E

16. Which of the following are B vitamins?
   a. folic acid **b.** insulin **c.** niacin
   d. riboflavin **e.** thiamine **f.** biotin
17. Identify the vitamin deficiency associated with
   a. scurvy
   b. rickets
   c. night blindness
18. Identify the deficiency disease associated with a diet lacking in
   a. vitamin B_1 (thiamine)
   b. niacin
   c. vitamin B_{12} (cyanocobalamin)

Solubility

19. Identify each vitamin as water-soluble or fat-soluble.
   a. vitamin A **b.** vitamin B_6
   c. vitamin B_{12} **d.** vitamin C
   e. vitamin K

20. Identify each vitamin as water-soluble or fat-soluble.
 a. ergocalciferol
 b. niacin
 c. riboflavin
 d. tocopherol

21. Identify each vitamin as water-soluble or fat-soluble.

 a.

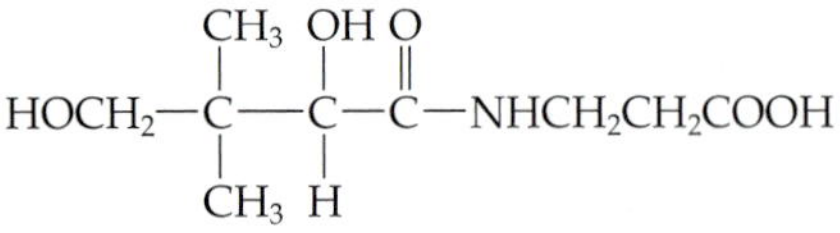

$$HOCH_2-\overset{\overset{\displaystyle CH_3}{|}}{\underset{\underset{\displaystyle CH_3}{|}}{C}}-\overset{\overset{\displaystyle OH}{|}}{\underset{\underset{\displaystyle H}{|}}{C}}-\overset{\overset{\displaystyle O}{||}}{C}-NHCH_2CH_2COOH$$

 b.

$$CH=CHC=CHCH=CHC=CHCH_2OH$$

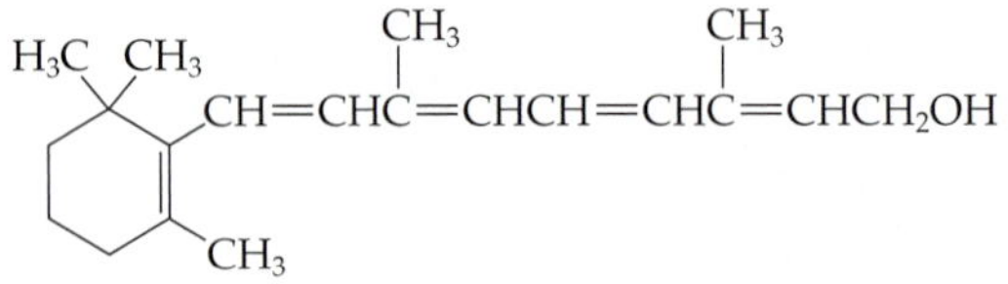

22. Each molecule of vitamin B_{12} has 63 carbon atoms. The molecular formula is $C_{63}H_{88}CoN_{14}O_{14}P$ and the molar mass is 1355 g/mol, and yet it is soluble in water. Explain.

Additional Problems

23. Name one function and one deficiency disease associated with
 a. vitamin A **b.** vitamin D
 c. vitamin E **d.** vitamin C

24. The phrases below refer to the structural changes that convert vitamins to coenzymes. Consult Table J.1 and identify:
 a. These vitamins are converted into coenzymes that catalyze oxidation-reduction reactions.
 b. This vitamin undergoes no further change (i.e., it is also a coenzyme).
 c. A pyrophosphate group is added to this vitamin.
 d. This vitamin is oxidized and phosphorylated.
 e. This vitamin is reduced.
 f. Methyl replaces cyanide in this vitamin.

25. Is vitamin C a single chemical compound? Does synthetic vitamin C differ from natural vitamin C? In what way(s) could *tablets* labeled natural vitamin C differ from those made with synthetic vitamin C?

23 Nucleic Acids and Protein Synthesis

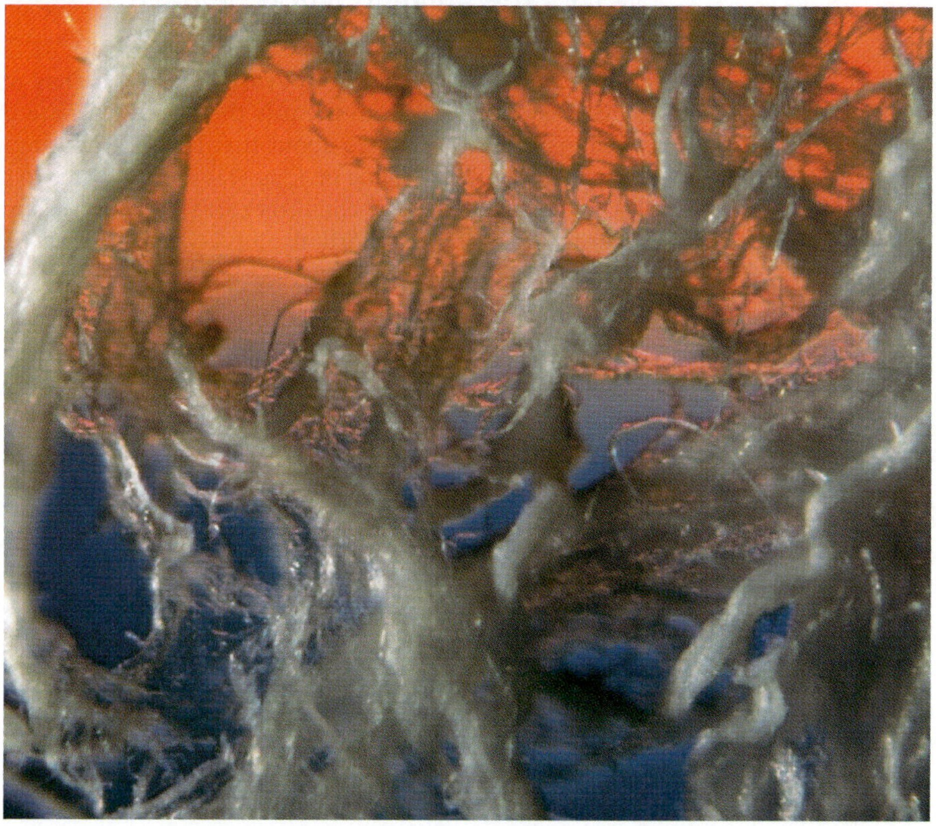

Artificial DNA synthesized from animal DNA. DNA is called the "blueprint of life" because it contains the information needed for the synthesis of all the body proteins.

In contrast to what many people believe, the complexity of the sciences increases as one proceeds from physics to chemistry to biology. Because the language of physics is mathematics, most people regard physics as the most difficult of the sciences. Yet physical phenomena can be described with mathematical precision because the relationships involved are comparatively simple. We can write an equation that accurately describes the behavior of gases or of subatomic particles. A functioning cell defies such analysis.

Nonetheless, the cell is slowly yielding its secrets. One of these secrets, perhaps the most important one, is the method by which the cell stores and transmits information on how to reproduce itself. Nucleic acids are the molecules that store the patterns of life. It is through nucleic acids that these patterns are passed from one generation to the next. Nucleic acids also control the synthesis of proteins, including the enzymes that mediate those biochemical reactions that make an organism what it is.

With understanding comes control. The biochemists and molecular biologists who are unraveling these mechanisms are also learning how to manipulate the structure of living matter. The repair of defective genes, the design of precise molecular medicines, and control—for better or worse—of our heredity may lie in the future. The twentieth century may be remembered as the nuclear age, but the twenty-first century could well become the age of molecular biology.

23.1 The Building Blocks: Sugars, Phosphates, and Bases

Nucleoproteins are found in every living cell. They are exceedingly complex, as might be expected from the role they play: they are the information and control centers of the cell. More about that later in the chapter. Let's look first at what nucleoproteins are made of. One way to find out is to take them apart.

Working carefully, chemists can separate nucleoproteins into a nucleic acid portion and a protein portion:

$$\text{Nucleoprotein} \longrightarrow \text{Nucleic acid} + \text{Protein}$$

The protein is highly basic. Hydrolysis reveals that it contains many units of the amino acids lysine and arginine:

$$\text{Protein} \xrightarrow{H_2O} \text{Lysine} + \text{Arginine} + \text{Other amino acids}$$

The nucleic acids can also be hydrolyzed. Controlled hydrolysis gives units called nucleotides. These units can be further hydrolyzed to phosphoric acid and compounds called nucleosides. Nucleosides can be hydrolyzed to the ultimate molecular constituents that are purine and pyrimidine bases (Chapter 17) and a pentose sugar (Chapter 19):

$$\text{Nucleic acids} \xrightarrow{H_2O} \text{Nucleotides} \xrightarrow{H_2O} \left\{\begin{matrix}\text{Nucleosides}\\ + \\ H_3PO_4\end{matrix}\right. \xrightarrow{H_2O} \left\{\begin{matrix}\text{Two purine bases}\\ + \\ \text{Two pyrimidine bases}\\ + \\ \text{A pentose sugar}\end{matrix}\right.$$

There are two kinds of nucleic acids. Each is a gigantic polymer that has nucleotides as the repeating units. **Deoxyribonucleic acid (DNA)** occurs in the cell nucleus. **Ribonucleic acid (RNA)** is found in all parts of the cell. The two nucleic acids differ only slightly in composition. Complete hydrolysis of DNA ultimately gives two purine bases, adenine and guanine, and two pyrimidine bases, cytosine and thymine. In RNA, however, the pyrimidine base thymine is replaced by another, uracil, and the sugar deoxyribose is replaced by ribose. These differences in composition are summarized in Table 23.1.

Table 23.1 Ultimate Hydrolysis Products of the Nucleic Acids DNA and RNA

	DNA	RNA
Purine bases	Adenine Guanine	Adenine Guanine
Pyrimidine bases	Cytosine Thymine	Cytosine Uracil
Pentose sugar	2-Deoxyribose	Ribose
Inorganic acid	Phosphoric acid	Phosphoric acid

Pyrimidine Bases

The **pyrimidine bases** in nucleic acids are substituted derivatives of the parent compound, pyrimidine. Pyrimidine is a heterocyclic six-member ring containing two nitrogen atoms in the ring. It does not occur free in nature, but its derivatives *uracil, thymine,* and *cytosine* occur in nucleic acids.

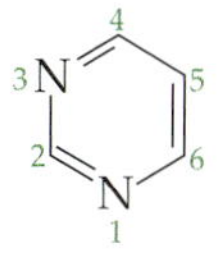

Pyrimidine

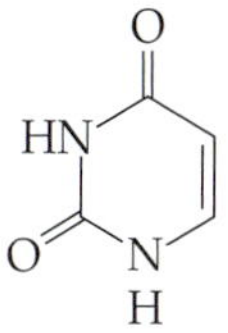

Uracil
(2,4-Dioxypyrimidine)

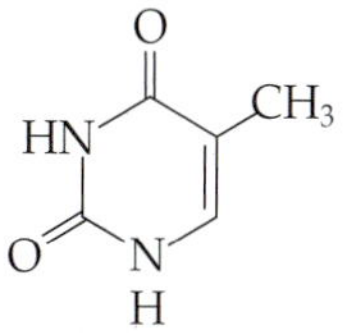

Thymine
(5-Methyl-2,4-dioxypyrimidine)

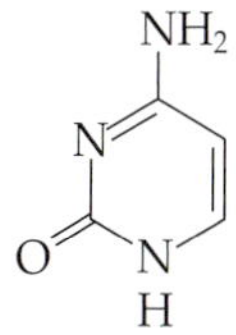

Cytosine
(4-Amino-2-oxypyrimidine)

Several other pyrimidine derivatives (called modified, or minor, bases) also are found in various nucleic acids. Among these are 5-methylcytosine and 5-hydroxymethylcytosine.

Purine Bases

The naturally occurring **purine bases** are derivatives of the parent compound purine, a heterocyclic amine consisting of a pyrimidine ring fused to an imidazole ring. Adenine and guanine are the major purine constituents of nucleic acids.

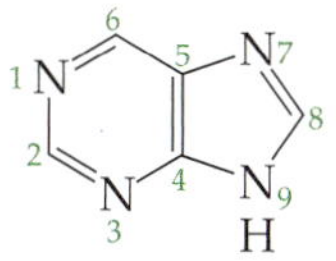

Purine

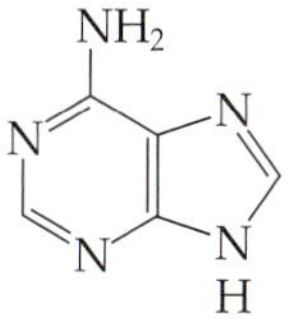

Adenine
(6-Aminopurine)

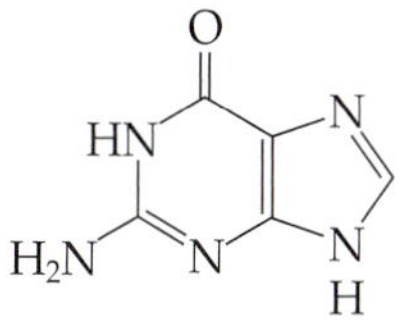

Guanine
(2-Amino-6-oxypurine)

Methylation is the most common form of purine modification, and methylated purines occur in varying amounts in nucleic acids. 6-Methyladenine and 2-methylguanine are two of the minor purine bases that occur in certain nucleic acids.

Collectively, purines and pyrimidines are referred to as *nitrogenous bases.*

Nucleosides

When a purine or pyrimidine base is combined with one of the pentose sugars, a compound called a **nucleoside** is formed. If the sugar is ribose, the compound is a *ribonucleoside.* If 2-deoxyribose is the sugar involved, the product is a *deoxyribonucleoside.*

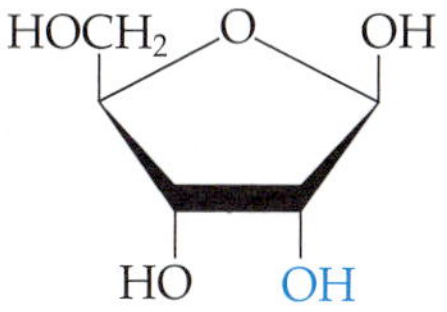

β-Ribose

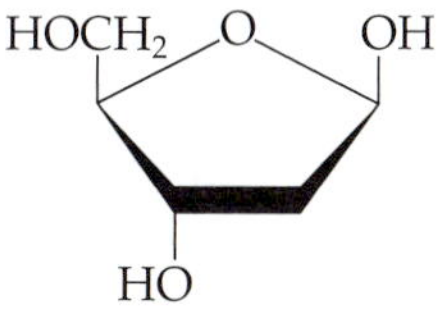

β-2-Deoxyribose

The bond joining the pentose to the nitrogen base is termed an *N-glycosyl linkage,* and it is always beta in naturally occurring nucleosides. The N-glycosyl linkage is formed between C-1′ of the sugar and N-1 of the pyrimidine base or N-9 of the purine base. A molecule of water is eliminated in the process. The following equation is given only to help you visualize the joining of the sugar to the base; nucleosides are *not* synthesized in this fashion in the cell.

The numbering convention is that atoms of the pentose ring are designated by primed numbers and atoms of the purine or pyrimidine ring are designated by unprimed numbers.

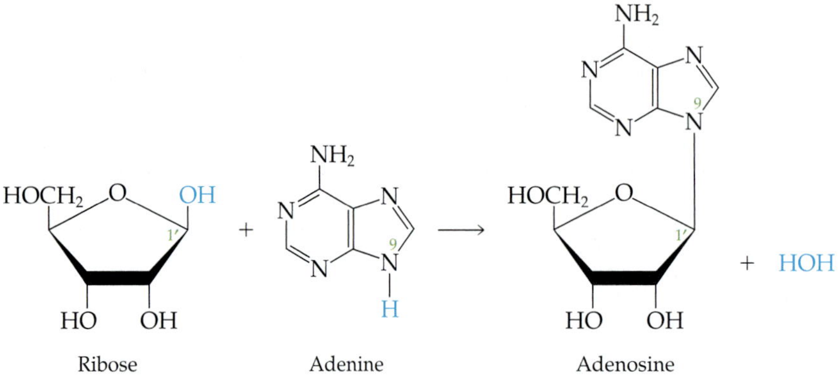

The common names of the ribonucleosides are derived from the names of the nitrogenous bases. The suffix *-osine* denotes purine nucleosides, and the suffix *-idine* is used for pyrimidine nucleosides. The prefix *deoxy-* is used if the base is combined with deoxyribose (deoxynucleosides)—deoxyadenosine, deoxyguanosine, deoxycytidine, and deoxythymidine. Structures and names of the major ribonucleosides and one of the deoxyribonucleosides are given in Figure 23.1.

FIGURE 23.1 ▶ The major pyrimidine and purine nucleosides.

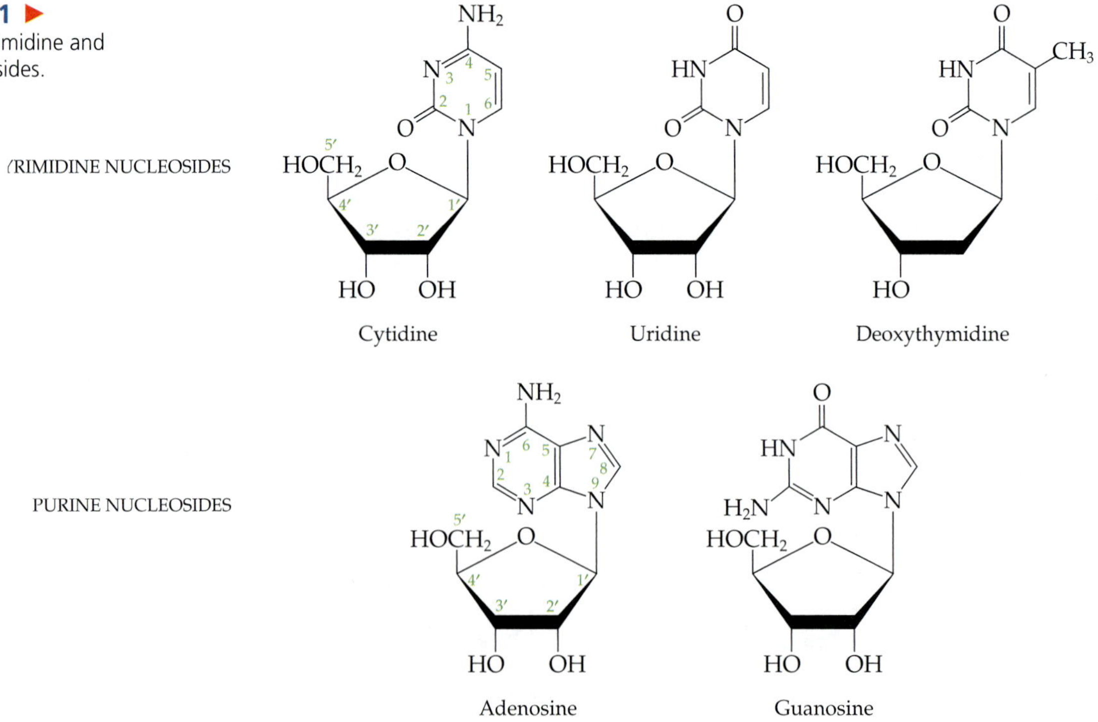

Some Pharmacological Nucleosides

Adenosine may serve as a chemical regulator throughout the body. Receptor sites for it have been identified. It appears that adenosine may regulate the function of neurons in the brain, dilate blood vessels in the heart, constrict bronchial tubes, and inhibit the aggregation of platelets. Caffeine may act as a stimulant by blocking adenosine receptors.

Several nucleoside derivatives have been used in medicine. One, puromycin (Figure 23.2a), is derived from adenosine. It is an antibiotic. First obtained from cultures of the fungus *Streptomyces alboniger*, puromycin is effective against protozoa and has shown some antitumor activity. Vidarabine (Figure 23.2b) is an antiviral drug. The base in vidarabine is adenine, but the sugar is arabinose, an isomer of ribose. 5-Fluorodeoxyuridine (see Section K.6) is an anticancer drug, and azidothymidine (see Section K.2) is a nucleoside used in the treatment of AIDS.

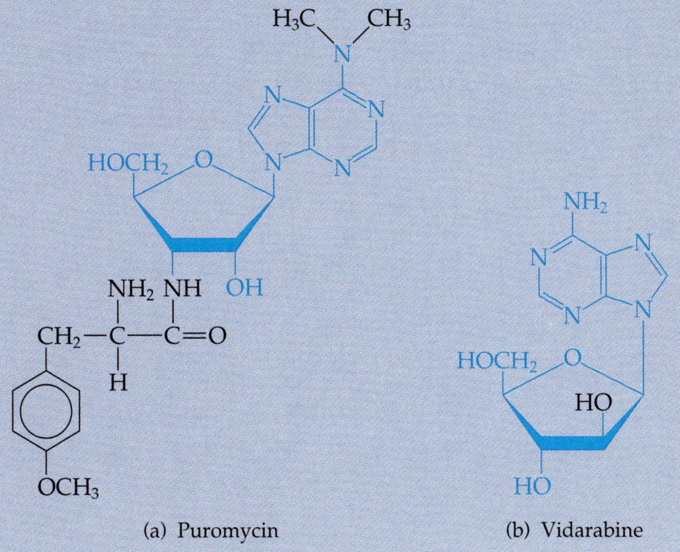

FIGURE 23.2
Puromycin (a) and vidarabine (b), two nucleoside derivatives used in medicine. See Figure K.2 for other antiviral drugs derived from nucleosides.

Nucleotides

The **nucleotides** are phosphate esters of the nucleosides and may be envisioned as resulting from the esterification of phosphoric acid with one of the free pentose hydroxyl groups. Note that again we are illustrating the combination of phosphate and a nucleoside. Nucleotides are *not* formed in this manner in the cell.

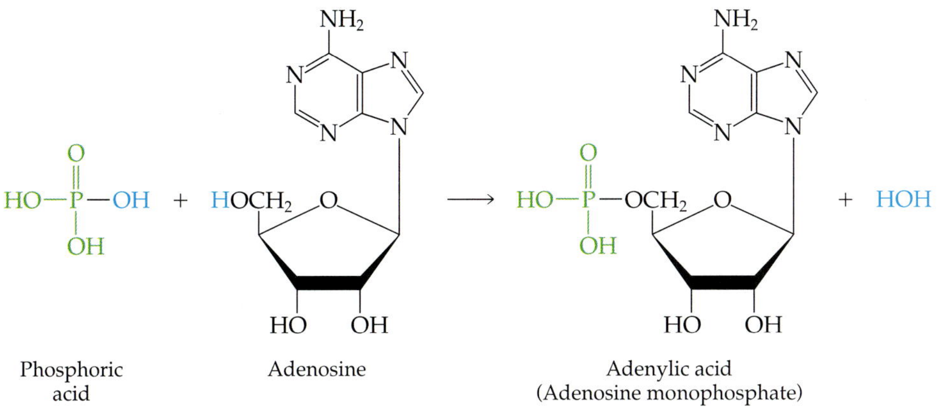

The nucleotides are named in two ways. In one system you drop the ending from the name of the corresponding nucleoside (either *-ine* or *-osine*) and add the ending *-ylic acid.* Thus, the nucleoside uridine, upon esterification with phosphoric acid, becomes uridylic acid. Similarly, guanosine becomes guanylic acid. In the other system, which is simpler and more frequently used, you use the nucleoside name as is and add the word *monophosphate.* Thus, adenosine, upon esterification, becomes adenosine monophosphate, often abbreviated AMP. The prefix *deoxy-* indicates that the sugar involved is deoxyribose rather than ribose. The names and structures of some nucleotides are given in Figure 23.3.

Nucleotides are the monomers from which DNA and RNA are synthesized. In addition, the nucleotides and some of their derivatives perform a variety of other functions in the cell. Adenosine diphosphate (ADP) and adenosine triphosphate (ATP) are involved in many metabolic and biosynthetic processes. We shall encounter them often in upcoming chapters. Structures of these nucleotide derivatives are shown in Figure 23.4. In addition, a cyclic 3′,5′-phosphate of adenosine occurs in which the phosphate group is bonded to two of the ribose carbons. The compound is adenosine 3′,5′-monophosphate (cyclic AMP) and, as we shall see in

FIGURE 23.3 ▶ The pyrimidine and purine nucleotides.

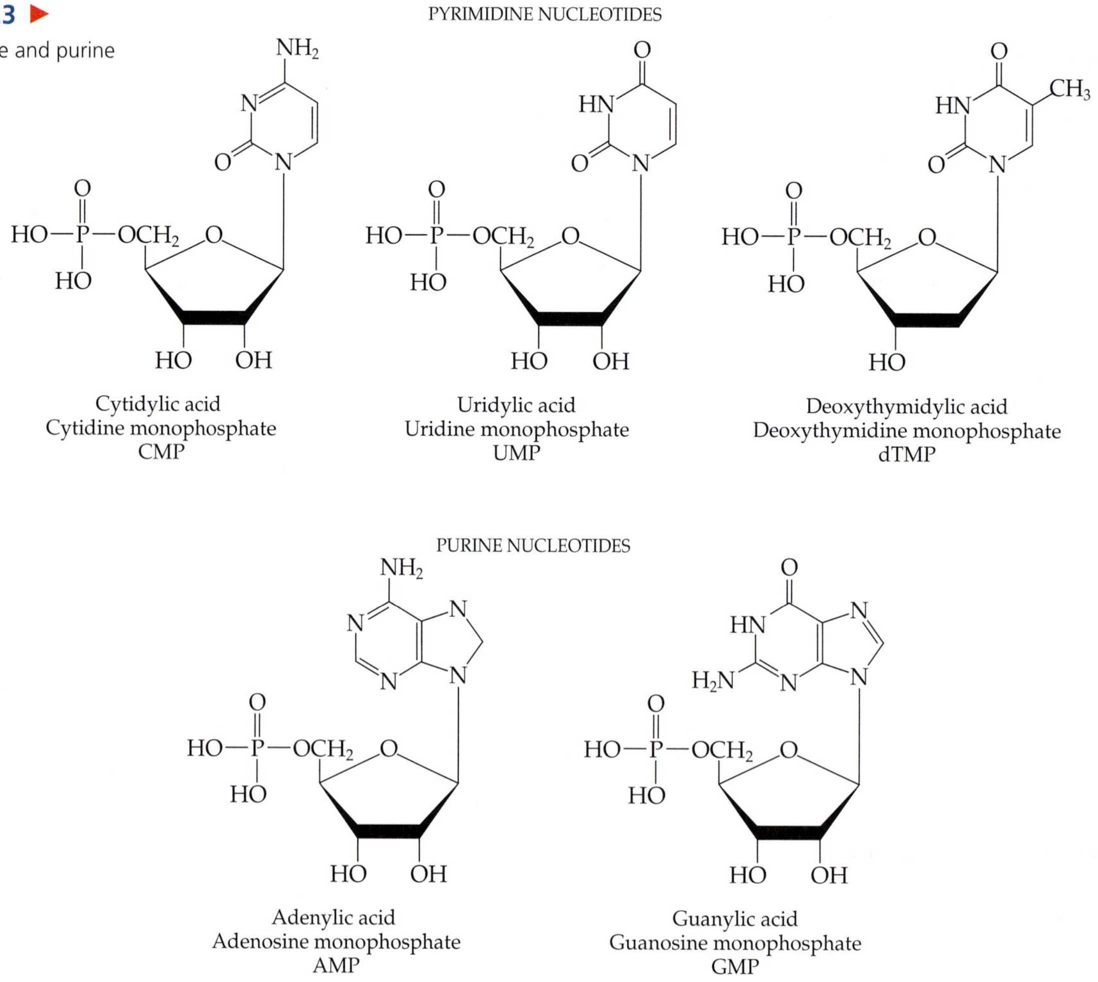

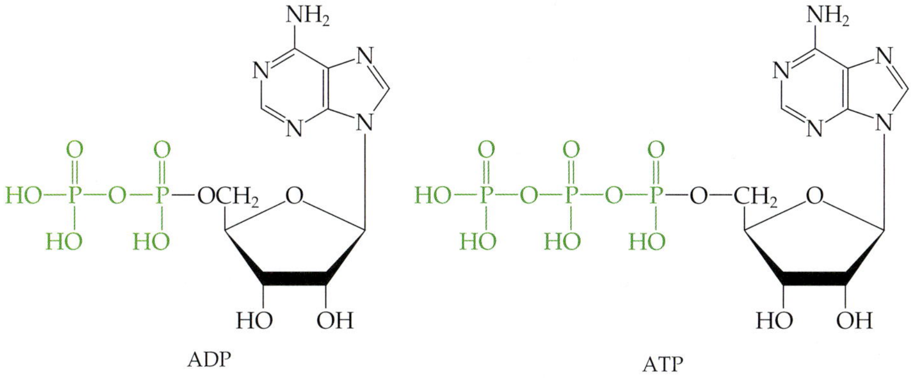

FIGURE 23.4 Structures of two important nucleotide derivatives.

Section 24.4, it plays a crucial role in metabolism. Certain other nucleotides are structural components of a number of important coenzymes. Refer to Table J.1 and notice that FAD, NAD^+, and coenzyme A are all adenine nucleotides.

23.2 The Base Sequence: Primary Structure of Nucleic Acids

Nucleotides are joined to one another through the phosphate group to form nucleic acid chains. The phosphate unit on one nucleotide forms an ester linkage to the hydroxyl group on the third carbon atom of the sugar unit in a second nucleotide. This unit is in turn joined to another nucleotide, and the process is repeated to build up the long nucleic acid chain (Figure 23.5). The backbone of the chain consists of alternating phosphate and sugar units. The purine and pyrimidine bases are branched off this backbone.

As we have seen, the sugar in DNA is 2-deoxyribose, and the one in RNA is ribose. The bases in DNA are adenine, guanine, cytosine, and thymine. Those in RNA are adenine, guanine, cytosine, and uracil. Note (in Figure 23.5) the one ionizable hydrogen on each phosphate unit. That is what makes these compounds nucleic *acids*. In solution or combined with basic proteins as nucleoproteins, the acid is ionized. Partial structures of DNA and RNA are shown in Figure 23.6.

Nucleic acids resemble proteins in one respect. To completely specify the primary structure of a nucleic acid, one must specify the sequence of bases. Unlike the proteins, which have 20 different amino acids, there are only four different bases in a nucleic acid. However, the molecular weight of a nucleic acid is often much greater than that of a protein, ranging into the billions for mammalian DNA. In the 1960s, base sequencing of nucleic acids was extremely laborious. For example, the primary structure of the nucleic acid called alanine transfer RNA, a molecule containing 77 nucleotide units, was determined in that decade. The work, done by Robert W. Holley and co-workers at Cornell University, took 7 years. Holley was rewarded with a share of the Nobel prize in 1968 for his part in the project.

The sequence of bases in short strands of nucleic acids is determined by using gel electrophoresis. Enzymes are used to cleave the nucleic acids at specific base sequences. The primary structure of each fragment is determined. Then overlapping parts are matched to give the base sequence of the whole strand. Now the work is largely automated, and sequences of thousands of base units per molecule have been determined.

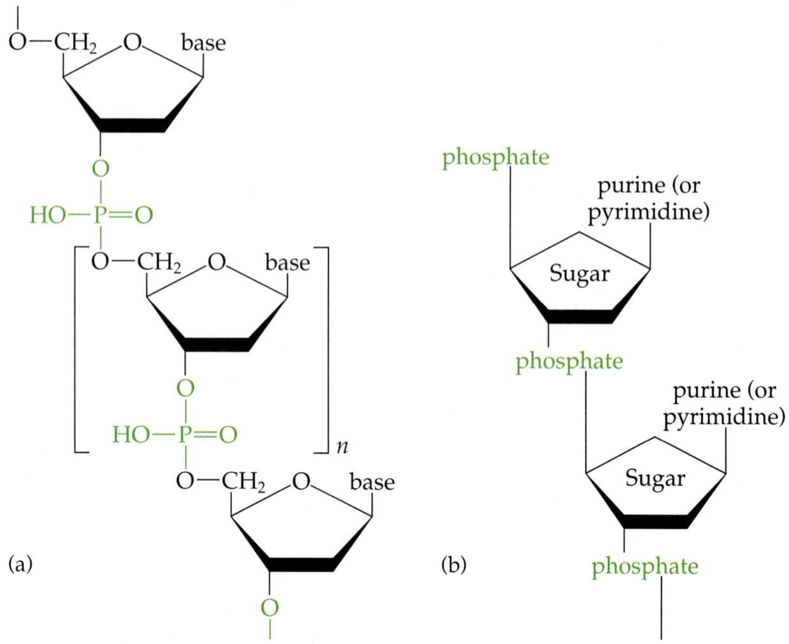

FIGURE 23.5 ▶
(a) The polymeric backbone of a nucleic acid, shown here for DNA. (b) A schematic representation.

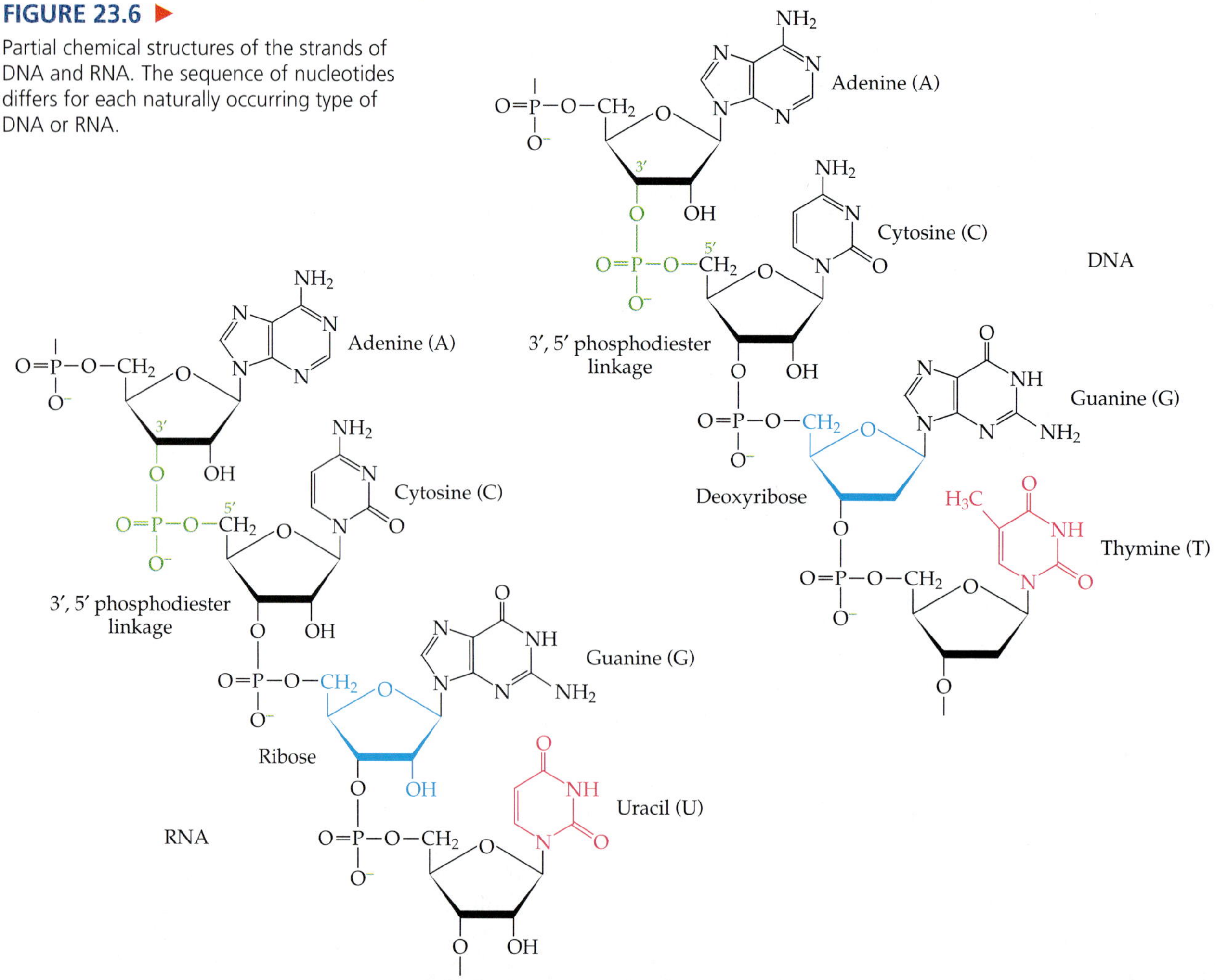

FIGURE 23.6 ▶
Partial chemical structures of the strands of DNA and RNA. The sequence of nucleotides differs for each naturally occurring type of DNA or RNA.

23.3 Base Pairing and the Double Helix: Secondary Structure of DNA

The shape of the gigantic DNA molecule was long a mystery. Early studies revealed no more than the fact that the structure exhibited a periodic pattern. A real breakthrough occurred in 1950, when Erwin Chargaff of Columbia University showed that the molar amount of adenine (A) in DNA is always equal to that of thymine (T). Similarly, the molar amount of guanine (G) is the same as that of cytosine (C). The bases must be paired, A to T and G to C. But how? The race was on, with an almost certain Nobel prize for the winner. Many illustrious scientists, including Linus Pauling, were working on the problem. However, at Cambridge University in 1953, two relative unknowns in the world of science announced that they had worked out the structure of DNA. Using data from the X-ray studies of Rosalind Franklin and Maurice Wilkins, which involved quite sophisticated chemistry, physics, and mathematics, and working with models not unlike a child's construction set, James D. Watson and Francis Crick determined that DNA must be composed of two helices wound about one another to form a **double helix.** The phosphate and sugar groups (the backbone of the nucleic acid polymer) form the outside of the structure, which is like a spiral staircase. The purine and pyrimidine bases are paired on the inside—with guanine always opposite cytosine and adenine always opposite thymine. These specific base pairs are referred to as **complementary bases.** In our staircase analogy, these base pairs are the stairsteps (Figure 23.7).

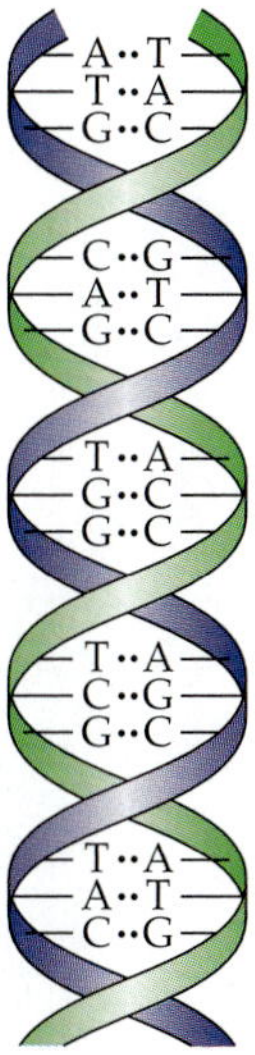

▲ **FIGURE 23.7**
The DNA double helix as portrayed by Watson and Crick.

This structure can explain how cells are able to divide and go on functioning, how genetic data are passed on to new generations, and even how proteins are built to required specifications. It all depends on the base pairing. Figure 23.8 shows other models of the DNA double helix.

Which brings up the still unanswered question, "Why do the bases pair in that precise pattern, always A to T and T to A, always G to C and C to G?" The answer is hydrogen bonding and a truly elegant molecular design. Figure 23.9 shows the two sets of base pairs. Notice two things. First, a pyrimidine is paired with a purine in each case, and the long dimensions of both pairs are identical (1.085 nm). If two pyrimidines were paired or two purines were paired, the two pyrimidines

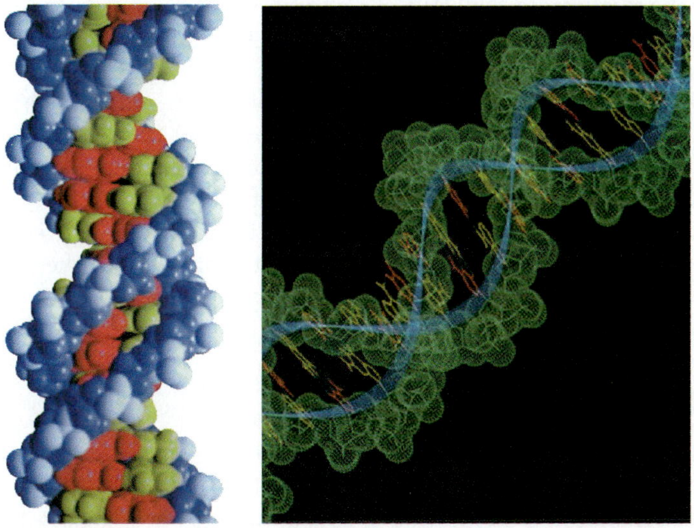

(a) (b)

◀ **FIGURE 23.8**
Two computer-generated models of the DNA double helix. (a) The blue and white atoms represent the sugar–phosphate chains that wrap around the outside of the helix. On the inside are the bases, shown in red and yellow. (b) A schematic illustration of the double helix, showing the hydrogen bonds that connect the complementary base pairs.

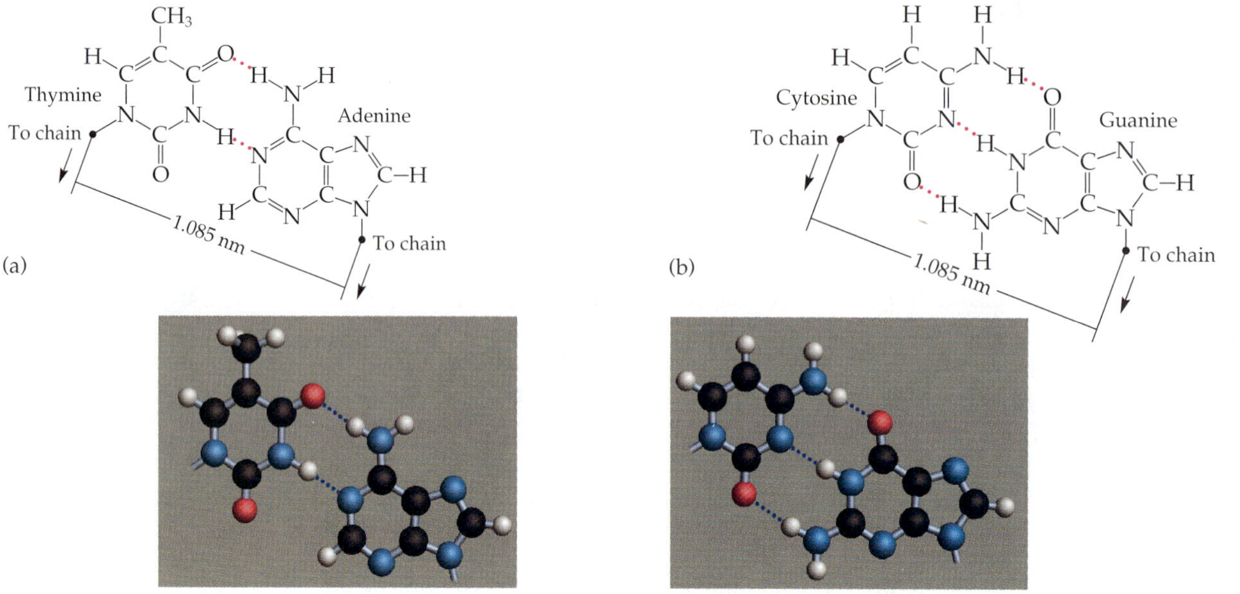

▲ **FIGURE 23.9**
Pairing of the complementary bases thymine and adenine (a) and of cytosine and guanine (b). The pairing involves hydrogen bonding as in DNA.

would take up less space than a purine and a pyrimidine, and the two purines would take up more space, as illustrated in Figure 23.10. If this were the situation, the structure of DNA would be like a staircase made with stairs of different widths. In order for the two strands of the double helix to fit neatly, a pyrimidine must always be paired with a purine.

The other thing you should notice in Figure 23.9 is that when guanine is paired with cytosine, three hydrogen bonds can be drawn between them, and between adenine and thymine there are two hydrogen bonds. It is the additive contribution of all these hydrogen bonds that imparts great stability to the DNA double helix.

There are about 10 base pairs per turn of the double helix. The acidic phosphate units are on the outside. In nucleoproteins, the highly basic proteins probably wrap around the double helix. Proton transfers from the phosphate units of DNA to the lysine and arginine side chains of the protein result in ionic charges. The protein is held to the nucleic acid, at least in part, by these salt linkages. It has been calculated that the total amount of DNA in a typical mammalian cell contains about 5.5×10^9 nucleotides. If all this DNA were stretched out end to end, it would extend more than 2 meters.

Watson and Crick received the Nobel prize in 1962 for discovering, as Crick put it, "the secret of life." It was not long after the development of their models that DNA was synthesized in the laboratory. In 1967, Arthur Kornberg of Stanford University carried out a test-tube synthesis of a single strand of DNA that was able to reproduce itself. Kornberg, of course, had to add the appropriate precursors, and he added the enzymes and cofactors essential to the process. Synthesis of life in a test tube? Hardly. A strand of DNA is still a long way from even the simplest functioning cell.

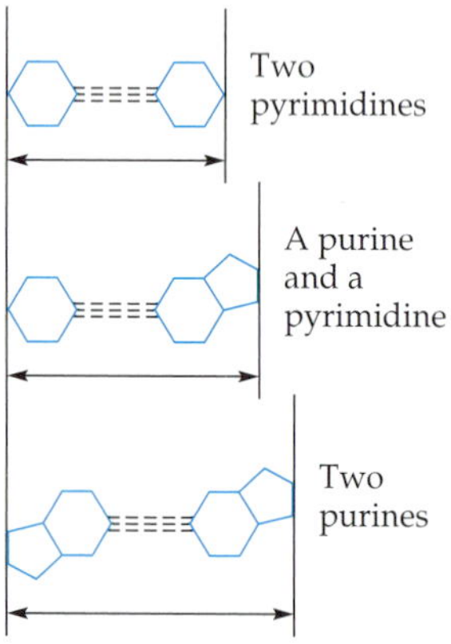

▲ **FIGURE 23.10**
Difference in widths of possible base pairs.

23.4 DNA: Self-Replication

Cats have kittens that grow up to be cats. Bears have cubs that grow up to be bears. How is it that each species reproduces after its own kind? How does a fertilized egg "know" that it should develop into a kangaroo and not a koala? The physical basis of heredity has been known for a long time. Most higher organisms reproduce sexually. A sperm cell from the male unites with an egg cell from the female. The fertilized egg so formed must carry all the information needed to make all the various cells, tissues, and organs necessary for the functioning of the new individual. For humans, that single cell must carry the information for the making of legs, liver, lungs, heart, head, hair, and hands—in short, all the instructions ever needed for growth and maintenance of the individual. In addition, if the species is to survive, information must be set aside in germ cells—either sperm or eggs—for the production of new individuals.

The hereditary material is found in the nuclei of all cells, concentrated in elongated, threadlike bodies called **chromosomes** (Figure 23.11). The number of chromosomes varies with the species. Human body cells have 46 chromosomes. The basic units of heredity, called **genes,** are arranged along the chromosomes in a linear fashion. During cell division, each chromosome produces an exact duplicate of itself. Sperm and egg cells carry only half the chromosomes of the body cells. Thus, in sexual reproduction, the entire complement of chromosomes is achieved only when the egg and sperm combine; a new individual receives half its hereditary material from each parent.

Chromosomes are fibers consisting of complex structures of DNA and proteins. Human chromosomes are composed of about 25% DNA and 75% protein. The DNA contains the genetic information; the proteins are a major factor in the regulation of gene expression.

Calling the unit of heredity a gene merely gives it a name. What are genes? What are they made of? The material of genes is nothing other than a distinct segment of a long DNA strand. (Some viruses carry genetic information in RNA; see Section K.1.) *Each gene codes for a specific polypeptide.* Transmission of genetic information involves the **self-replication** (copying or duplication) of the DNA strand.

Humans have about 100,000 different genes. Scientists have embarked on an ambitious endeavor, called the *human genome project,* to map the location of each of these genes on the 23 pairs of chromosomes. (The *genome* of an organism is its complete set of genes.)

The Watson–Crick double helix provides a ready model for genetic replication. If the two chains of the double helix are pulled apart and the hydrogen bonds between base pairs are broken, each chain can act as a *template,* or pattern, and direct the synthesis of a new DNA chain (Figure 23.12). In the cellular fluid surrounding the DNA are all the necessary nucleotides (monomers). It is simply a matter of a nucleotide with the proper base combining with its complementary base on the DNA strand. Keep in mind that adenine can pair only with thymine and guanine only with cytosine. Each base unit in the separated strand can pick up only a unit identical to the one that it had before. Each of the separating chains serves as a template for the formation of a new complementary chain.

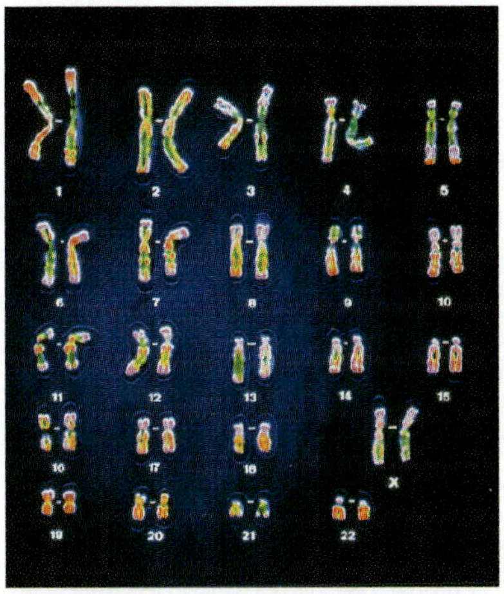

◄ **FIGURE 23.11**

Female chromosomes arranged in numbered homologous pairs. The male set differs from the female set only in the sex pair (bottom right); a male has an X and a Y instead of two Xs. The nucleus of each human body cell has 46 chromosomes, 23 from the mother and 23 from the father.

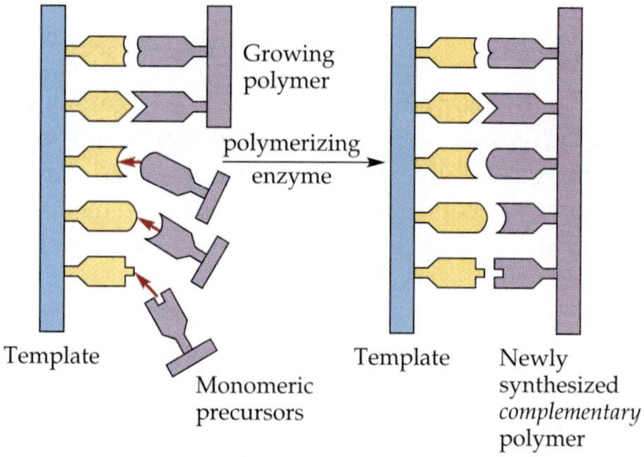

FIGURE 23.12 ▶

A schematic view of the formation of a complementary polymer upon a template surface.

As the nucleotides align, enzymes connect them to form the sugar-phosphate backbone of the new chain. In this way, each strand of the original DNA molecule forms a duplicate of its former partner. Whatever information was encoded in the original DNA double helix is now contained in each of the replicates. When the cell divides, each daughter cell gets one of the DNA molecules and all of the information that was available to the parent cell (Figure 23.13).

We keep saying that there is information encoded in the DNA molecule. DNA can be compared to a book containing directions for putting together a model airplane or for knitting a sweater. Knitting directions store information as words on paper. Letters of the alphabet are arranged in a certain way (e.g., "knit one, purl two"), and these words direct the knitter to perform a certain operation with needles and yarn. If all of the directions are correctly followed, the ball of yarn becomes a sweater.

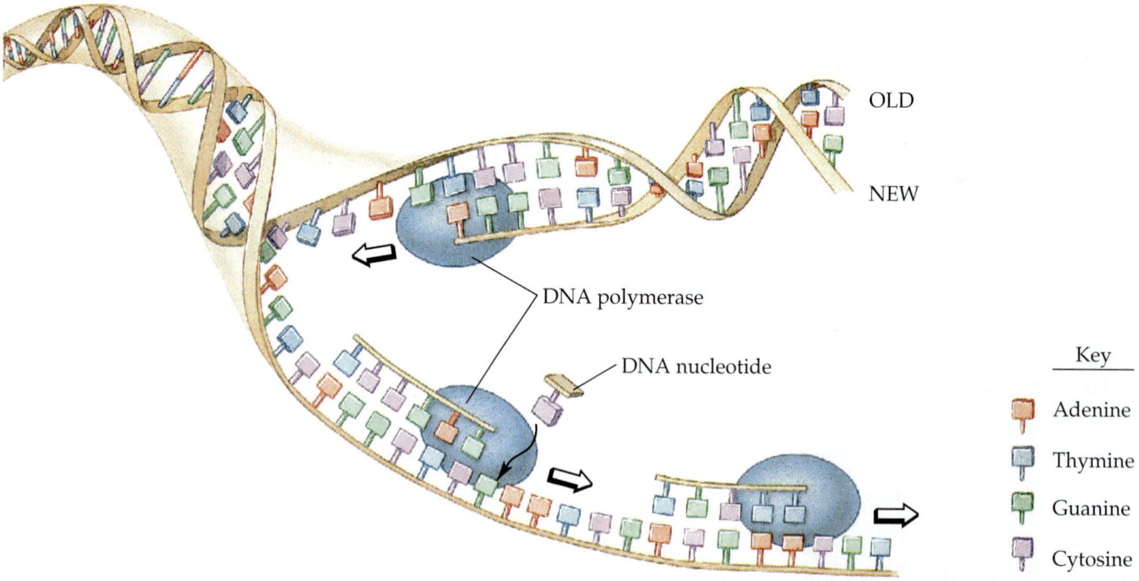

▲ FIGURE 23.13

A schematic diagram of DNA replication, which occurs by sequential "unzipping" of the double helix. The new nucleotides are brought into position by the enzyme DNA polymerase and phosphate bridges are formed, thus restoring the original double helix configuration. Each newly formed double helix consists of one old strand and one new strand, a process called semiconservative replication. (This representation is simplified; the nucleotides are actually triphosphate derivatives: ATP, TTP, GTP, and CTP.)

DNA Fingerprinting

In 1985, the British biologist Alec Jeffreys invented a technique and coined the term "DNA fingerprinting." Like fingerprints, a person's DNA is unique to that individual. Any cells—skin, blood, semen, saliva, etc.—can supply the necessary DNA sample. The DNA samples from a suspect are compared with the DNA obtained from evidence found at the crime scene. The technique is intricate and requires several chemical steps. In the final step, a "print" is represented as a series of horizontal bars resembling the bar codes imprinted on packaged goods sold in supermarkets.

DNA fingerprinting has been hailed as a major advance in criminal investigation. Several hundred criminal cases have been solved with this technology. Also, because children inherit half their DNA from each parent, DNA fingerprinting has been used to establish the parentage of a child of contested origin. The odds in favor of being right in such cases are said to be excellent—at least 100,000:1.

DNA fingerprinting received its greatest notoriety during the O. J. Simpson trial. That court case served to publicize the controversy regarding the use of DNA typing. Is the evidence from DNA analysis infallible? Do all individuals (aside from identical twins) truly have unique DNA fingerprints? What are the statistical probabilities that two DNA samples from different people will be different? And perhaps most significant, how careful are the people involved in the collecting, handling, and laboratory analysis of the samples from the crime scene?

How is information stored in DNA? It is the particular arrangement of bases along the DNA chain that encodes the directions for building an organism. Just as *saw* means one thing in English and *was* means another, the sequence of bases CGT means something, and TGC means something else. Although there are only four "letters"—the four bases—in the genetic code of DNA, their sequence along the long strands can vary so widely that there is an essentially unlimited information storage system. Even a tiny bacterium 3 μm long and 1 μm in diameter has 3 million base pairs. The genetic material of a human cell consists of 5 billion base pairs, and these pairs can specify 20 billion bits of information—enough information to fill 1000 books of 2000 pages each. Thus each cell can carry all the information it needs to determine all the hereditary characteristics of even the most complex organism. We shall see how this information is conveyed to the cell in Section 23.6.

James D. Watson (1928–) (right) and Francis H. C. Crick (1916–) (seated) proposed the double helix model of DNA in 1953. They were awarded the Nobel prize for this work in 1962.

23.5 RNA: The Different Ribonucleic Acids

A molecule of RNA consists of a single strand of the nucleic acid. Some internal (intramolecular) base pairing may occur in sections where the molecule folds back on itself. Because of this, portions of the molecule may exist in a double-helix form, but overall, RNA is considered a single-stranded molecule.

There are different kinds of RNA molecules in a cell, and all of them appear to be synthesized from DNA by a template mechanism analogous to DNA replication in many respects. To initiate RNA biosynthesis, the two strands of the DNA molecule begin to uncoil. This uncoiling occurs at specific sites, called *promoters,* on the DNA template. The nucleotides are attracted to the uncoiling region of the DNA molecule according to the rules of base pairing. Thymine in DNA calls for adenine in RNA, cytosine specifies guanine, guanine calls for cytosine, and adenine requires uracil. Recall from Section 23.1 that, in RNA molecules, uracil replaces DNA's thymine. Notice the similarity between the structures of these two bases. Because RNA is a complementary copy of information contained in DNA, RNA biosynthesis is referred to as **transcription.** We see later (Section 23.6) why this process is vital to all growth and development. Figures 23.14 and 23.15 depict the process schematically.

DNA Base	Complementary RNA Base
Adenine	Uracil
Thymine	Adenine
Cytosine	Guanine
Guanine	Cytosine

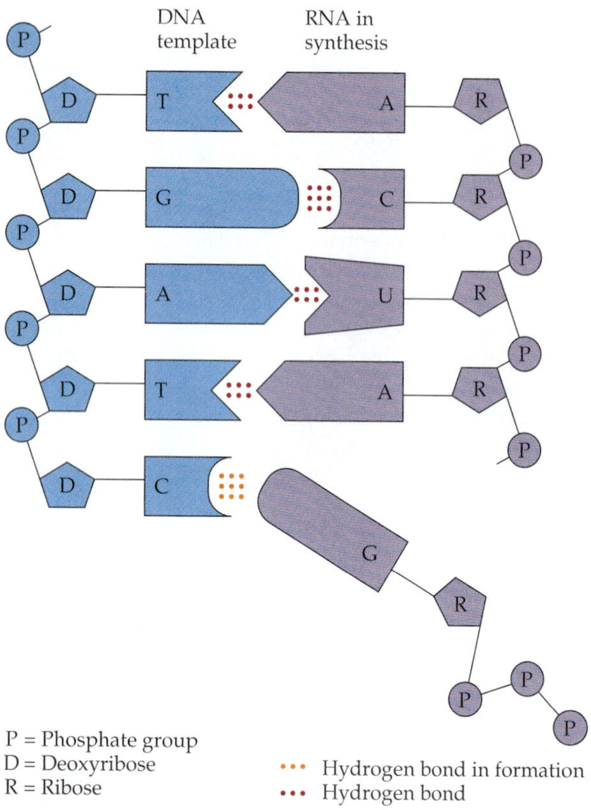

▲ **FIGURE 23.14**

Transcription of RNA on a DNA template. One strand of the DNA double helix serves as a template for alignment of RNA bases. Once a nucleotide is aligned, a phosphate bridge is formed, resulting in a single strand of RNA. The nucleotides in the new RNA molecule are joined together by the enzyme RNA polymerase.

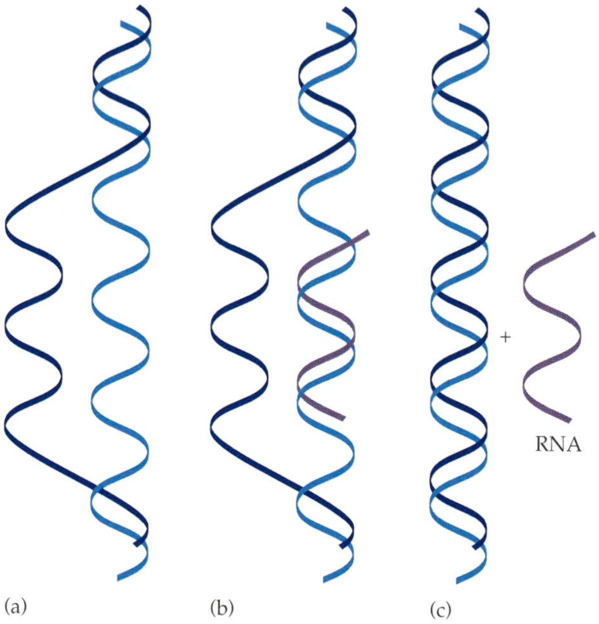

▲ **FIGURE 23.15**

(a) The DNA double helix is partly unwound. (b) An RNA is formed on the separated portion. (c) Hydrogen bonds are broken and the RNA is released.

Transcription differs from DNA replication in several ways. During transcription, in a given DNA region only one DNA strand serves as a template and is copied. In another region of the DNA, the other strand may serve as a template for the biosynthesis of a different RNA. Thus, RNA molecules are much shorter than DNA molecules. Also, RNA molecules do not remain hydrogen-bonded to DNA for any length of time. As soon as transcription is completed, the RNA is released and the DNA helix re-forms.

Three basic types of cellular RNAs are known to exist. The distinctions among them are made primarily on the basis of biochemical function. However, they also differ in molecular weight and in secondary structure (Table 23.2).

Table 23.2 RNA Molecules in *E. coli*

RNA Type	Relative Amount (%)	Average MW	Approx. Number of Nucleotides
Messenger RNA (mRNA)	5	4×10^5	1200
Ribosomal RNA (rRNA)	80	6×10^5	1800
Transfer RNA (tRNA)	15	3×10^4	70–90

Messenger RNA

Messenger RNA (mRNA) makes up only a few percent of the total amount of RNA within the cell. A molecule of mRNA exists for a relatively short time. Like proteins, it is continuously being degraded and resynthesized. The rate of mRNA degradation differs from species to species and also from one type of cell to another. In bacteria, one-half of the total mRNA is degraded every 2 min, whereas in rat liver the half-life is several days.

The molecular dimensions of the mRNA molecule vary according to the amount of genetic information the molecule is meant to encode. It is known, however, that there is very little intramolecular hydrogen bonding in mRNA and that the molecule exists in a fairly random coil. After transcription, which takes place in the nucleus, the mRNA passes into the cytoplasm, carrying the genetic message from DNA to the ribosomes, the sites of protein synthesis. In Section 23.6 we shall see how mRNA directly governs that synthesis.

Ribosomal RNA

Ribosomal RNA (rRNA) makes up 80% of the total cellular complement of ribonucleic acid. The **ribosome** is a cellular substructure that serves as the site for protein synthesis. Its composition is about 65% rRNA and 35% protein. The rRNA molecules and the proteins are bonded together by a large number of noncovalent forces, such as hydrogen bonds and hydrophobic interactions. Structurally, a ribosome is composed of two spherical particles of unequal size. The smaller of them has a distinct affinity for mRNA; the larger has an attraction for tRNA.

Ribosomes are extremely small particles visible only with the aid of an electron microscope. More often than not, they are seen as clusters known as *polyribosomes,* or *polysomes,* bound to the endoplasmic reticulum of animal and plant cells or to the cell membrane of microorganisms. When ribosomes occur in such aggregates, they are held together by strands of mRNA. On the average, five to eight ribosomes are simultaneously synthesizing the same polypeptide from the information in one mRNA strand (large proteins require long strands of mRNA, and as many as 100 individual ribosomes may be attached). The time required for the synthesis of an average-size polypeptide (~300 amino acids) is about 15 seconds in a bacterial cell and 2 or 3 minutes in a mammalian cell.

Transfer RNA

Transfer RNA (tRNA) is a relatively low-molecular-weight nucleic acid, soluble in solvents commonly used to isolate the higher-molecular-weight nucleic acids. It functions by attaching itself (with the aid of a specific enzyme) to a particular amino acid and carrying that amino acid to the site of protein synthesis at the precise moment specified by the genetic code. Each of the 20 amino acids found in proteins has at least one corresponding tRNA, and most amino acids have more than one. For example, there are two different tRNAs specific for the transfer of lysine, three for isoleucine, four for glycine, and six for serine. The existence of several tRNAs for the same amino acid is termed **multiplicity.**

A tRNA molecule has a "cloverleaf" structure (Figure 23.16). On one of the loops is a unique sequence of three bases that is different in the tRNAs for different amino acids. This triplet is called the **anticodon.** A specific amino acid becomes attached to the other end of the tRNA.

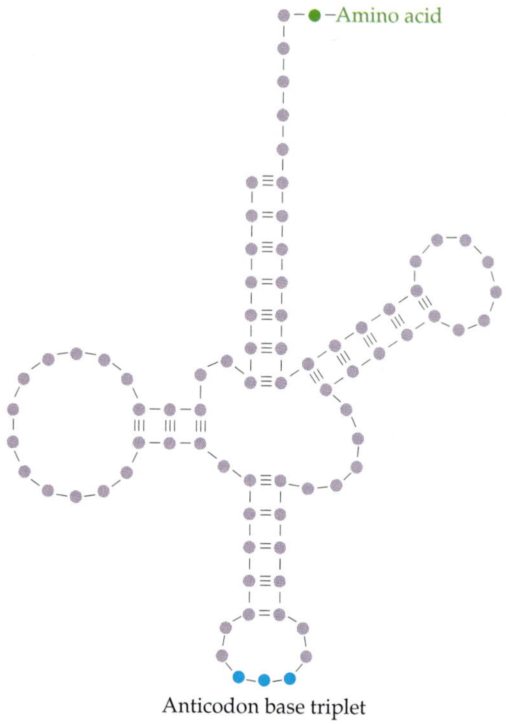

FIGURE 23.16 ▶

Cloverleaf diagram of a tRNA molecule. The molecule is a single chain, but it folds back on itself. Hydrogen bonds are formed, and large regions are characterized by base pairing.

23.6 Protein Synthesis

It has been estimated that about 5% of the DNA of higher organisms conveys the information for the synthesis of proteins. The remainder is involved in the regulation of protein synthesis, in the synthesis of tRNA and rRNA, and in maintaining the correct conformation of the DNA. See also "Genetic Regulation" (page 651).

Recall from Section 23.4 that a gene is the segment of a DNA molecule that codes for the biosynthesis of one polypeptide chain. (If a protein contains two or more polypeptide chains, each chain is coded by a different gene.) Each gene is a section of DNA that contains about 1000 to 2000 nucleotides. A human cell contains about 100,000 genes (although there is sufficient DNA to code for 80 to 100 times this number of genes).

Even if we accept the fact that DNA carries a message, there is still the problem of how this message is read. How is a particular sequence of bases along the DNA chain translated into a complex organism? The answer is that *DNA directs the synthesis of proteins.* Proteins serve as building materials and, most important, as enzymes. The mechanism by which the DNA blueprint is transformed into protein molecules involves the intermediacy of RNA molecules.

How can a molecule containing just four different monomeric units specify the sequence of the 20 amino acids that occur in proteins? If each nucleotide coded for one amino acid, then obviously the nucleic acids could code for only four of the 20 amino acids. Suppose we consider the nucleotides in groups of two. There are 4^2, or 16, combinations of pairs of the four distinct nucleotides. Such a code is more extensive but still inadequate. If, however, the nucleotides are considered in groups of three, there are 4^3, or 64, combinations. Here we have a code that is extensive enough to govern the primary structure of the protein molecule because it contains more than enough coding units to designate all 20 amino acids. Now we shall see how this code directs protein synthesis.

If the sequence of bases along the DNA strand determines the sequence of amino acids along the polypeptide chain, the information contained in the DNA must be conveyed from the nucleus to the site of protein synthesis. This is accomplished by orderly interactions between the nucleic acids and more than 100 enzymes. Recall that mRNA is made from a DNA template and so contains a base sequence that is complementary to that of the DNA upon whose surface it was synthesized. Once formed, the mRNA is transported across the nuclear membrane into the cytoplasm (and hence to the ribosomes), carrying with it the genetic instructions. *Each group of three bases along the mRNA strand now specifies a particular amino acid, and the sequence of these triplet groups dictates the sequence of the amino acids in the protein.* Because the code involves three bases per coding unit, it is referred to as a **triplet code.** The three-base coding unit is called a **codon.**

Now the cell faces the problem of lining up the amino acids according to the sequence called for by the mRNA and joining them together by means of peptide linkages. Because this process involves the transfer of the information encoded in the mRNA to the ultimate structure of the protein molecule, it is often referred to as **translation.**

Before the amino acids can be incorporated into a polypeptide chain, they must be activated. Activation occurs before each amino acid reacts with its particular tRNA carrier molecule. This crucial process requires certain "activation" enzymes and the participation of an ATP molecule (Figure 23.17). *Both the enzymes (aminoacyl tRNA synthetases) and the tRNAs are each highly specific for a particular amino acid.* The high degree of specificity of the synthetase enzymes is vital to the correct incorporation of amino acids into proteins. After the amino acid molecules have been activated and have reacted with the tRNA carriers, protein synthesis can take place.

When an amino acid is bound to its tRNA carrier, the combination is referred to as an amino-acyl tRNA complex.

Figure 23.18 depicts a schematic stepwise representation of this all-important process. After a certain codon of the mRNA strand has been "read" by a given ribosome, another ribosome may attach itself to the strand and begin to read it as the first ribosome moves on to read the next codon.

Thus in cells active in protein synthesis we find clusters of ribosomes connected by a single strand of mRNA (Figure 23.19). The amount of any particular protein in a cell depends on the balance between the rate at which it is synthesized (which is largely controlled by the rate at which its mRNA is synthesized in the nucleus) and the rate at which it is degraded. These events are summarized in Figure 23.20.

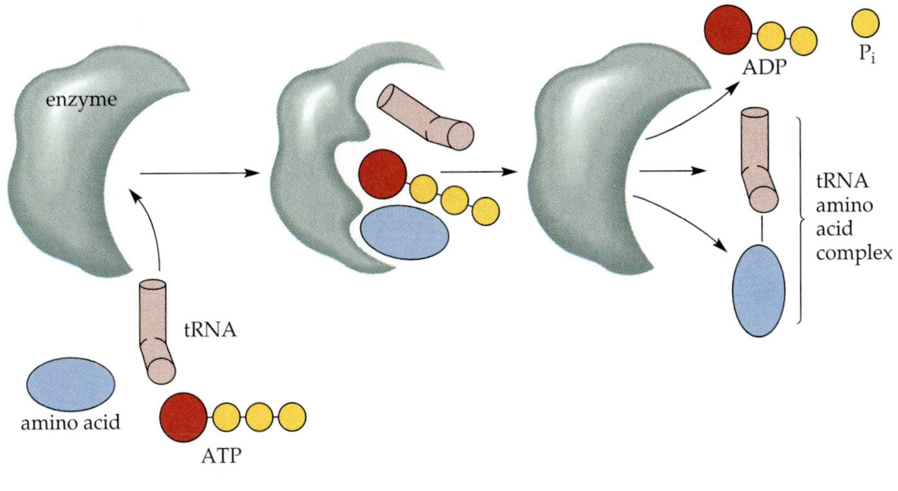

FIGURE 23.17

Each type of tRNA binds to a specific amino acid. Enzymes, one for each amino acid, catalyze the formation of the bond between the tRNA and its corresponding amino acid. ATP provides energy for the reaction.

FIGURE 23.18 ▶

The elongation steps in protein synthesis.

(a) Protein synthesis is already in progress at the ribosome. The growing polypeptide chain is bound to the peptidyl (P) site on the surface of the ribosome. At this point, the aminoacyl (A) site is vacant. The codon UUU is lined up above the A site. An activated tRNA molecule, which has the anticodon AAA, moves up to the ribosome. (The tetracyclines—Figure 22.19—block the binding of the aminoacyl tRNA to the A site on bacterial ribosomes, inhibiting protein synthesis and stopping bacterial growth. The tetracyclines do not bind to mammalian ribosomes and thus do not hinder protein synthesis in host cells.)

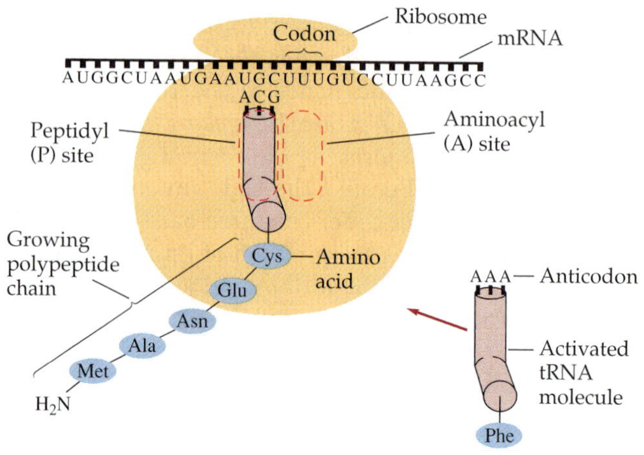

(a)

(b) The activated tRNA molecule is now bound to the ribosome at the A site. It is also bound to the mRNA molecule through base pairing of the codon and anticodon. The amino acid Phe is being incorporated into the polypeptide chain by the formation of a peptide linkage between the carboxyl group of Cys and the amino group of Phe. This reaction is catalyzed by the enzyme peptidyl transferase, a component of the ribosome. (Chloramphenicol acts by blocking the action of peptidyl transferase.)

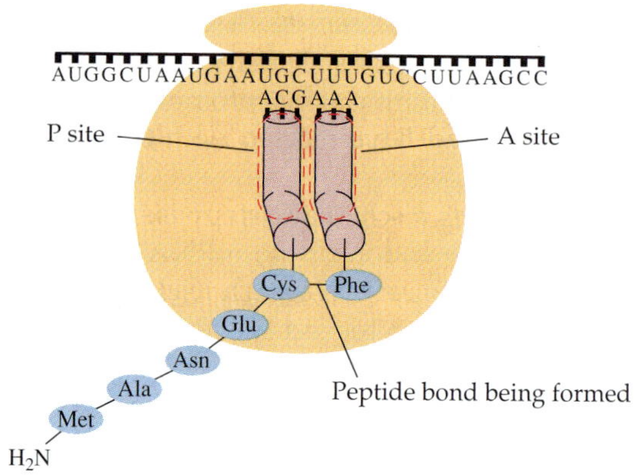

(b)

(c) The Cys–Phe linkage is now complete, and the growing polypeptide chain is now attached to the A site. The tRNA molecule has detached from the P site. It can now move off the ribosome and into the cytoplasm and pick up another amino acid.

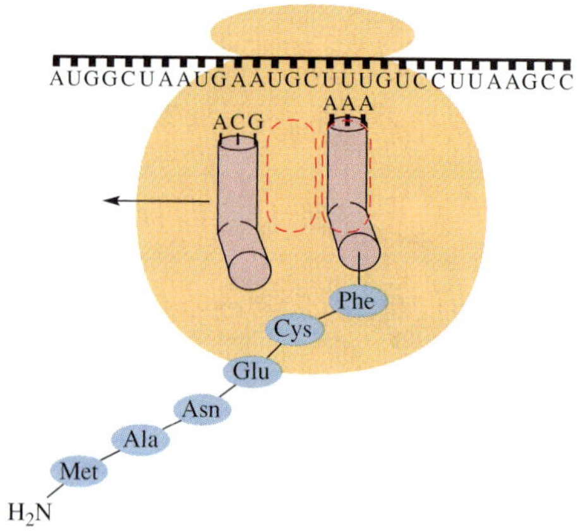

(c)

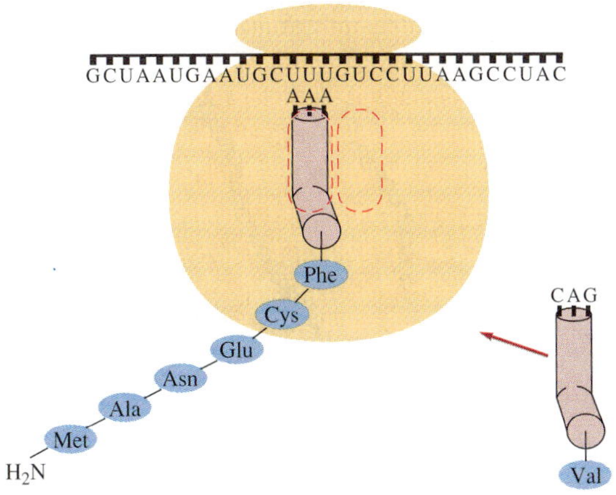

(d)

(d) The ribosome moves (translocates) to the right along the mRNA strand. The polypeptide chain and the tRNA molecule to which it is bound are shifted from the A site to the P site. This shift brings the next codon, GUC, into place over the A site. Notice that an activated tRNA molecule, containing the next amino acid to be attached to the chain, is moving into position on the surface of the ribosome. Its anticodon is CAG. (Erythromycin blocks the translocation reaction in bacteria.)

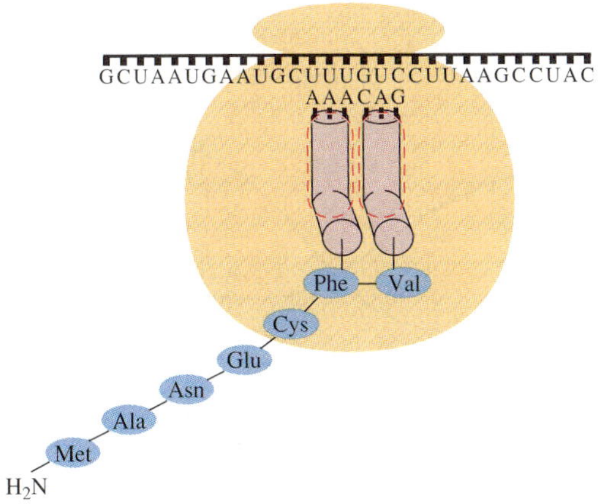

(e)

(e) The activated tRNA molecule, carrying the amino acid Val, is now in place on the ribosome. The peptide linkage between the carboxyl group of Phe and the amino group of Val is forming.

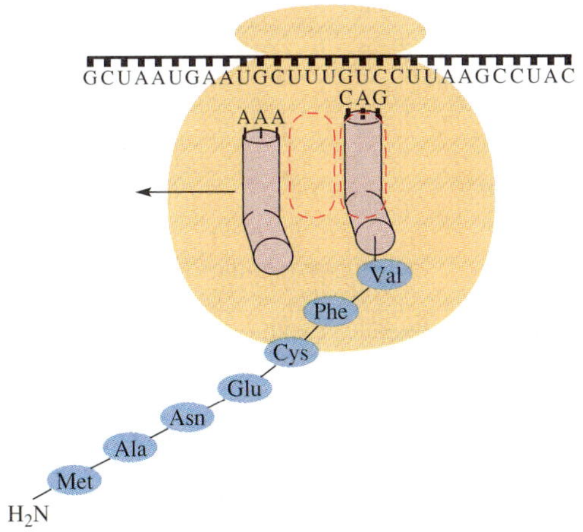

(f)

(f) The Phe–Val linkage is now complete, and the growing polypeptide chain is now attached, through a tRNA molecule, to the A site. The ribosome will translocate again, and the tRNA molecule plus the attached polypeptide chain will be in position at the P site. This process continues until the polypeptide chain is complete—that is, until one of the three termination codons appears at the A site. When the ribosome reaches the end of the message, both it and the polypeptide are released from the mRNA molecule.

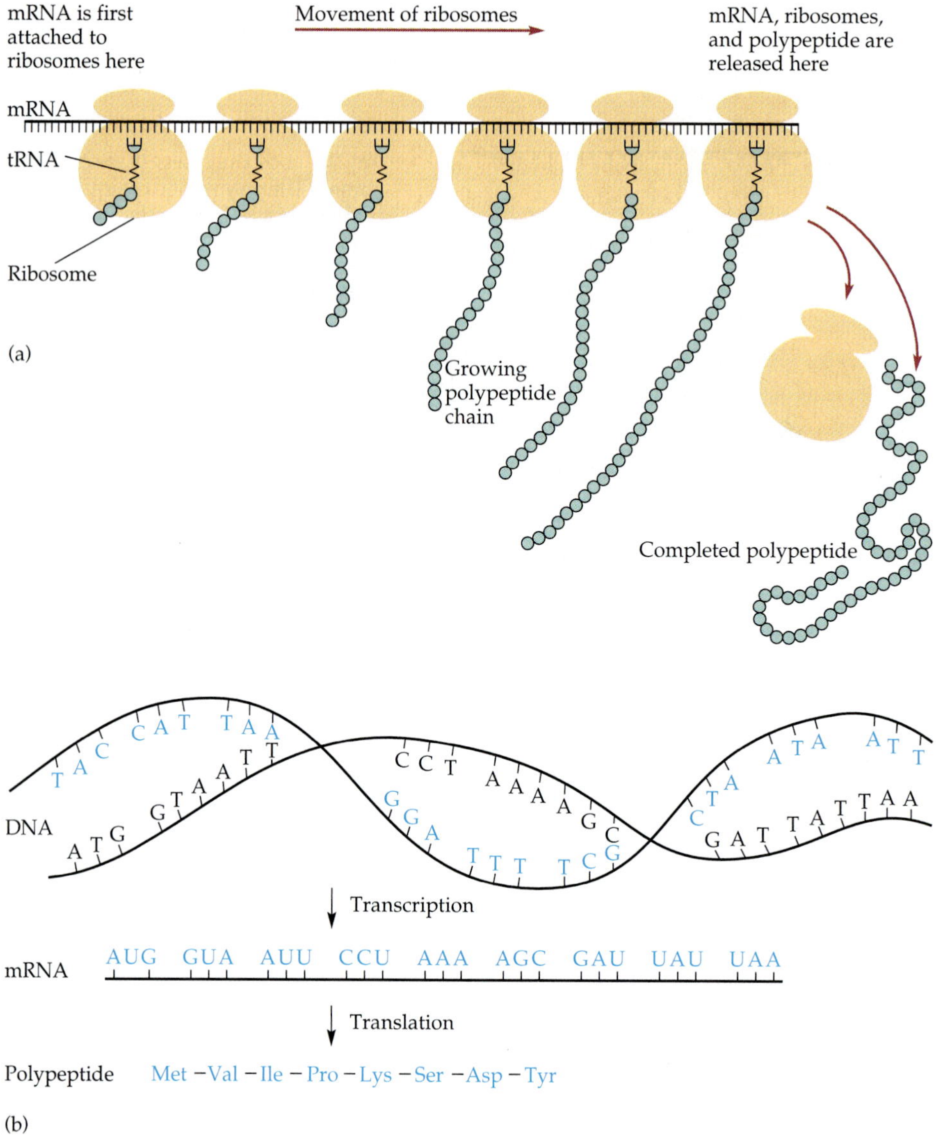

FIGURE 23.19 ▶

(a) Diagrammatic representation of a polysome, consisting of six ribosomes attached to one mRNA. The ribosomes have progressed various distances along the mRNA during translation, and each ribosome is associated with a progressively longer polypeptide chain. At the end of the message, the mRNA and ribosome separate, and the complete polypeptide is released. (b) The relationship between transcription (mRNA biosynthesis) and translation (polypeptide synthesis).

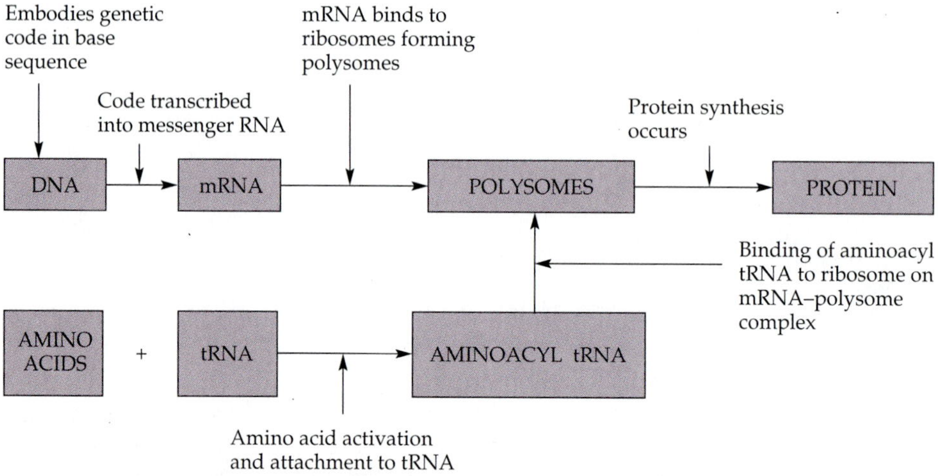

FIGURE 23.20 ▶

Outline of events in protein synthesis, from DNA transcription and amino acid activation to completed protein.

Genetic Regulation

It is essential at this point to note that the preceding discussions of DNA replication, RNA synthesis, and protein synthesis are correct but oversimplified. Several molecular components (including RNA primers) are required to initiate DNA replication. Additional proteins are required for the elongation and termination processes. Furthermore, ultraviolet radiation and certain chemicals are known to damage DNA by disrupting the sugar–phosphate backbone and/or by altering the purine and pyrimidine bases. DNA repair enzymes exist that can mend the backbone and correct the base sequence.

Also, the majority of plant and animal genes occur in pieces, spread out along the DNA. Parts of the gene are expressed (*exons*); the intervening regions (*introns*) are not expressed. Therefore, the RNAs transcribed from these genes have been found to be spliced—that is, the entire length of DNA is copied in a longer transcript that is cut once or several times and hooked back together to produce a shorter, functional messenger. In addition, protein synthesis is critically controlled at the level of transcription. We now know that there are structural genes, operator genes, promoter genes, regulatory genes, and repressor molecules (which are proteins), plus an involvement of GTP and cyclic AMP. Postribosomal modification of proteins occurs after synthesis on the mRNA–ribosome complex and involves many types of modifications (e.g., methylation or hydroxylation of amino acid side chains; breakage of peptide bonds to activate a zymogen or a hormone). An advanced biochemistry text will provide you with a detailed explanation of the fascinating studies in molecular genetics.

Genes have functional regions called *exons* interspersed with inactive portions called *introns.* During protein synthesis, introns are snipped out and not translated.

23.7 The Genetic Code

As we have stated, the sequence of bases on the mRNA directs the precise sequence of the amino acids for each protein. We have indicated that the codon, the unit that codes for a particular amino acid, consists of a group of three adjacent nucleotides on the mRNA. There are 64 possible triplet codons. Early experimenters were faced with the task of determining which codon (or perhaps codons) stood for each of the 20 amino acids. The cracking of the genetic code was the joint accomplishment of several well-known geneticists, notably H. Khorana, M. Nirenberg, P. Leder, and S. Ochoa (1961–1964). A genetic dictionary has been compiled and is given in Table 23.3. Of the 64 possible codons, 61 code for amino acids and three serve as signals for the termination of polypeptide synthesis (that is, as periods at the end of a sentence). Notice that only methionine (AUG) and tryptophan (UGG) have single codons. All other amino acids have two or more codons.

Further experimentation by Nirenberg threw much light on the nature of the genetic code. It now appears that:

1. The code is essentially universal—animal, plant, and bacterial cells use the same codons to specify each amino acid. (A few exceptions have been discovered.)
2. The code is degenerate—in all but two cases (methionine and tryptophan), more than one triplet codes for a given amino acid.
3. The first two bases of each codon are most significant; the third base often varies. This suggests that a change in the third base by a mutation may still permit the correct incorporation of a given amino acid into a protein (see Section 23.8). The third base is sometimes called the "wobble" base.

4. In general, codons with C or U as the second base specify the nonpolar amino acids, whereas codons with A or G as the second base specify the polar amino acids (see Table 21.1).
5. The code is continuous and nonoverlapping—there are no special signals, and adjacent codons do not overlap (except in the case of a few viruses that do have overlapping genes).
6. There are three codons that do not code for any amino acid. These are the termination codons; they are read by special proteins (called release factors) and signal the end of the translation process.
7. The codon AUG codes for methionine and is also the initiation codon. Thus methionine is the first amino acid in each newly synthesized polypeptide. This first amino acid is usually removed enzymatically before the polypeptide chain is completed; the vast majority of polypeptides do not begin with methionine.

23.8 Mutations and Genetic Diseases

Each step in the replication–transcription–translation process is subject to error. In replication alone, each time a human cell divides, a copy is made of 4 billion bases to make a new strand of DNA. There are perhaps one to two errors each time replication occurs. Most such errors are unimportant, but some have terrible consequences—genetic disease or even death may result.

We have seen that DNA directs the synthesis of proteins through the intermediary mRNA and that the sequence of bases in the DNA is critical and specific for the proper sequence of amino acids in proteins. On rare occasions, however, the base sequence in DNA may be modified either spontaneously (about 1 in 10 billion) or by exposure to heat, radiation, or certain chemicals. Any chemical or physical change that alters the sequence of bases in the DNA molecule is termed a **mutation.** The most common types of mutations are *substitution* (a different base is substituted), *insertion* (addition of a new base), and *deletion* (loss of a base). These changes within the DNA are called **point mutations** because the change occurs at a single nucleotide position (Figure 23.21).

The chemical and/or physical agents that cause mutations are termed **mutagens.** Examples of physical mutagens are ultraviolet and gamma radiation. They exert their mutagenic effects either directly or via free radicals induced by the

Table 23.3 The Genetic Code

First Base	Second Base: U		Second Base: C		Second Base: A		Second Base: G		Third Base
U	UUU	Phe	UCU	Ser	UAU	Tyr	UGU	Cys	U
	UUC	Phe	UCC	Ser	UAC	Tyr	UGC	Cys	C
	UUA	Leu	UCA	Ser	UAA	Termination	UGA	Termination	A
	UUG	Leu	UCG	Ser	UAG	Termination	UGG	Trp	G
C	CUU	Leu	CCU	Pro	CAU	His	CGU	Arg	U
	CUC	Leu	CCC	Pro	CAC	His	CGC	Arg	C
	CUA	Leu	CCA	Pro	CAA	Gln	CGA	Arg	A
	CUG	Leu	CCG	Pro	CAG	Gln	CGG	Arg	G
A	AUU	Ile	ACU	Thr	AAU	Asn	AGU	Ser	U
	AUC	Ile	ACC	Thr	AAC	Asn	AGC	Ser	C
	AUA	Ile	ACA	Thr	AAA	Lys	AGA	Arg	A
	AUG	Met	ACG	Thr	AAG	Lys	AGG	Arg	G
G	GUU	Val	GCU	Ala	GAU	Asp	GGU	Gly	U
	GUC	Val	GCC	Ala	GAC	Asp	GGC	Gly	C
	GUA	Val	GCA	Ala	GAA	Glu	GGA	Gly	A
	GUG	Val	GCG	Ala	GAG	Glu	GGG	Gly	G

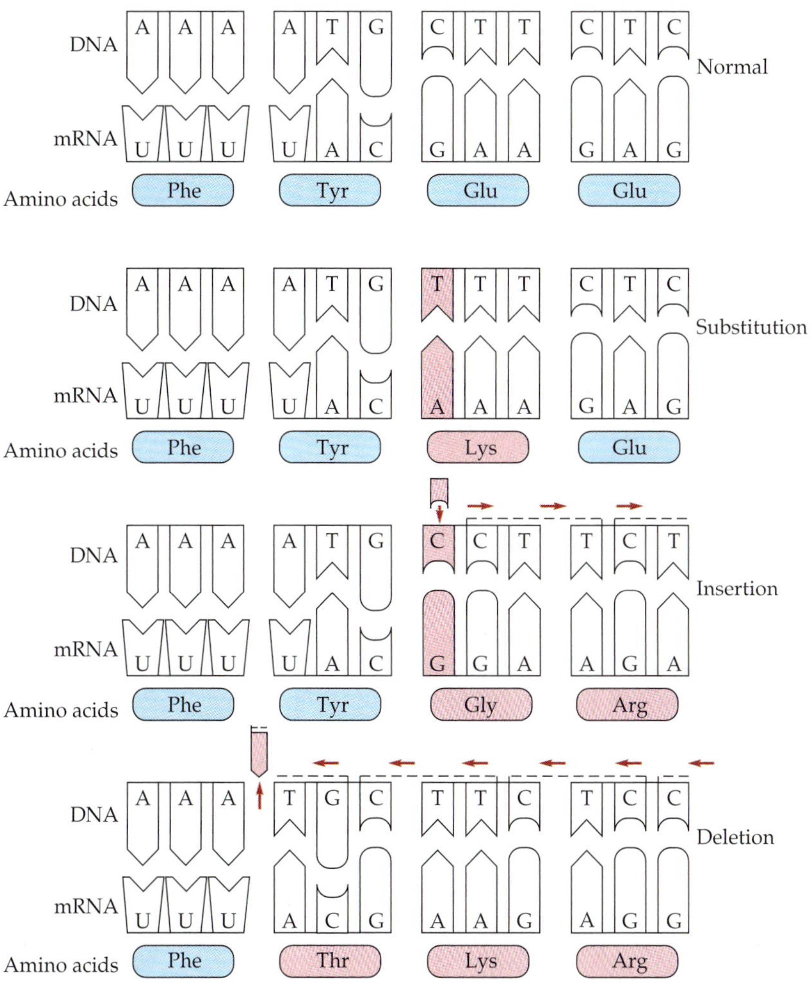

◀ **FIGURE 23.21**
Three types of point mutations.

radiation. Radiation and free radicals are known to cause covalent modification (often cross-linkage) of bases already incorporated into DNA. For example, upon exposure to UV light, two adjacent thymines on a DNA strand can become covalently linked, producing a thymine dimer (Figure 23.22). If not repaired, the dimer prevents formation of the double helix at the point at which it occurs. The genetic disease xeroderma pigmentosum is caused by a defective mechanism for the repair of pyrimidine dimers in DNA. (The enzyme that cuts out the damaged thymine dimers is not synthesized by the cells.) Individuals affected by this condition are abnormally sensitive to light and are more prone to skin cancer than normal individuals. An abnormal DNA is produced which apparently has no stop signal and results in the proliferation of cancer cells (see Special Topic K).

Among the chemical mutagens are two base analogs, 5-bromouracil and 2-aminopurine. They can be incorporated into the new DNA strand, but they exhibit faulty base pairing. 5-Bromouracil is incorporated into DNA in place of thymine, but it can base-pair with guanine (instead of adenine). 2-Aminopurine substitutes for adenine, yet it sometimes base-pairs with cytosine (instead of thymine). Hydroxylamine and nitrous acid are other chemical mutagens. Hydroxylamine (NH_2OH) deaminates cytosine, yielding a product that pairs with adenine instead of guanine. Nitrous acid (HNO_2) can convert cytosine to uracil, which can also

(a) (b)

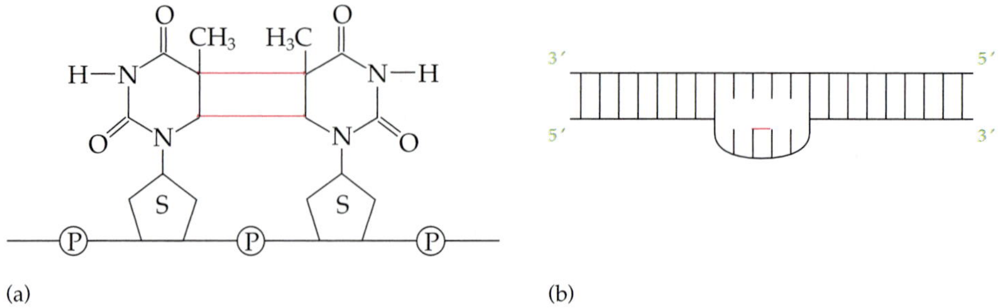

▲ **FIGURE 23.22**

(a) Structure of a DNA intrastrand thymine dimer. (b) Defect in the double strand produced by the thymine dimer. The dimerization is caused by ultraviolet light. This temporarily stops DNA replication, but the dimer can be excised by an enzyme and the strand can be repaired.

bond to adenine instead of guanine. That these compounds are both carcinogenic and mutagenic strongly indicates that disruption of DNA is fundamental to both processes.

Many mutations are perpetuated by replication and are thus inherited. Although it is possible for a gene mutation to be beneficial (thus permitting evolutionary advances through natural selection), most mutations are detrimental. If a point mutation occurs at a crucial position, the defective protein will lack biological activity and may result in the death of the cell. In such cases the altered DNA sequence is lost and will not be copied into daughter cells. Nonlethal mutations often lead to metabolic abnormalities or to hereditary diseases. Such diseases are called *inborn errors of metabolism* or **genetic diseases.** A partial listing of genetic diseases is presented in Table 23.4, and a few specific conditions are discussed here. In most cases the defective gene results in the failure to synthesize a particular enzyme.

Worldwide, about half a million babies are born with genetic defects each year.

Phenylketonuria (PKU) results when the enzyme phenylalanine hydroxylase is absent. A person with PKU cannot convert phenylalanine to tyrosine, which is the precursor of the neurotransmitters dopamine and norepinephrine as well as the skin pigment melanin (see Figure G.5).

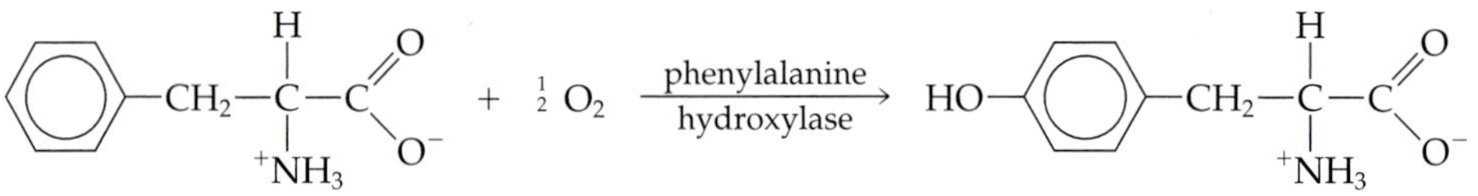

Phenylalanine $+ \frac{1}{2} O_2$ $\xrightarrow{\text{phenylalanine hydroxylase}}$ Tyrosine

In the absence of this step, phenylalanine accumulates, and the transamination (see Section 27.5) of phenylalanine to phenylpyruvate, normally a very minor process, becomes important.

Phenylalanine $\xrightarrow{\text{transaminase}}$ Phenylpyruvate

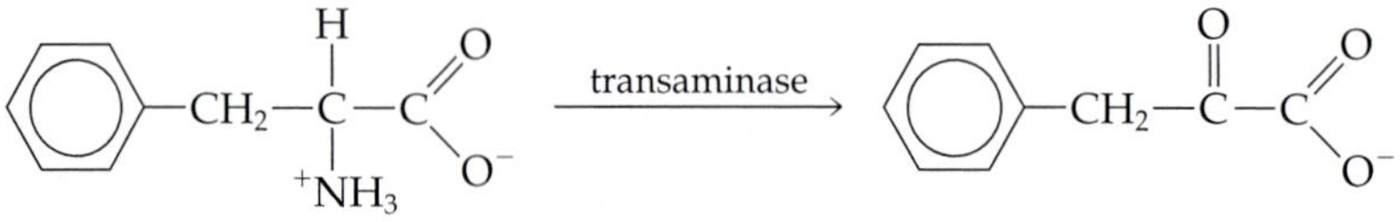

Table 23.4 A Partial Listing of Genetic Diseases in Humans and the Malfunctional or Deficient Protein or Enzyme

Disease	Responsible Protein or Enzyme
Acatalasia	Catalase (red blood cells)
Albinism	Tyrosinase
Alkaptonuria	Homogentistic acid oxidase
Cystathioninuria	Cystathionase
Fabry's disease	α-Galactosidase
Galactosemia	Galactose 1-phosphate uridyl transferase
Gaucher's disease	Glucocerebrosidase
Glycogen storage disease	Various types:
	α-Amylase
	Debranching enzyme
	Glucose 1-phosphatase
	Liver phosphorylase
	Muscle phosphofructokinase
	Muscle phosphorylase
Goiter	Iodotyrosine dehalogenase
Gout and Lesch–Nyhan syndrome	Hypoxanthine–guanine phosphoribosyl transferase
Hemolytic anemias	Various types:
	Glucose 6-phosphate dehydrogenase
	Glutathione reductase
	Phosphoglucose isomerase
	Pyruvate kinase
	Triose phosphate isomerase
Hemophilia	Antihemophilic factor (factor VIII)
Histidinemia	Histidase
Homocystinuria	Cystathionine synthetase
Hyperammonemia	Ornithine transcarbamylase
Hypophosphatasia	Alkaline phosphatase
Isovaleric acidemia	Isovaleryl-CoA dehydrogenase
Maple syrup urine disease	α-Keto acid decarboxylase
McArdle's syndrome	Muscle phosphorylase
Metachromatic leukodystrophy	Sphingolipid sulfatase
Methemoglobinemia	NADPH–methemoglobin reductase and NADH–methemoglobin reductase
Niemann–Pick disease	Sphingomyelinase
Phenylketonuria	Phenylalanine hydroxylase
Pulmonary emphysema	α-Globulin of blood
Sickle cell anemia	Hemoglobin
Tay–Sachs disease	Hexosaminidase A
Tyrosinemia	Hydroxyphenylpyruvate oxidase
Von Gierke's disease	Glucose 6-phosphatase
Wilson's disease	Ceruloplasmin (blood protein)

Excessive amounts of phenylpyruvate impair normal brain development, causing severe mental retardation (a mean IQ of 20)—about 1% of patients in mental institutions are phenylketonuric. The life span of untreated PKU individuals is significantly shorter than normal (75% are dead before age 30). The disease acquired its name from the high levels of this phenyl ketone in the urine. PKU may be diagnosed by assaying a sample of blood or urine for phenylalanine or one of its metabolites, and federal law requires that all newborns be tested (usually within the first two weeks). If the condition is detected, mental retardation can be prevented by giving the afflicted infant a diet containing little or no phenylalanine. Because phenylalanine is so prevalent in natural foods, the low-phenylalanine diet is composed of a synthetic protein substitute plus very small measured amounts of natural foods. The diet is maintained until the child is at least 3 years old, by which time brain development is completed. The incidence of PKU in newborns is about 1 in 15,000.

Another inborn error in the metabolism of tyrosine leads to *albinism*. Tyrosine serves as a precursor for the melanins, the pigments that color the skin, hair, and eyes. The absence of the enzyme tyrosinase prevents the occurrence of one of the reactions necessary for this conversion. The lack of pigmentation characteristic of albinism is the result.

Galactosemia results from a lack of the enzyme that catalyzes the formation of glucose from galactose. The blood galactose level is markedly elevated, and galactose is found in the urine. The baby experiences a lack of appetite, weight loss, diarrhea, and jaundice. The disease may result in impaired liver function, cataracts, mental retardation, and even death. If recognized in early infancy, the effects of galactosemia can be eliminated by removing milk and all other sources of galactose from the diet. As the children grow older, they normally develop an alternate pathway for metabolizing galactose, and thus the need to restrict milk is not permanent. The incidence of galactosemia in the United States is 1 in every 65,000 newborn babies.

There are several genetic diseases that are collectively categorized as *lipid-storage diseases*. As we shall learn in Chapter 26, lipids are constantly being synthesized and broken down in the body. If the enzymes that catalyze lipid decomposition are missing, the lipids tend to accumulate and cause a variety of medical problems. The enzymes that are responsible for lipid-storage diseases are known. The juncture at which the metabolic pathways go awry can be pinpointed. Unfortunately, however, no cure for these diseases has yet been developed. At present, genetic counseling of prospective parents who carry the defective gene is the only approach to control of the diseases.

In *Niemann–Pick disease*, a disease of infancy or early childhood, sphingomyelins accumulate in the brain, liver, and spleen because the enzyme sphingomyelinase is lacking. The accumulation of the sphingomyelins causes mental retardation and early death.

In *Gaucher's disease*, cerebrosides accumulate in the brain and cause severe mental retardation and death. Juvenile and adult forms of this disease are characterized by enlarged spleen and kidneys, hemorrhaging, mild anemia, and fragile bones. This disease is caused by the lack of a specific enzyme called glucocerebrosidase, which cleaves glucocerebrosides into glucose and sphingosine.

In the absence of the enzyme hexosaminidase A, gangliosides accumulate in brain tissue. The ganglion cells of the brain become greatly enlarged and nonfunctional. This effect, called *Tay–Sachs disease*, results in retardation of development, dementia, paralysis, and blindness. Death usually occurs before the age of 3. Tay–Sachs disease can be diagnosed by assaying the amniotic fluid (amniocentesis) for the enzyme. The absence of hexosaminidase A in the amniotic fluid allows for a recommendation for a therapeutic abortion because the disease is incurable. Genetic screening can identify Tay–Sachs carriers because they produce only half the normal amount of hexosaminidase A (although they do not exhibit symptoms of the disease). Tay–Sachs is most common in persons of Eastern European Jewish ancestry.

23.9 Genetic Engineering: Biotechnology

More than 3000 human diseases have a genetic component. Over the last decade or so, researchers have linked specific genes to specific diseases. Now the ability to use this information to diagnose and cure genetic diseases appears to be within our grasp. By determining the location of a gene on the DNA molecule, scientists have been able to identify and isolate genes having specific functions.

A gene is an elusive substance. There are approximately 10,000 genes on each human chromosome, and isolating the one defective gene that causes a particular genetic disease is a monumental task. One approach to gene location is to treat the DNA with enzymes called *restriction endonucleases.* This method yields segments of genetic material called *restriction fragment length polymorphisms* (RFLPs; pronounced "rif lips"). These segments are much easier to work with because they contain fewer genes. The RFLP pattern obtained upon enzyme treatment is an inherited one. RFLP patterns characteristic of certain families can be isolated. If the pattern of a relative matches that of a person with a genetic disease, that relative will probably develop the disease. Thus it is possible to identify and even predict the occurrence of a genetic disease.

A further hope of genetic engineering is that we will be able to introduce a functioning gene into a person's cells, thus correcting the action of a defective gene.

All living organisms (except some viruses) have DNA as their hereditary material. It should be possible, then, to place a gene from one organism into the genetic material of another. **Recombinant DNA technology** does just that.

By working backward from the amino acid sequence of the protein, scientists can work out the base sequence of the gene that codes for the protein. When isolated by the RFLP process, the gene is spliced into a special kind of bacterial DNA called a **plasmid.** The recombined plasmid is then inserted into the host organism (usually the bacterium *E. coli*). Figure 23.23 illustrates the production of recombinant DNA. The steps are as follows:

1. *Escherichia coli* bacteria are placed in a detergent solution to break open the cells.
2. The plasmids are separated from the chromosomal DNA by differential centrifugation.
3. Restriction enzymes (endonucleases) are used to cleave the plasmid at a specific short sequence in a way that creates overlapping, cohesive ("sticky") ends. Each restriction enzyme can recognize a specific sequence of four to six nucleotides in DNA. For example, the endonuclease designated as *Eco*RI cuts at the sequence GAATTC, and *Sma*I cuts at CCCGGG. More than 100 restriction enzymes, purified from different species of bacteria, are now commercially available.
4. *In vitro* combination of the same restriction enzyme with DNA from another organism (foreign DNA) or with synthetic DNA produces segments of DNA with cohesive ends that are complementary to those of the plasmid. [Because different restriction enzymes have different cleavage sites, a given strand of DNA can be separated into many different segments of varying lengths. It is therefore possible to insert almost any foreign gene(s) into *E. coli*.]

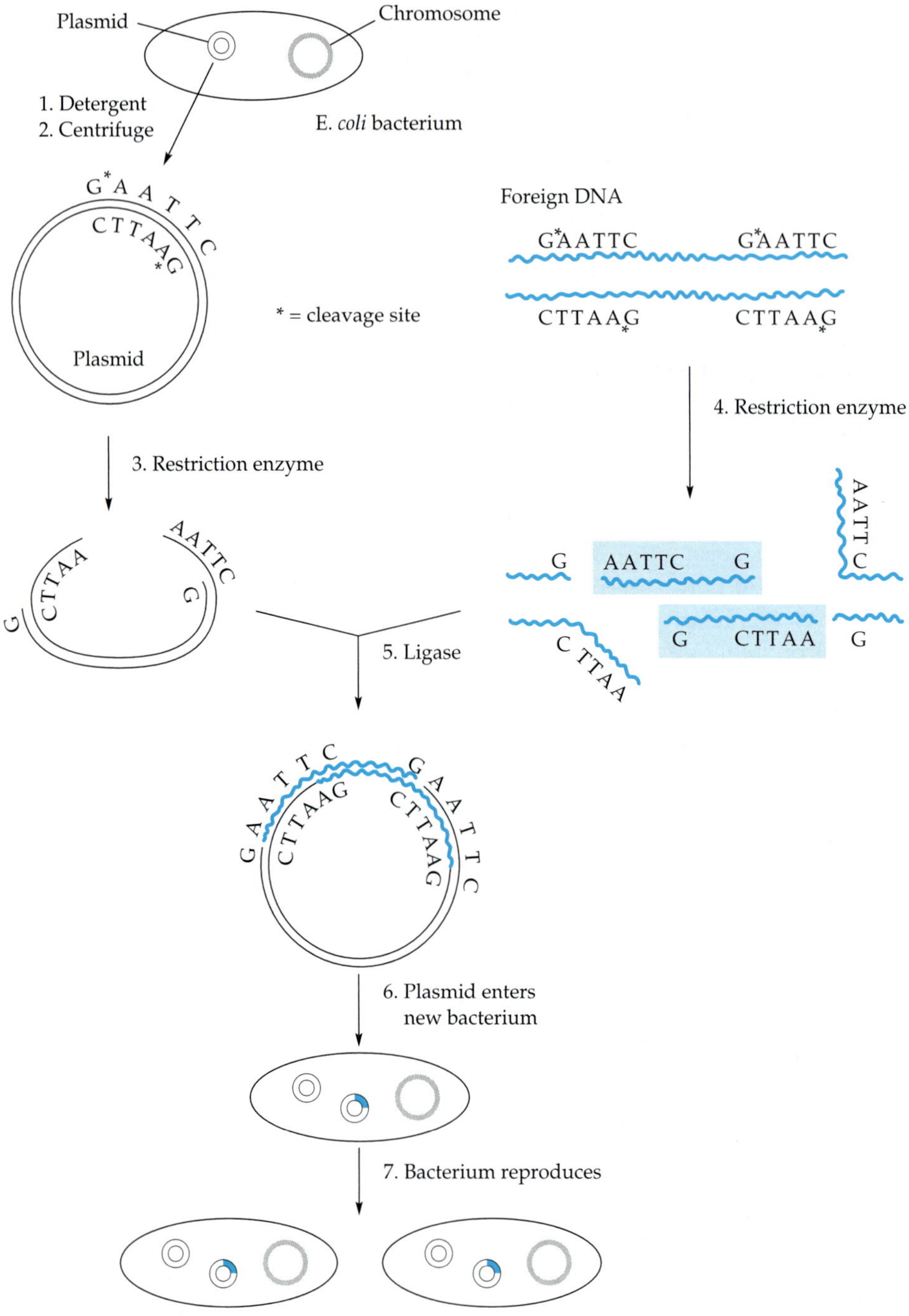

FIGURE 23.23 ▶ Recombinant DNA in *E. coli*.

5. The enzyme DNA ligase seals the foreign DNA segment into place in the plasmid.
6. The resealed plasmid is placed in a solution of calcium chloride containing *E. coli*. When the solution is heated, the bacterium cell membrane becomes permeable, allowing the plasmid to enter.
7. *Escherichia coli* reproduces by dividing (and thus doubling its population) at a rate of about once every 20 to 30 min. The new *E. coli* bacteria have characteristics dictated by their own genes as well as characteristics dictated by the genes transplanted from a different species.

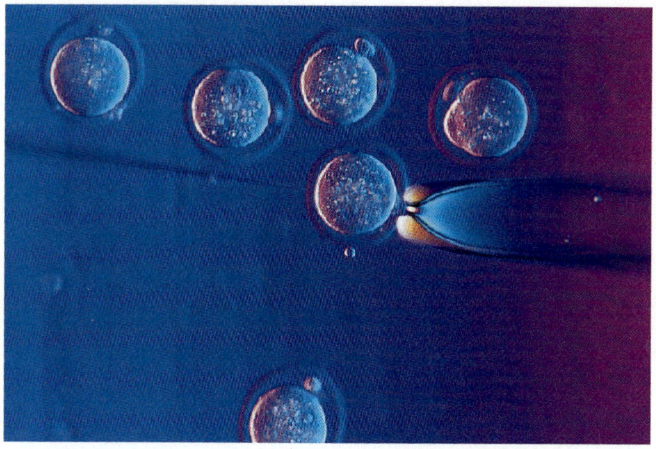

Under a powerful microscope, a tiny pipet is used to introduce foreign genes into animal cells.

The modified plasmids do all the things DNA does. They replicate, and they control the synthesis of proteins. Microorganisms multiply rapidly. Soon vast vats of these modified creatures are producing a protein specified by a "foreign" gene.

Many valuable materials, difficult to obtain in any other way, are now made using recombinant DNA technology. People with diabetes formerly had to use insulin from pigs or cattle. Now human insulin, a protein coded by human DNA, is being produced by the cell machinery of bacteria. All newly diagnosed insulin-dependent diabetics in the United States are now treated with human insulin produced by recombinant DNA technology. The hope for the future is that a functioning gene for insulin can be incorporated into the cells of insulin-dependent diabetics.

Human growth hormone, used to treat children who fail to grow properly, was formerly available only in tiny amounts obtained from cadavers. Now it is readily available through recombinant DNA technology. This technology also yields interferon, a promising anticancer agent. The gene for epidermal growth factor, which stimulates the growth of skin cells, has been cloned. It has been used to speed the healing of burns and other skin wounds. Scientists have even designed bacteria that "eat" the oil released in an oil spill, although success in actual spills has been minimal.

Proponents of recombinant DNA research are excited about its great potential benefits. Recombinant techniques are an enormous aid to scientists in mapping and sequencing genes and in determining the functions of different segments of an organism's DNA. The complete DNA sequences of more than 100 mammalian genes have been determined using recombinant DNA technology. An understanding of gene function and gene regulation is a primary goal of scientists working to cure cancer.

Human gene therapy has begun. Skin cancer patients have received injections of genetically altered white blood cells. The cells circulate in the blood seeking out malignant tissue. When one of these cells contacts a cancer cell, its foreign gene produces a toxic enzyme called tumor necrosis factor. Cancer cells get lethal doses, and the rest of the body is spared.

It is conceivable that recombinant DNA could lead to cures for genetic diseases. When appropriate genes are successfully inserted into *E. coli*, the bacteria can become miniature pharmaceutical factories, producing great quantities of insulin, clotting factor for hemophiliacs, missing enzymes, hormones, vitamins,

Many viruses readily penetrate human cells, where they use the cell machinery to replicate themselves (see Section K.1). A human gene can be cloned into a modified virus that will then transport that gene into the human cells, where it produces multiple copies of the human gene. The first human gene transplant was carried out in 1989. It was a demonstration project that proved the technique works, although it was not used to treat disease. The first successful human gene therapy was accomplished in 1990. Modified human white blood cells were infused into a young girl with a severe immune deficiency.

Circular plasmids isolated from *E. coli.* Genetically engineered plasmids are used as vectors for cloning DNA segments.

antibodies, vaccines, and so on. The production of DNA-recombinant molecules containing synthetic genes for the production of tissue plasminogen activator (TPA—a clot-dissolving enzyme that can rescue heart attack victims) has been accomplished in *E. coli.* Vaccines against hepatitis B (humans) and hoof and mouth disease (cattle) have been produced. In addition, it may be possible to breed human intestinal bacteria that can digest cellulose and to create new plants that can obtain their nitrogen directly from the air rather than from costly petroleum-based fertilizers.

Besides *E. coli,* scientists have used other bacteria as well as yeast and fungi in gene-splicing experiments. The enzyme rennin has been produced in fungi. Rennin is used commercially to coagulate milk into curds in the production of cheese. A bacterial plasmid has been used as a vector by plant molecular biologists. The bacterium is *Agrobacterium tumefaciens,* which can cause tumors in many plants. Scientists have succeeded in introducing genes for several foreign proteins (including animal protein) into plants by means of *A. tumefaciens,* at the same time eliminating its tumor-causing ability. One practical application would be to transfer the gene necessary for the synthesis of a deficient amino acid into a particular plant (e.g., the gene for methionine synthesis into soybeans).

A vector is any sort of delivery vehicle (e.g., plasmid, virus) for inserting genetic information into cells.

Concern over the potential for disaster in this type of research has lessened somewhat in recent years. Initially, scientists worried about the possibility of producing a deadly "artificial" organism. What if a gene that causes cancer were spliced into the DNA of a bacterium that normally inhabits our intestines? We would have no natural immunity against such an artificial organism. To protect against such a development, strict guidelines for recombinant DNA research have been instituted.

One of the products of recombinant DNA technology is a growth hormone for cows, bovine somatotropin (BST). When injected into cows, this hormone increases their milk production by 10 to 20%. The Food and Drug Administration, after 10 years of study, gave approval for the use of BST starting February 1994. Many milk producers are very positive about using BST, but some consumers are unhappy at the thought of having someone tamper with their milk. There is much controversy as to whether or not a hormone such as this should be used.

Another point of contention is that this research can be misused for political and social purposes. These techniques might be exploited for genetically engineered control of human behavior, even enhancing people's IQs. There are some people who believe that we should not meddle with evolution by creating new forms of life different from any that exist on Earth.

The new molecular genetics has already resulted in some impressive achievements. Its possibilities are mind-boggling—elimination of the genetic defects, a cure for cancer, a race of geniuses, and who knows what else? Knowledge gives power. It does not necessarily give wisdom. Who will decide what sorts of creatures the human species should be? The greatest problem we are likely to face in our use of bioengineering is that of choosing who is to play God with the new "secret of life."

Summary

The two types of *nitrogenous bases* most important in nucleic acids are **purine bases**—*adenine* and *guanine*—and **pyrimidine bases**—*cytosine, thymine,* and *uracil.*

One nitrogenous base combined with a five-carbon sugar forms a **nucleoside**—a ribonucleoside if the sugar is ribose and a deoxyribonucleoside if the sugar is 2-deoxyribose.

The combination of a nucleoside and a phosphoric acid group is a phosphate ester called a **nucleotide.** Polymer chains of nucleotide units are nucleic acids.

The nucleic acids in protein synthesis are **deoxyribonucleic acid (DNA)** and **ribonucleic acid (RNA).** DNA contains the nitrogenous bases adenine, cytosine, guanine, and thymine. The bases in RNA are adenine, cytosine, guanine, and uracil.

The sequence of bases in any nucleic acid defines the primary structure of the molecule. RNA is a single-chain nucleic acid. DNA comprises two nucleic-acid chains intertwined in a secondary structure called a double helix. The sugar-phosphate backbone forms the outside of the double helix, with the purine and pyrimidine bases inside. The pairing of bases, one from each chain, is always A–T and C–G. Adenine and thymine are one pair of **complementary bases,** and cytosine and guanine are another pair.

A cell's hereditary information is encoded in **chromosomes** in the cell nucleus. Each chromosome comprises many smaller hereditary units called **genes,** relatively short sections of DNA.

The transmission of genetic information from parent cell to offspring cell involves the **self-replication** of the parent cell's DNA. The double helix unwinds, and the hydrogen bonds joining complementary bases break so that there are two single strands of DNA, each a *template* for the new strand about to be synthesized.

For protein synthesis, three types of RNA are needed: *messenger RNA, ribosomal RNA,* and *transfer RNA.* All are made from DNA in the nucleus by a process is called **transcription.** The double helix uncoils just as in self-replication, and nucleotides base-pair to one DNA strand. Three of the pairings are as in DNA self-replication (C–G, G–C, T–A), but now, in RNA transcription, any template adenine calls for *uracil* on the RNA molecule (A–U). Once the RNA is formed, it breaks from the template and leaves the nucleus, and the DNA double helix reforms.

The **ribosome** is the site of protein synthesis. It is located outside the nucleus and is made of rRNA and protein. Clusters of ribosomes banded together by a strand of mRNA are *polysomes.*

Protein synthesis depends on **triplet codes,** the 64 possible ways the four nucleotides of DNA can combine in three-base sequences. Each three-base sequence on mRNA is a **codon.** The steps of protein synthesis are:

1. A mRNA strand moves from the nucleus to the cytoplasm and attaches to a ribosome. The nitrogenous bases on the mRNA are facing the ribosome interior, ready to be read by the ribosome.
2. A tRNA molecule, whose **anticodon** three-base sequence complements the mRNA codon now at the synthesis site on the ribosome, enters the ribosome. The anticodon part of this tRNA base-pairs with the mRNA *codon,* and the amino acid at the other end of the tRNA is attached to the protein chain being synthesized.
3. The bond holding the amino acid to the tRNA breaks; the anticodon part of the tRNA detaches from the mRNA codon; and the tRNA leaves the ribosome.
4. The ribosome moves along the mRNA strand to the next codon, and the process of adding another amino acid to the protein chain is repeated.

The sequence of nucleotides in an organism's DNA is the *genetic code* for that organism. The general term for any change in the code is **mutation,** and each specific change is a **point mutation.**

In **recombinant DNA technology,** a gene from one organism is inserted—"spliced"—into the DNA of a host organism. The host then reproduces large quantities of the altered DNA containing the foreign gene, and this DNA is harvested by humans and used in medical treatments and other applications.

Key Terms

anticodon (23.5)
chromosome (23.4)
codon (23.6)
complementary base (23.3)
deoxyribonucleic acid (DNA) (23.1)
double helix (23.3)
gene (23.4)
genetic disease (23.8)
multiplicity (23.5)
mutagen (23.8)
mutation (23.8)
nucleoside (23.1))
nucleotide (23.1)
plasmid (23.9)
point mutation (23.8)
purine base (23.1)
pyrimidine base (23.1)
recombinant DNA technology (23.9)
ribonucleic acid (RNA) (23.1)
ribosome (23.5)
self-replication (23.4)
transcription (23.5)
translation (23.6)
triplet code (23.6)

Review Questions

1. Explain what is meant by:
a. ribosome
b. template replication
c. multiplicity
d. complementary bases
e. genetic code
f. translocation
g. code degeneracy
h. mutagens
i. genetic disease
j. PKU
k. amniocentesis
l. recombinant DNA
m. plasmid
n. restriction enzyme

2. Distinguish between terms in each of the following pairs:
a. purine base and pyrimidine base
b. major base and minor base
c. ribose and deoxyribose
d. nucleoside and nucleotide
e. nucleotide and nucleic acid
f. DNA polymerase and DNA ligase
g. codon and anticodon
h. chromosomes and genes
i. transcription and translation
j. mutation and point mutation

3. Name the two kinds of nucleic acids. Which is concentrated in the nucleus of the cell?

4. How do DNA and mRNA differ in secondary structure?

5. Give the names of all the compounds that can be obtained from the complete hydrolysis of (**a**) DNA and (**b**) RNA.

6. The possibilities of hydrogen-bond formation are similar for uracil and thymine. Explain.

7. DNA and RNA are termed nucleic *acids*. What makes them acidic?

8. What constitutes the backbone of a DNA chain?

9. What are the differences among AMP, ADP, and ATP?

10. The primary structure of a protein is defined by the sequence of amino acids. What defines the primary structure of nucleic acids?

11. Why is it structurally important in the DNA double helix that a purine base always pair with a pyrimidine base?

12. What kind of intermolecular force is involved in base pairing?

13. Describe DNA replication.

14. Explain the role of mRNA in protein synthesis.

15. Explain the role of tRNA in protein synthesis.

16. Which nucleic acid(s) is (are) involved in transcription?

17. Which nucleic acid(s) is (are) involved in translation?

18. **a.** Which nucleic acid contains the codon?
b. Which nucleic acid contains the anticodon?

19. **a.** How many nucleotide units are present in a codon?
b. Why is it that a triplet of bases codes for each amino acid?

20. What is the relationship among chromosomes, genes, and DNA?

21. What is the basic process in recombinant DNA technology?

22. Discuss some applications of genetic engineering.

23. Give two examples of physical mutagens and four examples of chemical mutagens.

24. Name three genetic diseases, and indicate which enzyme is lacking in each.

Problems

Building Blocks

25. What is the sugar unit in RNA?

26. What is the sugar unit in DNA?

27. What are the major bases present in DNA?

28. What are the major bases present in RNA?

29. For each of the following, indicate whether the compound is a nucleoside, a nucleotide, or neither.

a.

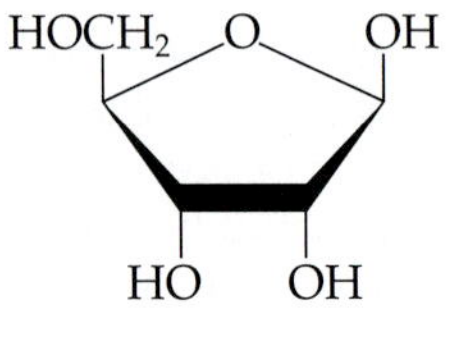

b. 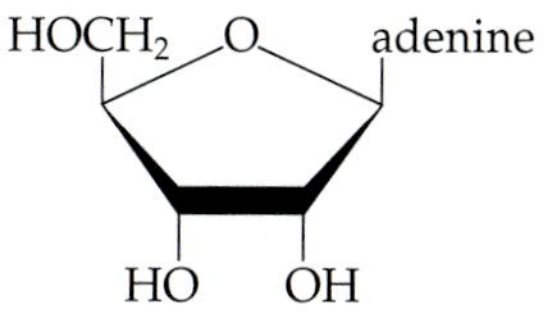

c. 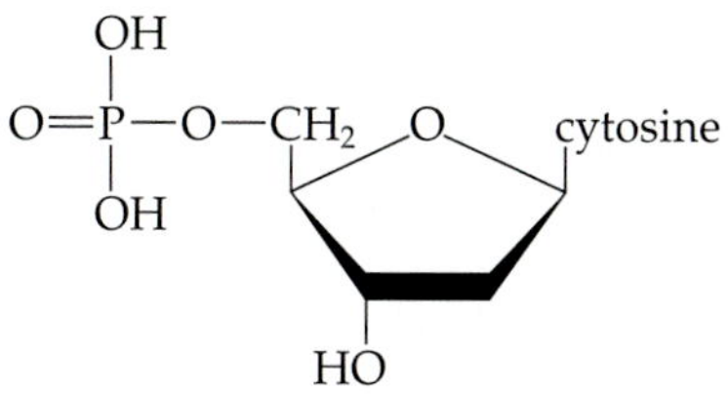

30. For each of the following, indicate whether the compound is a nucleoside, a nucleotide, or neither.

a. 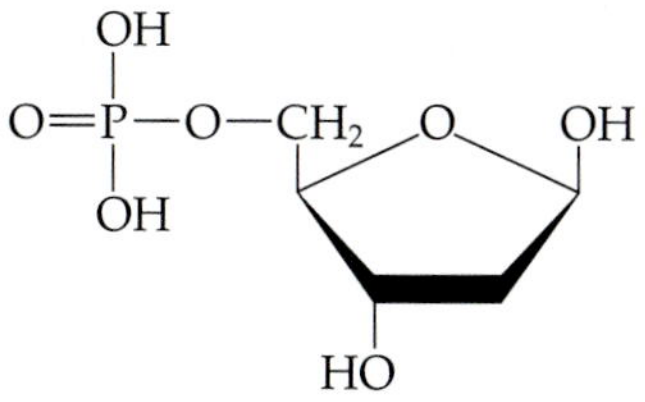

b. 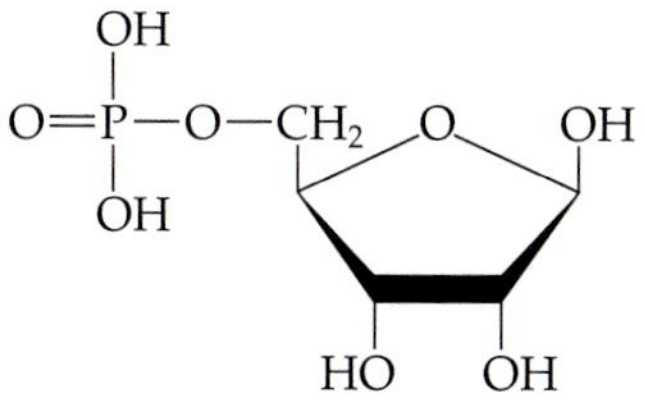

c. 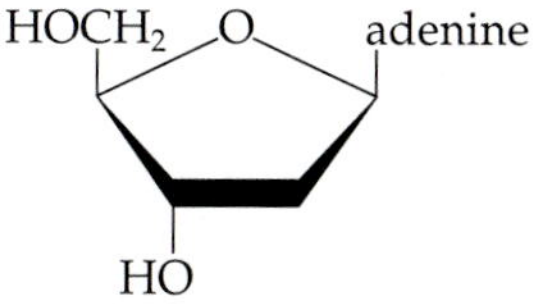

31. For each structure in Problem 29, indicate whether the sugar unit is ribose or deoxyribose.

32. For each structure in Problem 30, indicate whether the sugar unit is ribose or deoxyribose.

Structures and Names

33. Draw structural formulas for the following minor pyrimidine and purine bases.
a. 5-methylcytosine **b.** 5-hydroxymethylcytosine
c. 6-methyladenine **d.** 2-methyladenine

34. Draw structural formulas for the following minor pyrimidine and purine bases.
a. 6-oxypurine **b.** 4-thiouracil
c. 5,6-dihydrouracil **d.** 1-methylguanine

35. Using a schematic representation, show a length of nucleic acid polymer chain, indicating the positions of the sugar, phosphoric acid, and base units.

36. With the same sort of schematic representation you used for Problem 35, show the overall design of the double helix.

Base Pairing

37. In DNA, list the base that pairs with
a. cytosine **b.** adenine **c.** guanine **d.** thymine

38. In RNA, list the base that pairs with
a. adenine **b.** guanine **c.** uracil **d.** cytosine

39. The base sequence along one strand of DNA is ATTCG. What is the sequence of the complementary strand of DNA?

40. What sequence of bases appears in the mRNA molecule copied from the original DNA strand shown in Problem 39?

41. List the complementary triplets on tRNA for mRNA triplet
a. UUU **b.** CAU **c.** AGC **d.** CCG

42. List the complementary triplets on mRNA for tRNA triplet
a. UUG **b.** GAA **c.** UCC **d.** CAC

The Genetic Code

43. Using Table 23.3, identify the amino acids carried by the tRNA molecules in Problem 41.

44. Using Table 23.3, identify the amino acids carried by the tRNA molecules in Problem 42.

45. Refer to Table 23.3. List the amino acid sequence that results from the following base sequences on mRNA:
a. UUACCUCGA
b. GCGUCAUAA
c. CCCCCCCCC

46. If the DNA base sequence TTACTCTCA acted as a template for mRNA formation, what amino acid sequence would eventually be produced from the mRNA?

Additional Problems

47. In DNA replication, a parent DNA molecule produces two daughter molecules. What is the fate of each strand of the parent DNA double helix?

48. We say that DNA controls protein synthesis, and yet most DNA resides within the cell nucleus whereas protein synthesis occurs outside the nucleus. How does DNA exercise its control?

49. If the sequence of bases along an mRNA strand is UCCGAU, what was the sequence along the DNA template?

50. Show the replication of the DNA segment

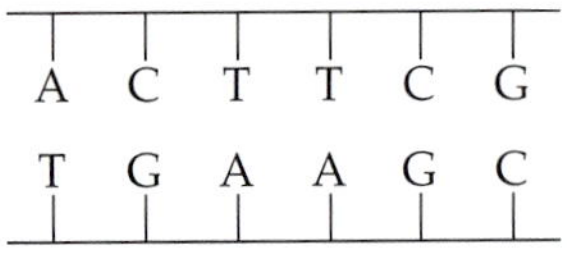

51. Write the RNA base sequence obtained upon transcription of the lower chain of the DNA segment

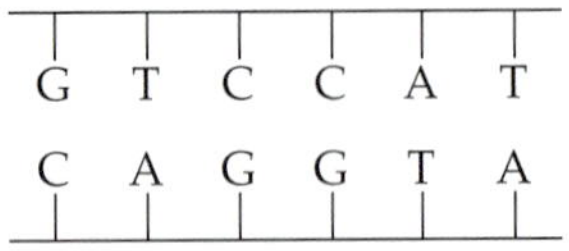

52. What are the two most important sites on tRNA molecules?

53. A hypothetical protein has a molecular weight of 60,000 daltons. Assume that the average molar mass of an amino acid is 120.

a. How many amino acids are present in the protein?

b. How many codons occur in the mRNA that codes for this protein?

c. How many nucleotide bases are found in the mRNA?

54. If the sequence of bases in a section of mRNA is AUGUACCACGGUACGCGGGUAUUGCUAGCC-GAUGGGUAA
what is the amino acid sequence in the polypeptide produced from this mRNA? (See Table 23.5.)

55. If the base sequence of a gene is TACGAATCTAGAATACTTCCAAAAGTATTTTGAT-ACATC
what is the amino acid sequence of the polypeptide synthesized?

56. The hormone somatostatin, produced in the pancreas, is composed of 14 amino acids. Somatostatin inhibits the release of a variety of hormones, including glucagon, insulin, and human growth hormone. It is believed to keep the pancreas's output of insulin and glucagon in a proper balance.

a. Given the following structure for somatostatin, postulate a base sequence in the mRNA that directs the synthesis of somatostatin. Include an initiation codon and a termination codon.

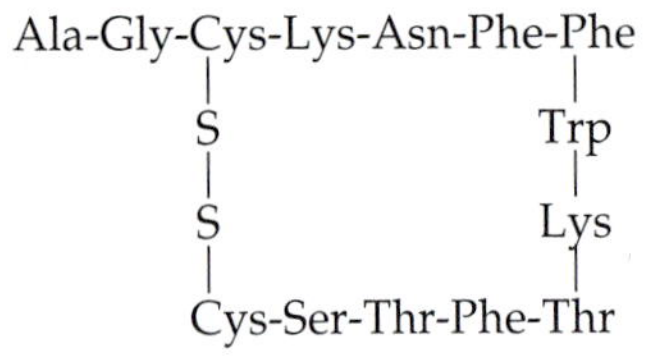

b. What is the base sequence of the DNA that codes for this mRNA?

57. Chemical bonds can be broken by ultraviolet radiation, gamma rays, certain chemical substances, and other means. What are the biological implications of these facts? (*Hint:* What effect might the breaking of chemical bonds have on DNA?)

58. Certain genes are implicated in cancer. Some of these *oncogenes* can be activated by single point mutations. Ordinarily, the DNA triplet GGT codes for the amino acid proline. What amino acid is formed when the DNA triplet is changed by mutation to (**a**) TGT, (**b**) GTT, and (**c**) CGT?

59. We shall see in Section 28.9 that sickle cell hemoglobin differs from normal hemoglobin as a result of the substitution of valine for glutamic acid as the sixth amino acid from the N-terminal end of the polypeptide chain.

a. What alteration in the base sequence of the DNA could have caused this substitution?

b. What is the resulting base sequence in the mRNA?

60. The following table contains just a few of the 200 or so known hemoglobin mutants. For each mutation, give the codon for the normal amino acid and that for the mutant amino acid.

Type	Chain	Position	Normal	Mutant
J	α	5	Ala	Asp
I	α	16	Lys	Glu
M	α	58	His	Tyr
D	β	121	Glu	Gln
K	β	136	Gly	Asp

61. For the DNA segment ACGTTAGCCCCAGCT, (**a**) Write the sequence of bases in the corresponding mRNA. (**b**) What is the amino acid sequence formed by translation? (**c**) What amino acid sequence results from (i) replacement of the red guanine by adenine, (ii) insertion of thymine immediately after the red guanine, and (iii) deletion of the red guanine?

62. Assume that a segment of a gene that codes for a particular enzyme has the base sequence TACGACG-TAACAAGC. (**a**) What effect results from a point mutation in which an adenine replaces the red guanine? (**b**) What effect results from a point mutation in which a thymine replaces the red adenine?

63. Following are the results of two point mutations. Which is likely to be more serious?

a. Valine is substituted for leucine.

b. Glutamic acid is substituted for leucine.

Special Topic K
Viruses and Cancer

A discussion of viruses seems particularly appropriate here because viruses are composed almost entirely of proteins and nucleic acids. The field of virology is a rapidly expanding one, and recent research efforts to establish connections between viruses and some cancers have yielded much important and exciting information.

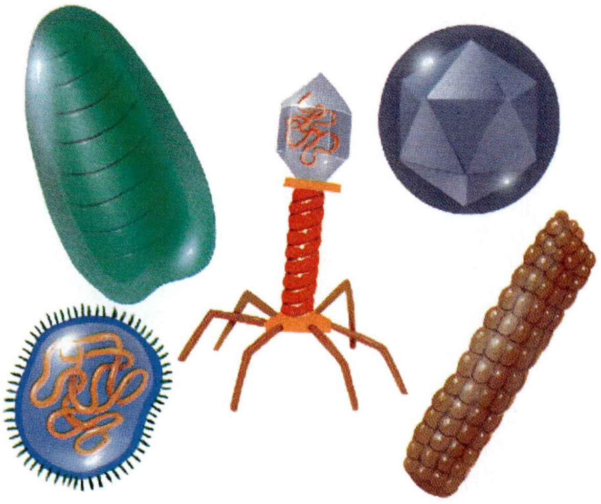

Viruses come in a variety of shapes that are determined by their protein coats. The rabies and herpes viruses have an extra coating derived from the membranes of the host cell.

K.1 The Nature of Viruses

Viruses are a unique group of infectious agents composed of a tightly packed central core of nucleic acids enclosed in one or more protein coats. They are divided into two main classes on the basis of nucleic acid content. Viruses contain either DNA or RNA, *but never both.* (Recall that the cells of organisms, from bacteria to humans, contain both kinds of nucleic acids.) Viruses differ from one another in both size[1] and shape. The influenza virus, for example, is about 10 times bigger than the polio virus. Viruses may be spherical, rod-shaped, or threadlike.

A *DNA virus* enters a host cell, where the viral DNA is replicated and directs the host cell to produce viral proteins. The viral proteins and viral DNA assemble into new viruses, which are released by the host cell. These new viruses can then invade other cells and continue the process. Figure K.1 depicts the life cycle of a *bacteriophage* (a virus that infects bacteria). Cell death and the production of new viruses account for the symptoms of viral infections. As Table K.1 indicates, a greater number of human diseases are of viral origin than of bacterial origin. Infectious diseases of viral origin (especially the common cold, influenza, and AIDS) are among the most significant health problems in our society.

Most *RNA viruses* use their nucleic acids in much the same way as do the DNA viruses. The virus penetrates a host cell, where the RNA strands are replicated and induce the synthesis of viral proteins. The new RNA strands and viral proteins are then assembled into new viruses. Some RNA viruses, called **retroviruses,** synthesize DNA in the host cell. This process is the opposite of the DNA-to-RNA transcription that normally occurs in cells. The synthesis of DNA from an RNA template is catalyzed

Table K.1 Human Infectious Diseases

Diseases of Bacterial Origin	Diseases of Viral Origin	
Cholera	AIDS	Measles
Diphtheria	Chicken pox	Meningitis
Dysentery	Cold sores	Mumps
Gonorrhea	Common cold	Pneumonia
Plague	Encephalitis	Polio
Syphilis	Gastroenteritis	Rabies
Tetanus	Genital herpes	Shingles
Tuberculosis	German measles	Smallpox
Typhoid fever	Hepatitis	Warts
Whooping cough	Influenza	Yellow fever

[1] Viruses are exceedingly small, much smaller than bacteria. The tobacco mosaic virus, for example, is approximately 300 nm long by 18 nm wide (300×10^{-9} m by 18 $\times 10^{-9}$ m) and has a molecular weight of about 40 million daltons. In general, animal viruses are larger than bacterial viruses. Most viruses are visible only under the electron microscope. (For comparison, the size of an average bacterial cell is 1500 nm long by 750 nm wide.)

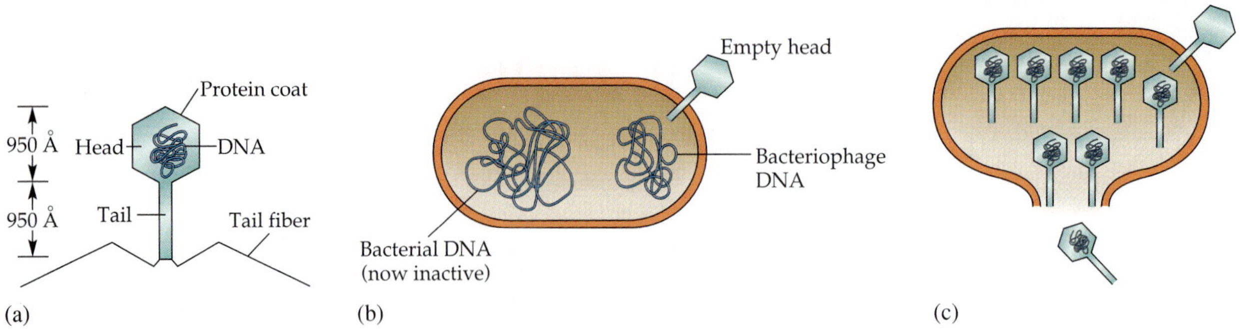

▲ **FIGURE K.1**

Life cycle of a bacteriophage. (a) Drawing of T2 bacteriophage. The bacteriophage DNA is located in the head of the particle, surrounded and protected by the protein coat. The end of the tail contains lysozyme, an enzyme that catalyzes the hydrolysis of the polysaccharide of the bacterial cell wall. (b) When the bacteriophage infects a bacterium, the phage DNA is injected into the bacterial cell. Synthesis of bacterial nucleic acids and proteins stops, and the cell machinery begins to produce phage DNA and protein. (c) As soon as all the phage constituents are synthesized, they begin to form new phage particles. The cell is filled with phage particles, and the bacterial cell wall is ruptured, releasing the phage progeny. As many as 150 to 300 progeny viruses are produced in 30 to 60 minutes from the infection of one bacterial cell by a single virus.

by the enzyme reverse transcriptase. The human immunodeficiency virus (HIV) that causes AIDS is perhaps the best-known retrovirus. The HIV invades and eventually destroys T cells, a group of white blood cells that normally help protect the body from infections. With the T cells destroyed, the AIDS victim succumbs to pneumonia or other infectious diseases.

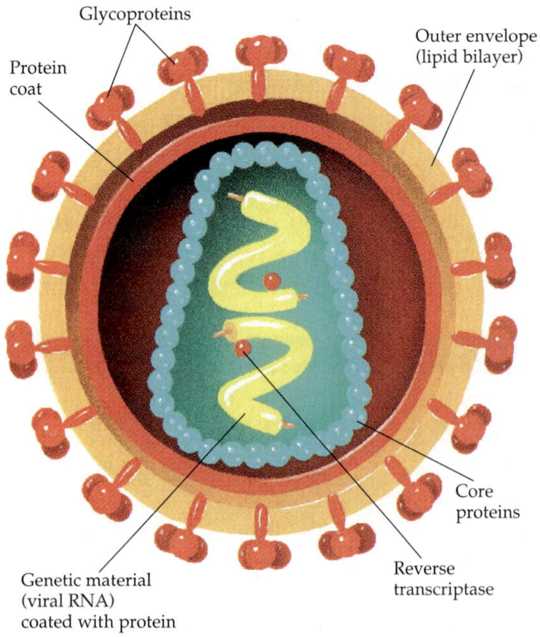

Cross section of the AIDS virus. The RNA on the inside is surrounded by a protein coat. The outer envelope is formed from the membrane of the host cell. Spikes of protein and carbohydrate project from the envelope and are used by the virus to attach to the host cell.

K.2 Antiviral Drugs

Scientists had to learn the normal biochemistry of cells before they could develop drugs that treat abnormal conditions caused by the invasion of viruses. Gertrude Elion and George Hitchings of Burroughs Wellcome Research Laboratories in North Carolina and James Black of Kings College in London did the basic biochemistry that led to the development of antiviral drugs and many of the anticancer drugs (Section K.6) and shared the 1988 Nobel prize in physiology or medicine. They determined the shapes of cell membrane receptors, and they learned how normal cells work. Then they and other scientists were able to design drugs to block receptors in infected cells. Today scientists use powerful computers to design molecules to fit receptors. Drug design often was hit or miss in its early decades. It is now becoming a more precise science.

Scientists have developed drugs that are effective against some viruses (Figure K.2). Amantadine helps prevent some influenza A infections. Acyclovir (Zovirax) controls flareups of the herpes viruses that cause genital sores, chicken pox, shingles, mononucleosis, and cold sores. Azidothymidine (Zidovudine—AZT) slows the onslaught of the AIDS virus, but the disease is still relentlessly fatal to those infected by the virus. AZT apparently acts by substituting for thymine. The reverse transcriptase enzyme incorporates AZT into the DNA chain, blocking the synthesis of the DNA. Unfortunately, the toxicity of AZT prevents its use in quantities

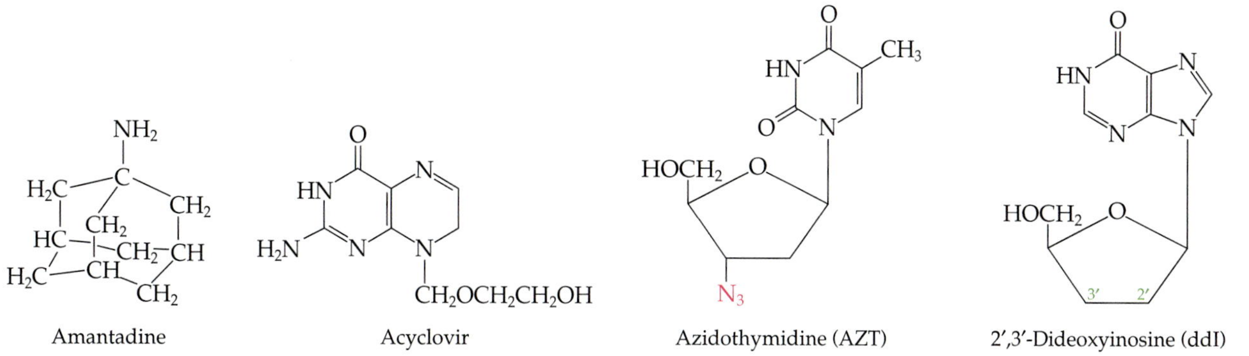

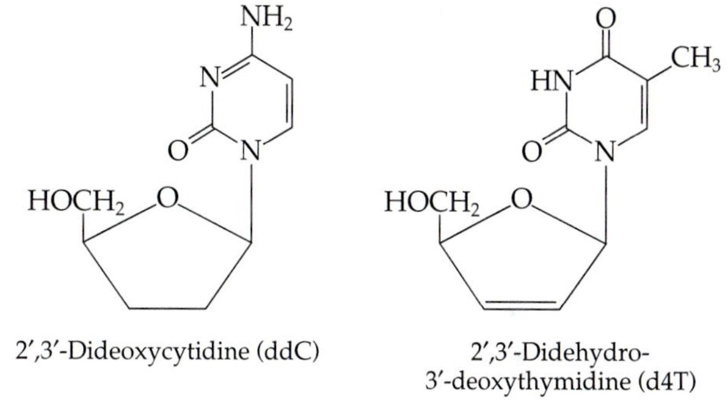

▲ **FIGURE K.2**

Some antiviral drugs.

sufficient to stop HIV replication completely. Other drugs similar to AZT are undergoing testing.

In subsequent years (1991–1994), three other nucleoside analogs, ddI, ddC, and d4T, were approved to treat AIDS. Each of them, in turn, was intended for HIV-infected individuals who no longer responded to or could not tolerate the other drugs. In the same manner as AZT, these three drugs work by inhibiting the replication of HIV.

K.3 Cancer and Its Causes

Carcinogens were mentioned often in earlier chapters. These substances cause the growth of tumors. A tumor is an abnormal growth of new tissue. Tumors may be either benign or malignant. **Benign tumors** are characterized by slow growth; they often regress spontaneously, and they do not invade neighboring tissues. **Malignant tumors** may grow slowly or rapidly, but their growth is generally irreversible. They often are called **cancers.** Malignant growths invade and destroy neighboring tissues. Actually, cancer is not a single disease. It is a catchall term for over a hundred different afflictions. Many are not even closely related to each other.

The World Health Organization estimates that 80 to 90% of cancer cases are caused by environmental factors, and 10 to 20% by genetic factors and (perhaps) viruses. Included most prominently among those "environmental" causes are cigarette smoking (40%), dietary factors (25 to 30%), and occupational exposure (10%).[2] That leaves 10 to 15% that may be caused by environmental *pollutants.* If you read the newspapers or watch television, you may get the idea that everything causes cancer. Even among those chemicals that have been suspect, however, many cannot be shown to be carcinogenic. Only about 30 chemical compounds have been identified as human carcinogens. Another 300 or so have been shown to cause cancer in laboratory animals. Some of the 300 are widely used, however.

Not all carcinogens are synthetic chemicals. Some, such as safrole in sassafras and the aflatoxins produced by molds on foods, occur naturally. Some researchers estimate that 99.99% of all carcinogens that we ingest are natural ones. Plants produce compounds to protect themselves from fungi, insects, and higher animals, including humans. Carcinogenic compounds are found in mushrooms, basil, celery, figs, mustard, pepper, fennel, parsnips, and citrus oils—almost everywhere a curious chemist looks. Carcinogens are also produced during cooking and as products of normal metabolism. With carcinogens virtually everywhere around

[2] There are 400,000 cancer deaths each year in the United States. Of these, 150,000 are related to cigarette smoking. Another 150,000 are related to diet.

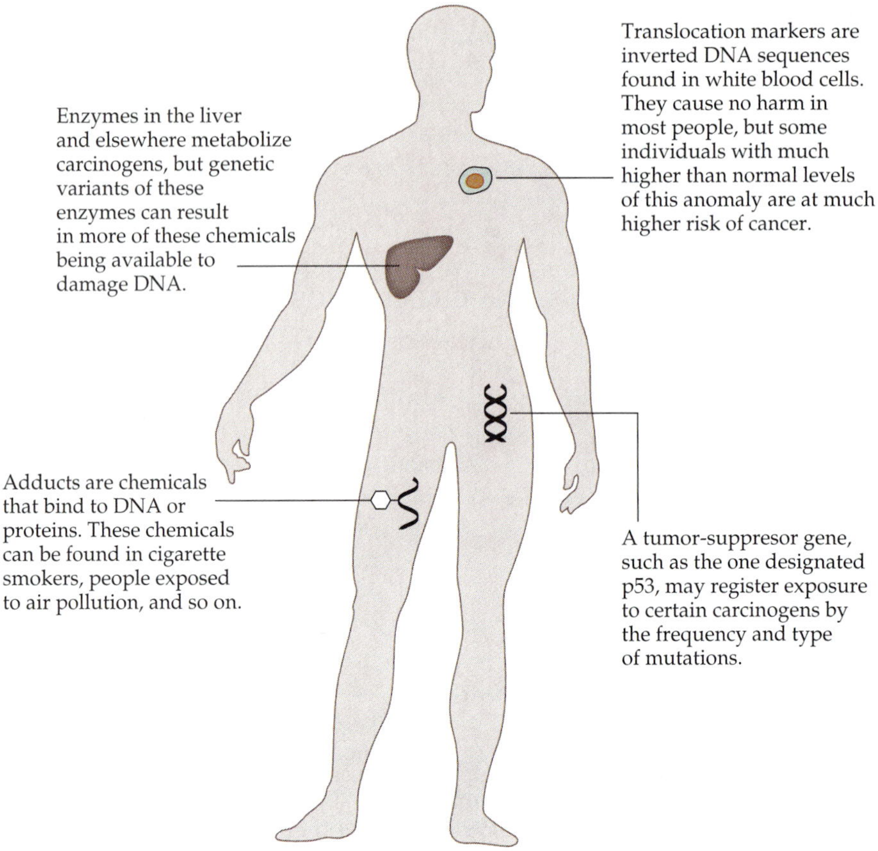

Enzymes in the liver and elsewhere metabolize carcinogens, but genetic variants of these enzymes can result in more of these chemicals being available to damage DNA.
Translocation markers are inverted DNA sequences found in white blood cells. They cause no harm in most people, but some individuals with much higher than normal levels of this anomaly are at much higher risk of cancer.
Adducts are chemicals that bind to DNA or proteins. These chemicals can be found in cigarette smokers, people exposed to air pollution, and so on.
A tumor-suppresor gene, such as the one designated p53, may register exposure to certain carcinogens by the frequency and type of mutations.

FIGURE K.3 ▶

Cancer risk at the molecular level.

us but not everyone developing some form of cancer, it must be that we must have some way of protecting ourselves from them.

How do chemicals cause cancer? Their mechanisms of action are probably as varied as their chemical structures. Some carcinogens chemically modify DNA, thus scrambling the code for replication and for the synthesis of proteins. For example, aflatoxin B is known to bind to guanine residues in DNA. Just how this initiates cancer, however, is not known for sure.

Researchers say that clues as to who is likely to get cancer can be found within the body's cells by measuring biological markers (Figure K.3). There is a genetic component to the development of many forms of cancer. Certain genes, called **oncogenes** (Greek *onkos,* mass), seem to trigger or sustain the processes that convert normal cells to cancerous ones. (Oncology is the branch of medicine that treats tumors.) Oncogenes arise from ordinary genes that regulate cell growth and cell division. They can be activated by chemical carcinogens, radiation, or perhaps some viruses. It seems that more than one oncogene must be turned on, perhaps at different stages of the process, before a cancer develops. We also have **suppressor genes** that ordinarily prevent the development of cancers. These genes must be inactivated before a cancer develops. Suppressor gene inactivation can occur through mutation, alteration, or loss. In all, 10 or 15 mutations may be required in a cell before it turns cancerous. There is hope that someday suppressor genes can be produced through genetic engineering and used in therapy.

K.4 Chemical Carcinogens

A variety of widely different chemical compounds are carcinogenic. We could not attempt to cover all the types of carcinogens here, and so we concentrate on a few major classes.

Some of the more notorious carcinogens are the polycyclic aromatic hydrocarbons (Section 13.17), of which 3,4-benzpyrene is perhaps the best known. Carcinogenic hydrocarbons are formed during the incomplete burning of nearly any organic material. They have been found in charcoal-grilled meats, cigarette smoke, automobile exhausts, coffee, burnt sugar, and many other materials. Not all polycyclic

The Development of Cancer

Cancer develops in a series of steps. Initially the cancer cells are restricted to a single location, called the *primary tumor.* All of the cells in the tumor are usually the daughter cells of a single malignant cell. At first the growth of the primary tumor simply distorts the tissue, and the basic tissue organization remains intact. Metastasis begins as tumor cells "break out" of the primary tumor and invade the surrounding tissue. When this invasion is followed by penetration of nearby blood vessels, the cancer cells begin circulating throughout the body.

Responding to cues that are as yet unknown, these cells later escape from the circulatory system and establish *secondary tumors* at other sites. These tumors are extremely active metabolically, and their presence stimulates the growth of blood vessels into the area. The increased circulatory supply provides additional nutrients and further accelerates tumor growth and metastasis. Death may occur because vital organs have been compressed, because nonfunctional cancer cells have killed or replaced the normal cells in vital organs, or because the voracious cancer cells have starved normal tissues of essential nutrients.

aromatic hydrocarbons are carcinogenic. There are strong correlations between carcinogenicity and certain molecular sizes and shapes. The mechanism of their action is currently under intense investigation. It already appears rather certain that the carcinogens are not the hydrocarbons themselves but the oxidation products formed in the liver.

Another important class of carcinogens is the aromatic amines. Two prominent ones are β-naphthylamine and benzidine:

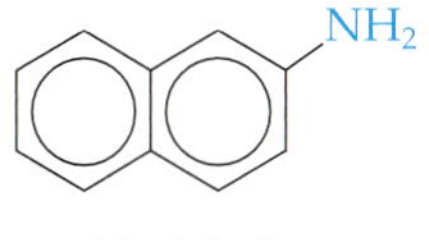

β-Naphthylamine

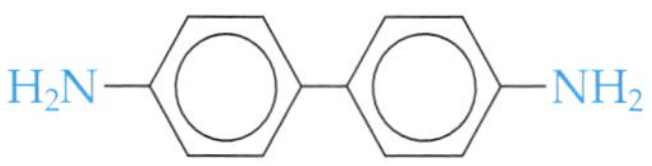

Benzidine

These compounds once were used widely in the dye industry. They were responsible for a high incidence of bladder cancer among the workers whose jobs brought them into prolonged contact with the compounds.

Several aminoazo dyes have been shown to be carcinogenic. An interesting example is 4-dimethylaminoazobenzene:

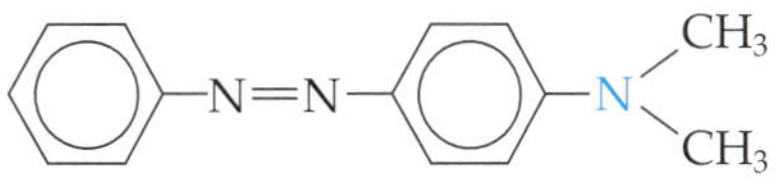

4-Dimethylaminoazobenzene

This compound is also known as "butter yellow." It was used widely as a coloring for butter and oleomargarine before its carcinogenicity became known.

Not all carcinogens are aromatic. Prominent among the aliphatic ones are dimethylnitrosoamine (Section 17.5) and vinyl chloride (Table 13.11). Others include three- and four-member heterocyclic rings containing nitrogen or oxygen. The epoxides

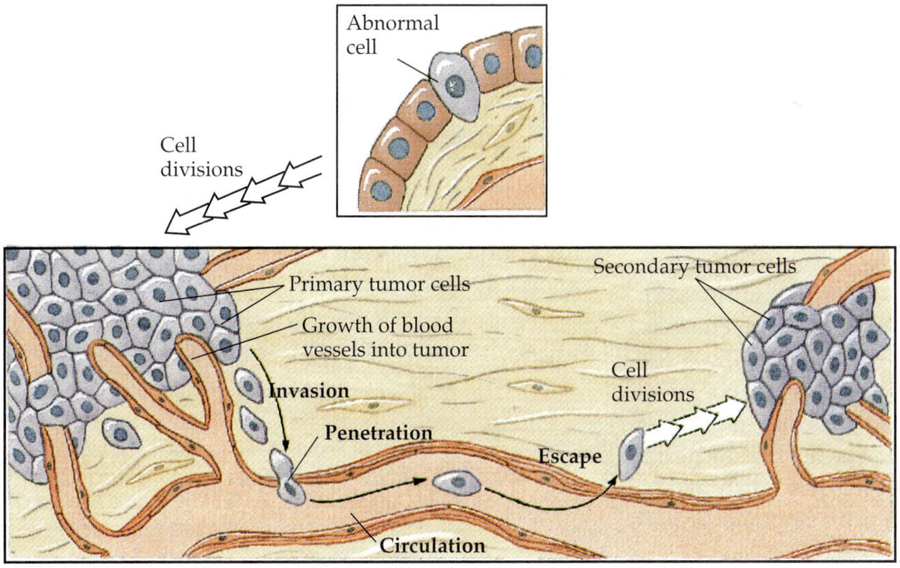

The development of cancer.

and derivatives of ethyleneimine are examples. Others are cyclic esters called lactones.

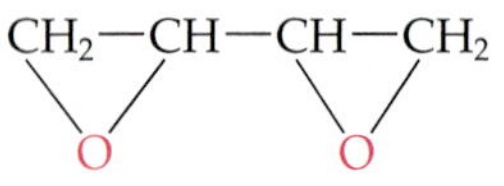

Bis(epoxy)butane

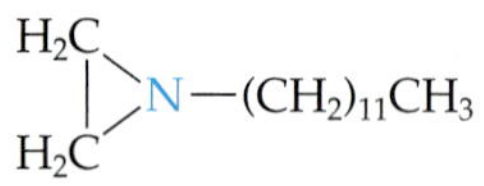

N-Laurylethyleneimine

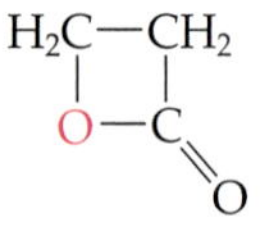

β-Propiolactone

Keep in mind that this list is not all-inclusive. Rather, its purpose is to give you an idea of the kinds of compounds that have tumor-inducing properties—and the list grows almost daily as research results are released.

K.5 Testing for Carcinogens

How do we know that a chemical causes cancer? Obviously, we can't experiment on humans to see what happens. That leaves us with no way to prove beyond doubt that a chemical does or does not cause cancer in humans. There are three ways, however, to gain evidence against a compound: bacterial screening for mutagenesis, animal tests, and epidemiological studies.

The quickest and cheapest way to find out whether or not a substance may be carcinogenic is to use the screening test developed by Bruce N. Ames of the University of California at Berkeley. The Ames test is a simple laboratory procedure that can be carried out in a petri dish. It assumes that most carcinogens are also mutagens, altering the genes in some way. (This usually seems to be the case. About 90% of the chemicals that appear on a list of either mutagens or carcinogens are found on the other list as well.)

The Ames test uses a special strain of Salmonella bacteria that have been modified so that they require histidine as an essential amino acid. The bacteria are placed in an agar medium containing all nutrients except histidine. Incubating the mixture in the presence of a mutagenic chemical causes the bacteria to mutate, so that they no longer require histidine and can grow like normal bacteria. Growth of bacterial colonies in the petri dish means that the chemical added was a mutagen, and probably also a carcinogen.

Chemicals suspected of being carcinogens can be tested on animals.[3] Tests using low dosages on millions of rats would cost too much, so tests usually are done by using large doses on 30 or so rats. An equal number of rats serve as controls. The control group is exposed to the same diet and environment as the experimental group, except that the control group does not get the suspected carcinogen. A higher incidence of cancer in the experimental animals than in the controls indicates that the compound is carcinogenic.

Animal tests are not conclusive. Humans usually are not exposed to comparable doses, and there may be a threshold below which a compound is not carcinogenic. Further, human metabolism is different from the metabolisms of the test animals. The carcinogen might be active in the rat but not in humans (or vice versa!).[4]

The best evidence that a substance causes cancer in humans comes from epidemiological studies. A population that has a higher-than-normal rate for a particular kind of cancer is studied for common factors in their background. It was this sort of study, for example, that showed that cigarette smoking causes lung cancer, that vinyl chloride causes a rare form of liver cancer, and that asbestos causes cancer of the lining of the pleural cavity (the body cavity that contains the lungs). These studies require sophisticated mathematical analyses. There is always the chance that some other (unknown) factor is involved in the carcinogenesis.

In the early 1990s, a series of epidemiological studies was carried out to determine whether or not the electric field surrounding high voltage power lines can trigger leukemia. The results of the study indicate that if there is an effect, it is quite small.

K.6 Chemicals Against Cancer

Chemists have designed molecules to relieve headache, cure infectious diseases, and prevent conception. Why can't they do something about cancer? They have done a lot, but much remains to be done. Treatment with drugs, radiation, and surgery has led to high rates of cures for some forms

[3] Animal studies cost about $1 million each and take 2 years.

[4] There is only a 70% correlation between the carcinogenesis of a chemical in rats and that in mice. The correlation between carcinogenesis in either rodent and that in humans probably is less.

of cancer (for example, one form of skin cancer). For other forms, such as lung cancer, the rate of cure is still quite low. Over 30 chemical substances are used widely in the treatment of cancer. That number will no doubt increase rapidly as our understanding of basic cell chemistry increases. We examine a few representative anticancer drugs here.

In cancer chemotherapy (Section 22.7), *antimetabolites* usually are compounds that inhibit the synthesis of nucleic acids. Rapidly dividing cells, characteristic of cancer, require large quantities of DNA. The anticancer metabolites block DNA synthesis and therefore block the increase of the number of cancer cells. Because cancer cells are undergoing rapid growth and cell division, they generally are affected to a greater extent than normal cells.[5]

The most widely used anticancer drug is cisplatin, a platinum-containing compound. Cisplatin binds to DNA and blocks its replication. Transplatin, an isomer of cisplatin, is ineffective, indicating that the *shape* of the molecule is all-important.

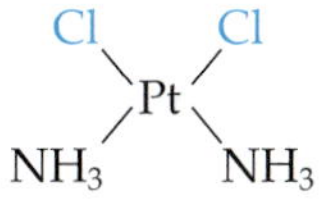

Cisplatin

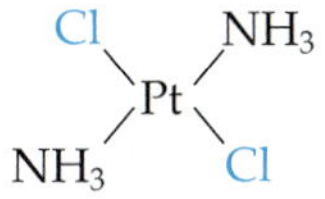

Transplatin

Two other prominent antimetabolites are 5-fluorouracil and its deoxyribose nucleoside, 5-fluorodeoxyuridine:

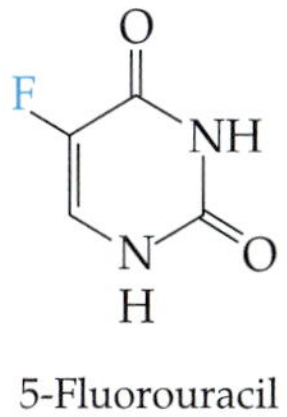

5-Fluorouracil

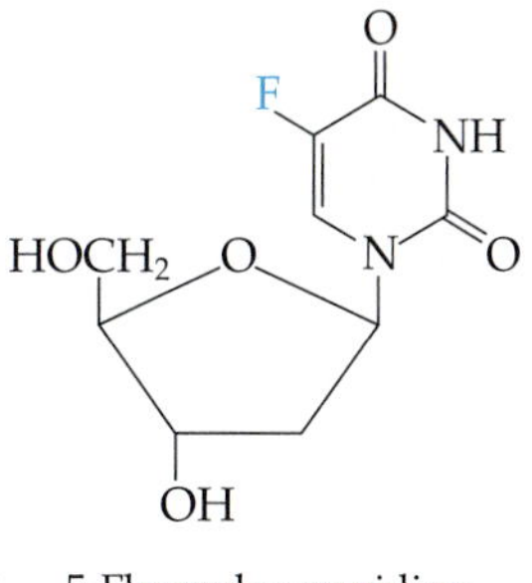

5-Fluorodeoxyuridine

In the body, both of these compounds can be incorporated into a nucleotide (Section 23.1). The fluorine-containing derivatives inhibit the formation of thymine-containing nucleotides required for DNA synthesis. Thus, both compounds slow the division of cancer cells. These compounds have been employed against a variety of cancers, especially those of the breast and digestive tract.

Another common antimetabolite is 6-mercaptopurine, which substitutes for adenine in a nucleotide:

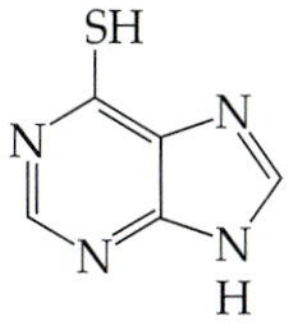

6-Mercaptopurine

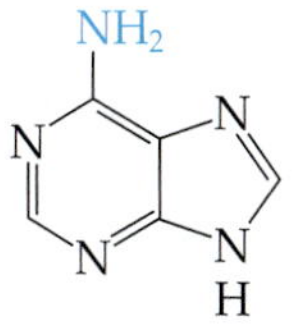

Adenine

The pseudonucleotide then inhibits the synthesis of nucleotides that incorporate adenine and guanine. Hence, DNA synthesis and cell division are allowed. 6-Mercaptopurine has been used in the treatment of leukemia.

Another antimetabolite, methotrexate, acts in a somewhat different manner. Note the similarity between its structure and that of folic acid:

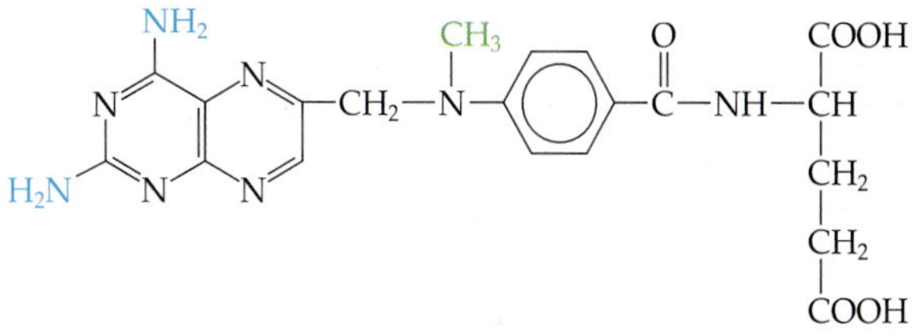

Methotrexate

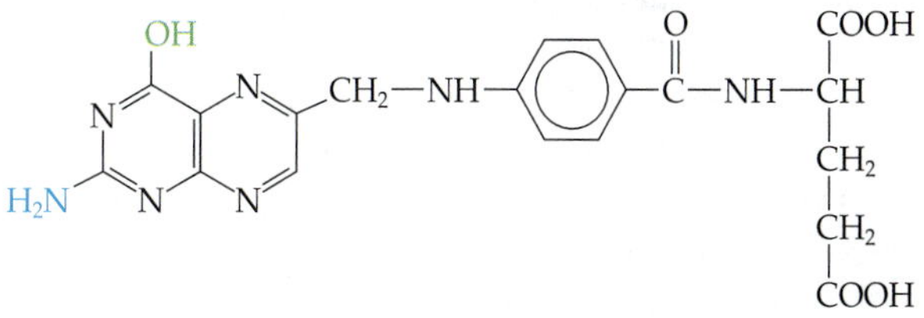

Folic acid

Like the pseudofolic acid formed from sulfanilamide (Section 22.7), methotrexate competes successfully with folic acid for an enzyme but cannot perform the growth-enhancing function of folic acid. Again, cell division is slowed, and cancer growth is retarded. Methotrexate is used frequently against leukemia.

K.7 Miscellaneous Anticancer Agents

There is a bewildering variety of anticancer agents that defy ready classification. Alkaloids from vinca plants have been shown to be effective against

[5] Cancer chemotherapy also affects noncancerous body cells that undergo rapid replacement, including those that line the digestive tract and those that produce hair. Side effects of the therapy include nausea and loss of hair. Eventually the normal cells are affected to such a degree that treatment must be discontinued.

leukemia and Hodgkin's disease. Actinomycin, a mixture of complex compounds obtained from the molds *Streptomyces antibioticus* and *Streptomyces parvus*, is used against Hodgkin's disease and other types of cancer. It is quite effective but extremely toxic. Actinomycin acts by binding to the double helix of DNA, thus blocking the replication of RNA on the DNA template. Protein synthesis is inhibited.

Sex hormones can be used against cancers of the reproductive system. For example, the female hormones estradiol (a natural hormone) and DES (a synthetic hormone) can be used against cancer of the prostate gland. Conversely, male hormones such as testosterone can be used against breast cancer. Such treatment often brings about a temporary cessation—or even a regression—in the growth of cancer cells.

The food additive butylated hydroxytoluene (BHT) has been shown to be anticarcinogenic in tests involving laboratory animals. Some people involved in cancer research speculate that the use of this additive as a preservative in foods may account for the declining rate of stomach cancer in the United States.

Similarly, there is some evidence that vitamin A may confer a resistance to some cancers. For example, persons suffering from vitamin A deficiencies exhibit a higher incidence of lung cancer. Vitamin C also may have an anticancer function. It has been shown to inhibit the formation of nitrosoamines (Section 17.5) under conditions similar to those in the human stomach. Dietary fiber is thought to protect against colon cancer.

Perhaps most notable is the protective value of a diet high in cruciferous vegetables (cabbage, brussels sprouts, broccoli, cauliflower, and kale). This kind of diet has been shown to reduce the incidence of cancer both in animal studies and in studies of human population groups. The chemical substances in these vegetables that act as anticarcinogens are not known. It seems quite likely, however, that by eating a balanced diet, including fresh fruit and vegetables, we also balance our carcinogens and anticarcinogens.

Chemotherapy is only a part of the treatment of cancer. Surgical removal of tumors and radiation treatment remain major weapons in the war on cancer. Modern management of cancers can involve surgery, radiation, and one or more anticancer drugs. Indeed, a combination of drugs is often considerably more effective than any one alone. It is unlikely that a single agent will be found to cure all cancers. Steady progress is being made, however. Rates of cure should improve as research progresses. Perhaps a greater hope lies in the prevention of cancer. Much active research is under way on the mechanisms of carcinogenesis. The more we learn about what causes cancer, the better equipped we shall be in trying to prevent it.

Key Terms

benign tumor (K.3)
cancer (K.3)
malignant tumor (K.3)
oncogene (K.3)
retrovirus (K.1)
suppressor gene (K.3)
virus (K.1)

Review Questions

1. What is the composition of viruses?
2. Distinguish a DNA virus from an RNA virus.
3. What are the shapes of viruses?
4. Compare the sizes of viruses and bacteria.
5. How does a DNA virus invade and destroy a cell?
6. How does an RNA virus invade and destroy a cell?
7. Name five diseases of viral origin and five diseases of bacterial origin.
8. List three antiviral drugs. Which, if any, cure viral diseases?
9. What is a tumor?
10. How are benign and malignant tumors different?
11. What is the single leading cause of cancer?
12. Name several natural carcinogens.
13. What are oncogenes? How are they involved in the development of cancer?
14. What are suppressor genes? How are they involved in the development of cancer?
15. List two major classes of anticancer drugs.
16. What is a mutagen?

Problems

Carcinogens

17. List some conditions under which carcinogenic hydrocarbons are formed.
18. Name two aromatic amines that are carcinogens.
19. Name two aliphatic carcinogens.

Anticancer Drugs

20. How does cisplatin work as an anticancer agent?
21. How does 5-fluorouracil act against cancer?
22. How does 6-mercaptopurine act against cancer?
23. What natural substance does methotrexate resemble? How does methotrexate act against cancer?

24 Carbohydrate Metabolism I

Weightlifting is an anaerobic activity. Energy is expended in short bursts.

Life requires energy. Living cells are inherently unstable and avoid falling apart only because of a continued input of energy. Living organisms are restricted to using certain forms of energy. Supplying a plant with heat energy by holding it in a flame will do little to prolong its life. On the other hand, a green plant is uniquely able to tap sunlight, the richest source of energy on Earth.

In our earlier discussion of carbohydrates, it was mentioned that human existence on this planet is directly dependent upon the plant kingdom. The animal world gets its foodstuffs, and hence its energy, from plant life (Figure 24.1). Plants obtain water, inorganic salts, and nitrogenous compounds from the soil, and carbon dioxide and oxygen from the atmosphere. With these raw materials, they are able to synthesize carbohydrates, lipids, and proteins. The energy required for these synthetic reactions is obtained from the sun—radiant energy (as sunlight) is the ultimate source of biological activity.

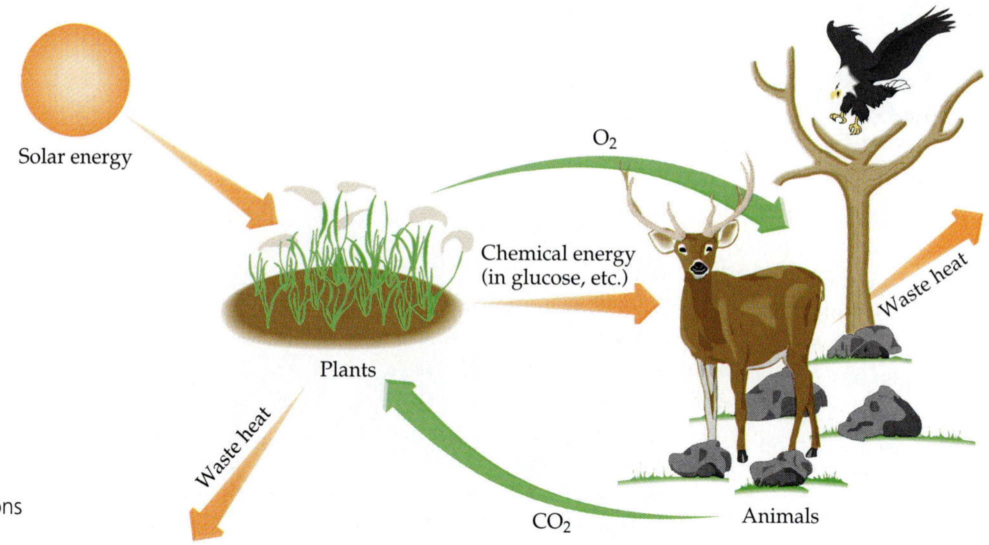

FIGURE 24.1 ▶ Some energy transformations in living systems.

The overall process by which glucose is formed from carbon dioxide and water at the expense of solar energy is termed **photosynthesis.** The general photosynthetic equation can be written as

$$6\,CO_2 + 6\,H_2O + 686{,}000\text{ cal} \xrightarrow{\text{sunlight}} \underset{\text{Glucose}}{C_6H_{12}O_6} + 6\,O_2$$

This synthesis is a distinguishing characteristic of green plants, and from the standpoint of human survival it represents the most important series of reactions (more than 100 enzyme-catalyzed steps) that occur on the surface of Earth. Notice that the formation of 1 mol of glucose requires the expenditure of 686,000 cal. Compared with the energy requirements of most other endothermic chemical reactions, this is a very large sum. (It is roughly equivalent to the heat required to raise the temperature of 7 quarts of water from 0 to 100 °C.) On the solar scale, however, it is a drop in the bucket.[1]

Animals cannot directly use the energy of sunlight. They must eat either plants or plant-eating animals in order to get carbohydrates, fats, and proteins and the chemical energy stored in them. Once digested and transported to the cell, a food molecule can be used in either of two ways. It can be used as a building block to make new cell parts or repair old ones, or it can be "burned" for energy.

The entire series of coordinated chemical reactions that keep cells alive is called **metabolism.** In general, metabolic reactions are divided into two classes. The breaking down of molecules to provide energy is **catabolism.** The process of building up the molecules of living systems is **anabolism.**

Any chemical compound involved in a metabolic reaction is a *metabolite.*

Carbohydrates are the primary metabolites of the animal kingdom. Recall that over half of the food that we ourselves consume is composed of carbohydrates.

[1] It has been estimated that on an average sunny day each square centimeter of Earth's surface receives about 1 cal of solar radiation every minute. Yet the energy intercepted by Earth is but a tiny part (2×10^{-7}%) of the total energy given off by the sun, and only a small fraction of this intercepted energy (~ 1%) is utilized in photosynthesis. The most important photosynthesizing organisms are the phytoplankton living in the world's oceans. Phytoplankton are the chief source of energy for aquatic organisms.

Carbohydrates serve as the chief fuel of biological systems, supplying living cells with usable energy. Like all other fuels, carbohydrates must be burned or oxidized if energy is to be released. The combustion (oxidation) process ultimately results in the conversion of the carbohydrate to carbon dioxide and water. The energy stored in the carbohydrate molecule during photosynthesis is released in the reaction:

$$C_6H_{12}O_6 \; + \; 6\,O_2 \; \longrightarrow \; 6\,CO_2 \; + \; 6\,H_2O \; + \; 686{,}000 \text{ cal}$$

This equation summarizes the biological combustion of foodstuff molecules by the cell (respiration). The term **respiration** is often used in a broader sense to include all metabolic processes by which gaseous oxygen is used to oxidize organic matter to carbon dioxide, water, and energy.

Both respiration and the combustion of the common fuels (wood, coal, gasoline) use oxygen from the air to break down complex organic substances to carbon dioxide and water. The energy released in the burning of wood is manifested entirely in the form of heat, but excess heat energy is useless and even injurious to the living cell. Organisms conserve almost half of the 686,000 cal by a series of stepwise reactions that liberate small amounts of usable energy, which is transmitted to the phosphate bond of ATP. The remainder of the energy is used to maintain body temperature. Before continuing, let us more closely examine the compound ATP.

Plant and animal cells exist in a *symbiotic* cycle—each requires the products of the other. Although H_2O and O_2 are abundant in the atmosphere, CO_2 is present only to the extent of 0.02 to 0.05%. It has been estimated that if animals were removed from the Earth, plants would consume all the atmospheric CO_2 in 1 to 2 years.

24.1 ATP: Universal Energy Currency

In Section 16.10, we discussed phosphate esters, and in Chapter 23, we learned of the nucleotide adenosine monophosphate (AMP). Probably the most important metabolic phosphate compound is adenosine triphosphate (ATP). It was first isolated from skeletal muscle tissue and has since been shown to occur in all types of plant and animal cells. The concentration of ATP in the cell varies from 0.5 to 2.5 mg/mL of cell fluid. ATP is a nucleoside triphosphate composed of adenine, ribose, and three phosphate groups (Figure 24.2).

The most significant feature of the ATP molecule is the phosphoric acid anhydride, or pyrophosphate, linkage:

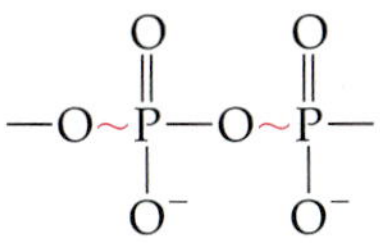

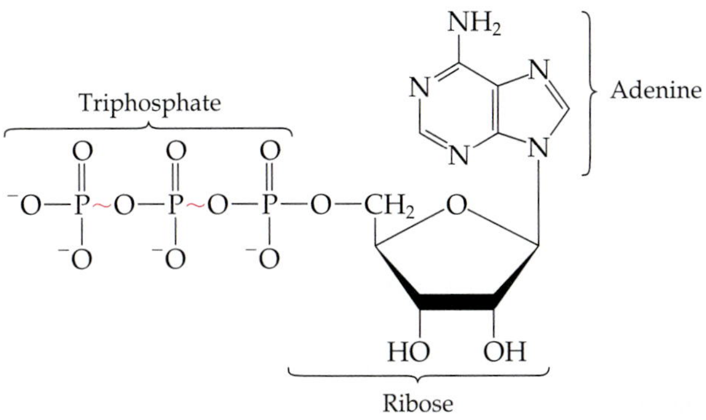

FIGURE 24.2
Adenosine triphosphate.

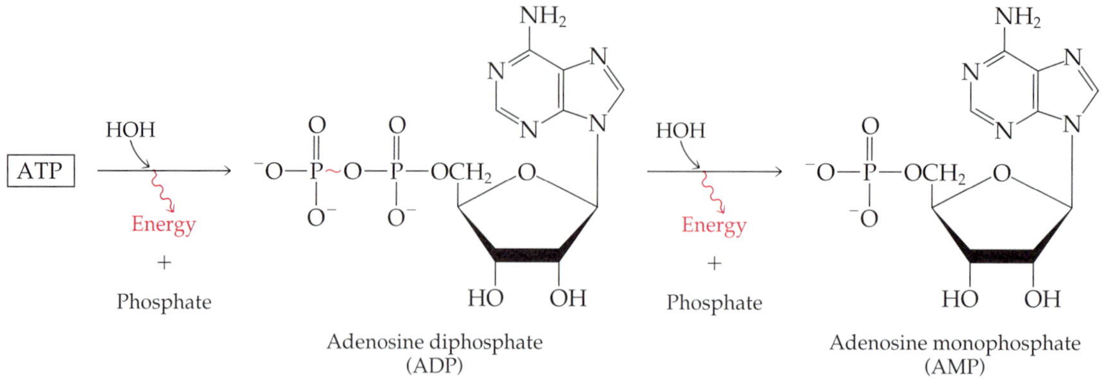

▲ **FIGURE 24.3**
The relationships among ATP, ADP, and AMP.

ATP is often called an *energy-rich compound.* Its pyrophosphate bonds are referred to as high-energy bonds and are sometimes symbolized by a squiggle bond (~). A *high-energy bond* is one that releases a relatively large amount of energy (>7000 cal/mol) when it is hydrolyzed. In this case, one of the driving forces for the reaction is to relieve the electron–electron repulsions associated with the negatively charged phosphate groups. It should be noted, however, that there is nothing special about the bonds themselves. The symbol ~ is just a device to focus attention on a portion of the molecule that undergoes reaction. Energy-rich compounds, then, are substances having particular structural features that yield high energies of hydrolysis, and for this reason they are able to supply energy for energy-requiring biochemical processes (Figure 24.3).[2]

The general equation for ATP hydrolysis is

$$\text{ATP} \xrightleftharpoons{H_2O} \text{ADP} + P_i + 7300\ \text{cal/mol}$$

P_i is the symbol for the inorganic phosphate anions $H_2PO_4^-$ and HPO_4^{2-} present in the intra- and extracellular fluids. About 1.2 g of phosphate is needed in the daily diet to replace the amount excreted in the urine.

The important feature of this biochemical reaction is its reversibility. The hydrolysis of ATP releases energy; its synthesis requires energy. [In a typical cell, an ATP molecule is consumed ("turned over") within 1 min after its formation.] Thus ATP is produced by those processes that supply energy to an organism (absorption of radiant energy of the sun in green plants and breakdown of foodstuffs in animals), and it is hydrolyzed by those processes that require energy (syntheses of carbohydrates, lipids, proteins; transmission of nerve impulses; muscle contraction). These couplings of ATP synthesis to processes that release energy and ATP breakdown to processes that require energy constitute one of the striking characteristics of living matter (Figure 24.4).

Because ATP is the principal medium of energy exchange in biological systems, it is referred to as the energy currency of the cell. However, it is not the

[2] The values in the literature for the energy released when ATP is hydrolyzed to ADP vary somewhat. This situation is due in part to the fact that reaction conditions (concentration, temperature, pH) have not always been the same in different laboratories and in part to the difficulties in obtaining exact values for the equilibrium constants. Most texts and research articles report values between −7000 and −8000 cal/mol for the free energy of hydrolysis of ATP at 25 °C and a pH of 7.0. We shall use a value of −7300 cal/mol throughout this text. However, this is only an approximate value because the concentrations of cellular reactants are not molar, as required for energy calculations.

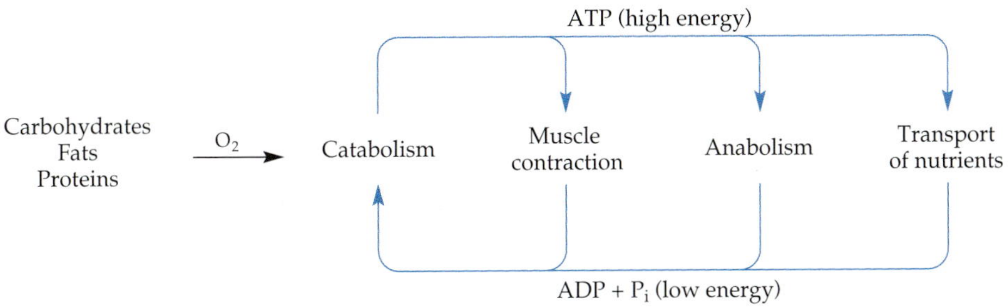

▲ **FIGURE 24.4**
ATP is called the energy currency of the cell. Energy stored in ATP, from catabolic reactions, can be used for mechanical work such as muscle contraction, for chemical synthesis (anabolism), and for transport of nutrients.

only high-energy compound. There are several other phosphate esters that provide energy for certain energy-requiring reactions. Table 24.1 lists a number of them. The use of these compounds will be illustrated in subsequent pages. Notice that the free energy of hydrolysis of ATP is approximately midway between those of the high-energy and the low-energy phosphate compounds.

Table 24.1 Standard Free Energies of Hydrolysis of Some Phosphate Compounds

Type	Example	$\Delta G°$ (cal/mol)
Acyl phosphate 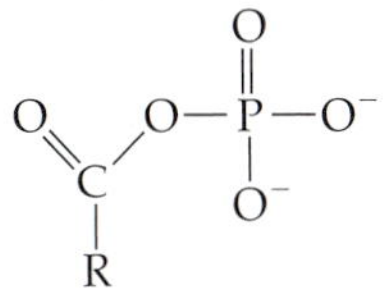	1,3-Bisphosphoglyceric acid Acetyl phosphate	−11,800 −10,300
Guanidine phosphate 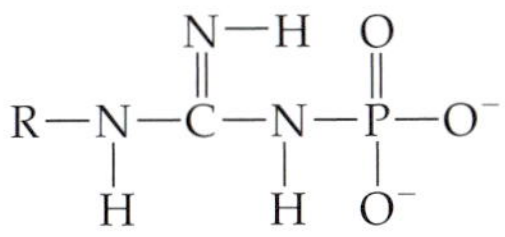	Creatine phosphate Arginine phosphate	−10,300 −7,700
Pyrophosphate 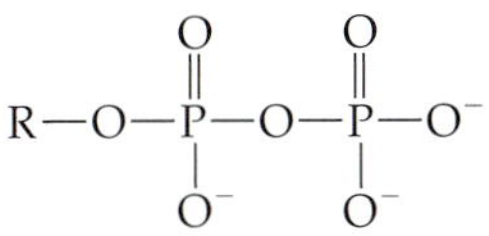	ATP ⟶ AMP + PP_i ATP ⟶ ADP + P_i ADP ⟶ AMP + P_i PP_i ⟶ 2 P_i	−7,700 −7,300 −7,300 −6,500
Sugar phosphate 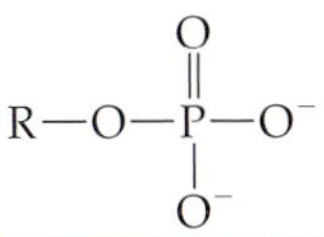	Glucose 1-phosphate Fructose 6-phosphate AMP ⟶ Adenosine + P_i Glucose 6-phosphate Glycerol 3-phosphate	−5,000 −3,800 −3,400 −3,300 −2,200

24.2 Digestion and Absorption of Carbohydrates

Digestion can be defined as a hydrolytic process whereby food molecules are broken down into simpler chemical units that can be absorbed by the body. In humans, digestion takes place in the digestive tract (Figure 24.5), and absorption occurs primarily in the small intestine. The digestive tract is a tunnel that runs *through* the body. Food in the digestive tract is in the tunnel, not in the body. The alternative name for the digestive tract, *alimentary canal*, perhaps conveys this image more clearly.

Food is sluiced through the canal by a flow of digestive juices and by physical pushes imparted by sections of the canal. Compounds that are changed during this journey into suitable forms are absorbed through the walls of the canal into the circulatory systems of the body. Those materials that can't be absorbed make their way through the entire length of the canal and out again. Without the process of digestion, very little of the food we eat would nourish us.

The average American consumes approximately 325 g of carbohydrates each day—about 160 g of starch, 120 g of sucrose, 30 g of lactose, 10 g of glucose, 5 g of fructose, and traces of maltose. A typical diet supplies about 50% of total body energy in the form of carbohydrate.

Starch is the principal carbohydrate ingested by humans. Its digestion begins in the mouth, where the food encounters **saliva,** a digestive fluid secreted by the salivary glands. Saliva is more than 99% water. It contains enzymes, inorganic ions, and a variety of organic molecules typical of other body fluids. In the mouth, food is chewed—that is, torn or crushed to a finer consistency. This reduction in the size of food particles facilitates digestion by providing greater surface area for enzymes to work on. Chewing also coats the food particles with mucin, a glycoprotein constituent of saliva. The mucin lubricates the food and makes it easier to swallow.

The secretion of saliva can be triggered by the sight, taste, smell, or even the thought of food. An average person produces about 1.5 L of saliva a day. Excessive flow may be caused by certain pathological conditions, such as mercury poisoning.

The principal digestive enzyme in the mouth is α-amylase, sometimes called ptyalin. This enzyme attacks the α-glycosidic linkages in starch more or less at random. The pH of saliva is about 6.8, the optimum pH for α-amylase. Cleavage

FIGURE 24.5 ▶

The human digestive tract.

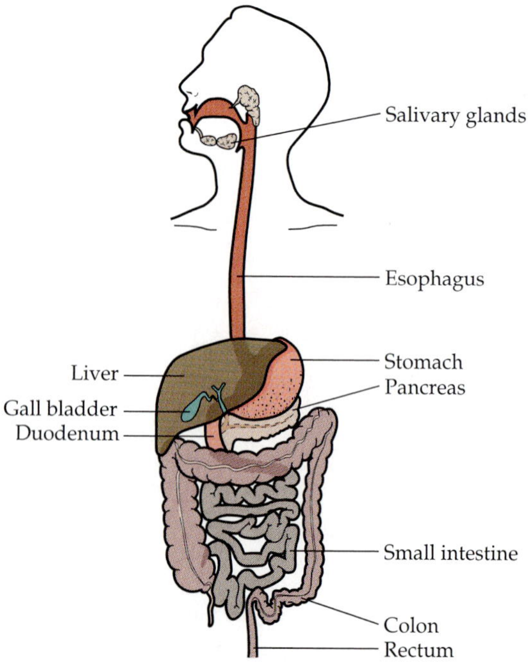

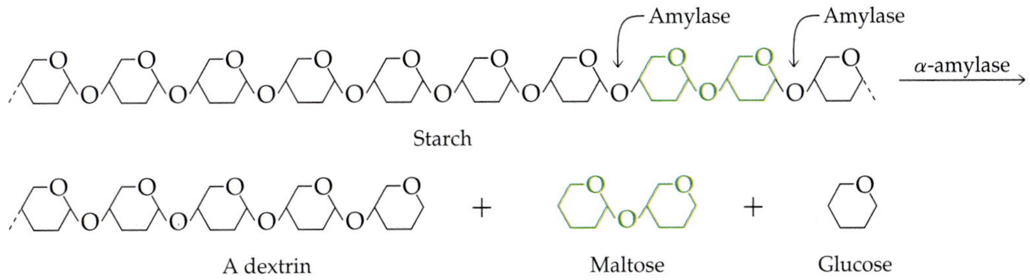

▲ **FIGURE 24.6**

A schematic representation of the hydrolysis of starch to dextrins, maltose, and glucose. See Chapter 19 for details of structure.

of the glycosidic linkages produces a mixture of dextrins, maltose, and glucose (Figure 24.6). Digestion of starch in the mouth is probably not necessary, except perhaps as a means of removing particles of starchy food lodged between the teeth.

α-Amylase continues to function as food passes through the esophagus, but it is quickly inactivated when it comes into contact with the acidic environment of the stomach. Very little carbohydrate digestion occurs in the stomach—acid-catalyzed hydrolysis proceeds too slowly at body temperature to be effective. The primary site of carbohydrate digestion is the small intestine, where another amylase, *amylopsin* (secreted from the pancreas), converts the remaining starch molecules, along with the dextrins, to maltose. Maltose is then cleaved into two glucose molecules by the enzyme maltase.

Disaccharides, such as sucrose and lactose, are not digested until they reach the small intestine, where they are acted upon by sucrase and lactase. The enzymes that catalyze the hydrolysis of disaccharides are termed *disaccharidases* and are located on the membranes of cells that line the inner surface of the small intestine. Ultimately, the complete hydrolysis of disaccharides and polysaccharides produces three monosaccharide units—glucose, fructose, and galactose. These monosaccharides are then absorbed through the wall of the small intestine into the bloodstream.

Absorption of most digested food takes place in the small intestine through the fingerlike projections, called *villi*, that line the inner surface (Figure 24.7). Each villus is richly supplied with a fine network of blood vessels and a central lymphatic vessel (Figure 24.8). Monosaccharides pass through the semipermeable membranous wall of each villus and are absorbed into the blood capillaries. However, the absorption is not via a simple process of osmosis or diffusion

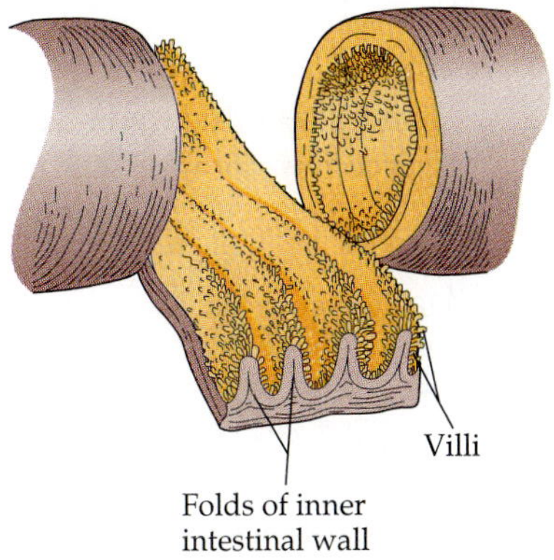

◀ **FIGURE 24.7**

A section of the small intestine opened to reveal the folds of the inner wall and the lining covered with villi.

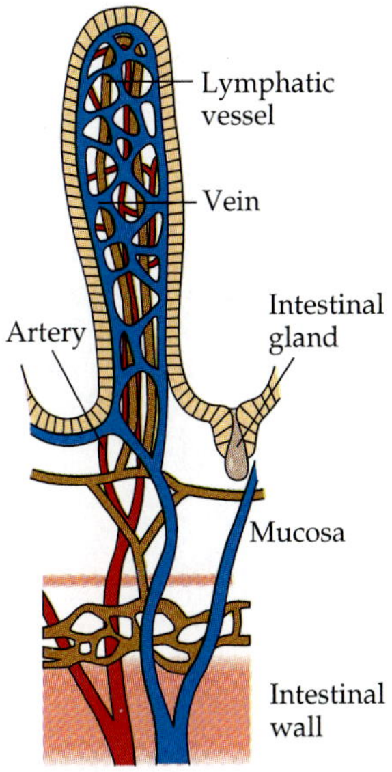

▲ **FIGURE 24.8**

Diagram of one of the intestinal villi. Each villus contains arteries, veins, and lymphatic vessels.

Any polysaccharides or disaccharides that escape hydrolysis by intestinal enzymes cannot be absorbed. Intestinal bacteria metabolize these carbohydrates to lactose, short-chain carboxylic acids, and gases (CO_2, CH_4, and H_2), a process that causes fluid secretion, increased intestinal mobility, and cramps.

through an inert membrane (i.e., *passive transport*). All cell membranes act selectively, a fact that implies that they play an active role in absorption. The passage of monosaccharides across the intestinal wall is an energy-requiring process, and so the term *active transport* is used to describe this type of absorption.

Following absorption, monosaccharides are carried by the portal vein to the liver. Here galactose and fructose are enzymatically converted to glucose or to phosphorylated intermediates that are metabolized by the liver. Some of the glucose passes into the general circulatory system to be transported to other tissues. Once in the tissues, this glucose may be oxidized to CO_2 and H_2O or converted to muscle glycogen, thus serving as a source of readily available energy for the muscles. The majority of absorbed glucose remains in the liver, however, and is either stored as glycogen (as a reservoir for the maintenance of the normal level of blood glucose) or converted to fat and exported to adipose tissue as lipoprotein complexes (the VLDLs; see Section 20.10). The average person has a sufficient amount of stored glycogen (in liver and muscle cells) to supply energy for about 18 h.

24.3 Blood Glucose

The concentration of glucose in the blood is referred to as the *blood sugar level.* Under normal circumstances, the blood sugar level remains remarkably constant at about 80 mg/100 mL of blood. However, because individuals differ in chemical makeup, a normal concentration of glucose may range from 70 to 100 mg/100 mL. (This amounts to a total of about 5 to 6 g of glucose, or 1 teaspoonful, in the entire body.) Soon after a meal, the blood sugar level may rise to 120 mg/100 mL, but it returns to the normal level within 2 h.

As we shall see in Section 24.4, regulation of blood sugar level is vital to the well-being of an individual. When the control mechanisms function improperly, there may be dire consequences. The condition of high blood sugar is called **hyperglycemia.** After severe starvation or vigorous exercise, the blood sugar concentration may fall below normal, leading to the condition called **hypoglycemia.** Neither condition, if temporary, is necessarily pathological, because the body has several methods of regulating the level of glucose in the blood.

If the concentration of a substance in the blood exceeds the renal threshold, the substance appears in the urine (see Section 28.12).

The kidneys may excrete excess glucose. The **renal threshold** value for glucose is fairly high, however, ranging from 150 to 170 mg of glucose per 100 mL of blood. In general, the kidneys are designed to conserve the glucose in the blood. Only when the blood glucose level goes well above an individual's normal range do the kidneys shunt some of the glucose into the urine. The liver also helps to regulate the blood sugar level by converting excess glucose to glycogen. Excess glucose may also be converted to fats for storage. And, of course, the glucose may be oxidized in the cells, producing energy.

Extreme hypoglycemia[3] can cause unconsciousness and lowered blood pressure and may result in death. Loss of consciousness is most likely due to the lack of glucose in the brain tissue, which has no capacity for glycogen storage and thus

[3] There is a great deal of controversy regarding hypoglycemia, but there are some clear-cut cases in which easily recognized symptoms are evident. These include general weakness, trembling, and rapid heartbeat. Severe cases can lead to delirium, coma, and even death. The treatment for hypoglycemia is simple: give the patient some sugar. If the person is unconscious, glucose solution can be administered intravenously. A common cause of hypoglycemia is an overdose of insulin. Diabetics generally carry candy to counteract the effects of excess insulin.

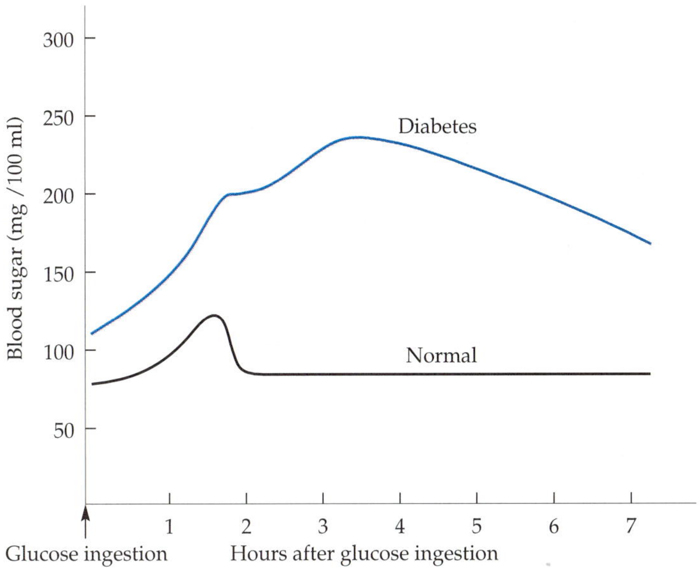

FIGURE 24.9

Glucose tolerance test results for a normal person and for a diabetic person.

is dependent upon a continuous supply of glucose for its energy requirements. The brain uses about 125 g of glucose per day. At rest, the total glucose requirement of all other tissues of the body (heart, liver, kidneys, muscles, etc.) is about 200 g/day. The total daily glucose requirements are normally met from the dietary intake of carbohydrates.

The major cause of hyperglycemia is diabetes mellitus (see Section 24.4). More than 10 million Americans have diabetes (either mild or severe), and this disease alone is the third leading cause of death (either outright or through side effects) in the United States. Diabetes is characterized by abnormal metabolism of carbohydrates, proteins, and lipids (see Section 26.7). Because a diabetic is unable to use glucose properly, excessive quantities accumulate in the blood and urine. Characteristic symptoms of diabetes are constant hunger, weight loss, extreme thirst, and frequent urination because the kidneys excrete large amounts of water in an attempt to remove excess sugar from the blood. High levels of glucose in the blood can cause various types of body damage. Among the many complications of diabetes are blindness, cardiovascular disease, gangrene, and kidney disease.

The **glucose tolerance test** is an important diagnostic test for diabetes mellitus. A patient's blood sugar level is determined after an overnight fast. Then a known amount (~75 g) of glucose is dissolved in about 400 mL of water and administered orally over a period of 5 to 10 min. Blood is drawn from the patient's fingertip at 30-min intervals after ingestion, and the blood sugar concentration is determined. After an initial rise, the blood sugar level falls rapidly in a normal individual. In a diabetic person, on the other hand, the increase in the blood sugar level is greater than normal and the level remains elevated for several hours (Figure 24.9).

24.4 Hormonal Regulation of Blood Sugar Level

The most important process in the maintenance of a constant blood glucose concentration is the synthesis and breakdown of glycogen in the liver. The liver is responsible for removing glucose from the blood when the concentration is too high (after a meal) and releasing it to the blood when the blood sugar level is too low (e.g., between meals). The activity of the liver, in this regard, is controlled by several hormones, including insulin, epinephrine (adrenaline), and glucagon.

Insulin, which is produced by the beta cells of the islets of Langerhans in the pancreas, *is the most important regulator of metabolism in the body.* It is released from the beta cells in response to a high blood sugar level. In general, insulin promotes anabolic reactions and inhibits catabolic reactions in the liver, muscle, and adipose tissue. (It has no effect on carbohydrate metabolism in the brain or kidneys.) Specifically, insulin performs the following functions:

1. Enhances **glycogenesis** (the formation of glycogen from glucose) in both the liver and muscle.
2. Promotes the entry of glucose into muscle, liver, and adipose tissue. (Note, however, that red blood cells and cells in the brain, kidneys, and intestinal tract do not require insulin for glucose uptake.)
3. Accelerates the conversion of glucose to fatty acids and hence the synthesis and storage of triglycerides in adipose tissue (see Section 26.3).
4. Inhibits the breakdown of glycogen and stored fat.
5. Promotes the transport of amino acids into cells and stimulates protein synthesis; inhibits intracellular degradation of proteins.
6. Suppresses **gluconeogenesis,** the synthesis of glucose from noncarbohydrate precursors (e.g., amino acids, glycerol, lactic acid). Gluconeogenesis occurs primarily in the liver and is particularly important in maintaining a constant blood sugar level during periods of starvation or strenuous exercise so that glucose can be supplied to brain tissue.

Hence, the principal role of insulin is to remove glucose rapidly from the blood, thus lowering the blood sugar level. The mechanism by which insulin accomplishes its prodigious tasks is largely unknown. Insulin binds to specific receptor proteins in the membranes of target cells. Just how the hormone–receptor interaction is coupled to the physiological responses is being investigated in many laboratories throughout the world.

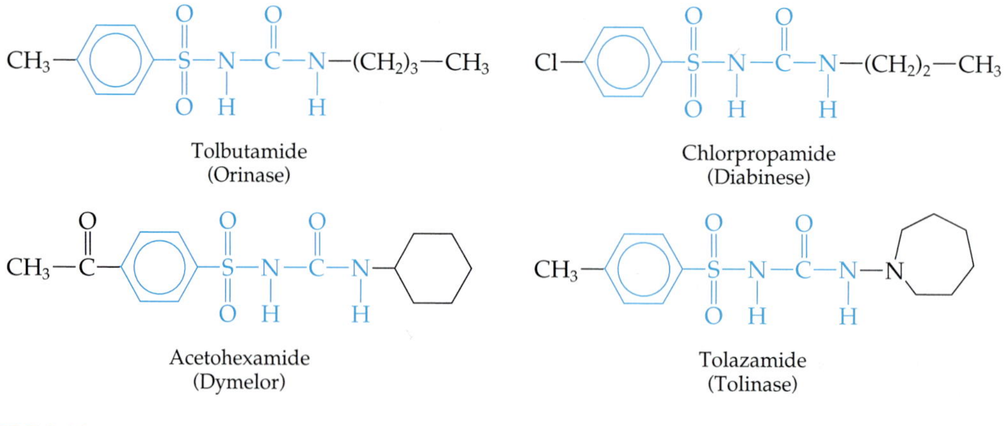

▲ **FIGURE 24.10**

Four first-generation antidiabetic drugs. These antidiabetic drugs are related to the sulfa drugs (see Figure 22.17). The replacement of the *p*-amino group on the benzene ring with other substituents accounts for the loss of antibacterial action.

Diabetes

If the pancreas does not secrete enough insulin and/or if there are insufficient (or defective) insulin receptors on the cell membranes, **diabetes mellitus** develops. Although medical science has made significant progress against this disease, it continues to be a major health threat. Consider a few of the serious complications:

- Diabetes is second only to trauma as the most frequent cause of lower limb amputations in the United States.
- It is the leading cause of blindness in adults older than 30.
- It is the leading cause of kidney failure.
- It doubles the risk of having a heart attack or stroke.

There are two types of diabetes. Insulin-dependent diabetics do not produce sufficient amounts of insulin to regulate their blood sugar levels. This type of diabetes develops early in life and is termed *Type I diabetes*. It is rapidly reversed by the administration of insulin (usually injected subcutaneously), and Type I diabetics can lead active lives provided they receive insulin as needed. Because insulin is a protein, it cannot be taken orally because it would be digested. Therefore, the Type I diabetic must be treated with daily injections of insulin. Limited success has been achieved by implanting insulin-producing cells from cadavers into adults suffering from Type I diabetes. In some patients, insulin production lasted from 3 to 12 weeks and significantly decreased the patient's need for injected insulin.

A current theory is that in Type I diabetics, insulin-producing cells of the pancreas are destroyed by the body's own immune system. These cells may have been altered by a viral infection (and thus appear as "foreign" to the immune system). Researchers have developed a simple blood test capable of predicting who will develop Type I diabetes several years before the disease becomes apparent. The blood test searches for antibodies that destroy the body's insulin-producing cells. The antibodies require several years to destroy enough islet cells to cause diabetes.

Type II (non-insulin-dependent) *diabetes* is by far the more common (about 95% of diabetic cases—about 10 million Americans), and it occurs late in life. Type II diabetics produce sufficient amounts of insulin, but either the beta cells do not secrete enough of it, or it is not utilized properly (because there is a lack of insulin-receptor proteins on the target cells, or the insulin-receptor proteins are defective). For these people, the disease can usually be controlled with a combination of diet and exercise alone, without insulin injections. Alternatively, there are oral antidiabetic drugs that stimulate the islet cells to secrete insulin (Figure 24.10). These compounds are effective only for persons who manufacture their own insulin but fail to release it in response to increased blood sugar level. Because these drugs have been shown to increase the risk of heart disease, their use has become controversial. Second-generation drugs, such as glyburide, are now available. They stimulate the release of insulin, as do the first-generation drugs, but they also increase the sensitivity of the cell receptors to the insulin. Like the earlier drugs, they must carry warnings about the increased risk of cardiovascular disease.

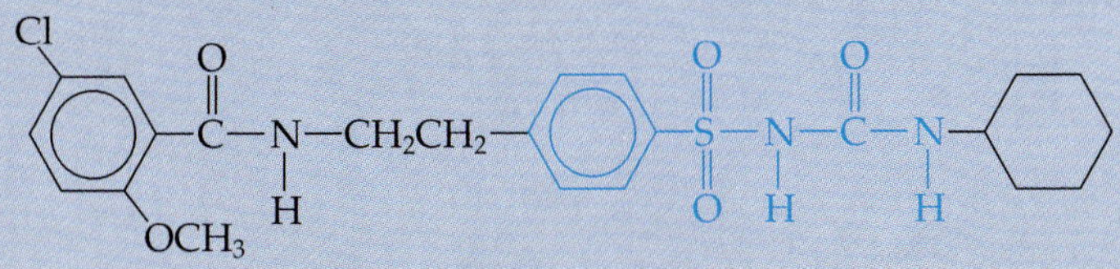

Glyburide, a second-generation antidiabetic drug

Diabetes mellitus is a metabolic disease that has been recognized for thousands of years. It is named from its symptoms—*diabetes mellitus* means excessive, sweet urine.

All other hormones that affect glucose metabolism act to raise the blood sugar level. Both epinephrine and glucagon exert their effects by binding to receptor proteins on the outside of the cell membrane (each binds to a different specific receptor). These receptor proteins are linked to the enzyme *adenyl cyclase,* which is bound to the inner membrane. When the outer receptor protein is unbound, the enzyme is not active. The binding of epinephrine or glucagon to their receptors causes conformational changes in other membrane proteins. It is these conformational changes that activate adenyl cyclase. The function of the enzyme is to catalyze the conversion of ATP to adenosine 3′,5′-monophosphate (cyclic AMP or cAMP). Adenyl cyclase is extremely efficient, and many cAMP molecules can be synthesized by a single activated enzyme:

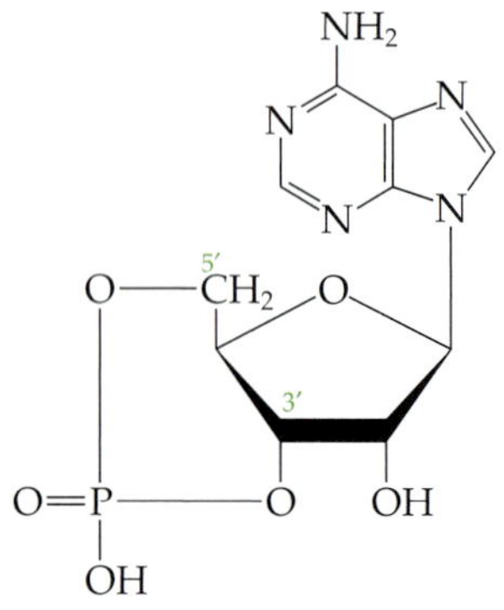

Adenosine 3′,5′-monophosphate (cAMP)

$$\text{ATP} \xrightarrow[\text{adenyl cyclase}]{\text{Mg}^{2+}} \text{cAMP} + \text{PP}_i$$

Cyclic AMP is often referred to as a *second messenger* because it transmits messages (delivered via the blood by the extracellular hormones—the primary messengers) from the cell membrane to enzymes within the cell (Figure 24.11). Cyclic AMP binds to and activates certain inactive enzyme precursors, thus beginning a cascade of cellular events that results in the stimulation of a wide range of catabolic processes and the inhibition of several anabolic reactions. (See an advanced biochemistry text for a discussion of the cAMP cascade.) It should be noted that cAMP serves as the second messenger for some neurotransmitters and several other hormones in addition to epinephrine and glucagon (Table 24.2).

Epinephrine (Section G.2) is secreted by the adrenal medulla in response to a low blood sugar level. It is also released during exercise or during periods of emotional stress, such as anger or fright, to provide the organism with additional energy. Epinephrine binds to its receptors (called *adrenergic receptors* after the hormone's other name, *adrenaline*), primarily on the membranes of muscle cells and to a lesser extent on the membranes of liver cells. It markedly stimulates **glycogenolysis,** the breakdown of glycogen to glucose. Epinephrine also promotes gluconeogenesis in the liver. Both of these processes increase the

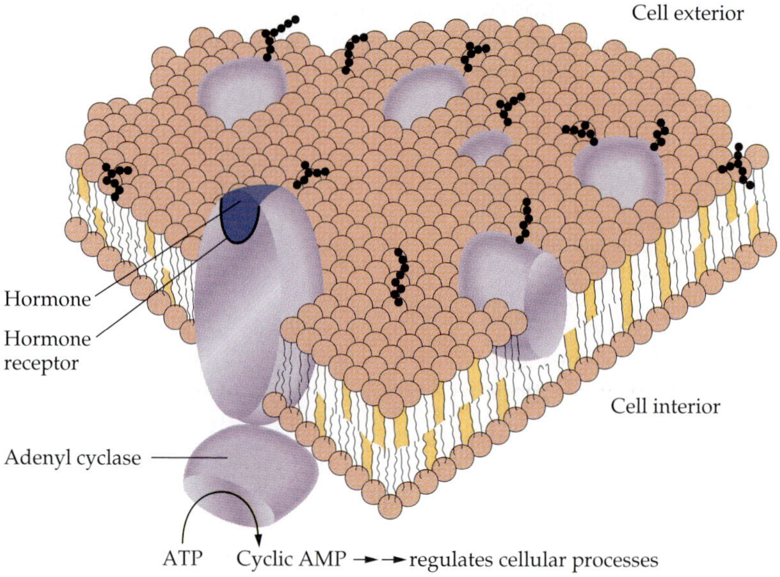

FIGURE 24.11 ▶ Binding of an extracellular hormone to a receptor protein activates adenyl cyclase inside the cell, and this enzyme then catalyzes the synthesis of cAMP.

Table 24.2 Hormones That Increase cAMP Levels

Tissue	Hormone	Principal Response
Bone	Parathyroid hormone	Calcium resorption
Muscle	Epinephrine	Glycogenolysis
Adipose	Epinephrine	Lipolysis
	Adrenocorticotrophic hormone	Lipolysis
	Glucagon	Lipolysis
Brain	Norepinephrine	Discharge of Purkinje cells
Thyroid	Thyroid-stimulating hormone	Thyroxine secretion
Heart	Epinephrine	Increased contractility
Liver	Epinephrine	Glycogenolysis
Kidney	Parathyroid hormone	Phosphate excretion
	Vasopressin	Water reabsorption
Adrenal	Adrenocorticotrophic hormone	Hydrocortisone secretion
Ovary	Luteinizing hormone	Progesterone secretion

amount of glucose in the blood, making that additional glucose available to the body tissues. Epinephrine increases the heart rate and raises the blood pressure.

Glucagon, like insulin, is a polypeptide (29 amino acids) hormone produced by the pancreas (the alpha cells of the islets of Langerhans), but glucagon is an antagonist of insulin. Like epinephrine, glucagon is secreted in response to a low blood sugar level, and its task is to increase the glucose concentration. Glucagon acts primarily on the liver (and adipose tissue) and not on skeletal muscle because there are no receptors for glucagon on muscle cell membranes. Glucagon stimulates gluconeogenesis and glycogenolysis in the liver, two processes that restore the blood glucose to its normal level. The concentration of glucagon in the blood of a diabetic is above normal, and this may be a significant contributor to the problems associated with diabetes.

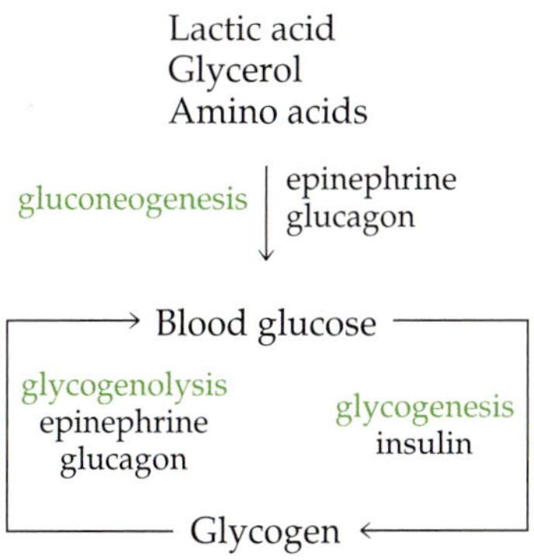

Glycogen metabolism is strongly influenced by the ratio of insulin to glucagon in the blood. Higher amounts of insulin lead to glycogen storage after a meal, whereas higher amounts of glucagon favor the breakdown of liver glycogen to add more glucose to the blood. It is the insulin/glucagon ratio that determines the outcome of carbohydrate metabolism. A high ratio leads to carbohydrate anabolism and storage, whereas a low ratio results in carbohydrate catabolism and utilization. The secretion of these hormones is directly governed by the blood sugar level.

24.5 Embden–Meyerhof Pathway

To fully appreciate the role of carbohydrates in living systems, it is necessary to understand their metabolic pathways. A **metabolic pathway** is a flow diagram that enables us to explain how an organism converts a given reactant to a desired end product.

In the 1930s, the German biochemists G. Embden and O. Meyerhof elucidated the sequence of reactions by which glycogen and glucose are degraded in the absence of oxygen (**anaerobic** conditions) to pyruvic acid. Embden and Meyerhof discovered that two apparently dissimilar processes—alcoholic fermentation in yeast and muscle contraction in animals—proceed by the same pathway as far as pyruvic acid is concerned. In the absence of oxygen, muscle cells (and certain other animal tissues) convert pyruvic acid to lactic acid, but under similar

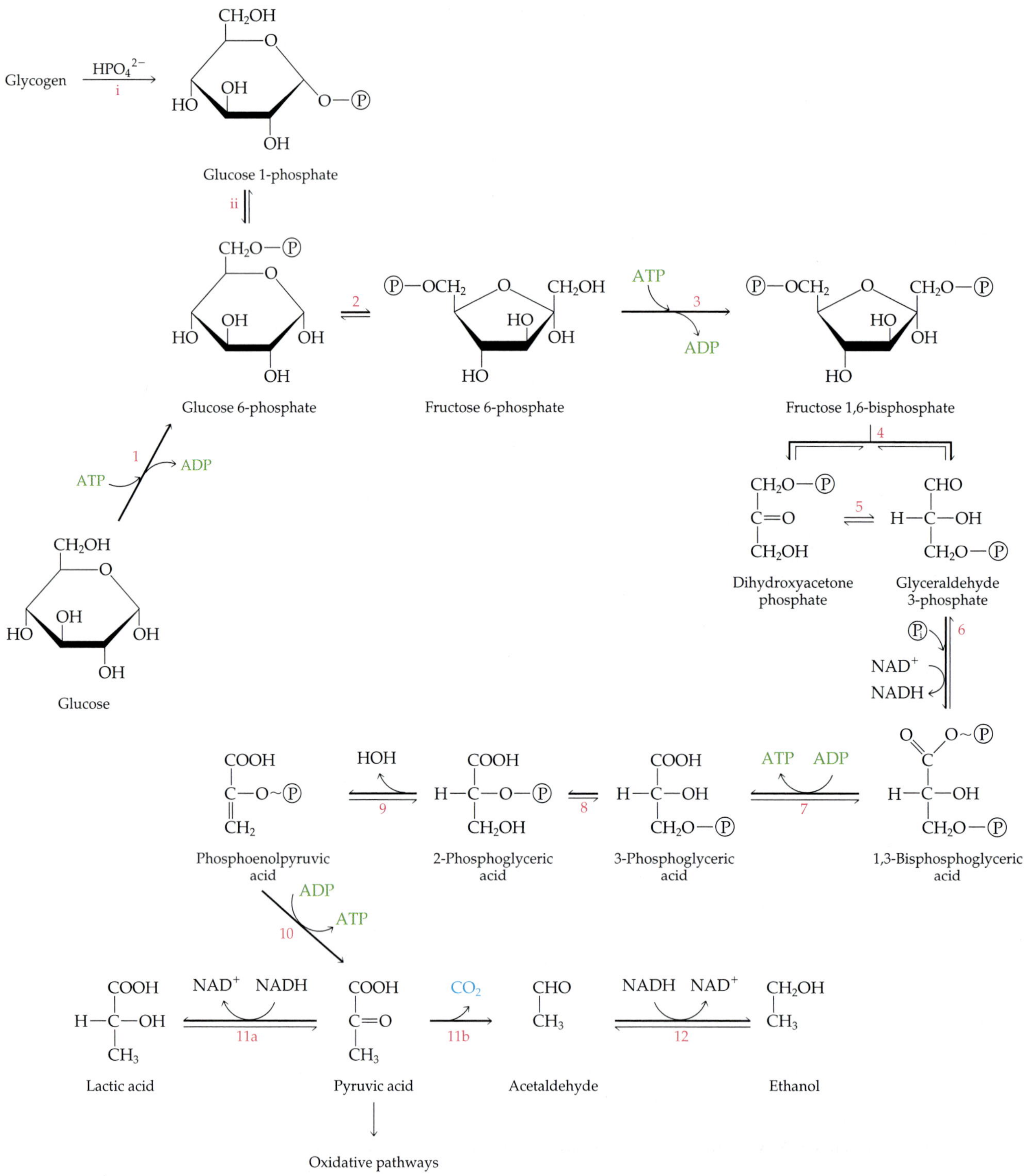

▲ **FIGURE 24.12**

Embden–Meyerhof pathway. Steps i and ii show the formation of glucose 6-phosphate from glycogen. Step 1 shows the formation of glucose 6-phosphate from glucose.

conditions the enzymes in yeast convert the pyruvic acid to ethyl alcohol and carbon dioxide.

In honor of these two scientists, the sequence of metabolic reactions is known as the **Embden–Meyerhof pathway** (Figure 24.12). Two other terms are used to distinguish the end products of the pathway. The anaerobic conversion of glucose to lactic acid and energy is referred to as **glycolysis** (glucose splitting), whereas the anaerobic conversion of glucose to ethyl alcohol, carbon dioxide, and energy is referred to as **fermentation:**

$$
C_6H_{12}O_6 \xrightarrow[\text{without oxygen}]{} 2\ CH_3-\overset{\overset{\displaystyle O}{\|}}{C}-\overset{\overset{\displaystyle O}{}}{C}\diagdown_{OH}
$$

Glucose · · · · · Pyruvic acid

$$
\xrightarrow[\text{oxygen}]{\text{without}}
\begin{cases}
\xrightarrow[\text{fermentation}]{\text{in yeast}} & 2\ C_2H_5OH\ +\ 2\ CO_2\ +\ \text{Energy} \\
& \text{Ethyl} \qquad\ \text{Carbon} \\
& \text{alcohol} \qquad \text{dioxide} \\[2mm]
\xrightarrow[\text{glycolysis}]{\text{in muscle}} & 2\ CH_3CHOHCOOH\ +\ \text{Energy} \\
& \text{Lactic acid}
\end{cases}
$$

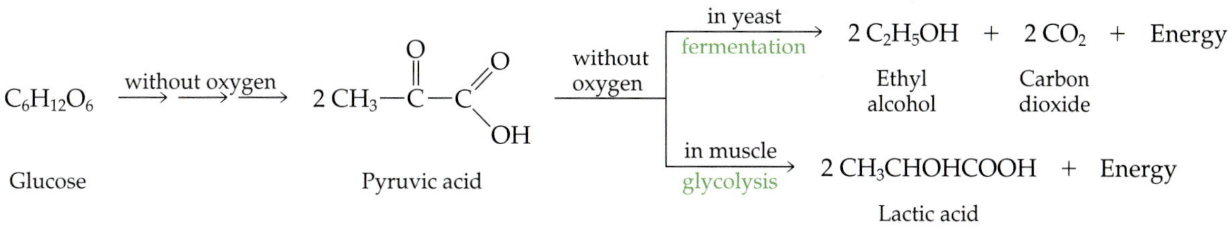

It is important to notice that the conversion of glucose to pyruvic acid represents an oxidation reaction ($C_6H_{12}O_6 \longrightarrow 2\ CH_3COCOOH\ +\ 4\ H^+\ +\ 4\ e^-$) that does not require oxygen. We shall see that the outstanding characteristic of the Embden–Meyerhof pathway is the utilization of the coenzyme NAD^+ as the electron acceptor (oxidizing agent). Thus glycolysis and fermentation differ only in the eventual fate of the pyruvic acid and hence in the means employed for the regeneration of the NAD^+.

The Embden–Meyerhof pathway is probably the best understood of all the metabolic pathways. More than a dozen enzyme molecules act in such a manner that the product of one enzyme-catalyzed reaction becomes the substrate of the next. All the reactions of the pathway occur in the cytoplasm, where the enzymes are in solution. The transfer of intermediates from one enzyme to the next occurs by diffusion. Because these enzymes are found in a soluble form in the cell, they were relatively easy to isolate and characterize.

Each reaction in Figure 24.12 has been assigned a number that corresponds to the discussion that follows. There is often a tendency to become lost in the complexity of this process and lose the overall perspective. We are interested in seeing how the principles of organic chemistry apply to these biochemical reactions, but it must always be kept in mind that *the central theme of metabolism is the extraction of chemical energy from foodstuff* and not merely the degradation of these molecules into simpler substances. In the discussion that follows, the full names of the pertinent enzymes are given in the body of the text. Only the general names are written above or below the arrows of the biochemical equations.

Step i The liver contains the only store of glycogen that can be converted to glucose and released into the blood for transport to all other tissues. Muscle glycogen, on the other hand, is used solely within the muscle. During stress and exercise, skeletal muscle cells use both the glucose transported in the blood and glucose obtained by glycogenolysis for energy production.

Liver and muscle cells contain enzymes called *phosphorylases* that catalyze the phosphorolytic cleavage of the α-1,4-glycosidic bonds at the nonreducing end of the glycogen chain to produce glucose 1-phosphate. The reaction is analogous to a hydrolytic cleavage, except that here inorganic phosphate takes the place of a water molecule. (Two other enzymes, a *transferase* and a *debranching enzyme*, are

McArdle's disease is a genetic disease in which there is a deficiency or absence of phosphorylase in muscle cells. Thus, stored glycogen cannot be used to provide energy. The patient has a limited capacity for exercise and suffers from painful muscle cramps.

required to hydrolytically split off glucose molecules at the branch points of glycogen.) Although this reaction is reversible, different enzymes and different reaction sequences are employed in the biosynthesis of glycogen:

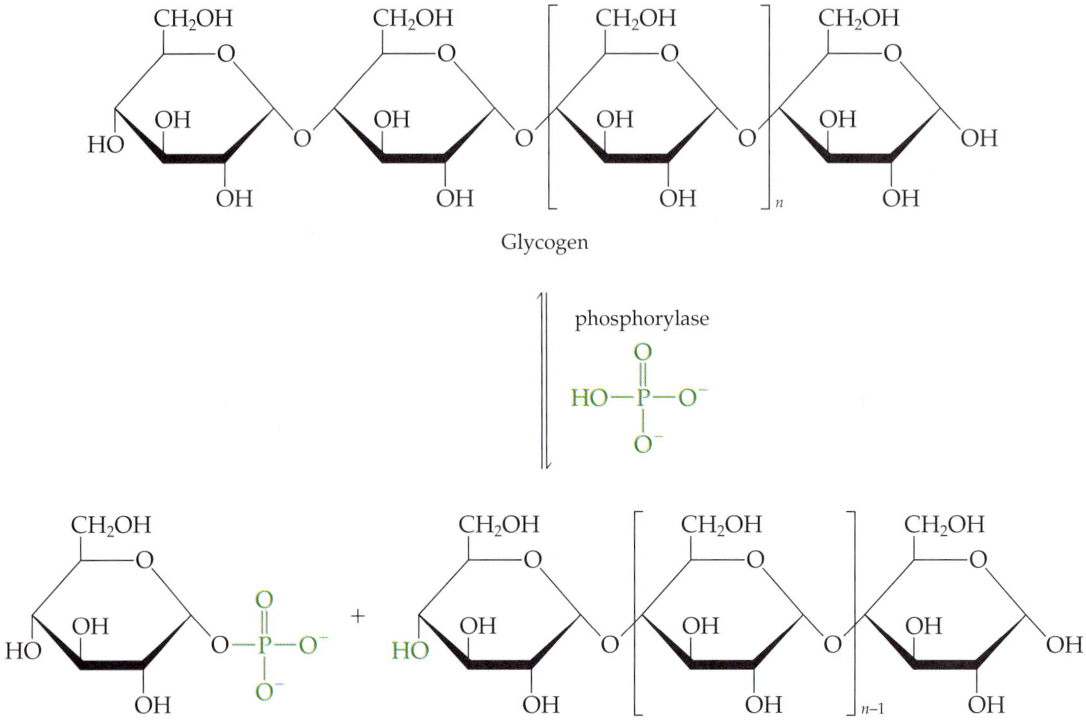

Glycogen

Glucose 1-phosphate

Step ii Glucose 1-phosphate and glucose 6-phosphate are readily interconvertible in the presence of Mg^{2+} ions and the enzyme *phosphoglucomutase.* (In general, a *mutase* is an enzyme that catalyzes the intramolecular transfer of a chemical group.) For simplicity, the isomerization reaction can be thought of as an intramolecular transfer of the phosphate group from C-1 to C-6 and, although this process is reversible, glucose 6-phosphate formation is favored:

The symbol Ⓟ is a shorthand notation for the phosphite group, PO_3^{2-}.

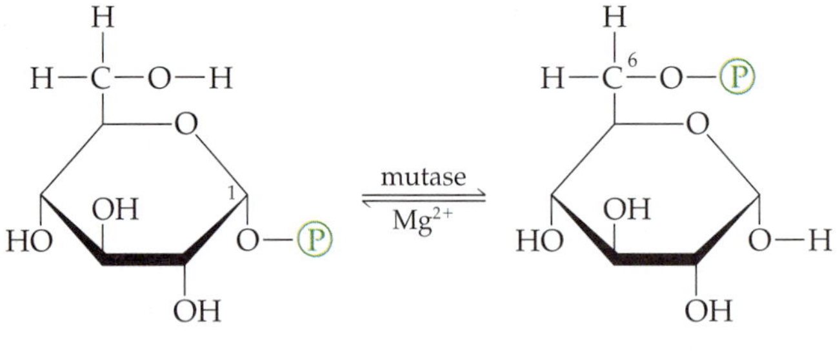

Glucose 1-phosphate

Glucose 6-phosphate

Step 1 The initial step, upon entry of glucose into yeast cells and most animal cells, is phosphorylation to glucose 6-phosphate. (With the exception of liver cells, there is very little free glucose inside most cells.) The phosphate donor in this reaction is ATP, and the enzyme, which requires Mg^{2+} ions for its activity, is *hexokinase:*

Metabolic Pathways of Glucose 6-Phosphate

The fate of glucose 6-phosphate is not the same in muscle cells as in the liver. Because the major role of glycogen in muscle is to provide energy for muscle contraction, the glucose 6-phosphate is immediately converted to fructose 6-phosphate. However, in the liver the chief function of glycogen is to serve as a reservoir of glucose molecules. The liver stores and releases glucose to meet the needs of other tissues. (The liver relies mainly on the oxidation of fatty acids for its own energy needs; see Chapter 26.)

When the blood sugar level decreases, glucose 6-phosphate in the liver is converted to glucose via a hydrolytic reaction catalyzed by glucose 6-phosphatase. (Muscle cells lack this enzyme and therefore cannot export glucose.)

$$\text{Glucose 6-phosphate} + \text{HOH} \xrightarrow[\text{(only in the liver)}]{\text{phosphatase}} \text{Glucose} + P_i$$

The glucose molecules then exit the liver and are transported by the blood for use by other tissues (primarily brain and skeletal muscle). Another genetic disease, von Gierke's disease, is caused by a lack of the enzyme glucose 6-phosphatase in liver cells. Hypoglycemia occurs because glucose cannot be formed from glucose 6-phosphate.

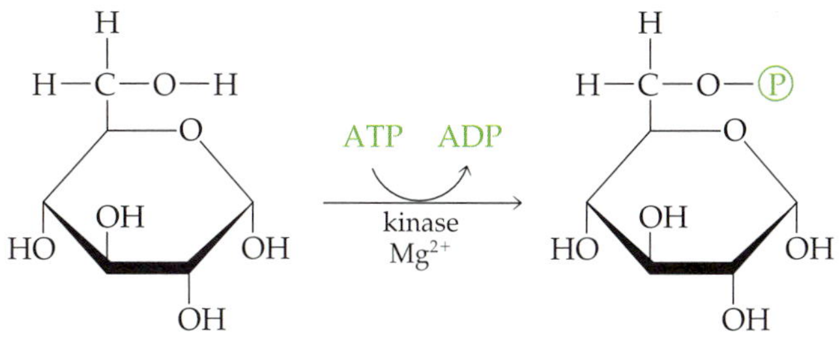

Glucose   Glucose 6-phosphate

The reaction is accompanied by an expenditure of energy because a molecule of ATP is being used rather than synthesized. However, this step is necessary for the activation of the glucose molecule. (We shall see that a second molecule of ATP is expended in Step 3. These two initiating molecules of ATP are recovered at a later stage of the pathway.) Step 1 is essentially irreversible in the cell. ATP is not formed to any appreciable extent by the reaction of ADP with a simple phosphate ester.

Step 2 Glucose 6-phosphate is isomerized to fructose 6-phosphate by the action of *phosphoglucose isomerase.* (In general, an *isomerase* catalyzes the interconversion of one isomeric form of a sugar to another isomeric form.) Phosphoglucose isomerase is highly specific for glucose 6-phosphate:

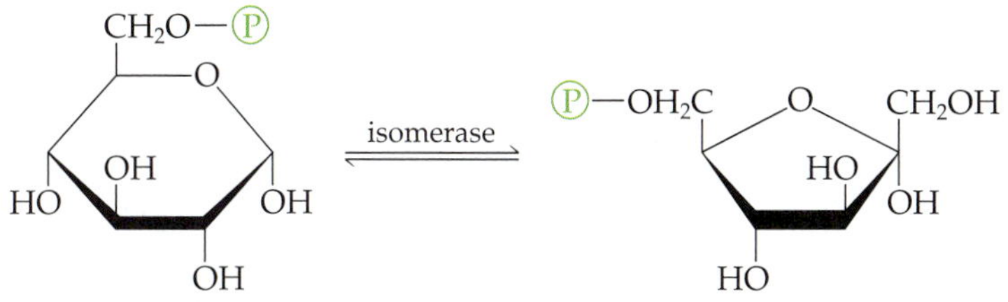

Glucose 6-phosphate   Fructose 6-phosphate

A mechanism for this aldose–ketose transformation involves the formation of an enediol intermediate and is best understood if the open-chain structures of the sugars are considered:

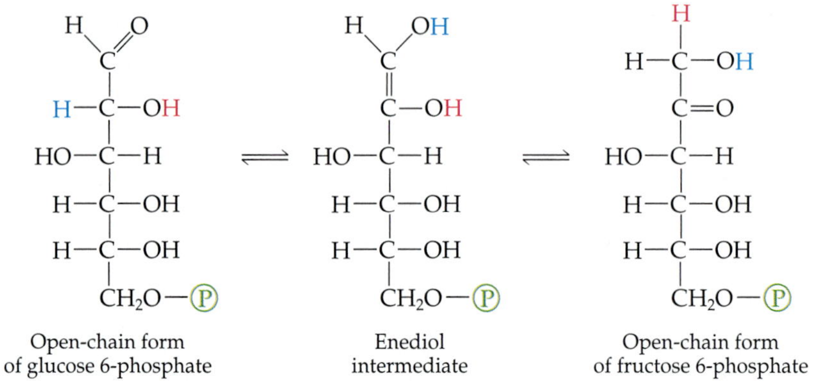

When a molecule contains two phosphate groups on different carbon atoms, the convention is to use the prefix *bis.* When the two phosphate groups are bonded to each other on the same carbon atom (for example, ADP), the prefix is *di.*

Step 3 Next follows another phosphorylation reaction, again involving ATP as the phosphate-group donor. The enzyme *phosphofructokinase*[4] is specific for fructose 6-phosphate and, like hexokinase, requires Mg^{2+} ions for activity. The reaction is irreversible and necessitates the expenditure of energy from a second molecule of ATP:

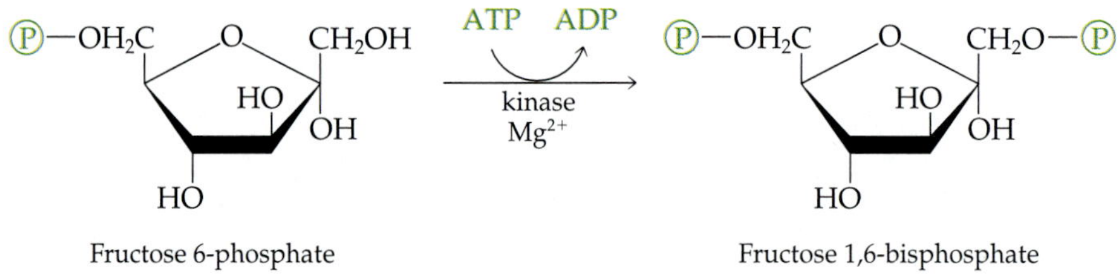

Step 4 Fructose 1,6-bisphosphate is enzymatically cleaved by *aldolase* into two molecules of triose phosphate. Again, a better understanding of the reaction is achieved if the fructose 1,6-bisphosphate is written in its open-chain form:

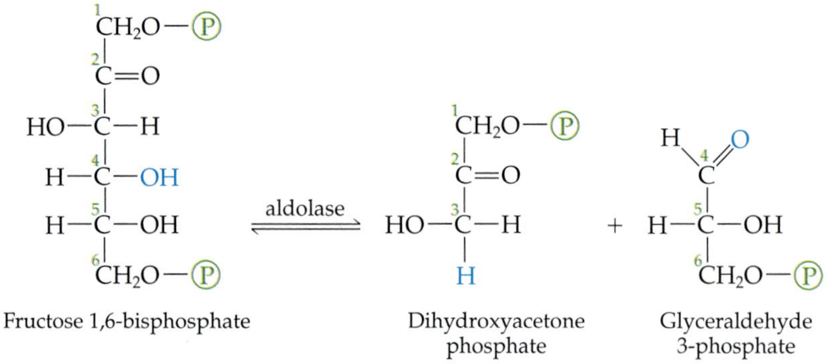

[4] Phosphofructokinase is an enzyme that is primarily responsible for controlling the rate of glycolysis. This enzyme is strictly regulated by the intracellular concentration of ATP. When there is a high concentration of ATP in the cell, ATP binds to the regulatory site, inhibiting the enzyme, and the pace of glycolysis decreases. When ATP is being used to provide energy, the regulatory site is vacant, the enzyme is active, and the glycolytic rate increases.

Step 5 The next step is concerned with the interconversion of the triose phosphates. This step is essential because only glyceraldehyde 3-phosphate can be metabolized by the body. If cells were unable to convert dihydroxyacetone phosphate to glyceraldehyde 3-phosphate, half of the energy stored in the original glucose molecule would be lost and nonmetabolizable ketotriose phosphate would accumulate in the cell. The enzyme for the isomerization reaction is *triose phosphate isomerase,* and the mechanism involves another enediol intermediate:

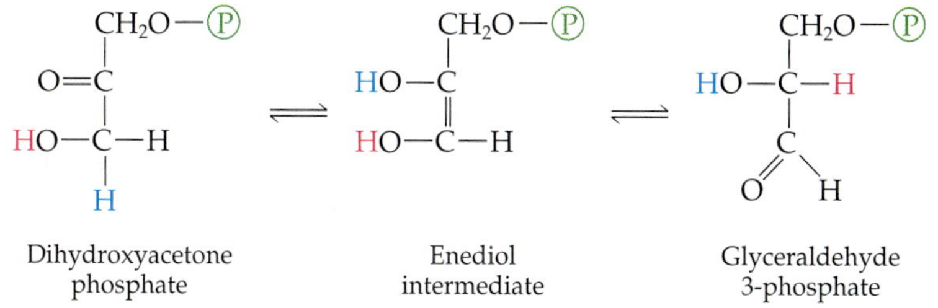

A summation of Steps 4 and 5 indicates that aldolase and triose phosphate isomerase have effectively accomplished the conversion of one molecule of fructose 1,6-bisphosphate to *two* molecules of glyceraldehyde 3-phosphate. All the remaining steps of glycolysis involve three-carbon compounds. Thus far, the pathway has required an energy input in the form of two molecules of ATP but has not yet released any of the energy stored in the glucose.

Step 6 We now come to the first oxidation–reduction reaction in the Embden–Meyerhof pathway. *Glyceraldehyde 3-phosphate dehydrogenase* contains the coenzyme NAD^+. In the process of oxidizing the aldehyde to a carboxylic acid, the coenzyme is reduced to NADH. Notice that the same enzyme also catalyzes a phosphorylation reaction in which inorganic phosphate is the phosphate donor. The energy released by the oxidation reaction is used for the subsequent phosphorylation reaction.[5]

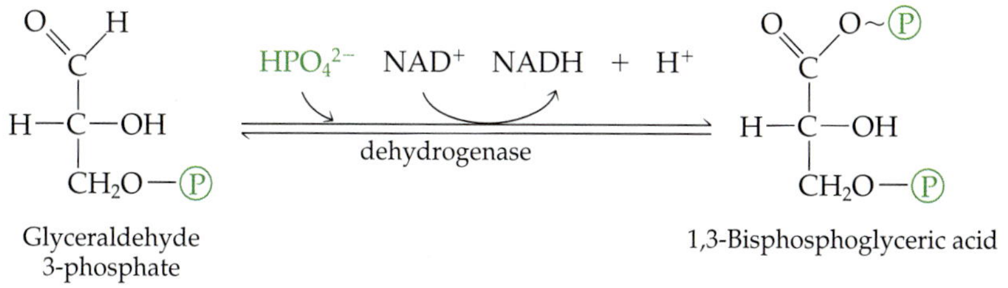

The reaction is more easily understood if it is considered in two steps: (a) oxidation and (b) phosphorylation. Note, however, that this presentation is an oversimplification that in no way implies the actual enzyme-catalyzed sequence of events.

[5] The poisonous effect of iodoacetic acid (see Table 22.2) in blocking glycolysis occurs at this step. The dehydrogenase enzyme contains free sulfhydryl groups at its active site. Iodoacetate irreversibly inhibits the enzyme by covalently bonding to these catalytic groups.

(a) Oxidation

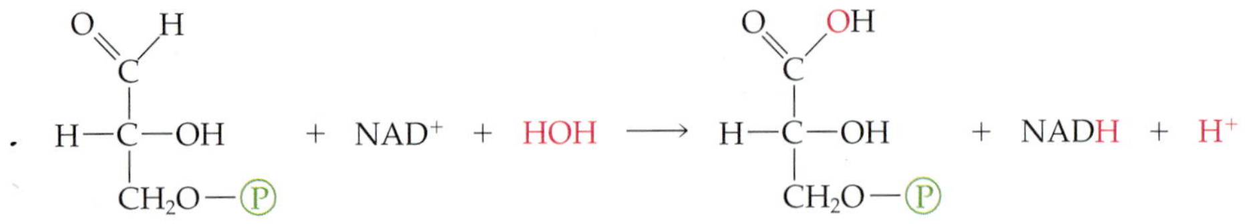

(b) Phosphorylation

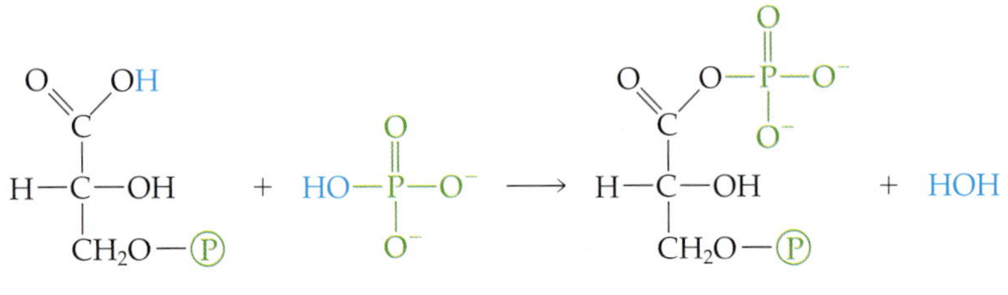

Step 7 1,3-Bisphosphoglyceric acid, the product of the reaction in Step 6, contains a high-energy acyl phosphate bond (see Table 24.1). It can transfer the phosphate group on C-1 directly to a molecule of ADP, thus forming a molecule of ATP. The enzyme that catalyzes the reaction is *phosphoglycerokinase,* which, like all other kinases, requires Mg^{2+} ions for activity. It is in this reaction that ATP is first produced in the pathway. Because the ATP is formed by a direct transfer of a phosphate group from a metabolite to ADP, the process is referred to as **substrate-level phosphorylation** to distinguish it from another ATP-synthesizing process we shall consider in Chapter 25.

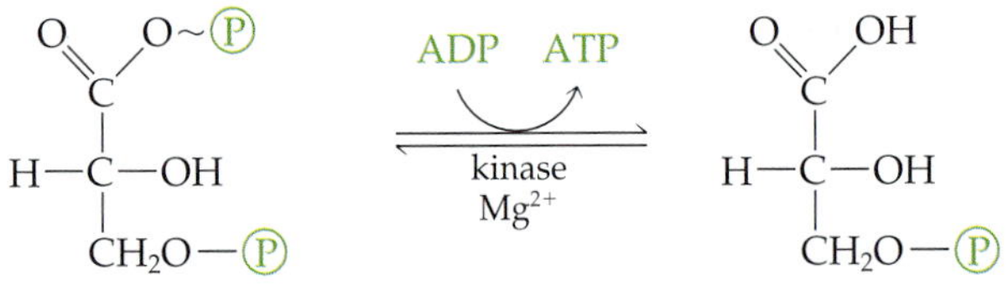

1,3-Bisphosphoglyceric acid — 3-Phosphoglyceric acid

Step 8 This step is similar to the intramolecular transfer of phosphate shown in Step ii. *Phosphoglyceromutase* catalyzes the exchange of a phosphate group from the hydroxyl group of C-3 to the hydroxyl group of C-2:

3-Phosphoglyceric acid — 2-Phosphoglyceric acid

Step 9 *Enolase,* which also requires Mg^{2+} ions for activity, catalyzes an alcohol dehydration reaction to produce phosphoenolpyruvic acid (PEP), a compound with a high-energy enol phosphate group:[6]

[6] Fluoride is an effective poison of the glycolytic pathway because of its inhibition of enolase (Table 22.2). Fluoride ions bind to the activator Mg^{2+} ions to form a magnesium fluorophosphate complex.

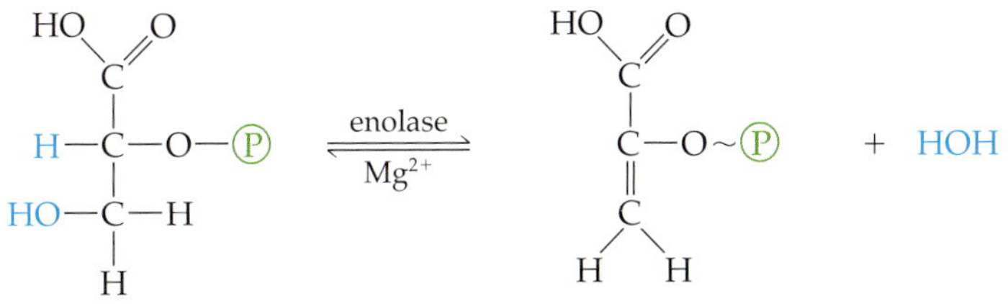

2-Phosphoglyceric acid

Phosphoenolpyruvic acid (PEP)

Step 10 This irreversible step is a second substrate-level phosphorylation. The phosphate group of PEP is transferred to ADP; one molecule of ATP is produced per molecule of PEP. It is likely that the reaction proceeds via the enol form of pyruvic acid, which is unstable and rearranges to pyruvic acid. *Pyruvate kinase* requires both Mg^{2+} and K^{+} ions for activity.

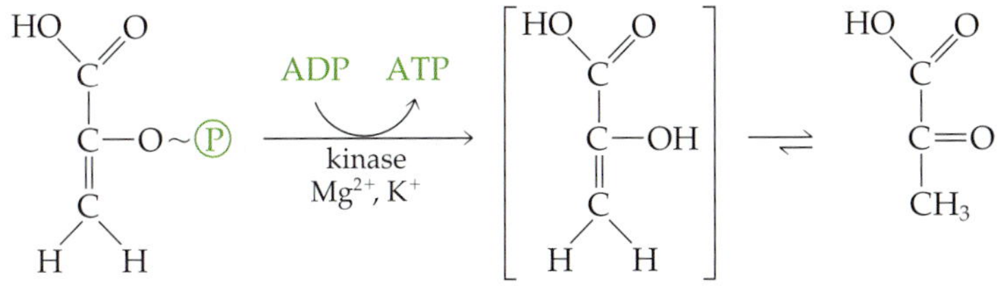

PEP

Enol form of pyruvic acid

Pyruvic acid

Steps 1 through 10 are identical for glycolysis and fermentation. Pyruvic acid, as we shall see, is the *crossroads* compound—its metabolic fate depends on the availability of oxygen, the organism under consideration, and the tissue involved.

24.6 Glycolysis

Under usual conditions and during moderate exercise, the respiration of muscle cells is **aerobic** (in the presence of oxygen; see Chapter 25). During strenuous exercise, however, the energy demand on the muscles is enormous, and the respiratory and circulatory systems are unable to deliver sufficient oxygen to these cells. (This condition, called **oxygen debt,** occurs when not enough oxygen is available for cellular activities.) As a result, the muscle cells must obtain energy via the anaerobic pathway.

Step 11a In the presence of NADH, *lactic acid dehydrogenase*[7] catalyzes the reduction of pyruvic acid to lactic acid (ketone to secondary alcohol):

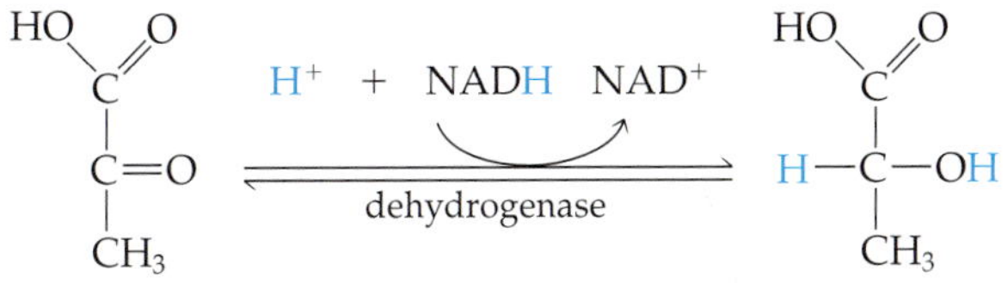

Pyruvic acid

Lactic acid

[7] This enzyme derives its name from its catalytic role in reversing this reaction. Enzymes catalyze both the forward and reverse reactions of an equilibrium mixture and may be named accordingly. An equally descriptive name for this enzyme is *pyruvic acid reductase* or *hydrogenase*.

This reaction is essential because it regenerates the NAD^+ needed in Step 6. If NAD^+ were not replenished, the cell's supply of this coenzyme could be swiftly depleted, the Embden–Meyerhof pathway would cease, and glyceraldehyde 3-phosphate would accumulate in the cell.

Lactic acid, then, is the end product of glycolysis.[8] If there were not some mechanism for its removal, it would accumulate in the muscle cells, raising the level of acidity. The generation of energy via the anaerobic pathway is therefore self-limiting. As lactic acid builds up in muscle tissue, the pH drops (from a resting value of 7.0 to about 6.5) and deactivates the enzymes required for glycolysis. The muscle's response to stimuli becomes weaker, and in extreme cases there may be no response at all. In this state the muscle is described as fatigued. Two processes act to maintain a proper level of lactic acid, and both require oxygen.

The Cori cycle is named in honor of Gerty and Carl Cori, who first described it. The Coris won a Nobel prize in 1947, the third husband/wife team to achieve this distinction (Marie and Pierre Curie were the first; Irène and Frédéric Joliot-Curie, the second).

1. Most (70 to 80%) of the lactic acid diffuses out of the muscle and is transported to the liver. There it may be oxidized to pyruvic acid and then to carbon dioxide and water (via the Krebs cycle; see Section 25.1) or converted back to glucose (gluconeogenesis). The anaerobic catabolism of glucose to lactic acid in muscle cells, the transport of the lactic acid via the blood to the liver, and the reconversion of lactic acid to glucose comprise the **Cori cycle** (Figure 24.13).
2. The 20 to 30% of the lactic acid that remains in the muscle cells can be reoxidized to pyruvic acid, which then enters the Krebs cycle and is further oxidized to carbon dioxide and water during those periods when the muscle cells receive an ample supply of oxygen.

FIGURE 24.13 ►
The Cori cycle.

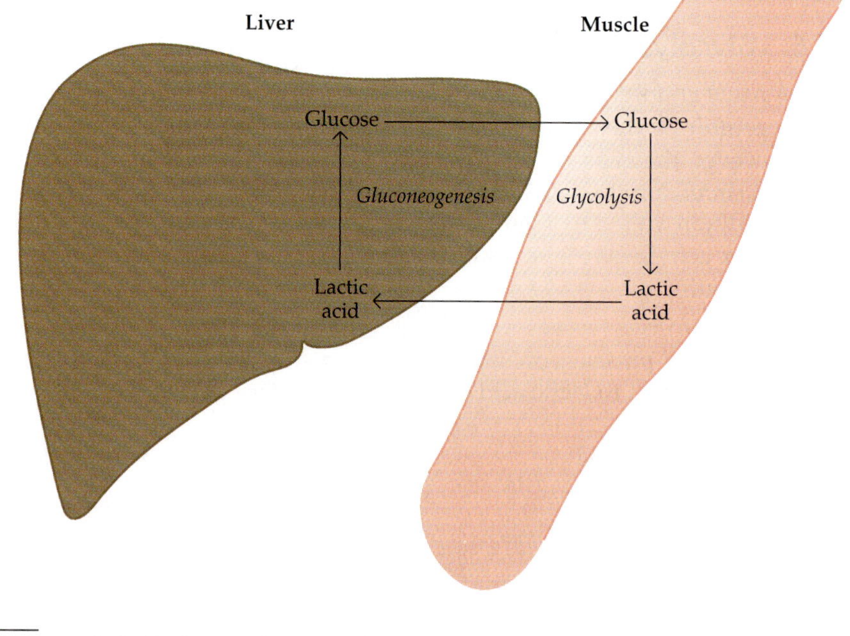

[8] The average individual, expending normal energy, produces about 120 g of lactic acid each day. (Vigorous exercise produces significantly more lactic acid.) It is estimated that one-third of this daily lactic acid is produced by tissues that are strictly anaerobic (i.e., erythrocytes, retina, renal medulla, epidermis)—tissues that do not have mitochondria.

We have seen that, when muscles use anaerobic pathways, they incur an oxygen debt. It is as if the body regards oxidation as the only proper source of energy for muscular activity and uses anaerobic metabolism as a temporary expedient. As soon as it can, the body oxidizes some of the resulting lactic acid back to pyruvic acid and ultimately to carbon dioxide and water. The energy released in the process is used to convert the rest of the lactic acid back to glucose and then to glycogen to be stored for future use.

When is this oxygen debt repaid? Just as soon as the very high level of muscular activity ceases. Sprinters, while running a 100-m dash, breathe in oxygen but still obtain only a fraction of the energy required for their intense muscular activity through aerobic processes. When the race is over, the sprinters continue to take in great gulps of air. This air is used to repay the oxygen debt incurred during the race. We continue to breathe hard even after we stop vigorous activity because our body chemistry is still catching up and needs more oxygen.

24.7 Fermentation

Yeast and many other organisms (including humans) are referred to as *facultative anaerobes,* which means they can function either anaerobically or aerobically, depending on circumstances. They utilize either an organic molecule or molecular oxygen as the terminal electron acceptor. In the absence of oxygen, yeasts metabolize glucose to carbon dioxide and ethanol, whereas in the presence of oxygen they metabolize glucose to carbon dioxide and water. (This is the reason air must be excluded from the fermentation vats in the production of alcoholic beverages.) *Strict anaerobes,* such as the bacteria that cause gangrene and botulism, can use only organic molecules as terminal electron acceptors in metabolism and are poisoned in the presence of oxygen.

Step 11b Yeasts and other microorganisms that can live in a limited supply of oxygen must also have some means of regenerating NAD^+. They do so by first decarboxylating pyruvic acid to acetaldehyde. This reaction, which is catalyzed by *pyruvic acid decarboxylase,* is responsible for CO_2 production in yeast (the reason yeast is used in certain baking processes). The enzyme requires the coenzyme thiamine pyrophosphate (TPP; see Table J.1) and Mg^{2+} ions. The decarboxylation reaction is irreversible:

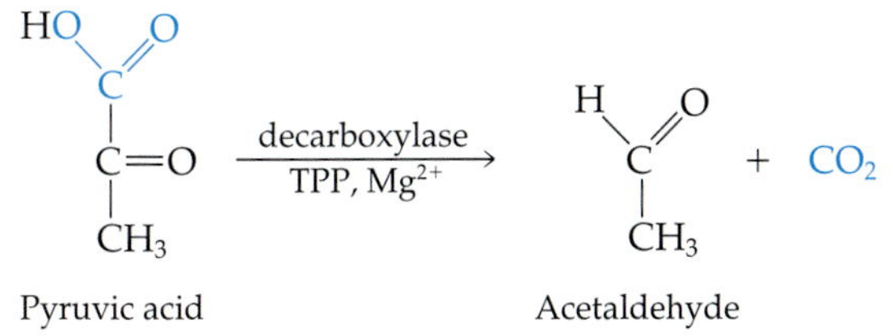

Step 12 The final step in fermentation is catalyzed by *alcohol dehydrogenase.* Acetaldehyde is reduced by NADH to ethyl alcohol, thus regenerating NAD^+. (When yeast is used in baking, the alcohol readily evaporates out of the oven.)

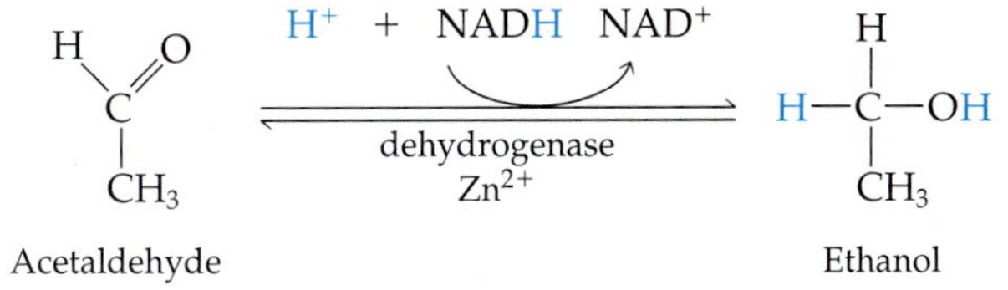

Alcohol Metabolism

About 3 g of ethyl alcohol is produced each day by microbial fermentation in the large intestine. For many people, this amount is insignificant compared with the amount ingested from alcoholic beverages. In humans, the principal route of metabolism of ingested alcohol is believed to be oxidation in the liver, first to acetaldehyde and subsequently to acetic acid. Most of the acetic acid is released by the liver and transported to other tissues, where it is converted to acetyl CoA. (As we shall see, the acetyl CoA is either metabolized to carbon dioxide and water or used in the biosynthesis of fat.)

$$CH_3CH_2OH \xrightarrow[\text{dehydrogenase}]{\text{alcohol}} CH_3C\overset{O}{\underset{H}{\diagup}} \xrightarrow[\text{dehydrogenase}]{\text{acetaldehyde}}$$

$$CH_3C\overset{O}{\underset{OH}{\diagup}} \xrightarrow[\text{synthetase}]{\text{acetyl CoA}} CH_3C\overset{O}{\underset{SCoA}{\diagup}}$$

In a pregnant woman, alcohol readily crosses the placental membrane and builds up in the fetus because the liver of the fetus does not contain the detoxifying enzymes. Therefore, children of alcoholic mothers have a high risk of *fetal alcohol syndrome* (FAS). FAS consists of facial deformities, growth deficiency, and mental retardation. It is believed to be the third most common cause of mental deficiency. (IQs average 35 to 40 points below normal.) Some researchers say that FAS can occur even if a pregnant woman drinks only moderately (one drink daily). In the United States, more than 2000 FAS children are born each year.

After heart disease and cancer, alcohol addiction is the third largest health problem in the United States. More than 200,000 people die of it each year (mainly as a result of cirrhosis of the liver, but also of cardiovascular disease and cancers of the mouth and larynx), and it is a factor in one of every 10 deaths. People who drive while intoxicated are responsible for about half of U.S. traffic fatalities. Alcohol-impaired driving is the leading cause of death and injury among those under 25 years of age.

One proven method in the treatment of chronic alcoholism is to administer a drug that inhibits the second step of alcohol metabolism. The drug disulfiram (Antabuse) successfully competes with acetaldehyde for the active site on the enzyme acetaldehyde dehydrogenase. As a result, when alcohol is ingested by an individual previously treated with disulfiram, the blood acetaldehyde concentrations rise five to 10 times higher than in an untreated individual. This effect is accompanied by severe discomfort. Characteristic physiological responses are an increase of the heartbeat and a reduction of the blood pressure. The individual experiences respiratory difficulties, nausea, sweating, vomiting, chest pains, and blurred vision. Once elicited, the effect lasts between 30 min and several hours. To avoid such discomfort, the patients must abstain from alcohol (and the treatments are continued until they can abstain voluntarily). Disulfiram should be administered only by a physician, and therapy is usually begun in a hospital.

$$CH_3CH_2 \diagdown \\ \underset{CH_3CH_2}{\diagup} N-\overset{S}{\overset{\|}{C}}-S-S-\overset{S}{\overset{\|}{C}}-N \diagdown \overset{CH_2CH_3}{\diagup}_{CH_2CH_3}$$

Disulfiram

More than two-thirds of the adult population in the United States drinks alcoholic beverages occasionally. The majority do so responsibly.

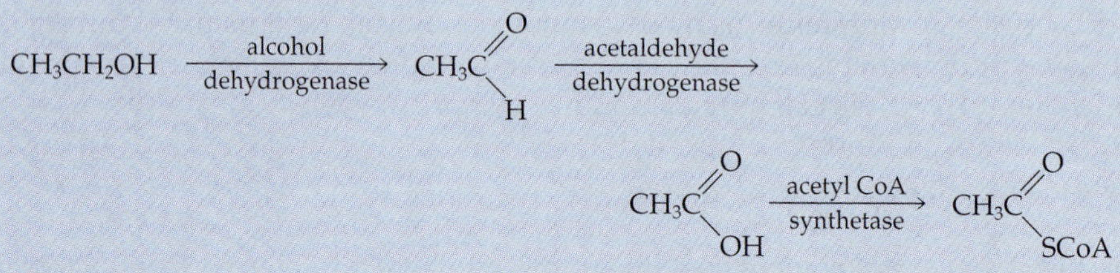

A syndrome is a set of conditions that occur together and are characteristic of a particular disease.

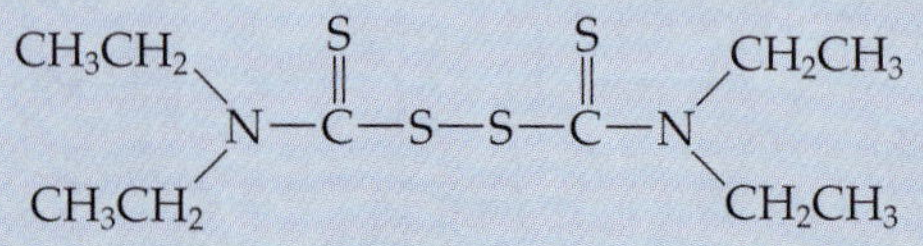

24.8 Reversal of Glycolysis and Fermentation

It has been mentioned that several degradative steps in the Embden–Meyerhof pathway are not reversible *in vivo*. (Single arrows are used in Figure 24.12 to indicate that the reactions in Steps i, 1, 3, 10, and 11b are irreversible.) This seems to be inconsistent with the fact that lactic acid and any other intermediates of the glycolytic sequence can be converted back to glucose and/or glycogen provided an energy source is available. Therefore, other reactions catalyzed by different enzymes must necessarily effect a reversal of Steps i, 1, 3, and 10.

The irreversibility of parts of a metabolic pathway is a common phenomenon—it will be encountered again in aerobic metabolism as well as in the metabolism of lipids and proteins. Catabolic and anabolic processes may have many of the same intermediates and many of the same enzymes, and yet their pathways are not identical. Similarly, more than one pathway may exist for the same process. The separation of degradative and synthetic pathways allows an organism to control one series of reactions without affecting the other. Because we are concerned primarily with reactions that yield energy and produce ATP, we shall not take up the anabolism of carbohydrates. These reactions are discussed in advanced biochemistry texts.

24.9 Bioenergetics of Glycolysis and Fermentation

The net energy yield in the form of high-energy phosphate bonds (moles of ATP) obtained from the anaerobic metabolism of each mole of glucose or glycogen can be easily calculated from Figure 24.12. One mole of ATP is expended in the initial phosphorylation of glucose (Step 1):

$$\text{Glucose} \xrightarrow{\text{ATP} \quad \text{ADP}} \text{Glucose 6-phosphate}$$

A second mole is consumed in the phosphorylation of fructose 6-phosphate (Step 3):

$$\text{Fructose 6-phosphate} \xrightarrow{\text{ATP} \quad \text{ADP}} \text{Fructose 1,6-bisphosphate}$$

Steps 4 and 5 are very significant, for it is in these reactions that each mole of six-carbon sugar is transformed into 2 mol of the triose phosphate glyceraldehyde 3-phosphate. In Step 7, each mole of 1,3-bisphosphoglyceric acid is converted to 1 mol of 3-phosphoglyceric acid; 1 mol of ATP is produced per mole of triose phosphate, which means *2 mol of ATP per mole of glucose:*

$$\text{1,3-Bisphosphoglyceric acid} \xrightarrow{\text{ADP} \quad \text{ATP}} \text{3-Phosphoglyceric acid}$$

In Step 10, 1 mol of ATP is generated for each mole of pyruvic acid formed from phosphoenolpyruvic acid; again, 2 mol of ATP is produced for each mole of glucose that entered the pathway:

$$\text{Phosphoenolpyruvic acid} \xrightarrow{\text{ADP} \quad \text{ATP}} \text{Pyruvic acid}$$

A summation of all these steps reveals that, for every mole of glucose degraded, 2 mol of ATP is initially consumed and 4 mol of ATP is ultimately produced. The net production of ATP is thus 2 mol per mole of glucose converted to lactic acid or to ethanol. If, however, glycogen is used as the source of glucose, Steps i and ii are operative instead of Step 1. It would then be necessary to expend only 1 mol of ATP (in Step 3) to produce fructose 1,6-bisphosphate, and the net yield of ATP would be 3 mol per mole of glucose 1-phosphate. Thus, the utilization of glycogen for anaerobic energy production represents a 50% increase in efficiency compared with the utilization of glucose:

Starting with Glucose (in Yeast)

$$C_6H_{12}O_6 + 2\,ADP + 2\,P_i \longrightarrow 2\,C_2H_5OH + 2\,CO_2 + 2\,ATP$$

Starting with Glycogen (in Muscle)

$$(C_6H_{12}O_6)_n + 3\,ADP + 3\,P_i \longrightarrow (C_6H_{12}O_6)_{n-1} + 2\,CH_3CHOHCOOH + 3\,ATP$$

Recall that about 7300 cal of free energy is conserved per mole of ATP produced (Section 24.1). Recall also that the total amount of energy that can theoretically be obtained from the complete oxidation of 1 mol of glucose is 686,000 cal. The energy derived from glucose by anaerobic metabolism, then, is only a minute amount of the total energy available—that is, either 2.1% [(2 × 7300)/686,000] or 3.2% [(3 × 7300)/686,000]. Thus, anaerobic cells extract only a small fraction of the total energy of the glucose molecule. Yet this amount is sufficient for their survival. Since they are not nearly as efficient as aerobic cells, it is necessary that anaerobic cells use more glucose per unit of time to accomplish the same amount of cellular work.

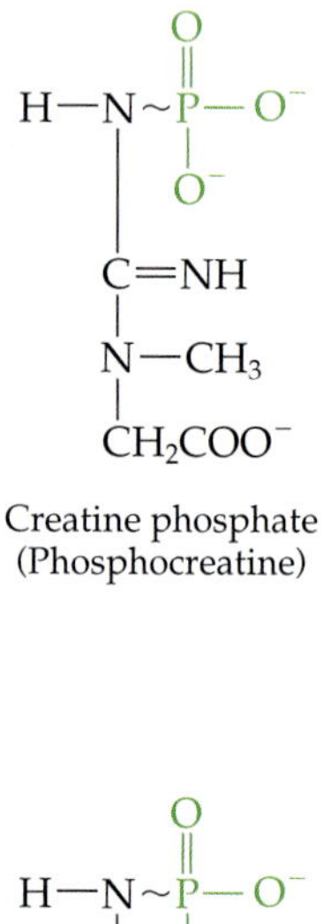

Creatine phosphate
(Phosphocreatine)

24.10 Storage of Chemical Energy

We now consider how muscle cells store the energy they extract from carbohydrates (and other compounds). In 1927, a compound called *creatine phosphate* was isolated from mammalian muscle. It was subsequently demonstrated to be the storage form of energy in the muscles of vertebrates and in nerve tissue. *Arginine phosphate* serves the same purpose in the muscles of invertebrates. These high-energy phosphate compounds, which serve as reservoirs of phosphate bond energy, are often called *phosphagens.* The high-energy bond of these particular phosphagens occurs between phosphorus and nitrogen rather than between phosphorus and oxygen.

The turnover rate of ATP is very high, and this precludes its use as a storage form of energy. A 70-kg man will hydrolyze and resynthesize about 70 kg of ATP per day. Therefore, the concentration of ATP in muscle is relatively low and cannot meet the demands of muscular exertion for more than 2 to 3 seconds. At rest, mammalian muscle contains four to six times as much creatine phosphate as ATP. As ATP is utilized, creatine phosphate, in the presence of creatine kinase, reacts with ADP to produce more ATP and creatine:

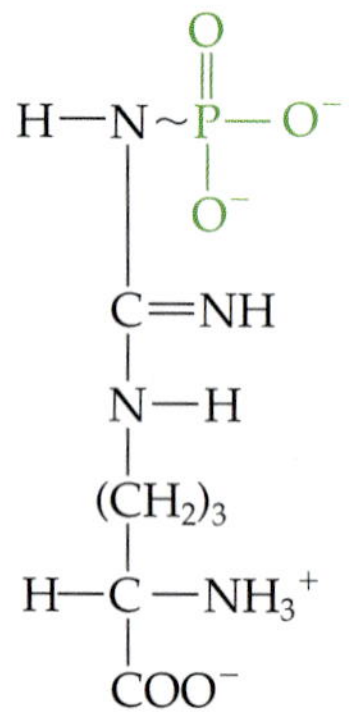

Arginine phosphate
(Phosphoarginine)

$$\text{Creatine phosphate} + ADP \underset{Mg^{2+}}{\overset{\text{kinase}}{\rightleftharpoons}} \text{Creatine} + ATP$$

The reaction is readily reversible. When muscular activity is required, the reaction proceeds to the right. When there is abundant ATP (from the catabolism of foodstuffs), the reaction proceeds to the left, and creatine phosphate is stored in the muscle cells. Because the concentration of creatine phosphate in the muscle is limited, this compound is useful only in generating a quick source of utilizable energy.

It has been estimated that the creatine phosphate in a cell can provide energy for about 20 s of strenuous exercise (long enough to sprint 200 m). Energy for prolonged activity is obtained from the anaerobic breakdown of glycogen synthesized during long periods of muscular inactivity. The glycogen stored in muscle cells is more readily available for energy production than is glucose from the blood. In the days just preceding a marathon, competitors "load up" on carbohydrates in order to maximize the amount of glycogen stored in muscle cells.[9]

The human body can store roughly 450 g of glucose as glycogen (about 350 g in all of the muscle cells and about 100 g in the liver). The liver glycogen reservoir allows us to eat intermittently because it provides the between-meal supply of glucose needed to maintain the blood sugar level. Stored glycogen, however, does not last long. After a 24-h fast, very little glycogen remains in the liver. We shall see in Chapter 26 that when carbohydrate intake exceeds the amount required for energy and for storage as glycogen, the excess is converted to fat and stored in adipose tissue. When the glycogen reserves are depleted, fatty acids provide the bulk of the energy for muscle cells.

Summary

Metabolism is the general term for all chemical reactions in living organisms. The two types of metabolism are **catabolism,** those reactions in which complex molecules are broken down to simpler ones with the concomitant release of energy, and **anabolism,** those reactions that consume energy in order to build complex molecules.

Oxidation of carbohydrates, a process called **respiration,** is the source of most of the energy used by cells. Living organisms obtain the energy they need for anabolism from *adenosine triphosphate* (*ATP*). Catabolic reactions move energy from food molecules to ATP molecules; anabolic reactions use the energy in ATP to create new compounds.

Digestion of carbohydrates begins in the mouth as saliva coats food particles and α-amylase breaks α-glycosidic linkages in the carbohydrate molecules. There is essentially no carbohydrate digestion in the stomach, and the saliva-coated particles pass through to the small intestine. Here amylopsin converts complex carbohydrate molecules (starches) to glucose. The glucose molecules then pass through the lining of the small intestine and into the bloodstream for transport to all the body cells.

Blood sugar level is the concentration of glucose in the blood. The condition in which this concentration is high is **hyperglycemia;** that in which this concentration is low is **hypoglycemia.**

The body regulates blood sugar level by controlling the amount of glycogen that is either synthesized or metabolized in the liver. When blood sugar levels are high, *insulin* secreted from the pancreas prompts liver and muscle cells to take excess glucose from the blood and convert it to glycogen (**glycogenesis,** the "birth of glycogen"). When blood sugar levels are low, little insulin is present in the blood, and liver cells convert glycogen to glucose. When blood sugar levels are really low, as when a person is running a marathon or starving, liver cells convert noncarbohydrate molecules to glucose (**gluconeogenesis,** the "birth of new glucose").

Epinephrine and *glucagon* also regulate glucose metabolism. Their action always raises blood sugar levels because they both promote gluconeogenesis and **glycogenolysis,** the breakdown of glycogen to glucose.

The series of reactions summarized in the respiration equation are known as the **Embden–Meyerhof pathway.**

[9] In carbohydrate loading, glycogen stores are first depleted by limiting intake of carbohydrates and training vigorously. Then, a few days before competition, the athlete cuts back on training and eats a diet high in carbohydrates. The presumption is that, under this regimen, the body will store more glycogen than usual. The benefits of carbohydrate loading, even for top athletes, are questionable. For casual athletes, the technique is probably of very little value.

These reactions **anaerobically** convert carbohydrate molecules (either glycogen or glucose) to pyruvic acid. In the process, *energy is extracted from glucose and transferred to ATP.* It is this energy, later released from the ATP, that allows all living organisms to function. The net energy yield in the pathway depends on whether the starting material is glucose or glycogen; that is, 2 mol ATP per mole of glucose and 3 mol ATP per mole of glucose units in glycogen.

The pyruvic acid produced in the Embden–Meyerhof pathway has two possible fates. In **glycolysis,** which takes place in animal cells, the pyruvic acid is converted to lactic acid. In **fermentation,** which takes place in yeast cells, it is converted to ethyl alcohol and carbon dioxide.

Key Terms

aerobic (24.6)
anabolism (Introduction)
anaerobic (24.5)
catabolism (Introduction)
Cori cycle (24.6)
diabetes mellitus (24.4)
digestion (24.2)
Embden–Meyerhof pathway (24.5)
fermentation (24.5)
gluconeogenesis (24.4)
glucose tolerance test (24.3)
glycogenesis (24.4)
glycogenolysis (24.4)
glycolysis (24.5)
hyperglycemia (24.3)
hypoglycemia (24.3)
metabolic pathway (24.5)
metabolism (Introduction)
oxygen debt (24.6)
photosynthesis (Introduction)
renal threshold (24.3)
respiration (Introduction)
saliva (24.2)
substrate-level phosphorylation (24.5)

Review Questions

1. Define or give an explanation for:
a. metabolite **b.** villi
c. blood sugar level **d.** renal threshold
e. gluconeogenesis **f.** glucose tolerance test
g. cyclic AMP **h.** adenyl cyclase
i. Ⓟ **j.** metabolic pathway
k. oxygen debt **l.** kinase
m. FAS **n.** Cori cycle
o. carbohydrate loading **p.** Antabuse

2. Distinguish between:
a. anabolism and catabolism
b. photosynthesis and respiration
c metabolism and digestion
d. active transport and passive transport
e. hypoglycemia and hyperglycemia
f. glycogenesis and glycogenolysis
g. anaerobic and aerobic
h. glycolysis and fermentation
i. a mutase and an isomerase
j. a phosphatase and a phosphorylase
k. facultative anaerobes and strict anaerobes

3. What is the general type of reaction used in digestion?
4. What is mucin? What is its function in saliva?
5. What are the end products of carbohydrate digestion?
6. In what section of the digestive tract is most carbohydrate digested?
7. If a cracker, which is rich in starch, is chewed for a long time, it begins to develop a sweet, sugary taste. Why?
8. Glucose appears in the urine of a diabetic patient. Does glucose in the urine always indicate diabetes mellitus? Explain.
9. What is the storage form of carbohydrate in the body?
10. In what tissues or organs are carbohydrates stored?
11. What happens to the monosaccharide galactose after it reaches the liver?
12. In glycolysis, how many molecules of pyruvic acid are produced from one molecule of glucose?

Problems

ATP: High-Energy Phosphates

13. What are the structural differences among ATP, ADP, and AMP?
14. Why is ATP referred to as the energy currency of the cell?
15. Referring to Table 24.1, indicate which compounds are high-energy phosphates:
a. ATP **b.** AMP **c.** ADP
16. Referring to Table 24.1, indicate which compounds are high-energy phosphates:
a. creatine phosphate **b.** glucose 1-phosphate
c. glucose 6-phosphate

Digestive Enzymes

17. Give the location of action and the function of
a. salivary amylase (ptyalin)

b. lactase

c. sucrase

18. Give the location of action and the function of
 a. pancreatic amylase (amylopsin)
 b. maltase

Blood Sugar Levels

19. What is the major factor in maintaining a constant blood sugar level?

20. What is the role of insulin in regulating blood sugar level?

21. How do epinephrine and glucagon act to raise blood sugar level?

22. Why can't insulin be taken orally?

23. In structure and purpose, how do oral drugs such as Orinase differ from insulin?

24. Briefly describe the two types of diabetes mellitus.

Cyclic AMP

25. When a hormone causes an increase in cAMP levels in the cells, where is the binding site for the hormone located?

26. Explain how the binding of a hormone at its receptor site can release cAMP in a target cell.

Biochemical Reactions

27. When aldolase catalyzes the cleavage of ketose mono- and bisphosphates, dihydroxyacetone phosphate is always one of the products. Complete the following catalytic reactions.

a.

$$
\begin{array}{c}
CH_2O-\textcircled{P} \\
| \\
C=O \\
| \\
HO-C-H \\
| \\
H-C-OH \\
| \\
H-C-OH \\
| \\
H-C-OH \\
| \\
CH_2O-\textcircled{P}
\end{array}
\quad \xrightleftharpoons{\text{aldolase}}
$$

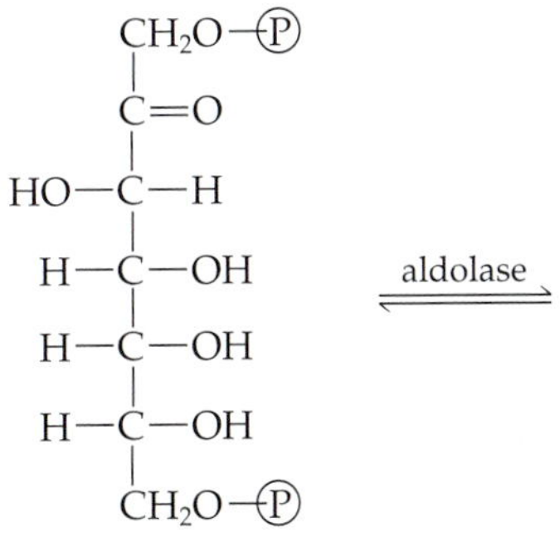

Sedoheptulose
1,7-bisphosphate

b. Fructose 1-phosphate $\xrightleftharpoons{\text{aldolase}}$

28. Draw a molecule of glucose, and label the carbon atoms in the 3 and 4 positions. **(a)** Which appears in lactic acid? **(b)** Which appears in ethanol or carbon dioxide?

29. In the Embden–Meyerhof pathway **(a)** which is the oxidative step? **(b)** What is the oxidizing agent?

30. In the Embden–Meyerhof pathway, **(a)** how is the reduced coenzyme reoxidized in yeast cells? **(b)** How is the reduced coenzyme reoxidized in muscle cells?

31. What is the fate of the lactic acid formed by muscular activity?

32. Write a balanced half-reaction for each conversion:
 a. Glucose \longrightarrow Pyruvic acid
 b. Pyruvic acid \longrightarrow Lactic acid
 c. Pyruvic acid \longrightarrow Ethanol + Carbon dioxide

33. Replace the question marks with the proper compounds:

 a. ? $\xrightarrow{\text{alcohol dehydrogenase}}$ Ethanol

 b. Fructose 1,6-bisphosphate $\xrightarrow{\text{aldolase}}$? + ?

 c. 2-Phosphoglyceric acid $\xrightarrow{?}$ Phosphoenolpyruvic acid

 d. ? $\xrightarrow{\text{glucose 6-phosphatase}}$ Glucose

 e. Glycogen $\xrightarrow{\text{phosphorylase}}$?

 f. Dihydroxyacetone phosphate $\xrightarrow{?}$ Glyceraldehyde 3-phosphate

 g. Glucose $\xrightarrow{\text{hexokinase}}$?

 h ? $\xrightarrow{\text{lactic acid dehydrogenase}}$ Lactic acid

34. Replace the question marks with the proper compounds:

 a. Fructose 6-phosphate $\xrightarrow{?}$ Fructose 1,6-bisphosphate

 b. ? $\xrightarrow{\text{phosphoglucose isomerase}}$ Fructose 6-phosphate

 c. Glyceraldehyde 3-phosphate $\xrightarrow{\substack{\text{glyceraldehyde} \\ \text{3-phosphate} \\ \text{dehydrogenase}}}$?

 d. 1,3-Bisphosphoglyceric acid $\xrightarrow{?}$ 3-Phosphoglyceric acid

 e. 3-Phosphoglyceric acid $\xrightarrow{\text{phosphoglyceromutase}}$?

 f. Pyruvic acid $\xrightarrow{?}$ Acetaldehyde

 g. ? $\xrightarrow{\text{pyruvate kinase}}$ Pyruvic acid

 h. Glucose 1-phosphate $\xrightarrow{\text{phosphoglucomutase}}$?

35. Refer to Problems 33 and 34, and select the equations (by number and letter) in which each process occurs:
 a. the expenditure of phosphate bond energy
 b. the formation of high-energy phosphate bonds
 c. isomerization reactions
 d. an oxidation reaction

36. Refer to Problems 33 and 34, and select the equations (by number and letter) in which each process occurs:
 a. reduction reactions
 b. a dehydration reaction
 c. a decarboxylation reaction

Additional Problems

37. List four phosphorylated and four nonphosphorylated metabolites of glycolysis and fermentation.

38. What is meant when glyburide is called a second-generation drug?

39. What critical role is played by both 1,3-bisphosphoglyceric acid and phosphoenolpyruvic acid (PEP) in the Embden–Meyerhof pathway?

40. Lactic acid dehydrogenase is not specific for pyruvic acid but will also catalyze the reduction of other keto acids. Write an equation for the enzymatic reduction of phenylpyruvic acid. Does this surprise you in light of what was said about the stereochemical specificity of this enzyme in Section 22.4? Comment.

41. How do muscle cells and liver cells differ in their metabolism of glucose 6-phosphate and glycogen?

42. The average adult consumes about 65 g of fructose daily (either as the free sugar or as part of sucrose). In the liver, fructose is first phosphorylated to fructose 1-phosphate, which is then split into dihydroxyacetone phosphate and glyceraldehyde. The latter compound is phosphorylated to glyceraldehyde 3-phosphate. Write out equations (using formulas) for these three steps, and give the specific names of the enzymes. Indicate which steps utilize ATP.

43. The alcohol we drink is detoxified by enzymes in the liver. Write out the sequence of reactions for the metabolism of alcohol.
 a. Which enzyme is inhibited by disulfiram?
 b. Does alcohol supply energy (calories) for the body? Explain.

44. Why is it more efficient for anaerobic cells to utilize glycogen as a source of energy than for them to use glucose?

45. Draw the structure of cyclic AMP.

46. Write the structural formula of creatine phosphate. What is its role in muscle contraction?

25 Carbohydrate Metabolism II

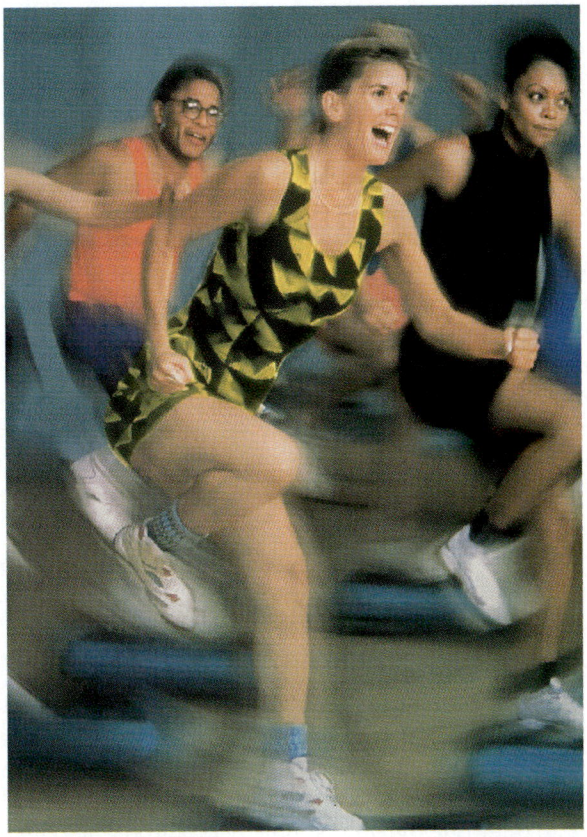

Prolonged exercise is fueled mainly by aerobic oxidation. Sustained aerobic activity enhances cardiovascular fitness.

In Chapter 24 we mentioned that pyruvic acid is a "crossroads" compound. All living cells metabolize glucose to pyruvic acid via the Embden–Meyerhof pathway. Cells that function anaerobically convert the pyruvic acid either to lactic acid or to ethanol. Recall (Section 24.9) that anaerobic metabolism accounts for only a small fraction of the total energy available from the glucose molecule. Cells that function aerobically, on the other hand, oxidize the pyruvic acid via a number of discrete enzymatic reactions to carbon dioxide and water:

$$C_6H_{12}O_6 \xrightarrow{\text{glycolysis}} 2\,CH_3CHOHCOOH \;+\; \sim 47{,}000\text{ cal}$$

$$2\,CH_3COCOOH \;+\; 5\,O_2 \longrightarrow 6\,CO_2 \;+\; 4\,H_2O \;+\; \sim 639{,}000\text{ cal}$$

From the standpoint of energy production, the oxidation of pyruvic acid is of considerable significance because it liberates most of the energy (93%) stored in the glucose molecule. Much of this energy is then conserved in a chemical form in the high-energy phosphate bonds of ATP.

25.1 The Krebs Cycle

The Krebs cycle is also known by two other names. It is called the *citric acid cycle* because citric acid is a most important intermediate. It is also known as the *tricarboxylic acid cycle* because four of the intermediate compounds have three carboxyl groups each.

A scheme for the complex series of reactions that brings about the oxidation of pyruvic acid to carbon dioxide and water was first proposed by Hans Krebs in 1937 (Nobel prize in physiology or medicine, 1953). Pyruvic acid is transported from the cytoplasm into the inner compartment of the mitochondria, cell components we discuss in Section 25.2. Within a mitochondrion, pyruvic acid is first decarboxylated to a two-carbon compound, which then enters a cyclic sequence of reactions known collectively as the **Krebs cycle.** The Krebs cycle produces ATP and provides metabolic intermediates for the biosynthesis of needed compounds. As we shall see in later chapters, the Krebs cycle is not restricted to the metabolism of carbohydrates. It also plays a vital role in the metabolism of lipids and proteins and thus occurs in almost all cells of higher animals.

Taken as a whole, the Krebs cycle may seem rather complex. All the reactions, however, are familiar types in organic chemistry: condensations, dehydrations, hydrations, oxidations, decarboxylations, and hydrolyses. The difference here is one of *in vitro* versus *in vivo conditions*. In a living organism, these reactions take place at constant temperature (37 °C in humans) and constant pH. Each is catalyzed by an enzyme. Every enzyme in the cycle has been identified, and the operation of this metabolic pathway *in vivo* has been completely verified by the use of isotopic tracers. All of the enzymes involved in the Krebs cycle are located within the mitochondria. Aerobic metabolism is thus separated from the reactions of glycolysis, which occur in the cytoplasm.

Figure 25.1 is a schematic outline of the Krebs cycle. Each reaction is numbered, and the individual steps of the sequence are now considered in detail.

Hans Krebs (1900–1981) discovered the cycle that bears his name. He also discovered the urea cycle (Chapter 27) and another cycle, the glyoxylate cycle, that is not discussed in this text.

Step 1 The pyruvic acid formed in the Embden–Meyerhof pathway is not an intermediate in the Krebs cycle. It must first be enzymatically decarboxylated and oxidized (*oxidative decarboxylation*) to yield the extremely important intermediate acetyl coenzyme A. The formation of acetyl CoA from pyruvic acid requires the sequential action of three enzymes and the participation of five coenzymes: coenzyme A, thiamine pyrophosphate (TPP), lipoic acid, NAD^+, and FAD (see Table J.1 for structures of the coenzymes):

$$CH_3-\overset{\overset{O}{\|}}{C}-\overset{\overset{O}{\|}}{C}-OH \;+\; CoASH \;+\; NAD^+ \;\xrightarrow[\text{TPP, FAD}]{\text{lipoic acid}}\; CH_3-\overset{\overset{O}{\|}}{C}-SCoA \;+\; NADH \;+\; CO_2 \;+\; H^+$$

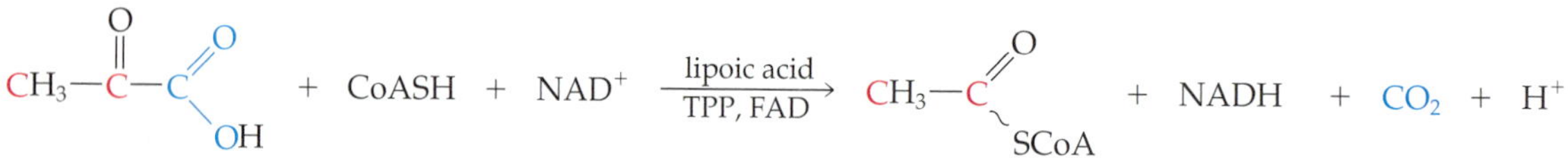

Pyruvic acid Coenzyme A Acetyl CoA

The name given to this multienzyme system is *pyruvic acid dehydrogenase complex*. The mechanism of the reaction is believed to involve four steps—decarboxylation, reductive acetylation, acetyl transfer, and electron transport. The initial decarboxylation step is irreversible. Therefore, the conversion of pyruvic acid to acetyl CoA is irreversible.

Step 2 Acetyl CoA is likewise not a true intermediate in the Krebs cycle. It enters the cycle by condensing with oxaloacetic acid, a four-carbon dicarboxylic acid, yielding citric acid. The reaction is highly exothermic because of the energy released by the subsequent hydrolysis of the thioester bond of acetyl CoA (7500

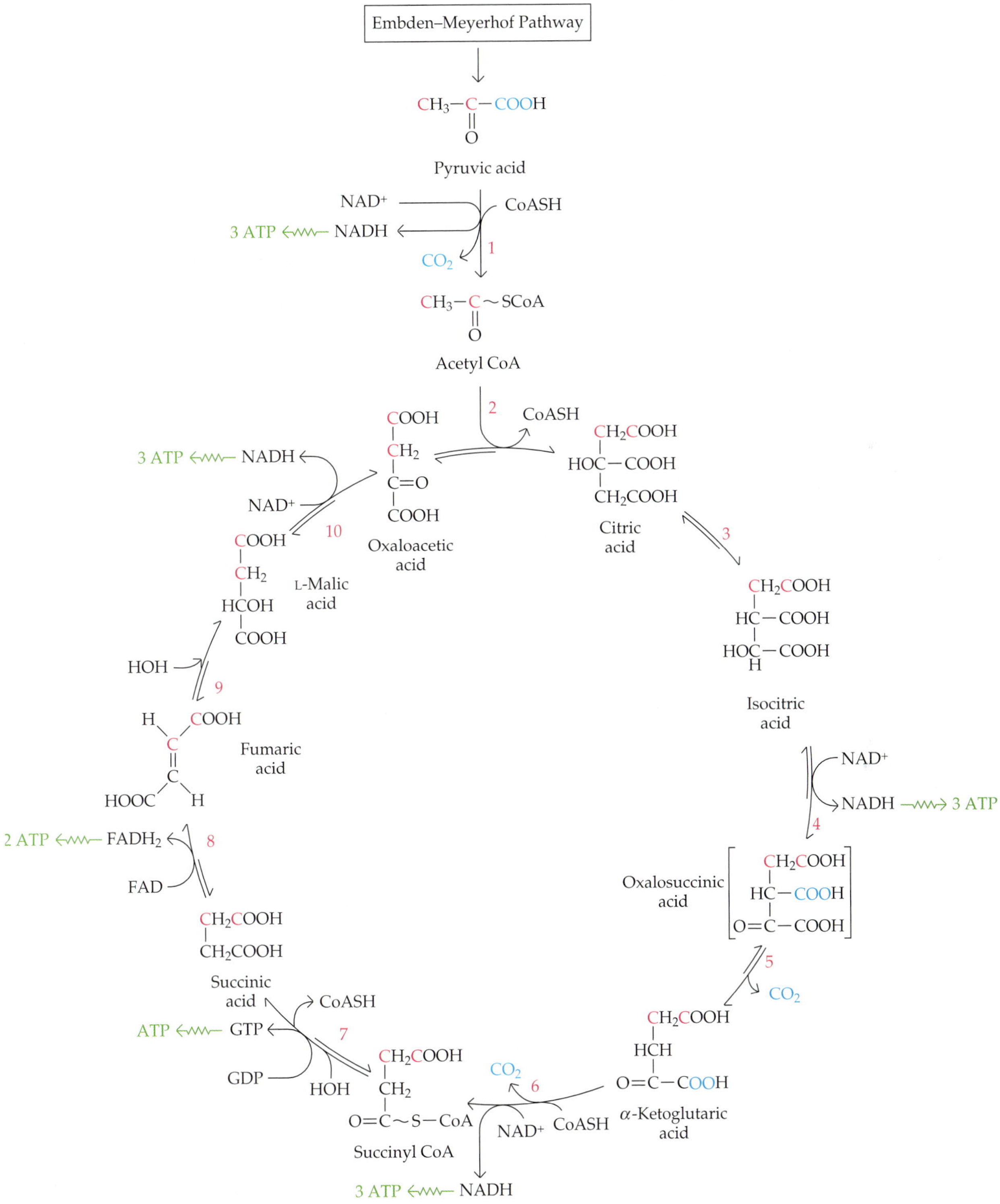

▲ **FIGURE 25.1**
The Krebs cycle.

cal/mol). The two carbon atoms that originate from acetyl CoA are shown in red here and in subsequent reactions. Note that this step regenerates coenzyme A:

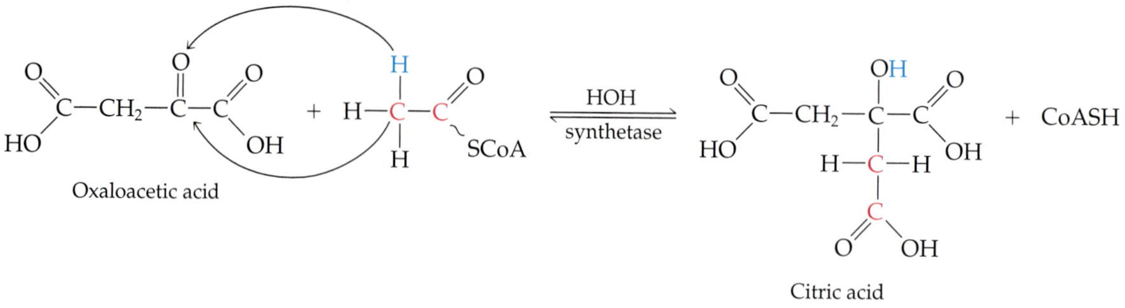

The Krebs cycle is regulated in part at this step. The reaction is catalyzed by *citric acid synthetase,* an enzyme that is inhibited by ATP and NADH. When the cell has sufficient ATP for its immediate needs, ATP molecules interact with the enzyme to reduce its affinity for acetyl CoA. The latter is then shunted to the biosynthesis of fatty acids, and the Krebs cycle slows down.

Step 3 The third step is sometimes considered to be two separate reactions. The single enzyme *aconitase* catalyzes successive dehydration and hydration reactions. The net result is the isomerization of citric acid to its less symmetrical isomer, isocitric acid. The intermediate is *cis*-aconitic acid, and it normally remains bound to the enzyme:

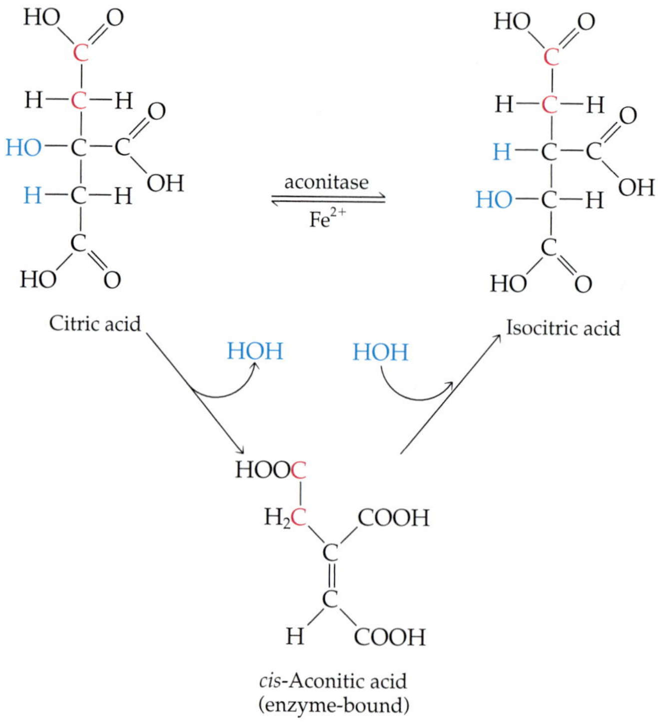

Steps 4 and 5 The next two steps are usually discussed together because experimental evidence indicates that the intermediate, oxalosuccinic acid, does not exist free but is firmly bound to the surface of the enzyme *isocitric acid dehydrogenase,* which catalyzes the oxidative decarboxylation of isocitric acid to α-ketoglutaric acid:

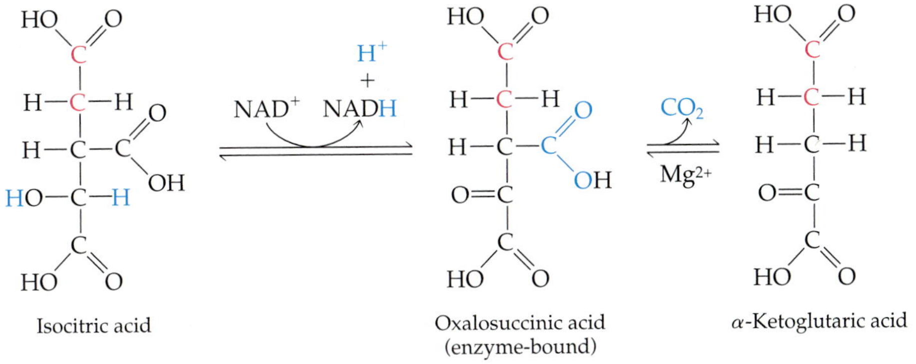

Step 6 This step is analogous to Step 1; it is catalyzed by another multienzyme system known as *α-ketoglutaric acid dehydrogenase complex,* which requires the same five coenzymes as pyruvic acid dehydrogenase:

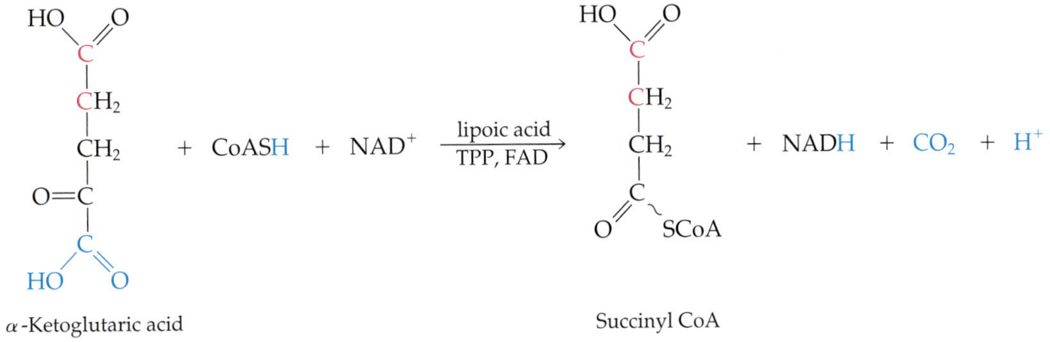

This is the only irreversible reaction in the Krebs cycle. As such, it prevents the cycle from operating in the reverse direction.

Step 7 In this reaction the energy released by the hydrolysis of the high-energy thioester bond of succinyl CoA (~8000 cal/mol) is used to form guanosine triphosphate (GTP) from guanosine diphosphate (GDP) and inorganic phosphate:

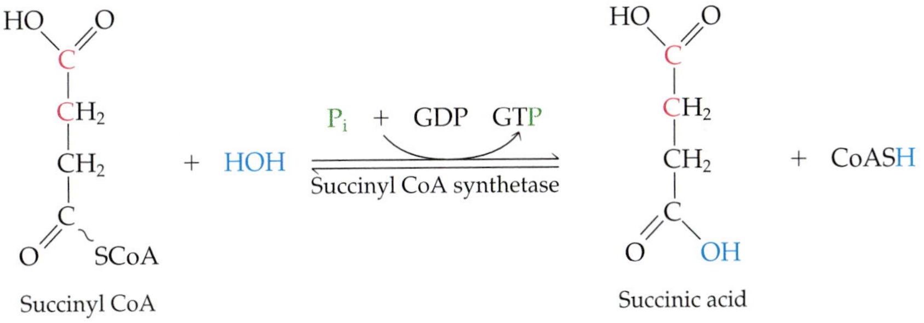

$$\text{GTP} + \text{ADP} \xrightleftharpoons{\text{kinase}} \text{GDP} + \text{ATP}$$

This reaction is significant because GTP has a higher free energy of hydrolysis than ATP and can readily transfer its terminal phosphate group to ADP to generate ATP in the presence of *nucleoside diphosphokinase.* Here we have another example of substrate-level phosphorylation. Step 7 is the only reaction in the Krebs cycle that directly involves a high-energy phosphate bond.

Step 8 *Succinic acid dehydrogenase* catalyzes the removal of two hydrogen atoms from succinic acid, thus forming fumaric acid:

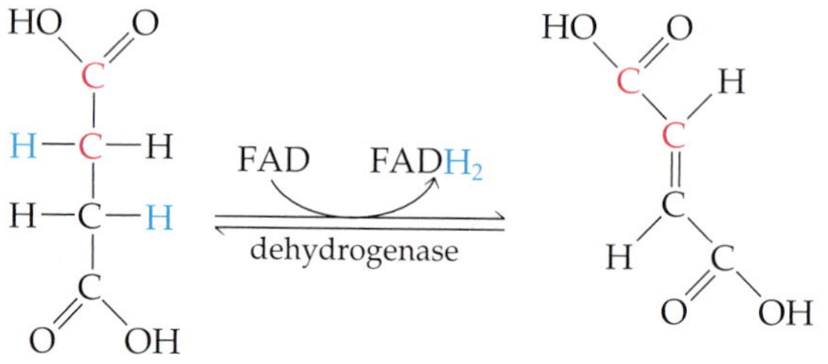

Succinic acid　　　　Fumaric acid

Recall that this enzyme is competitively inhibited by malonic acid (Section 22.6). This dehydrogenation reaction is the only one in the cycle that uses the coenzyme FAD rather than NAD^+. Succinic acid dehydrogenase is the only enzyme of the Krebs cycle located within the inner mitochondrial membrane. This permits electrons to be transferred directly into the electron transport chain when the $FADH_2$ is reoxidized (see footnote on page 712).

Step 9 The addition of a molecule of water across the double bond of fumaric acid to form L-malic acid is catalyzed by *fumarase.* This enzyme is highly stereospecific; trans addition occurs, so only the L isomer of malic acid is produced:

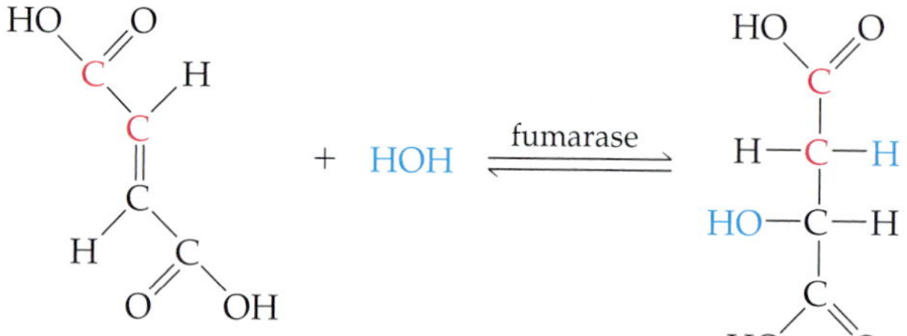

Fumaric acid　　　　L-Malic acid

Step 10 One revolution of the cycle is completed with the oxidation of L-malic acid to oxaloacetic acid, brought about by *malic acid dehydrogenase:*

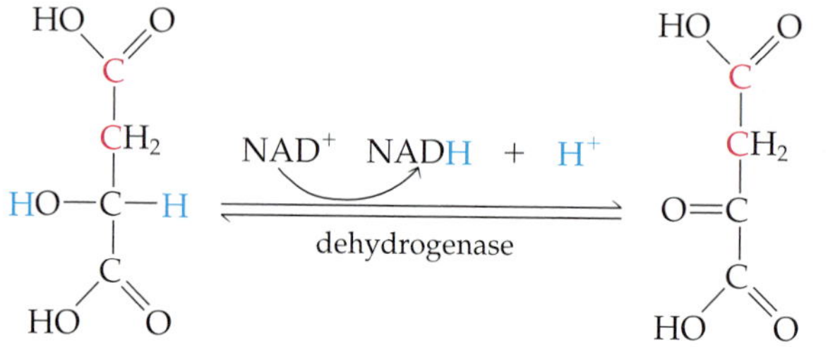

L-Malic acid　　　　Oxaloacetic acid

This is the fourth oxidation–reduction reaction that uses NAD^+ as the oxidizing agent. Oxaloacetic acid can accept an acetyl group from acetyl CoA, and the cycle is ready for another spin. Every time we go around the cycle, two carbons are fed into the system as acetyl CoA and two carbons are kicked out as carbon dioxide molecules.

25.2 Respiratory Chain: Electron-Transport Chain

We have stated that aerobic metabolism occurs only in the presence of molecular oxygen. It has also been mentioned that the major portion of solar energy stored in carbohydrates is conserved in the process. Yet nowhere in our discussion of the Krebs cycle has oxygen utilization or energy conservation been indicated. None of the intermediates of the cycle has been shown to be linked to phosphate groups, and no direct synthesis of ATP from ADP and inorganic phosphate has taken place. The Krebs cycle deals primarily with the fate of the carbon skeleton of pyruvic acid, describing the metabolites involved in its conversion to carbon dioxide. Two carbon atoms enter the cycle as acetyl CoA (Step 2), and two different carbon atoms exit the cycle as carbon dioxide (Steps 5 and 6). The coenzymes NAD^+ and FAD are reduced to NADH and $FADH_2$; no mechanism has yet been indicated for their regeneration. The reduced coenzymes must be reoxidized if the aerobic phase of carbohydrate metabolism is to continue.

All the enzymes and coenzymes necessary for the Krebs cycle and for the conservation of energy are localized in the **mitochondria,** small organelles often referred to as the "power plants" of the cell (Figure 25.2). Mitochondria are oval, dual-membrane structures. They may be randomly distributed throughout the cytoplasm, or they may be organized in regular rows or clusters. A cell may contain 100 to 1000 mitochondria, depending on its function, and the mitochondria can reproduce themselves if the energy requirements of the cell increase.

A mitochondrion has two lipid/protein membranes—an *outer membrane* and an *inner membrane* that is extensively folded into a series of internal ridges called *cristae.* Thus there are two compartments in mitochondria—the *intermembrane space* and the *matrix,* which is surrounded by the inner membrane. The outer

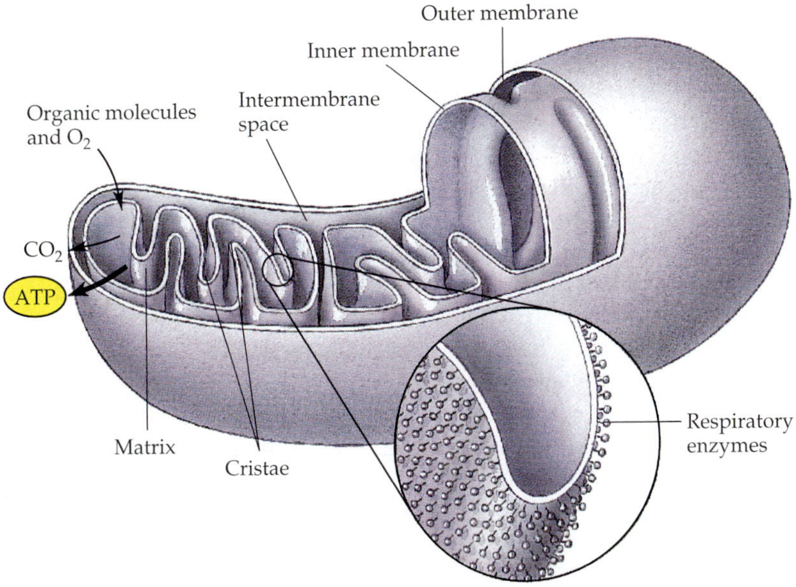

FIGURE 25.2

Cellular respiration occurs in the mitochondria. The subdivisions in a mitochondrion reflect the compartmentalized reactions that take place there. The inner compartment, which contains the soluble enzymes of the matrix, is separated from the intermembrane space by the inner membrane.

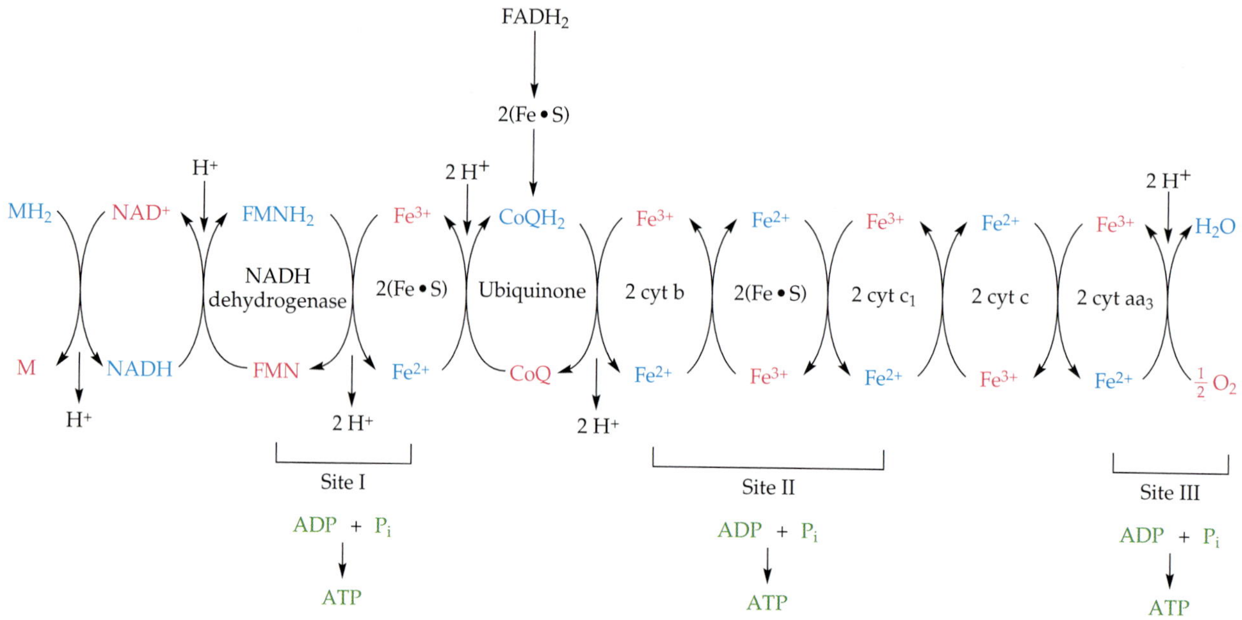

▲ **FIGURE 25.3**

A schematic diagram of the respiratory chain, showing the oxidized (magenta) and reduced (blue) forms of the carriers, proton (H^+) input to and output from the mitochondrial matrix, and the sites along the chain where oxidative phosphorylation occurs.

membrane is permeable, whereas the inner membrane is impermeable to most molecules and ions. (Water, oxygen, and carbon dioxide can freely penetrate both membranes.) The matrix contains all the enzymes of the Krebs cycle, with the exception of succinic acid dehydrogenase, which is embedded in the inner membrane. The enzymes that provide energy for the cell are also contained within the inner membrane; they are positioned in geometrically specific arrays capable of functioning as extremely efficient assembly lines. This sequence of highly organized oxidation–reduction enzymes is known as the **respiratory chain** or the **electron transport chain.** As we shall see, the operation of the chain is analogous to that of a bucket brigade.

Figure 25.3 illustrates the respiratory chain in the mitochondria. The sequence in which the electron carriers operate is determined by their respective reduction potentials. The **reduction potential** is a measure of the tendency of a substance to gain electrons compared with the standard hydrogen electrode (25 °C, 1 atm H_2, and 1 M H^+), which is arbitrarily assigned a value of 0.0 V. Because in biological systems we are usually interested in neutral solutions, a correction is made for changes in pH. At pH 7 the hydrogen electrode has a potential difference of −0.42 V when measured against the standard hydrogen electrode. The reduction potentials of some respiratory-chain intermediates are listed in Table 25.1.

For each molecule of pyruvic acid converted to carbon dioxide and water via the Krebs cycle, five dehydrogenation reactions occur. NAD^+ serves as the electron acceptor in four of these (Steps 1, 4, 6, and 10), and FAD is the oxidizing agent in the fifth (Step 8). In Figure 25.3, MH_2 symbolizes the reduced metabolites (pyruvic acid, isocitric acid, α-ketoglutaric acid, and malic acid) and M signifies the oxidized metabolites (acetyl CoA, oxalosuccinic acid, succinyl CoA, and oxaloacetic acid).

We now examine one of these conversions by writing the balanced half-reaction

Table 25.1 Standard Reduction Potentials for Some Respiratory Chain Intermediates

System (Oxidant/Reductant)	$E°$ at pH 7 (V)
$NAD^+/NADH$	−0.32
$FMN/FMNH_2$	−0.22
$CoQ/CoQH_2$	+0.04
Cyt b-Fe(III)/cyt b-Fe(II)	+0.07
Cyt c_1-Fe(III)/cyt c_1-Fe(II)	+0.23
Cyt c-Fe(III)/cyt c-Fe(II)	+0.25
Cyt aa_3-Fe(III)/cyt aa_3-Fe(II)	+0.29
$\frac{1}{2}O_2/H_2O$	+0.82

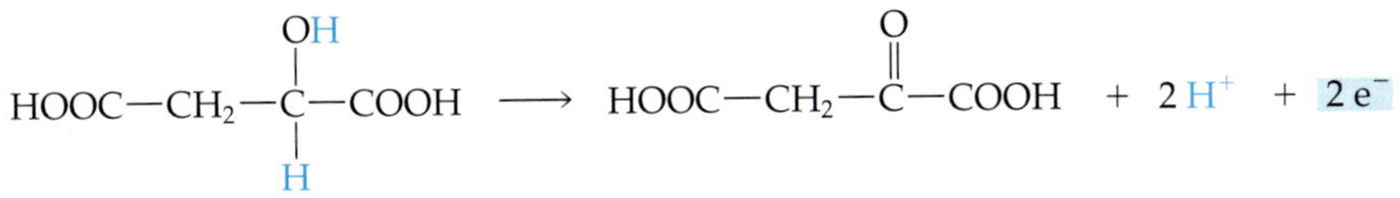

L-Malic acid   Oxaloacetic acid

It has been stated that the function of the coenzyme NAD^+ is to accept this pair of electrons. We can also write a balanced half-reaction for the reduction of the coenzyme; R— represents the remaining portion of the NAD^+ molecule (see Table J.1):

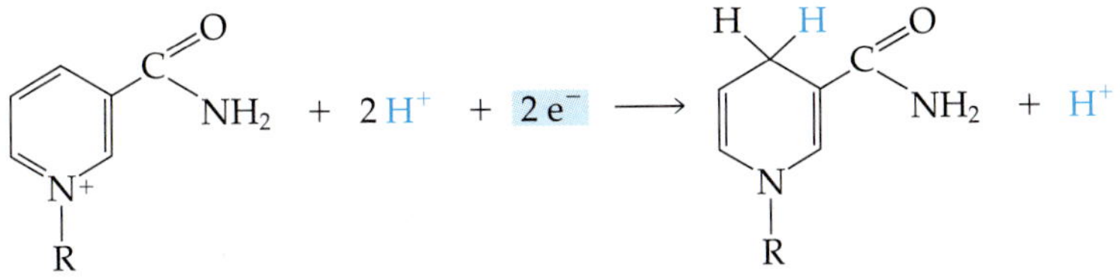

NAD^+ NADH

Two hydrogen ions and two electrons are removed from the substrate. The NAD^+ molecule accepts both electrons and one hydrogen ion. The other hydrogen ion is transported from the matrix, across the inner mitochondrial membrane, and into the intermembrane space.

In the next step of the respiratory chain, both electrons are passed on to an enzyme called *NADH dehydrogenase* whose coenzyme is FMN (see Table J.1). By passing the electrons along, NADH is reoxidized back to NAD^+ and FMN is reduced to $FMNH_2$. It is this step that accounts for the regeneration of NAD^+. Again we depict only the relevant portion of the FMN molecule:

$$\text{NADH} + \text{H}^+ \longrightarrow \text{NAD}^+ + 2\,\text{H}^+ + 2\,e^-$$

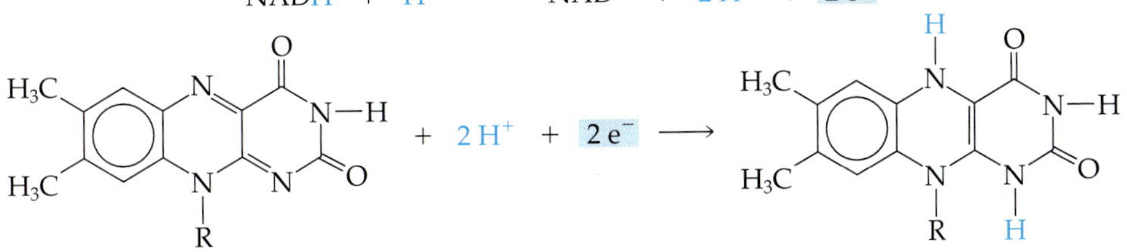

FMN $FMNH_2$

The electrons are then transferred from $FMNH_2$ to a series of iron–sulfur complexes (abbreviated Fe · S). A number of iron–sulfur proteins have been identified in the respiratory chain, complexed with other electron carriers. The iron atom in these proteins is in the Fe(III) form; by accepting an electron, it is reduced to the Fe(II) form (since each iron–sulfur complex can transfer only one electron, two molecules are needed to transport the two electrons and regenerate FMN):

$$FMNH_2 \longrightarrow FMN + 2\,H^+ + 2\,e^-$$

$$2\,Fe(III) \cdot S + 2\,e^- \longrightarrow 2\,Fe(II) \cdot S$$

The two electrons are next transferred from Fe(II) · S to coenzyme Q (CoQ), a quinone derivative containing a long isoprenoid side chain. Coenzyme Q is also called *ubiquinone* because it is ubiquitous in living systems. Several ubiquinones are known, differing only in the number of isoprene units in the side chain. The most common ubiquinone found in the mitochondria of animal tissues contains 10 isoprene units and is designated CoQ_{10}. The quinone ring is reversibly reducible to a hydroquinone, and so CoQ serves as the electron carrier between the flavin coenzymes and the cytochromes.[1] It is in this step that the oxidized form of the iron–sulfur complex is regenerated:

$$2\,Fe(II) \cdot S \longrightarrow 2\,Fe(III) \cdot S + 2\,e^-$$

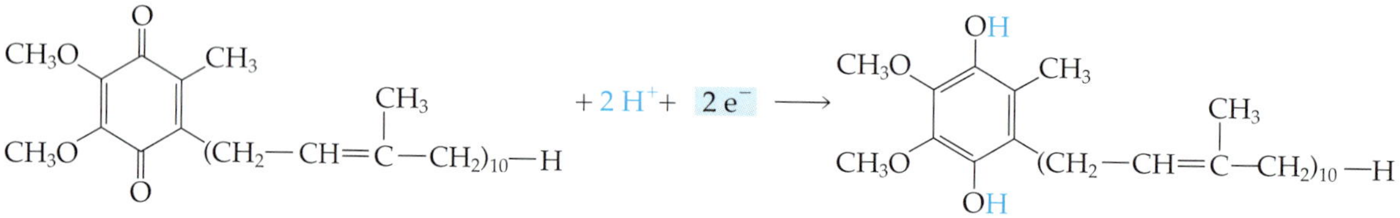

Coenzyme Q_{10} Coenzyme $Q_{10}H_2$

A series of compounds called **cytochromes** were probably the first entities to be associated with electron-transferring reactions. A number of these substances exist, and we have included four of them in Figure 25.3. (Notice that another Fe · S protein is located between two of the cytochromes.) The cytochromes are conjugated protein enzymes. Their prosthetic groups are iron porphyrins, which resemble heme (Figure 21.14), the pigment in hemoglobin and myoglobin. The various cytochromes differ with respect to (1) their protein constituents, (2) the manner in which the porphyrin is bound to the protein, and (3) the substituents on the periphery of the porphyrin ring. Such slight differences in structure bestow differ-

[1] Recall that we said the enzyme succinic acid dehydrogenase (which catalyzes the oxidation of succinic acid to fumaric acid—Step 8) contains the coenzyme FAD, which is reduced to $FADH_2$. The oxidized form of the coenzyme must be regenerated, and this is accomplished via the respiratory chain. The electrons from $FADH_2$ require a two-step transfer to enter the chain. First they are passed to an iron-sulfur protein,

$$FADH_2 \longrightarrow FAD + 2\,H^+ + 2\,e^-$$
$$2\,Fe(III) \cdot S + 2\,e^- \longrightarrow 2\,Fe(II) \cdot S$$

which in turn passes them on to CoQ_{10} (Figure 25.3):

$$2\,Fe(II) \cdot S \longrightarrow 2\,Fe(III) \cdot S + 2\,e^-$$
$$CoQ_{10} + 2H^+ + 2\,e^- \longrightarrow CoQ_{10}H_2$$

ences in reduction potential upon the different cytochromes. Like the iron in the Fe · S complexes, the characteristic feature of the cytochromes is the ability of their iron atoms to exist as either Fe(II) or Fe(III). Thus each cytochrome in its oxidized form, Fe(III), can accept one electron and be reduced to the Fe(II) form. This change in oxidation state is reversible, and the reduced form can donate its electron to the next cytochrome, and so on. Only the last cytochrome, cytochrome aa_3 (called *cytochrome oxidase*), has the ability to transfer electrons to molecular oxygen. Since the Fe(III)/Fe(II) system is only a one-electron exchange, two cytochrome b molecules are necessary to complete the oxidation of coenzyme $Q_{10}H_2$:

$$\mathrm{CoQ_{10}H_2} \longrightarrow \mathrm{CoQ_{10}} + 2\,\mathrm{H^+} + 2\,\mathrm{e^-}$$

$$2\,\text{Cyt b-Fe(III)} + 2\,\mathrm{e^-} \longrightarrow 2\,\text{Cyt b-Fe(II)}$$

Then

$$\text{Cyt b} \longrightarrow \text{Fe}\cdot\text{S} \longrightarrow \text{Cyt c}_1 \longrightarrow \text{Cyt c} \longrightarrow \text{Cyt aa}_3$$

In the final step, two molecules of the terminal electron carrier, cytochrome aa_3, pass their electrons to molecular oxygen, the ultimate electron acceptor. It has been estimated that about 95% of the oxygen used by cells reacts in this single process:

$$2\,\text{Cyt aa}_3\text{-Fe(II)} \longrightarrow 2\,\text{Cyt aa}_3\text{-Fe(III)} + 2\,\mathrm{e^-}$$

$$\tfrac{1}{2}\mathrm{O_2} + 2\,\mathrm{H^+} + 2\,\mathrm{e^-} \longrightarrow \mathrm{H_2O}$$

The combination of cyanide with the ferric ions of cytochrome aa_3 completely inhibits their reduction to the ferrous state, thus blocking the transfer of electrons to molecular oxygen. This accounts for the extreme toxicity of cyanide to living organisms—just 50 mg of HCN constitutes a lethal dose for a human.

Each intermediate compound in the respiratory chain is reduced by the addition of electrons in one reaction and is subsequently restored to its original form when it delivers the electrons to the next compound. Thus each pair of electrons removed from the substrates of the Krebs cycle ultimately reduces one atom of oxygen.

25.3 Oxidative Phosphorylation

The process whereby ATP synthesis is linked to oxygen consumption in the respiratory chain is referred to as **oxidative phosphorylation.** A model (the chemiosmotic hypothesis) linking ATP formation to the respiratory chain has been proposed by Peter Mitchell (Nobel prize, 1978; see an advanced biochemistry text for details). The energy required for ATP production results from the passage of an electron pair from NADH or $FADH_2$ to oxygen through a series of electron carriers. The electron-transport chain can be thought of as a biochemical "battery" because energy is obtained from its oxidation–reduction reactions.

Electron transport is tightly coupled to oxidative phosphorylation. The reduced forms of the coenzymes, NADH and $FADH_2$, are oxidized by the respiratory chain *only* if ADP is simultaneously phosphorylated to ATP. Within energy-utilizing cells, the turnover of ATP is very high. These cells, then, contain high levels of ADP, and they must consume large quantities of oxygen to continuously phosphorylate the ADP back to ATP. For example, about 20% of a resting adult's oxygen consumption occurs in brain tissue to supply energy to brain cells. (Much of this energy is used to maintain the concentration gradients across the membranes of the trillions of brain cells.) Resting skeletal muscles utilize about

30%, whereas during strenuous exercise these muscles account for almost 90% of the total oxygen consumption of the organism. The enzymes of oxidative phosphorylation are embedded in the inner mitochondrial membrane in association with the enzymes of the respiratory chain. The sites on the respiratory chain at which the oxidative phosphorylations are believed to occur are shown in Figure 25.3

From the data in Table 25.1, we can calculate the maximum amount of energy (E) made available when a pair of electrons travels from NADH to oxygen along the chain:

$$\begin{array}{lcll} & & & E \\ \text{NADH} & \longrightarrow & \text{NAD}^+ + \text{H}^+ + 2\,\text{e}^- & +0.32 \\ \tfrac{1}{2}\text{O}_2 + 2\,\text{H}^+ + 2\,\text{e}^- & \longrightarrow & \text{H}_2\text{O} & +0.82 \\ \hline \text{NADH} + \tfrac{1}{2}\text{O}_2 + \text{H}^+ & \longrightarrow & \text{NAD}^+ + \text{H}_2\text{O} & +1.14\ \text{V} \end{array}$$

Because the concentrations of the various compounds are unknown, reduction potentials yield only a rough estimate of the energy change for a reaction.

The energy change for the reaction can be obtained from the equation

$$\text{Energy change} = -nF\Delta E$$

where n is the number of electrons transferred and F is Faraday's constant (23,062 cal/V equivalent).

$$\begin{aligned} \text{Energy change} &= -(2)(23{,}062)(1.14) \\ &= -52{,}600\ \text{cal} \end{aligned}$$

This value of 52,600 cal represents a considerable amount of energy. If it were released all at once, much of it would be dissipated as heat and might damage the cell. Therefore, the respiratory chain delivers this energy in small increments to be used to phosphorylate ADP. It has been experimentally observed that three molecules of ATP are formed for every molecule of NADH oxidized in the chain (but only two ATPs are formed if the primary electron acceptor is FAD, as is the case when succinic acid serves as a substrate). The net equation for the respiratory chain is

$$\text{NADH} + \tfrac{1}{2}\text{O}_2 + \text{H}^+ + 3\,\text{ADP} + 3\,\text{P}_\text{i} \longrightarrow \text{NAD}^+ + \text{H}_2\text{O} + 3\,\text{ATP}$$

Recall that 7300 cal is required for the conversion of 1 mol of ADP to ATP. It can be determined that almost half of the energy released in the electron-transport chain is conserved in the formation of high-energy phosphate bonds:

$$\begin{aligned} \text{Energy conserved by respiratory chain} &= \frac{\text{energy conserved}}{\text{energy available}}(100\%) \\ &= \frac{(3)(7300)}{52{,}600}(100\%) = 42\% \end{aligned}$$

25.4 Energy Yield of Carbohydrate Metabolism

We now summarize the energy conserved (in ATP production) when one molecule of glucose is oxidized. Recall that, under aerobic conditions, the glycolytic sequence terminates with the formation of two molecules of pyruvic acid. Two molecules of ATP are obtained from substrate-level phosphorylation. In addition, since the pyruvic acid is not reduced to lactic acid, there are two molecules of NADH (from Step 6 of the Krebs cycle) that remain in the cytoplasm. We know that NAD^+ must be regenerated from NADH in order for glycolysis to continue. The problem, however, is that NADH cannot pass across the mitochondrial

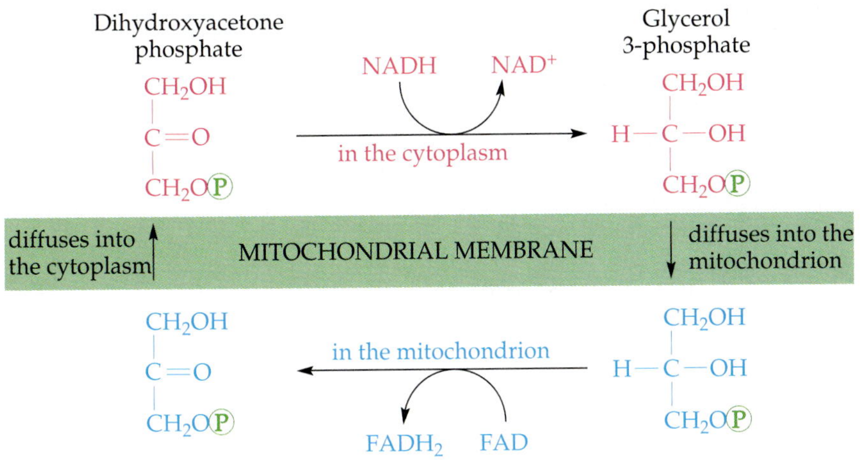

FIGURE 25.4
The glycerol phosphate shuttle.

membrane to be oxidized by the respiratory chain. Instead, electrons from NADH are transported across the membrane via a shuttle process involving glycerol 3-phosphate and dihydroxyacetone phosphate. These compounds can penetrate the outer mitochondrial membrane.

The first step in this shuttle (Figure 25.4) is the reduction of dihydroxyacetone phosphate by NADH to form glycerol 3-phosphate and regenerate the NAD^+. This reaction occurs in the cytoplasm. The glycerol 3-phosphate then enters the mitochondrion, where it is reoxidized to dihydroxyacetone phosphate, this time by a dehydrogenase that contains the coenzyme FAD instead of NAD^+. The dihydroxyacetone phosphate diffuses out of the mitochondrion and returns to the cytoplasm to complete the shuttle. The reduced form of the flavin coenzyme, $FADH_2$, is reoxidized as its electrons pass to the respiratory chain at the level of coenzyme Q_{10}. As a result, two molecules of ATP (instead of three) are formed every time a cytoplasm NADH molecule is reoxidized via the glycerol phosphate shuttle and the respiratory chain.[2]

The aerobic continuation of glycolysis yields 15 molecules of ATP from each molecule of pyruvic acid, which means 30 molecules of ATP per molecule of glucose. Table 25.2 lists the various reactions that result in ATP synthesis, the net equation for which is

$$C_6H_{12}O_6 + 6\,O_2 + 36\,ADP + 36\,P_i \longrightarrow 6\,CO_2 + 6\,H_2O + 36\,ATP$$

This equation summarizes the oxidation of 1 mol of glucose in certain aerobic cells (e.g., skeletal muscle, nerve). The energy released (686,000 cal) is coupled to the synthesis of 36 mol of ATP from ADP and inorganic phosphate. Assuming that 7300 cal is required for the synthesis of each mole of ATP, then $36 \times 7300 =$ 263,000 cal is conserved by the cell. The efficiency of conservation therefore is

$$\frac{263{,}000}{686{,}000}(100\%) = 38.3\%$$

This cellular recovery of energy compares favorably with the efficiency of any machine, and it represents a remarkable achievement on the part of the living

[2] The glycerol phosphate shuttle operates in most body cells. In certain tissues, however, particularly the liver and heart, another type of shuttle exists. This is the malic acid–aspartic acid shuttle. By a quite complex mechanism, it produces three molecules of ATP from the reoxidation of a cytoplasmic NADH molecule.

Table 25.2 Production of ATP During the Oxidation of One Molecule of Glucose to Carbon Dioxide and Water

Reaction	Type of Phosphorylation	Number of ATP Molecules Formed
Glucose $\longrightarrow$ 2 Pyruvic acid	Substrate level	2
Glyceraldehyde 3-phosphate $\xrightarrow[\text{FAD} \rightarrow \text{FADH}_2]{\text{NAD}^+ \rightarrow \text{NADH}}$ 1,3-Bisphosphoglyceric acid	Oxidative (via glycerol phosphate shuttle)	2×2
Pyruvic acid $\xrightarrow{\text{NAD}^+ \rightarrow \text{NADH}}$ Acetyl CoA	Oxidative	2×3
Isocitric acid $\xrightarrow{\text{NAD}^+ \rightarrow \text{NADH}}$ Oxalosuccinic acid	Oxidative	2×3
α-Ketoglutaric acid $\xrightarrow{\text{NAD}^+ \rightarrow \text{NADH}}$ Succinyl CoA	Oxidative	2×3
Succinyl CoA $\xrightarrow{\text{GDP} \rightarrow \text{GTP}}$ Succinic acid	Substrate level	2×1
Succinic acid $\xrightarrow{\text{FAD} \rightarrow \text{FADH}_2}$ Fumaric acid	Oxidative	2×2
Malic acid $\xrightarrow{\text{NAD}^+ \rightarrow \text{NADH}}$ Oxaloacetic acid	Oxidative	2×3
	Sum	36

organism. In comparison, automobiles are only about 5% efficient in utilizing the energy released in the combustion of gasoline.

What happens to the 62% of the energy that is not conserved? It is released as heat to the surroundings—that is, to the cell. It is this heat that maintains body temperature. If we are exercising strenuously and our metabolism speeds up to provide the necessary energy for muscle contraction, more heat is also produced. We begin to sweat to dissipate some of that heat. As the sweat evaporates, the excess heat is carried away from the body by the departing water vapor.

25.5 Muscle Power

Studies have shown that frequent exercise prolongs life and lowers the incidence of disease. Humans have about 600 muscles each. Exercise can make muscles stronger, more flexible, and more efficient in their use of oxygen. Strong muscles can do more work than weak ones. That is good, because the heart is an organ comprised mainly of muscle. A strong heart is a healthy heart. With regular exercise, resting pulse and blood pressure usually decline. After several months of an effective exercise program, pulse and blood pressure remain lower even *during* exercise. The net result, called the **training effect,** is that a person who exercises regularly is able to do more physical work with less strain.

People who expand their capacity to do more physical work under less strain often begin to think of doing more—faster and with more agility and accuracy—of whatever they do. These people become athletes. Exercise is an art, but it is increasingly also a science—a science in which chemistry plays a vital role.

The stimulation of muscle causes it to contract; that contraction is work and requires energy. The immediate source of energy for muscle contraction is ATP. The energy stored in this molecule powers the physical movement of muscle tissue. Two proteins, actin and myosin, play important roles in this process. Together they form a loose complex called **actomyosin,** the contractile protein of which muscles are made (Figure 25.5). When ATP is added to isolated actomyosin, the protein fibers contract. It seems likely that the same process occurs *in vivo*. Not only does myosin serve as part of the structural complex in muscles, it also acts as an enzyme for the removal of a phosphate group from ATP. Thus, it is directly involved in liberating the energy required for the contraction.

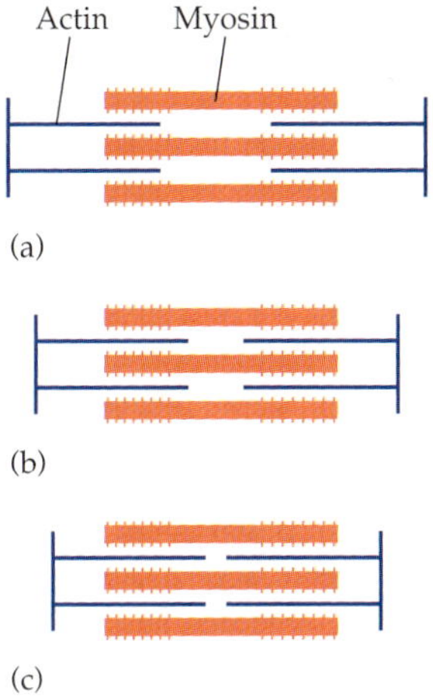

▲ **FIGURE 25.5**
Diagram of actomyosin complex in muscle.
(a) Extended muscle.
(b) Resting muscle.
(c) Partially contracted muscle.

In a resting person, muscle activity (including that of the heart muscle) accounts for only about 15 to 30% of the energy requirements of the body. Other activities, such as cell repair, transmission of nerve impulses, and even the maintenance of body temperature, account for the remaining energy needs. During intense physical activity, the energy requirements of muscle may be more than 200 times the resting level.

Fats are the major source of energy for sustained, low- or moderate-intensity activity (see Section 26.4). Prolonged exercise is fueled primarily by aerobic oxidation of both fats and carbohydrates. For example, long-distance runners derive only a small percentage of their energy needs from glycolysis. Aerobic oxidation (the Krebs cycle coupled to oxidative phosphorylation) provides most of the required ATP. Some glycogenolysis (hydrolysis of glycogen) and glycolysis (anaerobic metabolism of glucose) do occur, and after very long periods of this moderate muscle activity, lactic acid does build up and muscles do become fatigued. But because fats and some carbohydrates are supplying most of the energy through aerobic oxidation, the buildup of lactic acid takes much longer than during anaerobic activity.

In contrast with glycolysis, aerobic metabolism cannot be switched on quickly. At least one to two minutes of hard exercise must pass until the increase in breathing and heart rate ensures delivery of oxygen to muscle cells.

Muscle tissue seems to have been designed to provide for both short, intense bursts of activity and sustained, moderate levels of activity. Muscle fibers are divided into two categories: *slow-twitch* (Type I) and *fast twitch* (Type IIB). Table 25.3 lists some characteristics of these different types of muscle fibers. Type I fibers are called on during light or moderate activity. Their respiratory capacity is high, which means they can provide much energy via aerobic pathways, or, to put it

Table 25.3 A Comparison of Types of Muscle Fibers

	Type I	Type IIB[a]
Category	Slow twitch	Fast twitch
Color	Red	White
Respiratory capacity	High	Low
Myoglobin level	High	Low
Catalytic activity of actomyosin	Low	High
Capacity for glycogenolysis	Low	High
Number of mitochondria	High	Low

[a] There is a Type IIA fiber that resembles Type I in some respects and Type IIB in others. We discuss only the two types characterized in this table.

The New York Marathon is a test of the respiratory capacity of muscle.

another way, they are geared to oxidative phosphorylation. Notice that for Type I fibers, myoglobin levels are also high. Myoglobin (Section 21.12) is the heme-containing protein in muscle that transports oxygen (as hemoglobin does in the blood). Aerobic oxidation requires oxygen, and this muscle tissue is geared to supply high levels of oxygen. The capacity of Type I muscle fibers for glycogenolysis is low. This tissue is not geared to anaerobic generation of energy and does not require the hydrolysis of glycogen. The number of mitochondria in Type I muscle tissue is high, as we would expect, because oxidative phosphorylation takes place in the mitochondria. The catalytic activity of the actomyosin complex is low. Actomyosin is not only the structural unit in muscle that actually undergoes contraction; it is also responsible for catalyzing the hydrolysis of ATP to provide energy for the contraction. Low catalytic activity means that the energy is parceled out more slowly. That is not good if you want to lift 200 kg, but it is perfect for a long, sustained run.

Type IIB fibers have characteristics just the opposite of those of Type I fibers. Low respiratory capacity, low myoglobin levels, and fewer mitochondria all argue against aerobic oxidation. A high capacity for glycogenolysis and high catalytic activity of actomyosin allow this tissue to generate ATP rapidly via glycolysis and also to hydrolyze that ATP rapidly in intense muscle activity. Thus, this type of muscle tissue gives you the capacity to do short bursts of vigorous work. We say *bursts* because Type IIB muscle fatigues relatively quickly. A period of recovery in which lactic acid is cleared from the muscle is required between brief periods of activity.

The fields of sports medicine and exercise physiology have done much to increase our understanding of muscle action. Endurance training (e.g., running for long distances) increases the size and number of mitochondria in skeletal muscles. There is an increase in the level of enzymes required for the transport and oxidation of fatty acids (Section 26.4), for the Krebs cycle, and for oxidative phosphorylation. The increase in the mitochondrial enzymes is much greater for Type I fibers (used in prolonged, moderate-intensity activity, which is aerobic) than for Type IIB fibers (brief, intense activity, which is anaerobic). Endurance training also increases myoglobin levels in skeletal muscles, providing for faster oxygen transport, and stimulates additional capillaries to grow within the muscles, bringing additional oxygen-carrying blood to the muscles.

These changes can be observed after only one or two weeks of training. Muscle changes resulting from endurance training do not necessarily include a significant increase in muscle size,[3] in contrast to the effect of strength exercises such as weightlifting. Weightlifting develops fast-twitch muscles but does not result in the mitochondrial changes we have just described. The mitochondria of heart muscle, which is working constantly anyway, also undergo no change during endurance training.

Fast-twitch fibers increase in size and strength with repeated anaerobic exercise. Weight training does not increase respiratory capacity.

Athletes usually emphasize one type of training (anaerobic or aerobic) over the other. For example, an athlete training for a 60-m dash will do mainly anaerobic work, but one planning to run a 10-km race will do mainly aerobic training. Muscle-fiber type seems to be inherited. Research shows that world-class marathon runners may possess up to 80 to 90% slow-twitch fibers, compared with championship sprinters, who may have up to 70% fast-twitch muscle fibers. Some exceptions have been noted, however.

We hope we have made clear the central role ATP plays in many life-sustaining metabolic reactions. Our cells can make ATP, using the energy stored in foods, but only green plants can make the food. They also replenish the oxygen we breathe. Have you thanked a green plant lately?

Strength exercises build muscle mass but do not increase the respiratory capacity of muscles.

Summary

In cells that are operating aerobically, the pyruvic acid produced in the Embden–Meyerhof pathway is oxidized to carbon dioxide and water via the **Krebs cycle.** At the same time, ADP is converted to ATP, which means that energy originally in a glucose molecule (at the beginning of the Embden–Meyerhof pathway) becomes locked up in the high-energy phosphate bonds of ATP. This stored energy is later used to carry out the work needed to keep the organism functioning.

Pyruvic acid is produced in a cell's cytoplasm, but the

[3] Conversely, a muscle that is not used to any great extent tends to decrease in size and strength. There is a reduction in the number of mitochondria within each muscle cell and a reduction in the number of capillaries within the muscle fibers. This condition is called atrophy and is particularly noticeable in an individual upon removal of a cast after several weeks of limb immobilization.

Krebs cycle takes place in the organelles called **mitochondria** after the pyruvic acid has traveled from the cytoplasm to the interior of a mitochondrion.

The Krebs cycle describes the fate of the carbons in a pyruvic acid molecule but does not explain how the energy formerly in glucose ends up in ATP. The sequence describing this energy transfer takes place concomitantly with the carbon atom changes of the Krebs cycle and is called the **respiratory chain** (or **electron-transport chain**). The overall reaction is that molecular oxygen is reduced to water while ADP is converted to ATP.

The reactions of the respiratory chain are the $NAD^+/NADH$, $FAD/FADH_2$, and other conversions that are indicated alongside the reaction arrows in the Krebs-cycle diagram. Every time a carbon compound in the Krebs cycle is oxidized, a respiratory chain compound accepts the electrons lost in the oxidation (and so is reduced) and then passes them on to the next metabolite in the chain. At three sites along the chain, this transfer of electrons is accompanied by an ADP $\longrightarrow$ ATP **oxidative phosphorylation.** The reduction that goes along with this oxidation is that of molecular oxygen to water.

The energy yield of the combined Embden-Meyerhof–Krebs cycle pathway is 36 molecules of ATP for each glucose molecule. The energy locked up in the 36 ATP molecules is 38 percent of the energy contained in the glucose molecule.

Much of the energy in ATP is used for muscle contraction. Muscle fibers are made from two proteins, actin and myosin, combined into a complex called **actomyosin.** Actomyosin acts as an enzyme for the conversion ATP $\rightarrow$ ADP, and the energy released in this conversion allows a muscle to do work.

There are two main types of muscle fibers in humans. *Type I,* the *slow-twitch fibers,* are used to do light and moderate work. They operate under aerobic conditions. *Type IIB,* the *fast-twitch fibers,* are used in short burst of intense activity. They operate under anaerobic conditions, when the body cannot inhale oxygen fast enough to meet the needs of the muscle cells involved in the activity.

Key Terms

actomyosin (25.5)
cytochrome (25.2)
electron transport chain (25.2)
Krebs cycle (25.1)
mitochondria (25.2)
oxidative phosphorylation (25.3)
reduction potential (25.2)
respiratory chain (25.2)
training effect (25.5)

Review Questions

1. Define or describe:
   a. Krebs cycle
   b. oxidative decarboxylation
   c. mitochondria
   d. respiratory chain
   e. cytochromes
   f. coenzyme Q_{10}
2. What is the main function of the Krebs cycle?
3. What is GTP?
4. What is substrate-level phosphorylation?
5. What is oxidative phosphorylation?
6. List the component parts of the NAD^+ molecule. Which part is a B vitamin?
7. When NAD^+ and FAD act as oxidizing agents, which portion of each is reduced?
8. How is the energy of carbohydrate metabolism made available to the body cells?
9. Name the two shuttle systems that transport NADH from the cytoplasm into a mitochondrion.
10. a. What purpose do carbohydrates fulfill in metabolism?
    b. In what respect does carbohydrate metabolism resemble the burning of table sugar in a pan?
    c. In what way does it differ?

Problems

Krebs Cycle

11. Two carbon atoms are fed into the Krebs cycle as acetyl coenzyme A. In what form are two carbon atoms removed from the cycle?
12. What are the oxidizing agents most immediately involved in the Krebs cycle?

Respiratory Chain

13. How many molecules of ATP are formed by the oxidation of NADH in the respiratory chain?
14. How many molecules of ATP are formed by the oxidation of $FADH_2$ in the respiratory chain?
15. What is the electron acceptor at the end of the respira-

tory chain? To what product is this compound reduced?

16. What is the function of the cytochromes in the respiratory chain?

Biochemical Reactions

17. Replace the question marks with the proper compounds:

a. ? $\xrightarrow{\text{aconitase}}$ Isocitric acid

b. ? $\xrightarrow[\text{synthetase}]{\text{citric acid}}$ Citric acid

c. Fumaric acid $\xrightarrow{\text{fumarase}}$?

d. Isocitric acid $\xrightarrow{?}$ α-Ketoglutaric acid

e. α-Ketoglutaric acid $\xrightarrow{?}$ Succinyl CoA

18. Replace the question marks with the proper compounds:

a. Malic acid $\xrightarrow[\text{dehydrogenase}]{\text{malic acid}}$?

b. ? + ? $\xrightarrow[\text{diphosphokinase}]{\text{nucleoside}}$ GDP + ATP

c. Pyruvic acid $\xrightarrow{?}$ Acetyl CoA

d. Succinyl CoA $\xrightarrow[\text{synthetase}]{\text{succinyl CoA}}$?

e. Succinic acid $\xrightarrow[\text{dehydrogenase}]{\text{succinic acid}}$?

19. Refer to Problems 17 and 18 and select the equations (by number and letter) in which the following processes occur:
 a. isomerization b. hydration
 c. dehydration

20. Refer to Problems 17 and 18 and select the equations (by number and letter) in which the following processes occur:
 a. oxidation b. decarboxylation
 c. phosphorylation

21. Write balanced half-reactions for all the oxidation–reduction reactions in the Krebs cycle.

22. Write balanced half-reactions for all the oxidation–reduction reactions in the respiratory chain.

23. Write word equations for all the substrate-level phosphorylation reactions in glucose metabolism.

24. Write the net equation for the complete oxidation of 1 mol of glucose in a skeletal muscle cell respiring under aerobic conditions.

Muscle Metabolism

25. What is the function of the mitochondria?

26. What are the two functions of actomyosin?

27. Which energy reserves—carbohydrates or fats—are tapped in intense bursts of vigorous activity?

28. Which energy reserves—carbohydrates or fats—are mobilized to fuel prolonged low levels of activity?

29. Which type of metabolism—aerobic or anaerobic—is primarily responsible for providing energy for intense bursts of vigorous activity?

30. Which type of metabolism—aerobic or anaerobic—is primarily responsible for providing energy for prolonged low levels of activity?

31. Identify Type I and Type IIB muscle fibers as (**a**) fast twitch or slow twitch and (**b**) suited to aerobic oxidation or to anaerobic glycolysis.

32. Explain why high levels of myoglobin and mitochondria are appropriate for muscle tissue geared to aerobic oxidation.

33. Why does the high catalytic activity of actomyosin in Type IIB fibers suggest that these are the muscle fibers engaged in brief, intense physical activity?

34. Why can the muscle tissue that utilizes anaerobic glycolysis as its primary energy source be called on only for *brief* periods of intense activity?

35. Which type of muscle fiber is more affected by endurance training? What changes occur in the muscle tissue?

36. Birds use large, well-developed breast muscles for flying. Pheasants can fly 80 km/h, but only for short distances. Great blue herons can fly only about 35 km/h, but they can cruise great distances. What kind of fibers would each bird have in its breast muscles?

Additional Problems

37. If the methyl carbon atom of pyruvic acid is labeled, where does the label appear after the pyruvic acid goes through one turn of the Krebs cycle?

38. The complete oxidation of 1 mol of acetic acid in a calorimeter yields about 200 kcal.
 a. How much energy is stored as ATP when 1 mol of acetic acid is converted to acetyl CoA and metabolized via the Krebs cycle?
 b. What is the percent energy efficiency?

39. How many moles of ATP can be formed from each mole of lactic acid oxidized to CO_2 and H_2O?

40. Only two molecules of ATP are produced by the aerobic conversion of (**a**) succinic acid to fumaric acid and (**b**) glyceraldehyde 3-phosphate to 1,3-bisphosphoglyceric acid. Explain.

41. (**a**) How much ATP (in moles) is produced from 1 mol of glucose in a typical liver or heart cell? (**b**) What is the efficiency associated with this ATP production?

26 Lipid Metabolism

Lipids store energy and also insulate the organism against drastic changes in temperature.

Nearly all the energy required by the animal organism is generated by the oxidation of carbohydrates and lipids. Of all major nutrients, triglycerides are the richest energy source. The oxidation of 1 g of a typical fat or oil liberates about 9500 cal. (By comparison, the oxidation of an equal mass of carbohydrate liberates only about 4200 cal.) Lipid molecules contain a higher proportion of carbon–hydrogen bonds than carbohydrate molecules. Therefore, lipids have a greater capacity to combine with oxygen and consequently have a higher heat content. Whereas carbohydrates provide a readily available source of energy, lipids function as the principal *energy reserve.*

Men and women differ in their capacity to store lipids. The average percent body fat is about 16% for an adult male and 25% for an adult female. Male athletes in superb condition will have less than 7% body fat and females, less than 12% body fat. Americans consume about 100 to 125 g of lipids each day. This represents 34% of the daily calorie requirements. The National Cancer Institute, the American Heart Association, and most other health authorities recommend that Americans reduce their lipid intake and that not more than 25% of total calories be provided by lipids. Normally, triglycerides constitute about 98% of the total dietary lipids. As we shall see, however, significant quantities can be synthesized from carbohydrates when carbohydrate intake is high and lipid intake is low.

26.1 Fats as Fuels

A large percentage of any lipid molecule is saturated hydrocarbon (i.e., highly reduced). Thus fats may be thought of as analogous to combustible petroleum products. Carbohydrates, on the other hand, are analogous to alcohols (more oxidized), which are not nearly as effective as fuels.

The fat reserves in the average person provide sufficient energy to survive starvation for 30 days (given sufficient water). In comparison, glycogen in the liver (about 100 g) is depleted within 1 day. From the standpoint of efficiency of fuel storage, fats are far superior to carbohydrates. Glycogen, because of its many hydroxyl groups, is extremely hydrated. (About 2 g of water is bound to every gram of stored glycogen.) Thus in comparing stored mass of each fuel in the body, 1 g of fat contains more than six times the energy content of 1 g of hydrated glycogen.

The ability to store greater amounts of the more energy-efficient lipids is especially important for migrating birds and some terrestrial animals. The camel's hump, for example, is almost all adipose tissue (Section 26.3), and migratory birds rely on stored fats to supply the energy for long, sustained flight.

Example 26.1

One of McDonald's Big Mac sandwiches furnishes 541 kcal, of which 279 kcal comes from fat. What percentage of total calories is from fat?

Solution

Simply divide the calories from fat by the total calories. Then multiply by 100 to get percentage (parts per 100):

$$\%\text{ calories from fat } = \frac{279 \cancel{\text{kcal}}}{541 \cancel{\text{kcal}}} \times 100 = 51.6\%$$

Practice Exercise

One of Burger King's Whopper sandwiches furnishes 606 kcal, of which 288 kcal comes from fat. What percentage of total calories is from fat?

Example 26.2

The label on a macaroni and cheese dinner indicates that each 3/4-cup serving furnishes 290 kcal, 9 g of protein, 34 g of carbohydrate, and 13 g of fat. Calculate the percentage of calories from fat.

Solution

First, calculate the calories from each nutrient:

$$9\ \cancel{g} \times 4\ \text{kcal}/\cancel{g} = 36\ \text{kcal (from protein)}$$
$$34\ \cancel{g} \times 4\ \text{kcal}/\cancel{g} = 136\ \text{kcal (from carbohydrate)}$$
$$13\ \cancel{g} \times 9\ \text{kcal}/\cancel{g} = 117\ \text{kcal (from fat)}$$

Then calculate the percentage of calories from fat.

$$\frac{117\ \cancel{\text{kcal}}}{290\ \cancel{\text{kcal}}} \times 100 = 40.3\%$$

Practice Exercise

The label on a can of cream-style corn indicates that each one-cup serving furnishes 90 kcal, 2 g of protein, 22 g of carbohydrate, and 1 g of fat. Calculate the percentage of calories from fat.

Infants lacking essential fatty acids in their diet lose weight and develop **eczema,** an inflammatory skin disease characterized by scaly and crusty skin.

The nutritional aspects of lipids are still not completely understood, but we do know that they are not dietary necessities. An organism can survive on a lipid-free diet if carbohydrates and proteins are supplied as sources of metabolic energy. Certain lipids, however, are required for normal growth and development. These lipids supply the two fatty acids that the organism cannot synthesize. The unsaturated acids containing more than one double bond—linoleic and linolenic acids—are the **essential fatty acids.** Linoleic acid is used by the body to synthesize many of the other unsaturated fatty acids such as arachidonic acid. (Recall from Special Topic I, Section I.6, that arachidonic acid is required as a precursor for the biosynthesis of prostaglandins.) In addition, the essential fatty acids are incorporated into the structures of the membrane lipids (Section 20.9), and they are necessary for the efficient transport and metabolism of cholesterol. The average daily diet should contain about 2 to 3 g of the essential fatty acids.

In this chapter, we look at how our bodies store and metabolize fats. We also consider some of the problems associated with lipid storage and metabolism. Lipids offer one of the best illustrations of an old saying: Too much of a good thing can be bad.

26.2 Digestion and Absorption of Lipids

Lipids are not digested until they reach the upper portion of the small intestine. A hormone secreted in this region stimulates the gallbladder to discharge bile into the duodenum (the first 30 cm of the small intestine). In the context of lipid digestion, the principal constituents of bile are the bile salts (Section 20.10). These salts act as emulsifiers, disrupting some of the hydrophobic bonds holding the lipid molecules together. This emulsification is essential to lipid digestion. Bile salts act much like soap molecules. They break down large, water-insoluble lipids into smaller globules (micelles) and keep the smaller globules suspended in the aqueous digestive medium. The greatly increased surface area of the lipid particles and the opportunity afforded for more intimate contact with the lipases result in rapid digestion of the fats. Another hormone then promotes the secretion of the pancreatic juice. This juice contains the lipases that catalyze the digestion of triglycerides first to diglycerides and then to 2-monoglycerides and fatty acids. Phospholipids and cholesterol esters are also hydrolyzed into their component molecules.

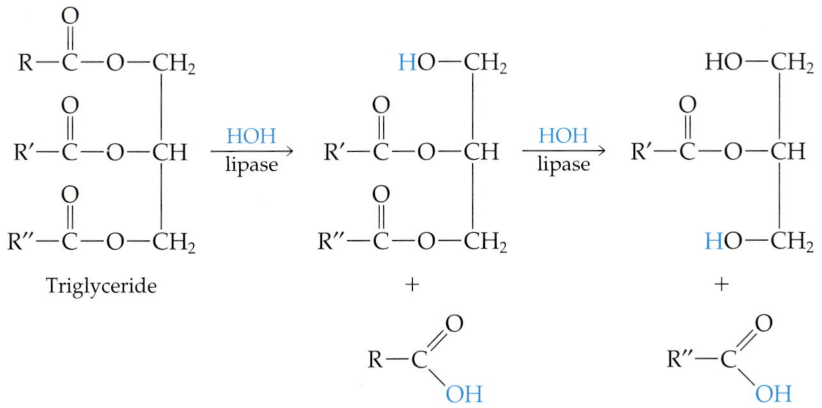

Biochemists do not agree upon the extent to which a lipid must be hydrolyzed in order to be absorbed. It is probable that, after emulsification, some lipids are absorbed directly through the intestinal membrane before being hydrolyzed. (Any lipid that is not absorbed is eliminated in the feces.) Once the monoglycerides, fatty acids, and free cholesterol pass into the cells of the intestinal epithelium, they are immediately resynthesized to triglycerides, phospholipids, or cholesterol esters. It has been shown experimentally that the glycerol needed to form these compounds is supplied by the metabolic pool within the intestinal cells and is not the glycerol obtained by lipid hydrolysis.

Lipids are, by definition, relatively insoluble in water.[1] Because blood is an aqueous solution, lipids are not soluble in blood. They can be efficiently transported by the blood, however, if they are first complexed with water-soluble proteins in the plasma. Such complexes are called **lipoproteins.** The triglycerides formed in the intestinal cells become associated with proteins and form lipoproteins called chylomicrons (recall Section 20.10). The chylomicrons are transported by the lymphatic system into the bloodstream.[2] Some of the triglycerides are carried to the liver, where they are modified and/or used to provide energy for liver functions. The remaining lipids (along with triglycerides synthesized in the liver) are either transported to specialized fat-storage cells (via VLDLs) or circulated through the blood as lipoprotein complexes (LDLs and HDLs). The circulating lipids are distributed to the various tissues to be incorporated into membrane lipids or oxidized for the generation of energy.

The concentration of lipids in the blood changes constantly. As lipids are absorbed from the digestive tract after a meal, the level in the blood increases. As blood lipids are removed to storage or are oxidized in certain tissues, the blood lipid level falls. On demand, the body can synthesize lipids from other foods or remove already-formed lipids from storage. Both activities result in an increase in

[1] There is one advantage to the low water solubility of fats. These compounds can be separated from other components of the blood by extraction with fat solvents such as toluene or methylene chloride. Such an extraction is always one of the first steps in any clinical procedure aimed at determining lipid levels.

[2] Lymph is tissue fluid (water and dissolved substances) that has entered a lymphatic capillary. Lymph vessels transport excess fluid from interstitial spaces and return it to the bloodstream. The blood system and the lymphatic system are separate, but there is a crossover point via the thoracic duct. After a high-fat meal, the lymph changes from a transparent yellow to a milky white because of the emulsified lipid.

Table 26.1 Range of Normal Levels of Lipids in Blood Plasma of a Fasting Person

Constituent	Concentration (mg/100 mL)
Free cholesterol	30–60
Cholesterol esters	75–150
Total cholesterol	120–250
Triglycerides	25–260
Lecithin	100–225
Sphingomyelin	10–47
Total phospholipids	150–250
Total lipids	400–700

the blood lipid level. Finally, the body excretes some lipids as a normal component of feces. These excreted lipids may be in the form of fats or soaps or fatty acids. Their removal via the intestine tends to lower the blood lipid level.

To standardize the conditions under which lipid concentrations are determined, lipids in the plasma are measured after fasting. This is the same procedure followed in the measurement of blood sugar levels. Typical *normal fasting levels* of lipids are given in Table 26.1. Abnormally high levels of triglycerides and cholesterol are thought by many medical practitioners to be involved in hardening of the arteries, a condition that may lead to rupture or blockage of vessels in the brain (a stroke) or in the heart (a heart attack; see Section 20.10).

26.3 Fat Depots

Fats are stored throughout the body. Principally, however, they are deposited in a special kind of connective tissue called **adipose tissue.** The cells making up adipose tissue are **fat cells,** and in them a large percentage of the cytoplasm is replaced by large droplets of triglycerides (approximately 90% of the mass of the fat cell; see Figure 26.1). Body locations containing large amounts of adipose tissue are called **fat depots.** Two such locations are just beneath the skin (subcutaneous fat depots) and in the abdominal area. Adults have approximately 30 to 40 billion fat cells that swell or shrink like a sponge depending on the amount of fat inside them. Adipose tissue is the only tissue in which free triglycerides occur in appreciable amounts. Elsewhere, in nonadipose cells or in the blood plasma, lipids are bound to proteins as lipoprotein complexes.

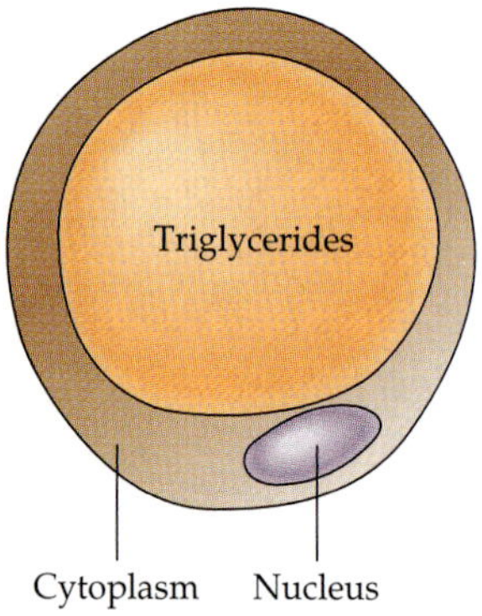

▲ **FIGURE 26.1**

A typical fat cell found in adipose tissue. Fat cells produce and store lipids; they are among the largest cells in the body. Notice that the nucleus and cytoplasm are displaced to the cell periphery by the large droplet of triglycerides.

Fat depots are found around vital organs, such as the heart, liver, kidneys, and spleen. There the adipose tissue serves as a protective cushion, helping prevent injury to the organs. Subcutaneous fat depots help insulate against sudden temperature changes. The adipose tissue acts just like the insulation in the walls of a house, trapping body heat and preventing it from escaping to the surroundings. In some ways, the tissue also acts like the furnace of a house. When the outside temperature drops, metabolic activity in the cells generates heat to compensate for the heat lost to the environment.

When food is taken into the body in excess of immediate needs, the excess is digested and shunted to storage areas. Digested lipids are tucked away in the adipose tissue; carbohydrates are converted to glycogen and stored, mainly in liver cells and muscle cells. When no more glycogen storage area is available, carbohydrates are converted to fats and stored in the depots. There is a continuous, dynamic change as molecules come and go from storage. If your weight is

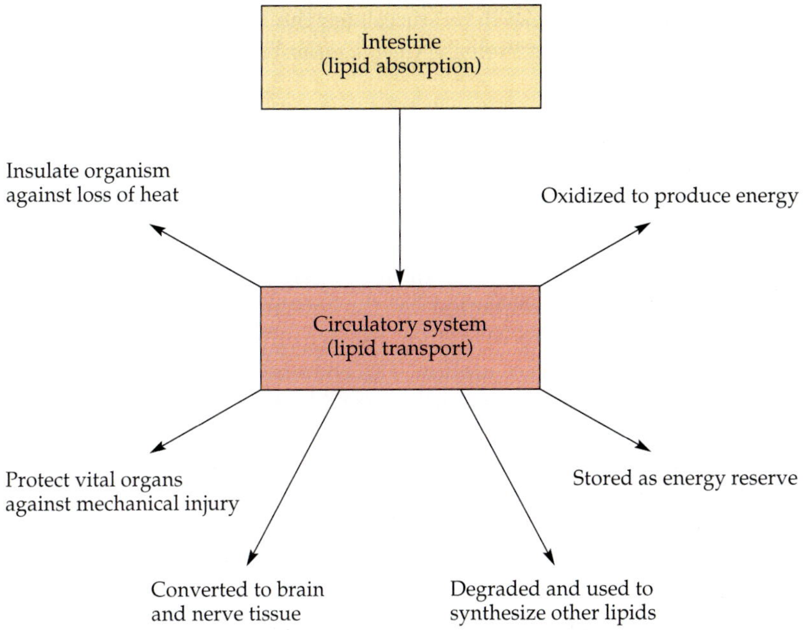

FIGURE 26.2
Metabolic fate of lipids.

Brown Fat

In addition to the normal white adipose tissue, some organisms contain another type of fatty tissue called **brown fat.** Its brown color is due to the presence of large numbers of mitochondria (stocked with red-brown cytochromes). In these specialized mitochondria, the reactions of the respiratory chain continue to take place, but no ATP is synthesized. The energy that would otherwise be used for ATP synthesis is released as heat, and the temperature of the organism rises. In essence, brown fat acts as a furnace to produce heat energy rather than stored energy.

Brown fat is particularly abundant in hibernating animals and in the necks and upper backs of newborn infants (as well as other mammals that are born hairless). These cells revive hibernating animals and maintain the body temperature of young animals. Adult humans have few brown fat cells because metabolic reactions usually generate more than enough heat so that body temperature can be regulated by the dissipation of this heat.

constant, there is an equilibrium in which the number of molecules arriving for storage is equal to the number being removed from storage. We shall discuss two situations in which there is an imbalance—obesity and starvation—in subsequent sections. The fate of lipids in the animal body is outlined in Figure 26.2.

26.4 Fatty Acid Oxidation

In order for the chemical energy in triglycerides to be released and so become available to the body, stored triglycerides must be cleaved to fatty acids and glycerol by the lipases in adipose tissue. Fat metabolism, like carbohydrate metabolism, is regulated by hormones. It is the same two hormones activated by low blood sugar levels (Section 24.4), epinephrine and glucagon, along with the neurotransmitter norepinephrine, that bind to receptor proteins in the cell membranes of adipose tissue. These hormones activate the enzyme *adenyl cyclase* (Section 24.4)

In humans, lipid breakdown occurs mainly at night. People on diets low in lipids are less able to sleep deeply. Some short-chain fatty acids and prostaglandins tend to induce sleep. In 1995, scientists at Scripps Research Institute found that *cis*-9-octadecenoamide,

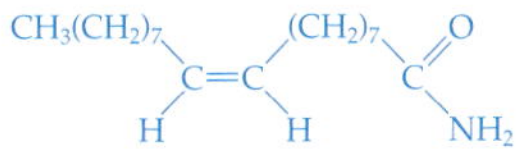

and related fatty amides induce sleep in rats. About 40 million people in the United States suffer from chronic sleep disorders and another 20 million to 30 million have occasional problems. Drugs developed from these compounds, which occur naturally in humans, would have a huge potential market.

to form cAMP from ATP. The cAMP then stimulates the hydrolysis (*lipolysis*) of triglycerides (by the lipases) and the release (*mobilization*) of fatty acids and glycerol from adipose tissue.

The glycerol obtained from fat hydrolysis in adipose tissue is transported to the liver, where it is converted to a carbohydrate derivative (see Section 26.6) and metabolized via the Embden–Meyerhof pathway. The fatty acids, bound to blood proteins, are transported by the blood to other tissues for oxidation. (Some fatty acids remain in the liver as fuel for liver cells.)

The fatty acids, which contain the bulk of lipid energy, are broken down in a series of sequential reactions accompanied by the gradual release of utilizable energy. Some of these reactions are oxidative and require the same coenzymes (NAD^+, FAD) used in carbohydrate oxidation. The enzymes involved in fatty acid catabolism are localized in the mitochondria along with the enzymes of the Krebs cycle, respiratory chain, and oxidative phosphorylation. This localization in the mitochondria is of the utmost importance because it provides for the efficient utilization of the energy stored in the fatty acid molecules.

At the beginning of the twentieth century, the German biochemist Franz Knoop showed that the breakdown of fatty acids occurs in a stepwise fashion with the removal of two carbon atoms at a time. Subsequent investigations have resulted in the separation and purification of the enzymes involved. The details of the degradation sequence are now fully understood. The reaction scheme is shown in Figure 26.3.

Step 1 The first phase of fatty acid metabolism occurs on the outer mitochondrial membrane. Fatty acids, like carbohydrates and amino acids, are relatively inert and must first be activated by conversion to an energy-rich fatty acid derivative of coenzyme A (called *fatty acyl CoA*). This activation is a two-step reaction catalyzed by *acyl CoA synthetase*. For each molecule of fatty acid activated, one molecule of coenzyme A and one molecule of ATP are used. First the fatty acid reacts with ATP to form a fatty acyl adenylate. Then the sulfhydryl group of coenzyme A attacks the acyl adenylate to yield fatty acyl CoA and AMP. Finally, the pyrophosphate formed in the initial reaction is hydrolyzed to phosphate ions by the enzyme *pyrophosphatase*. The net effect is the utilization of the two high-energy bonds in one ATP molecule to activate each molecule of fatty acid:

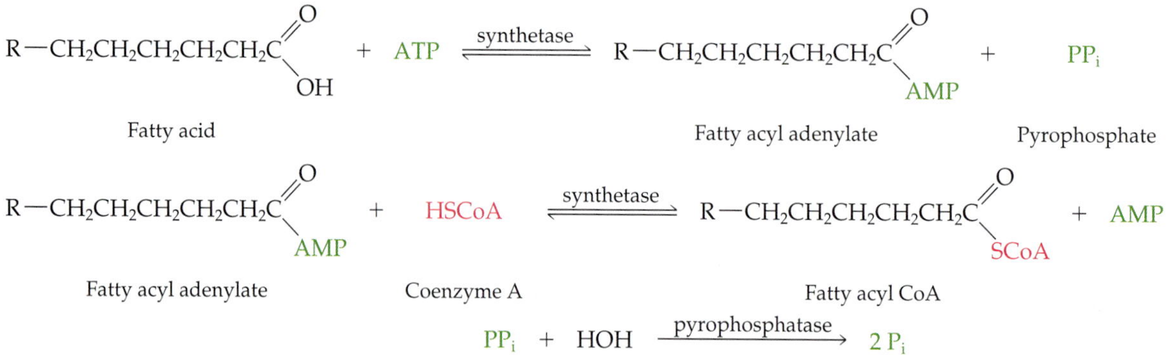

The fatty acyl CoA combines with a carrier molecule (carnitine; see Problem 53) and is transported into the mitochondrial matrix. Further metabolism of the fatty acyl CoA occurs entirely within the mitochondrial matrix via a sequence of reactions known as either β **oxidation** (because the β carbon undergoes successive

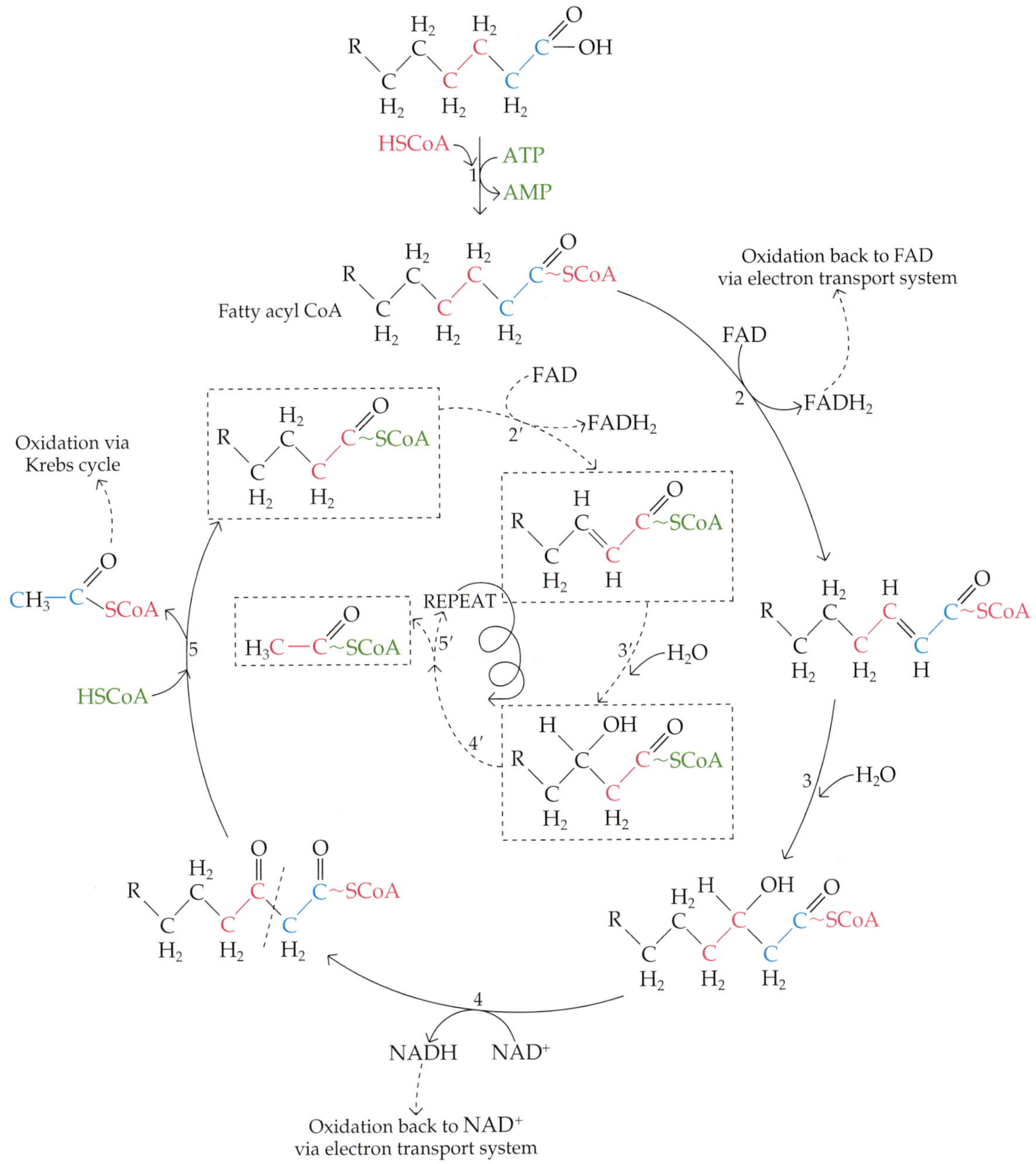

▲ **FIGURE 26.3**
Fatty acid oxidation.

oxidations) or the **fatty acid spiral** (because it involves the progressive removal of two-carbon units from the carboxyl end of the fatty acyl CoA).

Steps 2 to 4 of Figure 26.3 (dehydrogenation, hydration, dehydrogenation) are analogous to those involved in the conversion of succinic acid to oxaloacetic acid in the Krebs cycle (succinic ⟶ fumaric ⟶ malic ⟶ oxaloacetic). The functional group at the β carbon can be visualized as undergoing the following changes: alkane ⟶ alkene ⟶ secondary alcohol ⟶ ketone. The net effect is the introduction of a keto group beta to a carboxyl group.

Step 2 This first oxidation is catalyzed by *acyl CoA dehydrogenase*. The coenzyme FAD accepts two hydrogen atoms, one from the α carbon and one from the β carbon. The enzyme is stereospecific in that only the *trans*-alkene is obtained:

$$R-CH_2CH_2CH_2-\overset{\overset{H}{|}}{\underset{\underset{H}{|}}{C}}-\overset{\overset{H}{|}}{\underset{\underset{H}{|}}{C}}-C\overset{O}{\underset{SCoA}{}} + FAD \xrightarrow{dehydrogenase} R-CH_2CH_2CH_2-\overset{H}{\underset{}{C}}=\overset{}{\underset{H}{C}}-C\overset{O}{\underset{SCoA}{}} + FADH_2$$

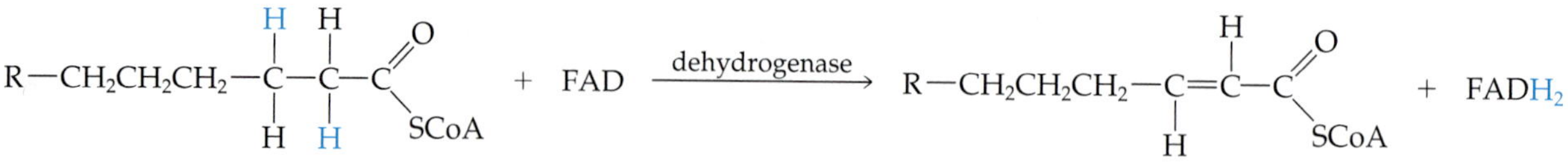

Fatty acyl CoA Enoyl CoA

Each molecule of $FADH_2$ reoxidized back to FAD via the electron transport chain supplies energy to form two molecules of ATP.

Step 3 As in the Krebs cycle, only the L isomer is formed when the stereospecific enzyme *enoyl CoA hydratase* adds water across the trans double bond:

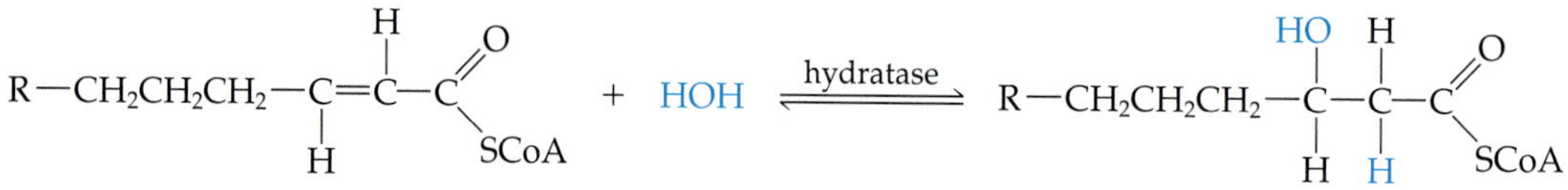

$$R-CH_2CH_2CH_2-\overset{H}{\underset{}{C}}=\overset{}{\underset{H}{C}}-C\overset{O}{\underset{SCoA}{}} + HOH \underset{}{\overset{hydratase}{\rightleftharpoons}} R-CH_2CH_2CH_2-\overset{\overset{HO}{|}}{\underset{\underset{H}{|}}{C}}-\overset{\overset{H}{|}}{\underset{\underset{H}{|}}{C}}-C\overset{O}{\underset{SCoA}{}}$$

Enoyl CoA L-β-Hydroxyacyl CoA

Step 4 The second oxidation is catalyzed by *β-hydroxyacyl CoA dehydrogenase*, which exhibits an absolute stereospecificity for the L isomer. Here the coenzyme NAD^+ is the hydrogen acceptor:

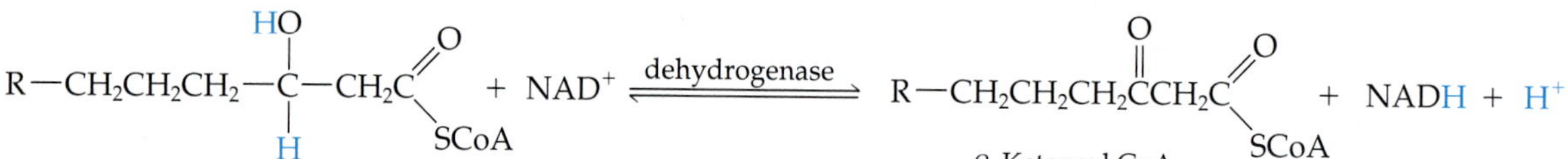

$$R-CH_2CH_2CH_2-\overset{\overset{HO}{|}}{\underset{\underset{H}{|}}{C}}-CH_2C\overset{O}{\underset{SCoA}{}} + NAD^+ \underset{}{\overset{dehydrogenase}{\rightleftharpoons}} R-CH_2CH_2CH_2\overset{O}{\overset{\|}{C}}CH_2C\overset{O}{\underset{SCoA}{}} + NADH + H^+$$

L-β-Hydroxyacyl CoA β-Ketoacyl CoA

The reoxidation of NADH to NAD^+ and the transport of hydrogen ions and electrons through the respiratory chain furnish three molecules of ATP.

Step 5 The final reaction is cleavage of the β-ketoacyl CoA by a molecule of coenzyme A. The products are acetyl CoA and a fatty acyl CoA whose chain length is shortened by two carbon atoms. The enzyme is *β-ketothiolase* (*thiolase*):

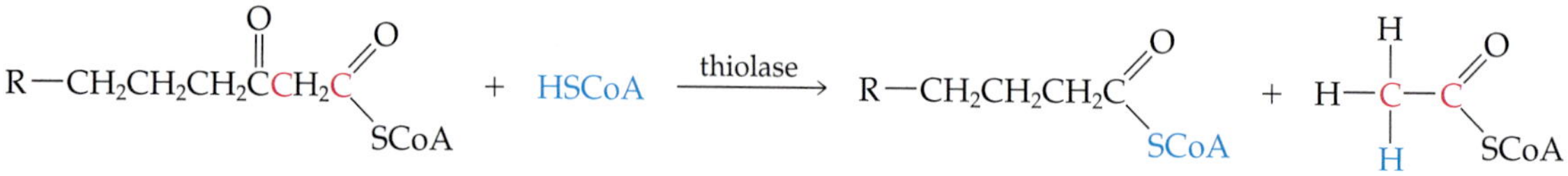

$$R-CH_2CH_2CH_2\overset{O}{\overset{\|}{C}}CH_2C\overset{O}{\underset{SCoA}{}} + HSCoA \xrightarrow{thiolase} R-CH_2CH_2CH_2C\overset{O}{\underset{SCoA}{}} + H-\overset{\overset{H}{|}}{\underset{\underset{H}{|}}{C}}-C\overset{O}{\underset{SCoA}{}}$$

β-Ketoacyl CoA Shortened fatty acyl CoA Acetyl CoA

The shortened fatty acyl CoA is then degraded further by repetition of steps 2 to 5; a molecule of acetyl CoA is liberated at each turn of the spiral. Normally, the spiral is repeated as many times as is necessary to break down a fatty acid containing an even number of carbon atoms, *n*, to *n*/2 molecules of acetyl CoA. Note that Step 1 in Figure 26.3 is necessary only at the beginning of the metabolism of each fatty acid molecule. Thus no further addition of ATP is necessary at subsequent turns of the spiral because the shortened fatty acids already contain the thiol ester. One molecule of ATP is sufficient to activate any fatty acid regardless of the number of carbon atoms in its hydrocarbon chain. (Unsaturated fatty acids, which are about 50% of the fatty acids found in humans are also incorporated into the β oxidation sequence. Two ancillary enzymes are required to convert them to normal substrates of the fatty acid spiral.) The overall equation for the β oxidation of palmitoyl CoA (16 carbons) is

$$CH_3(CH_2)_{14}C(=O)SCoA + 7\,FAD + 7\,NAD^+ + 7\,HSCoA + 7\,H_2O \longrightarrow$$

$$8\,CH_3C(=O)SCoA + 7\,FADH_2 + 7\,NADH + 7\,H^+$$

The acetyl CoA formed in the fatty acid spiral (plus the acetyl CoA obtained from glucose metabolism) is involved in a myriad of biochemical pathways. It may enter the Krebs cycle and be oxidized to produce energy, or it may be used as the starting material for biosynthesis of lipids (triglycerides, phospholipids, cholesterol, and other steroids). Acetyl CoA is also used in the formation of ketone bodies (Section 26.6). The various routes available to acetyl CoA are summarized in Figure 26.4.

When the body ingests more carbohydrate than it needs for energy and for glycogen synthesis, the excess is converted to fatty acids via the common intermediate, acetyl CoA. (In normal adults, roughly one-third of the total ingested carbohydrates may be converted to fatty acids and stored as fat.) *Fatty acids are built up from two-carbon units that come from acetyl CoA.* Although we do not discuss this biosynthetic pathway in this text, it should be pointed out that fatty acid synthesis is not simply the reverse of fatty acid breakdown. In humans, the liver is the major site of fatty acid synthesis. Synthesis occurs in the cytoplasm of liver cells, whereas fatty acid oxidation occurs in the mitochondria of various tissues. A completely different set of enzymes is involved. Synthesized fatty acids are used to provide immediate energy, or they are esterified with glycerol and stored as triglycerides in adipose tissue.

Finally, it is important to recall that the conversion of pyruvic acid to acetyl CoA is irreversible (Section 25.1) in mammals. Therefore the acetyl CoA molecules derived from fatty acids cannot be used to synthesize glucose. *Mammals can convert carbohydrates to lipids (via acetyl CoA). They cannot convert lipids to carbohydrates.*

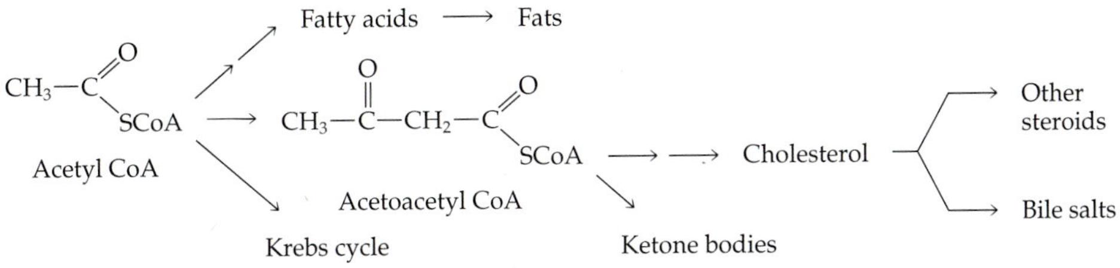

FIGURE 26.4 Acetyl coenzyme A plays a variety of roles in the chemistry of the cell.

Preferred Fuels of Various Tissues

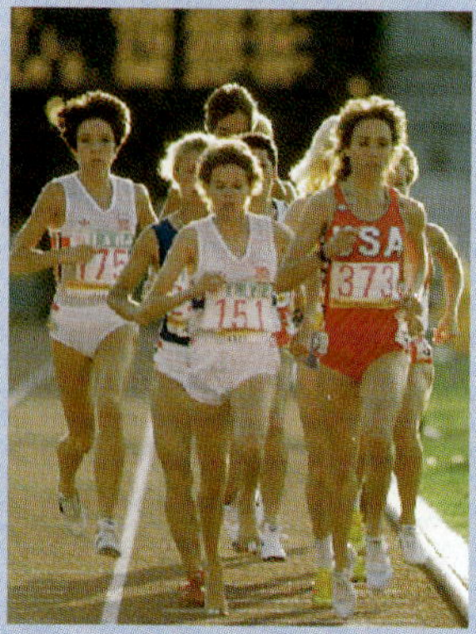

When athletes run a race, their initial fuel is mainly glucose derived from muscle glycogen. In longer races, liver glycogen and triglycerides in adipose tissue are broken down for fuel.

In a normal individual, certain organs use both glucose and fatty acids (in varying amounts) for their energy needs. The utilization rate of a specific fuel depends on the physiological state of the individual. For example, glucose (or glycogen) is the chief fuel for the immediate energy needs of active skeletal muscle (as in a sprint), whereas fatty acids are the major fuel for resting skeletal muscle. Fatty acids are the major fuel for cardiac muscle, although ketone bodies (see Section 26.7), glucose, and lactic acid are also used by the heart. During fasting and prolonged exercise, fatty acids are the preferred fuel for all types of muscles. The kidneys and adipose tissue use both glucose and fatty acids, whereas fatty acids are the preferred fuel of the liver. The brain uses glucose and ketone bodies (Section 26.7). It does not use fatty acids because they are bound to proteins and cannot diffuse across the blood–brain barrier. Red blood cells can use only glucose because they lack mitochondria and hence cannot obtain energy via oxidative phosphorylation.

26.5 Bioenergetics of Fatty Acid Oxidation

Note that the oxidation of fatty acids also produces large quantities of water, and it is this water that sustains migratory birds and animals (e.g., the camel) for long periods of time.

Let us again choose palmitic acid as a typical fatty acid of the human diet. Calculating its energy yield therefore instructs us about the yield of all other fatty acids. The combustion of 1 mol of palmitic acid releases a considerable amount of energy:

$$C_{16}H_{32}O_2 \quad + \quad 23\,O_2 \longrightarrow 16\,CO_2 \quad + \quad 16\,H_2O \quad + \quad 2340\text{ kcal}$$

The amount of energy that is made available to the cell and conserved in the form of ATP is readily calculable.

The breakdown by an organism of 1 mol of palmitic acid requires 1 mol of ATP, and 8 mol of acetyl CoA is formed. Recall from Figure 25.1 that each mole of acetyl CoA metabolized by the Krebs cycle yields 12 mol of ATP. Each turn of the fatty acid spiral produces 1 mol of NADH and 1 mol of $FADH_2$. Reoxidation of these compounds by the respiratory chain (within skeletal muscle cells) yields 3 and 2 mol of ATP, respectively. The complete degradation of 1 mol of palmitic acid requires seven turns of the spiral, and a total of 7 mol of $FADH_2$ and NADH is therefore formed. The energy calculations can be summarized as follows.

	Yield of ATP
1 mol of ATP is split to AMP and 2 P_i	−2
8 mol of acetyl CoA formed (8 × 12)	96
7 mol of $FADH_2$ formed (7 × 2)	14
7 mol of NADH formed (7 × 3)	21
Total moles of ATP	129

The number of turns of the spiral for the oxidation of a fatty acid containing *n* carbons is (*n*/2) − 1 because the final turn yields two acetyl CoA molecules.

The percentage of available energy that can theoretically be conserved by the cell in the form of ATP is therefore

$$\frac{\text{Energy conserved}}{\text{Total energy available}}(100\%) = \frac{(129)(7300)}{2{,}340{,}000}(100\%) = 40.2$$

The efficiency of fatty acid metabolism is comparable to that of carbohydrate metabolism (38%; see Section 25.4).

Recall we said that carbohydrates are used for the normal energy requirements of skeletal muscle, whereas lipids provide energy for prolonged activity. It has been estimated that after 4 h of sustained exercise, fatty acid oxidation supplies more than 60% of the muscle's energy demands.

26.6 Glycerol Metabolism

One molecule of glycerol is obtained from each molecule of triglyceride or phospholipid hydrolyzed. This glycerol is transported to the liver, where it is readily incorporated into the scheme of carbohydrate metabolism by conversion to dihydroxyacetone phosphate. This two-step process includes the phosphorylation of a primary hydroxyl group followed by the oxidation of a secondary hydroxyl group to a ketone:

$$
\begin{array}{ccc}
\underset{\text{Glycerol}}{\begin{array}{l} CH_2OH \\ | \\ CHOH \\ | \\ CH_2OH \end{array}}
& \xrightarrow[\text{kinase}]{ATP\quad ADP}
& \underset{\substack{\text{Glycerol} \\ \text{phosphate}}}{\begin{array}{l} CH_2OH \\ | \\ CHOH \\ | \\ CH_2O\text{\textcircled{P}} \end{array}}
\xrightarrow[\text{dehydrogenase}]{NAD^+\quad NADH\ +\ H^+}
& \underset{\substack{\text{Dihydroxyacetone} \\ \text{phosphate}}}{\begin{array}{l} CH_2OH \\ | \\ C{=}O \\ | \\ CH_2O\text{\textcircled{P}} \end{array}}
\end{array}
$$

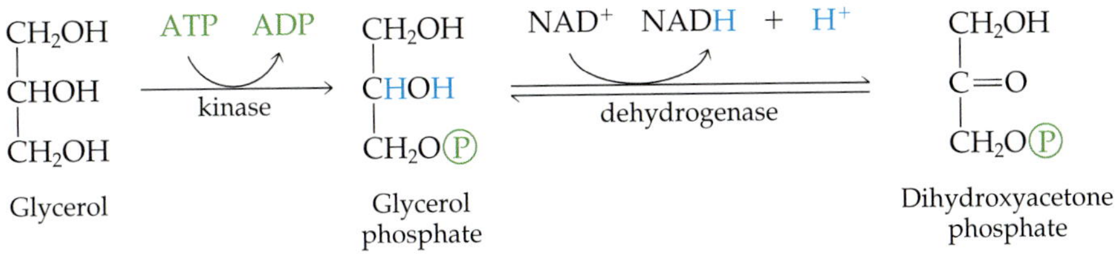

Dihydroxyacetone phosphate can be used to provide energy by conversion to pyruvic acid (via the Embden–Meyerhof pathway), or it can be transformed into glucose. Since glycerol is the starting compound for glucose synthesis, this is another important example of gluconeogenesis. Triglyceride catabolism is summarized in Figure 26.5.

Example 26.3

Calculate the moles of ATP produced by the oxidation of 1 mol of glyceryl stearate to CO_2 and H_2O in a muscle cell respiring aerobically.

$$
\begin{array}{l}
\overset{\displaystyle O}{\overset{\displaystyle \|}{C_{17}H_{35}C}}OCH_2 \\[4pt]
\overset{\displaystyle O}{\overset{\displaystyle \|}{C_{17}H_{35}C}}OCH \\[4pt]
\overset{\displaystyle O}{\overset{\displaystyle \|}{C_{17}H_{35}C}}OCH_2
\end{array}
\xrightarrow{\text{lipases}}\quad
3\ C_{17}H_{35}COOH \ + \
\underset{}{\begin{array}{l} CH_2OH \\ | \\ CHOH \\ | \\ CH_2OH \end{array}}
$$

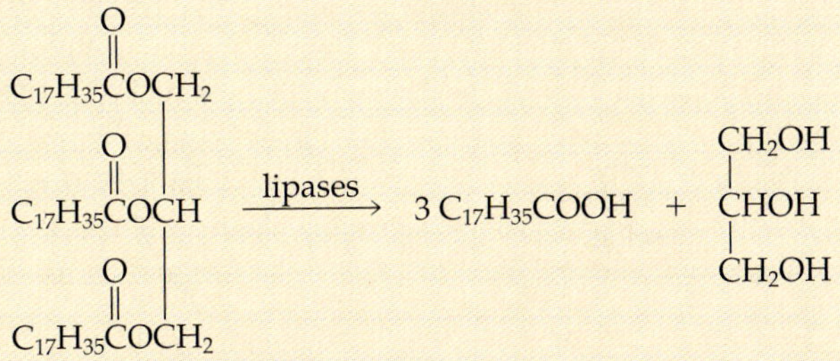

Solution

Each mole of stearic acid oxidized to CO_2 and H_2O yields 146 mol of ATP as follows:

1 mol of ATP used	-2
9 mol of acetyl CoA formed (9×12)	108
8 mol of $FADH_2$ formed (8×2)	16
8 mol of NADH formed (8×3)	$\underline{24}$
	146

Therefore, 3 mol of stearic acid yields $(3 \times 146) = 438$ mol of ATP.

One mole of glycerol, converted to dihydroxyacetone phosphate and metabolized via glycolysis and the Krebs cycle, yields 20 mol of ATP, obtained as follows. There is a net yield of 1 mol in the conversion of glycerol to dihydroxyacetone phosphate.

> The +2 ATP in the second step arises from the oxidation of NADH to NAD^+ via the glycerol phosphate shuttle and the respiratory chain.

$$
\begin{array}{c}
CH_2OH \\
| \\
CHOH \\
| \\
CH_2OH
\end{array}
\xrightarrow[\substack{-1\ ATP}]{ATP\quad ADP}
\begin{array}{c}
CH_2OH \\
| \\
CHOH \\
| \\
CH_2O\circled{P}
\end{array}
\xrightarrow[\substack{+2\ ATP}]{NAD^+\quad NADH\ +\ H^+}
\begin{array}{c}
CH_2OH \\
| \\
C{=}O \\
| \\
CH_2O\circled{P}
\end{array}
$$

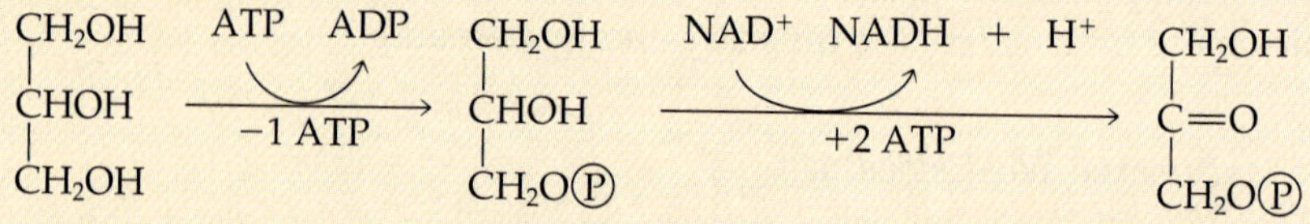

An additional 4 mol of ATP is formed by the subsequent conversion of dihydroxyacetone phosphate to pyruvic acid via the Embden–Meyerhof pathway. (Refer to Figure 24.12. Two mol of ATP in Step 6, one mol in Step 7, and one mol in Step 10.) Then 15 mol of ATP is produced by the oxidation of pyruvic acid by the Krebs cycle and electron transport chain.

$$\text{Total yield of ATP} = 438 + 20 = 458 \text{ mol}$$

Practice Exercise

Calculate the moles of ATP produced by the complete oxidation of 1 mol of the mixed triglyceride glyceryl lauropalmitomyristate. (See Chapter 20 for formulas of the fatty acids.)

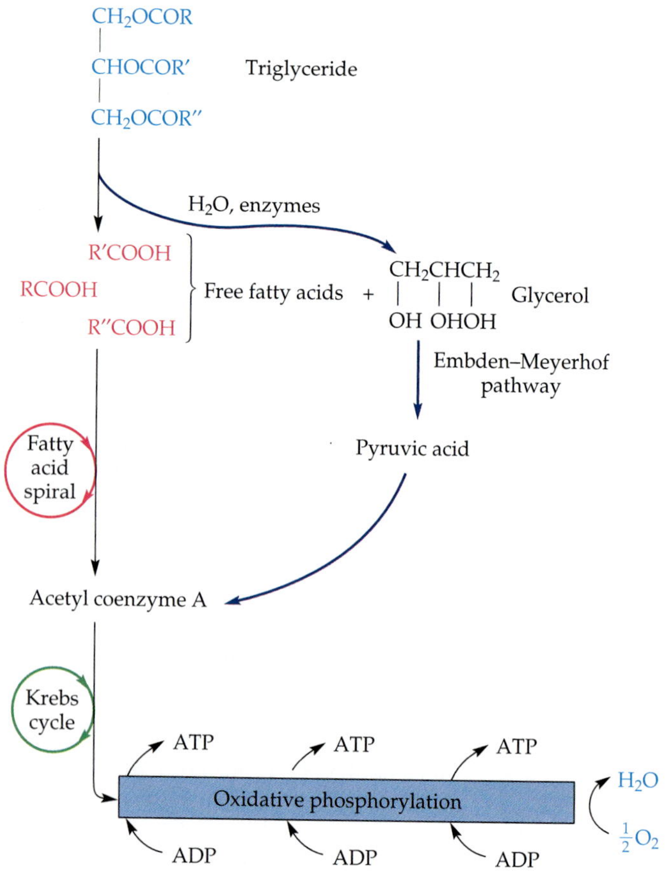

FIGURE 26.5 ▶

An outline showing how fat reserves (triglycerides) are used for the production of energy.

26.7 Ketosis

We previously mentioned some abnormal physiological conditions under which fatty acids provide the bulk of the energy for most body tissues. These conditions generally occur in conjunction with an impairment of carbohydrate metabolism (e.g., starvation, diabetes mellitus, or zero-carbohydrate "crash" diets). When cells don't receive sufficient amounts of carbohydrate, the rate of fatty acid oxidation increases and the production of acetyl CoA increases accordingly. The problem is that most of the acetyl CoA molecules cannot enter the Krebs cycle. Why? Because there is insufficient oxaloacetic acid (Step 2, Figure 25.1). When carbohydrate is lacking, oxaloacetic acid levels are reduced because oxaloacetic acid is used to synthesize glucose (via gluconeogenesis) and hence is not available to combine with acetyl CoA.

When the concentration of acetyl CoA in the liver mitochondria reaches a certain level, two acetyl CoA molecules combine, in a reversal of the final steps of the fatty acid spiral, to produce acetoacetyl CoA. The acetoacetyl CoA reacts with another molecule of acetyl CoA and with water to form β-hydroxy-β-methylglutaryl CoA, which is then cleaved to acetoacetic acid and acetyl CoA. Most of the acetoacetic acid is reduced to β-hydroxybutyric acid, and a small fraction of it is decarboxylated to carbon dioxide and acetone (Figure 26.6).

The acetone produced in the metabolism of excess acetyl CoA is transported to the kidneys and lungs. Being volatile, the acetone that reaches the lungs is expelled in the breath. The sweet smell of acetone, a characteristic of ketosis, is frequently noticed on the breath of severely diabetic patients.

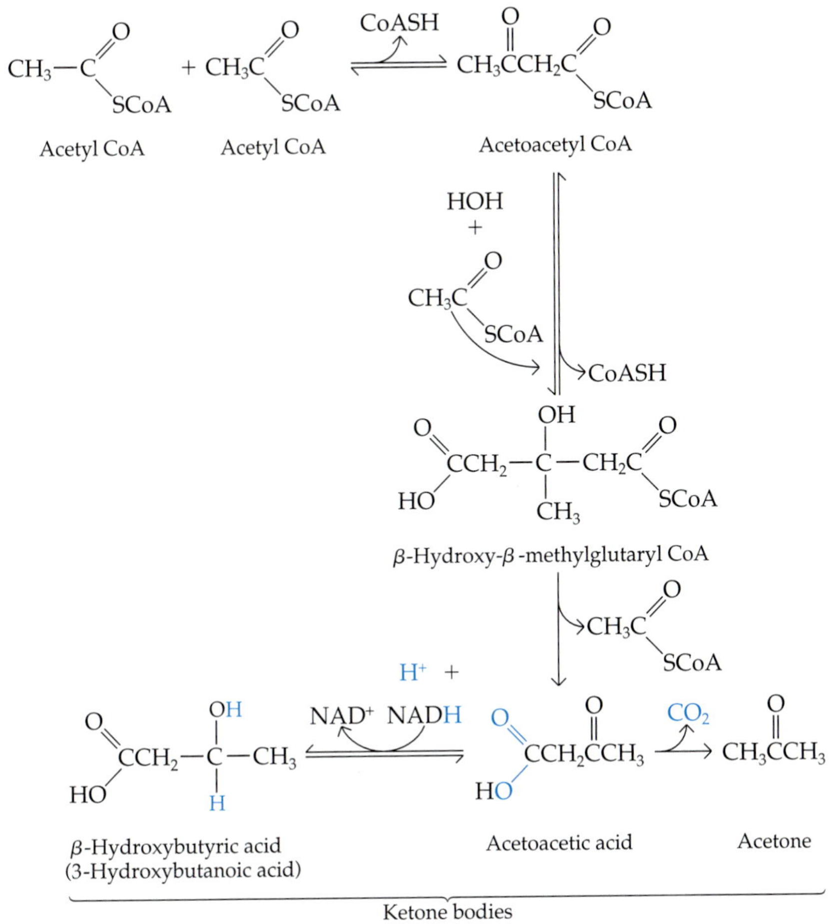

FIGURE 26.6
The formation of ketone bodies.

Acetoacetic acid, β-hydroxybutyric acid, and acetone are collectively referred to as **ketone bodies.** Normally the liver synthesizes small amounts of acetoacetic acid and β-hydroxybutyric acid and releases them into the blood for use as a metabolic fuel (to be converted back to acetyl CoA) by other aerobic tissues, including heart and brain. (During prolonged starvation, ketone bodies provide about 70% of the energy requirements of the brain.) The kidneys excrete about 20 mg of ketone bodies each day. Normally, blood levels are maintained at about 1 mg of ketone bodies per 100 mL of blood. However, when the rate of ketone body formation in the liver greatly exceeds the rate at which the tissues use them, excess ketone bodies accumulate in the blood (*ketonemia*—concentrations greater than 3 mg/100 mL) and in the urine (*ketonuria*). These conditions together are referred to as **ketosis.**

Because two of the three ketone bodies are acids, their presence in the blood in excessive amounts overwhelms the blood buffers and causes a marked decrease in blood pH (to 6.9 from a normal value of 7.4). This decrease in pH leads to a serious condition known as **acidosis.** Acidosis results in interference with the transport of oxygen by the hemoglobin molecule (see Section 28.9), and a fatal coma may result. In moderate to severe acidosis, "air hunger" sets in. Breathing becomes labored and very painful. The body also loses fluids as the kidneys eliminate large quantities of water trying to get rid of the acids. The person becomes dehydrated. The short oxygen supply and dehydration lead to depression. Even mild acidosis leads to lethargy, loss of appetite, and a generally run-down feeling. Untreated patients may go into diabetic coma. At that point, prompt treatment is necessary if the patient's life is to be saved. Any treatment that promotes the utilization of carbohydrates (for example, an injection of insulin) can alleviate both ketosis and acidosis.

Acidosis resulting from the production of acidic metabolites is sometimes called metabolic acidosis to distinguish it from respiratory acidosis (see Section 28.10).

26.8 Obesity, Exercise, and Diets

One of the anomalies of our time is that, while much of the world suffers from hunger and malnutrition, the major dietary problem in developed nations is obesity. **Obesity** is the condition in which an excessive amount of fat is deposited in the adipose tissue and the individual becomes overweight (20% over ideal weight for age and height). In rare cases, obesity is due to glandular malfunctions. The overwhelming majority of obese people are that way, however, because they eat more food than their bodies use as fuel. Adults need food primarily for the energy it can supply. Amounts eaten in excess of immediate energy requirements are stored as fat.

Obesity contributes to poor health, increased risk of cardiovascular disease and diabetes, and perhaps increased susceptibility to other diseases as well. An obese person has a shorter life expectancy than a person of average weight. One pound of adipose tissue is equivalent to about 3500 kcal. This tissue requires 200 miles of blood capillaries to serve its cells. Excess fat therefore puts extra strain on the heart—the heart has to work harder to supply blood to the extra tissue.

In our culture today, thin is in. Many formerly sedentary people have joined the ranks of joggers, walkers, tennis players, body-builders, and bikers in pursuit of the bulgeless body. About a million people per week are treated in weight-loss clubs and clinics. Countless others treat themselves with crash diets, dietary supplements, diet pills, and other programs. Diet books populate the bestseller lists. Many of these diet plans are simply fads. They are more likely to increase their creators' wealth than to decrease your weight. Many are grossly unbalanced in terms of necessary nutrients. An unbalanced diet, especially over an extended period of time, causes a variety of nutritional deficiencies, a decrease in resistance to disease, and a decline in general health.

The minerals, such as iron, calcium, and potassium, also are deficient in many crash diets. A deficiency of these minerals can interrupt the smooth function of nerve impulse transmission to muscle. This impairs athletic performance. Impulse transmission to vital organs also may be impaired in cases of severe mineral deficiency. Several deaths from cardiac arrest have resulted from variations of the "liquid protein diet."

Weight loss or gain is based on the law of conservation of energy. If we take in more calories than we use up, the excess calories are stored as fat. If we take in fewer calories than we need for our activities, our bodies burn some of the stored fat to make up for the deficit.

Total calories consumed is important, but even more important is the distribution of calories. Modern advice is that good nutrition should include generous portions of complex carbohydrates (starches), along with fruits and vegetables, modest portions of protein, and very small amounts of fats, oils, and simple carbohydrates (sugars). See the food guide pyramid in Figure 26.7.

How much fat is enough? The male body requires about 3% essential body fat, the female body 10 to 12%. It is difficult to measure percent body fat accurately. Skinfold calipers are quite inaccurate because they measure water retention as well as fat.

Instead of a direct measurement of fat, therefore, an index known as the body mass index (BMI) is commonly used. It is defined as follows:

$$\text{BMI} = \frac{700 \times \text{Body weight (lb)}}{[\text{Height (in.)}]^2}$$

In other words, for a person who is 5 ft 10 in. tall (70 in.) and weighs 180 lb, the body mass index is

$$\text{BMI} = \frac{700 \times 180}{70 \times 70} = 26$$

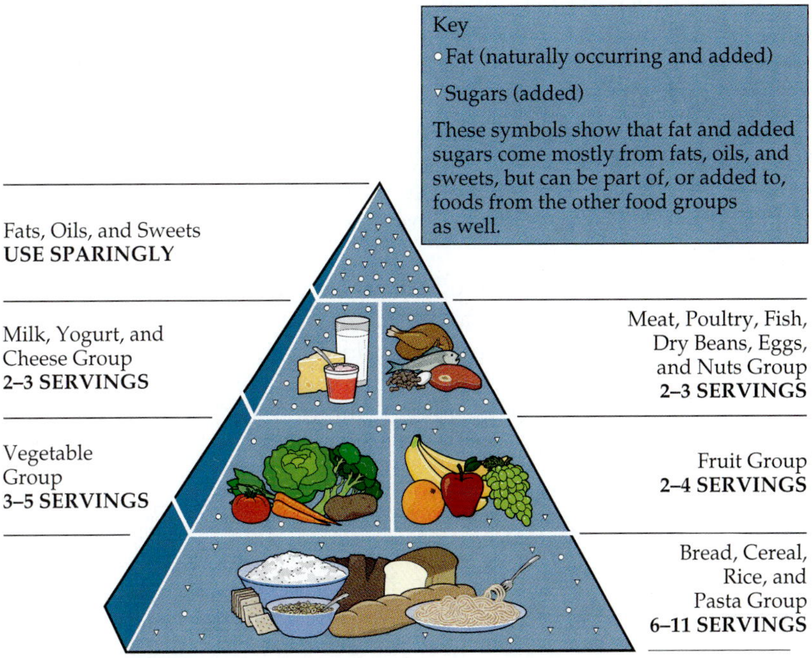

◄ **FIGURE 26.7**
The food guide pyramid. Start with bread, cereal, rice, and pasta, along with fruits and vegetables. Then add some servings from the milk and meat groups. Go easy on the fats, oils, and sweets.

Average BMI values for adults are between 20 and 25. Values greater than 27 may be associated with obesity as well as other related problems, such as high blood pressure or diabetes.

Example 26.4

What is the body mass index for a person who is 6 ft 2 in. tall and weighs 230 lb?

Solution

The person's height is $6 \times 12 + 2 = 72 + 2 = 74$ in.

$$\text{BMI} = \frac{700 \times 230}{74 \times 74} = \frac{161{,}000}{5500} = 29$$

Practice Exercise

What is the body mass index for a person who is 5 ft tall and weighs 150 lb?

It is possible to lose weight through dieting. One pound of fat is deposited in adipose tissue for every 3500 kcal in excess of the body's requirements. Therefore, if you reduce your intake by 100 kcal/day and keep your activity constant, you will burn off a pound of adipose tissue (3500 kcal) in 35 days. Unfortunately, people are seldom so patient, and they resort to more stringent diets. To achieve their goals more rapidly, they exclude certain foods and reduce the amounts of others. Such diets are harmful. Diets with fewer than 1200 kcal/day are likely to be deficient in necessary nutrients, particularly in B vitamins and iron. Finally, for most people, any weight lost through dieting is quickly regained when the dieter resumes old eating habits.

A moderately active person can calculate the calories needed each day to maintain proper weight by multiplying the desired weight (in pounds) by 15 kcal/lb. If you want to maintain a weight of 120 lb, you need (120 lb $\times$ 15 kcal/lb) = 1800 kcal per day.

According to one theory, hunger is regulated by the hypothalamus. When this gland senses that the level of fatty acids circulating in the bloodstream is low, it triggers the hunger mechanism. According to the **set-point theory,** each of us has a unique level at which the hypothalamus acts. Some of us must have a higher level of body fat than others to avoid constant hunger. This is consistent with the fact that obesity seems to be inherited. It is also rather grim news for those of us who would like to lose weight—we can do so only by being constantly hungry. There is some good news, however. It appears that the set point can be lowered through exercise.

People who do not increase their food intake when they begin an exercise program will lose weight. And, contrary to myth, exercise (up to 1 h/day) will not cause an increase in appetite. Most of the weight loss from exercise is due to the increase in metabolic rate during the activity, but the increased metabolic rate continues for several hours after completion of the exercise. Exercise helps us maintain both fitness and thinness.

Example 26.5

A fast game of tennis burns off about 10 kcal/min. How long would you have to play to burn off 1 lb of adipose tissue?

Solution

One pound of adipose tissue stores 3500 kcal of energy. To burn it off at 10 kcal/min requires

$$3500\,\cancel{\text{kcal}} \times \frac{1\text{ min}}{10\,\cancel{\text{kcal}}} = 350\text{ min}$$

Thus it takes 5 h and 50 min to burn off 1 lb of fat via a fast-paced game of tennis. (You don't have to do it all in one day.)

Practice Exercise

Walking a mile at a leisurely pace burns off about 100 kcal, and walking a mile at a brisk pace burns off about 175 kcal. How far do you have to walk (a) at a leisure pace and (b) at a brisk pace to burn off 1 lb of fat?

Most quick-weight-loss diets depend on factors other than fat metabolism to hook prospective customers. Many contain a *diuretic*, such as caffeine, to increase the output of urine. Much of the weight loss is water loss. Weight is regained when the body is rehydrated.

Other such diets depend on depleting the body's stores of glycogen. When carbohydrates are eliminated from the diet, the body draws on its glycogen reserves, depleting them in about 24 h. Recall that glycogen molecules have lots of hydroxyl groups that can form hydrogen bonds to water molecules. Each pound of glycogen carries with it about 2 lb of water held to it by these hydrogen bonds. Depleting the pound or so of glycogen results in a weight loss of about 3 lb (1 lb glycogen + 2 lb water). No fat is lost, and the weight is quickly regained when the dieter resumes eating carbohydrates.

If your normal energy expenditure is 2400 kcal/day, the most fat you could lose by total fasting for a day would be 0.69 lb (2400 kcal/day divided by 3500 kcal/lb adipose tissue). This assumes that your body would burn nothing but fat. It wouldn't. Recall that the brain runs on glucose, and if the glucose isn't supplied in the diet, it is obtained from protein (see Section 27.5). On any diet that restricts carbohydrate intake, you lose muscle mass as well as fat. When you gain the weight back (as 90% of all dieters do), you gain only fat. People who diet and then gain back the lost weight are replacing metabolically active tissue (muscle) with inactive fat. Weight loss becomes harder with each subsequent attempted diet.

On a weight-loss diet (without exercise), about 62% of the loss is fat. About 11% is protein (muscle tissue). The rest is water and a little glycogen.

Example 26.6

If you ordinarily expend 2200 kcal/day and you go on a diet of 1500 kcal/day, how long will it take to diet off 1 lb of fat?

Solution

You will use 700 kcal/day more than you consume. There are 3500 kcal in 1 lb of fat, so it will take

$$3500\,\cancel{\text{kcal}} \times \frac{1\text{ day}}{700\,\cancel{\text{kcal}}} = 5\text{ days}$$

(Keep in mind, however, that your weight loss will not all be fat. You will probably lose more than 1 lb in the five days, but it will be mostly water with some protein and a little glycogen.)

Practice Exercise

A person who expends 1800 kcal/day goes on a diet of 1200 kcal/day without a change in activities. How much fat will the person lose if the diet is followed for 3 weeks?

Any weight-loss program that promises a loss of more than a pound or two a week is likely to be dangerous quackery. The most sensible approach to weight loss is to (1) adhere to a balanced low-calorie diet that meets the RDA for essential

Table 26.2 Calories Expended in Exercise

Activity	Kcal per Hour*
Badminton, competitive singles	480
Basketball	360–660
Bicycling	
10 mph	420
11 mph	480
12 mph	600
13 mph	660
Calisthenics, heavy	600
Handball, competitive	660
Rope skipping, vigorous	800
Rowing machine	840
Running	
5 mph	600
6 mph	750
7 mph	870
8 mph	1020
9 mph	1130
10 mph	1285
Skating, ice or roller, rapid	700
Skiing, downhill, vigorous	600
Skiing, cross-country	
2.5 mph	560
4 mph	600
5 mph	700
8 mph	1020
Swimming, 25–50 yards per min	360–750
Walking	
Level road, 4 mph (fast)	420
Upstairs	600–1080
Uphill, 3.5 mph	480–900
Gardening, much lifting, stooping, digging	500
Mowing, pushing hand mower	450
Sawing hardwood	600
Shoveling, heavy	660
Wood chopping	560

* Caloric expenditure is based on a 150-lb person. There is a 10% increase in caloric expenditure for each 15 lb over this weight and a 10% decrease for each 15 lb under.

Adapted from E. L. Wynder, *The Book of Health: The American Health Foundation*. © 1981 Franklin Watts, Inc., New York. Used with permission.

nutrients and (2) engage in a reasonable, consistent, individualized exercise program. Table 26.2 provides the numbers of calories expended in several forms of exercise. The principles of weight loss are met by decreasing intake and increasing output.

All people, athletes included, should be aware of the possible risks of dieting. Vitamin deficiency often develops slowly over months or years. This problem can be corrected or avoided by careful nutritional planning or by taking vitamin supplements (although most nutritionists will argue that food is the best source of the body's nutritional needs). Some nutritionists suggest vitamin supplements for those who refuse (or are not able) to eat a balanced diet, but vitamin supplements should be taken only under the direction of a physician or registered dietitian.

Summary

The body needs, but cannot synthesize, two fatty acids, linoleic and linolenic. Because they must be present in the diet, they are called **essential fatty acids.**

Lipid digestion begins in the duodenum. First bile salts emulsify the lipid molecules, and then lipases hydrolyze them. The hydrolysis products pass into the intestine and are reconverted to triglycerides, phospholipids, and cholesterol esters. The lipids then combine with proteins to become **lipoproteins** and pass into the blood.

Body tissue that stores lipids is **adipose tissue,** and areas of the body containing large amounts of adipose tissue are **fat depots.** Triglycerides in adipose tissue are again hydrolyzed to fatty acids plus glycerol. The glycerol is carried to the liver, where it enters the Embden–Meyerhof pathway. The fatty acids are carried to all cells that can use them, and are oxidized via the **fatty acid spiral.**

In each turn of the spiral, the fatty acid is shortened by two carbons as one molecule of acetyl CoA is formed. When the chain has been shortened to four carbons, the final turn takes place because the two-carbon fragment resulting from this turn becomes an acetyl CoA molecule in step 5 of the spiral. Therefore the number of turns to oxidize a fatty acid completely is $(n/2)-1$, where n is the number of carbons in the acid. Because two acetyl CoA are formed in the final turn and one in each preceding turn, the number of acetyl CoA formed is $n/2$.

To activate a fatty acid for β-oxidation, the equivalent of two high energy phosphate bonds are consumed as the fatty acid combines with coenzyme A to form a fatty acyl CoA molecule. Each turn of the fatty acid spiral produces five ATP molecules. All the acetyl CoA formed during the spiral enters the Krebs cycle and yields 12 molecules of ATP. The net ATP yield of fatty oxidation is therefore

$(n/2) \times 12$ ATP from acetyl CoA
$[(n/2) - 1] \times 5$ ATP from all the turns of the spiral
MINUS the 2 high energy phosphate bonds required to activate the fatty acid

The efficiency of fatty acid catabolism in the human body is approximately 40%.

A human body not receiving sufficient carbohydrates for its immediate energy needs has a higher-than-normal rate of fatty acid oxidation. This elevated oxidation rate causes the acetyl CoA concentration in the liver to increase. The acetyl CoA produces acetoacetic acid, which is converted to β-hydroxybutyric acid and acetone. These three compounds are collectively known as **ketone bodies.** They accumulate in the blood in a condition known as **ketosis.** High blood levels of the two acidic ketone bodies overwhelm the blood buffers, so that the blood becomes too acidic, a condition called **acidosis.**

Key Terms

acidosis (26.7)
adipose tissue (26.3)
brown fat (26.3)
eczema (26.1)
essential fatty acid (26.1)
fat cell (26.3)
fat depot (26.3)
fatty acid spiral (26.4)
ketone body (26.7)
ketosis (26.7)
lipoprotein (26.2)
obesity (26.8)
β-oxidation (26.4)
set-point theory (26.8)

Review Questions

1. Define or explain:
   a. essential fatty acid
   b. eczema
   c. bile salts
   d. lipases
   e. chylomicrons
   f. lymph
   g. fat depots
   h. obesity
   i. lipolysis
   j. mobilization
   k. adipose tissue
   l. β oxidation
   m. ketonemia
   n. ketonuria
2. Compare the energy released when 1 g of carbohydrate and 1 g of lipid are oxidized completely in the body.
3. It has been said that 1 g of fat has more than six times the energy content of 1 g of glycogen. Explain.
4. What is a characteristic structural feature of the essential fatty acids?
5. What causes the flow of bile?
6. Describe the emulsifying action of bile salts. What function does emulsification serve?
7. What are the end products of fat digestion?
8. What are the functions of adipose tissue?
9. Epinephrine and insulin are usually associated with carbohydrate metabolism. What are the effects of these hormones on lipid metabolism?
10. What is the difference between the proteins that transport lipids (triglycerides, phospholipids, cholesterol) and the proteins that transport fatty acids?
11. List the chief fuel for:
    a. brain
    b. liver
    c. cardiac muscle
    d. kidneys
    e. resting skeletal muscle
    f. active skeletal muscle
    g. skeletal muscle (prolonged exercise)
    h. brain (prolonged starvation)
12. What is the principal cause of obesity?
13. Compare and contrast glycogen and adipose tissue as reserve energy sources.
14. List two ways to determine percent body fat. Describe a limitation of each method.
15. List some problems that result from low-calorie diets.
16. How much energy, in calories, is stored in 1.0 lb of adipose tissue?
17. Why does excess body fat strain the heart?
18. Describe the set-point theory of body weight.
19. Why does a diet that restricts carbohydrate intake lead to loss of muscle mass as well as loss of fat?
20. List two ways in which fad diets lead to a "quick weight loss." Why is this weight rapidly regained?
21. The ingestion of excess carbohydrates results in the deposition of fats in adipose tissue. Explain.
22. What is the significance of acetyl CoA in metabolism?
23. Why does a deficiency of carbohydrates in the diet lead to the formation of ketone bodies?
24. During a fast, which energy reserves are used first?
25. Which energy reserves supply the major part of the body's needs during a fast?
26. Why does acidosis result from ketosis?
27. Why does starvation result in acidosis?
28. How does lack of insulin lead to acidosis?

Problems

Fats as Fuels

29. A large double-decker hamburger provides 600 kcal of energy. How long would you have to walk to burn off that energy if 1 h of walking uses about 300 kcal?
30. One kilogram of fat tissue stores about 7700 kcal of energy. An average person burns about 40 kcal while walking 1 km. If that person walks 5 km a day, how much fat will be burned in 1 year?
31. How long would you have to run to burn off the 110 kcal in one glass of beer if 1 h of running burns off 1100 kcal?
32. How far would you have to run to burn off 5.0 kg of fat if your running burns off 100 kcal/km? (Assume 7700 kcal in 1.0 kg of fat.)

Digestion and Absorption of Lipids

33. Show, with equations, the chemical changes that triglycerides undergo during digestion.
34. Write the formulas for four fatty acids formed during the digestion of the lipids in peanuts (see Table 20.2).
35. What happens to the products of lipid digestion after they cross the intestinal wall?
36. a. In what form are lipids transported by the blood?
    b. To where are they transported?

Fatty Acid Oxidation

37. a. Why is the pathway for fatty acid oxidation called a spiral instead of a cycle?
    b. Why is it also called β oxidation?
38. What is the end product of the fatty acid spiral?
39. What types of reactions are involved in fatty acid metabolism?
40. How many molecules of acetyl CoA are produced in the metabolism of one molecule of palmitic acid?
41. If hexanoic acid with carbons 2, 4, and 6 as isotopic labels were metabolized to acetyl CoA and butyryl CoA via β oxidation, where would the labeled carbons appear?
42. How many turns of the fatty acid spiral are necessary to metabolize the following fatty acids?

a. myristic acid ($C_{13}H_{27}COOH$)
b. palmitoleic acid ($C_{15}H_{29}COOH$)
c. cerotic acid ($C_{25}H_{51}COOH$)

43. Why do most naturally occurring fatty acids contain even numbers of carbon atoms?

44. In which segment of fatty acid catabolism is the most energy made available?

45. How and where is $FADH_2$ from the fatty acid spiral converted back to FAD?

46. How many molecules of NADH are formed during the complete oxidation of one molecule of palmitic acid?

47. How many molecules of ATP are formed during the complete oxidation of one molecule of palmitic acid?

48. Calculate the total number of moles of ATP produced by the oxidation of 1 mol of glyceryl palmitate to CO_2 and H_2O in a liver cell respiring aerobically.

Additional Problems

49. Arrange the following in order of decreasing plasma lipid concentration: triglycerides, phospholipids, cholesterol esters.

50. Fat tissue has a density of about 0.90 g/mL, lean tissue a density of about 1.1 g/mL. Calculate the density of a person who has a body volume of 80 L and who weighs 85 kg. Is the person fat or lean?

51. Show, with equations, how the glycerol obtained from lipid hydrolysis is incorporated into the glycolytic pathway.

52. Use equations to show how acetone and β-hydroxybutyric acid are formed from acetoacetic acid.

53. Long-chain fatty acyl CoA molecules cannot traverse the inner mitochondrial membrane. A special transport mechanism that uses carnitine as the carrier molecule is required. The fatty acyl group is transferred from its thioester bond with CoA to the hydroxyl group of carnitine to form acyl carnitine plus CoA. Complete the following equation:

$$RCH_2CH_2CH_2CH_2CH_2C\overset{O}{\underset{SCoA}{\diagup}} \quad +$$

Fatty acyl CoA

$$CH_3-\overset{\pm}{N}(CH_3)(CH_3)-CH_2-\overset{H}{\underset{OH}{C}}-CH_2-C\overset{O}{\underset{OH}{\diagup}} \quad \overset{\text{acyl transferase}}{\rightleftharpoons}$$

Carnitine

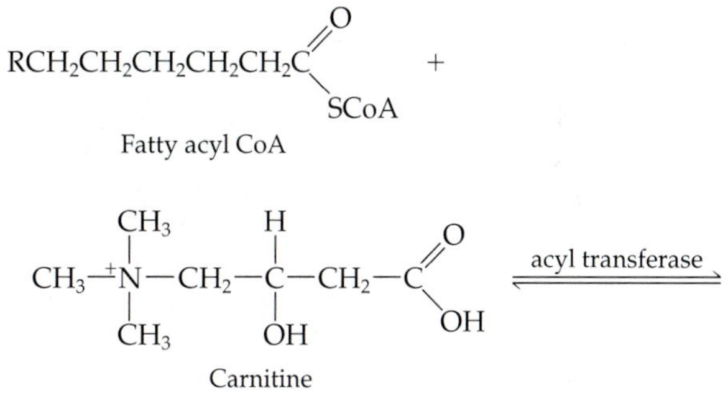

27 Protein Metabolism

Meats and vegetables supply the amino acids necessary to build body proteins.

Consider green plants. If proper amounts of nutrients—nitrates, phosphates, sulfur compounds, and so on—are available, green plants can make all the amino acids they need. From these amino acids the plants put together all the protein they need in order to function. We poor animals are not quite so versatile. We too can put together all the proteins we require, but only if we eat properly.

In Chapter 21 we indicated the essential nature of protein molecules and their vital functions in all living organisms. In Chapter 23 the biosynthesis of proteins was discussed in connection with the study of nucleic acids. In this chapter we deal primarily with the catabolic aspects of protein metabolism, with particular reference to the metabolic role of the amino acids.

27.1 Digestion and Absorption of Proteins

Like polysaccharides and lipids, intact proteins cannot normally be absorbed across intestinal membranes. They must first be hydrolyzed to their constituent amino acids. Protein digestion begins in the stomach, where the action of gastric juice hydrolyzes about 10% of the peptide bonds. Gastric juice is a mixture of substances secreted by the stomach. The chief components are water (more than 99%), mucin, inorganic ions, hydrochloric acid, and some enzymes. A protein hormone called *gastrin*, produced in the stomach, starts the flow of gastric juice. Flow can also be started by histamine. Indeed, it may well be that gastrin acts by releasing histamine (see Section 27.5), which in turn stimulates the secretion of gastric juice.

Occasionally, small polypeptides may be absorbed into the bloodstream. These foreign polypeptides act as *antigens*—they stimulate the formation of specific *antibodies* (see Section 28.5) and are probably responsible for the allergic reactions that some individuals develop for certain foods.

Hydrochloric acid is secreted by certain glands in the stomach lining. The pH of freshly secreted gastric juice is about 1.0, but the contents of the stomach may react with it, raising the pH to between 1.5 and 2.5. (The pain of a gastric ulcer is at least partially due to the irritation of the ulcerated tissue by the acidic gastric juice.) Hydrochloric acid is involved in the denaturation of food protein—it opens up the folds in the protein molecule to expose the chains to more efficient enzyme action.

The principal digestive component of gastric juice is *pepsinogen*, a zymogen produced in secretory cells located in the stomach wall. Pepsinogen is catalytically converted by hydrogen ions to its active form, *pepsin*. As mentioned in Section 22.2, the newly formed pepsin then acts as an autocatalyst in the activation of the remaining pepsinogen. Pepsin is an endopeptidase that catalyzes the hydrolysis of peptide linkages within the protein molecule. It has a fairly broad specificity but acts preferentially on linkages involving the aromatic amino acids tryptophan, tyrosine, and phenylalanine as well as methionine and leucine. The gastric juice of infants contains *rennin*, an enzyme having a specificity very similar to that of pepsin.

All enzymes that digest proteins (called proteolytic enzymes or *peptidases*) are secreted in the form of their inactive precursors, probably to protect tissues from digestion by their own enzymes. A mucous lining protects the digestive tract from hydrolysis by the proteolytic enzymes that function within the tract.

Food stays in the stomach for 2 to 5 h. It is broken down by pepsin and by the mechanical churning of the stomach into a thin, watery liquid called *chyme*. This material then passes, in small portions, into the duodenum. (Much coiled, the small intestine is about 7 m long in adult humans, but most of the enzymes of the intestinal juice are secreted from the duodenum.)

Protein digestion is completed in the small intestine by the action of digestive juices from the pancreas and from intestinal mucosal cells. The pancreatic juice passes through the pancreatic duct into the duodenum. This juice is sufficiently alkaline (pH ~7.5 to 8.5) to neutralize the acidic material passed on from the stomach. Pancreatic juice contains the zymogens trypsinogen and chymotrypsinogen.

The intestinal mucosal cells secrete the proteolytic enzyme *enteropeptidase*, which converts trypsinogen to trypsin (Figure 27.1). Trypsin then activates chymotrypsinogen to chymotrypsin. Both of these active enzymes are endopeptidases. Chymotrypsin preferentially attacks peptide bonds involving the carboxyl groups of the aromatic amino acids (phenylalanine, tryptophan, and tyrosine). Trypsin attacks peptide bonds involving the carboxyl groups of the basic amino acids (lysine and arginine). Pancreatic juice also contains the zymogen *procarboxypeptidase*, which is cleaved by trypsin to *carboxypeptidase*. The latter is an exopeptidase—it catalyzes the hydrolysis of the peptide linkages at the free carboxyl end of the peptide chain, resulting in the stepwise liberation of free amino acids from the carboxyl end of the polypeptide.

Two types of peptidases are secreted in intestinal juice: (1) an exopeptidase, *aminopeptidase*, which acts upon the peptide linkages of terminal amino acids

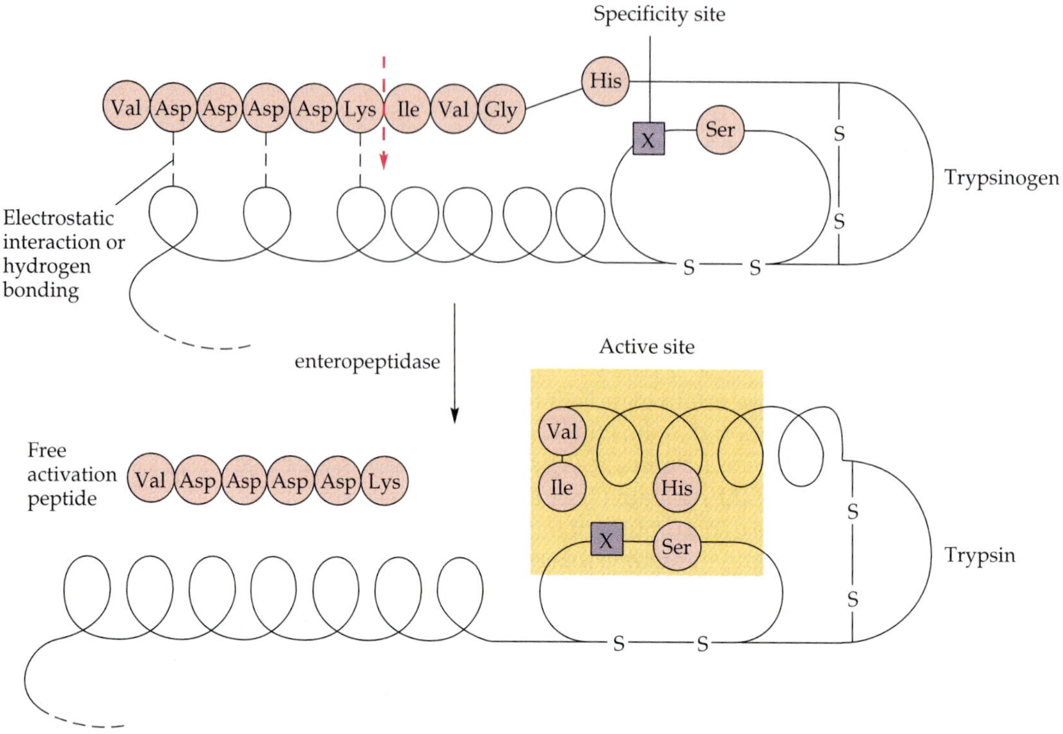

▲ **FIGURE 27.1**

Schematic representation of the structural changes in the activation of trypsinogen. Rupture of the lysyl–isoleucine bond (dashed arrow) in the N-terminal region leads to the liberation of the active peptide and causes the newly formed N-terminal region of the polypeptide chain to assume a more nearly helical configuration. This in turn permits a histidine and a serine side chain to come into juxtaposition so that these catalytic groups are properly aligned. The specificity site of the protein (X) is believed to preexist in the zymogen molecule.

possessing a free amino group, and (2) *dipeptidase* and *tripeptidase,* which cleave dipeptides and tripeptides. Figure 27.2 illustrates the specificity of protein-digesting enzymes.

The combined action of the proteolytic enzymes of the gastrointestinal tract results in the hydrolysis not only of the dietary (exogenous) proteins but also of endogenous proteins (the digestive enzymes, other secreted proteins, and dead epithelial cells). (The quantity of endogenous proteins hydrolyzed to amino acids is about 50 to 70 g each day.) The amino acids are actively transported (an energy-requiring process) across the intestinal wall into the circulatory system and are carried to the liver, the principal organ responsible for the degradation and synthesis of amino acids. The liver is also the site of synthesis of most blood

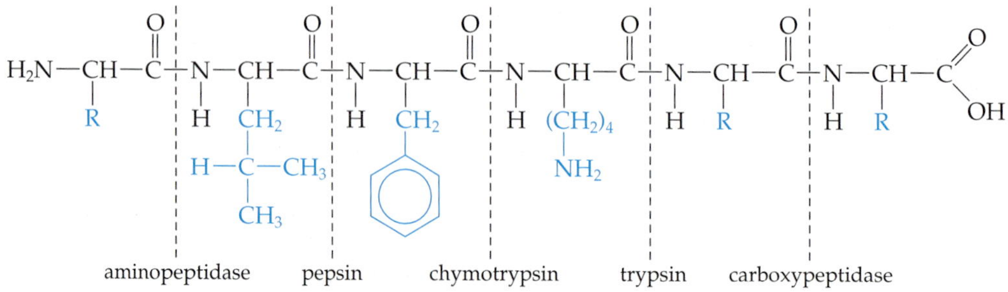

FIGURE 27.2 ►

Specificity of peptidase hydrolysis.

proteins, such as albumins, globulins, fibrinogen, and prothrombin (see Chapter 28). The amino acids synthesized in the liver and the amino acids obtained from protein digestion, along with the amino acids derived from the turnover of tissue proteins, are transported by the blood to all the cells of the body.

27.2 Nitrogen Balance

Cellular proteins are in a constant state of flux. Our bodies continuously take in proteins and break them down into their constituent amino acids. The liver and other body tissues take these amino acids and incorporate them into new proteins. Structural proteins, enzymes, and proteinlike hormones are continuously being degraded and rebuilt. Our proteins are in a state of dynamic equilibrium as "old" ones are hydrolyzed and "new" ones are synthesized.

The amount of protein synthesized or degraded per unit time gives the **turnover rate.** The protein turnover rate in a normal adult is about 300 g/day. The rate is different for proteins in different body tissues. Turnover rates represent the average residence time for a protein molecule in a tissue. They are usually expressed as half-lives. As with radioactive substances (Section 3.5), the half-life of a protein is the time interval required for one-half of the protein molecules in a given tissue to be replaced. Liver proteins and those in the blood plasma have rapid turnovers, with half-lives of 2 to 10 days. Muscle proteins have a more stable employment, with half-lives of about 6 months, and some collagen molecules may hang around for 3 years. The half-lives of enzymes vary widely from 10 minutes to 6 hours depending on their metabolic importance and the cell in which they function. Hair has no half-life because only synthesis (and not degradation) of hair protein occurs within the ectodermal (skin) cells.

A dietary intake of nitrogen is required to provide the raw materials for the biosynthesis of the various nitrogenous compounds. Nitrogen is lost in the continuous degradation of tissue protein and in the excretion of certain nitrogen-containing waste materials. Under normal conditions, an individual's intake of dietary nitrogen is equal to the amount of nitrogen lost in the feces, urine, and sweat. Such a condition is referred to as **nitrogen balance** or *nitrogen equilibrium.* Organisms are said to be in *positive* nitrogen balance (intake exceeds excretion) whenever tissue is being synthesized—for example, during periods of growth, pregnancy, and convalescence from disease. *Negative* nitrogen balance results from (1) an inadequate intake of protein (for example, fasting); (2) fever, infection, surgery, or a wasting disease; or (3) a diet that lacks, or is deficient in, any one of the essential amino acids. These factors accelerate breakdown of tissue protein (in an attempt to supply the missing amino acids), and nitrogen excretion exceeds intake.

27.3 The Essential Amino Acids

We said that higher plants and certain microorganisms are capable of synthesizing all of their amino acids from carbon dioxide, water, and inorganic salts. They obtain their required nitrogen either from soil nitrates or from atmospheric nitrogen (via nitrogen-fixing bacteria). Thus these organisms can grow on a medium that does not contain any preformed amino acids. Animals, however, can synthesize only about half of the amino acids they need in order to function properly. The remainder must be supplied in the diet. All of the amino acids required for the synthesis of a particular protein must be available to the cell at the time of protein synthesis. If just one amino acid is either absent or present in insufficient quantity,

Table 27.1 Essential and Nonessential Amino Acids for Humans

Essential	Nonessential
Lysine	Glycine
Leucine	Alanine
Isoleucine	Serine
Methionine	Tyrosine[a]
Threonine	Cysteine
Tryptophan	Aspartic acid
Valine	Asparagine
Phenylalanine	Glutamic acid
Histidine	Glutamine
Arginine[b]	Proline

[a] In the presence of adequate amounts of phenylalanine.
[b] Essential for growing children, not for adults.

From a health standpoint, "liquid protein" reducing diets are harmful to the individual because the protein is denatured and partially hydrolyzed gelatin. If this is the chief source of protein, the diet lacks the essential amino acid tryptophan.

the protein is not synthesized. For example, if a given protein requires four units of phenylalanine per molecule and 40 are available, only 10 protein molecules can be made, even though there may be enough of all the other amino acids to make a million protein molecules. Within these limits, proteins are synthesized as needed by our bodies.

An **essential amino acid** is one that cannot be synthesized by an organism, from the substrates ordinarily present in its diet, at a rate rapid enough to supply the normal requirements of protein biosynthesis. A list of essential amino acids for humans is given in Table 27.1. Notice that, as a rule, essential amino acids contain carbon chains or aromatic rings that are not intermediates in carbohydrate or lipid metabolism. The inability to synthesize these amino acids results not from a lack of the necessary nitrogen but rather from the animal's inability to manufacture the correct carbon skeleton. Supplied with phenylpyruvic acid, for example, an animal can readily synthesize the amino analog, phenylalanine. Lysine appears to be an exception because the entire preformed amino acid must be supplied. The general effect of a deficiency of one or more essential amino acids is to restrict growth and protein synthesis and produce a negative nitrogen balance.

A **complete protein** source supplies all the essential amino acids in the quantities needed for growth and repair of body tissues. Essential amino acids are best provided by animal protein. (Casein, the protein from milk, is especially beneficial because it is well balanced in its amino acid distribution.) Proteins that lack an adequate amount of one or more essential amino acids are termed **incomplete proteins.** Gelatin is an example of an incomplete animal protein because it is deficient in tryptophan. Most plant proteins are deficient in lysine and/or one other essential amino acid. Zein, the protein in corn, is deficient in lysine and tryptophan. People whose diets consist chiefly of corn may suffer from malnutrition even though the amount of calories supplied by the food is adequate. Protein from rice is short of lysine, methionine, and threonine. Wheat protein is lacking in lysine. Even the legumes, good nonanimal protein sources, are lacking in the essential amino acids methionine and tryptophan. People on vegetarian diets

Does Protein Build Muscles?

The pregame steak dinner consumed by football players in the past was based on a myth that protein builds muscles. If athletes want more muscle, says the myth, they should eat more protein. This is just not true. Although athletes do need the RDA for protein (based on grams of protein per kilogram of body weight), they do not need an excess. Protein consumed in amounts greater than the amount needed for synthesis and repair of tissue only makes the athlete fatter (due to excessive calorie intake), not more muscular.

Some extreme endurance athletes, such as triathletes and ultramarathoners, may need a bit more than the RDA for protein. Since nearly all Americans eat 50% more protein than they need, even these special athletes seldom need protein supplements.

Muscles are built through exercise, not through eating excess protein. When a muscle contracts against a resistance, creatine (Section 24.10) is released and stimulates the production of the protein myosin (Section 25.5), thus building more muscle tissue. If the exercise stops, the muscle begins to shrink after about 2 days. After about 2 months without exercise, muscle built through the exercise program is almost completely gone. (The muscle does *not* turn to fat, as some athletes believe. Former athletes often get fat, however, because they continue to take in the same number of calories and expend fewer.)

Amino Acid Supplements

Amino acids have been touted in articles and advertisements in newspapers and magazines as a cure-all for a variety of ailments including cold sores, depression, fatigue, insomnia, obesity, and pain. For example, there are claims that tryptophan and tyrosine cure depression and insomnia, and that leucine and phenylalanine are effective as pain relievers. However, the FDA states that it is dangerous for consumers to ingest large amounts of any one amino acid, and there is little evidence that they do any good. Animal studies have shown that amino acid imbalances can be created in the body by abnormal intakes of specific individual amino acids.

should consume a larger total quantity of protein than would be required from animal-protein diets and should include diverse vegetable protein sources to provide the proper amino acid requirements. The average individual requires about 50 to 60 g (about 2 oz) of protein in the daily diet. This is equivalent to the protein supplied by one quarter-pound hamburger or one regular cheeseburger. (In the industrialized countries, the average daily intake of proteins is about 100 g.)

Plants trap a small fraction of the solar energy that falls upon them. They use this energy to convert carbon dioxide, water, and mineral nutrients such as nitrates, phosphates, and sulfates to protein. Cattle eat the plant protein, digest it, and convert a small portion of it to animal protein. People eat this animal protein, digest it, and reassemble some of the amino acids into human protein. Some of the energy originally trapped by the green plants is lost as heat at every step of the food chain (Table 27.2). If people ate the plant protein directly, one highly inefficient step would be skipped. Extreme vegetarianism is dangerous, however. One would have to eat a wide variety of plant materials to be sure of getting enough of all the essential amino acids. Even then, an all-vegetable diet is likely to be lacking in vitamin B_{12} because this nutrient is not found in plants. Other nutrients scarce in all-plant diets include calcium, iron, riboflavin, and (for children not exposed to sunlight) vitamin D. A modified vegetarian diet that includes milk, eggs, cheese, and fish can provide excellent nutrition even with red meat totally excluded.[1]

Table 27.2 Efficiencies of Protein Conversions

Food	Efficiency of Production (%)*
Beef or veal	4.7
Pork	12.1
Chicken or turkey	18.2
Milk	22.7
Eggs	23.3

* Calculated by dividing the weight of edible protein by the weight of the protein feed required to produce it, then multiplying the result by 100.

27.4 The Chemistry of Starvation

When the human body is totally deprived of food, whether voluntarily or involuntarily, the condition is known as **starvation.** During total fasting, the body's glycogen stores are depleted rapidly and the body calls on its fat reserves. Fat is first obtained from around the kidneys and heart. Then it is removed from other locations. Ultimately even the bone marrow (which is also a fat storage depot) is depleted, and it becomes red and jellylike rather than white and firm.

In the early stages of a total fast, body protein is metabolized at a relatively rapid rate. The preferred energy source for brain cells is glucose. If none is available in the diet, the cells make it from amino acids obtained from the degradation

[1] It does appear that Americans are trying to reduce their consumption of red meat. In 1976 the average consumption of beef per person was 95 lb per year. By 1994 it had been reduced to 70 lb per person. This probably results from the widespread publication of the fact that a steady diet of red meat increases the rate of colon cancer.

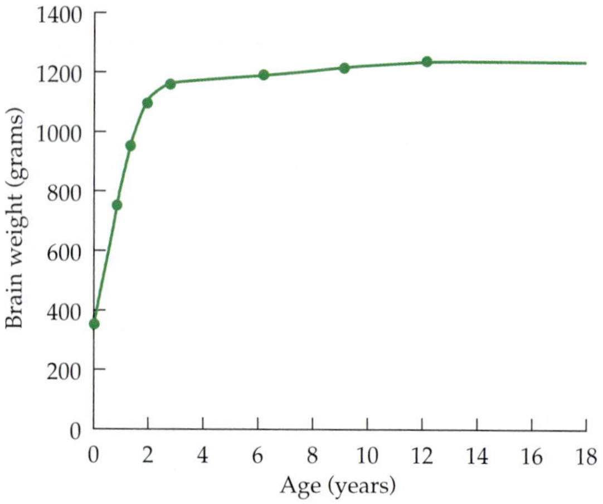

FIGURE 27.3 ▶ Growth of the human brain according to age.

of tissue protein (e.g., proteins from skeletal muscles and from membranes that line the digestive tract). A starving human may lose as much as 6% of her or his muscle mass per day. The loss of plasma proteins, especially albumins, occurs to an even greater extent. The nitrogenous metabolites from protein catabolism must be excreted through the kidneys. This requires large volumes of water. Starving people often die of dehydration.

After several weeks of starvation, the rate of protein breakdown slows considerably as the brain adjusts to using fatty acid metabolites for its energy source. When no more fat reserves remain, the body must again draw heavily on structural protein for its energy requirements. The emaciated appearance of a starving individual is due to depleted muscle protein.

Even low-carbohydrate diets high in complete proteins are hard on the body, which must rid itself of the nitrogen compounds—ammonia and urea—formed by the breakdown of proteins. This puts an increased stress on the liver, where the waste products are formed (see Section 27.7).

It is interesting to note that, contrary to a popular notion, fasting does not cleanse the body. Indeed, quite the reverse occurs. A shift to fat metabolism produces ketone bodies, and protein breakdown produces ammonia, urea, and other wastes. You can lose weight by fasting, but the process should be carefully monitored by a physician.

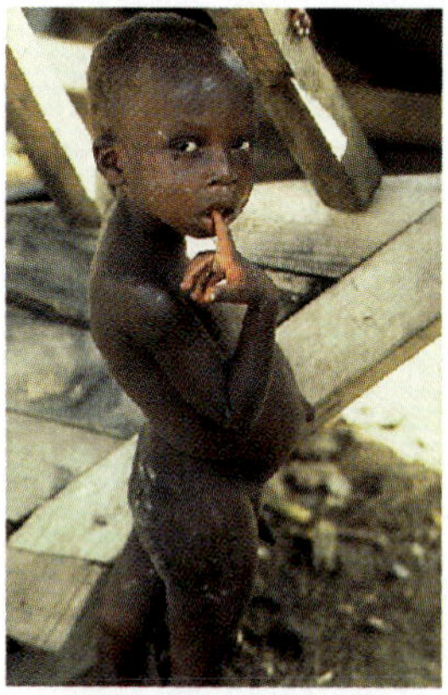

An extreme lack of proteins and vitamins causes a deficiency disease known as kwashiorkor.

Involuntary starvation is a serious problem in much of the world. Even so, starvation is seldom the sole cause of death. Weakened by starvation, the victims of starvation and malnutrition succumb to disease. Even usually minor diseases, such as chickenpox and measles, become life-threatening disorders. Barring disease, starvation alone would lead eventually to death from circulatory failure when the heart muscle became too weak to pump blood.

A similar situation occurs in the case of the protein deficiency disease kwashiorkor. Kwashiorkor results in extreme emaciation, bloated abdomen, mental apathy, diarrhea, lack of pigmentation of the skin and hair, and eventual death. This disease is at times prevalent in Latin America, Asia, and Africa, where corn or rice is the major food. It is said to be the most severe and widespread nutritional disorder among young children. Kwashiorkor can best be treated by the administration of adequate amounts of well-balanced protein. The problem, of course, is that animal protein is a scarce commodity in many of these areas.

Nutrition is especially important in a child's early years. Protein deficiency leads to both physical and mental retardation. The effect on a human's mental capacity is readily apparent from a consideration of Figure 27.3, which shows that the human brain reaches nearly full size by the age of 2 years.

27.5 Amino Acid Metabolism

Carbohydrates can be stored in the liver and muscles as glycogen. Fats can be placed on reserve in the fat depots. For proteins, however, there are no comparable storage facilities. Proteins are digested in the small intestine, and the resulting amino acids are absorbed directly into the bloodstream. Here they join a circulating pool of amino acids.

Members of the **amino acid pool** are there only on temporary assignment. New members are constantly added to the pool, and old ones are regularly withdrawn. The body can supply the pool by synthesizing amino acids from scratch or by breaking down tissue protein to obtain the constituent amino acids. It drains the pool to obtain raw materials for new protein synthesis or for the synthesis of other nitrogen-containing compounds such as heme. It can also use the amino acids as an energy source. The nitrogen stored in the pool cannot be recycled endlessly within the body, however. Each day some of the amino acids are catabolized, and the nitrogen is eliminated from the body as urea. This catabolism represents a net drain of material from the pool. It is to compensate for this drain that we require proteins in our diet.

The human body contains about 100 g of *free* (surplus) amino acids. Two of them, glutamic acid and glutamine, account for half of this total, whereas the essential amino acids constitute about 10 g. Most of the amino acids in the metabolic pool are used to synthesize any of the myriad of proteins necessary to the living organism. This is the major function of amino acids in the body. About 75% of the metabolized amino acids in the normal adult are used for the protein synthesis made necessary by the constant wear and tear on body proteins. However, amino acids also play an essential role in the anabolism of all other nitrogenous compounds.

Although the liver is the principal organ for amino acid metabolism, other tissues, such as kidney, intestine, muscle, and adipose tissue, are also involved. Generally, the first step in the breakdown of amino acids is the separation of the amino group from the carbon skeleton. The amino groups are then incorporated into almost all the other nonprotein nitrogen-containing compounds, such as neurotransmitters, hormones, nucleic acids, porphyrins, and creatine (Figure 27.4). A discussion of these specific metabolic pathways is beyond the scope of this text.

The carbon skeletons resulting from the deaminated amino acids are used to form either glucose (via gluconeogenesis) or fats (via acetyl CoA), or they are converted to a metabolic intermediate that can be oxidized by the Krebs cycle. Amino acid catabolism is particularly prevalent during hypoglycemia, fasting, and starvation. Because fatty acids cannot be converted to glucose, proteins must be hydrolyzed to amino acids that can provide the proper carbon skeleton to form the glucose needed by the brain. By the end of the overnight fast, gluconeogenesis is providing blood glucose from amino acids.

Transamination

Transamination is an exchange of functional groups between an amino compound and a keto compound. The α-amino group of any amino acid except lysine, threonine, proline, or hydroxyproline can be removed by transamination.

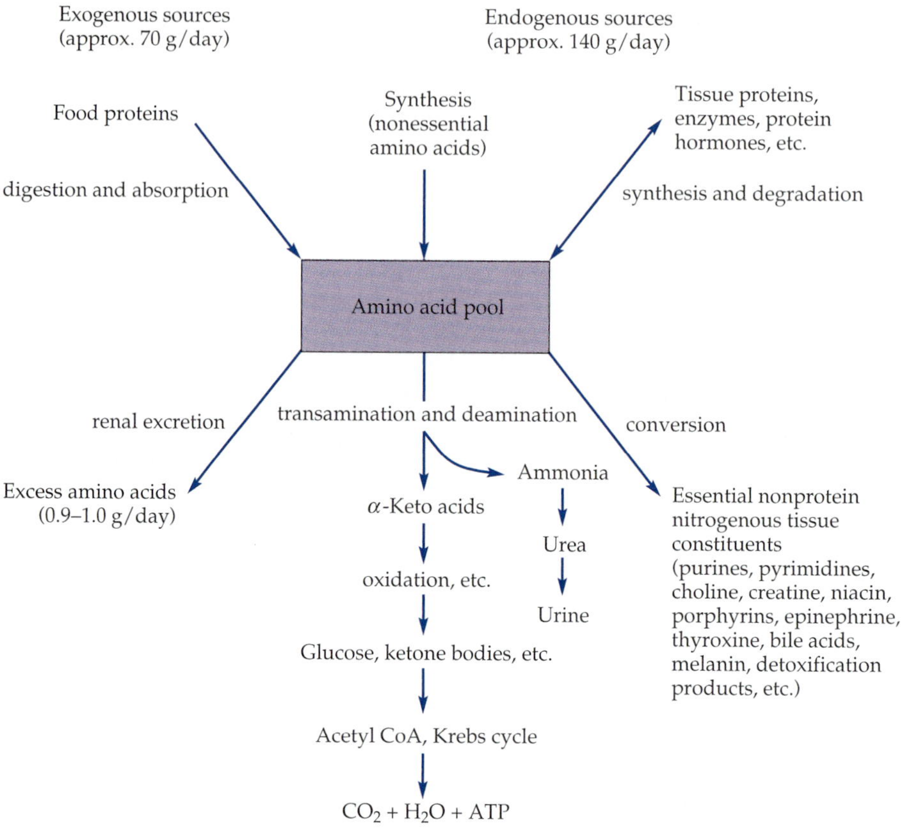

FIGURE 27.4 ▶ Scheme of general paths of amino acids in metabolism (average human).

The amino group is transferred to the keto carbon of one of three α-keto compounds—pyruvic acid, α-ketoglutaric acid, or oxaloacetic acid. It is probable that there are a large number of transaminases (also called aminotransferases), each catalyzing a reaction between a specific α-amino acid and some α-keto acid:

The three possible receptor compounds are converted to alanine, glutamic acid, and aspartic acid, respectively. These reactions are catalyzed by three transaminases. The enzymes, named for the product amino acid, are alanine transaminase, glutamic transaminase, and aspartic transaminase.

Alanine and aspartic acid are ultimately converted to glutamic acid through transamination reactions (Figure 27.5). These two transamination reactions are of special clinical interest. Normally, the blood contains a low concentration of transaminases. However, extensive tissue destruction is accompanied by rapid and striking increases in the blood transaminase levels. Glutamic-pyruvic transaminase (GPT) has a particularly high activity in the cytoplasm of the liver, and an elevated serum level of this enzyme is indicative of liver damage. Glutamic-oxaloacetic transaminase (GOT) is abundant in heart muscle (recall Section 22.7), and a sharp rise in the concentration of GOT in the blood is an indication of myocardial infarction.

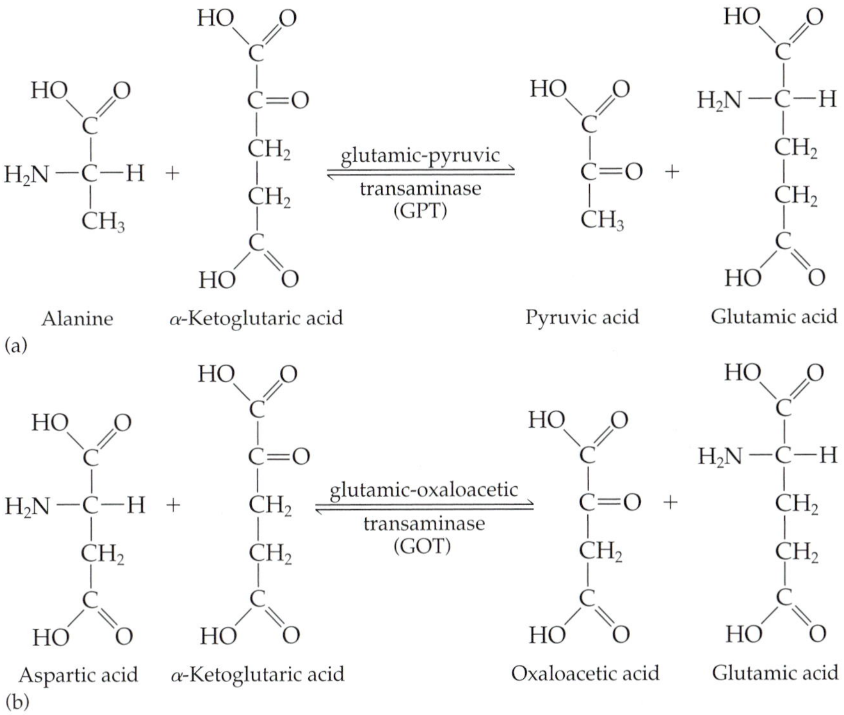

FIGURE 27.5 Transamination reactions catalyzed by (a) GPT and (b) GOT. Note that whichever the pathway, the final receptor for the amino group from several amino acids is α-ketoglutaric acid and the final product is glutamic acid. The reactions are reversible, in which case glutamic acid donates its amino group to α-keto acids.

Oxidative Deamination

We saw that in transamination reactions the α-amino groups of many amino acids are transferred to α-ketoglutaric acid to form glutamic acid. The latter compound then loses the amino group as ammonia and is oxidized back to α-ketoglutaric acid:

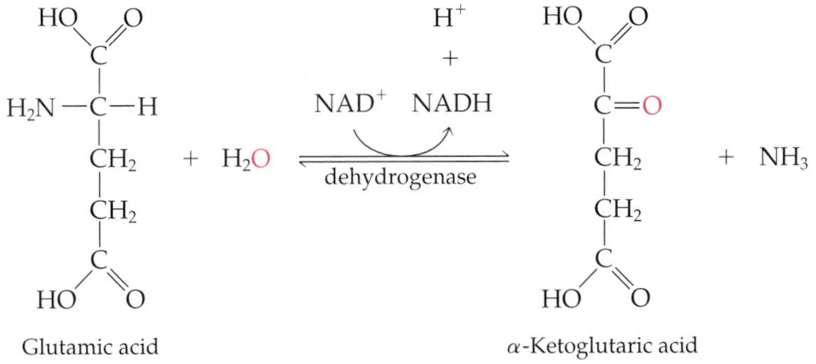

This process is termed **oxidative deamination,** and it occurs in the liver mitochondria. Glutamic acid dehydrogenase is unusual in that it uses either NAD^+ or $NADP^+$ as a coenzyme. (The reduced forms of these coenzymes, NADH and NADPH, are ultimately oxidized by the respiratory chain.) The ammonia formed from glutamic acid by oxidative deamination is converted to urea and excreted in the urine (see Section 27.7).

This equilibrium reaction is of central importance in linking protein metabolism and carbohydrate metabolism. The reverse reaction is significant in nitrogen metabolism because it is one of the few reactions in animals that can convert inorganic nitrogen (NH_3) to organic nitrogen (amino acids). The amino group in the glutamic acid can be passed on through transamination reactions, producing all the other cellular amino acids, provided the appropriate α-keto acids are available.

The Fate of the Carbon Skeleton

Any amino acid in the amino acid pool can make its way from the pool to the Krebs cycle, via a unique pathway. Most lose their amino groups through transamination. Once the amino group is gone, each carbon skeleton, incorporated into an α-keto acid, then follows its own special pathway. For example, phenylalanine undergoes a six-step reaction before it splits into fumaric acid and acetoacetic acid. The fumaric acid is an intermediate in the Krebs cycle, but acetoacetic acid must be converted to acetoacetyl CoA and then to acetyl CoA before it enters the cycle. Eventually the 20 or more individual pathways leading from the amino acid pool converge into five routes that join the Krebs cycle.

Those amino acids that can form any of the metabolites of carbohydrate metabolism can be converted to glucose (via gluconeogenesis) or oxidized to carbon dioxide, water, and energy; they are referred to as **glucogenic amino acids.** Amino acids that give rise to acetoacetyl CoA or acetyl CoA are precursors for fatty acid synthesis; they are called **ketogenic amino acids** (because they yield ketone bodies). Certain amino acids fall into both categories. Leucine is the only amino acid that is exclusively ketogenic. Table 27.3 classifies the amino acids with regard to the metabolic fate of their carbon skeletons. The products and metabolic routes of the 20 amino acids are summarized in Figure 27.6.

Table 27.3 Glucogenic and Ketogenic Amino Acids

Glucogenic		Ketogenic	Glucogenic and Ketogenic
Alanine	Glycine	Leucine	Isoleucine
Arginine	Histidine		Lysine
Asparagine	Methionine		Phenylalanine
Aspartic acid	Proline		Tyrosine
Cysteine	Serine		Tryptophan
Glutamic acid	Threonine		
Glutamine	Valine		

Decarboxylation

Several amino acid decarboxylases eliminate carbon dioxide from amino acids to form primary amines. Pyridoxal phosphate is the necessary coenzyme (Table J.1):

$$R-\underset{\underset{NH_2}{|}}{\overset{\overset{H}{|}}{C}}-\underset{OH}{\overset{O}{\overset{\|}{C}}} \xrightarrow[\text{decarboxylase}]{\text{pyridoxal phosphate,}} R-\underset{\underset{NH_2}{|}}{\overset{\overset{H}{|}}{C}}-H \; + \; CO_2$$

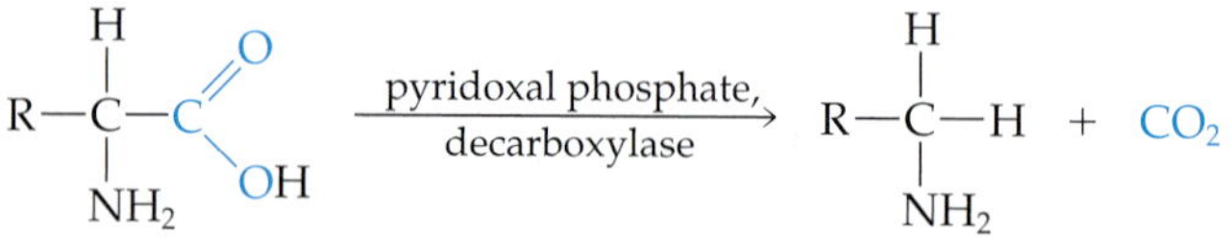

Decarboxylases are found primarily in microorganisms, but they are also found in some animal tissues. Intestinal bacteria are responsible for amino acid decarboxylation, and the foul smell of feces is due in part to the resulting amines. (Cadaverine and putrescine are formed upon decarboxylation of lysine and ornithine; see Figure 17.4).

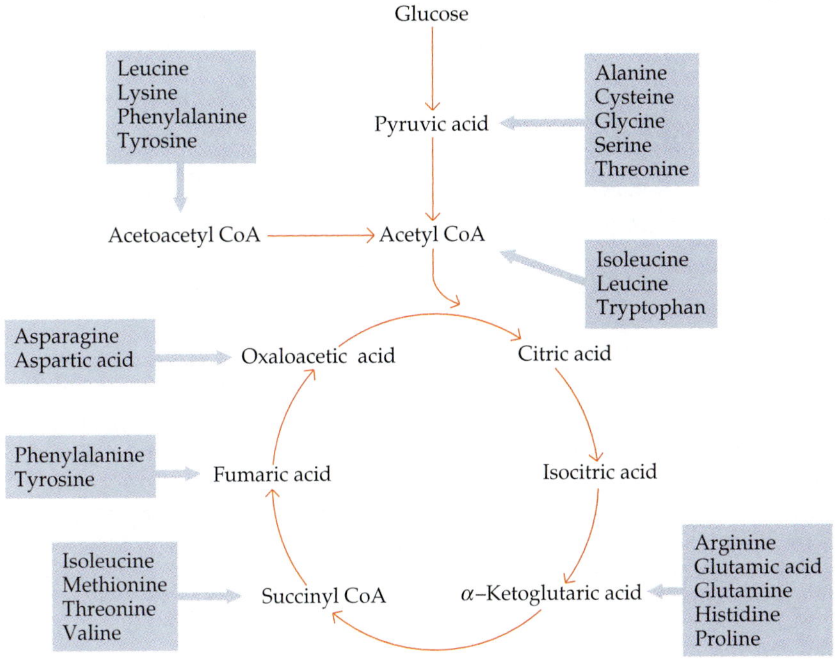

FIGURE 27.6
Fates of the carbon skeletons of amino acids.

Some of the amines produced as a result of decarboxylation have important physiological effects. Many people are allergic to pollen, dust, insect stings, and so on. By a mechanism not completely understood, the body responds to these foreign substances by decarboxylating histidine to *histamine* (allergic individuals synthesize abnormally high amounts of histidine):

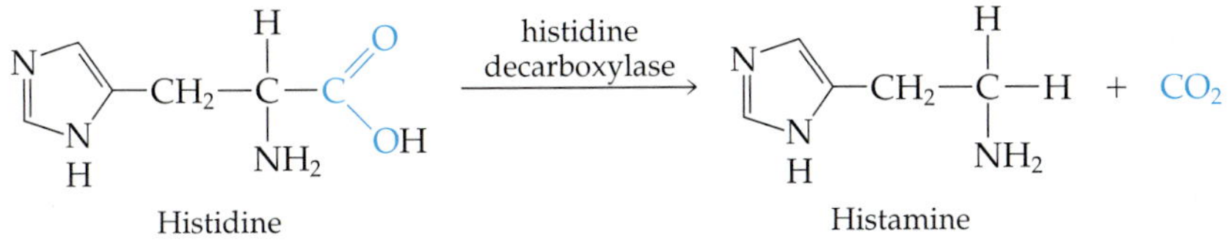

Histamine dilates blood vessels and thus initiates the inflammatory response (see Section I.2). The discomfort associated with hay fever and other allergies is due to the inflammation of the eyes, nose, and throat. Furthermore, the expansion of the blood capillaries causes a decrease in the blood pressure that, if severe enough, may induce shock. (Recall that prostaglandins are also released during an allergic reaction, and they increase the sensitivity to pain associated with inflammation; see Section I.6.)

Histamine is sometimes used in medicine. As little as 1 μg by injection causes a pronounced drop in blood pressure by dilating the blood vessels.

Besides histamine, several other neurotransmitters are synthesized by the decarboxylation of specific amino acids. Tyrosine is converted to tyramine by bacterial action:

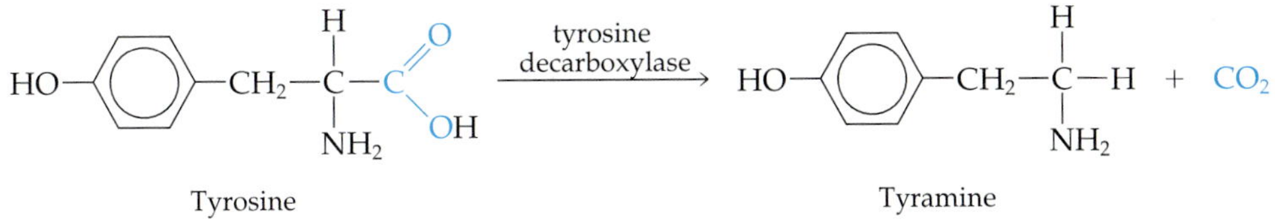

Antihistamines

Antihistamines are compounds structurally similar to histamine. They can occupy the receptor sites normally occupied by histamine (e.g., on smooth muscles that surround capillaries), thereby preventing the physiological changes produced by histamine (i.e., they are antagonists). Figure 27.7 illustrates some of the common antihistamines found in nasal decongestants, combination pain relievers, and hay fever preparations. The most common side effect of antihistamines is sedation, which accounts for their use as major ingredients in most over-the-counter sleeping pills. This effect can impair one's ability to operate machinery or drive a motor vehicle. Terfenadine (Seldane), an antihistamine approved for prescription use in 1985, does not cause drowsiness.

There are two types of receptors for histamine, designated as H_1 and H_2. The H_1 receptors are found in the walls of capillaries and in the smooth muscle of the respiratory tract. They affect the vascular changes (dilation and increased permeability of capillaries) and muscular changes (bronchoconstriction) associated with hay fever and asthma. H_1 receptors are blocked by the classical antihistamines (such as Benadryl), which relieve the symptoms of allergies. The H_2 receptors occur mainly in the wall of the stomach, and their activation causes an increased secretion of hydrochloric acid. Cimetidine blocks H_2 receptors and thus, by reducing acid secretion, is an effective drug for people with ulcers.

Antihistamines are not effective against colds, although they may temporarily relieve some cold symptoms.

Tyramine has a physiological action similar to, but weaker than, that of norepinephrine, which it resembles structurally.

Serotonin (5-hydroxytryptamine; see Section G.2) is formed by the action of a specific decarboxylase on 5-hydroxytryptophan:

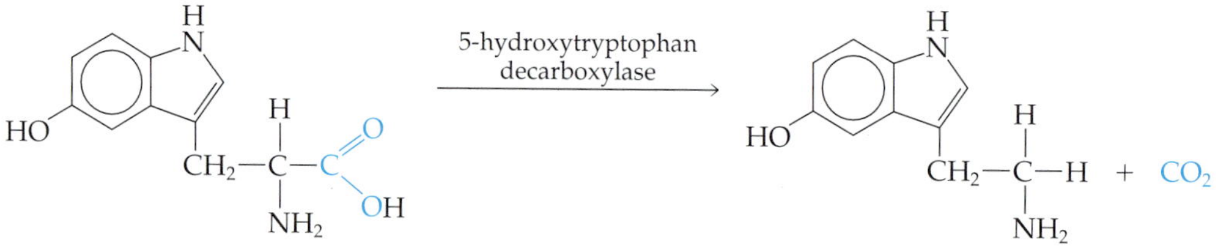

5-Hydroxytryptophan

Serotonin (5-Hydroxytryptamine)

The cell bodies of serotonin-containing neurons are located almost exclusively in the upper brain stem, from which axons project to other areas of the central nervous system. Serotonin constricts blood vessels, stimulates smooth muscle, and has a potent inhibitory effect on its postsynaptic neurons.

An important inhibitory neurotransmitter, GABA (γ-aminobutyric acid), is formed in the brain and spinal cord from the decarboxylation of glutamic acid:

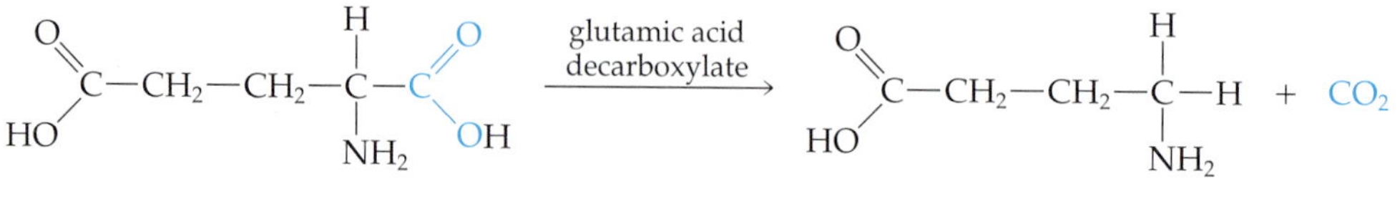

Glutamic acid

γ-Aminobutyric acid (GABA)

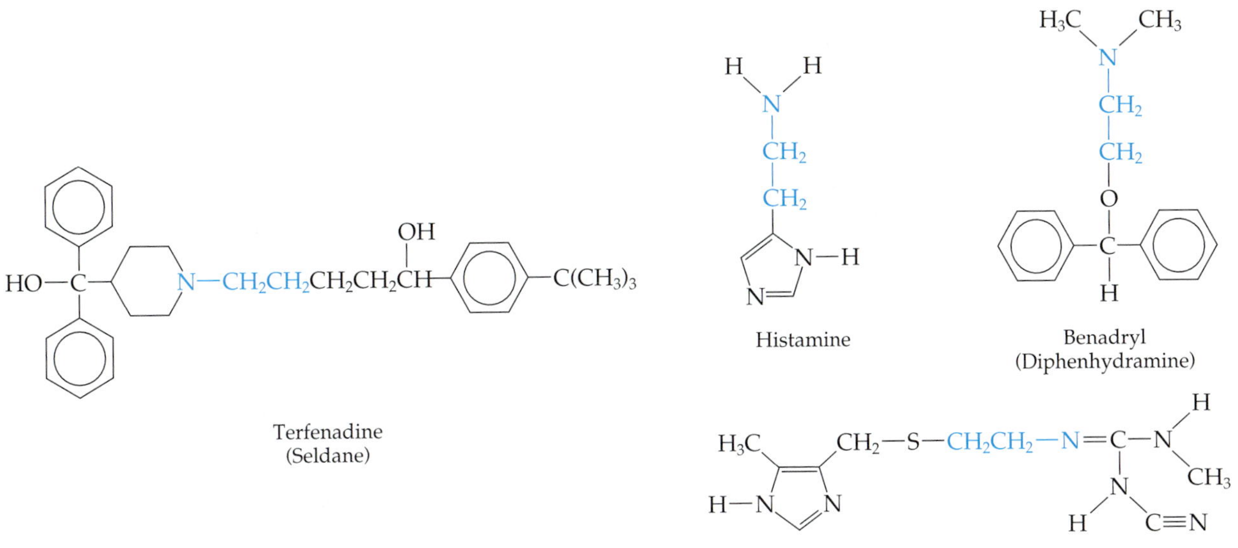

Terfenadine (Seldane)

Histamine

Benadryl (Diphenhydramine)

Cimetidine (Tagamet)

▲ **FIGURE 27.7**
The antihistamines act as competitive antagonists to histamine. Notice that the structural similarity to histamine is the substituted ethylamine moiety (blue).

GABA is thought to inhibit dopamine neurons in particular, as well as other neurons throughout the central nervous system.

Another decarboxylase catalyzes the decarboxylation of 3,4-dihydroxyphenylalanine (L-dopa) to dopamine:

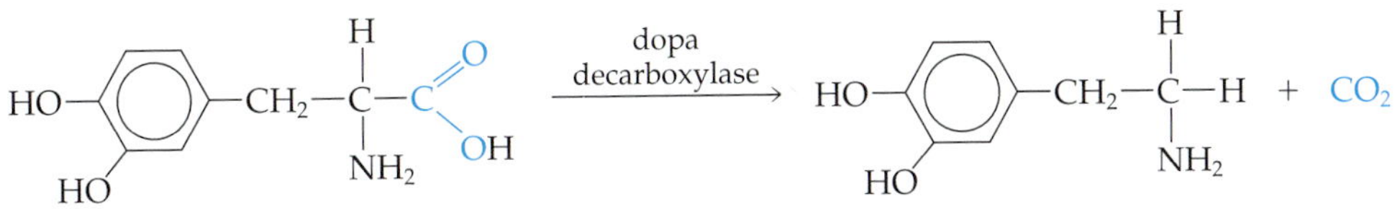

3,4-Dihydroxyphenylalanine (L-Dopa)

3,4-Dihydroxyphenylethylamine (Dopamine)

A deficiency of dopamine in the brain cells is a primary cause of Parkinson's disease, a disorder of the central nervous system that involves a progressive paralytic rigidity, tremors of the extremities, and unresponsiveness to external stimuli. Dopamine itself cannot be administered because it does not pass across the blood–brain barrier. A major breakthrough in the treatment of Parkinson's disease has been the use of L-dopa. (The L enantiomer is more effective and less toxic than the D form of the drug.) Large doses of L-dopa are administered orally; the drug is able to pass from the digestive system into the blood and then cross the blood–brain barrier. L-Dopa is decarboxylated to dopamine in the brain. (Incidentally, numerous studies have linked schizophrenia to an *overabundance* of dopamine in brain cells.)

27.6 Storage of Nitrogen

As mentioned earlier, excess protein is not stored to any appreciable extent in living organisms. The only significant way the body has to store small amounts of nitrogen is in the form of glutamine. This amide of glutamic acid is formed from glutamic acid and some of the ammonia produced in the oxidative deamination reactions described in the preceding section (this is another example of the conversion of inorganic nitrogen to organic nitrogen):

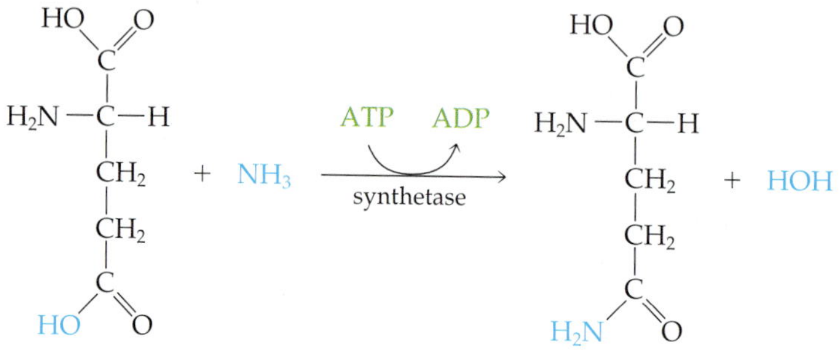

Glutamic acid
Glutamine

Glutamine is present in many tissues and in the blood and serves as a temporary storage and transport form of nitrogen. The formation of glutamine is the major method of disposing of ammonia from the brain. In the liver and kidneys, the amide group of glutamine can be donated to appropriate acceptor molecules in the biosynthesis of many nitrogen-containing compounds.

27.7 Excretion of Nitrogen

Whenever amino acids are used for energy production or for the synthesis of glucose or fat, an amino group is liberated in the form of ammonia, one type of *nitrogenous waste* the body must eliminate. Living organisms must have some mechanism for removing this ammonia from the cell because even low concentrations of it are poisonous. Levels of only 5 mg of ammonia per 100 mL of blood (hyperammonemia) are toxic to humans. The normal concentration of ammonia is about 1 to 3 μg/100 mL.

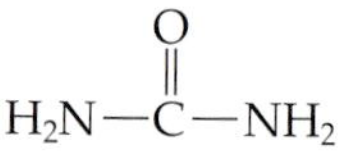
Urea

Organisms differ biochemically in the manner in which they excrete nitrogenous wastes. Most vertebrates and most adult amphibia excrete them as urea in the urine. Birds, reptiles, and insects convert them to uric acid. All marine organisms, from unicellular organisms to fish, excrete free ammonia. The ammonia is very soluble in water and is rapidly diluted in the aqueous environment.

In humans, 95% of nitrogenous wastes results from amino acid catabolism; the remaining 5% comes from the metabolism of other nitrogen-containing compounds (e.g., creatine, neurotransmitters, pyrimidines). The breakdown of purine bases in the human body results in the production of uric acid, and very small concentrations (about 0.5 g/day) of this acid are found in the urine and in the body fluids:

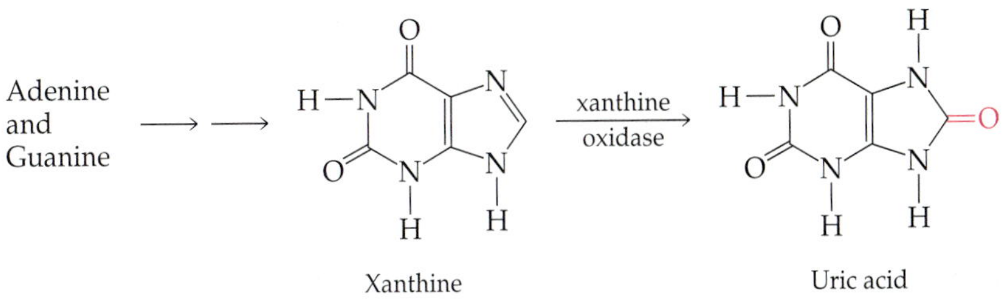

Under certain pathological conditions (impairment of purine metabolism), large quantities of uric acid are produced, and the plasma concentration of uric acid becomes abnormally high (>7 mg/100 mL). At physiological pH, the excess uric acid precipitates as the sparingly soluble monosodium salt. When such deposits occur in the joints of the body's digital regions, the surrounding tissue can become inflamed, causing the painful arthritic characteristics of **gout.** Crystals of monosodium urate may also accumulate in the kidneys (kidney stones) and cause extensive impairment of renal function. Kidney failure is another serious medical consequence of gout. (For some unknown reason, men suffer from gout to a much greater extent than women—only 5% of gout patients are women.)

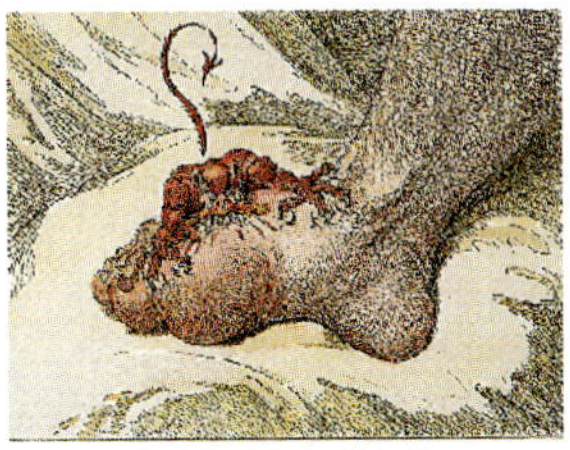
A foot afflicted with gout is shown in this old cartoon.

Another serious impairment of purine metabolism results from a genetic defect. In the *Lesch–Nyhan syndrome,* an enzyme necessary for the normal utilization of purines is absent. Affected children are mentally defective and exhibit a compulsive, aggressive behavior toward others (extreme hostility), as well as toward themselves (self-mutilation by chewing the tongue, lips, and fingers). The details of the relationship between the absence of the enzyme and the aberrant behavior is still unknown.

In mammals, the liver is the principal organ concerned with the formation of urea (by a series of reactions known as the **urea cycle;** see an advanced biochemistry text). Extensive liver damage (e.g., cirrhosis) or a genetic defect in any urea-cycle enzyme results in hyperammonemia and causes tremor, slurred speech, and blurred vision. A continuous rise in ammonia concentrations can lead to coma and death.

Once produced, urea is transported by the bloodstream to the kidneys and is eliminated in the urine. The kidneys filter about 100 L of blood each day. The normal individual excretes daily about 1.5 L of water, containing approximately 30 g of urea. This value varies greatly from day to day and can rise dramatically (to about 100 g of urea) upon ingestion of a high-protein diet. A long-term decrease in urea excretion is indicative of liver and/or kidney disease. Loss of nitrogen also occurs through the skin, mainly as sweat (about 2 g of urea) and via the feces (smaller amounts).

27.8 Relationships Among the Metabolic Pathways

A great variety of organic compounds can be derived from carbohydrates, lipids, and proteins. All organisms utilize the three major foodstuffs to form acetyl CoA or the metabolites of the Krebs cycle. These in turn supply energy upon subsequent oxidation by the cycle. A brief summary of the interrelationships of the metabolic pathways is given in Figure 27.8. We emphasize once again that in

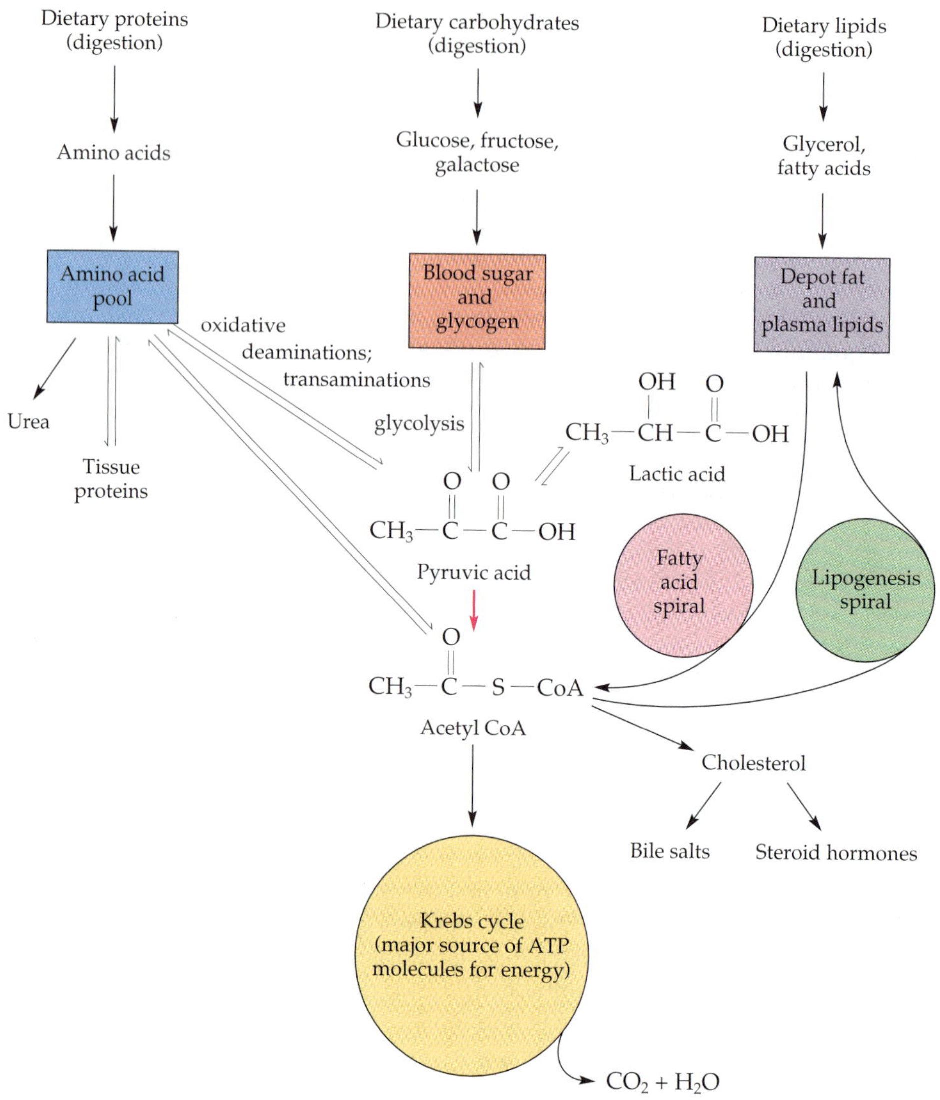

▲ **FIGURE 27.8**
Interrelationships of metabolic pathways.

mammals the conversion of pyruvic acid to acetyl CoA is irreversible. Thus, an acetyl CoA molecule derived from fatty acid degradation cannot be transformed directly to pyruvic acid. Consequently, mammals are unable to synthesize carbohydrates from lipids.

Summary

Protein digestion begins in the stomach as pepsinogen in gastric juice is converted to pepsin, the enzyme that hydrolyzes peptide bonds. The partially digested protein, called *chyme*, then passes to the duodenum, where the rest of protein digestion takes place. The resulting amino acids cross the intestinal wall into the blood and are carried to the liver, the site of most of the body's amino acid metabolism.

The body uses amino acids derived from food to make new tissues and other nitrogen-containing substances. At the same time, the body is breaking down old tissues and nitrogen-containing compounds. The relationship between these two processes, one of gain and one of loss, is called **nitrogen balance,** or *nitrogen equilibrium.* When nitrogen intake is greater than nitrogen loss and tissue is being built, the organism is in *positive nitrogen balance.* When nitrogen intake is less than nitrogen loss and body tissue is broken down in order to replenish the body's amino acid supply, the organism is in *negative nitrogen balance.*

Essential amino acids are those an organism needs but cannot synthesize for itself. These amino acids must be obtained from the diet. There are 10 essential amino acids for humans. A **complete protein** is one that contains all 10; an **incomplete protein** is one that lacks one or more.

Amino acids are not stored in the body the way carbohydrates and fats are. Instead, the amino acids obtained from food, the nonessential amino acids the body synthesizes, and those released as the body breaks down tissue all circulate in the blood in the **amino acid pool.** The body takes from this pool the amino acids it needs to synthesize (1) new proteins needed by the cells and (2) new nonprotein molecules that contain nitrogen.

The amino acids from the pool can be catabolized to provide energy. The carbon skeleton is sent to the Krebs cycle and the amino group is ultimately excreted in the urine. The first step in amino acid catabolism is separation of the amino group from the carbon skeleton. In a **transamination,** the amino acid gives its $-NH_2$ to pyruvic acid, α-ketoglutaric acid, or oxaloacetic acid. The products of this reaction are an α-keto acid containing the carbon skeleton of the original amino acid and a new amino acid. Transamination with pyruvic acid means the new amino acid is alanine, transamination with α-ketoglutaric acid means it is glutamic acid, and transamination with oxaloacetic acid means it is aspartic acid. Any alanine and aspartic acid formed in this way are ultimately converted to glutamic acid. The glutamic acid then undergoes **oxidative deamination** to yield α-ketoglutaric acid and ammonia.

Some amino acids are *decarboxylated* to primary amines.

A small percentage of the ammonia formed in amino acid deamination can react with glutamic acid to form glutamine. This substance can be stored in the body and so represents a way of storing excess nitrogen for a short time.

Key Terms

amino acid pool (27.5)
complete protein (27.3)
essential amino acid (27.3)
glucogenic amino acid (27.5)
gout (27.7)
incomplete protein (27.3)
ketogenic amino acid (27.5)
nitrogen balance (27.2)
oxidative deamination (27.5)
starvation (27.4)
transamination (27.5)
turnover rate (27.2)
urea cycle (27.7)

Review Questions

1. Define or explain:
- **a.** peptidase
- **b.** autocatalysis
- **c.** protein turnover
- **d.** nitrogen balance
- **e.** kwashiorkor
- **f.** starvation
- **g.** GABA
- **h.** Parkinson's disease
- **i.** L-dopa
- **j.** hyperammonemia
- **k.** gout
- **l.** urea

2. Distinguish between terms:
- **a.** pepsin and pepsinogen
- **b.** chymotrypsin and trypsin
- **c.** endopeptidase and exopeptidase
- **d.** positive and negative nitrogen balances
- **e.** essential and nonessential amino acids
- **f.** complete and incomplete proteins
- **g.** transamination and deamination
- **h.** GPT and GOT
- **i.** glucogenic and ketogenic amino acids
- **j.** histamine and antihistamine

3. What is the defense mechanism employed by the body to protect its tissues from the digestive action of the proteolytic enzymes?

4. In the treatment of diabetes, insulin is administered by injection rather than orally. Explain.

5. What are the products of protein digestion?

6. What is meant by the term "amino acid pool"?

7. Name three processes that add amino acids to the amino acid pool.

8. For what purposes are amino acids removed from the amino acid pool?

9. What happens to a body's nitrogen balance when the diet is lacking in one essential amino acid?

10. What happens to the nitrogen balance during fasting?

11. What happens to the nitrogen balance during recovery from a wasting illness?

12. Compare the turnover rates of enzymes and muscle proteins.

13. **a.** What is a unique structural feature of most of the essential amino acids?
 b. What is the effect of a deficiency of one or more essential amino acids?
14. What are three functions of amino acids in the body?
15. What is the significance of transamination in the metabolism of amino acids?

16. What is oxidative deamination?
17. What is the end product of purine metabolism?
18. What organs are responsible for (**a**) the synthesis of urea and (**b**) the excretion of urea?

Problems

Protein Digestion

19. Give the location of action and function of:
 a. pepsin **b.** trypsin
 c. chymotrypsin **d.** carboxypeptidase
 e. dipeptidase **f.** enteropeptidase
20. Indicate the expected products from the enzymatic action of pepsin, chymotrypsin, and trypsin on:
 a. Ala-Phe-Tyr **b.** Ile-Tyr-Ser
 c. Phe-Arg-Leu **d.** Thr-Glu-Lys
21. Indicate the location in a polypeptide chain cleaved by an aminopeptidase.
22. Indicate the location in a polypeptide chain cleaved by a carboxypeptidase.

Amino Acid Metabolism

23. The compound

$$CH_3CH_2CH(CH_3)\overset{\overset{\displaystyle O}{\|}}{C}-\overset{\overset{\displaystyle O}{\|}}{C}_{\diagdown OH}$$

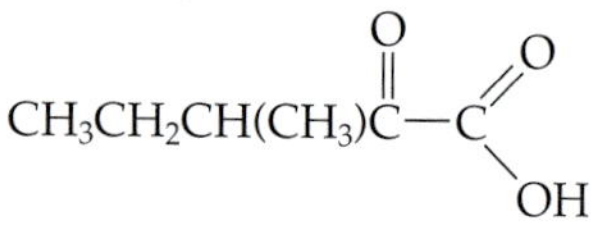

can supply the body with an essential amino acid. Show, with an equation, how this is possible.
24. Write the equation for transamination between phenylalanine and pyruvic acid.

25. What reaction is catalyzed by the enzyme GOT?
26. What reaction is catalyzed by the enzyme GPT?
27. What keto acid serves as a reactant in a transamination that produces alanine?
28. What product is formed by oxidative deamination of glutamic acid?
29. Write equations for the formation of each of the following compounds from an amino acid:
 a. pyruvic acid **b.** histamine
 c. α-ketoglutaric acid **d.** glutamine
30. Write equations for the formation of each of the following compounds from an amino acid:
 a. cadaverine **b.** oxaloacetic acid
 c. tyramine **d.** dopamine
31. Name each compound described:
 a. the amino acid that is exclusively ketogenic
 b. the principal amino-group acceptor in transamination
 c. the α-keto acid usually formed in oxidative deamination
32. Name each compound described:
 a. the compound decarboxylated to serotonin
 b. the compound decarboxylated to GABA
 c. the compound that serves as the temporary storage form of nitrogen in the body

Additional Problems

33. The RDA for protein is about 0.8 g per kilogram of body weight. How much protein is required each day by a 125-kg football player?
34. The animal body converts vegetable protein into animal protein. Explain.
35. **a.** Why are vegetarians instructed to eat a wide variety of vegetables?
 b. What is the best source of essential amino acids?
36. If 1 mol of alanine is converted to pyruvic acid and the pyruvic acid then metabolized via the Krebs cycle, how many moles of ATP are produced?

37. **a.** What toxic compound is formed in oxidative deamination?
 b. How does the body get rid of this compound?
38. If the essential amino acid leucine (2-amino-4-methyl-pentanoic acid) is lacking in the diet, a keto acid can substitute for it. Give the structure of the keto acid and rationalize the substitution.
39. **a.** How does Benadryl exert its effect?
 b. What is the difference between an H_1 receptor and an H_2 receptor?

28 Body Fluids

The testing of blood is one of the most important methods of diagnosing disease.

The processes of life occur for the most part in solution. The solvent for life processes is water, and the solutes are many—simple ions, small molecules, large molecules, and colloidal aggregates. These are all found in a variety of body fluids, and each in its own way is essential to life.

A 70-kg person has about 5 L of blood and about 10 L of interstitial fluid, the fluid that fills the space between cells. Fluid within the cells amounts to about 35 L. Each day the individual excretes an average of 1.5 L of urine, 0.2 L of water in the feces, and 1 L of water through the skin and lungs. All these losses are balanced by a combination of ingested water and water synthesized during metabolism.

You are indeed a solution—or rather, several solutions. However, the solutions of which you are composed are not static—they constantly renew themselves, with their compositions being quite dependent on the state of your health. Analysis of various body fluids represents one of the most powerful diagnostic techniques available to medical personnel.

28.1 Blood: An Introduction

Blood is the principal transport medium of the human body. It moves through a 100,000-km-long network of blood vessels. Some of these vessels are so small that blood cells have to line up to pass through. Blood carries (1) oxygen from the lungs to the tissues, (2) carbon dioxide from the tissues to the lungs, (3) nutrients from the intestines to the tissues, (4) metabolic wastes from the tissues to the excretory organs, (5) hormones from the endocrine glands to their target tissues, and (6) three major kinds of blood cells. In addition, blood helps maintain a fairly constant body temperature, an acid–base balance, an electrolyte balance, and a water balance. A rather remarkable substance, blood. And we've only mentioned a few highlights.

Blood consists of a straw-colored liquid portion (the plasma) and the formed elements (blood cells). It accounts for about one-twelfth of the body weight of the average individual. Moderate amounts lost through bleeding or blood donation are readily replaced. The volume of blood varies with body size. A 150-lb (68-kg) human has about 5 L of blood, roughly 55 to 60% of which is plasma. **Plasma** is an extremely complex solution, containing all of the biochemically significant compounds (carbohydrates, amino acids, proteins, lipoproteins, enzymes, hormones, vitamins) and inorganic ions. Although these compounds and ions are continuously entering and leaving the circulatory system, the overall composition of the plasma remains remarkably constant (a state of dynamic equilibrium exists). Approximately 90 to 92% of plasma is water; the remaining 8 to 10% consists of dissolved solids (mostly plasma proteins). Dispersed throughout the plasma are the **formed elements,** which account for about 40 to 45% of the total blood volume. The formed elements are the **erythrocytes** (red blood cells), **leukocytes** (white blood cells), and **thrombocytes** (platelets). The chemical analysis of blood samples is of great clinical significance. Increased or decreased amounts of certain substances may indicate a particular disease or condition.

28.2 Electrolytes in Plasma and Erythrocytes

A variety of ions are found in solution in both plasma and erythrocytes. These electrolytes play important roles in several life processes.

Sodium ions are found mainly in the plasma, and potassium ions are found chiefly in the erythrocytes. These positive ions do not migrate readily across cell membranes. Calcium and magnesium ions are the main dipositive ions in blood. (Calcium ions are necessary for blood clotting; if they are removed, the blood will not clot—see Section 28.8.) Calcium ions are not found in erythrocytes, but magnesium ions are. The metabolism of calcium and the metabolism of phosphorus are closely related, and plasma levels of the two usually vary in an inverse manner. Magnesium levels, on the other hand, sometimes parallel the variations in calcium levels and at other times parallel the variations in phosphorus levels.

The principal negative ions present in the plasma, in addition to inorganic phosphorus (as HPO_4^{2-}), are chloride, bicarbonate, and sulfate. Bicarbonate and hydrogen phosphate are involved in the acid–base balance of the blood (see Section 28.10). Table 28.1 lists some conditions that influence plasma levels of certain electrolytes as well as other blood constituents.

Table 28.1 Normal Composition of Blood

Constituent	Normal Range (mg/100 mL)[a]	Clinical Significance	
		Increased in	**Decreased in**
Calcium	8.5–10.3 (4.5–5.3 mEq/L)[b]	Hyperparathyroidism, Addison's disease, malignant bone tumor, hypervitaminosis D	Hypoparathyroidism, rickets, malnutrition, diarrhea, chronic kidney disease, celiac disease
Cholesterol, total	150–265	Diabetes mellitus, obstructive jaundice, hypothyroidism, pregnancy	Pernicious anemia, hemolytic jaundice, hyperthyroidism, tuberculosis
Uric acid	Male, 3–9 Female, 2.5–7.5	Gout, leukemia, pneumonia, liver and kidney disease	
Urea nitrogen	8–25	Mercury poisoning, acute glomerulonephritis, kidney disease	Pregnancy, low-protein diet, severe hepatic failure
Nonprotein nitrogen	15–35	Kidney disease, pregnancy, intestinal obstruction, congestive heart failure	Low-protein diet
Creatine	3–7	Nephritis, renal destruction, biliary obstruction, pregnancy	
Creatinine	0.7–1.5	Nephritis, chronic renal disease	
Glucose	70–100	Diabetes mellitus, hyperthyroidism, infections, pregnancy, emotional stress, after meals	Starvation, hyperinsulinism, Addison's disease, hypothyroidism, extensive hepatic damage
Chloride	96–106 mEq/L	Nephritis, anemia, urinary obstruction	Diabetes, diarrhea, pneumonia, vomiting, burns
Phosphate, inorganic	3–4.5	Hypoparathyroidism, Addison's disease, chronic nephritis	Hyperparathyroidism, diabetes mellitus
Sodium	136–145 mEq/L	Kidney disease, heart disease, pyloric obstruction	Vomiting, diarrhea, Addison's disease, myxedema, pneumonia, diabetes mellitus
Potassium	3.5–5 mEq/L	Addison's disease, oliguria, anuria, tissue breakdown	Vomiting, diarrhea
Carbon dioxide	Adults, 24–29 mEq/L Infants, 20–26 mEq/L	Tetany, vomiting, intestinal obstruction, respiratory disease	Acidosis, diarrhea, anesthesia, nephritis
Hemoglobin	Male, 14–18 g/100 mL Female, 12–16 g/100 mL	Polycythemia	Anemia

[a] Milligrams per 100 mL is also called milligram percent [b] mEq/L = milliequivalents per liter.

28.3 Proteins in the Plasma

The protein content of plasma is about the same as that of muscle and other tissues. More than 100 proteins have been identified in the plasma; most of them are synthesized in the liver. The normal concentration range of proteins is 7.0 to 8.0 g/100 mL of plasma. These proteins remain in the circulatory system and ordinarily are not used as sources of energy. At times of protein deprivation, the plasma protein concentration is maintained at the expense of tissue protein. Plasma proteins are grouped into three main classes on the basis of their solubility properties and methods of isolation.

Albumins

Albumins are the most abundant proteins (~55% by mass) in the plasma. Their major function is to maintain osmotic pressure (see Section 28.6) by controlling the water balance. They are also important for their buffering capacity and in the transport of fatty acids, certain metal ions, and many drugs (particularly nonpolar ones). Normal concentrations of albumins range from 3.5 to 4.5 g per 100 mL of plasma.

Globulins

Globulins account for about 40% by mass of the total plasma proteins. They have higher molecular weights than the albumins (150,000 as compared to 70,000). Three subclasses of globulins are recognized, and they differ from one another with respect to rate of movement in an electric field: *α-globulins* (0.7 to 1.5 g/100 mL of plasma), *β-globulins* (0.6 to 1.1 g/100 mL), and *γ-globulins* (0.7 to 1.5 g/100 mL). α-Globulins and β-globulins are synthesized in the liver. They form complexes (i.e., VLDLs, LDLs, and HDLs) with the water-insoluble lipids and transport these compounds through the aqueous media.

Unlike the other plasma proteins, the γ-globulins are synthesized by leukocytes in the lymph nodes. They combat certain infectious diseases (e.g., diphtheria, influenza, measles, mumps, typhoid) and are therefore classified as antibodies of *antibody* (see Section 28.5).

Fibrinogen

Fibrinogen is a large protein (MW 340,000) consisting of six polypeptide chains. It is synthesized in the liver and constitutes about 5% (0.3 to 0.4 g/100 mL of plasma) of the total blood protein. It functions in blood coagulation (see Section 28.8). The fibrinogen content of plasma increases when inflammatory or infectious conditions exist and during menstruation and pregnancy.

When fibrinogen and the formed elements are removed from the plasma (by a centrifuge), the liquid that remains is called the **serum.** The only distinction, therefore, between blood plasma and blood serum is the presence of fibrinogen in the plasma. *Because blood serum lacks fibrinogen, it is unable to clot* (see Section 28.8).

28.4 The Formed Elements

Apart from the respiratory function of hemoglobin in the erythrocytes (Section 28.9), the formed elements have specific roles that are not directly concerned with the general metabolic processes. This is in contrast to the plasma, which serves as the metabolic transport medium of the organism.

Erythrocytes

The erythrocytes are formed in the red bone marrow, and they are the most numerous of the formed elements. The blood of the average adult female contains about 5 million of these cells in every cubic millimeter (mm^3) of blood, and the value for the average adult male is 5.5 million/mm^3. (The volume of one drop of blood is about 100 mm^3, so there are about 500 million erythrocytes in each drop of blood.) Any condition that tends to lower the oxygen content of the blood causes an increase in the number of erythrocytes. Persons who live at high altitudes generally have higher erythrocyte counts than those who live at sea level. Conversely, increased barometric pressure results in a decrease in the erythrocyte count.

The term *hematocrit value* is applied to the volume (in percent) of packed erythrocytes in a sample of blood (usually 10 mL) that has been centrifuged under standard conditions. The cells are spun to the bottom of a centrifuge tube, and the supernatant liquid (the plasma) is drawn off the top. Usually about 45% of the formed-element component of a blood sample is erythrocytes, and so 45 is the normal hematocrit value. Variations from the normal value indicate the existence of certain pathological conditions. When **anemia** occurs, for instance, *the percentage of erythrocytes (and/or the percentage of hemoglobin) is abnormally low.* Anemia may result from (1) a decreased rate of erythrocyte production (aplastic anemia), (2) an increased rate of erythrocyte destruction (hemolytic anemia), or (3) an increased rate of erythrocyte loss (as in hemorrhaging). *Polycythemia* is the condition arising from an abnormally high percentage of erythrocytes.

Erythrocytes are disc-shaped cells that have a slight depression at the center, like a solid doughnut. Unlike most other cells, they contain neither mitochondria nor a nucleus. (The nucleus is lost during the development and maturation of the erythrocyte.) They cannot reproduce, have no aerobic metabolism, and are unable to synthesize carbohydrates, lipids, or proteins. They obtain all the energy they need from a metabolic pathway called the pentose phosphate shunt (see an advanced text) and from substrate-level phosphorylation in the Embden–Meyerhof pathway. The most significant components of the erythrocytes are the hemoglobin molecules. An individual's blood type is determined by specific short-chain polysaccharides that are bound to certain proteins (glycoproteins) on the membranes of the erythrocytes.

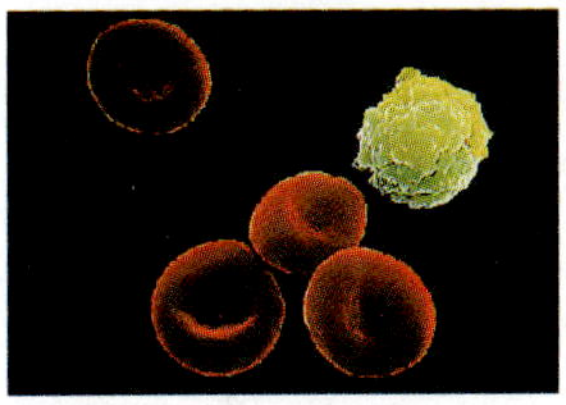

Erythrocytes are doughnut-shaped disks whereas leukocytes are roughly spherical. This scanning electron micrograph was made at 5200× magnification.

The major function of the erythrocytes is to transport oxygen from the lungs to the cells. Also, they assist in the transport of carbon dioxide from the tissues to the lungs. Erythrocytes in humans have a life span of about 4 months. (During this

Blood Doping

A radical attempt to improve athletic performance is called **blood doping.** A quantity of blood is withdrawn from the athlete, then stored for 6 weeks while the person's body manufactures replacement blood. Then, a few days before competition, the erythrocytes from the stored blood are returned to the athlete. This increased supply of erythrocytes enables the athlete's cardiovascular system to transport oxygen more efficiently. In controlled studies, blood-doped runners were able to improve their times by 9 s in an 8-km treadmill race, compared with a group of equally trained runners who were given merely a saline solution. Blood doping will never convert a mediocre runner to an outstanding one, and several sports groups have banned the practice.

time interval, each erythrocyte makes about 120,000 trips around the body.) During this period, there is no degradation or resynthesis of hemoglobin molecules in an erythrocyte. To maintain a constant level of erythrocytes, new ones are formed in the bone marrow at the same rate that old ones are eliminated by special tissues in the liver and spleen. It has been estimated that of the approximately 30 trillion erythrocytes in an average adult male, about 3 million are destroyed each second. Assuming that there are 300 million hemoglobin molecules in each erythrocyte, 900 trillion molecules of hemoglobin must be synthesized every second (by cells in the bone marrow) in order to maintain a constant supply.

Leukocytes

The composition of leukocytes resembles that of other tissue cells. They are nucleated, and they contain glucose, lipids, proteins, and other soluble organic substances and inorganic salts. Leukocytes constitute the body's primary defenders against foreign organisms (e.g., viruses, bacteria), and the blood is the vehicle that transports them to sites of infection.

The different varieties of leukocytes have specialized functions. *Lymphocytes* are involved in the synthesis and storage of antibodies (see Section 28.5). *Phagocytes (macrophages)* can leave the blood by squeezing between the endothelial cells that line the capillary wall. Phagocytes are attracted to sites of inflammation by chemicals released from injured tissue. They contain lysosomal enzymes, and their function is to engulf and digest the invading organisms.

Billions of leukocytes are produced each day in the bone marrow to replace the ones that die. On the average, there are about 7000 leukocytes per cubic millimeter of blood, but this value is subject to considerable variation. A higher-than-normal leukocyte count occurs during acute infections, such as *appendicitis* (16,000 to 20,000 per mm^3). High numbers of leukocytes may also appear during emotional disturbances and following vigorous exercise and/or excessive loss of body fluids. Viral diseases, such as chickenpox, influenza, measles, mumps, and polio, are accompanied by an abnormally low leukocyte count (<5000 per mm^3) because in fighting the viruses the leukocytes are killed faster than they can be produced. **Leukemia** is a cancer characterized by the uncontrolled production of leukocytes that fail to mature. Despite their numbers, these cells are unable to destroy invading pathogens, and the person has a lowered resistance to infections. Invariably some of the cancer cells *metastasize* (spread out) from the bone marrow or lymph nodes to other parts of the body. In these other tissues, the leukemic cells eventually crowd out the normal cells.

FIGURE 28.1 ▶
Relative sizes of the formed elements.

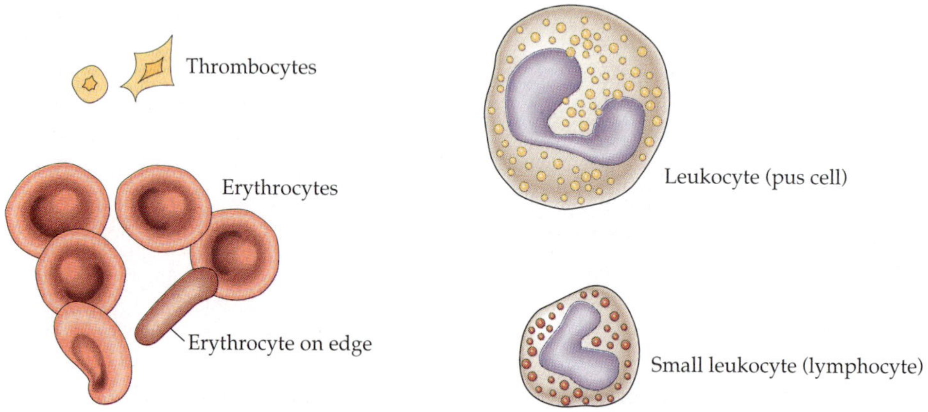

Blood Types

Blood is classified as type A, B, AB, or O depending on the presence or absence of specific glycoproteins (designated A and B) in the erythrocyte membrane. Individuals with type A blood have the A glycoprotein on their erythrocytes and have antibodies to the B glycoprotein in their plasma. If they are transfused with type B blood, the B antibodies attack the transfused erythrocytes, causing them to clump together and block small blood vessels. Type O blood cells lack both A and B glycoproteins, and antibodies to both A and B glycoproteins are present in the plasma. Type O blood can therefore be transfused into individuals with any blood type, but type O individuals can receive transfusions only of type O blood. Type AB blood cells have both glycoproteins and can be transfused only into another AB individual. But individuals with type AB blood can receive transfusions of other blood types, since their plasma lacks antibodies to either glycoprotein. Blood type is inherited.

Thrombocytes

There are about 250,000 thrombocytes in every cubic millimeter of blood. These small, nonnucleated cells contain proteins and relatively large amounts of phospholipids, mostly cephalin. They liberate species that are instrumental in the mechanism of blood clotting (see Section 28.8). An abnormally low thrombocyte count ($<$100,000 per mm^3) is related to a tendency to bleed.

Figure 28.1 shows the relative sizes of the various types of formed elements.

28.5 The Immune Response

As mentioned in Section 28.3, antibodies are gamma globulins. An **antibody** (also called an *immunoglobulin*) is a specialized protein synthesized by the body any time a foreign invader enters the body. Antibodies are associated with the sophisticated defense mechanism in the body termed the **immune response.** Whenever it is invaded by foreign substances, the body responds by synthesizing specific antibodies to fend off the invader. Because it triggers the synthesis of antibodies, the invader is called an **antigen** (it causes *anti*body *gen*eration). Each distinctive antigen causes a unique set of antibodies to be produced. The antibodies are designed to lock onto the particular antigen. They bind to it and (to oversimplify a bit) incapacitate it. In some instances the incapacitation involves the formation of a precipitate, with the antigen being tied up as part of an antibody–antigen clot.

When the body is invaded by pathogenic antigens (e.g., disease-causing bacteria or viruses), the immune response represents an effective defense mechanism. It is even possible to prime this mechanism. A **vaccine** containing a weakened form of the antigen (e.g., dead bacteria) will cause an individual to build up a certain level of antibodies for this particular antigen. If the same individual is then subjected to attack by a more virulent form of the antigen, the body responds much more rapidly and effectively to its presence. Having practiced on the dead invader, the system can easily handle the live one—it has gained an immunity to that particular organism.

Unfortunately, the body can't tell "good" antigens from "bad" ones. We want it to respond to a viral infection, but we wish it wouldn't attack purposely transplanted tissue. Nonetheless, the host body will respond to a donor skin graft, a kidney transplant, or a heart transplant as it does to a virus: by generating antibodies to attack it. When the body responds to foreign tissues in this way, the process is called *rejection*. This is a major problem associated with organ transplants. There are drugs that suppress the immune response, but when these are used on a transplant recipient to protect the transplanted organ, the patient also becomes more susceptible to infection. If you turn off the immune response, you turn off both the good and bad aspects of it.

There are a variety of diseases associated with the immune system. An overactive or misguided immune system can attack its own body organs, causing an **autoimmune disease.** It is thought that multiple sclerosis, some forms of arthritis, and some cases of diabetes are autoimmune diseases. The system senses something as foreign even though that something isn't. In the case of the three autoimmune diseases, the body attacks and destroys the myelin sheath of nerves (multiple sclerosis), the connective tissues in the joints (arthritis), or the insulin-producing cells of the pancreas (diabetes).

At the opposite end of the scale is *acquired immunodeficiency syndrome* (AIDS). In this deadly affliction, the immune system is destroyed by the HIV virus (Section K.1). The victims then succumb to a variety of infections or to a rare form of cancer.

Scientists have sought to enhance certain immune responses as weapons against diseases. By genetic engineering, researchers have cloned the genes for specific antibodies and placed them in cells that then produce *monoclonal antibodies*. In theory, monoclonal antibodies should attack and overwhelm specific antigens. In practice, they have been a bit disappointing, but they still hold considerable promise.

28.6 Osmotic Pressure

How does material get from the blood to the cells? Water, electrolytes, glucose, amino acids, and other materials all diffuse rapidly through pores in the capillary walls. This diffusion occurs in both directions—from the blood into the interstitial fluid outside the capillary and from the interstitial fluid back into the blood. Ultimately material makes its way from the interstitial space through cell walls and into (or out of) tissue cells.

The blood doesn't just sit in the capillaries while the diffusion takes place. It circulates—it continues to move through the vascular system. What keeps the blood moving is the heart, a pump that pushes the blood around the circulatory system. The pressure imparted to the blood by this pumping action falls off as the blood travels from the heart through the arteries, the capillaries, and the veins and finally back to the heart, where it gets a fresh push. Blood moves from the high-pressure arterial end of the capillaries to the low-pressure venous end.

The pressure that keeps the blood moving around the circulatory system also has a tendency to push the blood plasma out of the porous capillaries and into the interstitial space. In the absence of any countereffect, the rapid diffusion of material back and forth between blood and interstitial fluid would be accompanied by a slow but steady net loss of liquid from the blood to the interstitial space. Blood volume would drop, and tissue would swell with the extra fluid. There is, however, a countereffect—osmotic pressure.

When a living cell is placed in distilled water, water gradually flows through the semipermeable membrane into the cell. The environment outside the cell is 100% water, whereas the percentage of water in the internal environment is considerably lower owing to the presence of dissolved substances. Because the semipermeable membrane prevents these substances from leaving the cell, equalization of concentration can be attained only by the passage of the small water molecules into the cell. The diffusion of water from a dilute solution (high concentration of water) through a semipermeable membrane into a more concentrated solution (low concentration of water) is known as the process of *osmosis* (recall Section 9.10). **Osmotic pressure** is defined as the pressure required to prevent the occurrence of osmosis when two solutions of unequal concentrations are separated by a semipermeable membrane. The osmotic pressure depends solely on the concentration of solute particles (either ions or molecules) in the solutions involved.

The concentration of protein in the plasma far exceeds the concentration of protein in the interstitial fluid outside the blood vessels. This concentration gradient results in an osmotic pressure of about 25 mmHg. Therefore, if no external force were applied, liquids would be expected to diffuse from the interstitial fluid into the bloodstream. However, the pumping action of the heart creates the so-called *blood pressure* (see Section 28.7), and this pressure is greater at the arterial end of a capillary (~32 mm Hg) than at the venous end (~17 mmHg). Because the blood pressure at the arterial end is higher than the osmotic pressure, the natural tendency is reversed and there is a net flow *from* the capillary *into* the interstitial fluid. The fluid that leaves the capillary contains the dissolved nutrients, oxygen, hormones, and vitamins needed by the tissue cells. As the blood moves along the capillary branches, the blood pressure decreases until at the venous end the osmotic pressure is greater than the blood pressure and there is a net flow of fluid *from* the interstitial fluid *into* the capillary. The incoming fluid contains the metabolic waste products such as carbon dioxide and excess water (Figure 28.2). If we are in good health, everything balances nicely. To be sure, nutrients have diffused from blood to interstitial fluid and wastes have diffused from interstitial fluid to blood, but the volume of fluid in the two systems has remained constant.

Osmotic pressure, and hence the delicate balance of fluid exchange, are directly related to the concentration of albumins in the plasma. The normal half-life of albumins in the plasma is 20 days. If the albumin level is low, as might be the case from (1) malnutrition (low protein intake), (2) abnormal protein synthesis (liver disease), or (3) the loss of protein in the urine as a result of kidney disease (nephrosis), the osmotic pressure decreases. This results in a net efflux of fluids from the capillaries into the interstitial and cellular regions. This abnormal accumulation of fluids within the interstitial space produces noticeable swelling, particularly in the lower extremities. The condition so characterized is known as *edema*. Recall that children who suffer from the protein deficiency disease kwashiorkor (Section 27.4) characteristically have bloated abdomens. This

Edema can also result from the increased capillary permeability that accompanies an inflammatory reaction.

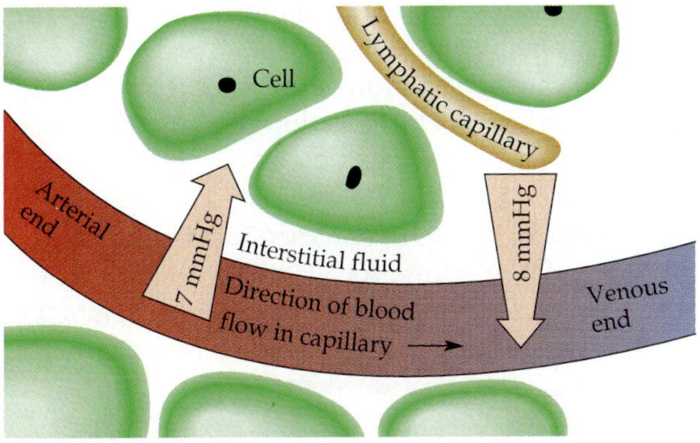

FIGURE 28.2

Oxygen and nutrients leave at the arterial ends of the capillaries. Carbon dioxide and other cellular waste products enter the venous ends of the capillaries.

swelling is caused by the accumulation of water, which leaves the blood because there are insufficient albumins to maintain the osmotic pressure of the blood.

The medical condition called **shock** also results from a loss of fluid from the vascular system. Capillary permeability increases, and albumins are lost to the interstitial fluid. The resulting outflow of fluid from the vascular system reduces blood volume enough to cause a dramatic decrease in blood pressure. There is a consequent decrease in oxygen-transporting capability with potentially fatal results. When someone goes into shock following a traumatic injury, it is because the delicately balanced transport system in the body has gone awry and could fail completely. Shock is a physiological, not a psychological, state. Treatment of the condition involves bringing the blood volume back up to normal levels.

28.7 Blood Pressure

Blood pressure is the force exerted by the blood against the inner walls of the arteries. The **systolic pressure** is the maximum pressure achieved during contraction of the heart ventricles. When the ventricles relax, the blood pressure drops, and the lowest pressure that remains in the arteries before the next ventricular contraction is called the **diastolic pressure.** The alternate expansion and contraction of the arterial walls can be felt as a *pulse* (when the artery is near the surface of the skin). Blood pressure measurements are reported as a *ratio of systolic pressure* (in mmHg) *to diastolic pressure* (in mmHg)—for example, 120/80.

The normal systolic and diastolic ranges for young adults are 100 to 120 and 60 to 80 mmHg, respectively. For older people, the corresponding normal ranges are 115 to 135 and 75 to 85 mmHg.

Blood pressure depends on several factors, including total blood volume, heart action, and action of the smooth muscles that surround the arteries. It is directly proportional to the volume of blood. Thus, if there is a loss of blood due to hemorrhaging, the blood pressure drops, but it returns to normal when the lost blood is replaced by a blood transfusion. Contraction of the smooth muscles in the walls of the arteries constricts the vessels (vasoconstriction), causing an increase in blood pressure. Conversely, dilation of the smooth muscles causes vasodilation and a decrease in blood pressure.

An approximate borderline reading for high blood pressure is 140/90.

One of the major health problems throughout the world is high blood pressure, or **hypertension.** Over 20% of the world's population is afflicted, and in the United States it is estimated that hypertension occurs in about 36 million people. To treat this disorder, it is necessary to attempt to counteract those factors that cause the increased pressure. Blood pressure can be reduced by any of the following procedures:

1. Administer diuretics (Section 21.6) and/or reduce sodium ion intake. Either step increases urine excretion and so decreases the blood volume.
2. Negate stimulating effects of epinephrine on the cardiac muscle with drugs (beta blockers) that specifically bind to epinephrine binding sites and/or reduce stress and therefore reduce epinephrine levels.
3. Administer vasodilators, drugs that cause relaxation of the smooth muscles in the arterial walls, and/or use drugs that block the action of smooth muscle vasoconstrictors.

If hypertension is not brought under control, the left ventricle must contract with a greater force against the increased arterial pressure. As a result of this increased workload, the heart muscle cells require additional oxygen, which cannot be delivered fast enough and in sufficient quantities. Heart cells die, and a myocardial infarction results.

28.8 Clotting of Blood

Blood clotting, or coagulation, is of the utmost importance to the organism. If such a mechanism did not exist, loss of blood would occur whenever a blood vessel was injured. The details of coagulation have been clarified considerably by research in recent years. The theory accounting for blood clotting has evolved from a two-step mechanism to the present model, which depicts a multistep cascade in which many protein factors are activated sequentially. Here, however, we restrict our discussion to the final two steps. (See an advanced biochemistry text for a more detailed explanation.)

When blood clots, the soluble plasma protein fibrinogen is converted to fibrin. Fibrin monomers undergo a polymerization reaction that results in the formation of insoluble needlelike threads. These threads enmesh the blood cells and effectively seal off the area where the blood vessel has been damaged (Figure 28.3).

The clotting mechanism becomes operative only when a tissue is cut or injured in some other way. Thrombocytes and damaged tissue cells are somehow activated and release a group of compounds collectively referred to as *thromboplastin.* In the presence of calcium ions and other cofactors, thromboplastin converts prothrombin to thrombin. Prothrombin is a zymogen (Section 22.2), and its activation is analogous to the activation of the various digestive enzymes. *Thrombin* is the actual clotting enzyme. It brings about the conversion of fibrinogen to fibrin by hydrolysis of two peptide fragments from the former. Blood coagulation can be summarized as

$$\text{Prothrombin} \xrightarrow[\text{Ca}^{2+}\text{, phospholipids, other factors}]{\text{thromboplastin}} \text{Thrombin}$$

$$\text{Fibrinogen} \xrightarrow{\text{thrombin}} \text{Fibrin}$$

Normally, blood takes 5 to 8 min to form a clot. After the tissue is repaired by the body, the fibrin clot is hydrolyzed into soluble components by the enzyme *plasmin* (which circulates in the blood as the zymogen *plasminogen*).

A number of substances, the *anticoagulants,* inhibit clotting by interfering with one or another of the reactions. Heparin is one of the principal anticoagulating agents. It is a polysaccharide rich in sulfate ester groups and is believed to block the catalytic activity of both thromboplastin and thrombin. Low concentrations of

Streptokinase, an enzyme isolated from a bacterium (and now available through recombinant DNA techniques), is used to treat myocardial infarctions. It functions by converting plasminogen to plasmin.

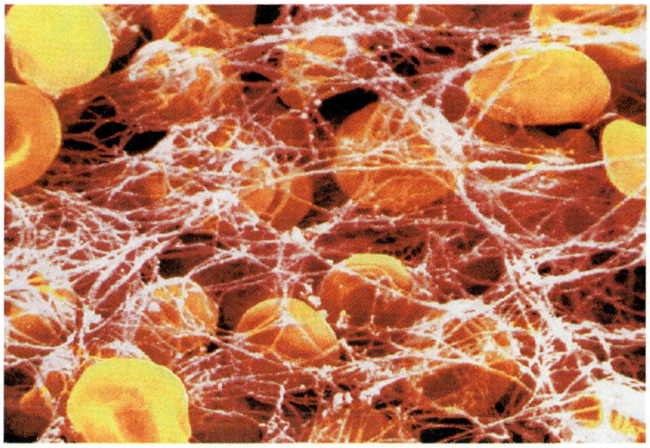

FIGURE 28.3
Scanning electron micrograph of erythrocytes enmeshed in fibrin fibrils—a part of a typical blood clot.

Hemophilia

Hemophilia is one of many inherited disorders characterized by inadequate production of clotting factors. The incidence of this condition in the general population is about 1 in 10,000, with males accounting for 80 to 90% of those affected. In hemophilia, production of a single clotting factor is reduced; the severity of the condition depends on the degree of reduction. In severe cases, extensive bleeding accompanies the slightest mechanical stress and hemorrhages occur spontaneously at joints and around muscles.

Transfusions of clotting factors can often reduce or control the symptoms of hemophilia, but plasma samples from many individuals must be pooled (combined) to obtain adequate amounts of clotting factors. Pooling is very expensive and increases the risk of infection with blood-borne infections such as hepatitis and AIDS. Gene-splicing techniques have been used to manufacture the clotting factor most often involved (factor VIII). This procedure should eventually provide a safer and cheaper method of treatment.

If one wishes to store blood plasma, an anticoagulant must be added to the freshly drawn blood to prevent clotting.

heparin are normally secreted into the circulatory system to prevent **thrombosis,** the formation of a clot within a blood vessel. Certain sodium salts are employed as anticoagulants when blood is collected for clinical purposes. The anions of these salts—citrate, oxalate, and fluoride—form strong complexes with calcium ions, thus preventing them from existing in the free ionic state. Without calcium ions in the plasma, blood will not clot.

Two other compounds, vitamin K and dicumarol, affect clotting. Vitamin K is a nutritional factor necessary for normal blood clotting (Section J.5). Animal blood deficient in vitamin K has a prolonged coagulation time because of a lack of prothrombin in the plasma. Vitamin K is a coenzyme in the oxygen-dependent carboxylation of the glutamic acid side chains of prothrombin to yield γ-carboxyglutamate residues. These carboxyglutamate groups must be present in prothrombin to enable it to bind to calcium ions during its conversion to thrombin. Dicumarol is believed to act as a metabolic antagonist of vitamin K. It prevents

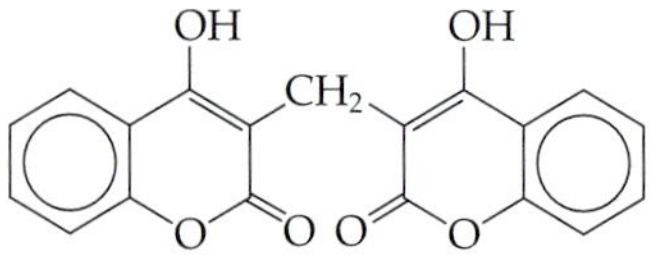

Dicumarol

Aspirin Against Thrombosis

Some doctors recommend that older persons and patients with histories of heart attacks and strokes should take aspirin regularly (one tablet every other day). Recall (Section I.6) that among its other physiological effects, aspirin inhibits the synthesis of prostaglandins and thromboxanes, thereby interfering with the aggregation of thrombocytes and diminishing the rate at which they release the blood-clotting factors (thromboplastin). Hence, aspirin prevents heart attacks by inhibiting thrombosis. The greatest danger from thrombosis is that the clot may become detached and travel through the blood to some vital organ, such as the heart or the brain. If the clot becomes lodged in and obstructs the blood vessels to these organs, their tissue cells are starved for oxygen and the cells die. If tissue death occurs in the brain, the condition is termed a *stroke*; if heart muscle tissue is destroyed, the condition is called a *coronary thrombosis* or a *mycardial infarction*. (If the clot, or a fragment of it, breaks loose and is carried away by the blood to be lodged in small blood vessels elsewhere (e.g., the lungs), it is called an *embolism*.

blood clotting either by repressing prothrombin formation or by inhibiting the enzyme for which vitamin K is a coenzyme. Dicumarol is frequently administered to patients who have suffered heart attacks caused by thrombosis, as a preventive measure against further clotting in the blood vessels.

Chemists have synthesized anticoagulants that have a greater potency than dicumarol. One of these is warfarin sodium (Coumadin),[1] which is unique in that it can be administered orally, intravenously, intramuscularly, or rectally. These drugs are usually administered before an operation or after heart attacks to minimize thrombosis.

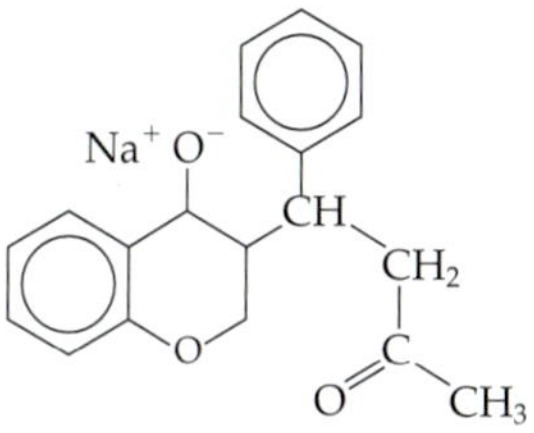

Warfarin sodium

28.9 Hemoglobin: Oxygen and Carbon Dioxide Transport

One of the main functions of blood is the transport of oxygen and carbon dioxide. The oxygen-carrying component is hemoglobin, which has a protein portion (globin) and a nonprotein portion, or prosthetic group, called heme (refer to Figures 21.13 and 21.14). A hemoglobin molecule contains four heme units, each with a central iron atom. Each iron atom is attached to the nitrogen atoms of four pyrrole rings.

As mentioned in Section 21.11, the protein portion of the hemoglobin molecule has four polypeptide chains that are not covalently bonded to one another. There are two identical α chains, each with 141 amino acid units, and two β chains, each with 146 amino acid units. The protein chains play an active role in oxygen transport. The fifth bonding site on the iron atoms is occupied by a nitrogen atom of a histidine that is the 87th amino acid unit in the α chain or the 92nd in the β chain.

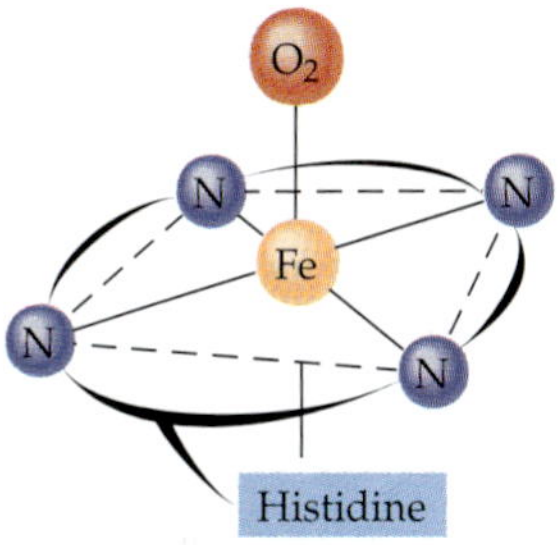

Schematic illustration of oxyhemoglobin.

Each hemoglobin molecule is capable of transporting four molecules of oxygen. Each oxygen molecule occupies the sixth bonding site of an iron in one of the heme units. Binding of one oxygen molecule to one heme facilitates the attachment of another oxygen to another heme because oxygenation changes the conformation of the chain. This conformation change seems also to induce changes in a neighboring heme–peptide unit, giving it a greater affinity for oxygen. Conversely, release of oxygen from one site facilitates release from a neighboring site. These factors enable hemoglobin to load and unload oxygen quite rapidly.

Normally, there is about 15 g of hemoglobin per 100 mL of blood. This amount of hemoglobin can combine with about 20 mL of gaseous oxygen (at STP). Without the hemoglobin, only 0.3 mL of gaseous oxygen could physically dissolve in 100 mL of plasma. Hemoglobin makes up about 90% of the total protein of an erythrocyte. The characteristic red color of blood is due entirely to the presence of hemoglobin or, more precisely, to the presence of the heme groups, which absorb strongly in the blue region of the spectrum (~400 nm).

The human body requires an enormous amount of oxygen for oxidative phosphorylation. The hemoglobin molecule is well suited to meet these demands because of its affinity for oxygen and because the attachment of oxygen to heme is readily reversible. In the alveoli of the lungs, hemoglobin comes into direct contact with a rich supply of oxygen (partial pressure of about 90 to 100 mmHg) and is converted to oxyhemoglobin. The oxyhemoglobin is carried by the arterial

---

[1] Warfarin is also employed as a rat poison. It is safe for use as a rodenticide because regular ingestion of massive doses is fatal to rodents (producing internal hemorrhaging) whereas a single, accidental ingestion by a child or pet is harmless.

circulation to the cells in which there is a low oxygen concentration (25 to 40 mmHg) and a relatively high concentration of carbon dioxide (~60 mmHg). The oxygen is given up to the cells. The resulting hemoglobin carries some of the carbon dioxide back to the lungs to be expelled, and more oxyhemoglobin is formed (see Figure 7.14).

Oxyhemoglobin does not transfer all of its oxygen to the tissue cells. Normally, every 100 mL of arterial blood combines with about 20 mL of oxygen. In the resting individual, the venous blood carries about 13 mL of oxygen per 100 mL of blood. Therefore, about 65% of the hemoglobin in venous blood is still combined with oxygen. When people are engaged in strenuous exercise, their oxygen demand is high and the percentage of oxyhemoglobin in the venous blood may fall to as low as 25%. Arterial blood is crimson; venous blood is a darker red, but it is not purple or blue.

What about people who live at high altitudes? In Leadville, Colorado, for example, where the altitude is 10,000 ft, the partial pressure of O_2 in the lungs is only about 68 mm Hg. Hemoglobin is only 90% saturated with O_2 at this pressure, meaning that less oxygen is available for delivery to tissues. People who climb suddenly from sea level to high altitude thus experience a feeling of oxygen deprivation, or *hypoxia*, as their bodies are unable to supply enough oxygen to the tissues. The body soon copes with the situation by producing more hemoglobin molecules.

Various chemical substances act as poisons by interfering with the transport of oxygen by hemoglobin. Oxidizing agents such as potassium ferricyanide can oxidize the iron of hemoglobin to the Fe(III) state. The same result is achieved *in vivo* by the action of nitrites and certain organic compounds (e.g., acetanilide, nitrobenzene, the sulfa drugs). The resulting compound, which contains iron in the Fe(III) oxidation state, is called *methemoglobin* and cannot transport oxygen. Small amounts of methemoglobin are normally present (about 0.3 g/100 mL of blood) in the erythrocytes, but appreciable amounts of this substance result in the pathological condition *methemoglobinemia*.[2]

Methemoglobin is brown. During cooking, red meat turns brown because of the oxidation of hemoglobin to methemoglobin. Dried bloodstains turn brown for the same reason.

Carbon dioxide (CO_2) is carried in the blood primarily as bicarbonate ion (HCO_3^-; 70%), or bound to hemoglobin (20%). Carbon dioxide binds most readily to hemoglobin that is not bound to oxygen (O_2), so as hemoglobin gives up its O_2 to the tissues, it becomes available to bind CO_2 picked up from the tissues.

Bicarbonate ions are formed when CO_2 diffuses into the blood from the tissues and reacts with water in the following reaction:

$$CO_2 + H_2O \longrightarrow HCO_3^- + H^+$$

The reaction producing HCO_3^- occurs primarily within the red blood cells, where it is catalyzed by carbonic anhydrase. Most of the HCO_3^- then diffuses back into the plasma. Carbonic anhydrase is also embedded in the walls of lung capillaries and catalyzes the reverse reaction, helping CO_2 diffuse out of the capillaries and into the alveoli (Figure 28.4).

[2] Salami and other preserved meat products contain nitrite salts as preservatives. (Recall that nitrites have been implicated in nitrosoamine formation; Section 17.5.) People who consume relatively large quantities of these foods have a tendency to develop methemoglobinemia. Also, nitrate ions are reduced to nitrite ions by microorganisms in the digestive tract. Concern exists over the high level of nitrates from fertilizer in the groundwater in some areas. Babies are particularly sensitive. Nitrites or nitrates cause methemoglobinemia and result in the blue baby syndrome.

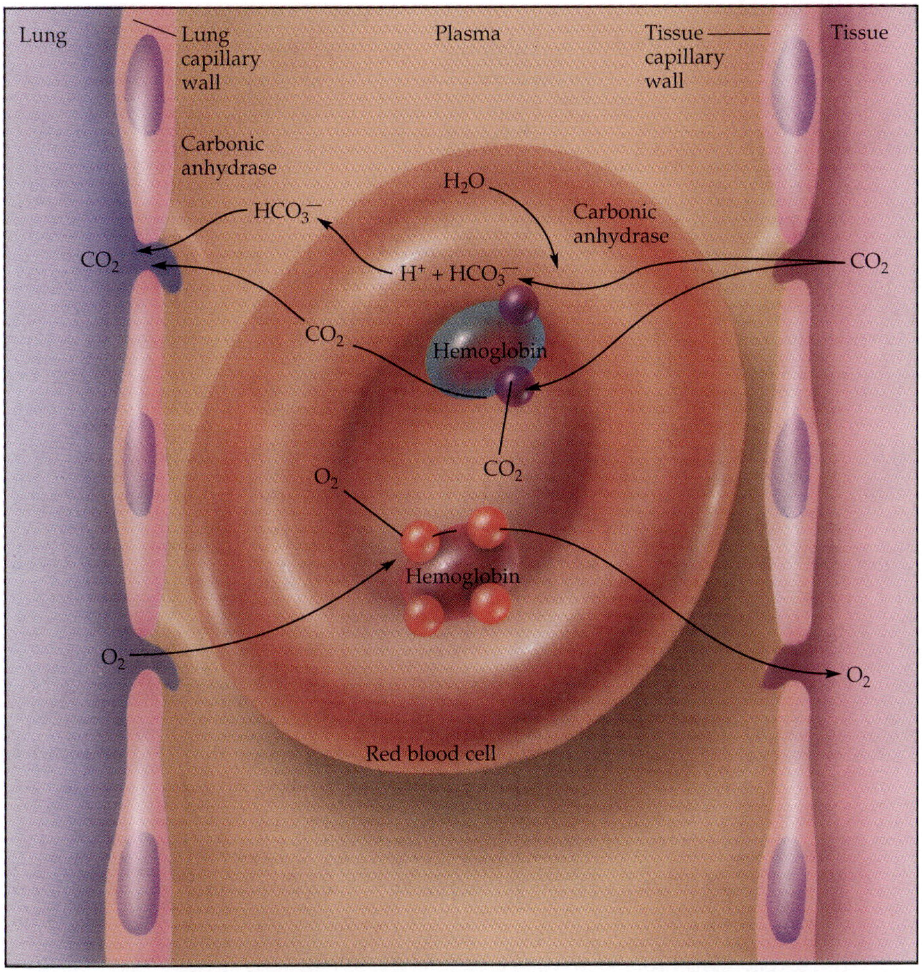

FIGURE 28.4

A schematic diagram that shows the exchange of gases between capillary blood and the air in the alveoli (left) and between capillary blood and body tissues (right).

Sickle Cell Anemia

There are a variety of abnormal hemoglobins, each responsible for a particular disease. Perhaps the most notorious is the one responsible for sickle cell anemia, an inherited disease that, if left untreated, may be fatal (due to infection, blood clots, or cardiac or renal failure). Sickle cell hemoglobin (HbS) differs from ordinary hemoglobin in only one amino acid—at the sixth position of the β chains, HbS has valine rather than glutamic acid:

Normal Hemoglobin

Val-His-Leu-Thr-Pro-Glu-Glu-Lys . . .
1 2 3 4 5 6 7 8

Sickle Cell Hemoglobin

Val-His-Leu-Thr-Pro-Val-Glu-Lys . . .

The change from a polar amino acid (Glu) to a nonpolar one (Val) reduces the overall charge on the hemoglobin molecule. If the altered hemoglobin molecule is fully oxygenated, there is no problem and it remains in solution. However, if the level of oxygenation decreases (e.g., at high altitudes or during vigorous physical exercise), the less soluble, deoxygenated hemoglobin molecules clump together, forming long, insoluble fibers, and force the erythrocytes to change from a round to a crescent, or sickle, shape (Figure 28.5). The abnormal cells become trapped in the capillaries and impair circulation. The resultant blockage of blood further decreases the oxygen supply to the affected areas of the body and increases the sickling of additional erythrocytes. The abnormal cells are subsequently destroyed by the spleen, and this leads to anemia (and subsequent tiredness and other permanent damage).

In the United States, about 10% (more than 2 million) of the African-American population are genetic carriers of the disease (heterozygotes), and 0.25% have the symptoms of sickle cell anemia (homozygotes). Genetic screening tests can determine whether prospective parents carry the sickle cell trait.

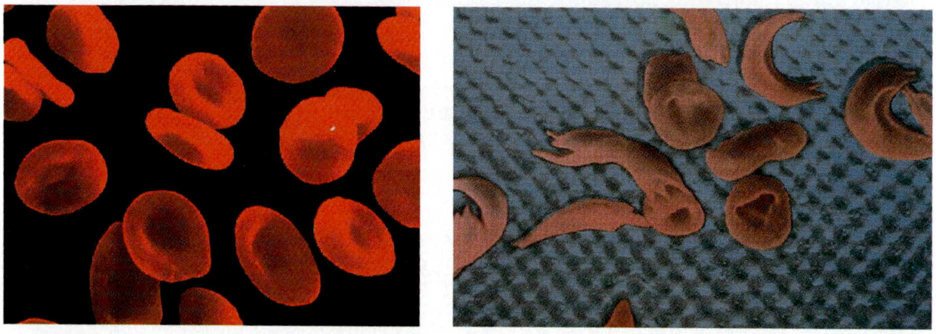

(a) (b)

FIGURE 28.5 ▶ Scanning electron micrographs of (a) normal and (b) sickled erythrocytes.

Metabolic Fate of Hemoglobin

When erythrocytes are destroyed, their hemoglobin molecules are completely catabolized. The porphyrin ring is first cleaved; then the globin and the iron are removed. The globin is digested, and its amino acids join the others in the metabolic pool. The iron is set free and incorporated into the iron-storage protein *ferritin*. This iron will be reused for the synthesis of new hemoglobin in the bone marrow. The porphyrin skeleton is of no further use to the body. It undergoes a series of degradation reactions that lead to the production of the bile pigments, chiefly *biliverdin and bilirubin*. These are colored substances that give the bile its yellow color. The degraded pigments are stored in the gallbladder and released into the small intestine. As they travel down the intestinal tract, they undergo additional transformations that result in a darkening of their color, thus accounting for the characteristic colors of feces and urine. The sometimes spectacular color changes observed in a bruise have a related cause. A bruise consists of blood released and trapped beneath the skin. As the blood is gradually broken down, a series of colored products are formed.

An average diet supplies about 12 to 15 mg of iron per day, only about 10% being absorbed. To maintain sufficient iron for hemoglobin synthesis, the body must retain the 20 to 25 mg of iron that is released each day.

An excess of bilirubin in the blood is responsible for the yellow color of the skin in jaundice (French *jaune*, yellow). Jaundice caused by an excess of bilirubin can arise from (1) infectious hepatitis, a condition during which the liver malfunctions and cannot remove sufficient bilirubin; (2) the obstruction of bile ducts by gallstones; and (3) an acceleration of erythrocyte destruction in the spleen (hemolytic jaundice). Jaundice occurs in a large percentage of newborn infants because of insufficient synthesis of the liver enzymes that decompose bilirubin. A common treatment of neonatal jaundice is to shine a special fluorescent light onto the baby's skin. The energy of the light is able to decompose some of the bilirubin just beneath the surface of the skin.

Carbon monoxide hemoglobin (CO-hemoglobin) is formed when hemoglobin combines with carbon monoxide. The Fe(II) ions in hemoglobin have a much greater affinity for carbon monoxide than they have for oxygen (by a factor of 200). Thus they preferentially combine with any carbon monoxide that is in the blood. CO-hemoglobin does not transport oxygen because all the iron-binding sites are tied up by the carbon monoxide molecules. If sufficiently large numbers of hemoglobin molecules become saturated with carbon monoxide (about 60%), death occurs because the blood fails to supply the brain with oxygen. All except the most severe cases of CO poisoning are reversible. The best antidote is the administration of pure oxygen. Artificial respiration may help if a tank of oxygen is not available. Because carbon monoxide poisoning impairs the blood's ability to transport oxygen, the heart has to work harder to supply oxygen to tissues. Chronic exposure, even to low levels of CO, as through cigarette smoking, puts an added strain on the heart and increases the chances of a heart attack.

28.10 Blood Buffers

Recall that the maintenance of pH within narrow limits is vital to the well-being of an organism. Any slight change in hydrogen ion concentration (± 0.2 to 0.4 pH unit) inhibits oxygen transport and alters the rates of the metabolic processes by decreasing the catalytic efficiency of enzymes. The blood plasma and the erythrocytes contain four buffering systems that maintain the pH of the blood between 7.35 and 7.45: (1) the bicarbonate pair, (2) the phosphate pair, (3) the plasma proteins, and (4) the hemoglobin of erythrocytes. In Section 11.6 we mentioned the buffering actions of the bicarbonate pair H_2CO_3/HCO_3^- and the phosphate pair $H_2PO_4^-/HPO_3^{2-}$. In Section 21.3 we indicated that the amino acids, in their zwitterionic forms, can neutralize small concentrations of either acids or bases. Since the plasma proteins and the globin of hemoglobin contain both acidic and basic amino acids, they tend to minimize changes in pH by combining with, or liberating, hydrogen ions. Thus they serve as excellent buffering agents over a wide range of pH values.

The principal buffer in tissue cells is the phosphate pair, whereas hemoglobin molecules are the most important buffers within the erythrocytes. The major buffer in the blood and interstitial fluid is the bicarbonate pair, owing to its intimate connection with respiration. Under normal conditions, the primary metabolic factor that tends to lower the pH is the continuous production of carbon dioxide and the acidic metabolites (acetoacetic acid, pyruvic acid, lactic acid, α-ketoglutaric acid, etc.). All of these compounds vary in concentration according to metabolic circumstances. When acids enter the blood, they are neutralized by bicarbonate ions, and the slightly dissociable carbonic acid is formed:

$$H^+ + HCO_3^- \rightleftharpoons H_2CO_3$$

It would seem that this reaction would alter the buffer ratio by decreasing the bicarbonate ion concentration and increasing the concentration of carbonic acid. The excess carbonic acid, however, is readily decomposed to water and carbon dioxide by carbonic anhydrase:

$$H_2CO_3 \xrightleftharpoons{\text{carbonic anhydrase}} H_2O + CO_2$$

The respiration rate is increased, and the carbon dioxide is eliminated at the lungs, thus preserving the proper buffer ratio.

Respiration rate is another factor that influences blood pH. *Respiratory acidosis* results from **hypoventilation,** a condition that arises when the rate of breathing is too slow. Hypoventilation is brought on by an obstruction to respiration (e.g., asthma, pneumonia, or pulmonary emphysema), by coronary attack, or by drugs that depress the brain's respiratory center (e.g., morphine, barbiturates). When the respiration rate is very low, carbon dioxide is not expelled from the lungs fast enough, and the H_2CO_3/CO_2 equilibrium is shifted to the left. This increased carbonic acid concentration results in a higher-than-normal H_2CO_3/HCO_3^- ratio and a subsequent decrease in the blood pH.

Hyperventilation, when the rate of breathing is too rapid, causes *respiratory alkalosis.* Hyperventilation arises during strenuous exercise, anxiety, crying, and hysteria. At high altitudes, hyperventilation may occur in response to low oxygen pressure. An increased rate of respiration accelerates the removal of carbon dioxide from the lungs, and the H_2CO_3/CO_2 equilibrium shifts to the right. The concentration of carbonic acid in the blood decreases, and the H_2CO_3/HCO_3^- ratio becomes lower than normal, with a subsequent increase in blood pH.

28.11 Lymph: A Secondary Transport System

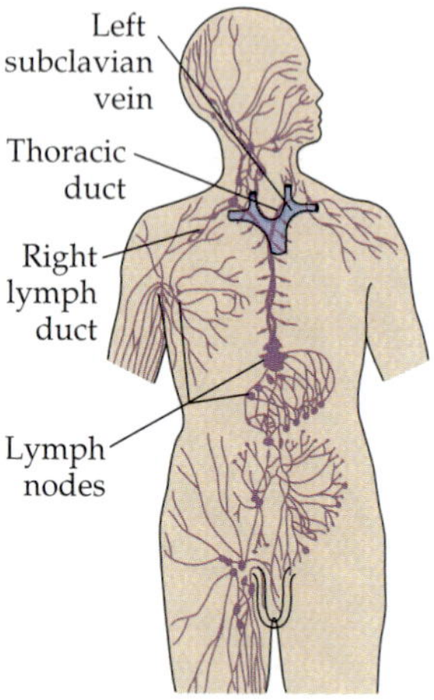

▲ **FIGURE 28.6**
The lymphatic system.

The lymphatic system also plays a role in transporting materials from one part of the body to another. This system, which returns components of the interstitial fluid to the bloodstream, is composed of veins and capillaries but no arteries (Figure 28.6). The capillaries are closed at one end. Interstitial fluid is absorbed into the lymph capillaries, in which it is called **lymph.** This lymph flows into larger and larger lymph veins. Eventually, two large lymph veins empty into veins of the blood circulatory system.

Lymph serves other functions. Fat absorbed from the intestine is picked up by lymph capillaries rather than blood capillaries. Lymph nodes serve as filters and as factories for the production of some forms of leukocytes. The leukocytes located there remove dead cells, bacteria, and other foreign elements from the lymph. It is in the lymph nodes that antibodies are synthesized. These nodes are lumpy enlargements in the lymph veins. Sometimes the nodes are so effective at filtering out bacteria from an infected area that they become swollen. Someone suffering from a sore throat may also exhibit swollen and tender lymph nodes in the neck area.

The role of lymph, although secondary to that of blood, is nonetheless an important one.

28.12 Urine: Formation and Composition

The kidneys remove metabolic waste products from the blood.[3] The functional units of the kidneys are called **nephrons** (Figure 28.7). Each nephron has a bulbous *Bowman's capsule,* which tails into a long, highly convoluted urinary tubule. The capsule surrounds a network of arterial capillaries called the *glomerulus.* One end of the glomerulus joins a small artery, and the other end forms a network of capillaries (purple in Figure 28.7) that surround the tubule. Finally

[3] Waste removal is only a secondary renal task, however. The principal role of the kidneys is to maintain stable concentrations of all the water and inorganic ions in the body.

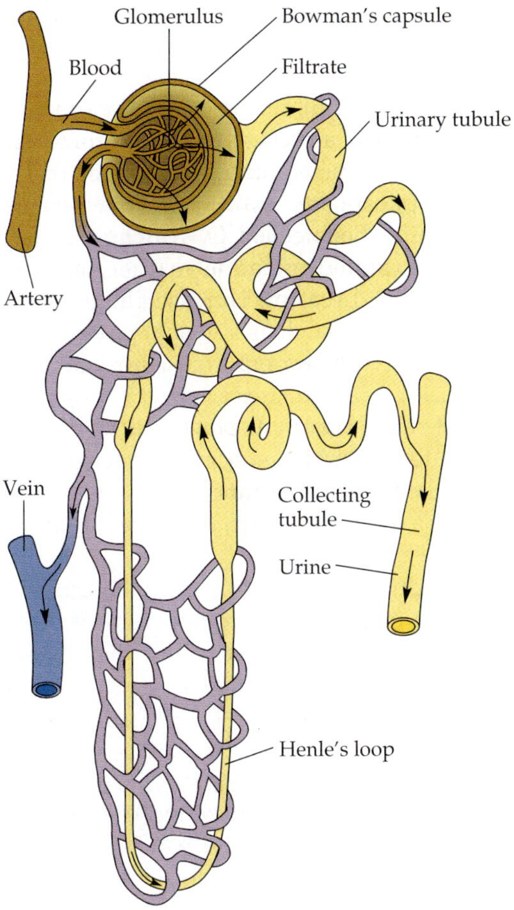

FIGURE 28.7

Structure of a nephron. Nephrons are the functional units of the kidney, where metabolic wastes are removed from the blood. The arrows indicate the path of fluids through the nephron. Each human kidney has about a million nephrons.

these tubule-surrounding capillaries join again and form a small vein. Blood flows from the artery into the glomerulus, then through the capillary network surrounding the tubule, and finally out the vein.

Most blood constituents except formed elements and large protein molecules filter through the capillary walls of the glomerulus and enter the Bowman's capsule. The glomerulus seems to act as a simple filter, with particle size being the main factor that decides what goes and what stays. The fluid that enters the Bowman's capsule and eventually collects in the tubule contains many valuable components as well as wastes. Consider water. About 170 L per day is filtered into the tubules. Most of this water—and valuable constituents such as glucose, amino acids, and salt—are reabsorbed by the blood through the walls of the tubules. Waste products such as urea, uric acid, and excess salts are passed on into collecting tubules and eventually excreted as **urine.** Knowing that healthy adults pass 1.1 to 1.5 L of urine each day, you can calculate that more than 99% of the water filtered through the glomerulus is reabsorbed through the tubules. We'd have quite a drinking problem if all that water weren't conserved.

The volume and composition of urine are quite variable. Its volume depends on liquid intake, amount of perspiration, presence of fever or diarrhea, and other factors. The composition varies with diet and state of health. Most substances have a renal threshold level. If the concentration of a substance in the blood exceeds this threshold value, the excess is not reabsorbed through the tubule but appears in the urine. The threshold level for glucose (blood sugar), for example, is

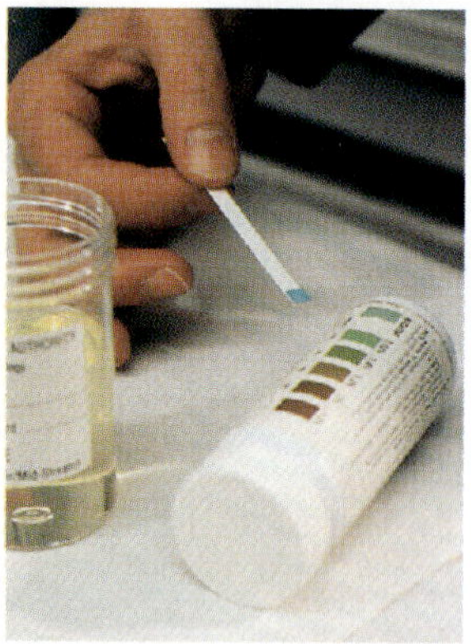

Urinalysis test strips. The reagent strip is dipped into the urine sample and then compared with a color chart for each constituent.

quite high (recall Section 24.3). Normally, nearly all glucose is reabsorbed. If the glucose level in the blood is too high, however (as happens in cases of uncontrolled diabetes), glucose shows up in the urine.

Some components of urine have relatively low solubility in water. When some condition (such as an increase in concentration or a change in pH) leads to the precipitation of these materials in the kidney, a stone is formed. *Kidney stones* (or renal calculi) usually consist of calcium phosphate [$Ca_3(PO_4)_2$], magnesium ammonium phosphate ($MgNH_4PO_4$), calcium carbonate ($CaCO_3$), calcium oxalate (CaC_2O_4), or a mixture of these compounds. Stone formation may accompany increased concentration of calcium ion caused by disease or increased ingestion of calcium ion. People who eat foods rich in oxalates (such as spinach) have a relatively high incidence of oxalate kidney stones. (What a great excuse for kids who don't want to eat their spinach!)

The kidney does far more than just get rid of wastes. It plays a vital role, for example, in maintaining the balance of water, electrolytes, acids and bases, and other components of the body fluid. When the kidneys malfunction, the body is in trouble. Analysis of the urine can often give a good indication of the health of the individual.

Wilhelm Kolff invented the artificial kidney at the Cleveland Clinic in the 1950s. Before that time, kidney failure meant death. There are now tens of thousands of people on dialysis. The number of dialysis patients and the costs—usually paid for by Medicare and Medicaid—are rising rapidly.

28.13 Sweat

It is interesting to note that female mosquitoes, seeking their meal of blood, find us by following a warm stream of air laden with carbon dioxide and lactic acid. We could foil them by not sweating and not breathing!

The skin is also an organ of excretion. Through it we lose water, electrolytes, nitrogenous wastes, and lipids. Water is lost directly through the skin and through the respiratory tract at a rate of about 700 mL per day. This water loss is called *insensible perspiration.* Perspiration from the 2.5 million sweat glands is called *sensible perspiration.* It is activated by a rise in blood temperature.

Sweat is about 99% water. It contains sodium ions, chloride ions, calcium ions, and smaller amounts of other minerals. A person working in a hot environment might lose 12 L of sweat per day, including 70 g of salt. Organic constituents of sweat include urea, lipids (body oils), creatinine, lactic acid, and pyruvic acid. Drugs such as morphine, nicotine, and alcohol also appear in sweat. The normal United States diet probably contains too much sodium chloride. Thus, it makes the most sense to replace the water component of lost sweat with pure water. Commercial "thirst quenchers" are quite popular with both serious and weekend athletes. These drinks, designed to replace the salts and water your body loses during long, sweat-provoking periods of exercise, may be too concentrated—a hazard that could lead to diarrhea. At best, the thirst quenchers are of marginal value.

Carbohydrate replacement drinks may be of some slight benefit to runners and cyclists who exercise vigorously for more than 2 h at a time. These drinks delay the onset of exhaustion by a few minutes.

If the air around us is not too humid, the water in sweat evaporates. Perspiration carries off not only wastes but also heat. It helps us keep cool on hot days. Each gram of water that evaporates absorbs 540 cal of heat from the body (Section 8.7). **Heat stroke** occurs when the heat regulation system fails. Without prompt medical attention, a victim of heat stroke may die.

28.14 Tears: The Chemistry of Crying

If ever you should cry over your chemistry grade, take consolation from the fact that this lacrimal fluid is responsible for maintaining the health of your eyes. Tears keep the eyes moist. The eyelids sweep the secretions of the lacrimal glands over the surface of the eye at regular intervals.

Tears are three-layered. There is an inner layer of mucus, then a layer of lacrimal secretions, and finally an outer layer of oily film that retards evaporation from the watery middle layer. Total normal secretion is about 1 g per day.

Chemically, tears have about the same composition as other body fluids. They have approximately the same salt content as blood plasma. Tears contain *lysozyme*, an enzyme that ruptures bacterial cell walls. This bactericidal action helps prevent eye infections.

Copious flows of tears may be caused by irritants, such as pepper, acid fumes, or a variety of chemical lachrymators. Tear gas, usually α-chloroacteophenone, is a specially designed eye irritant. The flow of tears can also be triggered by emotional upsets. Such crying is undoubtedly controlled by hormones. And if you feel like crying, go right ahead. Most psychologists say it's good for you. Indeed, many people have long believed that crying is beneficial. Richard Crashaw, an English poet of the seventeenth century, called tears "the ease of woe."

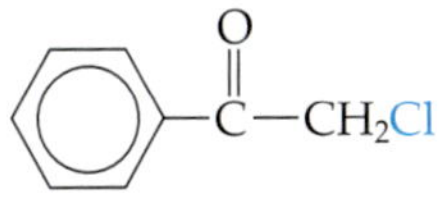

α-Chloroacetophenone

28.15 The Chemistry of Mother's Milk

Newborn mammals are nourished by a secretion from their mothers' mammary glands. Indeed, the presence of such glands in females characterizes the animal class Mammalia. The composition of milk varies from one species to another (Table 28.2). Amounts of fat and protein tend to be greater in marine mammals and those that live in cold climates.

Colloidal proteins and emulsified fats give milk its characteristic "milky" appearance. Casein is the precipitated protein from cow's milk. Humans beyond infancy often include some milk or milk products in their diet. This food is an excellent source of most nutrients. It includes all the essential amino acids. It also contains most vitamins and minerals. Cow's milk, however, is a poor source of iron, copper, and vitamin C. It is also short of vitamin D as it comes from the cow, so this vitamin is often added as a supplement to cow's milk intended for human consumption.

Cow's milk is sometimes given to infant humans as a substitute for their mothers' milk. Evidence now indicates that this may not always be a wise substitution. Human milk not only more closely matches the nutritional needs of human infants, but it also adds to the infant's immunological defenses against disease. Newborn infants are unable to make antibodies. They have *temporary passive immunity* because their blood contains antibodies made by the mother and passed across the placenta. Mother's milk produced within the first few days after childbirth is a rich source of gamma globulins. There is clear evidence that these immunoglobulins protect some newborn animals (ungulates such as cows, horses,

Table 28.2 Compositions of Milks (g/100 g)

Mammal	Fat	Protein	Lactose
Cow	4.4	3.8	4.5
Human	3.8	1.6	7.5
Goat	4.1	3.7	4.2
Reindeer	22.5	10.3	2.5
Porpoise	49.0	11.0	1.3

and sheep). The immunoglobins in mother's milk have a similar function in human infants.

And there you have it! Chemistry from molecules to mother's milk. Carbohydrate, lipid, and protein metabolism are all tied together. Indeed, in a living organism, everything is connected to everything else. Life is one huge, complicated set of chemical reactions—and, of course, a lot more. The whole of life is certainly much more than the sum of a set of chemical reactions.

We hope that we have enriched your life by helping you to learn something of the basis of chemistry and of the many ways chemistry touches your life every day. And we wish for you the proper reward for the many hours you have spent studying chemistry: the joy of success in your chosen profession.

Summary

Blood is 55 to 60% **plasma** and 40 to 45% **formed elements.** Plasma is a straw-colored liquid containing, in solution, all the compounds found in the body. The three types of formed elements are **erythrocytes** (red cells), **leukocytes** (white cells), and **thrombocytes** (platelets).

The principal ions in plasma are sodium, potassium, calcium, magnesium, phosphate, chloride, bicarbonate, and sulfate. The principal proteins are **albumins, globulins**, and **fibrinogen.** Blood minus fibrinogen and formed elements is **serum.**

The major functions of erythrocytes are to carry inhaled oxygen from lung to tissue cell and to carry carbon dioxide from tissue cell to lung to be exhaled. Leukocytes defend the body when foreign substances enter. Thrombocytes have a role in blood clotting.

Antibodies (also called immunoglobulins) are specialized proteins synthesized whenever a foreign substance enters the bloodstream. The invading substance is called an **antigen,** and its presence triggers the formation of a specific antibody, the overall mechanism being called the **immune response.**

The heart pumps the blood through the circulatory system. As the blood moves along, substances dissolved in it diffuse through the wall of the blood vessels, entering first the interstitial fluid and then tissue cells. The protein gradient from the blood to the interstitial space (much protein in blood, little in interstitial space) causes an osmotic pressure of about 25 mmHg. The blood pressure resulting from the heart's pumping results in a pressure of 32 mmHg at the arterial end of a capillary and about 17 mmHg at the venous end. The higher-than-osmotic pressure at the arterial end causes a net flow of liquid and dissolved nutrients in the direction capillary $\longrightarrow$ interstitial space. The lower-than-osmotic pressure at the venous end of the capillary causes a net flow of liquid and dissolved waste products in the direction interstitial space $\longrightarrow$ capillary. In this way, nutrients get from the blood to the rest of the body and metabolic wastes get from the rest of the body to the blood. These wastes are removed in the kidneys.

Systolic pressure is the maximum pressure the blood exerts on a capillary wall; it occurs when the heart is contracting. **Diastolic pressure** is the minimum pressure the blood exerts on a capillary wall; it occurs when the heart is relaxed. A blood pressure measurement is the ratio of systolic pressure to diastolic pressure.

A blood *clot* forms when a blood vessel is damaged. Thrombocytes and the damaged cells release *thromboplastin*, a substance that converts *prothrombin* to *thrombin*. This latter enzyme catalyzes the conversion of fibrinogen to *fibrin*, and the fibrin forms threads that seal off the damaged area.

Hemoglobin is made up of four protein chains (the globin) and four heme units, one on each chain. Each heme contains a central iron atom that bonds oxygen.

Buffers in the blood—mainly H_2CO_3/HCO_3^-, $H_2PO_4^-/HPO_4^{2-}$, and protein zwitterions—maintain the blood pH in the range 7.35 to 7.45.

A second transport system in the body is the *lymph system*. It contains veins and capillaries only, with the capillaries closed at one end. In this system, liquid flows in one direction only—out of the interstitial space and into the lymph vessels, where the fluid is known as **lymph.** The system carries the lymph to blood veins. *Lymph nodes* are tiny organs located at various places along the lymph-system pathway. They produce some leukocytes and filter dead cells, bacteria, and other foreign bodies from the lymph.

Waste products are removed from the blood as it flows through **nephrons,** the basic functional units that make up the kidneys. As the blood passes through the *glomerulus*, a clump of capillaries surrounded by the part of the nephron called *Bowman's capsule*, waste products leave the blood and enter the Bowman's capsule, then pass down the *urinary tubule* and out of the body in the urine. The kidneys regulate the water and electrolyte balance in the body.

Some water, electrolytes, nitrogenous wastes, and lipids pass through the skin in *sweat* and are in this way eliminated from the body.

Key Terms

albumin (28.3)
anemia (28.4)
antibody (28.5)
antigen (28.5)
autoimmune disease (28.5)
blood doping (28.4)
blood pressure (28.7)
diastolic pressure (28.7)
erythrocyte (28.1)
fibrinogen (28.3)
formed element (28.1)
globulin (28.3)
heat stroke (28.13)
hemophilia (28.8)
hypertension (28.7)
hyperventilation (28.10)
hypoventilation (28.10)
immune response (28.5)
leukemia (28.4)
leukocyte (28.1)
lymph (28.11)
nephron (28.12)
osmotic pressure (28.6)
plasma (28.1)
serum (28.3)
shock (28.6)
systolic pressure (28.7)
thrombocyte (28.1)
thrombosis (28.8)
urine (28.12)
vaccine (28.5)

Review Questions

1. Define or explain:
   a. lymph
   b. formed elements
   c. hematocrit value
   d. leukemia
   e. vaccine
   f. osmosis
   g. edema
   h. hypertension
   i. hemophilia
   j. anticoagulant
   k. thrombosis
   l. embolism
   m. methemoglobinemia
   n. jaundice
2. Distinguish between:
   a. aplastic anemia and hemolytic anemia
   b. anemia and polycythemia
   c. lymphocytes and phagocytes
   d. antibodies and antigens
   e. osmotic pressure and blood pressure
   f. systolic pressure and diastolic pressure
   g. vasodilation and vasoconstriction
   h. plasma and serum
   i. stroke and coronary thrombosis
   j. hypoventilation and hyperventilation
3. What are the three circulating fluids in the body?
4. List five functions of the blood.
5. Which body fluid cools the body? By what mechanism does this cooling occur?
6. Briefly describe the immune response.
7. Explain the processes involved when fluid is exchanged between the plasma and the cells (**a**) at the arterial end and (**b**) at the venous end of a capillary.
8. What determines the osmotic pressure in the blood vessels?
9. How are blood pressure measurements reported?
10. Cite three ways of treating hypertension.
11. Name three heme-containing compounds.
12. Name the two bile pigments formed from the porphyrin skeleton of hemoglobin.
13. What are the causes of jaundice?
14. a. Contrast the contents of arterial blood and venous blood.
    b. What are the colors of arterial blood and venous blood?
15. What is the normal pH range of the blood?
16. What is interstitial fluid?
17. What is shock?
18. List three functions of the lymphatic system.
19. Is cow's milk a "perfect" food? Is it a source of complete protein?
20. What advantage does human milk possess over cow's milk as a food for a human infant?

Problems

Blood

21. What are some of the nonprotein constituents of the blood?
22. Give the formulas for five inorganic ions present in the blood.
23. What are the functions of (**a**) albumins, (**b**) alpha and beta globulins, and (**c**) fibrinogen? Where are they synthesized?
24. What are the functions of the gamma globulins? Where are they synthesized?
25. Name the three formed elements of the blood, and give a function for each.
26. Compare erythrocytes to other cells.
    a. How are they similar? How do they differ?
    b. Where are erythrocytes formed?
    c. Where are they broken down?

27. Describe how each functions in blood clotting.
    a. prothrombin
    b. thrombin
    c. fibrinogen
    d. thromboplastin
    e. calcium ions
    f. fibrin
28. a. What is the role of vitamin K in blood coagulation?
    b. How does heparin act to prevent thrombosis?
    c. How does dicumarol act to prevent thrombosis?
    d. How do anions such as oxalates, fluorides, and citrates prevent blood coagulation?
    e. Write a structural formula for γ-carboxyglutamic acid.

Hemoglobin

29. Describe the chemical structure of hemoglobin.
30. Contrast the structural features of:
    a. heme and hemoglobin
    b. hemoglobin and myoglobin
    c. hemoglobin and oxyhemoglobin
    d. oxyhemoglobin and CO-hemoglobin
    e. oxyhemoglobin and methemoglobin
    f. hemoglobin and sickle cell hemoglobin
31. What is the oxidation state of iron in (a) hemoglobin, (b) oxyhemoglobin, (c) methemoglobin, and (d) CO-hemoglobin?
32. a. Why is carbon monoxide such a deadly poison?
    b. How is carbon monoxide poisoning treated?

Blood Buffers

33. Name the two blood buffer pairs. Why is the bicarbonate buffer pair effective in spite of the fact that the ratio of acid to anion is 1:10?
34. Illustrate, with equations, how blood buffers keep blood pH from changing when small amounts of acid or base are produced during metabolic reactions.

Urine

35. How is urine formed in the kidney?
36. List four factors that affect the volume of urine excreted.

Sweat and Tears

37. What is the difference between sensible and insensible perspiration?
38. List three inorganic and three organic constituents of sweat.
39. What are the three layers of tears?
40. What is the function of lysozyme in tears?

Additional Problems

41. List three organic components of urine.
42. What is the relationship between acidosis and oxygen transport?
43. The blood accounts for about 8.0% of the total body weight. Calculate the volume of blood your body contains (density of blood = 1.06 g/mL).
44. Which process results in the transfer of more solutes among fluids of the body: filtration or diffusion?
45. Use the following equilibrium to explain what causes (a) respiratory acidosis and (b) respiratory alkalosis:

$$H^+ + HCO_3^- \rightleftharpoons H_2O + CO_2$$

46. If the amount of colloidal protein in the interstitial fluid matched the amount in the blood plasma, would more or less fluid be likely to filter from the capillaries to the interstitial space?
47. When pneumonia blocks respiratory passageways and limits the transfer of carbon dioxide from blood to the atmosphere, blood CO_2 levels build up. In this situation, is the pH of the blood higher or lower than normal?
48. A common cause of death in young children is acidosis resulting from severe diarrhea. In severe diarrhea, large amounts of bicarbonate ion are excreted from the body. Why should this lead to acidosis?

Appendix I

The International System of Measurement

Measurement is discussed in Chapter 1. Conversions within a system of measurement and between systems are discussed in Special Topic A. Further discussion and additional tables are provided here.

The standard unit of length in the International System of Measurement is the **meter.** This distance was once meant to be 0.0000001 of the Earth's quadrant—that is, of the distance from the North Pole to the equator measured along a meridian. The quadrant, however, was difficult to measure accurately. Today, the meter is defined precisely as the distance that light travels in a vacuum during 1/299,792,458 of a second.

The primary unit of mass is the **kilogram** (1 kg = 1000 g). It is based on a standard platinum-iridium bar kept at the International Bureau of Weights and Measures. The **gram** is a more convenient unit for many chemical operations.

The derived SI unit of volume is the cubic meter. The unit more frequently employed in chemistry, however, is the cubic decimeter, which is often called a liter.

$$1 \text{ dm}^3 = 1 \text{ L} = 0.001 \text{ m}^3$$

Other SI units of length, mass, and volume are derived from these basic units by the use of prefixes (see Table 1.3, page 14).

Table I.1 Some Metric Units of Length

1 kilometer (km)	= 1000 meters (m)
1 meter (m)	= 100 centimeters (cm)
1 centimeter (cm)	= 10 millimeters (mm)
1 millimeter (mm)	= 1000 micrometers (μm)

Table I.2 Some Metric Units of Mass

1 kilogram (kg)	= 1000 grams (g)
1 gram (g)	= 1000 milligrams (mg)
1 milligram (mg)	= 1000 micrograms (μg)

Table I.3 Some Metric Units of Volume

1 liter (L)	= 1000 milliliters (mL)
1 milliliter (mL)	= 1000 microliters (μL)
1 milliliter (mL)	= 1 cubic centimeter (cm^3)

Table I.4 Some Intersystem Conversions

Length	Mass	Volume
1 mile (mi) = 1.61 kilometers (km)	1 pound (lb) = 454 grams (g)	1 U.S. quart (qt) = 0.946 liter (L)
1 yard (yd) = 0.914 meter (m)	1 ounce (oz) = 28.4 grams (g)	1 U.S. pint (pt) = 0.473 liter (L)
1 inch (in.) = 2.54 centimeters (cm)	1 pound (lb) = 0.454 kilogram (kg)	1 fluid ounce (fl oz) = 29.6 milliliters (mL)
	1 grain (gr) = 0.0648 gram (g)	1 gallon (gal) = 3.78 liters (L)
	1 carat (car) = 200 milligrams (mg)	

Table I.5 Some Conversion Units for Energy

1 calorie (cal)	= 4.184 joules (J)
1 British thermal unit (Btu)	= 1054.5 joules (J) = 252 calories (cal)
1 food "Calorie"	= 1 kilocalorie (kcal) = 1000 calories (cal) = 4184 joules (J)

Appendix II

Exponential Notation

Scientists often use numbers that are so large—or so small—that they boggle the mind. For example, light travels at 30,000,000,000 cm/s. There are 602,300,000,000,000,000,000,000 carbon atoms in 12 g of carbon. On the small side, the diameter of an atom is about 0.0000000001 m. The diameter of an atomic nucleus is about 0.000000000000001 m.

It is obviously difficult to keep track of the zeros in such quantities. Scientists find it convenient to express such numbers in *exponential form*.

A number is in *exponential form*—sometimes called *scientific notation*—when it is written as the product of a coefficient—usually with a value between 1 and 10—and a power of 10. For example,

$$4.18 \times 10^3 \quad \text{and} \quad 6.57 \times 10^{-4}$$

Expressing numbers in exponential form generally serves two purposes. (1) Very large or very small numbers can be written in a minimum of printed space and with a reduced chance of typographical error. (2) Explicit information is conveyed about the precision of measurements: the number of significant figures (Appendix III) in a measured quantity is stated unambiguously.

In the expression 10^n, n is the exponent of 10. The number 10 is said to be raised to the *nth power.* If n is a *positive* quantity, 10^n has a value *greater than 1.* If n is a *negative* quantity, 10^n has a value *less than 1.* We are interested mainly in cases where n is an integer. For example:

Positive Powers of 10

$10^0 = 1$

$10^1 = 10$

$10^2 = 10 \times 10 = 100$

$10^3 = 10 \times 10 \times 10 = 1000$

and so on

The power of 10 determines the number of zeros that follow the digit "1."

Negative Powers of 10

$10^0 = 1$

$10^{-1} = 1/10 = 0.1$

$10^{-2} = 1/(10 \times 10) = 0.01$

$10^{-3} = 1/(10 \times 10 \times 10) = 0.001$

and so on

The power of 10 determines the number of places to the right of the decimal point where the digit "1" appears.

To express 612,000 in exponential form,

$$612{,}000 = 6.12 \times 100{,}000 = 6.12 \times 10^5$$

To express 0.0000505 in exponential form,

$$0.0000505 = 5.05 \times 0.00001 = 5.05 \times 10^{-5}$$

A more direct approach to converting a number to exponential form is:

- Count the number of places the decimal point must be moved to produce a coefficient having a value between 1 and 10.
- The number of places counted then becomes the power of 10.
- The power of 10 is *positive* if the decimal point is moved to the left.

$$6\underbrace{1\,2\,0\,0\,0}_{5\;4\;3\;2\;1}. = 6.12 \times 10^5$$

- The power of 10 is *negative* if the decimal point is moved to the right.

$$0.\underbrace{0\,0\,0\,0\,5}_{1\;2\;3\;4\;5}0\,5 = 5.05 \times 10^{-5}$$

To convert a number from exponential form to conventional form, move the decimal point in the opposite direction. That is,

$$3.75 \times 10^6 = 3\underbrace{.750000}_{1\,2\,3\,4\,5\,6}$$

$$7.91 \times 10^{-5} = 0\underbrace{.00007}_{5\,4\,3\,2\,1}91$$

Most electronic calculators easily handle exponential numbers. A typical procedure is to enter the number, followed by the EXP key. The keystrokes required for the number 2.85×10^7 are

[2] [.] [8] [5] [EXP] [7]

and the result displayed is

[2.85^{07}]

For the number 1.67×10^{-5}, the key strokes are

[1] [.] [6] [7] [EXP] [5] [±]

and the result displayed is

[1.67^{-05}]

The keystrokes required on your calculator may be different from those shown here. Check the specific instructions in the manual supplied with the calculator.

Addition and Subtraction

To add or subtract numbers in exponential notation, it is necessary to express each quantity as *the same power of 10*. In calculations, this treats the power of 10 in the same way as a unit—it is simply "carried along." In the following calculation, each quantity is expressed in terms of the power 10^{-3}.

$$(3.22 \times 10^{-3}) + (7.3 \times 10^{-4}) + (4.8 \times 10^{-4})$$

$$= (3.22 \times 10^{-3}) + (0.73 \times 10^{-3}) + (0.48 \times 10^{-3})$$

$$= (3.22 + 0.73 + 0.48) \times 10^{-3}$$

$$= 4.43 \times 10^{-3}$$

Multiplication

To multiply numbers expressed in exponential form, *multiply* all coefficients to obtain the coefficient of the result, and *add* all exponents to obtain the power of 10 in the result.

$$0.0803 \times 0.0077 \times 455 = (8.03 \times 10^{-2}) \times (7.7 \times 10^{-3}) \times (4.55 \times 10^2)$$

$$= (8.03 \times 7.7 \times 4.55) \times 10^{(-2-3+2)}$$

$$= (2.8 \times 10^2) \times 10^{-3} = 2.8 \times 10^{-1}$$

Generally, an electronic calculator performs these operations automatically, and no intermediate results need be recorded.

Division

To divide two numbers in exponential form, *divide* the coefficients to obtain the coefficient of the result, and *subtract* the exponent in the denominator from the exponent in the numerator to obtain the power of 10. In the example below, multiplication and division are combined. First, the rule for multiplication is applied to the numerator and to the denominator, and then the rule for division is used.

$$\frac{0.015 \times 0.0088 \times 822}{0.092 \times 0.48} = \frac{(1.5 \times 10^{-2})(8.8 \times 10^{-3})(8.22 \times 10^{2})}{(9.2 \times 10^{-2})(4.8 \times 10^{-1})}$$

$$= \frac{1.1 \times 10^{-1}}{4.4 \times 10^{-2}} = 0.25 \times 10^{-1-(-2)} = 0.25 \times 10^{1}$$

$$= 2.5 \times 10^{-1} \times 10^{1} = 2.5 \times 10^{0} = 2.5$$

Raising a Number to a Power

To raise an exponential number to a given power, raise the coefficient to that power, and multiply the exponent by that power. For example, to raise 0.0066 to the *third* power, that is, to *cube* the number,

$$(0.0066)^3 = (6.6 \times 10^{-3})^3 = (6.6)^3 \times 10^{3 \times (-3)}$$

$$= (2.9 \times 10^2) \times 10^{-9} = 2.9 \times 10^{-7}$$

Extracting the Root of an Exponential Number

To extract the root of an exponential number means to raise the number to a *fractional* power—one-half power for a square root, one-third power for a cube root, and so on. Most calculators have keys designed for extracting square roots and cube roots. Thus, to extract the square root of 1.57×10^{-5}, enter the number 1.57×10^{-5} into an electronic calculator, and use the "$\sqrt{\ }$" key.

$$\sqrt{1.57 \times 10^{-5}} = 3.96 \times 10^{-3}$$

To extract the cube root of 3.18×10^{10}, enter the number 3.18×10^{10} into an electronic calculator, and use the "$\sqrt[3]{\ }$" key

$$\sqrt[3]{3.18 \times 10^{10}} = 3.17 \times 10^{3}$$

Problems

1. Express each of the following numbers in exponential form.
   - **a.** 0.000017 **b.** 19,000,000
   - **c.** 0.0034 **d.** 96,500
2. Express each of the following numbers in exponential form.
   - **a.** 4,500,000 **b.** 0.000108
   - **c.** 0.0341 **d.** 406,000
3. Carry out the following operations. Express the answers in exponential form.
   - **a.** $(4.5 \times 10^{13})(1.9 \times 10^{-5})$ **b.** $\dfrac{(9.3 \times 10^{9})}{(3.7 \times 10^{-7})}$
   - **c.** $\dfrac{(4.3 \times 10^{-7})}{(7.6 \times 10^{22})}$ **d.** $\sqrt{1.78 \times 10^{-5}}$
4. Carry out the following operations. Express the answers in exponential form.
   - **a.** $\dfrac{(2.1 \times 10^{5})}{(9.8 \times 10^{7})}$
   - **b.** $(6.2 \times 10^{-5})(4.1^{-5} \times 10^{-12})$
   - **c.** $(2.1 \times 10^{-6})^2$
   - **d.** $\dfrac{(4.6 \times 10^{-12})}{(2.1 \times 10^{3})}$

Appendix III

Significant Figures

Counting is usually exact: we can count exactly 18 students in a room. Measurements, on the other hand, are subject to error. One source of error is the measuring instruments themselves. A thermometer, for example, may consistently yield a result that is 2 °C too low. Other errors may result from the experimenter's lack of skill or care in using measuring instruments.

Suppose each of five students is asked to measure the length of a room with a meter stick. Table III.1 presents a possible set of results. The **precision** of a set of measurements refers to the degree of reproducibility among the set. The precision is good or high if each of the measurements is close to the average of the series. The precision is poor or low if there is a wide deviation from the average value. The precision of the data in Table III.1 is good; each measurement is within 0.005 m of the average value.

Table III.1 A Set of Measurements of the Length of a Room

Student	Length, m
1	14.157
2	14.150
3	14.153
4	14.159
5	14.156
Average:	14.155

The **accuracy** of a set of measurements refers to the closeness of the average of the set to the "correct" or most probable value. Measurements of high precision are more likely to be accurate than are those of poor precision, but even highly precise measurements are sometimes inaccurate. If the meter sticks used to obtain the data in Table III.1 were actually 1005 mm long, but still carried 1000-mm markings, the accuracy of the measurements would be rather poor, even though the precision remained high.

Look again at Table III.1. Notice that the five students agree on the first four digits (14.15); they differ only in the fifth digit. The last digit in a scientific measurement is usually regarded as uncertain; and all digits known with certainty, plus one of uncertain value, are called **significant figures.** The measurements in Table III.1 have five significant figures. In other words, we are certain that the length of the room is between 14.15 m and 14.16 m. Our best estimate, including the fifth uncertain digit, is the average value: 14.155 m.

It is rather easy to establish that 14.155 has five significant figures; we simply count the number of digits. In any measurement that is properly reported, all *nonzero* digits are significant. *Zeros* present problems because they can be used in two ways: to indicate a measured value or to position a decimal point.

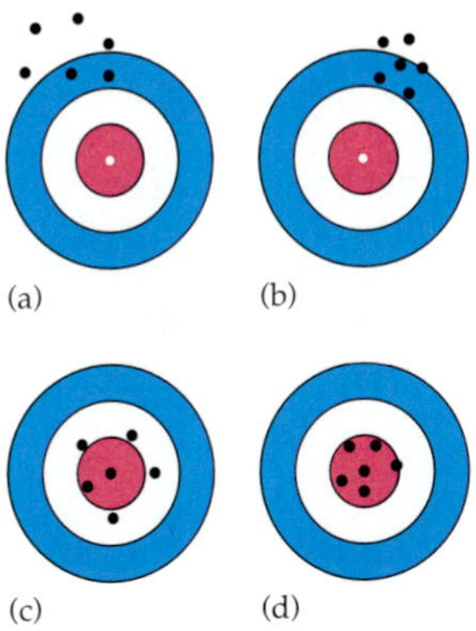

Comparing precision and accuracy. (a) Measurements of *low* accuracy and *low* precision are scattered and off-center; (b) those with a *low* accuracy and *high* precision form a tight off-center cluster; (c) those with *high* accuracy and *low* precision are evenly distributed but distant from the center; and (d) those with *high* accuracy and *high* precision are bunched in the center of the target.

- Zeros between two other digits are significant.
  Examples: 1107 (four significant figures); 50.002 (five).
- Zeros that precede the first nonzero digit are *not* significant.
  Examples: 0.000163 (three significant figures); 0.06801 (four).
- Zeros at the end of a number are significant if they are to the *right* of the decimal point.
  Examples: 0.2000 (four significant figures); 0.050120 (five).
- Zeros at the end of a number may or may not be significant if the number is written *without* a decimal point.
  Example: 400
  We do not know whether this number was measured to the nearest unit, ten, or hundred. To avoid this confusion, we use exponential notation (see Appendix II). In exponential notation, 400 would be recorded as 4×10^2 or 4.0×10^2 or 4.00×10^2 to indicate one, two, or three significant figures, respectively.

The concept of significant figures applies only to *measurements*—quantities subject to error. It does not apply to a quantity that is *counted,* such as the six faces of a cube or 12 items in a dozen. It also does not apply to *defined* quantities, such as 1 km = 1000 m. In this context, the numbers 6, 12, and 1000 are not limited to one, two, and three significant figures, respectively. In effect, each has an unlimited number of significant figures (6.000 . . . 12.000 . . . 1000.000 . . .) or, more correctly, each is an *exact* value.

III.1 Multiplication and Division

Not only must we recognize the relationship between the precision of a measurement and the number of significant figures used to express it, but we also have to obey this fundamental rule: *A calculated quantity can be no more precise than the data used in the calculation.* In multiplication and division, this leads to the general rule that answers should have no more significant figures than the factor with the *fewest* significant figures.

Example III.1

What is the area of a room, in square meters, that is 12.42 m long and 4.81 m wide?

Solution

The length of the room is expressed to four significant figures and the width to three. By whatever method we carry out the multiplication, we are limited to *three* significant figures in our answer.

$$12.42 \text{ m} \times 4.81 \text{ m} = 59.7 \text{ m}^2$$

The "answer" on this calculator is 59.7402, suggesting six significant figures. However, the rules for significant figures in multiplication tells us that in Example III.1 there can only be three. The answer is therefore 59.7 m².

Example III.2

For a laboratory experiment, a teacher wants to divide 453.6 g (1 lb) of sulfur among the 21 members of her class. What mass of sulfur, in grams, will each student receive?

Solution

Here we need to recognize that the number "21" is a counted number. It is not subject to significant figure rules. The answer should carry *four* significant figures, the same as in 453.6 g.

$$\frac{453.6 \text{ g}}{21} = 21.60 \text{ g}$$

Practice Exercise

Perform the indicated operations and give answers with the proper numbers of significant figures.

a. $73 \text{ m} \times 1.340 \text{ m} \times 0.41 \text{ m}$

b. $0.137 \text{ cm} \times 1.43 \text{ cm}$

c. $3.132 \text{ cm} \times 5.4 \text{ cm} \times 5.4 \text{ cm}$

d. $\dfrac{51.79 \text{ m}}{4.6 \text{ s}}$

e. $\dfrac{456.1 \text{ mi}}{7.13 \text{ h}}$

f. $\dfrac{305.5 \text{ mi}}{14.7 \text{ gal}}$

III.2 Addition and Subtraction

In addition or subtraction, the concern is not with the number of significant figures but with the number of digits to the right of the decimal point. If the quantities being added or subtracted have different numbers of digits to the right of the decimal point, find the one with the *fewest* such digits. The result of the addition or subtraction should contain the same number of digits to the right of *its* decimal point. The idea is this: if you are adding several lengths measured to the nearest *milli*meter and one measured only to the nearest *centi*meter, the total length cannot be stated to the nearest *milli*meter, no matter how precise the millimeter measurements are.

In addition and subtraction, it is often necessary to round numbers. The usual procedure is to round only the final result, using the following rules for rounding:

- If the leftmost digit dropped is 0, 1, 2, 3, or 4, leave the final digit *unchanged.*
  Example: 369.448 rounds to 369.4 if we want one decimal place.
- If the leftmost digit dropped is 5, 6, 7, 8, or 9, *increase* the final digit by *1.*
  Example: 538.768 rounds to 538.77 if we want two decimal places. (538.768 rounds to 538.8 if we want one decimal place.)

We apply these rules in Example III.3, and illustrate another point as well: in a calculation involving several steps, round only the final result. (This is quite easy to do using a hand-held calculator.)

Example III.3

Perform the following calculation and round the answer to the correct number of digits.

$$49.146 + 72.13 - 9.1434 = ?$$

Solution

In this calculation, we must add two numbers and, from their sum, subtract a third. We do this in two ways below. In both cases, we express the answer to two decimal places, the same number found in "72.13."

(a)

$$\begin{array}{r} 49.146 \\ +72.13 \\ \hline 121.276 \end{array} = \begin{array}{r} 121.28 \\ -9.1434 \\ \hline 112.1366 \end{array} = 112.14$$

(b)

$$\begin{array}{r} 49.146 \\ +72.13 \\ \hline 121.276 \\ -9.1434 \\ \hline 112.1326 \end{array} = 112.13$$

The preferred method is (b), where we do *not* round the intermediate result: 121.276. Note that if we use a hand-held calculator, there is no need to write down or otherwise take note of the intermediate result.

Practice Exercise

Perform the indicated operations and give answers with the proper numbers of digits.

a. 48.2 m + 3.82 m + 48.4394 m
b. 148 g + 2.39 g + 0.0124 g
c. 451 g − 15.46 g − 20.3 g
d. 15.436 L + 5.3 L − 6.24 L − 8.177 L

Appendix IV
Glossary

Absolute alcohol is 100% ethyl alcohol.

The **absolute scale** of temperatures has as the zero point the lowest temperature possible, or absolute zero.

An enzyme with **absolute specificity** acts on only one particular substrate.

An **acetal** is the product of a hemiacetal–alcohol reaction; it has two OR groups attached to the same carbon atom.

An **achiral** molecule is superimposable on its mirror image; it is *not* chiral.

An **acid** is (1) a compound that produces hydronium ions, H_3O^+, in water solution (Arrhenius theory) or (2) a proton donor (Brønsted–Lowry theory).

An **acid anhydride** is a substance that reacts with water to produce an acid; a nonmetal oxide.

The **acid ionization constant (K_a)** is the equilibrium constant describing equilibrium in the reversible ionization of a weak acid.

Acidosis is the condition that results when the pH of the blood falls below 7.35; oxygen transport is hindered.

Activation energy. *See* **energy of activation.**

The **active site** of an enzyme is that portion of the enzyme that binds to the substrate.

The **activity series** is a listing of metals in order of their abilities to displace one another from solutions of their ions or to displace H^+ as H_2 from acidic solutions.

Actomyosin is the contractile protein of which muscles are made; contains actin and myosin.

In **addition polymerization,** monomers add to one another to produce a polymeric product that contains all the atoms of the starting monomers. The monomers usually contain double bonds.

In an **addition reaction,** substituent groups join to hydrocarbon molecules at points of unsaturation—double or triple bonds. This type of reaction is typical of alkenes, alkynes, and carbonyl compounds.

Adipose tissue is connective tissue where fat is stored.

An **aerobic** process is one that requires the presence of oxygen.

An **agonist** is a molecule that fits and activates a specific receptor.

Albumins are globular proteins that are abundant in blood plasma.

An **alcohol** is an organic compound containing the —OH group on an aliphatic carbon atom.

An **aldehyde** is an organic compound with a *carbonyl* functional group that has a hydrogen atom attached and where the other group on the carbonyl carbon atom may be a hydrocarbon group or a second hydrogen atom.

An **aldose** is a sugar with an aldehyde group.

An **aliphatic** compound is an open-chain compound or a ring compound that has no aromatic groups.

An **alkali metal** is an element in Group 1A of the periodic table.

An **alkaline earth metal** is an element in Group 2A of the periodic table.

An **alkaloid** is a nitrogen-containing organic compound obtained from plants that has physiological properties.

Alkalosis is a physiological condition in which the pH of the blood rises above 7.45.

An **alkane** is a hydrocarbon with only single bonds; a saturated hydrocarbon.

An **alkene** is a hydrocarbon containing one or more carbon-to-carbon double bonds.

An **alkyl group** is a hydrocarbon group derived from an alkane by removal of a hydrogen atom.

An **alkyl halide** is a compound resulting from the replacement of a hydrogen atom of an alkane with a halogen atom.

An **alkyne** is a hydrocarbon whose molecules contain carbon-to-carbon triple bonds.

Allergen. *See* **Antigen.**

Allotropes are two or more forms of an element that differ in basic molecular structure. Diamond and graphite are allotropes of carbon.

The **alpha (α) form** of a cyclic monosaccharide has the OH on the hemiacetal carbon pointing down when the structure is drawn in the usual flat projection.

Alpha particles are identical to helium nuclei and are emitted by the nuclei of certain radioactive atoms as they undergo decay.

An **amide** is an organic compound with a *carbonyl* functional group that has an amino group or a substituted amino group attached.

An **amide linkage** is the bond between a carbonyl carbon atom and a nitrogen.

An **amine** is a compound that contains the elements carbon, hydrogen, and nitrogen; can be viewed as derived from ammonia by replacement of one, two, or three of the hydrogens by hydrocarbon groups.

The **amino acid pool** comprises the circulating amino acids obtained from both exogenous and endogenous sources.

Amylopectin is a form of starch with branched chains of glucose units.

Amylose is a form of starch with the glucose units joined in a continuous chain.

An **anabolic steroid** is a drug that aids in the building (anabolism) of body proteins and thus of muscle tissue.

Anabolism is the process of building up the molecules of living systems.

An **anaerobic process** is one that does not require the presence of oxygen.

An **analgesic** is a pain reliever.

An **androgen** is a male sex hormone.

Anemia is a condition characterized by abnormally low levels of erythrocytes.

An **anesthetic** is a substance that produces insensitivity to pain. *See also* **general anesthetic** and **local anesthetic.**

Anhydrous means "without water."

An **anion** is a negatively charged ion.

The **anode** is the electrode at which oxidation occurs.

An **antacid** is any basic substance used to neutralize stomach acid.

An **antagonist** is a drug that blocks the action of an agonist by blocking the receptor(s).

An **antibiotic** is a soluble substance, produced by a mold or bacterium, that is toxic to other microorganisms.

An **antibody** is a protein that binds to and destroys foreign substances.

An **anticoagulant** is a substance that inhibits the clotting of blood.

An **anticodon** is a sequence of three adjacent nucleotides on a tRNA molecule that is complementary to a codon on mRNA.

An **antidiuretic** is a water-conserving substance.

An **antigen** is a foreign substance that triggers the formation of antibodies.

An **antiinflammatory** substance inhibits inflammation.

An **antimetabolite** inhibits a significant metabolic reaction.

An **antineoplastic drug** inhibits the growth of cancer cells.

An **antioxidant** is a reducing agent used to prevent oxidation of foods and other products.

An **antipyretic** is a fever-reducing substance.

An **antiseptic** is a compound applied to living tissue to kill or prevent the growth of microorganisms.

An **apoenzyme** is the protein part of an enzyme.

An **aqueous** solution has water as the solvent.

An **aromatic compound** is an organic compound with a benzene-like structure. Resonance theory is used to describe the electronic structure.

An **artificial transformation** is the conversion of one element into another by artificial means.

An **aryl group** is a group derived from an aromatic hydrocarbon by removal of a hydrogen atom.

The **atmosphere** is the gaseous envelope surrounding the Earth (or other planet).

An **atmosphere** of pressure is equal to 760 mm Hg.

An **atom** is the smallest characteristic particle of an element.

Atomic mass. *See* **atomic weight.**

An **atomic mass unit (u)** is *exactly* one-twelfth the mass of an atom of carbon–12. The masses of the fundamental particles—electrons, protons, and neutrons—and of individual atoms are often expressed in these units.

The **atomic number (Z)** is the number of protons in the nucleus of an atom of an element.

The **atomic weight** of an element is the weighted average of the masses of the atoms of the naturally occurring isotopes of the element.

In an **autocatalytic reaction,** one of the *products* of the reaction acts as a catalyst for the reaction.

In an **autoimmune disease,** the immune system attacks its own body tissues.

Avogadro's hypothesis states that at a fixed temperature and pressure the volume of a gas is directly proportional to the amount of gas.

Avogadro's number is the number of particles of a substance in a mole of the substance; 6.02×10^{23}.

An **azeotrope** is a mixture of two or more liquids that boils at a constant temperature.

The B **complex** is the set of B vitamins; see Section J.6 for the list.

A **barbiturate** is a depressant anticonvulsant drug.

A **barometer** is an instrument used to measure atmospheric pressure.

A **base** is (1) a compound that produces hydroxide ions, OH^-, in water solution (Arrhenius theory) or (2) a proton acceptor (Brønsted–Lowry theory).

The **base ionization constant (K_b)** is the equilibrium constant describing equilibrium in the reversible ionization of a weak base.

A **basic anhydride** is a substance that reacts with water to produce a basic solution; a metal oxide.

A **benign tumor** is an abnormal growth of new tissue that grows slowly and does not spread to other tissues.

The **beta (β) form** of a cyclic monosaccharide has the OH on the hemiacetal carbon pointing up when the structure is drawn in the usual flat projection.

Beta particles are identical to electrons and are emitted by the nuclei of certain radioactive atoms as they undergo decay.

A **bilayer** is a layer two molecules thick. Phospholipids form bilayers with the polar heads sticking out in water and the nonpolar tails on the inside.

Bile is an alkaline fluid secreted by the liver that aids in the emulsification and digestion of fats.

Biochemistry is the study of all the substances and processes in living organisms.

Blood doping is a technique in which erythrocytes are removed and stored while the person's body makes more. The erythrocytes are subsequently returned to the bloodstream, increasing the person's capacity to transport oxygen to the tissues.

Blood pressure is the force exerted by the blood against the inner walls of the arteries.

A **body-centered cubic (bcc)** crystal structure has as its unit cell a cube with a structural unit at each corner and one at the center. *See also* **unit cell.**

The **boiling point** is the temperature at which the vapor pressure of a liquid becomes equal to atmospheric pressure.

A **bonding pair (BP)** is a pair of electrons shared by two atoms in a molecule.

Boyle's law states that for a given mass of a gas at constant temperature, volume varies inversely with pressure.

Brown fat contains many mitochondria that synthesize no ATP but instead release energy as heat.

A **buffer solution** is a solution containing a weak acid and its salt or a weak base and its salt. Small quantities of added acid are neutralized by one buffer component and small quantities of added base by the other. As a result, the solution pH is maintained nearly constant.

A **calorie (cal)** is the amount of energy needed to raise the temperature of 1 g of water by 1 °C (more precisely, from 14.5 to 15.5 °C). 1 cal = 4.184 J.

The food **Calorie** is 1000 calories (or 1 kcal); it is used to measure the energy contents of foods.

Cancer. *See* **malignant tumor.**

A **carbohydrate** is any of a group of compounds composed of carbon, hydrogen, and oxygen, including starches, sugars, and celluloses.

The **carboxyl group, —COOH,** is the functional group of the organic acids.

A **carboxylic acid** is an organic compound that contains the —COOH functional group.

Catabolism is the process of breaking down molecules to provide energy in living systems.

A **catalyst** is a substance that increases the rate of a reaction without itself being consumed in the reaction.

Catalytic reforming is a process of converting aliphatic hydrocarbons with low octane numbers into aromatic compounds with higher octane numbers.

A **cathode** is the electrode at which reduction occurs.

A **cathode ray** is a beam of electrons that travels from the cathode to the anode when an electric discharge is passed through an evacuated tube.

A **cation** is a positively charged ion.

A **cell nucleus** is a membrane-enclosed structure within a plant or animal cell that contains the genetic material.

The **Celsius** temperature scale defines the freezing point of water as 0 °C and the boiling point of water as 100 °C.

The **central nervous system** is the brain and spinal cord.

A **cephalin** is a phosphatide that contains ethanolamine as the amino alcohol.

A **cerebroside** is a glycolipid found in nerve tissue; it has a sugar unit (usually galactose or glucose), a fatty acid unit, and a sphingosine unit.

Charles's law states that for a given mass of gas at constant pressure, the volume varies directly with the temperature (on the absolute scale).

A **chemical bond** is a force that holds atoms together in compounds.

A **chemical equation** is a description of a chemical reaction that uses symbols and formulas to represent the elements and compounds involved in the reaction.

A **chemical property** describes how one substance reacts with other substances to produce new substances with altered compositions.

A **chemical symbol** is a representation of an element, made up of one or two letters derived from the English name of the element (or, sometimes, from the Latin name of the element or one of its compounds).

Chemistry is a study of the composition, structure, and properties of matter and of the changes that matter undergoes.

Chemotherapy is the use of chemicals to control or cure diseases.

A **chiral center** is an atom in a molecule that is attached to four different groups.

Chirality means "handedness"; a chiral molecule is *not* superimposable on its mirror image.

A **chlorofluorocarbon (CFC)** is a carbon compound that contains chlorine and fluorine.

A **chromosome** is a threadlike body in the cell nucleus that contains the hereditary materials.

The term **cis** is used to describe isomers in which two substituent groups are attached on the same side of a double bond in an organic molecule, or along the same edge of a square planar or octahedral complex ion. *See also* **geometric isomerism.**

A **codon** is a sequence of three adjacent nucleotides in mRNA that codes for one amino acid.

A **coenzyme** is an organic cofactor necessary for the function of an enzyme.

A **cofactor** is a substance necessary for the action of an enzyme. It may be a metal ion or a coenzyme.

A **colligative property** is a physical property of a solution that depends on the concentration of solute in the solution but not on the identity of the solute.

A **colloid** is a dispersion in which the dispersed matter has dimensions in the range from about 1 nm to 1000 nm.

The **combined gas law** is a combination of the three simple gas laws into one.

Combustion is the oxidation of a fuel by oxygen in an exothermic reaction; burning.

The **common ion effect** refers to the ability of ions from a strong electrolyte to repress the ionization of a weak acid or weak base.

A **competitive inhibitor** is one that blocks an enzymatic process by reversibly interacting with the enzyme at the active site.

Complementary bases are the base pairs adenine–thymine and guanine–cytosine.

A **complete protein** is one that supplies all the essential amino acids in the quantities needed for the growth and repair of body tissues.

A **compound** is a substance made up of atoms of two or more elements, with the different atoms joined in fixed proportions.

A **concentrated** solution contains a relatively large amount of solute in a given quantity of solvent.

Condensation is the conversion of a gas (vapor) to a liquid.

A **condensed structural formula** is an organic chemical formula that shows the atoms of hydrogen right next to the carbon atoms to which they are attached.

A **continuous spectrum** is a spectrum in which there is a continuous variation from one color to another.

A **conversion factor** is a numerical factor by which we multiply a quantity expressed in a certain unit to convert it to a quantity in another unit.

In the **Cori cycle,** glucose is oxidized to lactic acid in peripheral tissues, the lactic acid is transported to the liver where it is converted back to glucose, and the glucose is transported back to the peripheral tissues.

A **covalent bond** is a bond formed by a shared pair of electrons between atoms.

A **crystal** is a solid having plane surfaces, sharp edges, and a regular geometric shape.

In a **crystal lattice,** the fundamental units making up the crystal—atoms, ions, or molecules—are assembled in a regular, repeating manner extending in three dimensions through the crystal.

A **cytochrome** is an iron-containing globular protein that contains a heme group.

The **cytoplasm** is the portion of a plant or animal cell inside the cell membrane and external to the nucleus.

A **D sugar** has the OH group to the right on the chiral carbon farthest from the carbonyl group when the structure is drawn in a Fischer projection.

A **dalton** is a unit used by some biochemists that is the same as the atomic mass unit.

Dalton's law of partial pressures states that in a mixture of gases each gas expands to fill the container and exerts its own pressure, called a partial pressure, and that the total pressure of the mixture is the sum of the partial pressures.

A **deliquescent** substance is one that takes on water of hydration, and then continues to absorb water until the hydrated solid dissolves.

Denaturation is any process that alters the properties of a protein; it is accomplished by adding certain chemicals or by subjecting the protein to heat or radiation.

Denatured alcohol is ethanol to which some toxic or noxious substance has been added to render it unsuitable for drinking.

The **density (*d*)** of a sample of matter is its mass per unit volume—that is, the mass of the sample divided by its volume.

Deoxyribonucleic acid (DNA) is the type of nucleic acid found primarily in the nuclei of cells.

A **detergent** is a cleaning agent that has a water-soluble head and an oil-soluble tail.

A **dextrorotatory** substance rotates the plane of polarized light to the right.

Diabetes mellitus is a disease characterized by an abnormally high level of sugar in the blood.

Dialysis is a process that separates solvent and small molecules and ions from large ones by allowing the solvent and smaller particles to pass through a membrane that blocks the larger ones.

Diastereomers are stereoisomers that are not enantiomers.

Diastolic pressure is the minimum force exerted by the blood against the inner walls of the arteries between heartbeats.

Diffusion is the process by which one substance mixes with one or more other substances as a result of the movement of molecules.

Digestion is the hydrolytic process whereby food molecules are broken down into simpler chemical units.

A **dihydric alcohol** is one with two OH functional groups; a glycol.

A **dilute** solution is one that contains relatively little solute in a large quantity of solvent.

Dilution is a process of producing a more dilute solution from a more concentrated one by the addition of an appropriate quantity of solvent.

A **dimer** is a molecule formed from two monomer units.

A **dipeptide** is a compound composed of two amino acid units joined through a peptide bond.

A **dipole** is a molecule that has a positive end and a negative end.

Dipole forces are the attractive forces that exist among polar covalent substances.

A **diprotic acid** is an acid that can donate two protons per molecule.

A **disaccharide** is a carbohydrate 1 mol of which can be hydrolyzed to 2 mol of monosaccharide(s).

A **dispersion force** is an attractive force between an instantaneous dipole and an induced dipole.

Dissociation is the separation of the ions of an ionic substance as it dissolves in water.

A **dissociative anesthetic** is one that causes gross personality disorders, including hallucinations similar to those in near-death experiences.

Distillation is the boiling off of a volatile compound such as alcohol or water and then condensing its vapor. Solids and high-boiling compounds are left behind.

A **disulfide linkage** is a covalent linkage through two sulfur atoms.

A **diuretic** is a substance that increases the body's output of urine.

A **double bond** is a covalent linkage in which two atoms share *two* pairs of electrons between them.

A **double helix** is a two-stranded helical structure in which the two strands wrap around each other.

A **drug** is any substance that affects an individual in such a way as to bring about physiological, emotional, or behavioral change.

Dynamic equilibrium occurs when two opposing processes occur at the same rate, with the result that no net change occurs.

Eczema is an inflammatory skin disease characterized by scaly and crusty skin.

An **efflorescent** substance is a hydrate that gives up water to the atmosphere.

Electric current is the flow of electrons through a conductor or of ions through a solution or melt.

An **electrochemical cell** is a combination of two compartments in which metal electrodes are joined by a wire, and the solutions are brought into contact through a salt bridge or by other means.

An **electrode** is a metal strip or carbon rod dipped into a solution or molten compound to carry electricity to or from the liquid. *See also* **anode** and **cathode.**

Electrolysis is the decomposition of a compound by passing electricity through an ionic solution or a molten salt.

An **electrolyte** is a compound that, when melted or taken into solution, conducts an electric current.

The **electron** is a particle carrying the fundamental unit of negative electric charge. Electrons have a mass of 0.0005486 u and are found outside the nuclei of atoms.

Electron capture (E.C.) is a type of radioactive decay in which a nucleus absorbs an electron from the first or second energy level.

The **electron configuration** of an atom describes the distribution of electrons among the atomic orbitals in the atom.

Electron dot structure. *See* **Lewis structure.**

The **electronegativity** of an element is a measure of the tendency of its atoms in molecules to attract electrons to themselves.

Electron transport chain. *See* **respiratory chain.**

Electrophoresis is the process of separating a mixture by application of a buffer solution and an electric current.

An **element** is a substance composed of a single type of atom. Elements are the fundamental substances from which all material things are made.

The **Embden–Meyerhof pathway** is the sequence of metabolic reactions by which glucose is converted to either ethanol or lactic acid.

An **emulsifying agent** is a material that can stabilize an emulsion of two otherwise insoluble substances.

An **emulsion** is a colloid that has particles of liquid such as a fat or oil dispersed in water.

Enantiomers are mirror-image isomers; an enantiomer is not superimposable on its mirror image.

An **endorphin** is a naturally occurring peptide that bonds to the same receptor site as an opiate drug.

Endothermic describes a process or reaction in which, in a nonisolated system, heat is absorbed from the surroundings.

The **energy of activation** is the minimum energy needed to initiate a reaction.

An **energy level** is the state of an atom determined by the locations of its electrons among the various principal shells and subshells.

An **enkephalin** is a compound composed of a peptide chain of five amino acid units; a morphinelike substance produced by the body.

Enzymes are protein molecules that catalyze chemical reactions in living organisms.

An **enzyme–substrate complex** is the transitory complex formed when the substrate binds to the enzyme.

Equilibrium is a condition that is reached when two opposing processes occur at equal rates. As a result, the concentrations of the reacting species remain constant with time.

The **equilibrium constant (*K*)** is the constant that relates the concentrations of the species in an equilibrium.

The **equilibrium constant expression** is a particular ratio of concentrations of products to reactants in a chemical reaction at equilibrium. The expression has a constant value at a given temperature.

The **equivalence point** of a titration is the point at which two reactants have been introduced into a reaction mixture in their stoichiometric proportions.

Erythrocytes are red blood cells; they contain the hemoglobin that transports oxygen to the tissues.

An **essential amino acid** is an amino acid that is not produced in the body but must be included in the diet.

An **essential fatty acid** is a fatty acid that is not produced in the body but must be included in the diet.

An **ester** is a compound derived from a carboxylic acid and an alcohol. The —OH of the acid is replaced by an —OR group.

An **estrogen** is a female sex hormone that regulates the menstrual cycle.

An **ether** is an organic compound that has an oxygen atom between two hydrocarbon groups.

An **excited state** of an atom is a state in which one or more electrons have been promoted to an energy level higher than that in the ground state. *See also* **ground state.**

Exothermic describes a process or reaction in which, in a non-isolated system, heat is given off to the surroundings.

The term **expanded octet** refers to a situation in which the central atom in a Lewis structure is able to accommodate more than the usual octet of electrons in its valence shell. Expanded octets are encountered in molecules and polyatomic ions in which the central atom is a nonmetal of the third period or beyond.

A **face-centered cubic (fcc)** crystal structure has as its unit cell a cube with a structural unit at each corner and at the center of each face. *See also* **unit cell.**

The **Fahrenheit scale** is a temperature scale that defines the freezing point of water as 32 °F and the boiling point of water as 212 °F.

A **fat** is a triglyceride that is solid at room temperature.

A **fat cell** is found in adipose tissue; the cell contains large droplets of triglycerides.

A **fat depot** is a location in the body that has large amounts of adipose tissue.

A **fat-soluble vitamin** is a vitamin that dissolves in the fatty tissue of the body and is stored for future use; vitamins A, D, E, and K are fat-soluble vitamins.

The **fatty acid spiral** is a sequence of reactions by which two-carbon atom units are cleaved from the carboxyl end of a fatty acyl CoA molecule.

Fermentation is the process by which yeast produces alcohol.

Fibrinogen is the protein that is converted to fibrin by the enzyme thrombin in the clotting of blood.

A **fibrous protein** is a protein that is highly insoluble in water and that provides the structural elements for many animal tissues.

A **Fischer projection** is a representation of a molecule with one or more chiral centers based on the assumption that horizontal bonds project toward the viewer and vertical bonds project away from the viewer.

Formed elements are the cells in the blood: erythrocytes, leukocytes, and thrombocytes.

A **formula unit** is the simplest combination of atoms or ions consistent with the formula of a compound.

Formula weight is the mass of a formula unit relative to that of a carbon-12 atom; it is the sum of the weights of the atoms or ions represented by the formula.

A **free radical** is a highly reactive atom or molecular fragment that contains one or more unpaired electrons.

A **fuel cell** is a device in which chemical reactions are used to produce electricity directly from fuels and oxygen.

A **functional group** is an atom or group of atoms in an organic molecule that confers characteristic properties to the molecule as a whole.

A **gamma (γ) ray** is a form of electromagnetic radiation emitted by the nuclei of certain radioactive atoms as they undergo decay. Gamma rays are similar to X-rays but have higher energy and are more penetrating.

A **ganglioside** is a glycosphingolipid that contains several monosaccharide units and/or substituted sugars.

A **gas** is a substance that maintains neither shape nor volume.

A **gene** is a segment of a DNA molecule that codes for the biosynthesis of one polypeptide chain.

A **general anesthetic** is a substance that produces unconsciousness and insensitivity to pain.

A **genetic disease** is caused by an inherited mutation that causes metabolic abnormalities.

Geometric isomers (cis–trans isomers) have different configurations because of the presence of a rigid structure (such as a double bond or ring) in the molecule.

A **globular protein** is roughly spherical in shape and soluble in water as colloidal particles.

Globulins are proteins in the blood that are involved in fighting infectious diseases and form complexes that transport lipids.

A **glucogenic amino acid** is an amino acid that can be converted to pyruvic acid or other intermediates and then to glucose.

Gluconeogenesis is the synthesis of glucose from noncarbohydrate precursors.

The **glucose tolerance test** measures blood sugar levels after ingestion of glucose; it is a test for diabetes.

Glycogenesis is the formation of glycogen from glucose.

Glycogenolysis is the breakdown of glycogen to form glucose.

Glycol. *See* **dihydric alcohol.**

A **glycolipid** is any lipid with one or more carbohydrate units.

Glycolysis is the conversion of glucose to lactic acid. *See* **Embden–Meyerhof pathway.**

A **glycosidic linkage** is the bond that joins monosaccharide units; an acetal linkage.

Gout is an affliction caused by abnormal purine metabolism that produces excess uric acid that is deposited as urate salts in joints, particularly those of the big toe.

The **greenhouse effect** refers to the ability of $CO_2(g)$ and certain other gases to absorb and trap energy radiated by Earth's surface as infrared radiation.

The **ground state** of an atom is the state in which all electrons are in their lowest possible energy levels.

A **group** of the periodic table is a vertical column of elements having similar properties.

An enzyme with **group specificity** acts on only structurally similar molecules with the same functional group.

The **half-life** of a radioisotope is the period of time required for one-half of the atoms of the radioisotope to disintegrate.

A **half-reaction** is a portion of an oxidation–reduction reaction, representing either the oxidation process or the reduction process.

A **halogen** is an element in Group 7A of the periodic table.

A **halogenated hydrocarbon** is a hydrocarbon in which one or more hydrogen atoms have been replaced by a halogen atom.

The **heat index** is a measure of the discomfort caused by a combination of high temperature and high humidity.

A **hemiacetal** is formed when an alcohol molecule adds across the carbon–oxygen double bond of an aldehyde; it has an OH group and an OR group on the same carbon atom.

A **hemiketal** is formed when an alcohol molecule adds across the carbon–oxygen double bond of a ketone; it has an OH group and an OR group on the same carbon atom.

Hemophilia is a set of genetic diseases characterized by the inability of the blood to clot properly, leading to excessive bleeding.

The **Henderson–Hasselbalch equation** relates the pH of a solution of a weak acid and its salt to the pK_a of the weak acid and to the molar concentrations of the weak acid and its salt.

Henry's law states that the solubility of a gas is directly proportional to the pressure maintained in the gas above the solvent.

A **heterocyclic compound** is a cyclic compound in which one or more atoms in the ring is an element other than carbon.

A **heterogeneous mixture** is a mixture in which the composition and properties vary from one region to another.

A **holoenzyme** is a fully active enzyme consisting of an apoenzyme and a coenzyme.

A **homogeneous mixture** is a mixture throughout which the composition and properties remain constant.

A **homologous series** of compounds is a series in which successive compounds differ by one carbon atom and two hydrogen atoms. The properties of these compounds also vary in a regular manner.

A **hormone** is a chemical messenger that is secreted into the bloodstream by an endocrine gland.

A **hydrate** is a solid compound that incorporates water molecules into its basic structure.

A **hydrocarbon** is an organic compound that contains only carbon and hydrogen.

A **hydrogen bond** is a type of intermolecular force in which a hydrogen atom covalently bonded in one molecule is simultaneously attracted to a nonmetal atom in a neighboring molecule. Both the atom to which the hydrogen atom is bonded and the one to which it is attracted must be small atoms of high electronegativity—N, O, or F.

In a **hydrogenation reaction,** $H_2(g)$ is a reactant and H atoms are added to C atoms at a carbon-to-carbon double or triple bond.

A **hydrolase enzyme** is one that catalyzes a hydrolysis reaction.

Hydrolysis is the reaction of a substance with water; literally, a splitting by water.

A **hydronium ion (H_3O^+)** is a water molecule to which a hydrogen ion (H^+) has been added; the characteristic ion of an aqueous acid.

A **hydrophilic** substance has an affinity for water; literally, a water-loving substance.

A **hydrophobic** substance lacks an affinity for water; literally, a water-hating substance.

A **hydrophobic interaction** is an interaction of nonpolar groups on a protein as they turn toward the inside of the molecule and away from water molecules.

The **hydroxyl group** is the —OH group.

A **hygroscopic** substance is one that absorbs water vapor from the atmosphere to form a hydrate.

Hyperglycemia means the blood sugar concentration is above normal.

Hypertension is high blood pressure.

A **hypertonic solution** is a solution having an osmotic pressure greater than that of body fluids (blood, tears). A *hyper*tonic solution has a greater osmotic pressure than an *iso*tonic solution.

Hyperventilation is breathing too rapidly.

Hypoglycemia is a condition in which the blood sugar concentration is below normal.

A **hypotonic solution** is a solution having an osmotic pressure less than that of body fluids (blood, tears). A *hypo*tonic solution has a lower osmotic pressure than an *iso*tonic solution.

Hypoventilation is breathing too slowly.

The **ideal gas equation** (or **ideal gas law**) states that the volume of a gas is directly proportional to the amount of gas and its Kelvin temperature and inversely proportional to its pressure. Mathematically, it can be stated through the equation $PV = nRT$.

The **immune response** is the synthesis of antibodies in response to invading bacteria or viruses.

An **incomplete protein** is one that does not supply all the essential amino acids in the quantities needed for the growth and repair of body tissues.

The **induced-fit theory** states that the active site on an enzyme changes its shape somewhat to fit the substrate, much as a glove changes shape to fit a hand.

Inflammation is a tissue response to injury or stress.

Inorganic chemistry is the chemistry of the compounds of all the elements except carbon.

An **integral protein** is one that spans the lipid bilayer of the cell membrane.

Intermolecular forces are the forces of attraction between molecules.

The **International System of Units (SI)** is the measuring system used by scientists. It is based on seven base quantities and their multiples and submultiples.

Invert sugar is an equimolar mixture of glucose and fructose formed by the hydrolysis of sucrose.

The **iodine number** is the number of grams of iodine that are consumed by 100 g of fat or oil; an indication of the degree of unsaturation.

An **ion** is an electrically charged particle comprised of one or more atoms.

The **ion product of water,** K_W, is the product of the concentration of hydronium ion, $[H_3O^+]$, and the concentration of hydroxide ion, $[OH^-]$, in pure water or a water solution. At 25 °C, its value is 1.0×10^{-14}.

Ionic bonds are attractive forces between positive and negative ions, holding them together in a solid crystal.

Ionization of an acid or base is the formation of ions by the reaction of a molecular acid or base with water.

Ionizing radiation is radiation that causes the formation of ions from neutral particles.

An **irreversible inhibitor** is bound to an enzyme by a covalent bond and irreversibly destroys all activity.

The **isoelectric pH** is the pH value at which an amino acid or protein exists in an electrically neutral form.

An **isomerase enzyme** catalyzes the interconversion of isomers.

Isomerization is the conversion of a compound into one of its isomers.

Isomers are compounds that have the same molecular formula but different structural formulas.

An **isotonic solution** is a solution that has the same osmotic pressure as that of body fluids (blood, tears).

Isotopes are atoms that have the same number of protons—and the same atomic number—but different numbers of neutrons—and different mass numbers.

The **IUPAC system of nomenclature** is a systematic way of naming chemical substances so that each has a unique name.

The **kelvin (K)** is the unit of temperature on the Kelvin scale: a difference (interval) of one kelvin is the same as a difference of one degree on the Celsius scale.

A **ketogenic amino acid** is one that is degraded to ketone bodies.

A **ketone** is an organic compound whose molecules have a *carbonyl* functional group between two hydrocarbon groups.

A **ketone body** is acetoacetic acid, β-hydroxybutyric acid, or acetone.

A **ketose** is a sugar with a ketone group.

Ketosis is a condition characterized by elevated levels of ketone bodies in the blood.

The **kilogram (kg)** is the SI base unit of mass.

Kinetic energy is the energy of motion.

The **kinetic-molecular theory** is a model that uses the motion of molecules to explain the behavior of gases.

The **Krebs cycle** is the complex series of reactions by which acetyl CoA is oxidized to carbon dioxide and water.

An **L sugar** has the OH group on the left on the chiral center farthest from the carbonyl group when the structure is drawn in a Fischer projection.

Lactose intolerance is the inability to break down the sugar lactose; caused by a deficiency of lactase.

The **law of combining volumes** states that when gases measured at the same temperature and pressure are allowed to react, the volumes of gaseous reactants and products are in small whole-number ratios.

The **LD_{50}** of a substance is the dosage that is lethal to 50% of the population of test animals.

Le Châtelier's principle states that if a stress is applied to a system at equilibrium, the equilibrium shifts in the direction that will relieve the stress.

A **lecithin** is a phosphatide that contains choline as the amino-alcohol.

Leukemia is a cancer that is characterized by the uncontrolled production of leukocytes that fail to mature.

Leukocytes are white blood cells.

A **levorotatory** substance rotates the plane of polarized light to the left.

A **Lewis structure** is a representation of an element in which the chemical symbol stands for the core of the atom and dots placed around the symbol for its valence electrons.

A **ligase** is an enzyme that catalyzes the joining of two molecules.

An enzyme with **linkage specificity** acts on one kind of chemical bond—for example, an ester linkage.

A **lipase** is an enzyme that catalyzes the hydrolysis of fats.

Lipids are the components of biological systems that are insoluble in water but soluble in nonpolar or slightly polar solvents such as hexane and diethyl ether.

A **lipoprotein** is a protein–lipid complex that transports lipids in the bloodstream.

A **liquid** is a substance that assumes the shape of its container, flows readily, and maintains a fairly constant volume.

A **liter (L)** is a metric unit of volume equal to one cubic decimeter or 1000 cubic centimeters: $1 \text{ L} = 1 \text{ dm}^3 = 1000 \text{ cm}^3$.

A **local anesthetic** is a substance that produces insensitivity to pain yet leaves the patient conscious.

The **lock-and-key theory** of enzyme action holds that an enzyme and its substrate fit together like a lock and its key.

Lone pairs (LPs) are electron pairs assigned exclusively to one of the atoms in a Lewis structure. They are not shared, and hence are not involved in the chemical bonding. (Also called *nonbonding pairs.*)

A **lyase** is an enzyme, such as a decarboxylase, that catalyzes the nonhydrolytic cleavage of its substrate.

Lymph is interstitial fluid that has been absorbed into the lymph capillaries.

A **malignant tumor,** often called a cancer, grows and invades and destroys other tissues.

Marijuana is a preparation made from the leaves, flowers, seeds, and small stems of the *Cannabis* plant.

Markovnikov's rule states that when H—X adds across the double bond in an unsymmetrical alkene, the H goes on the carbon atom that already has the most hydrogen atoms attached. The X, which can be a halogen, OH, and so on, adds to the carbon atom at the other end of the double bond.

The **mass** of an object is a measure of the quantity of matter in the object.

The **mass number (A)** is the sum of the number of protons and neutrons in the nucleus of an atom.

Mass/volume percent is an expression of concentration in which the mass of the solute is divided by the volume of the solution and that quotient multiplied by 100%.

A **mechanism** is a series of individual steps in a chemical reaction that results in the net overall reaction.

The **melting point** of a solid is the temperature at which it melts—that is, comes into equilibrium with the liquid phase.

A **meso compound** is a stereoisomer that contains at least two chiral centers but has an internal symmetry plane.

A **metabolic pathway** is a flow chart that shows how a living organism converts a given starting material to an end product.

Metabolism is the set of all chemical reactions in living systems that break down large molecules for energy and component parts and build large molecules from component parts.

Metals are elements having a distinctive set of properties: luster, good heat and electrical conductivity, malleability, and ductility. Metal atoms generally have small numbers of valence electrons. Metals are found to the left of the stepped diagonal line in the periodic table.

The **meter (m)** is the SI base unit of length.

A **micelle** is a cluster of molecules that contain both polar and nonpolar groups. The polar heads stick out into water, and the nonpolar tails are turned to the inside.

A **millimeter of mercury (mmHg)** is a unit used to express gas pressure: 1 mmHg = 1/760 atm (exactly).

Miscible substances can be mixed in all proportions.

Mitochondria are subcellular units (organelles) that contain the enzymes necessary for the Krebs cycle and other metabolic pathways; they are the "power plants" of the cell.

A **mixed triglyceride** is an ester of glycerol with more than one type of fatty acid.

A **mixture** is a type of matter whose composition and properties may vary from one sample to another.

Molar heat of fusion is the quantity of heat that must be absorbed to melt 1 mol of a solid at a constant temperature.

Molar heat of vaporization is the quantity of heat that must be absorbed to vaporize 1 mol of a given liquid at a constant temperature.

The **molar mass** of a substance is the mass of 1 mol of the substance. It is numerically equal to the atomic weight, molecular weight, or formula weight, and is expressed in grams per mole.

The **molar volume of** a **gas** is the volume occupied by 1 mol of the gas at STP (22.4 L).

Molarity is an expression of the concentration of a solution in moles of solute per liter of solution.

A **mole (mol)** is the amount of substance that contains 6.02×10^{23} units of the substance.

The **molecular weight** is the average mass of a molecule of the substance relative to that of a carbon-12 atom; it is the sum of the masses of the atoms represented in the molecular formula.

A **molecule** is a discrete group of atoms held together by one or more shared pairs of electrons.

A **monohydric alcohol** is an alcohol with only one OH functional group.

A **monolayer** is a layer one molecule thick. Phospholipids form monolayers with the polar heads in water and the nonpolar tails sticking up into the air.

Monomers are small molecules that can be combined to make polymers.

A **monoprotic acid** is an acid that can donate only one proton per molecule.

A **monosaccharide** is a carbohydrate that cannot be hydrolyzed to simpler sugars.

Multiplicity is the existence of several tRNAs for the same amino acid.

A **mutagen** is any chemical or physical agent that causes mutations.

Mutarotation is the change in observed optical rotation of plane-polarized light as + and/or − forms of sugars move toward an equilibrium value.

A **mutation** is any chemical or physical change that alters the sequence of bases in DNA.

A **narcotic** is a drug that produces both narcosis (a profound stupor) and relief of pain.

A **nephron** is a basic unit of the kidney.

A **net ionic equation** is an equation that represents the actual molecules or ions that participate in a chemical reaction, eliminating all nonparticipating species ("spectator" ions).

A **neuron** is a nerve cell.

A **neurotransmitter** is a chemical that carries an impulse across the synapse from a nerve cell to a receptor on a receiving cell.

Neutralization is the reaction of an acid and a base to produce a salt and water.

A **neutron** is a fundamental particle, found in the nucleus of atoms, that has a mass of 1.0087 u and no electric charge.

Nitrogen balance is the state in which an individual's intake of dietary nitrogen is equal to the amount of nitrogen excreted.

The **noble gases** are the elements in Group 8A of the periodic table. They have the valence shell electron configuration ns^2np^6 (except helium, $1s^2$).

A **noncompetitive inhibitor** binds to an enzyme at a different site than does the substrate. By changing the shape of the enzyme molecule, the noncompetitive inhibitor blocks enzyme action.

A **nonelectrolyte** is a substance that exists exclusively or almost exclusively in molecular form, whether in the pure state or in solution.

Nonmetals are elements that lack metallic properties. They are generally poor conductors of heat and electricity and brittle when in the solid state, and they generally have larger numbers of valence electrons than do metals.

A **nonpolar covalent bond** is a covalent bond in which electrons are shared equally.

Nonpolarized light is ordinary light; light that is not polarized.

The **normal boiling point** of a liquid is the temperature at which the liquid boils when the prevailing atmospheric pressure is 1 atm.

Nuclear fission is the splitting of a large unstable nucleus into two lighter fragments and two or more neutrons. Mass destroyed in this process is converted to an equivalent quantity of energy, which is evolved.

Nuclear fusion is the joining together or fusing of lighter nuclei into a heavier one. In the process, some mass is converted to energy, which is evolved.

A **nucleic acid** is a polymer of nucleotides; the molecule of heredity.

A **nucleoside** is a combination of a purine or a pyrimidine base with a pentose sugar.

A **nucleotide** is a combination of a nucleoside and phosphoric acid; the monomer unit of nucleic acid.

The **nucleus** is the concentrated, positively charged matter at the center of an atom; composed of protons and neutrons. *See also* **cell nucleus.**

Obesity is a condition of excess body fat; an obese person is more than 20% above ideal body weight.

The **octane rating** measures the antiknock properties of a sample of gasoline on a scale that has isooctane at 100 and heptane at 0.

The **octet rule** states that most covalently bonded atoms represented by a Lewis structure have eight electrons in their valence shells.

A food **oil** is a triglyceride that is a liquid at room temperature.

An **oncogene** is a gene formed by mutation of a protooncogene that contributes to the development of a cancer.

An **optically active** substance is one that rotates a beam of polarized light.

The **optimum pH** for an enzyme is the pH at which the enzyme exhibits maximum activity.

The **optimum temperature** for an enzyme is the temperature at which the enzyme exerts its greatest catalytic effect. For many enzymes in the human body, the optimum temperature is body temperature (37 °C).

An **orbital** is a wave function that describes the space occupied by electrons with specific values for the main energy level, sublevel, and directional qualities.

Organic chemistry is the chemistry of compounds of carbon.

An **osmol** is the number of moles of a substance multiplied by the number of particles formed by each formula unit of solute.

Osmosis is the net flow of a solvent through a semipermeable membrane, from pure solvent into a solution or from a solution of a lower concentration into one of a higher concentration.

The **osmotic pressure** of a solution is the pressure that must be applied to the solution to prevent the flow of solvent molecules into the solution when the solution and pure solvent are separated by a semipermeable membrane.

Oxidation is a process in which the oxidation state of an element increases—that is, in which electrons are "lost."

The **oxidation state** of an element refers to the number of electrons transferred or shared in the formation of the chemical bonds in a substance.

Oxidative deamination is a reaction that removes the amino group of an amino acid as NH_3. The carbon atom that bore the amino group is oxidized to a carbonyl group.

Oxidative phosphorylation is the process that links ATP synthesis to oxygen consumption in the respiratory chain.

An **oxidizing agent** is a substance that causes oxidation and is itself reduced.

An **oxidoreductase** enzyme catalyzes a reaction in which one substance is oxidized and another is reduced.

An **oxygen debt** is an oxygen deficit resulting from anaerobic activity.

A **pascal (Pa)** is the basic unit of pressure in the SI system. It is a pressure of 1 newton per square meter, 1 N/m^2.

A **peptide** is a compound that has two or more amino acids joined through peptide bonds.

A **peptide bond** is the amide linkage that bonds amino acids in chains of peptides, polypeptides, and proteins.

Percent by volume is an expression of concentration in which the volume of the solute is divided by the volume of the solution and that quotient multiplied by 100%.

The **periodic table of the elements** is a tabular arrangement according to increasing atomic number that places elements having similar properties in the same vertical columns. (Mendeleev's original periodic table was arranged according to atomic weights, not atomic numbers.)

Periods are the horizontal rows of elements in the periodic table. In the modern table, the periods range in width from two members (first period) to 32 members (sixth and seventh periods).

A **peripheral protein** is embedded in only one face of the cell membrane.

A **petrochemical** is a synthetic substance made from petroleum or natural gas.

Petroleum is a naturally occurring liquid mixture consisting mainly of hydrocarbons.

The **pH** is the negative of the logarithm of the hydronium ion concentration in a solution: $pH = -\log [H_3O^+]$.

The **pH scale** is an exponential scale of acidity; a pH below 7 is acidic; exactly 7, neutral; above 7, basic.

A **phenol** is an organic compound with an OH functional group attached to a benzene ring or other aromatic group.

Pheromones are chemicals that are used for communication between members of the same species of insects.

A **phosphatide** is a phospholipid in which glycerol is esterified with two fatty acids and with phosphoric acid. The phosphoric acid is esterified with an amino alcohol.

A **phospholipid** is a phosphorus-containing lipid. Phospholipids are the most polar of lipids.

Photochemical isomerization is the conversion of a compound into one of its isomers by the action of light.

Photochemical smog is air that is polluted with oxides of nitrogen and unburned hydrocarbons, together with ozone and several other components produced by the action of sunlight.

Photosynthesis is the process by which green plants use solar energy to form glucose from carbon dioxide and water.

A **physical property** of a substance is a property that can be observed and specified without reference to any other substance and that does not produce changes in chemical composition.

Plasma is the noncellular part of blood.

A **plasmid** is a circular piece of DNA that is found outside the nucleus in bacteria.

The **pOH** is the negative of the logarithm of the hydroxide ion concentration in a solution: $pOH = -\log [OH^-]$.

A **point mutation** is a mutation that occurs at a single nucleotide position.

In a **polar covalent bond** between two atoms, electrons are drawn closer to the more electronegative atom, creating a separation of charge. One end of the bond is thought of as having a small negative charge, $\delta-$, and the other end, a small positive charge, $\delta+$.

A **polarimeter** is an instrument that is used to measure the optical activity of compounds.

Polarized light is light that vibrates in a single plane.

A **polyamide** is a condensation polymer in which the monomer units are joined by an amide linkage.

A **polyatomic ion** is an ion consisting of two or more atoms bonded together.

A **polyester** is a condensation polymer in which the monomer units are joined by an ester linkage.

A **polymer** is a giant molecule formed by the combination of smaller molecules (monomers) in a repeating pattern.

Polymerization is a type of reaction in which small repeating units (monomers) combine to form giant molecules (polymers).

A **polypeptide** is a polymer of amino acids, usually of lower molar mass than a protein.

A **polyprotic acid** is capable of donating more than one proton to an appropriate base.

A **polysaccharide** is a polymeric carbohydrate that can be hydrolyzed into many monosaccharide units.

A **positron** (β^+) is a positively charged particle having the same mass as β^- particles. Sometimes called "positive electrons," positrons are emitted by certain radioactive nuclei.

Potential energy is energy by virtue of position or composition.

A **precipitate** is an insoluble substance formed in a chemical reaction between ions in solution.

Pressure (*P*) is a force per unit area—that is, $P = F/A$.

A **primary (1°) alcohol** is one that bears the OH group on a carbon atom that is attached to only one other carbon atom.

A **primary (1°) amine** is one that has only one alkyl or aryl group on the nitrogen atom.

A **primary (1°) carbon atom** is one that is attached to only one other carbon atom.

The **primary structure** of a protein is its sequence of amino acids.

Products are the substances that are produced in a chemical reaction. Their formulas appear on the right side of a chemical equation.

A **proenzyme** is an inactive protein from which an enzyme is formed.

A **progestin** is a compound that mimics the action of progesterone.

A **prostaglandin** is a hormonelike compound, derived from arachidonic acid, that is involved in increased blood pressure, the contraction of smooth muscle, and other physiological processes.

A **protein** is a polymer of amino acids.

A **proton** is a particle carrying the fundamental unit of positive charge. Protons have a mass of 1.0073 u and are found in the nuclei of atoms.

A **proton acceptor** is a base: a substance that accepts H^+ (a proton).

A **proton donor** is an acid: a substance that gives up H^+ (a proton).

A **provitamin** is a substance that the body can convert into a vitamin.

A **purine base** is a heterocyclic compound with two fused rings. Adenine and guanine are examples.

A **pyrimidine base** is a heterocyclic compound with a single ring. Thymine, cytosine, and uracil are examples.

A **pyrogen** is a substance that causes a fever.

A **quaternary (4°) ammonium salt** has four alkyl or aryl groups on a nitrogen atom.

The **quaternary structure** of a protein is a specific arrangement of two or more polypeptide chains into larger units.

A **racemic mixture** is a mixture of enantiomers that is optically inactive because it contains equimolar amounts of molecules with opposite rotatory power.

The **rad** is a unit of absorbed radiation equal to 0.01 J/kg.

Radioactivity is the spontaneous emission of ionizing radiation by the atomic nuclei of certain isotopes.

A **radioisotope** is a radioactive isotope.

Reactants are the starting materials or substances consumed in a chemical reaction. Their formulas appear on the left side of a chemical equation.

Recombinant DNA technology is a set of techniques that incorporate genetic material from one organism into the DNA of another organism.

A **reducing agent** is a substance that causes reduction and is itself oxidized.

A **reducing sugar** is any carbohydrate that is capable of reducing Tollens's reagent.

Reduction is a process in which the oxidation state of an element decreases—that is, in which electrons are "gained."

The **reduction potential** is a measure of the tendency of a substance to reduce other substances.

Relative humidity is an expression of water vapor content as a percent of the maximum water vapor content possible.

The **rem** is a unit of ionizing radiation that produces the same damage to humans as 1 roentgen of high voltage X-rays.

The **renal threshold** is the level of a substance, such as glucose, in the blood above which the substance appears in the urine.

Representative elements are elements in which the subshell being filled is either an *s* or *p* subshell of the principal shell of highest principal quantum number (the outermost shell).

Resonance is a term used to describe a situation in which two or more plausible Lewis structures can be written to represent a species but in which the true structure cannot be written. The plausible structures are called *contributing structures.*

A **resonance hybrid** is a composite of the contributing resonance structures and is the true structure of a molecule that exhibits resonance.

Respiration is the process by which cells obtain energy by oxidizing organic molecules such as carbohydrates and lipids; oxygen is absorbed and carbon dioxide is given off.

The **respiratory chain** (also called electron transport chain) is the sequence of highly organized oxidation–reduction enzymes found in mitochondria.

A **retrovirus** has RNA as its genetic material. It synthesizes DNA in the host cell.

A **reversible reaction** can proceed in either the forward or the reverse direction, depending on conditions.

Reye's syndrome is a liver disorder occurring primarily in children and associated with the use of aspirin.

Ribonucleic acid (RNA) is the form of nucleic acid found mainly in the cytoplasm but present in all parts of the cell.

A **ribosome** is a cellular substructure that serves as the site for protein synthesis.

The **roentgen** is an exposure dose of X- or γ-radiation that produces ions with charges of 2.58×10^{-4} coulomb per kilogram of air.

Saliva is a secretion of the oral glands that aids in the swallowing and digestion of foods.

A **salt** is an ionic compound in which hydrogen atoms of an acid are replaced by metal ions.

A **salt linkage** is an interaction between an acidic side chain on one amino acid residue and a basic side chain on another; the resulting charges serve as ionic bonds between peptide chains or between two parts of the same chain.

Saponification is the alkaline hydrolysis of a triglyceride; literally, soapmaking.

A **saturated hydrocarbon** (alkane) has molecules that contain the maximum number of hydrogen atoms for the carbon atoms present. All bonds in the molecules are single covalent bonds.

A **saturated solution** contains the maximum amount of solute that can be dissolved in a particular quantity of solvent at equilibrium at a given temperature.

A **secondary (2°) alcohol** is one that bears the OH group on a carbon atom that is attached to two other carbon atoms.

A **secondary (2°) amine** is one that has two alkyl or aryl groups on the nitrogen atom.

A **secondary (2°) carbon atom** is one that is attached to two other carbon atoms.

The **secondary structure** of a protein is the arrangements of the polypeptide backbone, held together by hydrogen bonds. Two common examples are the alpha helix and pleated sheets.

Self-replication is the process by which a DNA molecule duplicates itself.

A **semipermeable** membrane is permeable to some solutes but not to others.

Serum is blood plasma from which fibrinogen has been removed.

The **set-point theory** holds that the hypothalamus monitors the level of circulating fatty acids in the blood; when the level is too low, the person is hungry. Each person has a unique set point.

Shock is a condition characterized by a loss of fluid from the vascular system.

A **simple cubic** crystal structure has as its unit cell a cube with a structural unit at each corner. *See also* **unit cell.**

A **simple triglyceride** is an ester of glycerol containing three identical fatty acids.

A **single bond** is a covalent linkage in which two atoms share one pair of electrons.

Smog (a contraction of the words "smoke" and "fog") is air that is visibly polluted.

A **solid** is a substance that maintains its shape and volume.

The **solubility product constant,** K_{sp}, describes the equilibrium that exists between a slightly soluble solute and its ions in a saturated solution.

A **solute** is a solution component that is dissolved in a solvent. A solution may have several solutes, which are generally present in lesser amounts than is the solvent.

A **solution** is a homogeneous mixture of two or more substances. The composition and properties are uniform throughout a solution.

A **solvent** is the solution component (usually present in the greatest amount) in which one or more solutes are dissolved to form the solution.

The **specific gravity** of a substance is the ratio of the mass of a given volume of the substance to that of an equal volume of water.

The **specific heat** of a substance is the quantity of heat required to raise the temperature of one gram of the substance by 1 °C (or 1 K).

The **specific rotation** of a substance is a physical property that is characteristic of the substance; it is a quantitative measure of the optical activity of the substance.

A **sphingolipid** is a phospholipid or glycolipid containing sphingosine rather than glycerol.

Sphingomyelin is a sphingolipid with units derived from sphingosine, phosphoric acid, choline, and a fatty acid.

The **standard conditions of temperature and pressure (STP)** for a gas are 273.15 K (0 °C) and 1 atm (760 mmHg).

Starvation is the voluntary or involuntary withholding of nutrition from the body.

An enzyme with **stereochemical specificity** acts on one stereoisomer but not on its enantiomer.

Stereoisomerism is a type of isomerism in which molecules have the same number and types of groups in the same order of attachment but differ in the arrangement of atoms in three-dimensional space.

A **steroid** is a lipid characterized by a structure with a particular arrangement of four fused rings.

A **stoichiometric factor** is a conversion factor relating molar amounts of two species involved in a chemical reaction (i.e., a reactant to a product, one reactant to another, etc.).

Straight-run gasoline is gasoline as it comes off the distilling column of a petroleum refinery.

A **strong acid** is an acid that is essentially completely ionized in solution. *See also* **acid.**

A **strong base** is a base that is essentially completely ionized in solution. *See also* **base.**

A **strong electrolyte** is a substance that exists exclusively or almost exclusively in ionic form in solution.

A **structural formula** is a chemical formula that shows how the atoms of a molecule are attached to one another.

Structural isomers have the same molecular formula, but they differ in the order of attachment of atoms and groups.

Sublimation is the direct passage of molecules from the solid state to the vapor state.

A **substance** is a type of matter having a definite or fixed composition and fixed properties that do not vary from one sample to another.

A **substrate** is the substance acted upon by an enzyme.

Substrate-level phosphorylation is the formation of ATP by direct transfer of a phosphate unit from a metabolite to ADP.

A **supersaturated solution** contains more solute than is present in a saturated solution, with the excess solute remaining in solution.

A **supressor gene** ordinarily inhibits the development of a cancer. When deactivated by a mutation, it allows the cancer to grow.

Surface tension is the amount of work required to extend a liquid surface.

A **synapse** is the gap between the axon end of a neuron and the receptor(s) on the receiving cell.

Synergism is the interaction of two or more substances to produce an effect that is greater than the sum of the separate effects of the substances.

Systolic pressure is the maximum force exerted by the blood against the inner walls of the arteries during a heartbeat.

A **tertiary (3°) alcohol** is one that bears the OH group on a carbon atom that is attached to three other carbon atoms.

A **tertiary (3°) amine** is one that has three alkyl or aryl groups on the nitrogen atom.

A **tertiary (3°) carbon atom** is one that is attached to three other carbon atoms.

The **tertiary structure** of a protein is the unique three-dimensional shape that results from the folding and bending of the protein backbone.

A **thrombocyte** is a blood platelet; thrombocytes are involved in clotting.

Thrombosis is the formation of a clot within a blood vessel.

The **tidal volume** is the volume of air inhaled and exhaled during normal breathing.

Titration is a laboratory procedure in which two reactants in solution are made to react in their stoichiometric proportions.

The **training effect** of regular exercise results in a lower pulse rate and lower blood pressure; it enables a person to do more physical work with less strain.

The term **trans** is used in organic chemistry to indicate geometric isomers in which two groups are attached to opposite sides of a double bond in a molecule.

Transamination is an exchange of functional groups between amino compounds and keto compounds.

Transcription is the process by which DNA directs the synthesis of RNA molecules during protein synthesis.

A **transferase** enzyme catalyzes the transfer of a group from one molecule to another.

Transition elements are elements in which the subshell being filled in the aufbau process is in a principal shell of less than the highest quantum number (an inner shell).

Translation is the process by which the information contained in an mRNA molecule is converted to a protein structure.

A **triglyceride** is an ester of glycerol containing three fatty acids; a triacylglycerol.

A **trihydric alcohol** is one with three OH functional groups.

A **tripeptide** is a compound that has three amino acid units joined through peptide bonds.

A **triple bond** is a covalent linkage in which two atoms share *three* pairs of electrons.

A **triplet code** is a set of three bases that codes for a particular amino acid in protein synthesis.

A **triprotic acid** is an acid that can donate three protons per molecule.

The **turnover rate** of an enzyme is the rate at which it converts substrate molecules to product.

The **Tyndall effect** is the scattering of a beam of light as it passes through a colloid; this makes a colloidal dispersion distinguishable from a true solution.

The **unit cell** of a crystal structure is the simplest parallelepiped that can be used to generate the entire crystalline lattice through straight-line displacements in all three dimensions.

The **universal gas constant (R)** is the numerical constant required to relate the pressure, volume, amount, and temperature of a gas in the ideal gas equation, $PV = nRT$. Its numerical value is $0.082057 \text{ L} \cdot \text{atm} \cdot \text{mol}^{-1} \cdot \text{K}^{-1}$.

An **unsaturated hydrocarbon** is a carbon–hydrogen compound having one or more multiple (double or triple) bonds between carbon atoms; an alkene or alkyne.

An **unsaturated solution** contains less of a solute in a given quantity of solution than is present in a saturated solution. It is a solution having a concentration less than the solubility limit.

The **urea cycle** is the metabolic pathway by which the nitrogen of amino acids is converted to urea.

Urine is the fluid formed in the kidneys and excreted through the urinary tract.

A **vaccine** is a preparation of a weakened antigen administered to cause the body to build up its immune system against an infectious disease organism.

Valence electrons are electrons with the highest principal quantum number. They are found in the outermost energy levels of atoms.

The **valence shell** is the outermost shell of electrons in an atom.

The **valence shell electron pair repulsion (VSEPR) theory** of chemical bonding describes the geometric shape of a molecule or polyatomic ion based on the mutual repulsions among electron groups surrounding the central atom(s) in the structure.

The **van't Hoff rule** is used to determine the maximum number of stereoisomers for a given structural formula; the number is 2^n, where n is the number of chiral centers in the molecule.

The **vapor pressure** of a liquid is the pressure exerted by the vapor in dynamic equilibrium with the liquid at a constant temperature.

Vaporization or **evaporation** refers to the conversion of a liquid to a gas (vapor).

A **virus** is a subcellular infectious agent that has a core of nucleic acids surrounded by proteins. It uses the host cell to replicate itself.

Viscosity is the resistance of a fluid (gas or liquid) to flow produced by intermolecular forces. The stronger the intermolecular forces, the more viscous the fluid.

The **vital capacity** is the maximum amount of air that can be forced from the lungs.

A **vitamin** is an organic compound that the body cannot produce in the amount required for good body health.

A **water-soluble vitamin** is one that is soluble in water; the B vitamins and vitamin C.

A **wax** is an ester formed from long-chain fatty acids and long-chain monohydroxy alcohols.

A **weak acid** is an acid that exists partly in ionic form and partly in molecular form in solution. *See also* **acid.**

A **weak base** is a base that exists partly in ionic form and partly in molecular form in solution. *See also* **base.**

A **weak electrolyte** is a substance that is present partly in molecular form and partly in ionic form in its solutions.

Weight measures the force of attraction between two objects and is related to the masses of the objects; a measure of the force of attraction of the Earth for an object.

An **X-ray** is a type of electromagnetic radiation produced by the impact of cathode rays (electrons) on a solid, such as on a dense metal anode (a target) in a cathode-ray tube.

A **zwitterion** is a molecule that has a positive charge on one atom and a negative charge on another; a dipolar ion.

Answers to Practice Exercises, Selected Review Questions, and Selected Problems

CHAPTER 1
PRACTICE EXERCISES
1.1 a. 1.00 kg; **b.** 475 lbs **1.2** Chemical: b; Physical: a,c
1.5 0.0163 g **1.6** 1.53 lb **1.7** 253 cm **1.8** 81.0 in **1.9** 310 K
1.10 218 °C **1.11** 173 °F **1.12** 66,700 cal = 66.7 kcal
1.13 5.14 °C **1.14** 19.3 g/cm³ **1.15** 66.0 g **1.16** 3.70 mL
1.17 1.39 g/mL

SPECIAL TOPIC A
PRACTICE EXERCISES
A.10 25.0 m/s **A.11** 13.6 g/mL **A.12** 5.44 mL

CHAPTER 2
PRACTICE EXERCISES
2.1 $^{90}_{42}Mo$ **2.2** 38 protons, 52 neutrons, 38 electrons **2.3** $^{90}_{37}X$ and $^{88}_{37}X$ are isotopes of the same element and $^{90}_{38}X$ and $^{93}_{38}X$ are isotopes of the same element. **2.4** 32
2.5

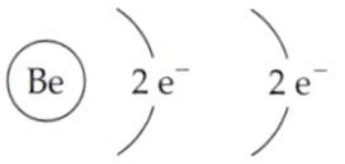

Be 2 e⁻ 2 e⁻

2.6

Al 2 e⁻ 8 e⁻ 3 e⁻

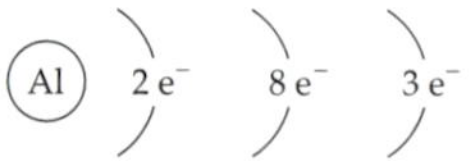

2.7 $1s^2 2s^2 2p^5$ **2.8** $1s^2 2s^2 2p^6 3s^2 3p^5$ **2.9** Rb: $5s^1$; Se: $4s^2 4p^4$

CHAPTER 3
PRACTICE EXERCISES
3.1 Californium −246 **3.2** Bromine −85 **3.3** 0.03825 mg
3.4 0.500 mg **3.5** 22,900 years old **3.6** a neutron
3.7 $^{188}_{79}Au \longrightarrow ^{0}_{-1}e + ^{188}_{78}Pt$

CHAPTER 4
PRACTICE EXERCISES
4.1 −2
4.2 a. :Är: **b.** ·Sr· **c.** :F̈: **d.** ·N̈· **e.** K· **f.** ·S̈:

4.3 Li· + :F̈· ⟶ Li⁺ + :F̈:⁻

4.4 ·Al· :Ö· Al³⁺ :Ö:²⁻
 ·Al· + :Ö· ⟶ Al³⁺ + :Ö:²⁻
 :Ö· :Ö:²⁻

4.5 MgBr₂ **4.6** Ca₃N₂ **4.7** lithium oxide **4.8** copper (II) bromide **4.9** iron (III) oxide
4.10 a. :B̈r· + ·B̈r: ⟶ :B̈r:B̈r:

b. H· + ·B̈r: ⟶ H:B̈r:

c. :Ï· + ·C̈l: ⟶ :Ï:C̈l:

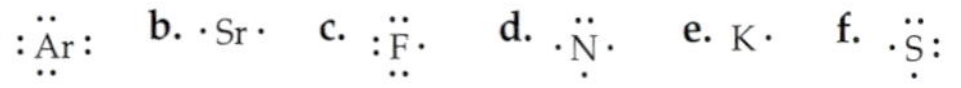

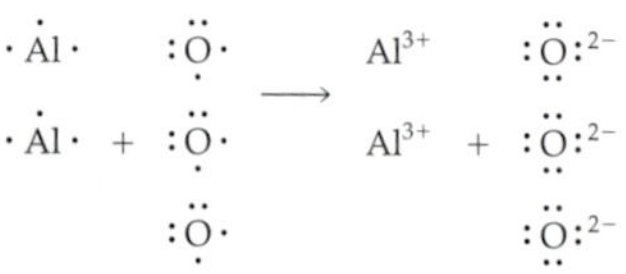

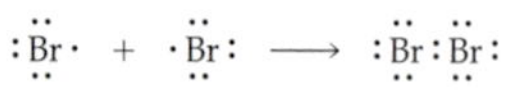

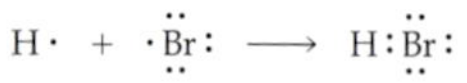

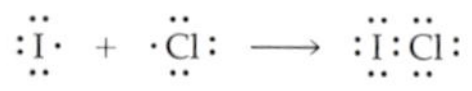

4.11 Bromine trifluoride; bromine pentafluoride **4.12** N₂O₅
4.13 P₄Se₃ **4.14** K₃PO₄ **4.15** Ca(CH₃CO₂)₂ or Ca(C₂H₃O₂)₂
4.16 Calcium carbonate **4.17** Potassium dichromate
4.18

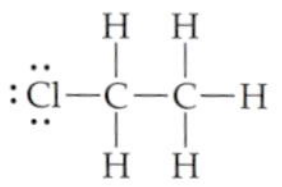

 H H
 | |
:Cl̈—C—C—H
 | |
 H H

4.19

 :Ö:
 ·F̈· ·F̈·

4.20

[H]⁺
|
H—P—H
|
H

4.21

:F̈:
|
N
‖
:Ö· ·Ö:

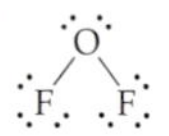

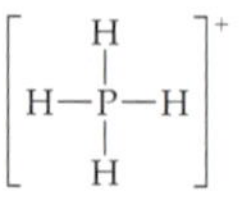

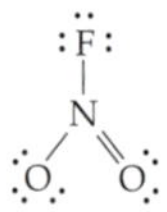

SPECIAL TOPIC B
PRACTICE EXERCISES
B.1 Linear **B.2** Pyramidal

CHAPTER 5
PRACTICE EXERCISES
5.1 $P_4 + 6H_2 \longrightarrow 4PH_3$ **5.2 a.** $3Mg + B_2O_3 \longrightarrow 2B + 3MgO$; **b.** $3NO_2 + H_2O \longrightarrow 2HNO_3 + NO$; **c.** $3H_2 + Fe_2O_3 \longrightarrow 2Fe + 3H_2O$ **5.3 a.** $2H_3PO_4 + 3Ca(OH)_2 \longrightarrow Ca_3(PO_4)_2 + 6H_2O$; **b.** $6CaO + P_4O_{10} \longrightarrow 2Ca_3(PO_4)_2$; **c.** $2Al(OH)_3 + 3H_2SO_4 \longrightarrow Al_2(SO_4)_3 + 6H_2O$ **5.4** 1.476 L CO_2 **5.5** 12.5 L O_2 **5.6 a.** 147.004 u; **b.** 98.960 u; **c.** 97.995 u
5.7 a. 138.206 u; **b.** 294.185 u; **c.** 342.224 u **5.8 a.** 1000. g H_2O; **b.** 0.756 g $C_4H_{10}O$; **c.** 73.7 g C_2H_6 **5.9 a.** 0.0664 mol Fe; **b.** 0.776 mol H_3PO_4; **c.** 2.84 mol C_4H_{10} **5.10** Molecular: 2 molecules of H_2S react with 3 molecules of O_2 to form 2 molecules of SO_2 and 2 molecules of H_2O. Molar: 2 moles of H_2S react with 3 moles of O_2 to form 2 moles of SO_2 and 2 moles of H_2O. Mass: 68.2 g of H_2S react with 96.0 g of O_2 to form 128.1 g of SO_2 and 36.0 g of H_2O.
5.11 a. 1.587 mol CO_2; **b.** 304.8 mol H_2O; **c.** 0.6060 mol CO_2
5.12 0.763 g O_2 **5.13** 0.967 g O_2 **5.14** The reaction will shift to the right, decreasing the concentration of CO.

CHAPTER 6
PRACTICE EXERCISES
6.1 $2Zn + O_2 \longrightarrow 2ZnO$ **6.2** $2PbS + 3O_2 \longrightarrow 2PbO + 2SO_2$ **6.3 a.** oxidation; **b.** oxidation; **c.** oxidation; **d.** oxidation
6.4 a. reduction; **b.** oxidation **6.5 a.** Al: +3, O: −2; **b.** P: 0; **c.** Na: +1, Mn: +7, O: −2; **d.** H: +1, O: −1; **e.** C: −2, H: +1, F: −1; **f.** C: +2, H: +1, Cl: −1 **6.6 a.** reduction; **b.** oxidation; **c.** neither; **d.** oxidation **6.7** Yes
6.8 a. Se + ⟨O₂⟩ ⟶ SeO₂

b. ⟨CH₃C≡N⟩ + 2 H₂ ⟶ CH₃CH₂NH₂

c. ⟨V₂O₅⟩ + 2 H₂ ⟶ V₂O₃ + 2 H₂O

d. 2K + ⟨Br₂⟩ ⟶ 2 K⁺ + 2 Br⁻

CHAPTER 7
PRACTICE EXERCISES
7.1 400 mmHg **7.2** 1800 mL (1.80 L) **7.3** 1.33 atm **7.4** 2.98 L
7.5 −167 °C **7.6** 9.82 g CO_2 **7.7** 40.9 mL **7.8** 0.670 L
7.9 0.0900 g/L **7.10** 5.0 L **7.11** 0.047 atm **7.12** 0.020 g H_2

CHAPTER 8
PRACTICE EXERCISES
8.1 a. KCl; **b.** HgS **8.2 a.** HCl; **b.** H_2S **8.3** NH_3: yes; CH_4: no; C_6H_5OH: yes; H_2S: no; H_2O_2: yes **8.4** 59.1 cal/g **8.5** 23.6 kcal
8.6 48.0 kcal **8.8** 180 cal **8.9** 10 kcal

CHAPTER 9
PRACTICE EXERCISES
9.1 0.968 M **9.22 a.** 9.00 M; **b.** 1.26 M; **c.** 0.274 M; **d.** 0.0242 M;

e. 0.123 M; **f.** 9.23 M **9.3 a.** 673 g KOH; **b.** 5.61 g KOH; **c.** 0.0561 g KOH; **d.** 4.63 g KOH **9.4** 0.030 L **9.5** 46.8% **9.6** Take 22.3 mL of acetic acid and add enough water to make 67.5 mL of solution. **9.7** 2.6% **9.8** Take 15 g of glucose and add it to 260 mL water. **9.9 a.** 1.5 osmol/L; **b.** 0.30 osmol/L; **c.** 1.32 osmol/L

CHAPTER 10

PRACTICE EXERCISES

10.1 H_3AsO_4 **10.2** HIO; $NaIO_3$; $NaIO_2$; NaIO **10.3** HNO_3 **10.4** KOH **10.5 a.** $Ca(OH)_2 + 2\,HCl \longrightarrow CaCl_2 + 2\,H_2O$; **b.** $Ca^{2+} + 2\,OH^- + 2\,H_3O^+ + 2\,Cl^- \longrightarrow Ca^{2+} + 2\,Cl^- + 4\,H_2O$; **c.** $2\,OH^- + 2\,H_3O^+ \longrightarrow 4\,H_2O$ or $OH^- + H_3O^+ \longrightarrow 2\,H_2O$

CHAPTER 11

PRACTICE EXERCISES

11.1 0.208 L **11.2** 0.0851 L **11.3** 0.1724 M **11.4** 0.4016 L **11.5** $[OH^-] = 0.025$ M; $[H_3O^+] = 4.0\times10^{-13}$ M **11.6** 9.00 **11.7** 8.57 **11.8** 1.6×10^{-11} M $= [H_3O^+]$; $[OH^-] = 6.2\times10^{-4}$ M **11.9** neutral **11.10** acidic **11.11** basic

SPECIAL TOPIC C

PRACTICE EXERCISES

C.1 a. $K = \dfrac{[H_2][I_2]}{[HI^2]}$; **b.** $K = \dfrac{[O_3]^2}{[O_2]^3}$; **c.** $K = \dfrac{[XeF_4]}{[Xe][F_2]^2}$

C.2 2.6×10^{-6} M **C.3** 9.1×10^{-6} M **C.4** 6.6×10^{-5} M **C.5** 3.7×10^{-5} M **C.6** 2.5×10^{-9} M **C.8** 3.2 **C.9** 2.94

CHAPTER 12

PRACTICE EXERCISES

12.1 Yes

CHAPTER 13

PRACTICE EXERCISES

13.3 a. 3-methylhexane; **b.** 2,4-dimethylpentane; **c.** 3-ethylhexane; **d.** 4-isopropylheptane

13.4 a. 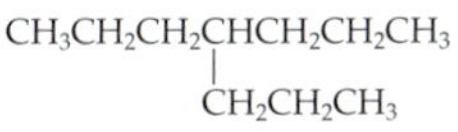 **b.** 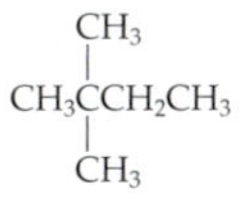

c. 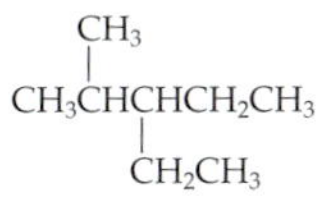 **d.** 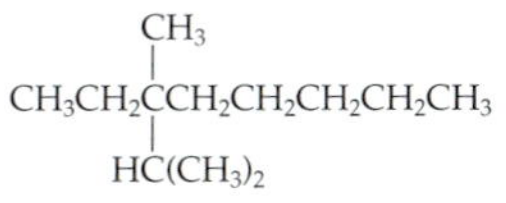

13.9 a. 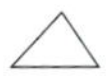 **b.** 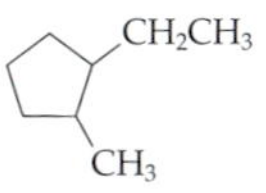

13.11 a. 4-ethyl-2-methyl-2-hexene; **b.** 1-methylcyclohexene

13.12 a. 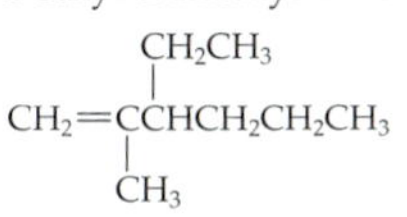 **b.** 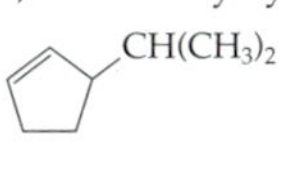

13.13 a. 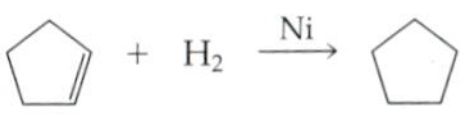

b. 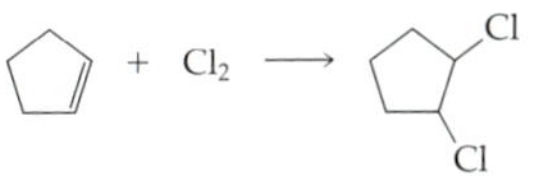

c. 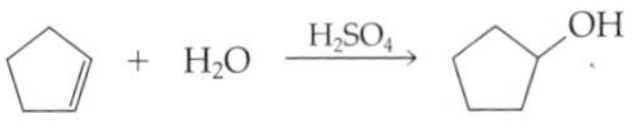

CHAPTER 14

PRACTICE EXERCISES

14.1 1-methylcyclopentanol

14.2 a. 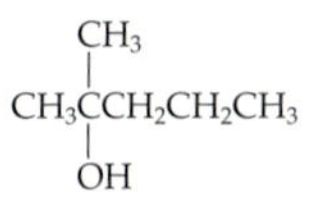 **b.** 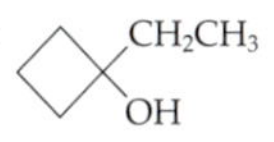

14.3 a. 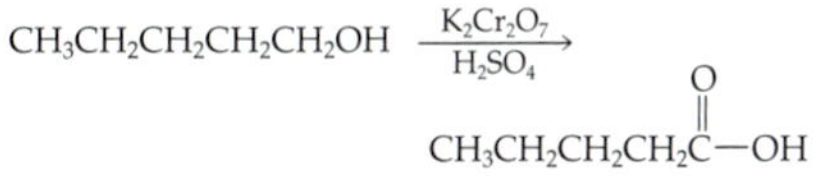

b. 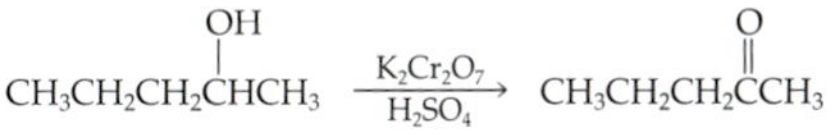

c. 

CHAPTER 15

PRACTICE EXERCISES

15.1 3,3-dimethylbutanal

15.2 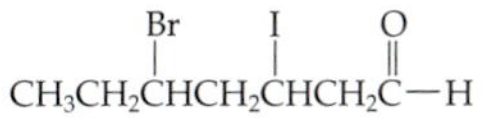

15.3 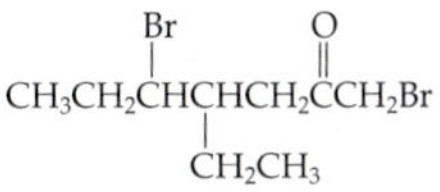

15.4 2-chlorocyclohexanone

15.6 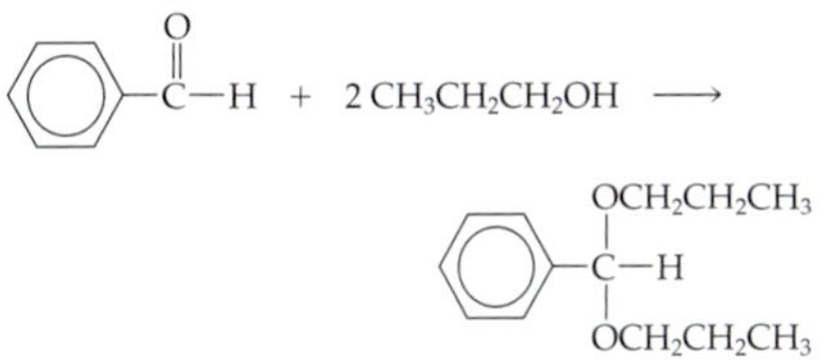

CHAPTER 16

PRACTICE EXERCISES

16.1 3-methylbutanoic acid

16.2 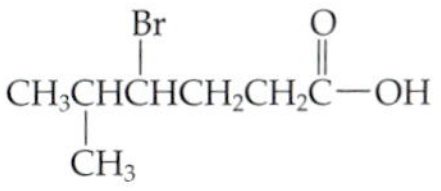

16.3 a. 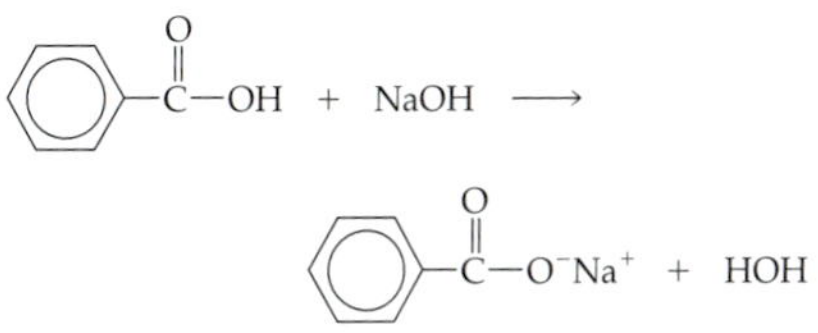

b. 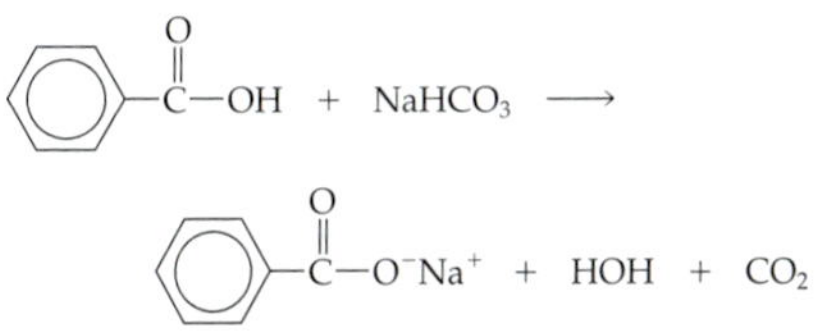

16.5 methyl benzoate

16.7 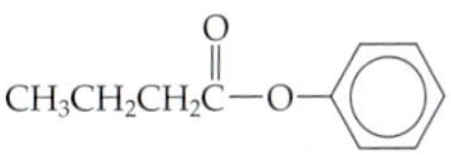

16.8
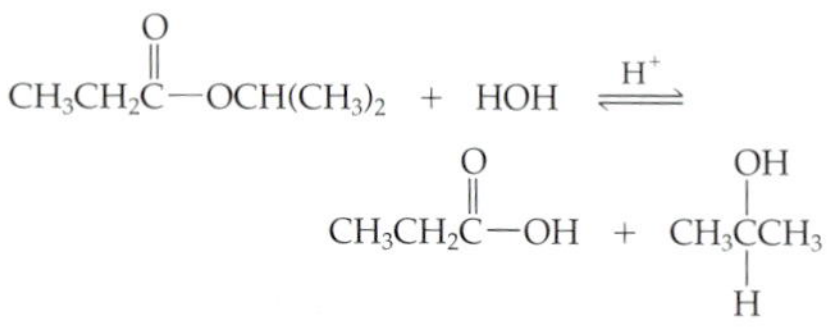

16.9
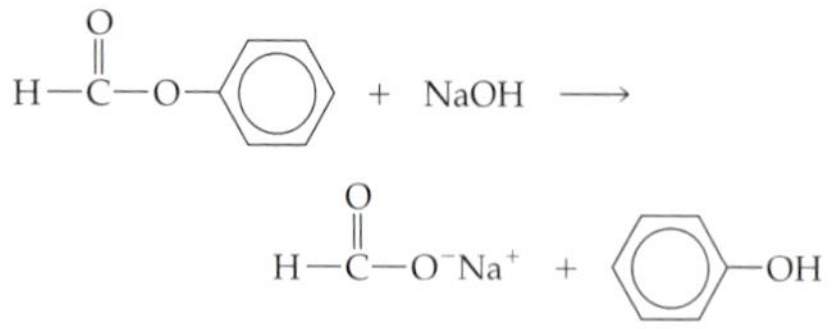

16.11

16.12
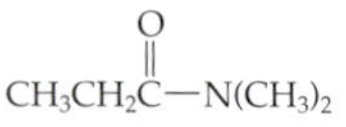

16.13
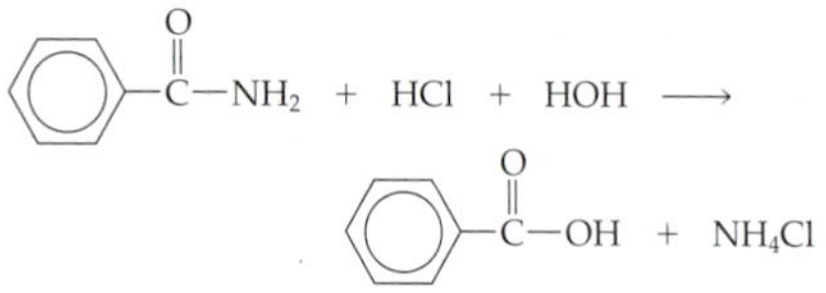

16.14
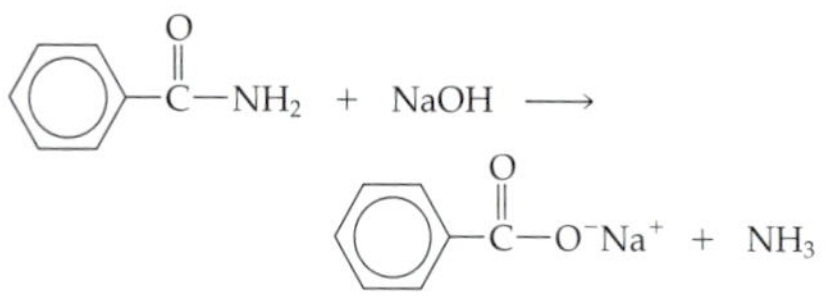

CHAPTER 17

PRACTICE EXERCISES

17.4 a. isopropylamine, primary; **b.** diethylmethylamine, tertiary; **c.** cyclopropylamine, primary **17.5** *p*-propylaniline

17.6
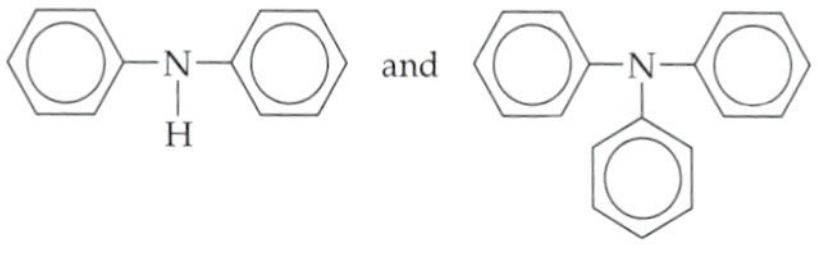

17.7
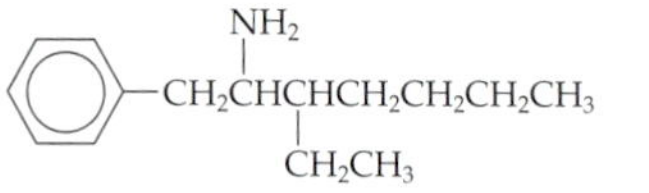

17.8 ethylammonium ion and tetraethylammonium ion

CHAPTER 18

PRACTICE EXERCISES

18.1
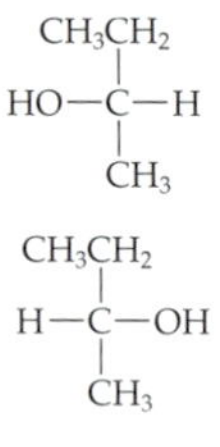

18.2
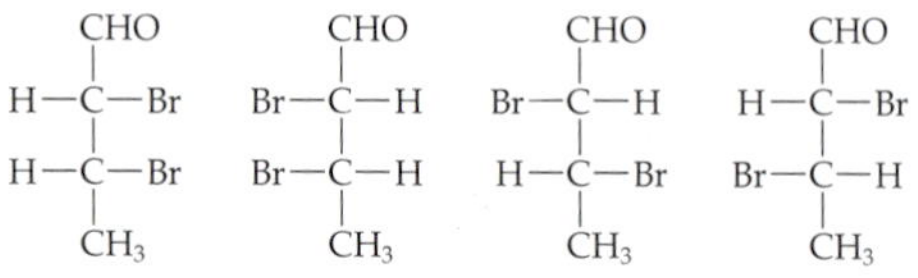

18.3
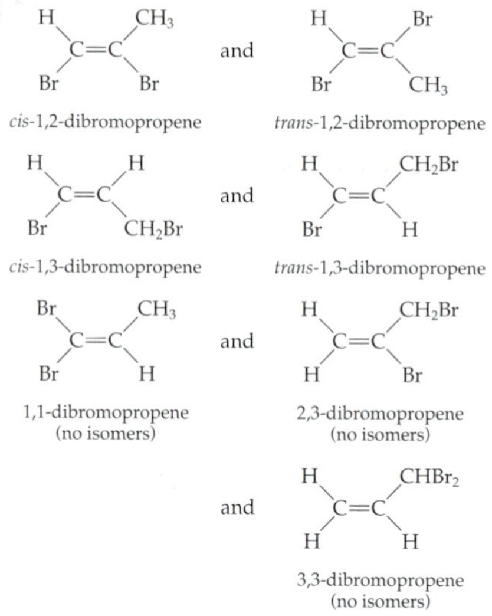
cis-1,2-dibromopropene
trans-1,2-dibromopropene
cis-1,3-dibromopropene
trans-1,3-dibromopropene
1,1-dibromopropene (no isomers)
2,3-dibromopropene (no isomers)
3,3-dibromopropene (no isomers)

CHAPTER 21

PRACTICE EXERCISES

(page 571)

a.
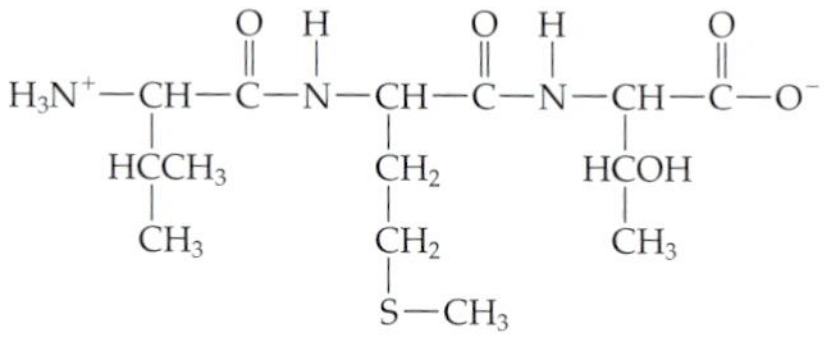

b.
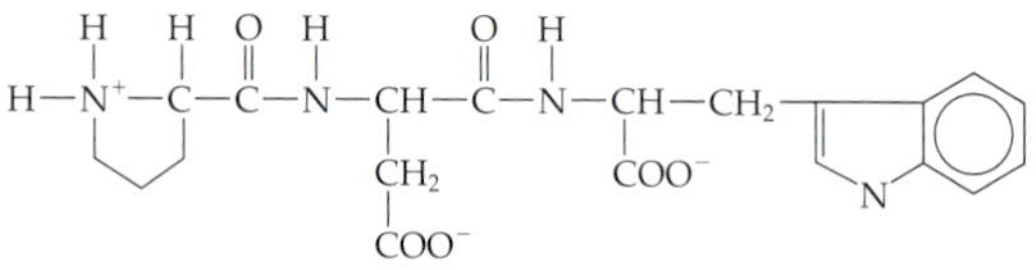

c.
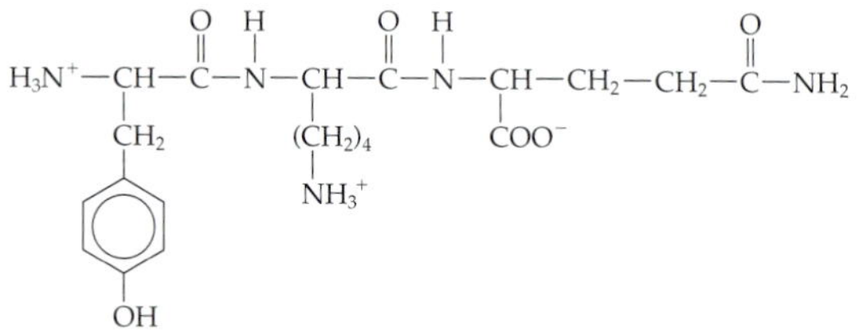

CHAPTER 26

PRACTICE EXERCISES

26.1 47.5% **26.2** 10% **26.3** 356 ATPs **26.4** 29 **26.5** 35 mi at a leisure pace, 20 mi at a brisk pace **26.6** 3.6 lb

---

CHAPTER 1

SELECTED PROBLEMS

3. a, c, and d **9.** Steam. **17.** Yes. Since the gravitational pull on both objects is equal, the mass will be proportional to the weight. **19. a.** Physical change; **b.** Chemical change; **c.** Chemical change **21.** Elements: a and c. Their composition is made up of only one type of fundamental substance. **23. a.** Substance; **b.** Mixture; **c.** Mixture **25. a.** homogeneous; **b.** heterogeneous (tea solution; chunks of ice) **27. a.** helium; **b.** nitrogen; **c.** fluorine; **d.** potassium; **e.** iron; **f.** copper **29. a.** H;

b. C; **c.** O; **d.** Zn; **e.** I; **f.** Hg **31. a.** 4.54 mg; **b.** 3.76 cm; **c.** 6.34 μg
33. 1000 mm, 100 cm **35. a.** L; **b.** equal in size **37. a.** 50,000 m;
b. 0.25 m **39. a.** 10 dL; **b.** 0.2 dL **41. a.** 15 g; **b.** 86 mg
43. a. 1500 mL; **b.** 0.018 L **45.** The sprinter **47.** The automobile. **49.** The diver on the 10 m platform. **51. a.** °C; **b.** Cal
53. a. 99 °F; **b.** −148 °F; **c.** 523 °F **55. a.** 37 °C; **b.** −15 °C;
c. −21 °C **57.** 1500 cal **59.** 140,000 cal **61.** 3.10 g/mL
63. 39.6 g **65.** 116,000 cm³ **67.** 5.23 g/cm³ **69.** 1.02

SPECIAL TOPIC A
SELECTED PROBLEMS
3. 4 (0.91 m) < 3 (1.04 m) < 1 (1.21 m) < 2 (1.91 m) **7. a.** 50,000 m;
b. 0.546 m; **c.** 98,500 g; **d.** 0.0479 L; **e.** 0.578 mg; **e.** 23.7 cm
9. a. 11.5 yd; **b.** 5.39 lb; **c.** 2.00 qt; **d.** 8.59 mi/hr **11. a.** 41.7 cm;
b. 3.94 L; **c.** 3.55 lb; **d.** 0.329 oz **13.** 145 kg **15.** 56 mi/hr
17. \$3.89/gal **19.** 0.66 g/mL **21.** 37.8 g **23.** 53.1 cm³

CHAPTER 2
SELECTED PROBLEMS
5. Proton: mass = 1 amu, charge = +1, location = nucleus;
Neutron: mass = 1 amu, charge = 0, location = nucleus; Electron:
mass = 0 amu, charge = −1, location = outside the nucleus
11. a. 11; **b.** 12 **15.** lithium-7 **23.** Yes. **25.** No. **27. a.** 20
each; **b.** 11 each; **c.** 9 each; **d.** 18 each **29. a.** 30 protons, 32
neutrons; **b.** 94 protons, 147 neutrons; **c.** 43 protons, 56 neutrons
d. 42 protons, 57 neutrons **31.** Isotope pairs: b **33.** 18

35. a.

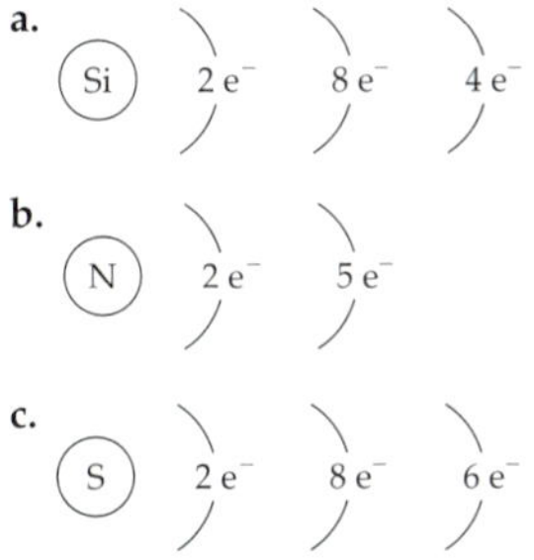

37. Si: $1s^2 2s^2 2p^6 3s^2 3p^2$; N: $1s^2 2s^2 2p^3$; S: $1s^2 2s^2 2p^6 3s^2 3p^4$
39. a. 4; **b.** 7; **c.** 13 **41. a.** Group: 4A, Period: 2, nonmetal;
b. Group: 2A, Period: 4, metal; **c.** Group: 2B, Period: 5, metal
d. Group: 7A, Period: 3, nonmetal; **e.** Group: 3A, Period: 2,
nonmetal; **f.** Group: 2A, Period: 6, metal; **g.** Group: 5A, Period: 6,
metal; **h.** Group: 7A, Period: 4, nonmetal **43. a.** Ga; **b.** Cu; **c.** I
45. b **47.** b, d, e **49.** Five electrons in a p subshell.

CHAPTER 3
SELECTED PROBLEMS
3. a. beta particle; **b.** alpha particle **5.** Mass number: no change;
Atomic number: no change **17.** Gamma **27. a.** alpha particle;
b. beta particle; **c.** neutron; **d.** deuterium
29. $^{209}_{82}\text{Pb} \longrightarrow \,^{0}_{-1}\text{e} + \,^{209}_{83}\text{B}$ **31.** $^{31}_{16}\text{S} \longrightarrow \,^{0}_{+1}\text{e} + \,^{31}_{15}\text{P}$
33. $^{87}_{35}\text{Br} \longrightarrow \,^{1}_{0}\text{n} + \,^{86}_{35}\text{Br}$ **35.** $^{24}_{12}\text{Mg} + \,^{1}_{0}\text{n} \longrightarrow \,^{1}_{1}\text{H} + \,^{24}_{11}\text{Na}$
37. a. $^{10}_{4}\text{Be}$; **b.** $^{1}_{1}\text{H}$; **c.** $^{4}_{2}\text{He}$ **39. a.** $^{2}_{1}\text{H} + \,^{2}_{1}\text{H} \longrightarrow \,^{3}_{2}\text{He} + \,^{1}_{0}\text{n}$;
b. $^{241}_{95}\text{Am} + \,^{4}_{2}\text{He} \longrightarrow \,^{243}_{97}\text{Bk} + 2\,^{1}_{0}\text{n}$; **c.** $^{121}_{51}\text{Sb} + \,^{4}_{2}\text{He} \longrightarrow \,^{124}_{53}\text{I} +$
$^{1}_{0}\text{n} \longrightarrow \,^{125}_{52}\text{Te} + \beta^+$ **41.** 24.12 days **43.** 335 hrs.

CHAPTER 4
SELECTED PROBLEMS
1. Noble gases **3.** A sodium ion and a neon atom both contain
10 electrons, however, sodium has one more proton (and a couple
more neutrons). **7. a.** 1; **b.** 4; **c.** 2; **d.** 1; **e.** 3; **f.** 1
13. a, c. ionic; **b.** polar covalent; **d.** nonpolar covalent

15. a. Na· **b.** :Ö: **c.** :F̈· **d.** Ȧl·

17. a. Ba: \longrightarrow 2e⁻ + Ba²⁺

b. :B̈r· + 1e⁻ \longrightarrow :B̈r:⁻

19. a. Ca: + :B̈r· + :B̈r· \longrightarrow Ca²⁺ + :B̈r:⁻ + :B̈r:⁻

b. Mg: + :S̈ \longrightarrow Mg²⁺ + :S̈:²⁻

21. a. Ca: Ca²⁺ **b.** :S̈ :S̈:²⁻

c. Rb· Rb⁺ **d.** :P̈· :P̈:³⁻

23. a. Na⁺ :F̈:⁻ **b.** K⁺ :C̈l:⁻ **c.** 2 Na⁺ :Ö:²⁻

d. Ca²⁺ 2:C̈l:⁻ **e.** Mg²⁺ 2:B̈r:⁻

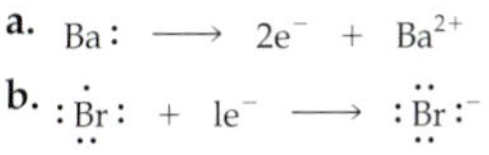

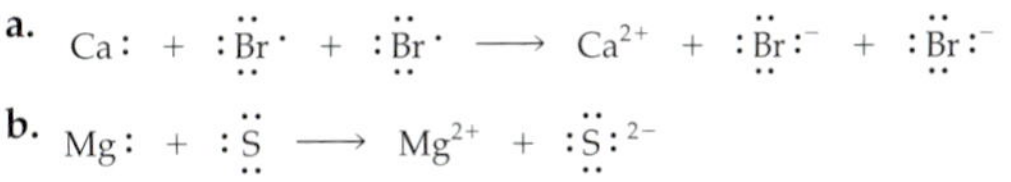

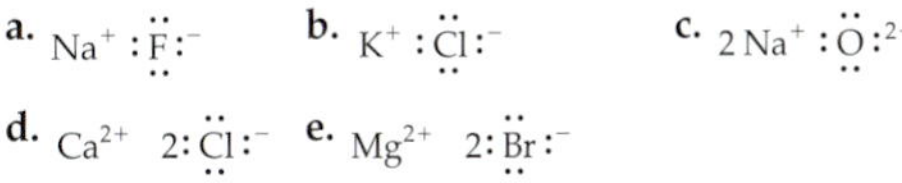

25. a. sodium ion; **b.** magnesium ion; **c.** aluminum ion; **d.** chloride
ion; **e.** oxide ion; **f.** nitride ion **27. a.** iron(III) ion (ferric);
b. copper(II) ion (cupric); **c.** silver ion **29. a.** Br⁻; **b.** Ca²⁺;
c. K⁺; **d.** Fe²⁺ **31. a.** sodium bromide; **b.** calcium chloride;
c. iron(II) chloride or ferrous chloride; **d.** lithium iodide; **e.** potassium sulfide; **f.** copper(I) bromide or cuprous bromide
33. a. carbonate ion; **b.** monohydrogen phosphate ion; **c.** permanganate ion; **d.** hydroxide ion **35. a.** NH₄⁺; **b.** HSO₄⁻; **c.** CN⁻;
d. NO₂⁻ **37. a.** MgSO₄; **b.** NaHCO₃; **c.** KNO₃; **d.** CaHPO₄
39. a. Fe₃(PO₄)₂; **b.** K₂Cr₂O₇; **c.** CuI; **d.** NH₄NO₂

41. :Ï:Ï: :represents bonding pair
 :represents lone pair

43. a. H:P̈:H (with H below) **b.** :F̈:C:F̈: (with :F̈: above and below)

45. a.
```
    H
    |    ..
H — C — O — H
    |    ..
    H
```
b.
```
  H
  |   ..
H—C=O
     ..
```
c. H—N̈—Ö—H (with H below N) **d.** H—N̈—N̈—H (with H below each N)

e.
```
F—C—F
    ‖
   :O:
```
f.
```
      ..
:C̈l—P—C̈l:
    |
   :C̈l:
```

47. a. N₂O; **b.** P₄S₃; **c.** PCl₅; **d.** SF₆ **49. a.** carbon disulfide;
b. dinitrogen tetrasulfide; **c.** phosphorus pentafluoride; **d.** disulfur decafluoride **51. a.** N; **b.** Cl; **c.** F **53. a.** B < N < F;
b. Ca < As < Br; **c.** Ga < C < O

55. a. :N̈=Ö: **b.** :Ï—Be—Ï: **c.**
```
    :C̈l:  :C̈l:
       \ /
 :C̈l—P—C̈l:
        |
      :C̈l:
```

57. a. :C̈l—Ö:⁻ **b.** :Ö—C̈l—Ö:⁻

c.
```
       :O:
        ‖        2−
H—Ö—P—Ö:
        |
       :O:
```
d.
```
      ..
:Ö—Br—Ö:⁻
      |
     :O:
```

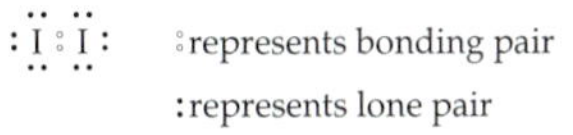

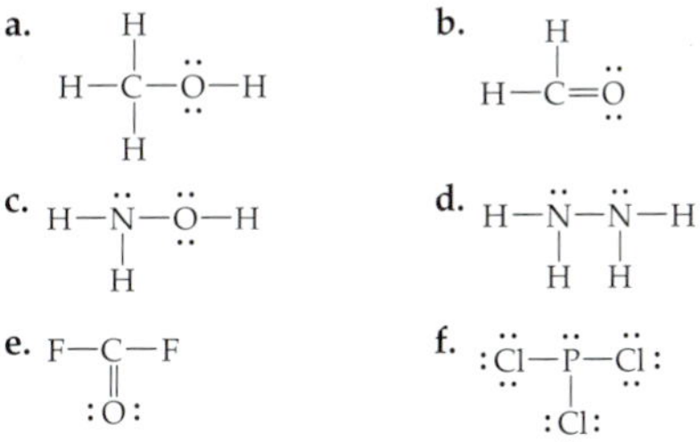

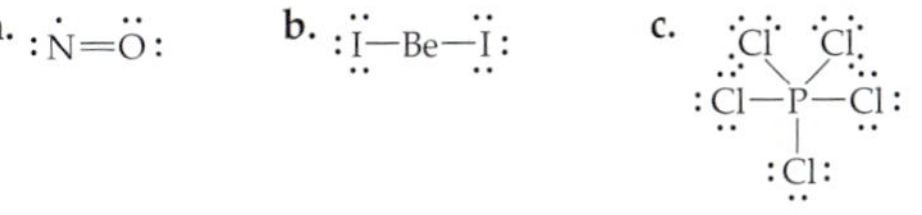

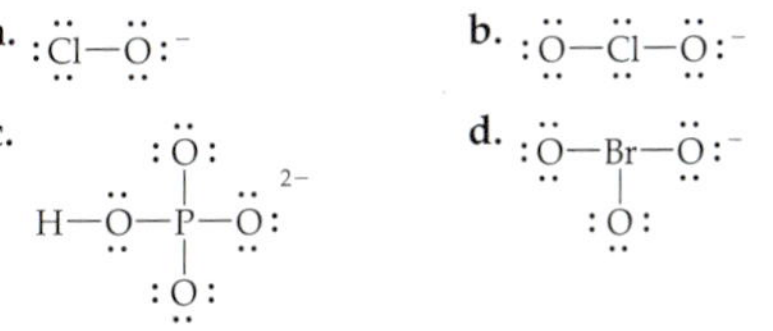

59. a. potassium nitrite; **b.** lithium cyanide; **c.** ammonium iodide;
d. sodium nitrate; **e.** potassium permanganate; **f.** calcium sulfate
61. a. sodium monohydrogen phosphate; **b.** ammonium phosphate; **c.** aluminum nitrate; **d.** ammonium nitrate

SPECIAL TOPIC B

SELECTED PROBLEMS

5. Polar covalent **11. a.** linear; **b.** triangular (trigonal planar) **13. a.** bent; **b.** tetrahedral **15. a.** pyrimidal; **b.** bent **17. a.** polar; **b.** polar; **c.** nonpolar **19. a.** $\underset{\delta+}{H}-\underset{\delta-}{O}$; **b.** $\underset{\delta-}{N}-\underset{\delta+}{Cl}$ **21.** Nonpolar. **23. a.** F—F, Cl—F, H—F; **b.** H—H, H—Br, H—F **25. a.** $F-F$, $\underset{\delta+}{Cl}-\underset{\delta-}{F}$, $\underset{\delta+}{H}-\underset{\delta-}{F}$; **b.** $H-H$, $\underset{\delta+}{H}-\underset{\delta-}{Br}$, $\underset{\delta+}{H}-\underset{\delta-}{F}$

CHAPTER 5

SELECTED PROBLEMS

5. The molecular weight of CO_2 is 44 u and the molar mass is 44 g/mol. **15.** The rate will increase. **21.** 22.4 L for all three **23. a.** 4; **b.** 4; **c.** 8; **d.** 6 **25. a.** 12; **b.** 8 **27. a, d, e.** balanced; **b, c.** not balanced **29. a.** $Cl_2O_5 + H_2O \longrightarrow 2\,HClO_3$; **b.** $V_2O_5 + 2\,H_2 \longrightarrow V_2O_3 + 2\,H_2O$; **c.** $4\,Al + 3\,O_2 \longrightarrow 2\,Al_2O_3$; **d.** $Sn + 2\,NaOH \longrightarrow Na_2SnO_2 + H_2$; **e.** $PCl_5 + 4\,H_2O \longrightarrow H_3PO_4 + 5\,HCl$ **31. a.** $Na_3P + 3\,H_2O \longrightarrow 3\,NaOH + PH_3$; **b.** $Cl_2O + H_2O \longrightarrow 2\,HClO$; **c.** $2\,CH_3OH + 3O_2 \longrightarrow 2\,CO_2 + 4\,H_2O$; **d.** $3\,Zn(OH)_2 + 2\,H_3PO_4 \longrightarrow Zn_3(PO_4)_2 + 6\,H_2O$; **e.** $C_3H_8 + 5\,O_2 \longrightarrow 3\,CO_2 + 4\,H_2O$ **33. a.** 156.9 u; **b.** 98.0 u; **c.** 294.2 u **35. a.** 0.435 g; **b.** 47.2 g; **c.** 45.0 g **37. a.** 1.57 mol; **b.** 0.0285 mol; **c.** 0.0600 mol; **d.** 0.0356 mol **39. a.** 2.65 L H_2O; **b.** 104.7 L O_2 **41. a.** 16.7 mol CO_2; **b.** 55.9 mol O_2 **43.** 2500 g NH_3 **45.** 11.3 g O_2 **47.** 2050 g **49.** 3600 g (balanced equation: $NH_3 + 2\,O_2 \longrightarrow HNO_3 + H_2O$) **51. a.** Equilibrium shifts to the left; **b.** Equilibrium shifts to the right; **c.** Equilibrium shifts to the right

Chapter 6

SELECTED PROBLEMS

13. a. $C + O_2 \longrightarrow CO_2$; **b.** $CH_4 + 2\,O_2 \longrightarrow CO_2 + 2\,H_2O$; **c.** $N_2 + O_2 \longrightarrow 2\,NO$; **d.** $C_3H_8 + 5\,O_2 \longrightarrow 3\,CO_2 + 4\,H_2O$ **15. a.** 0; **b.** +4; **c.** −2; **d.** +6 **17. a.** oxidation (Cl: +4 to +5); **b.** oxidation (Mn: +2 to +4); **c.** reduction (Br: +1 to 0) ; **d.** oxidation (Sb: -3 to 0) **19.** Some of the sulfurs in the H_2SO_4 are reduced to form SO_2 (+6 to +4).

21. a. $\underline{4\,Al} + 3\,\textcircled{O_2} \longrightarrow 2\,Al_2O_3$

b. $\underline{2\,SO_2} + \textcircled{O_2} \longrightarrow 2\,SO_3$

23. a. $\underline{Fe} + 2\,\textcircled{HCl} \longrightarrow FeCl_2 + H_2$

b. $\underline{CS_2} + \textcircled{O_2} \longrightarrow CO_2 + SO_2$

25. a. SO_2 is oxidized; HNO_3 is reduced. **b.** HI is oxidized; CrO_3 is reduced. **27.** I^- is oxidized; Cl_2 is reduced. **29.** Reduced. **31.** Acetylene is reduced; it gains hydrogens. (H_2 is oxidized to the +1 state.) **33.** NO_2^- is reduced; ascorbic acid is a reducing agent.

Chapter 7

SELECTED PROBLEMS

19. a. increase in pressure; **b.** increase in pressure; **c.** increase in pressure **35. a.** 749 mmHg; **b.** 1.12 atm; **c.** 0.949 atm **37.** 213 mmHg (8.39 in) **39. a.** 1090 mL; **b.** 2600 mmHg **41. a.** 9120 L; **b.** 19 hr **43.** 117 mL **45.** 160 °C **47. a.** 2.5 mol H_2; **b.** 2.23 mol SF_6; **c.** 1.66 mol CO_2; 5.0 g H_2 has the greatest number of molecules. **49.** 143 mL **51.** 2490 mmHg **53.** 0.489 m^3 **55.** 4.93 L **57.** 22.3 L **59.** 29.0 atm **61.** 0.078 mol Kr **63. a.** 1.25 g/L ; **b.** 3.48 g/L; **c.** 1.78 g/L; **d.** 1.25 g/L **65.** 710 mmHg **67.** 250 mmHg **69.** 95%

CHAPTER 8

SELECTED PROBLEMS

7. a. dispersion forces; **b.** dipolar forces; **c.** hydrogen bonds **29.** Ethane **31.** Xe **35.** 7200 cal/mol **37.** 96.8 cal **39.** 1.05 kcal **41.** 0.33 kcal **43.** 28.4 kcal **45.** CS_2; neither has a dipole moment, but CS_2 is smaller and has smaller dispersion forces. **47.** C_2H_5OH; C_2H_5OH molecules have the ability to hydrogen bond to one another, this creates a network of molecules held by intermolecular forces that would tend to stay in the liquid phase at a higher temperature. **49.** H_2S (lowest), H_2Se, H_2Te (highest); they are all relatively nonpolar molecules, the dispersion forces for the higher molecular weight compounds (H_2Se and H_2Te) would be greater.

CHAPTER 9

SELECTED PROBLEMS

11. 1 ppt < 1ppb < 1ppm < 1mg/dL < 1% **29. a.** 2.4 M; **b.** 0.700 M **31. a.** 0.907 M; **b.** 1.95 M **33. a.** 80.0 g NaOH; **b.** 7.65 g $C_6H_{12}O_6$ **35.** 0.208 L **37.** 2.05 L **39. a.** 4.83%; **b.** 5.09% **41. a.** 3.96%; **b.** 7.32% **43.** You would take 77.5 g NaCl and add it to 697.5 g of water. **45.** You would take 0.0400 L (40.0 mL) of acetic acid and add it to 1.96 L of water. **47.** 3.75 g of $MgSO_4$ dissolved in 250 mL of solution **49.** 100 mg/dL **51. a.** unsaturated; **b.** 38 °C **53.** Solute Particles: **a.** two; **b.** one; **c.** three. Osmol per one mole: **a.** two; **b.** one; **c.** three. **55. a.** 0.1 M $NaHCO_3$; **b.** 1 M NaCl **57. a.** 0.5 mol; **b.** 1.0 mol; **c.** 0.33 mol **59. a.** same; **b.** 2 osmol/L glucose ($C_6H_{12}O_6$)

CHAPTER 10

SELECTED PROBLEMS

7. Hydroxide ion (OH^-) **25.** Strong: b; Weak: c, d; Neither: a **27. a.** salt; **b.** strong base; **c.** salt; **d.** weak acid **29.** strong base **31.** weak acid **33.** an acid, weak **35. a.** HCl; **b.** H_2SO_4; **c.** H_2CO_3; **d.** LiOH; **e.** $Mg(OH)_2$; **f.** KOH **37. a.** sodium hydroxide; **b.** phosphoric acid; **c.** nitric acid; **d.** sulfurous acid; **e.** calcium hydroxide; **f.** hydrosulfuric acid **39.** Bromide ion; Hydrobromic acid **41.** Nitrite ion; Nitrous acid **43.** Oxalic acid **45. a.** $HI \longrightarrow H^+ + I^-$; **b.** $CH_3CH_2COOH \rightleftharpoons CH_3CH_2COO^- + H^+$; **c.** $HNO_2 \rightleftharpoons NO_2^- + H^+$; **d.** $H_2PO_4^- \rightleftharpoons HPO_4^{2-} + H^+$ **47. a.** $HNO_3 \rightleftharpoons NO_3^+ + H^+$; **b.** $KOH \longrightarrow K^+ + OH^-$; **c.** $HCOOH \rightleftharpoons HCOO^- + H^+$; **d.** $CH_3NH_2 + H_2O \rightleftharpoons CH_3NH_3^+ + OH^-$ **49.** H_2SO_4; an acid **51.** KOH; a base **53. a.** base; **b.** acid; **c.** acid **55.** $NaOH + HCl \longrightarrow NaCl + H_2O$ **57.** $Ca(OH)_2 + 2\,HCl \longrightarrow CaCl_2 + 2\,H_2O$ **59.** $H_3PO_4 + 3\,NaOH \longrightarrow Na_3PO_4 + 3\,H_2O$ **61.** $CO_3^{2-} + 2\,H_3O^+ \longrightarrow 3\,H_2O + CO_2$

CHAPTER 11

SELECTED PROBLEMS

5. Bicarbonate/carbonic acid buffer, phosphate buffer, and proteins **7.** decrease **19.** 0.167 L **21.** 0.481 L **23. a.** 0.417 L ; **b.** 0.00319 L **25.** 0.249 M HCl **27.** 0.0230 M $Ca(OH)_2$ **29.** 0.125 M HCl **31.** 20.8 mL H_2SO_4 **33. a.** 46.8 mL HCl; **b.** 11.9 mL HCl; **c.** 23.1 mL HCl **35. a.** 2; **b.** 4 **37. a.** 2; **b.** 3 **39. a.** 10.0; **b.** 10.0; **c.** 13.0 **41. a.** 2.48; **b.** 4.24; **c.** 3.09 **43.** 7.34 **45.** 11.7 **47.** 7.9×10^{-6} M **49.** $CH_3COO^- + H_2O \rightleftharpoons CH_3COOH + OH^-$ **51. a.** neutral; **b.** basic; **c.** (can't determine)

SPECIAL TOPIC C

SELECTED PROBLEMS

5. a. $K_a = \dfrac{[H^+][OCl^-]}{[HOCl]}$; **b.** $K_a = \dfrac{[H^+][C_6H_7O_6^-]}{[HC_6H_7O_6]}$;

c. $K_a = \frac{[H^+][HCO_2^-]}{[HCO_2H]}$

7. 4.00×10^{-7} **9. a.** 4.2×10^{-4} M; **b.** 3.5×10^{-3} M; **c.** 1.76×10^{-5} M **11. a.** $[OH^-] = 2.38 \times 10^{-11}$ M; **b.** $[OH^-] = 2.9 \times 10^{-12}$ M; **c.** $[OH^-] = 5.7 \times 10^{-10}$ M **13. a.** 6.71×10^{-4} M; **b.** 6.48×10^{-3} M; **c.** 8.60×10^{-6} M **15. a.** $[H^+] = 1.49 \times 10^{-11}$ M; **b.** $[H^+] = 1.54 \times 10^{-12}$ M; **c.** $[H^+] = 1.16 \times 10^{-9}$ M **17. a.** 6.2×10^{-10} M; **b.** 1.65×10^{-3} M; **c.** 4.62×10^{-5} M **19.** 1.80×10^{-4} M OH^-; 5.56×10^{-11} M H^+ **21.** 9.21 **23.** 4.20 **25.** 4.34 **27.** 4.24 **29.** 10.0

CHAPTER 12

SELECTED PROBLEMS

23. Strong: a,b,d,e; Weak: c **25. a.** $Ca + 2\,HCl \longrightarrow H_2 + CaCl_2$; **b.** $Ni + 2\,HCl \longrightarrow H_2 + NiCl_2$; **c.** $Mg + 2\,HNO_3 \longrightarrow H_2 + Mg(NO_3)_2$ **27. a.** $2\,Na + 2\,H_2O \longrightarrow H_2 + 2\,NaOH$; **b.** $Ba + 2\,H_2O \longrightarrow H_2 + Ba(OH)_2$ **29. a.** $Mg + Cu^{2+} \longrightarrow Mg^{2+} + Cu$; **b.** $Ag + Pb^{2+} \longrightarrow$ no reaction; **c.** $Fe + Zn^{2+} \longrightarrow$ no reaction; **d.** $2\,Al + 3\,Ni^{2+} \longrightarrow 2\,Al^{3+} + 3\,Ni$ **31.** Precipitation will occur. **33.** Precipitation will occur.

SPECIAL TOPIC D

SELECTED PROBLEMS

3. Helium is unreactive while hydrogen is combustible.
23. Low oxygen (O_2) availability

35. 

37. 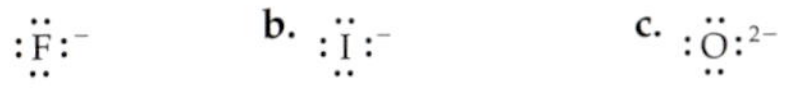

39. a. $4\,Li + O_2 \longrightarrow 2\,Li_2O$; **d.** $CaO + H_2O \longrightarrow Ca(OH)_2$; **c.** $S + O_2 \longrightarrow SO_2$

CHAPTER 13

SELECTED PROBLEMS

3. a. 2; **b.** 7; **c.** 4; **d.** 9 **7.** Saturated: c,d; unsaturated: a,b **9.** aromatic: a,d; aliphatic: b,c **19.** Organic: a, c; Inorganic: b **21. a.** NaOH; **b.** KCl **23. a.** same; **b.** same; **c.** isomers **25. a.** isomers; **b.** same; **c.** isomers

27. a. $CH_3CH_2CH_2CH_2CH_2CH_2CH_3$

b. 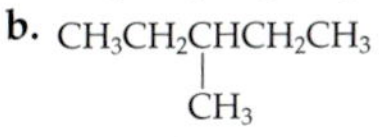 **c.** 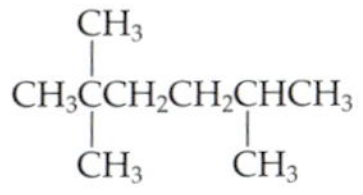

d. 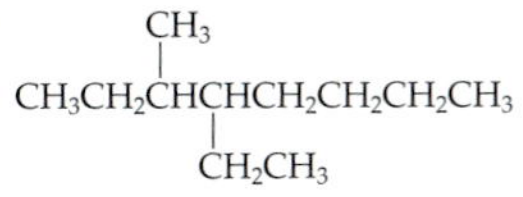

29. a. 3-methylpentane; **b.** 2,3-dimethylbutane

31. a. CH_3CH_2- **b.** 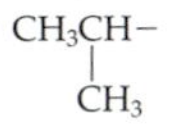

33. 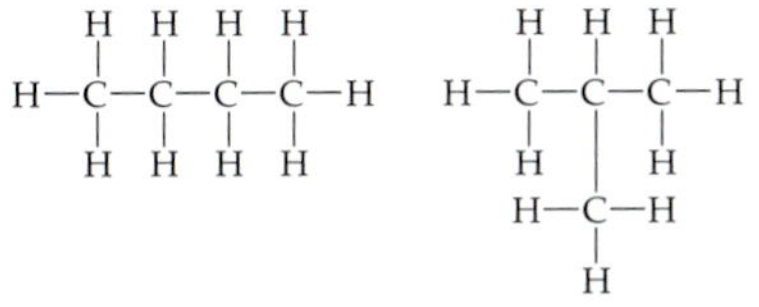

Butane Isobutane or methylpropane

35. a. methylcyclopropane; **b.** 1,2-diethyl-4-methylcyclopentane; **c.** cyclobutene; **d.** 3-ethylcyclohexene

37. a. 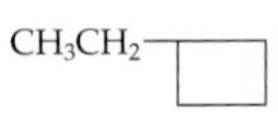 **b.** 

39. a. CH_3Cl; **b.** $CHCl_3$

41. 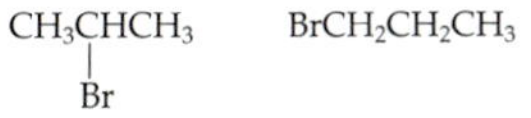

Isopropyl bromide 2-bromopropane Propyl bromide 1-bromopropane

43. a. $HC{\equiv}CH$ **b.** 

c. 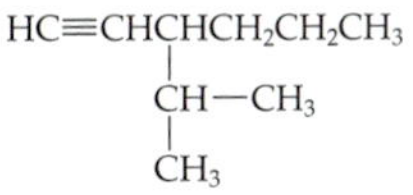 **d.** 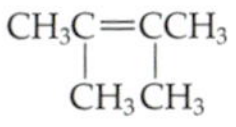

45. a. 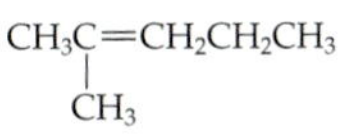 **b.** 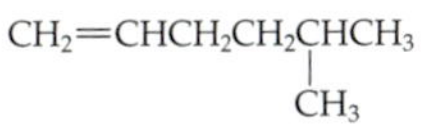

47. a. 2-methyl-1-pentene; **b.** 2-methyl-2-pentene; **c.** 2,5-dimethyl-2-hexene

49. a. 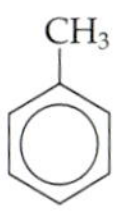 **b.** 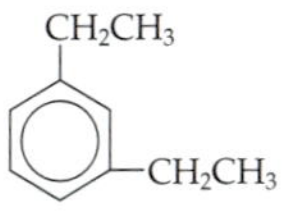 **c.** 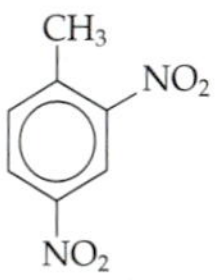

51. a. ethylbenzene; **b.** isopropylbenzene; **c.** 2-nitrotoluene; **d.** 3,5-dichlorotoluene **53. a.** pentane; **b.** $CH_3(CH_2)_4CH_3$; **c.** cyclohexane; **d.** $CH_3(CH_2)_7CH_3$

55. a. $(CH_3)_2CBrCH_2Br$; **b.** $CH_3CH(CH_3)CH_2CH_3$;

c. 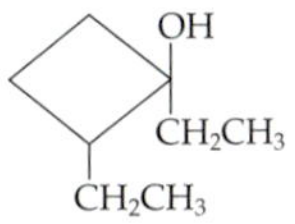

57. a. H_2, Ni; **b.** H_2O, H^+

59. a.  **b.** 

CHAPTER 14

SELECTED PROBLEMS

17. a. 1-hexanol; **b.** 2-hexanol **19. a.** 4,4-dichloro-2-butanol; **b.** 3,3-dibromo-2-methyl-2-butanol

21. a. $CH_3CH_2CHOHCH_2CH_2CH_3$; **b.** $CH_3CHOHC(CH_3)_2CH_3$

23. a. $CH_3CH_2CHOHCH(CH_3)CH(CH_3)CH_2CH_3$;

b. 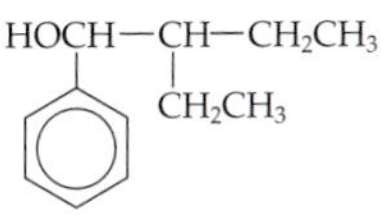

c. $CH_3CHOHCH_2OH$

25. a. dipropyl ether; **b.** diphenyl ether

27. a. $CH_3CH_2OCH_3$ **c.** 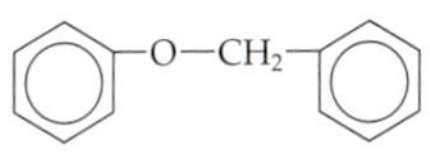

29. a. 2-nitrophenol; **b.** 4-bromophenol

31. a. 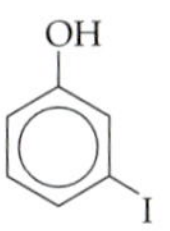 **b.** 

33. a. oxidation; **b.** dehydration; **c.** hydration

35.

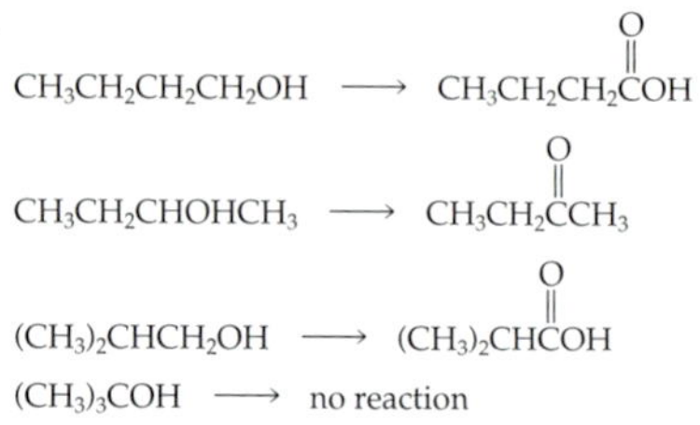

37.

39. **a.** $CH_3CHOHCH_2CH_3$ **b.**

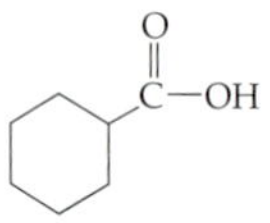

41. **a.** H^+, H_2O; **b.** $K_2Cr_2O_7$, H^+; **c.** conc. H_2SO_4, 140 °C, excess alcohol

43. **a.** $CH_3CH{=}CH_2$ **b.**

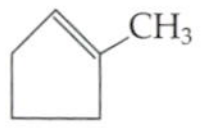

c.

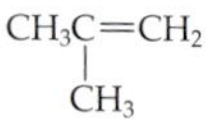

45. **a.**

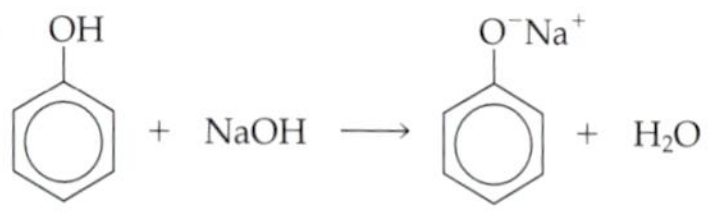

b. No reaction.

47. methanol, ethanol, 1-propanol 49. methanol, 1-butanol, 1-octanol

CHAPTER 15

SELECTED PROBLEMS

5. **a.** benzaldehyde; **b.** 3-hydroxypropanal; **c.** 4,4-dimethylpentanal; **d.** 2-chlorobenzaldehyde 7. **a.** 5-methyl-3-hexanone; **b.** cyclopentanone; **c.** 2-pentanone; **d.** 4-bromo-2,2-dimethyl-3-pentanone

9. **a.**

b.

c.

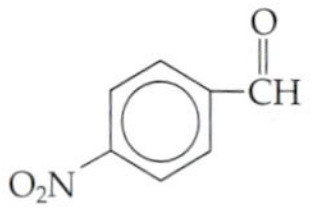

11. **a.**

b.

c.

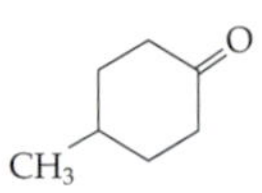

13. 2-propanol 15. acetaldehyde

17. **a.**

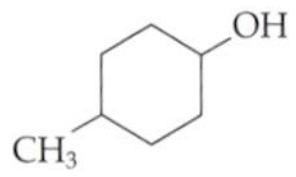

b. $(CH_3)_3CCH_2OH$ **c.** $HOCH_2CH_2CHBrCH_2CH_3$

19. **a.**

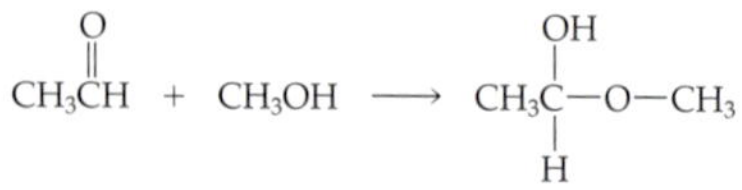

b.

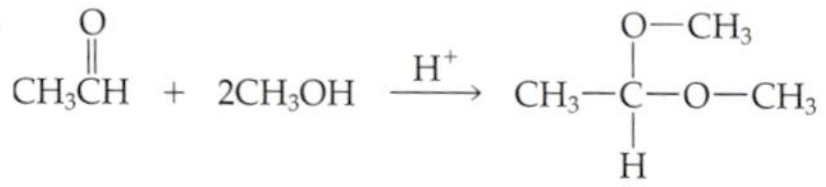

c.

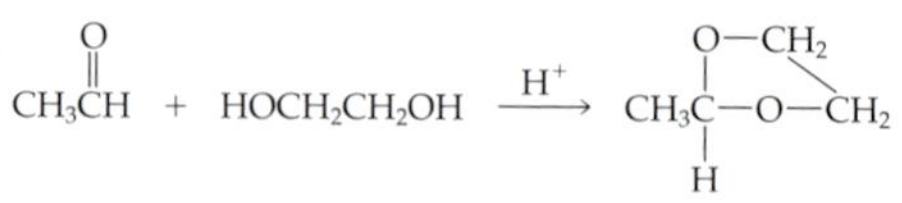

21. **a.**

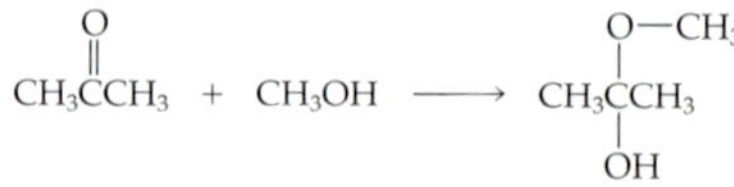

b.

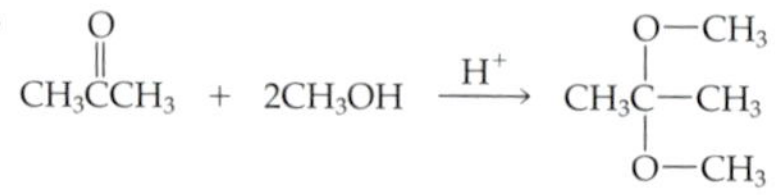

c.

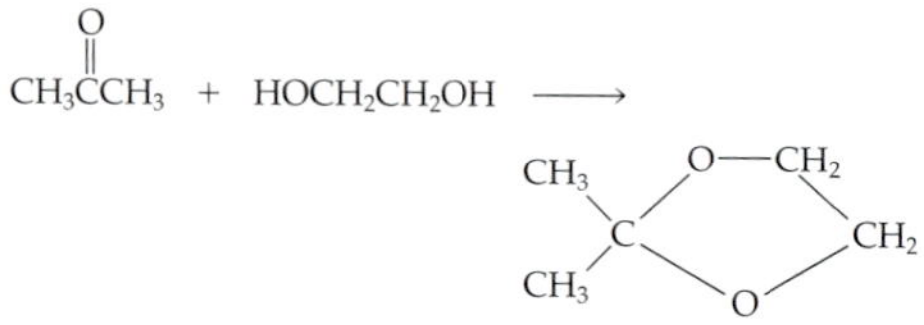

23. **a.** Yes; only pentanal would test positive; **b.** No; **c.** Yes, only pentanal would test positive; **d.** Yes, only pentanal would test positive; **e.** No 25. Phenylhydrazine gives an orange precipitate with 2-pentanone. 27. **a.** Ag^+, NH_3; **b.** CrO_3, HCl, pyridine, CH_2Cl_2; **c.** $2CH_3OH$, H^+ 29. a,d 31. c,e

CHAPTER 16

SELECTED PROBLEMS

3. short chain carboxylic acids; esters

5. **a.**

b.

c.

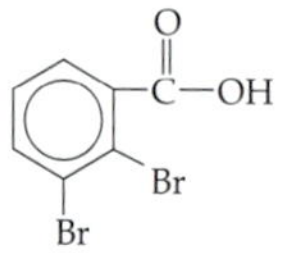

d.

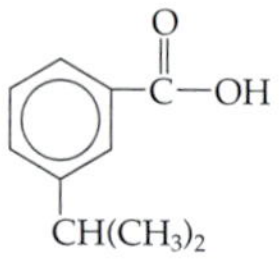

7. **a.**

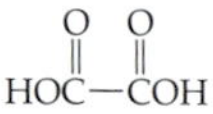

b.

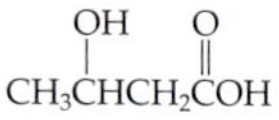

9. **a.** 3-methylbutanoic acid; **b.** 3,4,4-trimethylpentanoic acid; **c.** 4-hydroxybutanoic acid; **d.** 2,4-dimethylpentanoic acid

11. **a.**

b.

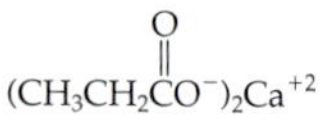

13. **a.**

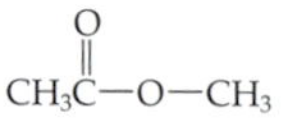

b.

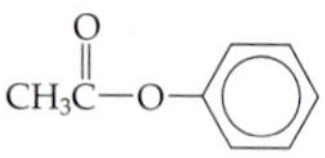

15. **a.**

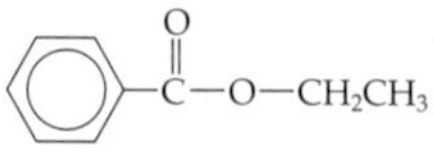

b.

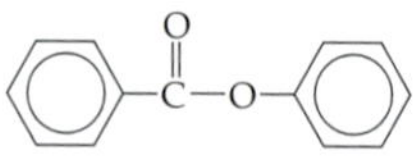

17. **a.** methyl benzoate; **b.** methyl formate (or methyl methanoate); **c.** ethyl propionate (or ethyl propanoate)

19. **a.**

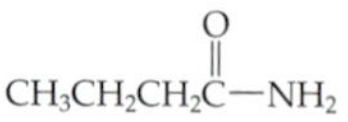

b.

c.

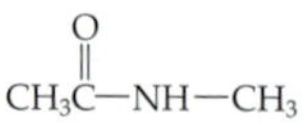

21. **a.** benzamide; **b.** 2-methylbutanamide; **c.** acetamide

23. II; Butanoic acid has the ability to form dimers which are hydrogen bonded to one another; the ether can't hydrogen bond.

25. I; The amide can hydrogen bond while the ester can't.

27. I; The acid can both hydrogen bond with water and ionize slightly in water. This helps it's solubility. The alkane is completely insoluble in water due to its lack of polarity.
29. I; The longer hydrocarbon chain of II will make it less able to mix with the polar water.
31. a.

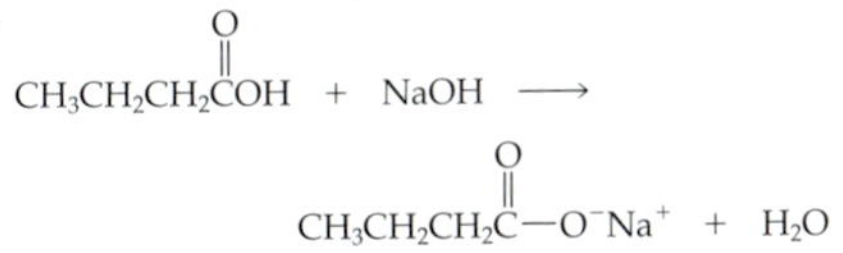

$$CH_3CH_2CH_2COH + NaOH \longrightarrow$$

$$CH_3CH_2CH_2C-O^-Na^+ + H_2O$$

b.

$$CH_3CH_2CH_2COH + NaHCO_3 \longrightarrow$$

$$CH_3CH_2CH_2C-O^-Na^+ + CO_2 + H_2O$$

33.

$$CH_3CO-CH_2CH_3 + H_2O \xrightarrow{H^+}$$

$$CH_3COH + HOCH_2CH_3$$

35.

$$\text{(benzene)}-C-NH_2 + H_2O \xrightarrow{H^+}$$

$$\text{(benzene)}-COH + NH_4^+$$

37. a.

$$CH_3CH_2C-O^-Na^+ + H_2O$$

b.

$$\text{(benzene)}\begin{matrix} COO^-Na^+ \\ COO^-Na^+ \end{matrix} + 2\,H_2O + 2\,CO_2$$

39. a.

$$\text{(benzene)}-C-O^-Na^+ + HOCH_2CH_2CH_3$$

b.

$$\text{(cyclohexane)}-O-CCH_3 + H_2O$$

41. a.

$$CH_3C-O-CH_2CH_2CH_3 + H_2O$$

b.

$$CH_3-O-CCH_2C-O-CH_3 + 2\,H_2O$$

43. a.

$$CH_3C-OH + NH_4Cl$$

b.

$$\text{(benzene)}-C-O^-Na^+ + HN(CH_3)_2$$

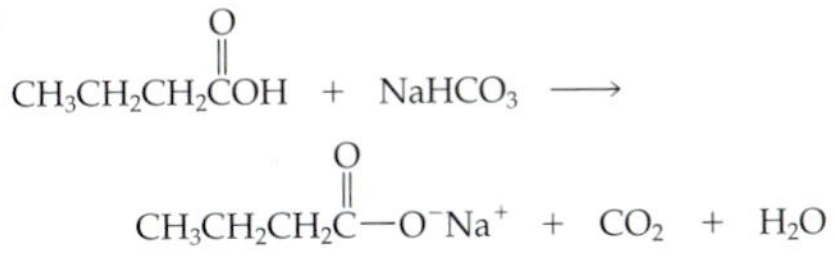

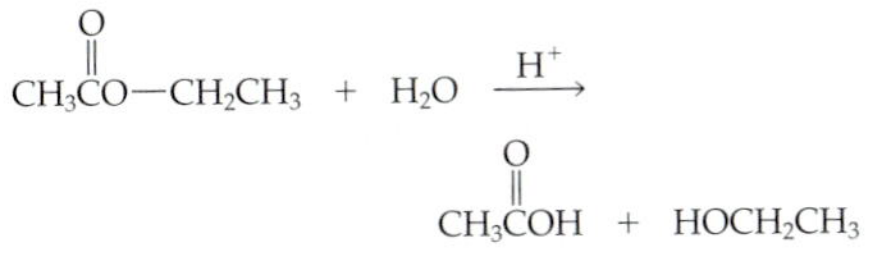

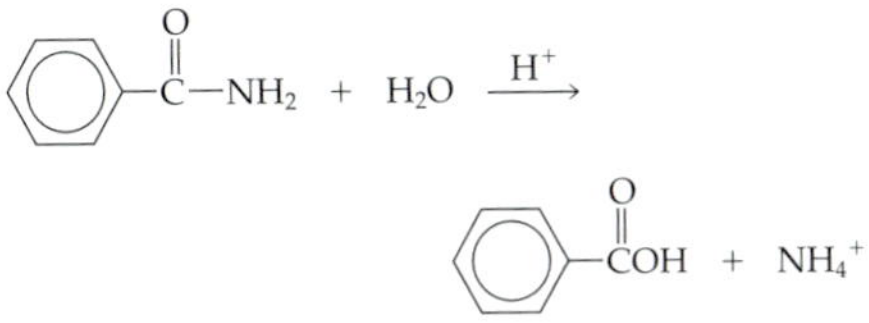

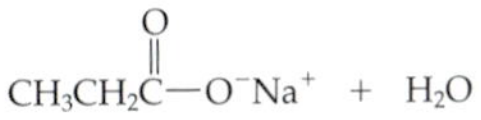

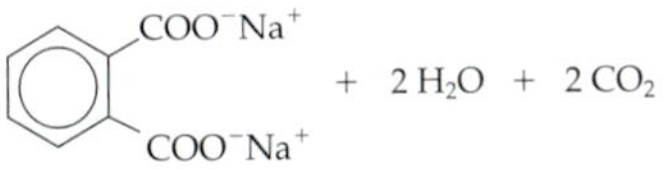

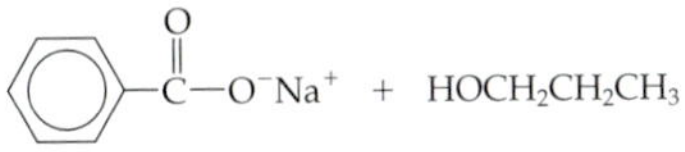

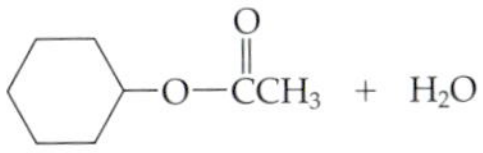

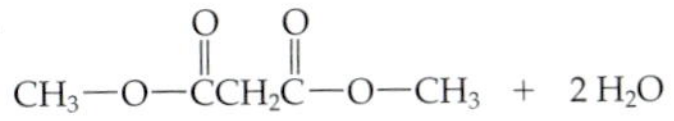

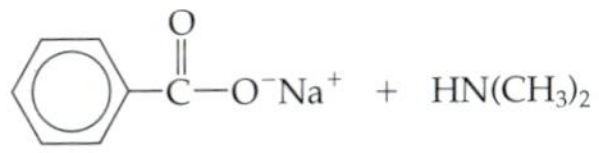

45. a. $K_2Cr_2O_7$, H^+; **b.** $K_2Cr_2O_7$, H^+; **c.** NaOH
47. a.

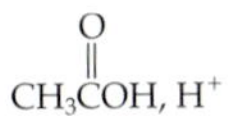

$$CH_3COH, H^+$$

b. LiOH

49. a.

$$CH_3CH_2O-P-OCH_2CH_3 \;(\text{with } OH)$$

b.

$$CH_3O-P-OH \;(\text{with } OH)$$

c.

$$HO-P-O-P-O-P-OH \;(\text{with } OH, OH, OH)$$

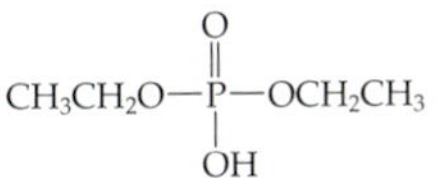

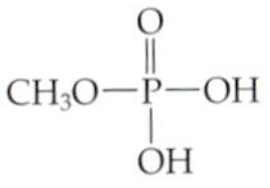

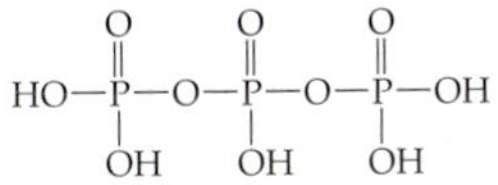

SPECIAL TOPIC F
SELECTED PROBLEMS
3. a.

$$\text{(benzene with COOH and } O-CCH_3\text{, } O)$$

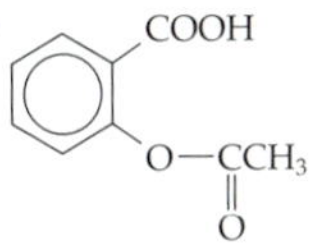

b. Different doses; different additives (such as buffering agents, acetaminophen, caffeine); **c.** All must contain acetylsalicylic acid.
15. Carboxylic acid **17. a.** $HOCH_3$, H^+; **b.** NaOH

CHAPTER 17
SELECTED PROBLEMS
3. acidic: c; basic: a; neutral: b,d **7. a.** amide; **b.** neither; **c.** both
9. a. alcohol (1°); **b.** amine (1°); **c.** alcohol (2°); **d.** amine (1°);
e. ether; **f.** phenol
11. a. CH_3NHCH_3

b. $CH_3CH_2N-CH_2CH_3$ (with CH_3)

c. (cyclobutane with OH and NH_2)

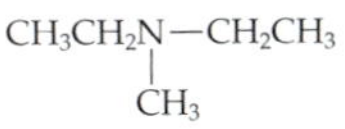

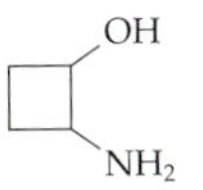

d. $HOCH_2CH_2NH_2$

13. a. (benzene with NH_2)

b. Br (benzene with NH_2)

c. (pyrimidine ring with N)

d. (benzene with $NHCH_2CH_3$)

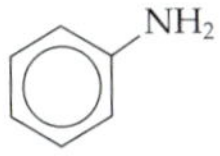

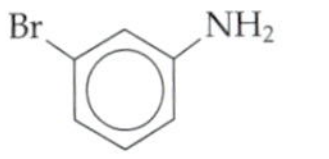

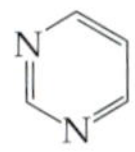

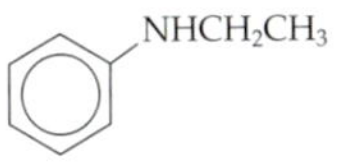

15. a. propylamine; **b.** methylisopropylamine; **c.** triethylamine;
d. 2-aminopentane
17. a. (benzene with $NH_3^+Br^-$)

b. $CH_3-N-CH_3^+Cl^-$ (with CH_3 top and CH_3 bottom)

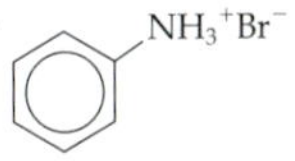

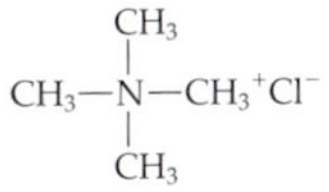

19. a. diethylammonium bromide; **b.** tetraethylammonium iodide
21. Butylamine. It can hydrogen bond (while pentane cannot).
23. Propylamine. Tertiary amines have no hydrogen bonded to the nitrogen; therefore they do not undergo hydrogen bonding.
25. $CH_3CH_2NH_2$. It has the ability to hydrogen bond with water.
27. NH_2 NH_2 NH_2 (on $CH_2CH_2CHCH_2CHCH_3$) — The more hydrogen bonding groups placed on a compound, the more its solubility in water.
29. $CH_3NH_3^+Br^-$ **31.** $[(CH_3)_3NH^+]_2SO_4^{-2}$
33.

$$CH_3(CH_2)_4C-NH(CH_2)_3CH_3$$

35.

$$\text{(benzene)}-C-NH-\text{(benzene)}$$

37.

$$CH_3CH_2NHCH_3 \text{ and } HOCCH_2CH_3$$

39.

$$\text{(benzene)}-NH_2 + HOCCH_2CH_3$$

41.

$$\text{(benzene)}-COH + NH_3$$

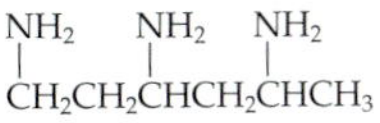

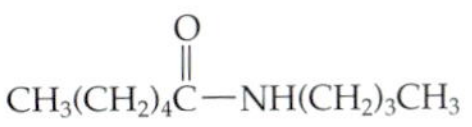

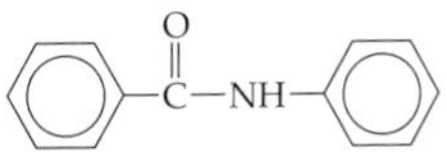

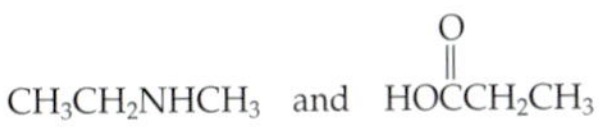

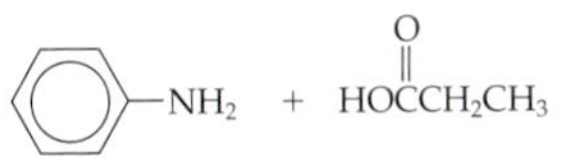

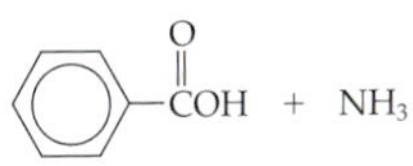

43. a. HCl; **b.** HNO_3

SPECIAL TOPIC G
SELECTED PROBLEMS
13. a. Barbiturates all contain a six-membered barbituric acid ring structure. **b.** Groups are added to a carbon located between a pair of carbonyl groups to alter the drug's properties such as effectiveness and length of duration. **15.** Amide, ketone, phenol, alcohol, amine **17.** Cocaine

CHAPTER 18
SELECTED PROBLEMS
3. Yes, no **7. a.** No; **b.** Yes; **c.** No. **9.** Enantiomers have the same physical and chemical properties. They rotate plane-polarized light the same number of degrees but differ in the direction (counterclockwise vs clockwise) the light is rotated. **11. a.** the carbon of the $-CH-$ group; **b.** none

13. a.
$$CH_3CHCH_2OH$$
$$\overset{|}{OH}$$

b.
$$CH_3CHCOOH$$
$$\overset{|}{NH_2}$$

c.
$$C_6H_5CH_2CHCH_3$$
$$\overset{|}{NH_2}$$

d.
$$CH_3CHCH_2CH_3$$
$$\overset{|}{Br}$$

e.
$$CH_3CHCHO$$
$$\overset{|}{OH}$$

f.
$$CH_3CH-CHCH_3$$
$$\overset{|}{OH}\ \overset{|}{OH}$$

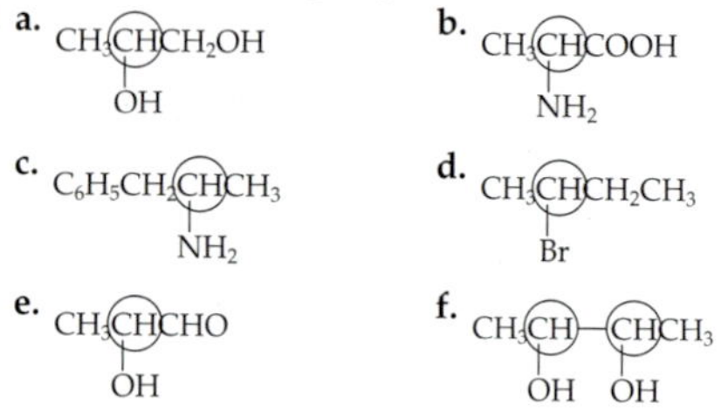

15. a. yes; **b.** no, **c.** yes; **d.** no

17. a.

$$\begin{array}{cc} CH_3 & CH_3 \\ | & | \\ H-C-OH & HO-C-H \\ | & | \\ CH_2CH_3 & CH_2CH_3 \end{array}$$

b.

$$\begin{array}{cc} COOH & COOH \\ | & | \\ H-C-CH_3 & CH_3-C-H \\ | & | \\ CH_3 & CH_3 \end{array}$$

19.

$$\begin{array}{cc} CHO & CHO \\ | & | \\ H-C-Cl & Cl-C-H \\ | & | \\ Cl-C-H & H-C-Cl \\ | & | \\ CH_3 & CH_3 \end{array}$$
Enantiomers

$$\begin{array}{cc} CHO & CHO \\ | & | \\ H-C-Cl & Cl-C-H \\ | & | \\ H-C-Cl & Cl-C-H \\ | & | \\ CH_3 & CH_3 \end{array}$$
Enantiomers

21. a,d **23. a.** 4; **b.** 8

25. a.

$$\begin{array}{cc} Br\ \ \ \ H & CH_3\ \ \ H \\ C=C & C=C \\ CH_3\ \ CH_2CH_3 & Br\ \ \ CH_2CH_3 \\ cis & trans \end{array}$$

b.

$$\begin{array}{cc} H\ \ \ \ H & CH_3CH_2\ \ H \\ C=C & C=C \\ CH_3CH_2\ \ CH_2CH_3 & H\ \ \ CH_2CH_3 \\ cis & trans \end{array}$$

c. none; **d.** none

e.

$$\begin{array}{cc} H\ \ \ \ H & CH_3\ \ \ H \\ C=C & C=C \\ CH_3\ \ \ COOH & H\ \ \ COOH \\ cis & trans \end{array}$$

f.

$$\begin{array}{cc} H\ \ \ \ H \\ C=C \\ CH_3\ \ \ CH(CH_3)_2 \\ cis \end{array}$$

$$\begin{array}{cc} CH_3\ \ \ H \\ C=C \\ H\ \ \ CH(CH_3)_2 \\ trans \end{array}$$

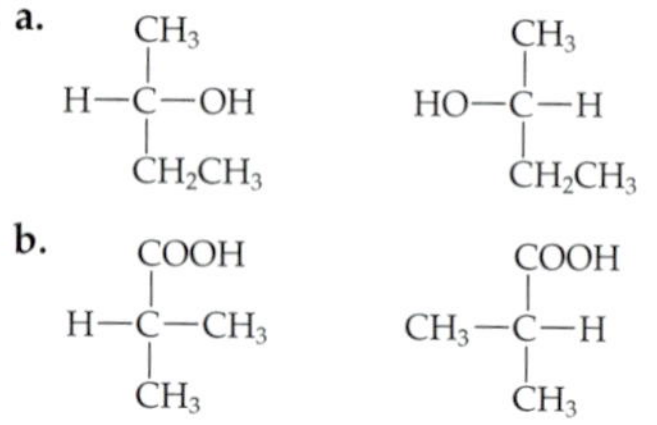

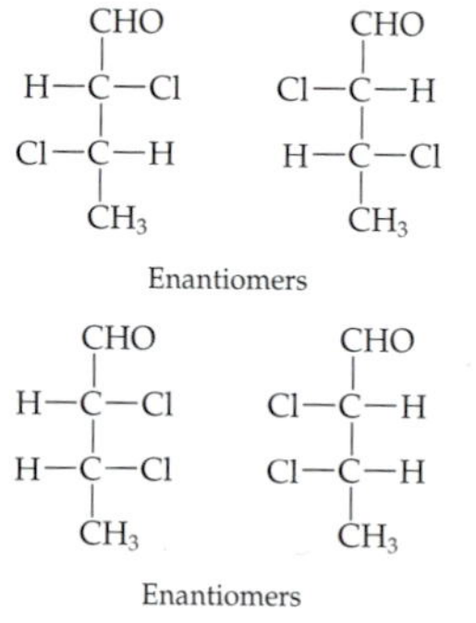

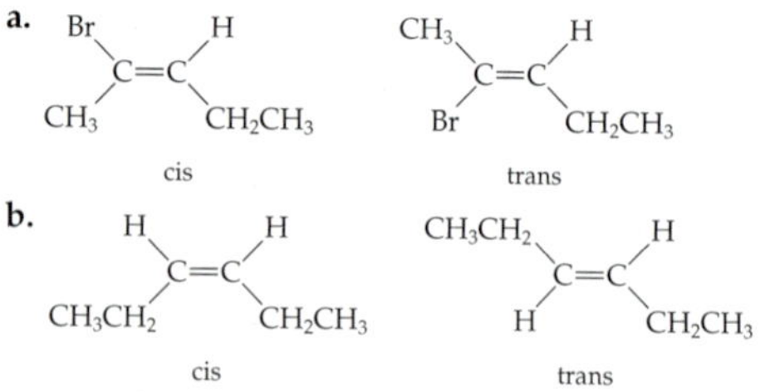

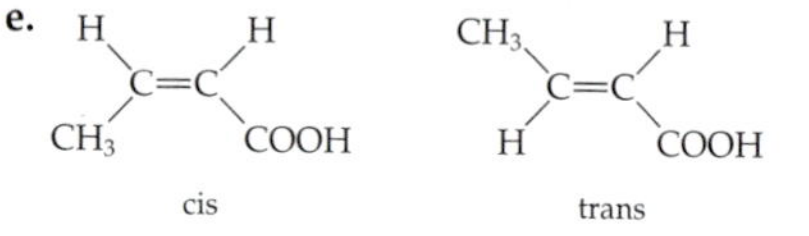

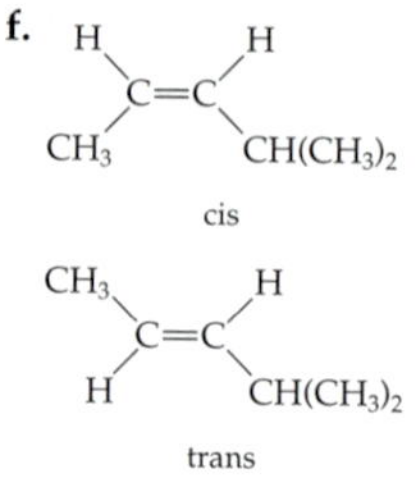

CHAPTER 19
SELECTED PROBLEMS
5. Yes. They are small molecules with plenty of hydrogen-bonding $-OH$ groups. **11. a.** glucose; **b.** lactose; **c.** glucose; **d.** fructose; **e.** sucrose; **f.** maltose **15. a.** D; **b.** L
17. Aldoses: a, b, c; Ketoses: d **19. a.** hexose; **b.** pentose

21.

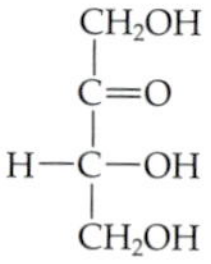

$$\begin{array}{c} CH_2OH \\ | \\ C=O \\ | \\ H-C-OH \\ | \\ CH_2OH \end{array}$$

23.

$$\begin{array}{cc} CHO & CHO \\ | & | \\ H-C-OH & HO-C-H \\ | & | \\ HO-C-H & HO-C-H \\ | & | \\ H-C-OH\ \ \text{D-Glucose} & H-C-OH\ \ \text{D-Mannose} \\ | & | \\ H-C-OH & H-C-OH \\ | & | \\ CH_2OH & CH_2OH \end{array}$$

25.

$$\begin{array}{c} CH_2OH \\ \\ OH \\ OH\ \ \ \ \ \ OH \\ OH \end{array}$$

27.

$$\begin{array}{c} CH_2OH \\ \\ OH\ \ \ \ \ OH \\ OH \\ OH \end{array}$$

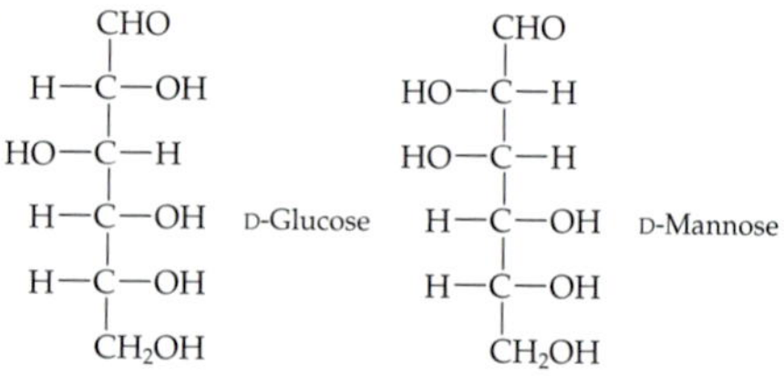

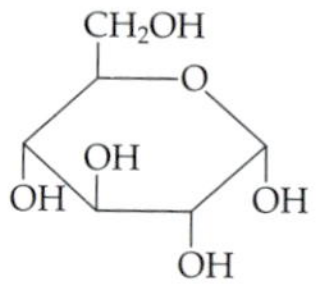

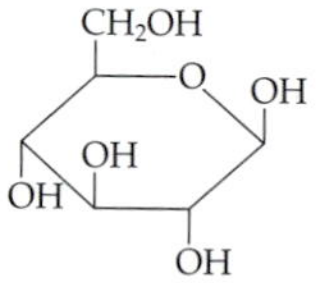

29. a. beta; **b.** alpha **31. a.** alpha; **b.** beta; both are reducing sugars **33. a.** beta; **b.** no; **c.** no **37.** a,b,c,and d will give positive tests **39. a.** glucose; **b.** glucose; **c.** glucose
41. a. $Ag(NH_3)_2^+$; **b.** CH_3OH, HCl

CHAPTER 20
SELECTED PROBLEMS
7. Corn oil. **11. a.** fatty acids and glycerol; **b.** salts of fatty acids and glycerol **19.** c,d **23.** Saturated: a,c; Unsaturated: b
25. b **27.** Glycerol: a; Sphingosine: b; neither: c
29. a. $CH_3(CH_2)_{14}COOH$; **b.** $CH_3(CH_2CH=CH)_3(CH_2)_7COOH$

c.

$$\begin{array}{l} O \\ \parallel \\ CH_2-O-C(CH_2)_7CH=CH(CH_2)_7CH_3 \\ | \ \ \ \ \ O \\ \ \ \ \ \ \ \parallel \\ CH-O-C(CH_2)_7CH=CH(CH_2)_7CH_3 \\ | \ \ \ \ \ O \\ \ \ \ \ \ \ \parallel \\ CH_2-O-C(CH_2)_7CH=CH(CH_2)_7CH_3 \end{array}$$

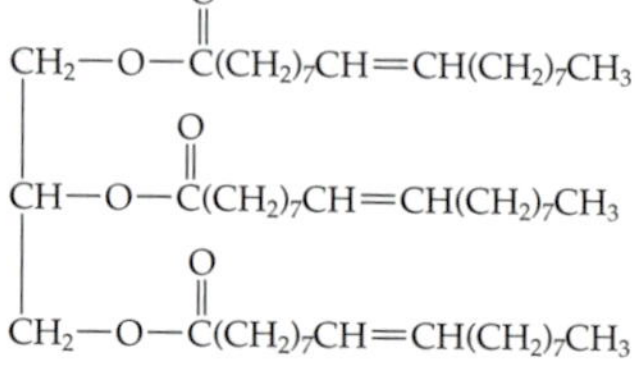

31. a.

$$C_{17}H_{29}\overset{O}{\overset{\|}{C}}-O-CH_2$$
$$C_{17}H_{29}\overset{O}{\overset{\|}{C}}-O-CH$$
$$C_{17}H_{29}\overset{O}{\overset{\|}{C}}-O-CH_2$$

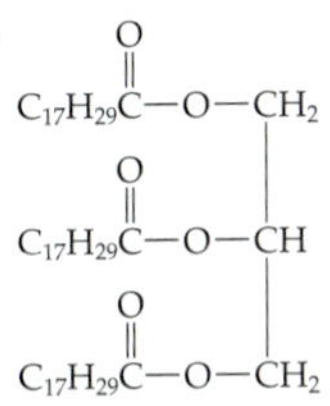

b. $C_{17}H_{33}COO^-Na^+$ **c.** $(C_{13}H_{27}COO^-)_2Ca^{+2}$

33.
$$\begin{array}{l} CH_2OH \qquad Na^+{}^-OOC(CH_2)_6CH_3 \\ CHOH \quad + \quad Na^+{}^-OOC(CH_2)_4CH_3 \\ CH_2OH \qquad Na^+{}^-OOC(CH_2)_8CH_3 \end{array}$$

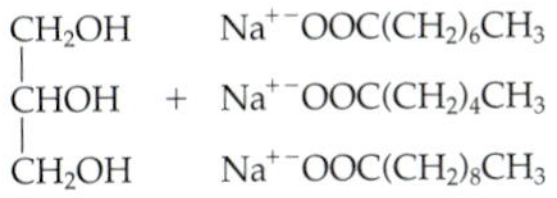

SPECIAL TOPIC I
SELECTED PROBLEMS
3. a. 2; **b.** 3; **c.** 1; **d.** 4 **5.** Pituitary **7.** Steroids

CHAPTER 21
SELECTED PROBLEMS
3. a. asparagine; **b.** glycine; **c.** hydroxyproline; **7.** Globular
17. a. $-CH_2$, **b.** $-(CH_2)_4NH_2$

$$-CH_2\overset{\displaystyle CH_2}{}$$
$$-CH_2$$

c.
$$-CH_2-\langle\bigcirc\rangle-OH$$

19. a. NH_2-CH_2-COOH **b.** $NH_2-CH-COOH$
$$\qquad\qquad\qquad\qquad\qquad\qquad\quad CH_3$$

c. $NH_2-CH-COOH$
$$\qquad\qquad CH$$
$$\qquad CH_3\;CH_3$$

21. $NH_3{}^+-CH-COO^-$
$$\qquad\quad CH_2$$
$$\qquad\quad CH_2 \qquad$$ Aspartic acid (also glutamic acid
$$\qquad\quad COOH \qquad$$ is an acidic amino acid)

23. a. proline; **b.** histidine, tryptophan; **c.** phenylalanine, tyrosine, tryptophan; **d.** citrulline, ornithine, dihydroxyphenylalanine, thyroxine, homocysteine, homoserine, β-alanine, γ-aminobutyric acid

25. a.
$$NH_3{}^+-CH_2-\overset{O}{\overset{\|}{C}}-NH-CH-COO^-$$
$$\qquad\qquad\qquad\qquad\qquad CH_3$$

b.
$$NH_3{}^+-CH-\overset{O}{\overset{\|}{C}}-NH-CH_2-COO^-$$
$$\qquad\quad CH_3$$

27.
$$NH_3{}^+-CH-\overset{O}{\overset{\|}{C}}-NH-CH-\overset{O}{\overset{\|}{C}}-NH-CH_2-COO^-$$
$$\qquad\quad CH_2OH \qquad\qquad CH_3$$

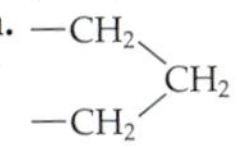

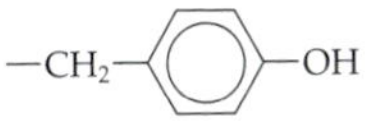

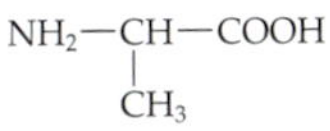

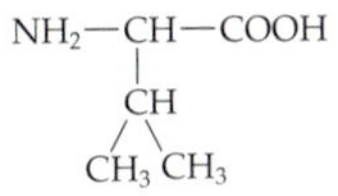

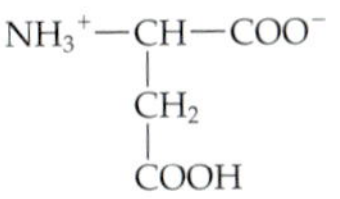

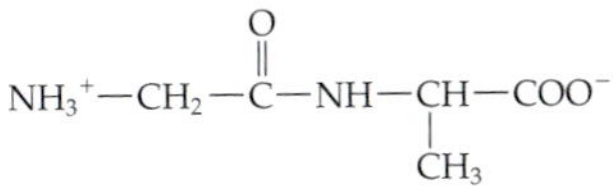

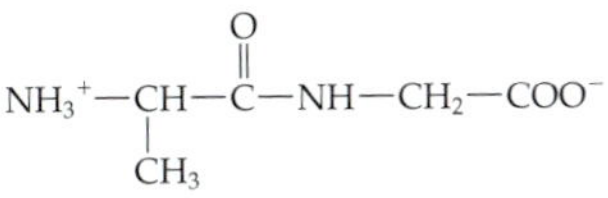

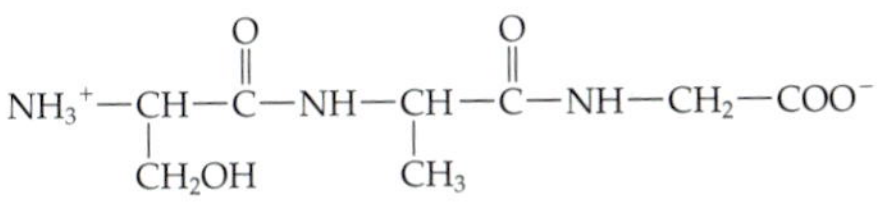

29. $H_2N-CH_2-COO^-Na^+$ **31. a.** low pH; **b.** high pH; **c.** isoelectric pH **33.** Silk has an arrangement of polypeptide chains in a manner where they run parallel to one another in a zigzag pattern. The chains are held together by hydrogen bonds. This secondary structure is called a β-pleated sheet.
35. Salt linkages, hydrogen bonds, disulfide linkages, and hydrophobic interactions. **37.** Globular: a; Fibrous: b,c
41. Alanine—stationary; histidine—to cathode; aspartic acid—to anode.

CHAPTER 22
SELECTED PROBLEMS
3. a. Maltose; **b.** cellulose; **c.** proteins (peptides); **d.** lipids
21. a. Lyases; **b.** peptidases; **c.** transferases; **23.** So that trypsin will not be active in the pancreas (where it's synthesized) and, thus, degrade important proteins in that tissue. **25.** It doubles.
27. Less active in both cases. **29.** It increases the rate. (More enzymes working at a given time on substrate). **31. a.** ethanol; **b.** Zn^{+2}; **c.** protein molecule without the Zn^{+2} ion
33. Yes. Yes. **35.** Lactase. Hydrolases. **37.** Penicillin-like compounds can be synthesized which resist cleavage by penicillase; a penicillase inhibitor (like clauvulinic acid) can be combined with the penicillin to prevent its (the penicillin) degradation.

SPECIAL TOPIC J
SELECTED PROBLEMS
11. Niacin **13.** Pantothenic acid **15.** Ascorbic acid—Vitamin C; Ergocalciferol—Vitamin D; Cyanocobalamin—Vitamin B_{12}; Retinol—Vitamin A; Tocopherol—Vitamin E **17.** Scurvy = vitamin C deficiency; Rickets = vitamin D deficiency; night blindness = vitamin A deficiency **19.** Water soluble: B_6, B_{12}, C; fat soluble: A, K **21. a.** water soluble; **b.** fat soluble

CHAPTER 23
SELECTED PROBLEMS
3. a. ribonucleic acid (RNA) and deoxyribonucleic acid (DNA); **b.** DNA **5. a.** adenine, guanine, thymine, cytosine, deoxyribose, phosphoric acid; **b.** adenine, guanine, uracil, cytosine, ribose, phosphoric acid **11.** A purine-pyrimidine pair fits well with the diameter of the helix. **17.** mRNA, tRNA and rRNA **23.** Physical mutagens: ultraviolet and gamma radiation; Chemical mutagens: 5-bromouracil, 2-aminopurine, hydroxylamine, nitrous acid **25.** ribose **27.** adenines, cytosines, guanines, and thymines **29. a.** neither; **b.** nucleoside; **c.** nucleotide **31.** Ribose: a, b Deoxyribose: c

33. a.

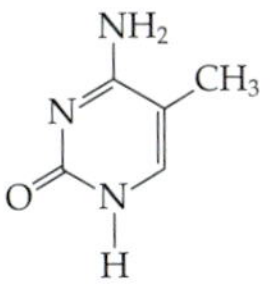

5-methylcytosine

b.

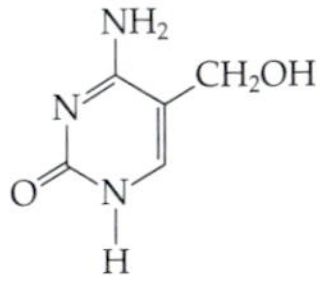

5-hydroxymethylcytosine

c.

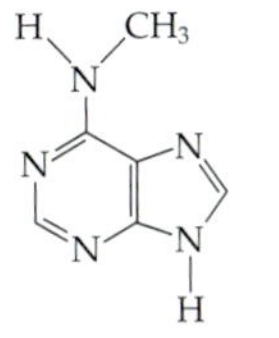

6-methyladenine

d.

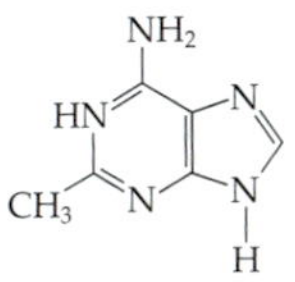

2-methyladenine

35.

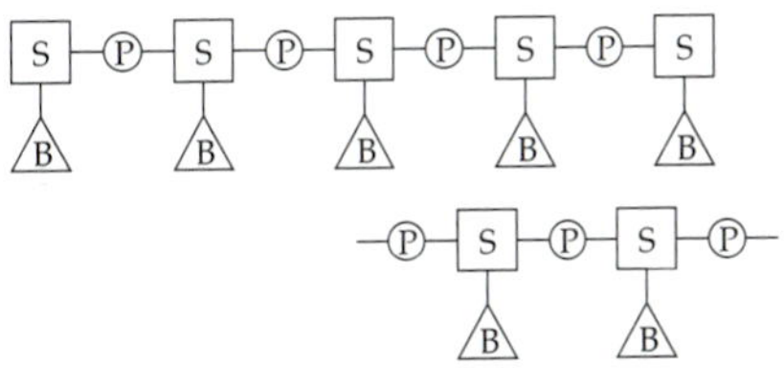

37. a. guanine; **b.** thymine; **c.** cytosine; **d.** adenine

39. . . .TAAGC . . . **41. a.** AAA; **b.** GUA; **c.** UCG; **d.** GGC **43. a.** Phe; **b.** His; **c.** Ser; **d.** Pro **45. a.** Leu-Pro-Gly; **b.** Ala-Ser; **c.** Pro-Pro-Pro

SPECIAL TOPIC K

SELECTED PROBLEMS

1. Nucleic acid (DNA or RNA) surrounded by a protein coat **17.** Charcoal-grilled meats, cigarette smoke, automobile exhausts, coffee, burnt sugar. **19.** Dimethylnitrosamine, vinyl chloride **21.** It is attached to sugar and phosphate groups and subsequently inhibits an enzyme involved in the formation of thymine nucleotides. **23.** Folic acid. It is a competitive inhibitor of an enzyme needed to transform folic acid into its active form. Without the active form of folic acid, thymine can not be synthesized and replication of the cell cannot occur.

CHAPTER 24

SELECTED PROBLEMS

3. hydrolysis **5.** Monosaccharides **9.** Glycogen **13.** AMP contains a single phosphate group attached to the ribose sugar, ADP has two phosphate groups hooked together onto the ribose sugar, ATP contains three phosphate groups. **15.** a,c

17.

Location	Function
a. mouth	—breaks down starch
b. intestine	—cleaves lactose to galactose and glucose
c. intestine	—cleaves sucrose to fructose and glucose

19. The synthesis and breakdown of glycogen in liver **21.** Both epinephrine and glucagon work on the liver to cause glycogenolysis, i.e., the breakdown of glycogen to glucose. The glucose produced enters the blood and, thus, raises blood glucose levels. **23.** Insulin is a protein while the oral antidiabetic drugs are small molecules, all with a para arrangement of the sulfonyl group to another substitutent. The oral antidiabetic drugs don't work like insulin; they stimulate release of insulin from the islet cells of the pancreas. **25.** On the outside of the cell membrane

27. a.

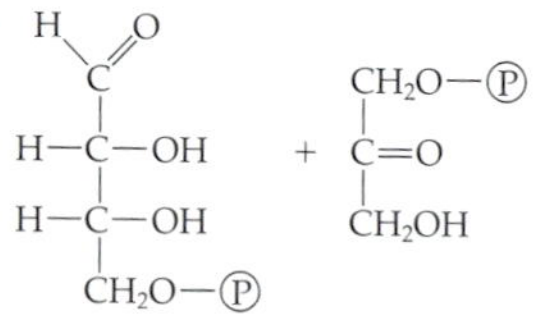

b.

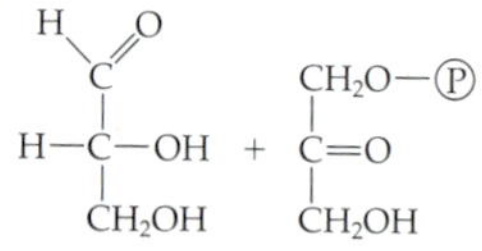

29. a. Step 6, glyceraldehyde 3-phosphate to 1,3-bisphosphoglyceric acid; **b.** NAD^+ **31.** 70–80% of the lactic acid diffuses out of the muscle into the bloodstream and is transported to the liver. There it may be oxidized to pyruvic acid and hence to CO_2 and H_2O (via the Krebs cycle) or it may be converted back into glucose (gluconeogenesis). The remaining 20–30% remains in the muscle cells where it can be reoxidized to pyruvic acid, which then enters the Krebs cycle and is further oxidized to CO_2 and H_2O (assuming the availability of oxygen in the muscle cells). **33. a.** acetaldehyde; **b.** dihydroxyacetone phosphate + glyceraldehyde 3-phosphate; **c.** enolase; **d.** glucose 6-phosphate; **e.** glucose 1-phosphate; **f.** triose phosphate isomerase; **g.** glucose 6-phosphate; **h.** pyruvic acid **35. a.** 33g, 34a; **b.** 34d, 34g; **c.** 33f, 34b; **d.** 34c

CHAPTER 25

SELECTED PROBLEMS

7. The nicotinamide and flavin groups **9.** The glycerol phosphate shuttle and the malic acid-aspartic acid shuttle **11.** CO_2 **13.** 3 **15.** O_2, H_2O **17. a.** citric acid; **b.** oxaloacetic acid + acetyl-CoA; **c.** L-malic acid **d.** isocitric acid dehydrogenase; **e.** α-ketoglutaric acid dehydrogenase complex **19. a.** 17a; **b.** 17a,17b,17c,18d; **c.** 17a

21.

$$CH_3\overset{O}{\overset{\|}{C}}COH + CoASH \longrightarrow CH_3\overset{O}{\overset{\|}{C}}SCoA + CO_2 + 2H^+ + 2e^-$$

$$HOOCCH_2CH(COOH)CHOHCOOH \longrightarrow HOOCCH_2CH(COOH)\overset{O}{\overset{\|}{C}}COOH + 2H^+ + 2e^-$$

$$HOOCCH_2CH_2\overset{O}{\overset{\|}{C}}COOH + CoASH \longrightarrow HOOCCH_2CH_2\overset{O}{\overset{\|}{C}}SCoA + CO_2 + 2H^+ + 2e^-$$

$$HOOCCH_2CH_2COOH \longrightarrow HOOCCH{=}CHCOOH + 2H^+ + 2e^-$$

$$HOOCCH_2CHOHCOOH \longrightarrow HOOCCH_2\overset{O}{\overset{\|}{C}}COOH + 2H^+ + 2e^-$$

23. 1,3-bisphosphoglyceric acid ⟶ 3-phosphoglyceric acid; phosphoenolpyruvic acid ⟶ pyruvic acid; succinyl CoA ⟶ succinic acid **25.** To carry out oxidative phosphorylation and, thus, produce ATPs. **27.** Carbohydrates **29.** Anaerobic **31.** Type I: Slow twitch, suited to aerobic oxidation; Type IIB: Fast twitch, suited to anaerobic glycolysis **33.** The high catalytic activity of actomyosin in Type IIB fibers suggests that the tissue can hydrolyze ATP at high rates and, thus, is important for bursts of vigorous physical activity. **35.** Type I (slow twitch). Endurance training increases the size and number of mitochondria in Type I muscle fibers, therefore aiding the individual in performing aerobic work.

CHAPTER 26

SELECTED PROBLEMS

7. Monoglycerides, fatty acids, and free cholesterol **11. a.** glucose; **b.** fatty acids; **c.** fatty acids; **d.** glucose and fatty acids; **e.** fatty acids; **f.** glucose; **g.** fatty acids; **h.** ketone bodies **25.** Fat reserves **29.** 2 hours **31.** 0.1 hr

33.

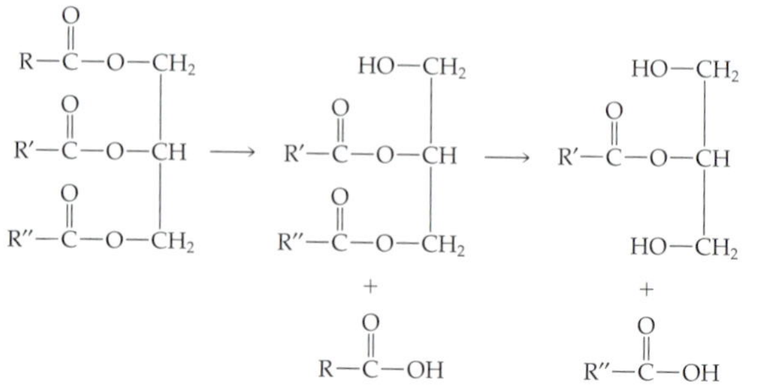

35. After the products of lipid digestion cross the intestinal wall, they are immediately resynthesized into glycerides, phospholipids, or cholesterol esters. These are then conjugated to proteins and transported to the blood via the lymphatic system.

37. a. Each turn of the spiral produces a fatty acyl CoA containing two fewer carbon atoms rather than reproducing the identical fatty acyl CoA. **b.** The carbon atom beta to the carboxyl group of the fatty acid undergoes successive oxidations. **39.** Dehydrogenation, hydration, thiolation **41.** Carbon 2 would appear in acetyl CoA. Carbons 4 and 6 would appear in the molecule of butyryl CoA. **43.** Fatty acid biosynthesis proceeds by the addition of two carbon units at a time to the growing hydrocarbon chain. Hence, most fatty acids are even numbered.

45. In the mitochondria during oxidative phosphorylation.

47. NADH produced = 31×3 ATP's/NADH = 93 ATP's
FADH$_2$ produced = 15×2 ATP's/FADH$_2$ = 30 ATP's
GTP's produced = 8×1 ATP/1GTP = 8 ATP's
131 ATP's

Since it costs 2 ATP's for initial fatty acid activation the net yield is 129 ATP's.

CHAPTER 27
SELECTED PROBLEMS
5. Amino acids **17.** Uric Acid

19.

Location	Function
a. stomach	—breaks down proteins
b. intestine	—breaks down proteins
c. intestine	—breaks down proteins
d. intestine	—breaks down proteins
e. intestine	—breaks down di- and tripeptides
f. intestine	—converts trypsinogen to trypsin

21. peptide linkage at the N-terminal end

23.

$$CH_3CH_2CH(CH_3)\overset{O}{\overset{\|}{C}}-\overset{O}{\overset{\|}{C}}OH \ + \ R-\underset{NH_2}{\overset{H}{\underset{|}{\overset{|}{C}}}}-COOH \ \xrightarrow{\text{transaminase}}$$

$$CH_3CH_2CH(CH_3)\underset{NH_2}{\overset{H}{\underset{|}{\overset{|}{C}}}}-\overset{O}{\overset{\|}{C}}-OH \ + \ R-\overset{O}{\overset{\|}{C}}-COOH$$

Isoleucine

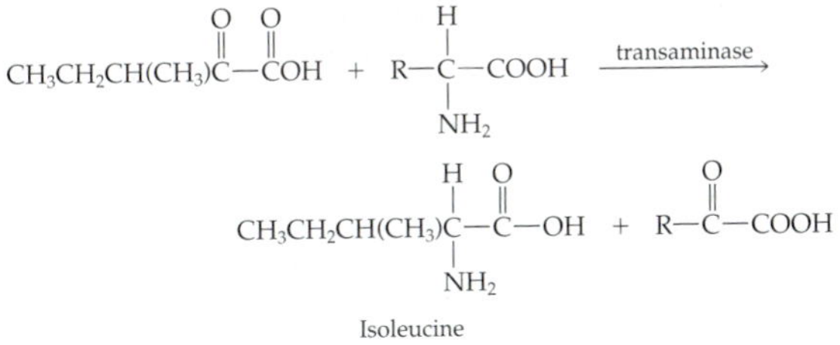

25. The transfer of an amine group from aspartic acid to α-ketoglutaric acid. **27.** Pyruvic acid

29. a.

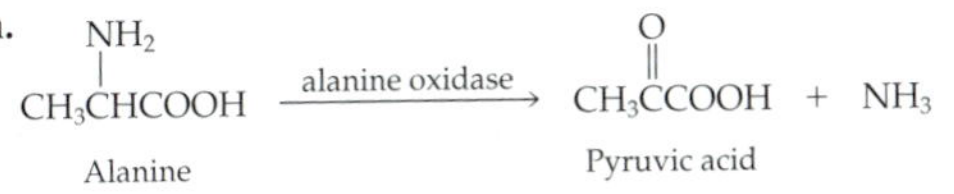

$$\underset{NH_2}{\overset{}{CH_3\underset{|}{CH}COOH}} \ \xrightarrow{\text{alanine oxidase}} \ CH_3\overset{O}{\overset{\|}{C}}COOH \ + \ NH_3$$

Alanine Pyruvic acid

b.

histidine decarboxylase

Histidine

Histamine

c.

$$HOOCCH_2CH_2\underset{NH_2}{\overset{NH_2}{\underset{|}{CH}}}COOH \ \xrightarrow{\text{oxidase}}$$

Glutamic acid

$$HOOCCH_2CH_2\overset{O}{\overset{\|}{C}}COOH \ + \ NH_3$$

α-ketoglutaric acid

d.

$$HOOCCH_2CH_2\underset{NH_2}{\overset{NH_2}{\underset{|}{CH}}}COOH \ + \ NH_3 \ \xrightarrow[\text{synthetase}]{\text{ATP} \quad \text{ADP}}$$

Glutamic acid

$$H_2N-\overset{O}{\overset{\|}{C}}CH_2CH_2\underset{NH_2}{\overset{NH_2}{\underset{|}{CH}}}COOH \ + \ H_2O \ + \ Pi$$

Glutamine

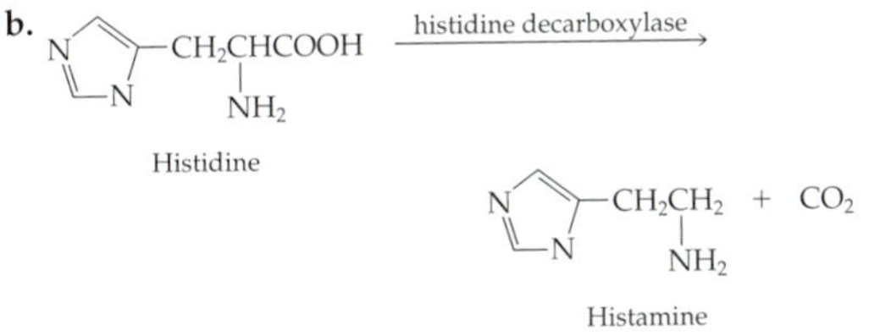

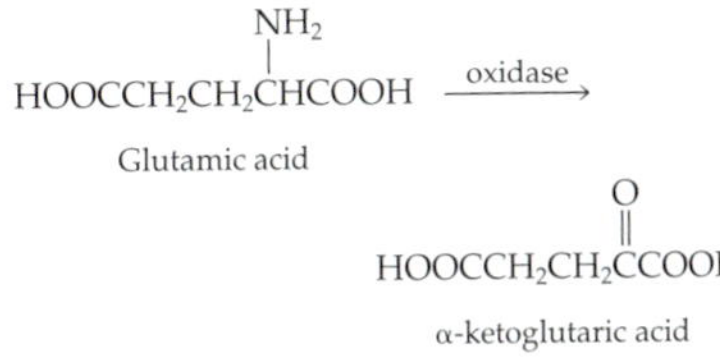

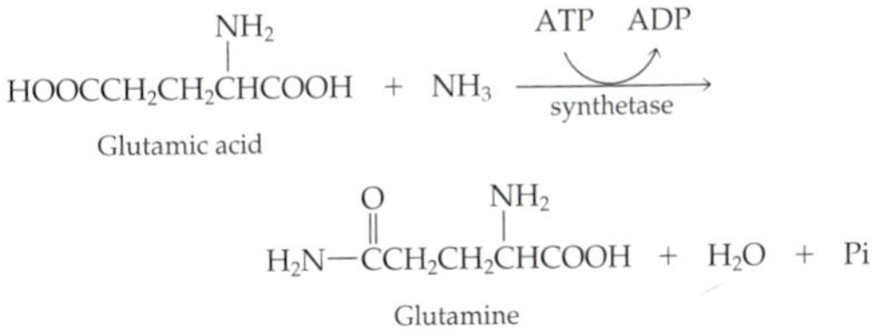

31. a. leucine; **b.** α-ketoglutaric acid; **c.** α-ketoglutaric acid

CHAPTER 28
SELECTED PROBLEMS
11. hemoglobin, cytochromes, catalase **15.** 7.35–7.45.
21. Carbohydrates, lipids, amino acids, hormones (nonprotein), vitamins, inorganic ions. **23. a.** Albumins maintain the osmotic balance and transport fatty acids. **b.** Globulins (alpha and beta) form complexes with lipids and transport them to all parts of the body. **c.** Fibrinogen functions in blood clotting. —They are all synthesized in the liver. **25.** Erythrocytes affect transportation of oxygen to the cells. Leukocytes destroy invading bacteria and other foreign substances. Thrombocytes liberate substances that are involved in blood clotting. **27. a.** Prothrombin is a globulin plasma protein produced by the liver. It is a zymogen that when activated by autoprothrombin C, catalyzes the conversion of fibrinogen into fibrin. **b.** Thrombin is the actual clotting enzyme that catalyzes the activation of fibrinogen. **c.** Fibrinogen is the soluble plasma protein that is converted by thrombin into the insoluble protein fibrin. **d.** Thromboplastin is a group of compounds released by blood platelets and damaged tissue. **e.** Calcium ions are necessary for the catalytic activity associated with both thromboplastin and autoprothrombin C. **f.** Fibrin is the insoluble protein endproduct of the blood-clotting cascade. Fibrin monomers polymerize, resulting in the formation of needle-like threads that enmesh to seal off the area where a blood vessel has been damaged. **29.** Hemoglobin is a conjugated protein with a molecular weight of about 68,000 daltons. Upon hydrolysis it yields a simple protein, globin, and four heme groups. The heme groups account for about 4% of the total molecular weight.
31. a. +2; **b.** +2; **c.** +3; **d.** +2 **33.** The bicarbonate pair, H_2CO_3/HCO_3^-, and the phosphate pair, $H_2PO_4^-/HPO_4^{2-}$. The proper buffer ratio is maintained by the decomposition of any excess carbonic acid to water and carbon dioxide. The carbon dioxide is removed from the equilibrium condition by its elimination at the lungs. **35.** In the kidneys, the glomerulus filters most components (except proteins and formed elements) into the kidney tubules. As this fluid passes through the tubules there is a selective re-uptake into the blood of important constituents that the body wishes to save. Waste products not desirable to the body are not reabsorbed and are excreted as urine. **37.** Insensible perspiration is water lost through the skin or respiratory tract while sensible perspiration is water lost via sweat glands.
39. An inner layer of mucus, then a layer of lacrimal secretions, and then, on the outside, an oily film.

Credits

Chapter 1

Page 1: Mark Harmel/FPG International. Page 3: Scala/Art Resource/Giovanni Stradano, The Alchemist. Studiolo, Palazzo Vecchio, Florence, Italy. Scala/Art Resource, N.Y. Page 5: NASA/Photo Researchers, Inc. Page 6: Carey Van Loon. Page 9: Bill W. Marsh/Photo Researchers, Inc. Page 12: Mark Burnett/Stock Boston. Page 15: Leonard Lessin/Peter Arnold, Inc. Page 15: Carey Van Loon/Carey Van Loon. Page 17: Joyce Photographics/Photo Researchers, Inc. Page 19: Richard Megna/Fundamental Photographs. Page 21: Frank Siteman/Stock Boston. Page 26: Daemmrich/Stock Boston.

Chapter 2

Page 31: Ken Eward/Science Source/Photo Researchers, Inc. Page 32: Bettmann. Page 33 (T): Science VU/Visuals Unlimited. Page 33 (B): Carey Van Loon. Page 34: Bettmann. Page 36: Bob Daemmrich/Stock Boston. Page 39 (All): Richard Megna/Fundamental Photographs. Page 40: NETTIS, JOSEPH/Photo Researchers, Inc. Page 46 (T): Novosti/Science Photo Library/Photo Researchers, Inc. Page 46 (B): Gary J. Shulfer/C. Marvin Lang. Stamp from the private collection of Professor C. M. Lang, photography by Gary J. Shulfer, University of Wisconsin, Stevens Point. "1957, Russia (Scott #1906) and 1969, Russia (Scott #3607);" Scott Standard Postage Stamp Catalogue, Scott Pub. Co., Sidney, Ohio. Page 48: Richard Megna/Fundamental Photographs. Page 54: Matt Meadows/Peter Arnold, Inc.

Chapter 3

Page 55: Science Photo Library/Photo Researchers, Inc. Page 60: Yoav Levy/Phototake NYC. Page 61: Wedgworth/Custom Medical Stock Photo. Page 65 (T): Bettmann. Page 65 (B): Ron Sherman/Stock Boston. Page 67: Atomic Energy Commission/FPG International. Page 70 (T): Martin Dohrn/Science Photo Library/Photo Researchers, Inc. Page 70 (B): DuPont Merck Pharmaceutical Company. Page 72 (T): Science Source/Photo Researchers, Inc. Page 72 (B): Science Photo Library/Custom Medical Stock Photo. Page 73: National Institutes of Health. Drs. M.J. Fulham and Giovanni Di Chiro, The Neuroimaging Section, NINDS, National Institutes of Health, Bethesda, Maryland. Page 74: Alexander Tsiaras/Stock Boston. Page 75: Peticolas/Megna/Fundamental Photographs.

Chapter 4

Page 80: Clive Freeman/Biosym Technologies/Science Photo Library/Photo Researchers, Inc. Page 83: Richard Megna/Fundamental Photographs. Page 84 (T): Carey Van Loon. Page 84 (B): Albert Copley/Visuals Unlimited. Page 85: Richard Megna/Fundamental Photographs. Page 94: Ken Eward/Science Source/Photo Researchers, Inc. / Tom Pantages/Phototake NYC.

Chapter 5

Page 113: Richard Megna/Fundamental Photographs. Page 118: Gary Shulfer/C. Marvin Lang. Page 122: Richard Megna/Fundamental Photographs. Page 127: Barry L. Runk/Grant Heilman Photography, Inc. Page 132: Novosti/Science Photo Library/Photo Researchers, Inc. Page 133: James King-Holmes/ICRF/Science Photo Library/Photo Researchers, Inc. Page 134: Michal Heron/Simon & Schuster/PH College.

Chapter 6

Page 144: Ray Ellis/Photo Researchers, Inc. Page 145 (TL): Richard Megna/Fundamental Photographs. Page 145 (T-a): Romilly Lockyer/The Image Bank. Page 145 (T-b): John Kaprielian/Science Source/Photo Researchers, Inc. Page 145 (T-c): Chris Jones/The Stock Market. Page 146 (L): Dan and Coco McCoy/Rainbow. Page 146 (C): David Madison/Bruce Coleman, Inc. Page 146 (R): Jose L. Pelaez/The Stock Market. Page 148: Royal Observatory, Edinburgh/AATB/Science Photo Library/Photo Researchers, Inc. Page 149 (TL): SuperStock, Inc. Page 149 (TR): Alan Pitcairn/Grant Heilman Photography, Inc. Page 153: Richard Megna/Fundamental Photographs. Page 154 (All): Richard Megna/Fundamental Photographs. Page 157 (All): Carey Van Loon. Page 159: Tomas Sennett/World Bank. Page 160 (T): Telegraph Colour Library/FPG International. Page 160 (BL): Craig Tuttle/The Stock Market. Page 160 (BR): Joseph Nettis/Photo Researchers, Inc.

Chapter 7

Page 164: Chase Swift/Westlight. Page 168: Mary Evans Picture Library/Photo Researchers, Inc. Page 172 (L): SuperStock, Inc. Page 172 (R): Science Photo Library/Custom Medical Stock Photo. Page 174 (All): Richard Megna/Fundamental Photographs. Page 176: Carey Van Loon. Page 180: Norbert Wu/Peter Arnold, Inc. Page 182: Kristen Brochmann/Fundamental Photographs.

Chapter 8

Page 190: Ralph A. Clevenger/Westlight. Page 199: Herman Eisenbeiss/Photo Researchers, Inc. Page 203: Carey Van Loon. Page 206: Carey Van Loon. Page 207 (C): Kim Heacox Photography/DRK Photo. Page 207 (B): Randy Brandon/Peter Arnold, Inc. Page 209: Jim Zuckerman/Westlight.

Chapter 9

Page 213: Stuart Westmorland/Tony Stone Images. Page 214: SuperStock, Inc. Page 220 (All): Richard Megna/Fundamental Photographs. Page 223: Richard Megna/Fundamental Photographs. Page 229 (T&C): Simon & Schuster/PH College. Page 229 (BL): Grant Heilman Photography, Inc. Page 229 (BR): Veronika Burmeister/Visuals Unlimited. Page 231 (All): Carey Van Loon. Page 233 (L): Stock Boston. Page 233 (R): John Lund/Tony Stone Images. Page 234: Jeff Greenberg/Visuals Unlimited.

Chapter 10

Page 239: John Curtis/The Stock Market. Page 240 (All): Fundamental Photographs. Page 241: Richard Megna/ Fundamental Photographs. Page 242: Richard Megna/ Fundamental Photographs. Page 243: Carey Van Loon. Page 250: Carey Van Loon. Page 251 (T): Lionel Delevingue/Phototake NYC. Page 251 (B): Ray Pfortner/Peter Arnold, Inc. Page 252: Dr. E.R. Degginger/Color-Pic, Inc. Page 253: Carey Van Loon. Page 254: Michael P. Gadomski/Photo Researchers, Inc.

Chapter 11

Page 258: L.S. Stepanowicz/Bruce Coleman, Inc. Page 261 (All): Carey Van Loon. Page 266 (T): Yoav Levy/Phototake NYC. Page 266 (B): Richard Megna/Fundamental Photographs. Page 267: Carey Van Loon. Page 268 (All): Richard Megna/Fundamental Photographs. Page 269: Richard Megna/Fundamental Photographs. Page 271: Dan McCoy/Rainbow.

Special Topic C

Page 276: Richard Megna/Fundamental Photographs. Page 280: Carey Van Loon.

Chapter 12

Page 283: John Coletti/Stock Boston. Page 284: Academy of Motion Picture Arts and Sciences. Page 285 (All): Fundamental Photographs. Page 290 (All): Carey Van Loon.

Special Topic D

Page 307 (T): Day Williams/Photo Researchers, Inc. Page 307 (B): Fundamental Photographs. Page 311: Geoff Tompkinson/Science Photo Library/Photo Researchers, Inc.

Chapter 13

Page 316: Ken Graham/Bruce Coleman, Inc. Page 317: Charles D. Winters/Photo Researchers, Inc. Page 327: Bruce Coleman, Inc. Page 331: Richard Megna/Fundamental Photographs. Page 332: NASA/Science Photo Library/Photo Researchers, Inc. Page 340 (All): Dr. E.R. Degginger/Color-Pic, Inc. Page 341: Phototake NYC. Page 342: Richard Megna/Fundamental Photographs. Page 344: Ray Pfortner/ Peter Arnold, Inc.

Special Topic E

Page 357: Dr. E.R. Degginger/Color-Pic, Inc.

Chapter 14

Page 361: Bill Banaszewski/Visuals Unlimited. Page 371: Richard Megna/Fundamental Photographs. Page 380: Bettmann. Page 383: SIU/Peter Arnold, Inc.

Chapter 15

Page 389: Michael Krasowitz/FPG International.

Chapter 16

Page 409: A.G.E. FotoStock/Westlight. Page 412: Richard Megna/Fundamental Photographs. Page 419: Wendell Metzen/ Bruce Coleman, Inc. Page 422: Amoco Fabrics & Fibers Co. Page 428: Dr. E.R. Degginger/Color-Pic, Inc.

Special Topic F

Page 439 (L): Dr. E.R. Degginger/Color-Pic, Inc. Page 439 (R): VU/Cabisco/Visuals Unlimited. Page 440: Drug Enforcement Administration. Page 441: Custom Medical Stock Photo.

Chapter 17

Page 446: Dr. David Scott/CNRI/Phototake NYC. Page 456: Jacques Louis David, 'The Death of Socrates'. Oil on Canvas, 51' X 77 1/4'. The Metropolitan Museum of Art, Wolfe Fund, 1931. Catharine Lorillard Wolfe Collection. Jacques Louis David/The Metropolitan Museum of Art.

Chapter 18

Page 473: Richard Megna/Fundamental Photographs. Page 483: Science Photo Library/Photo Researchers, Inc.

Chapter 19

Page 501: Jeffry Myers/FPG International. Page 502: Richard Megna/Fundamental Photographs. Page 513: Frank LaBua/ Simon & Schuster/PH College. Page 516: Dr. Jeremy Burgess/ Science Photo Library/Photo Researchers, Inc. Page 518: Don W. Fawcett/Visuals Unlimited. Page 519: Biophoto Associates/Photo Researchers, Inc.

Chapter 20

Page 525: Matt Meadows/Peter Arnold, Inc. Page 526: Stephen Frisch/Stock Boston. Page 532: Richard Megna/Fundamental Photographs. Page 542: SIU/Science Source/Photo Researchers, Inc. Page 546 (L): Custom Medical Stock Photo. Page 546 (R): Dan McCoy/Rainbow.

Chapter 21

Page 562: Photo Researchers, Inc. Page 563: Richard Megna/Fundamental Photographs. Page 578: Richard Megna/ Fundamental Photographs. Page 584: Ken Eward/Biografx/Photo Researchers, Inc. Page 585: CNRI/Science Photo Library/Photo Researchers, Inc..

Chapter 22

Page 595: Bucky Reeves/National Audubon Society/ Photo Researchers, Inc. Page 599: Clive Freeman, The Royal Institution/Science Photo Library/Photo Researchers, Inc. Page 612: Dr. Jeremy Burgess/Science Photo Library/Photo Researchers, Inc.

Special Topic J

Page 628: Thomas Hollyman/Photo Researchers, Inc.

Chapter 23

Page 631: Phil A. Harrington/Peter Arnold, Inc. Page 639: Will and Deni McIntyre/Photo Researchers, Inc. Page 643: A. Barrington Brown/Photo Researchers, Inc. Page 641: CNRI/Science Photo Library/Photo Researchers, Inc. Page 659: M. Baret/RHPHU/ Photo Researchers, Inc. Page 660: K.G. Murti/Visuals Unlimited.

Chapter 24

Page 673: Vandystadt/Photo Researchers, Inc.

Chapter 25

Page 703: Tom McCarthy/Rainbow. Page 704: Science Photo Library/Photo Researchers, Inc. Page 718: Louis Goldman/ Photo Researchers, Inc. Page 719: Ron Chapple/FPG International.

Chapter 26

Page 722: Johnny Johnson/DRK Photo. Page 732: Bill Ross/ Westlight.

Chapter 27

Page 744: Japack/Westlight. Page 750: Dagmar Fabricius/Stock Boston. Page 759: The Granger Collection.

Chapter 28

Page 763: Mark Burnett/Stock Boston. Page 767: Manfred Kage/Peter Arnold, Inc. Page 773: CNRI/Science Photo Library/ Photo Researchers, Inc. Page 778 (L): Dr. Gopal Murti/Science Photo Library/Custom Medical Stock Photo. Page 778 (R): Science Source/Photo Researchers, Inc. Page 782: Damien Lovegrove/ Science Photo Library/Photo Researchers, Inc.

Appendix

Page A-6: Fundamental Photographs.

Index

(f = figure, n = footnote or margin note, t = table)

D

T

Table of Atomic Weights (Based on Carbon-12)

Name	Symbol	Atomic Number	Relative Atomic Mass	Name	Symbol	Atomic Number	Relative Atomic Mass
Actinium	Ac	89	227.028	Mercury	Hg	80	200.59
Aluminum	Al	13	26.9815	Molybdenum	Mo	42	95.94
Americium	Am	95	(243)	Neodymium	Nd	60	144.24
Antimony	Sb	51	121.75	Neon	Ne	10	20.1797
Argon	Ar	18	39.948	Neptunium	Np	93	237.048
Arsenic	As	33	74.9216	Nickel	Ni	28	58.69
Astatine	At	85	(210)	Niobium	Nb	41	92.9064
Barium	Ba	56	137.327	Nitrogen	N	7	14.0067
Berkelium	Bk	97	(247)	Nobelium	No	102	(259)
Beryllium	Be	4	9.01218	Osmium	Os	76	190.23
Bismuth	Bi	83	208.980	Oxygen	O	8	15.9994
Boron	B	5	10.811	Palladium	Pd	46	106.42
Bromine	Br	35	79.904	Phosphorus	P	15	30.9738
Cadmium	Cd	48	112.411	Platinum	Pt	78	195.08
Calcium	Ca	20	40.078	Plutonium	Pu	94	(244)
Californium	Cf	98	(251)	Polonium	Po	84	(209)
Carbon	C	6	12.011	Potassium	K	19	39.0983
Cerium	Ce	58	140.115	Praseodymium	Pr	59	140.908
Cesium	Cs	55	132.905	Promethium	Pm	61	(145)
Chlorine	Cl	17	35.4527	Protactinium	Pa	91	231.036
Chromium	Cr	24	51.9961	Radium	Ra	88	226.025
Cobalt	Co	27	58.9332	Radon	Rn	86	(222)
Copper	Cu	29	63.546	Rhenium	Re	75	186.207
Curium	Cm	96	(247)	Rhodium	Rh	45	102.906
Dysprosium	Dy	66	162.50	Rubidium	Rb	37	85.4678
Einsteinium	Es	99	(252)	Ruthenium	Ru	44	101.07
Erbium	Er	68	167.26	Samarium	Sm	62	150.36
Europium	Eu	63	151.965	Scandium	Sc	21	44.9559
Fermium	Fm	100	(257)	Selenium	Se	34	78.96
Fluorine	F	9	18.9984	Silicon	Si	14	28.0855
Francium	Fr	87	(223)	Silver	Ag	47	107.868
Gadolinium	Gd	64	157.25	Sodium	Na	11	22.9898
Gallium	Ga	31	69.723	Strontium	Sr	38	87.62
Germanium	Ge	32	72.61	Sulfur	S	16	32.066
Gold	Au	79	196.967	Tantalum	Ta	73	180.948
Hafnium	Hf	72	178.49	Technetium	Tc	43	(98)
Helium	He	2	4.00260	Tellurium	Te	52	127.60
Holmium	Ho	67	164.930	Terbium	Tb	65	158.925
Hydrogen	H	1	1.00794	Thallium	Tl	81	204.383
Indium	In	49	114.818	Thorium	Th	90	232.038
Iodine	I	53	126.904	Thulium	Tm	69	168.934
Iridium	Ir	77	192.22	Tin	Sn	50	118.710
Iron	Fe	26	55.847	Titanium	Ti	22	47.88
Krypton	Kr	36	83.80	Tungsten	W	74	183.85
Lanthanum	La	57	138.906	Uranium	U	92	238.029
Lawrencium	Lr	103	(260)	Vanadium	V	23	50.9415
Lead	Pb	82	207.2	Xenon	Xe	54	131.29
Lithium	Li	3	6.941	Ytterbium	Yb	70	173.04
Lutetium	Lu	71	174.967	Yttrium	Y	39	88.9059
Magnesium	Mg	12	24.3050	Zinc	Zn	30	65.39
Manganese	Mn	25	54.9381	Zirconium	Zr	40	91.224
Mendelevium	Md	101	(258)				

Atomic masses in this table are relative to carbon-12 and limited to six significant figures, although some atomic masses are known more precisely. For certain radioactive elements the numbers listed (in parentheses) are the mass numbers of the most stable isotopes.